Lecture Notes in Computer Science 4557

Michael J. Smith Gavriel Salvendy (Eds.)

Human Interface and the Management of Information

Methods, Techniques and Tools in Information Design

Symposium on Human Interface 2007
Held as Part of HCI International 2007
Beijing, China, July 22-27, 2007
Proceedings, Part I

Springer

Volume Editors

Michael J. Smith
University of Wisconsin-Madison
Department of Industrial and Systems Engineering
2166 Engineering Centers Bldg., 1550 Engineering Drive, Madison, WI 53706, USA
E-mail: mjsmith@engr.wisc.edu

Gavriel Salvendy
Purdue University, Department of Industrial Engineering
Grissom Hall, 315 N. Grant St., West Lafayette, IN 47907-2023, USA
E-mail: salvendy@purdue.edu

Library of Congress Control Number: 2007930200

CR Subject Classification (1998): H.5, H.3, H.4, C.2, K.4.2, D.2

LNCS Sublibrary: SL 3 – Information Systems and Application, incl. Internet/Web and HCI

ISSN 0302-9743
ISBN-10 3-540-73344-2 Springer Berlin Heidelberg New York
ISBN-13 978-3-540-73344-7 Springer Berlin Heidelberg New York

Springer is a part of Springer Science+Business Media

springer.com

Printed in Germany

Typesetting: Camera-ready by author, data conversion by Scientific Publishing Services, Chennai, India
Printed on acid-free paper SPIN: 12083510 06/3180 5 4 3 2 1 0

Foreword

The 12th International Conference on Human-Computer Interaction, HCI International 2007, was held in Beijing, P.R. China, 22-27 July 2007, jointly with the Symposium on Human Interface (Japan) 2007, the 7th International Conference on Engineering Psychology and Cognitive Ergonomics, the 4th International Conference on Universal Access in Human-Computer Interaction, the 2nd International Conference on Virtual Reality, the 2nd International Conference on Usability and Internationalization, the 2nd International Conference on Online Communities and Social Computing, the 3rd International Conference on Augmented Cognition, and the 1st International Conference on Digital Human Modeling.

A total of 3403 individuals from academia, research institutes, industry and governmental agencies from 76 countries submitted contributions, and 1681 papers, judged to be of high scientific quality, were included in the program. These papers address the latest research and development efforts and highlight the human aspects of design and use of computing systems. The papers accepted for presentation thoroughly cover the entire field of Human-Computer Interaction, addressing major advances in knowledge and effective use of computers in a variety of application areas.

This volume, edited by Michael J. Smith and Gavriel Salvendy, contains papers in the thematic area of Human Interface and the Management of Information, addressing the following major topics:

- Design and Evaluation Methods and Techniques
- Visualizing Information
- Retrieval, Searching, Browsing and Navigation
- Development Methods and Techniques
- Advanced Interaction Technologies and Techniques

The remaining volumes of the HCI International 2007 proceedings are:

- Volume 1, LNCS 4550, Interaction Design and Usability, edited by Julie A. Jacko
- Volume 2, LNCS 4551, Interaction Platforms and Techniques, edited by Julie A. Jacko
- Volume 3, LNCS 4552, HCI Intelligent Multimodal Interaction Environments, edited by Julie A. Jacko
- Volume 4, LNCS 4553, HCI Applications and Services, edited by Julie A. Jacko
- Volume 5, LNCS 4554, Coping with Diversity in Universal Access, edited by Constantine Stephanidis
- Volume 6, LNCS 4555, Universal Access to Ambient Interaction, edited by Constantine Stephanidis
- Volume 7, LNCS 4556, Universal Access to Applications and Services, edited by Constantine Stephanidis

- Volume 9, LNCS 4558, Interacting in Information Environments, edited by Michael J. Smith and Gavriel Salvendy
- Volume 10, LNCS 4559, HCI and Culture, edited by Nuray Aykin
- Volume 11, LNCS 4560, Global and Local User Interfaces, edited by Nuray Aykin
- Volume 12, LNCS 4561, Digital Human Modeling, edited by Vincent G. Duffy
- Volume 13, LNAI 4562, Engineering Psychology and Cognitive Ergonomics, edited by Don Harris
- Volume 14, LNCS 4563, Virtual Reality, edited by Randall Shumaker
- Volume 15, LNCS 4564, Online Communities and Social Computing, edited by Douglas Schuler
- Volume 16, LNAI 4565, Foundations of Augmented Cognition 3rd Edition, edited by Dylan D. Schmorrow and Leah M. Reeves
- Volume 17, LNCS 4566, Ergonomics and Health Aspects of Work with Computers, edited by Marvin J. Dainoff

I would like to thank the Program Chairs and the members of the Program Boards of all Thematic Areas, listed below, for their contribution to the highest scientific quality and the overall success of the HCI International 2007 Conference.

Ergonomics and Health Aspects of Work with Computers

Program Chair: Marvin J. Dainoff

Arne Aaras, Norway
Pascale Carayon, USA
Barbara G.F. Cohen, USA
Wolfgang Friesdorf, Germany
Martin Helander, Singapore
Ben-Tzion Karsh, USA
Waldemar Karwowski, USA
Peter Kern, Germany
Danuta Koradecka, Poland
Kari Lindstrom, Finland
Holger Luczak, Germany
Aura C. Matias, Philippines
Kyung (Ken) Park, Korea
Michelle Robertson, USA
Steven L. Sauter, USA
Dominique L. Scapin, France
Michael J. Smith, USA
Naomi Swanson, USA
Peter Vink, The Netherlands
John Wilson, UK

Human Interface and the Management of Information

Program Chair: Michael J. Smith

Lajos Balint, Hungary
Gunilla Bradley, Sweden
Hans-Jörg Bullinger, Germany
Alan H.S. Chan, Hong Kong
Klaus-Peter Fähnrich, Germany
Michitaka Hirose, Japan
Yoshinori Horie, Japan
Robert Proctor, USA
Youngho Rhee, Korea
Anxo Cereijo Roibás, UK
Francois Sainfort, USA
Katsunori Shimohara, Japan
Tsutomu Tabe, Japan
Alvaro Taveira, USA

Human-Computer Interaction

Program Chair: Julie A. Jacko

Engineering Psychology and Cognitive Ergonomics

Program Chair: Don Harris

Universal Access in Human-Computer Interaction

Program Chair: Constantine Stephanidis

Jerzy Charytonowicz, Poland
Pier Luigi Emiliani, Italy
Michael Fairhurst, UK
Gerhard Fischer, USA
Jon Gunderson, USA
Andreas Holzinger, Austria
Arthur Karshmer, USA
Simeon Keates, USA
George Kouroupetroglou, Greece
Jonathan Lazar, USA
Seongil Lee, Korea
Ben Shneiderman, USA
Christian Stary, Austria
Hirotada Ueda, Japan
Jean Vanderdonckt, Belgium
Gregg Vanderheiden, USA
Gerhard Weber, Germany
Harald Weber, Germany
Toshiki Yamaoka, Japan
Mary Zajicek, UK
Panayiotis Zaphiris, UK

Virtual Reality

Program Chair: Randall Shumaker

Terry Allard, USA
Pat Banerjee, USA
Robert S. Kennedy, USA
Heidi Kroemker, Germany
Ben Lawson, USA
Ming Lin, USA
Bowen Loftin, USA
Holger Luczak, Germany
Annie Luciani, France
Gordon Mair, UK
Ulrich Neumann, USA
Albert "Skip" Rizzo, USA
Lawrence Rosenblum, USA
Dylan Schmorrow, USA
Kay Stanney, USA
Susumu Tachi, Japan
John Wilson, UK
Wei Zhang, P.R. China
Michael Zyda, USA

Usability and Internationalization

Program Chair: Nuray Aykin

Genevieve Bell, USA
Alan Chan, Hong Kong
Apala Lahiri Chavan, India
Jori Clarke, USA
Pierre-Henri Dejean, France
Susan Dray, USA
Paul Fu, USA
Emilie Gould, Canada
Sung H. Han, South Korea
Veikko Ikonen, Finland
Richard Ishida, UK
Esin Kiris, USA
Tobias Komischke, Germany
Masaaki Kurosu, Japan
James R. Lewis, USA
Rungtai Lin, Taiwan
Aaron Marcus, USA
Allen E. Milewski, USA
Patrick O'Sullivan, Ireland
Girish V. Prabhu, India
Kerstin Röse, Germany
Eunice Ratna Sari, Indonesia
Supriya Singh, Australia
Serengul Smith, UK
Denise Spacinsky, USA
Christian Sturm, Mexico
Adi B. Tedjasaputra, Singapore
Myung Hwan Yun, South Korea
Chen Zhao, P.R. China

Online Communities and Social Computing

Program Chair: Douglas Schuler

Chadia Abras, USA
Lecia Barker, USA
Amy Bruckman, USA
Peter van den Besselaar, The Netherlands
Peter Day, UK
Fiorella De Cindio, Italy
John Fung, P.R. China
Michael Gurstein, USA
Tom Horan, USA
Piet Kommers, The Netherlands
Jonathan Lazar, USA
Stefanie Lindstaedt, Austria
Diane Maloney-Krichmar, USA
Isaac Mao, P.R. China
Hideyuki Nakanishi, Japan
A. Ant Ozok, USA
Jennifer Preece, USA
Partha Pratim Sarker, Bangladesh
Gilson Schwartz, Brazil
Sergei Stafeev, Russia
F.F. Tusubira, Uganda
Cheng-Yen Wang, Taiwan

Augmented Cognition

Program Chair: Dylan D. Schmorrow

Kenneth Boff, USA
Joseph Cohn, USA
Blair Dickson, UK
Henry Girolamo, USA
Gerald Edelman, USA
Eric Horvitz, USA
Wilhelm Kincses, Germany
Amy Kruse, USA
Lee Kollmorgen, USA
Dennis McBride, USA
Jeffrey Morrison, USA
Denise Nicholson, USA
Dennis Proffitt, USA
Harry Shum, P.R. China
Kay Stanney, USA
Roy Stripling, USA
Michael Swetnam, USA
Robert Taylor, UK
John Wagner, USA

Digital Human Modeling

Program Chair: Vincent G. Duffy

Norm Badler, USA
Heiner Bubb, Germany
Don Chaffin, USA
Kathryn Cormican, Ireland
Andris Freivalds, USA
Ravindra Goonetilleke, Hong Kong
Anand Gramopadhye, USA
Sung H. Han, South Korea
Pheng Ann Heng, Hong Kong
Dewen Jin, P.R. China
Kang Li, USA
Zhizhong Li, P.R. China
Lizhuang Ma, P.R. China
Timo Maatta, Finland
J. Mark Porter, UK
Jim Potvin, Canada
Jean-Pierre Verriest, France
Zhaoqi Wang, P.R. China
Xiugan Yuan, P.R. China
Shao-Xiang Zhang, P.R. China
Xudong Zhang, USA

In addition to the members of the Program Boards above, I also wish to thank the following volunteer external reviewers: Kelly Hale, David Kobus, Amy Kruse, Cali Fidopiastis and Karl Van Orden from the USA, Mark Neerincx and Marc Grootjen from the Netherlands, Wilhelm Kincses from Germany, Ganesh Bhutkar and Mathura Prasad from India, Frederick Li from the UK, and Dimitris Grammenos, Angeliki Kastrinaki, Iosif Klironomos, Alexandros Mourouzis, and Stavroula Ntoa from Greece.

This conference could not have been possible without the continuous support and advise of the Conference Scientific Advisor, Gavriel Salvendy, as well as the dedicated work and outstanding efforts of the Communications Chair and Editor of HCI International News, Abbas Moallem, and of the members of the Organizational Board from P.R. China, Patrick Rau (Chair), Bo Chen, Xiaolan Fu, Zhibin Jiang, Congdong Li, Zhenjie Liu, Mowei Shen, Yuanchun Shi, Hui Su, Linyang Sun, Ming Po Tham, Ben Tsiang, Jian Wang, Guangyou Xu, Winnie Wanli Yang, Shuping Yi, Kan Zhang, and Wei Zho.

I would also like to thank for their contribution towards the organization of the HCI International 2007 Conference the members of the Human Computer Interaction Laboratory of ICS-FORTH, and in particular Margherita Antona, Maria Pitsoulaki, George Paparoulis, Maria Bouhli, Stavroula Ntoa and George Margetis.

Constantine Stephanidis
General Chair, HCI International 2007

HCI International 2009

The 13th International Conference on Human-Computer Interaction, HCI International 2009, will be held jointly with the affiliated Conferences in San Diego, California, USA, in the Town and Country Resort & Convention Center, 19-24 July 2009. It will cover a broad spectrum of themes related to Human Computer Interaction, including theoretical issues, methods, tools, processes and case studies in HCI design, as well as novel interaction techniques, interfaces and applications. The proceedings will be published by Springer. For more information, please visit the Conference website: http://www.hcii2009.org/

General Chair
Professor Constantine Stephanidis
ICS-FORTH and University of Crete
Heraklion, Crete, Greece
Email: program@hcii2009.org

Table of Contents

Part I: Design and Evaluation Methods and Techniques

Part II: Visualising Information

Part III: Retrieval, Searching, Browsing and Navigation

Part IV: Development Methods and Techniques

Part V: Advanced Interaction Technologies and Techniques

Part I

Design and Evaluation Methods and Techniques

Exporting Usability Knowledge into a Small-Sized Software Development Organization – A Pattern Approach

Kari-Pekka Aikio

Department of Information Processing Science
University of Oulu, Rakentajantie 3 P.O. Box 300, 90014, Oulu, Finland
kari-pekka.aikio@oulu.fi

Abstract. Frameworks and patterns on integrating usability activities into organizational work practices have been developed during the last years. However, usability and user-centered design activities remain fairly unknown to small-sized software development organizations. Empirical research on initiating usability activities in such organizational contexts is also limited to few cases. We present a case project in which our usability research team had six months to provide a small-sized software company with knowledge on how to improve usability of one of their products. Our approach is based on patterns of integration, on a selection of user-centered methods and on producing tailored usability artefacts.

Keywords: Integration, Knowledge, Patterns, Usability, User-Centered Design.

1 Introduction

Despite the growing trends in research that are steering organizations towards the use of an amalgamation of SE and HCI activities, most industry professionals have yet to follow suggestions from the academic and industrial research communities. For most part, software engineers and HCI practitioners continue to work separately. While collaboration between two groups does occur, it does not happen frequently enough or early enough in the software development life cycle [9]. User-centered design (UCD) -integration[1] literature identifies that some attempts at usability have resulted in a "one-time-only" phenomenon [2]. Strategic usability has been suggested as means to avoid usability pitfalls and the "one-time-only" – phenomenon. To make usability a part of the business solution, it has to be made uniform, manageable and systematic [15]. Latest development in UCD field of research has produced patterns and frameworks that provide usability practitioners and software developers with knowledge on how to integrate usability activities into existing organizational work practices. Although progress in the field continues, it still lacks empirical research that focuses on initiating usability work in small-sized enterprises.

This paper is organized into three sections. First section places our research into context by briefly outlining concepts and issues behind usability and software development

[1] Terms 'usability integration' and 'UCD-integration' are often used interchangeably.

M.J. Smith, G. Salvendy (Eds.): Human Interface, Part I, HCII 2007, LNCS 4557, pp. 3–11, 2007.

organization's size, strategic and tactical usability and patterns containing usability integration knowledge. Section two describes our case project. Section three consists of discussion, conclusions and recommendations for future work and research.

1.1 Usability and Software Development Organization's Size

According to Seffah et al. [16] studies indicate that usability integration is limited to large organizations. Seffah and Metzker [17] note that usability methods are relatively unknown and underused, and that they are inaccessible to common developers and small and medium-sized software development teams. Case study by Fellenz [7] describes how usability activities were introduced into a relatively small organization. However, that case does not explicate organization's size, but it is stated that a special department was established for conducting usability activities.

1.2 Strategic Usability

According to literature strategic usability is a broad and multifaceted concept. While focusing on usability, it also includes aspects such as business, organization, culture, people, processes, knowledge and change-agentry. [4], [12], [14], [13], [15], [18]. In 1997 Bloomer & al. [4] stated that usability can be successfully integrated into an organization by developing a strategy which leads to key usability benefits and supports overall business objectives. A decade later Venturi et al. [18] consider UCD being integrated (a) when UCD is brought in a timely way into the product life cycle, (b) UCD team is provided with the proper skills and experience, and (c) by means of a proper UCD infrastructure, (d) UCD is supported by management commitment, (e) when UCD awareness and culture are properly disseminated inside and outside of the organization and (f) the results of the UCD activities have impact on design decisions. Our case project describes initial steps of usability field work in a small-sized software development organization. At the other end of the spectrum is Venturi's definition of integrated user-centered design.

1.3 Field Tactics for Usability Practitioners and Software Engineers

Nielsen [14] clearly distinguishes strategic usability from tactical usability. According to him; "strategy doesn't work at the product level: that's where you need tactics like 'discount usability engineering'… ". Literature is rich on tactical knowledge on allocating and applying UCD-methods. Frameworks [8] and patterns [3] have been developed, but their approach in representing UCD-knowledge is different; a framework, such as Ferre's integration framework [8], provides criteria for evaluating the effectiveness and suitability of a given UCD-method, whereas patterns, such as Battle's patterns of integration, provide suggestions and practical UCD-knowledge. One clear distinction between these two sources of knowledge can be made; according to Ferre et al.[8], their framework is aimed for software developers and according to Battle, her patterns of integration is aimed for UCD-practitioners. Example framework and patterns both divide UCD-integration process into three phases respectively; 'initial or early', 'central or middle' and 'evolution or late'. In many ways they complement each other and together they provide a rich source for allocating UCD-activities. Our approach is mainly based on patterns and especially on Battle's pattern B (see below), but it has been augmented by Ferre's integration framework and Battle's pattern C.

1.4 Patterns of Integration

Patterns were originally introduced by Alexander [1] in late 1970's and ever since different types of patterns and pattern languages have emerged for software developers and usability practitioners. Plainly speaking, patterns aim to provide applicable knowledge to solving common problem situations.

Battle's patterns of integration [3] consists of four patterns designated A, B, C and D. Patterns A and B identify two sources of UCD-knowledge when initiating UCD work in an organization; (A) an internal usability group and (B) an external UCD-consultant. Pattern C describes the situation after UCD work is initiated and outcomes[2] of the previous phase function as input in the next. Pattern (D) describes the situation when UCD activities are fully integrated into all phases of product development lifecycle. Each pattern describes four parts of the process: (1) initial context, (2) problem, (3) solution and (4) resulting context. Pattern B (foot in the door – for external consultants) (see table 1.) resembles our case project in two aspects; (1) our position as external UCD-consultants and (2) case project took place during the late (or evolutionary) phases of the product's lifecycle.

Table 1. Battle's pattern B: foot in the door (for external consultants)

Pattern	**Area of Focus in a "Generic" Lifecycle**		
	Early	**Middle**	**Late**
(A): Foot in the door (for internal usability group)	**X**		
(B): Foot in the door (for external consultants)			**X**
(C): UCD focus on early definition and design	**X**		
(D): UCD in every phase	**X**	**X**	**X**

From external UCD-consultant's perspective pattern B's four parts look like this:

1. Initial context: the software development organization does not follow UCD-methods, but suspects that it has a usability problem. Company contacts a UCD/HCI consultant.
2. Problem: challenge is to sell UCD-services to the company and making a big impact while keeping costs low.
3. Solution: fix the immediate usability problems and help company in achieving quick "wins".
4. Resulting context: consultant's evaluation has led to short-term fixes and a long-term vision for the product. There is now potential to move to the next level (Pattern C).

While pattern B recommends conducting heuristic evaluation, usability testing and developing short-term and long-term recommendations, pattern C and Ferre's framework suggests application of methods, such as Personas [5], that produce outcomes with useful information for organization's future usability work. Our

---

[2] Jokela [11] has proposed an outcome-driven approach to usability integration process in 2002. The outcomes should be produced as a result of usability activities, in order to make usability design systematic [10].

approach to providing our partner software company with usability knowledge is a combination of patterns B and C.

2 Case Project

This research was conducted in a case project spanning six months, starting November 2005 and ending in April 2006. Project was funded by our partner software company. This section describes our case project as follows; composition of our usability team, description of our partner software company and its product, selection and application of UCD-methods and production of usability deliverables.

2.1 Usability Team

Our usability team consisted of four specially skilled usability researchers; one senior usability researcher with industrial experience, one with software engineering background, one focusing on psychological aspects of usability and one focusing on teaching usability and it's methodology. In addition our teams work was supported by a knowledge management researcher who followed and recorded our activities during the project.

2.2 A Small-Sized Software Development Company

Our partner company, according to definition by European Union commission [6], categorizes as a small-sized enterprise. Company, whose core business activity is web-based software development, contacted our department with interest in benefits of usability. Company had very practical motivation regarding this project. A sales manager stated following; "how can we improve our product's usability so that our new customers can learn to use it more efficiently? We are also interested in learning if usability can reduce training costs."

Stakeholders at the company consisted of a sales manager, a senior software designer, a software engineer and a product trainer. Company listed topics of interest such as usability knowledge, methods and processes regarding usability. Two issues came up in discussions with the company: (1) what advantages could usability provide for the next version of their product? (2) What could this kind of project provide for the company at the moment and in the future?

2.3 Product

Our usability activities focused on company's Content Management System (CMS) – product, which is a web-based publishing system. System is composed of several modules and it is being developed iteratively. During the project a new version of the system was under development and usability aspects would be taken into account in the next version of the product. End users of the product had so far not participated in the development process of the previous versions of the system.

2.4 Project Goals and Tasks

At the beginning of the project we discussed usability and its challenges with company's stakeholders. Topics included existing usability knowledge, training and the nature of usability work. The overall goal of the project was to make as big an impact as possible with our activities during the six months available.

Four tasks were set based on this discussion;

1. Identify current state of product's usability (conduct a requirements specification workshop, conduct usability tests and participate the training to product's use).
2. Identify company's long term goals regarding company's usability knowledge and short term goals regarding the outcomes of this project.
3. Design an implementation plan regarding steps and activities for this project.
4. Execute the plan.

2.4.1 Getting to Know the Product and Its Users

After the kick-off meeting our team participated a training session in which our team members were introduced to basic features of the product. In a two hour session we learned about the training company provided for its' clients and end-users. Learning some of the basic features ourselves helped our work in creating an understanding of what was considered essential tasks for new users of the product.

Company provided us access to their CMS-system so that we could familiarize ourselves with the product after the training. Having a demonstration environment similar to actual end use environment enabled us to conduct expert evaluations on the product over the Internet. Later we would use this environment to conduct usability tests from our usability laboratory.

One of the tasks of this case project was to provide our partner company with long term recommendations for future usability work. Although our approach to this project follows Battle's pattern B we estimated that our work would be incomplete if we did not analyze users of the product. User analysis is generally regarded as an activity best suited for initial cycles when requirements are elicited and analyzed [8]. Battle's pattern C and Ferre's framework recommend application of Personas – method [5]. Providing company stakeholders with knowledge on this method was considered essential for their future usability work.

We organized a workshop in which company's primary stakeholders and our team collaboratively identified main user groups and developed personas to represent these groups. At the beginning of the workshop our Personas-expert explained the method to our partners. After the briefing four teams were formed and they began to develop Personas descriptions based on identified user types. Altogether four personas were developed and assessed collaboratively. Finally one persona was chosen to represent a novice end user with basic training on using the product.

2.4.2 Usability Tests

A test scenario was developed representing several daily work tasks that our selected persona would encounter while using the product. Scenario and its tasks would be used in usability tests. After test scenario and test tasks were validated by company stakeholders a list of potential test users for usability tests was prepared.

Altogether four usability tests were conducted (one pilot test + three actual tests). All test users, except the pilot test, were end users of the product. Our usability testing laboratory facilitates a TechSmith Morae -software system, which enabled us to record test sessions at screen level and prepare a video run-through of our test findings.

Test participants were briefed to the test situation and they all filled a user profile questionnaire. Participants had five minutes to familiarize themselves with the equipment themselves before beginning the test. The basis for conducting usability tests was the application of a well recognized think aloud method. Each of the four tests took about one hour and all of them were recorded. Finally each test participant filled out an after-test questionnaire. Test findings were analyzed and combined to the test report. Test report and a test video run-through of selected findings were presented to company's stakeholders at the test report meeting.

2.5 Long-Term Recommendations

After conducting usability tests we began designing usability artefacts that would provide company with long term recommendations for future usability work. Two types of usability artefacts we considered valuable; a concept analysis document and a user interface design guideline document. Perhaps the most remarkable feature of these two documents is that they were tailored using examples from the product itself. Both documents were inspected at the project's end meeting and accepted as deliverables of the project.

2.6 Concept Analysis Document

In the concept analysis we wanted to pay attention how to avoid textual usability problems that may confuse and irritate users. However, such problems are usually cheap, and easy to correct and avoid. In the concept analysis we focused e.g. on terminology, which should be consistent and clear. Examples used in the analysis were from company's CMS product. We emphasized that a product `vocabulary' or `lexicon' should be created. Keeping terminology of the product more consistent between various modules of the whole product was considered helpful.

2.7 User Interface Design Guideline Document

Guideline document was developed to provide software developers with user interface design recommendations and suggestions by using sample pictures only of their own system. This feature of the guideline was very well received and appreciated by the designers. According to the sales manager, this guideline will ensure the basic quality of usability to the company and it is a concrete and important guide for developers. Sales manager also noted that the company will go through the guidelines with their business partner in cooperation as they negotiate UI design solutions.

2.8 Summary of Project Deliverables

During this research project we produced and delivered following usability artefacts for the company: (1) a test report on the current state of usability of their product for immediate (short-term) recommendations for developers, (2) a concept analysis document and (3) a user interface guideline for long-term product development. Personas –method was introduced to company stakeholders in a collaborative workshop session and four Persona descriptions were delivered to the company.

2.9 Post-project Interviews

After the project had officially ended a follow up interview was conducted for company's stakeholders. Results of that interview indicated that the Personas workshop session was perceived as the most satisfying task of the project. Stakeholders were also satisfied with deliverables we produced but whose production they did not directly influence. Most of all they appreciated the tailored usability artefacts that encapsulated practical design knowledge. Retrospective interviews were conducted for usability team members to identify successes and areas of improvement of our approach.

3 Discussion

Working as an external UCD consultant for a small-sized software company for a short period has its limitations. Short visits to company's site do not provide for thorough investigations on company's development processes and its personnel. Besides, the spectrum of UCD-integration is broad and mostly uncharted. Assuming that (a) basically all attempts at usability/UCD-integration begin with limited knowledge on usability and UCD on behalf of the development organization and (b) that Venturi's definition of UCD-integration works as criteria for a successful and complete UCD-integration, it is clear that there is a lot of ground to cover between those two points. Somewhere between points a and b each company will have make a decision for their source for UCD-knowledge (internal or external). Hiring external consultants periodically to evaluate the progress of products usability may work if knowledge is available how UCD-integration had advanced from point a towards point b.

Ideas behind concepts of usability and user-centered design are often difficult to communicate. Some basic ideas behind these concepts can be introduced during the short encounters between external usability experts and software company's stakeholders. If usability knowledge is intended to be transmitted through usability artefacts, such as usability test reports, concept analysis documents and design guideline documents, we recommend focusing on the quality of these artefacts. In other words; usability artefacts should be appear useful and usable for those who use them.

Patterns and frameworks do not clearly specify how it's approach suites organizations of different sizes. Quick usability fixes may work well with small companies and external usability consultants can guide companies with initial steps of usability work. Literature suggests more robust measures to managing UCD-activities, but such management requires resources. Future of usability work in small-sized enterprises depends not only on willingness to 'do' usability, but explicitly on organization's capability to continue and support usability work after the initial steps towards integrated user-centered design have been taken.

4 Conclusions

Our approach to exporting usability knowledge into a small-sized software development organization was based on knowledge embedded in patterns of integration. Patterns gave us insight into providing our partner company with usability knowledge that they would find useful and applicable. We accomplished this by focusing on the quality of the deliverables, namely the usability artefacts. Our partner software company expressed satisfaction on the outcomes of this case project and considered deliverables an important knowledge asset for their future usability work.

5 Recommendations for Future Work and Research

We recommend that further research on practical UCD-integration is conducted from perspectives of UCD-practitioners and software engineers. Focusing on collaboration between these two parties is important when UCD-knowledge is being exported over organizational boundaries. Development of organizational strategies and tactical guidelines on exporting and adopting UCD-knowledge is also recommended.

Acknowledgements

We thank Verkkoasema Oy for providing us an opportunity to study the initiation of usability work in an industrial setting.

References

1. Alexander, C., Ishikawa, S., Silverstein, M.: A pattern language: Towns, buildings, construction. Oxford University Press, Oxford (1977)
2. Aucella, A.F.: Ensuring Success with Usability Engineering. ACM 4, 19–22 (1997)
3. Battle, L.: Patterns of integration. In: Seffah, A., Gulliksen, J., Desmarais, M.C. (eds.) Human-Centered Software Engineering — Integrating Usability in the Software Development Lifecycle. Human-Computer Interaction Series edn, vol. 8, pp. 287–308. Springer, Heidelberg (2005)
4. Bloomer, S., Croft, R., Kieboom, H.: Strategic Usability: Introducing Usability into Organizations. CHI (1997)
5. Cooper, A.: The inmates are running the asylum: Why high-tech products drive us crazy and how to restore the sanity, 2nd edn. SAMS, USA (2004)
6. European Union Commission: Commission Recommendation of 06/05/2003, Concerning the Definition of Micro, Small and Medium-Sized Enterprises. 2007 (2003)
7. Fellenz, C.: Introducing Usability into Smaller Organizations. ACM 4, 29–33 (1997)
8. Ferré, X., Juristo, N., Moreno, A.: Framework for Integrating Usability Practices into the Software Process. Profes 1, 202–215 (2005)
9. Jerome, B., Kazman, R.: Surveying the solitudes: An investigation into the relationships between human computer interaction and software engineering in practice. In: Seffah, A., Gulliksen, J., Desmarais, M.C. (eds.) vol. 1, pp. 59–70. Springer, Netherlands (2005)

10. Jokela, T.: The KESSU Usability Design Process Model (Version 2.1. 2006), 22 (2004)
11. Jokela, T.: Making User-Centered Design Common Sense: Striving for an Unambiguous and Communicative UCD Process Model. NordiCHI, 19-23, 19–26 (2002)
12. Mayhew, D.J.: Business: Strategic Development of the Usability Engineering Function. Interactions 6, 27–34 (1999)
13. Miller, A.: Integrating Human Factors in Customer Support Systems Development using a Multi-Level Organizational Approach. CHI 13-18, 368–375 (1996)
14. Rosenbaum, S., Rohn, J., Humburg, J., et al.: What Makes Strategic Usability Fail? Lessons Learned from the Field. CHI. (1999)
15. Schaffer, E.: Institutionalization of usability: A step-by-step guide, Boston, USA, vol. 1. Addison-Wesley / Pearson Education Inc, London (2004)
16. Seffah, A., Gulliksen, J., Desmarais, M.C.: An introduction to human-centered software engineering: Integrating usability in the development process. In: Seffah, A., Gulliksen, J., Desmarais, M.C. (eds.) Human-Centered Software Engineering – Integrating Usability in the Development Process, vol. 8, pp. 3–14. Springer, Heidelberg (2005)
17. Seffah, A., Metzker, E.: The Obstacles and Myths of Usability and Software Engineering. Avoiding the Usability Pitfalls in Managing the Software Development Life Cycle. Communications of the ACM 47, 71–76 (2004)
18. Venturi, G., Troost, J., Jokela, T.: People, Organizations, and Processes: An Inquiry into the Adoption of User-Centered Design in Industry. International Journal of Human-Computer Interaction 21, 219–238 (2006)

Human Evaluation of Visual and Haptic Interaction

Hiroshi Ando[1,2], Yuichi Sakano[1,2], and Hirosh Ashida[1,3]

[1] National Institute of Information and Communication Technology
[2] ATR Cognitive Information Science Laboratories,
2-2-2 Hikaridai, Seika-cho, Soraku-gun,
Kyoto, 619-0288, Japan
[3] Gradualte School of Letters, Kyoto University,
Yoshida-honmachi, Sakyo,
Kyoto, 606-8501, Japan
h-ando@nict.go.jp, yuichi@nict.go.jp, ashida@bun.kyoto-u.ac.jp

Abstract. This paper describes psychophysical experiments which were conducted to evaluate the influence of haptic information on visual perception of three-dimensional (3D) interpretation of an object's shape in virtual environment. In particular, we investigated whether haptic information provided by a force-feedback device contributes to 3D interpretation of ambiguous visual patterns, which causes spontaneous alternations of bi-stable 3D percepts. The subjects' task was to report by pressing a key which 3D shape was perceived when they were touching the virtual haptic cube along a pre-defined trajectory on the cubic surface. To examine haptic influence on the bi-stable percepts, the duration of each percept was recorded. The results indicate that haptic information of the surface shape can impose a dynamic constraint on visual computation of 3D shapes. The evaluation methods and results could be used for developing human-machine interfaces that provide more natural and realistic sensation of 3D objects in the future.

Keywords: Human perception, Psychophysical experiments, Visual-haptic interaction, Virtual Environment.

1 Introduction

Vision and touch are fundamental sensory modalities for understanding three-dimensional (3D) shapes of objects. In our daily life, we often experience realistic sensation of 3D objects when we see, touch and manipulate them. To develop natural man-machine interfaces, there have been considerable efforts to develop virtual reality systems which produce pseudo-visual and pseudo-haptic sensation of 3D objects, including binocular stereo vision systems with polarized or liquid-crystal shutter glasses, and various types of force-feedback devices.

Nonetheless, it is not well known how humans perceive 3D objects through vision and touch. Does the human sensory system interpret visual and haptic information of 3D objects separately? Or, does the visual sensation enhance the haptic sensation, and vice versa? If there is an interaction between vision and touch, what are the amount and the way of interaction between them? Is there any method for objective measurement of such an interaction?

M.J. Smith, G. Salvendy (Eds.): Human Interface, Part I, HCII 2007, LNCS 4557, pp. 12–20, 2007.

If we can quantitatively evaluate the effect of interactions between touch and vision, the evaluated data could be used to develop better human interface systems. For instance, if visual information can enhance haptic perception of 3D shapes at certain spatial and temporal conditions, we could use the knowledge to develop a system which may reduce the complexity of haptic devices. If we find conditions where we obtain more natural sensation of 3D objects by adding haptic information to visual information, we could develop a more user-friendly interface by making use of such conditions.

The present study aims at proposing psychophysical methods to evaluate visual and haptic interactions. In particular, we investigated whether haptic information contributes to 3D interpretation of ambiguous visual patterns. The use of ambiguous visual patterns is a key idea for evaluating haptic influence on visual perception, since it seems much harder to make an empirical measurement of haptic influence on unambiguous visual patterns. The Necker cube is a well known example of ambiguous visual patterns, which causes spontaneous alternations of bi-stable 3D percepts [1]. To measure the amount of haptic influence on the perception of the Necker cube, a force-feedback device was used to generate a virtual haptic cube, which is consistent with either of the two visual interpretations. The subjects' task was to report by pressing a key which 3D shape was perceived while they were touching the virtual haptic cube along a pre-defined trajectory on the cubic surface. To examine haptic influence on the bi-stable percepts, the duration of each percept was recorded. We found that the duration of the visual percept consistent with the haptic information was much longer than that in the inconsistent case. We confirm that haptic information of the surface shape influences visual depth reversal of the Necker cube. The evaluation methods and results could be used for developing human-machine interfaces that provide more natural and realistic sensation of 3D objects in the future.

2 Psychophysical Experiments

The purpose of the psychophysical experiment was to examine whether haptic information interacts with visual information. If the haptic information does influence the visual information processing, we wished to quantitatively measure an amount of influence that the haptic information makes.

Visual and Haptic Stimuli

For visual stimulus, we used the Necker cube which causes spontaneous alternations of bi-stable 3D percepts, as shown in Fig. 1 (a).

To test haptic influence on the Necker cube perception, a force-feedback device, Phantom ™ [2] was used to generate a virtual haptic cube. The haptic cube was consistent with either of the two visual percepts, as shown in Fig. 1 (b). The main idea is that if the haptic information influences visual perception, we expect that subjects should more frequently perceive one of the two visual percepts, which is consistent with the haptic cube. To generate visual and haptic stimuli, we used Reachin API ™ software [3]. Human subjects monocularly observed the visual pattern on the CRT monitor through a half mirror, as shown in Fig. 2. Thus, the subjects felt as if they touched the visual pattern with the stylus of the force-feedback device.

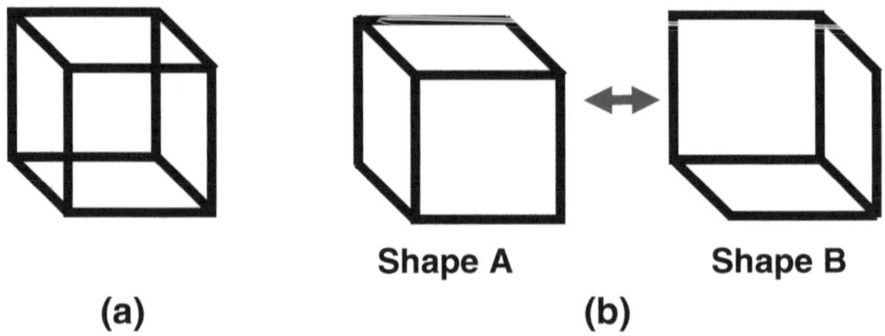

Fig. 1. (a) The Necker cube, (b) Bi-stable visual percepts; Shape A and Shape B

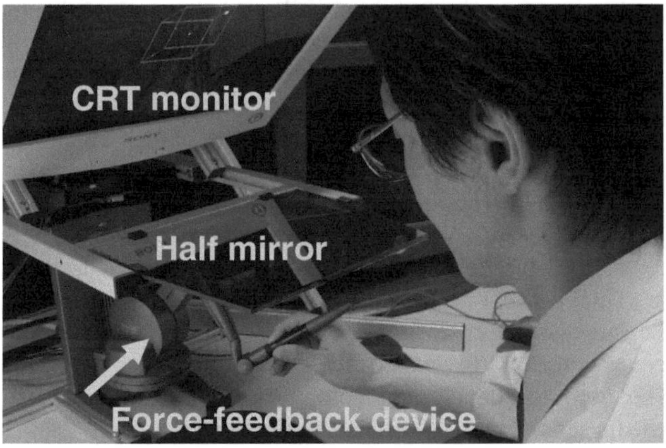

Fig. 2. Experimental apparatus: CRT monitor, a half mirror, and the force-feedback device, Phantom ™

Task

The subjects' task was to report by pressing a key which 3D shape was perceived while they were touching the virtual haptic cube along a pre-defined trajectory on the cubic surface, as shown in Fig. 3. While the subject perceived Shape A, he/she kept pressing a key for Shape A. When the percept changed to Shape B, he/she switched to press a key for Shape B. We allowed the subject to keep pressing another key, while he/she perceived an ambiguous pattern which could not be decided as Shape A or B. The subjects were instructed to use their right hand for touching the cube and to use their left hand for pressing the keys. The observation duration was 60 sec per condition. To examine the amount of haptic influence on the bi-stable visual percepts, the duration of each key press was recorded.

Human Subjects

10 Subjects participated in this experiment. All subjects have normal or corrected visual acuity. The subjects were instructed on the experimental procedures before starting the experiment. Each experiment took about one hour, including a few minutes of practice session.

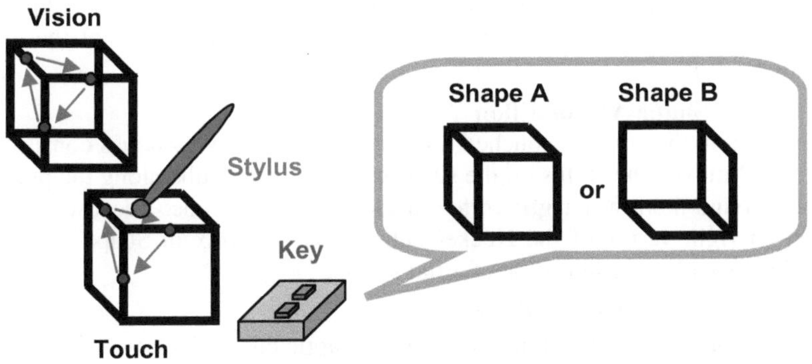

Fig. 3 The task, which required the subject to report whether he/she was perceiving Shape A or B by pressing a key while touching the virtual haptic cube

Conditions

We set following seven experimental conditions:

(1) FN (Fixation/ None) Condition

In this condition, the subjects fixated the eyes on a fixation point which was located in the middle of the Necker cube. No haptic feedback was provided to the subjects. Because this is the normal Necker cube condition, we expected spontaneous visual depth reversals during the observation of the ambiguous pattern.

(2) LA (Left/ Shape A) Condition

In this condition, the subjects touched a virtual haptic cube which was consistent with **Shape A** during the observation of the Necker cube. The subjects grabbed the stylus of the force-feedback device and moved it on the surface of the haptic cube along the pre-defined trajectory around the upper-**left** vertex inside the cube as shown in Fig. 3. A red dot was visually shown on the center of one side of the Necker cube for one second, and sequentially moved to the centers of two other sides. The subjects moved the stylus on the surface as the red dot visually moved. The tip of the stylus was visually shown with the Necker cube when the subjects performed the task.

(3) LB (Left / Shape B) Condition

In this condition, the subjects touched a virtual haptic cube which was consistent with **Shape B** and moved the stylus on the surface of the haptic cube along the pre-defined trajectory around the upper-**left** vertex inside the Necker cube. Because the location of the upper-left vertex of the Necker cube is not a vertex in Shape B, the haptic information provides the sensation of a flat surface. Subjects' task was the same as that of the LA condition.

(4) LN (Left / None) Condition

This is the control condition for the LA condition. When we pay visual attention to a vertex inside the Necker cube, it tends to be perceived in front. Because the subjects pay attention to the area around the upper-**left** vertex of the cube in the LA condition, even if they tend to perceive Shape A, the results could be attributed to the visual attention effect rather than the haptic effect. Therefore, we set the LN condition where the subjects move their eyes while observing the Necker cube as in the LA condition, but do **not** touch any haptic cube. The results of this condition thus show purely the effect of visual attention accompanied by the eye movement but not the effect of haptic information.

(5) RA (Right/ Shape A) Condition
In this condition, the subjects touched a virtual haptic cube which was consistent with **Shape A** and moved the stylus on the surface of the haptic cube along the pre-defined trajectory around the lower-**right** vertex inside the Necker cube. Because the location of the lower-right vertex of the Necker cube is not a vertex in Shape A, the haptic information provides the sensation of a flat surface.

(6) RB (Right / Shape B) Condition
In this condition, the subjects touched a virtual haptic cube which was consistent with **Shape B** and moved the stylus on the surface of the haptic cube along the pre-defined trajectory around the lower-**right** vertex inside the Necker cube. In this case, the haptic information provides the sensation of a protruded vertex.

(7) RN (Right / None) Condition
This is the control condition for the RB condition. In this condition, the subjects moved their eyes around the lower-**right** vertex of the Necker cube as in the RB condition, but did **not** touch any haptic cube. The results of this condition thus show purely the effect of visual attention accompanied by the eye movement but not the effect of haptic information, as in the LN condition.

3 Results

The main results are summarized in Fig.4. This figure shows the mean total duration of each percept (Shape A or B) for different conditions. The total duration of FN (Fixation) condition indicates that without haptic information the total duration of the visual percept A was about the same as that of the percept B. The duration of the percept A was slightly longer than that of the percept B since the viewpoint of Shape A may be more natural than that of Shape B.

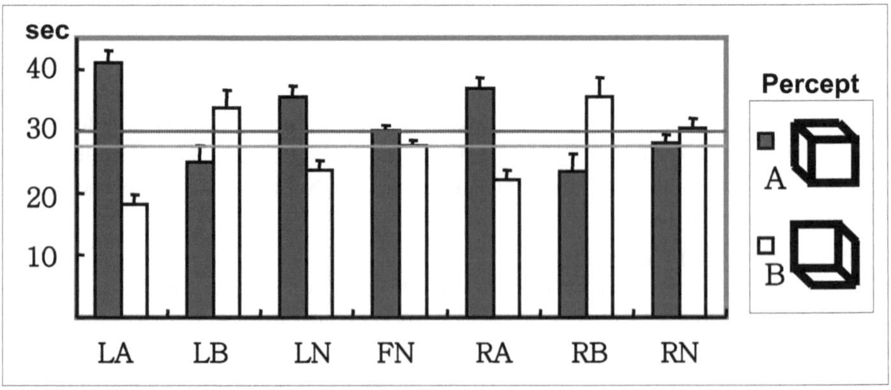

Fig. 4. The mean total duration of the percept A or B for different conditions

However, the total duration of the percept A was different from that of the percept B when the haptic information was added to the visual information. The duration of the visual percept consistent with the haptic information (i.e., the percept A in LA and RA conditions; the percept B in LB and RB conditions) was much longer than that in the inconsistent case (the percept B in LA and RA conditions; the percept A in LB and RB conditions).

The difference in perceptual duration found in LA and RB conditions may be attributed to visual attention accompanied by eye movements, since a corner of the Necker cube tends to be perceived in front when we pay attention to it. Nevertheless, the duration difference described above cannot be explained only by this visual attention effect, since the results of the control conditions (LN and RN) showed that the eye movement alone did not have as much effect. In the case of LN condition, for instance, the difference between the total duration of the percept A and that of the percept B is slightly over 10 seconds, whereas in the case of LA condition, the difference is nearly 25 seconds. Therefore, the difference between LA and LN conditions may indicate pure haptic influence on visual interpretation of the ambiguous pattern.

Furthermore, the data of LB and RA conditions also indicates that the eye movement cannot account for the duration difference, because the visual attention effect suggests that the duration of the percept A would be longer for LB condition and that the duration of the percept B would be longer for RA condition. The data shows the results contrary to the visual attention effect.

Therefore, the experimental data shown in Fig 4 suggests that the haptic information of the surface shape influences visual depth reversal of the Necker cube. The haptic influence, however, was not strong enough to totally suppress the visual percept which is inconsistent with the haptic cube.

Computational Analysis

To better understand the computational mechanisms of the visual and haptic interaction, we analyzed the distribution of the raw duration data. It is known that the duration data obtained by observing ambiguous visual figures can be well-fitted with the gamma distribution function [4]. The probability density function of the gamma function can be written as

$$P(t)=\frac{(t/\beta)^{\alpha}e^{-t/\beta}}{t\Gamma(\alpha)} \tag{1}$$

where (α,β) are the parameters that define the shape of the function, and the gamma function is given by

$$\Gamma(\alpha)=\int_0^t s^{\alpha-1}e^{-s}ds \tag{2}$$

Fig. 5 (a), (b) and (c) show the distributions of the duration data fitted by probability density function of the gamma function (1) in the case of the FN, LB, and LA conditions, respectively. In the FN condition where no haptic information was given to the subjects, the distribution function of the percept A is about the same as

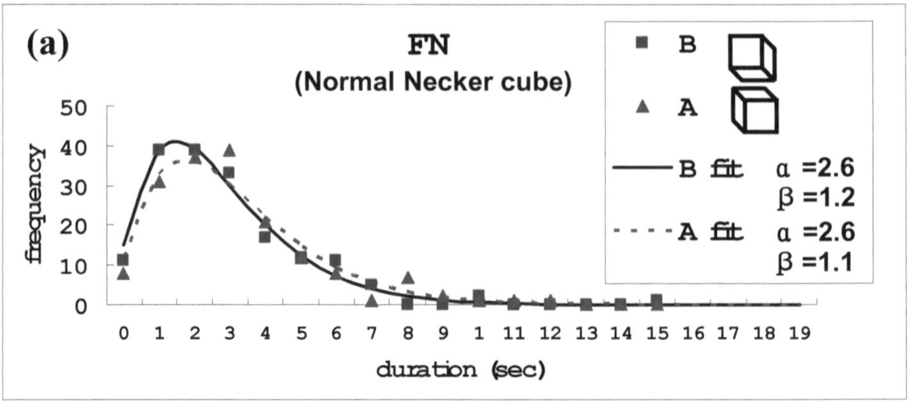

Fig. 5(a). The duration distribution of the FN condition fitted by the gamma probability function

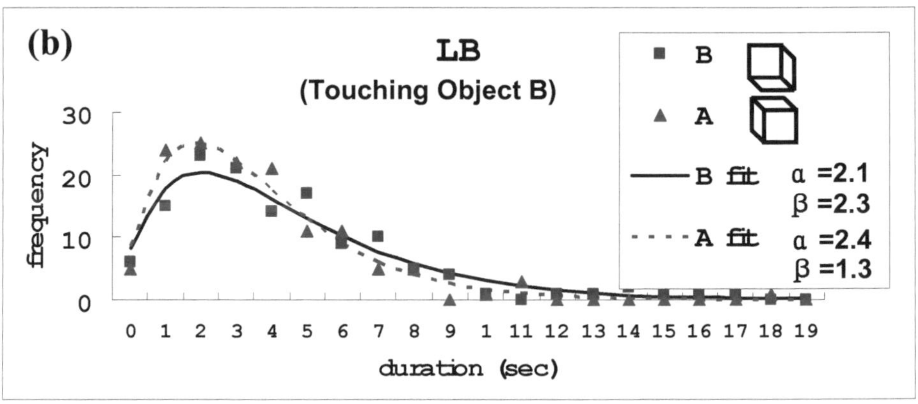

Fig. 5(b). The duration distribution of the LB condition fitted by the gamma probability function

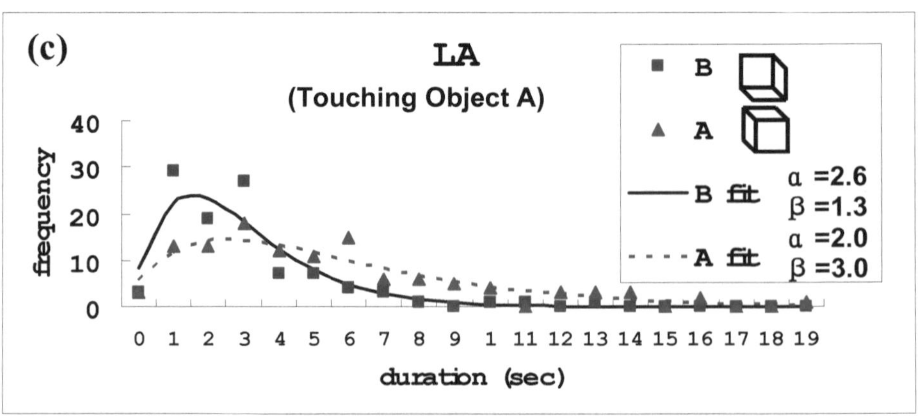

Fig. 5(c). The duration distribution of the LA condition fitted by the gamma probability function

that of the percept B. On the other hand, in the case of LB and LA conditions where the subjects obtained haptic information by touching virtual cubes, the distribution functions shift in the positive direction when their visual percepts were consistent with the 3D shapes they touched. These positive shifts in the duration distributions suggest that the haptic information can influence the visual interpretation of ambiguous patterns.

We further analyzed how different conditions affect the parameters (α,β) estimated by fitting the gamma distribution function (1). Fig .6 (a) and (b) show the estimated α and β for all conditions. As shown in the graphs, the parameter β varied significantly among different conditions, whereas α did not.

If α is an integer, the gamma distribution is derived from a Poisson process, where a random event occurs α times with an interval β. Because touching virtual objects appears to affect only β, haptic information may increase the temporal interval, but not the required number of neural events which cause perceptual alternation. This result suggests that the occurrence of depth alternation is visual per se and cannot be

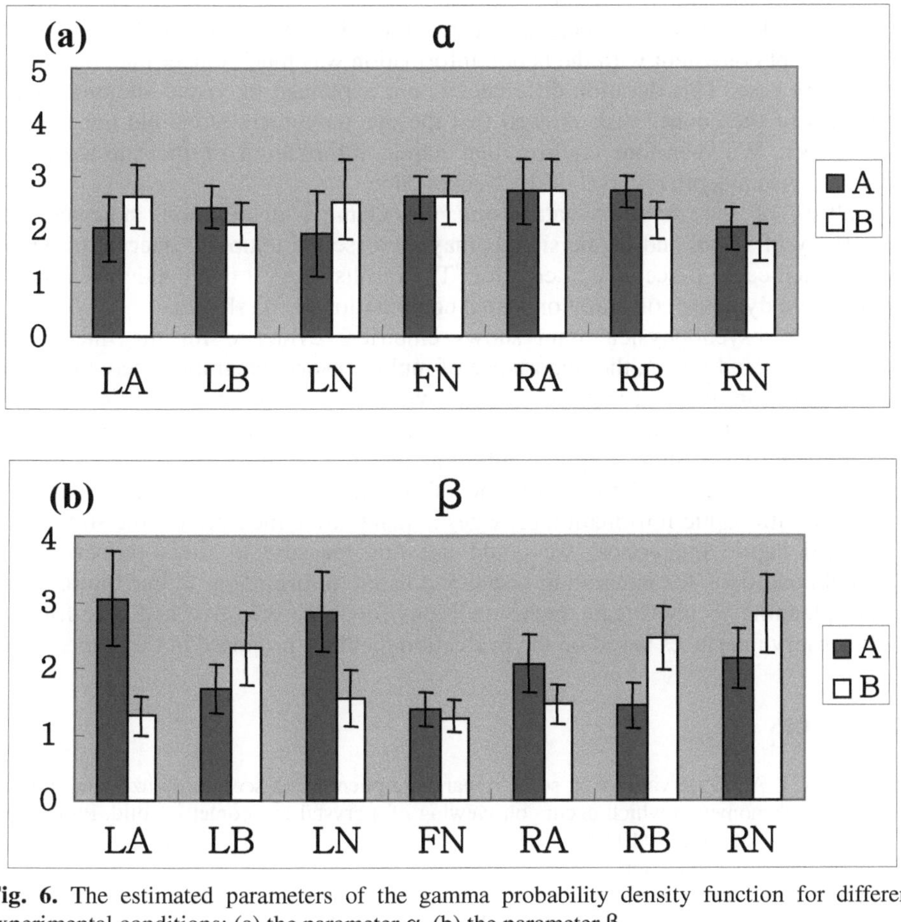

Fig. 6. The estimated parameters of the gamma probability density function for different experimental conditions: (a) the parameter α, (b) the parameter β

eliminated by the haptic information, but that the duration between the alternations could be controlled by the haptic information.

Furthermore, the graphs show that the estimated β of the LA, LB, RA, and RB conditions are all above the estimated β of the FN condition. This indicates that the interval between the perceptual alternations does not decrease even if the visual and haptic signals are inconsistent, whereas the consistency between the haptic and visual signals increases perceptual stability.

4 Discussion and Conclusions

In this paper, we evaluated visual and haptic interactions by human psychophysics using a visual display and a force-feedback device. We investigated whether haptic information contributes to 3D interpretation of an ambiguous visual pattern, the Necker cube. To measure the amount of haptic influence on the perception of the Necker cube, we generated a virtual haptic cube, which is consistent with either of the two visual interpretations. To examine haptic influence on the bi-stable percepts, the duration of each percept was recorded. We found that the duration of the visual percept consistent with the haptic information was much longer than that in the inconsistent case. This duration difference is not explained by visual attention, since the results of the control task showed that the eye movement alone did not have as much effect. We therefore confirm that haptic information of the surface shape influences visual depth reversal of the Necker cube.

Analysis of the duration data using the Gamma distribution indicates that consistency of visual and haptic signals may increase the temporal interval of neural events which cause perceptual alternation. The results suggest that haptic information can impose a dynamic constraint on visual computation of 3D shapes.

Since the psychophysical data shows empirical evidence for the interactions between touch and vision, the importance of multi-sensory integration increases when designing human-machine interfaces that provide more natural and realistic sensation of 3D objects. Furthermore, the evaluation methods proposed in this paper could be used for finding system requirements for developing effective human-machine interfaces. Because the duration of one of the bi-stable visual percepts which is consistent with haptic information provides a quantitative measure for the strength of visual and haptic interaction, we could use this measure to find optimal spatio-temporal conditions for integrating visual and haptic information. In our future work, we are planning to investigate such conditions for more effective and user-friendly multi-sensory interfaces based on the evaluation methods proposed in this paper.

References

1. Necker, L.A.: Observations on some remarkable phenomena seen in Switzerland; and an optical phenomenon which occurs on viewing of a crystal or geometric solid. Philosophy Magazine. 3, 329–337 (1932)
2. http://www.sensable.com/
3. http://www.reachin.se/
4. Borsellino, A., et al.: Reversal Time distribution in the perception of visual ambiguous stimuli. Kybernetik 10(3), 139–144 (1972)

Aporia in the Maps of the Hypermedia Systems

Francisco Cipolla-Ficarra

HCI Lab – F&F Multimedia Communic@tions Corp.,
Alaipo – Asociación Latina de Interacción Persona - Ordenador, and
Ainci – Asociación Internacional de Comunicación Interactiva
Via Pascoli, S. 15, C.P. 7 - 24121 Bergamo, Italy
ficarra@alaipo.com

Abstract. *Aporia* is a Greek word meaning helplessness or difficulty in dealing with, or finding out about, something. This investigation is aimed at determining the aporia existing in the present maps with satellite and traditional pictures. Moreover, we present a quality metric called *antinomy*, to eradicate the aporia in the design of interactive systems with cartographic contents. The present quality metrics can be applied to the digital maps. The importance of traditional iconography is also stressed in the maps and the need to carry out heuristic assessments before using a new iconography. We need to increase the quality in hypermedia systems, communicability and efficiency of the digital maps, quite aside from location, the multimedia technological support used in the presentation of the cartographic information into computers, palms, mobile phones, etc., and potential users of the interactive systems.

Keywords: Antinomy, Aporia, Cartography, Design, Evaluation, HCI, Hypermedia, Maps, Metrics, Quality, Semiotic.

1 Introduction

On many occasions researchers of semiotics have rejected the dichotomy between connotation and denotation because there are difficulties in reaching a clear distinction between the primary and secondary aspects of the meaning. For example, Roland Barthes usually speaks of denotation of the sign but not of connotation [1]. Umberto Eco retains the basic dichotomy, but since his theory of meaning rejects the dimension of reference, his concepts of denotation and connotation cover much of the traditional sphere of connotation. Denotation, according to Eco, is a "cultural unit, ... culturally recognized property of a possible referent" [2]. A connotation is also "a cultural unit" but it is "conveyed by its denotation and not necessarily corresponding to a culturally recognized property of the possible referent" [1]. Elsewhere, Eco defines a connotation as the "set of all cultural units which are institutionally associated in the receiver's mind with the signifier" [2]. The antinomy allows to establish a relation between primary and secondary meaning in a univocal way, for example: smoke and fire. However, it is necessary to establish a difference between natural and artificial signs. It is through semiotic convention that the sign exists, as long as a human group decides to use a thing as the vehicle of any other thing [2]. In

M.J. Smith, G. Salvendy (Eds.): Human Interface, Part I, HCII 2007, LNCS 4557, pp. 21–30, 2007.

cartography the use of the sky blue colour to indicate water and green to indicate areas of vegetation is a centuries-old convention, and this can easily be checked in any atlas printed on paper since the 15th century. The use of the current artificial signs in the hypermedia systems does not respect any kind of convention. In cartography the use of the sky blue colour to indicate water and green to indicate areas of vegetation is a centuries-old convention, and this can easily be checked in any atlas printed on paper since the 15th century. The use of the current artificial signs in the hypermedia systems does not respect any kind of convention. A classic example are the traffic signs for the viability of vehicles on streets, roads, highways, etc. which in many cases do not maintain that unique relationship in the same country. Consequently, to extrapolate these sign systems at world scale without previously carrying out an analysis, can entail negative consequences in the quality of the interactive cartographic systems.

The relationship between the icon and its meaning, strengthened by the text, does not always allow the elimination of the process of unlimited semiosis [1], which damages antinomy. In the interface design of hypermedia systems, some designers plump for guidelines in relation to the hardware and software that they use, i.e. Macintosh [3]. Others, in contrast, start to make icons taking as a reference the cultural environment where they are involved [4]. However, in both cases antinomy can be absent, because the problem of emulation and simulation of reality on the screens of the interactive systems, persists.

2 Maps: Evolution of the Emulation and Simulation of Reality

Currently cartography has reached a peak thanks to the new technologies, peak which is to be found in the intersection among of physical, social and digital domains. For example, in the human-computer interaction community explores the use of pervasive and/or ubiquitous computing technologies in inhabited environments [5], simulations for urban planning [6], helpful environment [7], designing effective digital systems inspired by, or included in, the physical world [8], etc. One of the classical problems in the evolution of maps has been the transition from spherical representation to plane representation. Once the main scales in cartography were determined, the next step was to analyze the different geometrical properties of the transformations. In the history of the evolution of the representation of maps we find three basic ways to depict planet Earth; cylindrical, conic and planar. Some examples of these models are: Lambert, 1772 (conic); Mercartor, 1569 (cylindrical), and Stereographic, Hipparchus, 160-125 B.C. (planar) [9]. Obviously, the Mercator map is an excellent example of map design for navigation purposes. This map replaced the previous model, known as gnomonic. Besides, Mercator introduced in his map, parallels and later on, meridians, which with the passing of the centuries helped to bring about a rectangular depiction of the globe. Several models and shapes were created throughout time. Some examples are: Sinusoidal (Cossin, 1570); Gall (Ortographic, 1855), and Robinson is the author of the ortophanic model (1963) –designed to look correct [9]. Only in the last half of the century has the map become more than a repository for information. Computerized cartography and modern photographic techniques have increased the density of information some 5,000-fold in the best of today's data maps compared to Halley's pioneering effort [10].

As we can see, a map is a schematic representation of reality that has evolved with the passing of time. It is not a natural language. It is a representation that has to be learned. The observer has to collect three abilities: the sight from above, the sense of direction and memory associated to 2D plan. The space-visual sense of direction requires a learning process in which the interactive systems can help, but we have to remember that feedback time must be immediate, i.e., a fraction of a second. Nevertheless, nowadays a certain slowness predominates in the moment of interaction with the image (Google Maps, TuttoCittà, ViaMichellin, etc.), even if the access and localization of the information required in the hyperbase management system is quick [11]. With the regard to the design of 3D, it has a widespread diffusion in the lines of that presented to youngsters and inexperienced users at the Simcity [12]. The success of interactive games lies in their diachronic evolution, which has developed from a cartography for the cities in 2D, to 3D as years have gone by. Besides, it enables the user to regulate the angle of vision. We introduce the notion of "bird vision" in quality measurement into interactive cartography. This belongs to the category of presentation and it refers to the angle of vision of the observer that is regarded as ideal for the correct fruition of the maps, 2D and 3D, where there can not be illumination effects such as shadows, brightness, reflections, etc. –see Appendix #1.

3 Communicability and Perception

The main problem lies in the presentation of the cartographic information, which is increased by the size of the screens. Besides, in design it is not possible to resort to transparencies as a solution in the use of icons, river names, mountains, etc. The transparencies can double the visual information [13]. The use of transparent menus can occupy the whole screen or part of it. Studies carried out by Harrison et al. [14] demonstrate the efficacy of this technique in interfaces, since the transparencies focus the user's attention on the interface by highlighting the slide than its background. For a correct use of transparencies it is advisable to avoid backgrounds with a very transparent cartography, since they can prompt visual confusion if it is the case that either transparent menus or navigation keys are used.

Another way to duplicate the information in a screen is by using diffuse images. A set of graphic elements maintains its clearest colours and the rest, which have been diffused, acquire a greyer shade. The difference between a transparent image and a diffused one in an interface is that the first can be watched from the background of the screen, whereas the second cannot. In digital cartography there is more information in the interface when transparencies are used than when diffuse is used. Furthermore, through the use of the images one can curb the antinomy in the visual perception of maps. Any map is a display for a reader, for a user. From it, one must gather information and ideas. Often one must use a map as the basis for making a decision. This theme is found throughout the history of human activity, and the maps which exist at any particular time and in any particular culture mirror the environmental concerns, activities and behaviours of those whose created and used them. The communicability of the maps in digital supports must take into account two usability principles: easiness to remember and easiness to use [15].

As a rule, the users of hypermedia systems tend to use that kind of cartography where straight horizontal lines prevail, whether they are real or unreal. In the real case, perception is strengthened when the parallels are present. In the second case, the cartographer manages to bend the geographical territory in such a way that it matches the imagined horizontal line. This last action tends to cut down the effect of memory distortions. Systematic errors in perception and memory present a challenge to theories of perception and memory. Users perceived and remembered curves in mountains and rivers in maps as more symmetric than they actually were. Symmetry, useful for detecting and recognizing figures, lead to distortions in maps and graphs figures alike. Sometimes, users were asked to sketch the curves of the graphs or the rivers of the maps, and other times, they were asked questions about the content of the maps. This was done to induce a natural comprehension attitude toward the figures, and to prevent subjects from simply memorizing line.

We then asked judges who knew nothing about the hypotheses to rate whether the drawn curves and rivers were more or less symmetric than the original ones. The remembered curves, whether in maps or graphs, were judged more symmetric than the originals. These errors in the directions of symmetry, however, apparently occur in perception, not in memory [16]. When attention was directed to the symmetry of the curves, remembered curves were drawn more symmetric than when attention was drawn to the asymmetry of the curve. Modern cartograms in frequent use include automobile strip maps, the distorted map used by many rapid transit, i.e., underground or subway. These examples are often much easier to understand or use than their geographically correct counterparts.

The visual variables described by Jacques Bertin are very important [17], for example, size and value are efficient for portraying quantitative data, while form and colour are more suited for qualitative data. Textures and directions can be used for both types of data. In the absence of any conceptual or meaningful factors, there are often perceptual factors that provide a frame of reference. For maps, there is an additional conceptual factor that is typically perfectly linked with the perceptually salient axes, namely the cardinal directions, north-south and east-west. Thus far, the evidence for alignment has come either from maps and environments, where both perceptual and conceptual factors suggest the horizontal and vertical as a reference frame, or from visual blobs, where perceptual factors suggest the horizontal and vertical [16].

Another factor that must be taken into account in order to increase the user's perception and attention in the cartography, and to decrease the aporia is the distribution of the maps and the navigation zones in the interface. According to Leonardo Da Vinci, when he defines “La Divina Proporzione”, that is, the proport-ional division of space in painting, since he sets down the areas that draw attention in a screen, there is a higher level of attention in the lower part of a rectangle than in its upper part. Besides, if two lines are drawn in the shape of a cross, you find that the right upper part has a higher perception and attention level than the lower one among Latin peoples [18].

4 Antinomy and Aporia in Cartography

In traditional paper cartography, the vector lines of the maps of a underground or/and subway are quickly assimilated by the population of a city. The colour, the letters and the numbers help the communicative process, reinforcing antinomy. Scales divides cartography in two big areas. The first one is the scale of 1:50.000 that uses the method called photogrametry, based on air frames and/or satellite images. In the case in which this scale becomes obsolete, there are other methods that are not considered in the present work because the aporia for the user of hypermedia systems is in the first group. In the classic structures of topography there's no problem, as the user has accustomed to bi-dimensional reading of cartographic information for many centuries.

The problem of aporia begins with digital urban cartography whose scale is of 1:1000, and in the symbology of thematic cartography above all. In urban cartography, the work in the real context is still necessary, i.e., go and measure the land in order to make a real analysis of the area, because from the sky it is difficult to define whether a path is practicable or not, whether a river is permanent or not, for example. In urban cartography the user can get his bearings quickly through the incorporation of cultural heritage, which must be included at the first level of visualization of the map and later on the kind of monument, building, etc. can be profiled. An excellent example is SimCity [12]. The tourist who visits a city for the first time and who, thanks to the 3D emulating maps, can get his bearings easily in the city. That is, that the emulation of the constructions in 3D favours antinomy.

Fig. 1. In the poster, the 3D emulations of the monuments help to locate in a rural area where the main sites of interest are to be found

Previous and next posters –figures 1 and 2, of tourist maps from the Italian region of Emilia-Romagna. The first mentions a gastronomic zone to taste a typical product named "culatello" (ham). In the said map it can be seen how aporia can appear in thematic symbols, especially if you do not know the language of the country, since icons are coupled with texts in Italian. Furthermore, there is no such thing as a universal symbol of cultural heritage in thematic cartography. In these maps the aporia can also appear in different categories: population (evolution, structure, density, distribution, etc.); economy (usage of the ground, industrial and commercial activity, etc.); physical geography (climate, hydrographic, geology, etc.); society (education, work, justice, etc); equipment and services (sports, health facilities, tourism, communication and transport, etc). In the international design of interfaces there are different analyses or studies [19], [20], [21], but a database has not been established, with free access and which takes into consideration the different cultural factors. In the second poster –figure 2, there are two maps of some castles in the Emilia-Romagna region.

Fig. 2. In the left zone there is an aerial version of the zone (satellite photography) which does not indicate, for example, latitude or longitude. However, in the map both the parallel and meridian lines are marked.

If we observe the satellite image of the zone with the current contents on-line and the navigation elements available to the user, it is very hard for him to find orientation because the whole environment is a rural plain. As a result, the green colour prevails in the satellite image without additional indications. In the off-line interactive systems, *In volo sull'Italia in 3D* [22], the use of transparent subtitles makes it possible to know in an efficient way the exact spot of the location seen from the satellite. Although the satellite images did not have the same quality as the current ones coming from NASA, as in the case of Google Maps, the interactive design was very good.

5 Towards a Quality Design for Digital Cartography Online

During the design of digital maps, the aporia can appear in the various stages:

1. Toponymy, i.e., the names of mountains, rivers, etc., that have to be written on the map. For example, in the Hispanic world the word Monaco refers to the Princedom of Monaco, whereas in Italian it means Monaco of Bavaria, that is to say, Munich.

2. Geographical information: into the information derived from the analysis of the image in a map, we have to incorporate the administrative or political division. For example, the provincial limits, the traffic markings, etc.
3. The cartographic style, that relates to the elements constituting the map: tourist information –if it is a thematic map, colour and typography used for titles, etc.

The present maps with pictures from the satellite allow different ways of above vision: classic or 2D, 3D simulation and mixed –or a combination of the two previous ones. In most on-line maps, in the categories of 3D simulation or mixed, the user has no opportunity to adjust the vision angle. The cartographic image is mainly vectorial and allows the same actions as a pdf format document: enlarge and reduce the area, emphasize phrases with a labeller, insert comments under the picture, etc. The indications in the maps tend to globalize in the tourist field. Nevertheless, if the traffic markings are bad in a country, this high degree of aporia is also in its thematic maps. The user orientates himself very well, where the 2D format streets can be accompanied with a 3D simulation of the main buildings.

In the Google Maps –beta online version, the interactive system easily finds and shows the city of Vancouver, but when you want to have access to the capital of British Columbia, that is, the city of Victoria, there is no such city in the database. Automatically other cities of the world bearing the same name in other countries appear on the left margin of the screen –see figure 3. However, the city of Victoria in Canada exists in the cartographic database of Google Maps.

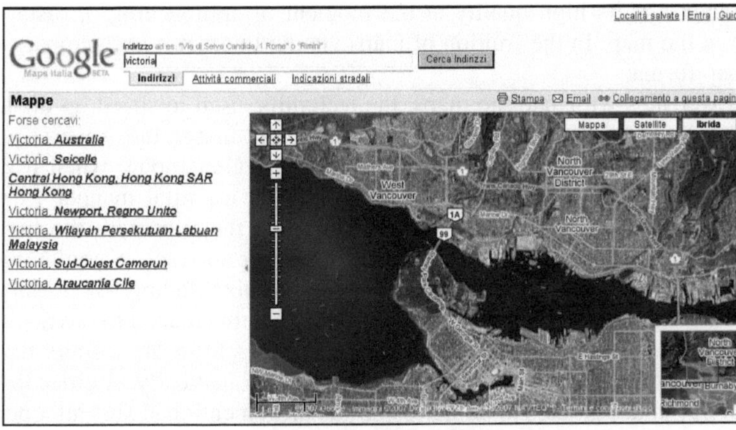

Fig. 3. The aporia appears in the accessibility of the information

An identical test has been carried out in the National Geographic on-line system and after visualizing the satellite map of Vancouver and searching for Victoria, the aforementioned city does show up quickly on the screen but in Hong Kong (China).

The use of the globe in the comic form is very positive from the standpoint of communicability, but its needs some changes to reduce aporia in the presentation of the information. For example, that globe with the shape of a comic should not generate shadows as in the Google Maps, and a semi-transparent version would be

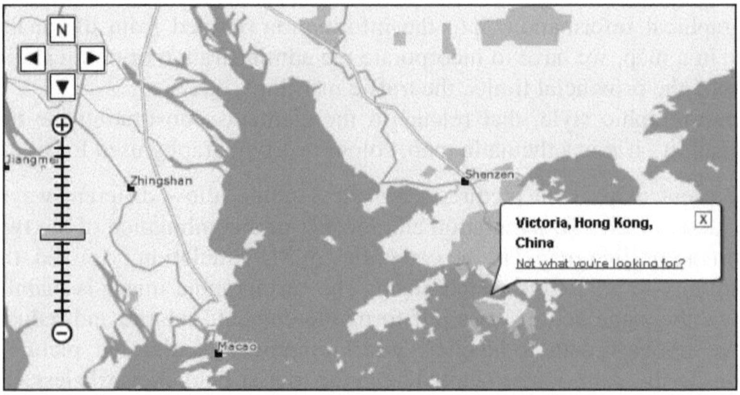

Fig. 4. A new search from the city of Vancouver to the city of Victoria, the globe in the comic form shape points at Hong Kong (http://plasma.nationalgeographic.com)

more convenient. In some cases, when the user chooses the mixed satellite option, that is, with the names of the main streets, roads, highways, etc., such a text is darkened by the shadow. The quality of the images in the maps slows down the presentation of the information in relation to the levels of approximation or distance from the zone of the map that is being watched at the moment, for instance, world, continent, country, city and street. In digital off-line cartography, the vectorial format prevails because of its high quality at the moment of approaching or distancing the perception of the map. In the fruition of maps via the Internet it is customary to resort to the bitmap format.

In some websites with satellite maps, the technique used to speed up the access to the map requested by the user is to divide the whole image into several parts. For example, it is the technique adopted by the National Geographic when it comes to presenting the map: it starts to generate the image in a partial manner from left to right. There is a loss of quality in the interaction of the user with the hypermedia system. However, from the standpoint of quality in the interactive communication it would be more convenient to use a transition effect of pixel fading which starts from a solid colour to the overall image of the map. The problems related to navigation in the system appear in the moment in which the user decides to locate a street that lies on the borders of the map. The system shifts the map automatically in order to locate it and point it out in the globe. Nevertheless, in the aforementioned shifting operation of the map, the original position gets lost and also the reference points, with which the user loses his bearings even more. The icons that mark the different points in the map do not allow the inference of the contents thereof. The user must select it and unfold it in order to see whether the contents are adjusted to what he is seeking. Moreover, if the listing happens to be extensive, the system inserts the pagination of the answer. As a consequence, this is hardly an ideal solution when the user does not know beforehand what he is trying to locate.

Perhaps it might work as a finder of addresses, phones, fax, etc. but in that case there are other databases on-line that do not insert a map for such activities. That is, to plan previously the information that must be shown, for example, to eliminate the option of watching non-existent highways in wide desert or rainforest regions, such as the Sahara or the Amazon. Leaving that space for another kind of theme, options such as can be the areas inhabited in those great regions (Sahara, i.e.). In the cartography of the cities, Anglo-Saxon users prefer maps with 3D perspectives to 2D maps. Latin European users prefer 2D maps, and Latin-American users remain indifferent in the face of a 2D or 3D map. The latter are also starting to value that cartography in the sightseeing areas they visit for the first time.

6 Conclusions

There is a fast diffusion of cartography in the new supports of multimedia information, however, a low quality in its fruition can be observed. Although the quality of cartographic pictures has increased in the last few years through the evolution of software, hardware procurement and image editing software of satellite data, etc. the design of interactive systems for this kind of material still remains superior in the hypermedia off-line systems compared with on-line ones. The main problem lies in the presentation of the information in the maps and the navigation zone in the screens of the different digital supports. On the one hand, there is a high amount of information in a tight space, on the other hand, there is no artificial or natural convention related to the iconography used for navigation. Accessibility in some cases is not only mistaken but also redundant. A way to solve this problem is to apply the criteria of richness of information according to the kind of user and the geographical context. The indications of the cultural heritage and the province, district limits etc., help orientation and location of what is being looked for in the satellite maps. However, the number of indications can either increase or decrease in relation to the zoom level the system offers to the user. In general terms, it is an option with several degrees of approximation and distancing. As a result, the user may get out of the map without being able to locate the information he is seeking, which exists in the database but can be found in a level that he/she has not visited. Another problem that fosters aporia in the interactive maps is that the pieces of information show in and off in relation to the zoom level, which is to some extent confusing to the inexpert user.

Referentes

1. Colapietro, V.: Semiotics. Paragon House, New York (1993)
2. Eco, U.: A Theory of Semiotics. Indiana University Press, Bloomington (1979)
3. Apple: Macintosh Human Interface Guidelines. Addison Wesley, Cupertino (1992)
4. Ben, C.: Notes from China: Handset Design. Interactions 13(4), 38–39 (2006)
5. Kjeldskov, J., Paay, J.: Public Pervasive Computing: Making the Invisible Visible. IEEE Computer 39(9), 60–65 (2006)
6. Davis, J., et al.: Simulations for Urban Planning: Designing for Human Values. IEEE Computer 39(9), 66–72 (2006)
7. Tate, A.: The Helpful Environment: Geographically Dispersed Intelligent Agents That Collaborate. IEEE Intelligent Systems 21(3), 57–61 (2006)
8. Cohen, P., McGee, D.: Tangible Multimodal Interfaces for Safety-Critical Applications. Communications of the ACM 47(1), 41–46 (2004)

9. Robinson, et al.: Elements of Cartography. John Wiley and Sons, New York (1995)
10. Tufte, E.: The Visual Display of Quantitative Information. Graphics Press, Cheshire (2001)
11. Schnase, J., et al.: Semantic Database Modeling: Survey, Applications, and Research Issues. ACM Transactions on Information Systems 11(1), 27–50 (1993)
12. SimCity 4 CD-ROM: Electronics Arts, Redwood (2003)
13. Harrison, B., Vicente, K.: An Experimental Evaluation of Transparent Menu Usage. In: CHI'96, Vancouver, pp. 391–398. ACM Press, New York (1996)
14. Harrison, B., et al.: An Experimental Evaluation of Transparent User Interface Tools and Information Content. In: ACM Symposium on User Interface Software and Technology (UIST '95), Pittsburgh, pp. 99–106. ACM Press, New York (1995)
15. Nielsen, J.: The Usability Engineering Life Cycle. IEEE Computer 25(3), 12–22 (1992)
16. Kaiser, M., Ellis, S.: Pictorial Communication in Virtual and Real Environments. Taylor & Francis, Washington (1993)
17. Bertin, J.: Sémiologie graphique: Les diagrammes, les réseaux, les cartes. EHESS, Paris (2005)
18. Cipolla-Ficarra, F.: Evaluation and Communication Techniques. In: Multimedia Product Design for on the Net University Education. Multimedia on the Net, pp. 151–165. Springer-Verlag, Heidelberg (1996)
19. Kellog, W., Thomas, J.: Cross-Cultural Perspective on Human-Computer Interaction. SIGCHI 25(2), 40–45 (1993)
20. Nielsen, J., del Galdo, E.: International User Interfaces. Wiley, New York (1996)
21. Shneiderman, B.: Designing the User Interface. Addison-Wesley, New York (1997)
22. In volo sull'Italia in 3D CD-ROM: Infmedia, Rome (2003)

Appendix 1: Quality Metric Antinomy

Table 1. Element, components/attributes and design categories: Content (CO), Dynamic (DY), Panchronic (PA), Presentation (PR) and Structure (ST)

Element	Components and/or attributes	Design categories
1) Bird Vision	Control of fruition and orientation	DY, PA, PR, ST
2) Geographical Information	Sense of direction and transparence	CO, PR
3) Perception and Memory	Maps structure, topology on the screen, inference and 3D constructions	PR
4) Toponymy	Relation between primary and secondary meaning: univocal, biunivocal, and semiosis unlimited. Languages and international iconography	CO, PR
5) Cartographic Style	Realism, emulation, simulation, and richness	CO, DY, PA, PR, ST

Model Based HMI Specification in an Automotive Context

Thomas Fleischmann

Elektrobit, Frauenweiherstr. 14,
91058 Erlangen, Germany
Phone: +49 (0) 9131 7701 288
Thomas.Fleischmann@elektrobit.com

Abstract. An overview of how a model based specification approach can be used in the domain of automotive human machine interface (HMI) development is presented. The common paper based specification approach is compared to a model based, tool supported process. Requirements from different stakeholders for such an approach are outlined. Intended audiences are all stakeholders involved in the creation of graphical user interfaces ranging from design, usability engineering, and prototyping to specification and final product realization.

Keywords: Model based, HMI Specification, Code generation, Domain Specific Language.

1 Introduction

The functionality contained within modern cars has increased drastically, and the car manufacturers (OEMs) face challenges in presenting the assistive systems and infotainment solutions to the driver in a usable way. Park- or lane assist and adaptive cruise control are just some of the examples. The following figure shows some more:

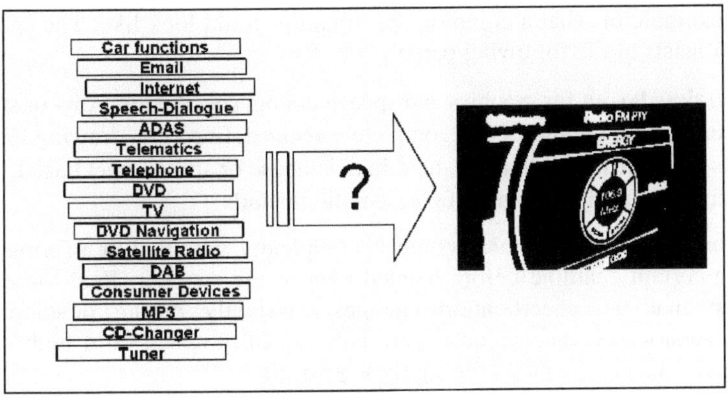

Fig. 1. Handling the complexity

M.J. Smith, G. Salvendy (Eds.): Human Interface, Part I, HCII 2007, LNCS 4557, pp. 31–39, 2007.

The number of features required forced the OEMs to shift away from the simple tuner and toward an integrated solution. The HMI became a brand identifier. Car manufacturers started to specify the look & feel of the HMI in order to get the supplier to build a custom system that fits into their brand image.

The common paper-based specification approach and its difficulties in the working process with the supplier will be described first to define the problem scope. Additional new social trends and as well technical features in the next automotive HMIs demand a change in the way these specifications are created.

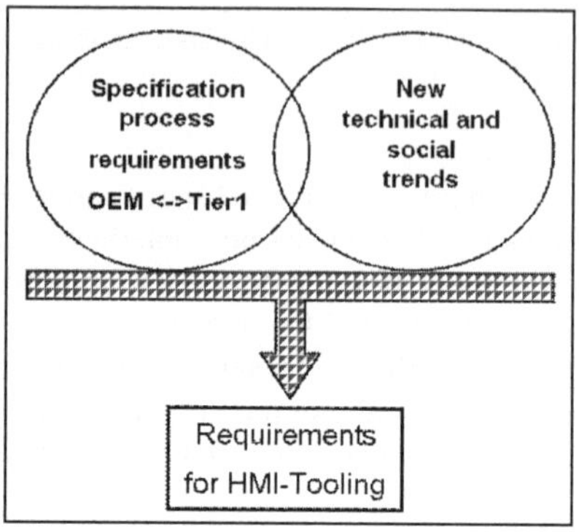

Fig. 2. Requirement input sources towards a solution

The solution for the problem lies in a domain specific and model based specification tool. Such a tool has to provide on one hand the technical features to allow specification of the latest multimodal user interface and on the other, to support the working process between OEM and Tier 1 supplier.

2 Paper Based Specification

It is still a common and widely used approach to specify an HMI on paper. Fig. 3 shows an example of what a common specification could look like. The specification typically consists of the following parts:

- Menu flow layout for graphics and speech dialog, specified in flow- or state charts
- Screen design, templates and complete screens defined in a drawing tool
- Texts for internationalization, held in a database or spreadsheet based
- Description of the software implementation details

This works well if the whole system is completely specified in a formal way and frozen at a certain deadline before handed over to the supplier. Real life experience has shown that the specification changes constantly during development and document versions become inconsistent. This typically results in a high number of very expensive change requests during these projects.

Another drawback is the missing simulation – final results can't be verified in a user clinic or by the top management before the system is a nearly finished.

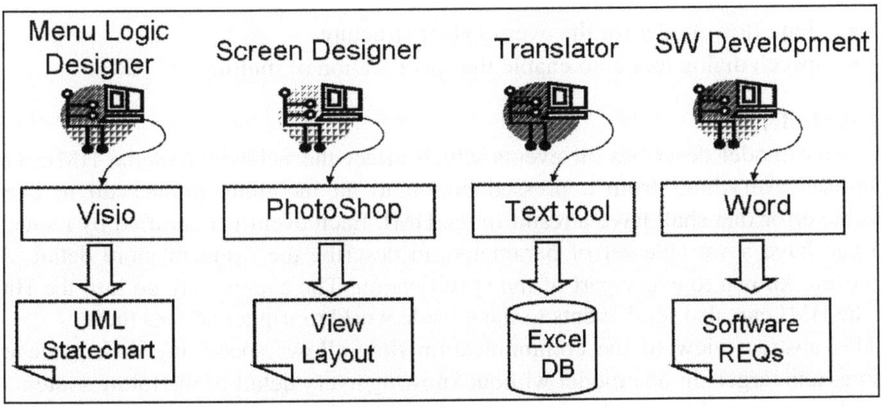

Fig. 3. Paper based specification

3 Solution - Model Based Specification

In order to create a consistent HMI system it is necessary to combine all involved specification elements into one model. Taking the speech dialog as example - the end user will notice that the graphic HMI part and the speech dialog are not synchronized or just loosely coupled if they are designed in two environments.

Fig. 4 shows the involved parties centered on the HMI model.

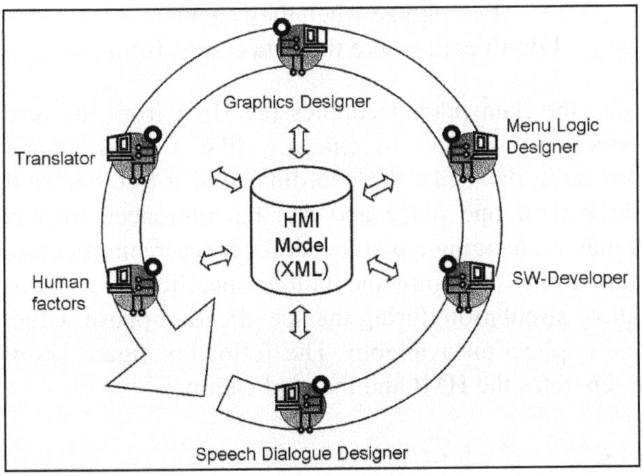

Fig. 4. Model based specification

The core parts of the model are:

- The event model to react to external triggers or events
- The data model to display and collect values from the applications
- Layout model for the screen definitions

- Menu flow model for the overall HMI structure
- Speech dialog model to enable the specification of multimodal HMIs

Event Model

The event model describes all events which affect the behaviour of the HMI. This comprises everything from a pressed button to an incoming phone call to every possible error that shall have a result in the HMI. Each event is identified by its name and can have a variable set of parameters to describe the event in more detail. All events are known to every part of the specification. The events may go into the HMI but the HMI can also send events to the outside world to trigger actions there.

This abstract view of the communication flow allows specifying different event sources and targets in one model without knowing every detail of the future system. A formal event model is also the first precondition for developing a simulation based on the specification.

Data Model

Specifying an HMI involves defining all data that the HMI requires from the system. If a value is displayed on the screen there has to be a provider for this data – the details of which software component this actually is does not matter while specifying the HMI. All data are kept inside the data pool component which is visible in the whole HMI and notifies interested components of data updates. To put it another way, this data abstraction follows the simple rule:

"I know the values exist;
I recognize when they change;
I don't care where the data comes from."

Following this rule, the data pool decouples the HMI from the rest of the system. Dynamic data which are subject to changes, like a tuner station list, must be distinguished from static data, like the coordinates of a title bar or the color model. All values are defined at one place and can be referenced throughout the whole system. This still makes changing a picture set for 500 screens an easier task.

The abstraction from the implementation specific data source is also one precondition to allow simulation during the specification phase without having a real tuner or telephone application available. The following figure shows how the data pool architecture separates the HMI and the application.

Layout Model

The layout model contains a formal description of each screen in the HMI. Each screen consists of multiple graphical control units called Widgets. The appearance of each Widget can be defined by a Widget-specific set of properties. Widgets should follow some basic rules. The first is that they should comply with the model-view-controller paradigm. This is nothing new to GUI developers. It simply means that graphical representation, behaviour and data must be kept separate to ensure flexibility. (e.g. A button can look completely different just by replacing its rendering component or work in a touch screen simply by replacing its controller part).

Secondly, the Widgets have to support inheritance to allow easy extension of the functionality of existing widgets.

The Widgets link different parts of the HMI model together. The most obvious link is a reference from for example, a label widget to a MP3-Text defined in the data pool. This slot would be provided by the MP3 application later in the target software. The Widget controller parts have to be able to act as event sinks or event sources into the system during a user's interaction with the Widget.

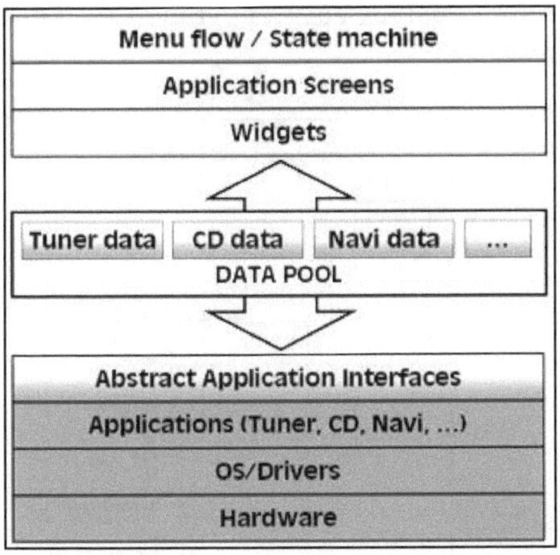

Fig. 5. HMI and applications separated by data pool

Menu Flow Model

The user's navigation through the whole HMI menu is specified in a menu flow model. A detailed description defines how each screen of the layout model can lead to the next screen. Illustrated UML state charts are used to define the menu logic. Each state without further child states can be linked to a screen from the layout model. A screen miniature is displayed in the diagram to allow a non expert in UML to easily follow the specified menu flow. The transitions between the states are triggered by events from the event model. Features provided by the UML state machine model like the history mechanism, conditional transitions or actions trigger when entering or leaving a state are supported as well.

Speech Dialog Model

The extension of the model with text-to-speech (TTS) prompts definitions and grammar rules will enable a full multimodal HMI specification and, if brought into a tool environment, an on-the fly simulation.

Currently many speech dialogs are created in independent tools decoupled from the graphical HMI model. Combining this in one model is a great step towards an integrated HMI specification. The resulting speech dialog model can be exported for different recognizer engines running in the device.

4 XML Export of a Domain Specific Language

Specifying an HMI model using these concepts will result in a domain specific language (DSL) which is very complex. Therefore a set of intelligent editors and tools is required to manipulate such a model. In order to allow others to setup automated processes using such a model, an export into a well documented data format is necessary. XML schemas fit this need very well. The figure below shows a part of the schema definition for a state definition from the menu flow model. Finally, the tool allows an export of the specified model into an XML file which is exactly compliant to the schema and therefore enables easy integration of the model data into a third party development process.

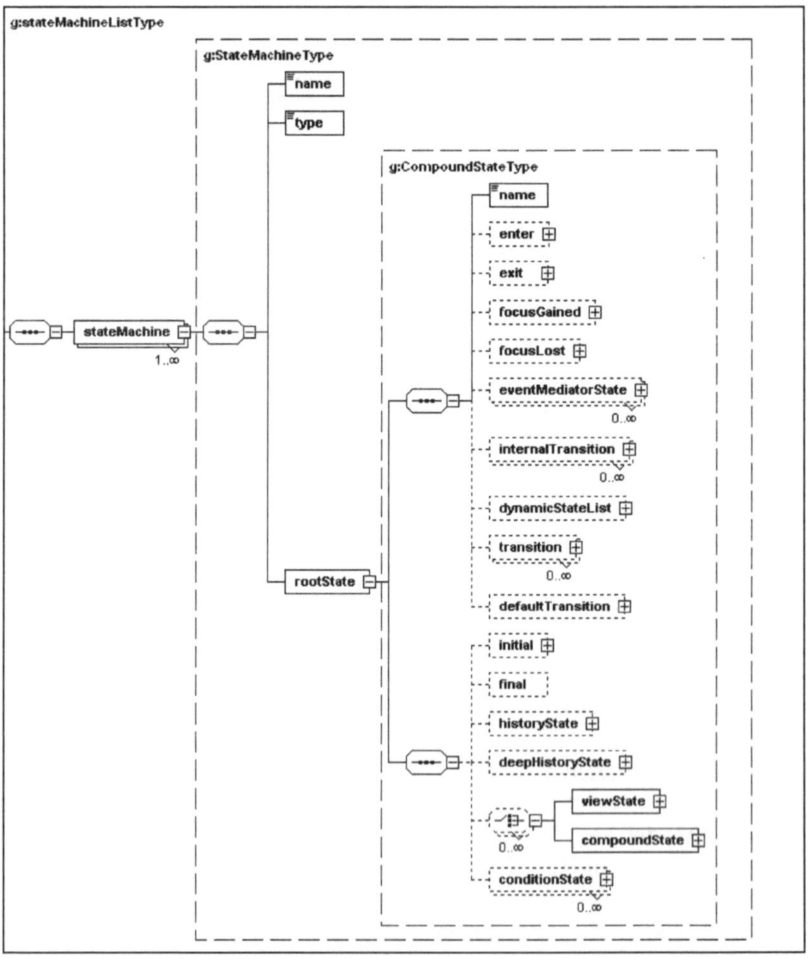

Fig. 6. XML Schema for a state definition

5 Tool Requirements

Some requirements for a tool, like simulation and automatic consistency checks, can already be derived from the above mentioned points. Some others are briefly outlined in the following section:

Rapid Prototyping
To try and test new HMI concepts, a rapid prototyping environment has to offer interfaces to integrate with new hardware and allow for a fast design of new HMIs. The reuse of elements from this prototyping into the actual specification is required to prevent "reinventing the wheel" during the final implementation phase.

Formal Specification
A model based tool must offer a way to create a formal specification – therefore it requires a formal element for every type of problem domain object. Multi-user access to this specification is a must in bigger projects.

Version Management
The specification will be given to the supplier at a certain point in time. If multiple parties start to change the model, an intelligent HMI tool has to provide version management, change logs and merge capabilities to support this as
Fig. **7** illustrates. The simulation model *V1.0* is handed over to a supplier who takes care of the realization. The supplier will extend the model into an implementation model *V1.2* which is used for code generation. The extensions made to create *V1.2* need to be merged with the next simulation model *V2.0*. Even though XML is used to store the model information, a simple text compare would not be of any help. In order to achieve a comparison on a logical level and to allow a conflict free merge a tool support for model comparison and merging is required.

Automatic Code Generation
If the whole system is specified in a formal model and this model can be exported in a machine readable format; the specification can be directly used to automatically generate the code for the embedded target system out of the specification. For speech dialogs the tool would directly provide the grammar, commands and prompts. For the graphic HMI it would provide all screen definitions and an executable state machine.

Test Support
Automated HMI testing as well as export functions for manual HMI test instructions and validation of specified behavior could be easily extracted from a machine readable HMI model and used by both OEM and Tier 1 developers. This covers the whole lifecycle from early usability testing to playback of regression tests in the maintenance phase of a project.

Independence
The OEMs have the need to stay independent from the supplier to a certain extent. Therefore Tier 1 suppliers can not create an HMI tool and offer it to an OEM. Only an

independent vendor can provide such a tool which provides open interfaces and support for standards to both the OEMs and Tier 1 developers.

Future Requirements

Future systems will include full graphics based cluster instruments, head-up displays and monitors for rear seat entertainment – all of this probably combined with speech dialogue systems with natural speech input. 3D graphics capabilities will have to be supported by an HMI tool as well as new input concepts like touch screens which are especially useful for Asian character input and handwriting recognition.

HMIs are getting increasingly complex while the society is getting older and the number of elderly people driving will increase. Future automotive HMIs will have to pay respect to this fact and HMIs will have to adapt to this user group.

Another challenge for the whole automotive industry lies in the seamless integration of various portable devices or portable navigation solutions. Initial attempts have been made to connect iPods and USB memory sticks but the different lifetimes of the car and the consumer industry will demand a flexible software architecture in the automobile and the tools to handle it.

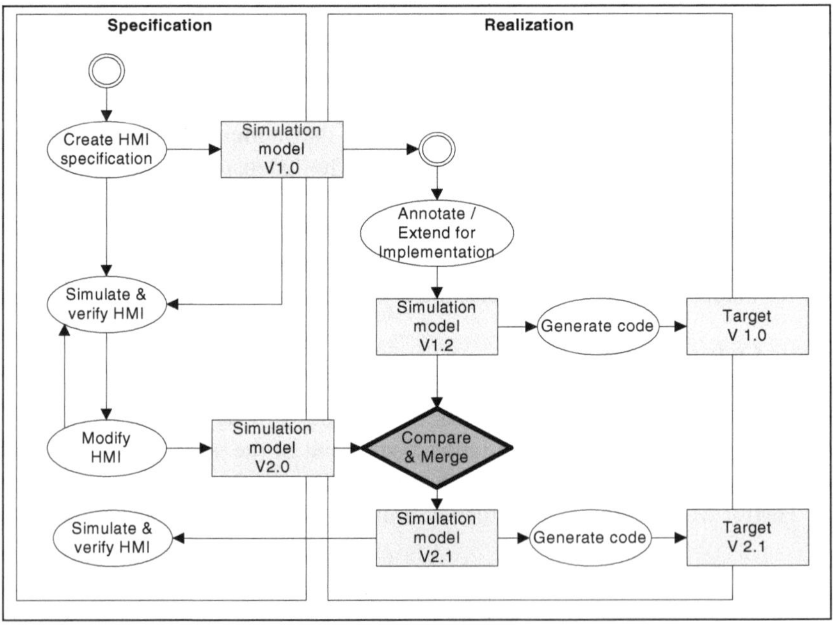

Fig. 7. Compare and merge of HMI versions

6 Status of the Tool

The basic HMI of the current Audi A6 model was successfully built by Elektrobit with a tool which implemented the techniques for model based HMI development described in this paper. Production start for the A6 was in early 2004. The automatic code generation supported by the tool was widely used during this project. The Audi HMI comprises roughly 500 screens to control tuner, CD, navigation and car setup functions.

Since 2005, Elektrobit has been providing the HMI tool **tresos® GUIDE** and its associated runtime software environment to the automotive market.

Reference

1. Dr. Holve, R.: A model-based Approach towards Human-Machine-Interfaces. ITS Telecommunication 2005, Brest, FRANCE (June 27-29, 2005)

Does Information Content Influence Perceived Informativeness? An Experiment in the Hypermedia

Yuan Gao

Anisfield School of Business, Ramapo College of New Jersey,
505 Ramapo Valley Road, Mahwah, NJ 07430 USA
ygao@ramapo.edu

Abstract. This paper reviews research in both information content and perceived informativeness in the literature, and examines the causal effect of two information content factors on perceived informativeness. A 2x2 factorial design was adopted in an experiment involving a hypothetical online retailer. Results from 120 surveys collected show strong support of the two hypotheses in the expected direction, i.e., both price and quality information had a significantly positive effect on perceived informativeness. Data also indicate that perceived informativeness is a significant predictor of visitor attitude toward the site and visitor intention to revisit.

Keywords: informativeness, content analysis, attitude toward a site.

1 Background

Previous consumer research has established a hierarchical model of advertising effects, spanning the spectrum from ad content to perception and attitude (Olney et al., 1991). A large body of research was devoted to studying the impact of executional factors on consumer attitude, e.g., ad content and format on advertising performance. This paper focuses on the potential influence of information content on perceived informativeness, a relationship that has attracted surprisingly few studies in the literature. An exploratory study is conducted to examine the influences of two representative content factors on perceived informativeness of a communication message.

Resnik and Stern (1977) view concrete information like price and quality as cues consumers can use to make intelligent decisions among alternative choices. Based on the presence or absence of content cues in a message, this methodology attempts to judge the amount of information communicated through an ad. Many articles subsequently published made use of this methodology in analyzing ad messages in various media, including magazine, TV, and newspaper advertising. Review of literature finds that print media are generally more informative than radio and TV advertising, and that informativeness differs across countries and product categories (Resnik & Stern, 1991; Abernethy & Franke, 1996). The content analytical approach has also been adopted in its

M.J. Smith, G. Salvendy (Eds.): Human Interface, Part I, HCII 2007, LNCS 4557, pp. 40–44, 2007.

general form. For example, Hwang et al. (2003) examined the functional components of 160 corporate Web sites in their ability to address a firm's message strategy.

This methodology attempts to explain information content through what is said in a message, without considering whether and how effective information is communicated to the consumers. On the other hand, informativeness is considered a perception. Research in marketing and advertising has focused on consumer perceptions of a communication message and how these perceptions influence advertising value and consumer attitude (Ducoffe, 1996; Chen & Wells, 1999). Informativeness of a commercial message is believed to be individual-specific and cannot be measured objectively. While asserting that concrete information helps consumers make intelligent comparisons and efficient purchase decisions, Resnik and Stern (1991) acknowledge that it would be unrealistic to create an infallible instrument to measure information because information is in the eye of the beholder.

Consumers consider information a major benefit of being exposed to advertising or any type of commercial messages. Information is considered one of the need-satisfying functions derived from media communications, according to the extended uses and gratifications theory (McQuail, 1983). If message content thought to be informative from the marketer's perspective is substantiated through consumer views, content analysis studies will possess more prescriptive power in message creation.

Among literature making such a connection, Aaker and Norris (1982) developed a list of 20 characteristic descriptors intended to explain a commercial's informativeness. Soley and Reid (1983) concluded that quality, components/content, price/value, and availability information affected perceived informativeness, while the total count of cues did not. Ylikoski (1994) found moderate support for the connection between the amount of informative claims and perceived informativeness in an experimental study using automobile advertisements.

The lack of theories in connecting content and perception may have been the main reason for the scarcity of research in this field. Nevertheless, no matter how different the perspectives of an advertiser and a consumer may be, to search for how they link to each other is to find the "focal point" where "the interests of buyers and sellers meet" (Ducoffe, 1995, p.12).

2 Informational Cues

Among the 14 informational cues in the content analytical scheme, price and quality seem to be most prevalent in their appearance in ad messages based on prior research. When a large number of commercial messages were sampled from magazines and newspapers, the majority of the 14 informational cues appeared significantly less frequently than price and quality (Stern et al., 1981; Abernethy, 1992). In Soley and Reid (1983), price and quality were among the four (the other two being components and availability) cues found significant in predicting informativeness. Price of a product indicates its relative worth or value inherent in a product. This cue was found in 37% of magazine advertising (Stern et al., 1981) and 68% of newspaper advertising

(Abernethy, 1992). It was also considered part of a sales incentive by Philport and Arbittier (1997) and was found to be one of the "product positioning" factors distinguishing brand communications by media. Price information is considered a major component of a buyer's search cost, along with product information (Bakos, 1997). The availability of price information represents substantial savings of time and money in obtaining intelligible and unknown, yet highly relevant information that consumers can use to evaluate competing products in their purchase decisions. Based on the above discussion, we propose the following hypothesis.

H1: Perceived informativeness is positively related to the presence of price information.

Product quality is another frequently found cue in advertising. This dimension communicates the message of the superiority of the product and the meaning of why a product is better than competitor offerings. It was found in 63.7% of newspaper ads (Abernethy, 1992). Quality is reflected in product characteristics that distinguish it from competing products based on an objective evaluation of "workmanship, engineering, durability, excellence of materials, structural superiority, superiority of personnel, attention to detail, or special services" (Stern et al., 1981, p.40).

Considering quality an intelligible and relevant ingredient within Aaker and Norris' (1982) definition of an informative commercial, we expect such statements to have a positive impact on perceived informativeness. We hypothesize a positive relationship between the presence of product quality information and perceived informativeness.

H2: Perceived informativeness is positively related to the presence of information on product quality.

In summary, we propose that product price and quality information impact the perceived informativeness of products. Though the original content analysis approach involved 14 informational cues, price and quality represent the two cues having a fairly high share of usage among advertisers (Abernethy, 1992; Stern, et al., 1981; Soley & Reid, 1983). More importantly, these are the two major categories of information consumers seek in their purchase decisions.

3 Methodology

This study adopted a 2x2 factorial design using a simple Web site with a limited number of pages showcasing some dorm furniture, which is likely of interest to the students. This study recruited 120 students through a gift incentive. Each participant was randomly assigned to a treatment group and asked to visit the assigned website (on local server) and complete a short questionnaire after about 10 minutes of visiting the site. The questionnaire includes manipulation checks and scale items on perceived informativeness adapted from Ducoffe (1996).

ANOVA was performed treating perceived informativeness as the dependent variable and manipulation of the presence of price and quality information as fixed factors in a 2x2 design. Results are shown in the following table:

Tests of Between-Subjects Effects

Dependent Variable: INFORMAT

Source	Type III Sum of Squares	df	Mean Square	F	Sig.	Eta Squared
Corrected Model	47.436[a]	3	15.812	7.584	.000	.164
Intercept	2437.506	1	2437.506	1169.049	.000	.910
MPRCE	21.534	1	21.534	10.328	.002	.082
MQLTY	24.150	1	24.150	11.583	.001	.091
MPRCE * MQLTY	1.752	1	1.752	.840	.361	.007
Error	241.864	116	2.085			
Total	2726.806	120				
Corrected Total	289.300	119				

a. R Squared = .164 (Adjusted R Squared = .142)

We note from the ANOVA analysis that those who were given price information perceived the site significantly more informative ($p< .01$, M=4.9306) than did those who were not (M=4.0833). Those who were given quality information perceived the site significantly more informative (M=4.9556) than did those who were not (M=4.0583). Thus both hypotheses H1 and H2 received support from the results above.

Additionally, using Pearson correlation coefficients, we found a significantly positive correlation of .797 ($p< .01$) between perceived informativeness and attitude toward the site. It indicates that perceived informativeness is a significant predictor of consumer attitude. It suggests that research in exploring content factors that contribute to perceived informativeness could potentially influence buyer attitude, and possibly subsequent behavior that may ultimately have an impact on the bottom line of a firm's sales and market share.

4 Discussion

This study used a 2x2 design to examine the potential effects of Web site informational cues on perceived informativeness. Unlike most observational research, this experimental approach enables the detection of a causal relationship between fixed factors and perceptual outcomes. The two most frequently included informational content cues by marketers, price and quality, turned out to have significant impact on perceived informativeness, at $p< .01$. This finding encourages further studies in examining the effects of other informational cues in Web sites through a similar approach or an expanded experiment in which more factors are manipulated.

The major limitation of this study is that only two content cues were tested. The incorporation of each additional content cue would require the doubling of the number of stimuli and thus the doubling of subjects needed to arrive at a meaningful conclusion. Due to our resource constraint, we started with the smallest factorial design, i.e., 2x2, in this exploratory study with the two most frequently appearing content cues, and planned on further exploration if results turn out to be encouraging.

We are indeed encouraged by the findings of this study. It represents an addition to current research that explores the effects of different combinations of elements in the hypermedia. Content elements of price and quality have largely been assumed to contribute to message informativeness, thus findings from this research verifies our suspicion all along and encourages the study of other content cues in future research.

Future research should replicate this study with other informational cues as fixed factors. To better represent each participant in more than a single level of treatment, a pair of Web sites can be employed in which a participant is exposed to a factor in one site but not the same factor in another. This could potentially reduce error variance.

Internet technology and e-commerce continue to grow. Consumers value information that helps them make better and more intelligent purchase decisions. This research further substantiates the argument that what makes a communications message valuable is no different in the new medium.

References

1. Aaker, D.A., Norris, D.: Characteristics of TV commercials perceived as informative. Journal of Advertising Research 22(2), 61–70 (1982)
2. Abernethy, A.M.: The information content of newspaper advertising. Journal of Current Issues and Research in Advertising 14(2), 63–68 (1992)
3. Abernethy, A.M., Franke, G.R.: The information content of advertising: a meta-analysis. Journal of Advertising 15(2), 1–17 (1996)
4. Bakos, J.Y.: Reducing buyer search costs: implications for electronic marketplaces. Management Science 43, 1676–1692 (1997)
5. Chen, Q., Wells, W.D.: Attitude toward the site. Journal of Advertising Research 39(5), 27–38 (1999)
6. Ducoffe, R.H.: How consumers assess the value of advertising. Journal of Current Issues and Research in Advertising 17(1), 1–18 (1995)
7. Ducoffe, R.H.: Advertising value and advertising on the Web. Journal of Advertising Research 36(5), 21–34 (1996)
8. Hwang, J., McMillan, S.J., Lee, G.: Corporate Web Sites as Advertising: An Analysis of Function, Audience, and Message Strategy. Journal of Interactive Advertising, 3(2), (2003), http://www.jiad.org/vol3/no2/mcmillan
9. McQuail, D.: Mass Communication Theory: An Introduction. Sage Publications, London (1983)
10. Olney, T.J., Holbrook, M.B., Batra, R.: Consumer responses to advertising: the effects of ad content, emotions, and attitude toward the ad on viewing time. Journal of Consumer Research 17, 440–453 (1991)
11. Philport, J.C., Arbittier, J.: Advertising: brand communications styles in established media and the Internet. Journal of Advertising Research 37(2), 68–76 (1997)
12. Resnik, A., Stern, B.L.: An analysis of information content in television advertising. Journal of Marketing 41(1), 50–53 (1977)
13. Resnik, A., Stern, B.L.: Information content in television advertising: a replication and extension. Journal of Advertising Research 31(2), 36–46 (1991)
14. Soley, L.C., Reid, L.N.: Is the perception of informativeness determined by the quantity or the type of information in advertising? Current Issues and Research in Advertising, 241–251 (1983)
15. Stern, B.L., Krugman, D.M., Resnik, A.: Magazine advertising: an analysis of its information content. Journal of Advertising Research 21(4), 39–44 (1981)
16. Ylikoski, T.: Cognitive effects of information content in advertising. Finnish Journal of Business Economics, vol. 2, (1994), Available online at http://www.hkkk.fi/̃teylikos/cognitive_effects.htm

Understanding Requirements of Ubiquitous Application in Context of Daily Life

Naotake Hirasawa[1], Tomonori Shibagaki[1], and Hideaki Kasai[2]

[1] Otaru University of Commerce Midori 3-5-21 Otaru, Hokkaido, Japan 047-8501
[2] NEC Software Hokkaido, Ltd, 28 Kita 8 Nishi 3 Kita-ku Sapporo, Hokkaido, Japan 060-0808
hirasawa@res.otaru-uc.ac.jp

Abstract. Progress of ubiquitous computing oriented ICT promotes development of information appliances with various functions. In spite of the progress, most of user cannot receive the benefit of them. In this paper as the reason of it, the problems related to increasing enormous amount of information contents were discussed. Based on the problems, there are new challenges for designing user experiences; interactions with vast amount of information contents and integrations ubiquitous computing applications into daily life. Analyzing these challenges from view of human-centered design could derive the necessity to services design based on ubiquitous technologies and validation of them.

Keywords: Ubiquitous computing, user experience, information content, information appliance.

1 Introduction

Recently "ubiquitous computing" is used in the context that micro-computers are embedded in various artifacts and user can receive various kinds of the benefits by their connecting each other. Japanese Government has published white paper which included life styles under ubiquitous computing [1]. It was called u-Japan policy whose life style seems to be the world through rose-colored glasses. Furthermore, for digital device manufactures ubiquitous-computing oriented information appliances are expected as new market break-through. However the progress of ICT like ubiquitous computing are not always welcomed for users. According to NRI report [2], most of Japanese people feel anxiety that they could not master the new appliances. Before receiving benefits from ubiquitous computing properly which are expected by computer scientists or engineers, some problems should be overcome.

For examples, we are facing the following problems. Firstly problem is that some people must arm with digital appliances. Progress of ICT promotes to succeed to implement various functionalities to small devices. Consequently we are surrounded by many appliances such as mobile phone, digital camera, USB memory, MP3 audio device, PDA and so on. Some businessmen must always have some devices among them.

M.J. Smith, G. Salvendy (Eds.): Human Interface, Part I, HCII 2007, LNCS 4557, pp. 45–50, 2007.

In contrast to above situation, some functions of computing tend to be concentrated on mobile phones. A lot of function does not always correlates to usability commonly. Actually some functions are not used at all for some people. Recently secure specification is strongly required. The stronger the security is the less easy the usage is. Consequently concentration of function does not yield conveniences of appliances.

Regarding information contents, amount of contents a person can handle in daily life is increasing rapidly based on advance of storage technology. At the same time, way of contents usage is getting various. As a result, users would be puzzled the usage and then the contents would be scattered like rubbish. Actually most of graphical contents by digital camera are said to remain unused.

Aside from the original concept of ubiquitous computing [3], its general trends would begin with device based concept embedding computer and anytime, anywhere network concept. Compared with developments of a large number of new digital devices, quite a few can be acceptable for user. As mentioned above, most of people in Japan are not necessary to have new devices in daily life. To bridge gaps between technological trend and user's necessity, these problems should be challenged in view of human-centered design. In this paper, a problem involving relations between user and information content is considered. As illustrating precedent cases, some implications of requirements for designing ubiquitous computing applications in the daily life are discussed from human-centered viewpoints.

2 Challenges to Vast Amount of Information Contents

Digital devices such as digital camera or hard disk recorder can promote to collect information contents. For examples, when we take photos with film camera, the number of photos per a film is limited from twelve to thirty six. On the contrary, we can take hundreds of photo with digital camera. Thus, amount of information contents and usage which user must handles is increasing rapidly as mentioned above. Thus, user has to manage vast amount of information contents as well as use the devices.

Furthermore the contents tend to be scattered in various devices. Regarding digital video, hard disk recorder can store it, personal computer can store it and mobile player device such as iPod can store it. The digital video has scattered in various devices at home. If he/she wants to watch a video, he/she must go to a device where the video might be stored and seek the video.

To solve the situation where information contents are scattered in various devices, there are some projects to set up platform for interoperable use of all digital contents at home. DLNA (Digital Living Network Alliance) [4] is a representative project for IT related leading industries (Fig.1). First DLNA project points out that we live in three digital islands at home; (1) the PC Internet world where PC and PC peripherals communicate, (2) the broadcast world of set-top boxes and traditional consumer electronics, (3) the mobile world of multimedia phones, PDAs, laptop computers and similar devices provides unparalleled connectivity and freedom of movement into and out of the home environment. User wants to access the information contents without regard to devices in these three domains. But the expectations have largely been unfulfilled under the present circumstances. To fulfill the expectations an interoperability framework is proposed, which will manage and distribute rich digital content to devices such as TVs and wireless monitors from devices such as digital stills cameras, camcorders and multimedia mobile phones. The framework is expected to define interoperable building blocks for devices and software infrastructure. It should cover physical media, network transports, media formats, streaming protocols and digital rights management mechanisms.

DLNA concept is itself excellent idea, however that accelerates the increase of information contents user can enjoy and must manage.

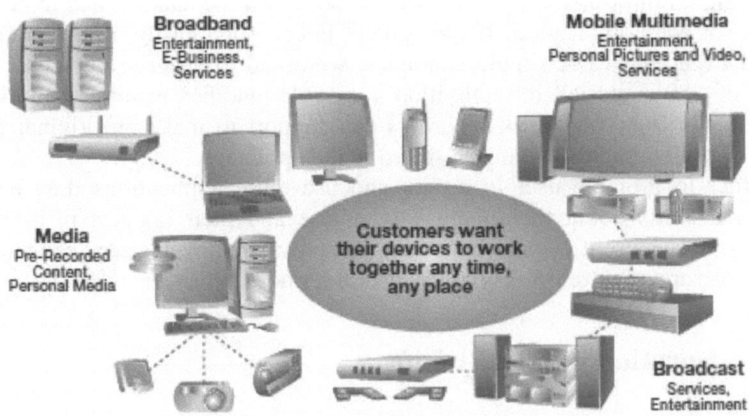

Fig. 1. DLNA Vision (Use Case Scenarios, White Paper, Digital Living Network Alliance 2004)

3 Contents Search in Daily Life

As mention above, it is definitely inevitable to interact with vast amount of information contents in daily life. Therefore some challenges in relation to the contents should be sorted out.

First, the effective information search should be invented to browse vast amount of contents. Technically in order to raise performance of information search the information should be restructured properly. If user interface based on the restructuring is implemented into devices, the system requires user to understand the structure. As a result it would be uneasy and often dull to search information in daily life. User interface of the contents search should be considered from user view so that search activity itself may be fun. XMB by Sony Inc is one of options among pleasant user interfaces.

Regarding original contents such as digital photos or digital movies, their search methods would be restricted. Because effective content attribute in searching is just only date or time. The contents category is limited to only time without adding new attribution by users themselves. Thus design of user interface for exploring contents would be restricted.

Second challenge is issue of development of new application using contents. In case that the information contents consist of analog data, it is very difficult to separate the contents from the media which store them. For instance, our experiences of hearing music with LP record are limited in living room or any room where the stereo player exists. If we need to copy music in LP record into cassette tape, we had to connect stereo player with tape recorder. After the contents were digitalized, it has become easy to separate them from media and easy to be moved between media. Now we can focus the contents themselves and invent various way of enjoyment. Not only in PC but also in hard disk recorder as home appliance, various types of digital contents such as digital movie, digital image and digital music file and so on are stored in a same device or in home network together.

Each digital information contents need to be easy to be handled and to be managed by each software application. If the content is compiled easily, user can enjoy new contents experience. For simple examples, we could experience ourselves "Best 25 music" or "80's music" through iPod's use. As another example, "x-Pict story" application in Sony hard disk recorders can support to make an original movie by compiling digital photos and music stored in the appliance.

In order to promote user to accept and use these applications they need to be understood easily about their main way of use and their impacts in user lifestyle context. In this sense it is quite important to represent enjoyable service images of the contents use and to develop easy to use user interfaces.

4 Incorporation into Daily Life

Many researches and developments related to technical issues based on ubiquitous computing and networking have been conducted. For example UbiLab [5] project is aiming to develop fundamental and application software for the ubiquitous computing environment. Ubila [6] project which consists of academic-industrial cooperation covers total system for supporting ubiquitous network and services, including core network, access network and ubiquitous end objects.

ICE-CREAM [7] project focused on the contents programs in end-user home environment. The objectives are to investigate the potential of new technologies for designing new concepts for interactive and enhanced broadcast, to create programs and environments in which users can interact with the content of the programs and compose personal programs to create personal flavor and emotion. The project investigated how to make compelling experiences for end users based on the possibilities of integrating technologies for interactive media, for example, DVB-MHP, MPEG-4, 3D graphics and Internet technologies. Technological options that address different levels of interactivity for end-users were investigated and implemented in prototypes, and supported by business frameworks.

Approach of these projects is typical technology centered although ICE-CREAM project conducted user testing. From technical viewpoint, ubiquitous computing can promise to offer immersive and exciting experience, however our daily life does not need always so stimulated. The point is whether user experiences base on those technologies can accustom users' way of life or not. The relevancy of new user experiences to our daily life should be considered to implement ubiquitous computing application.

In order to understand the relevancy between new technology experience and lifestyle habit, our research project conducted an experiment of how people merge Internet use in their habit of TV watching. The experiment was performed in situations which are close to informants' actual TV watching lifestyle in their living room.

As a result there were two patterns of browsing web site in watching TV; one pattern was that informants frequently browsed before and after watching TV. Before they decided the TV program they intend to watch, they repeated browsing and zapping by turns. Another pattern was that informant occasionally browsed and zapped at intervals to the end of test (Fig. 2).

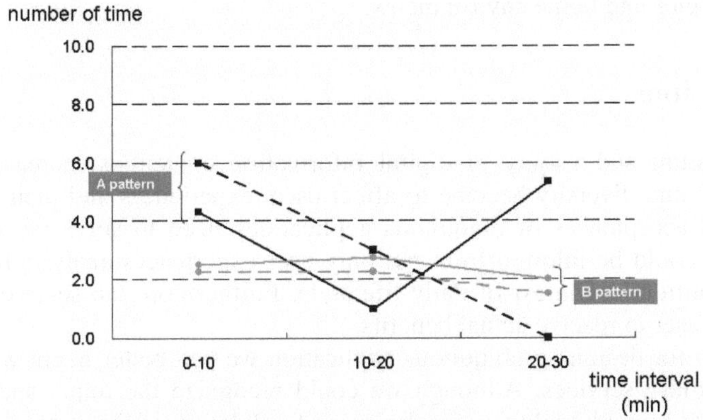

Fig. 2. Number of web access (full line) and zapping of TV channel (dot-line)

Watching TV programs are not always similar to browsing web site in view of cognitive interaction between user and contents, that is, the style of TV watching is mostly passive, whereas web site's one is active. This means it would be inconvenient to unite those two functions into a system such as TV with Web browsing or PC with TV tuner. But both of those activities work closely together although there are several patterns of them. This shows possibilities of new type of media environments including more than two devices that have those functions separately.

Through these results, the following implications could be shown at least.

First, the design scope of ubiquitous computing system should be extended to way of life context. It is difficult to specify requirements for the system only from user interaction with the system. The design must be begun with service definition which identifies relation between user and the information contents properly in the context of way of life.

Moreover, renewed additional criterions to present usability's criteria need to validate the service design. Actual installment of the ubiquitous computing system in the lifestyle context need not only fun or excitement viewpoint but also natural or piece and quite viewpoint. In particular habituation of using a system would be so important that the system might need to establish the way of life.

In order to integrate ubiquitous application into way of life and clarify the criterions for validation of new system, log data from actual human life style must be required. There have been some projects to monitor and log user activities. “KoKomemo” is one of logging systems with mobile phones with camera for user problems in town. It was developed in “Yaorozu” project [8] in Japan. The system can capture problems scenes in daily life.

But it is not sufficient to know total user context of lifestyle. PlaceLab [9] project conducts research by designing and building real living environments - "living labs" - that are used to study technology and design strategies in context. It is a residential condominium, designed to be a highly flexible and multi-disciplinary observational research facility for the scientific study of people and their interaction patterns with new technologies and home environments.

5 Conclusions

Recently amount and variety of digital information content is increasing rapidly. Their amount and diversity become to affect users experiences and their life style at home. Better acceptances of ubiquitous applications need to solve those problems. The solution could be inferred from not only user experiences analysis but also real services definition in context of daily life style. Furthermore the services should be validated for user to receive actual benefits.

Essentially for designing ubiquitous application we had better begin with building concepts of those services. Although we could recognize the importance, we must face undeveloped methodologies for design and validation of the services mentioned above. In parallel with technological projects, human-centered or multi- disciplinary projects need to be conducted to yield those methodologies especially in Japan.

References

1. The Ministry of Internal Affairs and Communications in Japan, u-Japan Policy (2004)
2. Nomura Research Institute, Cyber Life Observation (2002)
3. Weiser, M., Brown, S.J.: The Coming Age of Calm Technology, Beyond Calculation. In: Denning, P.J., Metcalfe, R.M. (eds.) The Next Fifty Years of Computing, pp. 75–85. Springer, Heidelberg (1997)
4. Overview and Vision White Paper, Digital Living Network Alliance (2004)
5. UbiLab, (2006), http://www.ht.sfc.keio.ac.jp/ubi-lab/
6. Ubilia, (2006), http://www.ubila.org/e/e_index.html
7. ICE-CREAM-phr-0404-01/Janse, The ICE-CREAM Project Final Report, Deliverable D22 (2004)
8. Yaorozu, (2005), http://www.8mg.jp/en/index.html
9. PlaceLab, (2004), http://architecture.mit.edu/house_n/placelab.html

Design for Confident Communication of Information in Public Spaces

Shigeyoshi Iizuka and Yurika Katagiri

NTT Cyber Solutions Laboratories, NTT Corporation
1-1 Hikarinooka Yokosuka-Shi Kanagawa Japan
{s.iizuka,katagiri.yurika}@lab.ntt.co.jp

Abstract. In a ubiquitous society, it is possible to use many kinds of information anytime, anywhere. Increasingly reliant on their mobile phones and PDAs, people often use information in public spaces. However, there are risks to entering highly confidential information such as personal data into a system. People have a strong awareness about the value placed on personal information, and worry about their information being leaked to people they don't want to see it. In public spaces people also worry about real-world leakage (i.e. by non-digital means). In spite of the security provided by law and technology, people cannot handle information with reassurance. Complete security means both communication security and physical security. The purpose of this research is to construct an environment where ubiquitous services can be used with reassurance. In this paper we describe our first experiment in this research area.

Keywords: Personal Space, Reassurance, Public Space.

1 Introduction

The spread of personal computers and the explosive growth of the Internet now make it possible for users to communicate a wide variety of information regardless of location or time of day. This means that even information of a highly confidential nature can be accessed and processed anywhere and anytime. As a result, even highly confidential information can potentially be handled any time, and any place. This situation bears risks, however, as the user's personal information could be leaked or made visible to parties unknown to the user. Of course, services that use confidential information are provided with robust security systems such as encryption, digital certification, and authentication technologies such as IC cards and biometric authentication. Therefore, we can say that these services are secure as far as their functions are concerned, but what about the environments in which these services are used? For example, have you ever found yourself worrying about people catching a glimpse of a display you are using on a busy sidewalk? In practice, it cannot be said that highly confidential information such as personal details can be used securely in public work environments.

Guidelines for designing work environments have previously been presented from a human-physiology perspective in the fields of architecture and human-factors

M.J. Smith, G. Salvendy (Eds.): Human Interface, Part I, HCII 2007, LNCS 4557, pp. 51–58, 2007.

engineering (Grandjean, 1989; Maruzen, 2001). Many of these guidelines, however, are based on the physiological characteristics of human beings, and there has been no research that takes account of the types of information handled in such an environment or of user reassurance. In short, these guidelines by themselves cannot clarify what format of work environment is best for providing users with a sense of reassurance. There is a need for guidelines and methods that can be applied to the design of public work environments that enable users to comfortably handle personal information. Therefore, we are researching and developing network architectures where IT equipment can be used in many different places and have initiated a study of secure space design technology because people need physical spaces where they can use these services securely (Iizuka, 2004; Goto, 2004; Iizuka, 2005a).

When taking action, humans feel reassurance based on three factors, reassurance by prior defense, reassurance due to grasp of the situation, and reassurance by subsequent compensation. Applying these factors to feelings of reassurance when a person is communicating information in a public space, it can be said that reassurance by prior defense takes the form of reassurance due to being physically prepared for the environment, reassurance due to grasp of the situation takes the form of reassurance obtained by knowing the surroundings during communication of information, and reassurance by subsequent compensation takes the form of keeping damage from predictable trouble to a minimum. Reassurance by subsequent compensation is considered to be a subject for insurance companies and other similar institutions, and reassurance by prior defense has already been advanced in research on secure space design technology, mentioned above. Actually, when people are communicating information in a public space, since there are many unknown people around them, the surroundings change moment-by-moment. To mitigate the problems this causes, a system that can always recognize the situation clearly, providing reassurance due to grasp of the situation, is needed for people to be able to work comfortably. So, in our research, paying attention to reassurance due to grasp of the situation, we decided to try providing the information to help users feel reassurance while communicating information in a public space. In this paper, we describe our approach to the research, the details of the experiment we conducted as the first trial, its results, and our future plans.

2 Approach

Some people want reassurance by prior defense when communicating information i.e., reassurance by secure space design technology. That is, users don't always want to communicate information relying only on their grasp of the surrounding situation, which depends on available information and judgment of whether the situation is reasonably secure. Prior defense means users want to concentrate on handling information without worrying about their surroundings. However, secure space design technology is still in the research stages. Even if this technology were in the utilization stage, it would still be difficult to apply it to all public spaces. Therefore, users must settle for reassurance due to grasp of the situation provided by presentation of situational information (Figure 1).

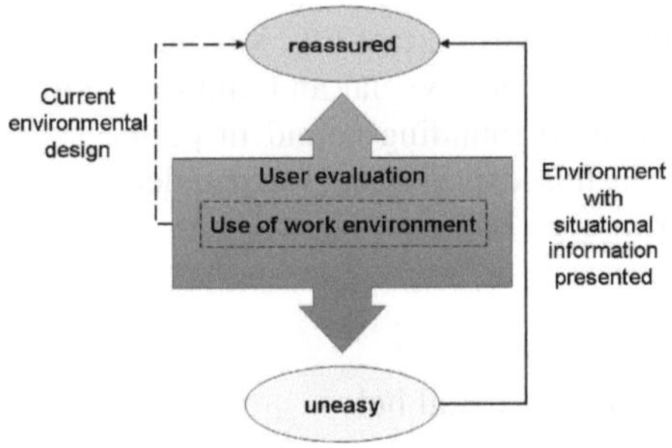

Fig. 1. Research design

The actual work environment is presented in Figure 2. Although the screen on which information is presented naturally faces the user, it also faces a person behind the user (Figure 3).

Fig. 2. Actual work environment

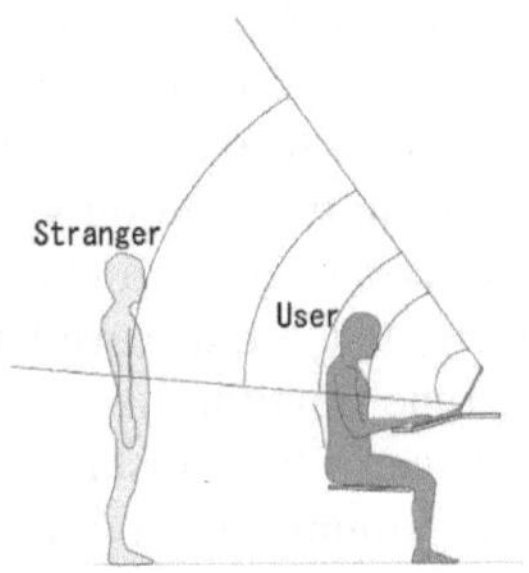

Fig. 3. Direction of screen

However, it is not easy for a user to know if someone is behind him. Therefore, we aim to reassure users about people behind them by presenting situational information.

3 Observation

We observed the effect (users' actions) that presentation of situational information (information about whether someone is behind the user) has on an information user. In this section, we describe the purpose and method and discuss the results.

3.1 Purpose

The first step in this research was investigation of the effect that presentation of situational information has on an information user. We assumed that uneasiness affects how accurately information is input so we decided to use the number of input mistakes as a metric of that effect.

Humans have five senses: hearing, tactility, vision, smell, and taste. Of these, it is thought that it is easiest to perceive visual and auditory information (Oyama, 2003). However, it is thought that surrounding sound in public space drowns out auditory information. Therefore, in this experiment, we decided to use visual information to provide situational information.

3.2 Method

The investigation is explained in detail below.

(1) Equipment

To present visual information, one can use text, pictures, or light. Since this was our first trial and we were concerned about cognitive speed, we decided to use light of two different colors. We wanted to observe whether information can be comfortably communicated and what effect it has on users by assigning one kind of information to each of these colors. We observed the effect this information had on users by assigning one color to safe and one color to unsafe situations. Since green is said to induce reassurance, and red to induce an uneasy feeling (Takahashi, 2002), in this investigation, we used green to signal situations where the user could communicate information with reassurance and red to signal situations where the user could not communicate information with reassurance. Specifically, we fabricated a display that shows circles of green and red light with a diameter of 11 mm on an LED (Figure 4). Considering it a place well within a user's field of view, we installed the display between the monitor and the keyboard of a normal PC set-up. We used the direction of a stranger's face, distance between a user and a stranger, and type of information being communicated by a user to determine what color to display. With this combination of factors, we made judgments about whether the situation was safe, and displayed either a red or a green light. We controlled the display from another room where we could see into the laboratory through a one-way mirror. We explained the meaning of the two colors to participants in advance.

Fig. 4. Experimental set-up

(2) Participants

We conducted the experiment with seven participants, all of whom were women who

- had experience communicating personal information (information that a user would not want others to see) on a PC,
- had experience operating PCs, and
- had used PCs to send email and shop on the Internet.

(3) Environment

In each trial there was a "user", someone inputting personal information and a "stranger", someone unknown to the user, in the room. The users input information on a PC equipped with one of our displays. Intentionally, in order that a stranger could see the PC monitor, and to make users aware that the stranger could see it, we used a 28-inch PC monitor. Moreover, in order that an experimenter might easily determine the distance between a stranger and a user, and to judge whether the stranger was in a position to see a display, the stranger was directed to stand in specified spots. Furthermore, to determine whether the pedestrian was looking at the screen (information), we asked the stranger to orient his body in one of two directions, one where he could see the information and the other where he could not.

(4) Procedure

In each trial there was one user and one stranger. The user inputted all information including person-specific and money-related information, preference- and behavior-related information, and present circumstances and history-related information into a form prepared beforehand. The stranger walked around at random, standing in different spots as directed. The user input information after a practice period. Trials were run with and without our display, each trial consisting of one information input. Then, the role of the user and stranger were changed and a trial was run using the same method. After the experiment, we asked participants to fill out a questionnaire and interviewed them all.

3.3 Results

We counted the number of typos for the color shown during each trial. We counted the number of typos per trial and compared the values for total number of typos for

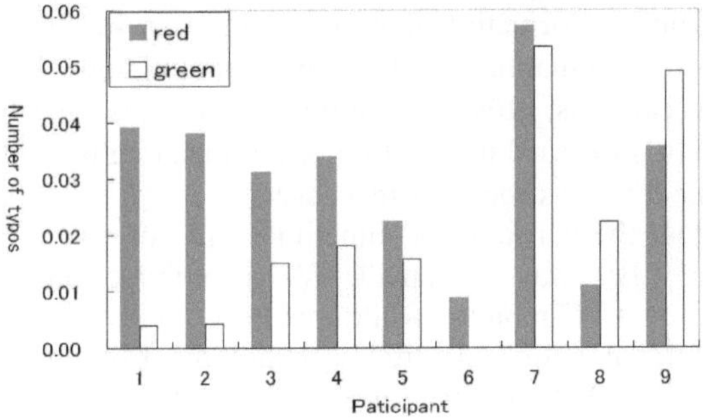

Fig. 5. Number of typos per trial for each participant

each time a color was presented. Figure 5 shows the number of typos per trial for each participant. As can be seen in the figure, there were two groups of participants: one with higher numbers of typos when the red light was presented and one with a higher number of typos when the green light was presented. With this in mind, we discuss the results of the questionnaires and interviews, as shown in Figure 6 below.

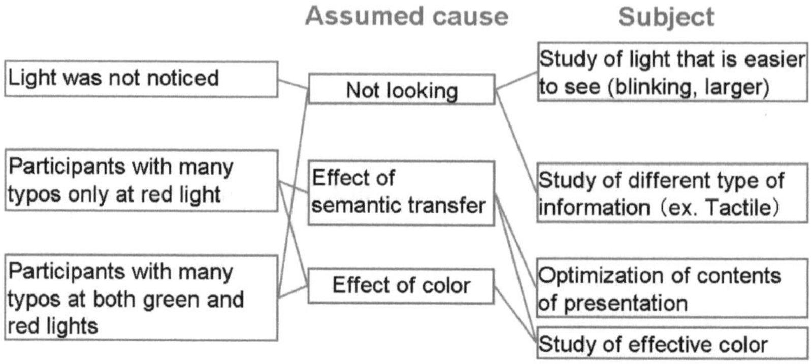

Fig. 6. Questionnaire and interview results

In the questionnaires and interviews the participants who made more typos when the red light was shown said that they were able to judge whether they could input information with reassurance (whether or not they personally felt a stranger's presence) based on the light that was shown. That is, the signals displayed enabled users to judge whether it was appropriate to communicate information based both on their own judgment and on objective information. As for the participants whose number of typos also increased when the green light was shown, we speculate that they became impatient because they were conscious of information about the inputting opportunity signaled by the green light, causing them to make typos. This suggests that the green light, which we thought would induce feelings of reassurance, does not always have a positive influence and may cause users to be impatient. Generally, the red and green lights were strongly associated with the corresponding message of a traffic light: red means "stop", and green means "go". Therefore, the supposition described above (green induces reassurance, and red an uneasy feeling) is not unalloyed semantic information that something can be done with reassurance. That is, the effect of the semantic transfer is not unambiguous. Red's common image induced users to be cautious, affecting the number of typos. Further study is needed about whether a color's external association or the meaning the color was intended to have in the experiment more strongly affects users.

We also noted that the participants sometimes did not notice the light. We think this is because the LED was too small. We also think that users may become myopically focused on a PC monitor while inputting information. It may also be that users sometimes did not notice the light because the LED was outside the user's immediate field of view. Although we installed the LED between the monitor and the keyboard thinking that it would be easily noticed, we now think that the effect was too small. To make the light more noticeable, we need to study ideas, such as changing the display location, using a blinking light, and changing the color of the light being shown.

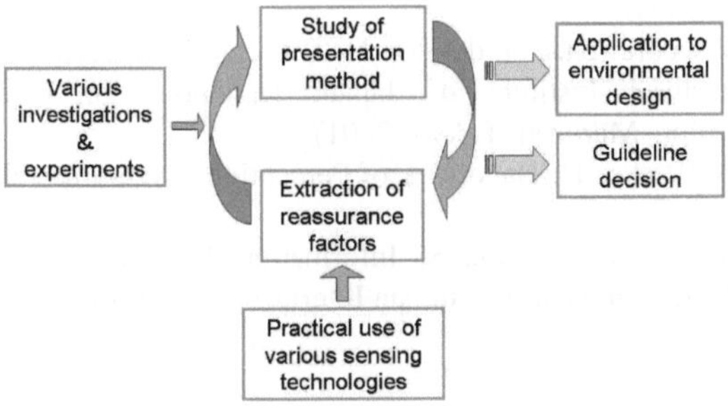

Fig. 7. Overall design of this research

4 Conclusion

We described our approach to designing provision of information intended to create feelings of reassurance when a user is inputting information in public spaces. The investigation was carried out as a first experiment on the effect of presenting two colors of light. We found that using or combining past research results about what affects people's reassurance won't necessarily create feelings of reassurance about communicating information in a public space. For example, different colors of light were found to have conflicting associations for users and to affect their feelings of reassurance in different ways. More research in this area is necessary. Therefore, we think it is necessary to investigate whether presenting a different kind of information (tactile for example) might work better. By compiling current and future results, we aim the type of information that best creates feelings of reassurance in a user communicating information in a public space. In the process, it may be necessary to identify the factors which give the user feelings of reassurance while communicating information in a public space. Using a varied approach that reflects knowledge and guidelines (Iizuka, 2005b) for secure space design technology based on research, (Figure 7) we aim to help create a public space in which users can communicate information with reassurance.

References

1. Goto, Y., Iizuka, S., Ogawa, K.: Research on Designing Environments with a High Degree of Reassurance, Architectural Institute of Japan, Summary of Technical Papers of Annual Meeting, Book E-1, Construction Planning, pp. 935–936 (2004)
2. Grandjean, E.: Ergonomics in Computerized Offices. Keigaku Shuppan, Tokyo (1989)

3. Iizuka, S., Nakajima, S., Ogawa, K., Goto, Y., Watanabe, A.: Reassurance When Using Public Terminals in Public Spaces. In: Proceedings of the 66th National Convention of Information Processing Society of Japan (IPSJ), 4-451-4-452 (2004)
4. Iizuka, S., Ogawa, K., Nakajima, S.: A study to develop public work environment design guidelines for handling personal information. In: Proceedings of HCI International 2005, incl. in CD-ROM (2005a)
5. Iizuka, S.: Pictorial secure space guidelines. NTT Technical Review 17(8), 46–49 (2005b)
6. Maruzen.: Architectural Institute of Japan, Handbook of Environmental Design, Comprehensive Edition. Maruzen, Tokyo (2001)
7. Oyama, T., Imai, S., Wake, T.: Handbook of Perception psychology. Seishin Shobo, Tokyo (2003)
8. Takahashi, K., Nakatani, M., Nishida, S.: Information Presentation from the standpoint of a sense of security. Proceedings of the Human Interface Symposium 2002, (2243) (2004)

Suggestion of Methods for Understanding User's Emotional Changes While Using a Product

Sang-Hoon Jeong

Department of Industrial Design, Mokwon University,
800 Doan-dong, Seo-gu, Daejeon 302-729, Republic of Korea
diasoul@mokwon.ac.kr

Abstract. The aim of this research is to suggest a most effective method for measuring user's emotions expressed while using a product. This study developed a tool that effectively measures the user's emotions expressed while using a product and that can complement the limitations of the psychological measuring method. The developed emotion logging software named VideoTAME, basically asks the subject to watch a recorded video clip of the user performing a specific task and to examine the emotional changes that had occurred. And a physiological measurement method that measures the user's emotions expressed during product use with VideoTAME and that is easily accessible in the design field was suggested. By upgrading VideoTAME and overcoming its limitations as a psychological measurement method, and also by using the physiological measurement method mentioned this study to measure the user's emotional changes, the correlation between the product usability and the user's emotion will be able to be defined more clearly.

Keywords: methods for measuring user's emotions, usability, user's emotions.

1 Introduction

Human emotion is the issue which is subjective, difficult to define and even more difficult to measure because it is personal. Cacioppo and Gardner concluded that "the measurement of emotion is a bustling research area" [2]. Generally, the methods for measuring human emotion are divided in two, one is the psychological way based on user's subjective evaluation and another is the physiological way based on physiological signals.

Traditionally, the measurement of emotion is usually dependent on the psychological method [10]. The most commonly used psychological method requires users to report their emotions with the use of a set of adjective interval scales (7-point scale, 5-point scale, etc.) such as semantic differential or verbal protocols. Mostly these emotions are measured in the form of an interview or a self-report. Lively research has been done on psychological measuring methods in areas such as psychology, linguistics, emotional engineering, etc. However, this method, based on subjective evaluation, has some limitations. When studying what users say about their emotional state, user's emotions tend to change according to time. Plus, precisely

M.J. Smith, G. Salvendy (Eds.): Human Interface, Part I, HCII 2007, LNCS 4557, pp. 59–67, 2007.

expressing emotions in words is not an easy task. And, because emotional experiences can be delicate, users sometimes (consciously or unconsciously) manipulate their reports. Most of all, since the subjective report is examined after the emotional part is experienced, the emotion expressed can be different before and after the examination. Rosenberg and Ekman showed that these kinds of self-report do get influenced from recent effects [12]. According to Scherer, self-reports are subject to distortions due to ego-defence tendencies and socially desirable effects. Moreover, relying on the subject's memory is also a significant limitation. There is always a time-lapse between the experienced emotion and the self-report. The bigger the time-lapse between experience and report, the more the results are distorted [13].

The physiological method, measuring physiological signals to use them as an objective index of human emotion, has been tried to measure the emotion more objectively. However, the method using the physiological signals also has problems. First, it is very burdensome and unnatural with recent technology. For example, a subject has to have sensors on his fingers to monitor the pulse and the electrodermal activity (EDA), and also has to have equipments on his head to measure an electroencephalogram (EEG) coming from the occipital and parietal lobe. Secondly, there are limitations in the stimulus to make subjects express their emotion. The stimulus is limited to one that makes subjects not need to move, such as visual or hearing equipments. Especially, to measure an EEG, the control of other waves has to be severe. It is still difficult to detect a pure EEG generated from emotion because eye blinking or tiny movements can make considerable noise. The third problem is an economic issue. High-costing equipments are required to sensor physiological signals and there are a lot of troublesome matters to furnish the testing environment. Fourth problem is the lack of consistency in the results of emotion research by measuring physiological signals. Even though it has consistent results about electrocardiogram (ECG) or EDA, it still requires an objective index about other signals. Since the stimulus and the physiological variables are different by researches, the research results by measuring methods are inconsistent [3]. The problems on these research methodologies are the difficulties of setting the framework for the emotion- specific reaction pattern and the relationship between emotion and autonomic nervous system reaction [11]. Therefore, it can be said that it is not easy to grasp human emotion just by physiological signals. Lastly, there are also difficulties in data analysis. It is not easy to approach the area from the design field since it requires propound knowledge of human physiological signals to analyze data from the tests.

Therefore, it is necessary to research on the methods for measuring user's emotion in the natural and accessible environment for the design field and for measuring user's emotion naturally expressed while using a product, and to solve problems of the two existing methods. The aim of this research is to suggest a most effective method for measuring user's emotion expressed while interacting with a product.

2 Development of a Tool for Measuring User's Emotions Expressed While Using a Product

If taking into consideration that the stimulus is not big, and the user has to keep moving while using a product, to measure the user's emotions, the psychological measuring method can be seen as more effective than the physiological measuring method. Because it is very difficult to overcome the limitations of the physiological measuring method mentioned above and to detect accurate physiological signals in these situations. In the previous research, I extracted a set of emotional words and representative emotions that can show changes in emotions of users while using a product and not words that were derived simply from the product's appearance. And I have come up with a suitable set of subjective evaluation scale for measuring user's emotions expressed while using a product [4, 5]. Through this, I suggest a more effective emotion measuring method that can complement the limitations of the psychological measuring method. The developed software named VideoTAME (Video Think Aloud for Measuring Emotion), basically asks the subject to watch a recorded video clip of the user performing a specific task and to examine the emotional changes that had occurred.

2.1 Objectives of Development and Structure of VideoTAME

To make up for the limitations of the previous emotion evaluation methods and taking into consideration the circumstances where the product would be used, the following objectives for development were derived to effectively measure user's emotions.

- Since emotion has to be measured after the emotion has been felt and experienced, emotion distortion should be minimized.
- The environment where the product is used should be as realistic as possible.
- User's emotion should be able to be measured in a natural environment where product is used.
- Environment should be easily accessible for the design field and should be able to sufficiently gather enough opinions from subjects.
- Collected data should be able to be transferred and utilized to other commonly used programs.
- Collected results should be databased and able to be taken out again to be used effectively.
- Suitable emotional words should be suggested to suit the various words users might come up with while using product.

The emotion evaluation tool developed based on the objectives listed above is composed largely of a Testing Module and an Analyzing Module. The Testing Module is composed of one part where the subject actually performs the experiment and the other part where the data is collected and recorded. The Analyzing Module is composed of a database part which can classify, store, and manage each experiment, and an analyzing part that plays and analyzes the collected data in various ways.

VideoTAME was developed using Microsoft Visual-Basic 6.0. A distinct characteristic is that it comes with 2 Windows Media Player engines. It is also compatible with various video clip formats such as AVI, MPG, WMV, MOV, etc. When the subject ends the session, the details of the experiment is recorded and saved in a text file format (*.txt). The saved file is then reworked as database in the Analyzing Module. The biggest characteristics of the Analyzing Module are that specific scenes can be captured from the video clip, and that all sorts of analyzed data can be directly converted into Microsoft Excel files by Excel Object 9.0. In addition to this, the subject can stop the video clip and then the selected emotional words automatically go into the sound folder. The designer designs it so that when it plays, the emotional words can come out as voice recording so that when the subject performs the task it can give a "think aloud" effect. The following is a rough overview of the development environment of VideoTAME:

- Development Language: Microsoft Visual-Basic 6.0
- Used Engine: Microsoft Windows Media Player 9.0, Excel Object 9.0
- Development Environment: CPU Intel® Pentium® 4 CPU 2.80GHz, 512MB, XP Service Pack2
- Operating Environment: CPU Intel® Pentium® 2 CPU 600Mhz, 128MB, WIN98 and above
- Size of Program: Testing Module (228KB), Analyzing Module(1.25MB)
- Program Structure: Testing Module (Video TAME.exe), Analyzing Module (VideoTAME- Analyzer.exe), guide.wav, Sound Folder

2.2 Testing Module of VideoTAME

This is the Testing Module where the subject performs the experiment by following the guiding message. The Testing Module starts as the subject selects the folder where the recorded file of the subject performing the task in the experiment room is located. The Testing Module is composed of four screens. An initial screen that shows basic information on general performance of experiment, a screen that shows the movie clip that the subject can see and evaluate emotional changes, a screen used after the experiment for emotion evaluation using the subjective evaluation scale and a screen with a brief survey.

In the part where the emotion is tested, the experiment subject examines his/her own emotional change by watching the video clip that was recorded in the experiment room when the task was performed. The video clip is composed of a part where the subject's facial expression can be seen and a part where the actual product can be seen. First of all, a task that can examine emotion should be selected, and then while playing the video clip, when there is visible emotional change the clip should be paused, then the subject should select all the emotions he/she felt from the list of the 32 emotional words. Color is added to the selected emotional words so that it can be easily distinguished.

Multiple answers can be selected for the emotional words, and after selected, subjects are guided to express how strong they felt in the selected emotional words by 5-point scale. The items of the scale are automatically changed according to the emotional word selected. At the end, the subject gets to briefly write down why emotional change had occurred.

Additional words can be selected while the clip is paused, and if there are none, the video clip can be resumed and the emotion evaluation test can be proceeded in the same method as above. Evaluations can be added for previous screens by rewinding the video clip, and revisions can be made for the previously selected emotional words. If all emotion evaluation of the performed task is completed, the screen goes to the next screen. In the survey that is shown at the end of the experiment, a general evaluation is given about the product applied by the subjective evaluation scale made in the previous section.

Finally, the experiment ends with some basic questions related to the subject (gender, age, academic background, occupation) and some questions related to the used product. For example, if the experiment was to evaluate emotion expressed while using a cellular phone, the following questions could be asked.

- How long have you been using a cellular phone?
- What previous types of cellular phone have you used? Please list all the company names.
- What type of cellular phone are you currently using? Please write down company name, model type, and purchased date.

2.3 Analyzing Module of VideoTAME

The Analyzing Module is where the designer or the researcher can play or analyze the collected data in various ways. The Analyzing Module starts as the designer selects the folder that contains the evaluation results of the Testing Module. The Analyzing Module is composed of two screens: A screen that allows the designer to examine the experiment results by playing the recorded video clip by each task, and a screen that variously analyzes the results that are exported to Microsoft Excel.

In the screen where the experiment results can be seen by each task, the designer can check the evaluated emotional change. The designer can easily recognize the part where the subject selected an emotional word by stopping the video clip because it automatically indicates it. Plus, the selected emotional words and the degree of it gets displayed at the bottom of the screen and lets the designer know through voice recording so that the designer can have a "think aloud" effect of the performed experiment. Other than this, the selected emotional words, the score given for the selected degree, the reason for choosing the specific word, and the average of the representative emotions can be identified through the selected emotional words. Lastly, when the analyzer finds something particular while playing the video clip, the situation can be input by clicking the "Description of Situation" button. The information that was input can be re-checked through the exported Microsoft Excel file. Through the exported Microsoft Excel file, variously analyzed information can be seen as follows.

•*Analyzed information per scene:* User's emotional change per situation can be analyzed by the following – the subject's task division, time that video clip was stopped to select emotional word, scene, information on explaining situation, emotional words and degree selected from subjects, reason for selecting those specific words, and the average of the representative emotions can be identified through the selected emotional words.

•*Analyzed Information per task:* User's emotions for each task can be analyzed by checking the average score of emotional words, number of times of emotional words selected, and the average score of representative emotions. The information on analysis per task applies not only to the subject but also to the whole group of subjects.

•*Information analyzing the whole experiment:* The emotional change occurred from the user using the product can be analyzed through the whole task, the number of times emotional words selected, it's average score and the average of the representative emotions can be identified through the selected emotional words. After the experiment, the results while using the product and after using the product can be compared showing the average scores of representative emotions and emotional words selected from the subjective evaluation scale. For analysis of the entire experiment, not only can the subject be analyzed but the whole group of subjects can be analyzed too.

•*Analysis of subject's basic information:* Information on subject's serial number, gender, age, academic background, occupation, etc. and answers to questions related to the experiment is displayed. These data can be used in various ways to analyze the results of experiment through cross-tabulation analysis.

2.4 Significance of the Development of VideoTAME

To summarize the significance of the development of VideoTAME would be as follows:

- Firstly, a more realistic product-using environment is suggested by letting the subject to examine his/her own emotional changes occurred while performing the task by actually watching the video clip that was recorded. Distortion on emotional measurement that Scherer had once pointed out as the limitations of psychological measurement [13] can also be minimized.
- Second, the emotions naturally expressed by users in a natural environment can be evaluated effectively by complementing the previously burdensome and unnatural equipments needed for the physiological measuring method.
- Thirdly, by easily analyzing and collecting various data on facial expressions and reasons on why the emotion was felt, it has become easier to be approached from the design field.
- Fourth, by comparing the situation while using the product and the emotional words selected in that situation, it suggests the possibility of analyzing the emotional change occurred from user, and the usability of the product.

3 Suggestion of Physiological Methods for Measuring User's Emotions Expressed While Using a Product

A physiological measurement method that measures the user's emotions expressed during product use with VideoTAME and that is easily accessible in the design field was suggested.

Almaden laboratory, IBM, made an emotion mouse to measure six basic emotions, such as happiness, surprise, anger, fear, sadness and disgust, by sensing ECG, skin temperature, photoplethysmographic (PPG) and EDA [1]. 'INNO 2000', the emotion mouse by BIOPIA Co. Ltd., Korean venture, came out on the market for the first time in the world, earlier than IBM.

The emotion mouse of this venture senses and analyzes user's PPG and EDA, applied in a lie detector, and then transmits them to computer. A personal computer, connected with the emotion mouse, grasps the data and shows them to user real time. Moreover, the data is stored for a day or a month to trace the changing state of stress. Kim et al.

who participated in the emotion mouse project, evaluated the reliability of physiological signs measured by INNO mouse. They compared EDA signs and PPG signs detected by the emotion mouse and those signs detected by MP 100 system (Biopac systems, Inc.), which is broadly used as a tool for analyzing out the physiological signs. At the result, it shows high correlation in both signs. Therefore, the physiological signs by INNO mouse have high reliability paralleled with the existing physiological signs-detecting tool and it can be used with the mediocre tools [6].

Levenson et al. founded that the EDA has obvious difference between a positive emotion and a negative emotion and it increases when experiencing a negative emotion [8]. The increase of a plasma volume means the vasodilation of peripheral arterial and the decrease of a plasma volume means the vasoconstriction of peripheral arterial. It is caused by suppression and activation of the sympathetic system. PPG, recording the change of blood velocity by sensing the change of photo volume, is generally used to record plasma volume. In Levenson's research, the emotion of sadness is able to be distinguished from other emotions because the change of plasma volume was bigger in sadness than in anger, fear and disgust [7].

And the Eyegaze, which can measure examinee's eyeball movements and the change of pupil size, has potential to be applied to the research of understanding user's emotion. The examinee can move relatively free with a head set for Eyegaze, but it is still a weak point to have burdensome equipment. Thus, this study uses equipment of cornea boundary-reflecting technique, measuring the eyeball movements by reflection angle of infrared rays shot on the cornea. With this equipment, an examinee doesn't need to put burdensome tool on and an examiner can gain precise data with relatively moderate prices. However, an examinee may have a mental burden because he should not move his head in a process of calibration. This study suggests the potential of the Eyegaze to measure the change of emotion by observing user's pupil. Partala et al. showed that pupil size was significantly larger after both negative and positive than neutral stimulation. And the results showed that the pupil size was significantly larger during negative highly arousing stimuli than during moderately arousing positive stimuli. The pupil size was also significantly larger after highly arousing negative stimuli than after moderately arousing neutral and positive stimuli [9].

This study suggests the method using the emotion mouse and the Eyegaze to measure user's emotion while using a product. An examinee performs several tasks with the emotion mouse through the simulator of a product on the computer monitor connected to the Eyegaze. While testing, the emotion mouse senses user's EDA and PPG and transmits the data to the computer. In addition, the Eyegaze can observe the change of pupil size. And a video camera records user's facial expression while testing. It suggests the potential to measure the change of emotion expressed while using a product by analyzing the physiological signs and facial expressions. INNO mouse, by Biopia Co.Ltd., measures examinee's emotion by detecting EDA and PPG while testing. Eyegaze Development System, by LC Technologies Inc., is used to observe the change of pupil size. It can record examinee's eye movement by using the cornea-boundary reflecting technique with infrared ray camera and observe the change of pupil size.

4 Conclusion

In this research, I have suggested methods for measuring a user's emotions in the natural and accessible environment of the design field.

First, I developed a tool that effectively measures the user's emotions expressed while using a product and that can complement the limitations of the psychological measuring method. In the Testing Module of the developed tool, the subject can view the recorded video clip of he/she performing some specific task, and examine the emotional changes occurred. In the Analyzing Module, results can be verified by running the total experiment clip by each task, and then the data can be sent to Microsoft Excel and be analyzed in various ways. The tool developed through this research, can be used to effectively measure a user's naturally expressed emotion while using a product.

And a physiological measurement method that measures the user's emotions expressed during product use with VideoTAME and that is easily accessible in the design field was suggested. This research suggested the method using the Emotion Mouse and the Eyegaze to measure user's emotion while using a product. While testing, the Emotion Mouse senses user's EDA and PPG and transmits the data to the computer. In addition, the Eyegaze can observe the change of pupil size. And a video camera records user's facial expression while testing.

By upgrading VideoTAME and overcoming its limitations as a psychological measurement method, and also by using the physiological measurement method mentioned above to measure the user's emotional changes, the correlation between the product usability and the user's emotion will be able to be defined more clearly. Through this, it is hoped that a basic framework for the development of interface design with consideration to the user's emotions will be illustrated.

References

1. Ark, W., Dryer, D.C., Lu, D.J.: The Emotion Mouse. In: Proceedings of HCI International '99 (the 8th International Conference on Human-Computer Interaction) on Human-Computer Interaction: Ergonomics and User Interfaces, vol. I, pp. 818–823. Lawrence Erlbaum Associates, Inc, Mahwah, NJ (1999)
2. Cacioppo, J.T., Gardner, W.L.: Emotion. Annual Review of Psychology 50, 191–214 (1999)
3. Cacioppo, J.T., Klein, D.J., Berntson, G.G., Hatfield, E.: The psychophysiology of Emotion. In: Lewis, M., Haviland, J.M. (eds.) Handbook of Emotions, pp. 119–142. The Guilford Press, New York (1993)
4. Jeong, S.H., Lee, K.P.: Conceptual Framework for Emotions in Usability of Products. Korean Journal of the Science of Emotion & Sensibility 8(1), 17–28 (2005)
5. Jeong, S.H., Lee, K.P.: Extraction of user's representative emotions expressed while using a product. Journal of Korean Society of Design Science 18(1), 69–80 (2005) (in Korean)
6. Kim, H., Heo, C.W., Choi, J.H.: Evaluation of Reliability of the Emotional Function Mouse. Journal of the Korean Society of Jungshin Science 5, 28–36 (2001) (in Korean)
7. Levenson Levenson, R.W.: Autonomic nervous system differences among emotions. Psychological Science 3, 23–27 (1992)

8. Levenson, R.W., Ekman, P., Friesen, W.V.: Voluntary facial action generates emotion-specific autonomic nervous system activity. Psychophysiology 27, 363–384 (1990)
9. Partala, T., Surakka, V.: Pupil size variation as an indication of affective processing. International Journal of Human-Computer Studies 59, 185–198 (2003)
10. Plutchik, R.: Emotions and life: perspectives from psychology, biology, and evolution, pp. 181–222. American Psychological Association, Washington (2003)
11. Prkachin, K.M., Williams-Avery, R.M., Zwaal, C., Mills, D.E.: Cardiovascular changes during induced emotion: an application of lang's theory of emotional imagery. Journal of Psychosomatic Research 47, 255–267 (1999)
12. Rosenberg, E.L., Ekman, P.: Coherence between expressive and experimental systems in emotion. Cognition and Emotion 8, 201–229 (1994)
13. Scherer, K.R.: Studying emotion empirically: issues and a paradigm for research. In: Scherer, K.R., Wallbott, H.G., Summerfield, A.B. (eds.) Experiencing emotion: a cross-cultural study, pp. 3–27. Cambridge University Press, Cambridge (1986)

Unconscious Transmission Services of Human Feelings

Mitsuhiko Karashima and Yuko Ishibashi

School of Information Science and Technology, Tokai University, 1117 Kita-Kaname, Hiratsuka, Japan
mitsuk@keyaki.cc.u-tokai.ac.jp, yuuuko105@hotmail.com

Abstract. This paper was focused on the next generation of ubiquitous services by using ubiquitous networks and devices. This paper was especially focused on the transmission services of feelings to others. In the paper, some conventional transmission services of feelings were introduced and a few of the unconscious transmission services of feelings to both specific targets and multiple targets were pointed out. The paper proposed two examples of the new services that would enable the users to transmit their feelings unconsciously to others by using ubiquitous networks and devices. These services were proposed through the PRESPE (Participatory Requirements Elicitation using Scenarios and Photo Essays) approach and called respectively the "aura transmission system" and "back scratcher system". Further discussion about the usefulness and possible negative influences on human nature or society by the new services was done, and the future research efforts of these services were described.

Keywords: ubiquitous services, transmission of feelings.

1 Introduction

Since the first advocating of "ubiquitous computing" by Mark Weiser at The Xerox Palo Alto Research Center around 1990, many researchers and governors have advocated "ubiquitous network systems", "ubiquitous interaction devices", and "ubiquitous services" for the future. For example, MIC (Ministry of Internal Affairs and Communications) Of Japan aims to achieve a "ubiquitous network society" (u-Japan) in which "anything and anyone" can easily access networks and freely transmit information "from anywhere at any time" by 2010. Ubiquitous network infrastructure has been developed and spread as in areas such as broadband (DSL or FTTH) network, public wireless LAN, and mobile phone internet in Japan. Many kinds of ubiquitous interaction devices, such as μ tip and RFID (Radio Frequency Identification) have also been developed. Many kinds of ubiquitous services using ubiquitous networks and devices, such as child monitoring systems by GPS mobile phone and user monitoring systems by home appliances, have been proposed.

In many proposed services, however, the users who the service developers centered on in the design processes of the proposed services were passive to the information

M.J. Smith, G. Salvendy (Eds.): Human Interface, Part I, HCII 2007, LNCS 4557, pp. 68–76, 2007.

from the services and could only acquire the information which was provided anytime and anywhere. There were few services by which the users could transmit the information anytime and anywhere (mobile emailing). The services where the users can acquire or transmit the subjective information are especially rare in Japan.

The blog hosting services and social networking services produce the virtual spaces in which the users can transmit their feelings as well as the objective information to the multiple targets. Some prototype services have also been developed in which the users could transmit their feelings to a specific person. IBM Japan Ltd. developed the "Kansei mail" system with the website, in which the mail sender could send the emotion of their emailing to the mail receiver by using special background animation with the text [1]. NEC Corp. developed "KOTOHANA" with two appliances in flower form and the internet, in which the user could transmit the emotion of their speech with the flower to the specific person who had the other flower by changing the color of the LED [2].

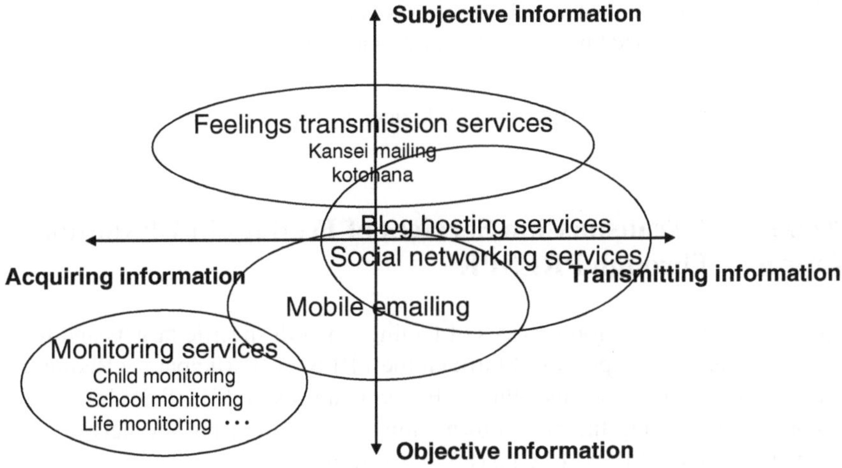

Fig. 1. Location of conventional ubiquitous services using ubiquitous networks and devices in two-dimensional space with Acquiring-Transmitting information axis and Subjective-Objective information axis

The development of these conventional transmission services of feelings revealed the reality of the feeling transmission services using ubiquitous networks and devices. Though in some cases persons might want to transmit the feelings intentionally, in other cases they might want to transmit the feelings unconsciously. These services, however, only targeted the communication between specific members and mainly the intentional transmission of feelings. The blog hosting services and social networking services also targeted the intentional transmission of feelings.

In this paper some examples of the active services are presented that would enable the users to transmit their feelings unconsciously to both specific targets and multiple

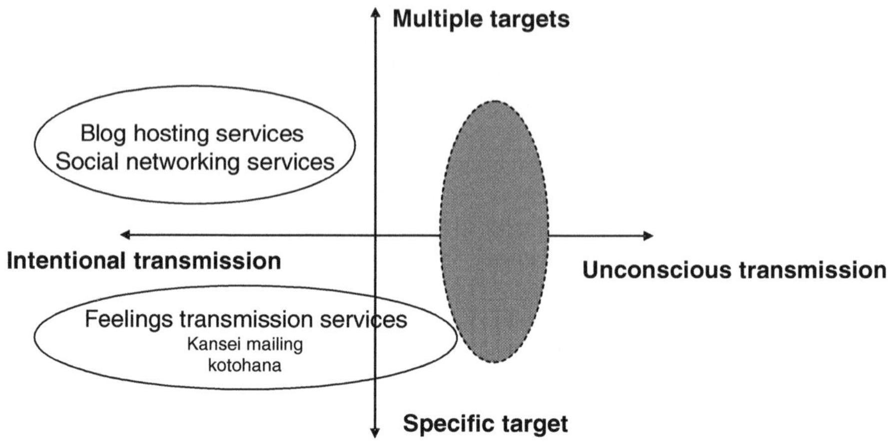

Fig. 2. Location of conventional ubiquitous services in two-dimensional space with Intentional-Unconscious targets axis and Specific-Multiple transmission axis

targets by using ubiquitous networks and devices as the next generation of ubiquitous services.

2 Proposal of Transmission Services of Feelings in Ubiquitous Networks Through PRESPE

The unconscious transmission services of feelings to both specific targets and multiple targets in this paper were proposed through the PRESPE (Participatory Requirements Elicitation using Scenarios and Photo Essays) approach [3]. Two brainstorming discussions for proposing the new information services by using the network system were done through the PRESPE. One (BS1) was done by the group of the universities' faculty members, students, and the designers in the collaborative workshop of ERGO-DESIGN MEETING and INFORMATION SOCIETY ERGONOMICS MEETING of JAPAN ERGONOMICS SOCIETY. The other (BS2) was done by the group of our university students in our laboratory.

2.1 PRESPE Procedure

The PRESPE was a participatory design approach introduced by Dr. Go, in which a group of major stakeholders worked together to produce and evaluate new designs of products and services. The PRESPE approach is consisted of two roles: coordinator and participants. The coordinator assigns a project theme and provides some support for the participants' activities. The participants work together in order to produce new products/services under the coordinator's control. The outline of the procedure of the PRESPE in this paper is shown in Table 1, which was slightly different from the original PRESPE procedure. Procedure (3), (4), and (5) are the brainstorming sessions.

Table 1. Outline of procedure of PRESPE

Procedure	Contents
(1)	One of the participants makes a photo-diary according to the requirement of the coordinator.
(2)	For the assigned theme, he/she creates photo-essays to reflect the personal experience with existing artifacts.
(3)	The participants choose the photo from the photo-diary which he/she made. They analyze the photo-essay, identify the concept behind it, and develop the design concept.
(4)	They envision the use scenarios and context of the newly designed product/service and translate scenes described in the scenarios into artifacts by making sketches of the use scenes.
(5)	They also conduct a claim analysis on the newly designed product/service to enumerate the potential tradeoffs.

2.2 PRESPE Conditions

In BS1, the participants were three different universities' faculty members, two different universities' students, and one designer. In BS2, the participants were three university students. In both brainstorming sessions, one of the participants who was a university student was required to take photos of one day from getting up to going to bed as the photo-diary and make the photo-essays for all photos. The assigned theme was "Something that makes me feel happy". The participants can propose the new service without technical constraints.

3 Proposed New Transmission Services of Feelings

3.1 Brainstorming Session 1

The participants chose the photo of Fig.3 from the photo-diary. The photo-essay of this photo is as follows. She felt happy that she read a book, did homework, and prepared something for a seminar at the lounge of the university. However she felt unhappy that she was obstructed by her friends when she was concentrating on doing something, while she welcomed the obstruction by her friends when she did something to kill time. If her feelings whether she wants to concentrate or not can be transmitted unconsciously to her friends, it will make her happier.

The participants analyzed the essay and they identified the keyword behind the essay. The keyword was "concentration". The participants conducted the use scenarios and context of the new service as follows.

One morning, a student is preparing something for the seminar. She has to make the report of the results of this week's research activities by one p.m.. Her friends find her. They try to talk to her, but the newly designed service makes them notice the feeling that she wants not to be obstructed by them. She can keep her concentration for the preparation of the seminar as they know her feeling from the new service and stop talking to her.

On the other hand, one afternoon, a student reads a book at the lounge to kill time because a class is cancelled. Her friends find her. The designed service makes them notice her feelings that she is bored and wants to talk with someone. She can enjoy talking with them as they know her feeling from the new service and talk to her.

Fig. 3. The photo chosen by the participants in BS1

(a) Scenario she is trying to concentrate

(b) Scenario she is bored

Fig. 4. Aura transmission system

Fig.4 illustrates the newly designed service by the participants which is named the "aura transmission system". This system aims at the unconscious presentation of the users' faint feelings, where it is mentally difficult to tell others in words either by text or voice in public spaces, e.g., "I don't want to be obstructed by any others", "I'm bored and want to talk with someone". The system is consisted of the sensors for gathering the physiological information related to autonomic nerve system, database system for analyzing the users' feelings from physiological information, representing devices of the users' feelings, and the ubiquitous network for data communication.

3.2 Brainstorming Session 2

The participants chose the photo of Fig.5 from the photo-diary. The photo-essay of this photo is as follows. She felt happy that she did cardio exercise by indoor bicycling while watching TV. However she felt unhappy that someone of her family visited her room when she did serious exercise, while she also felt unhappy that anyone of her family didn't visit her room in order to talk with her when she did relaxing exercise while watching TV and the TV program was boring. If her feelings whether she wants to be alone or not can be transmitted unconsciously to her family, it will make her happier.

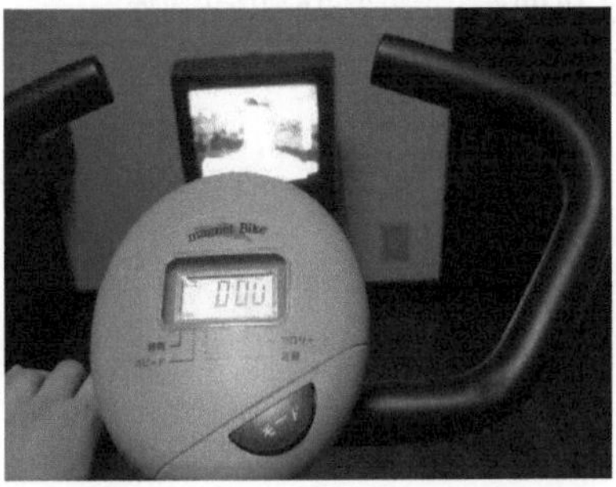

Fig. 5. The photo chosen by the participants in BS2

The participants analyzed the essay and they identified the keyword behind the essay. The keyword was "privacy". The participants conducted the use scenarios and context of the new service as follows.

One evening after dinner, a student does serious cardio exercise by indoor bicycling while watching TV. Her mother wants to visit her room and talk with her. Her mother tries to visit her room, but the newly designed service makes her mother notice the feeling that she doesn't want anyone to visit her room. She can keep on doing serious exercise alone as her mother knows her feeling from the new service and stops visiting her room.

On the other hand, one midnight a student does relaxing exercise while watching TV. She wants to talk with her mother because the TV program is boring. The newly designed service makes her mother notice the feeling that she wants to talk with her mother. She can talk with her mother during relaxing exercise as her mother knows her feeling from the new service and visits her room.

(a) Scenario she does serious exercise

(b) Scenario she does relaxing exercise

Fig. 6. Back scratcher system

Fig.6 illustrates the newly designed service by the participants which is named the "back scratcher system". This system aims at the unconscious presentation of the users' feelings, where it is troublesome to tell specific others in words either by text or voice in private spaces, e.g., "I want to talk with you now", "I don't want you to contact me now unless only briefly". The system is consisted of the sensors for gathering the physiological information related to autonomic nerve system, the video monitoring system for gathering the behavior information, database system for analyzing the users' feelings from the physiological and behavioral information, representing devices of the users' feelings, and the ubiquitous network for data communication.

4 Discussions

This paper proposed two newly designed services to transmit human feelings unconsciously to others through the PRESPE. The first service was designed from the keyword "concentration" and the second service was designed from the keyword “privacy”. Although these services were designed from different keywords and by different participants, both of them can be regarded as examples of the active services with which the users could transmit their feelings unconsciously to others by using ubiquitous networks and devices. It suggested that there are some potential needs of the services to transmit human feelings unconsciously to others as the next generation of the ubiquitous services.

On the other hand, in the claim analysis of both new services, the participants indicated that the users did not always want their feelings to be known to others. It revealed the necessity of the option in the services that the users could choose whether they used these transmission services of their feelings or not.

The participants also indicated that the others could not understand correctly the feelings because they could not understand the background of why the users had the feelings as these new services were only able to transmit the users’ feelings to others. These services should also transmit the information to the others that the others can understand the background of why the users have the feelings. Further research to clarify what kinds of information about the situations and the contexts are needed in order to enable the others to understand the reasons why the users have the feelings are in the field of the Ethno-methodology[4]. Future Ethno-methodological research will be needed in order to select what kinds of information about the situations and the contexts are transmitted to the others with the feelings in these services.

These new services might be useful for the users because their feelings are transmitted unconsciously to others without any troublesome effort. As the third indication, however, there might be the apprehension that it might have some negative influences on human nature or society that the users’ communication ability to transmit feelings to others is decreased by the transmission services of feelings such as the proposed services. Further research like the socio-technical design approach will be needed in order to clarify the influences of these new services on human nature or society[5] and it should be considered where these services are adopted even though these services can be developed technically wherever the people want.

5 Conclusion

Through the PRESPE, this paper presented two examples of the active services that would enable the users to transmit their feelings unconsciously to others by using ubiquitous networks and devices as the next generation of ubiquitous services. It suggested that there are some potential needs of the services to transmit human feelings unconsciously to others.

One was named the "aura transmission system" which aimed at the unconscious presentation of the users' faint feelings, where it was mentally difficult to tell others in words either by text or voice in public spaces. The other was named the "back scratcher system" which aimed at the unconscious presentation of the users' feelings, where it is troublesome to tell specific others in words either by text or voice in private spaces.

These new services might be useful for the users because their feelings were transmitted unconsciously to others without any troublesome effort. Some claims, however, which should be solved before the development of these services were indicated, and the directions of the future researches were also indicated in order to solve these claims.

References

1. Yamazaki, K., Furuta, K.: Kansei Interface Design for Network Computing. In: Proceedings of HCI International 2005 (2005) incl. in CD-ROM
2. NEC Cebit 2006, KOTOHANA (2006), http://www.nec-cebit.com/de/produkte/future-products/ kotohana.html
3. Go, K., Takamoto, Y., Carroll, J.M., Imamiya, A.: Envisioning Systems Using a Photo-Essay Technique and a Scenario-Based Inquiry. In: Proceedings of HCI International 2003, pp. 375–379 (2003)
4. Garfinkel, H.: Studies in Ethnomethodology. Polity Press (Blackwell Publishing), Oxford (1984)
5. Eason, K.: Understanding the Organizational Ramifications of Implementing information technology Systems. In: Helander, M., et al. (ed.) Handbook of Human-Computer Interaction, 2nd edn., pp. 1475–1495. Elsevier Science Publishers, Amsterdam (1997)

Do Beliefs About Hospital Technologies Predict Nurses' Perceptions of Their Ability to Provide Quality Care? A Study in Two Pediatric Hospitals

Ben-Tzion Karsh[1], Kamisha Escoto[2], Samuel Alper[1], Richard Holden[1], Matthew Scanlon[3], Kathleen Murkowski[4], Neal Patel[5], Theresa Shalaby[6], Judi Arnold[6], Rainu Kaushal[7], Kathleen Skibinski[8], and Roger Brown[9]

[1] Department of Industrial and Systems Engineering, University of Wisconsin-Madison, Madison, WI US
[2] Division of Health Services Research and Policy School of Public Health, University of Minnesota, Minneapolis, MN US
[3] Department of Pediatrics, Division of Critical Care, Medical College of Wisconsin, Milwaukee, WI US
[4] Children's Hospital of Wisconsin, Milwaukee, WI US
[5] Division of Pediatric Critical Care and Anesthesia, Department of Pediatrics, Vanderbilt Children's Hospital, Nashville, Tennessee US
[6] Vanderbilt Children's Hospital, Nashville, Tennessee US
[7] Department of Public Health, Weill Medical College, Cornell University, New York, New York US
[8] School of Pharmacy, University of Wisconsin-Madison, Madison, WI US
[9] School of Nursing, University of Wisconsin-Madison, Madison, WI US

Abstract. The purpose of this study was to test the hypothesis that nurse perceptions of technology they use in practice would affect their perception that they were able to provide high quality patient care. A survey assessing the variables was administered to 337 pediatric nurses from two academic freestanding pediatric hospitals in the US. Two separate equations were constructed, one to test whether technology perceptions affected individual quality of care and the other to test whether technology perceptions affected quality of care provided by the nursing unit. Nurse confidence in their ability to use hospital technology and their beliefs that the technologies were easy to use, useful, and fit their tasks are important predictors of nurse beliefs that they are able to provide quality care to their patients.

Keywords: quality of care, automation, information technology, self-efficacy.

1 Introduction

Nurses play a significant role in the quality of care patients receive [1] and spend the most time, among health care providers, with patients [2, 3]. Nurses have numerous demands on their time and practice in an environment with distractions and interruptions, often without supplies and resources readily available for patient care [4, 5]. Their ability to provide high quality care is increasingly difficult in such environments, and it is getting more complicated with automation.

M.J. Smith, G. Salvendy (Eds.): Human Interface, Part I, HCII 2007, LNCS 4557, pp. 77–83, 2007.

There are a wide variety of pressures on health care delivery organizations to adopt automation because of pressures from the government, purchasing groups and consumers [6-10]. This pressure seems to be having an effect. Recent estimates suggest that up to 40% of US hospitals are planning on implementing electronic order entry within the next five years [11]. Similarly, a 2002 survey found that 50% of the responding hospitals were considering implementing bar coding technology [12]. It is clear then, that the rapid pace of new technology introductions into healthcare delivery organizations will continue into the foreseeable future.

Perhaps the most talked about type of new healthcare automation is information technology (IT). The information technologies being implemented that can influence nursing care include electronic health records for viewing patient records, bar-coded medication administration systems to help ensure the five rights of medication administration, electronic nurse charting to help electronically capture patient vitals, smart IV pumps to improve the safety of intravenous medication administration, and electronic medication administration records which provide an electronic display of medication information. These technologies may help or hinder the provision of care depending on how well they support the nurses performance [13-17].

These technologies have the potential to greatly improve the quality of health care delivery. Some research suggests that the potential benefits, specifically patient safety benefits, can in fact be realized [9, 18-23]. However evidence is also emerging that these technologies, if not well designed, can lead to errors and patient harm [24-28]. This means more research is needed to understand how technologies can impact quality. As Bates warned, "the net effect (of information technologies) is . . . not entirely predictable, and it is vital to study the impact of these technologies" [18], p. 789). The purpose of this study was therefore to test the hypothesis that nurse perceptions of technology they use in practice would affect their perception that they were able to provide high quality patient care.

2 Methods

2.1 Study Design

A cross-sectional survey design was used to collect all data.

2.2 Sample

Full-time nurses from the pediatric intensive care unit, hematology-oncology unit, and a general medical/surgical unit of two freestanding academic pediatric hospitals were eligible to participate. Temporary nurses, float nurses, or nurses who did not directly provide patient care (e.g., nurses in educational positions) were not included in the target sample. The eligible sample from Hospital 1 was 203 nurses, and the eligible sample from Hospital 2 was 144 nurses. The response rates were 59.6% (n=121) and 54.2% (n=78) respectively. Of these, a majority were female (n = 193) and white (n = 191). 84 (42.2%) were between the ages of 18 and 29, 55 (27.6%) were between 30 and 39, 42 (21.1%) were between 40 and 49, and 18 (9.0%) were more than 50 years old. 170 (85.4%) had completed a college degree, and 19 (9.5%) had completed graduate or professional school.

2.3 Measures

All items were measured on 7-point Likert-type scales ranging from 0-6 and also included a "don't know" option. The anchors were 0 (Not at all) to 6 (A great deal). There were two outcome measures. The first, perceived individual ability to provide quality care, was a new 5-item measure developed based on components of nursing quality by Lynn and Moore[29] and Williams [30]. The second measured nurses' perceptions of the quality of care delivered on their unit. It too was a 5-item measure and was based on Shortell et al. [31].

The independent variables were chosen based on the Technology Acceptance Model [32-35], and were comprised of beliefs about current technologies used in the hospital. The wording of each question referred to the hospital's current technologies. Consistent with the Technology Acceptance Model, the beliefs measured were perceived ease of use, perceived usefulness, perceived compatibility with individual work style, subjective norms, and technology self-efficacy. Questions about ease of use, usefulness and compatibility were combined into a single 6-item measure of technology fit. Among the six items, two items measured technology usefulness and two measured ease of use based on Venkatesh et al. [36]. The two other items measured compatibility with work style and were from Moore and Benbasat [37]. Subjective norms was examined with two different measures. The first was a 2-item measure newly developed to measure nurses perceptions that patients or patients' families would want nurses to use the current hospital technologies. The second was a 2-item measure assessing the extent to which people important to the nurses would want them to use the hospital's technologies [38]. Finally, computer self-efficacy was measured using three items from Taylor and Todd [39].

Questionnaire items were evaluated using expert review and cognitive interviewing, the two most effective methods recommended for evaluating survey questions for content and comprehensibility [40-42].

2.4 Procedures

The study was publicized among the study units with a combination of informational in-services and presentations at staff meetings. Nurses were distributed a survey packet at the conclusion of the meetings or individually during work hours if they were not in attendance. Each survey packet contained a personalized cover letter, the employee survey, an informational sheet/consent form, a stamped business reply envelope, and a $5USD cash incentive. Nurses were instructed to fill out the survey on their own time, and place it in the U.S. mail using the enclosed business reply envelope. Reminder postcards were placed in all nurse mailboxes about one week after the initial distribution. Another employee survey was placed in non-respondent mailboxes approximately 7-10 days later. A final reminder postcard was distributed in mailboxes following another 7-10 days. Data were collected November-December 2005 and March-May 2006 at Hospital 1 and 2, respectively. Survey administration and data entry were conducted by the University of Wisconsin Survey Research Center. All data were double entered to ensure accuracy.

2.5 Analysis

Multiple regression was used. Two separate equations were constructed, one to test whether technology perceptions affected individual quality of care and the other to test whether technology perceptions affected quality of care provided by the nursing unit. Each model adjusted for hospital, unit, shift, average hours per week, job tenure, unit tenure, employer tenure and occupational tenure.

3 Results

The regression model for individual quality of care was significant ($F=5.12$, $p<0.001$). The adjusted R^2 for the model with the 8 covariates was 0.03. The adjusted R^2 for the model with the independent variables added was 0.24. Task-technology fit beliefs and technology self-efficacy beliefs both significantly predicted nurse perceptions of their individual ability to provide quality care (beta 3.93 and 4.17 respectively, $p < 0.001$). The regression model for unit quality of care was significant ($F=9.32$, $p<0.001$). The adjusted R^2 for the model with the 8 covariates was 0.10. The adjusted R^2 for the model with the four independent variables added was 0.31. Only task-technology fit beliefs predicted nurse perceptions of quality of care delivered by their unit (beta 5.31, $p < 0.001$).

4 Discussion

Nurse confidence in their ability to use hospital technology and their beliefs that the technologies fit (were easy to use, useful, and fit their work style) were important predictors of nurse beliefs that they are able to provide quality care to their patients. Technology fit also predicted perceptions that the nursing unit as a whole provided quality care. Neither measure of subjective norms was a significant predictor of the dependent variables. With the growing push for hospitals to rapidly implement health information technologies for improving quality of care and safety, care must be taken to ensure the designs fit the context. Nurse beliefs that technologies do not meet their needs could affect their behaviors with the technologies, which could further impact quality and safety.

Acknowledgments. This study was funded by a grant from the US Agency for Healthcare Research and Quality (1 R01 HS013610).

References

1. Institute of Medicine,: Patient Safety: Achieving a New Standard for Care. National Academies Press, Washington DC (2004)
2. Hall, L.M., Doran, D.: Nurse staffing, care delivery model, and patient care quality. Journal of Nursing Care. Quality 19, 27–33 (2004)

3. McNally, K.M., Page, M.A., Sunderland, B.V.: Failure-model and effects analysis in improving a drug distribution system. American Journal of Health System Pharmacist 54, 171–177 (1997)
4. Wolf, L., Potter, P., Sledge, J., Boxerman, S., Grayson, D., Evanoff, B.: Describing nurses' work: Combining quantitative and qualitative analysis. Human Factors 48, 5–14 (2006)
5. Ebright, P., Patterson, E., Chalko, B., Render, M.: Understanding the complexity of registered nurse work in acute care settings. Journal of Nursing Administration 33, 630–638 (2003)
6. Institute of Medicine: To Err is Human: Building a Safer Health System. National Academy Press, Washington DC (2000)
7. Institute of Medicine: Institute of Medicine: Crossing the Quality Chasm: A New Health System for the 21st Century. National Academy Press, Washington DC (2001)
8. Institute of Medicine: Preventing Medication Errors. National Academy Press, Washington DC (2007)
9. Shojania, K.G., Duncan, B.W., McDonald, K.M., Wachter, R.M. (eds.): Making Health Care Safer: A Critical Analysis of Patient Safety Practices. Agency for Healthcare Research and Quality, Rockville, MD (2001)
10. The Leapfrog Group: The Leapfrog Group for patient safety: rewarding higher standards. (Accessed March 6, 2004), Available at http://www.leapfroggroup.org
11. Merli, R., Stickler, M.: E-prescriptions promise patient safety, higher drug sales. (Accessed March 6, 2004), Available at http://www.kpmginsiders.com/display_analysis.asp?cs_id=96118
12. Institute for Safe Medication Practices: ISMP survey shows drug companies providing fewer unit dose packaged medications (March 6, 2002 ISMP Medication Safety Alert). (Accessed March 6, 2004) (2002), Available at http://www.ismp.org/msaarticles/cuttingprint.htm
13. Karsh, B.: Beyond usability for patient safety: designing effective technology implementation systems. Quality and Safety in Healthcare 13, 388–394 (2004)
14. Karsh, B.: A conceptual framework for understanding the safety implications of bar code enabled point of care (BPOC) systems. In (eds.): 11th International Conference on Human-Computer Interaction. Las Vegas (2005)
15. Karsh, B., Alper, S.J., Holden, R.J., Or, K.L.: A human factors engineering paradigm for patient safety – designing to support the performance of the health care professional Quality and Safety in Healthcare 15, i59–i65 (2006)
16. Karsh, B., Holden, R.: New technology implementation in health care. In: Carayon P. (eds.): Handbook of Human Factors and Ergonomics in Patient Safety (In Press, 2005)
17. Karsh, B., Scanlon, M.: Using bar code technology for medication administration: special human factors considerations for pediatric populations. In: Proceedings of the International Conference on Healthcare Systems Ergonomics and Patient Safety, pp. 135–138 (2005)
18. Bates, D.W.: Using information technology to reduce rates of medication errors in hospitals. British Medical Journal 320, 780–791 (2000)
19. Bates, D.W., Cohen, M., Leape, L.L., Overhage, J.M., Shabot, M.M., Sheridan, T.: Reducing the frequency of errors in medicine using information technology.[see comment]. Journal of the American Medical Informatics Association 8, 299–308 (2001)
20. Bates, D.W., Gawande, A.A.: Patient safety: Improving safety with information technology New England. Journal of Medicine 348, 2526–2534 (2003)

21. Kaushal, R., Barker, K.N., Bates, D.W.: How can information technology improve patient safety and reduce medication errors in children's health care? Archives of Pediatrics & Adolescent Medicine 155, 1002–1007 (2001)
22. Kaushal, R., Bates, D.W.: Information technology and medication safety: what is the benefit? Quality & Safety in Health Care. 11, 261–265 (2002)
23. Kuperman, G.J., Gibson, R.F.: Computer physician order entry: benefits, costs, and issues. Annals of Internal Medicine 139, 31–39 (2003)
24. Han, Y.Y., Carcillo, J.A., Venkataraman, S.T., Clark, R.S.B., Watson, R.S., Nguyen, T.C., Bayir, H., Orr, R.A.: Unexpected increased mortality after implementation of a commercially sold computerized physician order entry system Pediatrics, vol. 116, pp. 1506–1512 (2005)
25. Koppel, R., Metlay, J.P., Cohen, A., Abaluck, B., Localio, A.R., Kimmel, S.E., Strom, B.L.: Role of computerized physician order entry systems in facilitating medication errors. Journal of the American Medical Association 293, 1197–1203 (2005)
26. Nebeker, J.R., Hoffman, J.M., Weir, C.R., Bennett, C.L., Hurdle, J.F.: High rates of adverse drug events in a highly computerized hospital Archives of Internal Medicine, vol. 165, pp. 1111–1116 (2005)
27. Patterson, E.S., Rogers, M.L., Chapman, R.J., Render, M.L.: Compliance with intended use of bar code medication administration in acute and long-term care: An observational study Human Factors, vol. 48, pp. 15–22 (2006)
28. Thompson, D.A., Duling, L., Holzmueller, C.G., Dorman, T., Lubomski, L.H., Dickman, F., Fahey, M., Morlock, L.L., Wu, A.W., Pronovost, P.J.: Computerized physician order entry, a factor in medication errors: descriptive analysis of events in the intensive care unit safety reporting system. Journal of Clinical Outcomes Management 12, 407–412 (2005)
29. Lynn, M.R., Moore, K.: Relationship between traditional quality indicators and perceptions of care. Seminars for Nurse Managers 5, 187–193 (1997)
30. Williams, A.: The delivery of quality nursing care: A grounded theory study of the nurse's perspective. Journal of Advanced Nursing 27, 808–816 (1998)
31. Shortell, S.M., Zimmerman, J.E., Rousseau, D.M., Gillies, R.R., Wagner, D.P., Draper, E.A., Knaus, W.A., Duffy, J.: The Performance of Intensive-Care Units - Does Good Management Make a Difference Medical Care, vol. 32, pp. 508–525 (1994)
32. Davis, F.D.: Perceived usefulness, perceived ease of use, and user acceptance of information technology. Mis. Quarterly 13, 319–340 (1989)
33. Venkatesh, V.: Determinants of perceived ease of use: Integrating control, intrinsic motivation, and emotion into the technology acceptance model. Information Systems Research 11, 342–365 (2000)
34. Venkatesh, V., Davis, F.D.: A theoretical extension of the Technology Acceptance Model: Four longitudinal field studies. Management Science 46, 186–204 (2000)
35. Venkatesh, V., Morris, M.G., Davis, G.B., Davis, F.D.: User acceptance of information technology: Toward a unified view. Mis. Quarterly 27, 425–478 (2003)
36. Venkatesh, V., Speier, C., Morris, M.G.: User acceptance enablers in individual decision making about technology: Toward an integrated model. Decision Sciences 33, 297–316 (2002)
37. Moore, G.C., Benbasat, I.: Development of an instrument to measure the perceptions of adopting an information technolgy innovation. Information Systems Research 2, 192–222 (1991)
38. Venkatesh, V., Morris, M.G., Ackerman, P.L.: A longitudinal field investigation of gender differences in individual technology adoption decision-making processes. Organizational Behavior and Human Decision Processes 83, 33–60 (2000)

39. Taylor, S., Todd, P.: Assessing IT usage: the role of prior experience. Mis. Quarterly 19, 561–570 (1995)
40. Lessler, J.T., Forsyth, B.H.: A Coding System for Appraising Questionnaires. In: Schwarz, N., Sudman, S. (eds.) Answering Questions: Methodology for Determining Cognitive and Communicative Processes in Survey Research, pp. 259–292. Jossey-Bass Publishers, San Francisco, CA (1996)
41. Presser, S., Blair, J.: Survey Pretesting: Do Different Methods Produce Different Results? In: Marsden, P.V. (ed.) Sociological Methodology, pp. 73–104. Basil Blackwell, Oxford (1994)
42. Groves, R.M., Fowler, F.J., Couper, M.P., Lepkowski, J.M., Singer, E., Tourangeau, R.: Survey Methodology. Wiley-Interscience, Hoboken, NJ (2004)

Information Design for User's Reassurance in Public Space

Yurika Katagiri and Minoru Kobayashi

NTT Cyber Solutions Laboratories, NTT Corporation
1-1 Hikari-no-oka, Yokosuka-shi, Kanagawa, 239-0847 Japan
Katagiri.yurika@lab.ntt.co.jp, minoru@acm.org

Abstract. We are developing an intuitive information design that can increase the user's feeling of reassurance and security when handling private information in public space. We conducted an experiment to collect user responses to displays providing awareness of the activity behind them while they handled private information. Two approaches were examined: a simple two-color LED array and a mirror. The former yielded less distraction and faster response speeds while the latter was more intuitive and offered more reassurance.

Keywords: Reassurance, Public Space, Information Presentation.

1 Introduction

Many kinds of information are handled in public space because of the increased use of mobile terminal devices. However, risk becomes a concern when people enter highly confidential information like personal data; people have a strong awareness about the value of such information and worry about its leakage. Even though new laws and new technologies offer some relief, we still worry about leakage in real public space situations [1]. We are concerned with providing a sense of safety and reassurance to the usage of information systems and networks in public spaces.

Under the current technology, there is a space design technique that uses partitions in innovative manners to help users feel safe and reassured. If an environment is made using this space design technique, users can handle private information with little concern for what might be occurring around them. But the technique is still rather immature and difficult to apply it to all public spaces. We suggest replacing the partition technique with methods of dynamically presenting information about the user's environment. In public spaces the movement and flow of people is ever-changing thus private information usage situations change too. Providing with dynamic information of the environment a person can have consistent understanding and awareness of their specific situation.

The purpose of our research is to design and evaluate methods to present this awareness information. A user should easily feel reassured that others cannot see or access their private information when they are handling it in public spaces.

M.J. Smith, G. Salvendy (Eds.): Human Interface, Part I, HCII 2007, LNCS 4557, pp. 84–89, 2007.

2 Experiment

To use information with reassurance, a user must pass through processes of perceiving, understanding, and evaluating their surroundings. The most effective way to complete all processes is to present indicators of the current environment state. We considered two different presentation methods: one using a two-color LED array and another using a mirror. The LED array was used to present a rather abstract level of situation understanding. The mirror was used present the environment exactly as it is. Five trials (Table 1) were conducted using a mixture of these two methods and a setting where no information is provided at all.

A person's reassurance is mostly determined by qualitative factors, but we also had to consider the efficiency of their information use so we collected both subjective and numeric data.

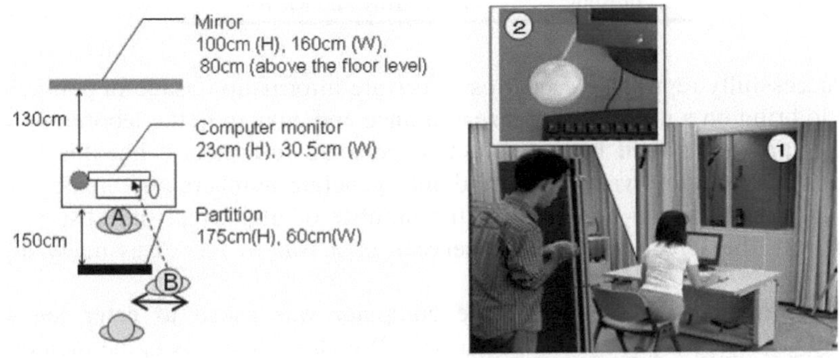

Fig. 1. The layout of laboratory (*left*) and the experimental setup (*right*)

2.1 Method

We attempted to create a scenario where an information user would be anxious about their data being captured by an observer standing behind the user. The mirror and the LED array were used to provide the user with awareness of the observer. The subjects, five men and fourteen women, were paid to participate in the experiment. One of the five men performed only as the observer in all trials. We used a man as the observer since a man's presence is sensed more heavily in personal space than a woman's [2].

The experimental setup is depicted in Figure 1. *A* is the information user and *B* is the observer. The information user was seated at a common office desk equipped with a keyboard and PC monitor (21 inch). Between the information user and an observer was a partition large enough to completely hide one man. The distance between the information user and the observer was set to 1.5 m as this is the radius around a person which, if a strange comes closer than this distance, he or she will feel uncomfortable [3]. The PC's numeric font was specified at 26 point font so that the observer could read the screen characters without difficulty. For the presentation method using a mirror, the mirror was large enough to provide full coverage of the region behind the information user (Figure 1). For the LED array method, the LED array (inset 2 in Figure 1) was placed on the desk next to the keyboard. It was 110 mm in diameter and each LED was 11 mm in diameter (inset 1 in Figure 1). The LED array was driven to illuminate either all red or all green LEDs. It is said that red color causes anxiety and green color provides reassurance [4]. LED color was switched by an experimenter according to whether the observer was in a position where he could view the PC screen (red) or not (green).

Table 1. Trials in the experiment

Observer	Information presentation
none	none
present	none
present	LED array
present	Mirror
present	LED array and Mirror

To successfully reproduce scenarios of private information usage in public spaces, we had to bring on a user's sense of reassurance and anxiety in the laboratory setting. The best approach would be to use actual personal information but this would be unethical. Our solution was to use randomly generate numbers and create a type of game; the user was told to minimize the number of digits that the observer could capture from the screen while the observer's goal was to record as many digits as possible.

The information user seated at the computer was asked to enter ten sets of randomly generated 8 digit numbers per trial. We chose 8 as this is the memory limit of most people [5]. The observer had to see, memorize, and write the numbers down. The request to enter ten 8 digit numbers per trial was made because we wanted to emphasize the difference in task time among each trial and the information users. The font size and input form size and position could not be changed but we allowed the information user to cover the PC monitor with their hand or body. The observer was instructed to periodically move from behind the partition to a specified side behind the information user. The total standing time in each position was the same for all information users.

In each trial, the information user was presented with a printed list containing the randomly generated ten sets of 8 digit numbers. The list could not be seen from any position of the observer. After each trial, the information user answered, in writing, a few questions about their impressions of the situation just experienced. Trial order was randomly set.

2.2 Results

After all 5 trials were completed the information users answered a final questionnaire asking which information presentation methods provided a stronger feeling of reassurance. For example, they were asked 'which arrangement, the LED array or mirror, provided a stronger feeling of reassurance?' Seven information users (participant IDs 12-18) felt more reassurance using the LED array, while the eleven other users (participants IDs 1-11) preferred the mirror. The following section classifies the information users according to their answer and discusses the time taken to finish each trial and the amount of data captured by the observer.

First, we discuss the eleven participants who preferred the mirror (data points marked by triangles in Figure 2). All of them had longer task times using the mirror method. A subgroup of five subjects (participant IDs 1, 2, 3, 4, 5) had the greatest number of digits captured by the observer when using the LED array. From the questionnaire results, a subgroup of eight subjects (participant IDs 1-6, 8, 11) did not care how long the task took. These subjects also said that the mirror was credible and steady. Overall, the mirror was preferred because it allowed a user to watch the observer and to have a good overall view of the observer's movement and other things in the environment. It was more important for them to enter the data slowly so that they may pay attention that the observer does not capture any of it.

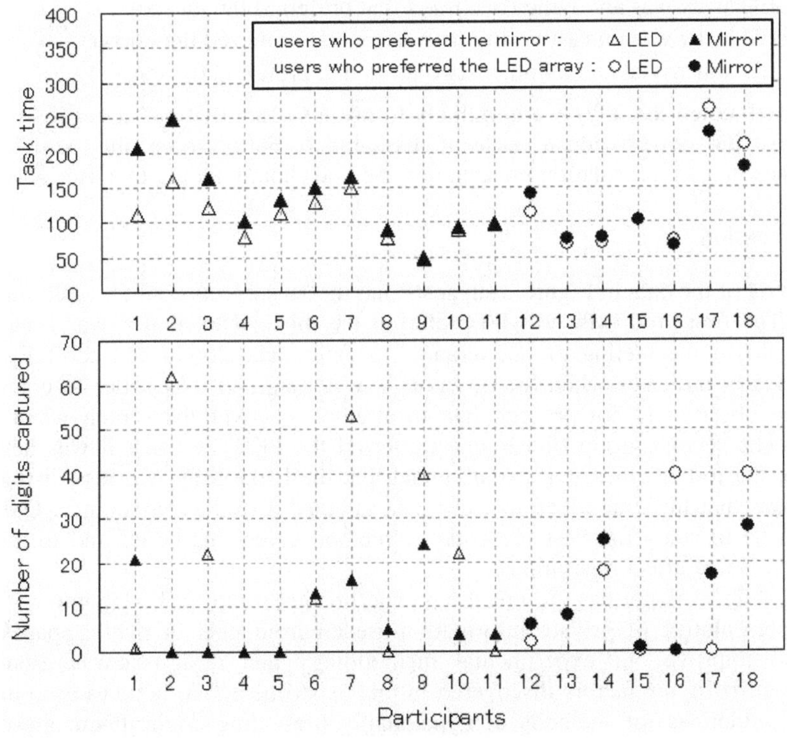

Fig. 2. Number of digits captured by behind people (*top*) and task time (*bottom*)

Next we turn to the seven participants who preferred the LED array. When using the LED array, more data was captured by the observer than when using the mirror. The task also took less time than when using the mirror. From the questionnaire, three users (participant IDs 14, 15, 18) felt that they accomplished their task more quickly given the LED array than when they used the mirror. Their opinion was that the fastest technique was correct and they commented that more data was captured by the observer when it took a long time to complete a trial. Six users said that the LED array was usable and it allowed them to concentrate on the task. The LED array was preferred because it eliminated the need to watch the observer.

To summarize, the two types of participants, classified according to their preferred information presentation method, had the following characteristics: those whose opinions as to which information presentation method enabled them to complete the task more quickly were wrong (those who preferred the mirror) and those whose opinions were correct (those who preferred the LED array). The former type felt that using the LED took less time than it actually did.

A negative image of a presentation method may impact the user's reassurance and resulting desired information presentation method to be used. In interviews after the trials we determined that:

- Mirrors are considered to be trivial and a bother (one user that preferred the LED).
- The LED array was annoying (two users that preferred the mirror).
- The LED array was untrustworthy (two users that preferred the mirror).
- The LED was trivial and a bother (two users that preferred the mirror).

These users noted that they were unlikely to use a system that had a negative image and/or that did not provide a sense of reassurance. Furthermore, they stressed the importance that an information presentation method should not affect a user's work.

2.3 Discussion

An analysis of the data in Figure 2 suggests that the 18 subjects could be split into two groups. The first group (those who preferred the mirror) felt that it was important, with regard to the feeling of reassurance, to enter data slowly and carefully. The mirror clearly indicates when the observer is watching and it is better to enter data when the observer is not present than to attempt to cover the screen when he is present. The second group (those who preferred the LED) felt that it was better to complete the job as quickly as possible and that the LED array was less distracting. We believe that the time taken to enter data affected a user's reassurance. Our next step will be to study the best presentation method given this belief and to identify other factors that affect reassurance.

Meanwhile, it is not easy to create an experimental setup in the laboratory that is truly representative of private information-use environments in public spaces. We intend to improve our experimental methodology and design new presentation methods utilizing the factors discovered in this experiment. We hope to use our data to form guidelines for methods of dynamically presenting environment awareness information for people handling private data.

3 Conclusion

This study is a first step towards information design that enables people to easily feel reassurance in public space; we focused on the reassurance possible when a situation can be assessed in an intuitive manner through the presentation of dynamic awareness information. We compared two approaches: the more abstract approach was an LED array while the more concrete approach was a mirror. Differences between the two presentation methods were categorized by trials using 18 subjects. Experiment results indicate that for some people reassurance is achieved by entering the data slowly and deliberately. Others preferred to work more rapidly. Regardless of the approach taken, the method used to present the information must be seen as reliable and non-intrusive.

References

1. Ministry of Internal Affairs and Communications, Japan: White Paper on Information and Communications in Japan, Gyosei 329 (2005)
2. Aono, A.: The Japanese journal of experimental social psychology. Japanese Journal of Experimental Social Psychology 42(2), 201–218 (2003)
3. Shozo,: SHIBUYA: A study of the shape of personal space. Bulletin of Yamanashi Medical College 2, 41–49 (1985)
4. Takahashi, K., Nakatani, M., Nishida, S.: Information Presentation from the standpoint of a sense of security. Human Interface Symposium 2002, pp. 289–292. Human Interface Society (2002)
5. Miller, G.A.: The magical number seven plus or minus two. Some limits on our capacity for processing information, Psychological Review 63, 81–97 (1956)

Use of Socio-technical Guidelines in Collaborative System Development

Hiroyuki Miki

Human Interface Laboratories, Oki Electric Ind. Co., Ltd.
1-16-8 Chuou, Warabi-shi, Saitama-ken 335-8510, Japan
hmiki@cf.netyou.jp

Abstract. The purpose of this research is to establish a design methodology for collaborative systems (artifacts for collaboration), which utilizes one of the social analyses called Ethnomethodology. So far, the author proposed guidelines, principles, and a model as components of the design methodology based on examples of collaborative assembly of an everyday product and remote instruction, which uses a movable laser pointer. This paper proposes a simple way to adapt previous results: it is the design matrix which can be used in the requirement acquisition phase, the design phase and the evaluation phase. The matrix supported its usefulness in the collaborative system development.

Keywords: guidelines, social analysis, Ethnomethodology, collaborative systems.

1 Introduction

As technology has become pervasive in society, some sociologists have begun to turn their attention to its influence at hospitals, air-traffic control towers, subway control rooms, and industries. This kind of social analysis (Ethnomethodology [1]) tries to understand how the evident characteristics of collaboration, easily grasped strategies and patterns of actions, are organized, by a detailed analysis of the links between these actions. It considers for instance the multiple roles of computers, instruments, paper, etc., situation dependent routines between individuals and groups, procedures, and accommodations that facilitate the organization of collaboration and their linkage.

However it was too profound for those in the scientific engineering field since the nature of the social analysis was highly specialized. It used particular terminologies, tried to analyze extremely narrow examples, and neglected its evaluation. It mainly discussed in detail the piled up improvisational behavior chains. Hence, the scientific engineers were required to modify the results before applying to the designs by translating the terminologies, abstracting the narrow examples, and adding the technological evaluations. These efforts were expected to be elaborated and organized as a design methodology.

In the following, section two briefly explains social issues and Ethnomethodology, and design guidelines based on Ethnomethodology. Section three describes the Socio-Technical guidelines (principles) which generalize previous design guidelines. Section four proposes the design matrix based on the Socio-Technical guidelines and the previous guidelines, followed by some discussions in section five.

M.J. Smith, G. Salvendy (Eds.): Human Interface, Part I, HCII 2007, LNCS 4557, pp. 90–97, 2007.

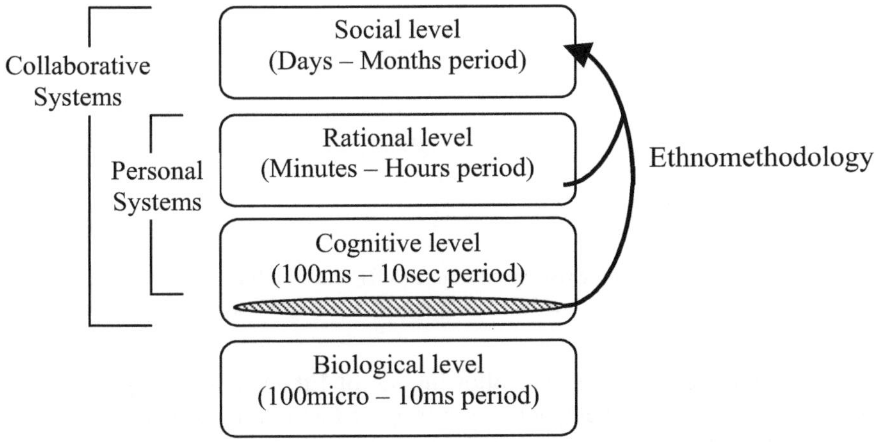

Fig. 1. Newell's time scale of human action [2] and Ethnomethodology

2 Social Issues and Ethnomethodology

Before talking about Socio-Technical guidelines, social issues are discussed briefly. Social (level) issues are such as leadership, power, organization, division of roles, and group performance. They have been studied in many research areas: Group Dynamics, Psychology, Sociology, Anthropology, Political Science, Education, Speech and Communication, and so forth.

Ethnomethodology is one of such research areas, yet it is very different from others. It describes social issues by explicating detailed human behaviors in a very short period of time (Fig.1). Other research areas usually ignore such details and explain social issues at the social level only in terms of abstract models and notions.

The main analytic characteristic of Ethnomethodology, namely explicating details, is well suited to the design of collaborative systems. It is because collaborative behaviors so much rely on interactive organizations of small actions.

For example, imagine a case where person A asks person B to take an object (Fig. 2 left). At first, A says to B "take this" and at the same time points to an object (such as a book on the table). Pointing by A connects with utterance "take this" by A. When B sees pointing by A to the book, B turns his body to the book and will say "ok". These body movement and utterance by B display his understanding of the utterance and the gesture by A. Afterwards, A draws away his hand, which was pointing at the book. By withdrawing the hand, A displays his understanding of the behaviors and understanding by B to all participants, including B [3]. As shown in this example, participants maintain and reorganize arrangements of bodies and tools interactively to monitor their pointing and other work [4].

Although these kind of interactive organizations are easy in the co-existing situation like Fig. 2 left, they are often difficult in distributed situations such as remote communications (Communicative Asymmetry [5]).

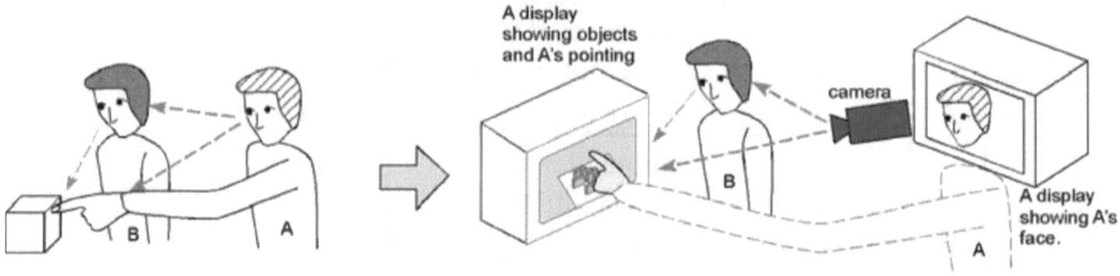

Fig. 2. Pointing in case of co-existing situation (left) and pointing using a system based on body metaphor (right)

To solve this design issue, the following set of Ethnomethodology-based design guidelines named "Body Metaphor" were proposed and verified so far (Fig. 2 right) [6]. With "Body Metaphor" design, instructions by an instructor are easily understood by an operator in the remote site without "Communicative Asymmetry". This situation is characterized as "Embodiment is achieved in this situation" in Ethnomethodology.

- Arrangement of bodies and tools requirement: The arrangement of bodies and tools should be appropriate for monitoring the other participants' behavior for both instructors and operators.
- Orientation of bodies requirement: Each participant should be able to turn his/her body to the other participants and shared artifacts.
- Gestural expression requirement: The instructor must be able to use freely not only verbal expressions, but also body movements and bodily expressions (gestures).
- Mutual monitoring requirement: Participants should be able to monitor each other's body arrangements, orientations, and gestures.
- Sequential organization requirement: Sequential and interactive organization of the arrangement of bodies, tools and gestural expressions must be possible.

Although "Body Metaphor" guidelines are useful especially in designing a remote instruction system, they tend to be restrictively focused on the instruction.

3 Socio-Technical Guidelines

To overcome the restrictive focus of "Body Metaphor", the author reorganized the guidelines and proposed more general guidelines (principles), namely Socio-Technical Guidelines, in the previous paper [7] based on Kirsh's information explicitness theory [8].

Relations between "Body Metaphor" notions and those of the proposed guidelines are illustrated in Fig. 3. Firstly, the figure shows that "Body Metaphor" is so conscious about "Locality" of Socio-Technical notions as to have three notions - arrangement, orientation, and mutual monitoring. Secondly, compared with action theory of

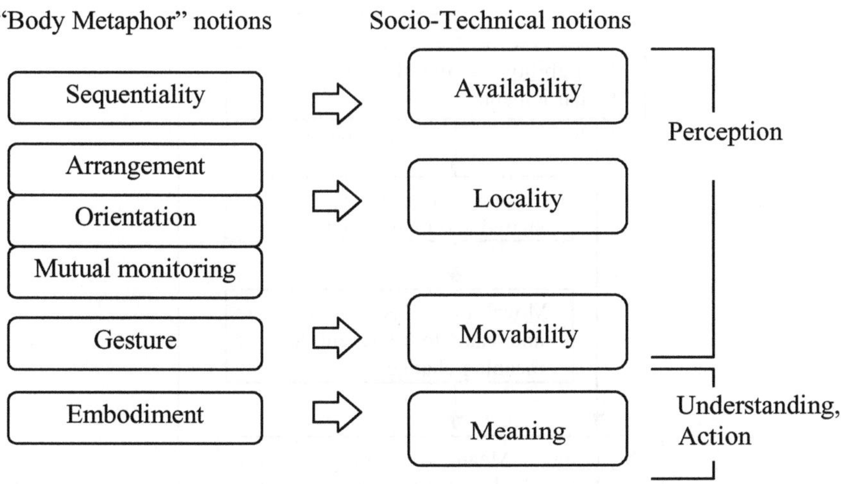

Fig. 3. Generalizations of "Body Metaphor" notions into Socio-Technical notions

cognitive engineering both "Body Metaphor" notions and Socio-Technical notions focus perception activities rather than understanding and action. This difference shows how both sets of notions inherit characteristics of Ethnomethodology. Seeing activity is not just a "snapshot" action but consisted of many accompanied pre/post and ongoing conducts.

In the Socio-Technical Guidelines, Kirsh's four concepts were expanded referring to key concepts of Ethnomethodology as follows [7].

- Availability: Required resource such as an artifact is instantly and easily referred and accessed.
- Locality: System organization and locality are necessary in order to establish "Availability". "Locality" must cover parallel actions since almost of all actions are normally in parallel motion. This consequently forms multiple localities.
- Movability: Normally communications will not be efficient for artifacts with complicated functions and architectures. In order to improve the communication, it is necessary to make physical motions (structured), explain with more words, and point with a finger and gesture (expression) [6]. If explanations are easy using such motions, then meanings can be modified easily, and the communication environment will be greatly improved.
- Meaning: "Locality" is about the relations between multiple actions and multiple localities. "Movability" is modification of meaning. Therefore, artifacts have several meanings in collaborative works. The principle of "Meaning" is to integrate other principles. The simplest way for integration is to select just one meaning.

Based on these Socio-Technical Guidelines, four steps explanatory model of collaborative use of artifacts (Fig. 4) and detailed guidelines were created [7].

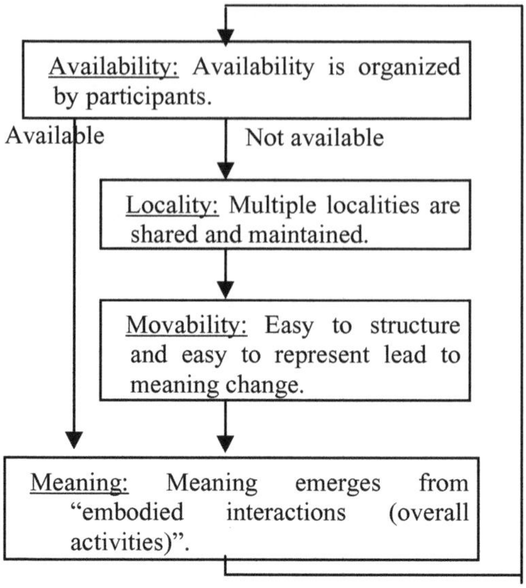

Fig. 4. Four steps explanatory model of collaborative use of artifacts based on four principles

4 Design Matrix: Use of Socio-Technical Guidelines

In the previous paper [7], use of the Socio-Technical guidelines and the model of Fig. 4 were explained according to simplified design phases – requirement acquisition, design, and evaluation.

In this paper, an easier way to use them is proposed, by using a design matrix. The matrix format is shown in Fig. 5. For usual engineers who are not familiar with Ethnomethodology, an approximate way can be better than detailed considerations and simulation which were explained in the previous paper.

In the column of the matrix, the four Socio-Technical notions are listed in chronological order accompanied by "Body Metaphor" notions except for "Meaning" cell. For "Meaning" cell, action theory notions are accompanied because action theory notions are more familiar to usual engineers and are more suited to consider cognitive issues.

In the row, two sides are considered in a remote collaboration setting: the user side and the operator side. Some requirements and specifications are for the user side or for the operator side only whilst there are common parts as well. As a result, there are three columns prepared (user side, operator side, and common of both sides), each of which is further divided into the individual factors (IFs) and Task/System/Environment factors (TFs) columns.

When using this matrix in the requirement acquisition phase, the engineer fills in each cell. In the row of IFs and TFs, their characteristics are described. For example, "Various users walking-up and standing" is written in the <characteristics> cell of IFs column of "User Side" and "Communication with voice and touch-panel screen" is written in its adjacent right cell of TFs column.

Remote collaboration settings / Socio-Technical notions	User Side		Operator Side		Common of both sides	
	Individual Factors (IFs)	Task/System/Env. Factors (TFs)	IFs	TFs	IFs	TFs
	<characteristics>	<characteristics>	<char.>	<char.>	<char.>	<char.>
Availability (Sequentiality)	<functional requirements>	<required specifications>	<func. req.>	<req. spec.>	<func. req.>	<req. spec.>
Locality (Arrangement, Orientation, Mutual monitoring)	Same as above (SA)	SA	SA	SA	SA	SA
Movability (Gesture)	SA	SA	SA	SA	SA	SA
Meaning (Understanding, Prediction/Projection)	SA	SA	SA	SA	SA	SA

Fig. 5. Format of design matrix based on Socio-Technical guidelines and "Body Metaphor" (IFs: Individual Factors, TFs: Task/System/Environment Factors)

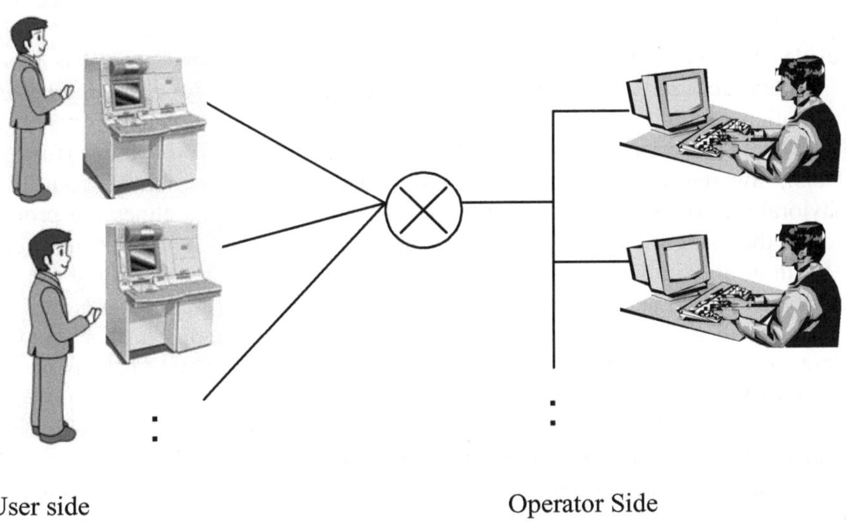

Fig. 6. Example of remote collaboration application

In a <functional requirements> cell, what the user can do and how the user feels are written for instance. In a <required specifications> cell, technological specifications are written to implement functional requirements of an adjacent left cell. For example, as <functional requirements> of "Locality" and the IFs of "User Side", "Able to figure out where to talk to" is written and "Speaker at almost the same height with user's eyes near the screen" is written in the adjacent right cell of it. All cells are written in the same way.

For the engineers in general, "Locality", "Movability", and "Prediction/ Projection" of "Meaning" tend to be out of account, so that the use of the matrix makes them realize to consider collaborative issues.

So far, the remote ticketing application and the remote consultation application were examined by the matrix and its usefulness was confirmed. In the applications, the user can be assisted by a remote operator when necessary (Fig. 6). The matrix is also used in both the design and the evaluation phases for better results.

5 Discussion

Two points are to be discussed in this section: they are comparisons with related researches and variations of the matrix use.

5.1 Related Researches

Recently several researchers have been involved in similar works to facilitate the usage of social and Ethnomethodological analysis. Such researchers are Martin et al. [9], Iqbal et al. [10], and Pinelle et al. [11]. Martin et al. [9] categorized the results from current social analysis and proposed guidelines with design notice points composed of the categorized cooperative work system. It can be utilized during the initial design process without participation of social researchers since it is more or less written abstractly. However, their guidelines do not support simulation of collaboration which is on the other hand supported by the approach of this paper.

Iqbal et al. [10] proposed three levels of analysis: plans and procedures (functional analysis), awareness of work (structural analysis), and distributed coordination (behavioral analysis). For each of three levels, around five guidelines are proposed. Although the categories (three levels) and guidelines are different from the proposal by Martin et al., similar disadvantages exist. The guidelines are mostly at the social level in Fig. 1 such as "Identify how individual work is oriented towards others". It can be difficult for the engineer to execute a guideline. In addition, their guidelines are not for design but for analysis. Their guidelines explain analysis steps of collaborative systems.

Proposal by Pinelle et al. [11] is also for analysis purpose, yet user activities can be simulated thanks to categories of the mechanics of collaboration.

Other than Ethnomethodology-oriented approaches, Situation Awareness approach by Endsley et al. [12] has more common points with the approach of this paper. Their approach has a systematic design and evaluation methodology, and deliberately separate perception, understanding, and prediction. Comparisons between their approach and the approach of this paper are future issues.

5.2 Variation of the Matrix Use

As a simple way to adapt the Socio-Technical guidelines, the proposed use of matrix does not cover detailed task analyses such as the goal-subgoal decomposition [12] or the scenario-based analyses [13]. However, when designing a large system or complex collaborative systems, such detailed analyses can be useful to differentiate many situations and goals which will make functional requirements and required specifications complex. With goal-subgoal decomposition, for instance, the matrix can apply to each of end (most detailed) subgoals. Similar technique was developed by Endsley et al. previously [12].

When there are existing long period of cultures and practices, field survey can be necessary before using the matrix [14].

6 Concluding Remarks

This paper proposed a simple way to adapt previous results, namely the design matrix which can be used in the requirement acquisition phase, the design phase and the evaluation phase. The initial use of the matrix supported its usefulness in the collaborative system development.

As a total design methodology for collaborative systems, further research is necessary in refining the matrix use and refining detailed design and evaluation techniques with the Socio-Technical guidelines and related guidelines.

References

1. Button, G.: Technology in Working Order, Routledge (1992)
2. Newell, A.: Unified Theories of Cognition. Harvard Univ. Press, Cambridge (1990)
3. Goodwin, C.: Pointing as situated practice, presented at the Max Planck Workshop on Pointing Gestures (1998)
4. Kendon, A.: Conducting Interaction. Cambridge Univ. Press, Cambridge (1990)
5. Gaver, W., Sellen, A., Heath, C., Luff, P.: One is not enough. In: Proc. INTERCHI'93, pp. 335–341. ACM Press, New York (1993)
6. Yamazaki, K., Yamazaki, A., Kuzuoka, H., Oyama, S., Kato, H., Suzuki, H., Miki, H.: GestureLaser and Gesturelaser Car. In: Proc. ECSCW'99, pp. 239–258. Kluwer, Dordrecht (1999)
7. Miki, H.: Design Methodology of Collaborative Systems Based on a Social Approach. In: Proc. of APCHI (2006)
8. Kirsh, D.: When is information explicitly represented? In: Hanson, P.P. (ed.) Information, Language, and Cognition, pp. 340–365. Oxford Univ. Press, New York (1991)
9. Martin, D., Sommerville, I.: Patterns of cooperative interaction. In: ACM Trans. on CHI, vol. 11(1), pp. 59–89. ACM, New York (2004)
10. Iqbal, R., Gatward, R., James, R.: A general Approach to Ethnographic Analysis for System Design. In: Proc. SIGDOC '05, pp. 34–40. ACM, New York (2005)
11. Pinelle, D., Gutwin, C., Greenberg, S.: Task Analysis for Groupware Usability Evaluation. ACM Trans. on CHI 10(4), 281–311 (2003)
12. Endsley, M.R., Bolte, B., Jones, D.G: Designing for Situation Awareness, Taylor & Francis (2003)
13. Carroll, J.M.: Making Use. MIT Press, Cambridge, MA (2000)
14. Beyer, H., Holtzblat, K.: Contextual Design. Morgan Kaufmann, San Francisco (1997)

Expert Systems Evaluation Proposal

Paula Miranda[1], Pedro Isaias[2], and Manuel Crisostomo[3]

[1] Escola Superior de Tecnologia de Setúbal,
Instituto Politécnico de Setúbal
Campus do IPS, Estefanilha
2910-761 Setúbal – Portugal
pmiranda@est.ips.pt
[2] Universidade Aberta,
Rua Fernão Lopes, 9, 1° Esq.
100-132 Lisboa – Portugal
pisaias@univ-ab.pt
[3] Faculdade de Ciências e Tecnologia,
Universidade de Coimbra, Polo II
3030 Coimbra – Portugal
mcris@deec.uc.pt

Abstract. It is well recognized that the evaluation of the expert systems is very important and difficult. This paper relates to the important issue of Expert System evaluation. A methodology for addressing the problem is presented. As evaluation model is proposed with the following main criteria: Usability; impact of the system in the behaviour and decisions of the user; Impact of the quality of the information provided to the user; financial impact; usefulness.

Keywords: Expert systems; evaluation; criteria; quality.

1 Introduction

In the current context of global, intense competitiveness, the systems' performance is a key competitiveness factor for manufacturing companies. The organizations have been challenged to keep pace with the rapid evolution of technologies. Built and managed knowledge to support decision in manufacturing processes are crucial for the organization.

The manufacturing processes are characterized by a continuous control, step by step, using decisions that require the required knowledge from the expert to relate actions and results.

The use of knowledge based systems, namely expert systems, to support the managers in the decision processes has evolved very quickly in the last several years. Moreover, it has been widely reported as a useful tool to enhance the performance of managers as they helped them to gain more knowledge, and consequently improve the quality of decisions.

A great deal of attention has been given to the use of expert systems in solving manufacturing problems. However, determining whether or not an expert system is justified to be applied in a particular manufacturing application has received very little attention or has almost been ignored.[1] [2]

Thus, it is fundamental to evaluate expert systems. Evaluation is the process of determining if a specific system is appropriate for its original requirements and objectives and also to assess if its performance satisfies users.

M.J. Smith, G. Salvendy (Eds.): Human Interface, Part I, HCII 2007, LNCS 4557, pp. 98–106, 2007.

The goal of this work is to propose a model for evaluating the performance of expert systems and presenting a case study of its implementation to an industry. The relevance of the topic is justified by the difficulties presented by the organization in evaluating expert systems.

So, an important question, related to the use of expert systems in organizations, is how to measure the results from its application. For this, it is necessary to identify the evaluation criteria and also how each criteria can be measured.

What to measure? Why to measure? How to measure? These are questions that are present at the organizational environment. The concern of evaluating results from the organization not only involves short term but also long term goals that can guarantee the sustainability of the previous defined strategy by the organization. It is important to point out that, independently of the methodology chosen by the organization, it has to consider that there are tangible results (essentially financial results) and intangible results (for instance users motivation) to be evaluated, both for the user of the system as well as for the organization.

From what has been exposed, it is necessary to propose an evaluating expert system performance model, composed by criteria (what needs to be measured) and metrics (what can be used to measure).

2 Methodology

In order to reach the main objective proposed by this study, the steps that will be followed are (see Figure 1):

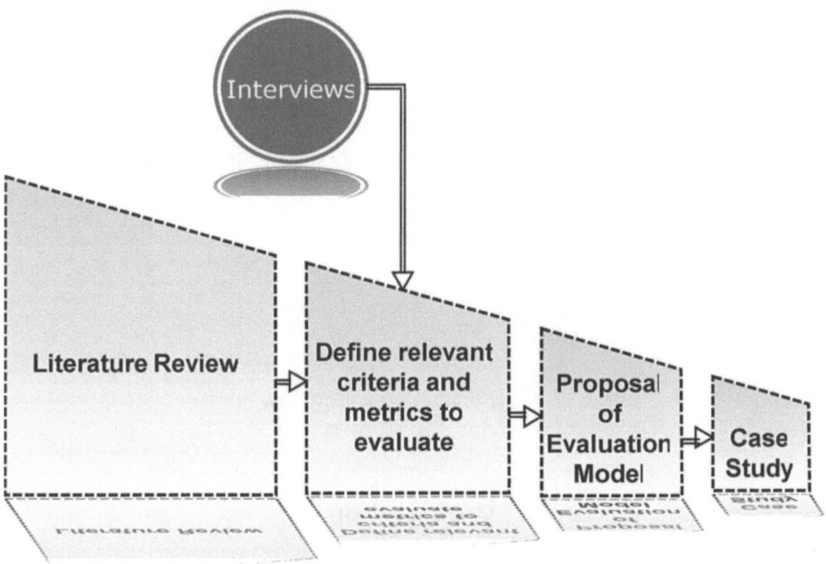

Fig. 1. Research Process

1. Literature review – this was used to build using literature references, an evaluation process to evaluate the performance of expert systems, grouping criteria and metrics. However it is important to say that there are no universally accepted standards for the evaluation of expert systems;
2. To find out from experts' interviews, about what are the relevant criteria and metrics to evaluate these systems;
3. Application to an organisation of the elaborated model – this will result in a case study. As part of the evaluation process, the assumptions made during the proposal of the model will be re-examined. This procedure will help to validate and test the applicability of the model in a specific organization and also to derive important conclusions regarding the model and the evaluation process.

3 Evaluation of the Performance

Expert Systems can be classified in several ways. One way is by the general areas they address. Thousands of expert systems are in use today in almost every industry and in every functional area [3].

How to evaluate the performance of an expert system becomes very important, because different domain has different requirements, different application background, different technical support, and different targets, it is very difficult to propose the same approach and criteria for evaluating different domain expert systems [4].

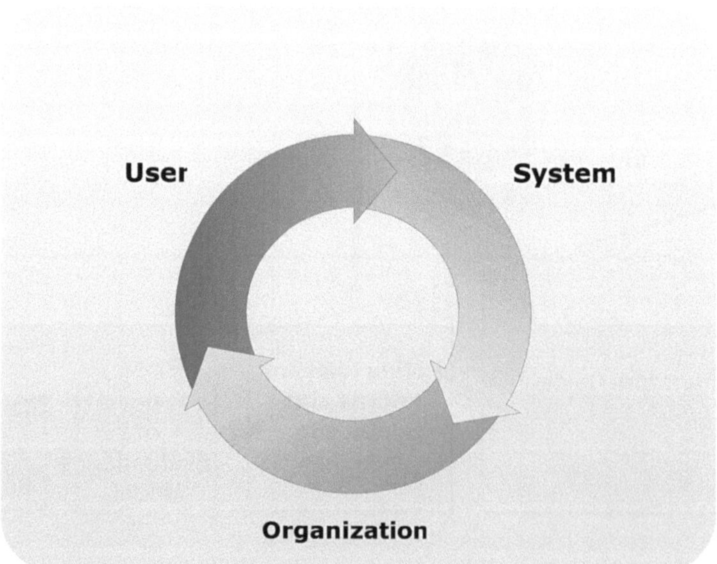

Fig. 2. Interactions between the entities involved in the evaluation of an expert system

Table 1. Performance Evaluation actor interrelations

System vs User	The system can provide the user with: ▪ Efficiency ▪ Efficacy ▪ Satisfaction
System vs Organization	The system can provide to the organization: ▪ Productivity ▪ Quality ▪ Decision taking, etc
User vs Organization	The user supported by the system contributes to improve the performance and to achieve the organization's goals

Several researchers point out as difficulties in the evaluation of an expert system the following:

- What characteristics should be evaluated?
- How to measure performance?

When one wishes to evaluate an expert system, one should have in consideration the evaluation as a global goal. An expert system cannot be seen as an isolated object but as an integrated entity, that is a product used by people, in a specific context, to reach certain goals. Therefore, in order to correctly assess and evaluate the performance of these type of systems, we have to analyse interactions between the three entities involved in the process: user, system and organization [5](see Figure 2).

4 Evaluation Model Proposal

Evaluation is the process of determining the likelihood that once deployed, the expert system will be used whenever appropriate. [6] Pertinent issues in evaluation are:

- Is the system user friendly and do the users accept the system?
- Does the system give correct results and is the logic of the system correct?
- Does the expert system offer an improvement over the practices it is intended to supplement?
- Is the system easy to learn and to become proficient on?
- Is the system useful as a training tool?
- Is the system in fact maintainable by the other than the developers?
- Can the system be used in the intended work environment?

As stated, after the literature review, one can conclude that there are no universally accepted standards for the evaluation of expert systems. However, there is a set of points that might be considered as essential in the evaluation of an information system, e.g. impact of the system at the organizational costs and benefits; impact f the system at the attitude of users (decisions, actions and behavioural changes); quality of information made available to users.

For the proposal of a model it is necessary to select evaluation criteria and specify how each criterion will be measured.

1. Usability is measures in relation to the system characteristics and to the degree of user satisfaction (see Table 2). The user friendliness of the system and acceptance of the system by the users should also be evaluated.

> *"Usability: the extent to which a product can be used by specified users to achieve a specified goals with effectiveness, efficiency and satisfaction in a specified context of use"* [7]

Table 2. Usability criteria proposed for Expert System Evaluation

Usability	➯ Easiness to Learn - Does the user understand easily the interface, its different paths and what he can do with it? - Are there any help to use instructions? (instructions should be provided as required and Access to help must always be available without being mandatory its reading) [8] ➯ Easiness to Use - Does the user know what to do and how to navigate with easiness? ➯ User Satisfaction - Reaction to interface, to the contents, to the document structure, to the interaction and navigation process and to available help [9]

2. Impact of the system in the behaviour and decisions of the user (see Table 3). The easiness of the users to learn how to operate the system influences their behaviour and the decisions that he or she has to make.

Table 3. Impact of the system in the behaviour and decisions of the user criteria proposed for Expert System Evaluation

Impact of the system in the behaviour and decisions of the user	- The system increases the users' confidence? - Does the system enable users to take decisions, establish compromises and provides guidance? - Does the system enable to improvements to decision makers performance? - Does the system change the nature of users' tasks and jobs in a matter that motivates continued use of the expert systems?

3. Impact of the quality of the information provided to the user (see Table 4).

Table 4. Impact of the quality of the information provided to the user criteria proposed for Expert System Evaluation

Impact of the quality of the information provided to the user	- The system provides adequate explanations? (how and why the decision was taken) - Correctness of the logic of the system (correctness, consistency and completeness of the inference rules) - Are the decisions made by the system correct compared to what the experts consider to be right? - Is the system adequate to replace the intelligence of the human expert? - Does the system change the nature of the users' tasks and jobs in a manner that motivates continued use of the expert system? - Is the level of knowledge sufficiently high? - Can the system be used as a training tool to transfer the knowledge of the expert to middle and top management and thus improve the performance of new employees? [10] - Is the representation of knowledge suitable? [11]

4. Financial impact - consideration of costs and benefits for the organization (see Table 5)

Table 5. Financial impact criteria proposed for Expert System Evaluation

Financial impact	- Does the system reduce cost? (does not require many experts; reduces routine tasks)

5. Usefulness - consideration of costs and benefits for the organization (see Table 6)

While there may be no universally accepted standards for field evaluation of expert systems, interviews to the experts can improve the process.

As stated, after the literature review and the experts' interviews it will be possible to propose a more focused evaluation model. The interview to experts is presented in appendix.

Table 6. Usefulness criteria proposed for Expert System Evaluation

Usefulness	- Is the system helpful to the user? - Is the system fast enough to satisfy the user? - Does the system respond on time? - optimisation of the operation - improvements to user performance

5 Conclusions

We can conclude that is difficult to evaluate the performance of an expert system especially when we intend to propose a standard model, because of its several applications. Moreover, not only it is necessary to select evaluation criteria but also to determine the minimum acceptable performance level for each criterion selected. In the formulation of the model it has been considered all the actors in the process: system, user and organization.

The paper describes the problem of the evaluation of expert systems and proposes a set of generic evaluation criteria. Implications of the evaluation and the evaluation criteria are discussed.

Nowadays the importance of quality in software development, and especially in expert systems development has greatly increased, and the risk of taking a wrong decision by using a non adequate decision making support system has decreased, but it is not inexistent yet[12].

References

1. Okogbaa, G.: Evaluation of Manufacturing Expert Systems: Framework and Model, Engineering Economist (1992)
2. Sørgaard, P.: Evaluating Expert System Prototypes, AI & Society, vol. 5(1). Springer, Heidelberg (1991)
3. Turban, E., Aronson, J.E.: Decision Support and Intelligent Systems, 6th edn. Prentice Hall, New Jersey (2000)
4. Yang, C., Kose, S., Phan, S., Kuo, P.: A Simulation-based Procedure for Expert System Evaluation. In: Proceedings of the IEA/AIE 13th International Conference, New Orleans (June 19-22, 2000)
5. Alves, J.: Avaliação do Desempenho de Sistemas Periciais, Tese de Mestrado, Universidade Nova de Lisboa, Instituto Superior de Economia e Gestão (2004)
6. Wentworth, J., Knaus, R., Aougab, H.: Verification, Validation, and Evaluation of Expert Systems, An FHWA Handbook, vol. I (1995)
7. ISO/DIS 9241-11, in European Usability Support Centre
8. Carvalho, A.: Testes de Usabilidade: exigência supérflua ou necessidade? In: Actas do 5° Congresso da Sociedade Portuguesa de Ciências da Educação., pp. 235–242. Sociedade Portuguesa de Ciências da Educação, Lisboa (2002)

9. Carvalho, A.: Avaliar a Usabilidade da Plataforma FLEXML: Descrição Dos Testes Realizados Com Utilizadores, VII Congresso Iberoamericano de Informática Educativa, pp. 197–206 (2004)
10. Cascante, P., Plaisent, M., Bernard, P., Maguiraga, L.: The Impact f Expert Decisions Support Systems of the Performance of New Employees. Information Resources Management Journal 14(4), 64–78 (2002)
11. Grier, A., Hsiao, R.: A Decision Support Expert System for Correction of Manufacturing Process Problems. In: Proceedings of the 3rd International Conference on Industrial Engineering Applications of Artificial Intelligence and Expert Systems. Charleston, South Carolina, United States. IEA/AIE, vol. 1, ACM Press, New York (1990)
12. Cojocariu, A., Munteanu, A., Sofran, O.: Verification, Validation and Evaluation of Expert Systems in Order to Develop a Safe in the Processo f Decison Making, Computational Economics 0510002, EconWPA (2005)

Appendix:

PART I

Below are enumerated some criteria that should be taken into consideration to evaluate expert systems.

Please rate ☒:
1 - Disagree 2- Slightly Agree 3- Agree 4- Agree much 5- Totally agree

1 – Usability

Easiness to Learn
☐ Disagree ☐ Slightly Agree ☐ Agree ☐ Agree much ☐ Totally Agree

Easiness to Use
☐ Disagree ☐ Slightly Agree ☐ Agree ☐ Agree much ☐ Totally Agree
Systems' user satisfaction
☐ Disagree ☐ Slightly Agree ☐ Agree ☐ Agree much ☐ Totally Agree

What additional aspects should be included in these criteria?

2 – Impact of the system in the behaviour and decisions of the user

Impact of the users in relation to the systems' use
☐ Disagree ☐ Slightly Agree ☐ Agree ☐ Agree much ☐ Totally Agree
Improvement of decision makers performance
☐ Disagree ☐ Slightly Agree ☐ Agree ☐ Agree much ☐ Totally Agree
Degree to influence decision making
☐ Disagree ☐ Slightly Agree ☐ Agree ☐ Agree much ☐ Totally Agree

What additional aspects should be included in these criteria?

3 – Impact of the quality of the information provided to the user

Information made available by the system enables improvements to quality in work

☐ Disagree ☐ Slightly Agree ☐ Agree ☐ Agree much ☐ Totally Agree

Correctness of the results given by the system

☐ Disagree ☐ Slightly Agree ☐ Agree ☐ Agree much ☐ Totally Agree

Adequateness of explanations provided

☐ Disagree ☐ Slightly Agree ☐ Agree ☐ Agree much ☐ Totally Agree

Adequateness of the system to replace the intelligence of the human expert

☐ Disagree ☐ Slightly Agree ☐ Agree ☐ Agree much ☐ Totally Agree

What additional aspects should be included in these criteria?

4 – Financial impact

Task reduction

☐ Disagree ☐ Slightly Agree ☐ Agree ☐ Agree much ☐ Totally Agree

Consideration of costs and benefits for the organization

☐ Disagree ☐ Slightly Agree ☐ Agree ☐ Agree much ☐ Totally Agree

What additional aspects should be included in these criteria?

5 – Usefulness

Systems help to the user

☐ Disagree ☐ Slightly Agree ☐ Agree ☐ Agree much ☐ Totally Agree

The system is fast enough to satisfy the user

☐ Disagree ☐ Slightly Agree ☐ Agree ☐ Agree much ☐ Totally Agree

System's response on time

☐ Disagree ☐ Slightly Agree ☐ Agree ☐ Agree much ☐ Totally Agree

Optimisation of the operation

☐ Disagree ☐ Slightly Agree ☐ Agree ☐ Agree much ☐ Totally Agree

What additional aspects should be included in these criteria?

PART II

1 – What are the key advantages of evaluating expert systems?

2 – What are the main difficulties to evaluate expert systems?

3 – In what type of situations should the expert systems be evaluated?

Evaluating Interfaces to Publicly Available Environmental Information

Peter Mooney and Adam C. Winstanley

National Center for Geocomptation,
National University of Ireland Maynooth,
Co. Kildare, Ireland
{peter.mooney,adam.winstanley}@nuim.ie

Abstract. In this paper we detail a number of recommendations regarding how to evaluate interfaces to systems providing public access to environmental data and information. The primary focus is based upon web and Internet-based interfaces. These recommendations will assist data providers and information managers to evaluate the effectiveness of their current data access interfaces. The paper also outlines some guidelines on how to ensure these software interfaces are accessible to all citizens.

Keywords: Environmental Information, Web-Based Interfaces, Accessibility, OGC Specifications.

1 Introduction

Given the high profile of environmental issues today it is not surprising that many citizens search for environmental information and data using the Internet [1]. The purpose of such searches are varied. Examples include scientists carrying out research, policy makers searching for documents on environmental legislation, or members of the public searching for information on the local environment in which they live. These different purposes and user categories mean that data providers must adopt a layered approach to environmental data distribution and information provision. Each category of user has different data requirements. These requirements are broadly classified as follows:

- Environmental scientists will require access to the highest resolution raw data. They will require that the data is distributed in the most suitable format for the thematic area. For example, large meteorological datasets should be distributed in the World Meteorological Organization (WMO) GRIB Format.
- Policy makers, strategic environmental assessment (SEA) reporters, and other researchers may not require the highest resolution data. However they still require aggregated data accompanied by graphical representations. They will require the data be distributed in accessible software formats that can be imported into popular statistical and data analysis software tools .

M.J. Smith, G. Salvendy (Eds.): Human Interface, Part I, HCII 2007, LNCS 4557, pp. 107–114, 2007.

– Public users (those not fitting into either of the previous categories) will require simple aggregated information. This information should possibly be accompanied by additional information on severe or unusual environmental events. For example, if Ozone air quality levels go above recommended public health protection thresholds. Simple statistical visualisations (trend charts, histograms, normal distributions) should be provided. If the underlying data has a distinct geospatial component visualisations through geographical maps should be considered.

This paper provides a brief overview discussion on the key components that we feel should be included when developing interfaces for access to publicly accessible environmental data. In this paper we define publicly accessible environmental data as data and information which can be accessed by any citizen without that citizen having to pay a data access fee. Our recommendations are based on our experience in management and distribution of environmental data arising from scientific research programmes in Ireland funded by the Irish Environmental Protection Agency.

2 Building the Interface to the Environmental Data

New user-interface metaphors to geospatial data, such as NASA Earthwind, Microsoft TerraEarth, and Google Earth, have increased expectation levels from the general public and scientific community toward availability of geospatial data and associated web-based 'geo-services'. Often mapping and Geographical Information Systems (GIS) services are often beyond the IT expenditures of most organisations or not viewed as fitting into the intended audience profile for the organisation [2]. Consequently environmental information is often distributed in difficult-to-use software formats or throught complicated user-interfaces requiring scientific expertise to navigate. This adversely affects the dissemination of environmental information. The general public will only support the sometimes inconvenient, expensive, and controversial actions that must be taken to protect the environment if they have access to information [3]. To understand why different user-interfaces must be considered it is necessary to summarise the different categories of environmental data.

2.1 Categories of Environmental Data

Environmental data and information has several interesting characteristics. These characteristics present challenges to user-interface designers when providing this environmental information to the public. These characteristics are summarised as follows:

– Cumulative environmental effects: Where environmental effects, such as climate change, must be viewed over very long timescales;
– Large Volumes of Data: Environmental data resulting from the monitoring of rapidly changing phenonemum usually contain large volumes of information

over short time periods. For example monitoring air pollution parameters is often carried out over five minute intervals on a continuous daily basis.
- Environmental data is often monitored over large geographical areas.
- Large number of parameters: Some environmental datasets contain a large number of inter-related variables. This is particularily evident in metrological datasets.
- Measurements at very low concentrations: In some environmental monitoring, such as water quality, some parameters are measured over a very small value scale.
- Public health information: Organisations that carry out air quality monitoring are required to highlight when various public health protection thresholds are exceeded.

The next two sections will describe various approaches to dealing with these categories of environmental data.

2.2 Distribution of Data Using Traditional Web Techniques

Using traditional web techniques involves using HTML in combination with graphical visualisations of the data and information in addition to linkages to download services for access to the raw environmental data. We recommend that this approach is best handled using a dynamic web-page generation approach. Users are presented with a simple web interface from which they select the parameters, time ranges, locations, or events they require information on. Graphical visualisations of these choices are then displayed, summary data or information is displayed, and links to the raw data used for the graphical visualisations. An example implementing these techniques is available on our greenhouse gas emissions inventory website `http://coe.epa.ie/ghg`. Figure 1 shows the number of unique public user requests to an information resource on our web-server. For this particular resource (greenhouse gas emissions inventory data) public users were must more interesed in the visualisations of the dataset using graphical means through histograms, distributions, and trend graphs. The document formats (PDF and MS Word) were reasonably popular. However it is difficult to assertain what these documents were used for after they were downloaded. The Comma Separated Values (CSV) files available for download with the raw data from the graphical visualisations provided to be the most popular raw-data format. This is probably due to the fact that the CSV format is readily imported into most popular statistical packages.

2.3 Distribution of Data Using OGC Web Map Services

Environmental data monitored over large environmental areas or data that contains distinctly local environmental interest benefit from be distributed in a map-based interfaces. The Open Geospatial Consortium (`http://www.opengis.org`) defines a Web Map Services (WMS) specifications. When OGC WMS are compared with traditional mapping processes the distribution of online mapping

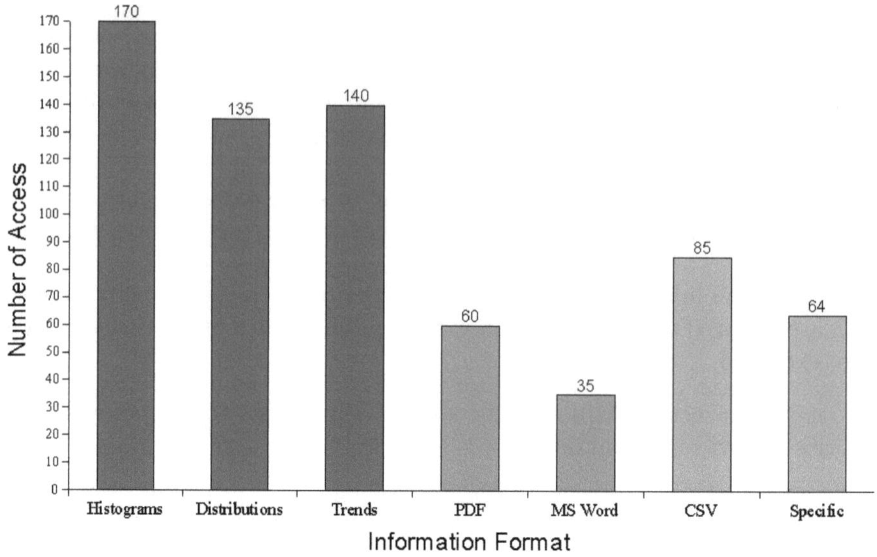

Fig. 1. Public Access to Various Information Formats

results are quicker, more easily customised, and more accessible to a greater number of people. No GIS or mapping skills are required on behalft of the end users. In line with the European INSPIRE Directive [4] the data management of the geospatial data is performed "as close to the source as possible". Providing OCG WMS access to the data allows all data and information be managed at one point. This reduces the possibility of duplication of data, and subsequent data quality issues as a result, while ensuring that the client user is always presented with the most current and available data. An example of ouput from an OGC WMS is displayed in Figure 2. This image was returned as the response to an WMS query. The user specified their options on a simple HTML web interface on a standard web-browser. The image depicts the Corine Land Cover Map for Europe 2000 (Wetlands Classification) layer with all towns in Ireland with population greater than $10,000$ (square dots). This is a static PNG image returned to the user. A user may draw the conclusions that the majority of large towns in Ireland are situated away from areas of wetland environments. The two layers of geospatial data which are integrated together in this graphic were specified by the user using a combination of checkboxes and drop-down-lists to select which *layers* they wished to see combined.

A well-known example of OGC WMS being used for environmental information is at the website of the Environment Agency UK (EA-UK). The EA-UK have a "What's in my backyard?" [5] section where public users can search by postal code or town name for environmental information on that area. The example shows triangles beside rivers which have water quality standard targets set for them.

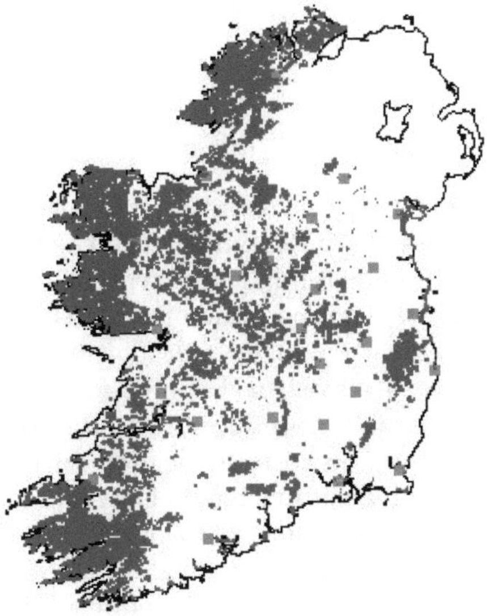

Fig. 2. An example of a simple static image returned from a OGC WMS query

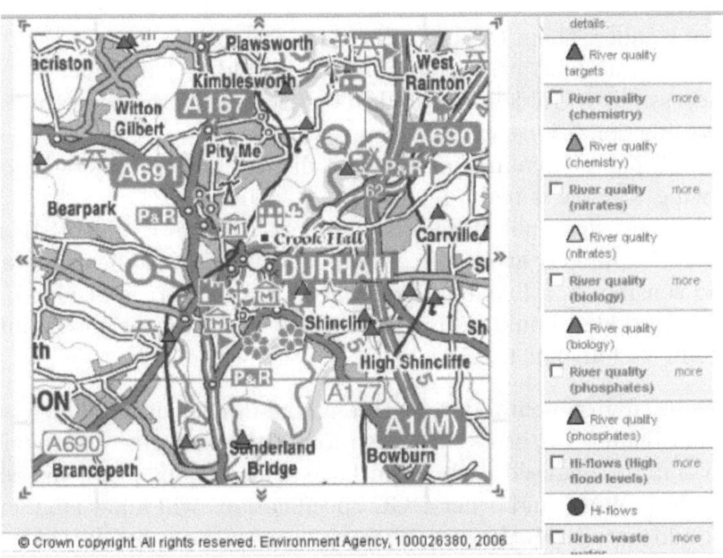

Fig. 3. An example map returned from a query for town of *Durham* on [5]

The list to the right-hand side of the map image allows users to enable the checkboxes of various other layers they wish to see on the map - for example river water quality in terms of biological status, or high flood level warnings. As in Figure 2 all of this functionality is available to the end user using at minimum a standard web-browser.

3 Conclusions

This paper has discussed in high level detail approaches to making environmental data and information accessible to the public. The key to successful distribution of these types of data is understanding the type of user audience. When the audience is spread across the wide spectrum of expertise in the general public the data access interfaces must cater equally well for all users. *Discovery* metadata about the data should be displayed in a simple format offering non-specialist users the opportunity to learn *who* produced the data, *where* it can be obtained, *how* it can be obtained and the costs associated with accessing the data (if there is a cost-recovery requirement involved). The ISO 19115 metadata standard provides for all of these discovery metadata items. The ISO 19115 metadata is stored in XML format and can be easily displayed in web interfaces and integrated into other data sources. There are a number of recommendations that we can make regarding web-based interfaces to publicly available environmental information:

1. Users should not be required to download any specialist software tools or packages to allow them to access the data or information. Current Internet browser software with basic configuration should be sufficient.
2. Web Accessibility Guidelines should be strictly adhered to[6]. In particular the following conditions should be available
   - All web pages (dynamic and static) display graceful degradation as the viewing window is resized
   - If the user does not have CSS support (or has it disabled) all pages providing publicly available content using CSS should be capable of being read sensibly in a linear fashion when CSS support is disabled.
   - All hyperlinks should be properly titled using the title attribute in the HTML `<a>` anchor tag.

Interfaces to environmental data with a global dimension provide a very wide range of functionality for users. We have found that interfaces to environmental data on a more localised scale often have some serious problems associated with them. OGC WMS and other OGC specifications will allow smaller organisations to build better interfaces to their data inventories quickly at a low cost of investment.

We are also investigating the use of user-logging information to help guide our development of more useful web-interfaces to our environmental data. In the web-applications serving out our data we carefully log the actions taken by visitors to the web site. User actions that are logged include which pages they visit (in chronological order), which datasets they view and/or subsquently download, or which graphical representations they view most frequently. This information is stored directly in a relational database where an more indepth analysis of the behaviour of users visiting our web sites can be performed offline. In combination with web-server log files this is a valuable tool towards finding out what public visitors find useful, discovering which data services are most popular, and also highlighting any barriers inadvertently placed before users as they search for environmental data and information from our web services.

3.1 Final Comments

Public concern for environmental issues will continue to increase. This places even greater emphasis on data providers and information managers to make environmental information available in a timely fashion, without any technological barriers, and in the most suitable information representations. Computing power is still doubling every 18 months[7] and many public users now have own home computers more powerful than the servers of yesterday. Portable media device proliferation is difficult to avoid on a dialy basis. Today PDAs, Cell Phones, portable music players etc. are web and Internet enabled. Increasing numbers of citizens are using these devices to access information on the Internet. We feel that data and information managers must move from thinking in terms of static HTML web sites with collections of links to data file downloads to innovative means of data distribution. This data and information should be thought of in terms in "pieces of information" and "information packages" which can be integrated with other information. Devices other than PCs or workstations running Internet browser software can then pos-sibly consume these pieces of information. Environmental data and information useful to better public understanding of their environment should made be available from systems with open data access interfaces (ie those implementing the OGC Specifications). This should allow citizens to choose any client software which implements the OGC Specifications to access these data services. We believe that at minimum for environmental data with geospatial components the OGC WMS specifications should be implemented allowing users to access to simple map images [8] of the data. Public access to the data and information should not incur costs on the end-user. Some cost recovery mechanism may implemented for very rare datasets or datasets which are products or very expensive monitoring programmes. Younger Internet users of today have been brought up on a 'web-diet' rich in exciting user interface experiences (gmail, flickr, youTube, etc). When these users become adult users and consumers of environmental and geospatial information in the future they will expect these types of easy-to-use and access user interfaces as standard.

Acknowledgments. This work is being carried out in the Environmental Research Center (ERC) of the Irish Environmental Protection Agency (EPA). It is a postdoctoral research fellowship (2002-CC-FS) funded by ERTDI and the National Development Plan

References

1. Schimak, G.: Environmental data management and monitoring system UWEDAT. Environmental Modelling And. Software 18(6), 573–580 (2003)
2. Maclachlan, J.C., Jerrett, M., Abernathy, T., Sears, M., Bunch, M.J.: Mapping health on the Internet: A new tool for environmental justice and public health research, Health And Place, Part Special Issue: Environmental Justice, Population Health, Critical Theory and GIS, March 2007, vol. 13(1), pp. 72–86 (2007)
3. Horinko, M.L.: Earth Observation Summit Conference, Keynote Address, Washington DC, USA (July 31, 2003) (Accessed February 07) http://yosemite.epa.gov/opa/admpress.nsf/
4. INSPIRE, Establishing An Infrastructure for Spatial Information in the European Community. PE-CONS 3685/2006, Brussels, Belguim (January 2007) (Accessed February 07) http://inspire.jrc.it
5. Environment Agency, What's in your backyard? The Environment Agency UK, (Accessed February 07) http://www.environment-ageny.gov.uk/maps/
6. Mooney, P., Winstanley, A.C.: Web Accessibility for Dynamically Generated Websites, IASTED Conference on Human Computer Interaction, Phoenix, Arizona, USA (November 2005)
7. Tuomi, I.: The Lives and Death of Moore's Law. First Monday, vol 7(11) (November 2002) (Accessed February 07) http://firstmonday.org/issues/issue7_11/tuomi/index.html
8. Schutzberg, A. OGC Newsletter, January 2006. (Accessed February 07) http://www.opengeospatial.org/pressroom/newsletters/200601/

Harder to Access, Better Performance? The Effects of Information Access Cost on Strategy and Performance

Phillip L. Morgan, Samuel M. Waldron, Sophia L. King, and John Patrick

School of Psychology
Park Place, Cardiff, CF10 3AT UK
MorganP4@Cardiff.ac.uk

Abstract. Whilst much effort is being invested into the design of human-computer interfaces that provide users with access to large amounts of information, there is less understanding of how human performance can be optimized in such interaction-driven contexts. We suggest one way to improve performance is to tailor the structure of the interface to the adaptive nature of cognition and orient behavior towards certain strategies, and away from others. Here, we show how small changes in Information Access Cost (IAC) – the time, physical and mental cost of accessing information – can effect powerful changes in cognitive processing strategies that subsequently affect performance. Increasing IAC on task-critical information was generally shown to encourage a more 'memory-intensive' processing strategy. This strategy not only improved future memory for task-relevant information, but also improved learning during problem solving, and planning. Implications for display designers in HCI are discussed.

Keywords: Information Access Cost (IAC), memory, problem solving, learning, interface design.

1 Introduction

Optimizing human performance is a key challenge in the design of human-computer interfaces. With rapid technological development, operators of complex systems, such as those found in the modern aircraft cockpit, now have access to an unprecedented volume of information [1]. To avoid information and cognitive overload that might impair performance [2], many approaches to cognitive engineering have attempted to provide design principles in order to organize and present information in the most accessible format (e.g., Ecological Interface Design [3], Configural Display Design [4], Proximity Compatibility Principle [5]). However, each of these approaches requires significant task analysis [6] before implementation, is often situation specific, and thus does not provide a generic design solution. Moreover, commensurate attention has not been paid to recent research conducted within the fields of cognitive and psychological science attending to the impact of small changes to information accessibility upon the deployment of adaptive cognition [7].

We suggest that a fruitful approach to interface design is to structure the interface to exploit the adaptive nature of cognition and orient operator behavior towards certain microstrategies and away from others. It is argued that boundary conditions

M.J. Smith, G. Salvendy (Eds.): Human Interface, Part I, HCII 2007, LNCS 4557, pp. 115–125, 2007.

and tradeoffs may exist associated with increasing the accessibility of information provided within an interface. Based upon recent experimental research examining the accessibility of information provided during interactive routine behavior [7], it is anticipated that under certain conditions, increasing the cost associated with accessing information from an interface may paradoxically prove beneficial.

Information Access Cost (IAC) – the time, physical, and mental cost associated with attaining information – was varied in the studies reported in this paper. Originally used by Ballard et al. [8] and recently by Gray et al. [7], IAC can be applied by covering important information with a mask and varying the cost of uncovering (e.g., a mouse movement and/or a time delay). Measures of strategy can be captured (e.g., the frequency with which a certain mask is removed and the time spent within the uncovered window) as well as typical performance measures (e.g., task completion time, errors).

1.1 Significance of Research Area

It is recognized that users perform tasks based on strategic adaptation to the structure of the task environment [9, 10] and the costs of performing certain operations [11]. In order to understand the link between interface design and human performance, it is important to appreciate the effect of the constraints of design upon the psychological processes selected to process information. Recent extensions of Anderson's work [9] on rational analysis have investigated how very small changes within the design of an interface can have surprisingly large consequences for the selection of microstrategies during interactive behavior. As noted by Gray & Boehm-Davis [12]:

> "Microstrategies develop in response to the fine-grained details of the interactive technology; that is, microstrategies focus on what most designers would regard as the mundane aspects of interface design: the ways in which subtle features of interactive technology influence the ways in which users perform tasks." (p. 322)

Instances whereby interface design encourages certain behavioral strategies, yet discourages others have recently been termed 'soft constraints' [13]. Of particular relevance to the current paper are soft constraints associated with small changes in IAC [13, 14]. Reliably, humans have been found to use memory more when IAC is high during routine copying [7, 14] and VCR programming [13, 15, 16]. When information is readily available within displays, cognitive processing strategies are reduced in favor of less cognitively demanding perceptual-motor strategies. Such 'interaction-intensive' strategies rely upon the interface as an external memory source [17].

Anderson and colleagues have argued that the human memory system operates to make most available the memories for information that are currently required [18]. Indeed, when analyzing the setting of goals during solution of the Tower of Hanoi, Anderson & Douglass [19] noted that since it was easy to reconstruct goals from the information provided within the external display, the price of memory failure was low. For this reason, they argued that there was little motivation to engage in goal rehearsal during externally supported solution of the Tower of Hanoi. With regards to display design, this suggests that when information is highly accessible within the interface, memory for visual-spatial information may be problematic due to a low probability of usage [20]. We, like others [13], predict an adaptation in processing strategy to one that is more memory-intensive when information presented within a display is harder to access, due to the time, physical and mental effort costs. Our novel prediction is that such soft constraints may be used within display design to harness the interface to promote learning and memory.

1.2 Goals and Objectives

The goals of this paper are to review our work on the powerful effects of different levels of IAC on task completion strategies, and their potentially beneficial effects on memory, problem solving and learning. We have the following two objectives:

1. To demonstrate that not only does variation in IAC affect the extent that a memory-intensive as opposed to a display-based strategy is used, but also that this will have an effect on the encoding and retention of visually presented information.
2. To identify the effect of IAC on the use of memory during more complex problem solving and learning.

Below we discuss evidence from investigations of the effect of IAC on strategies and performance during both simple and more complex tasks.

2 The Effect of IAC on Strategies and Performance in Simple Tasks

We utilized a simple routine copying task – the Blocks World Task (BWT) – to establish the effects of IAC on strategy and performance. Gray et al. [7], for example, demonstrated that when people complete the BWT, they made more use of memory by chunking a greater number of blocks per visit to the target window when IAC was higher. A novel aspect of our work is that we measured recall for target information, with the prediction that recall would be higher when the cost of accessing information increased. We also included measures of gaze recorded via an eye-tracker to explore qualitative changes in task strategies with varying level of IAC.

2.1 General Method

Apparatus and Procedure

The Blocks World Task (BWT)

In the BWT [8], participants attempt to copy a target pattern of colored blocks by selecting from a resource window palette of colored blocks and moving them into a workspace window (see Figure 1). No colored blocks were used twice. Participants completed 24 BWT problems: half of the participants completed a 'surprise memory' test after completing 12 BWT problems and the other half completed this test after completing all 24 BWT problems. During the surprise memory test, all masks were removed, the target and workspace windows were emptied, and participants had to recall the target pattern of the previous BWT problem by dragging blocks from the resources palette into the workspace window.

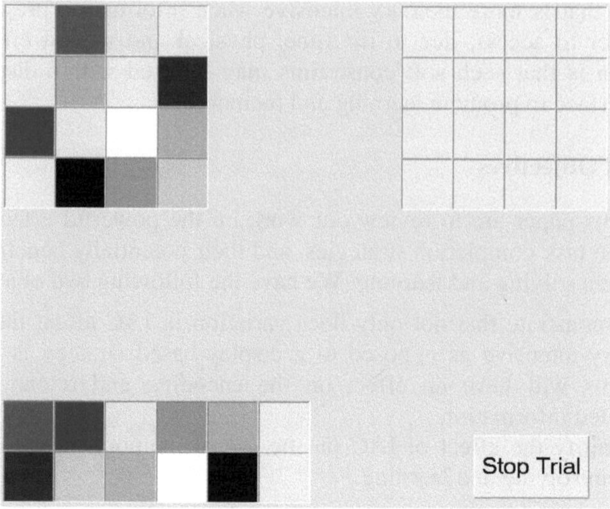

Fig. 1. Example of a Low IAC BWT start-state. (Target Window is top-left, Resource Window is bottom-left, and Workspace Window is top-right.).

Design

In the Low IAC condition, information was permanently available within all windows in the display. In both the Medium and High IAC conditions, grey masks covered all windows and each could only be uncovered by placing the mouse cursor over the mask. The mask reappeared as soon as the mouse cursor was moved out of the window. There was an additional one second time cost upon each uncovering of the target window in the High IAC condition.

A number of strategy measures were recorded throughout the experiment. First, the number of times participants viewed the workspace window was recorded using the eye-tracker. Second, the frequency of visits to the target window was recorded: data from the Low IAC condition came from the eye-tracker, and data for the Medium and High IAC conditions came from the Visual Basic program (in accordance with Gray & Fu [13, 16]). Third, the time participants spent viewing the target window was taken in the Low and Medium IAC conditions (using an eye-tracker) but not in the High IAC condition because of the possible contamination caused by the lock-out time.

In addition, the Visual Basic program recorded the number of moves made and the time taken to complete each trial.

2.2 Results and Discussion

Importantly, and in addition to Gray et al. [7], the introduction of a small IAC (operationalised by a mouse movement and/or a 1 second time cost) improved post-task recall. On the surprise memory test, more blocks were correctly recalled in the Medium and High IAC conditions (*ps* < .01) than in the Low IAC condition

($ps < .01$), although number of blocks recalled did not significantly differ between the Medium and High IAC conditions ($ps > .05$) (see Table 1). We suggest that a more developed cognitive representation was created in response to the higher cost of accessing information within the task environment and that this resulted in improved recall.

The improvement in recall in the higher cost conditions is further supported by the finding of differences in strategy between these conditions and the Low IAC condition, specifically, that even a 1 second delay and/or a mouse movement led to a strategy shift in the extent to which users relied on the display while completing the task. In accordance with Gray et al. [7], we found strategy differences across the different IAC conditions. The Low IAC trials were completed faster than the Medium and High IAC conditions ($ps < .01$), although there were no significant difference between the High and Medium IAC conditions. With regards to completion strategy, the target window was viewed more often in the Low IAC condition than in the Medium and High IAC conditions, and was uncovered more frequently in the Medium than in the High IAC condition ($ps < .05$). Furthermore, strategies involving copying blocks without any reference to the target window were significantly higher in both the High and Medium IAC conditions compared to the Low IAC condition ($ps < .001$).

Table 1. Effect of IAC on recall in a routine copying task

High IAC		Med IAC		Low IAC	
Mean	**SD**	**Mean**	**SD**	**Mean**	**SD**
9.00	1.78	9.00	4.37	4.70	1.95

3 The Effect of IAC on Strategy and Performance in More Complex Tasks

Our next objective was to examine whether the effects of IAC on producing powerful shifts in task completion strategies and performance during simple routine copying behavior could be generalized to more complex problem solving, including learning.

Two types of planning have been differentiated in the literature: display-based planning – conducted 'on-line' whilst solving the problem, and, memory-based planning – performed prior to executing a sequence of moves or actions. As one would expect, memory-based planning makes large demands upon memory, whereas display-based planning does not [21]. For this reason, we predicted that problem solving with low IAC would induce more display-based planning, whereas problem solving with high IAC would promote more memory-based planning.

Organization of memory is essential to learning during problem solving [22]. Therefore, if there is a greater reliance upon memory during planning due to increased IAC, learning might also improve. In contrast, reliance upon perceptually available information during display-based planning (which we predict would occur when problem solving with low IAC) might mask the mechanisms required for effective learning.

3.1 General Method

3.1.1 Apparatus, Procedure and Design

The Blocks Problem Solving Task

The Blocks Problem Solving Task (BPST) was developed at Cardiff University, and is similar to the eight-puzzle: participants recreate a goal-state pattern consisting of a number of colored blocks (number of blocks varied between some Experiments) in a 3 x 3 current-state window by moving one colored block at a time into one adjacent empty space (see Figure 2 for an example of a BPST). No colored blocks were used twice. Typically, participants had to complete between 6 and 12 BPST problems in each experiment.

IAC was manipulated in the same way as described in Section 2.1.2, except that the current-state window was always available irrespective of IAC condition and the IAC imposed in the High condition was increased to 2.5 seconds. Some of the dependent measures described in Section 2.1.2 were included, with the addition of qualitative data in the form of concurrent verbal protocols – recordings of participants' 'think-aloud' whilst solving problems – in order to collect evidence of planning strategy [23]. Unlike the experiments using the BWT in which recall the main dependent variable, problem solving was largely measured through the number of moves made in the current-state window per inspection of the goal-state window. Eye-tracking data are not reported here.

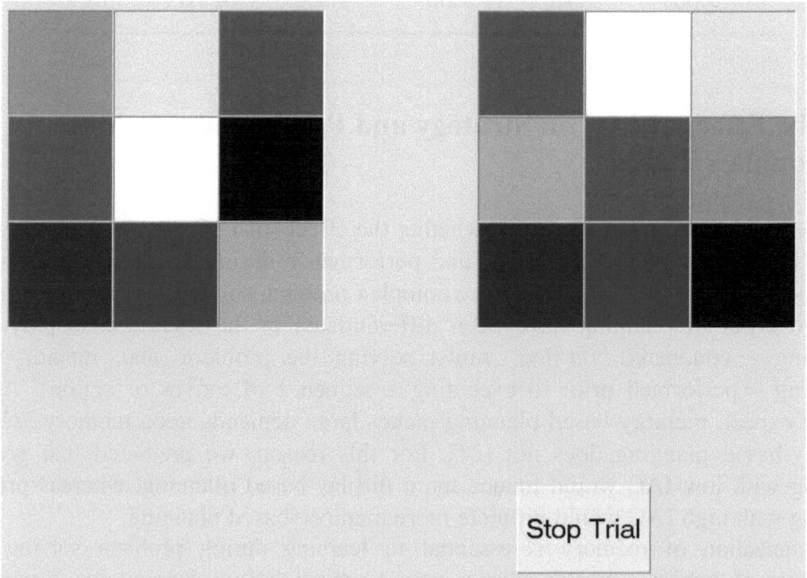

Fig. 2. Example of a Low IAC BPST start-state. (Goal-State Window is on the left, Current-State Window on the right.).

3.2 Results and Discussion

3.2.1 Effect of IAC on Problem Solving

Generally, small changes to IAC were shown to have the same effect on complex problem solving as they did on simple copying: participants chose to rely on memory more as IAC increased. For example, more blocks were moved per inspection of the goal-state in the High IAC condition compared to the Medium and Low IAC conditions ($ps < .05$) (see Table 2) and problems were completed faster in the Low IAC condition ($ps < .05$).

Table 2. Effect of IAC on number of problem solving moves per goal-state inspection

High IAC		Medium IAC		Low IAC	
Mean	**SD**	**Mean**	**SD**	**Mean**	**SD**
8.03	3.65	2.78	1.27	1.43	0.60

3.2.2 Effect of IAC on Planning

Concurrent verbal protocols were coded according to adjustments made to Ericsson's [24] coding scheme (designed for the eight-puzzle) in order to distinguish between largely memory or display-based episodes of planning, depending upon whether any moves were made during verbalization. If no move was executed during the verbalization, it was coded as a memory-based plan. If, however, a minimum of one move was made during verbalization, it was coded as a display-based plan. Using this coding scheme, qualitative evidence was found that high access cost promoted memory-intensive planning, whereas low access cost encouraged interaction-intensive planning (see Figure 3).

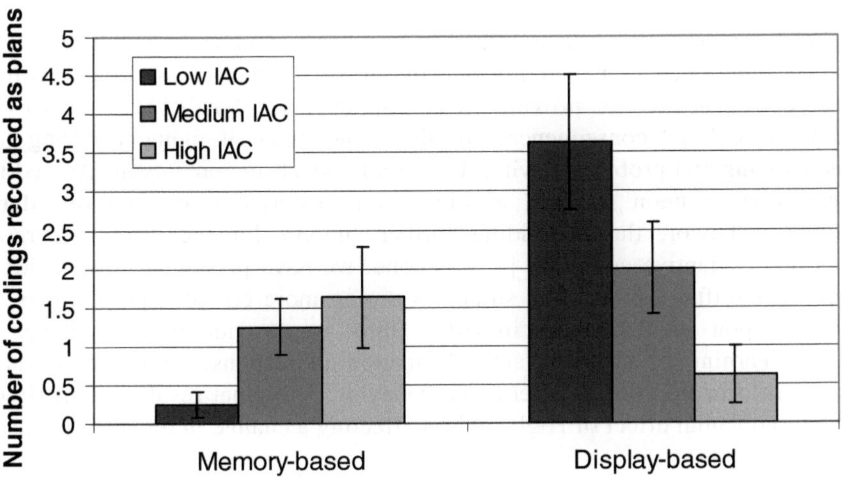

Fig. 3. Effect of IAC on frequency of display- and memory-based planning

3.2.3 Effect of IAC on Learning

The selection of a memory-based planning style (prompted by high access cost) led to more learning during repeated solution of an eight-puzzle-like BPST. Specifically, across trials 1 – 6, participants made fewer moves to completion in the Medium and High IAC conditions (*ps* < .05), but demonstrated no such improvement in the Low IAC condition (*p* > .05) (see Figure 4). These results support and expand on our findings from previous studies that a more memory-based interaction can lead to substantial benefits when learning is important.

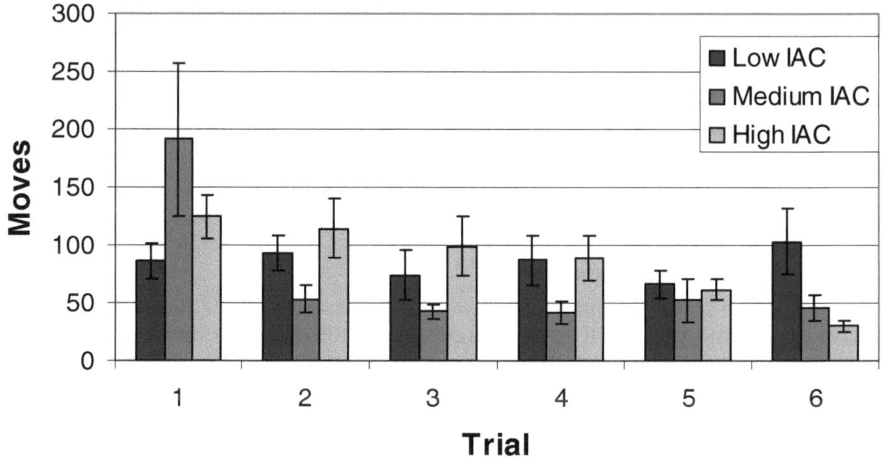

Fig. 4. The effect of IAC on problem solving moves over trials

4 General Discussion

First and foremost, our work presented in this paper demonstrates that externally supported human behavior is highly sensitive to small changes associated with the accessibility of information provided within the interface. Small IAC changes were found to have large consequences for the deployment of memory during both routine copying and problem solving. Essentially, small increments in IAC led to a stronger reliance upon internal, as opposed to external representations during interactive behavior, thus providing further support for the 'soft constraints' hypothesis of adaptive cognition [13]. Second, we have presented novel evidence for powerful shifts in processing strategies during more complex problem solving contingent upon small changes in IAC. Third, we provide novel evidence for improved learning of task-relevant information in response to higher costs on accessing information during interactive behavior. This final result was likely due to the consequential effect of High IAC on effecting a change in the adaptive use of human memory [18].

4.1 Implications for HCI Design

IAC is a subtle although powerful manipulation to force a change in task completion strategies that can have substantial effects on performance across a range of tasks and on a variety of performance measures. Our work is therefore relevant to display designers who wish to change or improve the user's strategy in response to changing task demands. For example, if a user becomes too passive, IAC could be introduced to increase awareness and improve performance.

Complex system interface designers should be aware of the behavioral repercussions caused by very small, or in Gray & Boehm-Davis's [12, p. 322] words "mundane" changes to the accessibility of information provided within an external display. For example, an increase in terms of milliseconds (Low to Medium IAC) had large consequences for use of memory during routine copying and more complex problem solving, even larger again when the cost was increased to either 1 or 2.5 seconds. Furthermore, it is anticipated that few system designers or engineers would envisage that actually *increasing* the cost associated with accessing task-relevant information during problem solving would *improve* learning.

Evidence from these studies has implications for supporting better operator behavior via increased information accessibility. However, we are not claiming that improved accessibility to information is not useful in certain circumstances, particularly when cognitive workload is already too high to support a strategy change in response to a higher IAC and when the time cost of introducing a higher IAC is impractical (e.g., in safety-critical situations). Indeed, the current findings do not detract from the many benefits observed when principles of information integration are put into practice [see 25, 26, 27 for useful reviews].

Designers of fusion displays need to effect a careful balance between reducing an operator's cognitive workload by making task critical information more available within the interface, whilst ensuring that the extraction of information is not made so effortless that it is either not processed or promptly forgotten. Increasing the cost of accessing information, even over a few seconds, will result in more cognitive processing that will provide extra resilience to system failure and forgetting of task critical information, especially in dynamic interactive situations. If memory is an important criterion of task performance, then small increases in IAC can be used to encourage a memory-intensive strategy that will improve recall for task-relevant information. If speed is an important criterion of task performance, then low IAC will facilitate this by leading to the adoption of interaction-intensive strategies, and use of the display as an external memory source. If access to goal-state information during problem solving is limited, then less display-based and more memory-based planning can be encouraged through increasing IAC. Finally, if it is necessary to learn a problem solving solution, this can be accelerated through a small increment in IAC.

Acknowledgements. This work was supported by a UK Ministry of Defence (MoD) Defence Technology Centre (DTC) Data Information Fusion (DIF) research grant # DTC/DG(R&T)/13/25/DIF.

References

1. Woods, D.D., Patterson, E.S., Roth, E.M.: Can we ever escape from data overload? A cognitive systems diagnosis. Cognition, Technology, & Work 4, 22–36 (2002)
2. Kirsh, D.: A few thoughts on cognitive overload. Intellectica 30, 19–52 (2000)
3. Rasmussen, J., Vicente, K.J.: Coping with human errors through system design: Implications for ecological interface design. International Journal of Man.-Machine Studies 31, 517–534 (1989)
4. Woods, D.D., Wise, J., Hanes, L.: An evaluation of nuclear power plant safety parameter display systems. In: Proceedings of the Human Factors 25th Annual Meeting, pp. 110–114. Human Factors Society, Santa Monica, CA (1981)
5. Carswell, C.M., Wickens, C.D.: Information integration and the object display: An interaction of task demands and display superiority. Ergonomics 30, 511–527 (1987)
6. Kirwan, B.B., Ainsworth, L.K.: A guide to task analysis. John Wiley & Sons, New York (1992)
7. Gray, W.D., Simms, C.R., Fu, W.-T., Schoelles, M.J.: The soft constraints hypothesis. A rational analysis approach to resource allocation for interactive behavior. Psychological Review 113, 461–482 (2006)
8. Ballard, D.H., Hayhoe, M.M., Pelz, J.B.: Memory representations in natural tasks. Journal of Cognitive Neuroscience 7(1), 66–80 (1995)
9. Anderson, J.R.: The adaptive character of thought. Erlbaum, Hillsdale, NJ (1990)
10. Payne, S.J., Howes, A., Reader, W.R.: Adaptively distributing cognition: A decision making perspective on human-computer interaction. Behaviour & Information Technology 20, 339–346 (2001)
11. O'Hara, K.P., Payne, S.J.: The effects of operator implementation cost on planfulness of problem solving and learning. Cognitive Psychology 35, 34–70 (1998)
12. Gray, W.D., Boehm-Davis, D.A.: Milliseconds matter: An introduction to microstrategies and their use in describing and predicting interactive behavior. Journal of Experimental Psychology: Applied 6, 322–335 (2000)
13. Gray, W.D., Fu, W.-T.: Soft constraints in interactive behavior: The case of ignoring perfect knowledge in-the-world for imperfect knowledge in-the-head. Cognitive Science 28, 359–383 (2004)
14. Fu, W.-t., Gray, W.D.: Memory versus perceptual-motor tradeoffs in a blocks world task. In: Proceedings of the Twenty-second Annual Conference of the Cognitive Science Society, pp. 154–159. Erlbaum, Hillsdale, NJ (2000)
15. Duggan, G.B., Payne, S.J.: Interleaving reading and acting while following procedural instructions. Journal of Experimental Psychology: Applied 7(4), 297–307 (2001)
16. Gray, W.D., Fu, W.-T.: Ignoring perfect knowledge in-the-world for imperfect knowledge in-the-head: Implications of rational analysis for interface design. CHI Letters 3(1), 112–119 (2001)
17. O'Regan, J.K.: Solving the 'real' mysteries of visual perception: The world as an outside memory. Canadian Journal of Psychology 46, 461–488 (1992)
18. Anderson, J.R., Schooler, L.J.: The adaptive nature of memory. In: Tulving, E., Craik, F.I.M. (eds.) Handbook of memory, pp. 557–570. Oxford University Press, New York (2000)
19. Anderson, J.R., Douglass, S.: Tower of Hanoi: Evidence for the cost of goal retrieval. Journal of Experimental Psychology: Learning, Memory, and Cognition 27, 1331–1346 (2001)

20. Anderson, J.R., Fincham, J.M., Douglass, S.: Practice and retention: A unifying analysis. Journal of Experimental Psychology: Learning, Memory, and Cognition 25, 1120–1136 (1999)
21. Davies, S.P.: Planning and problem solving in well-defined domains. In: Morris, R., Ward, G. (eds.) The cognitive psychology of planning, Psychology Press, Hove (2004)
22. Chase, W.G., Simon, H.A.: Perception in chess. Cognitive Psychology 4, 55–81 (1973)
23. Ericcson, K.A., Simon, H.A.: Protocol analysis: Verbal reports as data (Rev. ed). MIT Press, Cambridge, MA (1993)
24. Ericcson, K.A.: Instruction to verbalize as a means to study problem-solving processes with the 8-puzzle: A preliminary study. Report No. 458. Department of Psychology, University of Stockholm (1975)
25. Bennett, K.B., Flach, J.M.: Graphical displays: Implications for divided attention, focussed attention, and problem solving. Human Factors 35(1), 71–97 (1992)
26. Vicente, K.J.: Ecological interface design: Progress and challenges. Human Factors 44, 62–78 (2002)
27. Wickens, C.D., Hollands, J.G.: Engineering Psychology and Human Performance, 3rd edn. Prentice-Hall Inc, New Jersey (2000)

Measurement and Analysis of Performance of Human Perception for Information Communication Technology

Hidetoshi Nakayasu[1], Masao Nakagawa[2], and Hidehiko Hayashi[3]

[1] Institute of Intelligent Information and Communications Tech. Konan University, 8-9-1 Okamoto Higashinada-ku, Kobe 658-8601 Japan
nakaysu@konan-u.ac.jp
[2] Department Economics, Shiga University, 1-1-1 Bamba, Hikone Shiga 522-8522, Japan
[3] Advanced Information of Research and Education Center, Naruto University of Education, 748 Nakashima, Takashima, Naruto 772-8502
hhayashi@naruto-u.ac.jp

Abstract. An evaluation method of detection probability has been proposed by the authors for the measurement and evaluation of the industrial products based on the human performance by perception. As a practical application, the distributed parameters of psychometric curve were estimated by the experimental data concerning the three kinds of atribute of size, flatness degree and grayscale. As an example of experiments in the proposed methods, single madality and multiple dodalities sensory tests are applied to measure the perception performance of panels concerning the three atributes of FRP (fiber reinforced plastics) poroduct.

Keywords: Stimulus Threshold, Psychometric Curve, Paired Comparison, Statistical Estimation.

1 Introduction

Sine the psychometric curve method has been researched, the classical three of constant stimulation method, limit method and adjustment method are popular until now[1]. A sort capacity by human vision is very flexible; therefore such a soft information processing has been regarded unfit on the computer which is good at digital information processing. The outline of the objectives in this study is drawn in Figure 1. The two kinds of approaches for artificial and natural knowledge were illustrated in Figure 1. The former management is computerized such as artificial knowledge management. On the other hand, the latter is concerned with the human response by knowledge and skill of expert people. These two kinds of management processes must be integrated for the application to analysis, and evaluation in intelligence machine. To obtain the psychometric curve for detecting the deference on the modalities of figure, a prototype research was developed based on the constant stimulation method, has been proposed by the authors[2].

The aim of this work is to discuss the potentiality of identifying human perception and its application. In this paper, a procedure is presented for measurement of perceptive sensory function of various kinds of attribute in the inspection image data by the experiment of paired comparison under constant stimuli.

M.J. Smith, G. Salvendy (Eds.): Human Interface, Part I, HCII 2007, LNCS 4557, pp. 126–135, 2007.

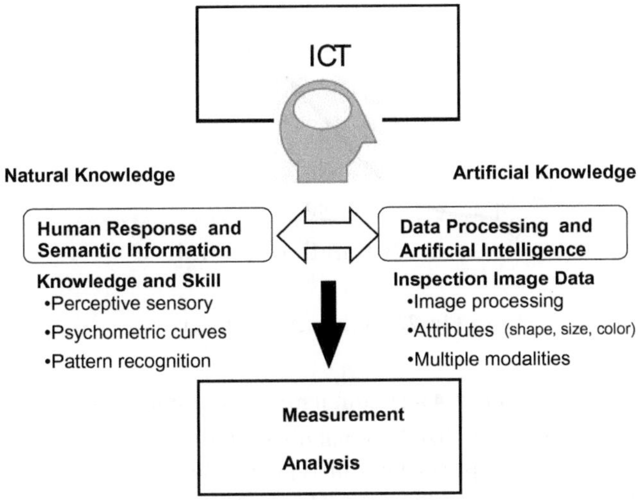

Fig. 1. Human perception for ICT

2 Measurement and Analysis of Perception

2.1 Measurement and Analysis of Single Modality

It is the problem of judging the existence of defect from inspected image by inspection machine. The problem is in the fact there is no way how to measure and analysis the degree of human performance to inspect it if the attribute of defect is more than the regulated value. The idea of this study is in the hypothesis that the reaction whether there is defect on the object image to be inspected is equal to the response of man who receives the perception stimulation. Such a response is shown as a probability curve that tells the reaction to stimulation strength as indicated by x-axis in Figure 2 when the amount of the feature in which perception stimulation intensity can be described by a variable amount of one dimension. This curve is called as psychometric curve.

Psychometric curve, as shown in Figure 2, is a continuous curve which is made from the detection probability as a function of stimulus strength, when the strength of stimulus is continuously varied over a range of values. The psychometric curve fits a continuous function $f(x)$ is such as:

$$P = f(x) \tag{1}$$

where

$$P = \Phi(y) = \int_{-\infty}^{y_i} \phi(y)dy, \quad y = \frac{x - \mu_x}{\sigma_x} \tag{2}$$

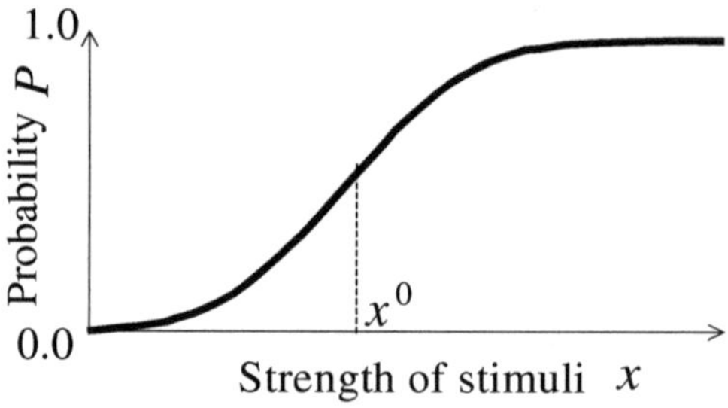

Fig. 2. Psychometric function

In Equation (2), $\Phi(\bullet)$ and $\phi(\bullet)$ is cumulative distribution function and probability density function of the standardized normal distribution.

In order to estimate the parameters in Eq.(2), the paired comparison method is used as a general method in the sensory test by which this curve is obtained[2]. It is a method of comparing a paired target sample, which is obtained by changing a standard sample and perception strength of stimulus x gradually, by which the strength of stimulus is set in constant value x^0 specified previously. In this experiment, a reactive category of the alternative of "x is bigger than x^0" or "x is smaller than x^0" is obtained, and the psychometric curve is identified by presuming the parameters of Eq.(2) by using the measurement appearance rate obtained as a result. The details of practical way to estimate the parameters statistically by Probit and Staircase method are shown in [3,4].

2.2 Measurement of Multiple Modalities

The probability obtained by psychometric curve by Eqs.(1),(2) mean the probability that one can feel the strength of stimuli of target object less than specified strength of stimuli such as threshold x^0 in Figure 2. This probability is also calculated as a probability that one can judge that the strength of stimuli of the target sample for paired comparison is less than that of standard sample. From this discussion, now suppose that this probability is defined by

$$P_i = P(E_i) \tag{3}$$

where $P(\bullet)$ is an occurrence probability of probability event $\bullet$. E_i means the probability event that a panel judges that the strength of the stimuli of the target sample is less than that of the threshold x_i^0 such as a standard sample for i-th attribute among multiple attributes to express the characteristic of interest. Thus the relationship

$$P(E_i) + P(F_i) = 1 \tag{4}$$

consists of E_i and F_i that are probabilistic events to be possible and impossible to detect the strength of stimuli x_i less than that of the threshold x_i^0 by means of judgement on i-th attribute.

On the other hand, now suppose that the detection probability P under the situation of multiple attributes are defined by

$$P = P(E_1 \cup E_2 \cdots \cup E_n) \tag{5}$$

where $P_1 \leq P_2 \leq \cdots \leq P_n$. In this condition, the confidence interval of detection probability P under multiple attributes can be estimated with upper and lower probability bounds P_L and P_U such as $P_L \leq P \leq P_U$ by the equations [5].

$$P_L = P_1 + (P_2 - P_{12}) + \sum_{k=3}^{n} \max\left\{P_k - \sum_{l=1}^{k-1} P_{lk}\,,\, 0\right\},$$

$$P_U = P_1 + \sum_{k=2}^{k_n} \min\{P_k - P_{lk}\},\ for\ l \in \{1,2,\cdots,k-1\} \tag{6}$$

In the equation, P_{lk} is the joint probability of the events E_l and E_k

$$P_{lk} = P(E_l \cap E_k) \tag{7}$$

that can be calculated by the integral

$$P_{kj} = \Phi(y_k^0, y_l^0) = \int_{-\infty}^{y_k^0} \int_{-\infty}^{y_l^0} \phi(y_k, y_l; \rho_{kl})\, dy_l\, dy_k \tag{8}$$

where $\phi(y_k, y_l; \rho_{kl})$ is the density function of two-dimensional standard normal distribution.

$$\phi(y_k, y_l; \rho_{kl}) = \frac{1}{2\pi\sigma_k\sigma_l\sqrt{1-\rho_{kl}^2}} \exp\left\{-\frac{1}{2\left(1-\rho_{kl}^2\right)}\left(y_k^2 - 2\rho_{kl} y_k y_l + y_l^2\right)\right\} \tag{9}$$

where

$$y_k = \frac{x_k - \mu_k}{\sigma_k},\ y_l = \frac{x_l - \mu_l}{\sigma_l},\ y_k^0 = \frac{x_k^0 - \mu_k}{\sigma_k},\ y_l^0 = \frac{x_l^0 - \mu_l}{\sigma_l},$$

$$\mu_{(.)} = \int_{-\infty}^{\infty} x_{(.)} \cdot f\left(x_{(.)}\right) dx_{(.)},\quad \sigma_{(.)}^2 = \int_{-\infty}^{\infty} \left(x_{(.)} - \mu_{(.)}\right)^2 f\left(x_{(.)}\right) dx_{(.)},$$

$$\sigma_{kl} = \int_{-\infty}^{\infty} \int_{-\infty}^{\infty} (x_k - \mu_k)(y_l - \mu_l) f(x_k, x_l)\, dx_k\, dx_l\,,$$

$$\rho_{kl} = \frac{\sigma_{kl}}{\sigma_k \cdot \sigma_l}$$

$$f(x_k) = \frac{1}{\sqrt{2\pi}\sigma_k} \exp\left\{-\frac{(x_k - \mu_k)^2}{2\sigma_k^2}\right\}$$

$$f\left(x_k,x_l;\mu_k,\mu_l,\sigma_k,\sigma_l;\rho_{kl}\right)=\frac{1}{2\pi\sigma_k\sigma_l\sqrt{1-\rho_{kl}^2}}\exp\left\{-\frac{1}{2\left(1-\rho_{kl}^2\right)}\left(y_k^2-2\rho_{kl}y_ky_l+y_l^2\right)\right\} \quad (10)$$

3 Practical Application

3.1 Design of Standard Sample for Sensory Tests

The mat FRP (fiber reinforced plastics) product defect inspection was assumed to be an exercise for the verification of the proposed method, and the perception psychometric experiment by which three attributes (size, aspect ratio, and grayscale) were to be controlled. Numeric conversion of stimulus strength of these three attributes applied correspondingly to the former report[3]. The mean value and the standard deviations of size x_{1j} (area), aspect ratio x_{2j} (ratio of the extended shaft and minor axis), and grayscale x_{3j} of the defect which exists on the surface of the FRP material which is obtained from the inspection image as shown in Figure 3 and Table 1.

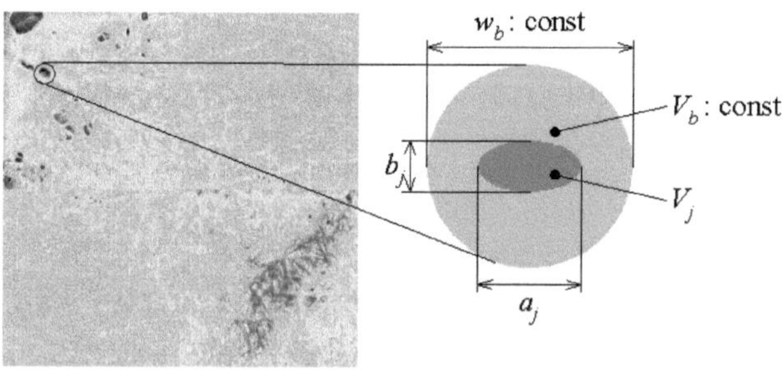

Fig. 3. Definition of dimensions from inspection image

The value of amount x_i^0, the feature of the standard sample, and the value of amount x_{ij} (i=1,2,3, j=1,2,...), the feature of the target sample were given by the following formulas. As to the amount of the feature x_1, which is the first attribute (size), the value that refers to the length of the long axis of an ellipse was used. The standard stimulus strength of standard sample as the amount of feature is set as follows:

$$x_1^0=\sqrt{\frac{4S}{\pi\mu_R}},\quad S=15\mu_A \quad (11)$$

$$x_{1j}=\sqrt{\frac{4S_j}{\pi\mu_R}},\quad S_j=15\{\mu_A+\sigma_A h_A(j-3.5)\} \quad (12)$$

Table 1. Statistics by inspection image data

	area size (mm^2)	aspect ratio	grayscale
mean	$\mu_A = 6.347$	$\mu_R = 0.328$	$\mu_V = 0.583$
S.D.	$\sigma_A = 19.113$	$\sigma_R = 0.197$	$\sigma_V = 0.422$

second attribute is flatness degree x_2 and is obtained by subtracting aspect ratio (of long axis and short axis) from 1 and is defined by

$$x_{2j}=(1-\mu_R)+\sigma_R h_R(j-3.5),\ x_2^0=(1-\mu_R) \quad (13)$$

where constant number μ_R, σ_R is the mean of aspect ratio and the standard deviation of aspect ratio respectively. In this case, the original data of aspect ratio is calculated by $1-\mu_R$ from the standard samples to strengthen stimulation according to the increase of stimulation strength. Finally, the strength of stimuli x_{3j}, x_3^0 of the degree of grayscale of object as the third attribute is quantified as

$$x_{3j}=\mu_V+\sigma_V h_V(j-3.5),\ x_3^0=\mu_V \quad (14)$$

where μ_V, σ_V are respectively the mean and the standard deviation of grayscale of object as shown in Table 1. The values of coefficients h_A, h_R, and h_V are decided by the previous experiments, and are 0.019, 0.055, and 0.028 respectively[3]. In this way, the strength of stimuli x_{ij}, x_i^0 (i=1,2,3, j=1,2,...) were set for the paired comparison, then they were presented as shown in Fig. 3.

Table 2 shows the value of stimulus strength of standard sample x_{1j}, x_{2j} and x_{3j} and target samples x_1^0 , x_2^0 and x_3^0 for single modality sensory test. Figures 4 and 5 show the examples of degree of stimuli for single and dual modalities sensory tests respectively.

Table 2. Stimuli strength of standard and target samples

i	x_i^0	x_{i1}	x_{i2}	x_{i3}	x_{i4}	x_{i5}	x_{i6}	x_{i7}
1	19.23	17.78	18.36	18.93	19.53	20.11	20.68	21.26
2	0.672	0.644	0.655	0.666	0.677	0.688	0.699	0.71
3	0.583	0.553	0.565	0.577	0.589	0.601	0.613	0.625

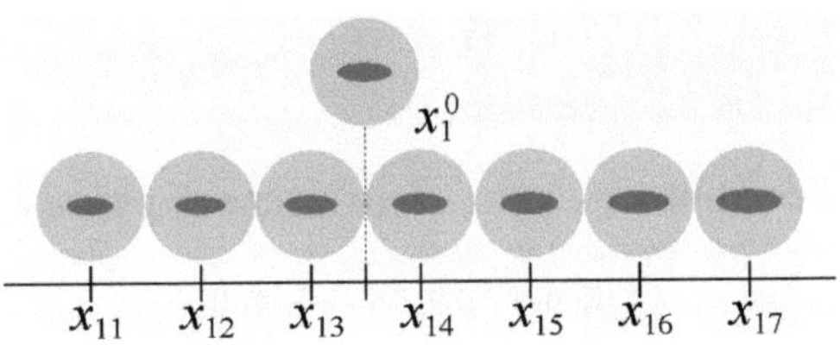

Fig 4. Examples of degree of stimuli (x_{1j}) for single modality sensory test

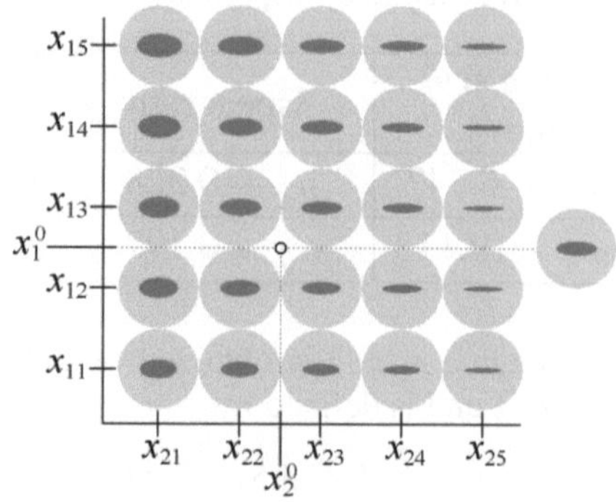

Fig. 5. Examples of degree of stimuli for dual modalities sensory test

3.2 Experimental Setup for Sensory Tests

In the experiment, the eyesight of the naked-eye or the correction made ten adults men and women of 20 years old were selected as subjects. The practical sensory tests are prepared for single and multiple modalities using the paired comparison test with standard sample and target sample that can be a pair of objects for comparison as presentation sample in Figure 6, because it requires the use in a pair of comparison experiment by the defect inspection. The perception experiment is an experiment that fixes the subject on chin rest as shown in Figure 6(a), presents 17-inch display forward of 60cm before the eyes, shows the subject stimulation images controlled by the stimulation generation program, and measures the reaction. Therefore, the program that is able to present each sample in a standard sample (left side in figure) and the target sample (right side in figure) as shown in Figure 6(b) has been developed. The program was developed by Visual Studio 6.0 (Visual Basic language) on Windows OS.

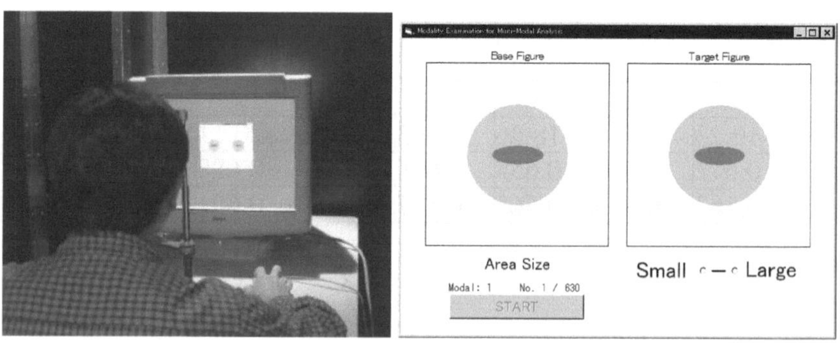

(a) apparent condition (b) scene of paired comparison test

Fig. 6. Experimental setup

3.3 Results and Discussions

3.3.1 Experimental Results of Single Modality Sensory Test

As an example of psychometric curve, Figure 7 illustrates the psychometric curve of subject A derived from the data measured by paired comparison test for single attribute of x_1. As can be seen Figure 7, a monotonically increasing function is derived from the ML estimation in the case of all experimental conditions.

Distributions of differences on characteristic of x_1, x_2 and x_3 among 10 subjects are shown in Figure 8. This distribution of PSE can be obtained by method of mean rank in due small order of PSE. From Figure 8, it is known that the mean values of PSE of x_1, x_2 and x_3 are 17.37, 0.647 and 0.548 respectively. Mean values of PSE become close the standard stimuli of x_1^0, x_2^0 and x_3^0. In addition, on the other hand, the coefficients of variation are 0.043, 0.018 and 0.020. Distributions of PSE indicate normality as can be seen to become near straight lines on the plane by normalized variable Y and strength of stimuli x_i.

It is shown in Figure 9 that the tendency of the results on PSE are strongly dependent on the values of threshold x_i^0.

3.3.2 Experimental Results of Multiple Modalities Sensory Test

From the result in experiments, the probabilities for dual and multiple attributes are larger than that of single attribute as shown in Figure 7. This tendency is same in all experimental conditions. Increasing amount of detection probability, however, is not same in different ways with the variation of correlation coefficient. In fact, the variation

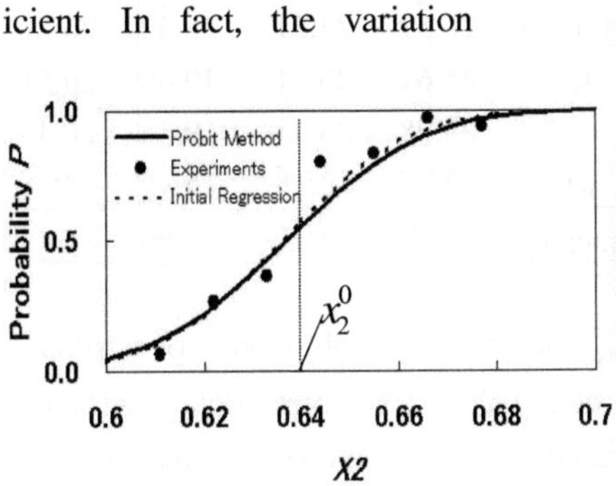

Fig. 7. Psychometric curves (Subject A)

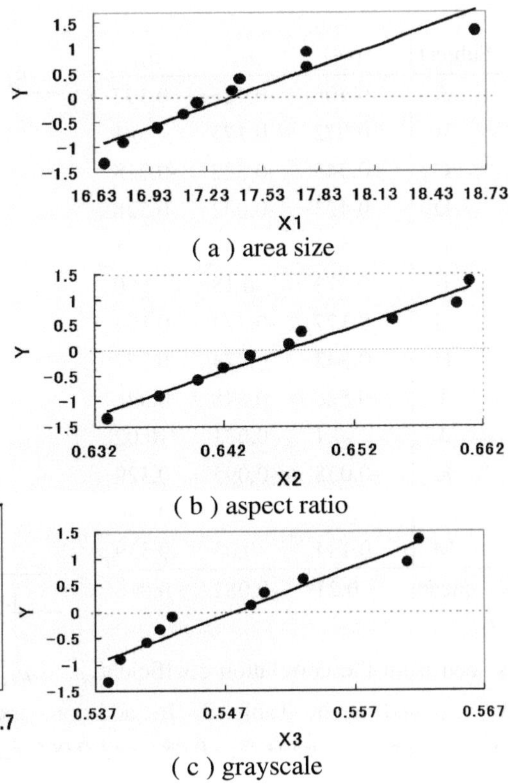

Fig. 8. Sample distribution of PSE

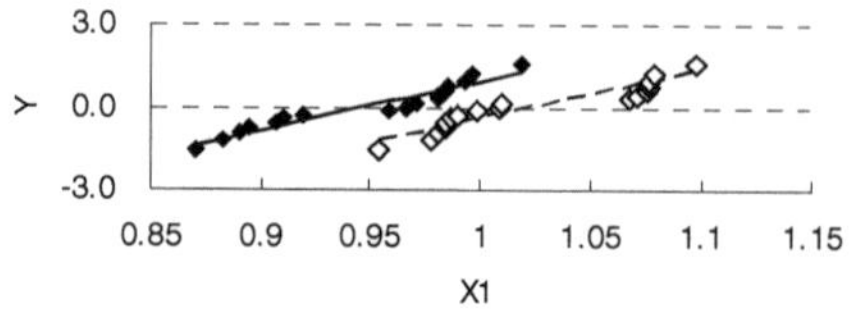

(a) area size

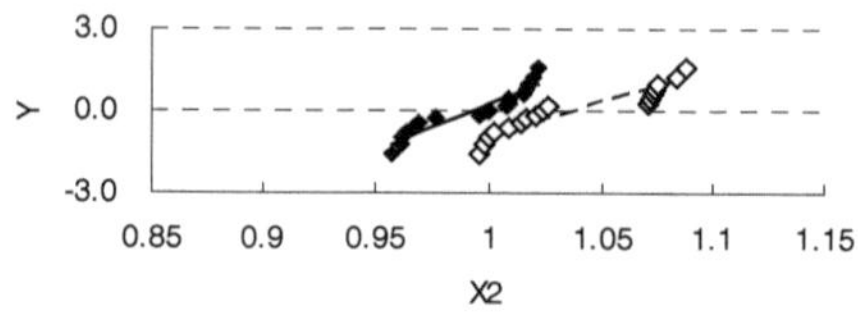

(b) aspect ratio

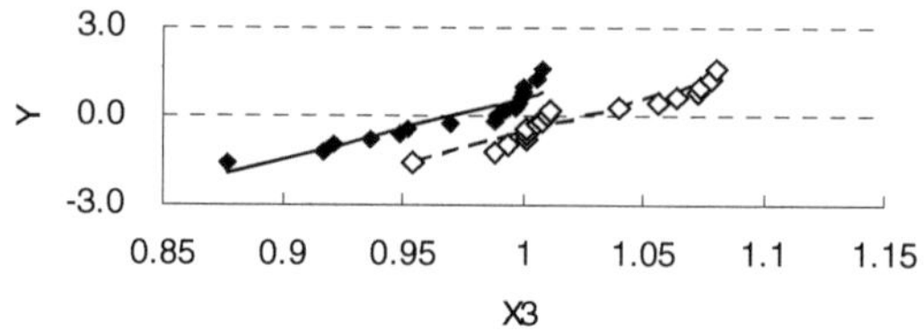

(c) grayscale

Fig. 9. Dependency on threshold on PSE distribution

Table 3. Correlation of coefficient

Subject	ρ_{12}	ρ_{23}	ρ_{31}
A	-0.406	-0.426	0.397
B	0.072	0.173	0.1
C	0.545	-0.522	-0.718
D	0.423	-0.042	0.528
E	0.532	-0.249	-0.315
F	0.573	-0.18	0
G	0.177	-0.373	-0.104
H	0.342	0.224	-0.272
I	0.528	0.548	0.048
J	-0.1	0.031	-0.02
K	-0.038	-0.095	0.329
L	-0.062	-0.09	0.346
M	0.144	-0.05	0.539
mean	0.21	-0.081	0.066

Table 4. Improving ratio

Subject	P12 (%)	P23 (%)	P31 (%)	PL (%)
A	31.65	32.01	18.5	32.15
B	23.85	22.23	23.41	23.85
C	15.83	33.74	37.76	37.76
D	18.06	25.67	16.14	25.67
E	16.07	29.01	30.11	30.11
F	15.29	27.88	25	27.88
G	22.17	31.08	26.65	31.08
H	19.45	21.41	29.38	29.38
I	16.14	15.78	24.23	24.23
J	26.59	24.51	25.32	26.59
K	25.61	26.51	19.66	26.51
L	25.98	26.43	19.38	26.43
M	22.7	25.8	15.94	25.8
mean	21.49	26.31	23.96	28.27

is seen from the correlation coefficient ρ_{12} , ρ_{23} and ρ_{31} for combinations of attributes x_1 , x_2 and x_3 in Table 3. In addition, mean values of correlation coefficient ρ_{12} , ρ_{23} and ρ_{31} are 0.21, –0.081 and 0.066 respectively. On the other hand, Table 4 shows improving ratio of detection probabilities for dual attributes: P12, P23, P31 respectively rise approximately 21%, 26% and 24% for mean in the case of dual attribute

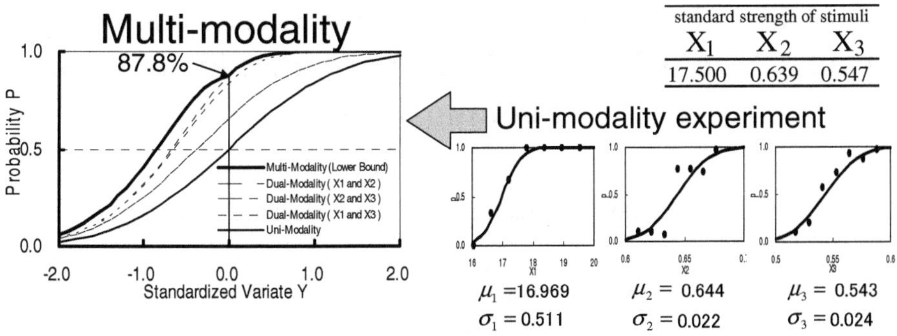

Fig. 10. Effect of multiple perceptions (subject C)

experiments. That is to say improving ratio of detection probabilities have a pronounced tendency to depend correlation coefficient. In addition, the improving ratio PL which is obtained by the approximation formula for the multiple attributes rises approximately 28% in comparison with the detection probability of past single attribute.

4 Conclusion

It was successfully accomplished in this study that the degree of human performance on visual perception of single and multiple modalities could be identified by using psychometric function and statistical estimation. As an application for practical use of proposed method, the way to produce the standard sample for sensory test was developed for detecting the human performance on the attribute of figure. It was verified that the sensory test for single modality and multiple modalities test enabled one can evaluate the human performance on visual perception by proposed method in this paper.

Acknowledgments. This research was supported in part of IICT, Konan University by MEXT ORC (2004-2008).

References

1. Ooyama, T., Imai, S., Wake, T. (eds.): Shinpen Kankaku Chikaku Handbook, Seishinshobo (in Japanese), pp. 19–41 (1994)
2. Hayashi, H., Nakagawa, M., Nakayasu, H.: Psychometric evaluation with correlation for multiple perceptions. In: Proceedings of the First World Congress of the International Federation for Systems Research, JAIST Press, CD-ROM, S3-5-2 (2005)
3. Nakagawa, M., Nakayasu, H.: Evaluation of detection probability for flaw by multiple modalities. Journal of the Society of Materials Science, Japan, (in Japanese) 54(1), 2–7 (2005)
4. Hayashi, H., Nakagawa, M., Nakayasu, H.: Evaluation Method of Detection Probability on Defect Inspection by Staircase Method Based on Psychometric Measurement. Journal of the Society of Materials Science, Japan, (in Japanese) 56(1), 50–55 (2005)
5. Madsen, H.O., Krenk, S., Lind, N.C.: Method of Structural Safety, Prentice Hall, pp. 116–120. Prentice Hall, Englewood Cliffs (1986)

Considerations on Efficient Touch Interfaces – How Display Size Influences the Performance in an Applied Pointing Task

Michael Oehl[1], Christine Sutter[2], and Martina Ziefle[2]

[1] University of Lueneburg, Institute of Experimental Business Psychology, Wilschenbrucher Weg 84, 21335 Lueneburg, Germany
oehl@uni-lueneburg.de

[2] RWTH Aachen University, Institute of Psychology, Jaegerstr. 17/19, 52066 Aachen, Germany
christine.sutter@post.rwth-aachen.de, martina.ziefle@psych.rwth-aachen.de

Abstract. The limited screen space in small technical devices imposes considerable usability challenges. On the one hand objects displayed on small screens should be big enough to be hit successfully, but also small enough to house several objects on the screen at the same time. However, findings up to now show that single pointing is more effective in a large display compared to a smaller display. In the present experiment this was also confirmed for an applied multidirectional serial pointing task. Especially in more difficult tasks, results point at a shift of the speed-accuracy tradeoff. In large displays a fast and comparably accurate execution is chosen in contrast to a very inaccurate and time-consuming style in small displays. From an ergonomic point of view the outcomes recommend an optimized balance of task difficulty and display size in small screen devices.

Keywords: Task difficulty, display size, pointing performance, small screen device, touch input.

1 Introduction

A broad user group increasingly uses mobile devices in business as well as private areas. The limited screen space in small technical devices imposes considerable usability challenges. On the one hand objects displayed on small screens should be big enough to be hit successfully, but also small enough to house several objects on the screen at the same time. Thus, the optimization of object size and display size is of crucial interest in modern device design. A standard research paradigm for the evaluation of the ergonomic design of hard- and software is given by Fitts' law [1]. It predicts movement time (MT) as a function of task difficulty (= ID, index of difficulty), determined by target distance (Amplitude) and target size (Width). The modification of the original law [1] by MacKenzie [2] is nowadays established in the ISO [3] with $MT = a + b \log_2 (A / W + 1)$. According to this function, bigger and nearer targets are easier to reach than smaller and farther away targets. A considerable

M.J. Smith, G. Salvendy (Eds.): Human Interface, Part I, HCII 2007, LNCS 4557, pp. 136 – 143, 2007.

amount of studies [e.g. 4-8] supported that the motor behavior operating an input device follows the same fundamental psychomotor principle as found for manual aiming movements. However, contrary to Fitts' law were findings of target size being of higher importance for the efficiency of aiming performance than target distance [5,6,9]. Sutter and Ziefle [5,6] showed that target size and target distance did not equally contribute to task difficulty as proposed by Fitts. But target size had a more powerful effect on movement time insofar that movement time was disproportional longer for smaller targets. Findings of Tränkle and Deutschmann [10] points at another restriction of Fitts' law as a design tool. The authors showed that pointing performance with a mouse was much more effective in a large display compared to a smaller display. Beyond the significant effect of task difficulty the impact of display size might be interpreted as a cognitive effect: The processing of larger space in which the movement has to be executed impose on a faster movement execution. Participants seemed to react more carefree and moved the mouse cursor much faster and snappier compared to the smaller display where movement space appeared more restricted. This was also observed for touch based interfaces operated with a stylus [11].

The discussion about task difficulty in modern applications might again be surveyed with this background in a more applied context. Displays in technical devices get smaller and smaller and their restricted movement space are contradictory to an efficient interaction. This study surveys an optimized design for touch interfaces in small screen devices with a stylus as input device and an applied pointing task. Data were analyzed according to two hypotheses: (1) The predicted effects of task difficulty on the basis of Fitts law [2] will be replicated ones more in this study. According to this, movement time should be prolonged in high task difficulties being composed of small target sizes and target distances being farther away. (2) The impact of display size and its interaction with task difficulty is considered. If assuming that display size acts as perceptual frame of reference, determining the speed of movement due to its ballistic nature or to cognitive effects, it can be deduced that larger display sizes should result in faster movements, accompanied by a lower accuracy of the pointing movement.

2 Method

The experiment was based on a two-factorial design with repeated measurements. The *independent variables* were the *display size* of the touch screen and the *difficulty of pointing task*.

Three display sizes were realized, covering a wide range of screen sizes present in real devices equipped with a touch interface (Figure 1; Display 1 = 6.00 x 8.00 cm, Display 2 = 12.00 x 16.00 cm and Display 3 = 18.00 x 24.00 cm). Difficulty of the pointing task was varied from 1.81 to 4.95 bits [2]. In dependence on display size, in maximum four target sizes (0.25; 0.50; 0.75 and 1.00 cm) and three target distances (2.50; 5.00 and 7.50 cm) were chosen. Due to the three different screen sizes and the spatial restrictions of the smallest display, respectively, the IDs to be realized differed across the display sizes. There were two IDs realized in all three display sizes, ID 2.58 bits and ID 3.46 bits. Thus, the research question at issue, the influence of display size on performance in an applied pointing task can be analyzed for these two IDs. Both IDs represent typical task difficulties for mobile devices. Exploiting the increasing display space with respect to the analysis of more IDs, five different IDs were realized for the medium display size (2.12; 2.58; 2.94; 3.46; 4.39 bits) and in total nine IDs were possible to be realized for the large display size (1.81; 2.12; 2.58; 2.94; 3.09; 3.46; 4.00; 4.39; 4.95 bits).

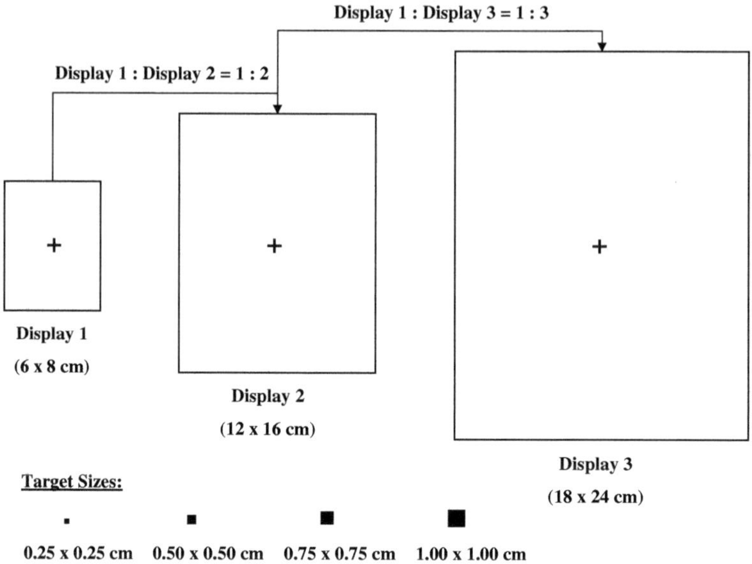

Fig. 1. Schematic illustration of display size and target size

Sixteen participants ($N = 16$) accomplished each 86 serial multidirectional pointing tasks on a high precision touch screen with a stylus. In each task, a sequence of nine square targets was presented directly one after another and users had to move the pen from a well-defined starting point inside the target area. Participants had to work on all display sizes and on all task difficulties (IDs). The order of display sizes was balanced over participants. The order of IDs within each display size was at random. To exclude confounding effects of movement direction and to realize the applied multidirectional pointing task, the target position was placed in eight different directions (0°, 45°, 90°, 135°, 180°, 225°, 270°, 315°) relative to the starting point of movement [3]. Time and spatial information of pointing movements (x-, y-coordinates) were recorded action-correlated with a logging and analyzing software tool [11]. A state of the art high precision 15'' (diagonal measurement) TFT touch screen (Iiyama AX 3819 UT) with a 1024 x 768 resolution and a "Touchlogic Twisted Nematic®" (RS 232C ELO Touchsystems) was used. The three different display sizes on the touch screen were created by the software program and represented simply a lighted area with the rest of the display kept dark and inactive. The stylus was a high precision professional touch input device for industrial applications (WES®).

Dependent variables were on the one hand time and accuracy of pointing as objective measures and on the other hand usability ratings in terms of task difficulty and their effort to complete the serial multidirectional pointing task.

The *time* measure of performance included the movement time to complete a whole task, which described the interval of time from correct pointing with the stylus on the onset point to final correct pointing on the last of all nine targets. For *accuracy*, the incorrect pointing was logged, defined as pointing error. This error occurred when participants did not hit the target itself, but pointed outside the target's boundaries.

Usability ratings were measured by *participants' judgments of task difficulty and effort* on a 4-point scale in a questionnaire (1 = low; 2 = reasonably low; 3 = reasonably high; 4 = high). In order to facilitate ratings for participants, the actual task difficulties (IDs) composed of target sizes and distances were operationalized in either smaller or bigger targets for each display size. In doing so, the smaller targets comprised the higher IDs whereas the bigger targets comprised the lower IDs.

3 Results

3.1 Effects of Task Difficulty

According to Fitts' law a regression analysis comprising all realized task difficulties (IDs) across the three display sizes revealed a high positive correlation ($r = .82$; $p < 0.001$; $R^2 = .77$) between task difficulty and movement time. Additionally, a high positive correlation ($r = .83$; $p < 0.001$; $R^2 = .72$) between task difficulty and pointing errors could be found.

Analyzing only the both IDs (2.58 and 3.46 bits) present in all three display sizes an analyses of variance (ANOVA) for repeated measurements showed the validity of Fitts' law for movement time in pointing tasks on all three display sizes (F (1,14) = 67.30; $p < 0.001$). Movement time increased significantly with increasing task difficulties (Figure 2). Again, a strong effect of task difficulty on pointing errors could be found (F (1,14) = 71.17; $p < 0.001$), i.e. the higher ID of 3.46 bits induced significantly more pointing errors than the lower ID of 2.58 bits (Figure 3).

Usability ratings of task difficulty and effort confirmed the described findings in pointing time and accuracy across all display sizes. For the subjective difficulty of smaller and bigger targets per display size, a non-parametric Friedman-Test revealed significant ranks ($\chi^2_{(5)} = 71.51$; $p < 0.001$). Generally, the smaller targets of each display size, which comprised actually the higer IDs, were rated as fairly "reasonably high = 3 of 4" in difficulty (small display: $M = 2.81$; medium display: $M = 2.69$; large display: $M = 3.13$). The smaller targets on the large display were significantly rated as most difficult of all realized targets, even compared with a non-parametric Wilcoxon-Test to the next difficult targets on the medium display ($Z = -2.11$; $p < 0.05$) and on the small display ($Z = -1.89$; $p = 0.059$). Amongst them, the objective most difficult pointing task of the whole study with an ID of 4.95 bits was present, but on the other both display sizes not. This finding confirms the quality of participants' usability ratings. Additionally, this is corroborated by participants' difficulty ratings of the bigger targets (lower IDs) on the three display sizes as fairly "low = 1 of 4" (small display: $M = 1.25$; medium display: $M = 1.25$; large display: $M = 1.38$). Comparing the average difficulty ratings for the higher IDs ($M = 2.88$) and lower IDs ($M = 1.29$) significant differences in ratings ($Z = -3.54$; $p < 0.001$) could be found.

A similar pattern appeared for the estimated task effort of smaller and bigger targets per display size. The non-parametric Friedman-Test revealed likewise significant ranks ($\chi^2_{(5)} = 62.78$; $p < 0.001$). The smaller targets of each display size were rated as fairly “reasonably high = 3 of 4” in effort (small display: $M = 2.69$; medium display: $M = 2.81$; large display: $M = 3.13$). Here again, the smaller targets on the large display were rated to be of highest effort. Participants rated the pointing effort of the bigger targets (lower IDs) on the three display sizes as fairly “low = 1 of 4” or “reasonably low = 2 of 4” (small display: $M = 1.50$; medium display: $M = 1.50$; large display: $M = 1.81$). Comparing the average effort ratings for the higher IDs ($M = 2.88$) and lower IDs ($M = 1.60$) a non-parametric Wilcoxon-Test showed significant differences in ratings ($Z = -3.54$; $p < 0.001$) just as well as for the usability ratings of task difficulty.

3.2 Effects of Display Size

The data for those both IDs (2.58 and 3.46 bits) present in all three display sizes were analyzed by analyses of variance (ANOVA) for repeated measurements. According to our second hypothesis a significant interaction between display size and task difficulty for pointing time (F (2,13) = 7.81; $p < 0.01$) and pointing errors (F (2,13) = 10.42; $p < 0.005$) could be found. From an ergonomic point of view this interaction is insightful, as it shows that the very same task difficulty can result in a different performance depending on how large or how small the display is.

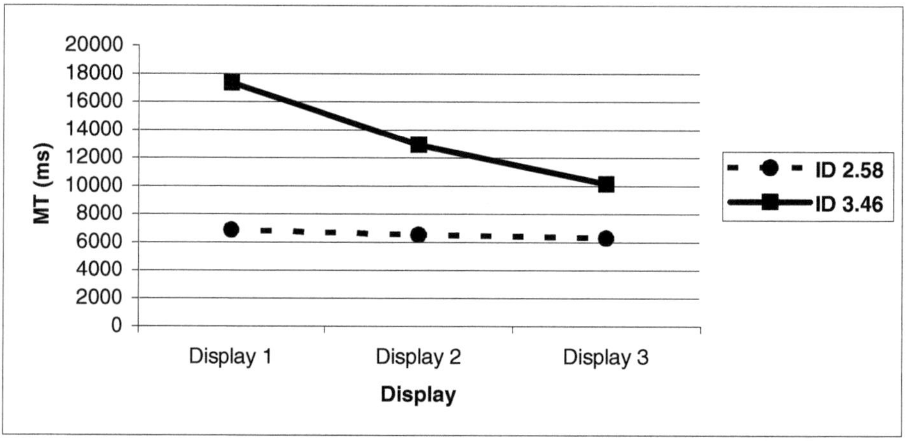

Fig. 2. Interaction effect for the display size (small = display 1; medium = display 2; large = display 3) and ID (2.58 and 3.46 bits): movement time (ms)

If ID is low (2.58 bits) pointing time was reduced clearly the larger the display size was (small display: $M = 6867.26$ ms; medium display: $M = 6531.67$ ms; large display: $M = 6295.44$ ms). The large display provided an observable gain of about 8% in pointing time compared to the same task difficulty in the small display. In direct contrast, for higher IDs (3.46 bits), pointing time has significantly increased for the small display ($M = 17326.16$ ms) when compared to the medium ($M = 12926.39$ ms)

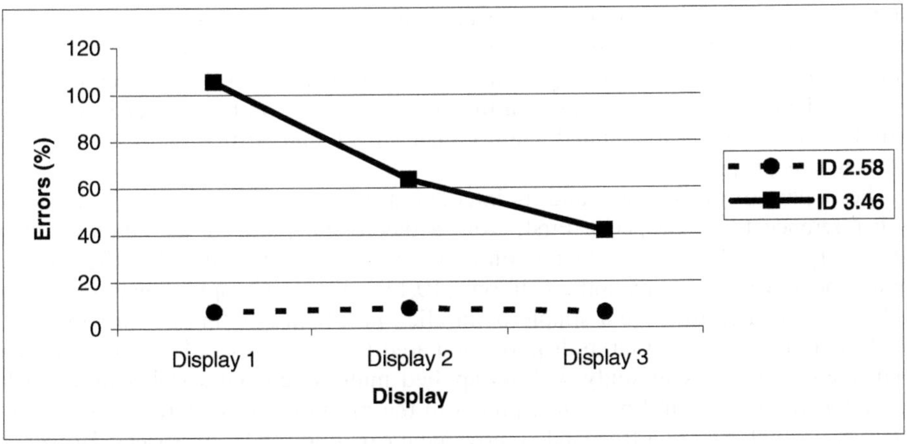

Fig. 3. Interaction effect for the display size (small = display 1; medium = display 2; large = display 3) and ID (2.58 and 3.46 bits): pointing error (%)

and large display (*M* = 10145.94 ms). The disadvantage of having a small display and a high task difficulty (ID of 3.46 bits) ranged at 41 % compared with the same task difficulty on the large display. This interaction is illustrated in Figure 2.

A similar pattern was found for pointing accuracy (Figure 3). While the percentage of errors did not differ due to display size for the lower ID 2.58 bits (small display: *M* = 7.11%; medium display: *M* = 8.56%; large display: *M* = 6.67%), a remarkable higher number of errors were revealed for the higher ID 3.46 bits depending on the display size: The lowest amount of pointing errors was found in the large display (*M* = 41.55%). It increased already for the medium display (*M* = 63.44 %) and mounted up to an error peak of 105.44% for the small display. An error rate of more than 100% means that more than nine errors occurred in the respective serial multidirectional pointing task. In this case, participants executed, on average, 1.05 errors per target until the whole serial pointing task of nine consecutive targets in the small display was finished properly.

A moderating influence of display size on participants' usability ratings like on the objective measures (pointing time and errors) could not be found.

4 Conclusions

The design of efficient touch interfaces is becoming a challenge in ergonomics. Displays in technical devices get smaller and smaller and their restricted movement space are contradictory to an efficient interaction. In literature the impact of display size on motor performance in combination with an increase of task difficulty in smaller displays is reported [10,11]. These results are theoretical unexpected, but can be interpreted in a cognitive or motor behavioral way [11]. Nevertheless, the practical implications are manifold towards the design of small screen interfaces.

The findings of the present experiment once more confirmed Fitts' law for pointing movements executed with a stylus. An increase of task difficulty in the applied multidirectional serial pointing task resulted in a lower pointing performance, i.e. movement time and pointing errors rose distinctly. These results stand in concordance with many other studies regarding input device performance in dependence on Fitts' law [e.g. 4-8]. Moreover, whenever in the present experiment the task execution was complicated users clearly stated their higher effort and difficulty of solving the pointing task properly.

The impact of display size and its interaction with task difficulty was considered with reference to two aspects. First, from a theoretical point of view this result is rather surprising. So far, the interaction between display size and task difficulty was found under strict experimental conditions [10,11]. When one rapid aimed movement had to be executed to solve a simple point-click task then pointing performance was much more effective in a large display compared to a smaller display. This was also confirmed in the present study with an applied multidirectional serial pointing task. The interaction was, until now, interpreted in the meaning of the perceived frame of reference, which allows a freehanded movement execution in larger spaces. However, the present results hint at different strategies of motor behavior. Especially regarding the more difficult pointing tasks (higher IDs), in large displays the speed-accuracy tradeoff [12] is shifted towards a fast and comparably accurate execution whereas in small screens a very inaccurate and time-consuming style is chosen. Thus, the cognitive processing of the frame of reference is very likely to affect the speed-accuracy tradeoff. However, this has to be addressed systematically in further research. Second, from an ergonomic point of view the interaction between display size and task difficulty is insightful. It shows that the very same task difficulty can result in a different performance depending on how large or how small the display is. The outcomes of this research allow ergonomic guidelines for an optimized usage and design of touch-based screen devices, assuring efficient and effective handling. It is therefore important to achieve an optimized balance of task difficulty and display size in small screen devices. That means that targets in small displays should be in easy reach for an effective interaction.

References

1. Fitts, P.M.: The information capacity of the human motor system in controlling the amplitude of movements. Journal of Experimental Psychology 47, 381–391 (1954)
2. MacKenzie, I.S.: Fitts law as a research and design tool in human-computer interaction. Human-Computer Interaction 7, 91–139 (1992)
3. ISO 9241-9:2000: Ergonomic requirements for office work with visual display terminals – Part 9: Requirements for non-keyboard input devices. International Organization for Standardization. Beuth, Berlin (2000)
4. Douglas, S.A., Mithal, A.K.: The ergonomics of computer pointing devices. London, Springer, Berlin Heidelberg
5. Sutter, C.: Charakteristische Nutzereigenschaften und ihr Einfluss auf die Bedienung von Notebook-Eingabegeräten: Eine arbeitspsychologische Evaluation von Touchpad und Trackpoint [Specific user characteristics and their impact on the usage of notebook-input devices: a work psychological evaluation of touchpad and trackpoint]. RWTH Aachen, Germany (2006), Online published PhD-Thesis, http://darwin.bth.rwth-aachen.de/opus3/volltexte/2006/1697/

6. Sutter, C., Ziefle, M.: Psychomotor efficiency in users of notebook input devices: Confirmation and restrictions of Fitts law as an evaluative tool for user-friendly design. In: Proceedings of the Human Factors and Ergonomics Society 48th Annual Meeting, pp. 523–528. Human Factors Society, Santa Monica, CA (2004)
7. Armbrüster, C., Sutter, C., Ziefle, M.: Target size and distance: Important factors for designing user interfaces for older notebook users. Work with Computing Systems, pp. 454–459. Damai Sciences, Kuala Lumpur (2004)
8. Armbrüster, C., Sutter, C., Ziefle, M.: Notebook input devices put to the age test: The usability of trackpoint and touchpad for middle-aged adults. Ergonomics (in press)
9. Sheridan, M.R.: A repraisal of Fitts law. Journal of Motor Behavior 11/3, 179–188 (1979)
10. Tränkle, U., Deutschmann, D.: Factors influencing speed and precision of cursor pointing using a mouse. Ergonomics 34/2, 161–174 (1991)
11. Oehl, M., Sutter, C., Ziefle, M.: Design of Touch Interfaces in Small Screen Devices: Effects of Display Size and Task Difficulty for Older and Younger Users. International Journal of Human-Computer Studies (under revision)
12. Pachella, R.: The interpretation of time in information-processing research. In: Kantowitz, B. (ed.) Human information processing: tutorials in performance and cognition, pp. 41–80. John Whiley and Sons, Hillsdale (1974)

Analysis and Evaluation of Recommendation Systems

Emiko Orimo[1,2], Hideki Koike[2], Toshiyuki Masui[3], and Akikazu Takeuchi[1]

[1] So-net Entertainment Corporation
[2] University of Electro-Communications
[3] Apple Computer Inc.
{emiko.orimo,akikazu.takeuchi}@so-net.co.jp, koike@is.uec.ac.jp, masui@pitecan.com

Abstract. Popular online services, such as Amazon.com, provide recommendations for users by using other users' rating scores for items. In this study, we describe three types of rating systems: score-rated, count-rated, and digital-rated. We hypothesize that digital-rated systems provide the most useful recommendations. Then we analyze the differences in the results of the rating when the granularity of the score changes. Finally, we visualize users by developing a 2-D visualization system that uses a multi-dimensional scaling method.

Keywords: recommendation system, rating algorithm, multi-dimensional scaling method, visualization.

1 Introduction

Popular online services, such as Amazon.com, provide a recommendation to users when a user opens a page describing an item or clicks a link to view its detail. Finding items from the recommendation list is accomplished by information retrieval using methods such as keyword search or browsing.

This recommendation method works well when the user has no exact target, but it lacks quantitative value. In general, calculation of recommended items utilizes a user's history of purchases or rating scores for items. For example, the user gives a score between 1 and 10 based on his or her evaluation of the item. Such rating scores, however, are not exact because people tend to give high scores such as 9 or 10. Ratings can be recognized as "interest in items." Thus, it might be thought that an item rated as a 9 or 10 by a user means that it is his or her favorite, but the rating is not definitive. Therefore, a binary rating such as "buy or not" or "listen or not" could provide a more useful recommendation. We hypothesize that such a binary rating makes it easier for users to rate items and also makes it easier for recommendation systems to perform calculations.

In this paper, we first observe existing rating systems. Then we analyze the difference between the results when the granularity of the rating scores changes. In order to analyze the results visually, we developed a 2-D visualization system that visualizes users who are making recommendations using a multi-dimensional scaling (MDS) method. MDS is widely used in various fields to analyze mutual relations among items. The quantification theory type III (QT-III) enables calculation of the

M.J. Smith, G. Salvendy (Eds.): Human Interface, Part I, HCII 2007, LNCS 4557, pp. 144–152, 2007.

"distance" between items. Using this distance, we can decide the geometrical position of each item so that similar items are placed physically near each other. For example, if user A answered that "Oasis" and "Beatles" are his or her favorite artists, "Oasis" and "Beatles" are near each other with respect to user A. In the same way, if user A and user B answered that "Oasis" is their favorite artist, users A and B are near each other with respect to "Oasis".

2 A Study of Rating Methods

In this study, we consider three types of rating methods for items. The first method gives regulated rating scores such as five stars. We call this a "score-rated type". The next method counts a user's actions such as history of purchase. We call this a "count-rated type". The third method expresses a user's interest in terms such as "1 or 0", meaning "I like it" or "I don't like it." We call this a "digital-rated type". In the following sections of this paper, we analyze the differences between the results obtained by each rating type.

2.1 Samples of Each Type of Rating

Score-Rated Type: Ratebeer.com. Ratebeer.com is a web service about beer. It has a huge amount of information about beer and also rating data by its users.

Fig. 1. Ratebeer.com

Once a user gives ratings about aroma, appearance, flavor, palate, and overall impression, Ratebeer.com converts them to official scores between 0.0 and 5.0. In this case, Ratebeer.com is categorized as a score-rated type.

Count-Rated Type : Last.fm. Last.fm is a web service related to music. This service stores users' histories of listening in real time. Using these histories, it provides recommended tracks and artists to each user. Last.fm is categorized as a count-rated type.

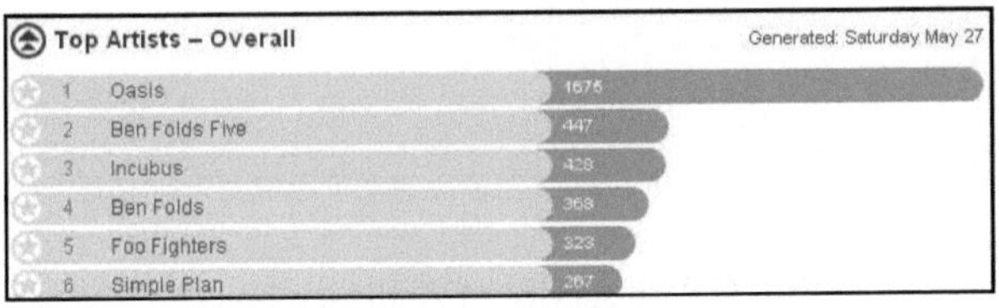

Fig. 2. Last.fm

Digital-Rated Type : Hondana.org. Hondana.org is an online bookshelf service. Its users can register any books they have. It does not require the users to rate books. Hondana.org can be categorized as a digital-rated type.

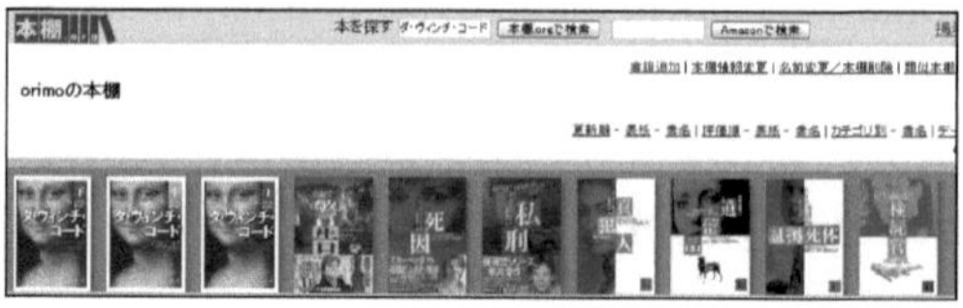

Fig. 3. Hondana.org

2.2 Creating a Data Set

Currently, there are many web services such as Flickr or del.icio.us that feed XML documents like RSS or Atom. In order to collect as much real rating data as easily as possible, we used the XML feeds and created ratings data from them.

Getting data from Web service. Service providers gather and use users' data for their own purposes. They also deliver the information as an XML document called an RSS feed. We can utilize this XML document in our applications.

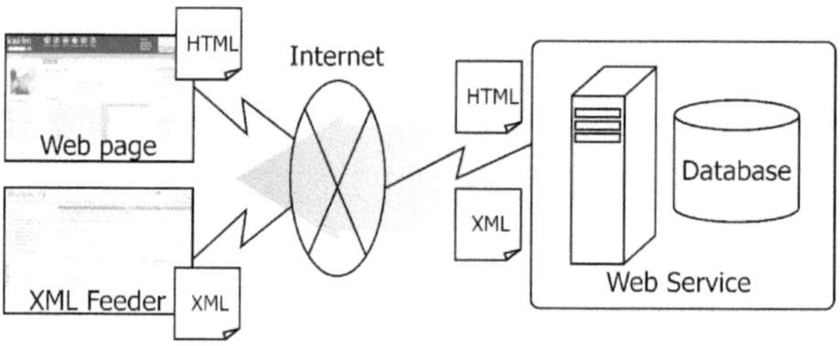

Fig. 4. Web service

Last.fm and Audioscrobbler.net. Audioscrobbler.net provides XML documents. Its data source is the users' listening habits at Last.fm. Using Audioscrobbler, it is possible to get such data as a profile, top artists, top albums, and top tracks for each user.

Fig. 5. Web service of Audioscrobbler

Figure 6 shows the XML about top tracks.


```
<topartists user="orimo">
<artist>
<name>Oasis</name>
<mbid/>
<playcount>1768</playcount>
<rank>1</rank>
<url>http://www.last.fm/music/Oasis</url>
</artist>
<artist>
<name>Ben Folds</name>
<mbid/>
<playcount>702</playcount>
<rank>2</rank>
<url>http://www.last.fm/music/Ben+Folds</url>
</artist>
                -snip-
</topartists>
```


Fig. 6. Users' Top Artists XML

3 Study of Each Data Type

We obtained top favorite artist data for 100 users from Audioscrobbler and conducted experiments described in the following sections. We converted these data to category data for the QT-III, which we called the original data set.

3.1 Original Data Set

In the original data set, it is very rare to see that two or more users listen to the same artist the same number of times. Consider, for example, user A who listened to a track by Oasis, for instance, 10 times and user B who listened to the same track 200 times.

Original Data

Track Data

track	orimo	koike	mifo
Oasis	429	0	40
LedZeppelin	0	598	20
Blur	18	22	23

Category Data

track	orimo	koike	mifo
Oasis40	0	0	1
Oasis429	1	0	0
LedZeppelin20	0	0	1
LedZeppelin598	0	1	0
Blur18	1	0	0
Blur22	0	1	0
Blur23	0	0	1

Normalized Data

Track Data

track	orimo	koike	mifo
Oasis	100	0	9
LedZeppelin	0	100	3
Blur	78	95	100

Category Data

track	orimo	koike	mifo
Oasis9	0	0	1
Oasis100	1	0	0
LedZeppelin3	0	0	1
LedZeppelin100	0	1	0
Blur88	1	0	0
Blur95	0	1	0
Blur100	0	0	1

Fig. 7. Original Data Set and Normalized Data Set

If A listened to a particular song 10 times and B listened to the same song 10 times, the result might be given as "A's listening habit is similar to that of B" since both listened to the song the same number of times. This is not appropriate because the total number of A's playing count is different from that of B.

It is, therefore, necessary to normalize the data. By setting max repeat count to 100, we normalized all repeat counts. For example, if user A listens to a song by Oasis 10 times and a song by Blur 5 times, the value of Oasis is set to 100 and that of Blur is set to 50. Using this normalized data, we calculate similarity of users' listening by QT-III.

3.2 Normalized Data Set

We converted the original data into a normalized data set, as shown in fig.7. Then we analyzed the difference in the result of the rating when the granularity of the score changed. In order to compare the results visually, we developed a 2-D visualization system that categorizes items by using MDS. Using this data set, we created a temporal data set of the count-rated type. Then we observed how users' positions changed on 2-D space.

3.3 Score-Rated Data Set

To create a temporal data set of the count-rated type, we converted the normalized data set as shown in fig.8. We call this result a score-rated data set. In ten grades of scoring, the user can give a score from 1 to 10. Since such fine granularity makes the rating complicated, few services use this rating. A system using five grades of scoring, as seen in YouTube.com, is more popular. In this case, users' positions are calculated and are plotted on a 2-D map as shown in fig.9. Both rating methods, however, have the problem that there is no exact rule for scoring, and this might reduce the reliability of the rating. In two grades of scoring, on the other hand, the user chooses "good" or "not good". In this case, users' position are calculated and plotted as shown in fig.10.

Scoring in Ten Grades

track	orimo	koike	mifo
Oasis1	0	0	1
Oasis10	1	0	0
LedZeppelin1	0	0	1
LedZeppelin10	0	1	0
Blur8	1	0	0
Blur9	0	1	0
Blur10	0	0	1

Scoring in Five Grades

track	orimo	koike	mifo
Oasis1	0	0	1
Oasis5	1	0	0
LedZeppelin1	0	0	1
LedZeppelin5	0	1	0
Blur4	1	0	0
Blur5	0	1	1

Scoring in Three Grades

track	orimo	koike	mifo
Oasis1	0	0	1
Oasis3	1	0	0
LedZeppelin1	0	0	1
LedZeppelin3	0	1	0
Blur3	1	1	1

Scoring in Two Grades

track	orimo	koike	mifo
Oasis1	0	0	1
Oasis2	1	0	0
LedZeppelin1	0	0	1
LedZeppelin2	0	1	0
Blur2	1	1	1

Fig. 8. Score-Rated Data Sets

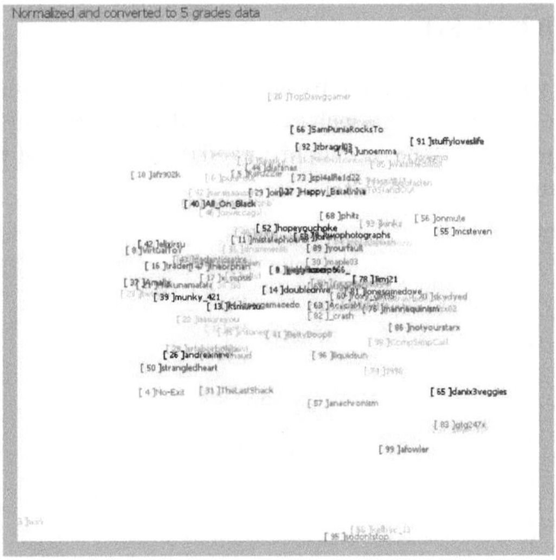

Fig. 9. Users' map using Scoring Data Set

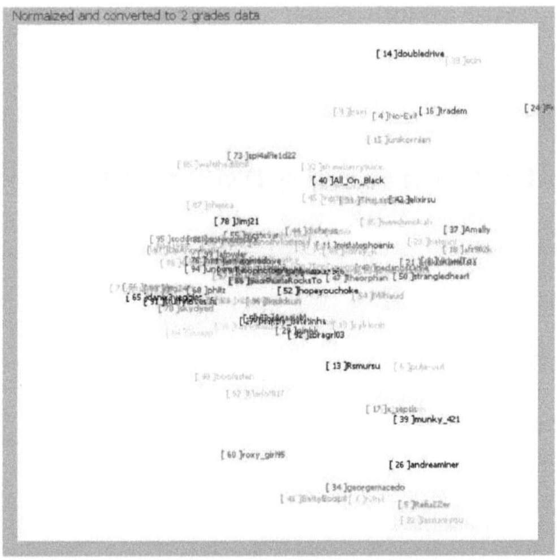

Fig. 10. User map using Scoring Two Grades data Set

3.4 Digital-Rated Data Set

In the next step of this study, we converted the original data set to a digital-rated data set. In this data set, we consider that "repeating artist A's track 100 times" and "repeating artist A's track 1 time" are the same. Scoring in digital is a rating based on selecting "yes" or "don't care". In this case, users' positions are calculated and plotted as shown in fig.11.

3.5 Study of Each Data Set

Using coordinate values calculated by QT-III, we drew a line graph (fig. 12) to compare changes of results from each score-rated data set. This shows that the ups and downs of the graph are almost synchronized, indicating that the granularity of the rating is not very important.

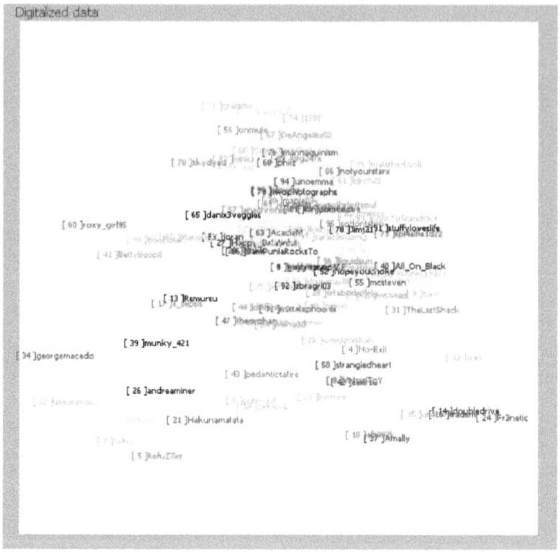

Fig. 11. User map using Digital-rated data Set

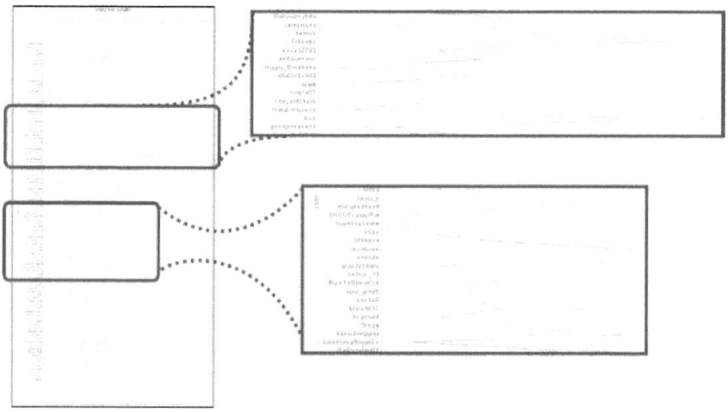

Fig. 12. Comparing results of Score-Rated data sets

Next, we compared the 2D maps generated by two-grade scoring and scoring in digital-rated scoring. The user "orimo" was set at the origin of coordinates and other users were placed by using the distances from "orimo" (fig. 13).

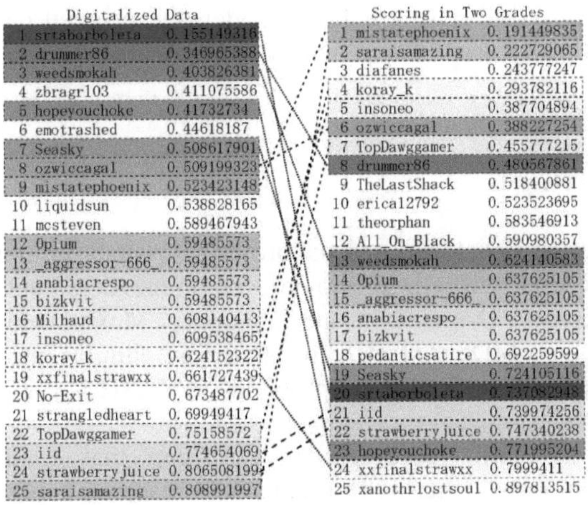

Fig. 13. Distances to Origin Point User

Since the top 25 similar users are almost the same, it could be said that both data sets could provide almost the same result. We also analyzed the reason why the distances between these users and "orimo" are close. Most similar users in the digitized data set were almost repeating the same artist's track. Most similar users in the two-grade scoring data set were repeating tracks of an artist similar to orimo's favorite. So it seems that both data sets apply to similar users.

4 Visualization System

We developed a visualization system using Apache Tomcat on Windows XP. The system was implemented using a Java Servlet and Java applets. We acquired XML

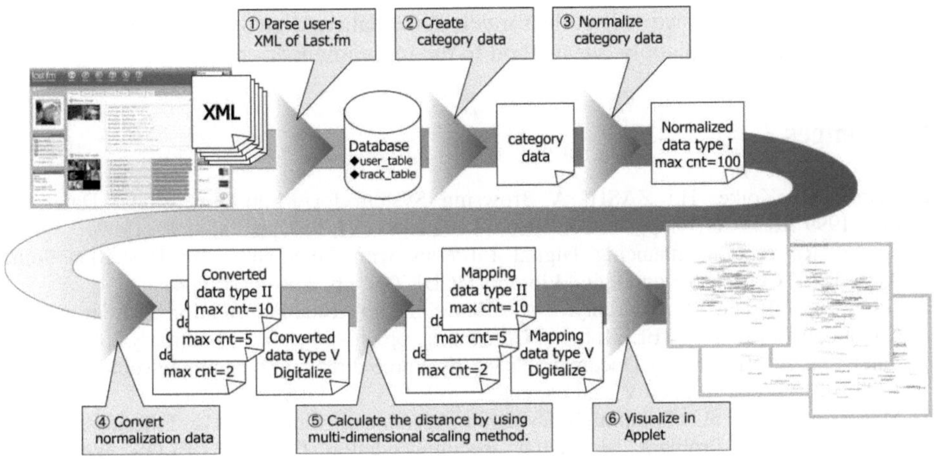

Fig. 14. Process of Calculating Mapping Data

data of Last.fm via HTTP. Using a Java Servlet, the original data are converted to normalized data. Finally, each user is visualized by using a Java applet. This process is shown in fig.14.

5 Related Work

The authors previously developed a visual browsing system for a movie database. The system, named ZASH, visualized movies, recommenders, actors, directors, and keywords on different 2-D planes in one 3-D space. Movies and recommenders are categorized by using MDS QT-III, and therefore similar movies are displayed physically near each other. One of the problems in ZASH is it requires users to give scores to the movies.

TechLens+ is a hybrid recommender algorithm that combines collaborative filtering and content-based filtering to recommend research papers to users. Through some experiments, it is shown that the algorithm gives a good recommendation. However, it also requires the users to give scores to papers.

Amazon.com uses recommendation algorithms to personalize the online store for each customer. The available selection radically changes based on customer interests. Amazon uses an algorithm called item-to-item collaborative filtering, but it also requires users to rate items.

6 Conclusion and Future Works

In this paper, we observed existing rating methods and identified that the granularity of the rating scores is not very important in calculating the similarities of users. In order to verify our hypothesis, we developed a system that collects a large data set from the Internet, normalizes the data, calculates the similarities of users by using MDS QT-III, and visualizes them on a 2D map. The experimental results support our hypothesis.

As a future project, we will collect much more data and analyze the similarities of users. Then we want to apply our method to the recommendation system.

References

1. Orimo, E., Koike, H.: ZASH: A Browsing System for Multi-Dimensional Data. In: Proc.1999 IEEE/CS Symposium on Visual Languages (VL'99), pp. 288–295 (1999)
2. Torres, R., et al.: Enhancing Digital Libraries with TechLens+. In: Proc.2004 Joint ACM/IEEE Conference on Digital Libraries(JCDL'04), pp. 228–236 (2004)
3. Greg, L., Brent, S., Jeremy, Y.: Amazon.com Recommendations. Item-to-Item Collaborative Filtering. IEEE Internet Computing, pp. 76–80 (January/February 2003)
4. Tatemura, J.: Visualizing Document Space by Force-directed Dynamic Layout. In: Proc. 1997 IEEE Symposium on Visual Languages (VL'97), pp. 119–120 (1997)
5. Maneeroj, S., et al.: Combining Dynamic Agents and Collaborative Filtering without Sparsity Rating Problem for Better Recommendation Quality. DELOS Workshop: Personalization and Recommender Systems in Digital Libraries (2001)

Collaborative Scenario Building: The Case of an 'Advertainment' Portal

Natalie Pang[1], Graeme Johanson[2], Sanxing Cao[3], Jianbo Liu[4], and Xin Zhang[5]

[1,2]Centre for Community Networking and Research, Monash University
[3] BCont Laboratory, Communication University of China; State Key Laboratory for Novel Software Technology, Nanjing University
[4]Information Engineering School, Communication University of China
[5]BCont Laboratory, Communication University of China
[1,2]{natalie.pang,graeme.johanson}@infotech.monash.edu.au,
[3-5]{c3x, ljb, zx}@cuc.edu.cn

Abstract. Based on the ongoing development of a portal intended for use during the upcoming Olympics event in 2008, the portal's main purpose is to allow volunteers, spectators, or any other participants of the Beijing Olympics to upload self-directed video clips and relevant advertisement clips associated with the event to the portal. Previous work has found dissimilar experiences in interacting with the portal – thus the need to use collaborative scenario building as part of the design process of the portal. The paper introduces the main purposes of the portal, and demonstrates how the intended outcomes of the portal and dialogues generated lead to the need for participatory design. The sessions building collaborative scenarios are also discussed, demonstrating how they can be used to guide stakeholder participation. Lessons learnt from the design exercise are discussed as a concluding note to the paper.

1 Introduction

The design of interactive systems has been found to require rich methodologies, with scenarios being one of them [2]. Scenarios specify characteristics of users and their tasks in given contexts. Other than raising usability requirements, they also help stakeholders of an information system undertake task analysis. There are numerous benefits of this method; such as helping designers consider characteristics of their users, their tasks, and the environments in which they will be carrying out those tasks. Usability issues can also be raised earlier rather than later, and realistic project timelines can be set using scenarios. Researchers have also found that this method promotes support amongst developers and can be used to aid in understanding requirements between users, developers, designers, and other stakeholders of an information or interactive system, while at the same time being used to generate frameworks for future evaluation studies – thus consistently employing a user-centred design approach [6], [13], [10], [3].

The fact that scenarios are also affordable and easy to use lead to several methods by which they can be built, such as cognitive walkthroughs [5], participant observations [9], user-interaction scenarios [8], or card-based methods [10]. Scenarios

M.J. Smith, G. Salvendy (Eds.): Human Interface, Part I, HCII 2007, LNCS 4557, pp. 153–162, 2007.

were also found to be improved by means of elaboration and collaboration with stakeholders [12]. Carroll [2] also reinforced this finding, arguing that generated scenarios of the various use of a system through participatory design helps to integrate many different kinds of knowledge and experience within the same context.

This paper is based on follow-up studies of earlier research conducted in 2005, when a hybrid approach was employed for the usability evaluation of the same portal, intended to be developed for use during the Olympics event in Beijing in 2008. In the pilot study it was not possible to include stakeholders in a participatory design experience due to geographic and communication difficulties. Since then researchers have engaged themselves with the project in an inclusive way, and this paper explores some of the lessons learnt in the participatory design exercise, with a particular focus on experiences gained through the collaborative building of scenarios.

2 Background

Scenarios are generally described as characteristics of users and their tasks in described and specified contexts. They are widely used in both academia and industry, for various applications such as design, evaluation, organizational planning, goal-setting, just to name a few. Because they are not proprietary, there are several methods and techniques that have been developed for the purpose of generating scenarios. Generation of scenarios is only one of the many benefits of this tool. Carroll [2] argues for this tool to be an integral element of user-centred strategies, which he described as scenario-based design and claims analysis.

According to Carroll's [2] work on the task-artifact cycle, design is explained as an iterative process; in other words a design of any system is never completely 'done'. Unlike traditional conceptions, design in this reasoning is never a linear process – with 'design scenarios at one point in time are the requirements scenarios at the next point in time' [14]. Technological contexts will always change, as how social contexts of use and the economic landscapes would also change and shape usage scenarios. Bonner and Porter [1] also found that users have consistently found it difficult to express and communicate ideas and concepts beyond their immediate experiences. It is therefore clear that the translation of user needs and feedback into practical design requirements cannot be satisfied in a linear fashion.

Giddens [7], in the construction of structuration theory, refers to as the 'duality of structure' – and in this case the recursive nature of design based on structuration theory is manifested in the properties of design as being created and changed by human actions. It also both supports and constrains such actions. Reinforcing this point, Carroll [2] highlights the value of maintaining a continuous and iterative view of situations and consequences of human work and activities, so as to encourage appreciation of various structural contexts and to understand how the dynamics of activities can shape such structures – gaining diverse usage perspectives and achieving effective design outcomes.

Learning can be facilitated through collaborative design, using design as a process of emergence rather than one-off purpose, and to propose as its appropriate 'design' methodology a reflexive process of participative, community and action based interpretations. Thanks to the growth of networks and the Internet, the potential for participation in collaborative design is more inclusive. Perhaps then, knowledge is not just held by 'experts', but whether tacit knowledge [4] is brought into the design interplay.

The use of design scenarios can usually be refined through collaboration with stakeholders. A related previous study [8] involved expert reviewers who undertook usability evaluation and constructed design scenarios using heuristics and scenario-based design. The latter method was found to be more effective in indicating causal relationships between system features and elements of a user interaction scenario. The method was found to be time-consuming which posed one of the major limitations to the project [8]. It was also found that both methods brought different benefits; when scenario-based design is conceived, as mentioned before, as a process of emergence and a reflexive process of participative, community and action based interpretations, it serves as an effective means to generate constructive dialogue and adopt inclusion quickly as one of the features of the project.

Technically a trial of a media content management platform, the portal in development, has been commissioned by the Humanistic Olympics Studies Centre for Beijing Olympics 2008, co-funded by the Chinese Ministry of Education, with support from China State Administration of Radio, Film and TV (SARFT). With the successful bid to hold the 2008 Olympics in Beijing, a team in CUC (Communication University of China), also a member of METIS Global Network (www.metis-global.net), the cross-cultural research organization in multimedia studies, began working on the project of producing an 'advertainment' portal for use in the Beijing Olympics.

The notion of 'advertainment' as the focus of content of the portal presents a portal born out of an age of convergence – as described by Price Waterhouse Coopers [11] -- to refer to the ability of different network platforms to implement different services and the merger of consumer devices. When it was first conceptualized, it was so named because the production of multimedia objects in the portal would suggest purposes of entertainment and advertising, whether or not they were commercial in nature.

The broad purpose of the portal is to allow volunteers, spectators, and other participants in the Beijing Olympics to upload self-directed and self-captured pictures, stories, video clips, and relevant advertisement clips associated with either the event itself or the lead-ups to it, and to interact with one another within the portal. Access to these resources is open to communities and facilitated through a web interface in the portal. Consistent with the theme of 'advertainment', access and use of the portal will also include business sponsors.

The theme of convergence of the portal has led to emphasis on the treatment of rich information objects, perceived to contain layers of meanings and constructions. For instance, each rich information object can belong to, or can be used by more than one person. Earlier studies have revealed that that different observers or users would evaluate the same content object based on diversified experience and knowledge, resulting in inconsistency in the interpretation of content features [8]. Users from diverse backgrounds and cultures, of various religions, and disparate social classes, for example, could view the same colours with very different sentiments. The approach to developing metadata for information objects and resources has been intricately designed to allow users to define their own tags to the multimedia objects which they create and share, while including them in the description model of the portal's infrastructure, to manage uncertainties in such metadata. This principle also supports the theme of convergence, allowing multiple layers of meanings to be constructed by members of the community. Rich multimedia objects such as videos and pictures have been found to warrant richer representations of meanings [8], which will be constructed by different users in the portal. As a medium of human action, the portal acted as a facilitator and repository where such meanings were exchanged, shared, and stored. At the same time, interactions also result in the shaping of the portal design, where humans act as the main agents.

Since the initiation of the project there have been various studies involving users in the last two years. One of the major challenges of this project is the multidisciplinary nature of researchers, and the fact that they are heavily dispersed geographically. It has been difficult to maintain continuous progress of the project. The researchers have been looking for ways to overcome communication difficulties and backlogs. Digital technologies and the Internet have proved to be very useful as collaborative design tools.

3 Collaborative Scenario Building

The use of scenarios is helpful in providing rich contexts about how and why users interact with any system – indicating usability pitfalls, identifying causal relationships between system features and user actions, or give rise to new functionalities that are missing. The approach of using scenarios in this study has been inspired by Carroll's scenario-based design [2]. In scenario-based design, contexts are generated through what is referred to as user interaction scenarios.

Carroll's work in scenario-based design does not stop at the generation of user interaction scenarios; claims analysis was later developed to enlarge the scope of the scenario-based design approach in order to provide greater focus in its analysis and outcomes. For the purpose of discussion the paper is focused on the generation of user interaction scenarios by various stakeholders. The claims analysis technique is used to aid in the analysis of these scenarios that have been generated collaboratively in the design exercise.

In earlier studies, it had not been possible to involve participation from various stakeholders [8]. Since the previous study the interfaces of the portal have undergone significant revisions, in response to feedback and also changes in user requirements. This study aims to build on previous findings through a participatory design approach to scenario building. According to Carroll [2], such an approach provides the opportunity to integrate many kinds of knowledge and experiences within the same context of use. In addition, this approach is also used as a guidepost for the inclusion of dialogue and participation between multidisciplinary researchers who are faced with distance and disciplinary challenges. Before explaining what this paper means by collaborative scenario building, it is necessary to outline the elements required to define a user interaction scenario. Table 1 defines these elements [12].

Table 1. Elements of a user interaction scenario [12]

Element	*Definition*
Setting	Situational details that motivate or explain goals, actions, and reactions of the actor(s)
Actors	Human(s) interacting with the computer or other setting elements; personal characteristics relevant to the scenario
Task goals	Effects on the situation that motivate actions carried out by actor(s)
Plans	Mental activity directed at converting a goal into a behavior
Evaluation	Mental activity directed at interpreting features of the situation
Actions	Observable behavior
Events	External actions or reactions produced by the computer or other features of the setting; some of these may be hidden to the actor(s) but important to scenario

These elements were used as a guide to generate scenarios in this study involving selected stakeholders of the interactive portal. An exercise was created and shared over the Internet amongst participants of the study. This is used as a tool to facilitate collaborative scenario building, but it is also explored for its feasibility as a technique for participatory design and redesign. As a start, a website was set up with detailed explanations of the method using examples from the previous study on the same method. This website functions as a learning experience for participants to be familiarized with the method. Selected interfaces that have been revised as a result of feedback from earlier studies [8] and changes in requirements were captured as well, to function as triggers for participants in the course of generating user interaction scenarios.

3.1 Data Collection and Analysis

In this paper the collection and analysis of design actions and statements is discussed mainly in terms of stakeholder perceptions in the ways they would interact with the portal. Attitudes of participants towards technologies, including the use of technologies such as digital cameras, mobile phones and the Web, to capture multimedia objects were also regarded as important to design actions. These attitudes were captured in warm up and introduction sessions for each group of stakeholders. Four groups of stakeholders were included in this study, namely: potential users, designers, developers, and researchers. The last group of stakeholders was selected because of its influence on the design and outcomes of the portal, which connects the portal to a larger picture of international research and development of interactive portals.

Workshops lasting one to three hours were conducted with each group of stakeholders, with facilitators trained in the method managing each workshop with participants. With all three workshops a warm-up session was carried out by the facilitator, using visual triggers and initial brainstorming discussion on the usefulness or functionalities of the portal. These results were recorded by facilitators.

After the warm-up sessions, participants were then guided into creating scenarios by working through the various elements of the user interaction scenarios. They were asked to think aloud whenever possible, and the facilitator also acted as a guide by asking specific questions relating to these elements. Scenarios were created around three main functionalities of the intended portal: login, view object, and the upload function. Figure 1 shows one example and the goal associated with this functionality. After the creation of these scenarios, discussions were also generated around the main portal interface. All interactions were transcribed by the facilitators in the workshops, and debriefs between researchers were also held face-to-face and online on the design experience and lessons learnt. Transcribed scenarios in Chinese were also translated.

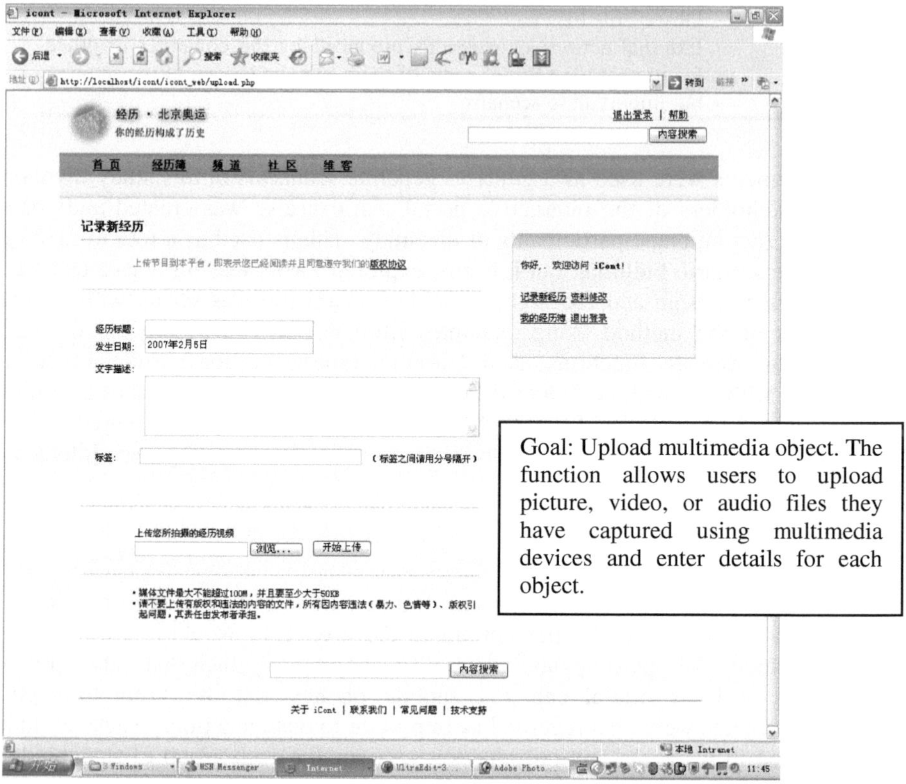

Fig. 1. Upload functionality and associated goal

3.2 Results and Discussion

The collaborative nature of the scenario-based design experience highlighted the importance of stakeholder roles. In addition to the purpose of building pragmatic usage scenarios for the portal, the study has been interested to study the impact of such roles on the outcomes of the design experience.

The building of each scenario involved at least one participant from each stakeholder group: users, designers, developers, and researchers. After the initial briefing participants worked separately on various workstations; building their own scenarios before coming together again for post-workshop discussions. The results were transcribed and analyzed with regards to stakeholder roles while building a list of functional and design requirements as contributed by various participants. Due to space constraints we are not able to display all results here. Table 2 shows selected extracts of one scenario using claims analysis [2], and grouped according to the four stakeholder roles: users, designers, developers, and researchers.

Table 2. An example of an analyzed interaction scenario, summarised according to roles

Function: View		
Role	*+ Claims*	*- Claims*
Users	Clear user options on the right menu trigger goals	Isolated user options (logout and help) and search function in white space on top – easy to miss
	Simple layout and frames gives clear indication of goal	Lack of control over multimedia object
	Annotations and associated comments increase interactivity	Post and reply comments function not clearly visible
Designers	Text annotations and comments provide context while user waits for multimedia object to load	User options in text - inadequate knowledge of the Chinese language would lead to confusion
	Annotations and associated comments increase interactivity	Post and reply comments function not clearly visible
	Clear user options on right frame trigger goals	Viewing pane is too small for larger clips
		No function to link/recommend associated clips
Developers	Viewing pane on left helps to trigger goal to upload clips	No tags/categories to browse related clips
	Buffer progress bar and playing progress bar aid in visibility of object/system status	
Researchers	Indication bars of page and clips help provide feedback on system status	Users used to reading from left to right may not find user options on the right usable
	Standardized control buttons of video aid users in understanding key functions	Table display (two options in a row instead of a list) of user options could be confusing
		Some functions not clearly visible, such as search

Not surprisingly, participants noted rather different potential difficulties with regards to interface scenarios. The actions and events noted by participants in their respective scenarios also differed significantly, with the designers and developers expressing greater expectations in the reactions of the interface upon the completion of certain actions. This is not unexpected, since the designers and developers involved in the study had been studying the portal for some time. These reactions were not noted by other participants in the users and researchers group. These participants, however, found more usability problems than the designers and developers, while more functional changes were suggested by the latter two groups. More than one participant also noted various factors influencing design process and outcomes, such as the level of experience with media technologies, time spent working on the portal, literacy, education, and attitudes also significantly influenced the degree of comfort with the interface and the kinds of changes that were suggested.

Such results demonstrated that there was no clear information processing by participants, and it is imperative that multiple stakeholders must be considered in design exercises. The collaborative scenario building method used in this study served as a technique to guide the involvement of multidisciplinary participants. Together with the visual triggers, stakeholder roles served as a ground to help participants create assumptions about the portal and expected usage scenarios. It must also be highlighted that past experiences of participants with regards to using technology and their involvement in the portal project further shaped their building of scenarios. The impacts of roles on generated scenarios were clear in the design process.

As earlier suggested, there were also other effects working to influence the developments of interaction scenarios. Past experiences with portal technologies, media types, and web interfaces were important factors as well as the levels of involvement in the project by participants. At the same time, cultural and demographical factors such as language literacy and educational levels were raised by participants. These factors appear to be shaping stakeholder roles. As it was not in the scope of the study to address such multifaceted influences on interaction scenarios, they have not been addressed at this point. However, it is of interest to frame such factors and take them into consideration in future design processes.

The collaborative scenario building exercise has been extremely helpful in demonstrating how the involvement of diverse participants can assist with pragmatic project outcomes. Simultaneously, it has also been beneficial in guiding stakeholder involvement and the generation of dialogue between stakeholders – an originally uphill task due to communication and disciplinary challenges.

While many benefits have been gained from this exercise, it is not without challenges. For one, scenario-based design can be time-consuming especially when it involves multiple stakeholder-participants. Because of the richness of the method, it was also found to be difficult to orient participants who were new to such exercises. The presence of facilitators was crucial to overcome such challenges.

The results provided evidence that past experiences and technological literacy had significant impacts on the design process and outcomes. Future improvements to the design method could include systematic consideration of such knowledge by participants. The challenge in this lies in the depth of inclusion of these factors and assessment of knowledge by different participants. These factors were implicitly revealed in the briefing and debriefing sessions in this study – which provided rich contexts to the design sessions but there needs to be more methodical integration of such knowledge – a working design framework – by participants in future collaborative design exercises.

Although participants worked independently to generate interaction scenarios, they were also able to discuss their results with each other during workshop debriefs. This was, however, not possible if they were in different workshops. Distances and time differences also posed hindrances to such collaborative discussions. Real-time recording, reading, and discussions of scenarios are desired to encourage greater interactivity and collaboration between participants. This will be looked into for future studies.

Other issues exist, challenging the effective development and usability of the portal. Huge quantities of heterogeneous media objects will make manual annotations and indexing of content objects almost impossible; yet the current media processing technologies, especially content feature extracting and retrieval technologies are not providing satisfactory performance for effective automatic annotations of media content objects. Moreover, in the intended open environment of the user-contributed content, the nature of user behaviors, content, and annotations are greatly diversified and thus difficult to predict and manage. Naturally, issues with digital rights management are also prevalent in the project.

4 Conclusions

The study has demonstrated how a collaborative design exercise can generate benefits for the development of an interactive portal. Other than achieving practical design outcomes, it was also useful as a guide for stakeholder involvement. Challenges remain, however, to both the design process and method.

Through the study, the importance of roles and other factors such as attitudes and past experiences on the design process and outcomes were also implicitly revealed. More work needs to be done to improve the depth and inclusion of knowledge by participants in the design process.

The two institutions engaged in this study had been working together since 2005 in multidisciplinary fields. Research teams from both institutions have been interested in the study of collaborative models to guide the design and development of interactive systems. While the Chinese team has been studying how such workflows can be integrated with technical frameworks of media management platforms, the team from Australia has been concerned about inclusive user studies and the development of methods and techniques to drive the development of local communities.

These research interests are also in congruence with the overall theme of the 2008 Beijing Olympics event – with the rich integration of event experiences through open access, the portal signifies the convergence of new media and humanity in the global community.

References

1. Bonner, J.V., Porter, J.M.: Envisioning Future Needs: From Pragmatics to Pleasure. In: Green, W.S., Jordan, P.W. (eds.) Pleasure with Products: Beyond Usability, pp. 151–161. Taylor & Francis, New York (2002)
2. Carroll, J.: Making use: Scenario based design of Human Computer Interaction, Manchester. MIT Press, Cambridge, MA (2000)
3. Faulkner, C.: The essence of human-computer interaction. Prentice Hall, New York (1998)
4. Giddens, A.: The constitution of society: Outline of the theory of structuration. University of California Press, Berkeley (1984)

5. Karat, J., Karat, C.M.: The evolution of user-centered focus in the human-computer interaction field. IBM Systems Journal 42(4), 532–541 (2003)
6. Pang, N., Schauder, D.: User-Centred Design and the Culture of Knowledge Creating Communities: A Theoretical Assessment. In: Burstein, F., Linger, H. (eds.) The Local and Global in Knowledge Management: Why Culture Matters, pp. 151–166. Australian Scholarly Publishing, Kew Melbourne (2006)
7. Pang, N., Schauder, D.: The Culture of Information Systems in Knowledge-Creating Contexts: Rethinking User-Centred Design. Special Issue of Informing Science Journal: A double helix relationship of use and redesign in Information Systems (forthcoming)
8. Pang, N., Cao, S., Schauder, D., Klein, R.: A Hybrid Approach in Applying Usability Evaluation: A Case Study of the MAMS platform in Olympics 2008. In: He, X., Hintz, T., Piccardi, M., Wu, Q., Huang, M., Tien, D. (eds.) Third International Conference on Information Technology and Applications, Sydney, vol. 1, pp. 82–87. IEEE Computer Society, Los Alamitos (2005)
9. Pang, N.: Digital Libraries as A Learning Environment Amongst Youths. In: Theng, Y.L., Foo, S. (eds.) Design and Usability of Digital Libraries: Case Studies in the Asia Pacific, pp. 254–268. IDEA Group Publishing, Hershey London (2005)
10. Preece, J., Sharp, H., Rogers, Y.: Interaction design: Beyond human-computer interaction. John Wiley & Sons, New York (2002)
11. Pricewaterhouse Coopers.: The Rise of Lifestyle Media: Achieving Success in the Digital Convergence Era. In: A Report by Technology Centre, Pricewaterhouse Coopers. Pricewaterhouse Coopers International Limited, New York (2006)
12. Rosson, M., Carroll, J.: Usability Engineering: Scenario-based development of HCI. Morgan Kaufmann Publishers, New York (2002)
13. Stowell, F.: Towards client-led development of information systems. Journal of Information Systems 1(3), 173–189 (1991)
14. Theng, Y.L., Goh, D.H.L, Lim, E.P., Liu, Z., Pang, N., Wong, P., Chua, L.C.: Intergenerational Partnerships in the Design of a Digital Library of Geography Examination Resources. In: Lim, E.-p., Foo, S.S.-B., Khoo, C., Chen, H., Fox, E., Urs, S.R., Costantino, T. (eds.) ICADL 2002. LNCS, vol. 2555, pp. 427–439. Springer, Heidelberg (2002)

A Case Study on Effective Application of Inquiry Methods to Find Out Mobile Phone's New Function

SangHyun Park[1] and YeonJi Kim[2]

[1,2] LG Electronics Inc. Mobile handset R&D Center
219-24, Kasan-dong, Kumchun-gu, Seoul, Korea
{sanghyun,yeonji_kim}@lge.com

Abstract. Recently, as the user-centered design has appeared to production design area as an important management philosophy, there has been various trials to reflect the user's needs in production in many aspects. However there are various kinds of related inquiry methods and each method has different characters according to the data traits, the participant's condition of inquiry, and the participant's number in needing inquiry. So it is very important to choose appropriate method fits in the inquiry purpose and the proceeding situation so that these inquiry methods are applied to mobile phone domain. Thereafter this study addresses the simple cases of nine inquiry methods which were applied for the purpose of finding out mobile phone needs/ new function, and then describes the result of this research as the following subjects. What significant data does each method have contrasted with the others? What methods can make out synergy effect when they match together? What methods can be omitted and are necessary to extract variable data from participants according to the purpose of inquiry?

Keywords: contextual design, mobile, user experience, latent need.

1 Introduction

There are many attempts to plan new production/ new function based on the factual user's needs. In the case of mobile phone, as the device role and the related technique is changing rapidly, it is getting more necessary to understand users with contextual information which is more vivid than other devices so that the new valuables can be offered to users. So mobile phone designers collect data on users with closest approach by user's inquiry[2]. This study addresses inquiry methods which were progressed to search needs from daily life information that is connected directly or indirectly with production on the purpose of finding out mobile phone's new function to offer value in daily life. It explains briefly about the method and process which applied in this research, and contemplates the meaning of the data from it by reviewing inquiry method which was applied according to the process phases. As the process of collecting user's whole daily life data is not only wide but also various in the available technique, there needs ,first of all, effective user's inquiry under this purpose. So this study will define the traits of inquiry method which has to be considered to apply effectively by case study.

M.J. Smith, G. Salvendy (Eds.): Human Interface, Part I, HCII 2007, LNCS 4557, pp. 163–170, 2007.

2 The Inquiry to Find Out Cell Phone's New Function

The purpose of this study was to find out the new function which can create new experience for mobile phone-users by supplying value that can make user's life comfortable. In other words, it tried to find out the uncomfortable things from information on daily life and needs which can be applied to mobile phone to make new mobile phone's function, so the inquiry was processed without confined to a special life area. That is, A general life style rather than a concrete action was initially examined to derive motivational level needs from UCPCD(User-Centered Product Concept Design)[3]. With regard to mobile application context model[2], all of components in this model need to be investigated, but this research mainly aimed to apply collected information including context to users and mobile phones. Thus, they can be functionalized into mobile phones only when they are based on the collection and generalization of users' life style information which is directly or indirectly related to mobile phones. Consequently, this research has shown what theme can improve users' life as follows.

2.1 User-Inquiry Process

As the purpose of this study is finding out the needs by contemplating whole daily life, it implemented wide inquiry and accordant various inquiry methods. The user-inquiry was implemented by four phases.

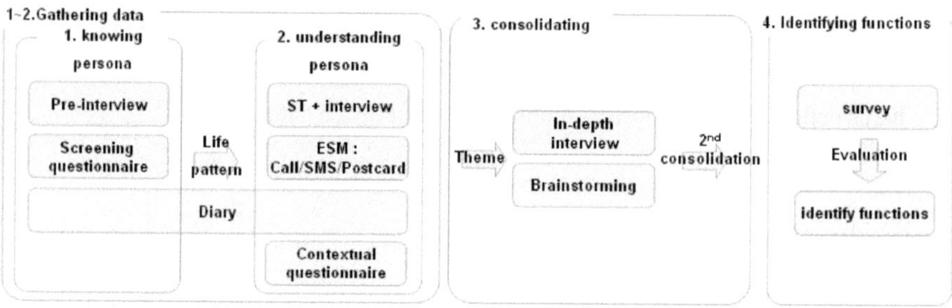

Fig. 1. User research process and applied methods

Phase one was to identify persona appropriateness of participants and the quality as participants to this purpose. Phase two was to collect data on daily life context and find out the needs. To access user experience, the viewpoints in [fig. 2] are considered[6], and the research methods are representatively used to collect data of each experience area. Experience sampling method by phone, like telephone interview, and Photo diary are selected for 'say' and observation + interview method for 'do'.

As the research condition of participants changed later on(for example, school vacation → beginning of school), various measures including applied disposable camera survey(questionnaires conducted to collect information of a participant'

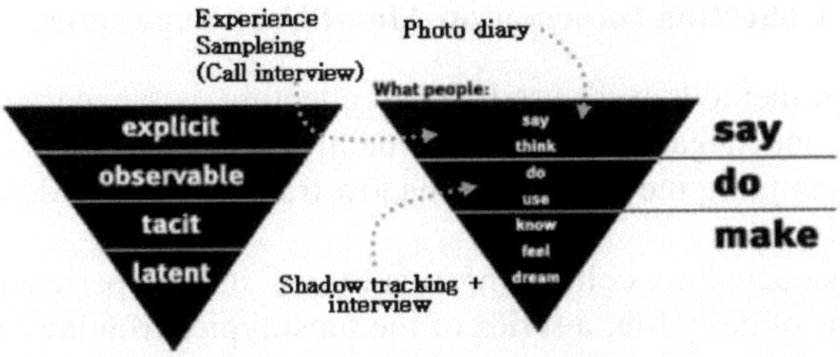

Fig. 2. Accessing user experience and inquiry methods

condition + self-photography), experience sampling method by SMS, and collection of information by postcard have been taken according to research conditions. Phase three was for consolidation; to find out the life theme and collect more related information based on life pattern from the collected information on phase one and two[1]. Phase four was to extent the new idea on function and identify it. The participants in this research offered appropriate information fit to this applied method by implementing all the user-inquiry phases. The number of co-working inquiry methods with participants were total nine and were progressed by the process like as figure 1. This study had tried to get valuable information through variety attempts for more than two months of research period and then differentiated the meaning and traits of the inquiry method and the things to be considered in the process of application. By the attempt, this study hoped to develop the effective application of the method which can be too complex in the course of inquiry which asks user's whole daily life, and described the things in detail to be considered in this session.

3 The Pros and Cons of Each Method

As a result of going through the process, it showed different information in the course of defining new function by the data from each method. Therefore, when we apply these methods to user research with purpose of finding out the new functions of mobile phone, the merit and the demerit were different according to each method.

3.1 Methods for Understanding Persona and Participants

Screening questionnaire conducted in phase one can help us figure out the characteristics of participants to provide us with useful information for the entire research in the long run. The pre-interview with the 1st round participants shows participants' personal information like the sphere of their life, acquaintances, environment and interests, and helps collect fundamental information to set the goals of future researches by collecting information of 5 areas of 'lenses model for better user understanding'. In this sense, the preliminary activities such as the 1st questionnaire and 2nd simple interview to figure out participants accurately can provide the researcher with good basic information when planning the research, and useful information to guide him in his investigation.

3.2 Methods for Collecting Information About User Experience

Seven investigation methods were used when collecting user experience information, and both positive and negative aspects of main research methods including photo diary, experience sampling method(Call), shadow tracking, and In-depth interview are summarized as follows.

Photo diary conducted to collect information of user experience through 'say' derived the patterns of daily life, a series of the most typical routines among daily life activities(for example, Get up at 8 am → Go to school at 10 am → Attend a class at 2 pm → Join a meeting at 7 pm → Go to sleep at 10 pm). These data will be very useful in giving more lucid explanation of an issue when key issues need to be examined to conduct contextual inquiry within a short period of time[8]. In addition, it can provide the researcher with a usual diary full of personal information including opinion about their hope/wish, philosophy and will, by collecting a large amount of personal emotional expression data. As people in their teens to thirty's make use of their blogs, participants can share daily updated information each other with ease and this helps make good progress in research. But information quality depends much on how sincerely participants attend the research, and we need to keep making sure that camera should be used as means to collect information.

The second method for 'say' is experience sampling method by phone. In this research, we made phone calls to participants at a point of time to collect information of their current circumstances and behavior at that moment and we had simple interviews with them to ask about the meaning and function of the mobile phones in their current circumstances. This method has advantage over other ones because it can help researchers to make up for insufficient information of viewpoints, places, events and relations among other information of their circumstances provided by participants, so this method can be made use of as "the joker" when conducting this research. However, we had better ask for participants' understanding in advance as they are connected with us at a point of time. Besides, they were asked to express mobile phones metaphorically, in order to give different definitions of the meaning and function of mobile phones. This was significant because this encouraged participants to suggest various answers by their creativity and expressive power. Thus, the items I have mentioned above resulted in constraints when screening participants.

Third, shadow tracking + interview has been done to figure out 'do'. A mobile phone has scenes of uses in every environment by nature, and as the circumstances keep changing while using it, the scope of investigation may expand and be complicated. So this research is focused on the main pattern of daily life from dairy data. This method has some advantages over other ones. First, it can collect the latest detailed information including behavior, looks, feeling, dialogues and interpersonal relations by examining participants in their daily life. Second, this method can help understand information of participants' circumstances derived with other methods, as it enables the researcher to experience their daily life. Third, it can make interpretation in the viewpoint different from the 1st person viewpoint like photo dairy, as it collects information of participants in the 3rd person viewpoint. But it has some difficulties in practical use. First, in this method the researcher is not a true shadow and this may result in forced interpretation. So we shared information even with those around participants in advance of this research for a period of preliminary preparation. Second, as participants were not equipped with any photographing/ recording device, we cannot exactly figure out how they use their mobile phones or what functions of mobile phones they use during shadow tracking. So this needs to be made up for by interviews. Third, as one of difficulties we meet in using examination method, family members' understanding is necessary for opening participants' private space(for example, their home).

In-depth interview used to access 'Make' aims to visualize latent needs, and these can be defined by key words of participants' life theme which is primarily consolidated based on the raw data derived with the methods I described above. The criteria for selecting life theme are the patterns of life, a series of anecdotes which repeatedly occur, frequent behavior, complaint in daily life, motive of abrupt change in emotion, the root of happiness, anger, sorrow and pleasure in life, and users' interests and needs for their mobile phones.

Time machine phone!!
"A mobile phone that makes you feel happy through your reminiscence"

Q) Please describe your memories in the past that make you feel happy whenever you think of them. (Image)

Q) Why do you feel happy when you think of those memories?

Q) Let's imagine there is a box of memories you can take out whenever you like to.
Q) What is it like?
Q) What color is it?

Q) Let's bring up memories from the box.
Q) what is the shape of your memory(human being? Animal? And so on..)?

Fig. 3. Example of life theme: time machine phone

Fig. 4. Images for Encouraging Mental hiccup

In short, when participants' information is derived additionally as for the life theme selected by the 1st consolidation, the function of a mobile phone to meet 5 layers of user experience factors in Jesse James Garrett[16] can be conceptualized. But in the participatory design to give shape to the concept of functions of a mobile phone, the consideration of so-called "Hiccoughs of thinking" factors determines how much amount of value of needs can be applied to a mobile phone(This is described with images in this research), and it may be difficult to ask participants to concretize the concept of a mobile phone taking the environment of a mobile phone into account.

4 Experience with Effective Application of Methods

When considering the traits of each method addressed in chapter three, there exists fair of useful methods which can be multiplied in those function when they combine together, and also there appeared inevitable inquiry method in application during this research session. The consideration for effective application of inquiry method was taken in three aspects throughout this case study.

First, it was matching between methods according to the degree of each method's coverage on inquiry duration and the amounts of detail data from specific context. In other words, in the case of this research, I used diary method which could catch out the whole daily life pattern, shadow tracking which could light the special situation, and experience sampling which could be applied to the situation that the diary and shadow tracking could not cover, and when these three methods are used together we can extract complete information.

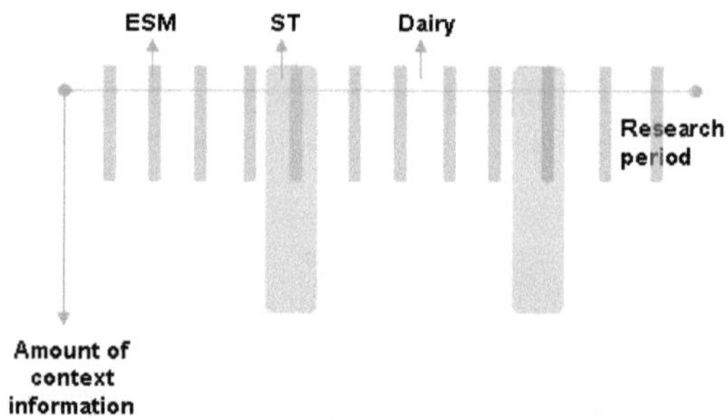

Fig. 5. A set of methods according to data property

Second viewpoint is intended to make an efficient application of methods by defining the features of the target information based on the features of methods when using each of them. When making an application of methods as above, various types of participants' hints that helped the analyst to make life theme out of them can be derived more efficiently. The types of significant information derived with main methods are classified as follows:

- Photo diary: life activity pattern, problems that repeatedly occur, activities that are repeatedly done in daily life, philosophy/life style, emotional expressions of daily life, interests expressed with photos
- Experience sampling method(Simple phone interview) : information of participants' circumstances, activity information(making up for diary information), Persona interests by defining the purpose of activities in circumstances, emotional status in circumstances, the function/meaning of a mobile phone by various points of time and circumstances in daily life, schedule information during research
- Observation(Shadow tracking): Usage behaviors of a mobile phone in circumstances, interests, the style of communication with those around them/subjects of discussion
- In-depth interview: opinion of life theme(user value-oriented), the role of theme as the concept of a mobile phone

Third viewpoint is for the methods which had to be implemented to heighten the quality of the whole inquiry by collecting necessary information. First, screening questionnaire should aim not only to examine the compatibility with persona, but also to see if a person got what it takes to be the participant in user research. In sort, this is because the participants in user research should be positive and critical and maintain their creativity with ripe experience in various activities to conduct the research successfully. Second, if the research is conducted to derive motivational level needs, activity pattern of persona should be figured out with photo dairy. This information will enable the researcher to outlook the entire aspects of a participant' life by activities, select the methods for researching best suited for his circumstances based on these data and make plans for making an application of them.

5 Conclusion

This study addressed the user research process used for finding out the new function of mobile phone, and investigated the traits of each method by comparing merits and demerits of applied various inquiry methods. By this I could arrange the items which have to be considered in applying methods to find out effective results from user research. We can take one more step toward latent needs based on users' circumstances and life by conducting this user research. But these needs can be reflected in products only when various information including market trend, technical trend, the identity of a product, and development environment are considered comprehensively after 2nd consolidation. It is presumed that these needs derived with this user research can be integrated when this information is taken into account and conceptualized collectively, and this will be done in further case studies.

References

1. Tamminen, S., Oulasvirta, A., Toiskallio, K., Kankainen, A.: Understanding Mobile Contexts. Mobile HCI 2003, pp. 17–31 (2003)
2. Ryan, C., Gonsalves, A.: The Effect of Context and Application Type on Mobile Usability: An Empirical Study. ACSC 2005, pp. 115–124 (2005)

3. Kankainen, A.: UCPCD: user-centered product concept design. Proceedings of the 2003 conference on Designing for user experiences, June 06-07, 2003, pp. 6–7. San Francisco, California (2003)
4. Paradise, J., Mynatt, E.D.: How do families remember: An initial analysis of diary and interview data. In: Proceedings of the HCI International Conference (HCII), Las Vegas, July 2005 (2005)
5. Turber, M.: Multi-disciplinary Teams in User Centered Development Processes. In Proceedings of the HCI International Conference (HCII), Las Vegas (July 2005)
6. Sanders, E., Dandavate, U.: Designing for Experiencing: New Tools. In: Proceedings of the first international Conference on Design & Emotion, Holland (1999)
7. Reimann, R.M., Bacon, E.: A Scenario-Based Approach to Creating Interaction Frameworks (2001)
8. Wixon, D.R., Ramey, J., Holtzblatt, K., Beyer, H., Hackos, J., Rosenbaum, S., Page, C., Laakso, S.A., Laakso, K.-P.: Usability in practice: field methods evolution and revolution (ed.), CHI 02, Minneapolis (MI), pp. 880-884 (2002)
9. Brown, B.A., Sellen, A., O'Hara, K.: A Dairy Study of Information Capture. in Working Life. In: Proceedings of the ACM 2000 Conference on Human. Factors in Computing Systems, pp. 438–445. ACM Press, The Hauge, The Netherlands (2000)
10. Park, S., Heo, J.: Usability questionnaire items suitable to online user survey. In: proceedings of the HCI 2006 (2006)
11. Kim, K.C., Kim, B.J.: A study on design process for public space by users behavioral characteristics. Journal of Korean Society of Design Science 17(1), 89–98 (2004)
12. Holtzblatt, K.: Customer-centered design for mobile applications. Personal and Ubiquitous Computing 9(4), 227–237 (2005)
13. Kantner, L., Rosenbaum, S., Sova, D.: Alternative Methods for Field Usability Research. In: Proceedings of SIGDOC 2003, October 2003, San Francisco, Calif (2003)
14. Tamminen, S., Oulasvirta, A., Toiskallio, K., Kankainen, A.: Understanding mobile contexts. Personal and Ubiquitous Computing 8(2), 2004 (2004)
15. Konkka, K., Koskinen, S.: Mobile User Experience: Challenges in International End-User Needs Research and Design. In: Proceedings of the HCI International Conference (HCII), July 2005, Las Vegas (2005)
16. Garrett, J.J.: The Elements of User Experience: User-Centered Design for the Web (2002)

An Experimental Examination of Customer Preferences on User Interface Design of Mobile Services

Heejun Park[1] and Seung Baek[2]

[1] Department of Information and Industrial Engineering, Yonsei University, 134 Shinchon-Dong, Seodaemoon-Gu, Seoul 129-749, Rep. of Korea
h.park@yonsei.ac.kr
[2] Division of Business Administration, Hanyang University, 17 Haengdang-Dong, Seongdong-Gu, Seoul 13-791, Rep. of Korea
sbaek@hanyang.ac.kr

Abstract. Designing mobile services is fundamentally different from designing online services. Not only are there differences in underlying technologies, but also in the way people use the services. If these differences are not taken into account, mobile services are likely to fail. However, it is quite difficult to provide a single-layered user interface in a small screen. This study aims at examining how design attributes of user interfaces of mobile services affect customer preferences. In order to explore customer preferences to each design attribute of mobile services, we measure and analyze customer's WTP (Willingness To Pay) toward their different interface designs. Ultimately, throughout the study, we try to answer how to design mobile services in small screen of mobile devices. In addition, we propose an optimal design solution that customers likely prefer.

Keywords: User Interface(UI), Willingness To Pay(WTP), Mobile Services.

1 Introduction

The Internet has provided an efficient and effective way of delivering information and services to millions of users who are connected to a wired network since it has been introduced. However, this wired network addresses two major constraints: time and place. These limitations have raised the issue of the mobile Internet, which enables users to access information from any place at any time. Recently, the population of using mobile Internet has been growing due to the widespread use of mobile devices and high-speed Internet services. Mobile technology, however, is still very much in its infancy regarding usability, display capabilities and computing performance. Thus, the main usage of mobile Internet is concentrated on ring/avatar downloads. For mobile Internet services to be propagated, they need to provide more intuitive and rich interface as well as contents of them. However, as the size of mobile devices has become smaller, it is quite difficult to provide intuitive and rich interface for users' conveniences [7].

This study aims at exploring customer preferences to user interface design attributes. We measure customer's WTP (Willingness To Pay) toward different interface design options. Ultimately, the study tries to answer how to design user interface of mobile services in small screen of mobile devices to satisfy users.

M.J. Smith, G. Salvendy (Eds.): Human Interface, Part I, HCII 2007, LNCS 4557, pp. 171 – 179, 2007.

2 User Interface of Mobile Services

The user interface and the size of mobile devices are the main concerns in designing mobile services. A well designed single-layered user interface will be more user friendly than the conventional one and it will be having edge over others. However, it is quite difficult to provide a single-layered user interface in a small screen. Designing mobile services is fundamentally different than designing online services in the Internet. Not only are there differences in underlying technologies, but also in the way people use services. If these differences are not taken into account, mobile services are likely to fail. If mobile services do not deliver users' needs with their convenience, these services will fail no matter how excellent the underlying technology is. Designing usable interfaces for tomorrow's mobile devices is not trivial but involves a series of challenges on Human Computer Interaction(HCI). Mobile services require efficient ways to record and access information under the circumstances that are often quite different from those where desktop computers are used. Displays on mobile devices are small, means of input are limited and use-contexts are very dynamic. The usability of mobile services consequently suffers from interfaces being very compact and cluttered with information, demanding the user's full attention. Many people say that mobile services are not yet attractive and usable for customers. There clearly is a lack of understanding of real user needs and how mobile internet can help users satisfy those needs.

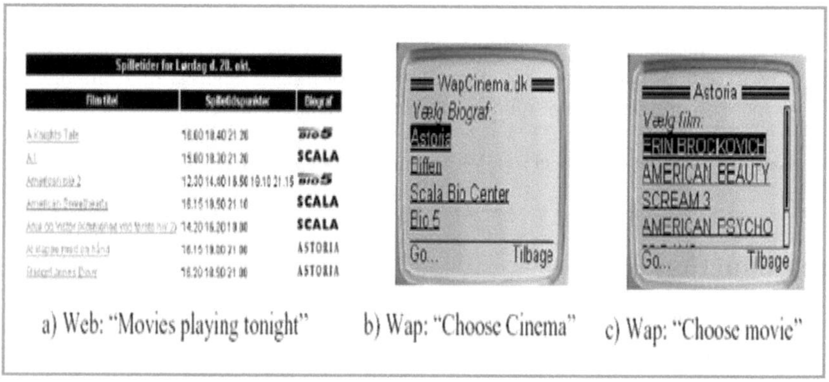

Fig. 1. Web Interface and Mobile Interface

Wireless Application Protocol (WAP) phones are a growing relevant part of the mobile market, and the number of WAP services offered is rapidly increasing. However, usability is crucial for these services which must be easily operated on small screens and keyboards. One of the reasons that many WAP sites have failed is that many service providers simply tried to carry their websites over to the mobile Internet. This shows that little thought has been given to the characteristics of mobile Internet and to the specific needs of people who use their mobile devices in real life. Understanding the fundamental differences between mobile and the web is essential for the success of mobile services [7]. Figure 1 compares a web interface with a mobile interface. Using the web site, users view all relevant information in one page. Whereas, accessing the services from WAP-based phones, users require a lot of clicking in a pre-defined sequence, due to the division of information into a large number of sub-pages. By following sequential menus, users can find specific mobile services. Generally, interfaces of WAP-based mobile services require minimal attentions and interactions. The usability of mobile services consequently suffers from interfaces being very compact and cluttered information. It is crucial to design the menus which can get the user's full attention, remaining simple and the required interaction minimal [9].

3 Willingness to Pay(WTR) for Mobile Services

Willingness to Pay(WTP) is defined as the maximum monetary amount that an individual would pay to obtain a good or service [1]. It is crucial in estimating the demand curves and determining optimal pricing schedules. The WTP has focused on cognitive structure of customer to specific tangible or intangible goods under experimental environment [8][10]. It is widely used to provide information to policy makers, regarding the economic value of non-market or non-pecuniary environmental assets. In this study, by measuring WTP for various interface designs, we try to investigate what features of interface design attributes affect customers' WTP [2][3].

To measure WTP, there are two methods: contingent method and conjoint method. In the contingent method, arbitrary goods or service are presented with specific prices, and then respondents take or reject the offers. While the contingent method asks respondents to evaluate real products, the conjoint method asks them to evaluate hypothetical products. The conjoint method allows researchers to presents respondents with a choice set that contains several alternatives that vary along several attributes, including price [5]. Respondents decide the rankings among the alternatives. Based on the rankings, researchers can analyze the preferences and utilities of respondents for specific goods/service.

4 Research Methodology and Hypothetical Products

In order to analyze the impacts of specific attributes on customers' preferences, this study employs conjoint analysis(shown on Figure 2) which shows relative importance of various utilities considered by consumers when they purchase service and suggests a service that consists of the most ideal combination of those utilities [5]. Analyzing relative impacts of interface design attributes, we are identifying the influencing factors that affect customers' WTP throughout focus group interview.

Its result shows four design attributes which are the information about service type, payment and recommendation to the use of specific mobile services shown in the menu and its structures and three attributes level for each design attribute as shown in Table 1.

With three attributes and four levels for each attributes, 81 (3*3*3*3) hypothetical products could be generated. All 81 possible hypothetical products are reduced to nine (shown in Table 1) based on a fractional factorial design due to the difficulty in surveying with all 81 questions [11]. In order to measure customer preferences to each of nine hypothetical products, the study employs the rank-order method, rather than the rating method. 116 respondents are asked to provide the maximum monetary amounts that they would pay to get these products. Based on the amounts, we investigate how much each menu design attribute of mobile services affects customer's WTP.

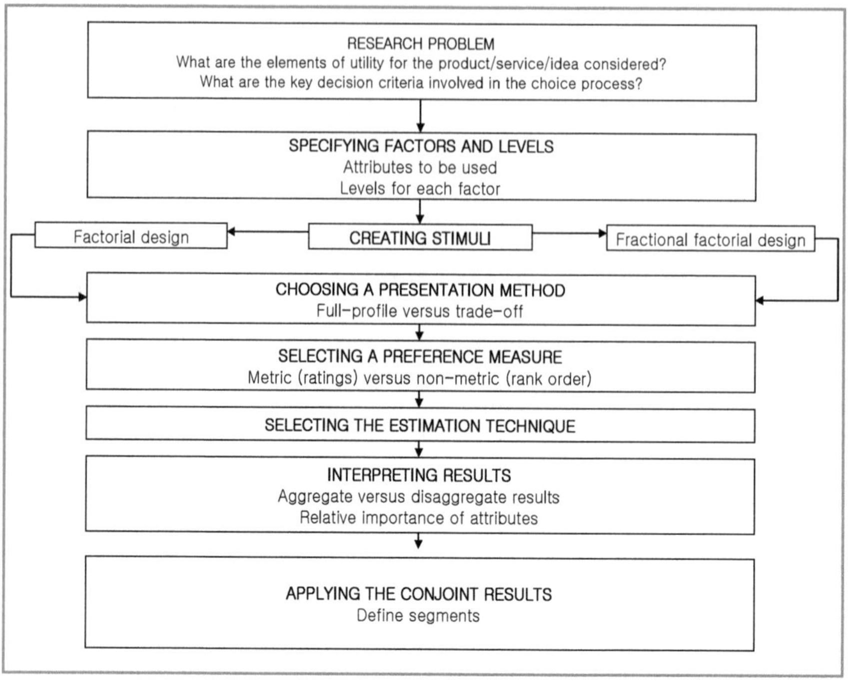

Fig. 2. Research model to conjoint analysis

Table 1. Hypothetical Products

Product	Mobile Service Type	Payment Information	Menu Position of Mobile Service	Recommendation Information
P1	No Information	No Information	First Line	User Evaluation Based Recommendation
P2	Entertainment	Coupon	Middle Line	User Evaluation Based Recommendation
P3	No Information	Cash Discount	Middle Line	No Information
P4	Information /Communication	Cash Discount	Second Line	User Evaluation Based Recommendation
P5	Information /Communication	No Information	Middle Line	Log Data Based Recommendation
P6	Entertainment	Cash Discount	First Line	Log Data Based Recommendation
P7	Entertainment	No Information	Second Line	No Information
P8	Information /Communication	Coupon	First Line	No Information
P9	No information	Coupon	Second Line	Log Data Based Recommendation

The focus of conjoint analysis is to explore customer preferences for product attribute levels and the customer benefits that may flow from the product attributes [6]. In addition, it allows us to measure part-worth functions at the individual level. Hence, if preference heterogeneity is present, by using cluster analysis, we can identify groups that have distinct preferences.

5 Results and Discussion

We interviewed 116 respondents consisting of 53 males and 63 females and most of them consist of 12-19 year olds in this study. In addition to portraying the impact of each level (entertainment, information / communication, no information et. al.) with the part-worth estimates, conjoint analysis can assess the relative importance of each design factor (information about mobile service type, payment of mobile service, position of mobile service, and recommendation to mobile service). Since the part-worth estimates are on a common scale, the greatest contribution to overall utility of preference, and hence the most important factor, is the factor with the highest range of part-worth.

The focus of conjoint analysis is to explore customer preferences for product attribute levels and the customer benefits that may flow from the product attributes. In addition, it allows us to measure part-worth functions at the individual level. Hence, if preference heterogeneity is present, by using cluster analysis, we can identify groups that have distinct preferences [5].

In this study, we examine the relative impacts of menu design attributes of mobile services on customer's WTP. We try to find an optimal menu design of mobile services that high-valued by users. Table 2 reports the means of the part-worth and relative importance for 116 subjects. The greater the relative importance, the more an attribute influences your preferences for mobile services. If the information about mobile service types is shown in menu, there is the highest possibility that customers purchase mobile services (27.51%). In other words, customers utilize information about mobile service types as critical information in purchasing of mobile services. The second influential menu design attribute is the payment information of mobile services shown in menu, (26.53%), the third is the positions of mobile services within menu (23.21%), and the last is recommendation information (22.75%). Based on the result, we can say that the understanding customers are the most important factor in terms of menu design. Many companies pay a lot of money to locate their services in higher positions in the menu, or use recommendation information to sell their services. Based on our study, we found that mobile customers tend to be more sensitive to services themselves, rather than the artificial information, such as recommendation information or menu positions.

In terms of part-worth for the mobile service types, customers prefer entertainment mobile services to other services. This result proves the actual usages of mobile Internet. Many users tend to use entertainment services, rather than other types of mobile services. In terms of payment information, customers prefer the mobiles service shown its cash discount information in the menu. When they use a specific mobile service, their decisions heavily depend on the payment information, especially on cash discount information. Regarding to recommendation information about mobile services, customers prefer the mobile service marked "Strongly Recommended" in the menu. Normally, service operators mark specific mobile services as "Strongly Recommended", when the services are popular and new. "Hit" marks are less attractive for customers than "No Information". In terms of menu position, customers are likely to purchase a mobile service in the middle line, rather than the second line or the first line. The relative importance of design attributes will guide the menu design of mobile services. The result shows that, for identical services, their different menu designs affect customer's purchasing power.

Table 2. Part-Worth and Relative Importance

Attribute	Part-Worth			Relative Importance (%)
	1	2	3	
Mobile Service Type	Entertainment	Information /Communication	No Information	27.51
	1.74	0.32	0.94	
Payment Information	Discount	Coupon	No Information	26.53
	1.54	0.96	0.50	
Menu Position	First Line	Second Line	Middle Line	23.21
	1.20	0.48	1.33	
Recommendation Information	Log Data Based Recommendation	User Evaluation Based	No Information	22.75
	0.85	1.16	0.99	

As result of this analysis Table 2 reports the means of the part-worth and relative importance for 116 subjects. In terms of part-worth for the mobile service types, customers prefer entertainment mobile services to other services. This result proves the actual usages of mobile Internet. Many users tend to use entertainment services, rather than other types of mobile services. In terms of payment information, customers prefer the mobiles service shown its cash discount information in the menu. When they use a specific mobile service, their decisions heavily depend on the payment information, especially on cash discount information. Regarding to recommendation information about mobile services, customers prefer the mobile service marked "Strongly Recommended" in the menu. Normally, service operators mark specific mobile services as "Strongly Recommended", when the services are popular and new. "Hit" marks are less attractive for customers than "No Information". In terms of menu position, customers are likely to purchase a mobile service in the middle line, rather than the second line or the first line. The relative importance of design attributes will guide the menu design of mobile services. The result shows that, for identical services, their different menu designs affect customer's purchasing power.

One of the most common uses of conjoint analysis is to group respondents with similar part-worths or importance values to identify segments. Because of the following reasons, conjoint analysis can be used as a useful method for implementing market segmentation and product positioning [4]. Based on 116 respondents, the study identifies three mobile service customer groups. Three clusters consist of 52, 28, and 36 respondents. In this study, we are to do cluster analysis as shown in Table 3.

In terms of mobile service types, respondents in three clusters have a similar preference structure. All of them give a significant preference to entertainment mobile services. However, regarding to payment information, menu position, and recommendation information, respondents in three clusters have different preferences. Respondents in cluster 1 prefer the mobile service shown cash discount information, located at the middle line, and marked as "Strongly Recommended" in the menu. Respondents in cluster 2 prefer the mobile service located at the second line in the menu position and marked as "Hit" in the menu. Unlike respondents in other two clusters, they do not seriously consider the payment information when they choose a specific mobile content. Respondents in cluster 3 prefer the mobile service shown coupon information and located at the first line in the menu. Unlike respondents in other two clusters, they do not seriously consider the recommendation information when they choose a specific mobile content. Respondents in cluster 2 and in cluster 3 have totally different preferences. While the cluster 2 regards the recommendation information as a critical factor in buying a mobile service, it does not seriously consider payment information in buying a mobile service. The cluster 3 seriously considers payment information, not payment information.

Table 3. Cluster Analysis

Attribute	**Item**	**Cluster**		
		Cluster 1	**Cluster 2**	**Cluster 3**
Mobile service type	Entertainment	**0.45**	**1.70**	**0.41**
	Information/communication	-0.87	-0.45	-0.57
	No information	0.42	-1.25	0.17
Payment information	Discount	**1.16**	-0.02	0.08
	Coupon	-0.81	-0.57	**1.47**
	No information	-0.35	**0.60**	-1.56
Menu position of mobile service	First line	-0.01	-0.17	**0.79**
	Second line	-1.13	**0.15**	-0.19
	Middle line	**1.15**	0.01	-0.59
Recommendation information about mobile service	Log data	-0.71	**0.35**	0.28
	User evaluation	**1.10**	-0.07	-1.01
	No information	-0.39	-0.27	**0.73**

Due to the small screens of mobile devices, the amount of information shown in the menu is minimal. Therefore, mobile service designers try to find out the condensed information that can catch customer's eyes. Some, like cluster 2, are sensitive to payment information shown in the menu. Others, like cluster 3, are sensitive to recommendation information shown in the menu. The result states the need to personalize the menu design for mobile services, as well as to personalize the mobile services.

One of the most common uses of conjoint analysis is to group respondents with similar part-worths or importance values to identify segments. Based on 116 respondents, the study identifies three mobile service customer groups as shown in Table 3. In terms of mobile service types, respondents in three clusters have a similar preference structure. All of them give a significant preference to entertainment mobile services. However, regarding to payment information, menu position, and recommendation information, respondents in three clusters have different preferences.

The results of this study provide a theoretical understanding of customers' preferences on user interface design of mobile internet services. The important user interface design attributes and the customer segments of mobile internet services revealed by this study

6 Conclusions and Recommendations

One often meets the argument that due to the complexity of the service model and the convergence of technologies and services expected in mobile services, it is very little relevant research available to help us understand the mobile services and their users. The main purpose of the study is to explore attributes that influence customers' preferences to mobile services. To explore the relative importance of the attributes, the study uses conjoint analysis. It finds that the most influential attribute is the mobile service types. This finding supports the current phenomenon that the main usage of mobile services is concentrated on entertainment services. Although this study tries to understand customers of mobile services empirically, it has very limited contributions to theoretical development. In addition, regarding to the relationships between attributes of mobile services and customers' preferences, it does not provide a theoretical explanation. However, the study helps us to understand mobile customers.

References

1. Brown, T.C., Gregory, R.: Why the WTA-WTP disparity matters. Ecological Economics 28, 323–335 (1999)
2. Cameron, T.A., James, M.D.: Estimating Willingness to Pay from Survey Data: An Alternative Pre-Test-Market Evaluation Procedure. Journal of Marketing Research 24, 389–395 (1987)
3. Gil, J.M., Gracia, A., Sanchez, M.: Market Segmentation and Willingness to Pay for Organic Products in Spain. International Food. and Agribusiness Management Review 3, 207–226 (2000)
4. Green, P.E., Krieger, A.M.: Segmenting Markets with Conjoint Analysis. Journal of Marketing, pp. 20–31 (1991)
5. Green, P.E., Srinivasan, V.: Conjoint Analysis in Consumer Research: Issues and Outlook. Journal of Consumer Research 5, 103–123 (1978)
6. Green, P.E., Srinivasan, V.: Conjoint Analysis in Marketing: New Developments with Implications for Research and Practice. Journal of Marketing 54, 3–19 (1990)
7. Hair, J.K., Rolph, J.R., Anderson, E., Tatham, R.L., Black, W.C.: Multivariate data analysis. Prentice Hall, Inc., Englewood Cliffs (1999)

8. Shefrin, H., Caldwell, D.: Determinants of the Magnitude of Willingness to Accept Relative to Willingness to Pay. Journal of Behavioral Decision Making 14, 87–105 (2001)
9. Walker, M.E., Morera, O.F., Vining, J., Orland, B.: Disparate WTA-WTP Disparities: The Influence of Human Versus Natural Causes. Journal of Behavioral Decision Making 12, 219–232 (1998)
10. Wertenbroch, K., Skiera, B.: Measuring Consumers Willingness to Pay at the Point of Purchase. Journal of Marketing Research, pp. 228–241 (2002)
11. Wittink, D.R., Cattin, P.: Commercial Use of Conjoint Analysis: An Update. Journal of Marketing 53, 81–86 (1989)

Basic Experimental Verification of Grasping Information Interface Concept, Grasping Force Increases in Precise Periods

Sigeru Sato[1], Muneo Kitajima[1], and Yukio Fukui[2]

[1] Institute for Human Science and Biomedical Engineering, AIST,
1-1, Higashi 1-chome, Tsukuba, 305-8566, Japan
[2] Department of Computer Science, University of Tsukuba,
1-1-1, Ten-no-dai, Tsukuba, 305-8573, Japan
{sato-s,kitajima.muneo}@aist.go.jp
fukui@cs.tsukuba.ac.jp

Abstract. In human-human communications, especially in face-to-face communication, sub-verbal and non-verbal messages have more importance than messages transported by words. On the other hand, in traditional man-machine interfaces, machines only understand pre-defined operations, and never understand operators' sub-verbal and non-verbal messages. This causes some usability problems in machine operations. We have proposed elsewhere that Grasping Information Interface Concept (GIIC) is the key idea to alleviate the above-mentioned communication gap between man and machine, and this paper verifies GIIC experimentally. GIIC regards grasping-and-moving as a fundamental hand operation necessary for performing tasks using modern man-machine interfaces, and behavioral measures associated with grasping, such as force, posture, etc., should have much potential in developing task-adaptive and operator-adaptive interfaces because it is known that how people grasp devices is dependent on the purpose of these tasks. GIIC provides machines with a communication channel to understand operators' intents more through the grasping information.

Keywords: silent messages, grasping force, pointing device, Fitts' law.

1 Introduction

1.1 Difference Between Human-Human Communication and Man-Machine Interfaces

In human-human communications, especially in face-to-face communication, it is known that most of information is communicated through sub-verbal and non-verbal messages. 55% of communications depends on non-verbal messages, such as facial emotional expressions, and 38% on sub-verbal messages such as vocal tone. They are called "silent messages." Only 7% of whole information is communicated by words [1]. On the other hand, in traditional man-machine interfaces, machines only understand pre-defined operations, and never understand operators' sub-verbal and non-verbal messages.

M.J. Smith, G. Salvendy (Eds.): Human Interface, Part I, HCII 2007, LNCS 4557, pp. 180–188, 2007.

This means that in man-machine interfaces, operators must express 100% of their intents in verbal channel which is used only 7% in human-human communication. This causes some usability problems in machine operations.

Table 1. Comparison of usage of communication channels in human-human communication and man-machine interface

Channel	Example	Human-human communication	Man-machine interface
Non-verbal	facial	55%	0%
Sub-verbal	vocal	38%	0%
Verbal	word, switch	7%	100%

1.2 Grasping Information Interface Concept (GIIC)

Grasping-and-moving is one of the fundamental hand operations necessary for performing tasks with modern man-machine interfaces. We have proposed Grasping Information Interface Concept (GIIC) in our previous papers that "grasping" could be used as an expression of operator's sub-verbal and non-verbal messages [2-5]. The key ideas came from Napier's observation and our characterization of pointing tasks. This paper verifies GIIC experimentally.

Grasping Expresses Operator's Intention. Napier's study indicates that a grasping posture is adaptive, and depends not only on the shape of the grasped object but also on the purpose of grasping [6]. This means that a grasping posture can be an indicator of the operator's intent. One can detect the operator's intent, through analysis of grasping form, grasping force and other grasping factors.

However, behavioral measures associated with grasping, such as force, posture, etc., have not been utilized for traditional interfaces. They can convey "silent messages" in man-machine interfaces.

This leads to a claim that we can identify the operator's intent if we can appropriately associate a set of features of a grasping posture such as grasping form, grasping force, etc. with a certain operator's intent. We call this claim as Grasping Information Interface Concept (GIIC). Though GIIC seems to be a promising idea for alleviating the communication gap between man and machine shown in Table 1, it has not been utilized for traditional interfaces.

Application of GIIC to the Designing of Remote Pointing Devices, Especially a New Computer Mouse. We applied GIIC to design a novel computer mouse for remote pointing tasks [2]. We suggest that a pointing task consists of two phases: approaching and positioning, which differ significantly in terms of the above-mentioned behavioral measures associated with grasping.

In approaching phase, the operator performs a rough task with faster and less accurate operation on the computer mouse, whereas in positioning phase, he/she performs a precise task with slower but more precise operation. The latter requires a higher value of stiffness and robustness in hand positions, which is caused by higher tensioning of hand and arm muscles, than the former does. Namely, the operator would grasp the mouse firmer for the precise task in positioning phase than he/she would do for the rough task in approaching phase. We argue that this is done intentionally. And thus operator's intention is represented by grasping information.

In precision tasks, a higher value of stiffness and robustness in hand positions are necessary than those required for rough tasks. Naturally, hand and arm muscles are tensioned to a higher degree in order to keep sufficient stiffness and robustness when in the hand position.

A remote pointing device consists of a pointing terminal and an operation terminal. A pointing terminal is driven and controlled with an operation terminal. There are many kinds of remote pointing terminals, such as industrial robot manipulators, cranes, surgery manipulators, etc., which are usually called as "pointing devices." A computer cursor on the screen is a type of remote pointing device. These pointing terminals are driven and controlled by operation terminals, such as, keyboards, joystick switches, levers, pedals, etc. The computer mouse is a type of operation terminal.

Grasping operation terminals such as joystick switches, levers, computer mouse, etc., are naturally expected to be grasped stronger in the positioning phase than in the approaching phase [2-5]. Therefore, by using grasping force as an indicator to make a real-time prediction on which phase a pointing task is in, automatic real-time adaptive gain adjustment can be realized without increasing complexity on the part of operators. This is the main argument of this paper, which we would further investigate by showing experiments for its verification in section 2 and its validation in section 3.

1.3 A Measurement Method to Validate GIIC: Easiness of a Pointing Task

A remote pointing device consists of a *pointing terminal* and an *operation terminal.* There are many kinds of remote pointing terminals, such as industrial robot manipulators, cranes, surgery manipulators, etc., which are usually called as "pointing devices." A computer cursor on the screen is a type of remote pointing device. These pointing terminals are driven and controlled by operation terminals, such as, keyboards, joystick, switches, levers, pedals, etc. Computer mouse is a type of operation terminal.

The time necessary for accomplishing pointing tasks is one of measures to evaluate the usability of remote pointing devices. The less time a pointing device requires for a user to accomplish a pointing task, the better the usability the device has. According to Fitts' law [7-8], the time T necessary for pointing a target of the size S located at a distance D is given by the following formula:

$$T = a + b \times \log_2 \left(D/S + 1 \right) \quad (1)$$

In this formula, the log term, $D/S+1$, is called Index of Difficulty (ID), and a and b are experimentally determined constants. The former, ID, defines the feature of a pointing task, namely, a pointing task becomes easier for a larger target, a larger S, located at a shorter distance, a smaller D; whereas the latter, especially b, is related to the usability of a pointing device. The smaller the estimate of b becomes, the shorter the pointing time results for a given task with a fixed value of ID.

2 An Experiment to Verify GIIC

To confirm experimentally the discussion above, an experiment was conducted. Differences in grasping forces in various conditions and phases of pointing tasks as described earlier were observed. The experimental task is remote pointing task on the computer screen. A mouse with a force sensor was used. The subjects were adults, skilled in computer usage and have no hand disabilities. Both male and female test subjects were used. The size of the test pool was 25.

Fig. 1. Appearance of the mouse used in the experiment

2.1 Apparatus

A mouse with a force sensor developed by [2] was used. Fig. 1 shows its appearance. To detect a grasping force, a thin, flexible pressure sensor was used as shown by the black part of the sidewall of the mouse body. Fig. 2 shows the grasping force sensor, constructed with a two-stripe electrode, printed on a flexible printed circuit sheet, and a pressure sensitive conductive rubber sheet. The amount of movement of the mouse was transferred to an A/D converter, then to a PC through a USB port. The amount of movement of a mouse cursor on the monitor was reduced by a predefined factor when the pointing task was judged to be in positioning phase.

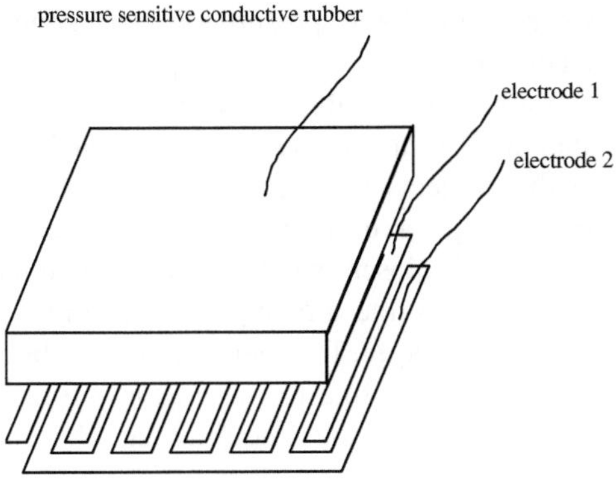

Fig. 2. Construction of grasping force sensor

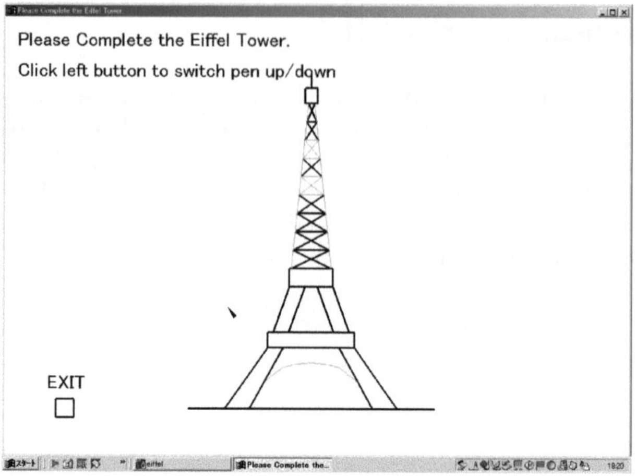

Fig. 3. The picture used in the experiment (The light gray lines are to be traced and drawn)

2.2 Experimental Tasks and Data Collection

The experimental task was to complete an illustration of Eiffel tower by adding some lines to a displayed picture shown in Fig. 3. There were both precise periods (pen-down to draw a line with an arrow-shaped cursor) and rough periods (pen-up to move the pen position with a plus-shaped cursor).

The subjects' operations were recorded every time a mouse movement event and/or a mouse button event were detected. Each record consisted of mouse position, mouse motion, cursor position, status of the left button, status of the phase detector, status of the cursor (arrow or plus), status of the drawn picture, and operation view.

2.3 Results

Fig. 4 and 5 show the trajectory of the cursor in one task and the corresponding grasping force, respectively. As shown in Fig. 5, grasping force changed significantly in a drawing task, and it looks like that there were two distinguishable periods with higher grasping force and lower grasping force, presumably corresponding to precise periods and rough periods, respectively.

The grasping force was varied between these precise and rough periods. In particular, the grasping force increased near the end of drawing each line. Especially, this point was conspicuous to the subjects who grasp the mouse loosely in rough periods. This suggests that the grasping force increases before the cursor reaches the goal or end point. By observing Fig.5 closer, we found another tendency that the grasping force increased near the end of a line-drawing. The grasping force increased before the cursor reaches the goal or end point. In order to examine this, we compared the average grasping force while drawing a line to the grasping force a half second before clicking mouse button to terminate the line-drawing. As shown by open

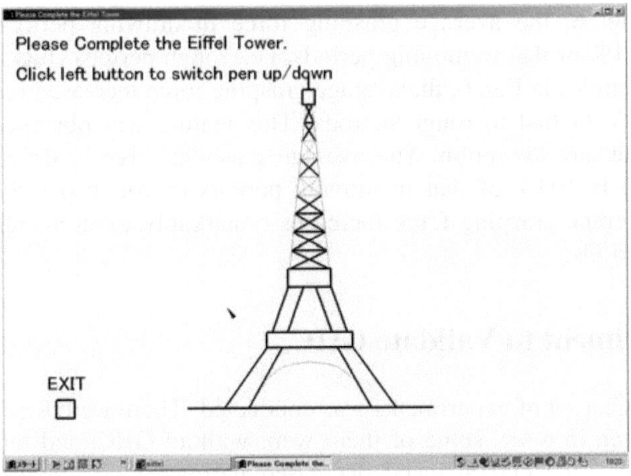

Fig. 4. Cursor trajectory in a drawing task

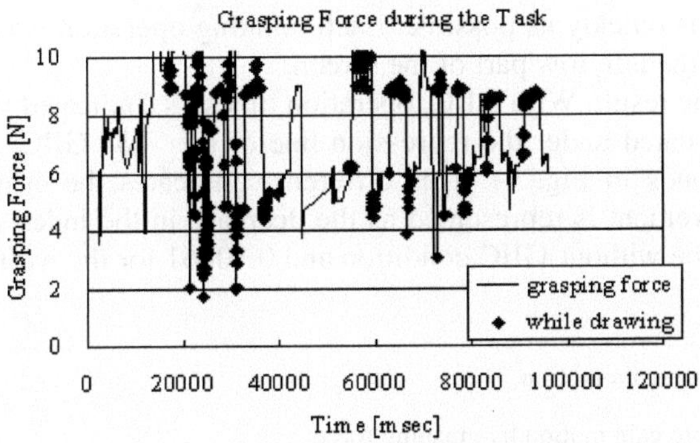

Fig. 5. Grasping force in a drawing task

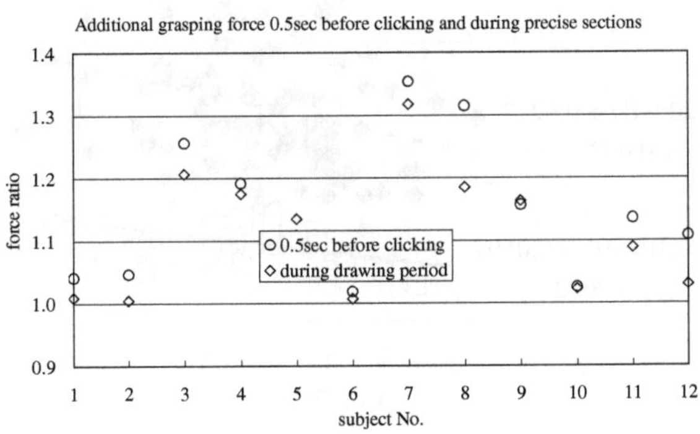

Fig. 6. The result of the experiment

diamonds in Fig. 6, the average grasping force in drawing periods, i.e., precise periods, was 111% of that in moving periods, i.e., rough periods. Just before clicking shown as open circles in Fig. 6, the average grasping force increased remarkably to an average of 114% of that in rough periods. This feature was observed in all of the subjects, without any exception. The average grasping force in drawing periods (= precise periods) is 103% of that in moving periods (= rough periods). Just before clicking, the average grasping force increases remarkably to an average of 114% of that in rough periods.

3 An Experiment to Validate GIIC

To validate GIIC, a set of experiments was conducted. There were 8 pointing tasks on a computer screen in a set, some of them were without GIIC, and others were with GIIC. With GIIC, the cursor speed decreased half when the mouse was grasped stronger than pre-defined threshold. The targets were displayed on the computer screen. The size of each target is randomly produced, ranging from four to 40 pixels. The positions of the targets are also randomized. The subjects are instructed to click all of the targets as quickly as possible. Each pointing operation was started from the starting square at the left, low part of the screen,

Fig. 7 shows the result. With GIIC, operation times *T*s (indicated with open squares in Fig. 7) were located under the regression line of the "non GIIC" group (indicated with filled diamonds in Fig. 7). This difference indicates the effect of GIIC. The amount of improvement is represented as the decrease in the index *b* in Fitts' law: it was 0.20081 for the without-GIIC condition and 0.10861 for the with-GIIC condition.

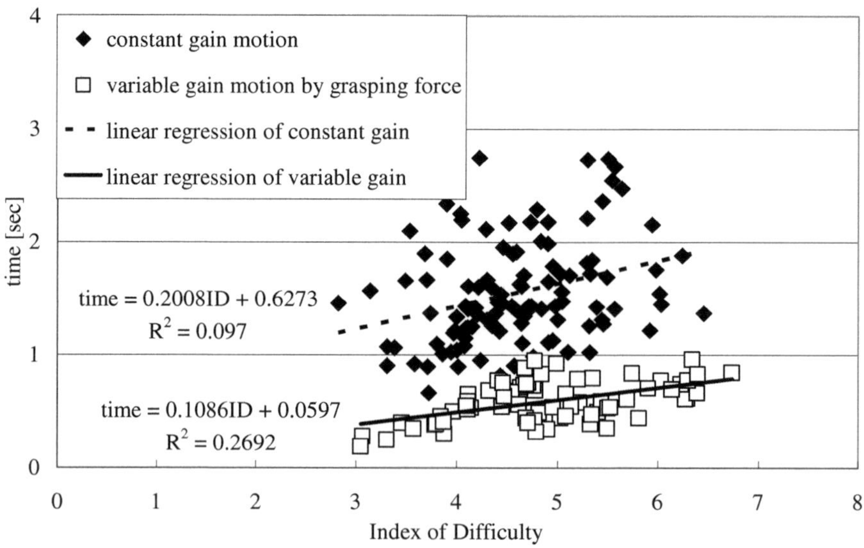

Fig. 7. An effect of GIIC

4 Discussion

The grasping force of the mouse is confirmed to be increased in precise periods, especially just before clicking.

As an example of GIIC, a new computer mouse with a real-time adaptive Mickey ratio adjuster is to be proposed as a typical remote pointing device. The grasping force of the mouse is confirmed to be increased in precise periods, especially just before clicking. The effect of real-time adaptive gain adjustment was also confirmed.

To apply the proposed concept to pointing tasks on the computer screen by mice, it is suggested that real-time gain (Mickey ratio) adjustment would be effective. Remote pointing tasks are separated into approaching and positioning phases. The use of real-time adaptive gain adjustment has advantages that make pointing tasks much easier than conventional pointing devices with a fixed gain. This is because, although there are two factors that decide the easiness of a remote pointing task, the distance D to the target and the size S of the target, they are not simultaneously effective. The distance D to the target is the major parameter in the approaching phase and the target size S is the major parameter in the positioning phase.

Because the purpose of real-time adaptive gain adjustment is to increase the easiness of pointing tasks, it must be added without any new gears or complexity.

With general pointing tasks, these two phases could not be known before the task is finished. Because the boundary between two phases exists only in the operator's mental or psychological processes, it is also difficult to detect.

GIIC has the potential of providing a natural usability with remote pointing devices by installing a real-time adaptive gain adjustment for grasped operation terminals. Grasping information has the potential to increase the usability of remote pointing devices.

Adapting the proposed concept, grasping information can be an indicator of the operator's intent. They include operators' "silent messages" for machines.

GIIC regards grasping-and-moving as a fundamental hand operation necessary for performing tasks using modern man-machine interfaces, and behavioral measures associated with grasping, such as force, posture, etc., should have much potential in developing task-adaptive and operator-adaptive interfaces because it is known that how people grasp devices is dependent on the purpose of these tasks. GIIC provides machines with a communication channel to understand operators' intents more through the grasping information.

References

1. Mehrabian, A.: Silent Messages. Wardworth, Belmont, CA (1971)
2. Sato, S., Kitajima, M., Fukui, Y.: A New Mouse with Cursor Speed Control by Grasping Force. Proc. HCI2001, New Orleans, Poster Sessions, pp. 291–292 (2001)
3. Sato, S., Kitajima, M., Fukui, Y.: A Mouse with Realtime Adaptive Mickey Ratio Adjustment by Grasping Force. Proc. UIST2002, Paris, pp. 53–54 (2002)

4. Sato, S., Kitajima, M., Fukui, Y., Proposal, A.: A Proposal of Grasping Information Interface for Remote Pointing Task. Proc. IEEE ISATP2003, Besançon, pp. 175–180 (2003)
5. Sato, S., Kitajima, M., Fukui, Y.: A Concept of Grasping Information Interface for Remote Pointing Task. Proc. IEEE ICRA2004, New Orleans, LA (2004)
6. Napier, J.R.: The prehensile movement of the human hand. Journal of Bone. and Joint Surgery 38-B, 903–912 (1956)
7. Fitts, P.M.: The Information capacity of the human motor system in controlling the amplitude of movement. Journal of Experimental Psychology 47, 381–391 (1954)
8. Fitts, P.M., Peterson, J.R.: Information capacity of discrete motor responses. Journal of Experimental Psychology 67, 103–112 (1964)

A Study of Information Flow Between Designers and Users Via Website Focused on Property of Hyper Links

Hidetsugu Suto[1], Hiroshi Kawakami[2], and Hisashi Handa[3]

[1] Department of Computer Science and System Engineering, Muroran Institute of Technology. 27-1, Muzumotocho, Muroran-shi, Hokkaido, 050-8585, Japan
suto@csse.muroran-it.ac.jp

[2] Graduate School of Informatics, Kyoto University. Yoshida-Honmachi, Kyoto 606-8501, Japan kawakami@i.kyoto-u.ac.jp

[3] Department of Information Technology, Okayama University. 3-1-1 Tsushima-naka, Okayama, 700-8530 Japan
handa@sdc.it.okayama-u.ac.jp

Abstract. The accessibility of information on web-space is considered to be focused on the properties of hyperlinks. Although the reasons to put in a hyperlink are multifarious, only one kind of link exists in HTML structure. As a result of this, users have to search for links by using syntactically techniques so as to retrieve information in web-space. In this paper, four kinds of hyper links are introduced into HTML, in order to improve the accessibility of information in web-space: External links, Internal links, Navigation links and Intra-unit links. Designers and users can share the image of the link target page by using these kinds of links.

1 Introduction

It is one of the things that annoy web designers, that what actions users take on a website are not clear. Users don't necessarily correctly understand the designers' intention, therefore the detailed design of a websites' structure is not always effective. On the other hand, users are also annoyed by guessing the designers' intention and searching for an item that should be clicked on when they encounter a website which is not well designed.

The most likely causes of these problems are the different expectations of links between users and the designer. The different expectations are caused by the fact that there is only one basic type of hyper link, although the reasons to put hyperlinks is multifarious[1]. As a result, the users cannot trace information by using "Meaning of link", which should be considered by the designer. Therefore, formal and meaningless symbol manipulations by using syntax theory technologies are forced on the user, such as generating "the label that it seem" from the keyword that it seem, though a search target is a page which has a semantic relationship with other pages[1]. In a word, the ability to understand the semantic structure of the website in the context of the page, and generating

M.J. Smith, G. Salvendy (Eds.): Human Interface, Part I, HCII 2007, LNCS 4557, pp. 189–198, 2007.

a label that seems most suitable for the required information is necessary to retrieve information from a website efficiently.

As technologies for retrieving information on a website, based on the meaning of words, semantic web and web ontology[2] are studied. However, the application of these technologies to general usages is difficult at the present stage because the concepts that need to be defined are vast[3]. Moreover, even if all of the necessary concepts can be defined, making the website users aware of them is not easy.

In this paper, the linked information is classified by the semantic upper and lower relation, from a current page. This classification doesn't depend on the content of an individual website, so to speak, it is meta-classification. Furthermore, information flows are analyzed by using a strict mathematical framework.

2 The Problem with Information Transmission Via Websites

2.1 Designers' Viewpoint

From a viewpoint of web accessibility, websites should be designed according to the semantic structure of information. Actually, some portal sites, e.g. Yahoo[4], are designed according to an original conceptual structure for accessibility. However, when thinking about the characteristics of web based media where everyone can easily send information, it is difficult to force all information senders to construct the appropriate structure.

The following items are enumerated as factors that make it difficult when constructing an appropriate layered structure.

1. Differences of mental modal caused by differences of individual experience.
2. Differences of knowledge of the information that needs to be sent.
3. A structure gradually collapses as a result of several staff updating at the same time.
4. Designers don't have a concept of the layered structure of information.

As for 1 and 2, the solutions are studied in the field of the above mentioned web ontology etc. In this paper, the aim is problem solving for items 3 and 4, without requesting designers to consider the layered structure of information, by introducing some loose restrictions based on a conceptual structure.

2.2 Users' Viewpoint

In general, website users spend a large amount of time looking for the item that should be clicked on. At this time, the users search while wondering "Is it correct?", causing a lot of mental stress to the user[5]. In other words, while searching a website, the operator feels numerous stresses when they can not understand the item that should be clicked, or after clicking, unexpected information appears. It is thought that these situations occur when the designers' intention is not transmitted to the users clearly.

In a website constructed by CMS (Contents Management System), e.g. XOOPS[6], ZOPE[7] or blogs, an item which uses a link that looks like a banner-type message. As a result, the user receives the impression that an institutional advertising page will appear. Therefore, novices are confused during the visit to the website because it is not understood intuitively where they need to click in order to read the contents of the website. At this time, the users cannot have the confidence in the type of information that is linked. Therefore, they feel hesitation in "the act so that a large mental cost is expended, i.e. click". Hence, if suitable link types for the character of linked information, which is linked, can be made, a reduction in the users' stress during information retrieval is expected.

3 Introducing "Loose Constraint"

The diversity of usage of link tag can be enumerated as a cause of making the users' forecasting of linked information difficult. However, it is not effective to strictly define the relationship between the link and the linked information, as already described in section 1. Hence, in this paper, loose restrictions are given to the relationship between a link and the linked information. First of all, clause of information in a website is discussed in order to provide restrictions.

3.1 Clause of Information

Pages of website are sets of information divided by a designer based on his/her own concept. For example, if the designer feels that one sentence became long too much, he/she divides it into two pages, on the other hand, if the contents are short, different information might be packed on one page by the designer. Therefore, guessing designers' intention is always required for website users.

This means that there may be improper delimitation of information on a website page. Hence, the concept of UNIT will be introduced as "a block of information" in a website. UNIT is a semantic delimitation of information. Two or more units might be included in one page, and one unit might be composed of two or more pages. For instance, two or more pages that include similar content, e.g. portfolio of work, commodity list and etc, are treated as a single UNIT. As a result, websites can be grasped as a layered structure of information that has UNITs as nodes.

3.2 Kinds of Links

Table 1 shows a relationship between an item on a page and a character of linked information from both user and designer viewpoints. In this table, "referring" means that there is information continuously being referred to. In general, designers consider the distinction between "link to an inside page" and "link to an outside page" comparatively strongly on account of the difference between "link to a file" and "link to a URL". Moreover, navigation links such as "Top", "Back", and "Next" are recognized as special items. On the other hand, users

Table 1. Relation between link items and linked information

	Property of link	Linked information
Designer	External	Related knowledge
	Internal	Subordinate/Superordinate/Single knowledge
	Navigation	Superordinate/Single knowledge
User	Back	Searching
	Forward	Referring/Searching
	Refer	Referring

Table 2. Relation between link items and linked information with the notion of UNIT

	Property of link	Linked information
Designer	External	Related knowledge
	Internal	Superordinate/Subordinate knowledge
	Intra-unit	Single knowledge
	Navigation	Superordinate knowledge
User	Back	Searching
	Forward (inside)	Searching
	Refer	Referring
	Forward (outside)	Leave site

have two modes, i.e. "searching" and "referring", during browsing. They consider the operations "back" and "next" because they exist in the button bar of a browser. Especially, the operation of "back" tends to be selected immediately after judging that a present page is useless[9].

The concept of four kinds of link types, i.e. "external link," "internal link," "intra-unit link," and "navigation link" are introduced based on the above discussion. Sharing the semantic structure of information in a website between designer and users can be established by specifying these links.

External links are used for connecting a current page and a page that is in another website. This type of link involves an exit of the website. As a rule, these types of links can be set to an arbitrary position in the website.

Internal links connect between UNITs in a website. This type of link is used when connecting an item with other contents in the website. They are chiefly used as a menu.

Intra-unit links are used when connecting an item with information in the same UNIT. In other words, the same kind of information as now referred to is linked by this type of link.

Navigation links are used to move freely in a website. For instance, buttons of "back" and "next" are classified as a navigation link. There is not a semantic relationship between a navigation link and the page which is linked. Breadcrumbs lists[5] are also classified as a navigation link.

Table 2 shows the relationship between the properties of the link and linked information, from the viewpoints of designers and users after introducing the concept of UNIT. From a designers viewpoint, the function of navigation links becomes clear because the connection is limited to a superordinate knowledge-unit.

Moreover, the role of internal links and intra-unit links are explicitly distinguished. "Forward" was classified into three kinds, i.e. "to inside of the website," "referring information" and "to outside of the website" from the users viewpoint. As a result, "the item that should be clicked when browsing" and "the item that should be clicked when referring" were clearly separated.

4 Analysing Information Flow

4.1 Channel Theory

Barwise and Seligman have proposed Channel Theory [10], which provides a mathematical framework for Dretske's "qualitative theory of information."

A classification $A = \langle tok(A), typ(A), \models_A \rangle$ consists of following three terms.

1. A set $tok(A)$ of objects to be classified.
2. A set $typ(A)$ of objects used to classify the tokens.
3. A binary relation $\models_A$ between $tok(A)$ and $typ(A)$ indicating the types to which tokens to be classified.

An infomorphism is a pair $\langle f^\wedge, f^\vee \rangle$ of functions. Given two classifications, A and B, an infomorphism from A to B written as $A \rightarrow B$ satisfies

$$f^\vee(b) \models_A \alpha \text{ iff } b \models_B f^\wedge(\alpha)$$

for $\forall \alpha \in typ(A)$, $\forall b \in tok(B)$, where $f^\wedge$ and $f^\vee$ are whole-part relationships.

An information channel $\mathcal{C} = \{f_i : A_i \rightarrow C\}_{i \in I}$ is an indexed family of infomorphisms with a common codomain C called the *core* of the channel. I is an index set.

4.2 Representation by Information Channel

In this section, the information flowing between a website designer and the user is represented by a model using Channel theory. Furthermore, the effectiveness of introducing four kinds of links, described in section 3, is discussed.

Classification of a Page. In a page, items which can be put on a link tag can be divided into {Banner, Menu, Icon, Word}. On the other hand, the character of the pages which are linked can be divided into {new, continuation of now referring, already referred}.

The Banner tag means transference to another website, so linked information is something new. A menu is used to connect to another page in the website. Pages linked by a menu are divided into two kinds, i.e. that which includes more detailed information and summarized information. An icon means a button which has the shape of an arrow, a house etc. It has the function of "Return," "Top," "Home," etc. An icon is used to freely move within a website. For example a right arrow is used to trace information one by one, and a left arrow is used to go back to information which has already been referred to. In this paper, the words "Back," "Next," "Top," etc. without graphic images are also regarded as icons. Links which are put on words in a document are used to change a topic to what the word (=keyword) means. Therefore, new information exists at the links included in the text.

From the consideration of the above-mentioned, items in a website and characters of the linked information are interpreted as token and type respectively. Suppose that "Banner," "Menu," "Icon," "Word" are defined as l_1, l_2, l_3, l_4, and "new," "referring," "referred" are defined as $\alpha_1, \alpha_2, \alpha_3$ respectively. The classification Pg of a page is defined as followings:

$$l_1 \models_{Pg} \alpha_1,\ l_2 \models_{Pg} \alpha_1,\ l_2 \models_{Pg} \alpha_3,$$
$$l_3 \models_{Pg} \alpha_2,\ l_3 \models_{Pg} \alpha_3,\ l_4 \models_{Pg} \alpha_1.$$

This classification can be represented by a Chu map[12] as follows, by regarding a token and a type as a column and a row respectively, and representing success of binomial relation by 1, 0.

$tok(Pg)$	α_1 New	α_2 Referring	α_3 Referred	$typ(Pg)$
l_1 Banner	1	0	0	
l_2 Menu	1	0	1	
l_3 Icon	0	1	1	
l_4 Word	1	0	0	

Classification of a User. The actions of a website user can be divided into m_x ={back, forward, refering}. These correspond to the basic functions of a web browser. Actions of "forward" include behavior that opens a new page by clicking on an item. The "referring" means an action that reads a series of information. Meanwhile, a user's status can be divided into β_x ={searching in a website, referring information}. As a result, token m_x is classified by type β_x as following.

$tok(Us)$	β_1 Searching	β_2 Referring	$typ(Us)$
m_1 Back	1	0	
m_2 Forward	1	1	
m_3 Refer	0	1	

Classification of a Designer. As described in Section 3.2, four kinds of links n_x ={external link, internal link, navigation link} are considered by a designer. As a consequence of the hierarchical structure of information in a website, the linked information can be divided into γ_x ={Related knowledge, Subordinate knowledge, Single knowledge, Superordinate knowledge}. As a result of this, token n_x is classified by type γ_x as following.

$tok(Ds)$	γ_1 Related	γ_2 Subordinate	γ_3 Single	γ_4 Superordinate	$typ(Ds)$
n_1 External link	1	0	0	0	
n_2 Internal link	0	1	1	1	
n_3 Navigation link	0	0	1	1	

Information Channel Between a Designer and a User. An information channel between classification Us and classification Ds must be considered in order to analyze the information that flows between a user and a designer.

Table 3 shows one of the information channels between Ds and Us via a classification Pg as a core. The left side of the second column of this table shows the infomorphism f_u from Us to Pg, and the right side of the second column of this table shows the infomorphism f_d from Ds to Pg. $\{f_u^{\wedge}(\beta_1) = \alpha_3 \mathrm{C} f_u^{\wedge}(\beta_2) = \alpha_1\}$ are represented by the combination of the row of the Chu map f_u and the row of the Chu map Pg, and $\{f_u^{\vee}(l_1) = m_3 \mathrm{C} f_u^{\vee}(l_2) = m_2 \mathrm{C} f_u^{\vee}(l_3) = m_1 \mathrm{C} f_u^{\vee}(l_4) = m_3\}$ are represented by the combination of the column of the Chu map f_u and the Chu map U_s.

Table 3. An information channel putting classification P_g as a core

	typ(Us) β_1	β_2				typ(Ds) γ_1	γ_2	γ_3	γ_4	
tok(Us) m_1	1	0				1	0	0	0	n_1 tok(Ds)
m_2	1	1	typ(Pg)			0	1	1	1	n_2
m_3	0	1	α_1	α_2	α_3	0	0	1	1	n_3
	0	1	1	0	0	1	0	0	0	l_1 tok(Pg)
	1	1	1	0	1	1	0	0	0	l_2
	1	0	0	1	1	0	1	1	1	l_3
	0	1	1	0	0	1	0	0	0	l_4
	f_u					f_d				

It is understood from $f_u^{\wedge}(\beta_2) = f_d^{\wedge}(\gamma_1) = \alpha_1$ that a website user can understand "the related information is traced when referring." However, a user cannot clearly understand the character of linked information in other situations.

4.3 Introducing "UNIT"

In this section, the loose restrictions are added to a hyperlink by introducing the concept of "intra-unit link" and "navigation link". These are described in section 3, and the changing of the information flow between a user, UNITs and the designer is discussed. In a UNIT, items which can use a link tag can be divided into l'_x ={Banner, Menu, Icon, Word} same as the page's case. The character of the information which is linked to items can be divided into α'_x ={new(outside of the website), new(inside of the website), continuation of now referring, already referred}. Therefore, token l'_x is classified by type α'_x as following.

tok(Ut)	α'_1 New(outside)	α'_2 New(inside)	α'_3 Referring	α'_4 Referred typ(Ut)
l'_1 Banner	1	0	0	0
l'_2 Menu	0	1	0	1
l'_3 Icon	0	0	1	1
l'_4 Word	0	1	0	0

By introducing the concept of UNIT, a conceptual model of users changes as shown in the lower part of table 2, and classification Us changes to Us' shown in following.

tok(Us')	β'_1 Searching	β'_2 Referring	β'_3 Leave site typ(Us')
m'_1 Back	1	0	0
m'_2 Forward (inside)	1	0	0
m'_3 Refer	0	1	0
m'_4 Forward (outside)	0	0	1

Table 4. An information channel putting classification U_t as a core

	$typ(Us')$ β'_1	β'_2	β'_3	$typ(Ut)$ α'_1	α'_2	α'_3	α'_4	$typ(Ds')$ γ'_1	γ'_2	γ'_3	γ'_4	
$tok(Us')$ m'_1	1	0	0					1	0	0	0	n'_1 $tok(Ds')$
m'_2	1	0	0					0	1	0	1	n'_2
m'_3	0	1	0					0	0	1	0	n'_3
m'_4	0	0	1					0	0	0	1	n'_4
	0	0	1	1	0	0	0	1	0	0	0	l'_1 $tok(Ut)$
	1	0	0	0	1	0	1	0	1	0	1	l'_2
	0	1	0	0	0	1	1	0	0	1	0	l'_3
	1	0	0	0	1	0	0	0	1	0	1	l'_4
	f'_u								f'_d			

Similarly, a conceptual model of designers changes as shown in the upper part of table 2, and classification Ds changes to Ds' shown in following.

$tok(Ds')$	γ_1 Related	γ_2 Subordinate	γ_3 Single	γ_4 Superordinate	$typ(Ds')$
n'_1 External link	1	0	0	0	
n'_2 Internal link	0	1	0	1	
n'_3 Intra-unit link	0	0	1	0	
n'_4 Navigation link	0	0	0	1	

Table 4 shows the information channel between classification Us' classification Ds' via a classification Ut as a core. It is understood that the following matters become clear from this table.

1. A single knowledge-unit is traced while referring. $(f'^{\wedge}_u(\beta'_2) = f'^{\wedge}_d(\gamma'_3) = \alpha'_3)$
2. Related information exists outside of a website. $(f'^{\wedge}_u(\beta'_3) = f'^{\wedge}_d(\gamma'_1) = \alpha'_1)$
3. Superordinate and subordinate knowledge-units can be traced while searching a website. $(f'^{\wedge}_u(\beta'_1) = f'^{\wedge}_d(\gamma'_2) = f'^{\wedge}_d(\gamma'_4) = \alpha'_2)$

5 An Experiment and Consideration

An experiment was carried out in order to verify the effectiveness of introducing four types of links.

Two websites, site A and site B were used for this experiment. These websites include a catalog of graduation works of Akita municipal junior college of arts and crafts[13]. The data of 604 works are contained in each catalog. The data is classified by course into eight groups, i.e. woodworking, textile A, textile B, Japanese lacquer, glass, ceramics, metalworking A and metalworking B, and also classified by the fiscal year. Site A and site B have quite the same hierarchical structure.

In site A, all items which are set with a link tag are displayed by navy color. On the other hand, site B is introduced with limited links. In site B, external links, internal links, navigation links and intra-unit links are displayed by dark cyan, navy, dark violet and orange red color respectively.

The testee did the following works on site A and site B.

1. Look at the photograph of the target work, and search the page in which the same image is included. (task 1)

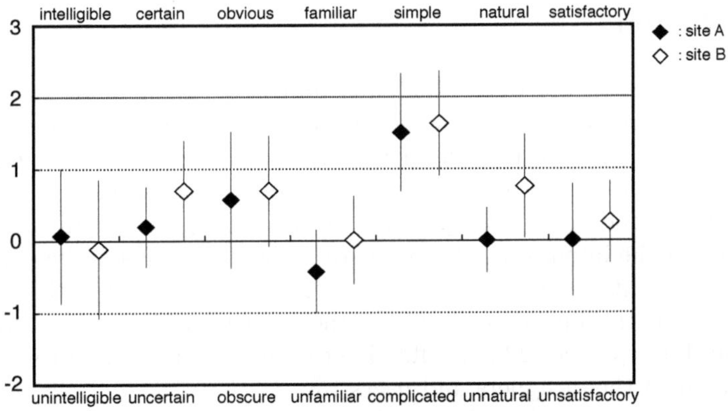

Fig. 1. A result of experimentation

2. Select one course from a specified group, and search the most favorite work in the course. (task 2)
3. Answer the questionnaire which was made according to SD method[14].

Work X of the woodworking course and work Y of the metalworking B course were selected as the search targets of task 1. X looks like a work of the ceramics course, and Y looks like it is made of wood at first glance. In task 2, the course groups were defined as followings. $F_1 = \{$woodworking, metalworking A, textile A, glass$\}$ $F_2 = \{$Japanese lacquer, metalworking B, textile B, ceramics$\}$

The test subjects were divided into four groups (G1 $\sim$ G4) and the experiment was carried out in the following order, in order to reduce the influence by context as much as possible. The subjects were 16 university students, aged from 18 to 23 years old, who are accustomed to the operations of a web browser.

	Step 1			Step 2		
Subject group	site	search target	course group	site	search target	course group
G1	A	X	F_1	B	Y	F_2
G2	A	Y	F_2	B	X	F_1
G3	B	X	F_1	A	Y	F_2
G4	B	Y	F_2	A	X	F_1

Fig 1 shows means and variances results of the experiment. The diamonds in the figure show means, upper and lower straight lines show variances. The black diamonds show data of site A, and white diamonds show data of site B respectively.

The "natural" has a significant difference between data of site A and site B (level of significance is 0.089). The differences are also slightly seen in "certainty" (level of significance is 0.102) and "familiarity" (level of significance is 0.142). This proves the consideration in section 4.3, i.e. a user comes to understand a designer's intention by introducing concept of UNIT.

As for "intelligibility," Site A reached a slightly higher value. It is thought that the cause of this is that the explanation of the four kinds of links gave an impression of difficulty to the subjects. It is expected that the value of "intelligibility" would be improved by increased user experience.

6 Conclusion

In this paper, the concept of "UNIT" and four types of links are introduced in order to give the restriction to a structure of website, called a loose restriction. Website designers come to be able inevitably to construct a semantic hierarchical structure by designing according to this restriction. Whereas conventional markup styles intend us to strictly define all the concepts. In addition, the users can understand the intention of the website designer by considering the kinds of links. As a result, it becomes easy to guess the semantic relationship between a linking page and the linked page, hence the reduction of stress in information retrieval can be expected. This is effective to a universal design of website from the standpoint of the designers and users[15].

The flow of information before and after the introduction of "UNIT" was analyzed by using the Channel theory in order to verify the effectiveness of the restriction. By analyzing the models, it has been understood that the item that should be clicked becomes clear in each case of referring information and searching in a website after introducing the concept of "UNIT." In addition, the experiment that used the sites where the restriction had been introduced was carried out in order to confirm the effectiveness of proposition of this paper.

References

1. Dreyfus, H.L.: On the internet (2001)
2. Antoniou, G., van Harmelen, F.: A Semantic Web Primer. Mit Press, Cambridge, MA (2004)
3. Matsuo, K.: Semantic Web is Next Big Thing? Trans. of Information Processing Society of Japan 45(5), 540–541 (2004)
4. http://www.yahoo.co.jp/
5. Krug, S.: A Common Sense Approach to Web Usability, Pearson Education, Inc., publishing (2000)
6. http://jp.xoops.org/
7. http://zope.jp/
8. Takeda, H., Ohmukai, I.: he State-of-Art and Prospects of Weblog - An Infrastructure of Semantic Web and Social Networking. Trans. of Information Processing Society of Japan 45(6), 586–593 (2004)
9. Spool, J.M., et al.: Web Site Usability: A Designer's Guide. Morgan Kaufmann Publishers, San Francisco (1999)
10. Barwise, J., Seligman, J.: Information flow. Cambrige University Press, Cambrige (1997)
11. Barwise, J., Perry, J.: Situations and attitudes, CSLI Publications (1999)
12. Gupta, V.: Chu Spaces: A Model of Concurrency, Ph. D Thesis, Comp. Sci Dept., Stanford Univ. (1994)
13. http://www.amcac.ac.jp/
14. Osgood, C.E., Suci, G.J., Tannenbaum, P.H.: The Measurement of Meaning. Univ. of Illinois Press (1957)
15. Suto, H.: Universality of Web Designing Focused on Property of Hyper Links. Proc. of the 32th SICE Symposium on Intelligent Systems, pp. 351–354 (2005)

Implementing the HCD Method into the Developing Process of a CPD System

Kevin C. Tseng[1,*], Tsai-hsuan Tsai[2], and Kun-chieh Wang[3]

[1] Chang Gung University, Taoyuan, Taiwan ROC
kevin@mail.cgu.edu.tw
[2] Chihlee Institute of Technology, Taipei, Taiwan ROC
dilystsai@gmail.com
[3] Ling Tung University, Taichung, Taiwan ROC
kcwang@mail.ltu.edu.tw

Abstract. This paper aims to investigate collaborative product development activities and highlights the importance of human perspectives for developing a collaborative product development (CPD) system. For CPD systems to be useful, it is necessary to effect usable and useful interaction. Therefore, the human-centred design (HCD) method had been implemented into the developing process of a CPD system. The modified HCD method specifies how the activities fit into the overall system development process. A proposed plan guides: (1) the design process activities described, (2) procedures for integrating these activities with other system development activities, (3) the individuals and the organisation(s) responsible for the design activities and the range of skills and viewpoints they provide and (4) effective procedures for establishing feedback and communication on design activities as they affect other design activities and methods for documenting these activities.

Keywords: collaborative product development, human-centred design, software engineering.

1 Introduction

Nowadays, the CPD systems are increasingly being seen as useful mechanisms for the sharing of design knowledge and the distributed design collaboration; particularly in the field of consumer goods production. Although many systems have been developed, few can be termed 'successful' (i.e. have a significant uptake by intended users, and used by them).

Industrial product development is an interdisciplinary activity requiring contributions from nearly all the functions of an organisation. Three functions are usually central to a product development project: marketing, design, and manufacturing. However, people in different functional departments have different disciplines. It is essential to assure the end system meets the target audiences' requirements, experience and expertise. A CPD system may thus be executed by a wide range of different types of users. These may be inexperienced users, unskilled people, users with average abilities and expert operatives.

M.J. Smith, G. Salvendy (Eds.): Human Interface, Part I, HCII 2007, LNCS 4557, pp. 199–207, 2007.

Documented reports of underperformance of ICT systems over several decades reveal that a major factor contributing to the disappointing outcome is the inadequate understanding of user requirements and thus failure to design new technologies to meet those requirements [1, 2]. Maguire [3] stated that the benefits of designing a useable system can be summed up as follows: increased productivity, reduced errors, reduced training and support, improved acceptance and enhanced reputation. Stamato [4] referred there are several factors that create the impetus for collaboration; however, technology is only one of the factors to affect collaborative product development. Instead, successful collaborative product development will have a balanced approach between users, organisations, and technology.

Human factors, also called Ergonomics, is a branch of applied science concerned both with the study of human behaviour in the context of technology and with the application of the knowledge gained from such study to real-world needs. For a collaborative product development system, it is centred on the characteristics and needs of team members in complex systems using advanced information and communication technology. Many poorly designed and unusable systems exist which users find difficult to learn and complicated to operate. These systems are likely to be underused, misused or fall into disuse with frustrated users maintaining their current working methods. The outcome is costly for an organisation using the system and wasteful for people learning the system. A global manufacturing organisation needs a CPD system that concerns all parameters during the product development life cycle. This requires the critical consideration of various tasks such as overall coordination, control, design rules, consistency, and data integrity to maximise the design efficiency and to minimise cost, time and resources waste. This can be achieved by the integration of different design areas through establishing a Web-based system, within an organisation in a consistent manner. Such integration should include a strategy for conflict resolution, to avoid disagreements within the different activities or areas. Research work in this area was conducted by several authors including Tseng et al. [5]. In addition, it is important early in the innovation cycle to identify and predict user needs and requirements. An effective application of theory requires a methodological framework which guides the process of the intervention, i.e., HOW, WHO, WHEN and WHERE. Hence, the HCD method has been implemented in achieving clear definition and prediction of needs, improved design and more effective involvement of end-users, as detailed in this paper.

This paper aims to investigate collaborative product development activities and highlights the importance of human perspectives for developing a CPD system. The HCD method systematically addresses phases of system design and development and considers the needs between users, organisation, and technology. The infrastructure of the HCD method is introduced, and then three development phases are presented accordingly for the development of the CPD system. The HCD method helps the system developer to identify and predict users' needs and requirements in the early design life cycle and guide the intervention, i.e., HOW, WHO, WHEN, and WHERE.

2 Challenges of a CPD System Design

For collaborative systems to be useful, it is necessary to effect usable and useful interaction. When designing these systems, several concerns at the theoretical and methodological content, and organisational and technical levels should be taken into consideration. Current problems faced to designing a CPD system have been addressed as follows:

- Insufficient knowledge, skill, and experience;
- Most CPD systems are designed by a third party such as a software engineer. They lack of sufficient knowledge, skill, and experience about the specific functions of the CPD system, the process for designing a CPD system, or the criteria for good design.
- Insufficient human-centred design guidelines;
- Most of the directions available to the CPD system are focused on outside the domain of software design in fields long involved in CPD research including human factors [6], human-computer interaction [7], cognitive science [8], graphic arts [9], and technical communications [10]. However, the guidelines from these disciplines apply more to software environments than CPD environments, where the user task is fundamentally different. CPD system, on the other hand, often promotes the user's acquisition of new knowledge, skills and attitudes needed for performance or for understanding. Users enter these environments to build their expertise. Their interactions with the environment may not produce visible results. The outcome of a CPD experience is often not reflected in the interface environment, but in the CPD user's mind. Guidelines for the design of a CPD user interface clearly must be based on set of assumptions different from those based on software users and tasks.
- Graphically Embellished Design.
- Many CPD interfaces are also poor examples of graphic design featuring ornate background fills, multiple bevelled information zones, and distracting animations that ultimately direct the user's attention to the elements that surround the core CPD information, rather than to the core information itself.

Research is needed to study, analyse and reflect on the use of technology. Interdisciplinary research should:-

- understand the factors that make the technology useful. What are the characteristics of collaborative systems that can be linked to measurable improvements in the user's and team's performance?;
- understand the end-user of persistent collaborative systems in relation to existing information technology for product development. The core functions of the existing CPD systems may interact with the proposed system;
- implement the HCD method that conducts the process of development of CPD system in human perspectives;
- understand user's essential properties which affect his/her interaction with computers;
- understand users' tasks and requirements and analyse what people do with computer systems and their interfaces;
- specify how the interface should function, how it should respond to the user and its appearance;

- design interfaces so that users' needs are fulfilled and the system matches users' characteristics;
- evaluate the properties of the developed system and their effect on people to ensure good quality.

Therefore, the human-centred design (HCD) method had been implemented into the developing process of a CPD system. This paper has attempted to illustrate the impact that HCD method has had on the development and take-up of CPD systems for consumer goods production, and can have in other sector too. It has shown that each issue considered to be relevant to the success/failure of the CPD system in consumer goods production can be addressed by the appropriate inclusion of users in the design and development process. It has also provided evidence that the use of HCD method, however simple, is a common feature of successful examples of these systems. While this is not conclusive, it seems to be useful evidence to support the effectiveness of the HCD method in this sector.

3 Human-Centred Design Method

The HCD (human-centred design) method (**Fig. 1.**) is a cycle of three flexible implementation phases:-

- planning and suggestions;
- implementation and improvement;
- check and evaluation.

Fig. 1. The Human-Centred Design Method

HCD method specifies how the activities fit into the overall system development process. A proposed plan guides: (1) the design process activities described, (2) procedures for integrating these activities with other system development activities, (3) the individuals and the organisation(s) responsible for the design activities and the range of skills and viewpoints they provide and (4) effective procedures for establishing feedback and communication on design activities as they affect other design activities and methods for documenting these activities.

Organisations may enter and move through the phases at different speeds and give different phases different emphasis depending on their individual circumstances, the availability of resources and the level of maturity of their ICT polices, strategies and programmes for collaborative product development.

The HCD method for developing a CPD system is used in five stages of the system development life cycle, as shown in authors' another paper [11]:

- to establish a management system for developing the CPD system as guidance to deepen and broaden existing management practices;
- to integrate existing organisational systems for collaborative product development.

To ensure compatibility with existing practice, the HCD method is modelled on a systematic framework [11] incorporated in the process of the system development. Organisations can either improve their CPD system performance or develop a new CPD system by using the HCD method as guidance to:-

- plan the human-centred design process for guiding the organisation to develop a CPD system based on human requirements during collaborative product development;
- understand and specify the context of use and the user and organisational requirements;
- assess the organisational systems for collaborative product development based on usability evaluation;
- set the priority for improving CPD system functions and applications.

Fig. 1. and **Table 1.** cover the outline of the three key phases of the HCD method for developing a CPD system. Each phase is divided into sub-phases; for instance, the planning and suggestion (PS) phase is divided into sub-phases PS1 to PS3. All the phase and its sub-phases are described in detailed tables. The tables contain:-

- key questions to focus on;
- a cross-check of how intended activities link to the organisation's chosen principles;
- who needs to be involved;
- what the key activities are;
- when they need to be implemented;
- a list of helpful resources;
- expected outcomes and outputs;
- key implementation issues of which the organisation may need to be aware.

Organisations may choose to establish a dedicated team to implement the development of the CPD system. The tables refer to such a team as the system development team. Similarly, an organisation may appoint a key individual to support the implementation process; this is referred to as the project manager.

Table 1. The Three Key Phases of the HCD Method for Developing the Proposed System

Management Phase	Purpose
Planning and Suggestion *Plan and suggest what ICT strategies/applications for collaborative product development need to be in their core processes and decision-making*	• Ascertaining the organisation's and end-users' needs for a CPD system for collaborative product development • Identifying and prioritising the organisation's key ICT issues and needs for collaborative product development • Analysing and managing strategies and processes to ensure the proposed system is integrating into current performance • Developing strategic plans to deliver the proposed system
Implementing and Improvement *Decide what needs to be done to implement/improve ICT strategies/applications for collaborative product development*	• Implementing the developed plan to design the proposed system • Implementing approval approach/methodology/system • Reviewing and improving existing ICT performance for collaborative product development
Check and Evaluation *Check that the proposed system meets user requirements and analysis results*	• Identifying, evaluating, categorising and managing the actions, impacts and outcomes of executing the proposed system • Establishing or reinforcing mechanisms for handling and responding to end users' feedback on reports • Monitoring the process of executing the proposed system • Communicating with current users' findings and recommendations to evaluate whole product life cycle • Taking preventive, corrective and innovative actions as appropriate • Consulting with end users on performance and future challenges

The outcomes and outputs column lists the likely output and outcomes of the recommended activities. It is not necessary to produce all the documents listed in the tables. Some organisations will favour producing all or most of the written output, to aid monitoring and assessment; others will prefer a less formal approach, or will be able to modify existing documents rather than create new ones.

3.1 Planning and Suggestion Phase

Table 2. reveals the planning and suggestion phase of what ICT strategies/ applications for collaborative product development are to be in their core processes and decision-making. Sub-phases are requirement list, analysis and management, and strategic planning.

Table 2. Planning and Suggestion Phase

Phase		Activity
PS1	Requirement list	• Ascertaining the organisation's and end-users' needs for a CPD system for collaborative product development • Identifying and prioritising the organisation's key ICT issues and needs for collaborative product development
PS2	Analysis and management	• Analysing and managing strategies and processes to ensure the proposed system integrates into current performance
PS3	Strategic planning	• Developing strategic plans to deliver the proposed system

3.2 Implementing and Improvement Phase

Table 3 reveals the implementing and improvement phase which decides what needs to be done to implement or improve ICT strategies and applications for collaborative product development. Sub-phases are implementing and managing programmes and improvement.

Table 3. Implementing and Improvement Phase

Phase		Activity
II1	Implementing and managing programmes	• Implementing the developed plan to design the proposed system • Implementing approval approach/methodology/system
II2	Improvement	• Reviewing and improving the proposed system for collaborative product development

3.3 Check and Evaluation Phase

Table 4 reveals the check and evaluation phase to check that the proposed system meets user requirements and analysis results. Sub-phases are action, impacts and outcomes and monitoring, measurement, auditing and feedback.

Table 4. Check and Evaluation Phase

Phase		Activity
CE1	Action, impacts and outcomes	• Identifying, evaluating, categorising and managing the actions, impacts and outcomes of executing the proposed system • Establishing or reinforcing mechanisms for handling and responding to end users' feedback on reports
CE2	Monitoring, measurement, auditing and feedback	• Monitoring the process of executing the proposed system • Communicating with current users' findings and recommendations to evaluate whole product life cycle • Taking preventive, corrective and innovative actions as appropriate • Consulting with end users on performance and future challenges

4 Implementing the HCD Method for Developing a CPD System

Some organisations may have well-developed ICT approaches or applications for CPD, but may need to clarify or revise their ICT strategic vision. For instance, some organisations may have a good understanding of, and a willingness to develop CPD system, but they are facing barriers during the implementation of some of the technologies. For these organisations, the check and evaluation phase would be a good benchmark to measure what functions are needed for collaborative product development.

For other organisations without well-integrated ICT approaches or applications for CPD, the planning or suggestions may be to help those organisations in their core processes and decision-making. The planning and suggestion phases would be a good place to start in these circumstances.

Other organisations may have urgent needs to implement ICT approaches or applications for CPD. The implementing and improvement phases would be a good way to implement or improve ICT strategies or applications for collaborative product development in those organisations.

5 Conclusions

This paper detailed the HCD method for the development of the CPD system. In order to design a system for collaborative product development, the HCD method was introduced, and then the three design phases were presented accordingly. The HCD method, which is described in details in this paper, by modelling the HCD method for designing the CPD system. The main point, described in this paper, contained the following features:

- The HCD method systematically divides the system development life cycle into three phases that cover not only the development procedures but also balances the needs between users, organisation and technology;

- Fundamentally, the HCD method considered the three central requirements, which are functional requirement, design requirement and technical requirement;
- The human-centred design method is implemented into the developing process a CPD system and management programme is listed for guiding the system developer to design the CPD system.

References

[1] Khalid, H.M.: Human factors of virtual collaboration in product design. presented at Proceedings of APCHI/ASEAN Ergonomics 2000 Conference, Amsterdam (2000)

[2] Regli, W.C., Hewett, T.: HCI Issues in Collaborative and Networked Engineering Design, presented at CHI '99 Workshop (1999)

[3] Maguire, M.: Methods to support human-centred design. International Journal of Human-Computer Studies 55, 587–634 (2001)

[4] Stamato, L.: Collaborating - Finding Common Ground for Multiparty Problems - Gray,B. Negotiation Journal-on the Process of Dispute Settlement 7, 265–277 (1991)

[5] Tseng, K., El-Ganzoury, W., Abdalla, H.: Integration of Non-networked Software Agents for Collaborative Product Development. Computer-Aided Design & Applications, vol. 2 (2005)

[6] Sanders, M.S., McCormick, E.J.: Human factors in engineering and design. McGraw-Hill, New York (1993)

[7] Schneiderman, B.: Designing the user interface: strategies for effective human-computer interaction. Addison Wesley Publishing Company, Reading, MA (1998)

[8] Norman, D.A.: Design of everyday things. Currency Doubleday, New York (1988)

[9] Mullet, K., Sano, D.: Designing visual interfaces: communication oriented techniques. Prentice Hall PTR, Englewood Cliffs, NJ (1995)

[10] Hackos, J.T., Redish, J.C.: User and task analysis for interface design. Wiley Computer Publishing, New York (1998)

[11] Tseng, K.C., Abdalla, H.: A novel approach to collaborative product design and development environment. Proceedings of the I MECH E Part. B Journal of Engineering Manufacture 220, 1997–2020 (2006)

Artificial Psychology

Zhiliang Wang

School of Information, University of Science and Technology Beijing,
Beijing, 100083, China
zhiliang_w@263.net

Abstract. This paper describes the basis content of Artificial Psychology as well as Affective Computing and Kansei Engineering and other technology that are correlated with the research. Also it reviews the research history of Artificial Psychology theory and practice, and the future research direction of the Artificial Psychology is indicated.

Keywords: Artificial Psychology, Kansei Engineering , Artificial Intelligence, Affective Computing.

1 Introduction

It is well known that on the one hand, research into Artificial Intelligence has been done to a very high level. But on the other hand, its purpose has been limited to imitating human intelligence talents such as judging, deducing, proving, explaining, identifying, sensing, planning, learning and problem-solving. The main task of these activities is how to present, acquire and utilize knowledge. This is just the initial stage in the field of virtual human being research. This is not nearly a wide-enough research range since human psychology and related activities include not only sensing, believing, memorizing and thinking, but also emotion, feelings, willingness, character and creativity. The theory of ARTIFICIAL PSYCHOLOGY(A.P.=Artificial Psychology in following) [1]is aimed at a comprehensive understanding of human psychology, (especially feelings, willingness, character, and creativity) by means of a systematic integration of psychology, brain science, information science, neuroscience, computing science and novel theories of automation science. This paper describes the basis content of Artificial Psychology based on Affective Computing and Kansei Engineering and other technology that are correlated with the research. Also it reviews the research history of Artificial Psychology theory and practice, and the future research direction of the Artificial Psychology is indicated.

2 Theoretical Research on Artificial Psychology

The theory of artificial psychology based on information science, is the comprehensive machine realization (computation and algorithms) of human psychological activities. Application of this theory includes development of robots with feelings, emotion and

M.J. Smith, G. Salvendy (Eds.): Human Interface, Part I, HCII 2007, LNCS 4557, pp. 208 – 217, 2007.

intelligence, human similar mechanics, and control patterns based on human brain. We know that the theory of human similar control is mainly the combination of weiner's "feedback" control theory and the artificial intelligence. It differs greatly from the control patterns of human brain because the control pattern of human brain is based on sense and the idea that sense decides behavior. But the current control system takes no account of the affective factors while making decision because it is not able to do that. There is a vast potential to apply artificial psychology theory to the design of more humanized commodities and to the development of market. The theory of Artificial psychology is the sophisticated stage of artificial intelligence. It is a new domain of automation and information science. Relationship between A.I(Artificial Intelligence) and A.P.(Artificial Psychology) is shown in fig.1. Its research will greatly promote the development of the theory of personification control, emotional robots, the design of humanized commodities and the market, and contribute to set up harmonious man-to-man and man-to-machine environment. The theory of Artificial psychology is an cross science. As shown in figure 2 its original basis is from brain science, psychology, physiology, ethics, neural science, anthropology, Kansei engineering, linguistics, aesthetics, jurisprudence, information science, computer science, automation science and artificial intelligence. The main applied areas are as follows: emotional control theory in robots, personification mechanics, the design of humanized commodity, sensible market development, Artificial Psychology Programming Languages, Artificial Creative Techniques, humanoid Techniques for Human Psychology Databases and Mathematical Models, Harmonic Interface for man-machine and IS multi-channel Interfaces.

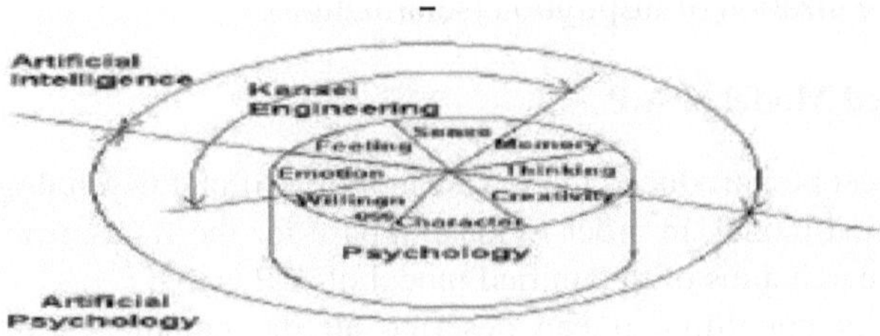

Fig. 1. Relationship Between A.I. and A.P.

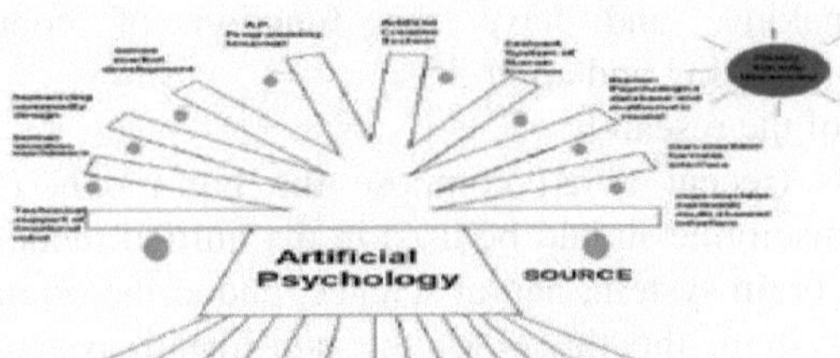

Fig. 2. Artificial Psychology - an Interdisciplinary Science

The research contents of the theory of artificial psychology are as follows:

1) To set up the theoretical system of Artificial Psychology(aims, rules, research content, application range, research methods), and especially to break the limitation of definitions, rules, and contents, make it conform to the moral rules of human beings which is not included in Artificial Intelligence.

2) To set up the theoretical system of Artificial Psychology by means of the theoretical achievements of A.I. and the relationships between A.I. and A.P.

3) To develop machine algorithms to repress bad emotions. It is decided by the rules of A.P.

4) To mathematically scale human psychological information whose main achievements were made by Japanese. Our national researchers also make some progress, such as the publication of <The analysis of fuzzy aggregate in psychology >, < measurement science of psychology>.

5) To set up a control mode of sense, feeling and emotion as the decision of behavior to imitate control mode of human brain. It is different from the control mode of A.I.

6) To probe the building-up of a programming language for Artificial Psychology. It is a challenging work. The programming language of A.I. is the presentation of knowledge and logical inference. In A.P., the programming language must be a kind of associative language whose character is associative inference, chaotic computation, divergent thinking and fuzzy induction.

7) Computation algorithm for emotion cultivation.

8) Machine realization of inspiration (scintillation).

2.1 The Unified Model of A.P

Mathematics must be introduced to the research of artificial psychology. Its premise is to build an unified model, in order to make a base for the future research and build a system. The research aims of the unified model of A.P is that :

Under a given condition, it can describe all the human emotion (Intelligence, imagination, study, memory, attention, consciousness, feeling and so on) or choose some small models from the big model. The small ones can not only describe one of psychology activities above, but also describe cognition—emotion, motivation—emotion—decision-making, and have the function of coordination, parallel, stratification, decision-making and control.

The general aims of the research:

The small models (recent aims) compose the big model (future aim). Much knowledge in multi-discipline should be used in the unified model. These knowledge include: psychology, brain system, neural science, endocrine science, physiology, the theory of complex system, the theory of the non-linear, system engineering, data structure, anthropology, behavioral science. Future aim: build big model recent aim: build feasible, useful and single-task, small model according to one of psychology activities.

The unified mathematical model we put forward preliminarily is as figure 3.
Its characters are:

1) The small models are subordinated to the unified big model.
2) Modular small models table the big model.
3) The unified big model can describe all of human psychology activities, while the small ones can describe certain one individually.
4) The inside of the small models is the progress of control, while the big one is the progress of the cooperative decision-making by small ones. The relationship between small ones is coordination, parallel, time-share, and coupling.
5) During the progress of model-building, we should use the knowledge of computer and system structure as technologic means, make use of the electronic circuit, and be on the basis of the theory of system, control, and intableation.

Here, we just put forward a concept of theory frame and will continue to do more research work about it. In another word, this is a very difficult but also very important project at all aspects of theory thought, computing methods and the special realization.

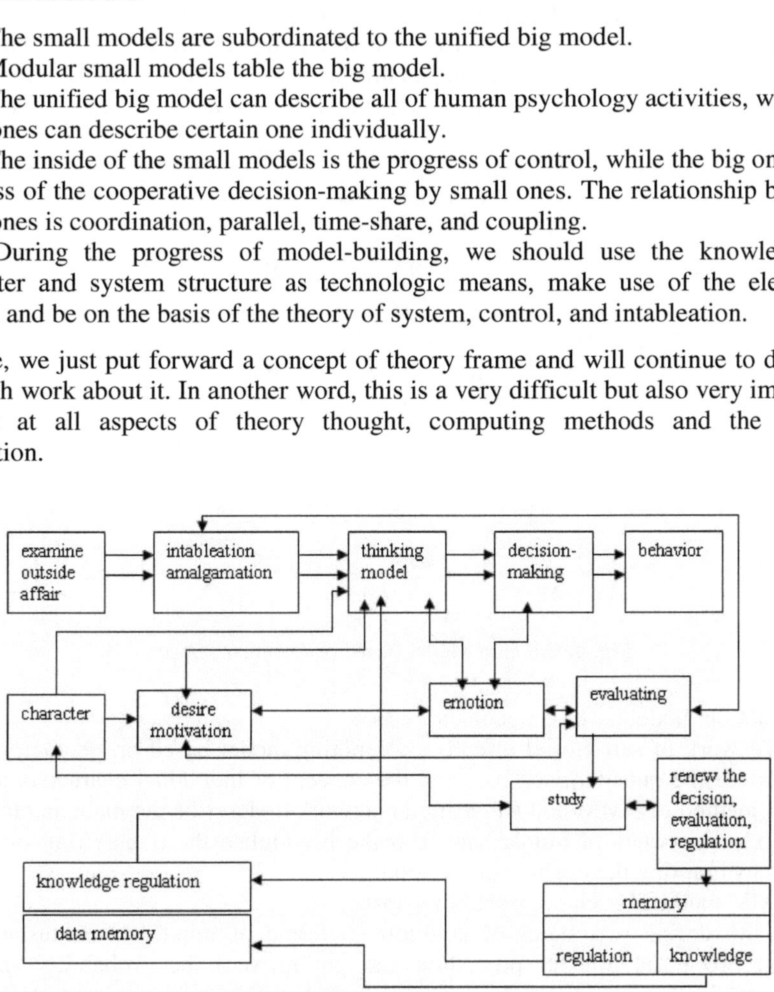

Fig. 3. The Unified Model of A.P

2.2 Affective Computing Model

The core of artificial psychology is that the effect of emotion at the psychology actives and the realization of the consciousness. The basic methods of artificial psychology are as follows: self-adjusting, positive emotion, coordinate work, energy balance, so the effective computing is the important content of artificial psychology. It has been considered an hard work and a challenge to establish effective model but we have done many jobs about it.

Based on emotion psychology, we define the mathematic space describing the emotion. In this space, we use mathematic theory to bring forward the effective computing methods which are easy to realize by computer and can simulate the producing, changing and transferring of the human emotion according to the rules of human emotion changing. We have bring forward and realized the emotion models which are based on geometry space, which are using HMM method based on probability space and which are the nonlinear dynamic model based on emotion dimensions. What mentioned above are the main research contents of artificial psychology.

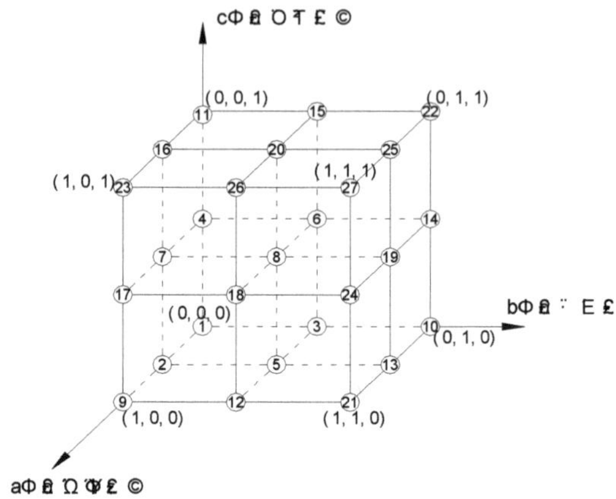

Fig. 4. Emotion Model Based on Geometry Space

1) Emotion model based on geometry space

On the work of self-closed effective computing model based on geometry space, we introduce the outside incentives and the concept of individual characters and use the joint method of HMM and BP artificial network to describe the main and the other psychology characters of human and to make it simulate the transferring of human emotion by ignoring the positive cross points.

2) HMM method based on probability space

First, we define two states of emotion — frame of mind and enthusiasm and accorded two basic transfer procedure and put forward the probability space of emotion states. Then we bring forward two models－One is based on Markov chain, another is based on HMM and emotion transfer model to simulate the two basic transfer procedure of emotion. We both define the emotion energy and emotion intension and emotion entropy to describe the emotion characters and emotion states. It is proved by computer simulation that these models can correctly describe the self-transfer procedure of emotions and the dynamic procedure of transferring and changing when outside incentives exist. They also can describe the emotion intension's changing rule under the influence of the outside incentives, present emotion state and personal character. They provide a new method for the theory research of effective computing and automatic creating.

3) nonlinear dynamic model based on emotion dimensions

According to basic emotion theory in emotion psychology and emotion dimension theory suggested by Wundt.W, we consider emotion procedure as a random dynamic procedure which controlled by a nonlinear dynamic equation. The general style of equation is as follow:

$$X=f(x, t)+g(x, t)u \tag{1}$$

The x represents the expectation value of emotion state and u represents the outside incentive.

We consider the machinery emotion procedure as a composed procedure of mood and enthusiasm. The mood procedure is used to simulate human mood and describe the steady state of machinery emotion. It is expressed by balance state in systematic dynamic equation when incentive equal zero. This balance state may express as isolated balance point or limit ring.

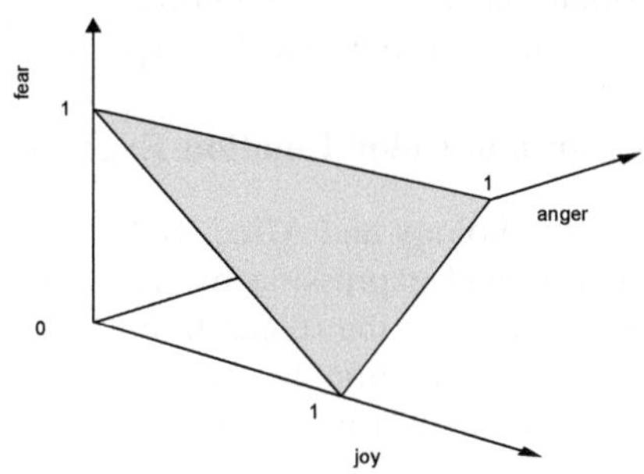

Fig. 5. The Identical Triangle in The Probability Space of Emotional State

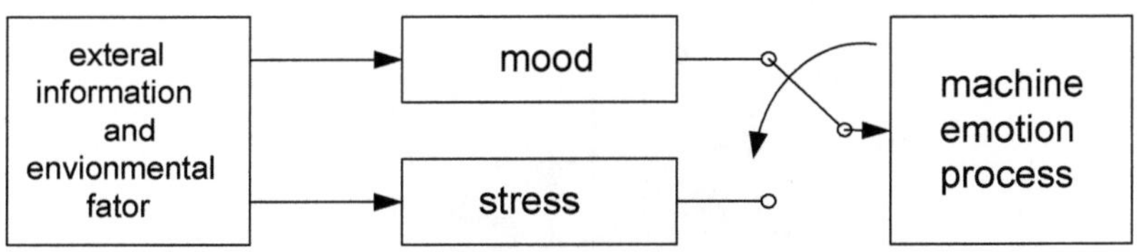

Fig. 6. Machinery Emotion Procedure

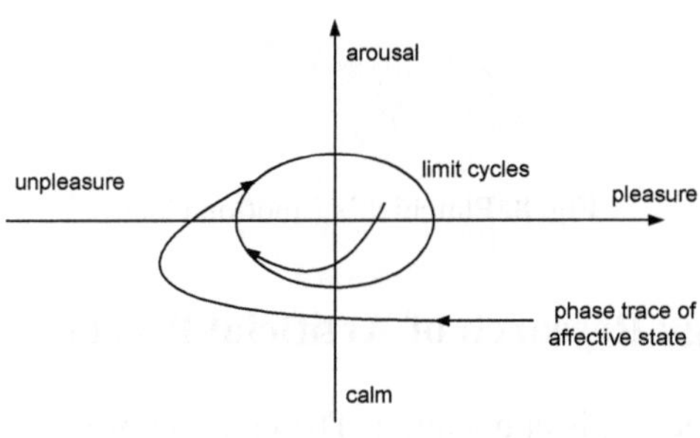

Fig. 7. The Transfer of The Machinery Emotion State

The enthusiasm procedure is used to simulate human enthusiasm and to discern the communicator's emotion through machine. The incentive is decided by environment. The enthusiasm procedure of machine is the respondent procedure of system to the environmental incentives.

The same scene and incentives may produce different emotion changing procedure because human's emotion is changefully so using random parameters to describe the changing procedure of emotion is appropriate. We consider the enthusiasm procedure as a random procedure using human mood as the initial state. The expectation value of the random procedure moves along the track of systematic dynamic equation. The rules of the machinery emotion procedure are that when no outside incentives exist, it express as the balance state of mood－systematic dynamic equation and when outside incentives exist and the balance state used as the initial state, it express as the random procedure which use the systematic response as the expectance curve.

2.3 The Correlation Research of Color-Emotion-Expression

In the research of Artificial Psychology and Affective Computing, it is very important that how to create the emotion and expression by means of machine. Here, we put forward that color creating emotion by the research of color theory and emotion cone theory of Plutchick; using basic emotion to create compound emotion, Figure 8 presents the emotion cone of Plutchick from Figure 8.

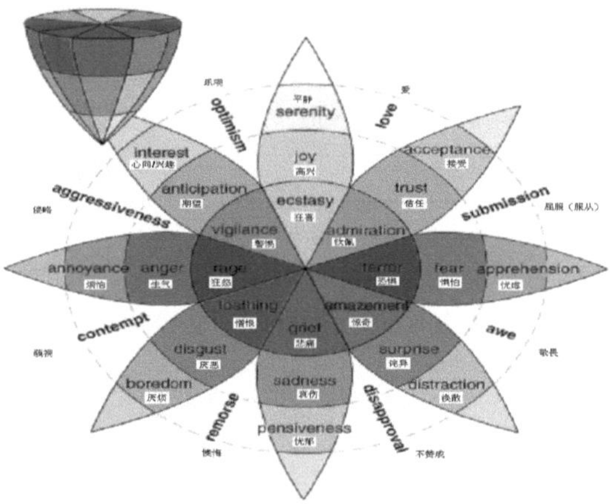

Fig. 8. Plutchick's Emotion Cone

3 The Technology Research of Artificial Psychology

The purpose of engineering is application. The creation of engineering is the creation of technology. So, the highest stage of the research development of Artificial Psychology is application. The most important point is to consider the human psychology and emotion component in the artificial systems, and make the artificial control systems and computer systems to adjust to human emotion , to realize harmony between human、man-computer and finally achieve the purpose of serving the people.

Our main works on application research about Artificial Psychology are as follows:

- **Research and Application of Personality Intelligent Fashion Shopping System Based on Network (Ma yun)** [2]

We use the method of integrating HSV color model algorithm and design theory of dress to derive its color eigenvalue, then improve features space, remedying the deficiency of original system by considering human vision system to color perceptual knowledge, dress variety and structure of dress, raising accuracy of clothing feature description. In order to overcome the problem that the fitness function of intelligent system is hard to express explicitly, improved IGA has been proposed firstly to implement on-line learning. Through IGA, human intuition and emotion is integrated into the evolution process to realize on-line retrieval by human-computer interaction. What is more, to deal with the problem that the user may tend to be tired arising from that the user has to evaluate a large number of individuals when the evolution time is too long, RBF neural networks is used for off-line learning to alleviate human fatigue.

Finally, a personality intelligent fashion shopping Website is established, and the experimental results demonstrates the effectiveness of our approach. Users think this system can express his or her kansei demands and be accord with human psychology characters.

- **Research on Modeling Artificial Emotion Based on HMM and Techniques Correlated with Virtual Human (Gu xuejing)** [3]

We realize an affective virtual human system based on the theory of artificial psychology. We build the dialogue engine for virtual human with the method of AI. We use frame structure to classify and store knowledge, this structure make data searching more quickly and make it easy to enlarge knowledge. It is more efficient for data searching and understanding by using the technology of pattern matching in our dialogue engine. The basic theory and method of HMM is studied before constructing the emotion model. We program the forward-backward arithmetic and Baum-Welch arithmetic by Visual Basic. It is the rational basis of the emotion model construction. Constructing emotion model is the key technology of virtual human system. Here, we present an artificial emotion model based on HMM. We get the parameter of matrix of emotion transfer probability and the vector of expression output probability. During modeling of emotion, we propose a definition of emotion entropy. We propose that emotion entropy is a scale to measure the stability of emotion and use it to restrict the initial emotion transfer probability matrix. It can help us measure one's character quantificationally. The results of emotion model test indicate that the emotional reaction of virtual human is according with human reaction. It means that this method of emotional modeling is feasible.

- **Study on Gait Feature Extraction and Human Identification Based on Computer Vision (Han hongzhe)** [4]

Gait recognition has recently received growing interest within the computer vision community. An efficient background updating algorithm based on Dynamic Intableation Window (DIW) is proposed. Updating decisions are made according to the pixel-wise Dynamic Intableation Window. Chromaticity distortion is measured in an effective way. The real time experiments have been done on a surveillance system in indoor environments as well as outdoor environments. Moreover, a new gait recognition method based on hidden Markov models (HMMs) and Fourier descriptors (FD) is put forward. The body contours are processed by Fourier descriptors. K-means clustering method is used to analysis the image sequence within a gait cycle, and gait is represented by key stances. The hidden Markov models are applied to model the gait, where the key stances are considered as analogues of states of the HMMs while the distance vector sequence is considered as the observed process. Finally, the experimental results demonstrate that our approach using linear discriminating analysis and support vector machine has a better recognition effect than other similar methods.

- **Research on Face Recognition Based on Kernel Function(Wang lijuan)** [5]

Because of the clustering character of skin color in the color space, skin color area can be detected based on YUV color space, and the face in video from complex background can be detected based on skin color density combined with horizontal and vertical gray projection. Then, the method of combined kernel function is proposed to recognize face based on the traditional Principal Component Analysis and Fisher Discriminating Analysis in this paper. By using kernel function, face features are extracted in the high-dimension space by using linear discriminating analysis to form nonlinear optimal features. Further more, focus on small sample leaning problem, we propose a one-to-rest method to extend the capability of support vector machine. Combined with nearest neighborhood classifier, the extracted face features are trained and classified. Experiments with ORL face-database and the images we collect from video show that when adopting the kernel method in the process of face recognition, the efficiency of feature extraction and the generalization ability of classification will be improved significantly and the real-time character and the recognition rate of the system is enhanced greatly.

- **Study and Implementation of Control and Communication System for Multi-Agent Robot System(Liang feng)** [6]

First, calculating model base on the agent abstract description is proposed. This model is composed of agent core and several function modules. Communication model, communication method, communication language and the functions of communication servers can be establish by this model. Secondly, the control system of the robot is designed and realized. The control system hardware makes up of servo control module, cable communication module, wireless communication module, behavior data storage module and power module. Software is the core of the control system. In this software, Function modules are designed as sub-agent, and scheduled by the agent core. This software can be easily replanted in different micro controllers and configured for controlling 1 to 24 degree-of-freedom robots. Ant colony Algorithm is chosen for the real-time schedule of the MARS after compared several common real-time schedule schemes, and optimized for real-time schedule application. Finally, this paper propose the realize steps about the improved Ant colony Algorithm.

- **A Teaching Assistant System Based on Affective Modeling (Meng xiuyan)** [7]

Based on Psychology and the theory of Artificial Psychology, a humanistic computer teaching system is presented in this paper. The core of this system is the affective interaction between teacher and student. An emotion-learning model is developed. Emotion space, four kinds of basic emotion and basic learning psychology are defined according to the Emotion psychology Theory in this model. The mapping between emotion and learning psychology is also established. The student's psychology can get through the expression being processed by affective model, and the psychology can be evaluated to get a value for teacher. Finally, this system was realized by using the recognition method that is based on digital image processing technology.

4 Conclusion

Artificial psychology is still a new conception and the research on it is sill in a primary stage, but it has an exciting perspective. For the future, our research is going to concentrate on the following fields:

1. Application of artificial psychology in cognitive teaching.
2. Research of human behavior and psychology components in intelligent transportation systems.
3. Novel algorithms in expression recognition.
4. Domestic affective robots.
5. Construction of affective computing model.
6. Construction of integral model of artificial psychology.
7. Theoretical study and intercross application in multidisciplinary field such as information science, psychology and brain science.

Acknowledgments. This work was supported by the National Science Foundation of China (No.60573059) and 973 National Basic Research Program of China (No.2006CB303100).

References

1. Wang, Z., Xie, L.: Artificial Psychology-an Attainable Scientific Research on the Human Brain. IPMM'99(KEYNOTE PAPER), Honolulu, USA, pp. 10–15 (ISTP) (July 1999)
2. yun, M.: Research and Application of Personality Intelligent Fashion Shopping System Based on Network. Master's Thesis of USTB? (February 2004)
3. xuejing, G.: Research on Modeling Artificial Emotion Based on HMM and Techniques Correlated with Virtual Human. Master's Thesis of USTB? (January 2003)
4. hongzhe, H.: Study on Gait Feature Extraction and Human Identification Based on Computer Vision. Doctor's Thesis of USTB? (December 2003)
5. lijuan, W.: Research on Face Recognition Based on Kernel Function. Master's Thesis of USTB? (December 2006)
6. feng, L.: Study and Implementation of Control and Communication System for Multi-Agent Robot System. Master's Thesis of USTB? (December 2006)
7. xiuyan, M., zhiliang, W., guojiang, W.: The research of a teaching assistant system based on artificial psychology. In: Proceedings of Affective Computing and Intelligent Interaction - First International Conference, Beijing, China, pp. 614–621 (2005)

Human-Friendly HCI Method for the Control of Home Appliance

Seung-Eun Yang[1], Jun-Hyeong Do[2], Hyoyoung Jang[2], and Zeungnam Bien[2]

[1] Systems Engineering & Integration Department KOMPSAT-3 Program Office, KARI, 45 Eoeun-Dong, Daejeon, 305-333, Korea
[2] School of Electrical Engineering & Computer Science Division of Electrical Engineering, KAIST, 373-1 Guseong-dong, Daejeon, 305-701, Korea
seyang@kari.re.kr, {jhdo,hyjang}@ctrsys.kaist.ac.kr, zbien@ee.kaist.ac.kr

Abstract. This paper describes a HCI system for the control of home appliance which is focused on human friendliness. This system utilizes two USB cameras to enable a user to select home appliance easily by hand pointing gesture. We propose two different methods of storing the three dimensional position of home appliance by user. Home appliance selection method and feedback for user's wrong pointing direction are also described in this paper. Because of the low cost installation, simple operation and interactive feedback, our proposed system enhances the usability and human-friendliness.

Keywords: Human-friendly interface, hand gesture, 3D position recognition.

1 Introduction

Recently, many studies of HCI have been conducted with the development of high technology. Those cutting edge technology drives people tend to find something is more convenient to use. Especially, user friendly systems are essential for elderly and disabled people who are suffer from operating home appliance which contains complex functions. In the case of remote control, user has to find proper remote control for each appliance, and too many buttons are cumbersome to use.

As an alternative way to control home appliance, instead of remote control, some research on voice control of home appliance were developed [1, 2]. However, it bothers user because it is required to keep a sensitive microphone nearby the user. And it is not suitable to express spatial information. In this point, gesture based interfaces are superior over interfaces based on speech recognition.

We have already developed the 'Soft Remote Control System,' which is an interface to control multiple home appliances based on hand gesture using three CCD cameras attached on the ceiling [3-5]. But it costs a lot of money and time because it needs a technician to install the camera and the camera itself is expensive. In addition, if no response appears from the system due to user's wrong pointing direction, it may make user confused.

M.J. Smith, G. Salvendy (Eds.): Human Interface, Part I, HCII 2007, LNCS 4557, pp. 218–226, 2007.

In order to improve human-friendliness in utilizing the 'Soft Remote Control System,' we propose a novel system with feedback capability using two USB cameras. In the proposed system, user may install the system easily by just putting two USB cameras wherever he/she wants. In this case, the distance between cameras is more than 50 centimeter and the user is in the viewing range of the camera. A feedback for the user's wrong pointing direction enables the user properly select what he/she wants operate.

The rest of this paper is organized as follows: Section 2 introduces overall configuration of the improved Soft Remote Control System which is used as a natural means to operate home appliance using hand gesture. Section 3 provides details of the initialization procedure of the system including home appliance position storing process. Especially, we provide two different way of initialization method to enhance the usability. In Section 4, home appliance selection and feedback method for user's wrong pointing direction will be discussed. In Section 5, we present the experimental results and discussion. Finally in Section 6, we conclude this paper.

2 Advanced Soft Remote Control System

In the conventional soft remote control system, 3 ceiling-mounted CCD cameras are used [1, 6]. Therefore, only expert can install the system and it takes a lot of time. On the other hand, arbitrary user can install the advanced system simply by putting two USB cameras wherever he/she wants to set them up. The proposed system finds relative position between two cameras through a pattern which is attached on the side of one camera. With the information, it calculates 3D positions of an object. For the user friendliness, the system use simple hand pointing gesture to calculate 3D positions of home appliances. Because of the diversity of pointing gesture depending on people, we define the pointing command as stretching user's hand toward an object so the center of the object is concealed by the hand.

When a user wants to control certain home appliance, he/she points the appliance with his/her hand to select it. But sometimes, user fails to select proper appliance. In this case, the system finds the nearest home appliance and let user know the direction and distance from the pointed position to the nearest appliance position. This concept of feedback enriches the application of HCI and enables user to select proper appliance.

Figure 1 shows flow chart of home appliance position storing and recognition respectively. For face/hand detection and tracking, we adopt a dynamic cascade structure using multimodal cues which is used in the previous system [7]. At first, the system recognizes a specific pattern automatically which is attached on the side of a camera by panning reference camera. The reference camera is left sided camera in the view point of user and becomes center of the three dimensional coordinate axis. Then, it calculates distance and angle between two cameras based on the result from the pattern analysis. With the information, it can calculate 3D positions of the user's face and hand.

For 3D position storing of the home appliance, we propose two different methods. In the first method, two directional vectors which are defined by the position of user's face and hand are used. In this case, user has to move from one place to other place to calculate 3D positions of home appliance. In the other method, two dimensional estimated position of each home appliance by user and one directional vector are used. In this case, user doesn't have to move two places but has burden to let the system know the approximate position of home appliance.

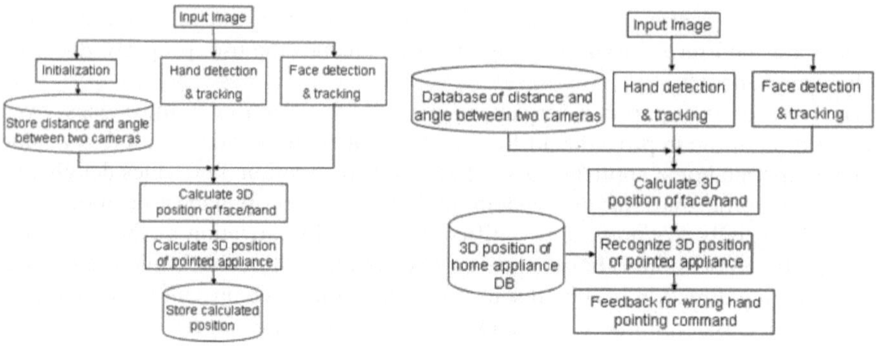

Fig. 1. Flow chart of home appliance position storing and home appliance selecting process

To enable user select home appliance, the system calculate directional vector which is defined by user's face and hand. The system select home appliance if the directional vector passes through the selection range of home appliance. If no appliance is selected, the system finds the nearest positioned appliance from the directional vector. Then let your know how to change his/her pointing direction to select the proper one.

3 Initialization Procedure of the System

3.1 Recognition of the Pattern

Because two USB cameras are placed in arbitrary position by user, the system needs information between two cameras to calculate 3D position of object. In this case, the system calculates distance and angle between two cameras using pattern. In order to

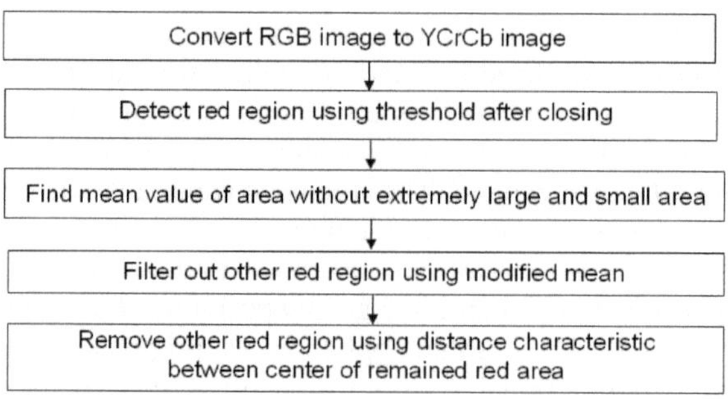

Fig. 2. Pattern recognition process

achieve the positional information in arbitrary environment, we utilize a pattern which consists of red colored circles as shown in Figure 5.

Figure 2 shows the pattern recognition procedure. Since RGB components are sensitive in light condition, we convert RGB image to YCrCb and discard Y component which contains luminance information. After split the image into Cr (red color information) and Cb (blue color information), detect red blobs using threshold. During the procedure, closing method is adopted to remove the fragments of red blobs.

In order to remove other red blobs that do not belong to the pattern, we use distances between each red blobs. Because the position of each red circle is fixed on the pattern, the length among each circle has common characteristics. As shown in Figure 3, we can easily discriminate between 9 circles of the pattern and other two red blobs using the characteristics.

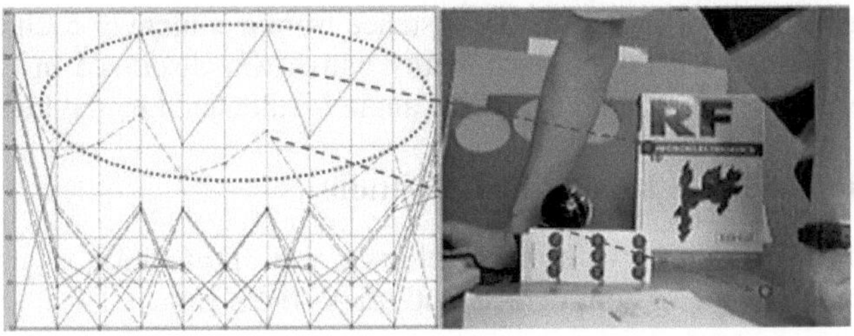

Fig. 3. Distance plot between centers of each red blob

3.2 Calculation of the Distance and Angle Between Two Cameras

After recognizing the pattern, we should calculate the distance and angle between two cameras. For this, we measured focal length of the camera using **GML MatLab Camera Calibration Toolbox** and **MATLAB™ (MATHWORKS, USA)** [8].

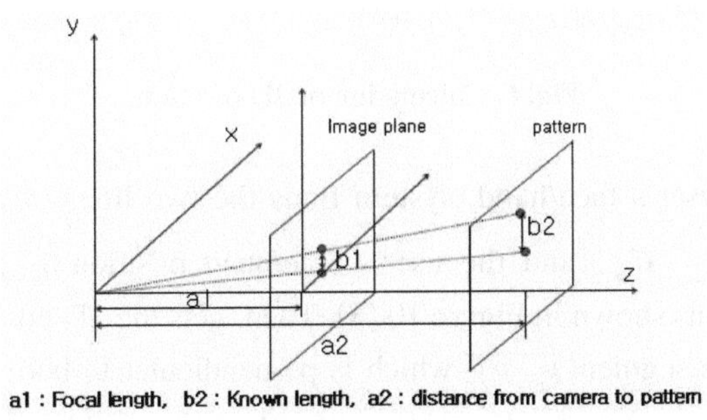

Fig. 4. Distance calculation from camera to pattern

Figure 4 shows the way to calculate the distance between reference camera and the pattern. Using focal length, we can calculate the length b1. And b2 is the distance between circles on the pattern, so, it is predefined value. Therefore, we can calculate the distance from the reference camera to the pattern by using proportionality.

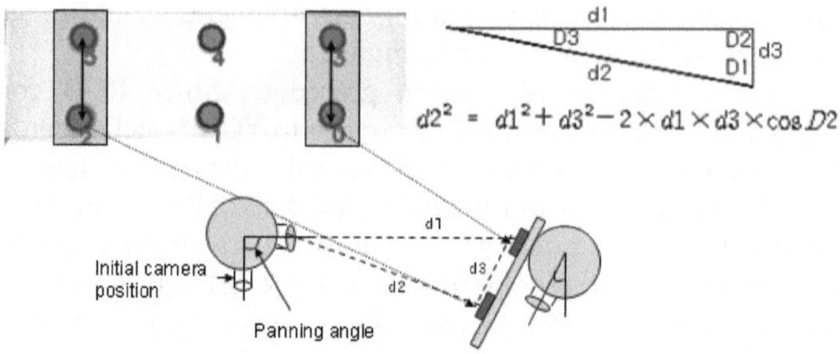

Fig. 5. Angle calculation using pattern

We can form a triangle using two columns on the pattern and the center point of reference camera shown in Figure 5. The distance from a camera to each column on the pattern can be calculated using proportionality which was discussed in Figure 4. If we know the three lengths of triangle, the inside angle is calculated by the law of cosines.

3.3 Calculation of Three Dimensional Positions

To calculate the 3D position of home appliance, the 3D position of user's face and hand is necessary. So, the system detects and tracks them first.

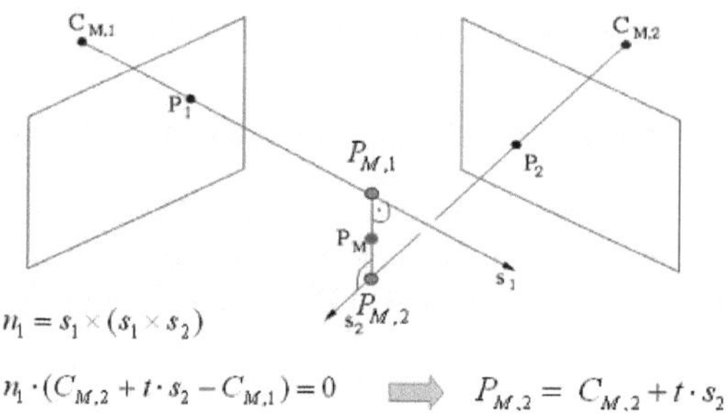

Fig. 6. Calculation of 3D position

After detecting user's face/hand, system finds the two lines, s_1 and s_2 from center of each camera $C_{M,1}$, $C_{M,2}$ and the user's face/hand position p_1, p_2 in each camera image, respectively as shown in Figure 10 [8]. Then, gets the 3D position by finding the midpoint of the line segment $\overline{P_{M,1}P_{M,2}}$ which is perpendicular to both s_1 and s_2. But in this case, if the two cameras are located too closely, error will be increased because of limited information from stereo image. So, for this system, we assume the minimum distance between cameras is 0.5m.

3.4 Two Different Methods for Home Appliance Storage

As mentioned in Section 2, we propose two different methods to store 3D position of home appliance. The first method is described in figure 7. To calculate 3D axis of a specific object, user should point to the object from different two positions. The system calculates 3D position of user's face and hand, then find directional vector. When user moves from position 1 to position 2, two USB cameras track user using pan-tilt function. In this case, to modify the relation about 3D axis, we defined translation matrix for panning and tilting angle.

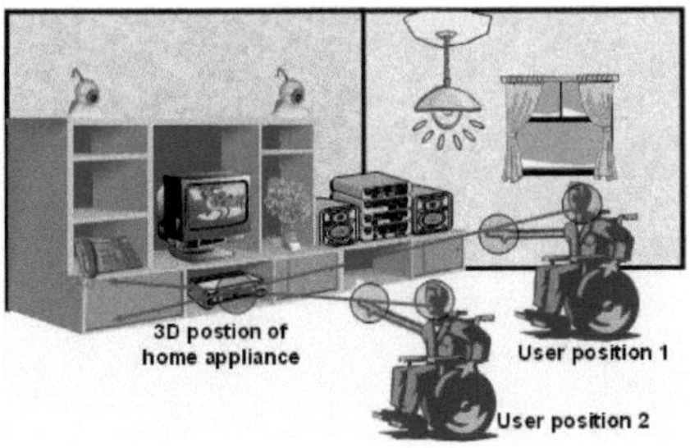

Fig. 7. Storing 3D position of home appliance by user's pointing command.

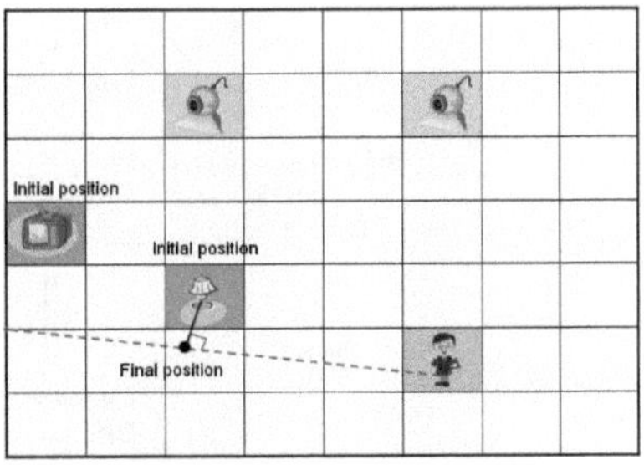

Fig. 8. Two dimensional grid plane which is used to store estimated position of home appliance.

But the first method gives burden to user to move one place to other place. To improve this incontinence, we devised a method which uses estimated position by user. As shown in figure 8, user places the relative position of appliances, camera and user on two-dimensional grid plane. When user points his/her hand to an object, the directional vector is mapped and displayed on the two dimensional grid plane and find a normal from the stored appliance position to the vector. The cross point which is described a dot in figure 8 becomes the 3D position of aimed appliance.

4 Home Appliance Selection and Feedback

To recognize the position of home appliance, system extends the directional vector which starts from user's face and ends at hand. The extension rate is determined by the length of user's face to each home appliance. At first, system calculates the distance between user and each appliance. And it detects the candidate position by extending directional vector for each extension rate. If the candidate position is included in the selection range, then the device is selected.

However, image data is very sensitive to environment and hand pointing command can't be done precisely every time. So, false recognition is inevitable. The most incident situation is that nothing is selected when user is pointing to a specific appliance. To reduce this problem and accommodate user interaction, the system provides feedback information. The feedback tells user the nearest positioned appliance from pointed 3D position by user. It also provides the information of direction and distance to select properly the nearest positioned appliance.

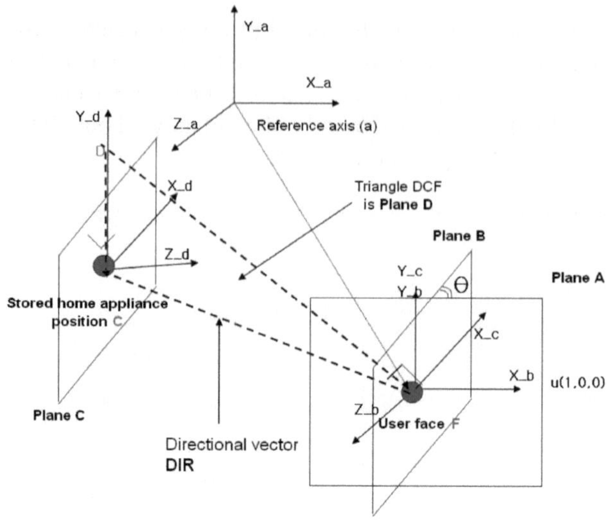

Fig. 9. Axis conversion from reference to user's view dependent axis

But the positional information should be transformed to an axis which is corresponds to user. Because the view point is changed according to user's seeing direction and position. Figure 9 shows the procedure of axis transformation where the axis_a is determined by the reference camera and axis_d is determined by user's view point. Every Y axis is parallel in the figure and we already know the directional vector from center of reference axis to stored home appliance position. So, the only information for axis transformation is the angle between axis_a and axis_d. The directional vector **DIR** is not always perpendicular to **Plane C**. But the imaginary plane DCF (**Plane D**) is always perpendicular both **Plane B** and **Plane C**. Because X_b and X_a are parallel, we only have to know the value of θ to formulate translation matrix between axis_a and axis_d by kinematics. Through calculating the angle $90^{o}+ \theta$ between unit directional vector **u(1,0,0)** and **Plane D**, it's possible to define translation matrix. Using the translation matrix, we can define user dependant axis from reference axis.

5 Experimental Result

To evaluate the performance of the system, we executed an experiment of storing home appliance position. The experimental results are shown in figure 10. User stored two home appliances by hand pointing command using different two methods in the same condition.

As we see in table 1, the first method looks superior to the second method. In the first method, user point to an object from different two places. In this case, user sees the object and points it directly. But in second method, user has to place each home appliance on a two dimensional grid plane. And there is no other tool to measure the distance and he/she has to place 3D positioned home appliance to 2D grid plane. Therefore, there is more chance of error.

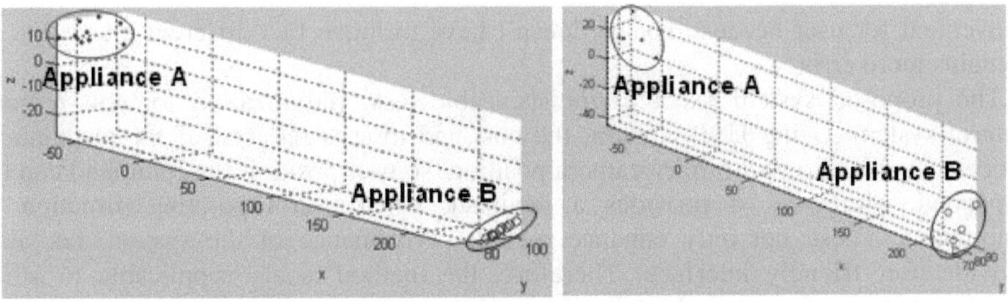

Fig. 10. Stored position of home appliance by the system

Table 1. Following table shows standard deviation of each axis of the stored home appliance position

	Standard deviation of appliance A			Standard deviation of appliance B		
	x	y	z	x	y	z
Method 1	10.04	7.06	2.80	1.816	4.345	1.944
Method 2	11.245	7.36	9.565	3.071	8.65	12.23

Table 2. Success rate of proper home appliance selection by user

	Appliance A	Appliance B
Success rate	96% (48/50)	88% (44/50)

Table 2 shows the recognition rate. User pointed 5 times to each appliance (Appliance A and Appliance B) respectively, and ten experiments were executed.

Table 3 shows the result of feedback. To evaluate the accuracy of feedback, we checked if the nearest positioned appliance is properly detected or not when no appliance is selected. And also, when the nearest positioned appliance is selected correctly, we observed whether the feedback direction is correct or not. Each experiment was done separately.

Table 3. Success rate of proper home appliance selection by user

	Nearest appliance detection	Feedback direction
Success rate	89% (89/100)	83% (83/100)

6 Concluding Remarks

In this paper, we proposed a method to store 3D position of a home appliance and recognize it by means of hand pointing command. For the storing the position of home appliance, we proposed two different methods. The first method gives burden user to move different place but more accurate. Otherwise, the second method is more convenient for user because he/she doesn't have to move two different places but it contains more error.

The proposed system executes indispensable task, initialization for soft remote control system. Using USB camera, the cost and time is reduced of the installation process. It also allows arbitrary camera position, so user position for command can be expanded. Moreover, it provides a feedback when user's pointing direction is ambiguous. These not only enhance overall performance of the system, but also provide user friendly interface. Therefore, the method is also applicable to other interface between human and intelligent system.

For the further study, we will focus on enhancing the robustness of the system by using user-adaptation technology and devising new pattern.

Acknowledgement. This Work is fully supported by the SRC/ERC program of MOST/KOSEF (Grant #R11-1999-008).

References

1. Jiang, H., Han, Z., Scuccess, P., Robidoux, S., Sun, Y.: Voice-activated environmental control system for persons with disabilities. In: Proc. of the IEEE 26th Annual Northeast Bioengineering Conference, pp. 167–169 (2000)
2. Lee, N.C., Keating, D.: Controllers for use by disabled people. Computing & Control Engineering Journal 5(3), 121–124 (1994)
3. Do, J.-H., Kim, J.-B., Park, K.-H., Bang, W.-C., Bien, Z.Z.: Soft Remote Control System using Hand Pointing Gesture. Int. Journal of Human-friendly Welfare Robotic Systems 3(1), 27–30 (2002)
4. Jung, J.-W., Do, J.-H., Kim, Y.-M., Suh, K.-S., Kim, D.-J., Bien, Z.: Advanced robotic residence for the elderly/the handicapped: realization and user evaluation. Proc. of the 9th Int. Conf. on Rehabilitation Robotics, pp. 492–495 (2005)
5. Dimitar, H., Stefanov, Z., Bien, Z., Bang, W.-C.: The Smart House for Older Persons and Persons With Physical Disabilities: Structure, Technology Arrangements, and Perspectives. IEEE Transactions on Neural Systems and Rehabilitation Engineering, vol. 12(2) (2004)
6. Do, J.-H., Jang, H., Jung, S.H., Jung, J., Bien, Z.: Soft Remote Control System in the Intelligent Sweet Home. In: Proc. Of IEEE/RSJ International Conference on Intelligent Robots and Systems, Edmonton, Canada, pp. 2193–2198 (2005)
7. Do, J.-H., Bien, Z.: A Dynamic Cascade Structure Using Multimodal Cues for Fast and Robust Face Detection in Videos. Pattern Recognition Letters (submitted 2005)
8. Kohler, M.: Vision Based Remote Control in Intelligent Home Environments. Proceedings of 3D Image Analysis and Synthesis, Erlangen, Germany, pp. 147–154 (1996)

A Framework for Enterprise Information Systems

Xi-Min Yang[1,2] and Chang-Sheng Xie[1]

[1] Department of Computer Science, Huazhong University of Science & Technology, Wuhan, China

[2] College of Computer Science, South-central University for Nationalities, Wuhan, China

yangximin@scuec.edu.cn, cs_xie@mail.hust.edu.cn

Abstract. In this paper, we present that Enterprise information systems (EIS) can be abstracted to a scenario in which users could schedule the task suites of controlled entities under enterprise security mechanism, and propose a novel Entity Driven Task Software Framework (EDTSF) for EIS. Based on existing hierarchical information system architectures, the EDTSF could clearly implements enterprise business partition, reduces coupling between data objects and tasks; it also increases flexibility and expansibility of the EIS by expanding service function of data object and reusing the scheduling mechanism at system level. We have noticed that little research has been done on the partition rules and methods of enterprise business. The EDTSF is a new software framework and its application results show that the EDTSF is an effective approach to analyzing, designing and implementing EIS.

Keywords: enterprise information system, framework, controlled entity, enterprise businesses partition.

1 Introduction

Enterprise information systems (EIS) is a computer application which contains a set of interrelated components that collect, storage, manipulate, and disseminate data and information, and provide feedback [1]. Today, enterprise requires that EIS must have higher sensitivity or dynamic adaptability in order to respond their demand changes over time. These demands include the adjustment of workflow, the alteration of management model, the evolution of user needs, or even if the redesign of process. Therefore, EIS should be updated quickly and reconfigured dynamically. Especially, EIS is increasingly apt to networking and integrating trends, what has gained more and more attention is how to protect enterprise's investments effectively and take full advantage of enterprise information resources.

Over the past decades, Research on EIS has made a number of important advancement with the evolvement of software engineering methodology and database techniques, such as CBSD (Component-Based Software Development)[2], SOA (Service Oriented Architecture)[3,4,5], AOIS (Agent-Oriented Information Systems)[6,7], and AUP (Agile Unified Process)[8,9]; But sometimes, developers would need to face the limited factors at every phase of the EIS lifecycle, which will be discussed in section 2. Currently, the development efficiency and software reuse

M.J. Smith, G. Salvendy (Eds.): Human Interface, Part I, HCII 2007, LNCS 4557, pp. 227–236, 2007.

are still two comparatively important aspects of the EIS research. According to the best of our knowledge, most of existing EIS development technologies focus on the research on the component description methods, the component composition technologies, and the connectors; but all of them are lacking in research on the partition rules and methods of enterprise business.

In this paper, we describe our approach to partitioning enterprise business. We pay much attention to how to apply the partition rules and methods of enterprise business to the hierarchical information system architecture; we also study approaches that can reduce the coupling of data and its processing procedures. As a matter of fact, all of (actual and logically defined) enterprise resources could be viewed as logic entities that are managed or controlled by operators. Therefore, we abstract EIS to a scenario in which users could schedule the task suites of controlled entities under enterprise security mechanism, and propose a novel Entity-Driven-Task Software Framework (EDTSF) for EIS of the existing hierarchical information system architectures. In EDTSF, it does not assign users to tasks for each application directly, instead, assigns users to entities and tasks to entities for each application. At the same time, data are only viewed as organizational form of enterprise information resource, and are degraded to the task's processing content. Any change in the user's duties and other requirements can be simply solved by assigning the user to another entity, reconfiguring the entity's tasks, and/or updating data processing procedures.

The remainder of this paper is organized as follows: Section 2 briefly analyzes some problems in the EIS development; Section 3 presents EDTSF, its architecture and framework development process; Section 4 compares EDTSF with other EIS development technologies; Section 5 gives a conclusion of this paper.

2 The Problem

Historically, the existing enterprise applications, especially those department applications that have been built at the different periods, usually are likely to be rather heterogeneous between them and even among one's components. The heterogeneities may exhibit themselves in the use of different techniques (i.e. programming languages, directive ideologies, and etc.), the availability on different hardware and operating system platforms and the use of different representations for the exchange of data[10~12]. So, the first task is the transplant of existing applications and their seamless integrating with each other. In the mean time, what must also be considered is how to furthest share and take full advantage of available enterprise resources, as well as how to protect enterprise's investments. This is also one of the reasons why the enterprise application integration (EAI) is a research hotspot at all times. Some of software development techniques (i.e. WebServices, SOA, ESB (Enterprise Service Bus) and Event Oriented Architectures) mainly focus on the approaches for exchanging data among applications, which could lead to increasing developer's effort in maintenance of data consistency and data integrality.

The methodology of CBSD has become one of major technologies used in today's EIS development and maintenance although it is yet to be mature. Most developers, especially primary ones, cannot gain practical advice in practice from CBSD except theoretic guidance. They must fill the gap with assumptions by themselves on the architecture when the Framework or Architecture was developing. While components can speed up the EIS project, the interface and communication mechanism in different components must be considered and translated. However, the more details are hidden in component or numbers of components were embedded in the projects, the more difficulties could be faced in the application maintaining and updating period. Furthermore, one that remains to be settled is how to implement the mapping from the problems of application domain to the realization of components, frameworks, and/or architectures [2,13~16].

Today, most of the EISs are generally built based on RDBMS and OO (Object-Oriented) technique. Application developers have been faced with a task of persistent object, including its storage approaches and ability to provide a transparent and robust applying method for developers. Aiming at the complexity of storage systems and the changeability of information organization structure, there are multiple ways that an object can be persisted, such as object/relational mapping and object-oriented databases. But relational databases lack many of the constructs necessary for true object persistence, and the difficulty in storing object-oriented data into a relational database, known as the object-relational impedance mismatch problem, still has not been completely solved. Object-oriented databases have advantages over relational databases because they are actually composed of objects instead of tables. Unfortunately, database schema migrations due to changes in class definitions can be costly, and support for ad-hoc queries is typically not as good as it is with relational databases[17]. Other options(i.e. approaches to embed SQL commands into objects, specialized data objects or data middleware, and etc.) probably result in tight coupling between data and object in which data are processed, or restrict their fitness, or affect the application system's performance with their complexities increasing.

When OO technique is used in the EIS development, enterprise business is often expressed as methods of objects. It means that the partition of enterprise business is achieved by aggregating methods into related objects. However, this processing method not only could affects the rationality of enterprise business representation, but also meets in turn the influence on object design, methods used in data expressing and data manipulating, and so on. For instance, it is not a good choice in digital hospital software to put the "register" method into patient-object, and is not conducive to the patient-object's persisting representation. To date, some approaches, including dynamic menu and role-based enterprise business partition [18,19], are expansion towards traditional methods in applications.

3 Entity Driven Task Information System Framework (EDTSF)

3.1 The Method and Definition

Enterprise employee can be regarded as relatively stable management units, and every management task can be regarded as a series of management and decision-making activities. These activities, in essence, are the rational usage or scheduling of enterprise resources under the direction and control of specific management ideas and models. All of the management objects, such as personnel, commodities, capital, information and so on, are generally called as the enterprise resources [19]. For example, in a hospital, the registrar's duty is management of registered evidence, and their work is to fill data items into registered evidence based on information what is patient provided. Therefore, all of actual and logically defined enterprise resources could be regarded as logic entities that are managed or controlled by operators. Thus, EIS could be abstracted to a scenario in which users could schedule the task suites of controlled entities under enterprise security mechanism. We then propose a novel framework for EIS named EDTSF based on this abstraction.

In EDTSF, enterprise information processing is performed through interactions among user, controlled-entity, task, and data-object. The first function of data-object is to describe the organization structure of enterprise information resources, which includes data items (name, type, length, etc.) and source (DBMS, tables or views, field list, agent-user info, etc.). The second function of data-object is to provide a consistent data accessing approach for other objects. The task is an all-encompassing procedure that collects, stores, manipulates, and disseminates information, and can result in properties change of controlled entity. To the goal, the task defines a set of data-object and the sequence of processing. One step is a single data processing function with format descriptions of input and output data. The controlled-entity is a real reflection to enterprise management object that could be an object container or an actual object. User's work is accomplished by gaining the management right of controlled-entity and scheduling tasks of controlled-entity. Finally, we define the EIS as following components:

- *U*, *E*, *T*, *D*, *P*, and *R* (users, controlled-entities, tasks, data-objects, processes, and rights, respectively);
- $UA \subseteq U \times E$, a many-to-many user to controlled-entity assignment relationship;
- $TA \subseteq T \times E$, a many-to-many task to controlled-entity assignment relationship;
- $PA \subseteq P \times T$, a many-to-many process to task assignment relationship;
- $DA \subseteq D \times E$, a many-to-many data-object to controlled-entity assignment relationship;
- $RA \subseteq U \times D \times R$, both user to data-object and data-object to right are many-to-many assignment relationships;
- $\forall p_i, \exists t_i e_i (t_i \in T \wedge e_i \in E \wedge (t_i, e_i) \in TA \wedge (p_i, t_i) \in PA), i = 1,2,3....$, ;
- $user: P \rightarrow U$, a function mapping each process p_i to the single user $user(p_i) \in \{u \mid (u, e_i) \in UA\}$; and
- $data\_objec\ t: P \rightarrow 2^D$, a function mapping each process p_i to a set of data objects $data\_objects(p_i) \subseteq \{d \mid (d, e_i) \in DA\}$ and process p_i has rights $\bigcup_{d \in data\_objects(p_i)} \{r \mid (user(p_i), d, r) \in RA\}$.

3.2 Architecture

The architecture of EDTSF, shown in figure 1, consists of application layer and presentation layer and data layer from the top to the bottom, and each layer can be subdivided relying on the actual applications.

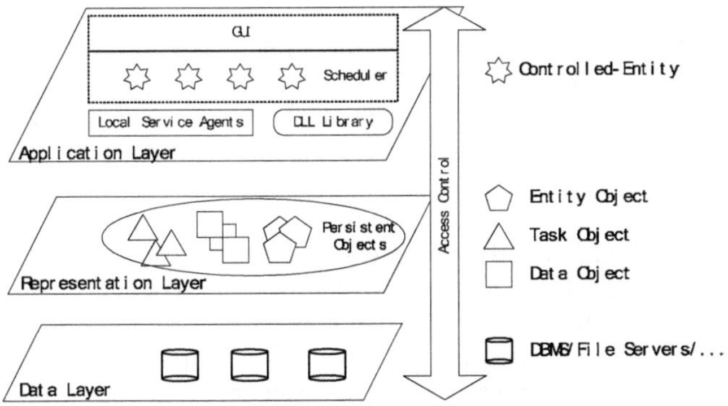

Fig. 1. The Architecture of EDTSF

Application layer: This layer consists of several function sub-layers: GUI, Controlled-entities centralized scheduler, local service agents and DLL library, represents all functions of the EIS. GUI is an interactive platform between user and application system. After a user logins, all controlled-entities managed by the user are shown in GUI's main window. Then the user can perform tasks by selecting one controlled-entity, scheduling its tasks, and browsing results. The Scheduler is a key component of application layer. After accepting user's instruction to execute one task of the selected controlled-entity, scheduler creates an instance of the controlled-entity and then sends a "start" message with one task's id to the controlled-entity. The local service agent is used for providing a uniform interface of presentation layer. The function of local service agent is transferring data access instructions of entities and tasks to presentation layer, and completing the inter-conversion of data format between presentation layer and objects in certain time. DLL library stores information of all data processing procedures, each procedure is relative to one processing step respectively. When receiving a request from the task object, DLL library automatically accomplishes finding and loading and running of a suitable procedure for the request.

Presentation layer: The key service of this layer is providing a presenting approach of persistence object for application layer. A group of associated or unassociated service objects, and coordinator objects, are expressed in this layer. For instance, data-object's duty is receiving requests from the application layer, creating connections to data sources, accessing data from these data sources, and converting data format to the objects required.

Data layer: This layer is a composition of various storage systems for enterprise, including local storage systems, remote storage systems, and others shared by existing applications or by cooperative organizations.

3.3 Development

The definition of controlled-entity and its task generally is stated as stability in EIS. Task's processing, workflow, and structure and format of data in task are easy to be changed. It is most important in EDTSF development to design and implement pervasive mechanism for tasks scheduling and approaches for presenting data-objects flexibly, namely the designs and implements of application layer and presentation layer. Figure 2 shows a simple development model of EDTSF we used in hospital information system project. A new object named operator is introduced for reducing quantity of controlled-entity. Here, the operator is a virtual object that can get the relative controlled-entities together according to partition rules of enterprise business duty. One operator can be assigned to different users and a user can be an owner of multi-operators. Relative to changeability of user duty, operator could keep stability between itself and controlled-entity. Therefore, administrator's efforts can be significantly alleviated. System configure is a tool in which administrator can define multifarious objects and set relationships between them and grant or revoke user's rights.

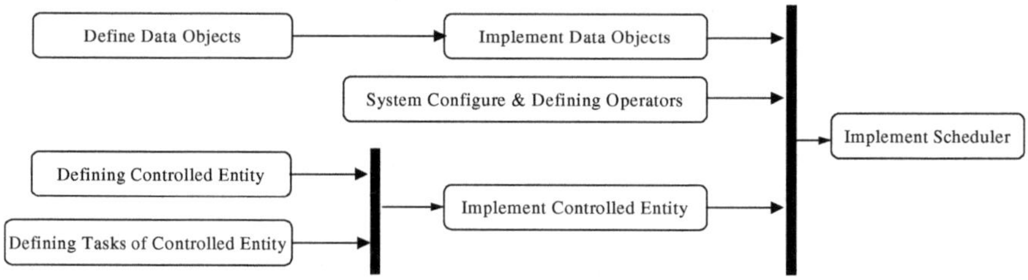

Fig. 2. Development Model of EDTSF

1) Definition of controlled entity

A clear characteristic of enterprise resources is that they would maintain stably in the certain time after they are vested in one department in enterprise, such as recipes and system users. At the other hand, one controlled-entity could also consist of other controlled-entities. To the goal, the concept of group is introduced, and one with the type of entity indicates itself being a controlled-entity, one with the type of group indicates itself being an entity group. So, we define controlled entity as follows:

```
Type = [Entity | Group];
CE = {ID, Name, Type, FGID, Icon, …, Description}.
```

Where, FGID is used to create hierarchical structure of controlled-entities, which is in response to organization form of enterprise resources. The value 0 indicates that the controlled-entity is a common entity, otherwise, it is one vested in a controlled-entity group.

2) Definition of data-object

The data-object's implement includes two missions. The first is the description of its data sources and each data's constituent parts. The question what should be considered in this part is the case of multi-data items combination. For the data what comprises multi-constituent parts, source of its each part and how to combine them all must be described. The second is to specialize data-object's applying properties, such as the form of output data (individual values or a dataset), the info of data item (i.e. name, display name, format, etc) and so on. Example 1 illustrates a definition of recipe data object based on XML.

Ex. 1. Partial definition of recipe data object based on XML

```
<DataObject ID=DO1011 name=Inpatient_recipe>
<Title>Inpatient Recipe</Title>
<Section name=Patient_info>
  <Item source=db_Patient Table=Inpatient_info>
   <Field=Patient_id>Inpatient ID</Field>
   <Field=Data_in format="yyyy-nn-dd hh:mm:ss">
    Enter Date
   </Field>
   ......
  </Item>
  <Item source=db_Patient table=Patient_info>
   <Field=Name>Name</Field>
   ......
  </Item>
</Section>
<Section name=Medicaments>
  ......
</Section>
<Section name=Physician_Info>
  ......
</Section>
<Section name=Finance_Info>
  ......
</Section>
</DataObject >
```

3) Implementation of controlled-entity

Known from the architecture of EDTSF, the key of implementing controlled-entity is task scheduling mechanism. A typical workflow of controlled-entity consists of three steps: data gaining, data processing and data storing, as shown in figure 3. The workflow's nonlinear execution would happen at time when data gaining and data storing need user's interactions to accomplish. Because EDTSF disperses these steps in different objects, the task scheduling mechanism is simplified to define the DLL exported function. For example, a exported function can be defined as follows:

```
Function Create(Task: TTask; OpID: Integer; UserID:
string): IStep;
```

Where Task is a task objects used by step process, OpID is a sequence number executed currently, UserID is user's identifier, and the function's result IStep is an interface definition. Methods the IStep interface included are:

```
IStep = interface(Iinterface)
   function RunStep: Boolean;
   function Abort: Boolean;
   ......
end;
```

Controlled-entity objects achieves its task scheduling by call methods of IStep interface implemented by task in relative event procedures.

4 Contributions

A kernel function in existing software architectures, such as SOA, CBSD and so on, is design enterprise business function module for reuse. Data representation middleware and Agent-based technologies are effective methods used to integrate generally heterogenic data source in enterprise. EDTSF, a hierarchical information system architecture employing entity driven task technology, combines the benefits of CBSD, SOA, AOIS and middleware technique in a manner that retain traceability and preserve correctness with respect to the original. EDTSF not only provides a method for implementing legible partition of enterprise businesses, but also decreases the coupling between data and data processing.

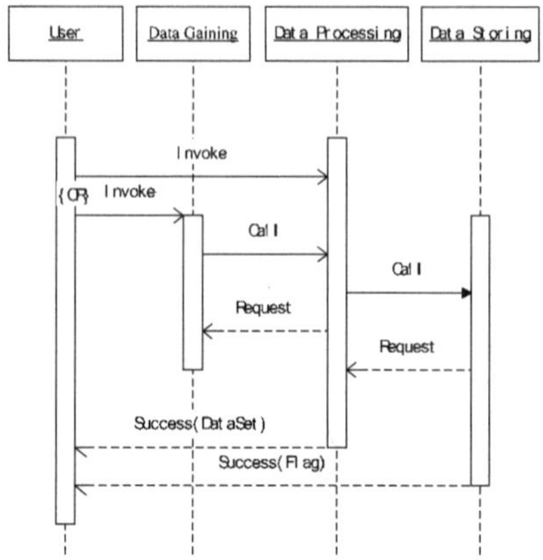

Fig. 3. Steps of a Single Task

Compared with other system developing technologies which we observed, the discussions on advantages of EDTSF as follows:

Using system configuration tool, EDTSF-based EIS can expediently plug-in/remove controlled-entities (include its tasks and the task's processes) and data-objects according to enterprise management demand change, such as business expansion and adjustment and so on. It means that EDTSF-based EIS can support enterprise business's plug and play. At the same time, application system's dynamic reconfiguration would be achieved too.

EDTSF has the universal flexibility because it implements reuse of the business partition approach and the task scheduling mechanism at system level, which is independent on the application environment. At the other hand, another one with distinction from component and service concepts in other technologies is the task object, which doesn't contain actual data it processed, only describes its required data and parses result back from data objects. The separation of task object and data-object decreases the coupling between them, and increases the responsibility of EDTSF in change on enterprise information structures and enterprise management demands. First, the recombination and the alteration of assignment for task object and data-object is easy-to-achieve. Second, if data description format doesn't change, the task object's modification cannot influence data-object, vice versa. If data description format changes, the only required to be modified is the parsing function that is global consistent. So, the application system would be continuously evolving tends to be prefect.

After introducing the concept of "controlled-entity", EDTSF could increase trustworthiness in enterprise business definitions, and make legible partition of enterprise business. In EDTSF, main form of the application system could be unified as a scenario in which controlled-entities are represented as a hierarchical structure based on the supply from enterprise resources management and the task scheduling approaches. Accordingly, EDTSF can reduce the complexity in traditional GUI such as MDI and SDI, and can support interface control technologies (i.e. dynamic menu, dynamically form loading, and so on) as well.

EDTSF can support the many kinds of development models and methods. The design and implement of task scheduling mechanism are simpler than the design and implement of connectors, which reduces the difficulty in developing EDTSF framework.

From the development processing view, other technologies referred in this paper realize the customization of existed components and services module at various levels, EDTSF realizes the customization of system configuration and its emphases is definition of controlled-entity, which include assignments from tasks to them, and data-objects. EDTSF is only an information system framework, and doesn't produce new system modules, but it may be used as an effective approach to exporting business and data components or service units. This is not the same as CBSD and SOA, which could bring new components or services unit in applications.

Besides the above benefits, the most important characteristic is that EDTSF would be used repetitiously after its development process had been completed, which would enable information system developers to pay much attention to analysis and design of enterprise businesses, and the Rapid Application Development (RAD) would also be archived.

5 Conclusions

In this paper, we abstract information systems to a scenario in which operator could schedule task suites of controlled-entities under enterprise security mechanism, and propose a novel entity driven task information development framework, named EDTSF, based on this abstraction. In EDTSF, the concept of "controlled-entity" is introduced to reflect the organization structure of enterprise resource and management duty in truth. EDTSF makes enterprise business partition more clearly, which is lacking in other information systems development technologies. EDTSF can speed up the development processing of application by archiving the reuse of task scheduling mechanism at system level. As an effective approached to information systems analyzing, designing, and implementing, EDTSF has been successfully used in some projects such as digital hospital software, student information system, and so on. In future, we will do research on the following aspects of EDTSF: 1) approaches to simplify express the data object; 2) optimization on business scheduling mechanism; 3) formalized definition of layer structure and; 4) access control technique.

References

1. Stair, R.M., Reynolds, G.W.: Principles of Information Systems (2005)
2. Feng, C., Qianxiang, W., et al.: An architecture-based approach for component-oriented development. In: Computer Software and Applications Conference, 2002. COMPSAC 2002. Proceedings. 26th Annual International, August 26-29, 2002, pp. 450–455 (2002)
3. Gold, N., Mohan, A., Knight, C., et al.: Understanding service-oriented software. Software(IEEE) 21(2), 71–77 (2004)

4. Chu, S.C.: From component-based to service oriented software architecture for healthcare. In: Enterprise networking and Computing in Healthcare Industry, HEALTHCOM 2005. Proceedings of 7th International Workshop on June 23-25, 2005, pp. 96–100 (2005)
5. Karhunen, H., Jantti, M., Eerola, A.: Service-oriented software engineering (SOSE) framework. In: Services Systems and Services Management, 2005. Proceedings of ICSSSM '05. 2005 International Conference, June 13-15, 2005, vol. 2, pp. 1199–1204 (2005)
6. Wagner, G.: Toward Agent-Oriented Information Systems. Technical report, Institute for Information, University of Leipzig (March 1999)
7. McDonald, J.T., Talbert, M.L., Deloach, S.A.: Heterogeneous DataBase Integration Using Agent-Oriented Information Systems
8. Stark, J.A., Crocker, R.: Trends in Software Process: The PSP and Agile Methods. IEEE Software, vol. 20(3), pp. 89–91
9. Alhir, S.S.: The Agile Unified Process (AUP), 2005, http://home.comcast.net/~salhir/TheAgileUnifiedProcess.PDF
10. Emmerich, W., Ellmer, E., Fieglein, H.: TIGRA - an architectural style for enterprise application modelling. In: 23rd international conference on Software engineering, Toronto, Ontario, Canada, ACM Press, New York (2001)
11. Majumdar, B., Dias, T., Mysore, U.: ESB: A bandwagon worth jumping on (May 2006), http://www.infosys.com/Technology/bandwagon-worth-jumping-on.pdf
12. Mosawi, A.A., Zhao, L., Macaulay, L.: A Model Driven Architecture for Enterprise Application Integration. In: Proceedings of the 39th Hawaii International Conference on System Sciences – 2006, IEEE, New York (2006)
13. Succi, G., Pedrycz, W., et al.: Package-oriented software engineering: a generic architecture. IT Professional 3(2), 29–36 (2001)
14. Xu, W., Yin, B.L., Li, Z.Y.: Research on the business component design of enterprise information system. Journal of Software 14(7), 1213–1220 (2003)
15. Kirk, D., Roper, M., Wood, M.: Defining the problems of framework reuse. In: Computer Software and Applications Conference, 2002. COMPSAC2002. Proceedings. 26th Annual International, August 26-29, 2002, pp. 623–626 (2002)
16. Chaudet, C., Greenwood, R.M, et al.: Architecture-driven software engineering: specifying, generating, and evolving component-based software systems. Software(IEEE) 147(6), 203–214 (2000)
17. Fontaine, A., Truong, A.-T., Manley, T.: A Survey of Strategies for Object Persistence. CSci 5802 - Spring (2006)
18. Ming-Hui, Y., Qi, F., Xue-Guang, C.: A Role Interactive Model Based on Method of information System Function Partition. Science & Technology Progress and Policy 9, 105–107 (2004)
19. Yue-ting, C., Xiao-dong, Z., Jin, D.: Research on the reconstructivity of agile supply chain management system. Journal of Tsinghua University 40(3), 68–71 (2000)

Information Behaviors of HCI Professionals: Design of Intuitive Reference System for Technologies

Eunkyung Yoo, Myunghyun Yoo, and Yongbeom Lee

Samsung Advanced Institute of Technology
P.O. Box 111 Suwon, 440-600, South Korea
{ek.yoo,mh.yoo,leey}@samsung.com

Abstract. Technology roadmaps are often referred for better decision making by HCI professionals who connect human factors with product development and innovation. We conducted user study that explore information seeking and tracing behaviors in using technology roadmap. The research revealed that HCI professionals exhibit distinctive patterns in using technology roadmap, depending on their expertise in technical knowledge and work experience. Finally, we designed new user interface of an interactive technology roadmap system based on the research findings. We demonstrated its usefulness in seeking task-dependent information, intuitiveness in information visualization, and easiness to use as a reference for technologies.

Keywords: Information behaviors, Interactive reference system, Interactive technology roadmap, User interface, Interaction design.

1 Introduction

HCI literature defines HCI professionals as bridge builders who deliver user-related information to the engineering process [2]. They need to consider tradeoffs such as technologies or business needs during user-centric design for new product development or innovation processes. A technology roadmap (TRM) is a tool HCI workers utilize for product decisions involving tradeoffs between human factors and technologies. Generic TRM is a time-based chart, comprising a number of layers that typically shows how a technology can be aligned to product, business strategy, and market opportunities [7]. A recent survey [7] confirms why most research on TRM has been concentrated in roadmapping process, instead of user interaction with TRM. It also answers why most TRM has static presentations even though it reflects diverse perspectives. These suggest that there is a need for study on user-centered interface design of TRM that is customized to user behaviors with respect information handling.

There are a number of studies on human information behaviors in various work contexts, [8] for example. However, there is little such literature in the context of HCI field. A recent article summarizes four information-seeking modes and its relation to the application design [5], but it mostly covers generic behaviors and web design.

In this paper, we focus on the information behaviors of the corporate HCI professionals and interactive visualization design. We describe user research on HCI

M.J. Smith, G. Salvendy (Eds.): Human Interface, Part I, HCII 2007, LNCS 4557, pp. 237–246, 2007.

workers, in terms of their needs and methods in seeking and tracing information on technologies. Based on the research findings, we propose a novel interaction design of interactive TRM system for HCI professionals. Finally, we present evaluation results on the usefulness, intuitiveness, and learning curve of our design.

2 Related Works

Our work builds on two different areas of related works. First is the design of interactive visualization for decision making tools, and the second is the studies on information seeking and tracing behaviors.

There are only a few examples of interactive visualization design for decision making tools. U.S navy developed a 3D system that visualizes undersea battlefield. This enables rapid and accurate decision making and planning. By tracking objects with consistent color scheme, the system supports easy navigation through a complex set of information [6]. In industry, Advanced Micro Devices Inc. [1] provides Flash-based interactive technology outlook which partially supports list of related products, and detailed information about selected items.

Previous research shows that there are various behavioral patterns in seeking and tracing information. A recent article establishes four different modes of approaching information [5]. Google-based intensive search, fact finding, and browsing behaviors for information gathering were also observed [4]. There are also diverse tracing methods for information re-use. With emergence of the web, people use functionalities of web browsers such as bookmark [3].

3 User Research

In this section, we present our in-depth user study on user tasks, information search patterns, and information tracing behaviors using TRM. The user sample is a representative set of HCI workers at Samsung. We employed a mock-workplace setting to obtain natural user behaviors with TRM. In the following, we first describe study methodologies and then present our research findings.

3.1 Methodologies

In this section, we describe our user study methods. We first delineate the user sample and the testing setup. Next we present the test procedures and analysis methods.

Our user sample consists of nine HCI professionals from six different business units, taking their expertise on TRM into account. They are mostly designers in emerging User Interface (UI) development and product UI planning, without engineering background. As shown in Table 1, participants have diverse range of work experience in HCI, ranging from 1 to 12 years. They also have at least ten years of web experience, and wide range of TRM experience from zero to 12 years. We divided their expertise level on TRM into novice (less than 3 years), intermediate (3 to 5 years), and expert (more than 6 years).

For test setup, we prepared a mock-workplace setup consisting of a computer and a printer. We surveyed four colleagues and found that computers are the predominant medium through which people access TRM. Thus we prepared HCI-related TRM in two different file formats - Acrobat PDF and Microsoft PowerPoint. One contained texts as well as images, while the other was text only. With pilot trials, we confirmed that indeed the two prepared digital formats comprise the majority of TRMs used at Samsung. We also found that in some cases people prefer to print the files for easier referral. To accommodate and encourage the most natural work behaviors using TRM, we allocated a printer at the mock-workplace setting.

The actual test consisted of interviews and observations while a participant used TRM. Each session took approximately an hour. For the interview and data analysis, we generated a set of open-ended questions. The question generation was dynamic in the sense that new questions were developed based on user responses. A representative set of major questions consistently employed during the tests are:

- To what degrees are you familiar with technical terms?
- What are the needs of your HCI related tasks for information on technologies?
- How do you find desired information with or without TRM?
- How do you keep the found technological information?
- How do you retrieve previously found information?

A facilitator led the session and two observers took notes, and the session was captured via digital audio recording and computer screen recording. Camtasia Studio, a screen recording software, was used to track down the participants' behaviors in using TRM. The observers recorded the participant's movement along TRM on prepared papers.

Features of the test data were extracted with respect to user expertise and job type, and then each user data was associated with appropriate stage of seeking and tracing process. Similarities between participants, information seeking and tracing patterns, as well as representative answers to test questions were formulated from the data. We analyzed how well the information needs, seeking and tracing behaviors observed in this test matched previous research findings.

Table 1. Summary description of participants

No	Job Title	Age	Exp. in HCI	Background	Experience on Roadmap	Years on Web
P1	Product UI planning	30	6	Industrial design	1 yr / Novice	10
P2	Product UI planning	33	1.5	Design planning	0 yr / Novice	13
P3	Product UI planning	35	12	Product design	12 yrs / Expert	13
P4	Emerging UI strategy	28	3	Graphic design	3 yrs / intermediate	10
P5	Product UI design	30	5	Cognitive science	4 yrs / intermediate	10
P6	Emerging UI design	32	6.5	Ergonomics	6.5 yrs / Expert	12
P7	Product UI planning	34	10	Industrial design	8 yrs / Expert	12
P8	Emerging UI planning	29	4	Product design	0 yr / Novice	8
P9	Emerging UI development	33	5	Computer science	5 yrs / intermediate	11

3.2 Findings

The result of this study suggests that an individual's level of technical expertise, type of HCI work, and web usage style are key influences on TRM usage behaviors. Specifically, the technical information-handling behaviors of the Samsung HCI professionals were highly dependent on their familiarity to technological terms on TRM, gained from work experiences. Another factor that influenced the TRM usage behavior was the web-surfing style in finding and tracing information.

Information Seeking. Our study shows that information seeking strategies are distinctively different by users' task and knowledge level on the relevant technology. This confirms Maurer's four modes of seeking information (2006).

Expertise. Expert users, with substantial TRM experience in HCI field, utilize TRM in order to acquire the possibility of technical implementation. They have clear project scope and are confident on what to find and how to approach the target information on TRM. As Maurer (2006) stated as 'known-item' mode and suggested several design approaches, expert users are aware of what they want, what words to use to describe it, and where to start to find it. Experts begin by pressing the short key combination for keyword search within TRM document, regardless of their file formats. When they find good matches, they start to explore details around the search results. In case there are no matches, they move onto the alternative step, 'Exploratory' mode. Experts' exploratory behavior is similar to that of intermediate users, but spends significantly less time. Moreover, expert users usually have better acquaintance to the technical terms, and they can better grasp what new terms mean. When confronted with highly technical terms, they employ a search engine to find out what it exactly means. To summarize, expert users have concrete methods for seeking target information with or without TRM.

Intermediate users have rough ideas of what they need to know but are not sure on what word to search for, as Maurer (2006) named this type of users as 'Exploratory'. Therefore, they begin by scanning through category, which is usually in the form of table of contents or bookmarks. When they encounter a seemingly-related terminology, they proceed to dissect the associated sub-categories and keep on narrowing down the categorical path. Once a relevant terminology is detected, the users expand the scope to neighboring contents. After exploring the neighborhood, they move on to unexplored categories and repeat the above steps until they find the target.

"According to my experiences in several types of technology roadmap, relevant information seems to be placed close to each other. Because I rarely find the target information at first stage, my strategy is to look for a keyword first, set it as an anchor and then explore further based on it." (Interview of Participant 9)

Novice users have vague ideas of what they need to know and where to start, due to lack of work experience and exposure to TRM. They exhibit similar patterns to the 'exploratory' mode in terms of browsing through the bigger categories and narrowing down to the details. However, novices spend more time and take larger number of steps, as they tend to examine the entire. They also spend more time to look for the meaning of unfamiliar terms.

All users search the internet for unfamiliar technological terms. They heavily rely on Google and Wikipedia. This is consistent with a recent study about search-based information gathering pattern with Google [4]. For more specific information, they look up corporate intranet.

Task. Information approach patterns are dependent on the users' tasks. The task of product UI planners is to generate new UI ideas for products to be released by certain timeline. For each targeted product, they consider industrial paradigm, market status, convergence issues, and so on. However, there is an important condition to meet – technologies should be mature enough for the production to meet the scheduled release. Therefore, on TRM, they look for a set of potential technologies that are feasible for implementation by the specified time.

On the other hand, the timeline of the emerging UI developers is more flexible. They usually look two to three years ahead for the advanced UI development. For example, when their project deadline is due at year 2009, they consider a series of technologies that will be usable for a UI development in 2009. See Fig. 1-D to compare the different timeline flexibility depending on the tasks.

Interestingly, both product UI planners and emerging UI developers emphasized the need for connecting with internal experts, since it is easier to reach and collaborate with them than those outside of their own organization.

"To get information about applicable technologies to implement in a product within 2-3 years, the best way is to have a meeting with engineers who develop those technologies. A few hour of discussion provides tacit knowledge within entire corporation on the technology, such as organizational strategies and vision." (Interview of Participant 7)

Information Tracing. Information tracing is defined as retrieving [5] what users have already seen. Results show that users have various methods to track the previously-seen information. Once finding desired information on TRM, 78% of the participants directly copy and paste the findings into their project documents. They claim that this method allows easier access to the findings. As for simple recording of the findings, users have several tactics to trace it. In our study, three users wrote down the list of technologies on their paper diary or on a digital file. Four users captured the screens and saved them with detailed file names, so that they can easily recognize the contents from the file name. Three of them utilized 'Snapshot Tool' of Acrobat (Reader) application to keep only the relevant parts. The other two users printed TRM out, highlighted the relevant technologies, and saved it in project folders.

As for detailed information of unfamiliar terms found on the web, all users first bookmark the website to keep their findings. They explain that bookmarking is their most usual behavior to store the information resources while surfing the internet. Similar to the keeping behavior on TRM mentioned above, most users recycle detailed information from the web in their documents or personal repositories for data storage.

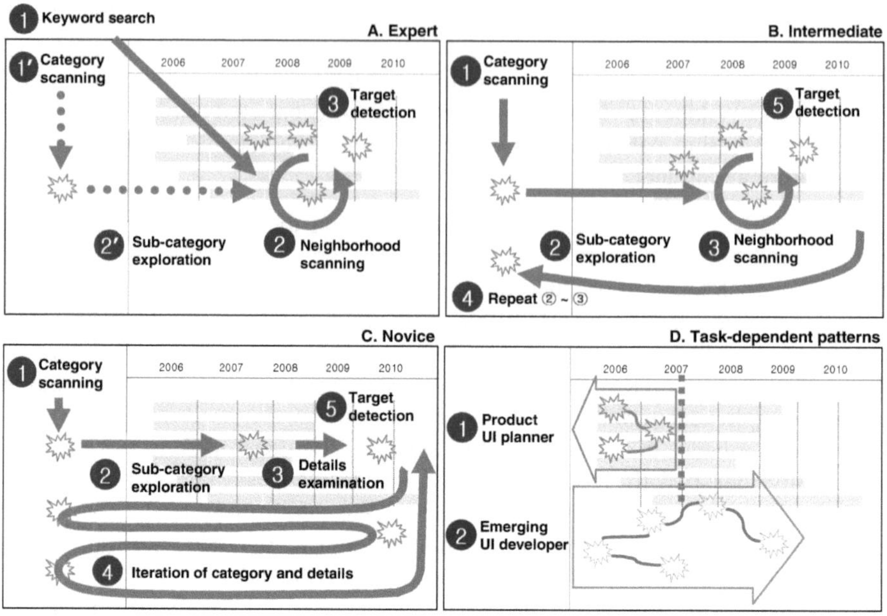

Fig. 1. Schematic diagrams on information-seeking patterns. Expertise-based patterns (A~C): expert (A), intermediate (B), and novice (C), and Task-dependent pattern (D).

4 Design of Interactive Reference System for Technologies

In this section we illustrate a novel user interface for interactive reference system on emerging UI technologies that we designed and implemented with Flash. We put together the test analysis results on HCI workers' information needs and behavioral patterns in intuitive ways. Reducing the learning curve for a new system is one of the most vital design factors for busy corporate users, considering the nature of the data-intensive system with enormous amount of technical information. In order to address this aspect, we adopted familiar terms from the web (e.g. bookmark, hyperlink), in naming our TRM functions. Our interaction designs are described with respect to the user needs on TRM and their behaviors in information seeking and tracing.

4.1 Expertise and Task-Focused Navigation

We propose three types of starting points - keyword search, overview explorer, and product navigator - for enhancing navigational experience and accuracy in finding the desired information. Each starting point reflects one of the primary information search methods of expert, intermediate, and novice or product planning users, as explained in Section 3.2. This multiple starting points provide users to choose their own approach to information seeking in TRM.

As for expert users, keyword search has the most potential for enhancing information seeking experience, by reducing the search time. Simple text input search returns the list of matched technologies. Matched keyword is highlighted in red, and the returns include not only the matched technological term but also related technologies within the same category, so that users can refer to them and explore further if necessary.

For the users in 'Exploratory' mode who begin the search by category scanning, there is the 'Overview' explorer that supports an easier exploration with assorted categorization. This supports hierarchical tree-search methods, with the first tier of primary categories shown initially, and lower-tier categories are shown once their corresponding higher tier category is selected.

The 'Product navigator' is designed for product UI planners or novice users who are unfamiliar with technical terms and need a browsing approach toward detailed technical information. It consists of sampled UI ideas in each product group. When a user selects a product idea, it shows the relevant list of technologies and their categories. Users can navigate further from the listed technologies.

4.2 Linkage of the Related Technologies

Our design provides links between related technologies, to support the search for a set of multiple technologies as well as the search patterns in which people seek related technologies from an anchor technology. The visualization of linkage between related technologies can be beneficial to those who are interested in further search. Selecting the icon 'R' in the label of a technology, which stands for 'relationship', displays lines that connect the anchor technology with the related ones.

4.3 Detailed Description

Our design also supports pop-up box containing details on the selected technology or category, to provide potential one-click answers. In particular, novice users can take the most benefit from this method, since such concise answers can satisfy the initial information need, as Maurer suggested (2006). This pop-up includes internal expert information, overview, technical specification, research examples from competitors or academia, patent strategies, images and diagrams, and other introductory information.

4.4 Bookmark

Bookmarks facilitate tracing and retrieving saved information. Descriptions of the bookmarks are automatically tagged, based on the stored content. Each icon has distinguishable design to indicate the saved content as an easy reminder for the users.

Each icon is automatically created depending on the elements in the saved content. Its variable features are layout, color, and tagged description. The layout is a miniature of the saved screen, so as to help users recognize previously seen content.

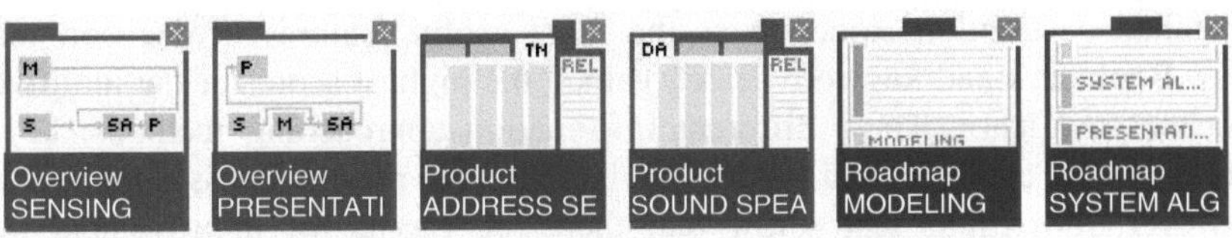

Fig. 2. Bookmark icon examples. Each color represents one of the starting points. Tags and inner images can differentiate different contents.

Each bookmark icon is automatically tagged for intuitive reminder, based on the stored content. TRM system creates a tag for each bookmark label using the associated category. For instance, when a user bookmark at the screen of sensing category in overview explorer, tagged icon is labeled as 'Overview-Sensing'.

Interaction with bookmark is done by a single click of mouse. One-click of the 'Bookmark' button at the top left of the screen generates a tagged icon with the current status of the screen, and places the icon at the bottom of the screen. Users can trace stored bookmarks by a simple click on the icons.

5 Evaluation

We evaluated the usefulness and intuitiveness of our proposed design as a technological reference system. We first present the evaluation process and then summarize the results.

Our design was evaluated via expert review by five corporate HCI professionals, in the form of a test questionnaire shown in Table 2. This questionnaire addresses the usefulness for their tasks, intuitiveness in searching for target information, and learning curve of the new interface. Seven-scale measure was used for the survey, and additional comments were obtained. The results are shown in Fig. 3.

Table 2. Expert review questionnaire

<table>
<tr><td colspan="2">[Overall evaluation]
1) Overall, the new interface of the technology roadmap is helpful in decision making for your usual task.
2) Overall, the new interface is easy to use in finding technological information.
3) Overall, I found the new interface of the technology roadmap easy to learn.</td></tr>
<tr><td colspan="2">[Detailed evaluation]
I found the following aspects are particularly:</td></tr>
<tr><td>4) useful for your task.</td><td rowspan="3">- bookmark (creation, retrieval, deletion)
- linkage between the related technologies
- keyword search
- overview browsing
- product navigation
- detailed description</td></tr>
<tr><td>5) intuitive to notice and understand the meanings.</td></tr>
<tr><td>6) easy to learn as new interface.</td></tr>
<tr><td colspan="2">[Comments]</td></tr>
</table>

On average, the participants voted 6.2 for usefulness, 5.8 for intuitiveness, and 5.7 for learning curve. In particular, bookmark and linkage between related technologies, and detailed description stood up as the most useful features in decision making process. Bookmark and keyword search gained the highest score for the intuitiveness and easiness to learn. Three different information approaching methods, keyword search, overview explorer, and product navigation were positively assessed by 72%.

However, the results also show that there are rooms for improvement on the proposed interaction design. There is a 28% gap between usefulness and the other two measures, intuitiveness and learning curve. According to the participants' comments, the 'R' icon is rather obscure and not very intuitive. Additionally, it does not solve the fundamental issue of text-intensive TRM. These results suggest the necessity of improvement in future studies.

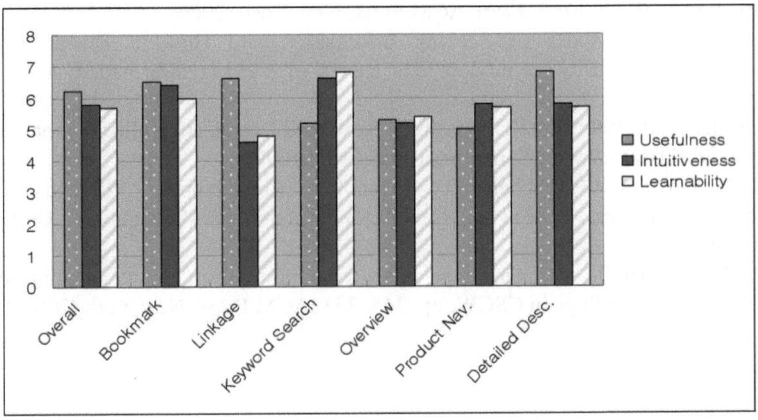

Fig. 3. Evaluation results in usefulness, intuitiveness, and learning curve of the new interface

6 Conclusion

We conducted user study with corporate HCI professionals to investigate their behaviors in seeking and tracing knowledge on technology roadmaps. Depending on the users' expertise level of technical knowledge and their task types, information seeking behavior was significantly different. Based on the research findings, we proposed an interaction design of an interactive technology roadmap system for HCI professionals. With new interaction methods, referring TRM can be more useful in HCI-related decision making. The new interactive reference system of the future UI technology overcomes the conventional static roadmaps by providing intuitive and easy-to-learn ways of information presentation.

In the future research, we need to develop intuitive methods to visualize unfamiliar but useful functions, such as icon and label that indicates the linkage to other technologies. We also need better visualization that covers text-intensive nature of TRM in order to improve the intuitiveness.

Acknowledgments. Thank for all the participants of the user research for this study.

References

1. Advanced Micro Devices, Inc.: Three Year Technology Outlook (n.d), http://www.amdcompare.com/prodoutlook/
2. Iivari, N.: Understanding the Work of an HCI Practitioner, Proceedings in the 4th Nordic conference NordiCHI (2006)

3. Jones, W., Dumais, S., Bruce, H.: Once Found, What Then? A Study of Keeping Behaviors in Personal Use of Web Information. In: Proceedings in ASIST 2002, Information Today Inc., pp. 391–402 (2002)
4. Kellar, M.: An Examination of User Behaviour during Web Information Tasks, Doctoral consortium in CHI, pp. 1763–1766. ACM Press, New York (2006)
5. Maurer, D.: Four Modes of Seeking Information and How to Design for Them, http://www.boxesandarrows.com/view/four_modes_of_seeking_information_and_how_to_design_for_them
6. Maxwell, D.: TALOSS: Three-Dimensional Advanced Localization Observation Submarine Software (2003), http://www.linuxjournal.com/article/6978
7. Phaal, R., Farrukh, C., Probert, D.: Technology Roadmapping: Linking Technology Resources to Business Objectives, Centre for Technology Management, University of Cambridge (2001)
8. Reddy, M., Dourish, P.: A finger on the pulse: temporal rhythms and information seeking in medical work. Proceedings in CSCW, pp. 344–353. ACM Press, New York (2002)

Part II

Visualising Information

3D World from 2D Photos

Takashi Aoki, Tomohiro Tanikawa, and Michitaka Hirose

The University of Tokyo
7-3-1 Hongo, Bunkyo-ku,
Tokyo, Japan
{takashi,tani,hirose}@cyber.t.u-tokyo.ac.jp

Abstract. A large number of the world's cultural heritage sites and landscapes have been lost over time due to the progress of urbanization. Digital archive projects that digitize these landscapes as virtual 3D worlds have become more popular. Although a large numbers of studies have been made on reconstructing 3D virtual worlds, the previous methods have been insufficient, because they require significant effort. In this study, we propose a new method of reconstructing a 3D virtual world only from photo images that requires little intervention. The idea is to reduce the learning curve of the software need and automate the method as much as possible so that we can digitize as many heritage sites as needed. In our approach, we first reconstruct 3D models from single 2D photos using an image based modeling and rendering(IBMR) technique. After reconstructing models from all the available photos, we connect the 3D models into one unified 3D virtual world. Specifically, we implemented a seamless connection algorithm that supports free viewpoint translation. And we demonstrated the reconstruction of part of a cultural heritage site based on our system.

Keywords: digital archive, image based modeling and rendering, occlusion interpolation, 3D model seamless connection.

1 Introduction

Time and the progress of urbanization are destroying a large number of cultural heritage sites and landscapes. In response, many researchers or enterprises have begun to archive them as digital data. These digital archive projects have become increasingly popular. These landscapes are digitized as computer graphics. In addition, today's virtual reality technology has made it possible to have a highly realistic experience of immersion in such landscapes using the archived data. Such systems are very useful for not only research or entertainment, but also educational purposes such as learning history. For example, Ando et. al. created a history learning system for elementary school students and demonstrated its effectiveness [1].

However, we are currently not able to create such highly realistic computer graphics without significant effort and it needs many highly skilled computer graphic designers and researchers. In particular, computer graphic designers have to master various 3D graphic tools and such training incurs significant costs. This is the reason why digital archiving could be expensive. Thus, these methods have only been adopted only for profitable business purposes such as video games or cinema films.

M.J. Smith, G. Salvendy (Eds.): Human Interface, Part I, HCII 2007, LNCS 4557, pp. 249–257, 2007.

Furthermore, the IBMR technique has recently become an important area of research in computer graphic. IBMR has enabled the representation of a photorealistic view without the use of a highly complicated geometric 3D model. The method simply requires a large number of photo images and a few 3D geometry data.

There are many kinds of IBMR technique today. However an ideal method has not been proposed yet. Current methods have both advantages and disadvantages. A simple IBMR technique is image morphing. Seitz et. al. presented view morphing, which is the most well-known image morphing method [2]. And another the well-known IBMR technique is the light-field rendering[3][4]. This technique is based on the idea that we are able to represent an arbitrary view if the recording of all light rays passing through all positions and undergoing all rotations is possible. In this approach, a free viewpoint image is computed by interpolation among a large number of viewpoint images that are taken from a camera array system. Although these methods require only images, they have several weaknesses, for example, they require large-scale equipment such as camera arrays. There is another IBMR technique that constructs a 3D geometric model from a video. This concept is called structure from motion(SfM). An example is the factorization method proposed by Tomasi ans Kanade[5]. This method enables computation of image feature points, the follow up of in every frame and the computation of the 3D positions of the image feature points.

On the other hand, Hoiem et. al proposed the "Automatic photo pop-up" method [6]. This is the starting point of our work. Their algorithm enables to automatically digitize a 3D model from a single 2D photo image. This is a revolutionary method because although a 2D photo image has no geometric information, the algorithm can infer 3D data of the image. Today's advanced computer vision techniques and pattern recognition algorithms made it possible to turn a 2D image into a 3D environment. However, the quality of these 3D models depends largely on the learning result of object recognition. As a result, this method has not been used universally.

In this paper, we present the reconstruction of a 3D model from a single 2D photo using the IBMR technique. After reconstructing the models from all the photos taken, we connect these 3D models into a unified virtual environment. Specifically, we applied the seamless-connection algorithm to support free viewpoint translation. To demonstrate the effectiveness of our method to create a landscape digital archive, we actually digitize the Japanese transportation museum which was closed on 15 May 2006. This is a famous Japanese historical building, however, because of the deterioration of the building, the museum is being transferred to a new building. We have taken about 40000 pictures to archive the interior and exterior of the building. In this paper, we demonstrate the reconstruction of a small part of the building based on the image data.

2 Constructing 3D World

We present the method of reconstructing a 3D model from 2D photos in this section. We first describe the assumptions of the virtual 3D world. Next, we present the details of our 3D reconstruction algorithm.

2.1 Assumptions

In our approach, we set up the 3D model coordinates as follows.

- The positive Y-axis is the up direction and is perpendicular to the ground.
- The Z-axis is mapped on the shooting aspect line and the negative direction is toward the front.
- The X-axis is from left to right.

We assume that 3D models follow the following:

- The 3D model is constructed with a single horizontal ground surface.
- Foreground objects stand on the ground surface perpendicularly.
- The ground surface is always the plane. (y=0)
- A foreground object may be constructed from one or more vertical surfaces.
- The Angles of view (horizontal and vertical) are known.
- The height of the shooting viewpoint is known.
- The vanishing line in the photo image is known.

The first two assumptions have an important role in our approach, because they limit the variability of the calculated depth to only one value. Using this technique, we can specify the depth of foreground object surfaces. Figure 1 shows the concept of the 3D model reconstructed by our method.

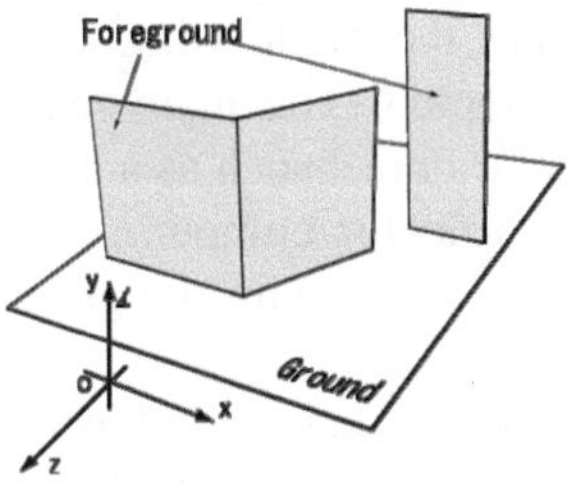

Fig. 1. Assumptions of 3D model

2.2 3D Reconstruction

We describe the method of 3D reconstruction from single 2D photos in this section. This approach requires two input images. One is a photo image and another is a segmented-region image which is created manually. Figure 2 shows an example of both images. From these images and parameters (angle of view, eye point height and vanishing line position), we compute the 3D geometry.

In particular, the segmented-region image is based on the following rules.

- The red region is the ground area in a photo image.
- The blue region is the sky area in a photo image.
- Other colors represent foreground objects in a photo image.
- A single surface in the 3D world is segmented with a single color.
- These colors do not represent metainformation such as the depth of a surface or other properties. They only represent an area in a image.

Fig. 2. The example of input images (left: photo image, right: segmented-region image)

Automatic image segmentation is currently a popular topic in computer vision, and much effort has been made to develop automatic image segmentation. However, images cannot be segmented with sufficient quality. Although the "Automatic Photo Pop-up" method has a full automated segmentation algorithm [6], it is not suitable for some situations. Thus, our approach requires little interaction during image segmentation.

To compute the 3D surface geometry, we first calculate the "feature edges" of each segmented region (shown in the left of Figure 3). The left and right edges are always vertical and located on the extreme left and right of the region respectively. The gradients of the top and bottom edges are computed by histogram analysis. As the right of shown in Figure 3, the sample data of this histogram are the gradients of line segments that are computed by subdividing the top or bottom curve. If the variance of the histogram is relatively small, the mode value is adopted for the edge gradient, otherwise the mean value is adopted. The main reason for this is to compute the precise feature edges of a irregular object such as a tree or a person.

Feature edges are important role for computing 3D geometry. In particular, the bottom edge represents a tangential line with the ground and is used to compute the depth of a surface.

Next, we compute the depth of a foreground object surface from the feature edges. In Figure 4, we show the 3D geometric relationship between a 2D photo and a 3D foreground object.

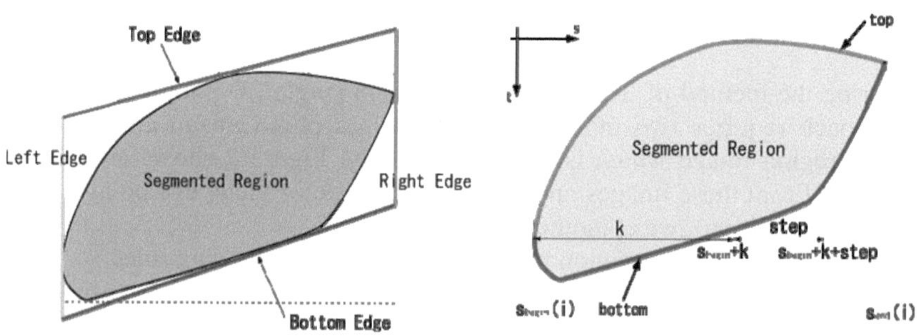

Fig. 3. Feature edges (left: concept, right: how to compute edge gradient)

As shown in Figure 4, we can calculate the depth of a foreground object surface Z_{3d}:

$$Z_{3d} = L \tan \psi \quad (1)$$

where L is the eye point height. Also θ , ϕ and ψ can be computed:

$$\theta = \arctan \frac{h}{(H/2)(1/\tan(\alpha_{Vaov}))}, \quad \phi = \arctan \frac{h_g}{(H/2)(1/\tan(\alpha_{Vaov}))}, \quad \psi = \frac{\pi}{2} - (\theta - \phi) \quad (2)$$

Where H is the photo image height and α_{Vaov} is the vertical angle of view. In the same way, other vector elements can be computed.

Furthermore, we can calculate texture map coordinates by projecting the 2D photo image to the 3D geometry model. However, this will not be described in this paper due to lack of space.

Finally, we create texture images for each object (the ground, sky and foreground objects).

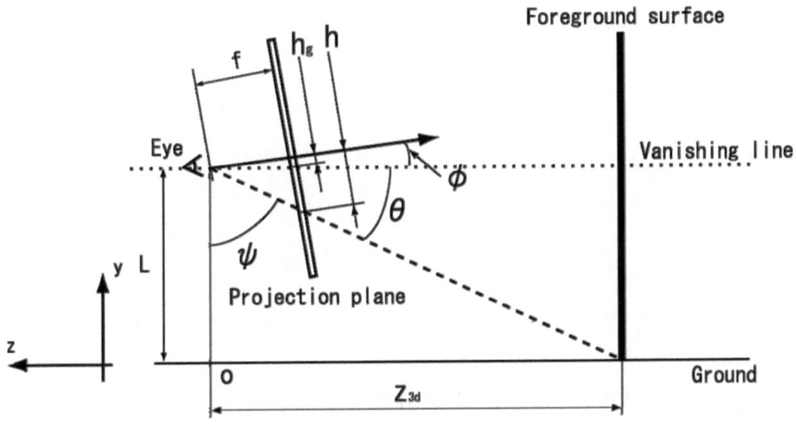

Fig. 4. Geometric relationship

2.3 Occluded-Area Interpolation

In our method, we apply texture image interpolation in accordance with the depth of the surface. A 3D model constructed by the method described in 2.2 has occluded texture image gaps. This approach fills these occluded gaps and makes it possible to support the free viewpoint translation and rotation in the virtual world.

There are two possible situations that we must consider before we fill the texture gap. One situation is when the texture image is occluded by a foreground object resulting in a texture gap. The other situation is when the texture image contains a gap or a hole. We employ an algorithm that can check whether the gap is the result of an occlusion. If it is the result of an occlusion, we employ the occluded-area interpolation algorithm to fill the gap.

This algorithm is simple. We first create a filling map that indicates the occluded area in a texture image. Figure 5 shows the concept of our algorithm. We scan a texture image over each coordinate (s and t). If there are colored pixels on a scan line and gaps between the extreme left and right, we indicate the gaps as interpolation candidates and we compute the depth value of each candidate pixel for regions i and j (see Figure 5). Comparing these two depth values, if the depth of region i is larger than that of region j, the candidate is occluded by an object in region j, otherwise it is not occluded and region i has a gap or a hole. By applying this algorithm to all the texture images, we can compute interpolation maps.

We fill the texture image gaps in accordance with the interpolation maps. We employ the back projection for lost pixels(BPLP) method on the eigenspace" as an interpolation algorithm [7]. This algorithm fills image gaps based on the local self-similarity of an image, i.e., one local region in an image is similar to another local region. The reason why we chose the BPLP method is that it only requires a source image as input information and takes relatively little time to carry out the process. In our approach, we must interpolate more than one texture image; thus, this algorithm suits our approach because the process is simple and relatively quick.

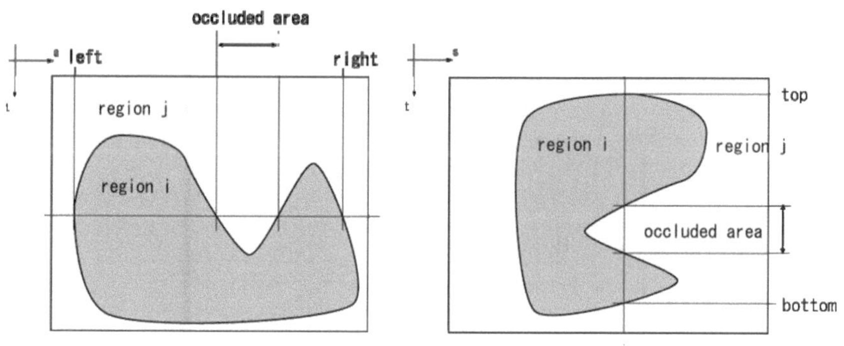

Fig. 5. Concept of interpolation map computation

2.4 Seamless Connection

A 3D model reconstructed from single 2D images has a small view-translation capacity. To solve this problem, we propose a 3D model connection method. Our approach is to switch seamlessly from one 3D model to another, similar to a picture-story show. By switching in accordance with the user's view-point position, we can represent a large 3D world.

Our method has 3 steps:

1. Link a 3D model and its metainformation (the position and rotation from where the photo is shot).
2. Search the 3D model from a database in accordance with the user's view position and rotation.
3. Seamless rendering.

We first link a 3D model and its metainformation (the shot position and rotation). This information is required for the searching at the next step. In our approach, we link the model by describing a 3D model data file path and its metainformation in one XML file.

Next, we search the displayed 3D models in accordance with the user's view information and 3D model metainformation. Our search method is based on an evaluation function. We calculate evaluation values for all 3D models and display a certain number of top scoring 3D models. The evaluation functions adopted here are

$$D_i = \cos\left(\frac{\sqrt{(x_e - x_{mi})^2 + (y_e - y_{mi})^2}}{P_{radius}} \cdot \frac{\pi}{2}\right) \tag{3}$$

$$R_i = \cos\left(\frac{1 - \cos(\theta_e - \theta_{mi})}{P_{angle}} \cdot \frac{\pi}{2}\right) \tag{4}$$

D_i evaluates the positional distance and R_i evaluates the rotational distance. (x_e, y_e) and θ_e are the user's view point position and rotation, respectively. And (x_{mi}, y_{mi}) and θ_{mi} are the 3D model (indexed i) position and rotation, respectively. P_{radius} and P_{angle} are parameters. These parameters depend on the 3D model dominant density in the virtual 3D environment. If the density of an area is high, these parameters should be set small; otherwise they should be set large. The output evaluation value is the product of D_i and R_i.

For rendering, we use the alpha-blending technique for connection. In this method, the blending ratio is computed by normalizing the evaluation values of the displayed 3D models. In addition, we edit the texture image alpha values:

$$\alpha(x, y) = A_x \sin\left(\frac{x}{W/2} \cdot \frac{\pi}{2}\right) \cdot A_y \sin\left(\frac{y}{H/2} \cdot \frac{\pi}{2}\right) \tag{5}$$

where W is the image width, H is the image height and A_x and A_y are the peak values of alpha. This operation makes it possible to form seamless boundaries, because the sin function is twice differentiable and people cannot recognize the differences in an overlapping area.

3 Experiments and Results

3.1 Occluded-Area Interpolation

Figure 6 shows the results of 3D reconstruction (without and with interpolation) from a single 2D photo. This demonstrates that our occlusion interpolation method is effective.

In our implementation, which uses OpenCV and Lapack libraries, it takes about 5 minutes to reconstruct a 3D model by including interpolation starting from a 3264x2448 input image on a Pentium 4 3.2 GHz computer. However, without interpolation, it takes about 10 seconds or less. This is because the interpolation method requires a principal component analysis(PCA) of a large number of dimensions.

Fig. 6. Result of 3D reconstruction (left: without interpolation, right: with interpolation)

3.2 3D Virtual World

Figure 7 shows images of the 3D virtual world. This demonstrates the effectiveness of our method at reconstructing a photorealistic 3D virtual world.

Fig. 7. Images of 3D virtual world showing seamless connection

4 Conclusion

In this paper, we proposed a method that can reconstruct a 3D virtual world only from photo images. By digitizing some parts of the Japanese transportation museum, we demonstrated the effectiveness of our method at digital archiving.

Our occluded-area interpolation algorithm is simple, but it can fill gaps in the texture image and makes it possible to remove the view translation and rotation limitations in 3D virtual worlds. Furthermore, our seamless connection method allows users to walk around large-space 3D virtual worlds.

One future aim of this research is to improve the 3D world quality. The 3D model reconstructed using our method cannot represent a curved surface and has a few computational errors under certain circumstances. In addition, there are some situations for which the BPLP interpolation method does not work well.

Another direction for future research will be its expansion into the use of the World Wide Web. This would allow the development of community spaces dedicated to photorealistic 3D virtual space creation to encourage the participation of the general public.

References

1. Ando, T., Yoshida, K., Tanikawa, T., Wang, Y., Yamashita, J., Kuzuoka, H., Hirose, M.: Proto-type Educational Contents by using Scalable VR System Historical Learning in Copan Ruins of Maya Civilization. In Trans. VRSJ, vol. 8(1) (2003)
2. Seitz, S.M., Dyer, C.R.: View Morphing. In: Proc. 23rd Annual Conf. Computer Graphics and Interactive Techniques, pp. 75–82 (1996)
3. Levoy, M., Hanrahan, P.: Light Field Rendering. In: Levoy, M., Hanrahan, P. (eds.) Proc. 23rd Annual Conf. Computer Graphics and Interactive Techniques, pp. 31–42 (1996)
4. Gortler, J,S., Grzeszczuk, R., Szeliski, R,, Cohen, F.M.: The Lumigraph. In: Proc. 23rd Annual Conf. Computer Graphics and Interactive Techniques, pp. 43–54 (1996)
5. Tomasi, C., Kanade, T.: Shape and Motion Without Depth. In: Proc. 3rd International Conf. Computer Vision, pp. 137–154 (1990)
6. Hoiem, D., Efros, A.A., Heber, M.: Automatic Photo Pop-up. In: Proc. ACM SIGGRAPH 2005, pp. 577–584 (2005)
7. Amano, T., Sato, Y.: Image Interpolation Using BPLP Method on the Eigenspace. in Trans. IEICE (D-II) J85(3), 457–465 (2002)

An Interactive Approach to Display Large Sets of Association Rules

Olivier Couturier[1], José Rouillard[2], and Vincent Chevrin[2]

[1] Centre de Recherche en Informatique de Lens (CRIL) – IUT de Lens, Rue de l'université, SP 18, F-62307 Lens Cedex, France
couturier@cril.univ-artois.fr
[2] Laboratoire Trigone/LIFL, CUEEP, Bâtiment B6, Cité Scientifique, F-59655, Villeneuve d'Ascq Cedex, France
{jose.rouillard,vincent.chevrin}@univ-lille1.fr

Abstract. Knowledge Discovery in Databases (KDD) is an active research domain. Due to the number of large databases, various data mining methods were developed. Those tools can generate a large amount of knowledge that needs more advanced tools to be explored. We focus on association rules mining such as "If Antecedent then Conclusion" and more particularly on rules visualization during the post processing stage in order to help expert's analysis. An association rule is mainly calculated depending on two user-specified metrics: *support* and *confidence*. All current representations present a common limitation which is effective on small data quantities. We introduced a new interactive approach which combines both a global representation (2D matrix) and a detailed representation (Fisheyes view) in order to display large sets of association rules.

Keywords: Knowledge Discovery in Databases (KDD), Human Computer Interaction (HCI), Visualization.

1 Introduction

In front of the increasing number of large databases, extracting useful information is a difficult and open problem. This is the goal of an active research domain: Knowledge Discovery in Databases (KDD). KDD techniques have been proposed and studied to help users to understand better and scrutinize huge amounts of collected and stored data [10]. KDD is a new hope for companies which use methods (e.g. statistical methods) that do not allow to tackle large amounts of data. Currently, commercial KDD tools panel develops quickly and some of these tools are marketed such as Purple Insight[1]. However, they are generally complex to use and they are not flexible depending on the user's problem. We focused on association rules mining (ARM) [1] but we oriented our work about the interaction between the user and a KDD process.

[1] http://www.purpleinsight.com/

M.J. Smith, G. Salvendy (Eds.): Human Interface, Part I, HCII 2007, LNCS 4557, pp. 258–267, 2007.

Precursory works are rather old because one of the first methods of mining correlations between Boolean values is the GUHA (General Unary Hypotheses Automaton) method [9]. Association rules interest was started three decades later thanks to the first large databases including commercial transactions [1]. The aim is to obtain rules such as "If *Antecedent* then *Conclusion*". This problem is also called *market basket analysis* and it is the starting point of ARM. In this case, each basket is relevant for one customer[2] depending on his needs and desires but if the supermarket tackles all baskets simultaneously, useful information can be extracted and exploited. All customers are different and buy different products in different quantities. However, market basket analysis consists in studing customer's behaviors as well as the factors which push them to carry out a kind of purchase. It allows to study what kind of products are bought together time with other products and consequently, to adapt a corresponding promotional campaign. The following simple example "If *Smoker* Then *Cholesterol* (75%)" means that a person who smokes has 75% of risk to have too much cholesterol. Although this method is initially planned for the great distribution sector, it can apply to other fields. The approach remains identical whatever the studied field: to propose models, tools and transdisciplinary methods in order to help the expert's analysis[3] in order to take the good decision.

Several works on Human Computer Interaction (HCI) are focused on Visual Information-Seeking Mantra illustrated by Shneiderman: *"Overview first, zoom and filter, then details on demand"* [22]. First of all, graphical tools must provide a global view of the system in order to deduce the main important points. The user must be able to determine the starting point of his analysis thanks to this global view. During his analysis, he can explore in-depth particular areas if he wishes so. Indeed, all details don't need to be displayed at the same time. Unfortunately, all current representations do not respect this crucial point, which is necessary in order to visualize large sets of knowledge. In addition, we need to display all metrics thanks to various colors pallets. Current tools still appear limited to display more than two metrics. Our purpose is how to obtain a representation adapted to visualization of large sets of association rules by jointly presenting a general view (the global) and a sight targeted on one or more particular elements (the detail)? This question constitutes the starting point of this work. The key to the success of a visualization prototype is a full compliance with such recommendations. We propose a hybrid visualization, which is composed of a 2D matrix to display an overview of our rules, and a fisheyes view to detail particular information. In the following, we will present our work corresponding to this research area.

This article is structured as follows: the second section introduces the framework of this work and presents in details ARM (Association rules mining) which constitutes the heart of our work. The third section describes our work which merges both HCI and KDD concepts based on human factors. This merging is crucial for

---

[2] Until the end of this paper, we consider one customer both man or woman

[3] In our work, the final user is a domain's expert. We will use independently the terms "expert" and "user".

relevant decisions support systems success. Finally, we conclude this article and propose some research ideas as perspectives.

2 Problem

This section formally introduces the association rules problem within a KDD process and it describes our motivations.

2.1 Knowledge Discovery in Databases (KDD)

In front of the increasing number of large databases, extracting useful information is a difficult and open problem. This is the goal of an active research domain: Knowledge Discovery in Databases (KDD). Nowadays, information which circulates in the whole world is mainly stored in digital form. Indeed, few years ago, a Berkeley University project estimated that in the world, the volume of annually generated data is equal to about one exa-byte[4] (i.e. 1 billion of gigabytes). Among these data, 99,997 % are available in digital form, according to [13] quoted in [18]. In commercial companies such as bank, insurances or distribution field, large amounts of customers' data are collected and they are not always exploited thereafter. How to make these data profitable in shorter running time? Indeed, current traditional request, as SQL (Structured Query Language) or OLAP[5] (On-Line Analytical Processing) are now

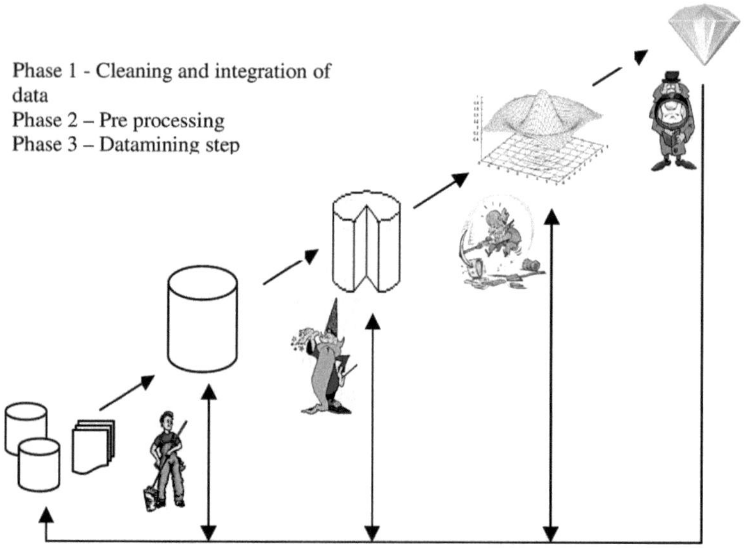

Fig. 1. KDD process

[4] 1 exa-byte (Eo) = 2^{60} bytes ; 1 zetta-byte (Zo) = 2^{70} bytes ; 1 yotta-byte (Yo) = 2^{80} bytes.

[5] Decision support software that allows the user to quickly analyze information that has been summarized into multidimensional views and hierarchies.

limited due to the increasing collection of large databases. To answer this problem, KDD is a new hope for companies which use methods (e.g. statistical methods) that do not allow to tackle large quantities of data (see Figure 1). Thanks to KDD techniques, large databases became rich and reliable sources for the generation and the validation of knowledge. Data mining is the main step of KDD process which consists in applying intelligent algorithms in order to obtain predictive models (or *patterns*). We focus on association rules mining which is a specific data mining task.

2.2 Association Rules Mining (ARM)

Association Rules Mining (ARM) [1] can be divided into two subproblems: the generation of the frequent itemsets lattice and the generation of association rules. The complexity of the first subproblem is exponential. Let $|I| = m$ the number of items, the search space to enumerate all possible frequent itemsets is equal to 2^m, and so exponential in m [1]. Let $I = \{a_1, a_2, \ldots, a_m\}$ be a set of items, and let $T = \{t_1, t_2, \ldots, t_n\}$ be a set of transactions establishing the database, where every transaction t_i is composed of a subset $X \subseteq I$ of items. A set of items $X \subseteq I$ is called *itemset*. A transaction t_i contains an itemset X in I, if $X \subseteq t_i$. Several ARM published papers are based on two main indices which are support and confidence [1]. The support of an itemset is the percentage of transactions in a database where this itemset is one subgroup. The confidence is the conditional probability that a transaction contains an itemset knowing that it contains another itemset. An itemset is frequent if support $(X) \geq$ *minsup*, where *minsup* is the user-specified minimum support. An association rule is strong if confidence(r) $\geq$ *minconf*, where *minconf* is the user-specified minimum confidence. Left part of an association rule is called *antecedent* and right part is called *conclusion*. Our motivations are described hereafter.

2.3 Motivations

The number of generated rules is a major problem on association rules mining. This number is too significant and leads to another problem called *Knowledge mining*. The human cycles spent in analyzing knowledge is the real bottleneck in data mining. This issue can limit the final user's expertise because of a strong cognitive activity. To solve it, visual data mining became an important research area. Indeed, extracting relevant information is very difficult when it is hidden in a large amount of data. Visual data mining attempts to improve the KDD process by offering adapted visualisation tools which allow to tackle various known problems. Those tools can use several kinds of visualization techniques which allow to simplify the acquisition of knowledge by the human mind. It can handle more data visually and extract relevant information quickly.

During the last few years, several graphical approaches were proposed to display association rules in order to help experts' analysis. The first works were done in a text-mode. Their efficiency is restricted to the database size. For instance, if an expert searches for particular information, he can occult some essential information for his analysis. To answer it, several works were proposed in order to present this rules set such as graphs and trees, 2D and 3D matrix or virtual reality. Currently, these graphical representations present advantages and drawbacks. One limitation of the aforementioned representations is to display all rules in the same screen space. Indeed, the user's visibility and understanding are proportionally reduced according to the number of generated rules. The common problem of the representations is that they are not simultaneously global and detailed. Indeed, global representations are quickly unreadable, whereas detailed representations do not present all information.

3 Visual Data Mining

The rise of KDD revealed new problems as *knowledge mining*. These large amounts of knowledge must be explored with specific advanced tools. Indeed, expertise requires an important cognitive work, *a fortiori*, a harmful waste of time for industrial. Extracting nuggets is a difficult task when relevant information is hidden in a large amount of data. In order to tackle this issue, visual data mining was conceived to propose visual tools adapted to several well-known KDD tasks. These tools contribute to the effectiveness of the processes implemented by giving understandable representations while facilitating interaction with experts. Visual data mining is present during all KDD process: upstream to apprehend the data and to carry out the first selections, during the mining, downstream to evaluate the obtained results and to display them. Visual tools became major components because of the increasing role of the expert within KDD process. Visual data mining integrates concepts resulting from various domains such as visual perception, cognitive psychology, visualization metaphors, information visualization, etc.

We focus on visualization during the post processing stage and we are interested by ARM. Independently of both context and task, ARM has a main drawback which is the high number of generated rules. Several works on filtering rules were proposed and a state of the art was presented in [4]. Although reducing the whole of generated rules significantly, this number remains however important. Expert must be able to easily interact with an environment of data mining in order to more easily understand the displayed results. This point is essential for the global performance of the system. Visual tools for association rules were proposed to reduce this cognitive analysis but they remain limited [4].

3.1 Visual Association Rules Mining

Various works already exist to help expert analysis in text-mode [16]. Several works on visual rules exploration were published [1], [2], [3], [25]. The main beliefs of our interactive ARM are described hereafter. All these tools use several methods which are textual, 2D or 3D way. The choice of one of them proves to be a difficult work. Moreover, their interpretations can vary according to the expert. Each one of these techniques presents advantages and drawbacks. It is necessary to take them into account for the initial choice of the representation. The effectiveness of these approaches is dependent on the input data files. These representations are understandable for small quantities of data but become complex when these quantities increase. Indeed, particular information can not be sufficiently perceptible in the mass. The common limitation of all the representations is that if they are global, they quickly become unreadable (size of the objects in 2D, occlusions in 3D) and if they are detailed, they do not provide an overall picture on these data to the expert.

3.2 Hybrid Representation of Association Rules

Data mining is effective if the tools are able to represent the great masses of results obtained. Moreover, various functions of interaction were proposed in HCI (overall picture, zooms, data filtering, visualizations of relations between posted graphic objects) in order to facilitate the task of the user who must know this information and decide the level of relevance of an element among others. The work presented in [4] showed that, even for experts of a field, tools facilitating research and navigation in large sets of information are unavoidable. We successively proposed several representations (summarized textual of decision rules, 2D visualizations, then colored 3D, see Figure 2) in order to reduce the cognitive effort of the expert. Among various known representations, allowing to apprehend a great mass of information, such as the hyperbolic lenses or trees [14], perspective walls [17], fish eye view (FEV) [8] or superpositions of transparent sights [12], the FEV constitutes one possible solution adapted for the fields that we studied (bank, health, etc).

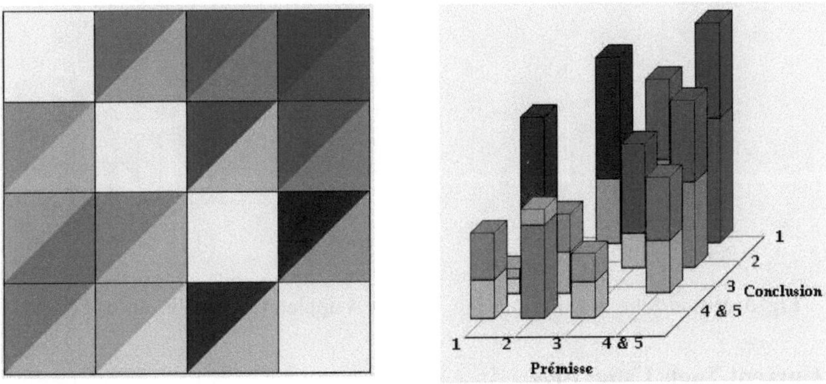

Fig. 2. 2D and 3D association rules visualization with *LARM*

In this paper, we present our study based on the continuation of the work started in [24] and [21] in order to interpret results in a visual way while preserving the context (see Figure 3b). For our study, we also tested *InfoVis* (see Figure 3c) [6]. However, several elements are not really adapted to our needs. For instance, with *InfoVis*, a rule is represented with the intersection of its metric. The number of metrics can not be higher than two. In our case, there can be much more. Consequently, legibility is reduce because of several rules can overlap (see Figure 3c). To answer this issue, we propose a representation in which the rules would be drawn in an allocated area (see Figure 3d). This kind of visualization makes it possible to represent several metrics in this area, thanks to various pallets of colors. In our example, we use two metrics and the allocated space is divided into two. With N metrics, the same space will be divided into N equal parts. Our assumption consists in supposing that we will have better results by hybridizing a semantically colored view [7] and a FEV. The user will be able directly to point the polygon (colored in a gradual way according to the value of one or more metric associated to the rule) of the FEV which appears to him most relevant according to its task.

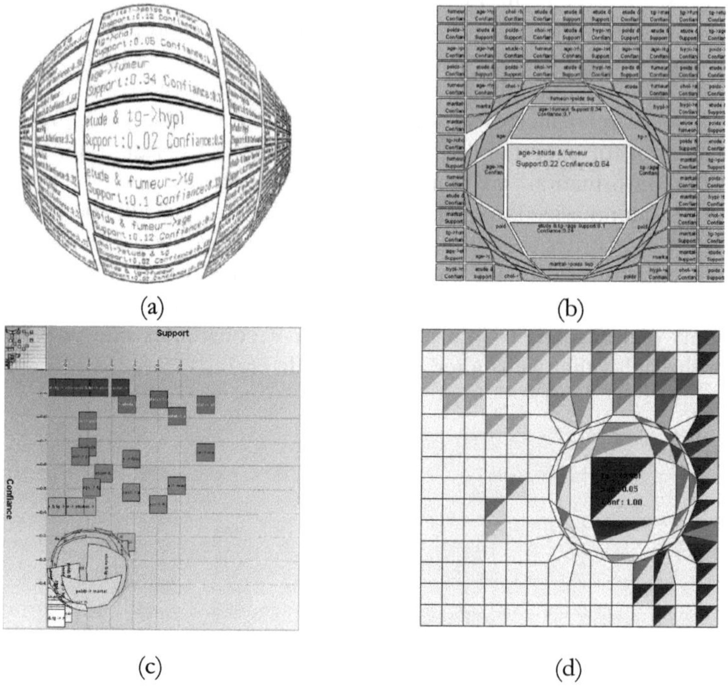

Fig. 3. Visualizations with (a) aiSee, (b) JAVA applet, (c) InfoVis, et (d) LARM

3.3 Current Tools Using FEV

We studied tools, allowing using our data files input and making it possible to visualize the rules using FEV. A system completely adapted to our problem does not exist, according to the literature related to this issue. Nevertheless, we studied the case of IDL (Interactive Language Dated) (www.rsinc.com) which is dedicated to the processing and the visualization of data (time series, images, cubes,...). IDL is based on compiled modules from the C language (there is a documentation that explain how to write such modules). It also makes it possible to write procedures and functions but also to make directed programming object. Since version 5, the programming of widgets became possible. The main aim of this language is to handle and display data with little investment in programming.

Then, we tested *aiSee* software (www.aisee.com), based on GDL (Graph Description Language) (near to IDL, but free). The *aiSee* software allows to read a data input file, (see Figure 4) resulting from a file (.gdl) and to display these data in

```
graph: {
    node: { title: "A" color: blue }
    node: { title: "B" color: red }
    edge: { source: "A" target: "B" }
    }
```

Fig. 4. Input format which is interpreted by GDL language

various forms, in particular in FEV. GDL describes a graph in term of nodes, edges, sub graphs and attributes. These attributes can be color, size, etc.

These two solutions seem adapted to our problems (see Figure 3a), but in real case, the following limitations appear: (a) Management of any field in a too generic way, however within the framework of the visualization of association rules, it is necessary to display several colors on the same node; (b) Difficult implementation for the developer. Indeed, the problems arising previously imply to code new modules in C language and that represents an important effort of implementation; (c) Usability reduced for the end-user. It should not be forgotten that an expert is a specialist within his field, but not necessarily within the tools (software) provided in order to achieve the task. It is essential to propose intuitive, simple and ergonomic interface. These languages appear to us more adapted to the scientists wishing to handle and display data, without need for producing code.

In order to start a footbridge between ECD and IHM, we conceived and developed our own display system exploiting a FEV (see Figure 3d). This implementation is carried out in JAVA and it is completely integrated into the LARM (Large Association Rules Mining) system which was presented in [5]. We thus proposed a display system of association rules allowing reducing the cognitive analysis while increasing the expert's efficiencies. According to our investigations, it seems that our approach is the only one able to propose a detailed and general sight simultaneously of association rules. Indeed, expert is in front of a general sight of the gradually colored rules and thanks to the FEV, he can obtain information on a highlight rule.

4 Conclusions and Further Works

After a study about current tools including FEV, we propose a first implementation which is integrated within an existing visualization rules platform (LARM). This approach can tackle large sets of rules in a same screen space. We show that our solution allowing to manage simultaneously both a global and a detailed representation. This is not yet the case in recent researches around KDD. This work is included in the *LARM* platform. We are testing it on both several banking and medical benchmarks. The first results are interesting and relevant. They show that human factors must play the first role at any time of the process during the decisional systems designing. Their efficiency depends on a good mix between KDD and HCI directly.

This work is a first outline to visualize large sets of association rules. However, it is necessary to evaluate our visualization system. In order to realize it, we will work in two times. Firstly, we wish to evaluate it thanks to the *discount usability testing* method [23]. The aim is to highlight the main advantages and the major drawbacks of our visualization method. Secondly, as soon as the first evaluation will be validated, we will evaluate in-depth our system thanks to experts intervention. Currently, we focus on the first point. In another way, we focus on clusters visualization. In our approach, the fisheyes view focus point is a detailed rule. We wish to use it to visualize a set of clusters thanks to a 3D representation. Finally, we wish to invest about a new active research domain: haptic system [19] which can be significant in human computer interaction. Indeed, haptic system can be profitable in a knowledge management data mining system in order to help KDD actors during the analysis. On a more global point of view, we are planning to integrate multimodal features to our system. In output, Text to Speech (TTS) and haptic feedback will be used to give information to the user. In input, Automatic Speech Recognition (ASR) will be an interesting modality in order to command the system (example: "zoom in", "zoom out", "save this subset on the disc", etc.) but also to interact more naturally with it, in a natural language manner (example: "what are the best combinations for the year 2006?"). Freehand manipulations on interactive surfaces are already used to retrieve geographical information, for instance [20]. It provides more friendly interaction with the system, and, according to us, nobody has adapted those techniques in a KDD context. We will propose different kinds of multimodality: first, exclusive multimodality will bring the opportunity to switch from a modality to another (speech instead of keyboard/mouse, for example), then we will propose synergic multimodality, in which the user could, for example, pronounce "Zoom 63%" while he/she will manipulate graphical object with the mouse.

Acknowledgments. This work has been partly supported by the "Centre National de la Recherche Scientifique" (CNRS), the "IUT de LENS" and the "Université d'Artois". Moreover, The authors are thankful to the MIAOU and EUCUE programs (French Nord Pas-de-Calais Region) and the UE funds (FEDER) for providing support for this research.

References

1. Agrawal, R., Mannila, H., Srikant, R., Toivonen, H., Toivonen, H., Verkamo, A.I.: Fast discovery of association rules, Advances in knowledge discovery and data mining, American Association for Artificial Intelligence, pp. 307–328 (1996)
2. ben Yahia, S., Mephu, N.E.: Emulating a cooperative behavior in a generic association rule visualization tool. In: Proceedings of the 16th IEEE International Conference on Tools with Artificial Intelligence (ICTAI'04), Boca Raton, Florida, USA (2004)
3. Blanchard, J., Guillet, F., Briand, H.: Exploratory Visualization for Association Rule Rummaging. In: Proceedings of the 4th International Workshop on Multimedia Data Mining MDM/KDD2003, Washington, DC, USA, pp. 107–114 (2003)
4. Couturier, O.: Contribution à la fouille de données : règles d'association et interactivité au sein d'un processus d'extraction de connaissances dans les données, PhD Thesis, Université d'Artois, CRIL, Lens, France (2005)
5. Couturier, O., Mephu, N.E., Noiret, B.: A formal approach to occlusion and optimization in association rules visualization. In: Proceedings of International Symposium of Visual Data Mining (VDM) of IEEE 9th International Conference on Information Visualization (IV@VDM'05), Poster, London, UK (2005)

6. Fekete, J.D.: The InfoVis Toolkit. In: Proceedings of the 10th IEEE Symposium on Information Visualization (InfoVis'04), pp. 167–174. IEEE Press, New York (2004)
7. Fekete, J.D., Plaisant, C.: Interactive Information Visualization of a Million Items. In: INFOVIS 2002. IEEE Symposium on Information Visualization, Boston, pp. 117–124 (2002)
8. Furnas, G.W.: Generalized Fisheye Views. Proceedings of ACM Conference CHI'86, ACM SIGCHI Bulletin 17(4), 16–23 (1986)
9. Hajek, P., Havel, I., Chytil, M.: The GUHA method of automatic hypotheses determination. In Computing (1), 293–308 (1966)
10. Han, J., Kamber, M.: Data Mining; concepts and techniques. Morgan Kauffman, San Francisco (2001)
11. Harisson, B.L., Vicente, K.J: An experimental evaluation of transparent menu usage. In: Proc of ACM Conference CHI'96, pp. 391–398. ACM Press, New York (1996)
12. Keim, D.: Visual Exploration of large data Sets. Communications of the ACM 44(8), 39–44 (2001)
13. Lamping, J., Rao, R., Pirolli, P.: A Focus+Context Technique Based on Hyperbolic Geometry for Visualizing Large Hierarchies. In: Proceedings ACM Conference on Human Factors in Computing Systems (CHI'95), Vancouver, Canada, pp. 401–408 (1995)
14. Liu, B., Hsu, W., Wang, K., Chen, S.: Visually aided exploration of interesting association rules. In: Zhong, N., Zhou, L. (eds.) Methodologies for Knowledge Discovery and Data Mining. LNCS (LNAI), vol. 1574, pp. 380–389. Springer, Heidelberg (1999)
15. Mackinlay, J.D., Robertson, G.G., Card, S.K.: Perspective Wall: Detail and Context Smoothly Integrated. In: Proc. ACM Conference CHI'91, pp. 173–179. ACM Press, New York (1991)
16. Nigay, L.: Modalité d'Interaction et Multimodalité, Habilitation à Diriger des Recherches, spécialité Informatique de l'Université Joseph Fourier - Grenoble I (2001)
17. Pietrzak, T., Martin, B., Pecci, I.: Information display by dragged haptic bumps. In: Proceedings of the 2nd International Conference on Enactive Interfaces, Genoa, Italy (2005)
18. Rekimoto, J.: SmartSkin: An Infrastructure for Freehand Manipulations on Interactive Surfaces CHI2002, Conference on Human Factors in Computing Systems took place in Minneapolis, Minnesota (April 20-25, 2002)
19. Rouillard, J.: Navigation versus dialogue sur le web, Une étude des préférences. IHM'99, Montpellier (1999)
20. Schneiderman, B.: The eyes have it: A task by data type taxonomy for information visualization. In: Proceedings of IEEE Symposiumon Visual Languages, Boulder, Colorado, USA, pp. 336–343 (1996)
21. Schneiderman, B., Plaisent, C.: Designing the user interface, 4th edn., International Edition, Boston, Addison-Wesley, Reading, MA (2005)
22. Vernier, F., et Nigay, L.: Représentations multiples d'une grande quantité d'information, IHM'97, Futuroscope de Poitiers, France (1997)
23. Wong, P.C., Whitney, P., Thomas, J.: Visualizing Association Rules for Text Mining. In: Proceedings of the 1999 IEEE Symposium on Information Visualization (INFOVIS'00), Salt Lake City, Utah, USA, pp. 120–128 (2000)

Integrating Sensor Data with System Information Via Interactive Visualizations

Jennie J. Gallimore[1], Elizabeth Matthews[2], Ron Cagle[2], Paul Faas[3], Jason Seyba[3], and Vaughan Whited[3]

[1] Wright State University. Dayton, OH 45435
{Jennie.Gallimore@Wright.edu}
[2] McAulay Brown, Beavercreek, OH 45430
{Elizabeth. Matthews,Ron.Cagle@macb.com}
[3] AFRL/HEAL,Wright Patterson AFB, OH, 45433-7604
{Paul.Faas,Jason.Seyba,Vaughan.Whited@wpafb.af.mil}

Abstract. Development of intuitive visualizations requires a systematic approach that includes a focus on the user. Creating interactive visualizations for complex systems often requires the integration of information from existing systems and sensor data to provide the operator with real-time information. The objective of this research was to fuse information from sensor technology with flightline maintenance information to support aircraft maintenance logistics. The research was conducted in two phases. A user-centered approach was used to design visualizations in each phase; however, in Phase II a cluster analysis technique was utilized to support the design. User feedback indicated that incorporating a technique to map data and decisions resulted in interactive visualizations that were well accepted by users and provided the important information needed for their decision making tasks.

1 Introduction

Creating interactive visualizations for complex systems often requires the integration of information from existing systems and sensor data to provide the operator with real-time information. The objective of this research was to investigate the fusion of information to provide decision makers with a shared vision to support collaborative work and provide insight into the use of resources (people, equipment, supplies, and information). The advantage of correctly fusing data from sensor technologies into complex operations is that decision makers will be more able to sense and then respond to issues before they affect the ability to fulfill a goal or mission, thus providing a more agile and responsive operation. The domain for this investigation was flightline maintenance. Logistics and sortie production mission success depends on effective use of resources, however, the flightline maintenance operational environment has limited cross echelon situational awareness. Personnel need a means to identify the impact of their logistics actions on operational capability.

Development of intuitive visualizations requires a systematic approach that includes a focus on the user, understanding what information is required to make decisions. Understanding there are no specific rules that are guaranteed to provide useful intuitive

M.J. Smith, G. Salvendy (Eds.): Human Interface, Part I, HCII 2007, LNCS 4557, pp. 268–277, 2007.

visualizations, there are several frameworks that focus on a user-centered approach such as the Work Centered Support Systems (WCSS) concept [1] and the Applied Cognitive Work Analysis (ACWA) [2]. These frameworks focus on performing up front cognitive work analysis (CWA) to understand user needs and transforming the data gathered into actual design of user-interfaces or visualizations. Even with these concepts, the difficulty lies in moving from information to design. Guidelines for design of visualizations are difficult because the visualizations are primarily domain and/or work dependent. Visualization design is both an art and a science, requiring creative ideas combined with knowledge of human capabilities and limitations related to both human cognition and visual perception. The challenge is in determining how to take the information and decisions and put them in the correct visualization so the information is readily available when needed and a shared vision is possible across users.

Other issues designers face are time and cost constraints, requiring quick development of concepts and rapid movement to a more detailed design before all requirements and information are completed. This often leads to the purchase and addition of commercial off-the-shelf technology (COTS) to existing systems with the intention that they will improve system performance. In some cases these stand-alone systems do not integrate well. In light of this fact, it is either not used or the user must duplicate their efforts across multiple systems. Radio frequency Identification (RFID) technology is a resource tracking technology example that is being inserted into systems without full consideration of how to integrate it into the work process.

1.1 Objectives and Approach

The purpose of the Smart Systems for Logistics Command and Control (SSLC2) research program was to investigate the integration of real time sensor information with existing information for flightline maintenance logistics support [3,4]. The project was divided into two phases in which the design and information from Phase I fed into the second phase. During both phases an interface for presenting the information was designed to integrate information from sensors with flightline information to support the way users conduct work. While both phases focused on a user-centered approach, it was determined that even within this approach it was difficult to determine how the information should be displayed visually. Therefore in Phase II we added a cluster analysis technique to support the design. This paper describes the interfaces developed in the two phases and user feedback.

2 Phase I Design Approach

Phase One's objective was to compare the use of COTS RFID technology (WhereNet) for supporting user tasks with a system in which sensor information was integrated with flightline information (SSLC2). This phase helped to gather additional system requirements for future phases. To develop the SSLC2 interface for this evaluation, a user-centered approach was followed. The first step consisted of systems analysis and CWA with data collected from subject matter experts (SMEs) at various Air Force bases through job shadowing and interviews. The second step focused on concept development and storyboard designs with SMEs participating as part of the design team. The third step developed use cases from the storyboards to create a simulation of the interface to allow for user testing. The focus of this phase was to support one decision; the fix/swap decision. The flightline information included was for only one maintenance day.

The SSLC2 general concept is illustrated in Figure 1 which presents the geographic view. The interface provides the user with aircraft status and location, daily flying schedules, personnel and equipment resources and their locations, tasks necessary to fix or swap aircraft, time to complete, time to move resources to the fix site, and types necessary to complete tasks, specific resources assigned to tasks, and a recommendation to fix or swap an aircraft. Users could also view the information using a flying schedule view. The left side of the screen presented the workflow process steps. The central portion of the screen provides the schedule and geographic views. The views can be changed from an overview of the entire flightline, to a problem view focusing on one aircraft. Users interacted with the system using a mouse and standard keyboard.

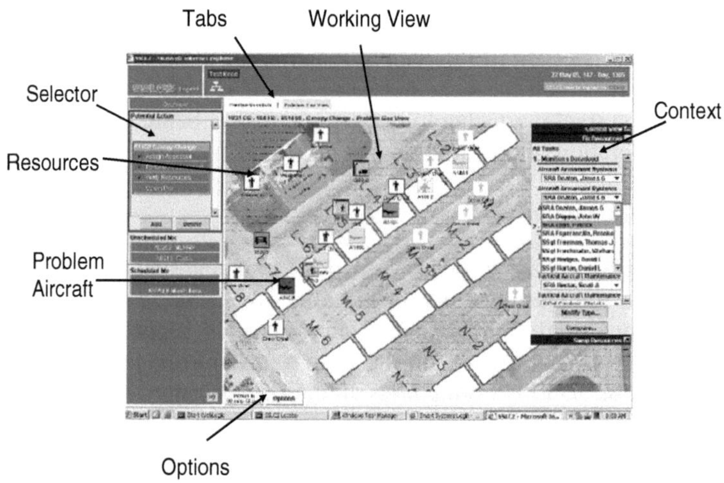

Fig. 1. Conceptual design phase I

2.1 User Feedback Phase I

Fifteen maintenance personnel from four airbases interacted with the interface in a simulation. All participants had production supervisor (ProSuper) and/or Expeditor experience. A great deal of data was collected on user feedback for this design and in-depth details can be found in the Phase I Final Report [5] and in Gallimore et al. [4].

In this section we describe some of the issues noted by users related to the interface. Participants made positive comments about the ability to see a flightline overview, both geographically and as a schedule. Prosupers and Expeditors are expected to multi-task and respond to more than one problem at a time. Participants suggested that a future system provide them the ability to multi-task and display information about every aircraft in order to support decision making.

SSLC2 also broke down each problem into a series of tasks to be completed. Each task listed associated equipment and personnel resources with their time to site and time to fix. Participants voiced their support for seeing the list of tasks and said they would use this feature. Participants also said they would prefer the ability to create or modify the tasks needed to fix problems. Participants indicated that having the entire listing of tasks and resources was overwhelming. Participants sometimes referred to the wrong task when trying to change a resource, and indicated too many steps were required to reallocate (up to eight). Participants wanted a click and point interface rather than dialog boxes and pull downs. Additional items and comments highlighting positive aspects of SSLC2 functionality included the ability to zoom in and out within the geographical views.

The results indicated that users preferred the integrated SSLC2 approach to the off-the-Shelf WhereNet system that only provided resource location information and no flightline information.

3 Phase II Design Approach

Since Prosupers and Expeditors make many decisions in addition to the fix/swap decision there are also significant collaborations that must take place on the flightline to maintain aircraft and prepare them for sorties. Multiple users need access to shared information and it must be provided beyond one day. Phase I was not meant to tackle all these issues, and feedback from Phase I was important for Phase II. However, it was determined that to incorporate multiple aircraft maintenance problems and to include all the data related to the flightline beyond one day, the interface would not be adequate. There were too many menus and steps to get to important information. The interface developed in Phase I had both positive and negative aspects.

The approach for Phase II was similar to that used in Phase I, however, an additional analysis tool, cluster analysis, was used for input into the conceptual design.. The conceptual design was presented to SMEs to refine concepts. A simulation of the user interface was developed to demonstrate concepts and receive feedback. This section details the approach and results of the cluster analysis, followed by a description of the user interface.

3.1 Cluster Analysis

For input to the design of the visualizations it is important to understand how the data elements map to the decisions. Not all information is needed for each type of decision. The goal is to provide users with easy access to data based on decisions they make. Current frameworks discuss the need for the CWA and understanding of the information and decisions to be applied to design, but moving from this information to design can still be difficult. How can the information and decisions be integrated and mapped to help designers?

In order to perform a mapping of what information was needed for different decisions and to determine their importance, a data matrix was created with the 31 decisions identified by ProSupers and Expeditors in columns and the 157 data elements they use on the flightline in rows (all decisions and data elements can be found in the Final Report [6].) Three SMEs were then asked to read each decision and rate the importance of the 157 data elements to that decision. They rated the information on a scale of 1 to 5; 1-not at all important, 2-somewhat important, 3-important, 4-very important, and 5-critical or extremely important. They also rated each data element to indicate whether the data should be visible most of the time.

The ratings were used as input for a hierarchical cluster analysis using Ward's method utilizing the JMP Professional statistics package (Version 5.0.1.2). The ratings were first averaged across the three SMEs. The cluster analysis was conducted across all decisions, which resulted in 28 clusters of the 157 data items. Clusters 27 and 28 all had data elements that had average rankings of 1 or below (not at all important.). The cluster analysis helped designers determine grouping of information and a general design concept.

3.2 Conceptual Design

The concept was to provide a direct tapping interface for use on a Tablet PC and provide users with global awareness of information and more detailed relevant information as needed without requiring users to continuously drill down through many levels. The visualization allows users to sort, rearrange, promote, focus, filter and spotlight data. Users must be able to make notes, highlight information, and collaborate with other users. Shniederman's concept of Overview first, zoom, filter, sort, and details on demand was followed [7]. The design is a tab-based interface to allow users to easily point to the needed information. The primary grouping of information on the tabs are: flying schedule, maintenance, personnel, equipment, facilities, fleet health, and additional resources (checklists, AF documents, lookup information). This Phase concentrated on all views except fleet health and additional resources. Figure 3 illustrates the general concept. The display is divided into four primary areas: A primary (content based) window, a geographic (geo) view, a detailed data view, and a message area. The information from the geo and detail window can switch places with the primary window so that a larger view can be seen. The information in the primary window always drives the information to be displayed in the detail and geo views, even when the windows change location. Because the window panes are limited in size it is necessary to allow the user to visualize the information without requiring them to continuously scroll. Therefore the concept calls for the use of zoomable user interface (ZUI) using space-scale techniques [8]. It is not possible to describe the entire interface in-depth, however, the basic views are briefly discussed.

Flying Schedule. The purpose of this view is to show the current day, week or month of scheduled sorties and all relevant information related to the aircraft and sorties (See Figure 2). The frame of reference is a time line with schedules. This view provides the user with the ability to choose the time frame by which to work; hours, day, week, etc. The user is able to filter and sort the order of the aircraft presented. If the user selects an aircraft icon on the left, specific details related to that aircraft are shown in the details pane, such as status, phase, turn, etc. The aircraft icon also provides

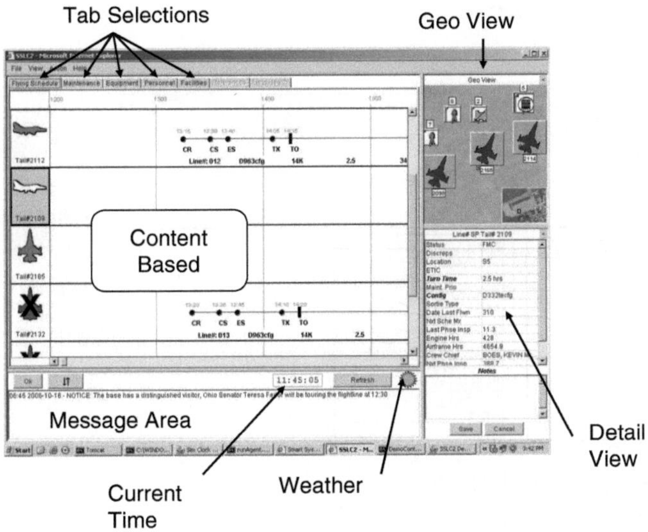

Fig. 2. Conceptual design phase II

information related to status via symbol design. If the user highlights the icon by selecting it, the location of the aircraft on the flightline is shown in the geo view.

Maintenance View. It was important to develop a maintenance visualization that could give users a great deal of information related to all the tasks and resources needed to perform a maintenance task. Monitoring and expediting unscheduled maintenance is a primary job of the Expeditor. The concept is to provide a global visualization providing the user with an overall impression and quick assessment of the tasks involved to fix aircraft, the flow of tasks, including those that can occur simultaneously, the ability to determine if there are any resource conflicts, and the ability to view multiple aircraft at the same time. Figure 3 illustrates the concept. The frame of reference is a vertical time scale and the ability to display multiple aircraft horizontally. The tasks required to complete a maintenance task are presented as a flow diagram, flowing vertically from top to bottom. Zooming functions allow users to select time frames. Each subtask is specified by a box divided into rows that provide quick details related to the tasks including resources assigned (e.g. type of personnel needed). To see more detail the user chooses a subtask to zoom in and see more detailed information. When the user puts the cursor over the subtask box, it expands in size so that all relevant information can be read. The other boxes are reduced in size but still visible to show the layout of the task sequence (i.e. fisheye). If the user selects a row on a box, the pertinent information is displayed in the details view. Completed tasks are highlighted (or grayed out). Areas of concern are highlighted (color coded) within each box when there is a potential conflict or if the task is running late. With this concept, users can glance at the tasks and quickly determine if there is an issue for them to concentrate their problem solving efforts on.

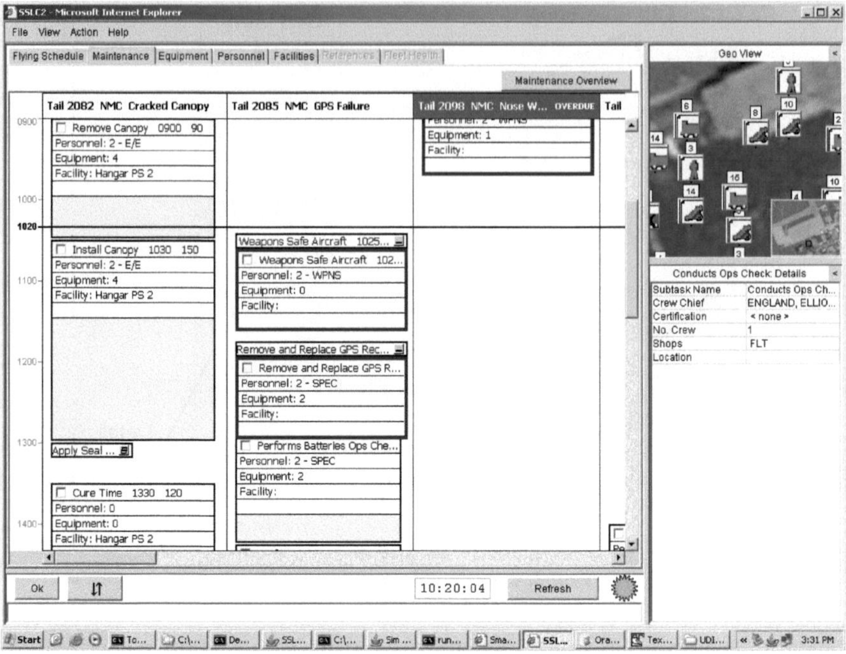

Fig. 3. Maintenance subtasks view

Equipment View. The equipment view is designed to provide the user with a global picture of resource availability as well as more detailed information if needed. The concept allows users to quickly glance at equipment to determine status (e.g. broken, in conflict, available). If they notice that many of their low density high demand resources are broken, they may consider how resource availability may affect their maintenance and plan accordingly. It also allows them to easily locate a piece of equipment. Equipment is displayed on the primary view using icons and coding to designate their status with details related to the equipment located in the details view when the equipment is selected. The use of RFID provides location of the equipment in the geo and details view.

Personnel and Facilities Views. Personnel are also resources. Expeditors indicated that finding people is a very time consuming task. The concept for the personnel view is similar to the equipment view; however, the frame of reference is a schedule with a time line. Active personnel tags on badges provide location as well as availability. The personnel view displays the person's name on the left and schedule on the right. Selecting the personnel icon on the left places detailed information related to that person in the details pane and shows the individual location on the geo view. The schedule indicates what tasks they are assigned and also conveys if they have been assigned to multiple tasks at the same time causing a conflict. Maintenance may occur at various facilities. The frame of reference for the facilities view is also a schedule view which shows the aircraft assigned to the facility and the time. Details related to the facility or schedules would be available in the details screen.

3.3 User Feedback

The simulation was presented to five personnel with Expeditor experience at the 180th FW Ohio ANG Toledo Express Airport, Swanton, OH. The subjects participated in a 1-hour simulation, were responsible for 8 aircraft, and information related to their sorties for a 1-week period. During the simulation aircraft maintenance problems occurred and the participant looked for issues, swapped aircraft, and changed resources. The simulation did not include the ability to zoom or filter due to time limitations; therefore users were required to scroll through the information. Zooming is a system requirement that will be included in the design. After completing the simulation they were asked to fill out a questionnaire to determine their opinions on the effectiveness of SSLC2 with respect to supporting decision making, situation awareness, collaboration, effectiveness of information integration, visualizations, and improving resource visibility and resource allocation. The questionnaire consisted of a 6-point rating scale (1-Not at all effective, 2-Not Effective, 3-Somewhat Not Effective, 4–Somewhat Effective, 5-Effective, 6–Extremely Effective).

Table 1 presents the focus of each of the 24 questions and average rating results. In general, the users provided ratings close to effective (5) for most questions. Often when ratings were closer to 4.6, participants mentioned the lack of zoom and filtering functions. Had these function been demonstrated it is very likely the ratings would be higher. Two questions related to messages and alerts averaged ratings of 4.6 and 4.25. Participants indicated they needed an audio alert to let them know when a message arrived. This is possible, but needs to be considered carefully since the users are often in a high noise environment. Some alerts may need audio while others of lower importance might not require audio. If an alert is very serious it can also be displayed on top of their working window.

With respect to the information and its placement, participants were asked if there was any information that should be added regarding sensed and flightline information. It was interesting to note that participants did not indicate any information was missing and only occasionally thought data should be located in a different place. Subjects were able to find most information very quickly. Even with limited time to interact with the system, the users learned the interface quickly and gave comfort ratings of 7-8 on a scale of 1-10 with 10 being the most comfortable.

4 Discussion

The use of cluster analysis was very helpful for developing constraints on how to integrate large amounts of data for arrangement in the interface. This data combined with an understanding of how users perform their work and what information is needed when they are making decisions helped in developing interactive visualizations to support users. While there was limited testing possible in Phase II due to resource constraints, the Phase II integrated visualizations received more positive comments than did the Phase I interface. The inability to include the zoom and filtering functionality with this complex interface affected user perceptions even though they were told during training the interface would include zooming and filtering in

Table 1. Average rating for questions related to Phase II SSLC2 Visualization

Question focus	Mean		Mean
Integration of information	5	Situation awareness maintenance tasks	4.8
Information organization	4.6	Situation awareness of resource status and location	5
Level of information detail	5	Situation awareness of facility availability	5
Visualization of flying schedule	4.6	Situation awareness of overall flightline maintenance	5
Visualization of unscheduled/ scheduled maintenance task list	5	Situation awareness of maintenance alerts	4.25
Visualization of unscheduled maintenance subtasks and resources	5.2	Information to predict how changes may affect maintenance tasks and resource use	5
Visualization of alerts and notifications	4.6	Providing a shared vision of flightline	5
Overall the visualization of the information presented	4.8	Enhancing collaboration among flightline personnel	4.4
Supporting decision making related to aircraft maintenance tasks	5.2	Improving resource allocation and utilization	4.8
Supporting resource allocation decisions	5	Supporting resource visibility	5.2
Managing information during decision making	4.8	Improvement over current practice	4.8
Coordinating resources to meet multiple objectives	5.2	Reduction in time to identify and make decisions	4.4
Average Rating Across all Questions	**4.86**		

the next phase. This was not unexpected as it is difficult for the users to conceptualize the functionality. This is another example of how constraints can affect design.

While the overall concept and navigation for the interface was easy and understandable, there is still a need to reduce the number of menu interactions and dialog boxes requiring keyboard inputs. Future phases should focus on easier ways to manipulate the data. Users indicated they would like to be able to tailor the interface for their use. For example, they would like to tailor the view of the windows to indicate if and when the different views (geo and detail) were available. They may also like to tailor the order of the information in the details screen and be able to turn on or off information in the views. For example, some users would like to have the job control number linked with Core Automated Maintenance System (CAMS) in the data, while others do not. Future phases should consider how the interface can be tailored and continue to test approaches.

5 Conclusions

Designing interactive visualizations that fuse information from sensors is important for a large number of applications. With large numbers of data elements and multiple decisions it is necessary to determine how that information can be grouped so that it supports user decisions and situation awareness. Using cluster analysis can provide designers with ways in which information should be grouped together and indicate what important data should always be visible or at top level versus data that can be accessed by drilling down. A user-centered approach is still necessary, but cluster analysis is another tool that can aid the designer.

Acknowledgements. This research is sponsored by the Air Force Research Laboratory (AFRL), HEAL, Wright Patterson AFB, Contract # FA8650-04-C-6404.

References

1. Eggleston, R.G., Young, M.J., Whitaker, R.D.: Work-Centered Support System Technology: A New Interface Client Technology for the Battlespace Infosphere. In: Proceedings of NEACON 2000, Dayton OH, IEEE 00CH37093 (October 10-12 (2000)
2. Gualtieri, J.W., Szymczak, S., Elm, W.: Cognitive Systems Engineering-Based Design: Alchemy or Engineering. In: Proceedings of the Human Factors and Ergonomics Society 49th Annual Meeting, pp. 254–258. Human Factors and Ergonomics Society, Santa Monica, CA (2005)
3. Gallimore, J.J., Maki, A., Faas, P., Seyba, J., Quill, L., Matthews, E.: The Need For A Human-in-the Loop Simulation Testbed for Logistics Decision Support Research. Summer Simulation Conference, pp. 84–89 (2005)
4. Gallimore, J.J., Quill, L., Cagle, R., Gruenke, J., Hosman, C., Matthews, E., Faas, P., Seyba, J., Young, I.: User Feedback on RFID and Integrated Flightline Data for Maintenance Decisions. Proceedings of the Institute of Industrial Engineers Annual Conference. May 2006, Orlando, CDROM (2006)
5. GRACAR Corporation, Smart Systems for Logistics Command and Control (SSLC2): Interim Final Report, Contract Number FA8650-04-C-6404. Dayton, OH (2006)
6. MacAulay Brown, Corp, Final Report, Smart Systems for Logistics Command and Control (SSLC2): Final Report, Contract Number FA8650-04-C-6404. Dayton, OH (2007)
7. Shneiderman, B.: The eyes have it: a task by data type taxonomy for information visualization. In: Proceedings of IEEE Workshop on Visual Language, Boulder, CO, pp. 336–343. IEEE Computer Society Press, Washington, DC (1996)
8. Furnas, G.W., Bederson, B.B.: Space-scale diagrams: Understanding multiscale interfaces. In: Katz, I.R., Mach, R., Marks, L., Rosson, M.B., Nielsen, J. (eds.) Proceedings of the ACM Conference on Human Factors in Computing Systems, pp. 234–241. ACM Press, New York (1995)

Fovea-Tablett®: A New Paradigm for the Interaction with Large Screens

Jürgen Geisler, Ralf Eck, Nils Rehfeld,
Elisabeth Peinsipp-Byma, Christian Schütz, and Sven Geggus

Fraunhofer-Institut für Informations- und Datenverarbeitung - IITB, Fraunhoferstr. 1,
76131 Karlsruhe, Germany
{juergen.geisler,ralf.eck,nils.rehfeld,
elisabeth.peinsipp-byma,christian.schuetz,
sven.geggus}@iitb.fraunhofer.de

Abstract. Today's desktop computers can be regarded as evolution of the typewriter. They are well suited for office applications but far from optimal for e. g. mechanical design, traditionally conducted on large drawing tables. Large screens are nowadays widely available for table-like computer workplaces. But their resolution is still too poor. To overcome this drawback the »fovea-approach« has been developed. A slim tablet PC (the so called Fovea-Tablett®) can be put deliberately on top of a large table screen. The position of the tablet in relation to the table screen is tracked and the screen content of the table display just below the tablet is displayed on the Fovea-Tablett®: just as if one would look through the tablet onto the table but with higher resolution. Whereas the table display is still good for overview, the Fovea-Tablett® brings highest resolution to the region, one is just focusing on.

1 Introduction

One basic idea of ubiquitous computing and ambient intelligence is to bring computer-based tools back to that, what already Aristoteles found roughly 2400 years ago: that »*every instrument is best made when intended for one and not for many uses*« [1]. The widespread desktop as well as the notebook computers can be regarded as the evolution of the typewriter: from an ergonomic point of view well suited for editing letters, setting up spread-sheets or preparing presentation slides, but in between overloaded by a vast amount of very different applications. So the classical drawing table nearly vanished as design workplace for engineers and became replaced by a desktop computer with one or two moderately large screens. But today's CAD[1] workbenches do not reach the ergonomically optimized size of the drawing table that allowed over the span, comfortably reachable by the two hands of the designer, a coherent view over the whole design object as well as a high resolved look onto that detail, the designer is currently working on (see Fig. 1, right hand). Nowadays the designer must either steadily fiddle with the zoom factor if he wants to change between overview and detail. Or he loads the overview on a separate screen what

[1] Computer Aided Design.

M.J. Smith, G. Salvendy (Eds.): Human Interface, Part I, HCII 2007, LNCS 4557, pp. 278–287, 2007.

dismembers the coherence of the object and causes additional cognitive workload to fit together the two views mentally. The same goes for the work with geographical maps for e. g. navigation planning (Fig. 1, left hand), planning of military operations or management of natural disaster relief activity. Before computers offered a rich functionality for map display and GIS[2]-supported geospatial reasoning, maps were laid out on a rather large table with a team of specialists grouped around it, performing their job by communication among each other logically *and* physically over the map. Today the teamwork is done mostly in front of the projection of a computer generated map: with the advantage of common view, but restricted possibility of common interaction with the geospatial information. And the detail work is conducted by single specialists in front of a common desktop workplace with the same problems as described above for the CAD.

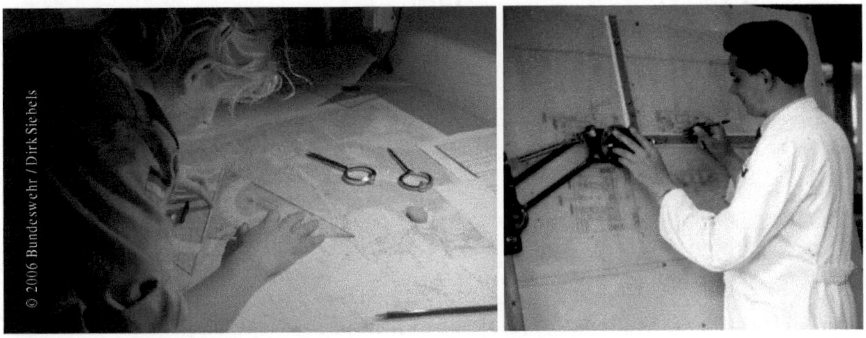

Fig. 1. Traditional work places for navigation planning (left hand) and mechanical design

Technical limitation in the production of wide-area computer screens to an affordable price retarded the construction of ergonomically optimized large screen computer-based workplaces for applications as discussed above. But in the most recent years one could observe a breakthrough of large flat panel displays based on LCD technology (Liquid Crystal Display), the most common computer display technology in between. Reaching a diagonal width of 50 inches and above they offer a much higher pixel density than the also large and since many years established plasma screens. The main driver for those large LCD screens is home television/cinema that opens a worldwide and highly attractive market. But are they also well suited for the future computer-based drawing or mapping table?

From Fig. 2 we can see, that for a viewing distance of 3 m (assumed as reasonable for home television) a resolution of more than 30 ppi (pixel per inch) is required to make the pixel lattice unperceivable. If we assume a 50 inch screen with 16:9 ratio and a common image size of 1.280 x 720 pixels (video standard HD 720), a resolution of 30 ppi will be achieved. This is sufficient for watching TV from the sofa in your living room. But even the image size of 1.920 x 1.080 (HD 1080) of a high end screen would only yield a resolution of 44 ppi. And this is still much less than enough if you take such a screen as drawing table and work on it from a distance of lets say 60 cm.

[2] Geographical Information System.

On that distance a resolution of at least 150 ppi is required in order to perceive sufficient detail and not to become distracted by the pixel lattice (see Fig. 2). To achieve this resolution a hypothetical image size of at least 6.600 x 3.700 would be necessary for a 50 inch 16:9 screen.

If we assume, that large flat panel displays with this image size will not be produced to a price affordable for the mass market in the near future, we see two choices. The first one is to assemble a large screen from a couple of smaller and higher resolving ones[3]. This »tile approach« is straightforward and, for vertical display walls and with rear projection instead of flat panel already realized with e. g. the HEyeWall [2]. Its drawbacks are the high technical effort to assemble the panels seamlessly and to provide a satisfying radiometric continuity over the whole display.

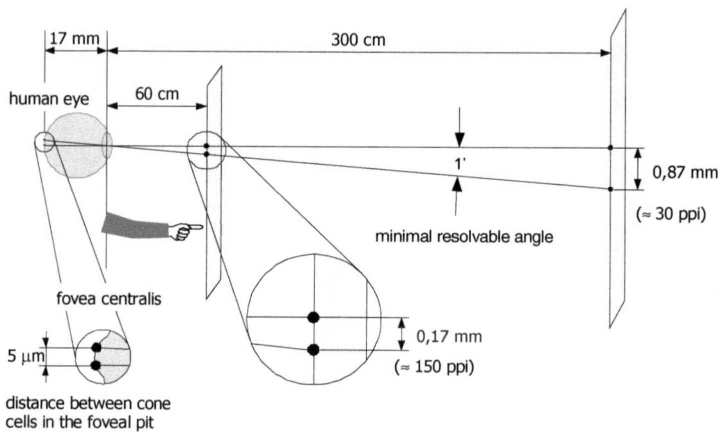

Fig. 2. Required ppi (pixel per inch) on a display derived from the least resolvable angle based on the distance of cone cells in the small foveal pit of the human eye; the distance of 60 cm is suited for touchable screens, the distance of 300 cm for e. g. home cinema

The second approach starts with remembering how the human (and other mammal) eye is designed. The so called fovea is the small area of the retina for sharp and detailed visual perception. The foveal pit (see also Fig. 2) covers only 1 ° of the visual angle but carries the highest density of the colour receipting cone cells (roughly 140.000 cells/mm^2). Outside the fovea this density falls quickly down to about 10 000 cones/mm^2. The eyes are steadily moving in order to focus sequentially small spots of interest whereas the retinal periphery is only taking care for the overview. Having in mind this very economic design of nature, we can combine a large flat panel LCD or a similar rear projection device with its poor resolution as display for the overview, and put a smaller, but much higher resolving display device, e. g. a tablet PC, on top of it for the detail view. The first step towards this idea has been undertaken by Baudisch ([3]) who fit a LCD screen into a large whiteboard, serving as screen for a front projection. While the large projected image showed the whole scene, e. g. a map, the

[3] E. g. 3 x 3 14" screens, each with an image size of 1 600 x 900 and a resulting resolution of 130 ppi, would be a big step towards the requirement for a drawing table.

LCD in the middle showed the respective cutout with higher resolution. With [4] a similar method was published but with two differently resolving projections instead of the combination of projection and flat panel. The disadvantage of those solutions is the fixed relation between the focus and the context view.

The »fovea approach«, described in this paper, overcomes the drawbacks of the fixed installation mentioned above, and allows by the use of mobile devices for the high resolution a deliberate shift of the point of interest, similar to the biological fovea, roaming across the scene to observe. The basic lines of this patented approach have first been published with [5], and, with emphasis on situation analysis as application area, in [6]. A similar solution is published with [7], with some differences explained later. The following sections describe the underlying method, and give an outlook for future research and development.

2 The »Fovea-Tablett®«

The equivalent of the foveal pit on a large screen workplace we do call »Fovea-Tablett®« (FT). An FT is a small, portable display unit, e. g. a tablet PC, with a rather high pixel density. It is simply placed on top of a much larger screen, here called the »overview table«. A measuring device determines its position and orientation with regard to the overview table. The measured position and rotation angle are wirelessly transmitted to the FT, which displays the graphics of the application (maps, technical drawings etc.) in such a way that the observer has the impression of looking through the FT onto the underlying image, but with a much higher resolution. Fig. 3 shows this arrangement with the so called »digital map table« of Fraunhofer IITB, the experimental system of a future workbench for the processing of geo-information. Fig. 4 gives an impression of the gain in visual acuity by the FT in relation to the overview table. Here the overview table, realized as a rear projection, shows a resolution of 22 ppi whereas the FT reaches 120 ppi.

2.1 Measurement of Position and Orientation

Measuring position and rotation of the FTs with respect to the overview table is carried out optically with a video camera. Therefore the FTs are equipped with a special visual marker, the so called MC-MXT (Multi-Cursor-MarkerXtrackT, see Fig. 5). Originally developed by Fraunhofer IITB for the purpose of high precision tracking of crash test dummies, those markers have an inner part of five points that determine the markers position. They are surrounded by a circular bar code giving each marker an identity and a defined rotary orientation. The so called Multi-Cursor-MarkerXtrackT image processing algorithm (MC-MXT tracker) detects and identifies the markers of the various FTs and calculates their position and rotation angle in real-time [8]. This kind of measurement distinguishes the FT approach from the ubiquitous graphics, published with [7], where an ultrasonic direction finder is used that can for one single device only determine the position but not the orientation.

Fig. 3. The »digital map table« of Fraunhofer IITB with its horizontal table for scene overview, two Fovea-Tabletts on top of this table illustrating scene details, and a vertical board displaying additional information.

Fig. 4. Gain of visual acuity with the Fovea-Tablett compared to the underlying overview table

Fig. 5. The coded MCT-MXT marker

Those coded markers can be slotted into one corner of the FT screen on a rather small area, e. g. 5 x 5 cm^2. But for our current experimental system as shown with Fig. 3 we went another way: while the frosted glass, where the overview image is projected on, is sufficiently transparent for objects laying directly on the glass, we fix a marker on the bottom of the FT and observe it with a video camera from below. In order to avoid interference with the projected image onto the overview table, the measuring camera operates at near infrared (0.78 – 1.5 μm) and we provide additional lighting in this spectral band from underneath. Fig. 6 shows the image taken by the IR camera positioned under the overview table with recognition results. The higher wavelength of infrared light and its diffusion through the frosted glass force to select a bigger marker in order to achieve sufficient measuring accuracy. The markers can get stuck deliberately underneath the FTs, e. g. as a self adhesive and easily removable tape. The measurement accuracy of the FT position is about ± 1 mm for the current arrangement and the orientation accuracy is better than ± 1 °. Measurements are taken with a frequency of approximately 20 Hz.

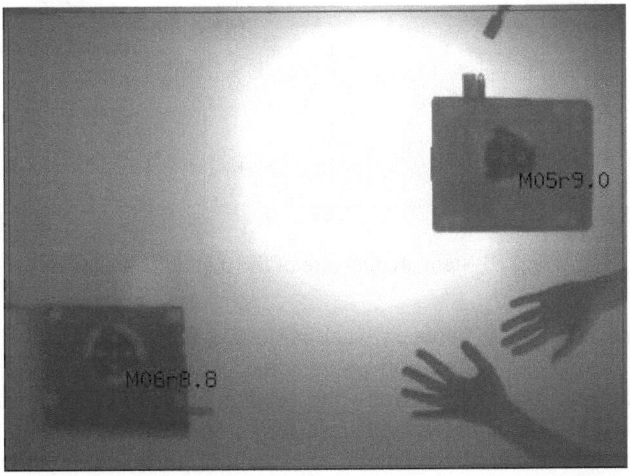

Fig. 6. Image taken by the IR-camera from underneath the table display

2.2 System Architecture

Fig. 7 gives a coarse view over the system architecture of the digital map table. The tracking camera is mounted above the overview table (or underneath it if rear projection is used). It is connected to the MC-MXT tracking server which detects the markers, determines their identity, position, and rotation angle from the camera image. The MC-MXT tracking server with the camera work independent from the application server as an embedded system with a TCP/IP socket interface using XML protocol.

The tracking information is passed over to the application server. Based on the data from the tracking server the application software calculates the coordinates of the views for the FTs. These data are passed to the FTs in dependency of their identities.

While tracking server and application server are connected by a wired LAN, the FTs and the application server communicate wirelessly, actually via Bluetooth. Bluetooth has been selected, because we decided for the current experimental system to run the application software not only on the server but also on the FTs (see, chapter 2.3). Therefore only position and orientation have to be transmitted for which a small bandwidth is sufficient. In the future the communication will move to wireless LAN.

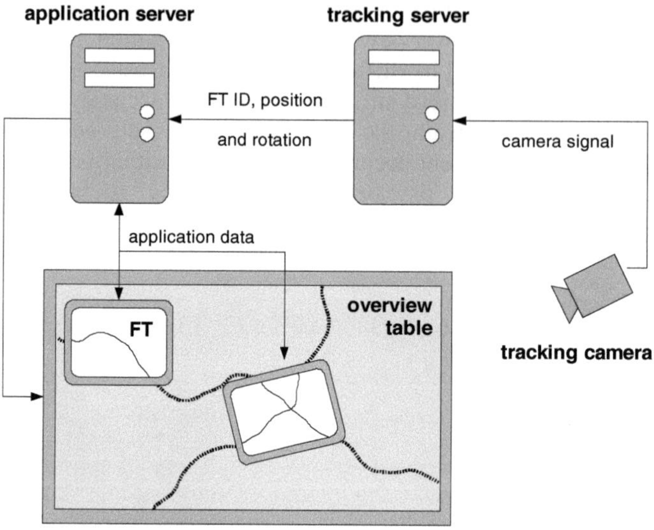

Fig. 7. System architecture of the digital map table

For reference calculation between work table and FTs the MC-MXT tracker must be calibrated. During the calibration process the coordinate transformation between the tracking camera images, which are the input for the MC-MXT tracker, and the viewer of the work table is calculated. In a first step, that has to be carried out only once for the whole table arrangement, the camera image gets referenced to the overview table. Therefore an calibration image with four defined cross points is projected onto the table. Each cross point is identified by its own colour. Four special MC-MXT calibration markers, only used within the calibration process, are marked with the same colors and laid on the projected cross points. So the camera »knows« the size of the overview table and the size of the markers. The second step has to be carried out for every Fovea-Tablett separately. For an FT calibration two images of the FT with the fixed identity marker are taken, whereby the FT has to be positioned on two defined regions of the work table. These regions are also indicated by the calibration image.

2.3 Interaction

At our experimental system, the digital map table, a GIS system as application software is running on the application server as well as on the FTs. So every tablet offers the full functionality of the application software at the user's hand. Various users can generate concurrently different logical views on their respective FT, e. g. showing different GIS layers, without interference between the users. It is also possible to take the FT away and work for a moment besides the table on a specific detail. Back at the table the FT will get instantly synchronized again with the overview image and will feed back the intermediate results to the application server.

Using regular tablet PCs, interaction with the FT is conducted with a digital pen, as Fig. 8 illustrates. A dedicated extra toolbar is overlaid that offers some functionality specific for the FT. The display can get decoupled from the overview so that the user can zoom or pan the image deliberately if he is for himself just diving into a specific detail. Afterwards he can quickly couple to the table with one pen click. All functionality that changes the view like zooming, panning or selection of map layers can also get activated from the FT including the overview. While the fovea approach is designed to working in teams, a clear management between those functionalities that affect the whole view has to be provided. If every team member could deliberately change the overview, confusion would be unavoidable. We decided that all manipulation concerning the overview is only allowed by one master FT, handled by the team leader. Other solutions are also thinkable but demand a higher discipline among the team members.

3 Discussion and Future Work

We have shown, that the fovea approach of combining large but low resolving with small, very mobile, and high resolving displays is a promising step towards ergonomic optimized work places that deal with extended objects like maps or technical drawings. While we have already gained experience with the first application area with our digital map table (for more detail see [6]), work for the area of future drawing tables is just in the beginning.

Fig. 9 illustrates the application of the digital map table for electronic circuit design. In the right part of Fig. 9 one can recognize, that the circuit details, like the

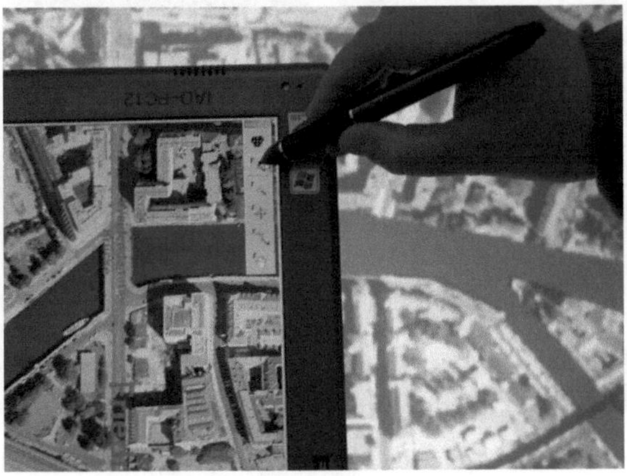

Fig. 8. Tool selection by pen

inscriptions are easily readable inside the FT but never on the overview table. Our first steps towards this electronic and mechanical design confirmed, that the interaction concept depends heavily on the selected application area. We see two general directions. The one is to use the Fovea-Tablett as a more or less simple detail viewer that shows exactly the underlying image but with high pixel density. Then the interaction with the pen offers the same but not more functionality as one would work with the mouse cursor in the respective area of the overview image. This direction has the advantage of being application independent and the demands on the device taken as Fovea-Tablett are rather low. But the space of interactions over the FT then is limited. - The second direction makes the Fovea-Tablett to an interaction device of its own that offers all of the functions the application software has, and is able to control the overview as well as its own detail view. This solution (the one we presented here with the digital map table) is much richer in the possibilities to interact and to use the specific advantages of the fovea approach. But it is also more specific with respect to the application software and demands separate installations on the fovea devices. Both directions have their own benefits and drawbacks and have both their own chance for a future use.

Finally it is necessary to allow also interactions directly on the overview table in order not to be forced to use the FT in every case. Beneath the usage of regular pen interfaces (as they are offered for writing on digital blackboard) we actually survey the chances of hand gestures, recognized by the same tracking camera as we use for the tablets. This is working already very well for simple image manipulation. The future research will concentrate on building an integrated architecture that harmonizes the interaction over the Fovea-Tabletts with those by hand gestures and others in order to come to a conclusive concept for the future work on large computer displays.

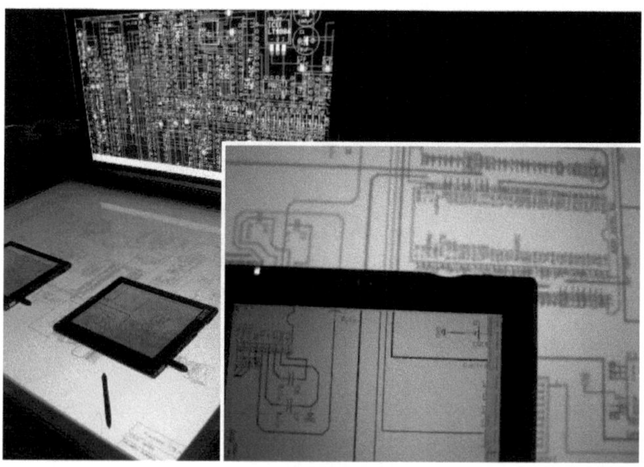

Fig. 9. Electronic circuit design: printed circuit board layout on the vertical screen, schematic diagram on the overview table with Fovea-Tabletts (detail view: right hand image)

Fig. 10. Manipulation of the overview image with camera tracked hand gestures (small image)

References

1. Everson, S. (ed.): Aristotle: The Politics, vol. 2. Cambridge University Press, Cambridge (1988)
2. Knöpfle, C., Stricker, D.: HEyeWall. Perfect Pictures for New Business Solutions, D. COMPUTER GRAPHIK topics, (1) (2004)
3. Baudisch, M.: Keeping in Context: A Comparative Evaluation of Focus plus Context Screens, Overviews, and Zooming. In: Proceedings of SIGCHI, pp. 259–266 (2002)
4. Ashdon, M., Robinson, P.: The Escritoire: A Personal Projected Display. Journal of WSCG, vol. 1(11) (2003)
5. Eck, R., Geisler, J., Rehfeld, N.: Vorrichtung zur visuellen Darstellung graphischer Detailinformation. German patent 10 2004 046 151.1, application on 23. 09. 2004
6. Peinsipp-Byma, E., Eck, R., Rehfeld, N., Geisler, J.: Situation Analysis at a Digital Situation Table with Fovea-Tablett. In: Proceedings of SPIE - Electronic Imaging, San José (to be published) (2007)
7. Sanneblad, J., Holmquist, L.E.: Ubiquitous Graphics: Combining Hand-held and Wall-size Displays to Interact with Large Images. AVI '06, Venezia, Italy (May 23-26, 2006)
8. Rehfeld, N.: Codierte Marker als Mess- und Interaktionskomponente für das Kindermuseum ZOOM in Wien, pp. 32–33. Fraunhofer IITB Jahresbericht, Karlsruhe (2001)

ZEUS – Zoomable Explorative User Interface for Searching and Object Presentation

Fredrik Gundelsweiler, Thomas Memmel, and Harald Reiterer

Human-Computer Interaction Lab
University of Konstanz, Universitätsstraße 10,
Box D 73, D- 78457 Konstanz, Germany
{gundelsw,memmel,reiterer}@uni-konstanz.de

Abstract. In this paper we describe a first version of ZEUS, a web application that combines browsing, searching and object presentation. With the zooming and panning based navigation concept of ZEUS and a hierarchical organization of the information space we try to solve the problems of information overload. It has to be evaluated if categorization, zooming and a full text search can minimize that the user gets lost in hyperspace. The concept of ZEUS is based on some thesis about human cognition, navigation and exploration which we hope to prove with evaluation and user testing of our application in the future.

Keywords: Zoomable User Interface, Interactive System, Complex Information Space, Usability, Navigation, Search, Browse.

1 Introduction

In this paper we describe a first version of ZEUS, a software application based on [18] that combines browsing, searching and object presentation which are usually developed separately in software systems. Some systems realize a good search, a nice object presentation or a well designed browsing but most fail to combine them in context of users' tasks. Literature research brought us to the conclusion that users are swapping between search and exploration mode while performing their tasks in information spaces [2], [20], [21]. This change of mode can occur at any time and is hard to predict because user tasks may change while detecting new information.

Software applications, especially for the web, are getting more and more complex with the progression of technological innovations, broadband internet and application programming interfaces like Adobe Flash. Not only commercial internet sites have to provide usable, efficient and effective search, browse and presentation mechanisms. Presenting a product on the web means to give the user an interactive experience including all product data and multi-media support (e.g. audio, video). It would be very nice for users to explore data in task context and several levels of detail. The buzzword "Web 2.0" represents not only principles like user contributed value and co-creation; it also animates to think over the application world. One possible future scenario is to have the web browser as main application handling all different user tasks from the search for information to the creation of an office document.

M.J. Smith, G. Salvendy (Eds.): Human Interface, Part I, HCII 2007, LNCS 4557, pp. 288–297, 2007.

2 Related Work

Operating systems like Windows Vista, Mac OSX Leopard or Sun's Looking Glass project [22] bring some innovation by introducing new search techniques like Spotlight, virtual desktop environments to swap different desktop views (OSX Leopard - Spaces) or windows in a 3d view that can be turned aside to save place on the desktop. Related to our work there are search applications like grokker [11], Liveplasma [12], DateLens [3] and DesignClicks [8] using interesting zoom-, browse and search-techniques in combination. Other systems like MuSE [9], the Attribute Explorer [25] or applications created with Piccolo [4] or Pad++ [5] realize filter concepts with immediate user feedback following or expanding the dynamic query approach [1]. Our ambition is to design a user interface (UI) which picks up good ideas and presents multiple ways a user system view could look like. We think that conventional WIMP (windows, icons, menus, pointers) systems won't solve the problems of usability and user experience we are faced with today. That's why we propose zoomable user interfaces (ZUIs) which can be used to take advantage of the natural way users navigate and orientate in space supporting their physical abilities and habits.

Raskin describes the zooming interface paradigm in [19]. He considers it as superior in comparison with conventional paradigms. "*The zooming interface paradigm can replace the browser, the desktop metaphor, and the traditional operating system.*" [19]. We took this paradigm as initial point for the development of our application. His argument is the better match between zooming navigation concepts and human cognition. Colin Ware supports this theory in [26] where he stated that humans move their bodies and thus their view mostly forward and backward but seldom sideward. Therefore a simple UI has to support just three degrees of freedom. Two degrees of freedom are needed for rotation (direction and pitch) which matches the movement of the head and one degree is for moving forward or backward. Transferred to the design of an UI this argument could be interpreted to realize a panning and zooming navigation. Panning the virtual information space complies moving the head and thus changing the visible area or aspect of interest. The zoom in and out is like moving forward and backward in space controlling the degree of interest of an object.

3 Technical Constraints

There are a lot of technical constraints when developing web applications. The main criterion is that the application has to run in a web browser. Especially when visualizing large datasets the bandwidth and data transfer rates have to be kept in mind. The application should make use of intelligent preloading of data which will probably be explored next by the user. Dependant on the complexity of the visual representation and number of data items the performance maybe an issue because the browser or used browser plug-ins do not support hardware rendering. Another issue is the diversity of browser implementations (Mozilla Firefox, Internet Explorer or Safari may interpret e.g. a java script in different ways). We thought about four approaches to web application development that could meet our requirements of rapid feedback, interactivity and data transfer.

One possible development approach is using DHTML in combination with AJAX [10]. This is working fine for most web browsers and no plug-in is needed. The disadvantage is that there are a lot of AJAX frameworks like Macao [13] or Mochi Kit [15] but no standard development environment which makes coding more comfortable. Additionally the UI and its interactive behaviour are not as easily implemented as with Adobe Flash/Flex 2 or Java. In contrary to the AJAX thin client solution, browser plug-ins can be used. Sun's java applets can be developed fast but are in need of the java plug-in and have an initial loading time on start-up. Adobe Flash/Flex 2 need the installation of the Flashplayer browser plug-in but have the advantage of easy-to-use development environments and standard support for vector graphics. Another interesting possibility comes from Microsoft with the Windows Presentation Foundation (WPF) [14]. Development of web applications is possible with the .net 3 environment and works with the Internet Explorer 7. WPF works with vector graphics and even supports 2d and 3d hardware rendering.

For development of the ZEUS prototype we have chosen Adobe Flash, the server side script language PHP as middleware and a free SQL database. With these components a comfortable development and prototyping of web applications is possible.

4 ZEUS - Theoretical Background

As we have seen a lot of technical constraints exist. Other problems appear when developing a concept for such a web application. One big problem is that the target user group is enormous and different users prefer different kinds of navigation, browsing, searching and object representation. One could reason that we need a system that supports all different navigation, interaction and visualization techniques. The user should be able to configure his preferred view. Following this approach arises two more problems. It is hard to decide which parts of the UI should be configurable. The second issue is the quantity of configurable options. Users are easily overextended if too many options are available. Apart from that it is essential for all options to be easy usable and accessible.

On the first glimpse the "Map View"-visualization of the grokker search engine [11] looks like our iNShOP application (see figures 1 and 2). That's why we want to point out the differences between both applications. Grokker builds up a hierarchy of the search results automatically (it can be searched in Yahoo, Wikipedia and Amazon Books). In addition to the outline view (an expandable list tree view) the search results can be explored in a hierarchical, zoomable "Map View". With a slider the user is able to narrow search results to the newest documents by date. In fact there are several important differences between the applications. A first technical and maybe negligible point is that grokker is written in the java programming language and therefore needs the java web plug-in to be installed while ZEUS is realized with Adobe Flash and needs the Flashplayer browser plug-in which is widespread among internet users. ZEUS doesn't create its categories automatically. They have to be part of the database connected to ZEUS. With the web 2.0 one could think about to involve user in the categorization process by enabling them to create their own categories and change the assigned categories of data items. This approach generates a more comprehensible categorization than a computer system could calculate automatically. Apart from that the main interaction concept of grokker is like the one of ZEUS. A problem with grokker could be the depth of the hierarchy. While ZEUS makes the depth visible to the user by the category combo boxes for each level, the grokker-hierarchy is not visible to the user. ZEUS is more flexible with the categorization because users can change and select a proper category for each level.

Another difference is the main concept. Grokker is a web search engine that tries to get away from the conventional list presentation of search results. In contrast the focus of ZEUS lies on browsing the information space and on the presentation of simple and complex multimedia data which has to be supported by a full text search. A stringent semantic zoom concept realized in ZEUS is the visualization of search results in different detail depths. Grokker doesn't visualize the details of a result in the "Map View" but in a detail area on the right side of the application. In ZEUS the user can zoom in to an item to view all of the available information directly in the visualization. If the item consists of a lot of information a further zoom in on the item is possible to explore all of its information areas.

A feature of ZEUS that grokker does not have is the change of visualizations for the category tiles. The user can chose between round or square category representation using grokker but no other kind of visualization like a scatterplot or a treemap. A change of the category tiles enables the user to use different visualizations for different data item collections. One concept of Zeus that is not yet implemented is the collection of important data items. The user then can create own category tiles to store individual collections of relevant information.

4.1 Information and Navigation Overload

Information Overload or *Information Pollution* is the term expressing the flood of information rolling over today's internet users. The term *Information Overload* was coined 1970 by Alvin Toffler in [23]. Individuals that are confronted with too much information accomplish their tasks less efficient. The same problem may occur if users can't understand the organization of a presented information space in a search application. This can be the outcome of a bad information organization and presentation or *Information Overload* as described in [24]. If the UI presents too much information at once, users aren't able to process it correctly. They overlook information, categorize relevant information as irrelevant or even feel interfered. The same problem can occur with an "overloaded" navigation. In this case the user is confronted with too many navigation options. Difficulties with the logical organization of the information space have direct effect on the orientation and decision-making ability of the users in the digital world. Too many links and possible ways alienate users so that they can't decide how to reach their user goals. Our thesis is that we can solve these problems by a hierarchical information structure and zooming/panning enriched navigation supported by a full text search.

4.2 Navigation for Searching and Browsing

With ZEUS we develop a ZUI that supports searching and browsing in large information spaces with regard to user tasks, goals and activities. Therefore our system combines search and filter techniques with zooming interaction techniques to narrow search results. The wide range of internet application users with varying computer and domain experience makes it necessary to develop adaptable navigation and visualization concepts. Novices and experts shall be able to solve their tasks by using different strategies.

Users are mixing two basic approaches while trying to find the needed information. These two main strategies are on the one hand the directed search for known items and on the other hand the exploration (e.g. by browsing) of different items of interest. The used terms for searching and browsing are slightly different in literature. In [21] the term "teleporting" is used when a user directly jumps to the needed information. The browsing strategy is described by the term "orienteering". The user narrows the information space by "*a series of steps (e.g., selecting links) based on prior and contextual information to hone in on the target*" [21]. Our opinion is that we can support both strategies best with a UI that includes an always visible full text search and a flexible hierarchical visualization of the information space. In this way the user can change his strategy from searching (or teleporting) to browsing (or orienteering) or the other way round at any time. Concerning the browsing strategy we think that navigation by animated panning and zooming supports the collection of information and the orienteering tasks of users in the best way. It is easier for users to move visually through an information space to explore its contents than to navigate through a hyperlinked collection of sites which organization is invisible to the user. This animated exploration mode has an additional advantage regarding the search function. Is the search result (e.g. triggered by a full text search) one item or a collection of items, the user view can be focused on the relevant item or item group by automatically zooming and panning to its position.

5 ZEUS – User Interface Design and Concept

In the following we present the idea of ZEUS in form of a sample application called iNShOP that is connected to a music database. The application consists of a main area visualizing the objects and categories, and a filter area for search and choosing categories (see figure 1). Filter and category operations are triggered by selecting category attributes in the combo boxes. Selecting "Music type" as first category level organizes the results like in figure 1 with the categories "Drum'n'Base", "Classic", "Pop", "Dance/Electro" and the other eight. Our thesis is that this mix of hierarchical and flat data representation enables a quick switch between search and exploration mode and therefore a more efficient and effective work.

Tiles are the main visual components used to organize the information space and visualize the data items. There are two different types of tiles. Category tiles organize

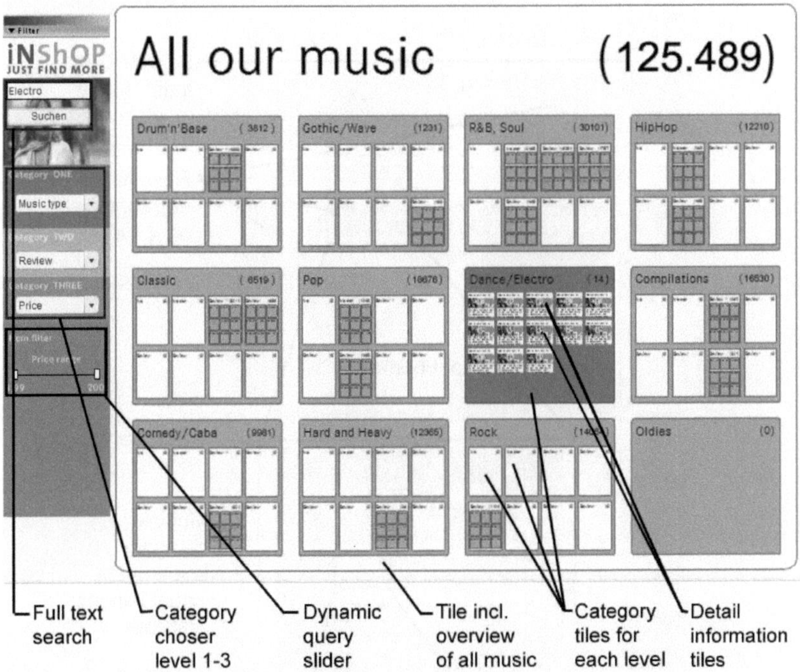

Fig. 1. ZEUS main overview by example of a virtual music store we called iNShOP

the information space in groups on different hierarchy levels. They can include further category tiles or information tiles to visualize the data items in that level. An information tile visualizes one item and can include text, images and multimedia objects like video and sound. Selecting a category in a combo box initiates a recalculation of the tile organization with a redraw of tiles.

For the system decision weather to further categorize the information space or visualize the included information tiles, we can set an item threshold. That means if e.g. less than fifty items (threshold=50) fall in the category "electro", they are shown in the "electro" tile and the information hierarchy is not expanded. If more than fifty items fall in the "electro" category the visual hierarchy is expanded and further category tiles (e.g. price or review points) are created within the "electro" tile. Additionally we included a price slider to narrow the price range of visualized items.

5.1 Main Navigation

The main navigation of ZEUS concerns all visible areas except the information tiles. Semantic and geometric zooming is used to show relevant information at different aspects and/or degrees of interest. Figure 2 shows a short scenario how a search for an artist called "Moby" could look like.

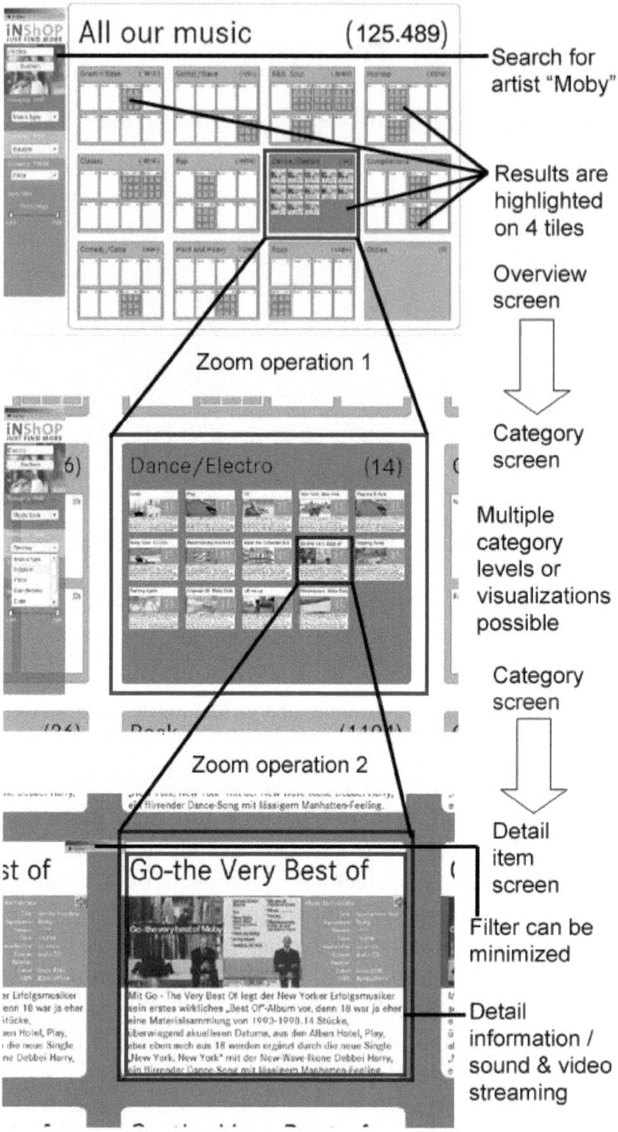

Fig. 2. ZEUS from overview to detail view by zoom in

First the user enters "Moby" as search query into the search field located in the movable and resizable filter area (best visible figure 1 left). The system processes the query and highlights the results. In this case there are results for "Moby" on four tiles ("Drum'n'Bass", "HipHop", "Compilations" and "Dance / Electro"). Now the user is able to change the category combo boxes to let the system build up a hierarchy matching his interests. Maybe on the first level he could be interested in the review rating, so he would chose "review" for the first level combo box. For simplicity we assume he is interested in the "Music type" categorization on the first level because he wants to see which albums are in the category "Dance/Electro". On the first glimpse the user sees that fourteen albums of the artist fall into this category. With a click on the "Dance/Electro" tile the user triggers zoom operation one and the tile enlarges to screen fitting size. Zoom and pan actions are clarified to the user by an animation [6]. Now that the category tile is larger, the included albums are better readable. In future there should be an area where the user can select different visualizations for the category tile view. Let's assume that the user is interested in the album with the title "GO – the very best of". A click on that tile triggers a further zoom into the detail information tile (figure 2 bottom and figure 3). Now the user can see all album details. The way back to the category overview is triggered by clicking on the category tile area visible between the detail information tiles. So it is possible to navigate back to the main overview at any time.

5.2 Navigation and Visualization Within Information and Category Tiles

Navigation within an information tile should follow some straightforward rules so that the main navigation concept still remains consistent.

Fig. 3. ZEUS detail item view showing all data of an item e.g. an audio cd

In a detail view title, artist, price, format and so on are shown. Additionally there media streaming can be included. The category tiles shall be able to change their look and feel. In contrary to the systems in the related work section different visualizations can be integrated into our "meta"-system. It will be possible to integrate various visualizations like a table, scatterplot, treemap or others. Each category tile will be able to present its contents independently by other visualizations. Additionally different databases or XML sources can be connected to ZEUS. For each data type dependent on the kind of data other basic templates are available to visualize the data objects. Let's imagine we have music data objects like in iNShOP. The basic template can show title, label, band, other detail information, album image and play an example sound stream. An easy way to add video support would be to adapt a basic template. A play button can trigger the video stream to start and has to be included in our adapted template. Additionally the visual overall representation can be different from the basic template. We propose to use templates for each kind of data domain (e.g. sound compact discs, books, cars or computers).

6 Discussion of Results and Future Work

In summary, we have developed a first version of a nested ZUI that combines search and browse techniques to support users in their search modes. New in ZEUS are the combination of searching and browsing in one UI, the idea of switching the visualization for item groups and the connection of data sources whose results are visualized by different adaptable templates. Our assumption is that users are generally not very familiar with zoom navigation but it may be more effective and efficient than conventional navigation, especially coupled with a full text search and a hierarchical information organization. The main thesis that has to be verified is that human cognition is better supported by zooming and panning than by any other navigation style when exploring an information space visually. A technical issue is runtime rendering without hardware support. Therefore zooming operations with more than approximately 200 items (depending on hardware and complexity of items) are too slow and the system gets stuck. Maybe a change of the development environment to Adobe Flex or Microsoft WPF can avoid this. Future work will be the implementation of visualizations contained in the category tiles and further improvement of the UI. Additional search and exploration techniques like scent trails [16] or navigation by query [7] can be implemented. At the end we would like to test different application versions (which differ in database, search techniques, browsing techniques, with and without zooming and panning) against each other.

References

1. Ahlberg, C., Shneiderman, B.: Visual Information Seeking: Tight Coupling of Dynamic Query Filters With Starfield Displays. In: Ahlberg, C. (ed.) Proceedings of Human Factors in Computing Systems, pp. S313–S317. ACM Press, New York (1994)
2. Bates, M.J.: Toward an Integrated Model of Information Seeking and Searching. The Fourth International Conference on Information Needs, Seeking and Use in Different Contexts, Lisbon, Portugal (September 11-13, 2002)
3. Bederson, B.B., Clamage, A., Czerwinski, M.P., Robertson, G.G.: Datelens: A fisheye calendar interface for pdas. ACM Transactions onComputer-Human Interaction 11(1), 90–119 (2004)
4. Bederson, B., Grosjean, J., Meyer, J.: Toolkit Design for Interactive Structured Graphics IEEE Transactions on Software Engineering (TSE). HCIL-2003-01, CS-TR-4432, UMIACS-TR-2003-03 (January 2003)
5. Bederson, B.B., Hollan, J.D.: PAD++: A zooming graphical user interface for exploring alternate interface physics. In: UIST 94: 7th ACM Symposium on User Interface Software and Technology, pp. 17–27. ACM press, New York (1994)
6. Bederson, B.B., Meyer, J., Good, L.: Jazz: An extensible zoomable user interface graphics toolkit in Java. UIST'00, ACMSymposium on User Interface Software and Technology, CHI Lett. 2(2), 171–180 (2000)

7. Cunliffe, D., Taylor, C., Tudhope, D.: Query-based Navigation in Semantically Indexed Hypermedia. In: Proceedings of the eighth ACM conference on Hypertext, Southampton, United Kingdom, pp. 87–95. ACM Press, New York (1997)
8. Designklicks visual presentation of art 2007, URL: (last visited: 09.02.2007) http://designklicks.spiegel.de/
9. Furnas, G.W., Zhang. X.: MuSE: A Multiscale Editor. In: Proceedings of ACM UIST'98, pp. 107–116 (1998)
10. Garrett, J.J.: Ajax: A New Approach to Web Applications. Adaptive Path LLC (February 18, 2005), Url: (last visited: 09.02.2007) http://www.adaptivepath.com/publications/essays/archives/000385.php
11. Grokker seeking system. URL: (last visited: 09.02.2007) http://www.grokker.com/
12. Liveplasma seeking system. URL: (last visited: 09.02.2007) http://www.liveplasma.com/
13. Macao Ajax Framework. Url: (last visited: 09.02.2007) http://macao.sourceforge.net/
14. Microsoft Windows Presentation Foundation (WPF) at msdn. Url: (last visited: 09.02.2007) http://msdn2.microsoft.com/en-us/netframework/aa663326.aspx
15. Mochi Kit Ajax Framework. Url: (last visited: 09.02.2007) http://mochikit.com/
16. Olston, C., Chi, E.H.: ScentTrails: Integrating Browsing and Searching on the Web. ACM Transactions on Computer Human Interaction 10(3), 177–197 (2003)
17. Perlin, K., Fox, D.: Pad: an alternative approach to the computer interface. In: proceedings of Computer graphics and interactive techniques (1993)
18. Ken, P., Jon, M.: Nested user interface components. In: Proceedings of the 12th annual ACM symposium on User interface software and technology, Asheville, North Carolina, United States, November 07-10, 1999, pp. 11–18. (1999)
19. Raskin, J. (ed.): The Humane Interface. New Directions for Designing Interactive Systems. Addison-Wesley Verlag, Reading, MA (2000)
20. Rose, D.E.: Reconciling information-seeking behavior with search user interfaces for the Web. JASIST 57(6), 797–799 (2006)
21. Schaffer, E., Straub, K.: The answer you're searching for is...browse. In: UI Design Update Newsletter (January 2005), Url: (last visited: 12.02.2007) http://www.humanfactors.com/downloads/jan052.htm
22. Sun Microsystem: Project Looking Glass. Url: (last visited: 12.02.2007) http://www.sun.com/software/looking_glass/
23. Toffler, A.: Future Shock. Random House Inc., Reissue Edition (September 1984)
24. Turetken, O., Sharda, R.: Development of a fisheye-based information search processing aid (FISPA) for managing information overload in the web environment (ISSN: 0167-9236). In: Decision Support Systems, June 2004, vol. 37(3), pp. 415–434. Elsevier Science Publishers B.V, Amsterdam, Netherlands (2004)
25. Tweedie, L., Spence, B., Williams, D., Bhogal, R.: The Attribute Explorer. In: CHI'94, Boston, Massachusetts USA, April 24-28, 1994, ACM, New York (1994)
26. Ware, C.: Information Visualization: Perception for Design. Morgan Kaufmann, San Fransisco, Kalifornien (2004)
27. Allison, W., James, L., Michael, S.: Goal-directed zoom. In: CHI 98 conference summary on Human factors in computing systems, Los Angeles, California, United States, April 18-23, 1998, pp. 305–306 (1998)

Folksonomy-Based Collaborative Tagging System for Classifying Visualized Information in Design Practice

Hyun-oh Jung, Min-shik Son, and Kun-pyo Lee

Human Centered Interaction Design Laboratory, Department of Industrial Design, Korea Advanced Institute of Science and Technology, Daejeon, Republic of Korea
{brianjung,uniquex,kplee}@kaist.ac.kr

Abstract. The aim of this research is to suggest folksonomy-based collaborative tagging system for supporting designers in group who interpret visualized information such as images through grouping, labeling and classifying for design inspiration. We performed field observation and preliminary studies to examine how designers interpret visualized information in group work. We found that traditional classification methods have some problems like lack of surface and time consuming. Based on this research, we developed PC based group work application, named I-VIDI. By implementing I-VIDI based on functional requirements, we have showed how I-VIDI reduces problems found from current image classification methods such as KJ clustering and MDS. In future case study, we plan to conduct extensive user research to evaluate the system further as well as adding more functions which can be usefully applied to collaborative design work.

Keywords: Collaborative Tagging, Image Classification, Information Visualization, Information Organizer, Folksonomy.

1 Introduction

1.1 Research Background

Because designers should collect a lot of data and analyze them to develop a new concept, collaborative work became important issue.[1]

Current typical methods for information structuralizing in design field are KJ clustering and image map generation via multidimensional scaling. For visualizing information, designers often use diagrams. To analyze visual information and get insights, designers assign keywords on images and classify them.[2]

However, these methods have some problem like limitation of surface and time efficiency as amount of information increased. To solve these problems and support collaborative design problem solving, this research focused on developing tool that support collaborative visual information analyzing. As a start point of this research, collaborative tagging has similar concept to this project that tagging on visual information and share[3], and we expect this aspect can be applied effectively.

---

[1] Marvin E. Shaw: Group structure and behavior of individuals in small group, Journal of Psychology (1954) 139-149.

[2] Gert, Passman: Design with Precedents (2005).

[3] http://www.wikipedia.org/wiki/folksonomy

M.J. Smith, G. Salvendy (Eds.): Human Interface, Part I, HCII 2007, LNCS 4557, pp. 298–306, 2007.

1.2 Research Objectives and Methodology

Aim of this research is finding limitations of current methods which used by designers to analyzing visual information, and develop tool that solve the problems and support visual information based collaborative design work. By these objectives, the research carried out by following process described in Fig. 1.

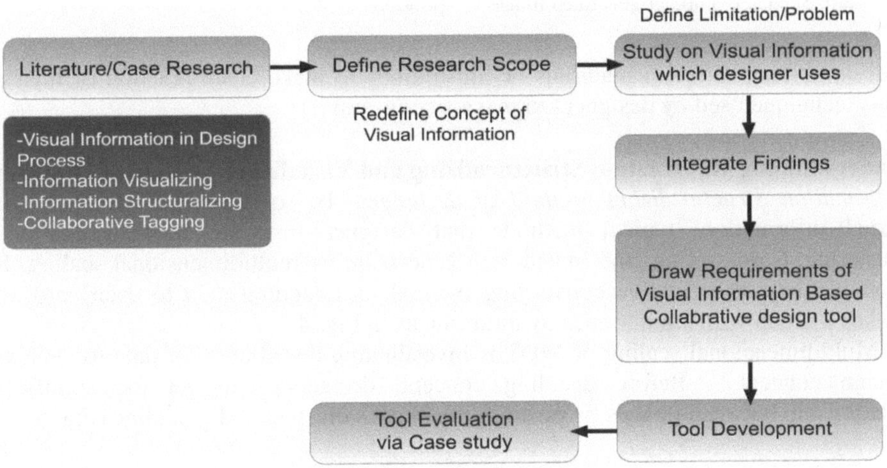

Fig. 1. Research Process

2 Study on Theoretical Background

2.1 Visual Information Based Design and Collaborative Tagging

Understanding Visual Information in Design Process. In this research, we studied on type and concept of visual information used in design process, and tried to redefine the scope of visual information in design field.

Definition of Visual Information in Design. In design field, "Visual Information" means visual artifacts that can be processed by design purpose, which consisted with line, shape, texture, color, or volume, the basic elements of visual objects.[4]

Images as Visual sources in Form reference phase. Various image collection or notes become collage or one big image and can be used as source of design idea to visualizing context and evaluate them.[5]

Image collage as Source of Design Inspiration. Collage is regarded as a work of visual arts made from an assemblage of different forms, thus creating a new whole[6]

---

[4] Kolli, R. Pasman: Some consideration for designing a user environment for creative ideation (1993) 72-77.

[5] Gert, Passman, Design with Precedents (2005) 78-81.

[6] http://en.wikipedia.org/wiki/Collage

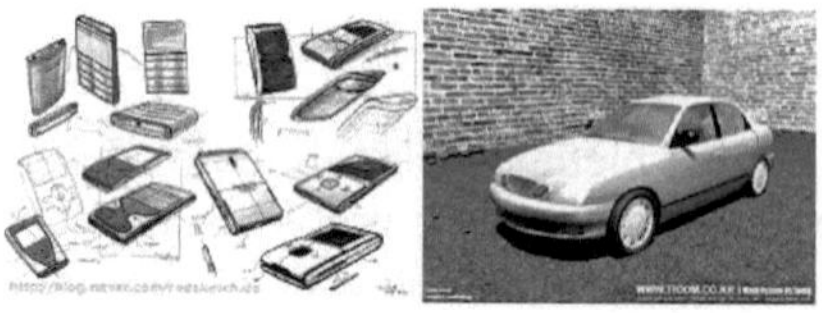

Fig. 2(left). Example of visual information: Sketch, Computer generated image

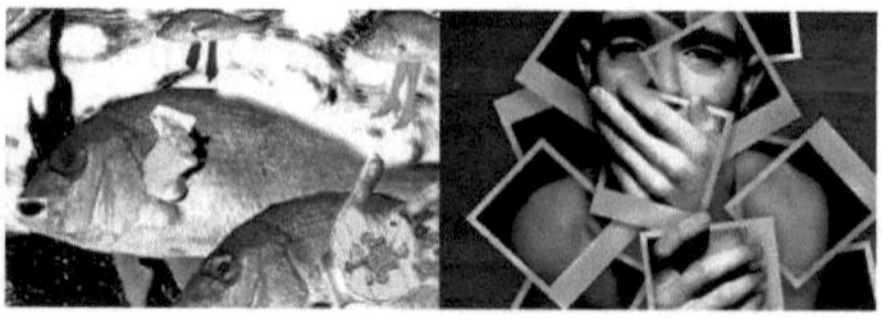

Fig. 3(right). Collage: Source of design inspiration

that combining different materials or illustrations to make chain reaction of images. This technique used by designers to make a suggestion.

Understanding Information Structuralizing and Visualizing Method of Designer. *Information Structuralizing Method of designers.* To solve design problems with visual information, typical methods that designers use are KJ(Kawakita Jiro) clustering, Card sorting, and image map generation by multidimensional scaling. KJ method, the representative converging method, is assigning data to each card and makes a group with similar ones by intuition, as in Fig. 4.[7]

Multidimensional scaling or MDS is investigating distribution of data to deciding design concept [8]. Before deciding concept, designers can get the significant relationship between images by position the images on specified guideline (Fig. 5).

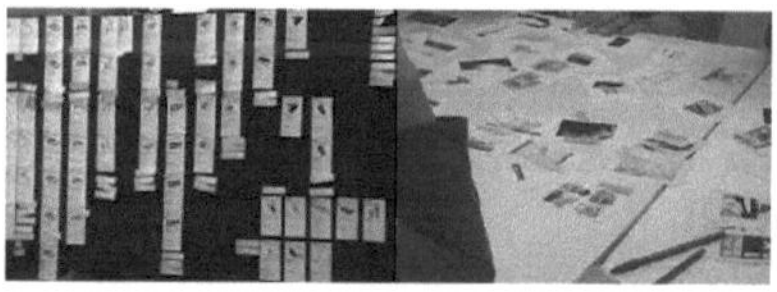

Fig. 4(left). Information clustering by KJ method

Fig. 5(right). Image positioning by MDS

Information Visualizing Method of designers. Popular method to visualizing information in design field is graphic organizer such as diagram. Diagram is way to express the abstract relationship between information using standardized shape[9].

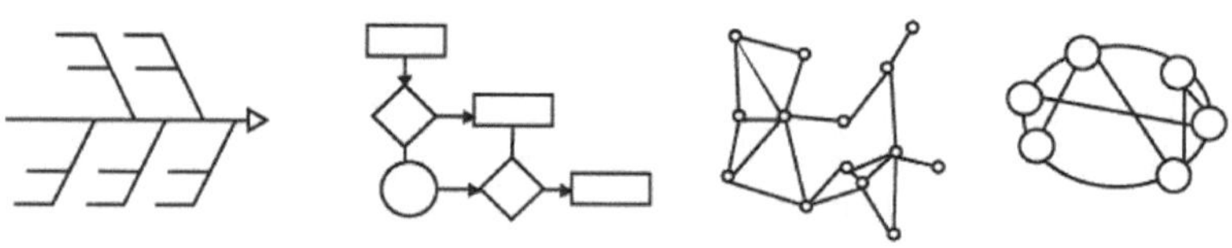

Fig. 6. Diagrams for visualizing information

7 http://www.mycoted.com/KJ-Method

8 William Write: From Function to Centex to Form: Precedents and Focus Shifts in the Form Creation Process, In Proceedings of C&C, April 1215 (2005) 195-204.

9 Collin Ware, Information Visualization, Perception for Design (2005).

Collaborative Tagging as Information Assigning and Analyzing. Tag is keyword that represents the web contents in subject or category. It can be expressed freely as a word, two-character word[10]. Tagging method is popular in web site like Flickr.com[11].

2.2 Redefinition of Concept of Visual Information for Research

Based on result of literature research and case research, we redefine the visual information in design field as following:

- Design precedence image as form factor reference data in design process
- Collage as source of design inspiration in concept decision
- Array and distribution as a result of KJ clustering and MDS
- Tag information of images
- Visualized information as a result of diagrams

3 Requirement Deductions for Tool Development

3.1 Understanding Requirement Through Field Research

To find the way how people use visual information, we visited educational space that teaching product styling. Participants of field interview are 20 people of men and women in undergraduate course 3rd and 4th year of Kookmin University[12]. To investigate their workspace, we took a photo of their individual/public studio. Also we visited laboratories where using visual information. Field research and interview performed on 21st~22nd July, 2006.

Practical Usage of Visual Data in Collaborative Work Environment. Product precedence images, images for trend map, visual data for concept development are main visual information for cooperative environment.

Practical Method to grant information to image data. Most popular way to assigning information to images is direct annotation on images with text memo.

Sharing Visual data. They bring the source itself like magazines to others.

Practical Usage of visual data to solve design problem. They used image as reference of product shape, color and concept deduction.

Problems of current image data structuralizing method. The biggest problem is it takes too much time to structuralizing images using KJ clustering, road mapping, and so on.

3.2 Study on Information Structuralizing Method in Design Field

The usage and limitation of current information structuralizing methods in design field are like following Table 1.

---

[10] http://www.wikipedia.org/wiki/Tag
[11] http://www.flickr.com/photos/tags/
[12] http://english.kookmin.ac.kr

Table 1. Usage and limitation of information structuralizing methods

Method name	Usage in design field	Limitations
KJ Method	Clustering relative image	Insufficient surface for a lot of images
	Assign image as keyword on image cluster	Hard to reveal the relationship between groups
MDS	Find trend between images	Difficult to positioning images on various conditions due to the fixed axes.
	Visualize image distribution on specified conditions	Difficult to measure scale value between images
Card Sorting	Hierarchical Information grouping	Difficult to reveal the relationship between groups
	Simulating web site structure	Unable to change standard

3.3 Study on Information Visualizing Method in Design Field

Based on literature research, flow chart, concept mapping, matrix and webbing are typical diagram method to visualizing information. There characteristics are describing, comparing/contracting, classifying, sequencing, casual arranging, and decision making. By purpose of information visualizing, usage of diagram method is distinguished.

3.4 Study on Collaborative Tagging

Folksonomy, a part of collaborative tagging, is compound word of Folk(people)+order+nomos(law) and it means "Categorizing by people"[13]. Folksonomy is commonly used in website like Flickr, Del.icio.us, and blog services. Table 3 shows comparing between taxonomy and folksonomy, the collaborative tagging.

Table 2. Comparing between taxonomy and folksonomy

Aspect	Taxonomy	Folksonomy
Information structure	Hierarchical tree	Radial shape network
System Characteristic	Systematic, Static, One-way	Non-systematic, Dynamic, Interactive
Information generation	Make category before enter information	Enter information first, and classify
Expansion	unable	Possible
Categorizing cost	High	Low

[13] http://www.wikipedia.org/wiki/Taxonomy

3.5 Deduction of Requirements

Based on result of field research, we draw requirements of tool through process described in Fig. 7. Tool requirements are as following:

- Clustering a lot of images quickly, and overcome limit of surface.
- Tag on specified are of image
- Variable axes composition ability in MDS.
- Fluent image sharing between participants.
- Enable remote conference, not only local.
- Separate individual work space from public one.

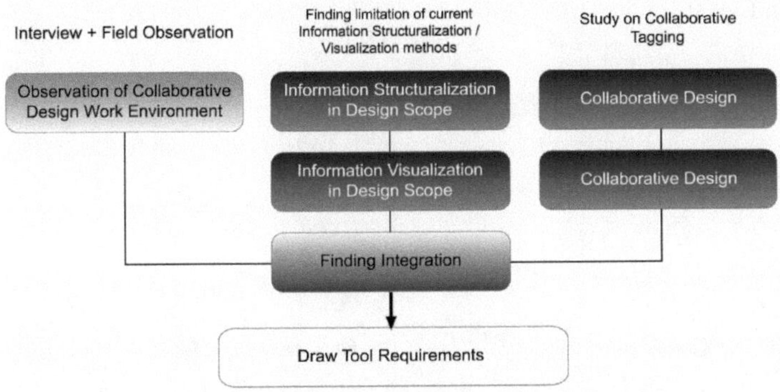

Fig. 7. Process of Requirements deduction

4 Development of I-VIDI Tool (Interpretation of Visualized Information for Design Inspiration)

4.1 Schema and Organization of I-VIDI

System schema of I-VIDI is described in Fig. 9.

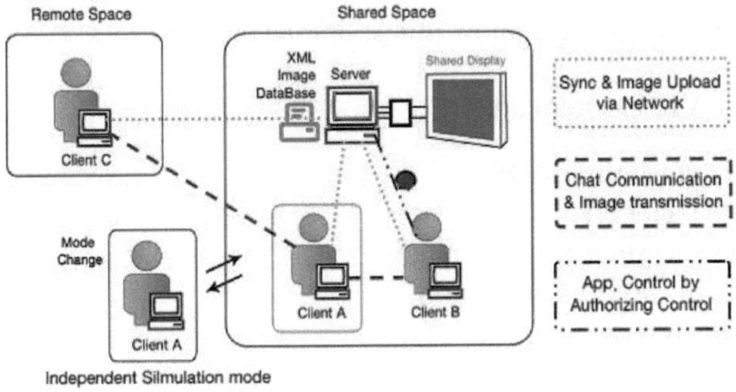

Fig. 8. I-VIDI system schema

I-VIDI is PC based application developed with Flash action script 2.0 and Visual C++. FSCommand is used to connect flash and Visual C++ application, and communication is established via socket. I-VIDI designed to support local and remote conference in shard display. Every client screen synchronized via network connection. Also, members are allowed simulate work independently within connected session. Image data which participants upload to server is saved as XML data. I-VIDI is designed to support collaborative work in design concept generation phase of product design.

4.3 Functions of I-VIDI.[14]

I-VIDI application consists with server and client. Whole layout of each application is described in Fig. 9.

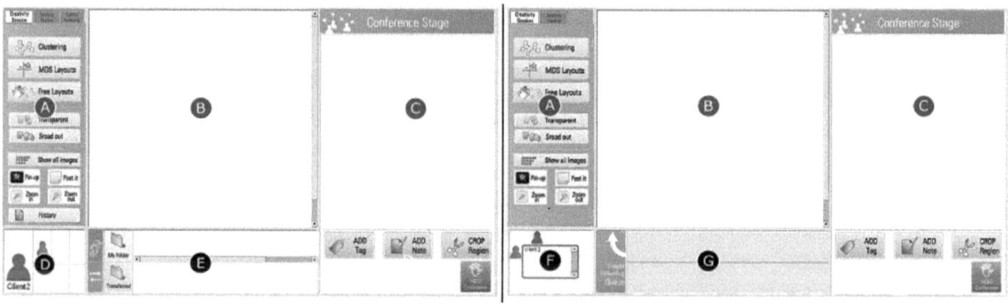

Fig. 9. I-VIDI client(left) and server(right) layout

- A(UI panel): Controls to using I-VIDI system
- B(Image Stage): Display area for KJ clustering or MDS results.
- C(Conference Stage): Shared area for keyword assign, direct annotating
- D(Client communication panel): Show client connection information. Image sharing channel. Client only.
- E(File management panel): Show image data
- F(Client Log-in status panel): Show client status and control authorize. Server only.
- G(Image uploading queue): Show image upload status

Key features of I-VIDI system are:

- Automated KJ clustering/MDS
- Image share via network
- Make a tag on specified area of image
- Two types of Keyword visualization
- Task history browsing

[14] I-VIDI demo video is available at http://storm541.dothome.co.kr/ividi.wmv

5 I-VIDI Evaluation Through Case Study

5.1 Plan and Progression of Case Study

Purpose of case study is to observe how the participants use I-VIDI system and measure usability and satisfaction of application. Participants for case study are a team consists with 4 men and women students who have experience of styling based design, and age range is 20s. Given task is generate concept of mobile music player for 20s women. This case study performed 2 times. After each session, we conducted quantitative evaluation on usability and satisfaction with 7-point Likert scale debriefing questions.

5.2 Result Analysis and Conclusion

According to the case study participants' answer, I-VIDI contributes to increase satisfaction of conference session. Especially in image arrangement, collaborative tagging and KJ clustering aspects, participants gave high score. However, image browsing and annotating on specified area function got relatively low score.
As a conclusion of case study, the benefits of using I-VIDI in conference are as following:

- Help collective intelligence via collaborative tagging
- Increase efficiency of conference in terms of time and cost
- Support drawing concrete concept

6 Conclusion and Further Studies

In contrast with taxonomy based classification, folksonomy way can assign multiple categories on data by tag. Namely, classification system can be selected by user dynamically. We suggested I-VIDI, collaborative visual information tagging system too through this research, and we proved I-VIDI is improved conference environment in efficiency, usability and satisfaction aspect by case study and debriefing.

Finally, through the case study, we set the research direction and improvement for further work:

6.1 Technical Limitations and Problems

Because more than one people involved in I-VIDI conference environment, authority control is important. And when more than 8 people involved, network slowed down due to the communication overload.

6.2 Expansion to Integrated Tool for Whole Design Process

By adding sketching or idea visualization functions, I-VIDI can become integrated design solution aiding tool.

References

1. Colin, W.: Information Visualization, Perception for Design (2005)
2. Gert, P.: Design with Precedents (2005)
3. Keller. A: For Inspiration only, Designer interaction with informal collections of visual material (2005)
4. Buzan, T., Buzan, B.: The mind map book. Millennium Edition, pp. 144–155. BBC Worldwide Ltd, London (2000)
5. Crowley, T., Baker, E., et al.: Mmconf: an infrastructure for building shared applications. In: Proceedings of the Conference on Computer Supported Cooperative Work, ACM Press, New York (1990)
6. Eppler, M., Burkhard, R.: Knowledge Visualization, Encyclopedia of Knowledge Management, Idea Group (2005) (to appear)
7. Eckrt, C.M., Stacery, M.M: Sources of inspiration: A language of design. design studeis 21(5), 523–538 (2000)
8. Greenberg,: Computer-supported Cooperative Work and Groupware. Academic Press, San Diego (1991)
9. Koyama, K.: The KJ editor: a supporting tool for idea generation (in Japanese), Thesis for Master of Engineering, Toyohashi University of Technology (1988)
10. Kolli, R.: Pasman Some consideration for designing a user environment for creative ideation. In: Proceedings of the interface '93, pp. 72–77 (1993)
11. Noguchi, H.: How do constraints Facilitate designer's Thinking? In: Proceedings of 4th International Round Table Conference on Computational Models of Creative Design, pp. 265–275 (1998)
12. Nagia, Y.: How does designer think with drawing in design process? In: Proceedings of 5th Asia Design Conference, International Symposium on Design Science, pp. 77–97 (2001)
13. Osborn, A.F.: Applied Imagination, Scriber's.,New York, 58 (1953)
14. Greenberg, S.: Sharing views and interactions with single-user applications. In: Proceedings of the Conference on Office Information Systems, pp. 227–237 (1990)
15. Greenberg, S., Marwood, D.: Real Time Groupware as a Distributed System: Concurrency Control and its Effect on the Interface. In: Proceedings of the ACM Conference on Computer Supported Cooperative Work, pp. 207–217 (1994)
16. Tatar, D, Foster, G., Bobrow, D.: Design for conversation: Lessons from Cognoter. In: Computer Supported Collaborative Work, pp. 55–79. Academic Press, San Diego (1991)
17. Viégas, F.B.: Visualizing Email Content: Portraying Relationships from Conversational Histories, CHI (2006)
18. Wang. W., Wang, H., Dai, G., Wang, H.: Visualization of Large Hierarchical Data by Circle Packing, CHI (2006)
19. Wright, W.: From Function to Centex to Form: Precedents and Focus Shifts in the Form Creation Process. In: Proceedings of C&C, London, United Kingdom, April 12-15, 2005, pp. 195–204 (2005)
20. Wattenberg, M.: Visual Exploration of Multivariate Graphs, CHI (2006)
21. Wright, W.D., Schroh, P., Proulx, A., Skaburskis, B.: Cort: Oculus Info Inc. The Sandbox for Analysis - Concepts and Methods, CHI (2006)
22. Chen, Y.: Image Categorization by Learning and Reasoning with Regions. Journal of Machine Learning Research 5, 913–939 (2005)
23. Yukari, N.: How Designer Transform Keywords into Visual Images, C&C'02, pp. 118–112 (2002)

Interactive Product Visualization for an In-Store Sales Support System for the Clothing Retail

Karim Khakzar[1], Rainer Blum[1], Jörn Kohlhammer[2], Arnulph Fuhrmann[2], Angela Maier[3], and Axel Maier[3]

[1] Fulda University of Applied Sciences, Department of Applied Computer Sciences, Marquardstr. 35, 36039 Fulda, Germany
{Karim.Khakzar,Rainer.Blum}@informatik.hs-fulda.de
[2] Fraunhofer Institute for Computer Graphics (IGD), Fraunhoferstraße 5, 64283 Darmstadt, Germany
{Joern.Kohlhammer,Arnulph.Fuhrmann}@igd.fraunhofer.de
[3] Reutlingen University, School of Textiles and Design, Alteburgstr. 150, 72762 Reutlingen, Germany
{Angela.Maier,Axel.Maier}@reutlingen-university.de

Abstract. The development of an in-store sales support system that focuses on the "virtual try-on" of clothing is the aim of the research project "IntExMa". Based on sophisticated virtual reality technology, the interactive system provides visualization of made-to-measure shirts in combination with digital customer counterparts. The system is intended for seamless integration into existing processes at the point-of-sale and for the support of the collaborative consultation process between salesperson and customer. This paper describes the various system parts stemming from different research disciplines and their integration under the goal of high usability in an everyday setting.

Keywords: Sales process support, usability, virtual try-on, physically-based simulation, product visualization.

1 Introduction

Computer-supported visualization of clothing has been an important topic in the textile business for many years. One recurring idea in this respect has been the support of clothing sales by a virtual try-on of garments: A digital representation of a human being (avatar) is dressed in digitally represented clothing with the aim to provide a statement of fit and optical appearance of the respective garments in reality.

Rather different technical approaches have been taken to implement this promising concept. A typical representative of the technically more lightweight solutions is, for example, My Virtual Model Inc. [1]. This service offers avatars with limited customization options and does not consider individual body measurements. It can be classified as "pseudo-3D", as the dressed avatar can be viewed from four different sides. Some well-known apparel distributors (Lands' End®, Hennes & Mauritz®, Adidas® etc.) employ the system on their websites, though it can only give a rather vague impression of the clothing's fit and optical appearance for an individual customer.

M.J. Smith, G. Salvendy (Eds.): Human Interface, Part I, HCII 2007, LNCS 4557, pp. 307–316, 2007.

A major advancement in recent years was the realistic physical simulation of clothing and the according three-dimensional visualization on highly individual digital humans. The representative research project Virtual Try-On [2] was able to demonstrate a complete process chain: automated measuring of the customer via body scanning, generation of individual avatars, realistic simulation of correct fit, and authentic visualization of textile characteristics and drapery of the clothing.

High-end approaches like this suffer from limitations concerning the required processing time and the continuous automation between the different steps. This conflicts with the requirements of an interactive system and complicates the practicable integration into existing sales processes. At the moment, solutions for a high-end "virtual try-on" are a topic primarily for CAD applications in the textile sector, where garment designers are enabled to create virtual prototypes.

In this paper, intermediate results of the research project "IntExMa (Interactive Expert System for Made-to-measure Clothing)" are presented, aiming at the development of an effective "virtual try-on" system for a sales support that is point-of-sale compatible. The basic concept is to provide added value by supporting in-store, "real world" sales processes with e-commerce technology ("virtual shopping"). Here, we especially look at shops of small- and medium-sized enterprises that provide specialist products and services. The target setting is characterized by the interplay between customers and sales staff, where the intended system provides information, visualizes product data, and supports a well-founded buying decision. Sales assistants shall not be replaced, but shall rather be supported naturally and unobtrusively in their interaction with the customers.

The aim to enrich the shopping process at the point-of-sale with information technology does not intend to overload the shops with sophisticated and expensive state-of-the-art equipment, especially when it comes to hardware. On the contrary, it is intended to keep the solution affordable, space-saving, and easy to integrate in existing store environments and processes.

The particular sales context addressed in this work is made-to-measure shirts. Though provided with examples of the available array of fabrics and other product components, e.g. collars, and some made-up shirts, customers are usually not able to preview the shirt with all their specifications. Virtual product representations can enable customers to get a clearer understanding of their selected combination as a complete shirt. Furthermore, information technology has the potential to support the selection process and to improve access to a large range of design alternatives.

From the user's perspective, the main system components are a large-sized three-dimensional product visualization and separate user interfaces for customers and sales staff (see Fig. 1). During consultation, the sales assistant enters the customers' body measures into the system. From that data a three-dimensional figure is automatically generated and dressed in the desired made-to-measure shirt, following the metaphor of a "virtual mirror". As the shirt representations are generated individually from real sewing patterns, their drape is simulated with according physical parameters, and the texturing and lighting is optimized for a high degree of realism. The system provides a proper evaluation of fit and optical appearance of the later product for the individual customer. The possibility to easily compare different shirt configurations further supports the buying decision.

The main contribution of the reported work lies in the exemplary realization of a system that is able to visualize and make accessible a comprehensive data space of product and customer data (individual made-to-measure shirts and virtual figurines) in an effective, integrated, and easy-to-use way.

We demonstrate a usability-driven implementation of a complex system concept. This concept implies the demand for an efficient system that reacts to customer input in a responsive way. Thus, state-of-the-art technology from different research disciplines is made accessible with rather basic user interface components to provide intuitive functionality that is easy to use by non-specialists in an everyday setting.

Fig. 1. IntExMa prototype: customized interaction device (middle) and 3D product visualization with GUI on top (right)

2 Method

In order to ensure high usability and user acceptance for the system, it was decided to adopt a user-centered design process according to ISO 13407 [3]. This implies the steps: i) context of use analysis, ii) requirements specification, iii) production of design solutions, iv) evaluation of design results with the two end-user groups, salespersons and customers, and (v) feedback into the next cycle. Repeatedly, different user interface alternatives are considered during the design phases, implemented as rapid prototypes, and tested in the project labs. The most promising alternatives find their way into the superordinated system.

Test participants were always matched to the demographic profile of shirt shop customers and sales staff. The evaluation phase of the first cycle took place with real customers and salespersons in the real made-to-measure tailor shop of one of the project partners in Hamburg, Germany.

A user-centered design process is naturally very adequate for a system's front-end or its user interface, respectively. In parallel, the advancement of the different back-end software components has to be aligned with the test cycles. The requirements for the avatar generation, for the 3D-enabled garments including pre-positioning and physical simulation, and for the realistic visualization have been quite stable from the start and were hardly influenced by user test results. Nevertheless, in order to have fully functional versions ready to be deployed for the different prototype releases and the user tests, the work plan was partitioned into incremental packages.

During the project, two cycles are conducted including a subsequent finalization phase for incorporation of the final user test results.

3 Design and Implementation

3.1 Avatar Generation

One basic requirement for the implementation of the "virtual mirror" metaphor and an effective checking of fit and appearance of the clothing, is the support of realistic and individual digital humans. Their measurements and proportions must be optimized for a convincing product illustration result, in combination with the simulation component.

The area of digital humans has been the subject of research and development for approximately twenty years. In [4], Magnenat-Thalmann and Thalmann provide a comprehensive overview of the plenitude of methods. Many techniques can not be completely automated, while many others require complex technical equipment. In contrast, this project's approach is the use of simplistic technique with a minimum number of measures and without the need for body scanners. For the context at hand, the method "parametric deformation" was chosen: Provided with sparse input data it can deliver impressive results and does not call for special equipment in the shop.

Parametric deformation is based on the linear transformation of vectors. In our case, a complete 3D human figure consisting of a polygon mesh is required as a base model. Starting from this basis, all deformations need to be created only by relocating existing vertices - the so-called morph targets. For a specific deformation one or several relocations are applied concurrently, each weighted with a specific value.

As the resulting avatars must comply with the given body measurements, we apply a measuring algorithm. Based on the polygon mesh it calculates the sum of distances between a predefined number of vertices in a specified order. This concept covers girth and length measures, likewise.

The deformation process is performed in discrete steps - morph actions alternate with the measuring of the according measuring section. This sequence stops as soon as the specified dimension is achieved or when it satisfies a predefined tolerance value, respectively. The initial deformation value is predicted based on the assumption of a linear dependency between two test measurements that are initially calculated.

Morph targets may also cover aspects of the human body appearance like age and gender that go beyond body measurements. Here, a fuzzy-style curve can be defined as input filter in order to achieve nonlinear mappings. This solution also supports color per vertex as well as texture mapping. Several tools have been developed for the preparation of a "generation" of avatars – base avatar plus morph targets – e.g. for measuring the section definition, the generation of real-world to model scaling factors, and the graphical analysis of the target quality.

The quality of the results highly depends on the quality and quantity of the input data. The more detailed the base avatar is and the more morph targets and measuring sections (and measurements) are provided, the more precise the reproduction of the real person will be. Currently, we model the avatars with the help of standard software for digital human modeling and we are using 10 morph targets.

3.2 3D Enabled Garments

Apart from the realistic generation of the avatar, another step is the generation of realistic garments. To perform this step, initial patterns with the following properties are required: i) the patterns must be generated to be made-to-measure; ii) according to the components selected by the customer the corresponding patterns have to be generated, e.g. specific collar, cuff, pocket etc.; iii) the generation of the patterns have to take place automatically, i.e. without human interaction. In addition, the interface to the 3D simulation has to be supplied with information for the 3D representation and simulation, e.g. seam information, orientation in space, information for the pre-positioning, information on buttons, pleats, darts, logos, and a few more. This information is also important for a realistic 3D representation.

To generate such patterns and information it is necessary to have the underlying garment constructions. Due to requirement ii), one sees easily that there are many possibilities of combinations. Requirement iii) includes, that there must be a description of the garment construction, which can be handled in an easy way and which is flexible concerning the variants. The descriptions of the pattern constructions are stored in XML format. To describe the constructions in XML, an XML schema definition was set up that defines construction steps. To ensure that all combinations can be generated, the separate steps are designed in a way that they can be parameterized by measure data and fitting descriptors. Each step may or may not be executed according to the selected combination of components.

The XML schema also covers the elements for the 3D simulation, especially seam information, button information, information on texture, and material properties which are necessary for a realistic 3D representation. To define this information in the construction, they are again built as steps with the advantages aforementioned. Thus, this information can depend on the fitting. For instance, it is possible that there is the same pattern, but with different seam information.

Different pattern constructions can be stored in separate files. This keeps the descriptions of the constructions easy to handle to some extent. A method has been developed for the selection of the files. This selection is also defined in XML. It may happen that the construction in one file depends on a different construction in another file. In this case, the execution sequence of the constructions is important. The pattern generator recognizes these dependencies.

The whole pattern generation process happens as follows. The fitting parameters are passed from the user interface to the pattern generator. In the first step the corresponding files are chosen. Usually, there will be a set of files for one garment. Due to the possible dependencies described in the previous paragraph, it is then necessary to sort the files in the right sequence. For this, a graph representation of the sequence is used. In the next step, the constructions are executed, which results in a set of patterns with all the descriptions that are required for the simulation. In the last step, all the information is put together in one XML document that is passed to the 3D simulation.

3.3 Pre-positioning and Physical Simulation

The simulation of the real-world process of dressing can be very complex, e.g. the simulation of a person that tries to get in a sweater would involve complex movements of the arms. However, there is a much easier way to dress virtual humans, which was proposed by Volino and Magnenat-Thalmann [5]. There, the single cloth patterns are positioned manually around a virtual human body. But time-critical applications like the proposed interactive product visualization require automated and faster mechanisms. In this context, a pre-positioning by user interaction must be replaced by an automatic pre-positioning.

Therefore, a novel approach was proposed by Fuhrmann et al. [6], where the virtual cloth patterns are positioned automatically on bounding surfaces that enclose the body segments, e.g. torso, arms and legs. The resulting pattern positions serve as initial positions in the following physically-based sewing process. It is also possible to pre-position several garments simultaneously by computing a series of bounding surfaces lying upon each other. Also, pockets of a shirt can be handled as patterns and pre-positioned over a shirt. Accessories like buttons are also processed during the pre-positioning.

We use a particle system for the realistic physically-based simulation of the triangulated garment patterns. The movement of each particle is controlled by Newton's laws of motion. Internal cloth forces can be very large due to the fact that cloth strongly resists against stretch. The standard approach to cope with this is to use an implicit time integration method [7]. These methods allow large time steps for the simulation but unfortunately require considerable computational efforts per time step. We follow another approach of Fuhrmann et al. [8], which avoids large forces and can be computed very efficiently. The key idea is to replace internal forces by geometric constraints, which can be parameterized for different cloth behavior. Besides efficiency this method has the advantage that it is extremely stable, and is therefore ideal for interactive systems. This is particularly useful if the user of the system is not a simulation expert but a salesperson.

During the simulation, the patterns are sewn together along their seaming lines. After sewing, gravity is activated and the garment is put into its final shape. The simulation is stopped when the particles reach a stable state.

The physically-based simulation also has to handle self-collisions and collisions between the human body and the cloth. Numerous methods have been proposed in

Fig. 2. An avatar dressed with a shirt. The close-ups show details like the collar, the 3D buttons and seam threads. The image in the lower right was rendered without the avatar and self-shadowing to visualize the cloth thickness and the seam allowance of the cuff.

recent years, cf. [9] for an overview. Our system solves the problem by testing only particles against the surface of the body and each other. This approach saves a lot of computations compared to a full triangle-triangle intersection test. Distances between particles and the human body are rapidly computed with a signed distance field [5].

3.4 Realistic Visualization

The simulated garments and the virtual human are finally visualized. In order to create realistic images, a photo of a real environment is used for lighting the virtual scene. One can think of specific light situations like a sunny landscape or even the real shop in which the customer is standing. By using several 360° photos of the environment the complete dynamic range of real light can be captured. This results in a high-dynamic-range (HDR) environment map, which is used to light the virtual scene.

Besides correct lighting, shadows and self-shadowing must be considered to create realistic images. We employ structured importance sampling [10] to transform the HDR-image into a set of directional light sources. For the rendering of shadows cast by directional light sources we use numerous shadow maps [11]. Although shadow maps work in image space and are often prone to aliasing artifacts, they are a good choice, since the effect of aliasing decreases by increasing the number of lights.

Further important aspects of realistic visualization include details like the seam allowances, seam threads and imprints. We create additional geometry for the seam allowances automatically from the input data, where the width of each seam allowance is specified. In contrast to a thickening of the cloth, which can be done by simply extruding the geometry, the seam allowances are modeled as a stripe of geometry with constant width. This stripe is then attached to the border of the patterns and folded around (see Fig. 2).

The seam threads and imprints are visualized with additional textures. These textures are partially transparent and rendered after the original pattern texture. For the seam threads, extra sets of texture coordinates are computed in order to be able to visualize highly detailed seam threads.

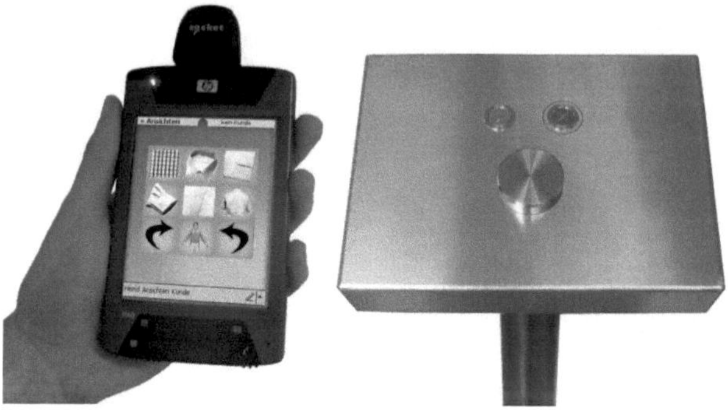

Fig. 3. PDA (3D interaction mask) and an interaction device with knob and buttons

3.5 User Interfaces

The basic interaction concept of the IntExMa prototype is oriented at a collective system usage by customer and salesperson. While the latter has access to the entire functionality of the system, only selected functions are offered to the customers. This aspect results from interim usability tests, where only the shop personnel interacted with the system in the sense of maximized customer service and decreased complexity for the customer. However, the great majority of the customer test group members asked for independent access to at least some of the system's functions.

As a consequence, two different user interfaces are offered: a PDA for the sales staff and a special interaction device for the customers. The interaction device consists of only two buttons and a rotating knob (see Fig. 1 and 3). These two specific interaction means were chosen based on a sustained program of research conducted during "IntExMa" and a previous project [12].

Via the PDA, the salesperson has access to the electronic product catalog with capacious ordering functions, the customer database, and the 3D interaction functionality. Concerning the latter, the PDA works like a remote control. If the actual shirt configuration is altered on the PDA, the changes are directly transferred to the 'virtual try-on' and a new simulation is initiated. In case of exchanging the cloth, only the texture is exchanged to save the simulation time. As soon as the customer's body measurements are entered, the according avatar is generated and directly displayed and dressed. Normally, the virtual figures are only shown completely vested, but alternatively the pre-positioning and the simulation (see above) can be visualized. Furthermore, the salesperson can turn the avatar around its vertical body axis and zoom in to details of the shirt. A bar-code scanner is integrated into the PDA. It allows to scan real products in the shop resulting in direct alteration of the shirt configuration in the PDA and the 3D scene.

Concerning the user interface design of the PDA, we tried to adopt existing expertise in the form of design guidelines. Results of an accordant analysis indicated that handheld design is only sparsely covered yet, especially when it comes to business process support. Our current PDA design is based on the analysis, interpretation and amendment of established guidelines for desktop user interfaces, e.g. ISO 9241 and some of the rare mobile design guides, interpreted for the present context of use. The results of this work, a rather complete set of principles, exemplary rules and according concrete examples have led to a separate publication [13].

The interaction device hardware was custom-designed, as no commercially available device could be found, that provides the desired form factor, simplicity and robustness of construction. The associated 2D graphical user interface (GUI) is overlaid on top of the 3D product visualization (see Fig. 1). The aim was to exploit the available display area as widely as possible without covering the relevant 3D content.

With the green confirmation button the user can invoke the GUI and the product catalog, navigate to different levels of its hierarchical structure, and choose the desired shirt components, which are again directly illustrated in the "virtual try-on". The currently configured product detail is at the same time automatically zoomed in within the 3D scene. The red abort button allows going backwards in the navigation path and, having reached the top level, allows fading out the GUI. Browsing the different entries inside a menu level is achieved with the rotary knob. With its help, the avatar can also be rotated, presumed that the GUI is "closed" (faded out). Both, PDA and interaction device, operate on the same data and can be used concurrently.

4 Results and Conclusion

The results of this work can be split up into technical and usability-related findings. From the technical point of view, the project has successfully achieved to integrate sophisticated state-of-the-art technology from different research disciplines (virtual reality, textile engineering, physics, and mathematics). The result is a rather complete and self-contained system consequently oriented at the demands of the point-of-sale. Room for improvement can nevertheless be identified: reduce overall computing time for improved interactivity, allow further customization of avatars for better individualization (though in the limits of available customer data), and reduce required manual effort for digital content production.

Concerning usability, we demonstrate an example how the context of use of a particular area, sales process support at the point-of-sale, shapes a technically complex system concept. The user testing conducted so far indicates the promising potential of the employed user interface concepts: Though being rather "unspectacular", they successfully support our goal, to provide intuitive, effective and easy to use information access in an everyday setting. Also, the distribution of interaction possibilities between customer and sale staff is appreciated. It balances the service demands and the "play instinct" of the customers. Also, the customers like the idea of being supported by the salespersons in this way. The latter esteem the system's abilities to depict complex information, which enhances their customer consultation.

Our approach to a "virtual try-on" for made-to-measure clothing in the shop was rated to provide added-value to made-to-measure shirt shopping - by customers, sales staff and shop owners. Further detailed findings for the current prototype will result from the scheduled next evaluation phase.

Acknowledgments. This research was carried out during the project IntExMa and financially supported by the German Federal Ministry of Education and Research.

References

1. My Virtual Model Inc, last retrieved 15/02/07 http://www.mvm.com/
2. Virtual Try-On, (last retrieved 15/02/07), http://www.human-solutions.com/virtualtryon/index.shtml
3. ISO 13407, Human-centred design processes for interactive systems (1999)
4. Magnenat-Thalmann, N., Thalmann, D.: Handbook of Virtual Humans. John Wiley and Sons Ltd, New York (2004)
5. Volino, P., Magnenat-Thalmann, N.: Developing simulation techniques for an interactive clothing system. In: Proceedings of Virtual Systems and MultiMedia '97, pp. 109–118 (1997)
6. Fuhrmann, A., Sobotka, G., Gross, C.: Distance fields for rapid collision detection in physically based modeling. In: Proceedings of GraphiCon 2003 (2003)
7. Baraff, D., Witkin, A.: Large steps in cloth simulation. In: Cohen, M. (ed.) SIGGRAPH 98 Conference Proceedings, Annual Conference Series, Orlando, FL, USA. ACM SIGGRAPH, pp. 43–54 (1998)
8. Fuhrmann, A., Gross, C., Luckas, V.: Interactive animation of cloth including self collision detection. Journal of WSCG 11(1), 141–148 (2003)
9. Teschner, M., Kimmerle, S., Heidelberger, B., Zachmann, G., Raghupathi, L., Fuhrmann, A., Cani, M.-P., Faure, F., Magnenat-Thalmann, N., Strasser, W., Volino, P.: Collision Detection for Deformable Objects. State of the Art Report, Eurographics EG'04, August 2004 (2004)
10. Agarwal, S., Ramamoorthi, R., Belongie, S., Jensen, H.: Structured importance sampling of environment maps. In Transactions on Graphics 22, 605–612 (2003)
11. Williams, L.: Casting curved shadows on curved surfaces. In: Computer Graphics (SIGGRAPH 78 Proceedings), pp. 270–274 (1978)
12. Blum, R., Häberling, S., Khakzar, K., Westerman, St.: User Interfaces for an In-store Sales Process Supporting System. In: Proceedings of CIS^{2}E 06, International Joint Conferences on Computer, Information, and Systems Sciences, and Engineering (to appear)
13. Blum, R., Khakzar, K.: Design Guidelines for PDA User Interfaces in the Context of Retail Sales Support. In: Proceedings of HCII, 12th International Conference on Human-Computer Interaction (to appear, 2007)

A Visualization Solution for the Analysis and Identification of Workforce Expertise

Cheryl Kieliszewski[1], Jie Cui[2], Amit Behal[1], Ana Lelescu[1], and Takeisha Hubbard[3]

[1] IBM Almaden Research Center, 650 Harry Road
San Jose, California 95120 USA
{cher,abehal,lelescu}@us.ibm.com
[2] IBM China Research Lab, Beijing, China
ciujie@cn.ibm.com
[3] Texas A & M University, College Station, TX
hubbake1@cs.tamu.edu

Abstract. Keeping sight of the enterprise's workforce strengthens the entire business by helping to avoid poor decision-making and lowering the risk of failure in problem-solving. It is critical for large-scale, global enterprises to have capabilities to quickly identify subject matter experts (SMEs) to staff teams or to resolve domain-specific problems. This requires timely understanding of the kinds of experience and expertise of the people in the firm for any given set of skills. Fortunately, a large portion of the information that is needed to identify SMEs and knowledge communities is embedded in many structured and unstructured data sources. Mining and understanding this information requires non-linear processes to interact with automated tools; along with visualizations of different interrelated data to enable exploration and discovery. This paper describes a visualization solution coupled with an interactive information analytics technique to facilitate the discovery and identification of workforce experience and knowledge community capacity.

Keywords: information visualization, workforce management, unstructured data.

1 Introduction

Often, in today's global business environment, a team of consultants with different kinds of expertise and different levels of experience must quickly be brought together in response to a potential client engagement or to address a client issue. This is especially true of any kind of outsourcing engagements, where one company is contracted to run a piece of another company's business. Fast and dynamic team definition capability is critical to the success of addressing these business needs. One key component to this capability is the ability to quickly identify the appropriate persons with the appropriate experience for assignment to the team. However, for many large scale companies, this may be the equivalent of finding a needle (the appropriate person) in a haystack (amongst 100s or 1000s of people resources).

In large-scale geographically dispersed firms, there is also a great need to understand the skills of the organization as a whole. This means more than just identifying experts in specific domains to lead others or address difficult problems. It

M.J. Smith, G. Salvendy (Eds.): Human Interface, Part I, HCII 2007, LNCS 4557, pp. 317–326, 2007.

is the understanding of the overall skills, experience and expertise landscape within the firm. This information can be used to intelligently choose client engagements and to evolve the workforce to meet future market or strategic needs. On the other hand, losing sight of the knowledge community weakens the entire business, leading to poor decision-making which may result in failures in problem-solving.

Both of these goals require evaluating knowledge worker experience and expertise. How can this be accomplished? In the context of large enterprises, vast volumes of experience information are embedded in both structured and unstructured data sources. Identification of individual employee's skills can be determined by mining structured databases, such as a resume database or an employee skills database. In recent years, searching and sorting through structured data has become a relatively automated process. However, additional information is often required to determine one's experience around a given skill or a particular role.

Unstructured data sources, such as white papers, presentations, proposals, reports or manuals provide evidence of demonstrated knowledge and experience of an individual and teams. In addition, these sources are often more up-to-date than structured sources. However, the programmatic analysis of unstructured sources is not 100% accurate, so the analysis becomes a semi-automatic process that requires a human to be highly engaged in the analytics process. The analytic tool design needs an additional element to ensure interactivity and appropriate feedback to make sense of the analytic landscape.

The primary scenario we used to develop our expertise locator solution was to provide assistance to a human resource analyst in finding a subject-matter expert (SME). Where, the identification of a SME is imperative to the resolution of a critical situation in a client engagement. This scenario could be extended to be team creation, where a team of people with specific experiences may need to be identified and assembled to meet the needs of a particular client engagement.

From the idea of team creation, a secondary scenario for the project was to provide a tool for long-term workforce planning that would provide analysts with a snapshot of the distribution of the organization's experiential make-up. This concept includes understanding skill and experience overlap and gaps with the ability to interact with and validate the automated analytic output.

In general, there are two pieces to this problem to understand an experience landscape: (1) proper data warehousing and analytical technologies and (2) the human analytic process and technology interaction. This paper focuses on the facilitation of the human analytics process through improved data visualization that support discovery and insight of knowledge community experience in order to find expert resources and evaluate experience capacity.

2 How Is Expertise Found?

To gain a better understanding of how human resource analysts identify experts, we interviewed three analysts whose job it is to find information and people that can solve impending business problems. One of the most interesting aspects of their jobs is that these analysts are like internal head-hunters; they are the people that organizations turn to for help in finding an expert after having exhausted all of their own search leads.

Our interviews confirmed our hypothesis that the analysis and insights process is not linear. That is, analysts move back and forth between and through information, depending on what is discovered during the exploratory process. Intuitively, this probably isn't a surprise to anyone, but has been a difficult behavior to understand and mimic [e.g., 1].

However, what we found was that the analysts performed their searches for information or people within a relatively structured process. For the most part, this process is defined by what they are asked to find (i.e., information or people) and the focus of the databases they have access to. For example, if an analyst is asked to find an expert who has in-depth knowledge of a particular product or process, she would focus her search by first weeding-out any inconsequential information with respect to the request. Then, she would determine relationships that are of consequence and may facilitate the search (e.g., a particular client relationship or a certain geographical region).

Once this groundwork is complete, she would then methodically search for communities of experience in people-oriented databases, such as directories, knowledge repositories, or discussion forums, around the relationships of consequence. She would then hone in on individuals with particular expertise. As can be imagined, this is a highly iterative process where, although automated queries against databases are being performed, the analyst must manually sort through and make sense of the results to identify either an individual or community of expertise who will fulfill the request.

3 Scenario and Our Approach

Our team was challenged with creating a technology solution that would facilitate the iterative process of searching, sorting and analyzing unstructured information to more efficiently locate knowledge worker experience and expertise. The primary scenario for development of this expertise locator solution was to afford fast and accurate identification of a SME to assist in resolving a critical situation. This scenario was based on the project requirements and the interview findings.

As mentioned earlier, identification of individual employee skills can often be determined by mining structured databases. For example, if an employer wants to know who has experience with SAP software applications for assignment to a project, a resume database or an employee skills database can be queried to find such individuals. However, additional information is often required to determine one's level of experience and expertise for a given skill or a particular role. This information often resides in unstructured sources such as white papers, presentations, proposals, and other business documents that are not as easily mined and examined. The general problem being solved by our solution is the facilitation of the human analytics process, through improved data visualizations in support of discovery and insight in the domain of workforce management.

In order to reach the goals of an analysis, the visualization design of the supporting analytics output must be flexible, yet be able to present a significant amount of information in an easily digestible manner. The presentation must also keep the user reminded of the analysis goal while working with the analytical details [2].

To date, skills and expertise location products tend to focus on improved search and analytical processing technology but are lacking in their portrayal of information to facilitate understanding. For example, data visualizations tend to be asynchronous displays of 2-dimensional information that the resource analyst has to page through. The visualizations are helpful, but this type of interaction requires the user to manually review, recall and associate visually disparate information—an inefficient process to aid discovery and insight.

A more efficient method is to structure the interaction to progressively sequence the type and amount of information within context of the discovery and insight activities. To incorporate this method for expertise location, output from a text analytics platform was used to analyze a corpus of business documents to identify SMEs and the experiences that they posses. (Note: detailed description of the analytics platform is beyond the scope of this discussion [see 3].)

However, as stated in the interview findings, analysis is an iterative process where identification of a solution to a problem is focused (and refocused) depending on what the information landscape is that is being explored. With that said, our design strategy was to provide a flexible interaction solution that:

1. provides context around detailed inquiries;
2. allows for a non-linear work process that affords switching between tasks;
3. allows dynamic modification of search results to focus or refocus the query; and
4. implements visualization techniques that support progressive exploration to facilitate discovery and validation.

This system would surface the various mappings between experience and people resources provided by the analytic platform, to facilitate the semi-automated identification of expertise. Where, the final goal was to demonstrate the value of visualization techniques in enhancing the efficiency and effectiveness of identifying knowledge worker experience and determining expertise.

4 Our Solution

Our solution design had to consider (1) the kind of problems that might need to be solved and how to facilitate a query; (2) structure of the search parameters; (3) review and editing of the search results; (4) display of the results at different levels of detail; (5) flexible user interaction between searching, editing and reviewing the results; and (6) implementation as an AJAX enabled interactive web-based user interface. Based on these considerations and text analytics platform constraints, the user interface was designed to include four basic elements: search and context, landscapes, progressively sequenced graphical presentation and all within an interactive tabbed framework.

4.1 Search and Context

Based on the interview findings, we thought it was important that the analyst be provided some degree of problem context while performing an analysis. It was decided that problem context would be linked to the search parameters. Queries to the underlying data were initiated by the analyst as a key term search. If desired, the analyst was also able to include structured field filters in the query to focus the range of returned results. Recall, the analyst would often try to determine relationships of consequence to focus her search and analysis. The structured filters afforded a small degree of relationship selection and focus. Once the search results were returned, the parameters remained in sight to aid in recall of problem context.

4.2 Interactive Tabbed Framework

Output from the analytics platform was in the form of editable landscapes with weighted categories that provided a measure of relatedness to the query parameters. These were used to populate the interactive tabbed framework for simple and complex insight.

The entire user interface was designed within a simple tabbed framework. This was done so that the analyst could easily modify previously viewed information during the analysis process. Where, edits performed in one tab would be reflected in all other tabs. One thing to keep in mind is that all of the results are only driven at the user interface level and are not persisted at this level of the technology; any edits to the result set through the user interface would instantiate a recalculate via the text analytics platform and produce altered output (Figure 1).

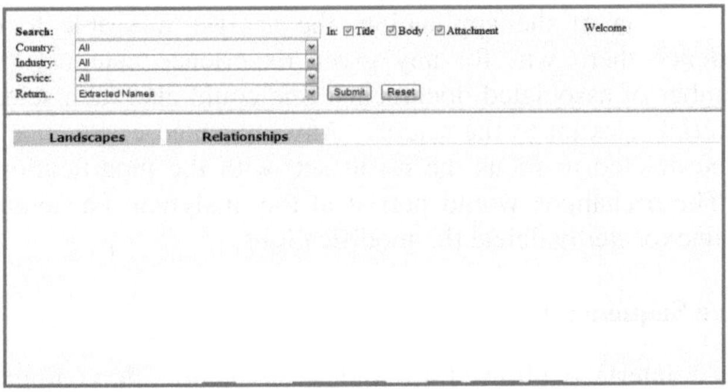

Fig. 1. Illustration of basic user interface layout

Output from the analytics platform was in the form of Landscapes and Relationships. The tabbed framework was designed to first provide the analyst with a Landscape of the output based on 1-dimensional taxonomy classifications. Two taxonomies were generated via the text analytics platform (for this particular application), each being an individual dimension of the overall information landscape. Information was categorized either within an experience dimension or a people dimension under Landscapes. In addition, each category linked to the source documentation in a side panel, which is included in every tab. This provided a path to evidence and afforded simple insight into concepts found in the documents along with interactive editing to refine the output results (Figure 2).

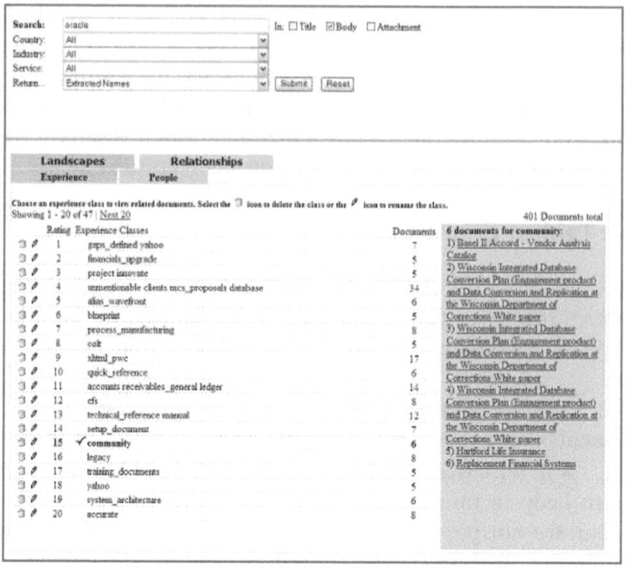

Fig. 2. Landscape dimensions with 1-dimensional output

In the first iteration of the application, the analyst was able to determine the amount of evidence there was for any given experience category (or person) by viewing the number of associated documents. She could also then select a document and review material relevant to the search. Any unwanted experience categories (or people) could be deleted to focus the result set, with the modifications reflected in Relationships. These changes would persist at the analytical backend, so switching tabs would not unexpectedly delete the modifications.

4.3 Progressive Sequencing

Since the tabbed interface allowed for quick navigation throughout the analytics process, we needed progressive sequencing to show the context of the user's actions, while displaying more information. Just as we kept the search parameters once we displayed the results, we applied the same idea within tabs by keeping a view of the user's selections when navigating the results.

Once the analyst was satisfied with the output in the Landscapes, she could then explore the relationships between experience and people by selecting the Relationships tab and perusing the visualizations. A cross-landscape analysis of experience and people resulted in 2-dimensional matrix output that could be used to populate any number of visualization types.

The Relationships tab includes four sub-tabs—each of which provided a different view of the matrix to create maps of the experience field. The intention in providing different views was to provide multiple perspectives of the same information. In addition, different levels of granularity were supported within each view to support discovery and complex insight to identify hidden patterns and relationships. The progressive sequencing concept was applied in each tab, in that the display of the coarse information was linked and visually alongside the more granular information.

The four views that composed Relationships were: (1) co-occurrence table, (2) experience affinity, (3) people affinity, and (4) network diagram. The co-occurrence table was a sortable representation of the raw 2-dimensional matrix. The experience and people affinity allowed users to navigate through one dimension and drill down via relationships into the other dimension. The network diagram represented the relationships as an interactive graph of nodes and edges and provided a bird's-eye view of the relationships.

The co-occurrence table allowed the analyst to quickly identify a set of experts for a group of experiences (Figure 3). Both the rows and columns were sortable so that the analyst could easily view (a) the top people with a particular experience or (b) experiences of an individual person. Also, the cells were color coded categorically as having a high, medium or low affinity score. So, our analyst could quickly gauge the strength of the relationship represented by each cell.

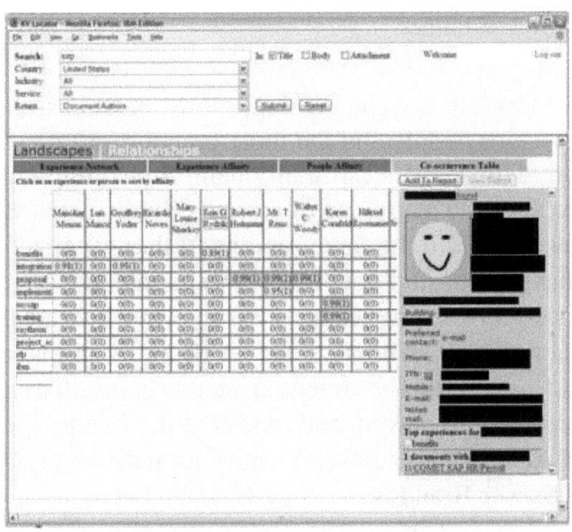

Fig. 3. Example of co-occurrence table

A dossier was generated and composed based on experience-people relationships and by accessing a directory repository and including contact information, if applicable. The information that populated the dossier depended on the interest of the analyst—identifying experience first or identifying people first. If the analyst was more interested in identifying experience, the dossier would populate with a list of the top people and supporting documents. If the analyst was more interested in identifying a person, the dossier would populate with contact information, top experiences and supporting documentation.

The experience and people affinity tabs were symmetric representations of the data output (Figure 4). In general, the affinity results provided the inherent similarity between experience and people. As with the co-occurrence table, the affinity results were displayed categorically as high, medium and low. The progressive sequencing

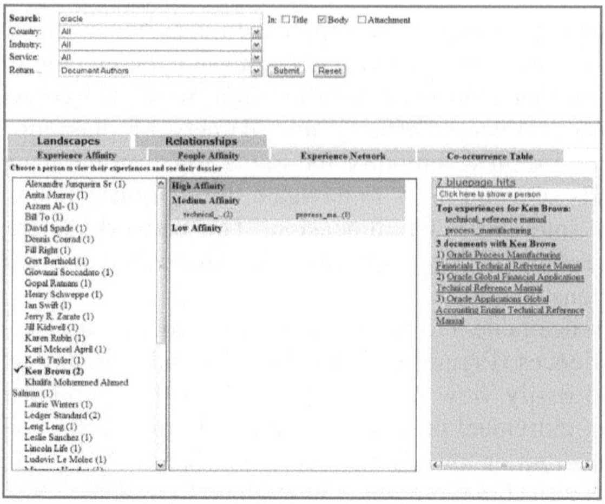

Fig. 4. Progressive sequencing in People Affinity, coarse-to-granular information movement from left-to-right in the linked frames

within the view was from left to right. Where, more coarse information was provided as individual experience or people and the most granular information was provided in the composite dossier. This view is especially helpful in finding the top people with a particular experience(s) or the aggregate of an individual's experience.

By selecting elements within the three regions of the visualization, the view would update to illustrate the relationship between found people, experience affinities for the person and the dossier. For example, let's assume that our analyst wanted to find someone with experience troubleshooting Server Brand X and is located in the United Kingdom. She would have initially created a search using this information as query parameters. Then received, reviewed, and modified the Landscape results. From there she would select the Relationships tab to identify an individual (or set of individuals) with high affinity to Server Brand X.

If we use the experience affinity view as an example, our analyst would progress through the visualization by selecting Server Brand X in the left-most frame, which would update the center affinity frame and dossier frames with pertinent information. The context of that selection is displayed by a check mark along with the initial search parameters. Then, our analyst can select and explore further on an individual of interest. These actions also update the dossier by adding contact information and shortens the list of displayed documents to just those relating to that person and experience combination.

The fourth sub-tab within Relationships is the network view (Figure 5). Placing the people and experiences as nodes in a network graph allows the analyst to see the relationships between people and experiences; and is a way to visualize the whole organization's experience based on the search parameters. Where, this view allows the analyst to see the relative abundance of experience in the organization and the gaps in experience within the organization. Similar to the other views, selecting a

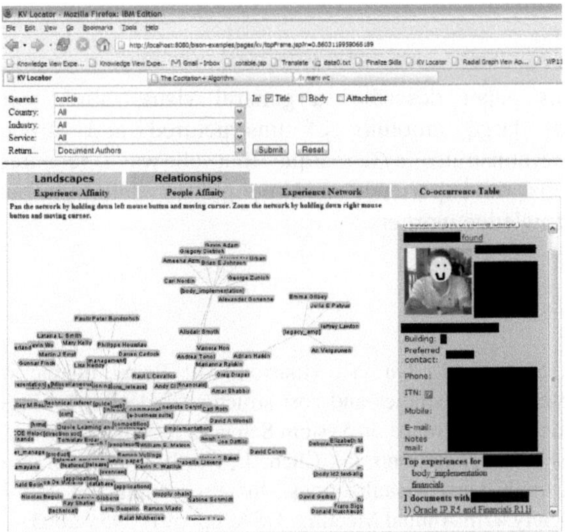

Fig. 5. Network view

node in the network will display the appropriate dossier to the right. In addition, the network graph is a unique view in that it can represent more than two dimensions, which the co-occurrence table and affinity views do not allow.

4.4 Discovery and Insight Activities

We recognize that this type of interaction model and information sequencing is not a new idea with respect to data visualization—as evidenced by a history of visualization references [e.g., 4, 5]. However, the visualizations coupled with the text analytics platform is a new approach within the domain of workforce management and expertise location.

The novelty of this design is that individual visualizations were systematically linked in different combinations for synchronous data display within context of the problem space. Through the overall interaction framework, the resource analyst is able to search for a desired experience type, examine and refine the landscapes, view relationships between experience categories and people, iterate on this sequence to refine the investigation, and then contact the SME to validate knowledge or availability. In addition, the analyst can also check the overall knowledge distribution of the organization around a certain topic to identify overlap or gaps.

The framework allowed for the exploration of both 1-dimensional data for simple insight and multi-dimensional data for complex insight to identify hidden patterns and relationships in maps of cross-landscape analysis. Data presentation was provided using visualizations techniques that analysts were already accustom with as tables, tree-diagrams, network diagrams, affinity diagrams, and a composite dossier. Where, the dossier provided a detailed profile and contact information of the found expert for further investigation and decision-making.

5 Summary

In summary, this paper describes a general visualization solution that affords understanding of large amounts of unstructured and structured data. More specifically, it demonstrates a technique that allows a resource analyst to view synchronous maps of experience and people to determine expertise within a community of knowledge workers.

References

1. Russell, D.M., Slaney, M., Qu, Y., Houston, M.: Being literate with large document collections: Observational studies and cost structure tradeoffs. In: Proceedings of the 39th Hawaii International Conference on System Sciences (2006)
2. Ryu, Y.S., Yost, B., Convertiono, G., Chen, J., North, C.: Exploring congnitive strategies for integrating multi-view visualizations. In: Proceedings of the Human Factors and Ergonomics Society 47th Annual Meeting, USA, pp. 591–595 (2003)
3. Behal, A., Chen, Y., Kieliszewski, C., Lelescu, A., He, B., Cui, J., Kreulen, J., Maximilien, M., Rhodes, J., Spangler, S.: Understanding complex IT environments using information analytics and visualization. 1st Symposium on Computer Human Interaction for Management of Information Technology, March 30-31, 2007, pp. 30–31. Cambridge University Press, Cambridge, MA (2007), http://chimit.cs.tufts.edu
4. Bederson, B.B., Shneiderman, B. (eds.): The craft of information visualization: Readings and reflections. Morgan Kaufmann, San Francisco, CA (2003)
5. Card, S.K., Mackinlay, J.D., Shneiderman, B. (eds.): Readings in information visualization: Using vision to think. Academic Press, San Diego, CA (1999)

The Study of Past Working History Visualization for Supporting Trial and Error Approach in Data Mining

Kunihiro Nishimura[1] and Michitaka Hirose[2]

[1] Research Center for Advanced Science and Technology, The University of Tokyo
4-6-1, Komaga, Meguro-ku, Tokyo, 153-8904, Japan
[2] Graduate School of Information Science and Technology, The University of Tokyo
7-3-1, Hongo, Bunkyo-ku, Tokyo, 113-8656, Japan
{kuni,hirose}@cyber.rcast.u-tokyo.ac.jp

Abstract. Scientists in the data mining field are constantly faced with the challenge of finding useful information from a huge amount of information. We have to analyze the data until we can get the appropriate information. We have to select one part of the data, compare them against each other, or arrange them in certain order. This approach is also known the trial and error approach. A trial and error approach requires the users' judgment, for example, to correctly set certain parameters; it is an approach that place importance not only to the end result, but also to the process in achieving the end result. In this paper, we propose visualization methods to visualize past working history for supporting trial and error approach in data mining. We use our methods to visualize web browsing logs and data browsing logs in genome science fields.

Keywords: Information Visualization, Past Working History, Web Visualization.

1 Introduction

We usually have to shift through a vast amount of information that is available on the internet or from knowledge databases when we want to find a specific answer to our question. If we know the target keyword, we can perform the search and find the solution almost immediately. However, if the subject is more ambiguous, we might not have a target keyword immediately. We have to perform the search by specifying several target keywords, check the result and refine the query string iteratively. In short, we have to do the trial and error approach.

When conducting a research by trial and error, we tend to forget previous trials. However, there is a possibility that the previous trials' results are better than the current result. Also, there are times when we do the same trial because we forgot that it has already been conducted in the past. In this study, we developed visualization techniques to support trial and error approach for data mining.

Our proposed visualization method visualizes the working process as well as the results. Because the users are able to glance at the parameters of trial and errors in the past, they are able to deduct the relationship between the trial result and the parameters used in that trial.

M.J. Smith, G. Salvendy (Eds.): Human Interface, Part I, HCII 2007, LNCS 4557, pp. 327–334, 2007.

2 Information Visualization for Supporting Data Mining

Information visualization is the use of computer-supported, interactive, visual representations of abstract data to amplify cognition [1]. Visualization with spatial structures is called scientific visualization. On the other hand, information visualization is a term applied to visualizing information without spatial structures. For example, web networks, documents, and directory structures are targets of information visualization [2]. With adequate information visualization, it is easy to carry the information to the users. Information visualization is an effective method to grasp a complete view of the data.

In data mining, we want to grasp, understand, and interpret the target data. We manipulate, compare, calculate the data, and get the information we want. In order to optimize the search result, data mining process often requires user interaction. When the data mining result are visualized, it is easier for users to understand the result. Thus, we believe that the visualization of data with interaction support can enhance users' data mining process.

Users do trial and error approach in data mining when they want to get specific information in the data. Visualization of the manipulated result is called "visualization of trial". Visualization methods of trial often include displaying various kinds of graphs. Thus, there are a lot of visualization methods of trial. During data mining process, users manipulate of the data (this is called a trial), and check the result by visualization of the trial. Then users interpret them and decide next manipulation. Users repeat this trial (manipulation, check the visualization result, and interpretation and decision) until they can get suitable information.

In addition, working process is as important as trials in data analysis. A final result is often based on the previous trials. Relationship between current trial results and previous results are important because trial and error approach often depend on accumulation of the user's previous trials' result. That is, it is important to display the "result" of trials but also "process" which means "past working history" and context of analysis in data mining.

In this paper, we focus on visualization of past working history for supporting trial and error approach in data mining.

3 Visualization of Past Working History

As we have stated in previous section, it is important to visualize working result but also to visualize working history. In this section, we discuss visualization of past working history.

Past working history consists of time and work process. When data is visualized and a user manipulates it, data has spatiality. Thus, past working history has three spatial axes: time, work process, and spatiality. Past working history generates automatically according to user's data analysis. We can get this history as logs of interactions.

Visualization of past working history offers three advantages for the user:

1. The ability to access previous state/process and manipulate data at that state
2. The accessibility to understand the relationship between current state and the procedures taken to achieve it
3. It can act as a work memory support for the user

We propose visualization of past working history because of the three advantages. Our concept of past working history visualization is shown in Figure 1. Data is visualized in a point that a user manipulates the data as a trial. The user can interact with the visualized result and can manipulate it. According to the user's analysis, the data is visualized and past working history is accumulated. The data mining process is visualized like a road. The road represents analysis process and a road branch represents a process branch. Branches of road indicate user's trials histories. When the user takes a bird's eye view, the user can get the whole view of the analysis. The user can access the working history and interact with them.

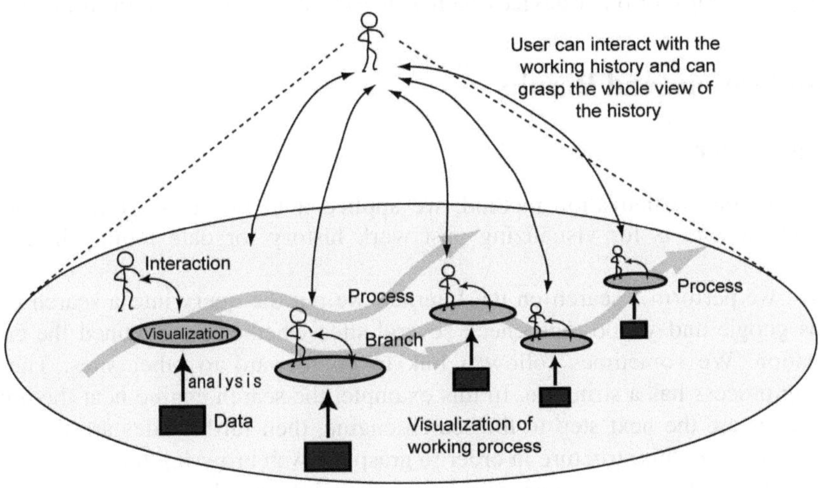

Fig. 1. Concept of visualization of past working history. Past working history is visualized like a road and connected each visualized result. Users can grasp the whole view of the history as bird's eye view and interact with them.

To visualize past work history, there are two kinds of visualization methods. One is visualization with three axes written above. The other is visualization without any axes. This visualization method includes footprints as a browser history and markers as memo. In this paper, to visualize past work history efficiently, we have developed three categories for visualization method using the three axes:

1. Past Working History Visualization with Time Axis
2. Past Working History Visualization with Process Axis
3. Past Working History Visualization with Spatial Axis

Time axis is generated automatically. As time naturally flows, the work process is generated by users' analysis. Spatial axis is generated by user's interactions. We use these axes for visualization of past working history. When data has several states and the user do trial and error approach, we define the state position and the history is visualized that are taken one axis as time and the other axis as state. When data has hierarchic structure, we use this structure in the visualization. For example, a user conduct a clustering method manually, data is divided from top to bottom. The clustering process can generate a dendrogram (tree diagram) that indicates data structure. We use this structure as history and add functions on the dendrogram. When the user interacts with visualized objects such as slide-bars that enable the user to set parameters, slide-bars remain as the user set. This slide-bars status is a spatial history. When the user transforms the location of visualized objects, the previous position is a spatial history. We also visualize these kinds of spatial histories.

To visualize the history, various kinds of log data are required in the analysis. Log data should include time, analysis state, parameters, thumbnails, and relationships between multiple states. There are many ways to get log data. For example, by logging application's data parameter, web browser history, logger software which takes logs, OS logs, or other devices such as RFID cards which contains usage logs.

4 Applications and Results

4.1 Applications

To evaluate our visualization method, we applied it to two tasks. First is for web browsing. Second is for visualizing past work history for data mining in genome science.

When we perform a search on the Internet, we put the query into a search engine such as google and yahoo, and check several sites whether we obtained the correct information. We sometimes follow a link to go forward to other sites. The web browsing process has a structure. In this example, the search engine is at the top and several sites are the next step to the search engine, then further sites are the tertiary step. We visualize this structure in order to grasp the web browsing process.

In the field of genome science, researchers are facing huge experimental data. Technology of sequencing and microarray enable us to get large biological data at a time. Using these high-throughput technologies, experimental data are accumulated rapidly and genome researchers have to analyze data and interpret them biologically and medically. Genome data consists of patients, expression levels of genes, copy number of genomes, sequence of genomes, and so on. Genome researchers observe and check the data in whole chromosomes level, or in one chromosomal level, then zoom up into a chromosomal band level. Finally, they could also zoom up to a genome sequence level. In this search process, they want to find abnormal regions from the data and to grasp the relationship between abnormal regions. Consequently they will compare among different patients because they want to learn if the abnormal regions are isolated cases or common among many individuals.

Thus we apply our visualization methods to the field of genome science and visualize their searching and checking data histories to support their analysis.

4.2 Visualization with Time Axis

We applied visualization method using time axis for web browsing history. We took Firefox browser's history and captured the page screenshots using ImageMagic library. Visualization is based on OpenGL.

We visualize the web browser history in two and three dimensional spaces. Z direction in three dimensional visualization is the time axis and Y direction in two dimensional visualization is also time axis. We distribute web site screenshots according to the history in time series. The same domain web site is located in the same position in a space. The result of visualization with time axis for web browser history is shown in Figure 2.

Fig. 2. (upper left) Visualization with time axis in two dimensional spaces. X axis represent domains and y axis represents time. Web thumbnails are presented. (upper right) Visualization with time axis in three dimensional spaces. Domains are distributed in 2 dimensionally and Z axis represents time. (lower) Visualization with time axis in three dimensional spaces. Z axis represents time and X axis represents domains. Web thumbnails are stood using X and Y dimension.

We also visualize genome science application history. Genome science researchers look the data in whole view level and in each chromosome level. We have developed genome copy number viewer for supporting genome researchers. The viewer enables them to present both whole chromosome view and each chromosome view. User can compare patients' data using the viewer. We have obtained logs of this viewer and visualized them at a same time. We took one axis as time and others as the relationship between whole and part views. That is, data structure is distributed two dimensionally. When the user searched for data from chromosome 1 to chromosome Y, the history visualized a pattern. The result of visualization is shown in Figure 3. The result reveals their analysis process that they check the whole view and see the details view, then they go back to whole view and repeated this cycle.

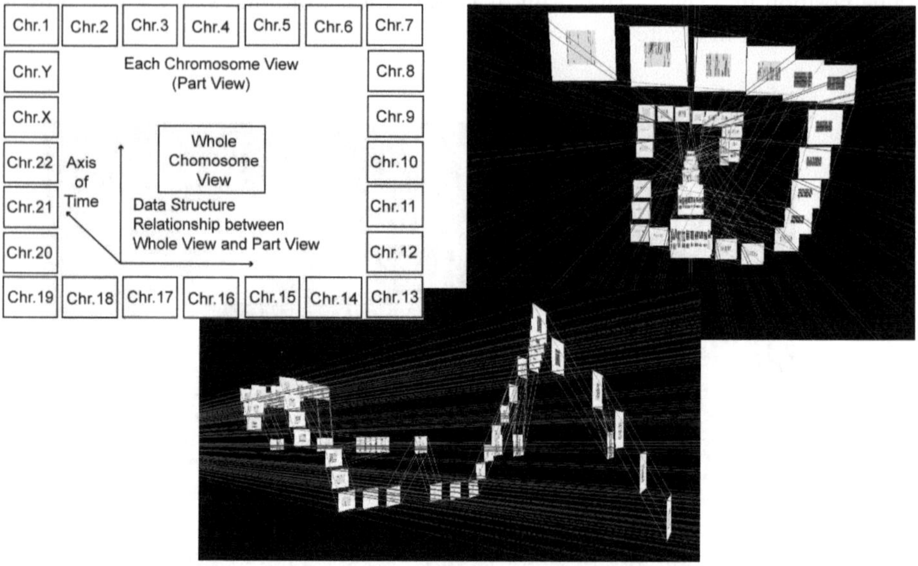

Fig. 3. (upper left) Distribution of data. Data structure is located in two dimensional spaces and the other axis indicates time. Each chromosome view is located in a rectangle. Center of the view is whole chromosome view. (upper right) Visualization with time axis and data structure. There are 24 locations for each chromosome view and 1 center location for whole chromosome view. When a user see and check the data according to chromosome number, each chromosome visualization results are distributed spirally. (lower) The same view from different angle. Time axis is visualized from right to left. Each views which visualized as thumbnails are connected each other, thus a user can grasp transitions of the work history. The user can tell that user checked the data whole view and each view to-and fro.

4.3 Visualization with Process Axis

We applied visualization with process axis for browsing application. We took the interaction history from our original application for browsing genome data.

We have developed the graph viewer for genome data. The viewer is for analyzing genome copy number and genome researchers want to detect abnormal copy number regions. They compare the candidate's abnormal copy number regions that are detected by one patient data with other patients' data to check whether those regions are common. The viewer supports this analysis. It presents graphs that indicate genome copy number for each patient. Graphs are distributed in tile fashion. A user can select a graph to check the detail, the graph will be zoomed in. Then user click the graph again, the view will be zoomed up again. Then, the user can find an abnormal region and click it. The viewer extracts the same region for all patients and distributed in a line. It provides the user a comparison view among patients. Then, the user can know which one is common abnormal region. We got interaction history in this analysis. The viewer outputs the state and the thumbnail at the time when there is any interaction automatically. When the user analyzes the data, the interaction history is accumulated automatically.

The analysis has a hierarchy. There are five steps: whole view, selection of one view, zoom view, the selection of an abnormal region in the view, and comparison among patients. We visualize this process as an axis with a dendrogram (tree structure) that shows relationship between whole and part views. Figure 4 shows the viewer and history visualization result.

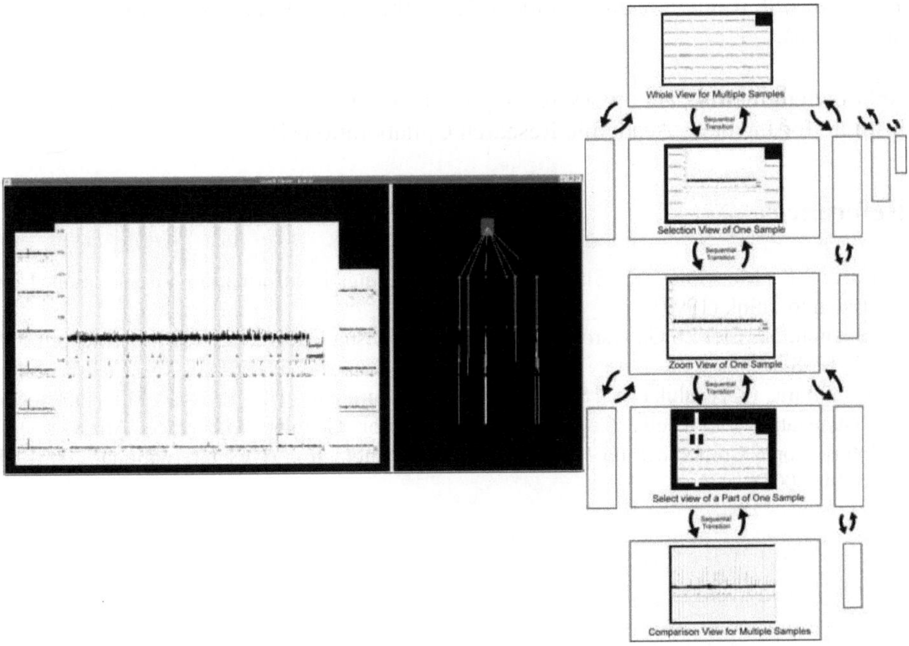

Fig. 4. (left) Distribution with process axis. The left window indicates a viewer and right window indicates visualization of work history. Work history is visualized as a dendrogram which is tree structure. (right) Work process has a hierarchal structure and it can be visualized as a dendrogram. This figure indicates the work process in an image. There are 5 levels. A user can see whole view for multiple samples. Then the user can select one sample. The selected view is zooming up. The user can select one region where the user feels abnormal. The selected region can be compared among multiple samples in the last view. This process is visualized as a dendrogram in right image.

4.4 Visualization with Spatial Axis

When a user interact with virtual objects and move them spatially, we can visualize the interaction history with spatial axis. One example is the slide bar. A user can set parameters by manipulating slide bars. It is implemented with three dimensional positioning sensors. The other example is moving histories. When a user move or fly in the virtual environment, the trajectory indicates history. We visualize the history with spatial in a virtual environments.

5 Summary and Discussion

In this paper, we proposed three visualization methods. We introduced two applications for our visualization methods as results.

The results indicate that history is useful to grasp the process. However, past history visualization has limitation because it depends on the amount of history data. There is still room for improvement to further incorporate and utilize history of interactions.

Acknowledgments. This work is supported in part by Core Project from Microsoft Institute for Japanese Academic Research Collaboration (IJARC).

References

1. Card, S.K., Mackinlay, K., Shneiderman, B.: Readings in Information Visualization: Using Vision to Think (1999)
2. Gershon, N., Eick, S.G., Card, S.K.: Information visualization. In. ACM interactions 5, 9–15 (1998)
3. Nishimura, K., Ishikawa, S., Hirota, K., Aburatani, H., Hirose, M.: Information Visualization of Allelic Copy Number Data for Genome Imbalance Analysis. 11th International Conference on Human - Computer Interaction (HCI International 2005) CD-ROM (2005)

Towards a Metrics-Based Framework for Assessing Comprehension of Software Visualization Systems

Harkirat Kaur Padda, Ahmed Seffah, and Sudhir Mudur

Department of Computer Science & Software Engineering, Concordia University, Montreal, Canada
{padda,seffah,mudur}@encs.concordia.ca

Abstract. Despite the burgeoning interest shown in visualizations by diverse disciplines, there yet remains the unresolved question concerning comprehension. Is the concept that is being communicated through the visual easily grasped and clearly interpreted? Given the vast variety of users and their visualization goals, it is difficult for one to decide on the effectiveness of different visualization tools/ techniques in a context independent fashion. To capture the true gains of visualizations, we need a systematic framework that can effectively tell us about actual quantifiable benefits of these visual representations to the intended audience. In this paper, we present our research methodology to establish a metrics-based framework for comprehension measurement in the domain of software visualization systems. We also propose an innovative way of evaluating a visualization technique by encapsulating it in a visualization pattern where it is seen as a solution to the visualization problem in a specific context.

Keywords: Software visualizations, comprehension, measurement, metrics, cognition, perception, GUI, patterns etc.

1 Introduction

There are many visualization tools/techniques available today, and many more continue to be introduced as computer-based information processing pervades different domains. Visualization is often seen as a way to help people gain insight about large, related and complex information or artifacts. Despite this apparent proliferation, various researchers have seen many shortcomings in existing visualizations tools which tend to considerably reduce their overall value to users. A detailed study of existing literature shows that possible shortcomings include: 'navigational problems', 'improper context', 'lack of evaluation', 'ineffective imagery', cognitive overload', and interaction difficulty'. With these drawbacks, the utility of the visualization systems is questionable and this is the point where we need to seek measurement. The success of any visualization system relies on its support for providing 'user insights' to understand underlying artifact represented through the visual. If the visualization system does not achieve this objective, it is of little use and suffers from poor quality. Clearly, comprehension is the most important aspect that determines the quality of any visualization system. For visualization systems, we define comprehension as the degree to which information represented through visuals can be grasped and interpreted correctly in a specified context.

M.J. Smith, G. Salvendy (Eds.): Human Interface, Part I, HCII 2007, LNCS 4557, pp. 335–344, 2007.

Existing visualization systems, with sheer volume of information, place high cognitive load on the users. They provide little help to interpret the meanings of different visualizations being displayed. Gleaning from the literature, one can see that no matter how efficient a visualization tool/technique may be, or how well motivated from theory it is, if it does not convey information effectively then its' usefulness is questionable. We need to study and answer research questions like- how well the visualization system's intent is met through visuals and interaction techniques; how well the user's intent is met by the visualization system; whether these representations are really effective in terms of achieving their major goal of providing 'user insights' for which they were developed and how can we measure whether the visualization has been appropriately comprehended by intended users. Clearly, what is needed is a framework which enables us to systematically carry out studies for measuring the comprehension aspects of visualization tools/ techniques.

To empirically assess the value of visualizations in a practical sense, we need to study their usefulness in a particular field of their usage; which in our case is software visualizations. In the domain of software visualizations, visualization technologies provide graphical abstractions of the huge source code and assert to ease in perception of this invisible entity by giving it altogether a different aspect than that of a source code. In addition, given the vast variety of users (software engineers in our case) and their distinct visualization goals, it is difficult for one to decide on the effectiveness of different software visualization tools/techniques in a context independent fashion. This notion of 'context of use' has become a pandemic in almost all measures of HCI field; usability itself is not independent of this criteria. We plan to deal logically with this influential factor of any evaluation mechanism by conducting an empirical study with five comparable software visualization tools. Ensuring the effectiveness of a software visualization tool/technique involves understanding of how users use it. This is being simulated in our research by using controlled experiment approach with the help of typical users of these tools. Unlike other researchers who have conducted empirical studies to informally judge the strengths and weaknesses of their tools, our goal is to objectively quantify the overall effectiveness in terms of supported user's comprehension. The software visualization tools (i.e. SA4J [3], SHriMP [10], Structure 101[14], Surveyor [15], and VizzAnalyzer [16]), which we are exploring are the static software visualization tools. These are mainly used in academia and industry during software maintenance purposes to extract the structure of any software system. We investigate the issue of evaluating a visualization tool/ technique by encapsulating it in a pattern format describing the applicable 'context of use' that is appropriate for it. The context of use is defined in detail by studying the actual needs, characteristics of software maintainers and matching their needs with the capabilities of different visualization tools/techniques.

Our proposed metrics-based framework will integrate both quantitative and qualitative measures of users' comprehension which are derived from human's cognitive and perceptual capabilities along with the interface features of any visualization system. Our evaluation criterion is based on three well-defined principles for effective visual communication in HCI – principle of organization, principle of economize and principle of communication proposed by Marcus [5] along with Norman's Cognitive Principles [7]. Like other standard software engineering models (i.e. McCall, Boehm, GQM etc.), the proposed framework incorporates a hierarchical decomposition of users' comprehension factors, their associated criteria, metrics along with their interpretation, as well as overall evaluation context for comprehension assessment in the form of visualization patterns.

The rest of this paper is organized in various sections as follows:
Section 2 describes the general background and related research in the domain of software visualizations, section 3 along with subsections is a detailed explanation of our research methodology towards the establishment of the metrics-based framework, and finally in section 4, we conclude by enumerating the benefits of our approach and the current stage in the development of this framework.

2 Background and Related Work

The two main facets of software visualizations (SV), static and dynamic, are intended to support many different software development activities; and the developers of these tools claim that their usage improves the productivity of their users especially the software maintainers. Without measurement or evaluation in some form, it is very difficult to realize the true value of such visualization tools to the software community. There is still little progress in the evaluation of software visualizations, as most research effort is being spent on the development of yet more novel visualization techniques, ideas and technological innovations rather than judging the effectiveness/usefulness of the currently available SV tools/techniques. In short, the field of empirical investigation of software visualization tools/ techniques is rather immature and only a few researchers have worked informally to characterize and assess the usefulness of these SV tools/techniques. In the following paragraphs, we briefly summarize various related studies conducted by other researchers in the domain of software visualizations.

Bassil and Keller [1] conducted an online survey of software visualization tools using a questionnaire approach. The questionnaire was designed using existing taxonomies to extract a list of properties of software visualization tools. The objective of the study was twofold - to assess the functional, practical and cognitive aspects of visualization tools that users' desire, and to evaluate support of code analysis in various existing tools that users' use in their environment. The authors recognized a total of 34 functional aspects along with 13 different practical properties of software visualization tools. They also summarized the cognitive aspects of visualization tools in terms of various usability elements like 'ease of use', 'effectiveness', and 'degree of satisfaction' etc.

Knight and Munro [4] briefly discuss two main perspectives that should be taken into account when deciding whether or not visualization is effective. These are - the suitability for the tasks that the visualization is intended to support, and the suitability of representation, metaphor and mapping based on the underlying data. They also highlight that domain and data structures have a considerable affect on the effectiveness of any visualization.

Marcus et al. [6] conducted a usability study to assess the effectiveness of a software visualization tool named sv3D. The aim of the study was to determine the usefulness and improvement of sv3D as a new technology to support program comprehension. The source program was a documentation software application which was rendered using 3D metaphor of poly cylinders and containers. A total of 35 participants participated in usability study. The participants were divided into two groups: one group answered the questions using sv3D tool and other group responded the questions using tabular data with metrics and source code utilizing the search features in Visual Studio.NET. The answers were analyzed and compared to judge the effectiveness of sv3D tool.

Pacione et al. [8] conducted an empirical evaluation of five dynamic visualization tools. The aim of their study was to compare the performance of these tools in general software comprehension and specific reverse engineering tasks. The performance of the tools was judged by conducting a case study with a drawing editor. The evaluation was carried out by a single user who had the knowledge of the drawing editor and dynamic visualization tools. The tools were compared based on four categories – extraction technique, analysis technique, presentation technique, and abstraction level. The questionnaire was divided into two sections- large scale questions expressing the course of a software comprehension effort, and small-scale questions resembling the course of a specific reverse engineering effort.

Storey et al. have performed a number of experiments with software visualization tools [11, 12, and 13]. In these studies, their primary objectives were to: compare the effectiveness of their tool on five usability dimensions, observe different strategies used by participants while comprehending program under study; how the tools were supporting these set of preferred strategies; devise a framework for describing, comparing and understanding visualization tools that provide awareness of human activities in software development and provide feedback for tool developers and researchers. Their framework has five key dimensions: Intent (to capture the general purpose and motivation that led to the design of visualization), Information (data sources that a tool uses to extract relevant information), Presentation (how the tool presents the extracted and derived information to users), Interaction (refers to interactivity of the tools), Effectiveness (determines if the proposed approach is feasible and if the tool has been evaluated, deployed).

3 Research Methodology Towards Establishing a Framework

As we have seen in the previous section, many researchers have applied different strategies for empirical investigation of SV tools, but without a unified measurement framework. Our goal is to provide a measurement framework that can objectively quantify the overall effectiveness in terms of supported user's comprehension. To achieve this objective, we are proposing a hierarchical framework to properly investigate the issue of comprehension evaluation as shown in Fig. 1. The three bottom layers are similar to the other well-known software engineering models for measuring the software quality. However, the top three layers are the fundamental layers that are needed before conducting any empirical investigation of the SV tools with their users.

Our research methodology is a step by step refinement of these layers, and is described in the following sections.

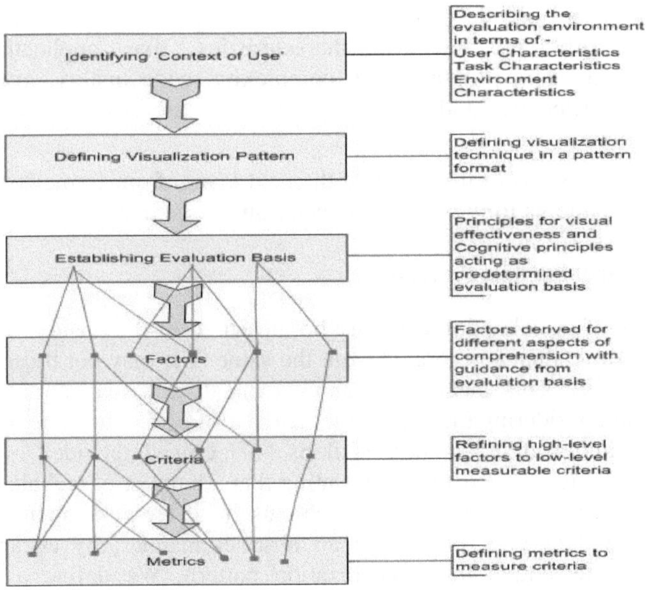

Fig. 1. Research Framework

3.1 Identifying Context of Use

To conduct an evaluation of any SV tool, the foremost step is to identify the 'context of use' that captures the boundaries of evaluation. This is done so that the elements which may influence the evaluation are appropriately summed up. The context of an experiment can be described on three basic dimensions - user characteristics, task characteristics and environment characteristics as briefly described below:

- User characteristics - In our research with static software visualization tools, our users are software engineers. So, the characteristics that have significant impact on their performance are - age, gender, spatial-ability, education, experience (application domain knowledge, programming language expertise, visualization tools expertise) etc.
- Task characteristics - Tasks selected for an evaluation should be representative of what the users (software engineers) do with the static software visualization systems and must be manageable, suitable for a laboratory evaluation. Based on a thorough literature survey, we have identified the needs for software maintenance and have compared the maintenance needs with the tasks supported by current software visualization tools. This is done because the users' tasks that these SV tools support are linked to or are derivable from the typical and elementary information needs of software maintainers. We have developed a catalogue of software maintenance tasks that should be supported by any static software visualization tool. The other characteristics of tasks, like- task type, task size and complexity, task time and cost constraints are also summed up appropriately.
- Environment characteristics - The environment for the experiment is described in detail by determining the appropriate software/ hardware, social components for it.

In our framework, we describe in detail the software part of environment by enumerating different software characteristics like application domain, programming domain, program size, complexity, code quality, availability etc. Hardware characteristics and social characteristics are also studied in detail for experimental purposes.

This 'context of use' is further fed into the next layer of our framework to describe the visualization patterns for visualization techniques used in an experiment.

3.2 Defining Visualization Pattern

The true quality of visualization can only be measured in the context of a particular purpose, as the same image generated from the same data may not be appropriate for another purpose [9]. This means we cannot evaluate a visualization technique in isolation without considering the applicable 'context of use'. A technique can be good in one context and bad in another. Wilkins [17] coined the idea of formalizing visualization techniques into patterns by stating that a number of techniques are being reused to solve recurring visualization problems in different domains. That is, we have to evaluate their effectiveness in an abstract manner; by encapsulating the visualization tool/technique in a visualization pattern. We define a visualization pattern as a visualization problem that occurs in a certain context and for which visualization technique can be a solution.

A visualization problem could be solved by a number of different techniques. Consequently, there are many different patterns that could be derived for the same problem. For the tools under our investigation, we are proposing a number of different visualization patterns that have common visualization problem and context as shown in Fig. 2 and Fig. 3. All these visualization patterns are to solve a common visualization

Title	**Radial Tree**
Context	The display consists of a number of software objects (packages, classes and interfaces) and their inter-relationships or structural dependencies in the source code
Problem	How to display large hierarchical tree structures showing dependencies among software objects?
Forces	▪ To visualize and navigate large trees in a radial space ▪ Focus + context viewing allowing enormous trees to fit within fixed space of a computer screen ▪ Zoom in on a particular item while keeping in view of the neighborhood context
Solution	Use a Radial Tree representation [3] 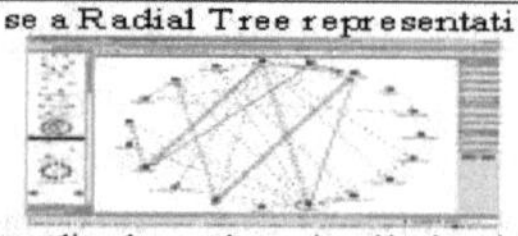 To display the detailed picture of relationships between application objects and detailed map of dependencies and dependents of every package, class or interface in an application. The idea is to display different software objects and their relationships in a radial fashion, where the object nodes are placed round the circle and their relationships are shown with directed lines emanating from the source to destination node.
Examples	Sunburst, RadViz
Related pattern	Pyramid/Skeleton View, Tree/Graph, Cone Tree, TreeMap, Explorer etc.

Fig. 2. Radial Tree Pattern

Title	**Pyramid or Skeleton View**
Context	The display consists of a number of software objects (packages, classes and interfaces) and their inter-relationships or structural dependencies in the source code
Problem	How to display large hierarchical tree structures showing dependencies among software objects?
Forces	▪ To visualize and navigate large software structures within fixed space of a computer screen ▪ Overview of entire structure of software system in the form of pyramid of small squares ▪ Details on demand by providing a 'data tip' to access the detailed information about any object under selection ▪ View relationships among items. Users can select an item and then highlight the dependent items
Solution	Use a Skeleton View [3] 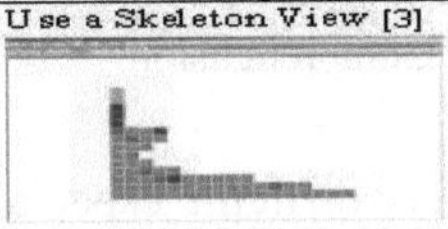 The basic idea here is to represent the software as a pyramid of dependencies- the entities with only outgoing dependencies on the bottom, those with only incoming dependencies on the top. Each square corresponds to either one object (class/ interface/ package) or one tangle. This layered view of the system is constructed by putting objects (class/ interface/ package) that do not depend on anything at the bottom of the visualization. The objects that are dependent on the lowest layer appear in the layer above, and so on. In this view, a stable system should have a normal pyramid shape. An unstable system may look like an upside down pyramid shape. Tangles, dependents of an entity etc. can be specifically colored.
Examples	Icicle plot
Related pattern	Radial Tree, Tree/Graph, Cone Tree, TreeMap, Explorer etc.

Fig. 3. Pyramid/Skeleton View Pattern

problem of displaying large hierarchical tree structures of dependencies among different software objects. The contributing forces may vary in each pattern; however, to distinguish pattern from one another solution is entirely different in each pattern.

3.3 Defining the Evaluation Basis

Our evaluation criterion is based on the work of two eminent researchers (Aaron Marcus and Donald A. Norman) in HCI community who have their expertise in the field of visual communication and human cognition. We believe that the basic principles proposed by them in their respective areas are the fundamental evaluation objectives that contribute to the overall comprehension of any visualization system regardless of its domain. The three principles proposed by Marcus [5] on visual communication i.e. 'Principle of Organization', 'Principle of Economize' and 'Principle of Communication' along with the criteria proposed by Norman cognitive principles [7] on 'Affordances', 'Mapping', 'Constraints' etc. are the building blocks of our evaluation basis. These guiding principles and criteria are studied further in detail in order to determine their affect on human comprehension. The next layer in our hierarchical framework is actually derived from this evaluation basis.

3.4 Determining the Comprehension Factors and Criteria

For any visualization system, data rendered in visual form is perceived or interacted upon by the user of that system as shown in Fig. 4.

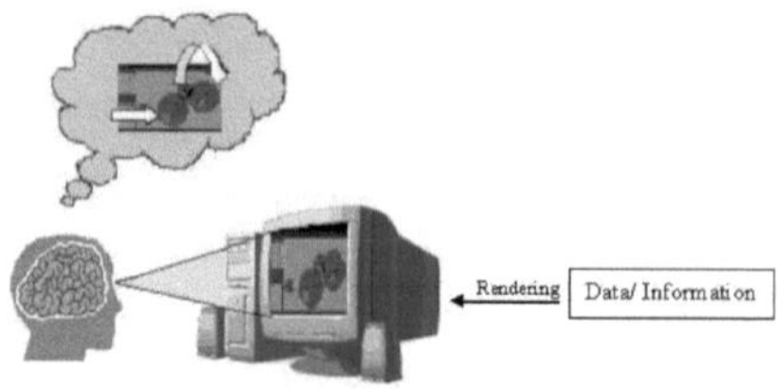

Fig. 4. Aspects of Comprehension

The issue of comprehension evaluation is difficult especially with visualization systems where many different aspects like human cognition, perception, information structure and visualization interface play different roles and affect one another. Brief explanations of each of these aspects are as under:

- Information structure: The information structure has a profound affect on user comprehension. Sometimes the data that is rendered is not perfect by itself due to many causes like: corruption of data, incompleteness, inconsistency, information complexity, uncertainty, imperfect presentation etc. [2]. The net affect is that the visual, which is used to represent it, is not easy to comprehend.
- Visualization interface: Naturally, the interface is a crucial part of any visualization system, as it essentially forms the link between the user and the visualization itself. An easily understandable UI helps the user to interpret the visualization and perform the correct operations. Based on the Marcus's [5] principle of organization and principle of economize, we have derived a set of factors/ criteria for visualization interface comprehension like consistency, navigability, screen layout, simplicity, clarity/intuitiveness, distinctiveness, emphasis etc.
- Perception: Perception is an integral part of any visualization and perceptual features of visualizations like color, orientation, contrast, position, size etc. along with the criteria proposed by Norman [7] like affordances, metaphors/ symbolism, familiarity can interfere with successful comprehension.
- Cognition: In order to judge the degree of comprehension, it is also necessary to study the human information processing or the cognition of information in human mind. The visualization tool/technique should provide an ergonomic design that matches the cognitive capabilities of the user. To ensure that, Marcus' third principle of communication and Norman's Cognitive Principles guide us to have a number of factors/criteria like legibility, readability, multiple views, naturalness of interaction/mapping etc.

Currently, all the factors in these four aspects of comprehension excluding "information structure" are being studied further in detail to determine their appropriateness to the SV tools. If needed, they can be further decomposed into measurable criteria.

3.5 Determining the Metrics

The lowest level of our framework is a collection of metrics and measures to quantify the related criteria and is currently under implementation. We will be adopting the metrics proposed by various researchers and will define new metrics. We are devising a set of questions to be asked in a controlled experiment. The questionnaire will incorporate qualitative and quantitative aspects of visualizations which will feed the data for our subjective and objective metrics respectively. Metrics interpretation will be an integrated part of our evaluation framework.

4 Conclusion

In addition to the general measurement benefits, our framework will be a reusable solution that could be applied to other domains in real-world settings to measure comprehension/effectiveness of any visualization system. Specifically, we expect the following benefits:

1. Prior attention to the most important visual design principles for understanding what factors of the visualization system can influence users' comprehension.
2. Provide a flexible hierarchy of the factors and criteria, so that evaluators could select those that are most appropriate according to their evaluation objectives.
3. Appropriate documentation of the test environment, in terms of the 'context of use' and encapsulation template for visualization techniques in terms of the patterns, for better analysis.
4. Data collection efforts will be concentrated, since the required data elements will be already defined.
5. Data interpretation will be more efficient and effectively tied to selected objectives.

Currently, we are refining our repository of factors/criteria and are working on a set of metrics that can be applied for the controlled experiment. We hope that our framework will help and guide other researchers in different domains as well.

References

1. Bassil, S., Keller, R.K.: Software Visualization Tools: Survey and Analysis. In: Proc. of 9th Intl. Workshop on Program Comprehension. Toronto, Canada, pp. 7–17 (2001)
2. Gershon, N.: Visualization of an Imperfect World. IEEE Computer Graphics and Applications 18(4), 43–45 (1998)
3. Iskold, A., Kogan, D., Begic, G.: Structural Analysis for Java (SA4J). An alphaworks Java technology from IBM, (2004), [Accessed October 28, 2006], Available from: < http://www.alphaworks.ibm.com/tech/sa4j >
4. Knight, C., Munro, M.: Visualisations; Functionality and Interaction. In: Alexandrov, V.N., Dongarra, J., Juliano, B.A., Renner, R.S., Tan, C.J.K. (eds.) Computational Science - ICCS 2001. LNCS, vol. 2074, pp. 470–475. Springer, Heidelberg (2001)
5. Marcus, A.: Principles of Effective Visual Communication for Graphical User Interface Design. In: Human Computer Interaction –Towards the year 2000, 2nd edn., pp. 425–441. Morgan Kaufmann, San Francisco California (1995)
6. Marcus, A., Comorski, D., Sergeyev, A.: Supporting the Evolution of a Software Visualization Tool through Usability Studies. In: Proc. of 13th Intl Workshop on Program Comprehension (2005)
7. Norman, D.A.: The Design of Everyday Things. Doubleday, New York (1990)
8. Pacione, M.J., Roper, M., Wood, M.: A Comparative Evaluation of Dynamic Visualisation Tools. In: Proc. of 10th Working Conf. on Reverse Engg., pp. 1095–1350 (2003)

9. Rushmeier, H., Botts, M., Uselton, S., Walton, J., Watkins, H., Watson, D.: Panel: Metrics and Benchmarks for Visualization. In: Proc. of 6th IEEE Visualization Conference, 422 (1995)
10. SHriMP. The CHISEL Group. University of Victoria, BC, Canada. [Accessed November 2007,2006] Available from: <http://www.thechiselgroup.org/shrimp/manual>
11. Storey, M.A.D., Wong, K., Fong, P., Hooper, D., Hopkins, K., Müller, H.A.: On Designing an Experiment to Evaluate a Reverse Engineering Tool. In: IEEE Proc. of 3rd Working Conf. on Reverse Engg., pp. 31–40, Los Alamitos, CA (1996)
12. Storey, M.A.D., Wong, K., Müller, H.A.: How Do Program Understanding Tools Affect How Programmers Understand Programs? In: Proc. of 4th Working Conf. on Reverse Engg., Amsterdam, Holland, pp. 12–21 (1997)
13. Storey, M.A.D., Čubranić, D., German, D.M.: On the Use of Visualization to Support Awareness of Human Activities in Software Development: A Survey and a Framework. In: Proc. of ACM symposium on Software visualization, St. Louis, Missouri, pp. 193–202 (2005)
14. Structure101. Headway Software Technologies. [Accessed November 08, 2006], Available from: < http://www.headwaysoft.com/index.php >
15. Surveyor, Lexient Corporation. [Accessed November 2006], Available from: < http://www.lexientcorp.com/codeanalyzer/products.htm >
16. VizzAnalyzer, ARiSA Group, Växjö University, Sweden. [Accessed October 28, 2006], Available from: < http://www.arisa.se/index_projects.html >
17. Wilkins, B.M.: A Pattern Supported Methodology for Visualisation Design. Doctoral dissertation, University of Birmingham, UK (2003)

Facilitating Visual Queries in the TreeMap Using Distortion Techniques

Kang Shi, Pourang Irani, and Pak Ching Li

Department of Computer Science, University of Manitoba
Winnipeg, Manitoba, R3T 2N2, Canada
shikang@gmail.com, {Irani,pkli}@cs.umanitoba.ca

Abstract. TreeMap is one a common space-filling visualization technique to display large hierarchies in a limited display space. In TreeMaps, highlighting techniques are widely used to depict search results from visual queries. To improve visualizing the queries results in the TreeMap, we designed a continuous animated multi-distortion algorithm based on fisheye and continuous zooming techniques. To evaluate the effectiveness of the new algorithm, we conducted an experiment to compare the distortion technique to the traditional highlighting methods used in TreeMaps. The results suggest that the multi-distortion technique is only effective with small result sets but not as effective as simple highlighting for large search result sets.

Keywords: TreeMap, distortion, searching, visual query, fisheye, continuous zooming, visualization.

1 Introduction

Search techniques are widely used in many areas of interaction such as searching in file systems (hierarchical data). Generally, a search action includes three steps: users submit search criteria to the system, the system performs the search in the database, and finally the results of the search are represented to the users. In most cases, such as search in Windows Explorer, the search results are presented in plain text to the users. However, this type of representation is not effective and has some disadvantages. One disadvantage is that the large quantities of search results can not be represented in a limited display space. The users need to scan and scroll and only can view a part of the result set. Another disadvantage is that users have difficulty in knowing the priority and the relationship between two or more search results. In addition, users also find it difficult to know the relationship between a search result and a non-result.

In this paper we introduce a visualization technique to represent search results in hierarchical data. Hierarchical data, typically represented in the form of a tree, are widely used in a variety of areas such as a file system. However, large hierarchies require complex interaction models as they typically exceed the available display space. Space-filling visualization, such as TreeMaps [4] or Sunburst [5], is one approach to resolving this problem. TreeMaps make efficient use of the display area

M.J. Smith, G. Salvendy (Eds.): Human Interface, Part I, HCII 2007, LNCS 4557, pp. 345–353, 2007.

and provide limited structural information. In the TreeMap, the display area is divided into nested rectangular regions to map an entire hierarchy of nodes and their children. Each node uses an amount of space relative to the weight of the item being represented in the hierarchy.

To facilitate a search results visualization technique in the TreeMap, we applied animated fisheye and semantic zooming techniques to build a multi-distortion algorithm. With this distortion algorithm, users get the information about the priority of each result and can visually identify the relations between the results. Furthermore, some result nodes that are not visible in the TreeMap could be expanded to a visible size. We conducted an experiment to evaluate the effectiveness of the distortion technique. Results show that the distortion technique is effective only under a few conditions, but highlighting is still quite effective for presenting search nodes in TreeMaps.

2 Related Work

This paper is primarily inspired by the concept of continuous semantic zooming (CSZ) developed by Schaffer et al [2]. This technique is characterized by two distinct but interrelated components: continuous zooming [1] and presentation of semantic content at various stages of the zoom operation. CSZ manages a 2D display by recursively breaking it up into smaller areas. A region of interest becomes the focus and as the continuous zoom is applied successive layers of a display "open up". At each level of the operation the technique enhances continuity through smooth transitions between views and maintains location constraints to reduce the user's sense of spatial disorientation. The amount of detail shown in parts of the display is controlled by pruning the display and presenting items of non interest in summary form. The approach we have implemented is similar in concept to the continuous semantic zoom: smooth transition between zoom levels is applied and content visibility is increased as the nodes enlarge.

To implement the animated distortion, Shi et al [3] developed a uni-distortion algorithm to continuously change the size of a single node in the TreeMap. The algorithm is similar in concept to the continuous semantic zoom: smooth transition between zoom levels is applied and content visibility is increased as the nodes enlarge. Compared to the traditional method of browsing file contents in the TreeMap, the uni-distortion algorithm allows users to open nodes and view the contents without opening multiple layers of the hierarchy. While a target node in the TreeMap is selected, the uni-distortion algorithm executes in three steps: computing the neighbour nodes of the target node, changing the size of the neighbour nodes, and distorting neighbour nodes.

The distortion algorithm in the TreeMap can also be applied to the representation of search results. For instance, after viewing the TreeMap, users may typically be interested in searching for items with specific content (for example, files with certain keywords). Typically this type of interaction will result in obtaining a set of result items with each item having a degree of priority. In this case, the distortion algorithm can be applied to multiple nodes simultaneously. The multi-distortion algorithm is an extension of the uni-distortion algorithm with two new rules.

3 Algorithm

The multi-distortion algorithm is expanded from the uni-distortion algorithm. Some major differences between the multi-distortion technique and the uni-distortion technique are as follows. In the multi-distortion algorithm, a target node cannot become a neighbour of other target nodes, and one node cannot become a neighbour of two target nodes. Furthermore, additional mappings are used in the multi-distortion algorithm. The distortion method has one primary attribute: the amount of distortion (amplitude). This attribute can be used to distinguish the levels of significance for each result. Therefore, significance is assigned to the amplitude of the distortion. For instance, a distortion with large amplitude can imply higher priority content than a distortion with small amplitude. The behavior of the multi-distortion algorithm is depicted in figure 1. Figure 1 (a) is the initial state of the TreeMap. The highlighted nodes are the search result nodes which will get distorted. Figure 1 (b) to (d) shows that the size of the nodes increased due to the distortion. The amplitude of the distortion indicates the priority of the node. Larger amplitude represents higher priority. Figure 1 (e) to (f) show that the size of nodes decrease because of the distortion. After state (f), the TreeMap returns back to the original state of the TreeMap (a).

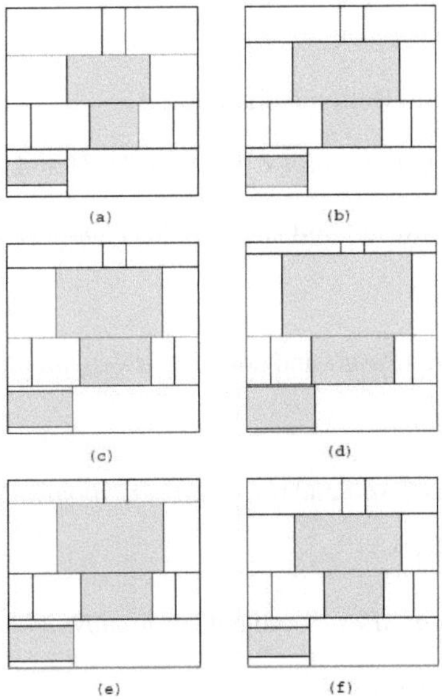

Fig. 1. The behavior of the multi-distortion algorithm. Figure (a) is the original state, figure (b)-(d) indicate the size of each node increased, figure (e)-(f) indicate the size of each node decreased. After state (f), the state of TreeMap goes back to (a).

Algorithm 1 describes the distortion of one node in the multi-distortion technique. It starts by determining the neighbours of the node of result (node A). If the neighbours are not other result nodes or neighbours of other result nodes, it decreases the size of these nodes. The size of A increase at the same time to provide the distortion effect. Then the entire TreeMap is redrawn. This process is repeated until an external event stops the distortion process or the node has reached a fixed maximum width or height.

Algorithm 2 describes the final multi-distortion algorithm. A1, A2, ..., An are search result nodes which should be distorted. Algorithm 11 is applied one by one on these result nodes to generate multiple distortions.

Algorithm 1.

```
OneNodeinMultiDistortAlgorithm(A)
global gLeftNeighbour, gRightNeighbour
global gTopNeighbour, gBottomNeighbour
bool isResult  \ \Whether a node is a search result
bool isNeighbour  \ \Whether a node is a neighbour of another search result
gLeftNeighbour = ComputeLeftNeighbour(A)
gRightNeighbour = ComputeRightNeighbour(A)
gTopNeighbour = ComputeTopNeighbour(A)
gBottomNeighbour = ComputeBottomNeighbour(A)
while DISTORTING = true and A.width < MAX_WIDTH and A.height < MAX_HEIGHT
do {
    if not gLeftNeighbour.isResult and not gLeftNeighbour.isNeighbour
    then DistortLeft(amount)
    if not gRightNeighbour.isResult and not gRightNeighbour.isNeighbour
    then DistortRight(amount)
    if not gTopNeighbour.isResult and not gTopNeighbour.isNeighbour
    then DistortTop(amount)
    if not gBottomNeighbour.isResult and not gBottomNeighbour.isNeighbour
    then DistortBottom(amount)
    RedrawTreeMap(ROOT)
    Sleep(sleep int erval)
```

Algorithm 2.

MultiDistortAlgorithm($A1, A2, \ldots, An$)
OneNodeinMultiDistortAlgorithm($A1$)
OneNodeinMultiDistortAlgorithm($A2$)
...
OneNodeinMultiDistortAlgorithm(An)

4 Experiments

To evaluate the effectiveness of the multi-distortion algorithm, we designed this experiment to compare the distortion technique to the conventional method of showing search results (highlight) in the TreeMap. We defined priority as how often search keywords occur in a node, or whether all search keywords occur in a node. Priority in the distortion technique is mapped onto the amplitude of the animated distortions. In the highlight approach priority is indicated by the level of saturation. We anticipated the following effects from our experimental data:

Hypothesis 1: With small results set (number of results ≤ 5), the distortion technique is more effective than the highlight technique (speed and accuracy).

Hypothesis 2: With large results set ($5 <$ number of results ≤ 10) the highlight technique is more effective than the distortion technique (speed and accuracy).

4.1 Method

Twelve students participated in this experiment. Half of the subjects were assigned to one condition: Distortion first, and half of them were assigned to the other condition: Highlight first. Subjects were familiar with the concept of searching in windows file systems. All subjects were familiar with the highlight technique and the TreeMap, but none had any experience using distortion to represent search results in the TreeMap.

One hierarchy containing one thousand files was used for this experiment. Two different types of search keywords were used for the experiment: long and short. Using the short search keywords, the search would generate 3, 4, or 5 search results (small set). Using the long search keywords, the search would generate 6, 8, or 10 search results (large set). Six short search keywords and six long search keywords were used in the experiment. To reduce learning effects, I used two sets of search keywords (Set A and Set B) which would generate the same numbers of search results but in entirely different positions in the TreeMap. Half the subjects started the experiment with the Distortion method and the other half started with the Highlight method. After completing the tasks using one set of keywords, the subjects switched to use the other set of keywords with the other method.

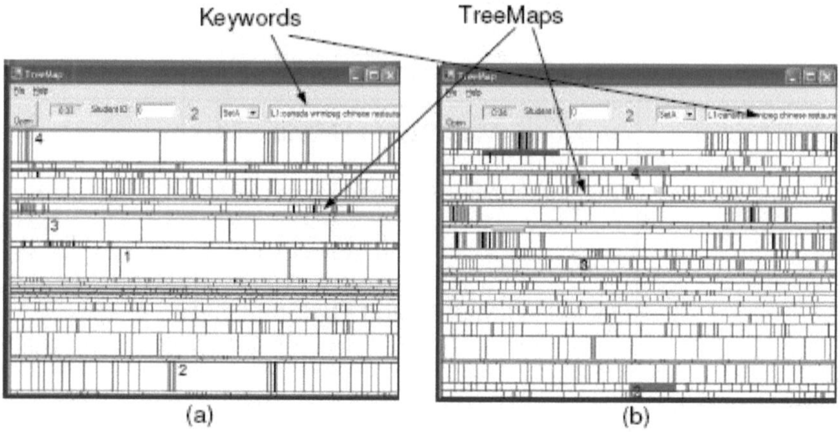

Fig. 2. Interface used in the experiment. Users are required to identify all search results and their priorities with (a) the distortion method, and (b) the highlight method.

As shown in Figure 2, this experiment consisted of two types of representations: distortion and highlight. In the distortion method, all search results were animated in and out using distortion algorithms until the subjects identified all results or the subjects withdrew from the task. The amplitude of the distortion represented the priority of each result. A result with larger amplitude has a higher priority. In the highlight method, all search results were filled by a color. Subjects were required to identify all the results or could withdraw from the task. The saturation of the color represented the priority of the result. A result with higher saturation has a higher priority.

4.2 Procedure

In this experiment, the task of the subjects is to identify the priority of the search results from highest to lowest. Before starting the experiment, each subject got familiarized with both representations. Each participant performed the tasks with six different short keywords and six different long keywords using both methods. The 12 trials were executed in the following sequence S1, S2, S3, ..., S6, L1, L2, L3, ..., and L6 (S represents the small sets of search results and L represents the large sets of search results). No time limit was set for the tasks, and the subjects were free to finish the trial if they could not identify all search results.

We recorded the time to execute the task, the number of results identified by the subjects, and whether the subject identified the correct priority for each search result in all conditions. In summary, the whole experiment involved: 12 subjects×2 main conditions×2 type of search result sets×6 trials = 288 trials in total.

4.3 Results

To test hypotheses 1 and 2 mentioned at the beginning of this section, we recorded the time it took the participants to select the items that are in the result set. Accuracy in

Table 1. The average completion times and average error rates for small and large result sets with both methods (standard deviations are in parentheses)

		Time	Error Rate
Distortion	Small Set	5.01 (1.47) sec	4.48%
	Large Set	7.06 (2.83) sec	7.43%
Highlight	Small Set	2.43 (0.66) sec	3.49%
	Large Set	3.30 (1.11) sec	2.65%

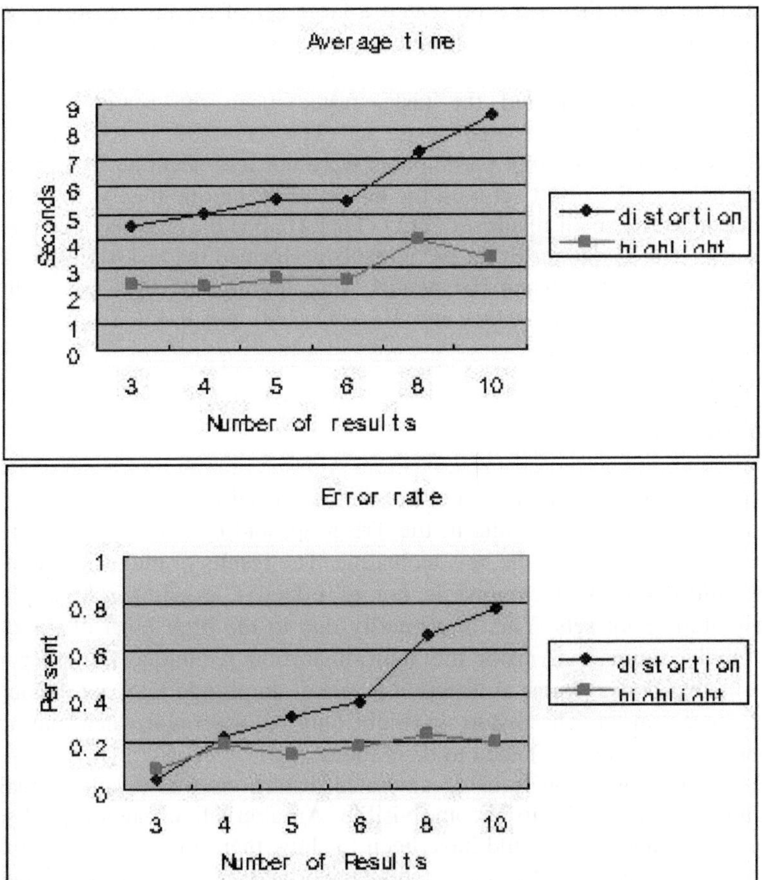

Fig. 3. Average time and average error rate with distortion and highlight methods

selecting the order of importance was also recorded, i.e. to get a perfect score the participant would need to select the items from the highest priority to the lowest priority in the correct order.

Time to locate search items and accuracy in determining the correct priority were analyzed using a repeated measure uni-variate ANOVA and a paired sample T Test. The results are summarized in Tables 1, and the average time and average error rate are described in Figure 3. An alpha level of 0.05 was used for all statistical tests. The main effect of representation type on time to complete the task and the main effect of representation type on accuracy were significant, $F(1, 11) = 44.61$, $p < 0.001$ and $F(1, 11) = 19.25$, $p = 0.001$ respectively. Participants performed significantly faster and more accurately when the search results were highlighted than when they are presented using distortion.

The main effect of the size of the search result on time to complete the task was significant, $F(1, 11) = 11.68$, $p < 0.001$. Participants performed significantly faster when they were presented with a small set of search results (less than five items in the result set) than when they were presented a large set of results. However, the main effect of the size of the search result on accuracy was not significant $F(1, 11) = 3.63$, $p = 0.083$.

The participants completed the tasks faster when the search results where highlighted vs. distorted in a small result set ($T(1, 11)=6.88$, $p < 0.001$). However, The analysis also suggests that participants were not less accurate in identifying the priority of items in the result set with the distortion technique than with the highlight representation in the small result set ($T(1, 11)=1.16$, $p = 0.27$). This does not support Hypothesis 1. In the large result set participants performed the task significantly faster and were more accurate when the results were highlighted than when they were distorted ($T(1, 11)=5.27$, $p < 0.001$ and $T(1, 11)=7.06$, $p < 0.001$, respectively). This is a support of hypotheses 2.

5 Conclusion

In this paper we introduced a new multi-distortion visualization technique to facilitate the visualization of search results in the TreeMap, and conducted an experiment to evaluate the effectiveness of the new technique. The results of the experiment suggest that the multi-distortion technique is not as effective as simple highlighting for showing search result sets. This is primarily due to the high level of distractibility created by the animations from the multi-distortion technique. Another possible reason is having nodes distort at different levels of amplitude may not enhance focus and user concentration. The distortion might only be beneficial when one node or a few nodes (less than 3 nodes) need to be in focus.

However, the motivation in using animated distortions was to allow small nodes that are not clearly visible, to become visible. An alternative method to display the distortions simultaneously would have been to show them in series starting from the highest priority to the lowest priority. This would have allowed users to locate nodes that would have not been visible otherwise but still provide a fair indication of items in the tree. We are pursuing this research track in the near future.

References

1. Bederson, B.B., Hollan, J.D., Perlin, K., Meyer, J., Bacon, D., Furnas, G.W.: Pad++: A zoomable graphical sketchpad for exploring alternate interface physics. Journal of Visual Languages and Computing 7(1), 3–32 (1996)
2. Schaffer, D., Zuo, Z., Greenberg, S., Bartram, L., Dill, J., Dubs, S., Roseman, M.: Navigating Hierarchically Clustered Networks through Fisheye and Full-Zoom Methods. ACM TOCHI 3(2), 162–188 (1996)
3. Shi, K., Irani, P.P., Li, P.C.: An Evaluation of Content Browsing Techniques for Hierarchical Space-Filling Visualizations. IEEE InfoVis, pp. 81–88 (2005)
4. Shneiderman, B.: Tree visualization with TreeMaps: a 2D Space-Filling Approach. ACM TOG 11(1), 92–99 (1990)
5. Stasko, J.T., Zhang, E.: Focus+Context Display and Navigation Techniques for Enhancing Radial, Space-Filling Hierarchy Visualizations. IEEE Infovis, pp. 57–65 (2000)

ActiveScrollbar: A Scroll Bar with Direct Scale Ratio Control

Hongzhi Song[1], Yu Qi[1], Lei Xiao[1], Tonglin Zhu[1], and Edwin P. Curran[2]

[1] HCI Group, College of Informatics, South China Agricultural University, Guangzhou, 510642, China
{hz.song,yuliangqi,lein xiao,tlzhu}@scau.edu.cn
[2] School of Computing and Mathematics, Faculty of Engineering, University of Ulster at Jordanstown, Northern Ireland, BT37 0QB, UK
ep.curran@ulster.ac.uk

Abstract. Scroll bar is one of the most frequently used components of graphical user interfaces (GUIs). It is generally considered to provide overview + detail functionality. The scale ratio of a scroll bar refers to the rate of the document dimension to the proportion being displayed. It is usually determined by the length of the document and the height of the display window if considering the case of a vertical scrollbar. The user has no direct control to the scale ratio through the scroll bar, which is inconvenient when dealing with long documents in overview and navigation tasks. This inconvenience is more prominent for small display devices. This paper presents a novel GUI component, ActiveScrollbar, that enhances the standard scroll bar by providing direct scale ratio control without consuming more screen space. This component was expected to be more useful in hand held devices.

Keywords: Information Navigation, Information Overview, GUI Component, ScrollBar, Overview + Detail.

1 Introduction

Scroll bar is one of the most frequently used GUI components. It is generally considered to provide overview + detail functionality. The position of the scroll thumb inside the scroll trough denotes the current position of the detailed view within the overall document. The length of the scroll thumb also portrays the proportion of the document currently shown.

The minimum scale ratio for scroll bars is 1:1, and they are often removed from the display at this ratio because there is no purpose to display an extra widget with the entire document shown in the detail view. Their maximum proportional scale ratio is on the order of 100:1, but it can be set higher if we don't care so much of the proportion. For example, scroll bars can be used to navigate through documents that are several hundreds of pages long. At high scale ratios the length of the scroll thumb does not accurately reflect the proportion of the information space shown in the detail region. To do so truthfully could cause the thumb to shrink to less than one pixel, this will create obvious usability problems in acquiring the thumb. For this reason, a minimum thumb size of approximately ten pixels is used, which reduces the problems

M.J. Smith, G. Salvendy (Eds.): Human Interface, Part I, HCII 2007, LNCS 4557, pp. 354–358, 2007.

of target acquisition but introduces problems of precise document control because small thumb movements causes large "jerky" document movements [1]. At high scale ratios, users must find alternative techniques to smoothly move their documents, such as the scroll bar's end arrows, the cursor keys, or rate based scrolling etc.

Traditional scroll bars do not allow direct configuration of the scale ratio, scale ratio control is either hidden from view or performed through a separate widget. For example many software packages provide magnifying glass like icons to control the zooming function. This is not easy to execute, especially for novice users. In standard desktop interfaces, length of the scroll thumb immediately adapts to changes in zoom level within the detail region. There is no reason why the reverse control could not also be implemented, allowing the zoom level to be controlled directly by manipulating the length of the scroll thumb [1]. This paper presents ActiveScrollbar, A scroll bar with direct scale ratio control that enhances the functionality of the traditional scroll bar.

2 Related Work

Traditional scroll bars typically encode only spatial information. However several researchers had experimented with variants that additionally portray semantic information. Value bars [2] embellished the scroll trough with dashed lines, the thickness of which depended on numerical values extracted from columns in formatted data files, to depict the location of interesting or outlier data. Hilland Hollan [3] also used similar techniques to convey the read and edit history of document regions. Masui [4] used a scroll bar variant for browsing and filtering large lists of data. Byrd [5] evaluated a related technique with inconclusive results but positive subjective responses from participants. McCrickard [6] did an evaluation of the traditional scroll bar and its variants. Bederson et al. [7] used visual demarcations in the scroll bar trough to convey the presence of search matches across several months of appointment data in the PDA calendar Date-Lens. Range-slider [8] is similar to ActiveScrollbar, but it was targeted to control the number of records being displayed from a database rather than a generalized GUI component.

3 Objectives

The aim of this work was to enhance the functionality of the standard scroll bar in dealing with long documents or long components of GUIs. To pursue this aim, four objectives were set up as below:

1. A GUI component was to be designed which should be compatible with standard scroll bars.
2. The component should help the user more in gaining an overview of long documents or long components.
3. The user should have direct control to the scale ratio of the displayed document through this component.
4. The component should be space saving, and it was expected not to take more screen that the standard scroll bar.

4 Methods

Two more arrows were provided on both ends of the scroll thumb of a standard scroll bar, see Figure1. These two arrows control the scale ratio of the displayed document, but they also affect the current position of the detailed view within the overall document because adjusting them will change the length of the scroll thumb. The user performs the function by dragging these two arrows. Dragging the top arrow up increases the height of the scroll thumb, and dragging it down decreases the height. Dragging the bottom arrow up and down performs the reverse function.

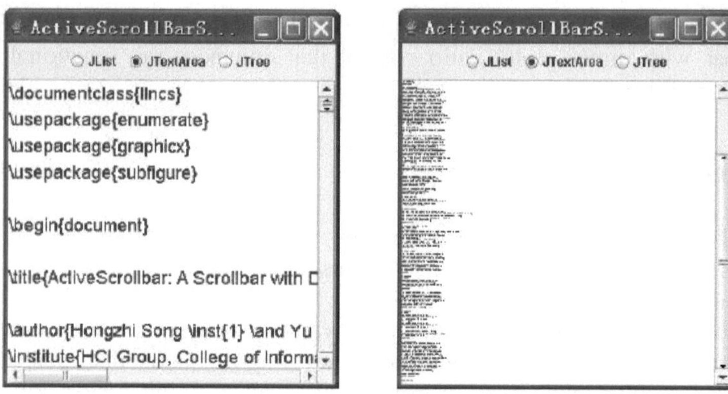

(a) Maximum scale ratio for detail. (b)Minimum scale ratio for overview.

Fig. 1. ActiveScrollbar used to control the scale ratio of a long document

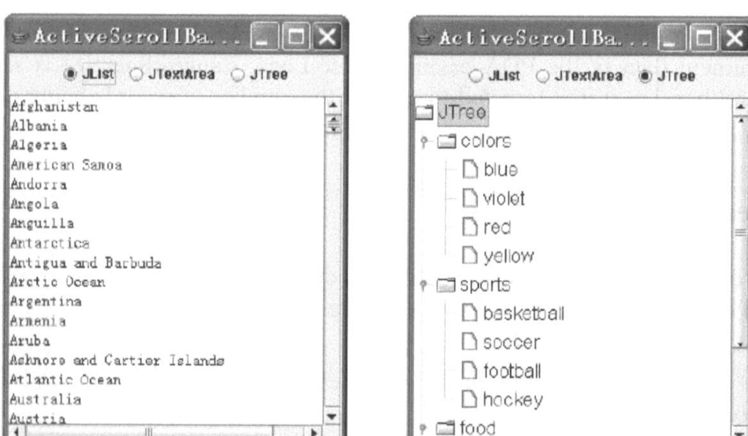

(a) ActiveScrollbar used on a list. (b) ActiveScrollbar used on a tree.

Fig. 2. ActiveScrollbar used to control the scale ratio of long GUI components

The functions of ActiveScrollbar were exemplified by relating the scale ratio to the font size of the displayed document. Increasing the font size means raising the scale ratio, and decreasing the font size means reducing the scale ratio. Three use cases for controlling the scale ratio were illustrated, one was a document in Figure1, another one was a list component in Figure 2(a), and the third one was a tree component in Figure 2(b).

5 User Test

A couple of users were invited to try ActiveScrollbar, and they were all experienced computer users. The trial was subjective, and the goal was mainly to find user preference of the new widget comparing to its traditional peer. The users showed their interest in ActiveScrollbar. Some of them thought ActiveScrollbar was more direct in manipulation. Most users thought that it was easy to use.

A problem was also found. Although the function was thought to be good, but it is not intuitive enough, the users did not know what it is for until it was tried, or they were told to try dragging the small arrows beforehand. Perhaps a more intuitive icon than the two arrows needs to be designed.

6 Implementation

ActiveScrollbar was developed in Java. It subclasses the Java *JScrollBar* component in the Swing GUI toolkit, so it can be used as a replacement for *JScrollBar*. The example used to generate Figure 1 is the content of this paper prepared in LaTeX format, Figure 2(a) is the list of countries, and Figure 2(b) is the default example of the Java *JTree* component. The widget responds to user interaction immediately, and the scale ratio changes smoothly when dragging the mouse on the two small arrows.

7 Discussion

This work produced a scrolling control component, ActiveScrollbar, that is fully compatible with the standard scroll bar, but having direct control to the scale ratio of the displayed document. It can be generally used as a substitute to the standard scroll bar. The scale ratio is programmable by the practitioner. For example the controlled variable can be mapped to the page size or the font size of the displayed document. If not mapped to any property the component will degrade to a standard scroll bar.

8 Conclusion

Traditional scroll bars do not allow direct manipulation of the scale ratio, but the length of the scroll thumb immediately adapts to changes in zoom level within the detail region. There is no reason why the reverse control could not also be implemented, allowing the zoom level to be controlled directly by manipulating the length of the scroll thumb. Although such a widget is expected, it has not been provided yet. Some related work produced useful widgets for this purpose, but they were not designed for general use in everyday user interfaces. ActiveScrollbar implemented the user control of the scale ratio on scroll bars so that zooming control components may be removed from GUIs, and more screen space would be saved. It was expected to be more useful in dealing with long documents or long components such as GUI lists, trees or tables etc. It was also expected to be more useful in hand held devices.

Acknowledgements

This work was jointly sponsored by the National Science Foundation of Guangdong Province of China under grant number 06300433, and Talent Introducing Fund from South China Agricultural University under grant number 2005K099. The feed back and suggestions to improve Active Scrollbar from voluntary users are appreciated.

References

1. Cockburn, A., Karlson, A., Bederson, B.B.: A review offocus and context interfaces. HCIL Tech Report 2006-09, Department of Computer Science, University of Maryland, College Park, MD 20742, USA (2006)
2. Chimera, R.: Valuebars: An information visualization and navigation tool. In: Proc. CHI'92: Human Factorsin Computing Systems, Monterey, California, USA, pp. 293–294. ACM Press, New York (1992)
3. Hill, W.C., D., H.J., Wroblewski, D., McCandless, T.: Editwear and readwear. In: Proc.CHI'92: Human Factorsin Computing Systems, Monterey, California, USA, pp. 3–9. ACM Press, New York (1992)
4. Masui, T.: LensBar: Visualization for browsing and filtering large lists of data. In: Proc. IEEE Symposiumon Information Visualization '98, North Carolina, USA, pp. 113–120. IEEE Computer Society Press, Los Alamitos (1998)
5. Byrd, D.: Ascrollbar-based visualization for document navigation. In: Proc. ACM Fourth International Conferenceon Digital Libraries, Berkeley, CA, USA, pp. 122–129. ACM Press, New York (1999)
6. McCrickard, D., Catrambone, R.: Beyond the scrollbar: An evolution and evaluation of alter nativen avigation techniques. In: Proc. IEEE Symposiumon Visual Languages '99, Tokyo, Japan, pp. 270–277. IEEE Computer Society Press, Los Alamitos (1999)
7. Bederson, B.B., Clamage, A., Czerwinski, M.P., Robertson, G.G.: DateLens: A fish eye calendar interface for PDAs. ACM Transactionson Computer-Human Interaction 11(1), 90–119 (2004)
8. Ahlberg, C., Shneiderman, B.: The Alphaslider: Acompact and rapid selector. In: Proc. ACM Conferenceon Human Factorsin Software, CHI'94, Boston, MA, USA, pp. 365–371. ACM Press, New York (1994)

Communication Analysis of Visual Support System That Uses Line Drawing Expression

Shunichi Yonemura[1], Tohru Yoshida[2], Yukio Tokunaga[2], and Jun Ohya[3]

[1] NTT Cyber Solutions Laboratories
[2] Shibaura Institute of Technology
[3] GITS, Waseda University

Abstract. This paper proposes a system that automatically deforms the images transmitted in both directions so as to achieve two goals: easing the concerns of users and ensuring well enough support. We examine here the effect of visual information quality on the effectiveness and impression of collaboration between a novice user and an operator.

Keywords: Line drawing, Deformation image, Technical Support, Communication Analysis.

1 Introduction

Various communications services are emerging with the spread of the broadband network. This is complicating system operation, and call centers are handling a lot more inquiries. Most calls are made by users with little technical skill and the questions are raised often. Such novices are best supported by verbal explanations by experts in conjunction with visual information such as pictures of computer screens. Video telephone systems are becoming more prevalent and are a logical infrastructure on which to build a novice support service. Unfortunately, there is a lot of resistance to such systems since most users feel uncomfortable with showing their faces and dwellings to strangers, the experts. Key problems are the loss of privacy and the feeling of breakdown in security. This paper proposes a system that automatically deforms the images transmitted in both directions so as to achieve two goals: easing the concerns of users and ensuring well enough support. We examine here the effect of visual information quality on the effectiveness and impression of collaboration between a novice user and an operator.

2 Deformation Image by Line Drawing Expression

Figure 1 shows the process of image deformation. An original image is input from the Web camera of a PC, and binarized. Edges are then extracted. Finally, block sampling

M.J. Smith, G. Salvendy (Eds.): Human Interface, Part I, HCII 2007, LNCS 4557, pp. 359–365, 2007.

yields the deformed image. The extracted sub blocks are transmitted to the other PC. Figure 2 shows the configuration of the prototype system. In the prototype system, a PC with a Web camera was set at both the user's side and the expert's side; they were connected by LAN. The output of the Web camera was deformed by the local PC and then transmitted to the other PC. DirectX of Microsoft was used in developing the deformation program.

Fig. 1. Outline of image deformation processing

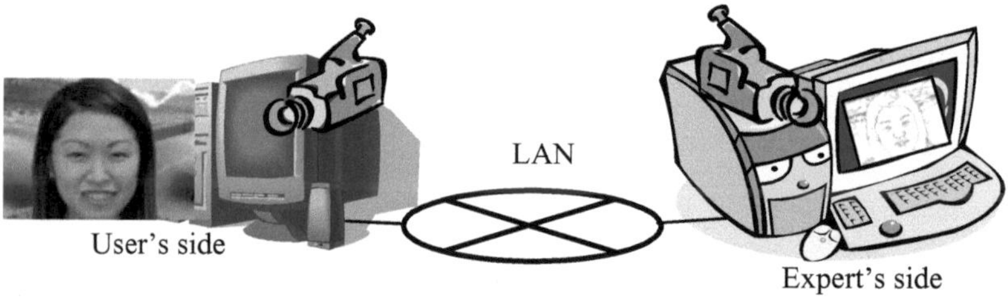

Fig. 2. Configuration of the prototype system

3 Experiment

(1) Participants

Ten women (28 to 49 years old) were chosen as the novice users. All had some experience with browsing the Web and communicating via E-mail; none had experience in setting up networks and/or developing programs. The participants were randomly divided into two equal groups; one used the deformed images and the other used the original images. One woman that had work experience at a call center was used as the expert in all trials.

(2)Equipment

The novice had two PCs. The first PC was a desktop PC connected with the Internet. The second PC was a note book type PC. The task assigned to the novice was to activate a mail program on the note book. Two images were displayed on the screen of the desktop PC, see Figure 3. The larger(smaller) image was sent from the other side (own side). The two desktop PCs were directly connected to realize interactive teleconferences between the expert and the novice.

(3)Procedure
The novice's task was to setup a mail system on the note book PC imagining that she were at home. The novice was directed to access the call center and to execute the instructions from the operator. The expert ran the same mail program and could copy the screen to the novice's desktop PC. The novice was instructed to complete the task as rapidly as possible. In the experiment, the conversation was divided into six stages following the work flow;

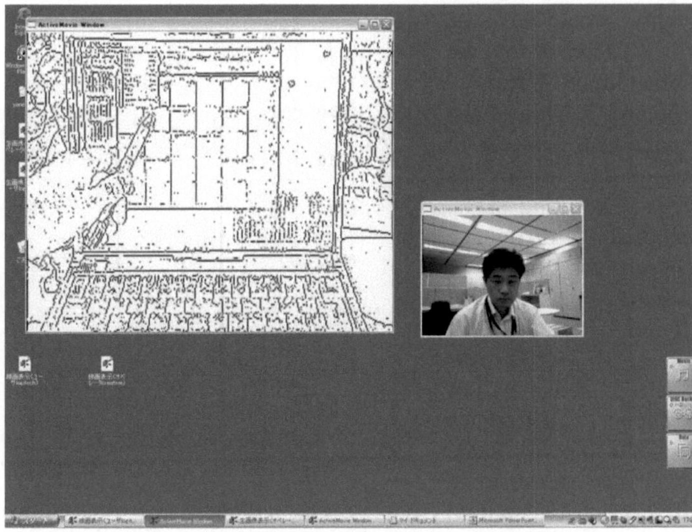

Fig. 3. An example of screen displayed on desk-top PC

- Stage 1. Circumstance verification; The participants contacted the operator and explained the condition of the PC.
- Stage 2. Starting up the mail application; The participants had to launch the mail application.
- Stage 3. Set up the mail parameters; The participants had to set parameters into the mail application. It was at this stage that the participants encountered the problem.
- Stage 4. Error analysis; The participants had to specify the source of the trouble and fix the problem following to the indication of the operator.
- Stage 5. Trouble shooting A; The participants had to activate "Receive HTML mail" button.
- Stage 6. Trouble shooting A; The participants had to activate "Permit image download" button.

4 Results

Figure 4 shows the average task completion time of the two groups and the range of times. The original image group took 15.8 minutes, while the deformed image group took 21 minutes. However, no significant difference was detected (F(1,8)=2.60, p=0.146>0.05) between the groups. That is, the deformed image group roughly matched the work efficiency of the original image group.

Figure 5 shows the average task completion time of the two groups for each working steps. Remarkable difference is seen in stage 3. However, no significant difference(F(1,8)=2.238, p=0.173 > 0.05) was detected. Tendency of significant difference was detected in stage five. However, absolute difference between two groups was small. Therefore, the result shows that remarkable differences were not seen between deformed image group and original image group in terms of task completion time.

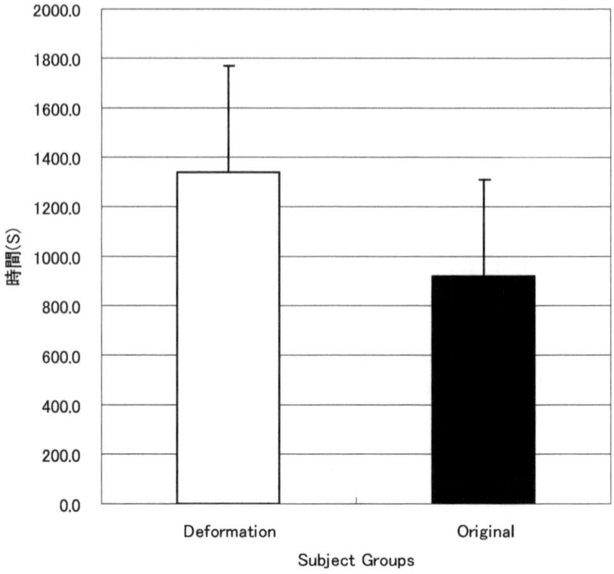

Fig. 4. Average task completion time for each group

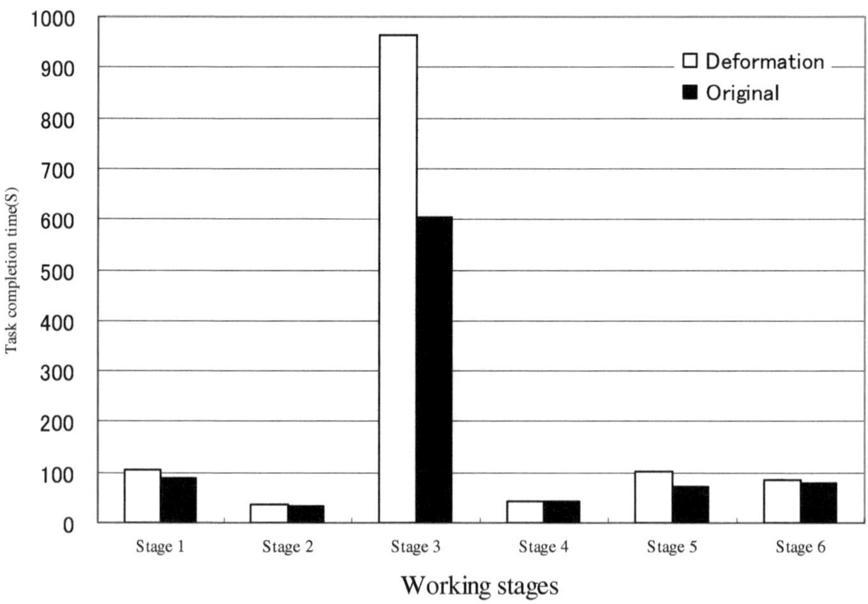

Fig. 5. The average task completion time of the two groups for each working steps

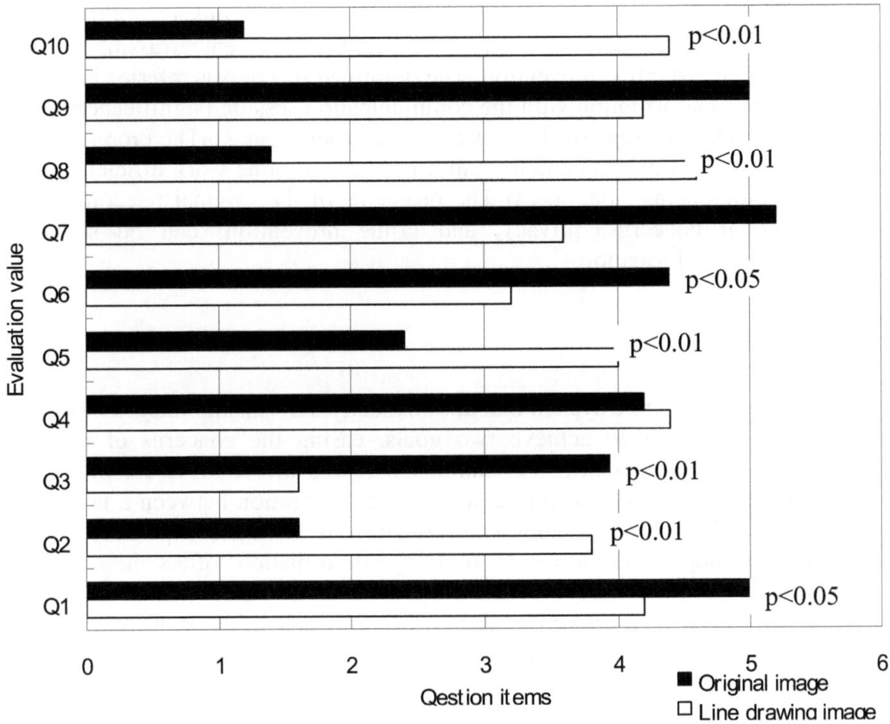

Fig. 6. An average rating value of the participant to each question item

5 Discussion

The subjects answered a questionnaire on their impressions of the support provided. Each items of the questionnaire were following:

- Q1 Easiness: It is easy to talk with using this image.
- Q2 Tension: I am highly tensioned with using this image. (inv)
- Q3 Sense of incongruity: I have a sense of incongruity with using this image. (inv)
- Q4 Light-hearted: I am light-hearted with using this image.
- Q5 Embarrassment: I am embarrassed with using this image. (inv)
- Q6 Irritation: I am irritated with using this image. (inv)
- Q7 Satisfaction: I am satisfied with using this image.
- Q8 Privacy: I feel the problem of privacy in using this image. (inv)
- Q9 Sense of security: I feel the sense of security in using this image.
- Q10 Crime prevention: I feel the problem of crime prevention in using this image. (inv)

A Lickert scale, range of 1 to 7, was used to assess ten items such as ease of use, concern, and privacy. In the item to which the (inv) mark adhered, the evaluation value was made inverse. As shown in figure 6, it became clear (significance level of 1%) that the deformed image was superior to the original image in terms of less personal concerns (reduced tension, less incongruity, less embarrassment), better privacy, and enhanced crime prevention. The deformed image was inferior in terms of lower ease of use and irritation with the communication session (significance level of 5%). We draw the following conclusions from the experiment. (1)The proposed visual support system (based on image deformation) offers the same work efficiency as the original teleconferencing system. (2) The problems of the original teleconferencing system (personal concerns, privacy, and crime prevention) can be solved by introducing image deformation.

6 Conlusions

In this paper, we proposed a system that automatically deforms the images transmitted in both directions so as to achieve two goals: easing the concerns of users and ensuring good enough support. We examine here the effect of visual information quality on the effectiveness and impression of a collaboration between a novice user and an operator. We draw the following conclusions from the experiment. (1)The proposed visual support system (based on image deformation) offers the same work efficiency as the original teleconferencing system. (2) The problems of the original teleconferencing system (personal concerns, privacy, and crime prevention) can be solved by introducing image deformation.

References

1. Fish, R.S., Kraut, R.E., Chalfonte, B.L.: The VideoWindow System in Informal Communications. In: CSCW 90 Proceedings, pp. 1–11 (1990)
2. Ishii, H., Kobayashi, M.: ClearBoard: A Seamless Medium for Shared Drawing and Conversation with Eye Contact. In: CHI '92, pp. 525–532 (1992)
3. Rutter, D.R., Stephenson, G.M.: The role of visual communication in synchronising conversation. European Journal of Social Psychology 7, 29–37 (1977)
4. Rutter, D.R., Stephenson, G.M., Dewey, M.E.: Visual Communication and the content and style of conversation. British Journal of Social Psychology 20, 41–52 (1981)
5. Harada, E.: Cognitive interface of communication media: What kind of task is 'dialogue', IPSJ SIG Technical report, 16-2, pp. 9–16 (in Japanese) (1994)
6. Harada, E.: L'esprit d'aujourdhui, vol. 306, pp. 75–83 (in Japanese) (1993)
7. Harada, E., Nanbu, M.: Understanding interaction from a personal view: A psychological analysis of the cognitive tasks within remote conversation, The 14th Annual conference of JSAI, S2-02, pp. 60–61 (in Japanese) (2000)
8. Veinott, E.S., Olson, J., Olson, G.M., Fu, X.: Video Helps Remote Work: Speakers Who Need to Negotiate Common Ground Benefit from Seeing Each Other, CHI99, pp. 302–309 (1999)
9. Jones, O.D., Monk, A., Watts, L.: Some advantages of video conferencing over high-quality audio conferencing: fluency and awareness of attentional focus. International Journal of Human-Computer Studies 49, 21–58 (1998)
10. Sellen, A.J.: Speech Patterns in Video-Mediated Conversations, CHI92, pp. 49–59 (1992)

11. Kraut, R.E., Fussell, S.R., Siegel, J.: Visual Information as a Conversational Resource in Collaborative Physical Tasks. Human-Computer Interaction 18, 13–49 (2003)
12. Fussell, S.R., Setlock, L.D., Kraut, R.E.: Effects of Head-Mounted and Scene-Oriented Video Systems on Remote Collaboration on Physical Tasks. CHI2003 5(1), 513–520 (2003)
13. Norman, D.A.: Cognitive engineering, User Centered System Design. Lawrence Erlbaum Associates, Mahwah, NJ (1986)
14. Ryo, O., Keiko, K.: Does the education in a Department of Education promote the power of information use. Research Report of JSET Conferences, vol. 3-2, pp. 57–60 (2003) (in Japanese)
15. Shunichi, Y., Masaru, M., Momoko, N., Katsuhiko, O.: Communication support for novice users using visual information in collaborative trouble shooting. The. Japanese Journal of Ergonomics 42(5), 329–336 (2006) (in Japanese)
16. Oishi, T., Tokunaga, Y., Yonemura, S., Oya, J.: Visual Communication System Based on Edge Abstract. IPSJ technical reports, 2005-HI-112, pp. 51–56 (2005) (in Japanese)
17. Shunichi, Y., Masaru, M., Momoko, N.: Impact of visual communication on the effective collaboration of a novice user and an operator (in Japanese). Technical Report of IEICE HCS2004-61(2005-3), pp. 7–12 (2005)

Integrating Data Quality Data into Decision-Making Process: An Information Visualization Approach

Bin Zhu, G. Shankar, and Yu Cai

Information Systems Department
School of Management
Boston University
{bzhu,gshankar,ycai}@bu.edu

Abstract. Poor data quality (DQ) has been a troubling issue within many organizations. Thus it is important for a decision-maker to be aware of the data quality of the information based on which he/she makes a decision. However, the mental integration of DQ metadata in to a decision process poses a significant cognitive challenge. A decision-maker could be overloaded and may not effectively take advantage of the DQ metadata [2]. This paper thus proposes a visualization approach to reduce the amount of cognitive resource required for such mental integration.

Keywords: Information visualization, data quality data.

1 Introduction

Data quality (DQ) management is becoming a critical activity in organizations [1, 5]. It could lead to increased cost, poor execution of organizational strategies, and poor decision-making [5]. To manage quality effectively, it is crucial to provide decision-makers with DQ metadata, *data that describes the quality of the data used in the decision-process*. In the form of a [0, 1] ratio or as a percentage representing the quality, DQ metadata permits decision-makers to gauge data quality in the context of the decision-task [4].

However, the mental integration of DQ metadata in to a decision process poses a significant cognitive challenge. A decision-maker could be overloaded and may not effectively take advantage of the DQ metadata [2]. This paper thus proposes a visualization approach to reduce the amount of cognitive resource required for such mental integration. We developed a prototype system, SPIDEV (Special Purpose interface for Interactive Decision Evaluation using Visualization), that not only facilitates the easy integration of DQ metadata into decision-making process, but also enables decision-makers to incorporate contextual factors for better decision outcomes.

This study focuses on supporting individual analytical decision tasks [3]. Analytical decision-making is a computational approach applied to well-structured decision tasks and usually involves a mathematical model [3]. A decision-maker

M.J. Smith, G. Salvendy (Eds.): Human Interface, Part I, HCII 2007, LNCS 4557, pp. 366–369, 2007.

interacts with the model to answer his/her "what-if" questions and seeks an optimal solution. An information visualization system thus could be applied to visualize the decision outcomes of the different decision alternatives. Decision makers could then easily identify a decision alternative leading to a desired decision outcome by browsing visual representations of possible decision outcomes.

2 The Visualization Approach Proposed

The SPIDEV system visualizes decision outcomes that incorporate DQ metadata to help identify optimal solutions. Figure 1 displays the interface of the prototype system. The visualization displayed adopts the spoke chart. The spokes are used to visualize both the data and its associated quality. Each spoke represents a decision variable in a decision task (i.e. price, size, or quantity). The length of the spoke represents the importance of that decision variable. To make the display more comparable, the importance of all decision variables is standardized to fit in the wheel. Such standardization is achieved according to the importance and value of the decision variable and is also determined by the data quality of the data associated with each decision variable. The DQ metadata, defined as a percentage, is also represented along with the regular data in the spoke chart. It is visualized by the length of the colored part of the spoke. The color code also indicates the quality of the data for each decision variable. For instance, color green represents higher data quality than color red.

Since we implemented all the decision rules at the backend of this visualization, a decision-maker could immediately see the decision outcomes of different decision alternatives when using the sliders in the interface. And the "web" created by linking the points on the spokes shows the solution space. A decision-maker could choose to maximize the overall value of the decision outcome, measured by the area of the solution space. He/she could also maximize the importance of decision variables with high data quality (in color green) to get as much overall value as possible while simultaneously ensuring the quality of decision outcome.

3 Summary

The prototype system presented in this short paper proposes an approach that allows easy integration of DQ data into decision-making process. It also provides decision-makers the flexibility to customize their decision criteria for the decision context. Our future research includes a series of empirical studies to validate the performance of the system. Such studies include measuring its effect on decision outcomes (with and without SPIDEV), the reduction in differential performance caused by the task complexity and individual differences, and measuring the increased usage of DQ metadata. Future work also includes understanding the decision circumstances under which SPIDEV is likely to be more effective.

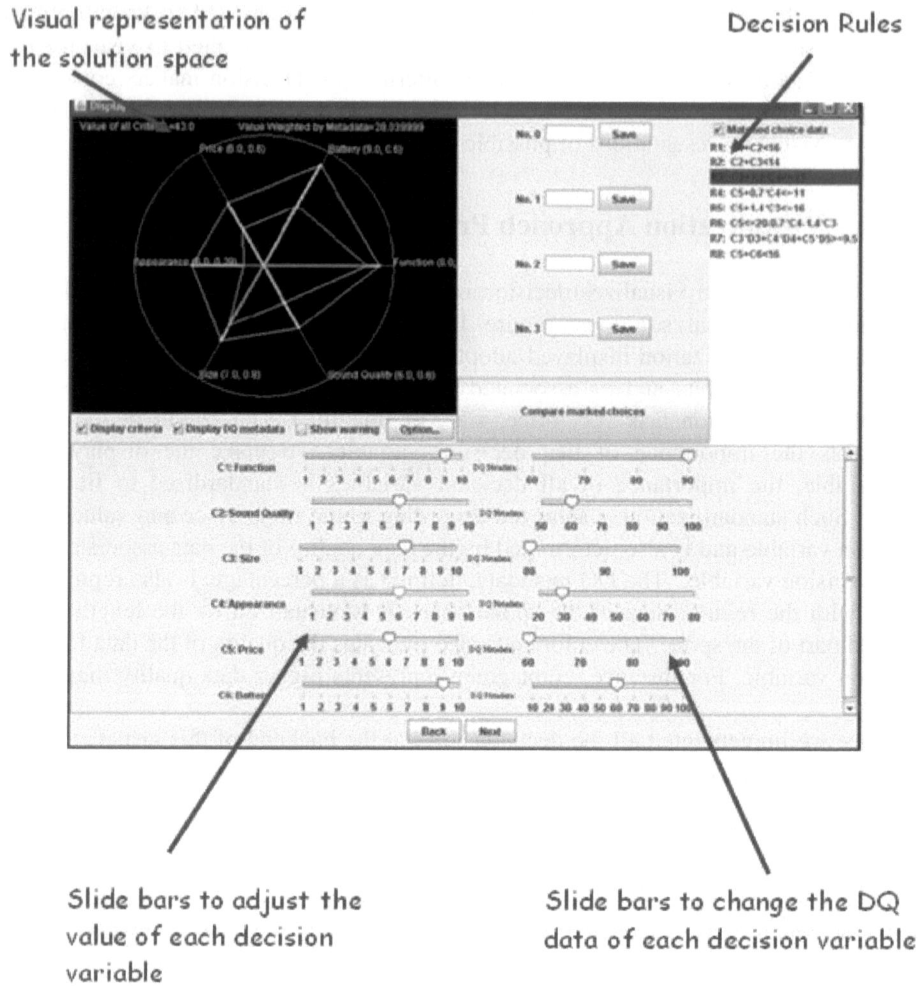

Fig. 1. The Interface of the SPIDEV Prototype System

References

1. Ballou, D.P., Wang, R.Y., Pazer, H., Tayi, G.K.: Modeling information manufacturing systems to determine information product quality. Management Science 44(4), 462–484 (1998)
2. Chengalur-Smith, I.N., Ballou, D., Pazer, H.: The impact of data quality information on decision making: an exploratory analysis. IEEE Transactions on Knowledge and Data. Engineering 11(6), 853–864 (1999)
3. Nutt, P.C.: Selecting tactics to implement strategic plans. Strategic Management Journal 10(2), 145–161

4. Shankaranarayanan, G., Ziad, M., Wang, R.Y.: Managing Date Quality in Dynamic Decision Environment: An Information Product Approach. Journal of Database Management 14(4), 14–32 (2003)
5. Wang, R.Y.: A Product Perspective on Total Data Quality Management. Communications of the ACM 41(2), 58–65 (1998)

Part III

Retrieval, Searching, Browsing and Navigation

HCI and Information Search: Capturing Task and Searcher Characteristics Through 'User Ability to Specify Information Need'

Naresh Kumar Agarwal and Danny C.C. Poo

School of Computing, National University of Singapore
3 Science Drive 2, Singapore 117543
{naresh,dpoo}@comp.nus.edu.sg

Abstract. Human-Computer Interaction (HCI) is all about the way in which people interact with computer systems. This paper focuses on the cognitive aspects of HCI when a user is searching for information, so as to facilitate effective user interactions with vast amounts of available information. Search engines provide a 'one-size-fits-all' model, which do not adequately cater to the differing needs of searchers at different points in time (continuously changing situations in time/space, as per sense-making theory). We posit that from a system designer's point of view, capturing the 'user ability to specify his information need' will help operationalize task/searcher characteristics (hence the user need) and help the designer provide better interfaces for search that fit the needs of the user and lead to search efficacy and searcher satisfaction. The study should advance HCI for search through greater understanding of user needs, enhance search interfaces and lead to theory development.

Keywords: Human-computer interaction for search, sense-making theory, task-technology fit, information searching and retrieval.

1 Introduction

Human-Computer Interaction (HCI) is all about the way in which people (individuals, groups or organizations) interact with computer systems [1]. HCI involves *physical* aspects of the interaction (are the keys spaced right?), *perceptual* aspects (is the text color easy to see against the background?), *cognitive* aspects (will these menu names be understood?), and *social* aspects (will people trust each other on this auction site?) [1]. This paper focuses on the cognitive aspects of human-computer interaction when a user is searching for and retrieving information, so as to facilitate effective user interactions with the vast amount of information available information today.

1.1 Limitations of Current Search Interfaces

Search engines have become the gatekeepers to online information by helping people find information [2]. However, no single search engine indexes the entire WWW (maximum coverage estimated to be 69.6% as of 2005 [3]). Search engines don't perform well when two search words have the same meaning or a single word has

M.J. Smith, G. Salvendy (Eds.): Human Interface, Part I, HCII 2007, LNCS 4557, pp. 373–382, 2007.

multiple meanings [4,2]. Effective query formulation is possible only when the users are already familiar with the research topic and can see the subtle differences in its vocabulary [5] i.e. it depends on the user's ability to specify information need (which in turn depends on the characteristics of the task and the searcher). Also, the interests of the users vary with time (Dervin [6], as per her sense-making approach, calls them continuously changing *situations in time/space*) and cannot be represented by a fixed set [7] or a 'one-size-fits-all' model widely prevalent in the search engines of today. Thus, there is a lack of 'fit' between the information systems available for search and the task needs of different searchers or of the same searcher at different times or different stages of the information search process.

Knowledge discovery can happen if information systems are designed which take into the account the searcher's differing abilities to specify information need (depending on the complexity of the task, prior domain knowledge, etc.) and provide technology features that match the task and searcher characteristics reflected from this ability. Designers of interfaces for search must strive to design interfaces that capture the unique context of the searcher (his task and searcher characteristics, which define his need). While this may be easier said, this is a humongous task, as the needs of a searcher reside in his head and change continuously (sense-making theory). Different tools and technology characteristics have been developed to facilitate different aspects of searching behavior. However, the efficacy of these tools may depend on the user's differing ability to specify their information need.

In this paper, we describe how from a system designers point of view, capturing the user ability to specify his information need, will help operationalize task and searcher characteristics (hence the user need) and help a system designer provide better interfaces for search, which fit the needs of the user and lead to search efficacy and searcher satisfaction. The study should advance HCI for search through greater understanding of user needs, enhance search interfaces and lead to theory development.

In the next section, we discuss calls for collaboration between the person-centric information seeking studies and the system-centric information retrieval studies. We also look at the two theories that provide the theoretical basis for this study. In Section 3, we look at task, searcher characteristics and need, and explain the difficulty in capturing a user's information need. In Section 4, we look at the searcher's ability to specify information need, which can be used to operationalize task and searcher characteristics. Section 5 concludes the paper and mentions future work to be done.

2 Need to Bridge 'System-Centric' and 'User-Centric' Studies

The tradition of research into information seeking considers information seeking from a systems perspective and information users as passive, situation independent receivers of objective information [8]. Yet it has been often accepted that information needs and information seeking processes depend on user's tasks [9,10,11]. Thus, many studies have investigated the relationships of various types of tasks and information seeking behavior [12].

Since the early 1980's, there has been a movement away from system-centeric studies to person or user-centered studies [13]. Person-centered research offers understanding of information seeking and use within the various contexts of people's lives. Important meta-theories (e.g. Dervin's sense-making [14]) and models (e.g. Wilson's model of information seeking [15], Information Search Process [16], etc.) have been developed. New ways of looking at information seeking have emerged, such as Savolainen's [17] work on Everyday Life Information Seeking. However, transfer of concepts across user studies and information retrieval/information systems remains problematic and insufficient [18]. Kuhlthau [18] has called for collaboration between the insights of user studies and the innovations of information retrieval and information systems. Ingwersen and Jarvelin [19,20] have also concluded that Information Retrieval research needs extension toward more context and Information Seeking research needs extension towards task and technology. Taking Kuhlthau's [18], Ingwersen's and Jarvelin's [19,20] call, this study endeavors to work towards a 'fit' between the needs of the information searcher (user studies/information seeking) and the services provided by the information system (information retrieval) through the user ability to specify his information need.

The study draws its theoretical basis from two theories – the Task-Technology Fit Theory and the Sense-Making Theory.

2.1 Task-Technology Fit Theory

The ability of IT to support a task is expressed by task-technology fit (TTF), which implies matching the capabilities of the technology to the demands of the task [21]. As shown in Fig. 1, a match between the characteristics of the task and the searcher (including his needs) and the characteristics of the technology will lead to a task-technology fit, which will result in increased search precision and a higher level of searcher satisfaction.

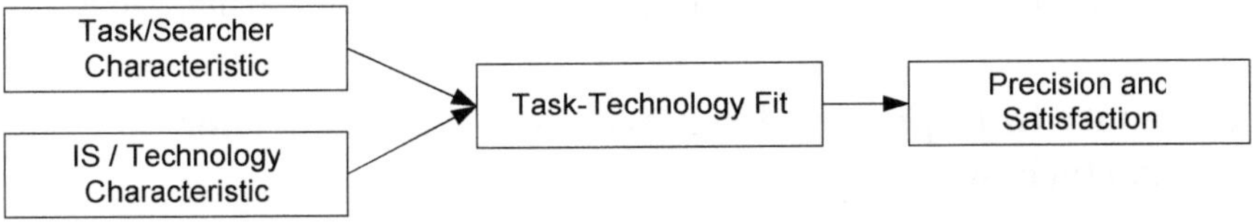

Fig. 1. TTF model adapted for Information Search and Retrieval

Kuhlthau [22,18], in his study of the information search process of lawyers, observed that the lawyers' experiences and expectations in complex information seeking tasks were not easily accommodated by the information system available to them. Similarly, Van Ossenbruggen and Hardman [23] observed that though general purpose search engines are widely prevalent, their focus on providing a *one-size-fits all* model for search leads to inadequate search results. Thus, it is imperative that the characteristics of the information system match the needs of the searcher and his search task.

2.2 Sense-Making Paradigm

Using Brenda Dervin's [6] sense-making paradigm, we propose that task and searcher characteristics can be operationalized as the searcher's ability to specify information need. According to the paradigm, the need of a searcher is unique only in a *situation in time-space* and keeps changing. The theory proposes that information is not 'something that exists apart from human behavioural activity.' Rather, it is 'created at a specific moment in time-space by one or more humans' [6 p.63]. Unlike other approaches to information seeking that see information as something 'out there' that is transmitted to people (as Dervin says, an information 'brick' that is put into a human 'bucket'), sense-making sees information as construed internally in order to address gaps or discontinuities [24,15].

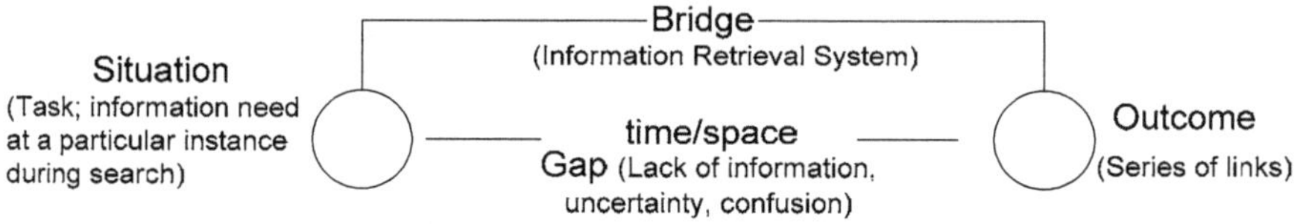

Fig. 2. Dervin's model (adapted from [15])

Sense-making is implemented in terms of four constituent elements: a *situation* in time and space, which defines the context in which information problems arise; a *gap*, which identifies the difference between the contextual situation and the desired situation (e.g. uncertainty); an *outcome*, that is, the consequences of the sense-making approach, and a *bridge*, i.e. some means of closing the gap between situation and outcome [15] (see Fig. 2). Tools are needed to support continuous sense-making. An effective way for the information system to accommodate the context of the search (searcher and task characteristics) is to understand how well the searcher is able to specify his need before the IS, and develop technology features pertaining to that.

3 Difficulty in Capturing User Need and Task/Searcher Characteristics

Having looked at the two major theories underlying this study, let us investigate the difficulty in trying to capture the user need and his unique search context, based on the characteristics of the searcher and the task at hand.

3.1 Task

Kim and Allen [25] cite a number of empirical studies that have supported the premise that user's search performance and/or patterns differ depending on the task. Saracevic and Kantor [26] have found that the specificity and complexity (*broad* and *specific* questions) of search task have an impact on search performance. Several studies have used task with different levels of specificity to investigate the impact of tasks on search behavior – *closed* and *open* tasks [27], *general* and *specific* tasks [28], *topical* and *factual* tasks [29] and *subject* and *known-item* searches [30].

In the course of promoting a seven-step strategy to web searchers, Pffafenberger [31] divided tasks based on the amount of information needed for a topic into three types of questions: 1) *Finding specific information* 2) *Collecting a few sources of high quality information* 3) *Collecting everything on a topic* i.e. conducting an exhaustive search to retrieve all available material. Spool *et al.* [32] defined four types of questions: 1) *Simple fact questions*, simplest type with only one correct answer 2) *Judgment questions*, where user must locate as well as analyze potential answers 3) *Comparison of fact questions*, where the user researches two or more questions to arrive at an answer 4) *Comparison of judgment questions*, situations involving comparisons and judgments [33]. Kuhlthau [16] defines a 6-stage information search process and the feelings and thoughts common to each stage. A successive search process is implicit in Kuhlthau's analysis of the search activity [15].

Thus, different characteristics of task lead to different information need and questions that need answering. The tasks might also differ depending on which stage of the information seeking person the searcher is at.

3.2 Searcher Characteristics

Prior research has investigated different aspects of the searcher during information search:

- **Individual differences.** Aspects such as the user's cognitive abilities, cognitive style and problem-solving style [25]
- **Information goal.** Limberg [34] found that within the same assignment, the goals of fact-finding, getting a right answer or analyzing and synthesizing resulted in quite different outcomes. Information goal might explain the different approaches to information seeking by individuals with the same or similar task. Kuhlthau [22] found in her study that the expert (who was seeking to add value to the client's knowledge) had quite different goals in information seeking than the novice (who was looking for the right answer).
- **System knowledge.** System knowledge is the searcher's familiarity/expertise with the Information System and searching techniques [35].
- **Domain knowledge.** Domain knowledge is the searcher's knowledge of the search subject or topic. Miura, Fujihara and Yamashita [36] concluded that task-related domain-specific knowledge has a much greater impact on various stages of their retrieval behaviour (compared to knowledge relevant to search engines or browsing i.e. system knowledge).

The above characteristics, as well as factors like experience and prior success with search play an important role in determining the context and need of the search.

3.3 Information Need

'Need for information consists of the process of perceiving a difference between an ideal state of knowledge and the actual state of knowledge' [37 p.463]. Need involves reaching a desired goal [24]. Harter [38] argues that to talk about an individual's *information need* is virtually the same as describing his or her 'current psychological state', because needs shift stochastically as each relevant piece of information is encountered. One bit of knowledge may raise questions, lead to another fact or a new conclusion, and so forth, which changes one's knowledge state and hence what one finds relevant and worth seeking [24].

As per Dervin's [6] sense-making approach, new knowledge from query results (outcome) may raise questions, lead to another fact or a new conclusion (situation in time/space), which changes one's knowledge state (state). Dervin's use of *time/space* underlines the ever-changing nature of information need.

Belkin *et al.* [9]'s concept of need is that of an 'anomalous state of knowledge' (ASK). An ASK exists when a person recognizes that there is an anomaly (gap or uncertainty) in their state of knowledge regarding a situation or topic. He may then try to address the anomaly by requesting for information. He will then judge if the anomaly has been resolved. If it is not resolved, another ASK may be generated, or the motivation to address it may be exhausted. Case [24] interestingly points out that the searcher always 'gives up' eventually, because there is always more that could be known regarding a topic. The question of 'when' is determined by available resources and the searcher's level of motivation. Thus, the searcher's need is a complex context consisting of the perceived work task or interest as well as perceptions and interpretations of: knowledge gap or ASK and relevance; uncertainty and other emotional states; the potential sources for the solution (if any) of the work task or interest; the intentionality i.e. goals, purposes, motivation, etc.; information preferences, strategies, pressures (costs, time); self i.e. own capabilities, health, experiences; systematic and interactive features and information objects [39].

From the discussion on information need above, it is evident that from the information retrieval system's perspective, it is extremely difficult to pin-point the user's information need at a particular time/space and to provide compatible technology features to address this need. 'Typically, the cost of acquiring full context is simply too high, compared to the benefits, let alone possible privacy issues' [40].

However, even though the searcher's information need cannot be adequately captured by the information system, what is possible to capture is the searcher's ability to specify information need. It is also possible to be unaware of one's true need [41]. At such times, a searcher is often looking for the right questions or the right keyword to ask, rather than the answer to the need right away.

4 Ability to Specify Information Need

The searcher's **ability to specify information need** is defined as *the state resulting from task and searcher's characteristics which determines how effectively the searcher is able to interact with the information system to retrieve information pertaining to the need.*

The same person may exhibit different abilities to specify information need depending on the complexity/unstructuredness of the task he is working on and the domain knowledge he has in the area of search. However, two different people will have different abilities to specify information need depending on their individual

Table 1. Effect of Task and Searcher Characteristics on the ability to specify information need

Task Characteristics	Searcher Characteristics
Leading to 'low' ability to specify information need	
Task. *Broad* questions [26]; *open* tasks [27]; *general* tasks [28]; *topical* tasks [29]; *subject* searches [30]; *judgment* questions [32] **Information Search Process.** *Initiation / exploration phases* of the search process [16]	**Individual Differences.** Has *low level of cognitive ability* [25]; has a cognitive style with *high field dependence* [25] **Information Goal.** Information *goal of the novice* [22] **System Knowledge.** Has *low system knowledge* [35] **Domain Knowledge.** Has *low domain knowledge* [36]; knows *fuzzy or incomplete information* [33]
Leading to 'high' ability to specify information need	
Task. *Specific* questions [26]; *closed* tasks [27]; *specific* tasks [28]; *factual* tasks [29]; *known-item* searches [30]; *simple fact* questions [32] **Information Search Process.** *Formulation / collection phases* of the search process [16]	**Individual Differences.** Has *high level of cognitive ability* [25]; has a cognitive style with *high field independence* [25] **Information Goal.** Information *goal of the expert* [22] **System Knowledge.** Has *high system knowledge* [35] **Domain Knowledge.** Has *high domain knowledge* [36]; knows *accurate or precise information* [33]

differences (e.g. cognitive ability) and experience in using the search system. Different tasks and searcher characteristics can be classified as leading to low or high ability to specify information need (see Table 1).

Table 1 shows examples of task characteristics (including characteristics of the information search process) and searcher characteristics (individual differences, information goal, levels of system and domain knowledge) which determine a searcher's ability to specify information need. E.g. using the classification of tasks by Spool *et al.* [32], a searcher dealing with judgment questions will have lower clarity of the task and a lower ability to specify information need, compared to when he is dealing with simple fact questions. Similarly, a student starting to work on a new project (with little domain knowledge) will have a lower ability to specify his information need appropriately, compared to another student who has been working on the project for 2 years.

Topical well-defined requests on content (only) is just one approach to document retrieval (more suited for a user with high ability to specify information need) albeit the most popular in IR research. A user with low ability to define information need may define requests vaguely. Also, requests may be non-topical (e.g. by journal or genre) and/or non-content based [42].

5 Conclusion and Future Work

In this study, we have proposed the 'searcher ability to specify his information need' as a solution to capturing the task/searcher characteristics. As per Dervin [6]'s sense-making theory, the needs of searchers change continuously (a unique situation in time/space) and is very difficult to pin-point specifically from an IS perspective, because it resides in the searcher's head. However, operationalizing these complex constructs as the searcher's ability to specify information need is possible. Technology characteristics can be built to match this ability. This will help designers develop better interfaces for search, which will lead to greater human-computer interaction and search efficacy.

The study answers Kuhlthau [18]'s call for collaboration between the person and system-centered aspects of information seeking/retrieval. It also takes on the calls of Ingwersen and Jarvelin [19,20] by providing a way to include context and task in developing technology interfaces for search. The study will contribute to theory development in the field of HCI pertaining to information search and retrieval. The importance of this effort is highlighted by recent efforts such as the Information Retrieval in Context (IRiX) workshop of ACM SIGIR and the HARD track of the Text REtrieval Conference (TREC), that seek to leverage additional information about the searcher and his context to improve search accuracy.

From the practitioner's perspective, the study will be useful for developers of information systems for search – knowledge providers, content providers as well as designers of next-generation web search engines. As the organizers of IRiX 2005 mention in the preface of their proceedings, 'The underlying hypothesis (and belief) is that by taking context into account, the next generation of retrieval engines dependent on models of context can be created, designed and developed delivering performance exceeding that of out-of-context engines.' [43].

Future work on the study will include designing research studies utilizing the 'user ability to specify information need'. The relationship between user ability to specify need and technology characteristics will be investigated to test for fit.

References

1. Dix, A.: Human-Computer Interaction and Web Design. In: Proctor, R.W., Vu, K.-P.L. (eds.) Handbook of Human Factors in Web Design, pp. 28–47. Lawrence Erlbaum, Mahwah, New Jersey (2005)
2. Morahan-Martin, J., Anderson, C.D.: Information and Misinformation Online: Recommendations for Facilitating Accurate Mental Health Information Retrieval and Evaluation. CyberPsychology & Behavior 3(5), 731–746 (2000)
3. Sullivan, D.: New Estimate Puts Web Size At 11.5 Billion Pages & Compares Search Engine Coverage. SearchEngineWatch (May 17, 2005). Retrieved Feb 14, 2007 from http://blog.searchenginewatch.com/blog/050517-075657
4. Deerwester, S., Dumais, S.T., Furnas, G.W., Landauer, T.K., Harshman, R.: Indexing by latent semantic analysis. J. Society for Info. Sc. 41(6), 391–407 (1990)
5. Belkin, N.J.: Helping people find what they don't know. Communications of the ACM 43(8), 58–61 (2000)
6. Dervin, B.: From the mind's eye of the user: The sense-making qualitative-quantitative methodology. In: Glazier, J., Powell, R. (eds.) Qualitative research in information management. Libraries Unlimited, Englewood, CO, pp. 61–84 (1992)
7. Narayanan, S., Koppaka, L., Edala, N., Loritz, D., Daley, R.: Adaptive Interface for Personalizing Information Seeking. CyberPsychology & Behavior 7(6), 683–688 (2004)
8. Dervin, B., Nilan, M.: Information needs and uses. In: Williams, M.E. (ed.) Annual review of Info. Sc. & Tech. 21, pp. 3–33. Knowledge Industry Publ., White Plains, NY (1986)

9. Belkin, N.J., Oddy, R., Brooks, H.: ASK for information retrieval. J. Documentation. 38(2), 61–71 (1982)
10. Ingwersen, P.: Information retrieval interaction. Taylor Graham, London (1992)
11. Mick, C.K., Lindsey, G.N., Callahan, D.: Toward usable user studies. JASIS. 31(5), 347–365 (1980)
12. Bystrom, K., Jarvelin, K.: Task complexity affects information seeking and use. Information Processing and Management. 31(2), 191–213 (1995)
13. Wilson, T.D.: Information needs and uses: Fifty years of progress? In: Vickery, B.C. (ed.) Fifty years of information progress: A J. Documentation review, Aslib, London, pp. 15–52 (1994)
14. Dervin, B.: An overview of sense-making research: concepts, methods and results to date. In: Int. Comm. Asso. Annual Meeting (1983)
15. Wilson, T.D.: Models in information behavior research. J. Documentation. 55(3), 249–270 (1999)
16. Kuhlthau, C.C.: Inside the search process: Information seeking from the user's perspective. JASIS 42(5), 361–371 (1991)
17. Savolainen, R.: Everyday life information seeking: approaching information seeking in the context of 'way of life'. LIBRES. 17(3), 259–294 (1995)
18. Kuhlthau, C.C.: Towards collaboration between information seeking and information retrieval. Information Research. 10, 2 (2005)
19. Ingwersen, P., Jarvelin, K.: The Turn: Integration of Information Seeking and Retrieval in Context Series. The Information Retrieval Series. 18(XIV), 448 (2005)
20. Jarvelin, K., Ingwersen, P.: Information seeking research needs extension towards tasks and technology. Information Research 10(1), 17 (2004)
21. Goodhue, D.L., Thompson, R.L.: Task-technology fit and individual performance. MIS Quarterly 19(2), 213–236 (1995)
22. Kuhlthau, C.C.: Seeking meaning: a process approach to library and information services, 2nd edn. Libraries Unlimited, Westport, CT (2004)
23. van Ossenbruggen, J., Hardman, L.: Smart Style on the Semantic Web. Semantic Web Workshop, WWW (2002)
24. Case, D.O.: Looking for information: A Survey of Research on Information Seeking, Needs, and Behavior. Academic Press, Elsevier, San Diego, CA (2002)
25. Kim, K.-S., Allen, B.: Cognitive and task influences on Web searching behavior. JASIST. 53(2), 109–119 (2002)
26. Saracevic, T., Kantor, P.: A study of information seeking and retrieving II. Users, questions, and effectiveness. JASIS 39(3), 177–196 (1988)
27. Marchionini, G.: Information seeking strategies of novices using a full-text electronic encyclopedia. JASIS 40(1), 54–66 (1989)
28. Qiu, L.: Analytical searching vs. browsing in hypertext information retrieval systems. Canadian J. Info. and Lib. Sc. 18(4), 1–13 (1993)
29. Kim, K.-S.: Users, tasks and the Web: Their impact on the information-seeking behavior. In: Proc. 21st national online meeting, pp. 189–198 (2000)
30. Matthews, J.R., Lawrence, G.S., Ferguson, D.K.: Using online catalogs: A nation wide survey. Neal-Schumann, New York (1983)
31. Pfaffenberger, B.: Web search strategies. MIS Press, New York (1996)
32. Spool, J., Scanlon, T., Snyder, C., DeAngelo, T.: Web site usability: A designer's guide. Morgan Kaufmann, San Francisco (1999)
33. Bhavani, S.K., Drabenstott, K., Radev, D.: Toward a Unified Framework of IR Tasks and Strategies. Proc. ASIS Annual Meeting 38, 340–354 (2001)

34. Limberg, L.: Information use for learning purposes. In: Vakkari, P., Savolainen, R., Dervin, B. (eds.) Proc. ISIC 1996, pp. 275–289. Taylor Graham, London (1997)
35. Wildemuth, B.M.: The Effects of Domain Knowledge on Search Tactic Formulation. JASIST. 55(3), 246–258 (2003)
36. Miura, A., Fujihara, N., Yamashita, K.: Retrieving information on the World Wide Web: Effects of domain specific knowledge. AI and Society. 20(2), 221–231 (2006)
37. van de Wijngaert, L.: A policy capturing study of media choice: The effect information [sic] of needs and user characteristics on media choice. In: Wilson, T.D., Allen, D.K. (eds.) Information behavior: Proc. ISIC 1998, Taylor Graham, London, pp. 463–478 (1999)
38. Harter, S.P.: Psychological relevance and information science. JASIS 43, 602–615 (1992)
39. Ingwersen, P., Jarvelin, K.: Information Retrieval in Context. In: Proc. SIGIR IRiX Workshop, pp. 6–9 (2004)
40. Hawking, D., Paris, C., Wilkinson, R., Wu, M.: Context in Enterprise Search and Delivery. In: Proc. SIGIR 2005 IRix Workshop, pp. 14–16 (2005)
41. Green, A.: What do we mean by user needs? British J. Academic Librarianship 5, 65–78 (1990)
42. Jarvelin, K., Ingwersen, P.: Extending Information Seeking and Retrieval Research toward Context. In: Proc. SIGIR 2004 IRiX Workshop, pp. 11–15 (2004)
43. Ingwersen, P., Jarvelin, K., Belkin, N.: Proc. SIGIR IRiX Workshop (2005)

Hierarchical Image Gathering Technique for Browsing Surveillance Camera Images

Wataru Akutsu, Tadasuke Furuya, Hiroko Nakamura Miyamura, and Takafumi Saito

Graduate School of Bio-Applications and Systems Engineering,
Tokyo University of Agriculture and Technology
wataru@vc.base.tuat.ac.jp

Abstract. We propose an image gathering and display method for efficient browsing of surveillance camera images. The proposed method requires large cost to inspect lengthy image sequences taken by a surveillance camera. The proposed method involves generating a still image by gathering the moving parts from image sequences captured by a fixed camera. The gathered images are generated for several intervals of time and are displayed hierarchically. A user can easily browse the scene by observing the images with moving parts. Since detection and recognition of the target objects are performed by a human operator, efficient and reliable browsing can be established.

Keywords: gathered images, temp oral resolutions, hierarchical display.

1 Introduction

In recent years, the surveillance camera has been used extensively in crime prevention, and surveillance cameras have become ubiquitous. As such, the costs associated with monitoring and/or browsing captured images has increased greatly. For efficient use of surveillance cameras, it is necessary to reduce these costs. Surveillance cameras are used in various situations. For example, the view angle may be fixed or variable (i.e., panned or zoomed). The amount of traffic at the capture location may be significant or slight. The image may be monitored in real-time or recorded for browsing afterward. In the present study, we focus on the situation with fixed view angle cameras set up in areas having a low volume of traffic, for browsing afterward. Such a situation is not special, and several low-cost surveillance camera systems may be used in this situation, especially at night.

In such a situation, a large number of images are captured and stored, but most of these images are of the background only. Usually, users only want to view scenes containing moving objects, especially humans. Therefore, it is necessary to browse these images efficiently. As naive browsing methods, a large number of images can be checked simultaneously by displaying several thumbnail-size images, or viewed in a shorter time by fast-forwarding through video sequences. However, both methods still have enormous costs and are not practical for lengthy image sequences. In order to search, extract, and track humans automatically in surveillance camera images, various techniques have been proposed and developed. However, it is difficult to

M.J. Smith, G. Salvendy (Eds.): Human Interface, Part I, HCII 2007, LNCS 4557, pp. 383–390, 2007.

recognize humans without error in various situations. Since many objects such as small animals, shaking trees, and lighting changes, could be detected as moving objects, and it is sometimes difficult to recognize whether a detected object is human. A number of methods recognize humans well, but require precise parameter settings for each situation.

For efficient image browsing, we propose a new display method with image gathering. A gathered image is a still image that contains all moving objects in all of the frames during a given period of time. The gathered image is a kind of photographic playback, and the user can easily determine whether human(s) exist in the image.

In the present paper, we also propose a hierarchical image gathering and display method. The gathered images are generated for several intervals of time length, and are displayed hierarchically. A user can easily browse the scene by looking at the moving parts of the images. Since detection and recognition of the target objects are performed by a human operator, efficient and reliable browsing can be established.

2 Related Research

2.1 Automatic Recognition Methods

Several techniques have been proposed for efficient browsing of surveillance camera images. One such method is automatic human detection and recognition by image processing. Background subtraction is a simple method by which to detect moving objects[1,2]. Techniques for distinguishing human have also been studied extensively. These techniques are based on object size[3] or pattern training and neural-network learning of human behavior[4,5]. In addition, techniques for detecting the position of humans[6] and contour definition of moving objects[7,8,9] have been proposed. The major problem in automatic detection and recognition is detection leakage and misrecognition. In addition, a number of methods require precise parameter setting for each situation. However, perfect detection and recognition of humans in various situations is difficult.

2.2 Hierarchical Display

Multi-resolution video[10], proposed by Finkelstein,*et al.*, efficiently displays a video sequence. This technique uses the hierarchical change of spatial and temporal resolutions. We applied the hierarchical change of temporal resolutions to a surveillance camera image. However, since the multi-resolution video uses an image that smoothes each pixel in the time direction of the axis, moving objects that appear only for a short time become thin and disappear in low time resolutions. Our method , on the other hand, extracts the moving parts and displayed them without smoothing.

3 Generation of the Gathered Image

A gathered image is a still image that contains all moving objects in all of the frames during a given period of time. The generation strategy of the gathered image is as follows. The intensity of an image rapidly changes when a moving object comes into the view. Therefore, if the intensity change is examined during the time period at a fixed pixel position, a moving object at a pixel can be detected by examining the frame with the maximum intensity change. By applying this process to all pixels, the gathered image of moving objects can be obtained (Fig. 1). For better results, approximately one frame per second is used for gathering. To reduce noise, the average intensity of a small rectangular block is used instead of the single pixel intensity. The average intensity of the small rectangular block is as follows:

$$L_n(Q_i) = \frac{\sum_{(x,y)\in Q_i} |B_{n+1}(x,y) - B_n(x,y)| + |B_n(x,y) - B_{n-1}(x,y)|}{2S}$$

Q_i : ith block

$B_n(x,y)$: luminance of picture element (x,y) in the nth frame

S : number of pixels in each Q_i

The resulting image includes almost all moving objects, unless multiple objects are located in the same position. The user can easily determine whether human(s) are present during the time period. If only background is present in a gathered image, it is no longer necessary to check the corresponding time period. The image might be complicated if too many moving objects exist (Fig. 2). In such cases, we need the gathered images for shorter time periods. In addition, we can easily confirm the presence of human(s) in such images.

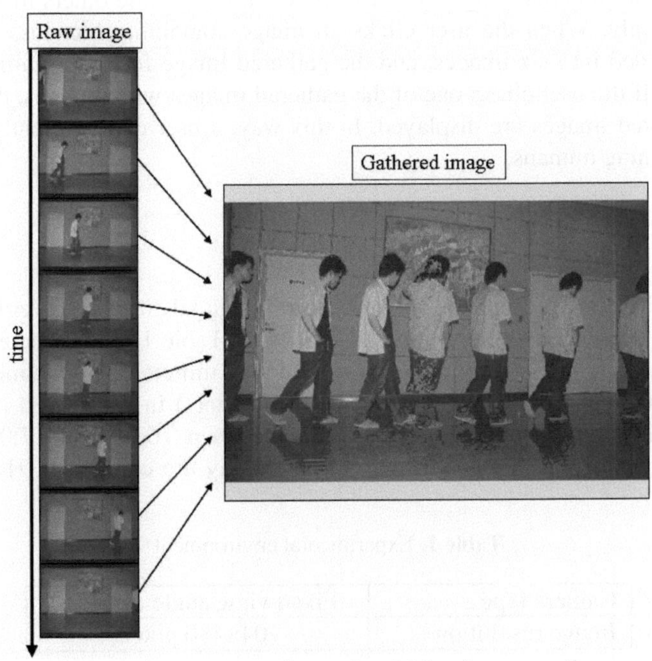

Fig. 1. This figure shows a time sequence image in which a human walks across the captured area. Information from several row images is gathered into one still image. As a result, the user can easily confirm the behavior of the subject in the image.

(a) Sample1 (b) Sample2

Fig. 2. The gathered image is obfuscated when too many moving objects exist

4 Hierarchical Image Gathering and Display Method

For easier management of the gathered images for various time periods, we propose a hierarchical image gathering and display method. The gathered images are generated for several intervals of time length and are displayed hierarchically. When browsing the images captured for a twelve-hour period, for example, the period is divided into twelve periods, and the gathered images for each one-hour period are displayed. Some of the one-hour gathered images might include humans, while others may contain the background only. When the user clicks an image containing humans, the one-hour period is divided into six images, and the gathered image for each ten-minute period is displayed. If the user clicks one of the gathered images with humans, then, ten one-minute gathered images are displayed. In this way, a user can efficiently browse the scenes containing humans.

5 Experiment

We applied the proposed method to a one-hour image taken by a surveillance camera (Fig. 3). The experimental environment is shown in Table 1. The gathered image was hierarchically managed in three stages (one hour, ten minutes, and one minute.)

First, the user can confirm the presence of human(s) in a gathered image of one hour. This shows that human(s) are present between 16:00 and 17:00. Next, six gathered images of ten minutes in the next hierarchy are confirmed. Here, a human

Table 1. Experimental environment

Camera type	Fixed view angle cameras
Image resolution	704×480 pixels
Serial shoot duration	1/second
Block size	2x2 pixels

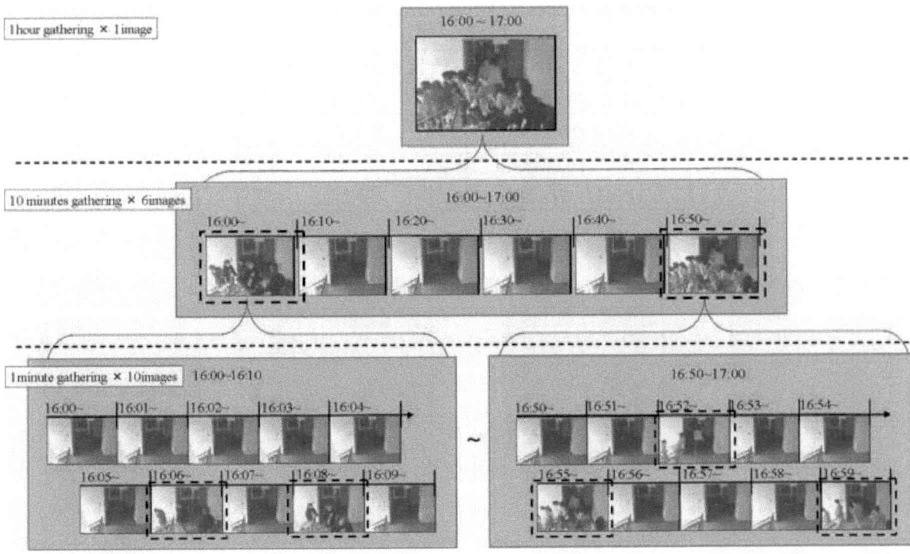

Fig. 3. Hierarchical image gathering and display. This figure shows that the user can easily select only images in which human(s) appear. As a result, effective browsing is established.

can be confirmed to be present from 16:00 to 16:10 and from 16:50 to 16:60. Similarly, only gathered images of human(s) are confirmed by the image of the subordinate position hierarchy. As a result, the presence of human(s) between 16:00 and 17:00 can be confirmed at 16:05, 16:06, 16:08, 16:52, 16:55, and 16:59.

6 Discussion

6.1 Browsing Efficiency of Hierarchical Display

A user can easily browse only necessary parts, i.e., times at which human(s) appear, by hierarchically managing the gathered image. In the experiment, 3,600 images can be confirmed through a number of gathered images. Efficient browsing of a large number of images is established. When you manage the image far into the future, only the chosen part need be manage, and an improvement in the efficiency of image management is achieved. Moreover, equivalent results are obtained in other environments.

6.2 Appropriate Block Size for Consolidating Image

Gathered images of each block size (2×2, 4×4, 8×8, 16×16 pixels) are generated. It is predicted that the outline of the object will be preserved when the block size is small. However, the influence of the block noise is considered, and the processing time of the gathered image increased. The image quality and processing time have a trade-off

Table 2. Processing time for each block size

Block size	Processing time(s)
2×2 pixels	96.44
4×4 pixels	93.77
8×8 pixels	92.42
16×16 pixels	89.66

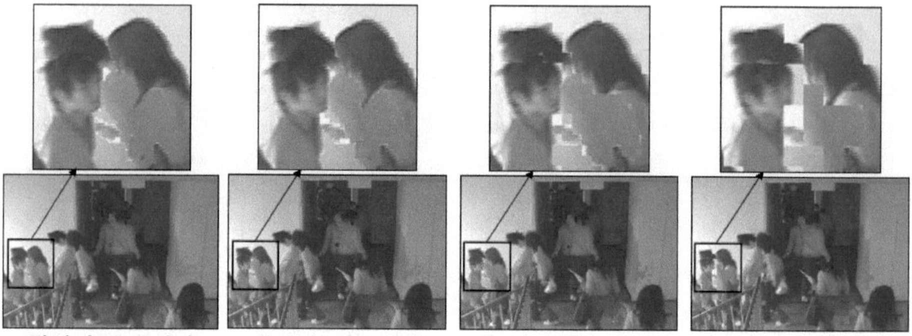

Fig. 4. This figure shows gathered images in which the block size changes. The top images are magnified areas of the lower images.

relationship. Then, the size of the block is changed to generate the gathered images, and the influence of the change is investigated. The processing time for each block up to five minutes (300 pieces) is shown in Table 2.

The influence of the noise is negligible in the minimum block size of 2x2 pixels. The processing time of the minimum block size is 12.1% larger than that of the maximum block size. However, the processing time for the minimum block size is only approximately 32% of the captured time interval, and thus, real-time processing is also possible. Considering the clarity (Fig. 4) of the gathered image and the processing time, 2x2 pixels is the best block size in the proposed method.

6.3 Situation with Rapid Illumination Change

There is an environment in which a sensor light operates when a human approaches (Fig. 5). In such an environment, the luminance value of the entire image changes rapidly. Therefore, there is a possibility that human(s) information may be overwritten by information on lighting. We therefore conduct an experiment in such an environment. The gathered image of a time period that contains lighting is generated (Fig. 6). Most of the time, humans appear on the gathered image, although human(s) information might not appear, depending on the human(s) color. However, information of lighting is plainly shown. As a result, the human(s) appearance is not overlooked by browsing the hierarchy below.

Fig. 5. This figure shows the environment with lighting. The part in which a rapid luminance changes is enclosed by a red frame. The part in which the human appeared is enclosed by a blue frame.

Fig. 6. Gathered image of time period that contains lighting. Information of lighting can be obtained, although there is a part in which the person disappears, depending on their color.

7 Conclusion

In the present study, we proposed an image gathering and display method for efficient browsing of surveillance camera images. The greatest advantages of the proposed method, compared to automatic recognition methods, are its robustness and reliability. There is no need for threshold setting in the proposed method. If small animals or other meaningless moving objects appear during a time period, the gathered image does include these objects. However, the user can easily recognize that the moving objects are meaningless and can discard them. The obtained experimental results show that the proposed method is also robust for changing lighting conditions. Therefore, the proposed method is effective for use in various situations.

References

1. Cucchiara, R., Grana, C., Piccardi, M., Prati, A., Sirotti, S.: Improving Shadow Suppression in Moving Object Detection with HSV Color Information. IEEE Intelligent Transportation Systems Conference Proceedings, pp. 334–339 (2001)
2. Haritaoglu, I., Harwood, D., Davis, L.: W4: Real-Time Surveillance of People and TheirActivities. IEEE Transactions on Pattern Analysis and Machine Intelligence 22(8), 809–830 (2000)
3. Jacinto, C., Jorge, M.: Performance Evaluation of Object Detection Algorithms for Video Survillance. IEEE Transaction on Multimedia 8(4), 761–774 (2006)
4. Gian, F., Lucio, M., Carlo, R.: Automatic Detection and Indexing of Video-Event Shots for Surveillance Application. IEEE Transction on Multimedia 4(4), 459–471 (2002)
5. Hu, W., Tan, T., Wang, L., Maybankl, S.: A Survey on Visual Surveillance of Object Motion and Behaviors. IEEE Transaction on CyberneticsPart C: Application and Reviews, vol. 34(3) (2004)
6. Hayashi, K., Haga, T., Seki, M., Sasakawa, K.: Detecting Human Position and Action in Surveillance Scenes. IPSJ CVIM14, 47(9), 12–20 (2006)
7. Kass, M., Witkim, A., Terzopoulos, D.: Snakes, Active Contour Models. Int. J. Computer Vision 1.1(4), 321–331 (1988)
8. Sethian, J.: Level set Methods, 1st edn. Cambridge University Press, Level set Methods (1996)
9. Osher, S., Sethian, J.A.: Fronts Propagating with Curvature Dependent Speed, Algorithm based on Hamilton-Javobi formation. J. Computational Physics 79, 12–49 (1998)
10. Finkelstein, A., Jacobs, C.E., Salesin, D.H.: Multiresolution Video. In: Proceedings of SIGGRAPH 96, pp. 281–290 (1996)

Self-help Troubleshooting by Q-KE-CLD Based on a Fuzzy Bayes Model

Pilsung Choe[1], Mark R. Lehto[1], and Jan Allebach[2]

[1] School of Industrial Engineering, Purdue University, 1287 Grissom Hall, West Lafayette, In 47907-1287, USA
[2] School of Electrical and Computer Engineering, Purdue University, 1285 Electrical Engineering Building, West Lafayette, In 47907-1285, USA
{pchoe,lehto,allebach}@purdue.edu

Abstract. The previous study [9], [10] showed the fuzzy Bayes model successfully predicted print defects with a 50% hit rate at the first top prediction and an 80% hit rate within the top five predictions. However, the previous study was limited to English. In this study, Korean and English descriptions in predicting print defects by Korean subjects were evaluated based on fuzzy Bayes models. For the study, Korean descriptions were collected in Korea, and Bayes models were developed and evaluated. The result shows that Korean subjects much more accurately predicted print defects when they used Korean descriptions than English descriptions. Afterwards, English descriptions by US subjects will be collected, and both Korean and English lexicon data will be compared. Finally, the study will investigate a Korean-English cross language diagnosis (Q-KE-CLD) system to identify print defects based on the fuzzy Bayes model.

Keywords: fuzzy Bayes model, diagnosis, troubleshooting, Q-KE-CLD, cross language diagnosis, translation, CLIR, cross language information retrieval, parsing, lexical analysis, search.

1 Introduction

Due to the rapid spread of internet technology, troubleshooting using Web sites is becoming more popular because of its advantages in terms of economy and efficiency [2], [5], [6]. However, troubleshooting on Web sites written in English is often very difficult for users whose native language is not English, because of the difficulty of understanding English as well as the complexity of the troubleshooting process to navigate Web sites. In this respect, query-translation based cross language diagnosis (Q-CLD) using the fuzzy Bayes model can be a feasible solution for non-native English speaking users (NNEU) to troubleshoot on Web sites written in English because it avoids the process of navigating the complex Web sites written in English. The fuzzy Bayes model has been proven to be a good tool for identifying print quality issues [9], [10]. The result showed the performance was a 50% hit rate with the top one prediction, 70% with top three predictions, and 80% with top five predictions [9], [10]. However, the study was limited to English so that the result can not be directly

M.J. Smith, G. Salvendy (Eds.): Human Interface, Part I, HCII 2007, LNCS 4557, pp. 391–400, 2007.

applied for NNEU to predict print defects, because the English descriptions used by NNEU may be different from those of native English speaking users. In addition, entering English queries is sometimes challenging for NNEU, although they use English in many situations such as reading, writing, and communicating. They often feel they could use their own languages as queries to find information they need, even though the information is written in English.

As a practical application, this study investigates a Korean-English cross language diagnosis (Q-KE-CLD) based on a fuzzy Bayes model to identify print quality defects. The original idea of cross language diagnosis (CLD) came from cross language information retrieval (CLIR), which is defined as finding relevant documents in a second language using a query expressed in a first language [1], [3], [11], [12]. Likewise, Q-KE-CLD is defined as finding relevant documents to diagnose problems in English using a query expressed in Korean. In order to investigate the Q-KE-CLD, four prediction cases are compared: (1) Korean predictions based on a Korean training set by Korean subjects; (2) English predictions based on an English training set by Korean subjects; (3) English predictions based on an English training set by US subjects; and (4) Korean predictions (which need translations to English) based on an English training set by Korean subjects. Case (1) or Case (3) is expected to provide an upper bound of the prediction accuracy, and case (2) is assumed to give a lower bound due to the deficiency of language ability of Korean subjects. In this aspect, case (4) is expected to have better prediction accuracy than case (3). The four cases are evaluated in terms of prediction accuracy, user performance, and user preference. In this paper, cases (1) and (2) are analyzed, and cases (3) and (4) will be conducted as a following study. Finally, the Q-KE-CLD system will be implemented and evaluated to determine whether the system can be a possible solution for localizing print quality troubleshooting Web sites [2], [5], [6].

2 Experiment for Model Development and Validation

In order to develop and validate fuzzy Bayes models to identify print defects, an experiment was conducted in Korea for three weeks from November 21 to December 8, 2006. The experiment was intended to collect English and Korean description data by Korean subjects that describe 32 representative print defects of laser printers manufactured by one company. The Bayes models obtained from the experiment will be implemented in a query search function on Websites for self-help print quality troubleshooting.

2.1 Subjects

A total of 42 Korean subjects applied for the experiment. However, only 40 subjects participated in the experiment because two subjects were found to be colorblind from the Ishihara color test [4]. The data from the first 30 subjects were used to develop fuzzy Bayes models, and the data from the last ten subjects were used to evaluate the fuzzy Bayes models. The subjects were composed of 37 males and 3 females, all of which are undergraduate, graduate students, or a post doctorial researcher in engineering or business schools from five universities in Korea (Hanyang University at Seoul, Hanyang Univeristy at Ansan, POSTECH, Pukyong National University, Kumoh Natioal Institute of Technology). All subjects had TOEIC (Test of English for International Communication) scores greater than or equal to 550 (Average: 702), which was intended to qualify subjects who were assumed to read information written in English on the Web sites.

2.2 Procedure

The experiments were conducted under fluorescent illumination in five different laboratories located at five universities, which are similar to a normal office using laser printers. As a formal procedure, subjects signed a consent form approving that they would participate the experiment. Prior to the experiment, a colorblindness test

Table 1. Thirty Two Print Defects

Defect	Defect description	Background Color	Orienta-tion	Pattern
D_1	Horizontal lines repeating every 1.4 inches	C	H	P_1
D_2	Horizontal lines repeating every 1.4 inches	M	H	P_1
D_3	Horizontal lines repeating every 1.4 inches	Y	H	P_1
D_4	Horizontal lines repeating every 1.4 inches	K	H	P_1
D_5	Horizontal lines repeating every 3.7 inches	C	H	P_2
D_6	Horizontal lines repeating every 3.7 inches	M	H	P_2
D_7	Horizontal lines repeating every 3.7 inches	Y	H	P_2
D_8	Horizontal lines repeating every 3.7 inches	K	H	P_2
D_9	Broad fuzzy horizontal bands repeating every 1.4 inches	C	H	P_3
D_{10}	Broad fuzzy horizontal bands repeating every 1.4 inches	M	H	P_3
D_{11}	Broad fuzzy horizontal bands repeating every 1.4 inches	Y	H	P_3
D_{12}	Broad fuzzy horizontal bands repeating every 1.4 inches	K	H	P_3
D_{13}	White vertical lines or bands	C	V	P_4
D_{14}	White vertical lines or bands	M	V	P_4
D_{15}	White vertical lines or bands	Y	V	P_4
D_{16}	White vertical lines or bands	K	V	P_4
D_{17}	Vertical lines or line segments	C	V	P_5
D_{18}	Vertical lines or line segments	M	V	P_5
D_{19}	Vertical lines or line segments	Y	V	P_5
D_{20}	Vertical lines or line segments	K	V	P_5
D_{21}	Fine vertical lines	C	V	P_6
D_{22}	Fine vertical lines	M	V	P_6
D_{23}	Fine vertical lines	Y	V	P_6
D_{24}	Fine vertical lines	K	V	P_6
D_{25}	Missing color	C	X	P_7
D_{26}	Missing color	M	X	P_7
D_{27}	Missing color	Y	X	P_7
D_{28}	Missing color	K	X	P_7
D_{29}	Tinted background	C	X	P_8
D_{30}	Tinted background	M	X	P_8
D_{31}	Tinted background	Y	X	P_8
D_{32}	Tinted background	K	X	P_8

C: cyan, M: magenta, Y: yellow, K: black; H: horizontal, V: vertical, X: others; P_i: pattern i was conducted because print defects in the experiment were closely related to colors. The Ishihara color test, which consists of a series of pictures of colored spots, was used to screen the colorblind subjects. A pretest questionnaire which was designed to measure the subjects' general knowledge about printers, print defects, and troubleshooting on the Web sites, was provided to the subjects at the beginning of the experiment. During the experiment, a pair of printouts and an answer sheet were provided to the subjects in random order to eliminate learning effects. One of each pair had no print defect and the other had a defect so that the subjects could compare the two pages and describe the defect. As shown in Table 1 (Refer to www.hp.com/cpso-support-new/pq/4600/home.html), there were a total of 32 print defects which had 8 different print defect patterns for four different colors, cyan (C), magenta (M), yellow (Y), and black (K). Among these print defects, "Horizontal lines repeating every 1.4 inches (P_1)", "Horizontal lines repeating every 3.7 inches (P_2)", and "Broad fuzzy horizontal bands repeating every 1.4 inches (P_3)" for any color are classified into horizontal print defects (H). "White vertical lines or bands (P_4)", "Vertical lines or line segments (P_5)", and "Fine vertical lines (P_6)" for any color are classified into vertical print defects (V). "Missing color (P_7)" and "Tinted background (P_8)" for any color are classified into others (X) on the print quality troubleshooting Web site. After comparing a pair of printouts, the subjects described on the answer sheet how they perceived the print defect in Korean or English. After describing all defects in one language, they followed the same process with another set of printouts in the other language in random orders from subject to subject. From the experiment, 1280 descriptions in English and 1280 descriptions in Korean were obtained. Out of each set of 1280 data, 960 were used for training and 320 were used for predictions.

3 Model Development

3.1 Parsing

Parsing is the process of using grammar to assign a syntactic analysis to a string of words, that is, the process of recognizing a sentence and simultaneously building a representation of its structure [11], [13]. The most common and easiest methods to implement parsing are based on keywords by chunking that are used to build sets of attribute values [9], [13]. For this study, shallow parsing analysis [9], [13] was conducted using the Textminer [7] which Purdue University has developed. Every word in the descriptions was separated by a space as shown in Figure 1.

ID	word	Tpred
15	yellow	t
15	line	t
15	empty	t
16	black	t

(a) English word list example

ID	word	Tpred
92	회색	t
92	바탕	t
93	파란	t
93	프린트	t

(b) Korean word list example

Fig. 1. Word list examples

From the word lists, some non-essential words not helpful to identify print defects such as “is”, “are”, “and”, and “of” in English, and "거", "것", and "한다" in Korean were removed for a convenient analysis.

3.2 Lexical Analysis

Lexical analysis is used to divide a sentence into keywords and recognize the same word in its various possible forms [9]. In this study, only morphological similarity was considered and semantic similarity was not considered. For example, “line and lines”, band and bands”, “dark and darkness”, or “add and added” are categorized into the same morphs. On the contrary, “error and fail” in English, or "붉은 and 빨간" in Korean are semantically similar, but morphologically different. Therefore, these words are categorized into different morphs. A total of 203 morphs in English and 304 morphs in Korean were created from the analysis. The result shows that Korean subjects used more morphs in Korean than those in English to describe all the print defects.

3.3 Fuzzy Bayes Model

The fuzzy Bayes model has been studied as a very useful technique to index new annotations to be added to an annotated bibliography [8] or to assign index terms or links to documents in a hypertext system [14]. The relation between descriptions and possible categories of print defects can be quantified using the expression.

$$P(D_i \mid w) = MAX_j\, P(D_i \mid w_j) = MAX_j \frac{P(w_j \mid D_i)P(D_i)}{P(w_j)},\ (i=1,2,\ldots,32 \text{ and } j=1,2,\ldots,q) \quad (1)$$

where q is the number of words in the description. $P(D_i/w)$ refers to the probability of a category of print defect I_i given the word present in the description. $P(w_j/D_i)$ is the probability of word w_j given a category of print defect D_i. $P(D_i)$ is the probability of a category of print defect I_i. $P(w_j)$ is probability of a particular word w_j in the description. The operator Max_j then selects the maximum probability calculated over the j words. The model calculated the probability of all keywords being included in certain categories of print defects as shown in Figure 2. The keywords include single words (Figure 2a) and word combinations (Figure 2b) in the word list. Paired words, triplet words, and quadruplet words were considered for the word combinations because the previous study [9], [10] insisted that word combinations greater than word quadruplets were not helpful in identifying print defect categories. In addition, the keywords for the analysis were included in case that the frequency of single words or any word in the word combinations is greater than four. With those rules, description data for both Korean and English training set was reduced to 955. Analysis of the descriptions showed that certain keywords were strongly associated with certain categories of print defects. The strength of these associations was described using mean and Likelihood Ratios as shown in the Figure 2. The likelihood Ratio (LR) for keywords was calculated by taking the ratio of the probability of word w_j given the print defect category D_i divided by the probability of word w_j. That is, $LR(D_i/ w_j) = P(w_j/D_i)/P(w_j)$, and $P(D_i/ w_j) = LR(D_i/ w_j) \times P(D_i)$ from equation (1).

Consequently, the LR is proportional to the Bayes estimator which also can be calculated by taking the ratio of frequency within a defect category divided by total frequency within the whole data set of a keyword. Taking Figure 2a as an example, the word "black" appeared 53 times out of the total 955 descriptions, and 7 times out of 30 of D_{32}. Thus, the likelihood ratio (LR) for the "black" is calculated as follows:

$$LR(D_{32}/black) = P(black/ D_{32})/P(black) = (7/30)/(53/955) = 4.20$$
$$P(D_{32}/black) = LR(D_{32}/black) \times P(D_{32}) = 4.20 \times (30/955) = 7/53 = 0.13.$$

The higher the LR is, the higher the accuracy in identifying print defects.

word	defect	frequency	Total	mean	LR
a	8	3	14	0,21428571429	7,05665024630542
all	28	2	5	0,4	12,7333333333333
black	32	7	53	0,1320754717	4,20440251572327
black	4	9	53	0,16981132075	5,59206245933637

(a) Example for English single word

word	defect	frequency	Total	mean	LR
빨간&가로&개	6	2	4	0,5	15,9666666666667
빨간&가로&개	2	2	4	0,5	15,9666666666667
빨간&가로&선	6	2	4	0,5	15,9666666666667
빨간&가로&줄	2	7	15	0,46666667	14,9022222222222

(b) Example for Korean word combination

Fig. 2. Examples of keywords and Bayes estimators

4 Model Validation

4.1 Model Performance Within Top Five Predictions

As mentioned before, 320 additional descriptions were used to evaluate a fuzzy Bayes model developed by 955 descriptions. Applying each description to the Bayes model, LR was calculated for all keywords of English and Korean and sorted according to the LR. Table 2 shows an example of LR listed in descending order for one description, "added some lines of different color" which was a description of D_{23} (Yellow fine vertical lines). The first keyword of the description is a paired word, "ink & yellow", and it predicts D_{31} (Yellow tinted background) with LR = 12.73, which is wrong. The second keyword of the description is a word triplet, "line & vertical & yellow", and it predicts D_{23} (Yellow fine vertical lines) with LR = 11.50, which is correct.

By the same process, 320 descriptions for prediction were analyzed in the aspect of perdition accuracy. Table 3 shows that the English model successfully predicted 26%, 46%, 54%, 60%, and 66%, and Korean model successfully predicted 33%, 54%, 66%, 72%, and 78% within the first, the second, the third, the fourth, and the fifth predictions. Overall percentages of correct prediction rates were much decreased compared with those of previous study [9], [10], 50%, 66, 70%, 77%, and 81%. The main two reasons for the difference are as follows: (1) The number of print defects was increased from 16 in the previous study to 32 of this study; (2) there were many similarities among the 32 defects in this study. For example, D_1, D_2, D_3, and D_4 are exactly the same defects except for their colors. Print defects, D_1, and D_5 are very similar except for the distance between lines. However, subjects often failed to notice some critical factors such as the color and the distance.

Another result is that there was significant difference between English and Korean predictions, which showed that there were some problems when Korean subjects used English queries to find information and which supports the need for research into a Q-KE-CLD (query translation based Korean-English cross language diagnosis) system. The main reason for the difference between English and Korean predictions by Korean subjects was that when they described print defects in English, they tried to make the descriptions simple due to their lack of ability to describe them in English.

Table 2. Fuzzy Bayes Estimation Based on LR

Keyword	Prediction	Frequency	Total	Mean	LR
ink&yellow	31	2	5	0.16	12.73
line&vertical&yellow	23	13	36	0.30	11.50
vertical&yellow	23	14	41	0.28	10.87
line&vertical&yellow	15	12	36	0.30	10.61
vertical&yellow	15	13	41	0.28	10.09
line&vertical&yellow	19	9	36	0.30	7.96
vertical&yellow	19	10	41	0.28	7.76
line&yellow	3	24	111	0.17	6.88
line&yellow	7	24	111	0.17	6.88
line&yellow	23	19	111	0.17	5.45
line&yellow	15	17	111	0.17	4.88
yellow	3	30	208	0.11	4.59

Table 3. Model Accuracies within Top Five Predictions

Prediction Language	Correct Prediction	Top	Top Two	Top Three	Top Four	Top Five
English Predictions	Number (out of 320)	84	146	172	193	210
	Ratio	26%	46%	54%	60%	66%
Korean Predictions	Number (out of 320)	104	173	211	229	249
	Ratio	33%	54%	66%	72%	78%
Previous Study (Leman, 2003)	Number (out of 64)	32	42	45	49	52
	Ratio	50%	66%	70%	77%	81%

As mentioned previously, 203 morphs for English descriptions and 304 morphs for Korean descriptions were used. That is to say, when they described print defects in English, they missed many critical features in the print defects such as color, defect orientation, thickness, and distance. Therefore, in the case of English predictions, there were relatively fewer keywords to identify the print defects.

4.2 Model Performance By Print Defect Features

As mentioned before, colors, orientation, and patterns of the defects are the main features used to identify print defects. In this respect, the data was also re-classified based on the main features for both English and Korean descriptions. Figure 3 shows the prediction of defect orientation in English using the Textminer [7] as an example of predicting the main defect features.

ID	category	prediction	accuracy	predictor
1004	H	H	1	horizon
1005	V	X	0	dot&white
1006	V	V	1	stripe&white
1007	V	V	1	two&white

Fig. 3. Predicting defect orientations by English

Table 4 summarizes the accuracies of the first top predictions when the fuzzy Bayes model predicted colors, orientations, and patterns of the print defects with both Korean and English descriptions. As shown in Table 4, in general, the prediction accuracy of each of the main features was much higher than the accuracy of print defects of English (26%) and Korean (33%). Moreover, any accuracy in predicting the main features was higher than the accuracy (50%) in the previous study [10]. Especially, the prediction accuracy for defect orientations is over 90% for both English (90%) and Korean (94%). In the case of defect colors, Korean predictions (87%) were much more accurate than English prediction (55%), which means that Korean subjects were much worse at describing features of defect colors when they used English. In the case of defect patterns, English predictions (61%) were a bit higher than Korean predictions (57%), but rather close. As expected, multiplying all

Table 4. Model Accuracies by Main Features of Defects

Defect features	English prediction	Korean prediction
Colors (C, M, Y, K)	55% (177 out of 320)	87% (278 out of 320)
Orientations (H, V, X)	90% (288 out of 320)	94% (302 out of 320)
Patterns ($P_1, P_2, \dots, P_7$)	61% (194 out of 320)	57% (181 out of 320)
Multiplying three accuracies	30% (0.55 x 0.9 x 0.61)	47% (0.87 x 0.94 x 0.57)
Defects ($D_1, D_2, \dots, D_{32}$)	26% (84 out of 320)	33% (104 out of 320)

accuracies for the main features (30% for English and 47% for Korean) was much higher than the accuracy for print defects because there were many other factors influencing the prediction accuracy such as line thickness, distance between lines, and number of lines.

5 Conclusions

The fuzzy Bayes model correctly predicted 26%, 46%, 54%, 60%, and 66% for English descriptions and 33%, 54%, 66%, 72%, and 78% for Korean descriptions within the first, the second, the third, the fourth, and the fifth predictions by Korean subjects. The percentage of correct prediction of Korean descriptions was much higher than that of English because Korean subjects tried to make English descriptions simpler due to their reluctance to use English, though the subject selection was restricted to individuals who have TOEIC scores over 550. In addition, a result shows that accuracies of predicting colors, orientations, and patterns of print defects were 55%, 90%, 61% respectively by English predictions, and 87%, 94%, 67% by Korean predictions, which were much higher than the accuracies respectively (26% by English, 33% by Korean) of predicting print defects.

6 Further Study

As mentioned before, the objective of this study is to investigate a Q-KE-CLD system. Therefore English description data by US subjects will be collected and a fuzzy Bayes model from the data will be developed. English descriptions by Korean subjects will be evaluated based on the fuzzy Bayes model developed with English descriptions by US subjects. In addition, Korean descriptions will be translated into English and evaluated based on the fuzzy Bayes model developed with English descriptions by US subjects. Finally, the Q-KE-CLD system will be developed and evaluated based on the fuzzy Bayes models from data obtained by Korean and US subjects.

Acknowledgments. The authors would like to thank all professors - In-Jae Jeong, Min-Yong Park (Hanyang University, Seoul), Dongmin Shin, Chang Hee Han, Jong Soo Kim, Sun Hur, Seon Eung Ahn (Hanyang University, Ansan), Sung Ho Han (POSTECH, Pohang), Sang-Ho Kim, Wook-Gee Lee (Kumoh National Institute of Technology, Gumi) and Pyung-Hoi Koo (Pukyong National University, Busan) - for their help in the collection of the Korean description data.

References

1. Chen, J.: The construction, use, and evaluation of a lexical knowledge base for English-Chinese cross-language information retrieval. Doctoral dissertation, Syracuse University (2003)
2. Choe, P., Kim, C., Lehto, M.R., Lehto, X., Allebach, J.: Evaluating and improving a self-help technical support Web site: Use of focus group interviews. International Journal of Human-Computer Interaction 21(3), 333–354 (2006)

3. Diekema, A.R.: Translation events in cross-language information retrieval: Lexical ambiguity, lexical holes, vocabulary mismatch, and correct translations. Doctoral dissertation, Syracuse University (2003)
4. Ishihara, S.: 1996. Ishihara's tests for colour-defficiency. 38 plates edition (2003)
5. Park, H.J., Santos, H., Kim, C., Choe, P., Kumontoy, R., Low, K., Oldenberger, K., Ortiz, M., Lehto, X., Lehto, M.: A Web-based self-diagnosis tool to solve print quality issues. Proceedings of NIP22 22nd International Conference on Digital Printing Technologies, Denver, CO, September 17-22, 2006, pp. 465–471 (2006)
6. Kim, C., Choe, P., Lehto, M.R., Allebach, J.: Development of a web-based interactive self-help troubleshooting tool for print quality problems. In: International Conference on Human-Computer Interaction, Las Vegas, Nevada USA, July 22-27, 2005, pp. 22–27 (2005)
7. Lehto, M.R.: TextMiner Manual (Ann Arbor, MI: Consumer Research,Inc.) (2004)
8. Lehto, M.R.: Warnings and safety instructions: electronic hypertext version 2.0. Fuller technical publications, Ann Arbor, MI (1994)
9. Leman, S.: Decision support system to predict print quality. Master thesis, Purdue University (2001)
10. Leman, S., Lehto, M.: Interactive decision support system to predict print quality. Ergonomics 46(1-3), 52–67 (2003)
11. Mitkov, R.: The Oxford handbook of computational linguistics. Oxford University Press, Oxford, New York (2003)
12. Salton, G.: Experiments in multi-lingual information retrieval. Information Processing Letters 2(1), 6–11 (1973)
13. Smith, G.W.: Computers and human language. Oxford University Press, Oxford, New York (1991)
14. Zhu, W.: Decision support for indexing and retrieval of information in hypertext systems. International Journal of Human-Computer Interaction 11(4), 349–371 (1999)

A Treemap-Based Result Interface for Search Engine Users

Shixian Chu, Jinfeng Chen, Zonghuan Wu, Chee-Hung Henry Chu, and Vijay Raghavan

The Center for Advanced Computer Studies, University of Louisiana at Lafayette
P.O. Box 44330, Lafayette, LA 70504-4330, U.S.A.
{sxc1825,jxc5466,zwu,cice,raghavan}@cacs.louisiana.edu

Abstract. Search engines and metasearch engines on the World Wide Web typically display representative information such as the result document titles and snippets in a one-dimensional ranked list. Alternatively, a few others such as Clusty and Kartoo cluster their results so that users can, to a certain extent, interact with documents, keywords and/or clusters. The number of documents that can be effectively presented in one screen is usually limited to between 10 and 20. In this paper, we propose a method based on treemap visualization that substantially improve the information compactness. In addition, it provides a few unique post-search interaction methods that enable a user to manage a large number of results. An online prototype system is built. Various experiments are designed and done to evaluate the efficiency, effectiveness and usability.

Keywords: Information visualization, web search interface.

1 Introduction

Search engines and metasearch engines such as Google (www.google.com), Yahoo! (www.yahoo.com),Windows Live (www.live.com), and many others essentially have the same one-dimensional rank list-based search result interface. Taking Google as an example, in a result page for the query "treeemap", the information of 10 out of more than 1.5 million results, including the title, url and snippet, were displayed on the first page (Fig. 1). Only the first 6 or 7 results can be seen in a 1280 by 800 sized window of the Web browser, if the user does not click the mouse to scroll down. Obviously, the user has to click the "next page" button to get more results, if the first 10 results cannot satisfy him or her.

A number of other search engines, such as Clusty (www.clusty.com) (Fig. 2), Kartoo (www.kartoo.com), and Ask (www.ask.com), cluster the results into different categories before presenting them to the users. The clustering helps users if they need to narrow their search to a more specific domain. Fundamentally, however, there are still only a small number of results that can be seen within one screen.

The current search engine result display methods have a few common limitations, such as:

- The presentation often makes it hard for users to find desired information, especially if it is not within the top-ranked search results.

M.J. Smith, G. Salvendy (Eds.): Human Interface, Part I, HCII 2007, LNCS 4557, pp. 401–410, 2007.

- Users may not be able to collect enough relevant materials from the first few screens of search results, while it is a tedious, and sometimes impractical, job to go through pages of scrolled lists.Many returned results scattered in the list may be not of interest to the user while he/she still has to scan through them.

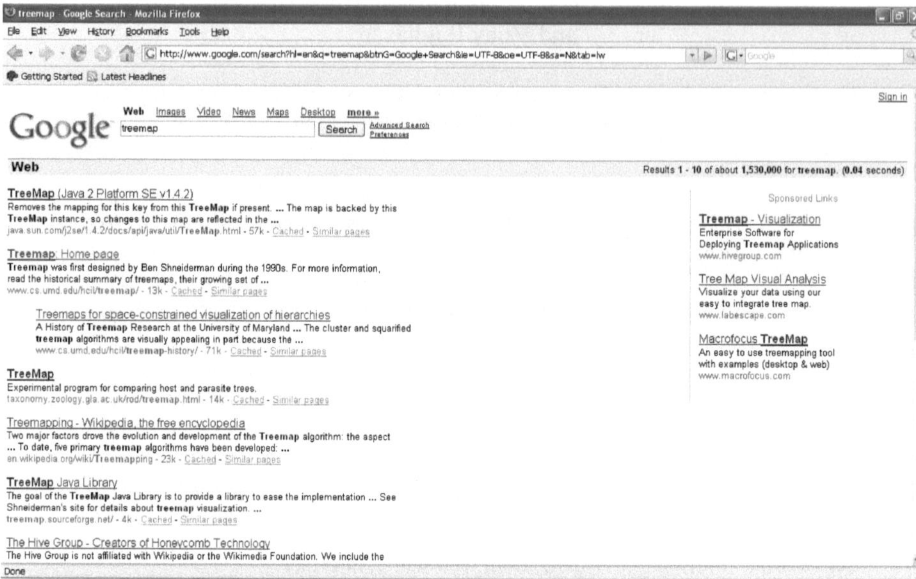

Fig. 1. A screen shot of Google's search result page

These limitations make it difficult for a user to manage large amounts of information returned from search engines, though such information is already retrieved and available for use.

In this paper, we present a treemap-based search result visualization solution for search engine users. Treemap is a space-constrained visualization of hierarchical structures. In our work, such a structure is used to represent search results and the relation between phrases and documents. Thus, hundreds of results can be displayed in one screen. Furthermore, a few useful interaction methods that cannot be realized in traditional search engine interfaces can be applied to further empower users' capability to manage a large number of search results.

In addition to providing a general solution for applying treemap in search engine result visualization, our other contributions in this paper include that, we propose two treemap algorithms to enhance user experiences; we also provide our study on performance and usability; finally, we have implemented a Web-based prototype system for interested users to have first-hand experience.

In section 2, we briefly go over treemap, the concept and related prior arts. In section 3, we explain our method in detail. In section 4, we compare our result visualization methods with a few other major search engines, including Google, Clusty, Kartoo. Finally, we conclude in section 5.

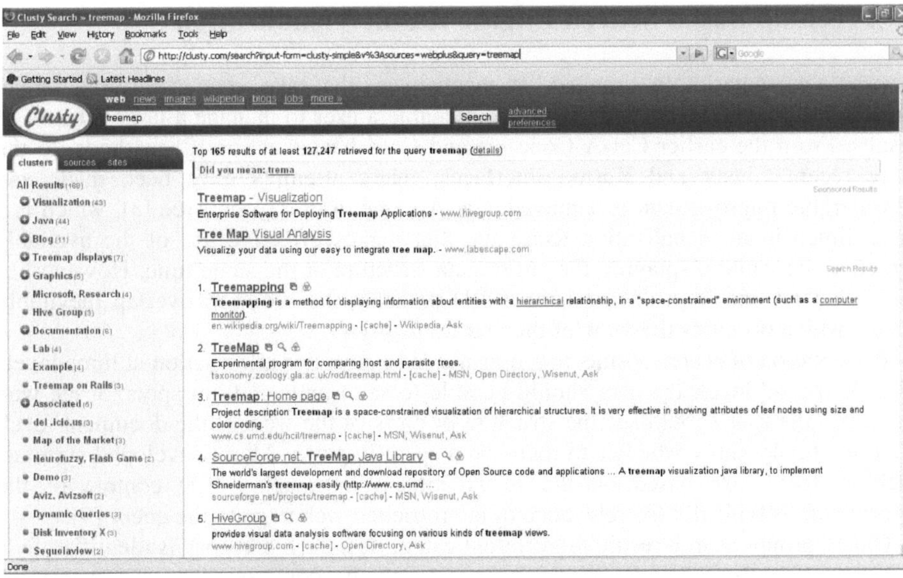

Fig. 2. A screen shot of Clusty's search result page

2 Treemap and Related Work

Treemap is a space-constrained visualization of hierarchical structures. It turns a tree into a planar space-filling map. Typically, in a two-dimensional space, each node is represented by a rectangle; child nodes are arranged to be nested in their parent node's rectangle. The arrangement, size, and color of each rectangle represent different attributes of the node. A treemap can enable users to compare nodes and sub-trees even at varying depth in the tree, and help them spot patterns and exceptions. A lot of studies have been done since Ben Shneiderman invented the treemap in early 1990s. For example, several treemap algorithms, from using rectangles to other polygons and circles, from two dimensional to three dimensional, have been designed; treemap visualization has been integrated into research works ranging from data mining to bio-informatics; powerful open source software has been developed. It has been successfully applied to many industries, from inventory control, project management, to financial analysis and e-commerce. A comprehensive list of publications and applications of treemap is maintained at www.cs.umd.edu/hcil/treemap.

Essentially, treemap visualization helps display an enormous amount of information and reveals underlying patterns in the data. In the context of the Web, treemap has been applied to visualize:

- online news: Newsmap at www.marumushi.com/apps/newsmap/Newsmap,
- websites: del.icio.us being the most popular treemap; and
- the Usenet: Netscan at netscan.research.microsoft.com/treemap/ Netscan.

In this paper, we apply it to search engines for visualization of their retrieved results.

3 Treemap Based Search Engine Result Visualization

Beyond a presentation in a ranked list, mechanisms that take advantage of the human capacity to process visual information can enable a user to manage a large amount of results. From the earlier Cat-A-Cone [1] and Card Visualization [2] methods, to the more recent Clusty and Kartoo interfaces, many attempts have been made and considerable improvement is achieved. Cat-A-Cone used a Cone Tree [3], which is a three-dimensional visualization technique that makes effective use of the available screen space while displaying the entire data structure at the same time. However, as the 3-D cone tree is projected onto a 2-D plane in the browser, overlap inevitably occurs which obscures the view of the rear items [4].

Presentation of search results to a user needs to consider information at three levels [5]. At the set level, the user should be able to see trends and hot spots; at the web site level, the user should see the structure or parts of the web; at the document level, the user should know whether to focus on a specific page. Many developed systems, such as [6-8], are based on the focus+context approach. The context of the information is typically the relevance of the retrieved web page to the query [9].

Clusty groups search results in different categories; each category is identified by a key phrase as the label. By clicking on a label, the corresponding documents are displayed. In some sense, Clusty method can be seen as an extension of Card visualization. Kartoo utilizes a graphic interface and a proprietary algorithm. A few result documents and key phrases are shown in a "contour map". When the mouse hovers over an object (i.e. a document or a key phrase), curve lines are drawn to link related objects. However, it appears that this method can only manage a few tens of records and is not scalable.

3.1 A General Tree-Map Based Solution

The basic idea of our method is: Firstly, search result documents will be clustered into hierarchical groups (based on content, title and/or snippet). The hierarchical cluster structure is essentially a tree with each node being a cluster or a sub-cluster. Each node has its "name", which is a key phrase extracted from the documents within the cluster. Secondly, we map the tree into clustered treemap. Each cluster in the treemap represents a cluster of documents. Each document is represented by a block in the treemap cluster, with a name (the title of the document, for example). The size of a block is used to indicate result relevancy. The bigger the block, the more relevant (to the query) the corresponding documents. Features of blocks, including colors and brightness, are used to indicate the file type, domain, date, location, size and other document properties. Finally, such a treemap will be shown to users. Users can also conveniently customize the block feature-document property mapping through additional functions provided at the interface.

Comparing to existing result displaying/visualizing techniques, the advantages for our method are as follows.

1. It allows users to comfortably take advantage of natural intuition, spatial cues and perception at the interface. Logical cluster structure of the search results can be expressed clearly; Relationships among phrases and docs are also clear; Comprehensive information and trend can be presented—for example, with just a

glance, the user can get the sense of how many results are there in each domain or how many document are located in a given country.

2. Also, our method can meet the two stringent requirements of the online web search:
   a. Information compact — hundreds of results can be shown in one result page;
   b. Network and computation efficient — In our implementation, the size of a result page with two hundred search results is around 75k to 100k bytes. In average, one Google result web page of 10 search results is around 32k bytes. A Clusty result page of 20 clustered search results is about 150k bytes. Also, the treemap algorithm does not require intensive computation.
3. It is interesting.

4 Experiments and Evaluations

There are many aspects of a visualization system that can be tested for its performance. In [10], the factors considered for the success of visual structures are: (i) target user group, (ii) type and size of result sets, and (iii) the nature of task. By varying these factors, the performance of an interface can be judged by: (i) the task completeness or effectiveness, (ii) the task performance time or efficiency, and (iii) the user subjective acceptance or satisfaction. A scenario-based evaluation of the interface used in [11] varies the user group to measure the time to complete a task, the perceived precision, and users' subjective reactions. We have designed several tests to evaluate the performance of search result visualization methods. More specifically, the following aspects were assessed:

- Time to find the most relevant result
- Capability of providing the macro-view of search results
- Maximum Topological distance
- The Index of ease of use
- The Interestingness Index

Each of the following sub-section addresses one of the aspects listed above. In most tests, we compare our treemap-based method with those used by a few other major search engines (Table 1). Note that our search engine gets results information and uses cluster information from SRC MSN; it subsequently uses the centralized treemap to visualize the organized information. In our experiments, we assumed that the quality and coverage of these five search engines are virtually equal to each other, for the ease of comparison.

Table 1. Search engines and their URLs used in our experiments

Search Engine	URL
Google	http://www.google.com
SRC	http://rwsm.directtaps.net/
Kartoo	http://www.kartoo.com
clusty	http://www.clusty.com
Treemap-based	http://lincweb.cacs.louisiana.edu:8080/treemap

4.1 Time to Find the Most Relevant Result

In this test, we compare how much time is needed for a user to find his/her desired results. There are two types of queries that a user can send, viz. implicit and explicit.

Implicit Queries refer to queries that a user's intentions cannot be easily identified from the query itself or a user does not know what to anticipate from search results. To guide testers to send proper implicit queries, we adopted the description at [12] and divide a query into the following 5 sub-categories:

1) A query that has Ambiguity. For example, "Java" is an ambiguous query as it may mean a programming language or coffee.

2) A query that has many sub-topics. Many query terms contains sub-topics, such as a query one of our testers sent: "Argentina", its sub-topics include its history, travel, business, and so on.

3) A query for finding unknown fact of peoples. As an example, one of our tester sent "bill Clinton".

4) A query for finding unknown relationship of peoples. When a query is two person's name, find out their relationships. For example, "Bush Blair". (Note: None of our testers chose such kind of queries)

5) A query that is a question, to find out the possible answers. For example, one of the testers sent queries such as "smallest animal" and "fastest animal".

Explicit Queries refer to queries that have key terms that can hopefully pinpoint the specific desired result documents. For example, a test sent a query "Distinctive image features from scale-invariant key points" to find the paper that has the query string as its title.

Testers were required to send different implicit queries and explicit queries to each search engine, and record the time it takes to find desired information from results. If it takes the user more than 3 minutes to find satisfying information, please mark time to be infinite

Table 2. The number of clicks it takes to find satisfactory results from a search engine

	# of Clicks	Google (# of queries)	Clusty (# of queries)	SRC (# of queries)	Kartoo (# of queries)	Treemap (# of queries)
Implicit Queries (32 queries in total)	1	*13*	*1*	*0*	*11*	*26*
	2	*5*	*21*	*19*	2	2
	3	2	*7*	7	*10*	2
	4+	*8*	2	*4*	*4*	*1*
	NSRF*	*4*	*1*	2	5	*1*
Explicit Queries (12 queries in total)	1	*12*	*12*	*12*	*10*	*12*
	2	*0*	*0*	*0*	*1*	*0*
	3	*0*	*0*	*0*	*0*	*0*
	4+	*0*	*0*	*0*	*1*	*0*
	NSRF*	*0*	*0*	*0*	*0*	*0*

* NSRF = No Satisfactory Result Found

Table 3. The time (in seconds) it takes to find satisfactory results from a search engine

	Time	Google	Clusty	SRC	Kartoo	Treemap
Implicit Queries (32 queries in total)	<5s	*10*	*9*	*12*	*6*	*16*
	5s~10s	*6*	*7*	*7*	*5*	*8*
	10s~60s	*7*	*12*	*9*	*11*	*4*
	60s~180s	*5*	*3*	*2*	*4*	*2*
	NSRF*	*4*	*1*	*2*	*5*	*1*
Explicit Queries (12 queries in total)	<2s	*7*	*2*	*0*	*0*	*0*
	2s~5s	*5*	*8*	*11*	*2*	*9*
	5s~10s	*0*	*2*	*1*	*4*	*2*
	10s~180s	*0*	*0*	*0*	*6*	*1*
	NSRF*	*0*	*0*	*0*	*0*	*0*

* NSRF = No Satisfactory Result Found

Due to complexity of the test, and limitations on time and resource, we could only perform the test at a relative small scale. Six (6) volunteers have participated in this test. Each one was given time to practice using the five search engines until they are familiar with the interfaces, each of them was asked to send 5~6 implicit queries and 2 explicit queries to each search engine. The time and the number of mouse clicks between that a query is sent and that a satisfactory result is found, are recorded.

From tables 2 and 3, we could clearly see the advantage of treemap method when the search query is implicit. In 26 out of 32 queries, only 1 click is needed to get the satisfactory results, while the best in the other four competitors only can do this in 13 queries (Google). Similarly, in 26 out of 32 queries, it took less than 10 seconds to find results while the record of the best in 4 competitors is 19 queries (SRC).

For explicit queries, Google has advantages in terms of number of clicks and time needed over others. However, the performance of treemap method is still comparable with the other three search engines.

4.2 Capability of Providing the Macro-view of Search Results

As mentioned in previous sections, most current search engines can only effectively display a limited number of search results, while our treemap-based method can display hundreds of results on the computer screen. As more information is available, it provides a better basis for a user to get the macro view of his/her search. In Table 4, we summarize the capability of the search engines in terms of the macro view of search results.

Google lists results in a ranked list. The other four search engines cluster results in different categories. Such categorization often helps a user to understand the search better. From the different sizes of clusters, the different color code of documents based on the document publication date, document type, domain of the website, or other features, Treemap has a unique property, viz. color coding, which can clarify the distribution based on some feature of results. Among all search engines, only through the treemap-based method can a user can get the answers of questions such as in which domains the searched topic was most popular (Figure 3), or in which years the searched topic was most active, with just at a glance of the result page.

4.3 The Topological Distance Between Related Result Documents

Clustering help similar documents be close to each other on the screen. When information in one result document is not sufficient, the user can easily locates the related document in the neighborhood. Among the five methods, treemap displays related documents in a cluster in the most compact manner. We borrow the concept of *topological distance* to illustrate this feature.

Take the first cluster in Fig 3, which is labeled by the keyword "industrial", as an example; it has 10 documents d_0, d_{12}, d_{24}, …, d_{96}. In the treemap visualization, any two documents are separated by essentially more than 2 documents. For example, between d_0 and d_{96} , there are d_{12}, d_{80} or d_{60} and d_{71}. So, we can say the maximum topological distance between two related documents is 2. However, in case of SRC or Clusty, the clustered results are lined in a ranked list. Thus, the maximum topological distance between two related documents is 8, as between d_0 and d_{96} , there are the other 8 documents in the clusters listed. Since there is no clustering, Google is worse in this aspect. Usually, Google display 10 results a page, that virtually prevent the d_0 on page 1 and d_{96} on page 9 from appearing together. Even though Google could be customized to show 100 documents in a page, between d_0 and d_{96} , there will be some other 95 documents listed. We note that Katoo was not included in the comparison because its proprietary atypical visualization method makes it difficult to apply the topological distance to its interface.

4.4 Ease of Use and Interestingness

After they conducted experiments described in 4.1, the 6 volunteers were asked to rank the five search engine result visualization methods based on their degree of ease of use and that of interestingness. In both cases, we use a 7-point scale: 1 means very easy to use/very interesting and 7 means very difficult to use/very boring.

Table 4. Macro-view of search results

	Google	Clusty	SRC	Kartoo	Treemap
#Results shown on a screen	<10	<10	<10	20 ~30	>100
Are results categorized based on content?	No	Yes	Yes	Yes	Yes
Can a distribution of common features be shown?	No	No	No	No	Yes

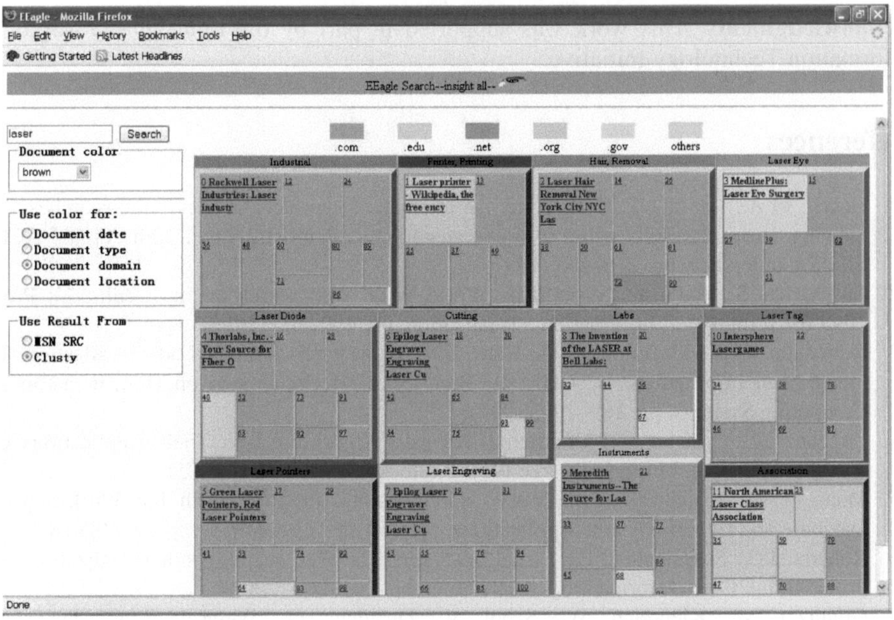

Fig. 3. A screen shot of treemap search result page for the query "laser"

Table 5. Average subjective scores of ease of use and interestingness of search engines

Search Engine	Ease of Use (score: rank)	Interestingness (score : rank)
Treemap	1.83 : 2	2.67 : 1
Google	1.5 : 1	4.5 : 5
SRC	3.16 : 3	4: 4
Clusty	3.33 : 4	3.83 : 3
Kartoo	4.66 : 5	2.83: 2

The average scores are listed in Table 5, which shows that the treemap is in the second place, considered to be quite easy to use, only slightly less so than Google but better than others. As for interestingness, treemap is considered to be the most interesting method among all five search engines.

5 Conclusions

In this paper, we proposed 1) a treemap-based result visualization method for Web search engines, and 2) a set of methods to evaluate the performance of the interface. Our tests show that, in most cases, the treemap method outperforms that of other search engines in information compactness, macro-view capability, efficiency of finding results for implicit query, ease of use and interestingness.

Acknowledgments. This work was supported in part by the Louisiana Governor's Information Technology Initiative.

References

1. Hearst, M.A., Karadi, C.: Cat-a-cone: An interactive interface for specifying searches and viewing retrieval results using a large category hierarchy. In: Proc. 20th Annual ACM SIGIR Conference, pp. 246–255 (1997)
2. Mukherjea, S., Hara, Y.: Visualizing World Wide Web search engine results. In: Proc. 1999 IEEE Int. Conf. Information Visualization, pp. 400–405 (1999)
3. Robertson, G.G., Mackinlay, J.D., Card, S.K.: Cone trees: animated 3d visualizations of hierarchical information. In: CHI '91: Proc. SIGCHI Conference on Human factors in Computing Systems, pp. 189–194 (1991)
4. Engdahl, B.: Ordered and Unordered Treemap Algorithms and Their Applications on Handheld Devices, Master's Degree Project, Stockholm, Sweden (2005)
5. Mann, T.M.: Visualization of WWW-search results. In: Proc. 10th Int. Workshop on Database and Expert Systems Applications, pp. 264–268 (1999)
6. Roberts, J.C., Suvanaphen, E.: Visual bracketing for Web search result visualization. In: Proc. 7th Int. Conf. Information Visualization, pp. 264–269 (2003)
7. Granitzer, M., Kienreich, W., Sabol, V., Dosinger, G.: WebRat: Supporting agile knowledge retrieval through dynamic, incremental clustering and automatic labeling of Web search result sets. In: Proc. 12th IEEE Int. Workshops on Enabling Technologies: Infrastructure for Collaborative Enterprises, pp. 290–301 (2003)
8. Cellary, W., Wiza, W., Walczak, K.: Visualizing Web Search Results in 3D. IEEE Computer 37(5), 87–89 (2004)
9. Nguyen, T., Zhang, J.: A novel visualization model for Web search results. IEEE Trans. Visualization and Computer Graphics 12(5), 981–988 (2006)
10. Mann, T.M., Reiterer, H.: Evaluation of different visualizations of Web search results. Proc. 11th Int. Workshop on Database and Expert Systems Applications, pp. 586–590 (2000)
11. Hoeber, O., Yang, X.D.: A comparative user study of Web search interfaces: HotMap, Concept Highlighter, and Google. In: Proc. 2006 IEEE/WIC/ACM Int. Conf. Web Intelligence, pp. 866–874 (2006)
12. accessed on 02/15/2007, http://rwsm.directtaps.net/usage.htm

Development of an Approach for Optimizing the Accuracy of Classifying Claims Narratives Using a Machine Learning Tool (TEXTMINER[4])

Helen L. Corns[1], Helen R. Marucci[1,3], and Mark R. Lehto[2]

[1] Liberty Mutual Research Institute for Safety, 71 Frankland Road, Hopkinton, MA 01748, USA

[2] School of Industrial Engineering, Purdue University, 1287 Grissom Hall, West Lafayette, In 47907-1287, USA

[3] Department of Work Environment, UMASS Lowell, 1 University Avenue, Lowell, MA 01854, USA

Abstract. This paper demonstrates a successful application of a Fuzzy Bayes machine-learning tool for classifying large amounts of narrative text, involving the use of ROC curves to identify optimum prediction threshold values at which to filter predictions for manual review. Different thresholds were used for different categories to optimize results and effectively minimize resources necessary for manual coding of a large number of claims narratives randomly extracted from a large U.S. insurer. The results indicated that utilizing a computer approach with strategic assignment of manual narratives filtered out approximately 15% of the narratives for manual review and resulted in a final accuracy at the two-digit classification level of 81%..

Keywords: Textmining, Occupational Safety, Workers Compensation, BLS, Textminer.

1 Introduction

Narrative text analysis has many different applications across many disciplines. One important application allows for the analysis of qualitative data, routinely collected in administrative data systems. In previous studies [1], [2], [3], a method was introduced to classify narratives using Textminer[4] software developed based on Fuzzy Bayes Logic.

In this study an approach was developed for using the Textminer tool to discriminate claims narratives from a large US insurer into approximately 20 two-digit event categories. The approach attempted to maximize accuracy while minimizing resource requirements of manual coding. The objective of this study was to demonstrate that a computerized approach that classifies some narratives with high confidence and then selects out narratives for manual review is very effective for large administrative databases.

2 Methods

This Textminer method uses a Fuzzy Bayes algorithm to maximize the probability of a classification category given the word, word combinations or word sequences in the

M.J. Smith, G. Salvendy (Eds.): Human Interface, Part I, HCII 2007, LNCS 4557, pp. 411–416, 2007.

accident description. A computer program was developed by one of the authors (ML) to calculate probabilities and run the Fuzzy Bayes algorithm [4]. A random sample of approximately 12,000 claims narratives were obtained from a large US insurer. Two manual coders assigned classifications according to the Bureau of Labor Statistics (BLS) Standard Occupational Injury and Illness Classification (SOIIC) two-digit event protocol. Only narratives where both coders agreed on the final classification were used in this analysis. This yielded a dataset of 10,389 narratives with gold standard codes. The computer learned (assigned probabilities) from 7,389 manually coded narratives and then, using the Fuzzy Bayes algorithm, assigned codes to 3,000 new narratives.

The method had three main steps. First, Textminer was used to assign narratives into one-digit classifications. The Fuzzy Bayes model using single, paired, three, and four sequenced words as predictors was used. The narratives were then placed into separate categorical databases based on this first assignment. Textminer was run a second time on each categorical database to evaluate the more specific code (2nd digit).

The accuracy and strengths (probability of the classification given the words in the narrative) for each two digit assignment were evaluated to decide what threshold values would maximize on accuracy. Receiver Operating Characteristic (ROC) curves were constructed to help determine the optimum threshold. The optimum threshold

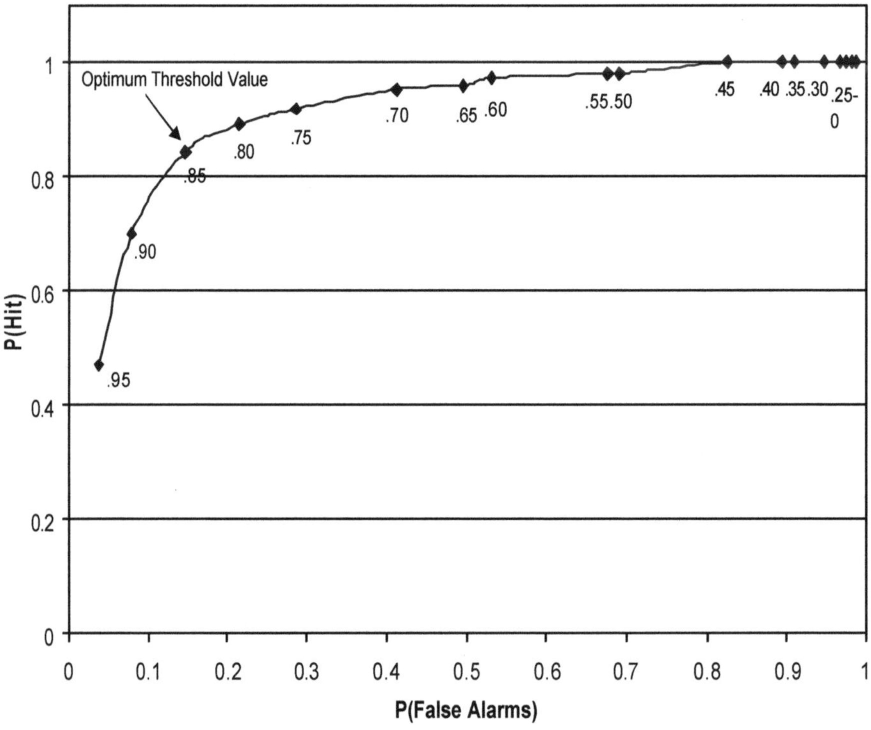

Fig. 1. Receiver Operating Characteristic (ROC) Curve: Fall to Lower Level (n=159)

Table 1. Threshold values assigned to each two digit classification

BLS Two Digit Event	N	Threshold	P(hits) [1]	P(FA) [2]
Contact:				
Contact unspecified	18	0.55	0.14	0.00
Struck against	163	0.85	0.40	0.08
Struck by	272	0.85	0.73	0.25
Caught in-between	100	0.80	0.77	0.09
Rubbed/abraded by vibration	4	0.55	---	0.02
Falls:				
Falls unspecified	2	0.70	---	---
Fall to lower level	159	0.85	0.84	0.15
Jump to lower level	10	0.85	0.75	0.00
Fall on same level	299	0.85	0.94	0.56
Bodily Reaction (BR) & Exertion:				
BR & exertion unspec	16	0.50	0.07	0.00
Bodily reaction	218	0.80	0.49	0.08
Slip, trip without a fall	162	0.90	0.75	0.03
Overexertion	460	0.85	0.88	0.32
Repetitive motion	98	0.80	0.82	0.05
Bodily conditions	10	0.80	0.67	0.00
Exposure to harmful environment:				
Contact with electric current	34	0.85	0.97	0.02
Contact with temp. extremes	108	0.85	0.90	0.11
Exposure to caustic,noxious sub	123	0.80	0.94	0.09
Exposure to noise	36	0.80	1.00	0.02
Exposure to traumatic event	60	0.85	0.96	0.04
Transportation Accident:				
Transportation accident unspecified	2	0.85	---	---
Highway accident	205	0.85	0.99	0.32
Non highway accident	74	0.85	0.68	0.06
Pedestrian struck by vehicle	89	0.80	0.82	0.09
Fires & Explosions:				
Fire - unintended	8	0.70	---	---
Explosion	13	0.70	1.00	0.00
Assaults & Violent Acts:				
Assaults & violent acts by person	111	0.70	1.00	1.00
Assaults by animals	6	0.70	1.00	0.00
Non classifiable	140	0.90	1.00	0.93
Total	3000			

[1] The probability that given manual coder assignment into a category, the program also correctly assigned the classification to that category.

[2] The probability that given that the true prediction is in another category the program incorrectly assigned the classification to that category.

was selected where a more strict threshold (i.e. going from .85 to .9) gained more hits than false alarms (i.e. (difference in hits between .85 to .9) /(difference in false alarms between .85 and .9 was greater than 1)). This optimized threshold will be at some point on the convex hull of the ROC curve.

An example of utilizing ROC curves to determine optimum threshold values is shown in Figure 1. The threshold values that were used for each category are shown in Table 1.

If the probability of a category given the maximum predictor did not meet the threshold value the narrative was filtered out for manual review. This process was then repeated for each two-digit event classification. Using this approach a total of 454 narratives (15%) would be filtered out for manual review

Table 2. Improved sensitivity and specificity using maximum thresholds for selection of narratives for manual review

BLS Two Digit Event	N	All narratives classified with Textminer only		Textminer & Manual coding below threshold		
		Sensitivity	Specificity	Sensitivity	Specificity	N(%) manually coded
Contact:						
Contact unspecified	18	0.00	1.00	0.61	1.00	11(61%)
Struck against	163	0.47	0.99	0.69	1.00	59(36%)
Struck by	272	0.72	0.95	0.80	0.98	63(23%)
Caught in-between	100	0.80	0.99	0.92	0.99	16(16%)
Rubbed/abraded by vibration	4	0.00	1.00	0.75	1.00	3(75%)
Falls:						
Falls unspecified	2	0.00	1.00	0.00	1.00	0(0%)
Fall to lower level	159	0.73	0.98	0.77	0.99	12(8%)
Jump to lower level	10	0.20	1.00	0.30	1.00	3(30%)
Fall on same level	299	0.86	0.96	0.88	0.96	16(5%)
Bodily Reaction (BR) & Exertion:						
BR & exertion unspec	16	0.00	1.00	0.38	1.00	6(38%)
Bodily reaction	218	0.41	0.99	0.64	0.99	69(32%)
Slip, trip without a fall	162	0.53	0.98	0.61	0.99	22(14%)
Overexertion	460	0.87	0.92	0.91	0.95	55(12%)
Repetitive motion	98	0.79	0.99	0.82	0.99	8(8%)
Bodily conditions	10	0.50	1.00	1.00	1.00	4(40%)

Table 2. *{Continued}*

BLS Two Digit Event	N	All narratives classified with Textminer Sensitivity	Specificity	Manual coding below threshold Sensitivity	Specificity	N(%) manually coded
Exposure to harmful environment:						
Contact with electric current	34	0.94	1.00	0.97	1.00	2(6%)
Contact w temp extremes	108	0.94	0.99	0.97	1.00	10(9%)
Exposure to caustic,noxious sub	123	0.88	0.99	0.92	0.99	8(7%)
Exposure to noise	36	0.92	1.00	0.97	1.00	2(6%)
Exposure to traumatic event	60	0.88	1.00	0.90	1.00	1(2%)
Transportation Accident:						
Transportation Accident Unspecified	2	0.00	1.00	0.00	1.00	0(0%)
Highway Accident	205	0.99	0.98	0.99	0.99	2(1%)
Non Highway accident	74	0.49	0.99	0.74	1.00	19(26%)
Pedestrian struck by vehicle	89	0.61	1.00	0.70	1.00	8(9%)
Fires & Explosions:						
Fire - unintended	8	0.00	1.00	0.13	1.00	1(13%)
Explosion	13	0.08	1.00	0.38	1.00	3(23%)
Assaults & Violent Acts:						
Assaults & Violent Acts by person	111	0.77	1.00	0.85	1.00	9(8%)
Assaults by animals	6	0.00	1.00	0.33	1.00	2(33%)
Non Classifiable	140	0.29	1.00	0.54	1.00	40(29%)
Overall	**3000**	**0.71**	**0.97**	**0.81**	**0.98**	**454(15%)**

3 Results

The resulting accuracy of each two-digit classification is shown in table 2. Results are compared for Textminer alone to classify all narratives, and using Textminer complemented with selecting out the more difficult narratives (those predicted at a lower threshold) for manual review. The overall gain in accuracy was 10% with many categories improving dramatically ([Bodily Reaction] sensitivity improved 22% and [Non-highway accident] sensitivity improved 26%).

4 Conclusions

Results indicate that utilizing a computer approach, with strategic assignment of manual narratives, filtered out approximately 15% of the narratives for manual review. This resulted in a final accuracy at the two-digit classification level of 81%. While some specific categories have lower predicted strengths and are more difficult to code, these are likely to be filtered out for manual review, thus maintaining a high accuracy even in these categories (Table 2). In order to maintain high accuracy requirements, manual coding may never be totally replaced, however, strategically assigning manual coders the more difficult narratives, while allowing the computer to classify common accident narrative allows for a more efficient and cost effective utilization of resources. As an example of the potential impact, in a study requiring classification of approximately 85,000 claims narratives, it is estimated that the burden of 3500 person-hours of coding could be reduced by approximately 500 person-hours.

References

1. Wellman, H., Lehto, M., Sorock, G.: Computerized coding of injury narrative data from the National Health Interview Survey. Accid. Anal. Prev 36, 165–171 (2004)
2. Lehto, M., Sorock, G.: Machine learning of motor vehicle accident categories from narrative data. Methods Info Med. 35(4-5), 309–316 (1996)
3. Sorock, G., Ranney, T., Lehto, M.: Motor vehicle crashes in roadway construction work zones: an analysis using narrative text from insurance claims. Accid. Anal. Prev. 28, 131–138 (1996)
4. Lehto, M.R.: TextMiner Manual (Ann Arbor, MI: ConsumerResearch, Inc.) (2004)

The Interface of VISTO, a New Vector Image Search Tool

Tania Di Mascio[1], Luigi Laura[2], and Valeria Mirabella[2]

[1] Dipartimento di Ingegneria Elettrica e dell'Informazione
Università di L'Aquila Monteluco di Roio, I-67040, L'Aquila - Italy
tania@ing.univaq.it
[2] Dipartimento di Informatica e Sistemistica
Università di Roma "La Sapienza", via Salaria 113, I-00198, Roma – Italy
{laura,mirabel}@dis.uniroma1.it

Abstract. We discuss the interface of VISTO, a Content Based Image Retrieval system for vector images in SVG (Scalable Vector Graphics) format. The system includes different engines and one common graphical interface. In fact, to support the different requirements of different application domains, the system offers a variety of engines. Most notably, due to its modular architecture, the system allows users to add engines at runtime; the interface provides support for newly added engines, including parameter tuning. The interface is designed for two classes of users: application domain users and researchers in the field of multimedia. Application domain users can use both query-by-sketch and query-by-example to search collections; researcher users can test, tune, and compare engines, and they can design datasets to be used in batch mode. VISTO is an open source project developed in Java; it uses advanced features of the language, such as the Core Reflection API, to dynamically adapt the interface to the available engines. The system is being validated in different application domains, including the production of 2D animation, the Sign Writing Language, and the BLISS Language.

Keywords: CBIR, vector images, SVG format, user interface, query formulation.

1 Introduction

Nowadays there are several Content Based Image Retrieval (CBIR) tools available (see, e.g., [1, 5, 9, 19, 12, 17]); anyway almost all of them are dealing with raster images. However, starting from the system described [7], we developed an open source CBIR system for vector images (VISTO) in the SVG format; the SVG format [21], currently being developed by the W3C, is the standard for vector images on the Web. Vector images would be conveniently used on the Web, allowing faster download speeds, thanks to reduced size compared to raster images, as well as client-side scaling, thus avoiding new images to be sent. VISTO (Vector Image Search TOol) is a highly modular Java application, available from [23] under the GPL license. In this paper we detail the design issues of the VISTO interface; it is designed for two classes of users: *application domain users* and *researchers in the field of*

M.J. Smith, G. Salvendy (Eds.): Human Interface, Part I, HCII 2007, LNCS 4557, pp. 417–426, 2007.

multimedia users. Application domain users can use both query-by-sketch and query-by-example to search collections; researcher users can test, tune, and compare engines, and they can design datasets to be used in batch mode. In order to assist researchers in the multimedia field, the interface should provide easy support for the following task:

- *Tuning an engine*. The retrieval effectiveness of an engine can sensibly vary according to some parameters (*Precision* and *Recall*), whose optimal value can depend on the underlying document collection, see [8]. The interface should provide the researcher to ability plotting and comparing different results.
- *Definition of test-sets*. To evaluate an engine it is necessary to test it against users building test-sets, which include documents, queries and answers. The system should provide researchers a way to define, store and recall test-sets.
- *Recording testing sessions*. A test session is usually done against on or more standard test-sets. The interface should allow researchers to start and stop test sessions, providing him the statistics for each session, including plots of averages.

We would also like to point out that, thanks to the high modularity of the VISTO architecture, it is possible with few efforts to use the interface for a different image format: one should only provide an appropriate image rendered and similarity engines for the new format.

2 Related Work

The research area concerning CBIR systems for images described by the vector model is quite new. In fact, as pointed out in [9] *"to the best of authors' knowledge, there are no proposals (neither theoretical, nor applicative) that try to solve CBIR when images are represented in a vector model"*. A first system has been presented in [7], with its interface detailed in [6], but, since it is a proprietary system, neither the application nor the sources are available for research purposes; furthermore it has been designed for a proprietary format of vector images. VISTO is probably the first open source, freely available, to deal with CBIR of vector images in an open standard such as SVG.

However, we find appropriate to give a summary on the state of the art concerning existing CBIR systems for raster images. The goal of this section is to highlight the fact that, notwithstanding we stem with vector images, VISTO has many characteristics in common with the existing systems, while it improves some of their aspects concerning the user interface and the interaction with it. We have collected the majority of the systems with demos and studied their main components: the engine system and the user interface. In this paper we focus on aspects related to the user interface. For a complete survey on CBIR systems we refer to [22]. We defined the terms web-like and windows-like to categorize the *interface type*. While the web-like interface (see, e.g., Blobworld [5] and MARS [19]) strictly follows the web environment, then the design of a Windows™ interface (see, e.g., ImageFinder [3] and Netra [17]) reproduce the classical windows application. Most of the systems studied support a web-like interface. This is motivated by the fact that the great majority of the CBIR systems realized up to now is dedicated to the search of images in the Web. The kind of supported *query formulation* is mainly the query-by-image, where the user aims at extracting images of a given category (in other words the retrieval system should discriminate among categories). The query-by-example response to a query typically formulated by an example drawing; using this kind of query formulation the user aims at extracting similar images (partially identical image, or a partially identical object in the image). Examples of systems supporting both query-by-example and query-by-image are, among others, AltaVista Photofinder [1], Amore [18], and PictoSeek [10]. A *draw window* represents the possibility to draw an image as input of the search process. This characteristic is given only by two of the considered systems: DrewSearch [15], and C-bird [16]. Again the application domain motivates this aspect; in fact, since the search of photos is the main goal, the necessity to draw is not very important. Besides the use of raster images, the studied systems have other common characteristics, among others:

- the lack of a module for the visualization of graphics concerning the results of the performed operations;
- the lack of a software module dedicated to tune the system.

All these aspects highlight the necessity to have a complete interaction whit users; the interaction with users is the main aspect that we considered in the designing process of VISTO interface. In what follows we summarize the main characteristics of the system proposed in this paper and compare them with systems studied:

- contrarily to traditional CBIR, we decide dealing with vector images;
- differently from many of CBIR, VISTO gives the possibility to set several engines;
- differently from most of the systems in the literature, the interface of VISTO has a windows-like interface; this choice was motivated by the fact that at this stage, VISTO is a desktop application and not a web application;
- differently from many of CBIR, VISTO allows to perform both query-by-image and query-by-example;
- contrarily to traditional CBIR, the VISTO interface has a dedicated software for the vectorial draw; in fact for example in the production of 2D the presence of a draw area is so basic for the cartoonist in the searching process of scenes already existing in the database.

3 The VISTO Architecture

VISTO is an open source system developed in Java. It has a highly modular architecture, with the following main components:

- *the engine* - it can be seen as black boxes giving in input two SVG images output a similarity value; the engines are implemented as single Java classes;
- *the SVG parser* - it extracts features from the SVG images; note that some engine may have their SVG parser built-in;
- *the interface* - it is detailed in the next section;
- *the SVG rendered* - it is called by the interface to draw the SVG images;

- *the indexing modules* - they allow storing, computing and retrieving image features in order to speed up on-line answer time;
- *the statistics module* - it is used to record test sessions and to provide charts and reports.

At the moment, we use some open source libraries: Batik [4] that provides SVG utilities used both by the parser and the rendered, JFreeReport [13] that is used by the statistics module. As we previously mentioned, the use of Java Core Reflection API [12] allows engines to be seen as plug-in from the interface, and therefore they can be added and removed at runtime without re-compiling the whole application. This allows researchers, witch want studying the effectiveness of a particular engine, to implement only one Java class.

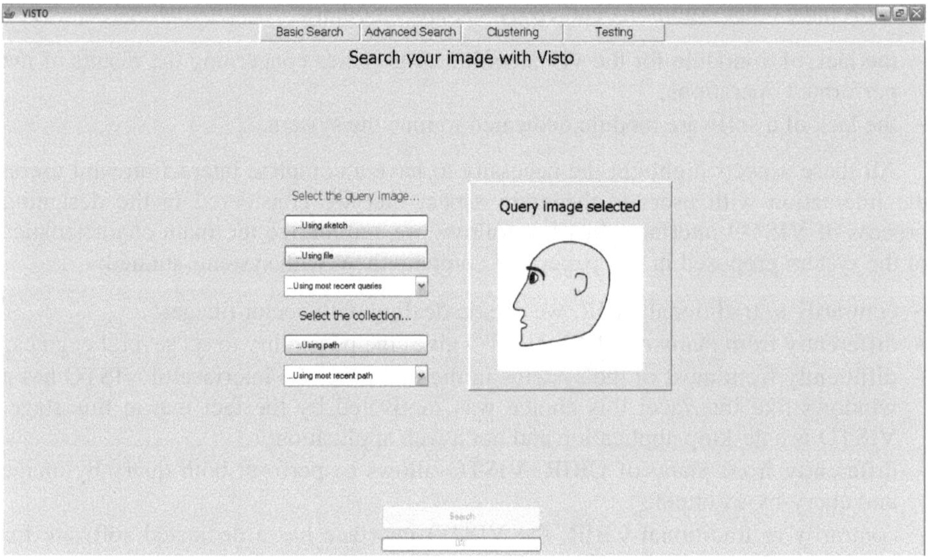

Fig. 1. VISTO – the Basic Search Tab - the query-selection window

Engines currently available in VISTO follow the ideas from [7]: an image is considered as an inertial system and the center of mass is used as selected point. The inertial system is obtained by discretizing the image, and associating material points with basic elements obtained by the discretization process. Vectors of descriptors are built from invariant moments of the inertial system, while similarity between any two images is computed as the similarity between the corresponding descriptor vectors. The moment sets supported by VISTO are Hu's moment set, Zernike's moment set, and Bamieh's moment set. The retrieval efficiency of moment invariants for retrieval by primitive features is documented in the literature for raster-based CBIR systems (see, e.g. [2, 11, 14, 24]); our preliminary tests with VISTO [8] confirm the

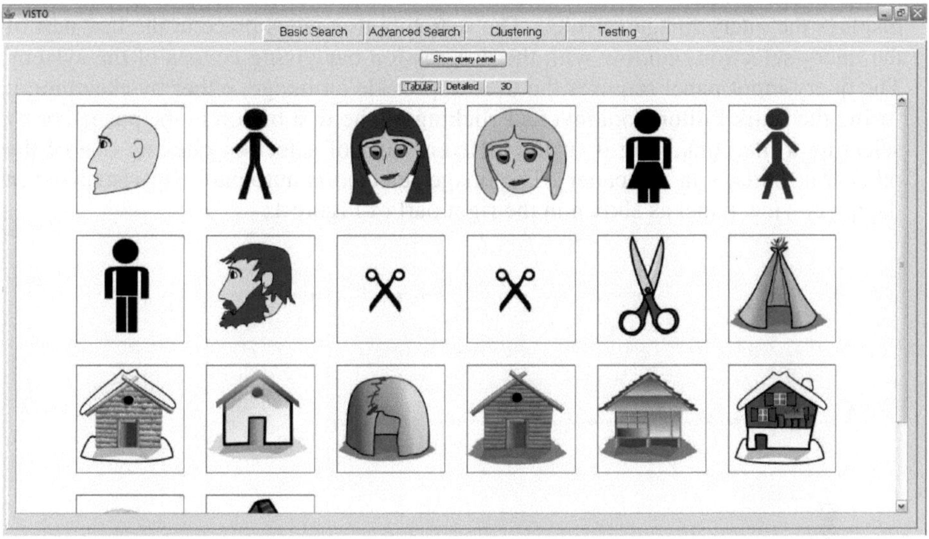

Fig. 2. VISTO – the Basic Search Tab - the query-result window – "Tabular" visualization

experimental results of [7] for vector images. We are also experimenting an engine based on the Latent Semantic Indexing (LSI) [20]; we are at an early stage but it seems that also in this field LSI can be an effective retrieval technique (see also [24]).

4 The VISTO Interface

The VISTO *interface type* is windows-like; it supports both kind of *query formulation*, query-by-example and query-by-image; moreover, it supports a *draw window* to draw images as query input. Result images may be selected as target images in a new search, in an incremental *querying process*.

As mentioned in the introduction, the VISTO interface was designed for two classes of users: it provides a Basic Mode for application domain users, and an Advanced Mode for researchers in the field of multimedia. The *"Basic Search Tab"* supports the Basic Mode, the *"Advanced Search Tab"*, the *"Testing Tab"*, and the *"Clustering Tab"* support the Advanced Mode. While the *"Basic Search Tab"* and the *"Advanced Search Tab"* are composed of two windows, the query-selection window and the query-result window, the *"Clustering Tab"* and the *"Testing Tab"* have just one window. Hereinafter we describe main characteristics of different tabs.

"Basic Search Tab" - Designed to handle user's input actions, the query-selection window is always displayed, differently from the query-result window that is invoked when the results of the query are ready to visualize. In particular:

- the query-selection window of the *"Basic Search Tab"* is composed of two panels (see Figure 1); the query-input panel, in the left part of this window, that accepts user input actions, and the query-view panel, in the right part of this window, that displays the query image selected. The search button takes place in the low part of the query-selection window with the progression bar giving echoes of the system. The query-input panel requires the user to provide an image, either by sketching it (using the Batik Editor Tool invoked clicking on the first button in the panel), or by selecting a file containing it (using different kind of selection clicking one of the other four buttons in the panel). The image selected is automatically visualized in the query-view panel as shown in the right part of Figure 1;

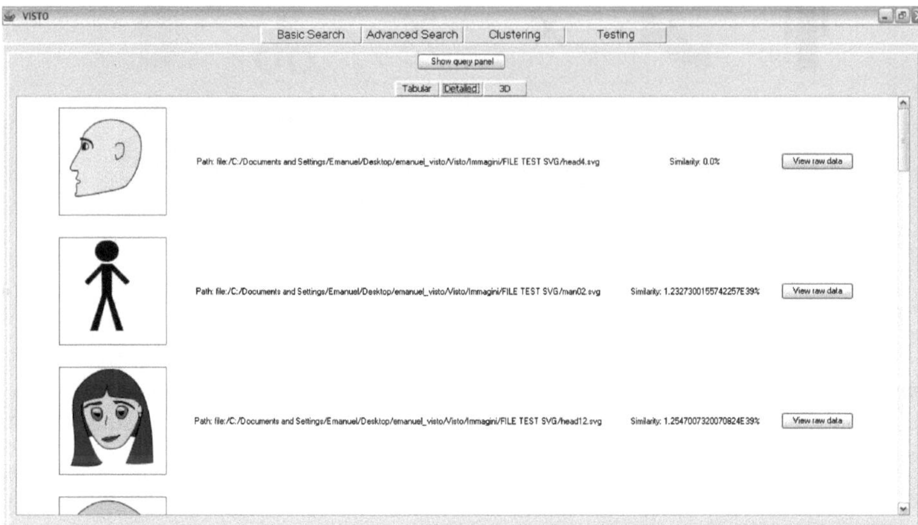

Fig. 3. VISTO – the Basic Search Tab - the query-result window – "Detailed" visualization

– the query-result window of the *"Basic Search Tab"*, shown in Figures 2 and 3 is composed of two panels; the result-view panel that displays the images retrieved ranked by similarity and the re-query-input panel that allows users to increment querying process. In order to better use the display space, the re-query-input panel is visualized only clicking on the *"Show query panel"* button displayed in the high part of the query-result window, and the result-view panel use tabs to support different type of visualization. When displayed, the re-query-input panel has the same query-input panel form of the query-selection window; the *"Tabular"* visualization (Figure 2), the *"Detailed"* visualization (Figure 3) and the *"3D"* visualization are the different supported types of result visualization. Users can also just point and click on individual image result to be provided as target image in a new search; in this case, the re-query-input panel is automatically showed in the high part of this window and the selected image is also automatically shown in the same panel;

"Advanced Search Tab" - Also in this case, the query-selection window is always displayed, and the query-result window is invoked when the results of the query are ready to visualize. In particular:

– the query-selection window of the *"Advanced Search Tab"* is composed of four panels (see Figure 4); in the high part of this window, there are the query-input panel that accepts user input actions, the query-view panel that displays the query image selected, and the search-engine panel that contains engines supported by VISTO; users can select an engine browsing different tabs of this panel. The fourth panel, containing the selected engine setting parameters, the available indexing and the folder indexing is automatically zoomed on when users click on the *"Hide other parameters"* button displayed in the center of this window. The search button and the progression bar take place in the low part of this window.

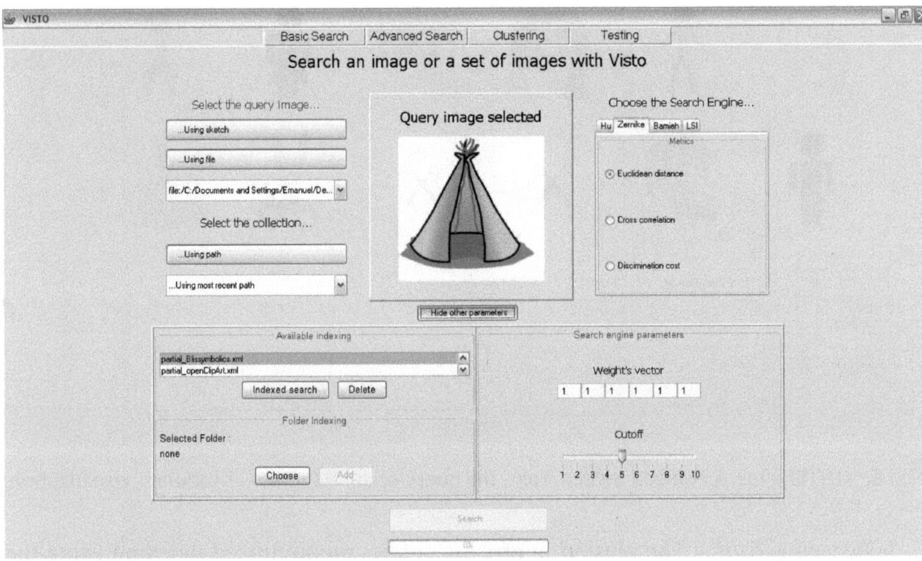

Fig. 4. VISTO – the Advanced Search Tab - the query-selection window

– the query-result window of the *"Advanced Search Tab"* shown in Figure 5, is composed of three panels; the re-query-input panel (visualized clicking on the *"Show query panel"* button) that allows users to increment querying process, the result-view panel that displays query results in tabular, detailed and 3D visualization type, and the analysis-panel (visualized clicking on the *"Hide statistics"* button) that provides an initial indication on the search effectiveness, by displaying in chart form, figures about Precision and Recall obtained with the selected engine. Objects, their positions and the philosophy followed in the design process are the same as for this query result window as for the query-result window of the *"Basic Search Tab"* (see Figures 2 and 5).

"Testing Tab" - This tab is dedicated to researches in the field of multimedia to favor an in depth analysis of the tool effectiveness, designing datasets, which include documents, queries and answers, to be used in batch mode. It is composed of two

panels as shown in Figure 6.: the test-input panel that supports users in selecting a new image file to add it in the test set and the test-view that shows the path of selected image's files added in the test set. The *"Add to query list"* button, in the low part allows the adding operation. It is worth noting that only after a testing session made using this tab, the *"Hide statistics"* button in the query-result window of the *"Advanced Search Tab"* is able to work!

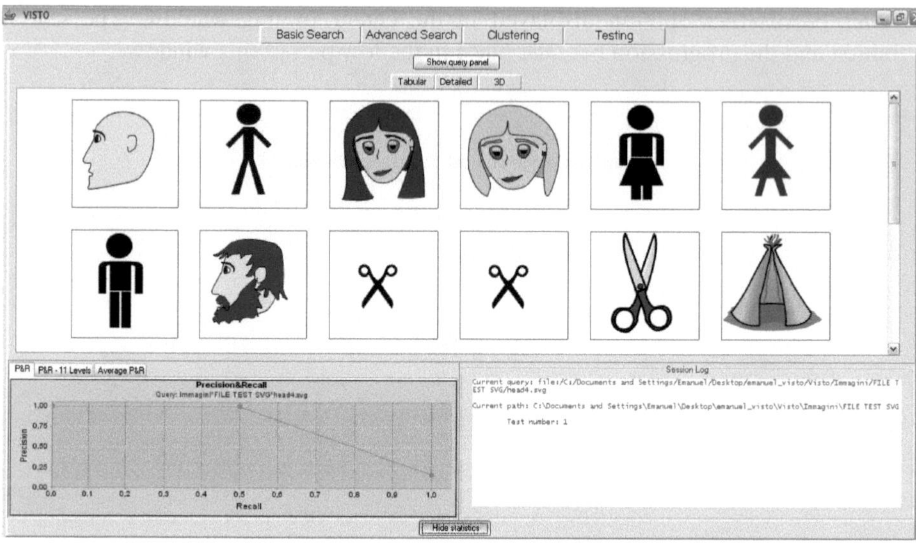

Fig. 5. VISTO – the Advanced Search Tab - the query-result window – "Tabular" visualization

"Clustering Tab" - The clustering process creates an optimized indexing using the cluster algorithm [25]. The objects of this tab are spatially organized according to the same philosophy used in the design process of the others tabs. It supports researchers in the clustering creation process used in the test sessions. Since in this paper we deal with interface of VISTO, the clustering process is not here described; see [25] for details.

Finally, it is worth noting that the:

- VISTO interface follows the Visual Design laws in order to guide users in performing tasks in a natural and usable way.
- As objects in basic mode dedicated tabs as objects in advanced mode dedicated tabs are positioned obtaining a correct balance between the full and the empty space.
- As objects in basic mode dedicated tabs as objects in advanced mode dedicated tabs are positioned in respecting to the central symmetry.
- The use of Hide/Show buttons optimizes the visualization of query results and the balance of display space.
- Different visualization types of query results allow users to a friendly browsing.

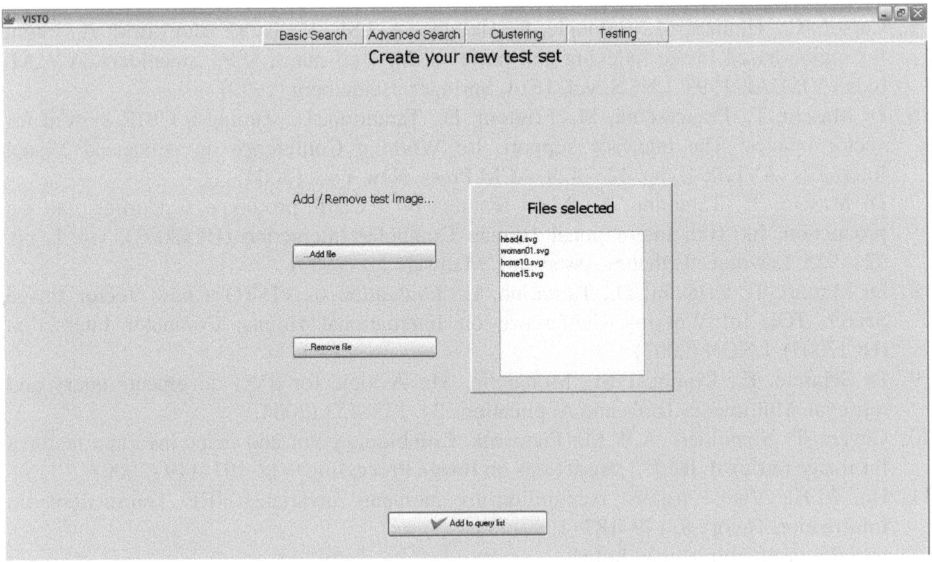

Fig. 6. VISTO – the Testing Tab

5 Conclusions

In this paper we presented VISTO, a CBIR system for images in the SVG format. VISTO is available under the GPL license and can be downloaded from [23]. The interface supports both application domain users and researchers in the multimedia users in a natural and friendly way. Most notably, engines can be added at run-time, and the interface provides support for parameter tuning of even the newly added engines; furthermore the interface allows the testing of several engines all together; reports, comparison graphs, Precision versus Recall plots representing standard measure of effectiveness are produced. In order to speed up the answer time of on-line queries, an indexing module, controlled by the *"Clustering Tab"* of the VISTO interface, is able to compute, store and recall features extraction from image collections. With few efforts VISTO can be used for other image formats different from SVG, and even for raster images; it only needs the appropriate image rendered and similarity engine(s).

References

1. http://www.altavista.com/sites/search/simage
2. Babu, M.M., Kankanhalli, M., Lee, W.F.: Shape measures for content based image retrieval: a comparison. Information Processing and Management 33(3), 319–337 (1997)
3. Baeza-Yates, R., Ribeiro-Neto, R.: Modern Information Retrieval. Addison Wesley, Reading (1999)
4. Batik SVG Toolkit, (2006), URL: http://xml.apache.org/batik/.

5. Carson, C., Thomas, M., Belongie, S., Hellerstein, J.M., Malik, J.: Blobworld: A system for region-based image indexing and retrieval. In: Huijsmans, D.P., Smeulders, A.W.M. (eds.) VISUAL 1999. LNCS, vol. 1614, Springer, Heidelberg (1999)
6. Di Mascio, T., Francesconi, M., Frigioni, D., Tarantino, L.: Tuning a CBIR system for vector images: The interface support. In: Working Conference on Advanced Visual Interfaces (AVI2004), pp. 425–428. ACM Press, New York (2004)
7. Di Mascio, T., Tarantino, L.: Main features of a CBIR prototype supporting cartoon production. In: 10th International Human Computer Interaction (HCI2003), vol. 1, pp. 921–925. Lawrence Erlbaum Associates, Mahwah, NJ (2003)
8. Di Mascio, T., Frigioni, D., Tarantino, L.: Evaluation of VISTO a new Vector Image Search TOl. In: Working Conference on International Human Computer Interaction (HCI2007), LNCN (2007)
9. Di Sciascio, E., Donini, F.M., Mongiello, M.: A logic for SVG documents query and retrieval. Multimedia Tools and Applications 24, 125–153 (2004)
10. Gevers, T., Smeulders, A.W.M.: Pictoseek: Combining color and shape invariant features for image retrieval. IEEE Transactions on Image Processing 9(1), 102–119 (2000)
11. Hu, M.K.: Visual pattern recognition by moments invariants. IRE Transactions on Information Theory 8, 179–187 (1997)
12. http://attrasoft.com/abm34.html
13. JFreeChart, (2006), URL http://www.jfree.org/jfreechart/
14. Kapur, D., Lakshman, Y.N., Saxena, T.: Computing invariants using elimination methods. In: IEEE International Conference on Computer Vision (1995)
15. Lew, M., Lempinen, K., Huijmans, N.: Webcrawling using sketches. In: Proceedings of the 2nd International Conference on Visual Information Systems, pp. 77–84 (1997)
16. Li, Z.N., Zaiane, O.R., Yan, B.: C-bird: Content-based image retrieval from image repositories using chromaticity and recognition kernel. Technical report, School of Computing Science, Simon Fraser University (February 1998)
17. Ma, W.-Y., Manjunath, B.S.: Netra: A toolbox for navigating large image databases. Multimedia Systems 7(3), 184–198 (1999)
18. Mukherjea, S., Hirata, K., Hara, Y.: Amore: A world wide web image retrieval engine. The. WWW Journal 2(3), 115–132 (1999)
19. Ortega, M., Rui, Y., Chakrabarti, K., Mehrotra, S., Huang, T.S.: Supporting similarity queries in MARS. In: Proceedings of 5th ACM International Multimedia Conference, pp. 403–413 (1997)
20. Papadimitriou, C., Raghavan, P., Tamaki, H., Vempala, S.: Latent semantic indexing: A probabilistic analysis. J. Comput. Systems Sciences (61), 217–235 (2000)
21. Scalable Vector Graphics (SVG), W3 Consortium. http://www.w3.org/Graphics/SVG
22. Veltkamp, R.C., Tanase, M.: A survey of CBIR systems. In: Content-based image and video retrieval, pp. 47–101. Kluwer Academic Publishers, Dordrecht (2002)
23. VISTO: a Vector Image Search TOol (2006), URL: http://www.linus.ing.univaq.it/ visto
24. Yang, L., Algregtsen, F.: Fast computation of invariant geometric moments: a new method giving correct results. In: IEEE International Conference on Pattern Recognition, pp. 201–204 (1994)
25. http://www.elet.polimi.it/upload/matteucc/Clustering/tutorial_html/

Using Autobiographic Information to Retrieve Real and Electronic Documents

Daniel Gonçalves, Tiago Guerreiro, Renata Marin, and Joaquim A. Jorge

Dep. Eng[a]. Informática, IST
Av. Rovisco Pais, 1000 Lisboa, Portugal
{daniel.goncalves,tjvg,jaj}@inesc-id.pt, renata.marin@gmail.com

Abstract. Current systems force users to store their documents in hierarchic filesystems. However, users remember their documents not in terms of ad-hoc categories, but of their contents, why they were written, etc. Describing documents using such autobiographic information would be a simpler and more effective way for users to retrieve them. Furthermore, we found that users mention printed versions of documents when retrieving them. However, current systems don't regard the printed and electronic versions of a document as facets of the same entity. To solve these problems we developed an interface that allows users to browse their autobiographic information, gathered by a special-purpose monitoring subsystem, to retrieve their documents. We bridged the gap between the real and electronic worlds by maintaining the association between paper and electronic documents resorting to RFID technology. Tests show our interface allows users to efficiently retrieve their documents.

Keywords: personal information management, document retrieval, RFID.

1 Introduction

It is easy for users that have tried to find an electronic document stored somewhere in their computers to remember how ineffective and frustrating traditional retrieval mechanisms can be. All files containing documents are stored somewhere in the hierarchic file system and have a file name. Unfortunately, apart from that information, there is little else that can be used to find the correct document. The file system's hierarchy could help but it is often hard for users to classify all their documents into categories in that hierarchy. A document might seem to belong to more than a folder, or none of the existing ones might apply. These and other problems make it difficult to properly classify documents. Indeed, Thomas Malone found that many users would rather not classify their documents at all, preferring to store them in roughly unsorted piles [5]. But, as we have already discussed, a good classification into the hierarchy is crucial, as it is one of the few cues users have available at a later time to help them find their documents.

Compounding on the classification problem are the large numbers of documents that users must deal with nowadays, on a daily basis. Not only have computers left schools, labs and the enterprise to become commonplace on most citizen's homes, but also each user now performs a large and growing number of tasks with them. From

M.J. Smith, G. Salvendy (Eds.): Human Interface, Part I, HCII 2007, LNCS 4557, pp. 427–436, 2007.

writing letters to paying taxes, computers are increasingly common. For each task one or more documents are often produced. Thus, and considering that most users are now non computer-savvy, better and easier ways to organize and retrieve personal documents must be developed.

While research on this area has been taking place for decades, it concerns itself mostly with the retrieval of text-based documents. Keyword search is common, even in recent systems such as Google Desktop. This reflects the reality of some years ago. Today, with the lowering costs of storage and bandwidth, multimedia formats are gaining more and more importance, as attested by popularity of web sites as Flickr and YouTube. Many users now store hundreds or thousands of digital photos, music files, and video clips. These should be amenable to retrieval in a way similar to that of text-based documents.

Finally, many works in the field of Information Retrieval try to help users to retrieve any kind of text-based documents, from news articles to scientific papers in digital libraries. Such works don't give enough relevance to *personal* documents. The retrieval of a user's own documents, unlike the general case of document searching on the web or libraries, has a different but very specific nature. Users looking for personal documents are not tying to find any documents on a given subject, or some document they have some reference to. Instead, they are trying to find *their own documents*, with which they have intimately interacted in the past, having written or read them for a reason, at some point in time. Some particular tasks might have been performed to complete that document, or they had to send it to a friend or colleague for revision. In short, users remember about their documents a wealth of autobiographic information they cannot associate to documents in the general case. Remembering this autobiographic information is easier than recalling arbitrary values imposed upon them by the computer, such as the location in the filesystem. Thus, using this information would undoubtedly be more natural and easier. Furthermore, users are able to recall a large amount of such information and consequently can provide the retrieval system with a large number of hints of where to find a document they seek [3].

It is this ease of using autobiographic information to store and retrieve documents that makes users sometimes resort to their email tools to perform those tasks, even if they are not directly supported by the tools [8]. This is due to the fact that email messages are perhaps the type of document the users most frequently deal with that have autobiographic information associated to them. The message's sender or receiver, its date, subject, etc. can be used to finding them more easily than it would be otherwise possible.

1.1 The Gap Between Real and Electronic Documents

One particular aspect of autobiographic information has been for the most part neglected in previous solutions: real documents. It often occurs that users trying to find a specific electronic document remember they printed it to give to some one or to review it. They might even have a printed copy of the document with them. However, this information is for the most part useless. The reason for this is that once a document has been printed, nothing connects it to its electronic version. The computer is incapable of regarding the two different versions of the document, the printed and the electronic one, as different facets of a same entity. This would make possible two different scenarios: retrieving a paper document from its electronic version and retrieving an electronic document from its print version. A way to bridge the gap between the real and electronic worlds was, thus, sorely needed.

Some previous works try to accomplish this in different ways. Video-Based Document Tracking [4] tries to solve the physical/virtual document association problem with the help of an overhead camera, used to track physical documents on a desktop and link them to the corresponding virtual documents. The system detects changes in a document stack, and uses this information to establish the desk's contents. The movement of documents is tracked with a video camera. The video is analyzed with computer vision techniques to connect the document with his virtual copy on the disk. It relies on the visual pattern of the first page of a document to identify it, which is clearly not enough when several documents with similar first pages are present at the same time. Also this technique can only recognize documents if their first page is facing upwards and not occluded by some other document or object. As such, this and other similar approaches (such as using barcodes) are ill suited for document retrieval, not only because of their inaccuracy, but also because they don't support the users' work practices: users don't store their documents side-by-side, face up, placing them in piles instead.

Other solutions, aiming at a more natural and accurate linking between virtual and real documents are based on RFID (Radio Frequency Identification) technology. Historically, the roots of the RFID technology can be traced back to the World War II. It was used by the British to distinguish their aircrafts from the enemy's. RFID is a generic term for technologies that use waves to automatically identify objects. Each tag is identified with a unique serial number. A microchip is attached to an antenna that enables the RFID to transmit the identification to a reader. This reader converts the radio waves reflected back from the RFID tag into digital information that can be passed on to computers [6]. Recently, the growing popularity of RFID technology has made readers and tags less expensive, to the point where tagging each document is feasible.

One of the earliest works that tried to bridge physical and virtual worlds through the use of RFID technology was Want et al. [7]. It tries to connect the physical objects with its virtual representation using various types of tags.

More recently, the association between virtual representations and real objects has been extended to encompass personal items that accompany the users on their everyday tasks [2]. This work tries to help people not to forget important objects. The different relevant objects are tagged. RFID readers are present at each of the locations where the user usually spends some time at. Users are given a mini personal server that communicates with their watches. Whenever a user passes close to a reader without an important object (that he could have forgotten) the watch reminds the user of if. This work shows how RFID technology can be employed not only to help users to initiate the retrieval of objects they remember, but also how the computer itself can identify a user's needs and act proactively to provide the required objects.

Abu Safiya et al. [1] developed a project that more directly addresses the information retrieval problem. The concept of Document Database (where all electronic documents are indexed) is extended to allow the representation of printed documents and their physical location. With this, a company can manage all its documents, becoming possible to know at all times where a specific document is or which documents are in a specific room. The RFID readers must be located in strategic positions, places where documents usually accumulate. While interesting, this work is a large scale project, trying to deal with an organization's entire document collection. Thus, only important documents are tagged identified. It is mostly a localization-based project and the retrieved information is quite poor. Also, the retrieval process is unidirectional: a user can't retrieve a virtual document from its real replica. A proper way to handle *personal* documents was still to be developed.

1.2 Our Approach

To solve the aforementioned problems, we developed a system that automatically collects a wide range of autobiographical information. Provisions were made to allow it to semi-automatically associate printed and electronic documents, with the help of RFID tags. This information is stored interrelated with the remaining autobiographic data, forming a coherent whole.

A custom-designed interface can be used to browse that information in search of personal documents. User tests show this retrieval to be simple and effective.

In the following section, we describe the monitoring system, responsible for collecting all relevant autobiographic information. Next, we will explain the browsing interface, to which a discussion of the user tests performed using it will ensue. Finally, we will conclude pointing to relevant directions for future work.

2 Collecting the Autobiographic Information

To be able to build an interface that allows users to describe the autobiographic information they remember about their documents to retrieve them, it is necessary to somehow collect all that information beforehand so that the correct documents can be identified. There is a wide range of potentially relevant information [3]. As such, the users won't be willing to enter it all by hand, either using a special purpose interface or annotating their documents. Thus, it is necessary to automatically and continuously collect it, in an effective but unobtrusive way.

To that end, we created a monitoring system that monitors the users' actions at the computer and stores all relevant information. A problem we faced when designing this system is the large degree of variability that can be found in terms of software applications, configurations and services employed by users. It soon became apparent that a single monolithic system would never satisfy the needs of all users. We decided to build a plugin-based system instead. Each plug-in is responsible for collecting data from a different source and can be independently configured to match each users' particular details. Also, it will be possible to enable only those plugins that make sense in any given system.

The overall architecture of the system can be found in Fig. 1. There, we can see that the different monitoring plugins are managed by the monitoring system, configurable with the help of its graphic user interface. This monitoring system stores all data in a special-purpose knowledge-base. The data therein can then be used by the retrieval interface to allow users to quickly and effectively find their documents.

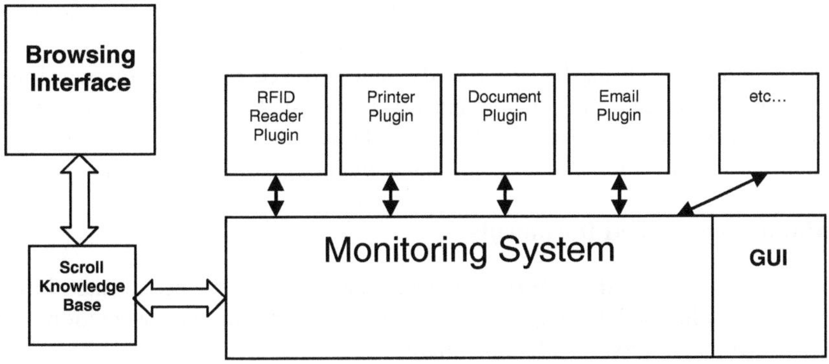

Fig. 1. Overall architecture of the monitoring system

Overall, we created plugins to collect data from the following sources:

- All of the users' documents stored in their hard drives
- All changes performed on those documents, in real time
- All emails already present in the users' email clients (currently, Mozilla Thunderbird, Microsoft Outlook and Microsoft Outlook Express are supported)
- All email messages sent and received by the user
- All programs ran by the users
- The users' agenda (Microsoft Outlook and Palm Desktop)
- The web pages visited by the users
- All documents printed by the users
- All paper documents located by RFID readers

The challenges were many, concerning mainly poorly documented formats, improperly followed standards and the wide range of encodings, formats and software versions that can be found. In the end, we were able to obtain adequate behavior for all plugins.

Each plugin collects as much information from the relevant source as possible. For instance, the email plugins record not only an email's subject, but also when it was received, to whom it was sent, whether it was a reply of another message, any documents attached to it, etc. All meta-information that can be gleaned from the documents themselves is extracted. For instance, all data in an mp3 file's ID3 tags is collected. Also, text-based documents are analyzed, tokenized, stemmed, and relevant keywords selected using the tf-idf algorithm.

All this data is stored in the *Scroll Knowledge Base*. This is a special-purpose knowledge base developed especially to store autobiographic data. It is based on the RDF standard. It abstracts from it by providing facilities to handle high-level concepts, such as classes and their properties, in easy ways. An RDF Schema dubbed QUILL was created to provide enough expressiveness to allow the representation of all relevant information pertaining a given document (author, dates, title, keywords, etc.), and all events that influence it (being sent by email as an attachment, for instance).

Although each plugin works separately from the others, an effort was made to ensure all the information stored in the knowledge base is tightly coherent. For instance, if a document that had previously been indexed by the Document Plugin is found to be sent as an email attachment by the Email Plugin, instead of creating a duplicated entry for that document in the knowledge base the existing one is annotated with the new information. The same happens for other similar cases.

2.1 Monitoring Printed Documents

Of the different plugins that were developed, one that merits special attention is the printer monitor plugin. This plugin is responsible for creating the association between the digital and paper versions of a document.

To accomplish this, we rely on an RFID infrastructure composed by a fixed RFID reader (with a 30-40 cm range) capable of reading multiple tags simultaneously, connected to a PC. The relatively short range is important, as we want to know precisely where our documents are (a desk or a shelf, for instance). Our RFID tags are passive (with no power source), as we only need to receive the tag ID. Each printed document will have its own tag, which identifies it. The printing plugin associates the tag's ID with the corresponding virtual document.

Fig. 2. Tagged documents on the user's desk

Whenever a document is printed, the operating system generates an event. Our plugin intercepts that event to perform the association. From the event, we are able to get the name of the file being printed. The user is supposed to fetch the printed document from the printer and stick an adhesive RFID tag to it. Then, the document can simply be placed in the user's desk, where the RFID reader is placed. The plugin will detect the new tag and automatically associate it to the electronic document that was printed (Fig. 2). From that moment onwards, both versions will be treated by the system as the same document.

The only problem that sometimes hinders this process is that we can get the file name from the print job event, but not the entire path. Thus, it is necessary to inspect the Scroll knowledge base, where all the users' documents are indexed, to identify all files with that name. If there is only one such file, the association is automatic. If the query returns more than one result, several different files with the same file-name, the users are asked to choose between them. All different possibilities will be displayed in a list, sorted by modification date, since it is likely that a file being printed has been recently modified. However, a study we performed shows that over 81% of names identify a single document, with nearly 96% of documents sharing their name with, at the most, two others. Thus, in the general case, the need to choose the correct document is not a problem.

3 Browsing Interface

To allow the users to take advantage of all the autobiographic information stored in the knowledge base, a browsing application was created. This application, depicted in Fig. 3, allows the users to see and inter-relate all different information factoids, regardless of their source. It is based on the concept of *views*. Just as there were different plugins in the monitoring system to monitor different data sources, there are different views in the interface to display information about different kinds of entity in the knowledge base. These views are not fixed or hard-coded. Rather, they are automatically built from the data in the knowledge base, allowing the system to be quickly and easily extended, if necessary. At this moment, the interface supports the following views:

- *Document View*, displaying all information regarding a document;
- *Email View,* that shows all pertinent data about an email message;
- *Person View,* in which everything about a person is presented;
- *Date View,* where everything that took place around a certain moment in time can be seen.

Each view displays all information to which an object of the appropriate type is connected. For instance, the Document View shows all properties of a document, such as its title, authors, creation and modification date, keywords, etc. All views also show other elements of other views with which the displayed element is connected (all emails a certain document was attached to, for instance).

It is possible to navigate between the views by double-clicking on a specific element in the interface. For instance, clicking on the text field with a document's creation date will transfer the user to the View Date for the appropriate time period. It is thus possible to navigate throughout the different information in the knowledge base in search of the appropriate document. Furthermore, it is possible to enter constraints in the interface. For instance, by entering the "doc" string in the field corresponding to the document's extension, in the document view, and then pressing the "Add" button, we will constrain the system to display information only about "doc" files. At all times, every document that satisfies the active constraints is displayed in the list at the right side of the interface. The user can click on each of those to see their information in the Document View. Finally, a thumbnail of the document being displayed in the Document View is shown in the top right corner of the interface. In this way, we take advantage of the users' associative memories and help them recognize the documents they seek.

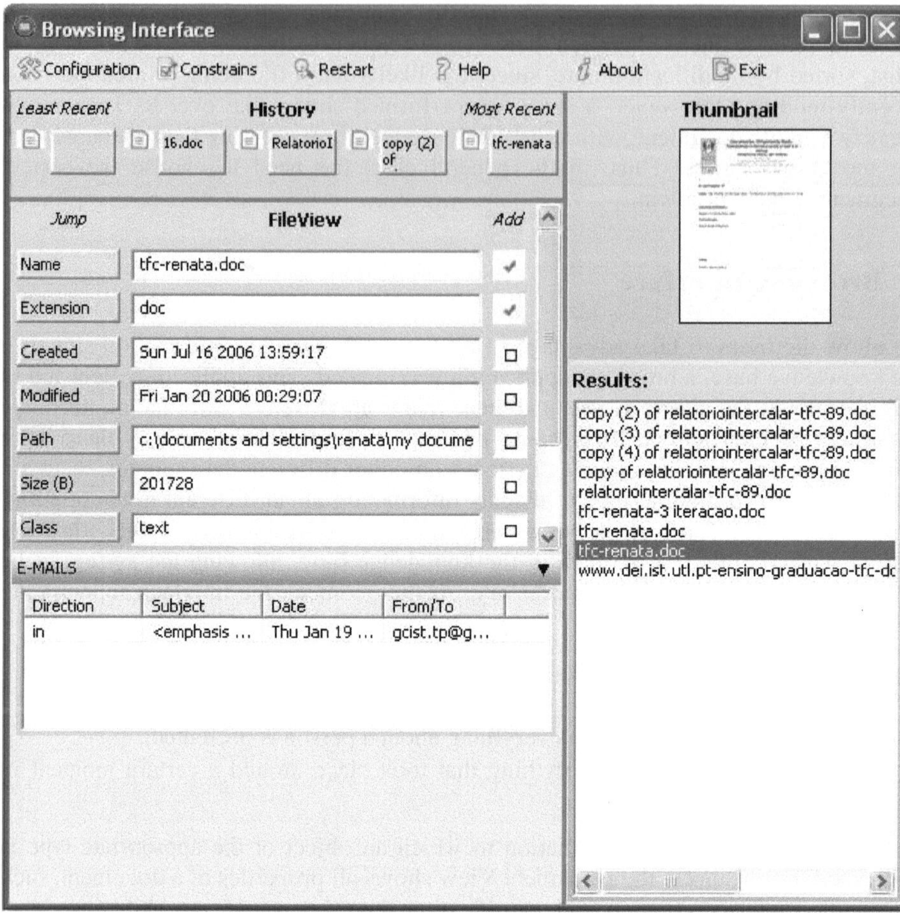

Fig. 3. Browsing Interface

4 Evaluation

After explaining to the users how the system works, they were asked to print out two documents and associate their printed and electronic versions. Both the time it took to do so and the steps they had to were recorded. Six users participated in this part of the study. We found that it is fairly easy for users to establish the connection between both versions of a document. It took them, on average, 3.7 steps to do so. There was statistically significant difference between the documents regarding the number of steps. However, performing those steps was faster for the second document (49.6s vs. 97.3s for the first one), showing that there is a learning effect and that proficient users might get even lower times.

The times it took to retrieve an electronic document from its printed counterpart and vice-versa, using the browsing interface, were also measured. Finding a real document from its electronic counterpart took, on average, 36.3 seconds. This time includes finding the electronic document using the browsing interface and shows this process to be efficient. Retrieving an electronic document from its printed counterpart was faster, taking only 23s on average.

The browsing interface was also evaluated. Twenty users were asked to retrieve two documents: a recent one (created up to two weeks ago), an old one (created over six months ago). The order in which the documents were considered varied from user to user, to prevent unduly biases to the results. The users' own documents (and other autobiographic information sources such as emails) were indexed and used in the tests, as only for those documents would the users recall relevant autobiographic information.

The time to retrieve the recent document was, on average, 77s, while for the old document it was of 92.6s. This was to be expected, as users had a worse time remembering relevant information about older documents. Nevertheless, all documents were found which in itself is encouraging. More interestingly, 89% of recent documents and 72% of old ones were found in less than two minutes.

The number of steps it took to find a document was also greater for older documents than for more recent ones (6.2 and 5.5, respectively). Again, this reflects the need to perform a more extensive search when looking for older documents.

While the times we presented might seem a little high at first, if we consider that the retrieval was performed by novice users, and that the times include all stages of retrieval, including opening the target document to verify it is the correct one, it becomes clear that our approach produces results which are better than many traditional approaches.

5 Conclusions

While users remember a wealth of autobiographic information about their personal documents, current document retrieval tools don't make it possible to use that information. Furthermore, users often associate to their electronic documents printed versions of those documents. Sometimes they can even have the printed document in front of them and still be no closer to find its electronic counterpart.

To solve these problems, we developed a system that continuously monitors the user's actions at the computer. All documents created and modified are indexed. A plugin-based architecture allows the system to collect data from different sources in an integrated fashion, from the emails sent and received by the user to the web pages visited. This allows a knowledge base of the autobiographic information to be constructed without the explicit need for user intervention, which few would be willing to give. In particular, the monitoring system is able to bridge the gap between real and electronic documents using RFID technology. To navigate the information in the knowledge base, a special-purpose browsing tool was created. It allows users to explore all relevant information until a document is found.

User tests show that both the association between real and electronic documents and the retrieval of a document of any of those kinds can be done quickly and with little effort.

In the future, it would be interesting to extend this approach to other real-world entities besides documents. Also, while we have brought together electronic and real documents when the former are printed, scanned documents should also be handled. Finally, as technology prices continue to drop, an extended usage scenario with more readers should be considered.

Acknowledgments. The authors would like to thank all users that participated in the studies described in this paper. This work was funded by Project BIRD, FCT POSC/EIA/59022/2004.

References

1. AbuSafiya, M., Mazumdar, S.: Accommodating Paper in Document Databases, DocEng'04
2. Borriello, G., Brunette, B., Hall, M., Hartung, C., Tangney, C.: Reminding about Tagged Objects using Passive RFIDs. In: Davies, N., Mynatt, E.D., Siio, I. (eds.) UbiComp 2004. LNCS, vol. 3205, Springer, Heidelberg (2004)
3. Gonçalves, D., Jorge, J.: Tell Me a Story: Issues on the Design of Document Retrieval Systems. In: Bastide, R., Palanque, P., Roth, J. (eds.) Engineering Human Computer Interaction and Interactive Systems. LNCS, vol. 3425, Springer, Heidelberg (2005)
4. Kim, J., Seitz, S.M., Agrawala, M.: Video-Based Document Tracking: Unifying Your Physical and Electronic Desktops, UIST '04
5. Malone, T. (ed.): How do People Organize their Desks? Implications for the Design of Office Information Systems. ACM Transactions on Office Information Systems, vol. 1(1), pp. 99–112. ACM Press, New York (1983)
6. RFID Journal – The History of RFID Technology http://www.rfidjournal.com/article/articleview/1338/1/129
7. Want, R., Fishkin, K., Gujar, A., Harrison, B.: Bridging Physical and Virtual Worlds with Electronic Tags. ACM Conference on Human Factors in Computing Systems, Pittsburgh, PA, May 1999, pp. 370–377 (1999)
8. Whittaker, S., Sidner, C.: Email over-load exploring personal information management of email. In: Conference proceedings on Human factors in computing systems, pp. 276–283. ACM Press, New York (1996)

A Video Digest and Delivery System: "ChocoParaTV"

Kota Hidaka[1], Naoya Miyashita[1], Masaru Fujikawa[1], Masahiro Yuguchi[2], Takashi Satou[1], and Katsuhiko Ogawa[1]

[1] NTT Cyber Solutions Laboratories , NTT Corporation,
1-1 Hikarinooka, Yokosuka-Shi, Kanagawa, Japan
{hidaka.kota,miyashita.naoya,fujikawa.masaru,
satou.takashi,ogawa.katsuhiko}@lab.ntt.co.jp
[2] Enterprise Business Division, NTT Communications Corporation,
2-5-1 Kouraku, Bunkyo-Ku, Tokyo, Japan
masahiro.yuguchi@ntt.com

Abstract. This paper proposes a new video digest and delivery system, "ChocoParaTV", for applications such as blogs that include video content, and CGM sites. The video digests are generated based on automatic extraction of emphasized speech portions. To deliver the digest information and the digest to audiences, we also propose a new video digest description: CH-RSS; it can generate digested contents dynamically to match the user's requested digest time. With the ChocoParaTV browser, the users first view the digest and then visit the original web site if they want to watch the original content. The effectiveness of the proposed system was confirmed by a field experiment; about 40% of the users accessed the original web sites through the ChocoParaTV browser digests. This indicates that our method is very effective in making video content attractive to the audience.

Keywords: video, digest, RSS, blog, CGM, speech parameter, quick view.

1 Introduction

Advances in broadband network technologies offer the opportunity to distribute multimedia contents. Blogs have become popular for disseminating information and many more sites now include video content. CGM (Consumer Generated Media) also offers integrated multimedia communication. The question we have to ask here is whether an efficient method can be provided to enable users to grasp video content given that the amount of sites that include video will increase rapidly. This paper proposes a new video digest and delivery system, called "ChocoParaTV" ('Choco,' in Japanese, means 'a little' and 'Para' is an imitative word for turning over pages), for applications such as blogs that include video content, and CGM sites.

2 Objectives and Related Work

Figure 1 overviews ChocoParaTV. The ChocoParaTV server starts collecting contents by using RSS, already registered, from blogs and CGM sites, when new videos are

M.J. Smith, G. Salvendy (Eds.): Human Interface, Part I, HCII 2007, LNCS 4557, pp. 437–445, 2007.

submitted to the sites. After generating a digest, the server delivers the digest information and the content to audiences. They use the ChocoParaTV browser to review the digest video and can then simply access the original site if desired.

Our technical objectives are as follows.

(1) Create a video digest automatically.
(2) Deliver the digest information.
(3) Browse videos quickly and passively.

Various studies [1-3] have demonstrated a number of ways by which users can confirm contents effectively such as news, dramas, clear speech contents, and natural speech contents on a definite theme. However, little is known about their feasibility when applied to contents such as blogs and CGM sites. In this paper, we apply a multimedia content digest technique based on the automatic extraction of emphasized speech [4-6] to ChocoParaTV in order to realize objective (1).

For objective (2), we employ RSS. RSS stands for Rich Site Summary or Really Simple Syndication; it describes the title, date and summary of articles on a Web site. By using an "RSS reader" [7], users can get newly submitted items of Web sites. However, the basic RSS reader is inappropriate for viewing video contents, because RSS does not support an effective way of rapidly understanding audio visual data. In this paper, we propose a new video digest description that can generate digested contents dynamically to match the user's requested digest time. We define the description as "Content Highlight RSS: CH-RSS".

To achieve objective (3), we propose the "ChocoParaTV browser" which allows digests to be viewed in the shortest possible time while still allowing good comprehension. By using the ChocoParaTV browser, an audience can get new video content passively because the ChocoParaTV server automatically delivers video digests.

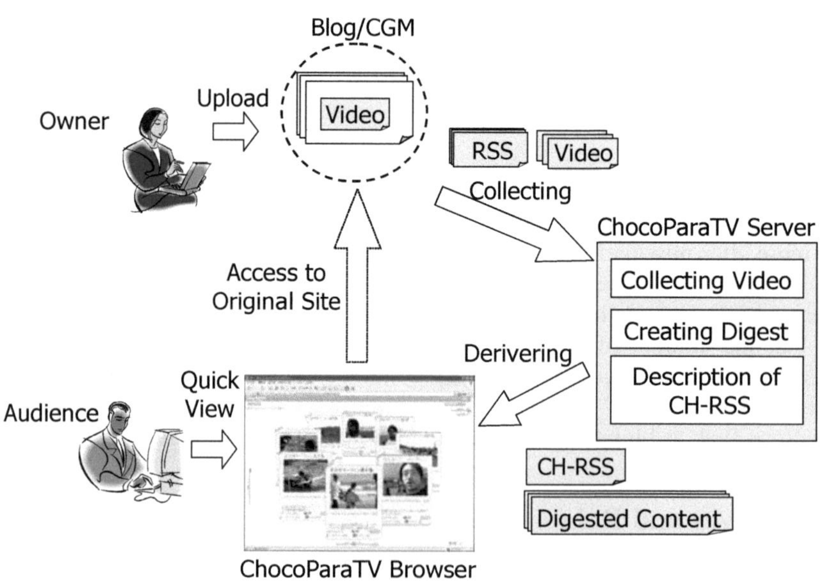

Fig. 1. The overview of ChocoParaTV

The following section describes the proposed digest method. Section 4 and Section 5 describe CH-RSS and ChocoParaTV browser, respectively. The following section gives the results of a field experiment. The final section gives our conclusion and future work.

3 Digest Method

In this section, we briefly summarize the proposed digest method since the method is fully described in our previous work [4-6]. Figure 2 shows the flow of the proposed video digest process. Given a video content, the method extracts speech information and calculates prosodic parameters such as pitch, power, and speaking rate. From these prosodic parameters, the "degree of emphasis" is analyzed by referring to an emphasis probability table, which is prepared in advance via a learning procedure. Since the durations of emphasized speech segments are quite short, the extracted portions do not exhibit semantic continuity. We can define the portions of speech that can be understood semantically as "prosodic phrases". Prosodic phrases with the higher degree of emphasis are assigned the higher priority. When a user requests a digest of a video, the method selects as many prosodic phases as will fit within the user specified digest time.

Figure 3 shows a schematic of our digest method. (a) Speech signal is separated from video content and analyzed to estimate (b) the degree of emphasis, and to extract (c) prosodic phrases. All prosodic phrases are sorted in descending order of their degrees of emphasis. The degree of emphasis of a prosodic phrase is defined by the highest value in the analysis windows. As shown in the figure, the degrees of emphasis of prosodic phrases I, II, III, and IV are P, Q, R, and S respectively. (d) shows the extraction of two prosodic phrases, i.e., I and III , and (e) shows 'digest time'. The digest is generated by concatenating the video segments corresponding to the extracted prosodic phrase.

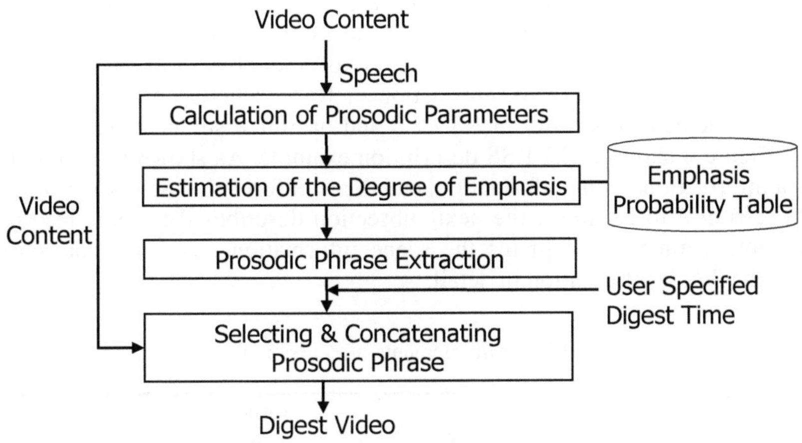

Fig. 2. Flow of the proposed video digesting process

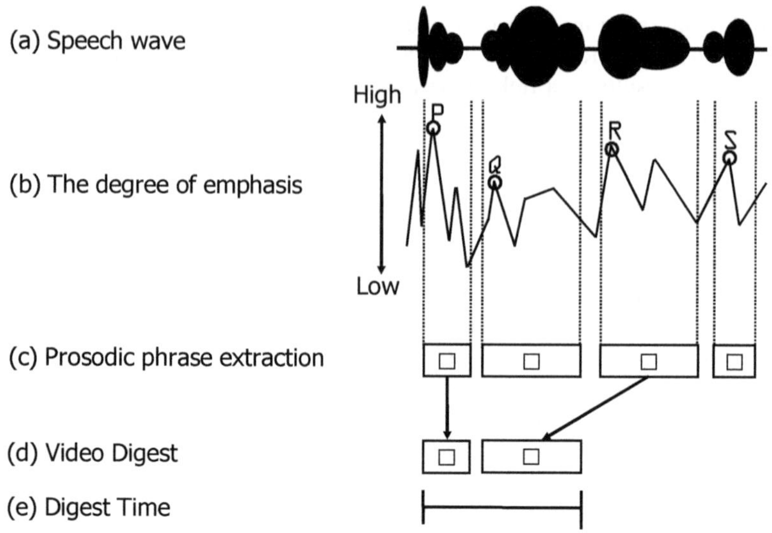

Fig. 3. The schematic diagram of the used digest method

4 Proposed CH-RSS Description Method

4.1 Outline of CH-RSS

CH-RSS has three layers: (1) media information, (2) scene information and (3) digest information. The objectives of these layers are to search by keywords, to seek and play scenes, and to generate digests, respectively.

In creating the CH-RSS description method, we need to address the problem of the RSS variants popular among blog sites. There are a few versions of RSS: RSS0.9x, RSS1.0 and RSS2.0. We use RSS2.0 because it is easy to extend new tags and elements and is widely used in Podcasting and other multimedia delivery services. Since we focus attention on the satisfaction of the objectives mentioned above without decreasing the effective information of RSS, <chrss:item> is defined as an optional sub-element of <item> in RSS 2.0 and has three sub-elements as shown in Table 1. Figure 4 shows a CH-RSS description example. As shown in Figure 4, parts of media information, scene information, and digest information are indicated.

With reference to Figure 4, the next subsection describes the media information. The subsection after that explains the scene information, and the final subsection describes the digest information in detail.

Table 1. Sub-elements of <chrss:item>

Sub-element	Layer	Objective
<chrss:media>	Media information	Search and play
<chrss:playlist>	Scene information	Seeking and playing position
<chrss:summary>	Digest information	video digest

```
<?xml version="1.0" encoding="utf-8"?>
<rss version="2.0" xmlns:chrss="http://xxx/chrss/">
<channel>
 <title>blog</title>
 <item>
  <title>Trip</title>
  <chrss:item src="http://xxx.com/xxx.avi">
```

Media Information

```
<chrss:media>
 <chrss:info what="abc" who="def" where="ghi"/>
 <chrss:video format="avi" compress="mpeg4"
           bitrate="2M" width="320" height="240"
           duration="180.0"/>
 <chrss:audio format="wav" compress="pcm"
           bits="16" sampling="11025" channel="2"/>
</chrss:media>
```

Scene Information

```
<chrss:playlist>
 <chrss:scene id="s0" from="0.000" to="0.800"/>
 <chrss:scene id="s1" from="0.800" to="13.830"/>
 ...
</chrss:playlist>
```

Digest Information

```
<chrss:summarylist>
 <chrss:summary id="d15" primary="true"
           duration="15" href="http:/.../d15.avi">
  <chrss:scene ref="s1"/>
  <chrss:scene ref="s6"/>
 </chrss:summary>
</chrss:summarylist>
  </chrss:item>
 </item>
</channel>
</rss>
```

Fig. 4. CH-RSS description example

4.2 Media Information

The media information layer <chrss:media> consists three sub-elements, i.e. <chrss:info>, <chrss:video>, and <chrss:audio>. While <chrss:info> indicates entities such as the title, author photographer and the location where the video was shot, for searching, <chrss:video> and <chrss:audio> indicate the video and audio formats, compression scheme, and sampling rates, for media playback.

4.3 Scene Information

The scene information layer <chrss:playlist> has a sub-element <chrss:scene> which divides the video into separate scenes, and indicates the starting and ending times of each scene. This is used when you want to seek and play the original video from the beginning of a certain scene. As shown in Figure 4, *id* means arbitrary ID number, and the starting and ending times are those of the prosodic phrase, respectively.

Figure 5 (a) shows a schematic diagram of the scene information layer. The horizontal axis represents time, and S0, S1, S2, S3, S4, and S5 are ID numbers of the different scenes described by <chrss:scene>.

4.4 Digest Information

The digest information layer <chrss:summarylist> contains information about how the different scenes, into which the video was divided by the scene information layer, were condensed to suit the arbitrary digest time.

There is a sub-element <chrss:summary> in <chrss:summarylist>. As shown in

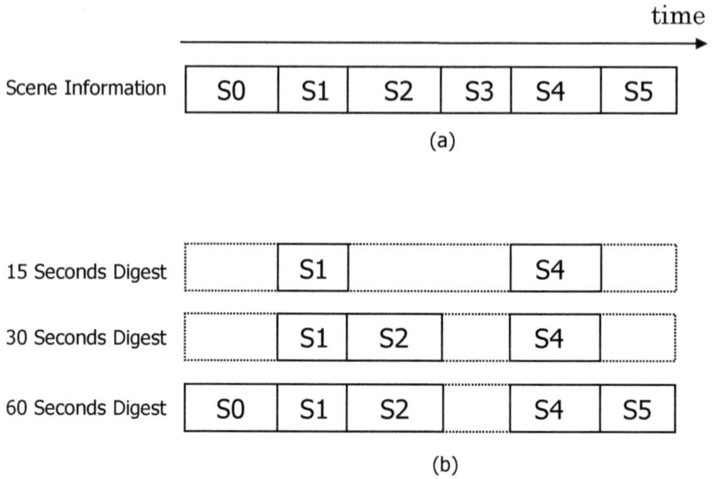

Fig. 5. Schematic diagram of (a) the scene information and (b) the digest information

Figure 4, *primary="true"* means the digest prior to all others, and *duration* indicates digest time. When a video digest is generated in advance, the digest is referred to by *href*.

Figure 5 (b) shows a schematic diagram of the digest information. <chrss:summary> has a sub-element <chrss:scene> which refers to the ID number described in 4.3 by *ref*. Since the degree of emphasis of each scene is calculated and the priority sequence of the scenes is determined mentioned in Section 3, multiple digests can be generated such as a 15 second digest, a 30 second digest and a 60 second digest as shown in Figure 5 (b). According to <chrss:summary>, an owner/audience specified digest delivery can be performed.

5 ChocoParaTV Browser

ChocoParaTV browser provides a bird's eye view of a video; users can watch digests as shown in Figure 6 (i). Since the ChocoParaTV browser is based on Flash, it is not necessary to install any special applications. ChocoParaTV browser watches for updates of CH-RSS on the ChocoParaTV server regularly, and acquires the newest digest information. When a video digest is generated in advance, ChocoParaTV browser downloads the digest, if not, it skips to play position for digesting according

(i) Overview of ChocoParaTV Browser

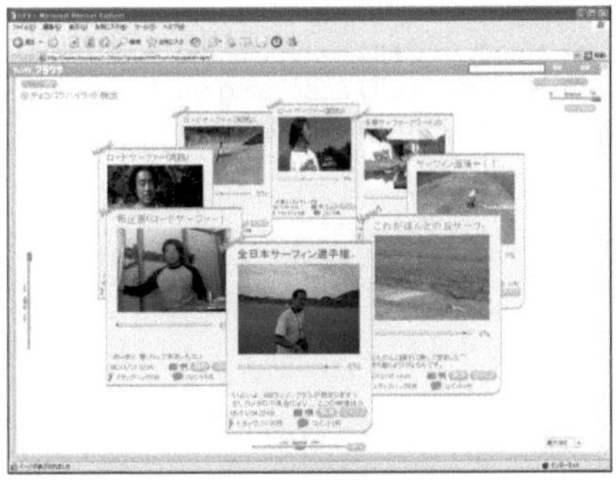

(ii) Digest Display Window

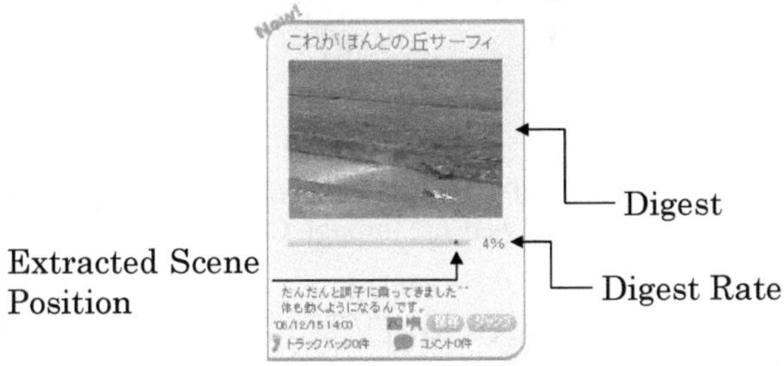

Fig. 6. ChocoParaTV browser

to <chrss:summary>, mentioned in 4.4. Even while a user is watching another video using the ChocoParaTV browser, the user can be notified, passively, of the availability of the latest video.

Figure 6 (ii) shows a digest display window. Extracted scene position is known by CH-RSS, and digest rate is defined as a ratio of digest time to total duration of the original content. By clicking the digest display window, a user can access the original web site easily. Thus, the user first views the digest and then visits the original web site if desired to watch the original content.

6 Experiment and Result

A key performance attribute of "ChocoParaTV" is how effective it is in getting users to access the original web sites. To assess this metric, we conducted a field experiment as follows. The period of the experiment was from 12/5/2005 to 3/31/2006. We opened the site (http://www.chocopara.tv) on the Internet to the public. According to site records, users could activate ChocoParaTV browser easily.

Over the period of the experiment, about 7,100 UUs (UU: Unique Users) accessed the site, and about 4% of the UUs booted ChocoParaTV browser. Figure 7 shows (α)

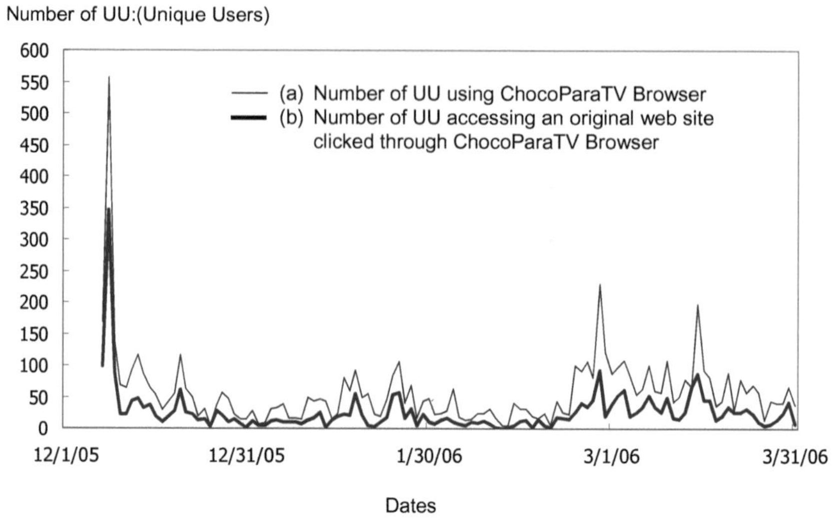

Fig. 7. (α) the profile of number of UU using ChocoParaTV browser, and (β) the profile of number of UU accessing an original web site through ChocoParaTV browser

the profile of the UUs using ChocoParaTV browser, and (β) the profile of the UUs who accessed the original web site through ChocoParaTV browser. We estimated that 43% of the users accessed the original sites through ChocoParaTV browser.

7 Conclusions

This paper has proposed a new video digest and delivery system: "ChocoParaTV", for applications such as blogs that include video content, and CGM sites.

ChocoParaTV server collects contents by using RSS from blogs and CGM sites as new videos are submitted.

The video digests are generated by the automatic extraction of emphasized speech portions. The proposed method uses prosodic parameters such as pitch, power, and speaking rate to estimate the degree of emphasis. By selecting a sufficient number of emphasized prosodic phrases, the method can produce a video digest to satisfy the user set digest time.

For delivering the digest information and the digest to audiences, we proposed a new video digest description: CH-RSS; it can generate digested contents dynamically that match the user's requested digest time. We define CH-RSS as an optional sub-element of RSS 2.0; it realizes the functionality needed to support video search and play, skipping to play position, and digest generation.

To browse videos quickly and passively, we proposed ChocoParaTV browser. ChocoParaTV browser provides a bird's eye view of a video, i.e., users can watch the digests on their screen. By using ChocoParaTV browser, the users first view the digest and then visit the original web site if desired to watch the original content.

The effectiveness of the proposed system was confirmed by a field experiment; about 40% of the ChocoParaTV browser users accessed the original web sites through the digests. This indicates that our method is a very effective at attracting the audience to video content. For ChocoParaTV borwser, the question of how to determine the arrangement of digest display windows is still open, and is left as future work.

Acknowledgments. We would like to thank Ken-ichiro Shimokura, Project Manager of NTT Cyber Solutions Laboratories, for encouraging us to write this article. We have had helpful discussions with the office support staff of Megumi Machiguchi, Kaori Takei, and Harumi Matsuura.

References

1. Lienhart, R., Pfeiffer, S., Effelsberg, W.: Video Abstracting. Communicaton, ACM 40(12), 55–62 (1997)
2. He, L., Sanocki, L., Gupta, A., Grudin, J.: Auto-Summarization of Audio-Video Presentations, Proc. ACM Multimedia 1999. pp. 489–498 (1999)
3. Hori, C., Furui, S.: A New Approach to Automatic Speech Summarization. IEEE Trans. Multimedia 5(3), 368–378 (2003)
4. Hidaka, K., Mizuno, O., Nakajima, S.: A New Speech Content Summarization Technique based on Speech Emphasis Extraction. Proc. ASJ Fall Conf. 2002, pp. 99–100 (2002)
5. Hidaka, K., Machiguchi, M., Takeuchi, J., Mizuno, O., Nakajima, S.: A New Multimedia Content Summarization Technique based on Automatic Speech Emphasis Extraction. IPSJ Symposium Series, 2003 7, 17–24 (2003)
6. Hidaka, H., Takeuchi, J., Matsuura, M., Ibaraki, H., Nakajima, S.: A Study on Efficient Video Preview Method based on Automatic Speech Emphasis Estimation. Journal Inst. Image Elec. Eng. Japan 34(5), 505–511 (2005)
7. goo RSS Reader: http://reader.goo.ne.jp/

Extraction of Anchor-Related Text and Its Evaluation by User Studies

Bui Quang Hung, Masanori Otsubo, Yoshinori Hijikata, and Shogo Nishida

Department of Systems Innovation, Graduate School of Engineering Science, Osaka University, 1-3 Machikaneyama, Toyonaka, Osaka 560-8531, Japan
{bqhung.otsubo,hijikata,nishida}@nishilab.sys.es.osaka-u.ac.jp

Abstract. Semantic Text Portion (STP) is a text portion in the original page which is semantically related to the anchor pointing to the target page. STPs may include the facts and the people's opinions about the target pages. STPs can be used for various upper-level applications such as automatic summarization and document categorization. In this paper, we concentrate on extracting STPs. We conduct a survey of STP to see the positions of STPs in original pages and find out HTML tags which can divide STPs from the other text portions in original pages. We then develop a method for extracting STPs based on the result of the survey. The experimental results show that our method achieves high performance.

Keywords: user study, text mining, web mining, semantic text portion, link structure, anchor.

1 Introduction

In the field of Web mining, many researchers come to focus on the link structure. When there is a link from a web page to another one, the former is called the original page and the latter is called the target page. One of the most important characteristics of the link structure is that the text portions around the anchors in the original pages describe the target pages [4]. Henzinger, in her survey on the link structure analysis [7], explains that this characteristic originates from the following human factor. Many authors of original pages create links because they think the links are useful for the readers. A link from an original page to a target page can be seemed as a recommendation about the target page by the author of the original page. The author also writes some texts around the anchor to explain the target page to the readers from his own viewpoint. These text portions are semantically related to the target page. We give the following definition about this kind of text portions.

Definition 1. *Semantic Text Portion (STP) in an original page is a text portion which is semantically related to the anchor pointing to the target page.*

We realize that there are two types of STP from the viewpoint of its locations. One type exists around the anchor. The other type exists in the upper-level structure of the original page. We call the former type *the Local Semantic Portion (LSP).* We call the latter type *the Upper-level Semantic Portion (USP).* STPs can be used for many applications such as automatic summarization [1, 2, 6] and document categorization [5, 9, 11, 12, 13, 14].

M.J. Smith, G. Salvendy (Eds.): Human Interface, Part I, HCII 2007, LNCS 4557, pp. 446–455, 2007.

Researchers proposed various methods for extracting STPs. These methods are anchor-text method, fixed-window method, sentence-based method, paragraph-based method, and list-based method. The anchor-text method is the simplest one. It extracts the text portion between the tags <A> and </A> of the anchor. The fixed-window method extracts the anchor text and the pre-determined number of words around the anchor. The sentence-based method extracts one or more sentences around the anchor. The paragraph-based method extracts the paragraph which begins with the anchor followed by texts. The list-based method extracts the list item which includes the anchor.

These methods are too simple to extract all the STPs in one original page. The problem of extracting STPs is that they locate in various kinds of location like the text around the anchor, the page title, the list title, the first row of the table and so on. Therefore the previous methods cannot extract STPs in high precision and especially in high recall.

Our approach to solve this problem is as follows. We conduct a survey of STPs to see which kinds of text portion in an original page are related to the anchor. We hope that we will find out some HTML tags which can semantically divide STPs from the other text portions in original pages. Based on the result of the survey, we develop a method for extracting STPs. Our method represents an original page by a DOM tree to analyze its document structure. DOM (Document Object Model) is an API to access any parts of a Web page which is standardized by W3C [17]. Our method then extracts STPs by using specific HTML tags which are found in the survey.

The most serious shortcoming in the previous researches is that they did not survey where STPs are written in an original page and did not evaluate their methods from the viewpoint of extracting STPs. They only proposed their simple methods and used the text portions extracted by their methods for upper-level applications. They did not consider whether the extracted text portion itself is semantically related to the target page or not. In our research, we conducted a deep survey of the location of STPs and evaluated our method from the viewpoint of extracting STPs by inviting three evaluators. We made a dataset which consists of more than 1000 real original pages for the survey and a dataset which consists of 200 real original pages for the evaluation. The evaluators judged which text portions are real STPs in those pages. We decided on the text part which is a real STP by the majority vote. In the evaluation, we compared the texts extracted by our method to the real STPs given by evaluators. We then compared our method to the previous methods in extracting STPs. The experimental results showed that our method can achieve high precision and also the highest recall among the previous methods.

In brief, the contributions of this paper are as follows:

- We deeply survey the locations of STP in original pages for the first time.
- We propose a method for extracting STPs from the result of the survey.
- We evaluate extracted text portions by using real STPs given by evaluators for the first time.

The rest of this paper is organized as follows. Section 2 discusses the survey of STP and Section 3 explains our method for extracting STPs. In Section 4, we evaluate STPs extracted by our method and compare our method to other methods. Section 5 provides some concluding remarks and directions for future research.

2 Survey of STP

2.1 Dataset and Survey Method

We prepared 1108 real original pages in our survey. We invited three evaluators to give us the right answer of USPs. For each original page in the dataset, we show the three evaluators its content and the anchor pointing to its target page. Then we ask them to judge which text portions are semantically related to the anchor. We define a real STP as the text portion which is judged to be semantically related to the anchor by at least two evaluators.

2.2 Survey of LSP

A. Positions of LSPs in original pages

Through the survey, we realized that LSPs are located in one of the following five places: table, list (ordered and un-ordered list), definition list, paragraph, or DIV object. Table 1 shows the number of LSPs in each place in 1108 original pages.

Table 1. Number of LSPs in each place

Position	Total
Paragraph	320
Ordered and un-ordered list	354
Definition list	56
Table	339
DIV	39

B. HTMLs tags for dividing LSPs from the other text portions

This subsection explains the result of survey about what kind of HTML tag can divide the LSP from the other text portions in each place.

a) When the LSP is in a paragraph. From here, we call the paragraph which has the anchor to the target page the *current paragraph.* There are the following two cases. ***Case 1:*** *The LSP covers the whole paragraph.* There is only one anchor in the paragraph or there is no line feeder tag in the paragraph. <P> tag and </P> tag can divide the LSP from the other text portions. ***Case 2:*** *The LSP is one part of the paragraph.* There are several anchors and several line feeder tags in the paragraph. There are four sub-cases as shown in Fig. 1. In Sub-case 1 and Sub-case 2, the LSP is divided by the line feeder tag before the anchor and the line feeder tag before the next anchor. In Sub-case 3 and Sub-case 4, the LSP is divided by the line feeder tag after the previous anchor and the line feeder tag after the anchor.

b) When the LSP is in a DIV object. There are the following two cases. ***Case 1:*** *The LSP covers the whole DIV object.* The LSP is divided from the other text portions by the <DIV> tag and </DIV> tag. ***Case 2:*** *The LSP is one part of the DIV object.* The LSP is divided from the other text portions by line feeder tags like a)-Case 2.

c) When the LSP is in a list. After here, we call the list item which includes the anchor to the target page the *current list item.* We call the list which has the current list item the *current list.* There are the following two cases. ***Case 1:*** *The LSP covers the whole current list item.* The LSP is divided from the other text portions by the <LI> tag and </LI> tag. ***Case 2:*** *The LSP covers one part of the current list item.* The LSP is divided from the other text portions by line feeder tags like a)-Case 2.

d) When the LSP is in a definition list. There are the following two cases (see Fig. 2). ***Case 1:*** *The LSP covers the definition term including the anchor and the definition description of the definition term.* The LSP is divided from the other text portions by the <DT> tag before the anchor and the </DD> tag after the anchor. ***Case 2:*** *The LSP is a part of the definition description.* There are several anchors and several line feeder tags in the definition description. The line feeder tag before the anchor and the line feeder tag before the next anchor can divide the LSP from the other text portions.

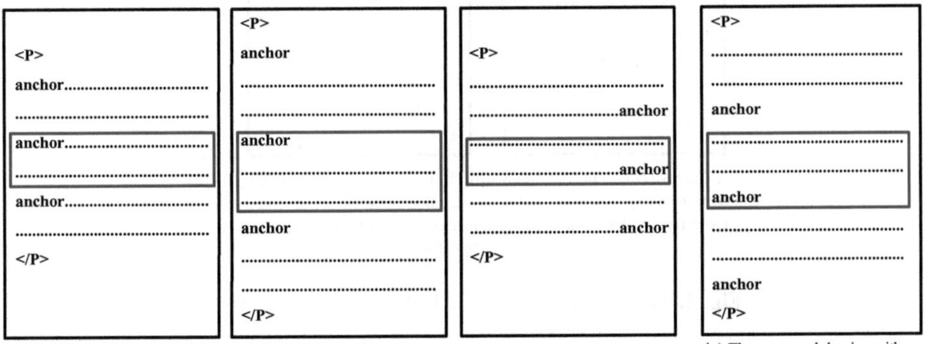

i)The paragraph begins with an anchor followed by texts. There is no line feeder tag between the anchor and texts

ii) The paragraph begins with an anchor followed by texts. There is one or more line feeder tag(s) between the anchor and texts

iii) The paragraph begins with texts. There is no line feeder tag between the anchor and texts

iv) The paragraph begins with texts. There is one or more line feeder tag(s) between the anchor and texts

Fig. 1. Four sub-cased when the LSP is one part of a paragraph

e) When the LSP is in a table: There are the following five cases (see Fig. 3). After here, we call the cell, where the anchor to the target page exists, the *current cell.* We call the row which has the current cell the *current row.* We call the table which has the current row the *current table.* ***Case 1:*** *The LSP covers the whole current cell.* The LSP is divided by <TD> tag and </TD> tag. ***Case 2:*** *The LSP is a part of the current cell.* The LSP is divided from the other text portions by line feeder tags like a)-Case 2. ***Case 3:*** *The LSP covers several cells (not all cells) in the current row.* There are the following two sub-cases. ***Sub-case 1:*** The current row begins with an anchor. The <TD> tag before the anchor and the </TD> tag before the next anchor divide the LSP from the other text portions. ***Sub-case 2:*** The current row begins with texts. The <TD> tag after the previous anchor and the </TD> tag after the anchor can divide the LSP from the other text portions. ***Case 4:*** The LSP covers the current row. The <TR>

tag and </TR> tag of the current row can divide the LSP from the other text portions. ***Case 5:*** The LSP covers several rows of the table. There are the following two sub-cases. ***Sub-case 1:*** The table begins with an anchor. the <TR> tag before the anchor and the </TR> tag before the next anchor can divide the LSP from the other text portions. ***Sub-case 2:*** The table begins with texts. The <TR> tag after the previous anchor and the </TR> tag after the anchor can divide the LSP from the other text portions.

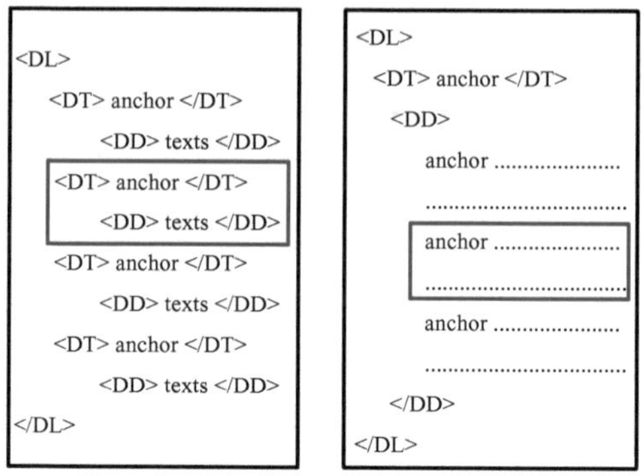

i) The LSP covers the definition term including the anchor and the definition description.

ii) The LSP is a part of the definition description and there are several anchors and several line feeder tags in the definition description.

Fig. 2. Two cases in which the LSP is in a definition list

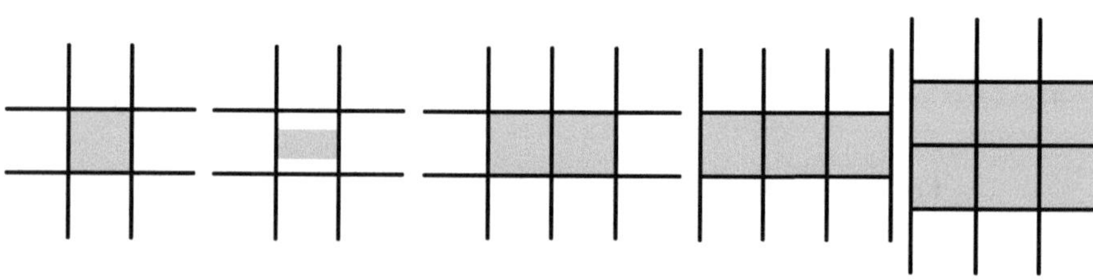

i) The LSP covers the current cell

ii) The LSP is one part of the current cell

iii) The LSP covers several cells of the current row

iv) The LSP covers the current row

v) The LSP covers several rows

Fig. 3. Five cases in which the LSP is in a table

2.3 Survey of USP

Table 2 shows the survey result. The left column shows the type of upper-level object which is related to the anchor, the center column shows the number of pages which has each type of upper-level object, and the right column shows the HTML tags which can divide the USP from other text portions.

Table 2. Result of the survey of USP

Upper-level object	Total	HTML tags used for extracting
Page title	1097	<Title> and </Title>
H1	326	<H1> and </H1>
H2	209	<H2> and </H2>
H3	153	<H3> and </H3>
H4	18	<H4> and </H4>
H5	26	<H5> and </H5>
H6	7	<H6> and </H6>
Table Header	6	<TH> and </TH>
The first row of the current table	48	<TR> and </TR>
The first row of an upper-level table	82	<TR> and </TR>
The text portion at the top of the current list	64	line feeder tags
Another row of the current table	46	cannot extract
Another row of an upper-level table	167	cannot extract
Another table	278	cannot extract
Another list	36	cannot extract
Another paragraph	372	cannot extract

3 Extraction of STP

3.1 Extraction of LSP

Firstly, our method represents an original page by a DOM tree. It then identifies which location (paragraph, list item, definition list, table, or <DIV> object) the anchor to the target page belongs to. After that, the method extracts the LSP from the identified location. The detail of the method is as follows.

The method identifies which location the anchor belongs to according to the type of the parent tag as follows: <P>: the anchor is in a paragraph; <LI>: the anchor is in a list item, <DT> or <DD>: the anchor is in a definition list; <TD>: the anchor is in a cell table; <DIV>: the anchor is a DIV object.

Then the method extracts the LSP from each location as follows:

a) If the anchor is in a paragraph, list item, definition object (<DD>) or DIV object. The method checks the number of line feeder tags in the parent object to the anchor.

- If there is no line feeder tag, it then extracts the whole texts of the object.
- If there is one or more line feeder tag(s), it then checks the number of anchors in the object. If there is only one anchor, it then extracts the whole text of the object. If there are several anchors, it then checks whether the object begins with an anchor or texts. If the object begins with an anchor, it extracts the text portion between the line feeder tag before the anchor and the line feeder tag before the next anchor. If the object begins with texts, it extracts the text portion between the line feeder tag after the previous anchor and the line feeder tag after the anchor.

b) If the anchor is in a cell of a table. The method tries to expand to nearby cells by following the left and right directions from the current cell. It repeats this expansion until it meets a cell which includes a different anchor. If it can expand to all cells of the current row which includes the current cell, it tries to expand to nearby rows by following the up and down directions. It repeats this expansion until it meets a row which includes a different anchor. There are the following four cases in the result of this expansion:

■ ***Case 1:*** *The method cannot expand to any other cells.* The method extracts the LSP from the current cell by the same method as in a).

■ ***Case 2:*** *The method can expand to other cells but it cannot expand to all cells of the current row.* The method then checks whether the current row begins with an anchor or texts. If it begins with an anchor, the method extracts the text portion between the <TD> tag before the anchor and the </TD> tag before the next anchor. If it begins with texts, the method extracts the text portion between the <TD> tag after the previous anchor and the </TD> tag after the anchor.

■ ***Case 3:*** *The method can expand to all cells of the current row.* It extracts the whole texts in the current row.

■ ***Case 4:*** *The method can expand to other rows of the table.* It checks whether the table begins with an anchor or texts. If it begins with an anchor, the method extracts the text portion between the <TR> tag before the anchor and the </TR> tag before the next anchor. If it begins with texts, the method extracts the text portion between the <TR> tag after the previous anchor and the </TR> tag after the anchor.

c) If the anchor is in <DT> object of a definition list. It extracts the whole texts of the <DT> object and the whole texts of its <DD> object.

3.2 Extraction of USP

Our method extracts USPs as follows:

■ It extracts the page title and all the upper headers from H1 to H6. If there are several headers at the same level, it extracts the nearest one to the anchor.

■ It checks whether the anchor is in a table. If the anchor is in a table, it checks whether a table header exists.

- If a table header exists, the method extracts the table header.

- If a table header does not exist, the method checks whether or not the first row of the current table satisfies at least one of the following two conditions. (1) The number of its cells is smaller than the number of cells in the other rows. (2) There is no anchor in it while there are anchor(s) in all the other rows of the current table. If the first row of the current table satisfies at least one condition, the method extracts the first row of the current table. If the first row of the current table does not satisfy any condition, the method checks whether or not the first row of the upper-level table (if it exists) satisfies at least one of the above two conditions. If it satisfies at least one condition, the method extracts it. If it does not satisfy, the method continues to check the first row of the upper-level table of the previous upper-level table (if it exists). The method repeats this process until it finds out the first row which satisfies at least one condition or there is no more upper-level table.

■ The method checks whether the anchor is in a list item. If it is in a list item, the method checks whether there is a text portion at the top of the list. If there is a text portion and its number of words is smaller than a threshold a, the method extracts this text portion. We set a as 20 because in our survey of USPs, there is no list title which has the number of words which is greater than 19.

4 Evaluation

The dataset we prepared for our experiments contains 200 original pages. We evaluated our method from three criteria: *precision*, *recall*, and *average number of words*. We also compared our method to other conventional methods in extracting STPs.

4.1 Evaluation of Our Method

Table 3 shows the experimental results. Experimental results show that our method extracts LSPs in high precision (97.01%) and in high recall (93.94%). The number of words in the texts extracted as LSPs (20.36 words) is quite similar to the average number of words of the real LSPs (21.07 words). From this result, we can see that our method can identify the positions of LSPs in original pages accurately.

Our method extracts USPs in 89.43% precision and in 74.35% recall. The average number of words in the texts extracted as USPs (8.54 words) is almost same as the average number of the real USPs (9.35 words). These precision and recall are smaller than the precision and the recall in extracting LSPs. This can be explained as follows. Based on the result of the survey of USP, our method extracts the page title, the headers (H1~H6), the first row of the current table, the first row of the upper-level table, and the text portion at the top of the current list. However, in some original pages, these text portions are not related to the anchor. For example, some authors put the same name (in most cases, the name of the web site) to all pages. Some authors use headers or tables not for structuring the content of the document but for decorating the web page or creating the layout for the web page. This is why our method extracts noise keywords. Our method cannot extract STPs in another paragraph, in another row of the current row (not the first row), in a row of the upper-level table (not the fist row), in another table or in another list. This means that there are STPs which exist in popular places and our method cannot extract.

Our method extracts both LSPs and USPs in 94.08% precision and in 85.03% recall. The average number of words of extracted texts is 28.89. This is almost same as the average number of the real STPs (30.43 words). We do not know this precision and recall is high among other existing methods. The next subsection compares our method to the existing methods.

Table 3. Evaluation of our method for extracting STPs

	Precision (%)	Recall (%)	Avg. number of words of extracted texts	Avg. number of words of real STPs
LSPs	97.01	93.94	20.36	21.07
USPs	89.43	74.35	8.54	9.35
both of LSPs and USPs	94.08	85.03	28.89	30.43

4.2 Comparison of Our Method to Other Methods in Extracting Both LSPs and USPs

Our method extracts both the text portion around the anchor and the text portion in the upper-level structure in the original page. Therefore, it achieves the highest recall (85.03%) compared to the other methods. The precision of our method is 94.08%. It is lower than the precision of the anchor-text method (100%) and the sentence-based method with Option 1 (100%). Because anchor text and the sentence, which includes the anchor, are always related to the anchor. We can see that these methods take the safest measure to extract STPs. Therefore their recalls are low. From this result, we can see that our method brings a good balance between precision and recall.

Table 4. Comparison of our method to other methods in extracting both LSPs and USPs

Method	Precision (%)	Recall (%)	Avg. number of words of extracted texts
Our method	94.08	85.03	28.89
Anchor-text method	100	19.37	3.43
Fixed-window method (50 words around the anchor)	29.52	52.78	51.38
Sentence-based method (3 sentences around the anchor)	60.1	51.03	25.54
Paragraph-based method	100	18.32	7.16
List-based method	84.24	19.78	7.92
Object-based method	70.95	50.28	367.45
Extracting all upper-level objects	13.01	39.98	1081.71
Roy's Method	84.17	20.06	5.89

*Average number of words of the real STPs is 30.43.

5 Conclusion and Future Work

This paper concentrates on extracting a semantic text portion (STP) from an original page. STP is a text part which is related to the anchor to the target page. Firstly, we found two types of STP: local semantic portion (LSP) and upper-level semantic portion (USP). We conducted a survey for each type of STP by using 1108 real original pages to find HTML tags which can semantically divide STPs from the other text portions in original pages. We invited three evaluators to participate in our survey to judge which text portions in an original page are STPs. We then developed a method for extracting STPs based on the result of the survey. Our method represents an original page by a DOM tree to analyze its document structure. It then extracts STPs by using specific set of HTML tags which are found in the survey. We then conducted experiments to evaluate our method and compare it to the previous methods in extracting STPs. We evaluated the texts extracted by each method by comparing them to the real STPs given by three evaluators. The experimental results showed that our method achieves high precision and the highest recall compared to the previous methods.

The shortcoming of our survey and our extraction method is that they just consider the relevance to the anchor but do not consider the type of relevance. We found in the survey that STPs are either facts or people's opinions (evaluation and categorization).

In some applications, we should select the type of STPs. In the summarization, the user may want to see only the people's evaluation. In the categorization, the user may want to see a categorization created by a user group from their viewpoints. We will study STPs by considering whether they are facts, people's evaluation or people's categorization as a future work.

References

1. Delort, J., Meunier, B.B., Rifqi, M.: Enhanced Web Document Summarization Using Hyperlinks. Proc. 14th ACM Conference on Hypertext and Hypermedia (HT'03), pp. 208–215 (2003)
2. Amitay, E.: Using common hypertext links to identify the best phrasal description of target web documents. In: Proc. Post-Conference Workshop on Hypertext Information Retrieval for the Web (SIGIR'98), pp. 271–276 (1998)
3. Open Directory http://dmoz.org/
4. Davison, B.D.: Topical Locality in the Web. In: Proc. 23rd Annual International Conference on Research and Development in Information Retrieval (SIGIR2000), pp. 272–279 (2001)
5. Roy, S., Joshi, S., Krishnapuram, R.: Automatic Categorization of Websites based on Source Type. Proc. 15th ACM Conference on Hypertext & Hypermedia, pp. 38–39 (2004)
6. Amitay, E., Paris, C.: Automatically summarizing web sites: Is there a way around it? Proc. ACM 9th International Conference on Information and Knowledge Management, pp. 173–179 (2000)
7. Henzinger, M.: Link Analysis in Web Information Retrieval. IEEE Data. Engineering Bulletin 23(3), 3–8 (2000)
8. Google, http://www.google.com/
9. Otsubo, M., Hung, B.Q., Hijikata, Y., Nishida, S.: A Basic Study on Web Page Classification Method by Anchor-Related Text. Proc. SICE Annual Conference, pp. 3622–3625 (2005)
10. Chakrabarti, S., Dom, B., Gibson, D., Keinberg, J., Raghavan, P., Rajagopalan, S.: Automatic Resource list Compilation by Analyzing Hyperlink Structure and Associated Text. Proc. 7th International World Wide Web Conference, pp. 65–74 (1998)
11. Glover, E.J., Tsioutsiouliklis, K., Lawrence, S., Pennock, D.M., Flake, G.W.: Using web structure for classifying and describing web pages. Proc. 11th International World Wide Web Conference, pp. 562–569 (2002)
12. Attardi, G., Di Marco, S., Salvi, D.: Categorisation by context. Journal of Universal Computer Science 4(9), 719–736 (1998)
13. Furnkranz, J.: Exploiting Structural Information for Text Classification on the WWW. In: Hand, D.J., Kok, J.N., Berthold, M.R. (eds.) Advances in Intelligent Data Analysis. LNCS, vol. 1642, Springer, Heidelberg (1999)
14. Blum, T.M.: Combining Labeled and Unlabeled Data with Co-Training. Proc. 11th Annual Conference on Computational Learning Theory, pp. 92–100 (1998)
15. Kleinberg, J.M.: Authoritative Sources in a Hyperlinked Environment. Journal of the ACM (JACM) 45(5), 604–632 (1999)
16. Brin, S., Page, L.: The Anatomy of a Large-Scale Hypertextual Web Search Engine. Proc. 7th International Conference on World Wide Web, pp. 107–117 (1998)
17. http://www.w3.org/DOM/

Rough Ontology: Extension of Ontologies by Rough Sets

Syohei Ishizu, Andreas Gehrmann, Yoshimitsu Nagai, and Yusei Inukai

Aoyama Gakuin University,
5-15-1 Fuchinobe, Sagamihara City, Japan 229-8558
ishizu@ise.aoyama.ac.jp, gehrmann@yhc.att.ne.jp,
nagai@ise.aoyama.ac.jp, 5703014@ise.aoyama.ac.jp

Abstract. Ontology is widely used in the areas of knowledge engineering, web based data mining, etc. In rough set theory, accuracy of approximation of set and a concept of granularity are introduced. Rough set theory is very useful to define dependency among attributes and extract decision rules from the set. One of our main aims of this paper is to propose a concept of rough ontology. A concept of rough ontology is extended concept of rough set, and it enables us to use flexible information system in the form of ontology. And rough ontology is useful to introduce concepts rough set theory in to ontology. In this paper we formulate a concept of rough ontology, which is extended concept of rough set theory. We define upper and lower approximation, accuracy of approximation of preference, concept of granularity of preference. And we also show the property of rough ontology.

Keywords: Rough set theory, Ontology, OWL, information system, accuracy.

1 Introduction

With the development of the semantic web, methods for knowledge extraction and knowledge representation are becoming of great importance. However, the existing concepts of an ontology provides no means to capture incomplete knowledge about instances in a domain. Rough set theory is a natural choice for dealing with incomplete information. Incorporating Rough set theory into existing ontology concepts provide possibilities to quantify the degree of accuracy of knowledge.

Rough set theory was proposed by Pawlak [1,2] as an extension of the set theory in order to represent incomplete knowledge. It is useful to represent imperfect knowledge, to mine data, and to analyze attributes dependency. On the other hand ontologies are introduced for complete knowledge representation [9]. In this paper we propose a concept of a rough ontology, which utilizes ideas of both approaches for representation of incomplete knowledge.

In defining concepts usually two approaches are relevant, extensional definition and intensional definition of concepts used in an ontology. An extensional definition of a concept formulates its meaning by specifying its extension, that is, every object that falls under the definition of the concept is directly identified. An intentional definition gives the meaning of a term by giving all the properties required of something that falls under that definition; the necessary and sufficient conditions for belonging

M.J. Smith, G. Salvendy (Eds.): Human Interface, Part I, HCII 2007, LNCS 4557, pp. 456–462, 2007.

to the set being defined. We apply rough sets to grasp incomplete information on extential definition of concepts.

2 Concepts of Approximation and Accuracy of Approximation in Rough Set Theory

In this section, we overview the concepts of rough set theory and ontology, and the next section we define the concepts of rough preference. Relationships among the concept of rough set theory and rough preference will be shown in Table 1 in section 3. An information system is defined as 4-tuple $<U, Q, V, f>$ where U is finite set of objects, Q is finite set of attributes, $V = \bigcup\{V_q \mid q \in Q\}$ and $V_q = \{f(x,q) \mid x \in U\}$ is range of the attribute q, and $f : U \times Q \rightarrow V$ is called information function such that $f(x,q) \in V_q$ for every $q \in Q$ and $x \in X$. For every $P \subset Q$, $IND(P) = \{(x,y) \in U \times U \mid \forall q \in Q : f(x,q) = f(y,q)\}$ denotes equivalence relation. For any $P \subset Q$ and $x \in U$, $[x]_P = \{y \mid \forall q \in P : f(x,q) = f(y,q)\}$ denotes an equivalence class. For any $Y \subset U$ and $P \subset Q$, P-upper approximation $P^*(Y)$ of Y, and P-lower approximation $P_*(Y)$ of Y are defined as follows.

$$P^*(Y) = \bigcup\{x \in U \mid [x]_P \subset Y\}$$
$$P_*(Y) = \bigcup\{x \in U \mid [x]_P \cap Y \neq \phi\}$$

For every set $Y \subset U$, accuracy of approximation of Y by P is defined as follows.

$$\alpha_P(Y) = card(P_*(Y)) / card(P^*(Y))$$

A set of attributes $P \subset Q$ depends on a set of attributes $P' \subset Q$, denoted $P \rightarrow P'$, if $IND(P) \subset IND(P')$

3 Concepts of Accuracy of Approximation in Rough Ontology and Rules Generation

3.1 Ontology Information System

In this section we introduce the concepts of rough ontology. In Rough set theory an Information system of rough set is based on the decision table, which is theoretically same as relational database. Ontology has more flexible information structure. One of our main aims of this paper is to propose a concept of rough ontology. A concept of rough ontology is extended concept of rough set, and it is enable us to use flexible information system by the form of ontology. And rough ontology is useful to introduce concepts of accuracy of approximation of class, a concept of granularity, defining dependency among attributes, and extraction of decision rules.

Ontology information system is defined as $<U, Q, C, Dom, Range, rel>$. Where U is a finite set of individuals, Q is a finite set of property names. Let C be a finite class of subsets of U. This means if $c \in C$, then $c \subset U$. Each property name has domain, range, and relation, e.g., $Dom : Q \to C$, $Range : Q \to C$, $rel(p) \subset Dom(p) \times Range(p)$ for any $p \in Q$. Note $rel(p)$ may not be function, but relation.

Ontologies are frequently represented by RDF expression. RDF expression of the takes a form of a set of triples $RDF \subset U \times Q \times U$, where $(x, p, y) \in RDF \leftrightarrow (x, y) \in rel(p)$.

In an ontology, class hierarchy is important concept. Class hierarchy is easily represented by the form of $< C, \subseteq >$, where $\subseteq$ is set theoretical inclusion.

In order to compare usual rough set theory with ontology system directly, we introduce some conditions on rough ontology

1) Every property has same domain.
2) Individuals are divided by domain and union of the ranges of properties.
3) Every relation is functional.

Under the conditions, the following notations are introduced. Since every property has same domain, so $U = Dom(q)$ where $q \in Q$. Let $V_q = Range(q)$ where $q \in Q$, then V_q is finite set of individuals. Let $V = \bigcup\{V_q \mid q \in Q\}$. Let $f : U \times Q \to V$ be defined as $f(x, q) = \{y \mid (x, y) \in rel(q)\}$. Since every relation is functional, then f is function. Then $<U, Q, V, f>$ can be regarded as information system.

Conversely rough ontology is free from these conditions, and the conditions 1), 2), 3) may characterize flexible features of ontology information system comparing with information system in rough set theory. In ontology information system, domain of properties may not same. There may be some individuals, which involved in both range and domain of properties. For some $p \in Q$, $rel(p)$ may not be function, and $rel(p)(x)$ may be null or multiple values.

Since usual ontology information system may not satisfy the conditions 1), 2), 3), we can not apply rough set theory directly to ontology system. We need the new rough set theory extended to ontology system. From the definition of ontology system, we introduce ontology information system. From the nature of relation rel, i.e. rel is not functional, we introduce extended information function. $\tilde{f} : U \times Q \to Pow(U)$ next as follows. $\tilde{f}(x, q) = \{y \mid (x, y) \in rel(q)\}$. Where $Pow(U)$ denote power set of U. We define ontology information systems as $<U, Q, C, \tilde{f}>$.

3.2 Rough Ontology and Accuracy of Approximation

Next we introduce extended concepts of rough ontology and accuracy of approximation by a set of properties. $IND(P)$ for any $P \subset Q$, for any $P \subset Q$ and $x \in U$, $[x]_P$, for any $Y \subset U$ and $P \subset Q$, P-upper approximation $P^*(Y)$ of Y , and P-lower approximation $P_*(Y)$ of Y . For every $P \subset Q$, $IND(P) = \{(x, y) \in U \times U \mid \forall q \in Q : \tilde{f}(x,q) = \tilde{f}(y,q)\}$ denotes equivalence relation. For any $P \subset Q$ and $x \in U$, $[x]_P = \{y \mid \forall q \in P : \tilde{f}(x,q) = \tilde{f}(y,q)\}$ denotes an equivalence class.

For any $Y \subset U$ and $P \subset Q$, P-upper approximation $P^*(Y)$ of Y , and P-lower approximation $P_*(Y)$ of Y are defined as follows.

$$P^*(Y) = \bigcup\{x \in U \mid [x]_P \subset Y\},\ P_*(Y) = \bigcup\{x \in U \mid [x]_P \cap Y \neq \phi\}$$

For every set $Y \subset U$, accuracy of approximation of Y by P is defined as follows.

$$\alpha_P(Y) = card(P_*(Y)) / card(P^*(Y))$$

A set of attributes $P \subset Q$ depends on a set of attributes $P' \subset Q$, denoted $P \rightarrow P'$, if $IND(P) \subset IND(P')$. We extended the concepts of rough set theory, and proposed ontology system and rough ontology, and accuracy of approximation. Table 1 shows the relationships among the concepts of rough set and rough ontology. As shown in Table 1, each concept of rough set theory is naturally extended to rough ontology.

3.3 Rule Generation

Rule generation is one of the important roles in rough set theory. Let Y be $Y \subset U$. If $\alpha_P(Y) = 1$, then rules can be generated and represented by attributes only P . Now we consider the rules of ontology information system. First we define the condition of rule represented by P as $\varphi_P = \{(q, X_q) \mid q \in Q\}$. A set of rules represented by P is defined as R_P . We also define set of individuals which satisfies the condition φ_P as $U_{\varphi_P} = \{x \in X \mid \forall q \in Q : \tilde{f}(x,q) = X_q\}$. Next we define the concept of applicability of rules to Y .

Definition 1
Let Y be a sub set of U . Let R_P be a set of rules represented by P . If the following condition is satisfied then R_P is called applicable to Y .

(1) $\forall \varphi_P \in R_P : U_{\varphi_P} \subset Y$

(2) $\forall y \in Y \cap U : \exists \varphi_P \in R_P : y \in U_{\varphi_P}$

Proposition 1

Let Y be a sub set of U, and R_P .is a set of rules. R_P is applicable to Y, if and only if $\alpha_P(Y) = 1$.

Table 1. Relationships among the concepts of rough set and rough ontology

Rough set	Rough ontology
Information system $<U, Q, V, f>$ $f : U \times Q \to V$ $V = \bigcup\{V_q \mid q \in Q\}$ $V_q = \{f(x,q) \mid x \in U\}$	**Ontology system** $<U, Q, C, Dom, Range, rel>$ **Ontological information system** $<U, Q, C, \tilde{f}>$ $\tilde{f} : U \times Q \to Pow(U)$
Indiscernibility relation $IND(P) = \{(x, y) \in U \times U \mid$ $\forall q \in P : f(x,q) = f(y,q)\}$	**Indiscernibility relation** $IND(P) = \{(x, y) \in U \times U \mid$ $\forall q \in P : \tilde{f}(x,q) = \tilde{f}(y,q)\}$
Equivalence class $[x]_P = \{y \mid \forall q \in P :$ $f(x,q) = f(y,q)\}$	**Equivalence class** $[x]_P = \{y \mid \forall q \in P :$ $\tilde{f}(x,q) = \tilde{f}(y,q)\}$
Rough set $Y \subset U$ **Upper and lower approximation** $P^*(Y) = \bigcup\{x \in U \mid [x]_P \subset Y\}$ $P_*(Y) = \bigcup\{x \in U \mid [x]_P \cap Y \neq \phi\}$	**Rough ontology** $Y \subset U$ **Upper and lower approximation** $P^*(Y) = \bigcup\{x \in U \mid [x]_P \subset Y\}$ $P_*(Y) = \bigcup\{x \in U \mid [x]_P \cap Y \neq \phi\}$
Accuracy and dependency $\alpha_P(Y) = card(P_*(Y)) / card(P^*(Y))$ $IND(P) \subset IND(P')$	**Accuracy and dependency** $\alpha_P(Y) = card(P_*(Y)) / card(P^*(Y))$ $IND(P) \subset IND(P')$

Table 2. Ontology information system for sneakers and decision attributes

	Color (q_1)	Model (q_2)	Closing mechanism (q_3)	Material (q_4)	Decision attributes(q_5)
sneaker1(x_1)	white	low	lace	Textile	o
sneaker2(x_2)	black	high	lace, Velcro fastener	textile, leather	o
sneaker3(x_3)	colorful	low	ace	textile	×

Table 2. *(continued)*

sneaker4(x_4)	black	low	-	Textile	×
sneaker5(x_5)	white	high	lace	textile, leather	×
Sneaker6(x_6)	colorful	low	-	leather	o

Table 3. Analysis of accuracy of approximation

P	Accuracy of approximation
q_1, q_2, q_3, q_4	1.00
q_1, q_2, q_3	1.00
q_1, q_2, q_4	1.00
q_1, q_3, q_4	1.00
q_2, q_3, q_4	0.67

P	Accuracy of approximation
q_1, q_2	0.67
q_1, q_3	0.67
q_1, q_4	1.00
q_2, q_3	0.33
q_2, q_4	0.33
q_3, q_4	0.17

P	Accuracy of approximation
q_1	0.00
q_2	0.00
q_3	0.17
q_4	0.17

Table 4. Decision rules

P	attribute1(q_1)	attribute2(q_4)
q_1, q_4	white	textile
	black	textile, leather
	colorful	leather

(a) Decision rules for selection good "o"

P	attribute1(q_1)	attribute2(q_4)
q_1, q_4	white	textile, leather
	black	textile
	colorful	textile

(b) Decision rules for selection bad "×"

In Tables 2-4, we show a simple example of rough ontology for sneakers. We show ontology information system (Table 2), the accuracy of approximation (Table 3) and resulted decision rules (Table 4).

4 Summary

One of our main aims of this paper is to present concepts of rough ontology which is an extended concept of from rough set theory. We generalize the concept of an ontology by using rough set theory to represent incomplete knowledge about concepts given by extentional definitions to the concepts of ontology information system, rough ontology, and accuracy of approximation. We define a set of rules which applicable to Y, and show the properties of rough ontology. Proposition 1 shows that if a set of rules R_P is applicable to Y, if and only if $\alpha_P(Y) = 1$. We also show an example, which illustrates the concepts of rough ontology. By the example we show how to generate rules is demonstrated.

References

1. Gehrmann, A., Ishizu, S.: Representation of Evaluation Spaces over Attributes Structures - a practical example-. In: Proc. of SCI 2001. Information Systems, vol. 2, pp. 139–145 (2001)
2. Gehrmann, A., Ishizu, S.: Minimal evaluation structure for inconsistent multi-attributes decision-making – selection and addition method of evaluation attributes, vol. 14(1), pp. 85–98 (2005)
3. Ishizu, S.: An Attributes Structure on Evaluation Space. Advances in Systems Study II, 33–38 (1995)
4. Ishizu, S., Amakasu, H.: Basic Concepts and Algorithms for Construction of Evaluation Structures. JASMIN 12(2), 21–36 (2003)
5. Klir, G.J.: Fuzzy Measure Theory. Kluwer Academic, Plenum Publishers (1993)
6. Nicoletti, M.D.C., Uchoa, J.Q., Baptistini, M.T.Z.: Rough relation properties. Int. J. Appl. Math. Comput. Sci. 11(3), 621–635 (2001)
7. Pawlak, Z., Slowinski, R.: Rough set approach to multi-attribute decision analysis. European J. of OR. (72), 443–459 (1994)
8. Pawlak, Z.: Rough set. Int J. Inf. Comp.Sci. 11(5), 341–356 (1982)
9. Wolterink, T.: What is an Ontology? http://referaat.cs.utwente.nl/new/papers.php?confid=1

Selecting Target Word Using Contexonym Comparison Method

Hyungsuk Ji[1], Bertrand Gaiffe[2], and Hyunseung Choo[1]

[1] School of Information and Communication Engineering
Sungkyunkwan University, Korea
[2] Loria, France
jihyungsuk@skku.edu, Bertrand.Gaiffe@loria.fr,
choo@ece.skku.ac.kr

Abstract. For an advanced next-generation Human-Computer Interaction (HCI) interface, combining natural language processing (NLP) is an inevitable choice as human language is the most common and sophisticated communication device. Among various topics in NLP, word sense related topic is one of the most challenging areas. In this paper we present a method adopted from ACOM that automatically generate bilingual lexicon using aligned parallel corpus. The results of the test on the predefined test set for the English and French Bibles show the method correctly produce target words with 70% correct ratio. Besides, the proposed method generates target words that reflect contextual relationship between source and target words such as *garrison* and *Philistins.*

Keywords: Bilingual lexicon, bitext, parallel corpus, ACOM.

1 Introduction

For an advanced next-generation Human-Computer Interaction (HCI) interface, combining natural language processing (NLP) is an inevitable choice, as human language is the most common and sophisticated communication device. Among various topics in NLP, word sense related topic is one of the most challenging areas, since a word has manifold meanings and identifying it requires complicated process [1]. Despite the advances achieved in NLP research, understanding a word's meaning is still a difficult task and indirect way to grasp it such as word sense discrimination [2] or target word selection for a given source word [3] have been conducted in the area.

Automatic bilingual lexicon construction using aligned parallel text (or bitext) is one of the most practical areas in machine translation (MT) [4-8]. Gale and Church proposed character level mapping technique [9], Melamed SIMR technique [10], Gaussier flow network models [11], Smadja collocation study [12], Fung and Ngai ontology related study [13], Dorr, Levin interlingual MT [14]. In general stop words list is used to skip frequent functional words such as *of, if, then*, etc., but for some studies like microstucture [16], these words are important to identify and align texts. However it is difficult to apply such methods for languages with very different structures. For these languages, a model that considers structural feature was proposed [17].

Constructing automatic bilingual lexicons from bitext assumes in general the existence of a systematic corresponding word or words for a given word. From this

M.J. Smith, G. Salvendy (Eds.): Human Interface, Part I, HCII 2007, LNCS 4557, pp. 463–470, 2007.

point of view, the failure of finding the corresponding pair is considered as erroneous and incorrect bilingual lexicon construction. But if the correct corresponding target word in language *A* for a source word in language *B* is not systematically used in some context, the failure of finding the 'correct' target word should not be considered as such.

In this paper we present a method for automatic bilingual lexicon construction that is theoretically similar to one that has been used for automatic extraction of contextually related words in monolingual corpus. In the next section, theoretic position for the current study is presented followed by the introduction of the method and the results.

2 Backgroud

Like other domains in NLP, grabbing a word's meaning in a specific context is not a trivial task, since identifying a word's meaning requires not only linguistic analysis but also extra-linguistic information. The latter, often referred to as a pragmatic issue, poses a difficulty in NLP tasks especially in machine translation. The difficulty is related to the fact that a word's meaning cannot be fully represented by one pre-defined one in a dictionary or thesaurus. For example even though the word *W1* in a language *L1* has an equivalent target word *W2* in the language *L2* in a bilingual dictionary, an examination should be carried out whether or not *W1* is really equivalent to *W2* in various respects such as equivalence in usage, in connotational sense, register, etc. An extreme view concerning a word's meaning states that a perfect synonym or a perfect target word for a source word does not exist. While this view shows one correct aspect about a word's meaning, it does not explain the possibility that a number of languages could describe or state certain concrete common situation.

For this reason, we hypothesize that given a specific context, a word's meaning represents a specific 'one meaning', and not all possible meanings, and thus it is possible in general to find an appropriate target word in a different language. We hypothesize also that a well prepared bitext (parallel text) does show this equivalence in a very sophisticated manner and by analyzing bitexts, well-suited source-target words matching could be found.

In this study, we adopt Automatic Contexonym Organizing Model (ACOM) [18-20] that produces contexonyms (contextually related words) from a given monolingual corpus. The original ACOM model inspects a monolingual corpus and produces and organizes contexonyms for a given word and in order to find a proper source-target sets of words in parallel corpus, ACOM's method was modified for the use in cross-lingual context.

A parallel text is a document that has precise translation pairs for a given language. We use the Bible as a bitext in this study. The Bible is translated in thousands languages, many of which the proper dictionaries does not exists yet. In this sense, the general method to find a proper source-target word is important to a language where no dictionary is ever made. Another reason to use the Bible as a bitext source lies in the fact that the corresponding verse-to-verse structure is well designed for the require processes in the study. It is true that the different version of the Bible has different systems of numbering and structuring verses. In general, however, the matching is nearly perfect compared to other bitexts where identifying corresponding sentences is not a trivial task. Finally, the Bible has different versions according to the degree of formal and informal language. This is very precious source to identify a word in a required situation which is not always apparent in many documents.

3 Method

The general method used in the current study to construct automatic bilingual lexicon is as follows: (1) making aligned parallel texts from the corpora in different languages (2) making cross lingual contexonym databases for each language (3) selecting contexonyms in target language for a given source word.

In this study the Bible text was used for bitext and the first step involves only some trivial tasks: as the biblical text is precisely translated and verified, the verses between the source text and the target text match nearly perfectly and only some adjustment of exceptional non corresponding verses across different versions of the Bible is needed. In case the verse *m* in the Bible *A* corresponds to the merged verse *m-n* in the Bible *B*, the verses *m* and *n* in *A* were merged.

Some hypotheses were made prior to processing the steps.

Hypothesis 1. If the portion *p* in the source texts and *q* in the target text express or depict the same notion or event then there exist the sub-portions *p2* and *q2* in each text which comprise bilingual lexicon.

This hypothesis states that if a bitext is a translation of a certain language then there should exist corresponding portion of segments between the texts in each language. The segment can be a word but this does not mean that every word in text *A* has its corresponding target word in text *B*. Instead of this strong assumption, we assume that there could be no corresponding word in target language for a source word:

Hypothesis 2. Unlike word to word correspondence in bilingual dictionaries, it is allowed that the corresponding target word for a source word does not exist.

This hypothesis requires in turn that in such a case there should be a method that can provide some information in target language for a given source word. We consider the current method that we explore might answer to this requirement.

We propose that a model that is able to produce the information on contextually related sense of a word can be implemented on the automatic bilingual lexicon study to enrich both semantic and pragmatic value of a source word. We use ACOM for this purpose, which has been reported to produce contexonyms for a word in the same language [18-20]. We modify the method that was used in ACOM to implement it for bitext environment. The procedures for the production of cross lingual contexonyms are as follows:

Step 1. The co-occurrence of the target words (not the source words) for a given source word are counted and stored in the database. For the word *W* in the language *L1*, the co-occurred words *V1*, *V2*, *V3*, ... in language *L2* and not the words in language *L1* are counted.[1] The co-occurrence database for the language *L2* is constructed in the same fashion.

[1] For the comparison of two corpora in the same language, *L1* and *L2* represent simply different corpora.

Step 2. For the given word *W*, the first candidate contexonyms (refers to [20] for more explanation on this term) of the target words are selected by the frequency criteria, that is, the portion of α ($\alpha \leq 1$) of target words are selected that co-occur the most frequently with *W*.
Step 2-2. Mutual information method is applied also to compare the results.
Step 3. The candidate cross-language contexonyms are examined whether they are valid contexonyms or not. This is done by examining whether or not the candidate contexonyms of the contexonyms (children of the child) have the original source word. This procedure requires the other database that is composed of target language entries and source words' contexonyms.
Step 3-2. In choosing valid contexonyms, the extra examination is applied on whether or not the stop words list.

4 Test on Pre-defined Word Set

In word sense related study, the available number of words to be tested surpasses easily 10,000. Since the results of this number of words cannot be published in general paper, some authors let their models be tested on Internet as were done in Latent Semantic Analysis [21] and ACOM. In this study we propose as an alternative using a pre-defined word set for the publication of NLP research. The following number set might be reusable for NLP related experiments:

Ji's number set: 10, 100, 200, 300, 400, 500, 600, 700, 800, 900, 1000, 2000, 3000, 4000, 5000, 6000, 7000, 8000, 9000, 10000, 100000

This number was used in the current study to select test set of words in a word list file which is sorted in descending order of the word frequency. That is, the 10th, 100th, 200th, 300th, … , 100000th the most frequent words in the source text are selected for the test. Since words with very low frequencies are positioned at the latter half of such database,[2] the sparse numbers for the latter part of the number set might be a reasonable choice. This kind of pre-defined number set has several advantages over the traditional randomly generated test set: there exists a number of random number generation methods and a myriad of test sets can be produced by changing the factors, giving too much flexibility to the test of the model. If the proposed number set, say Ji's number set, is accepted in the community, a rigorous 'random sample' is assured since the test set of words is pre-defined regardless of methods to be applied to, as the frequency of each word in a given corpus is constantly fixed.

5 Results

Table 1 shows the results of the test set of words chosen using Ji's number set for English (KJV[3]) and the French Bible (Segond). For the word *for*, *than* and *being*, stop

[2] For the database constructed on the corpus KJV, the word positioned at 46% of the frequency-based sorted database has the frequency 3 (i.e. the word appears only 3 times), and the word positioned at 69% of the database has the frequency as low as 1.

[3] King James Version.

Table 1. Result for the source words test set generated by Ji's number set. The most probable target word is listed first. For the mutual information method, first 5 the most probable words are listed.

Source word	Target words (Frequency cutoff)	Target words (Mutual Information)
for	pour (8)	car, pour, toi, c, nous
Father+father	Père+père	Père+père, mère, mon, son, notre
than	que	vaut, mieux, plus, grand, que
being	étant (17)	étant, Christ, été, était, Jésus
places	lieux	hauts, lieux, parfums, villes, Juda
darkness	ténèbres, lumière, obscurité, ombre,	ténèbres, obscurité, lumière, ombre, nuit, jour, mort
husband	mari, elle, femme,	mari, mariée, liée, femme, elle
bound	lié, chaines, lier, liés, lia,	liée, lié, lier, chaînes, prison
company	troupe, multitude, toute, Koré,	troupe, Koré, multitude, foule, assemblée
Levi	Lévi, Kehath, Merari, Siméon	Lévi, Kehath, Merari, Siméon, noms
hurt	mal, malheur	mal, malheur, aucun, paix, faire
visions	vision, visions	vision, eue, visions, Daniel, avais
garrison	Philistins, poste, Jonathan, portait,	poste, dent, portait, hébreux, Jonathan, forteresse, Philistins
departeth	violence, esprit, détourne, pleine, va[4]	écumer, amant, arrogant, violemment, appui, arrache
looks	crains, rebelles, famille, hautains, humilié, discours, orgueilleux, regards	abaisses, diamant, scorpions, quoiqu, hautains, visages
therefrom	détourna, pêcher	appliquez, commit, Nebath, détourna, pêcher
unequal	voie, droite,	droites, plutot, voie, droite, voies
furnaces	muraille, fours, tour	opposite, reparées, fours, chœur, Haschub
tend	précipitation, diligent, agit, arrive, projets	diligent, précipitation, mènent, disette, projets
outrageous	résistera, cruelle, fureur, jalousie, impétueuse, colère	impétueuse, cruelle, résistera, jalousie, fureur, colère

[4] '*s*' is found when no stop words filtering is applied that is the part of the corresponding word *s'en va*.

words list filtering was not applied in 'frequency cutoff method since the words themselves are stop words. The stop words list was generated by counting tokens in the British National Corpus (BNC) and the Gutenberg Project corpus. In Table 1, most words have correct corresponding target words. For certain words, the corresponding words are not correct. For example for the English word *garrison*, the French word *Philistins* was proposed as a corresponding word.

At first this looks totally erroneous, but a close investigation on the original text gives an explanation to the results. The English word *garrison* appears 19 times in KJV and in the verses having the word *garrison*, the word *Philistines* also appears 12 times, whose correct corresponding French pair is *Philistins*. This shows that there exists a strong bound between the two words. In the French Bible Segond, the corresponding pairs of *garrison* are *garnison*, *troupes* and *poste* and the verse "kept the city of the damascenes with a garrison" has its pair "faisait garder la ville des Damascéniens" where no direct corresponding French noun is found. Since the word *Philistins* and *garrison* is very closely related in the biblical context the result may be considered as reflecting this contextual relationship.

The English word *departeth* corresponds to several French expressions in the target corpus such as *s'en va*, *détourne*, *se détourne*, *s'éloigne*, *se retire de*, *est infidèle*[5] whereas the frequency of *departeth* is only 8 in the given corpus. This low frequency together with various corresponding target words makes it virtually impossible to find the correct matching pair. In some cases such contextual relationship is even systematic. For example, the word *unequal* has *droite* for its related word. This is due to the fact that in French *unequal* is systematically translated "*pas droite* (*not right*)". In fact this antonym correlation in statistical approach for corpus study is a well known problem.

Overall, 14 correct words out of 20 source words were produced. For the frequency cutoff method, the correct words *pour* and *étant* were found in the 8th and 17th position respectively when stop words list filtering was not applied, whereas for mutual information method the correct target words were found. The word *looks* appears only 5 times in KJV and that is why the corresponding words *regards* is found in the 8th position.

6 Conclusion

Using ACOM which was originally developed to generate and organize contextually related words, we proposed a flexible automatic bilingual lexicon generation method. For most source words in the test set, correct target words were selected. Besides, as shown in the results, the obtained target words reflect valuable contextual relationship with a given source word. These results might be of much value especially in constructing intelligent and enriched bilingual lexicon. In combination with other automatic bilingual lexicon generation method, our proposed method can contribute to the construction of enriched bilingual lexicons.

---

[5] "a wife treacherously departeth from her husband" and "une femme est infidèle à son amant".

Acknowledgments. This research was supported by the Ministry of Information and Communication, Korea under the Information Technology Research Center support program supervised by the Institute of Information Technology Assessment, IITA-2006-(C1090-0603-0046).

References

1. Brown, P.F., Pietra, S.D., Pietra, V.D., Mercer, R.L.: Word Sense Disambiguation using Statistical Methods. In: Proceedings of the 29th Annual Meeting of the Association for Computational Linguistics, ACL, pp. 264–270 (1991)
2. Schütze, H.: Automatic word sense discrimination. Computational Linguistics 24(1), 97–123 (1998)
3. Dagan, I., Itai, A.: Word sense disambiguation using a second language monolingual corpus. Computational Linguistics 20(4), 563–596 (1994)
4. Brown, P.F., Lai, J.C., Mercer, R.L.: Aligning sentences in parallel corpora. In: Proceedings of the 29th annual meeting on Association for Computational Linguistics, Morristown, NJ, USA, Association for Computational Linguistics, pp. 169–176 (1991)
5. Brown, P.F., Pietra, V.J.D., Pietra, S.A.D., Mercer, R.L.: The mathematics of statistical machine translation: parameter estimation. Comput. Linguist. 19(2), 263–311 (1993)
6. Fung, P.: A pattern matching method for finding noun and proper noun translations from noisy parallel corpora. In: Proceedings of the 33rd annual meeting on Association for Computational Linguistics, Morristown, NJ, USA, Association for Computational Linguistics, pp. 236–243 (1995)
7. Resnik, P., Melamed, I.D.: Semi-automatic acquisition of domain-specific translation lexicons. In: Proceedings of the fifth conference on Applied natural language processing, pp. 340–347. Morgan Kaufmann Publishers Inc, San Francisco, CA (1997)
8. Wu, D., Xia, X.: Large-scale automatic extraction of an English-Chinese translation lexicon. Machine Translation 9(3–4), 285–313 (1994)
9. Gale, W.A., Church, K.W.: A program for aligning sentences in bilingual corpora. Comput. Linguist. 19(1), 75–102 (1993)
10. Melamed, I.D.: A portable algorithm for mapping bitext correspondence. In: Proceedings of the 35th annual meeting on Association for Computational Linguistics, Morristown, NJ, USA, Association for Computational Linguistics, pp. 305–312 (1997)
11. Éric Gaussier: Flow network models for word alignment and terminology extraction from bilingual corpora. In: Proceedings of the 17th international conference on Computational linguistics, Morristown, NJ, USA, Association for Computational Linguistics, Selecting Target Word Using Contexonym Comparison Method 7, pp. 444–450 (1998)
12. Smadja, F., McKeown, K.R., Hatzivassiloglou, V.: Translating collocations for bilingual lexicons: a statistical approach. Comput. Linguist. 22(1), 1–38 (1996)
13. Carpuat, M., Fung, P., Ngai, G.: Aligning word senses using bilingual corpora. ACM Transactions on Asian Language Information Processing (TALIP) 5(2), 89–120 (2006)
14. Dorr, B.J.: The use of lexical semantics in interlingual machine translation. Machine Translation 7(3), 135–193 (1992)
15. Levin, L., Nirenburg, S.: The correct place of lexical semantics in interlingual MT. In: Proceedings of the 15th conference on Computational linguistics, Morristown, NJ, USA, Association for Computational Linguistics, pp. 349–355 (1994)

16. Moulin, B., Rousseau, D.: Knowledge acquisition from prescriptive texts. In: IEA/AIE '90: Proceedings of the 3rd international conference on Industrial and engineering applications of artificial intelligence and expert systems, pp. 1112–1121. ACM Press, New York (1990)
17. Kim, S., Yoon, J., Song, M.: Structural feature selection for english-korean statistical machine translation. In: Proceedings of the 18th conference on Computational linguistics, Morristown, NJ, USA, Association for Computational Linguistics, pp. 439–445 (2000)
18. Ji, H., Ploux, S.: Automatic contexonym organizing model. In: Proceedings of the 25th annual meeting of the Cognitive Science Society, pp. 622–627 (2003)
19. Ji, H., Ploux, S., Wehrli, E.: Lexical knowledge representation with contexonyms. In: Proceedings of the 9th Machine Translation, pp. 194–201 (2003)
20. Ji, H.: A Computational Model for Word Sense Representation Using Contextual Relations. PhD thesis, INPG, Grenoble, France (2004)
21. Foltz, P.W., Kintsch, W., Landauer, T.K.: The measurement of textual coherence with latent semantic analysis. Discourse Processes 25, 285–307 (1998)

Distance-Based Bloom Filter for an Efficient Search in Mobile Ad Hoc Networks

Byungryong Kim[1] and Kichang Kim[2]

[1] Department of Computer and Science Engineering, Inha Univ., 253, YongHyun-Dong, Nam-Ku, Incheon, 402-751 Korea
doolyn@inha.ac.kr
[2] School of Information and Communication Engineering, Inha Univ., 253, YongHyun-Dong, Nam-Ku, Incheon, 402-751 Korea
kchang@inha.ac.kr

Abstract. This study proposes a keyword search technique to reduce the traffic caused when applying Distributed Hash Table(DHT) base P2P systems as middleware for ubiquitous computing application in order to ensure effective search in Mobile Ad-hoc Network (MANET). For query with many keywords, nodes should be visited as much as the number of keywords and many posting files(inverted list) should be sent to the next node and these posting files bring about traffic. To solve this problem firstly the size of inverted list nothing to do with search result was diminished using distance of the hash values obtained by hashing keywords within document; secondly the positive false rate of bloom filter can be lowered since the size of posting file to be hashed can be reduced through distance-based bloom filter technique and by doing this the size of inverted list to be sent can be reduced as well. This technique is based on Plaxton's approach and the proposed technique was evaluated through a series of tests.

1 Introduction

In mobile ad-hoc environment every node has a wireless interface and network is voluntarily organized based on high mobility and limited resources. Each node shares or provides service provided by other node and resources demanded in many applications. Due to limited communication bandwidth, limited communication area, mobility, and the cut-off of power, a mode to reduce power consumption, each node has limited permanent connection with other nodes. This property is largely identical to voluntary organization, resource share or cooperative operation on which P2P systems are based. Therefore P2P technique can be applied as middleware for the application in mobile ad-hoc environment. This study proposes an effective keyword base search technique for the search applications in structured mobile P2P network based on Distributed Hash Table(DHT) among many P2P techniques. Currently there are two types of P2P file share and search method in large. One is Gnutella[1,2] where nodes save and mange each data and that used in unstructured P2P networks including Google[3], Yahoo[4], KaZaA[5], etc. The other is DHT where each node manages posting file mapping which of document has the keyword. In DHT node corresponding to the keyword manages posting file of the keyword. DHT provides far more

M.J. Smith, G. Salvendy (Eds.): Human Interface, Part I, HCII 2007, LNCS 4557, pp. 471–479, 2007.

effective search service than with broadcasting method and is the foundation of structured P2P systems such as Chord[6], CAN[7], Tapestry[8], Pastry[9], etc.

In case of local indexing that posting file is saved in local it leads to high cost of search to broadcast search query to every node. Reversely global indexing that posting file is managed by specific node provides more effective DHT base search of $O(log\text{N})$ than local indexing and each node manages and is responsible for posting file about keyword. However this also leads to high search cost because when search including n keywords is requested posting file should be sent to perform JOIN operations. Accordingly if this technique is used as middleware as it is in mobile ad-hoc environment, considerable consumption will be allowed in network traffic. The aim of this study is to reduce all the possible traffic.

Approx. 71.5% of query requesting search includes at least two keywords[10]. Which means in 71.5% of query in structured p2p systems search is performed only by visiting at least two nodes. In this case to obtain final result value JOIN operations should be performed and to perform the operations great amount of posting files must be sent. Therefore if possible it is important to distinguish URL list nothing to do with correct URL list corresponding query. If not possible, the size of URL list(posting file) should be minimized. Therefore this study proposes efficient keyword search method enabling search applications when applying structured P2P systems to establish communication infrastructure in mobile ad-hoc network. Proposed technique determines which node will be firstly visited by distance of keywords and makes possible to distinguish URL list nothing to do with query through distance value saved in node. With this technique it is possible to have small storage overhead and the search cost can be effectively decreased without any additional communication cost.

Rest of this paper is composed as follows: in chapter 2 we understand the basic operations of structured/unstructured p2p systems and search techniques for many keywords proposed earlier. Chapter 3 explains technique proposed in this study in detail. Chapter 4 explains distance-based bloom filter method combining bloom filter technique[11,12,13] to enhance proposed technique and evaluates the performance of technique through test. The last chapter 5 gives a conclusion.

2 Related Works

Unstructured p2p systems retrieve needed data by broadcasting search query to neighboring main server or neighboring node but this model is weak in that lots of traffics occur by flooding and to solve this point hybrid p2p model is presented. Using search query as super peer effectively diminishes flooding packets. Structured p2p systems are based on Distributed Hash Table(DHT). DHT is able to manage data and node data effectively by placing data hashed value through hashing function as SHA-1 and address information(ip address and port number) of node in the same hash space(i.e. Chord, Pastry, CAN, Plaxton mesh[14]). This study is based on DHT and Plaxton's approach is used to explain the proposed technique.

Fig. 1 shows the keyword search in DHT base Overlay network. Each node places hash value of keyword included in Data and node ID in the same hash range. In fig. 1 node A is responsible for keyword a's posting file(URL list of Data including keyword a) and node B is responsible for keyword b's posting file. Let's assume that node

C requests the data search including keyword a and b to node A. In step 2 node A should send the whole posting file for which the node is responsible to node B and in step 3 node B performs intersection(JOIN) operations using the keyword a's posting file and the posting file of keyword b for which the node is responsible and sends the result to the initial node A. In step 2 six URLs in total are sent but the total search results are only two. Ultimately performing query including lots of keywords generates a great deal of traffic and this necessarily results in a big load to mobile environment.

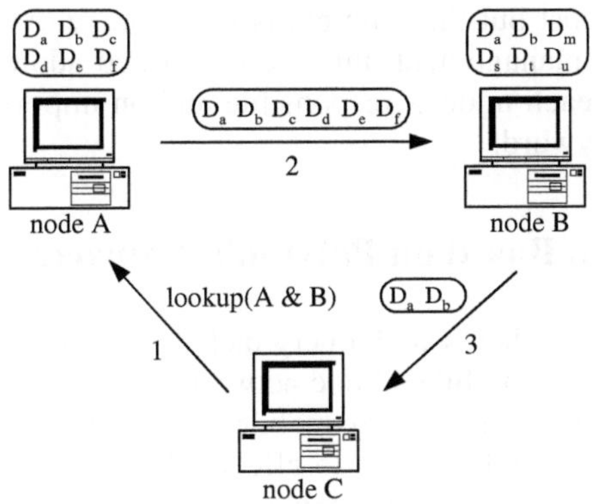

Fig. 1. A Simple approach to keyword search(a AND b)

To solve this problem many techniques have been proposed. A Hybrid partition strategy, Multi-Level Partitioning(MLP)[15] was intended to solve the problem by sending query to only reasonable number of peers to get correct result by dividing each node into logical k number of groups and sub groups. There are also bloom filter[10] to compress the intermediate findings (the intermediate findings before performing JOIN operations) and method to cache the former search results. Keyword fusion[16] provides effective search by each node's managing keyword fusion, a group of common keywords residing in the whole network. SCALLOP[17] was proposed to solve hotspot problem using balanced lookup tree. By evenly dispersing search request hotspot is avoided. KSS[18,19] proposed technique to create reverse list for term-set. mSearch[20] gives a solution with hybrid-index using multicast tree. CAN base pSearch[21] enables search to be effective visiting small number of nodes and sending small amount of data for search. In spite of all these many studies problems in multi-keyword search are not fully resolved yet.

MLP should perform grouping of each node and manage it by keeping dividing the groups into sub-groups again that this requires communication cost for the maintenance of each group. Bloom filter may have greatly varying hit rate according to the number of hash function and the size of bit vector. Also in case of search including many keywords, the search process is complicated. With keyword fusion, the more found common keyword, the lower search cost required but some amount of time is needed to find the common-keyword. Whenever common-keyword is found also the communication cost to be added to fusion dictionary managed by each node is incurred. As time goes by the amount of common keyword increases so that the size of fusion dictionary and partial keyword set increase accordingly and the storage overhead increases because all the combinations must be re-hashed and saved to each node additionally. SCALLOP avoids hot-spot and performs even lookup request but additional storage cost is required for lookup-table. pSearch tries to perform document search in the closest relation to query but it does not guarantee perfect search. In KSS the more the keyword included in document, the more the combinations of keywords so it increases insert and storage overhead and mSearch also generates storage overhead to save multicast tree. Hybrid-indexing[22] requires very small amount of cost but the problem is that storage overhead of each node is huge since saving and management must be performed additionally on reverse list of keyword for which each node is responsible and on important keyword of documents including the keyword.

3 Keyword Search Based on Palxton's Approach

As explained in chapter 2 when search query includes lots of keywords it should visit many mobile agents. Under mobile ad-hoc network, due to the high mobility of nodes and the limit of resources, visiting many mobile agents needs high search cost. Therefore this study proposes search method to effectively diminish posting list to be sent by each node in JOIN computing. To explain the proposed technique prefix based routing algorithms, plaxton's approach is applied.

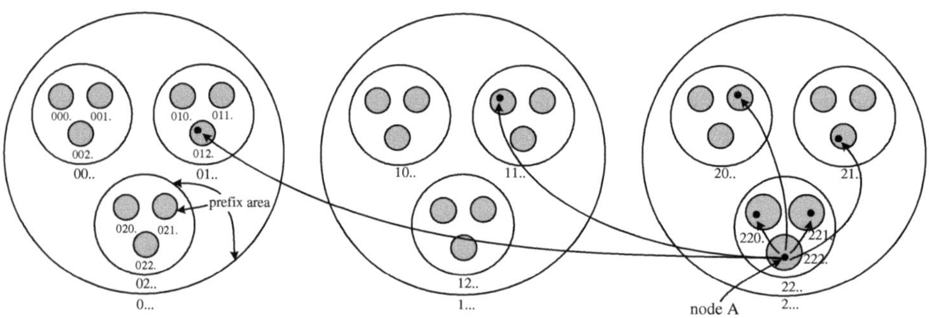

Fig. 2. A Prefix based routing in Plaxton's approach

Prefix based routing algorithms, plaxton's approach applies address data of each node and data(contents) shared by each node as the same hash limit. Hash limit has b value of antilogarithm and node and data have n-digit number as ID. Data (key-value Pairs) is saved to the next node in numeric order, i.e. when base is 3 key 0002 is saved to node 0002(there is no 0003). In the end key is managed under the same prefix group and in the same prefix group it is saved to the next node in numeric order. Fig. 2 shows the plaxton mesh base network when base is 3.

In fig. 2 node included in a prefix area at least knows one of nodes in other group (small word theory), i.e. 222 with node A. Nodes in the group topologically know the neighboring nodes such as 220, 221., 21.. , 0…, 1… In other words n nodes have these data as routing table therefore logarithmic link count is b*n= b*$O(log$N).

Platon's approach uses prefix based routing algorithms. In each step routing is done as the closest ID with the ID which is intended to find. For example when the destination is 0003 and b=3, the search at node 2221 is shown on Fig. 3.

start:	2 2 2 1	
step 1:	0 1 1 2	
step 2:	0 0 1 2	
step 3:	0 0 0 1	
step 4:	0 0 0 2	-- hit

Fig. 3. A Prefix based routing search when destination is 0003 and b is 3

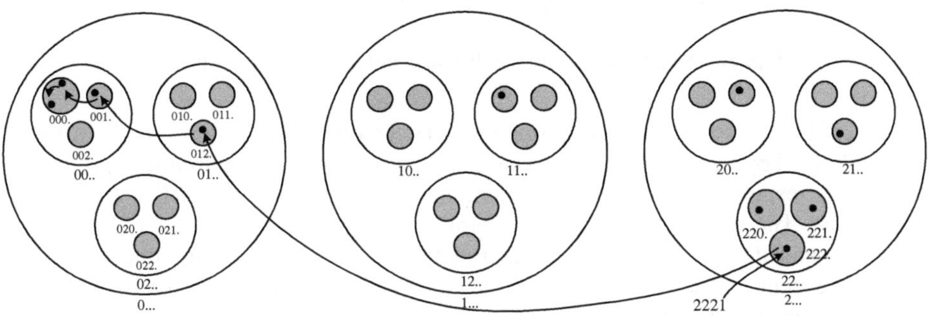

Fig. 4. 4 steps to find key 0003 at node 2221 in prefix based routing

Fig. 3 shows the case when base is 3. Since prefix of node requesting search is 0, finding index with prefix 0 from routing table, search query is sent to the node (id 0012) within the index. Reaching the node 0002 after repeatedly passing through the process finalizes the search. In fig. 4 in order to reach the destination the maximum $O(log\text{N})$ steps are needed. In this DHT base mobile P2P network, processing search query including many keywords generates as much as $O(log\text{N})$ * the number of keywords as explained on fig. 1 in order to send data from one node to another. In addition problem to be solved is the huge amount of data(posting list) sent from one node to another and many of these data are discarded. Hence if possible posting list with high possibility to be the final search result should be sent and posting list nothing to do with the final search result should not be sent from the beginning.

To solve this problem in the proposed method the node firstly receiving search query filters posting list not matching to the final search result using distance, the difference of hash values among keywords within search query. Each node passes through the following process to publish posting list.

In fig. 5 step *p1* hashes each keyword within data shared by node with function like SHA-1 and makes it to m-bit identifier. Keywords w_0, w_1, and w_2 are hashed into 0210, 2001, and 0112 each. In step *p2* these hashed values are arranged in descending order. The order of arranged keywords is now w_2, w_0, w_1 from w_0, w_1, w_2. In step *p3* arranged keywords are grouped in pairs, the difference of hash values between the pairs is computed and the difference is the distance between two keywords. In the end the distance between w_2 and w_0 is 0021 and the distance between w_0 and w_1 is 1022. Final posting list is created in step *p4* adding these computed distance. However the distance of the smallest keyword becomes -1. In fig.4 each posting list created this way is saved to node 0112(Data's URL, w_2, 0000), node 0211(Data's URL, w_0, 0021) and node 2002(Data's URL, w_1, 1022). Node 0112 in fig. 4 is responsible for URL of data including keyword w_2, node 0211 for w_0, and node 2002 for keyword w_1.

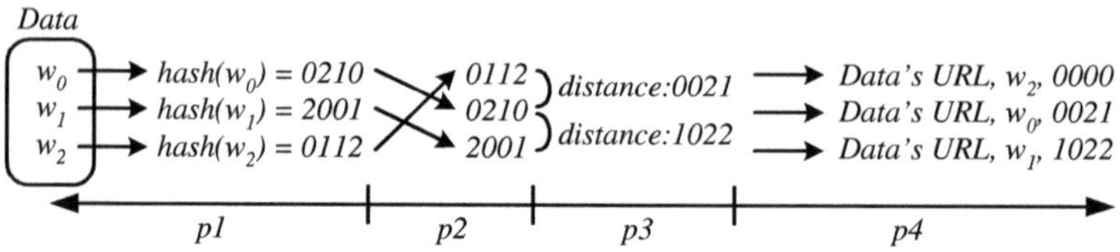

Fig. 5. 4 Steps to compute the distance and add this to posting list

Generally in case of performing search query Search(w_0 & w_1 & w_2) node 0211 responsible for keyword w_0 is firstly visited and all the URL list including w_0 is sent to node 2002 responsible for w_1. JOIN operations are performed on URL list including this and w_1. The result value of JOIN operations is again sent to node 0112 responsible for w_2 and JOIN operations are performed again. The final result value performed this way is sent to the node requesting search. However proposed search algorithms here are different from this and the process is as follows.

Every keyword included in search query is hashed at the first node requesting search. And then visit is performed on from node responsible for keyword having the largest value by hashed value. Therefore node 2002 responsible for keyword w_1 is firstly visited. Node 2002 sends posting list only to node 0211, responsible for the second largest keyword excluding list with distance larger than 1002, the difference between w_0, the second largest keyword and w_1. At node 2002 URL with distance larger than 1022 is excluded for the following reason.

Since all the keywords included in Data are arranged in descending order by hashed value, distance larger than w1 and 1022 means that the data does never include w_0. But if smaller than 1022 it means that the data may or may not include w_0 so that it should not be excluded. Likewise also at node 0211 responsible for node w0, intermediate search list is sent to node responsible for w_2, excluding URL with distance larger than 0021. Accordingly many URL lists are discarded by distance value. Consequently low traffic is used that in mobile ad-hoc environment with high mobility and limited resources this helps perform effective search.

As studied by now, URL list without specific keyword can be easily removed by applying distance to posting list. Test results show that traffic is reduced by approx. 40% with 2 keywords included in query and traffic is reduced by 29% on average with maximum 5 keywords. Test results will be examined in chapter 4. In this study bloom filter technique was applied to enhance the proposed algorithms. Chapter 4 explains and tests how the distance of keywords included in data can be applied to bloom filter technique.

4 Applying Distance to Bloom Filter

In an example applying bloom filter as shown on [10], the filter is made by hashing all the keywords included in data shared by nodes or the filter is created by hashing posting list, URL list including each keyword. In the end the size of bloom filter should be increased to reduce positive false and if not the positive false gets high. At the same time sending traffic has no choice but to increase. But with the proposed distance technology since the number of keywords to be hashed is considerably reduced and then positive false rate can be reduced, sending traffic can be lower than when using bloom filter only. Fig. 6 is search flow using bloom filter when applying distance.

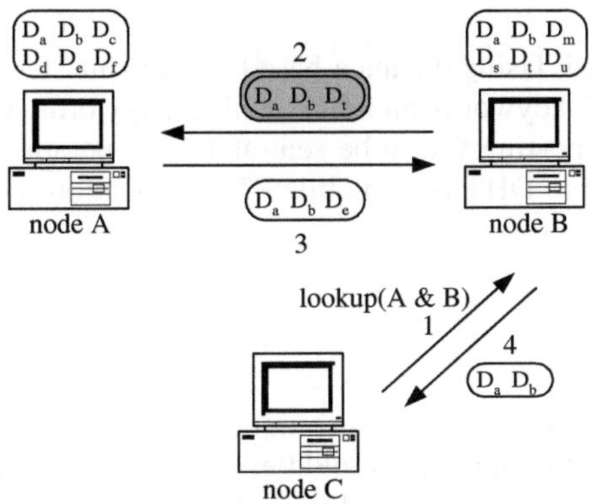

Fig. 6. Search process flow with bloom filter applied to distance

In fig. 6 node A is responsible for posting list(D_a, D_b, D_c, D_d, D_e, D_f) of data including keyword A and node B is responsible for posting list of data including keyword B. Search(A & B) begins at node I. Because the hash value of B is bigger than the hash value of A, search query is firstly sent to node B. In case of [10] using all the posting lists bloom filter is created but applying distance can exclude great amount of URL list. In fig. 6 hashing only D_a, D_b, and D_t makes bloom filter. Only 3 URL lists, not 6 lists, are used so the correctness is improved. Node A sends D_a, D_b, and D_e, matched URL lists through received bloom filter to node B. Node B performs JOIN operations and search is completed sending the resulting values, D_a, D_b to node C.

In step 2, since bloom filter with high correctness is sent by distance in step 3 the amount of URL lists irrelevant to search results can be diminished much more. Fig.7 shows the test results on the performance of Plaxton's approach, distance, distance-based bloom filter. Chord-Simulator[23] was used in the test and the test was performed by making Chord simulator modified by adding distance process, multi keyword process routine, and bloom filter making routine to original Chord simulator.

According to the test results when the number of keywords included in query was 2 with the application of distance, traffic was reduced by 40% and when the 3 keywords are included in query the performance was improved by approx. 35%. With 2 to 5 keywords the traffic was reduced by 29% on average. With bloom filter the rate was

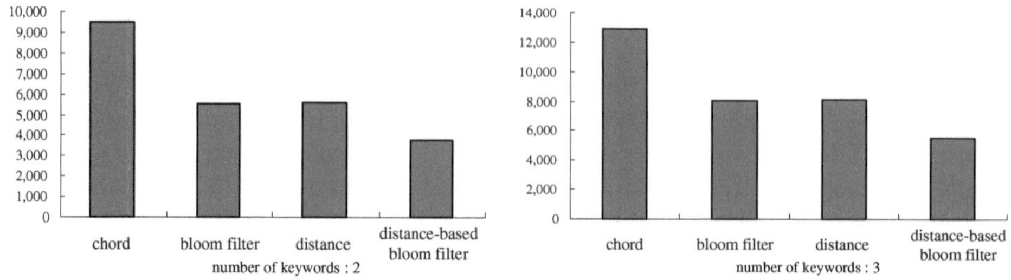

Fig. 6. Comparison of posting lists to be sent at the intermediate stage when keyword included in search query is 2, 3

similar or a bit increased. Using distance-based bloom filter technique, the traffic was reduced by 61% with 2 keywords and 57% with 3 keywords. Consequently with the decrease in the size of posting file to be sent at JOIN operations at each node, it can pay a great contribution to DHT base mobile P2P application service in mobile ad-hoc network.

5 Conclusion

This study proposed keyword search technique to decrease the traffic caused when DHT base P2P systems are applied as middleware for ubiquitous computing applications in order for effective search in Mobile Ad-hoc Network(MANET). Keywords included in document were hashed and using the distance between hash values URL list(inverted list) not included in document can be removed. When the number of keywords is 5 at maximum the traffic was declined by 29% on average. Distance-based bloom Filter technique was proposed to enhance the performance. Since the size of inverted list to be hashed can be reduced due to distance, positive false rate of bloom filter was lowered. Finally through the distance-based bloom filter large amount of inverted lists irrelevant to search was decreased.

Acknowledgements. This work was supported by INHA UNIVERSITY Research Grant.

References

1. Gnutella. http://gnutella.wego.com
2. The Gnutella Protocol Specification v0.41 Document Revision 1.2. http://rfc-gnutella.sourceforge.net/developer/stable/index.html/
3. http://www.Google.com
4. http://www.Yahoo.com
5. KaZaA media dekstop, http://www.kazaa.com/.
6. Stoica, I., Morris, R., Karger, D.R., Kaashoek, M.F., Balakrishnan, H.: Chord: A scalable peer-to-peer lookup service for internet applications. Proc. of the ACM SIGCOMM, pp. 149–160 (2001)

7. Ratnasamy, S., Francis, P., Handley, M., Karp, R., Shenker, S.: A scalable content-addressable network. In: SIGCOMM'01 (2001)
8. Hildrum, K., Kubiatowicz, J.D., Rao, S., Zhao, B.Y.: Distributed object location in a dynamic network. In: Proc. of the 14th ACM Symp. on Parallel Algorithms and Architectures (SPAA), August 2002, ACM Press, New York (2002)
9. Rowstron, A., Druschel, P.: Pastry: Scalable, decentralized object location, and routing for large-scale peer-to-peer systems. In: Guerraoui, R. (ed.) Middleware 2001. LNCS, vol. 2218, pp. 329–350. Springer, Heidelberg (2001)
10. Reynolds, P., Vahdat, A.: Efficient Peer-to-Peer Keyword Searching. In: Endler, M., Schmidt, D.C. (eds.) Middleware 2003. LNCS, vol. 2672, Springer, Heidelberg (2003)
11. Broder, A., Mitzenmacher, M.: Network Applications of Bloom Filters: A Survey. In: Proc. of the 40th Annual Allerton Conference on Communication, Control, and Computing, pp. 636–646 (2002)
12. Dillinger, P.C., Manolios, P.: Bloom Filters in Probabilistic Verification. In: Hu, A.J., Martin, A.K. (eds.) FMCAD 2004. LNCS, vol. 3312, Springer, Heidelberg (2004)
13. Dillinger, P.C., Manolios, P.: Fast and Accurate Bitstate Verification for SPIN. In: Graf, S., Mounier, L. (eds.) Model Checking Software. LNCS, vol. 2989, pp. 57–75. Springer, Heidelberg (2004)
14. Plaxton, C.G., Rajaraman, R., Richa, A.W.: Accessing nearby copies of replicated objects in a distributed environment. In: ACM Symposium on Parallel Algorithms and Architectures, pp. 311–320 (1997)
15. Shi, S., Yang, G., Wang, D., Yu, J., Qu, S., Chen, M.: Making peer-to-peer keyword searching feasible using multi-level partitioning. In: Voelker, G.M., Shenker, S. (eds.) IPTPS 2004. LNCS, vol. 3279, p. 2. Springer, Heidelberg (2005)
16. Liu, L., Ryu, K.D., Lee, K.-W.: Keyword fusion to support efficient keyword-based search in peer-to-peer file sharing. In: CCGRID 2004, pp. 269–276 (2004)
17. Chou, J.C.Y., Huang, T.-Y., Huang, K.-L., Chen, T.-Y.: SCALLOP: A Scalable and Load-Balanced Peer-to-Peer Lookup Protocol, pp. 419–433
18. Gnawali, O.D.: A Keyword-Set Search System for Peer-to-Peer Networks, Master's Thesis, Department of Computer Science, Massachusetts Institute of Technology (June 2002)
19. Liang, Z., Fu-tai, Z., Fan-yuan, M.: KRBKSS: A Keyword Relationship Based Keyword-set Search System for Peer-to-peer Networks, Journal of Zhejiang University Science, vol. 6A(6) (2005)
20. Gulati, A., Ranjan, S.: Efficient Keyword Search using Multicast Trees in Structured p2p Networks, submitted to Middleware (2005)
21. Tang, C., Mahalingam, M., Xu, Z.: psearch: Information retrieval in structured overlays (October 20, 2002)
22. Tang, C., Dwarkadas, S.: Hybrid Global-Local Indexing for Efficient Peer-to-Peer Information Retrieval. In: Proceedings of the Symposium on Networked Systems Design and Implementation (NSDI) (June 2004)
23. The Chord Simulator, http://pdos.csail.mit.edu/chord/sim.html and http://cvs.pdos.csail.mit.edu/cvs/~checkout~/sfsnet/

Integrated Physically Based Manipulation and Decision-Making Tree for Navigation to Support Design Rationale

Ji-Hyun Lee and Tian-Chiu Li

Graduate School of Computational Design, National Yunlin University of Science & Technology, Touliu, Yunlin 640, Taiwan
{jihyun,g9434714}@yuntech.edu.tw

Abstract. We explore the idea of integrated physically based manipulation and decision-making tree for navigation to support design rationale. It introduces a way to support user participation. This paper also proposes and validates a framework which combines several techniques to aid user to make decisions. We hope this research can assist untrained users to quickly feel comfortable and satisfied for a computer aid system by rational way. This paper focuses on a system prototype for the client customization process in apartment plan design to exemplify our concepts.

Keywords: Rational, Navigation, Decision-making tree, Fuzzy.

1 Introduction

It places importance on user participation in contributing to the success of a generative system has been a long held theoretical belief [10]. There are two fundamental issues in user participation: First, most users have little knowledge about design professionals. Second, the users are novices who may not easily manipulate the software the designers use. Research prototypes and applications address the first question by rational knowledge-intensive systems. When users need the knowledge of design for rational, the system should guide them to the goal through vast information space. To search and visualize the vast information space, building decision-making trees is necessary for navigation. Navigation is the cognitive theory that incorporates landmarks as general orientation points and be able to display design spaces at arbitrary levels of details and allow users to identify their abstraction settings. Decision-making tree can provide a navigational manner that allows designers to access desired information in a generative design system easily. In addition, decision-making tree can be operated by user to enable knowledge management and reuse. It can help users to generate their alternatives and provide reasoning by causal relationships. Physically-based manipulation is introduced to address the second question, because it can let people understand their decisions easily by interacting design elements intuitively and iteratively in the schematic design layout process. This paper proposes a 3D generative system using navigation concept to integrate physically-based manipulation and rational decision-makings to help apartment buyer to participate the layout design through information space.

M.J. Smith, G. Salvendy (Eds.): Human Interface, Part I, HCII 2007, LNCS 4557, pp. 480–489, 2007.

This paper also proposes and validates a framework which combines all techniques we mention above to aid user to make decisions. The objective is to provide the user with effective and appropriate decision-making representations and inference mechanisms to support the design rationale. We focus on a system prototype for the client customization process in an apartment plan design.

2 Literature Review

In the last several decades there has been a tremendous wave of interest in the relationship between rational and design process. In the past, design matter usually depends on professional designer's experience, ability, and preference. This kind of design always causes that design process has a lot of contradiction and irrationality.

Relatively, rationality brings about comprehension between profession designer and client. Rational is a good understanding of the ideal process will help us to know how to proceed and having a standard procedure. It becomes much easier to measure the progress that a project is making. We also can compare the project's achievements with those that the ideal process calls for. Decision making is an important part for rational design. A series correct decision is more approachable for rational. Gordon [6] introduced the generic information-processing model integrating views of naturalistic decision making. It helps us to know people's cognition and decision making in real-world. In this model, information enters the model and then is processed at one of three levels: automatic skill-based processing, intuitive rule-based processing, and analytical knowledge-based processing.

Parnas and Clements [8] also descript why irrationality will happen:

(1). In most cases the people who commission the building of a design case do not know exactly what they want and are unable to tell us all that they know.
(2). Many of the details only become known to us as we progress in the implementation, the resulting design may be one that would not result from a rational design process.
(3). Experience shows that human beings are unable to comprehend fully the plethora of details that must be taken into account in order to design and complete and correct project.
(4). All but the most projects are subject to change for external reasons.
(5). Human errors can only be avoided if one can avoid the use of humans.
(6). We are often burdened by preconceived design ideas, ideas that we invented, acquired on related projects.
(7). Often we are encouraged, for economic reasons, to use design component that was developed for some other project.

The central purpose of this study was to investigate how to approach rational design. For catching people's requirement, Fuzzy logic is a rule-base type of control that uses concepts from fuzzy set theory, and it has emerged as a criterion of intelligent control capable of dealing with complex and ill-defined problems [5]. The process of designing fuzzy logic systems is explained by the following development steps [5]:

(1). **Fuzzification:** defining linguistic variables and types of membership functions to describe a situation.
(2). **Inference:** formulating the proposition logic such as IF-THEN statements to make fuzzy rules for inference.
(3). **Defuzzification:** obtaining the crisp results.

For having a steady rational design process, it builds a knowledge base system is necessary. Knowledge base system supports the user's actions or controls and prompts the user's intention so that the user can design drawings according to the standard or the best way [11]. The rule of knowledge in knowledge base system can include fuzzy theory and expert knowledge to infer recommend of design for user. But, knowledge base system has self contradiction and redundancy information problems [11].

Navigation is a way which can look for useful material in large information spaces. Navigable space has become a new tool of labor and a common way to visualize and work with any data, since it can be used to represent both physical spaces and abstract information spaces [7]. Many researchers have suggested that imposing a spatial structure to information spaces (regardless of their inherent structures) could make the task of seeking information a more effective and pleasing activity [3]. A decision making cone tree (Fig.1), it is one of the best known 3D graph layout techniques in information visualization [4], supports user to present and record user's motive and history path. It also navigators predict goal for design process.

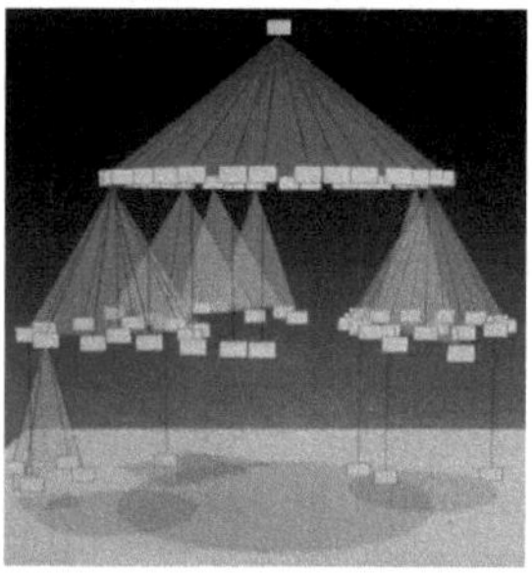

Fig. 2. Cone tree (Image courtesy of Palo Alto Research Center; Brian Tramontana, photographer.)

In order to advance navigation, one of the geometric models, known as direct manipulation (DM), however, can support design integration in terms of continuous representation of the objects and actions of interest; physical actions or presses of labeled buttons instead of complex syntax; and rapid incremental reversible operations whose effect on the object of interest is immediately visible [9].

Physically based manipulation, a concept of DM, therefore, can be recognized as a promising approach to implement interactive computational tools to solve the user participation problems because it allows people to communicate their ideas easily by interacting intuitively and iteratively with the design elements.

3 Developing Framework for Rational Design

This research attempts to approach rational design. So, knowing human's decision making process is helpful for creating a module to help user making decision.

Based on the Gordon's model, we want to develop and match our conceptual framework of the information processing procedures with the generic model, each of which has different level of details. We show the modified framework in Fig. 2.

3.1 Framework

Automatic processing and intuitive rule-based processing. Automatic processing, which is at the lowest level of the model, directly triggers an action response. Reflex is activated in this process. In a sense, this is not even really decision making or problem solving [6]. For intuitive rule-based processing, the cues trigger retrieval of appropriate cue-action rules from long-term memory. This research tries to integrate techniques from processing in order to aid people who can make decision easily. Navigation support will travel the information space to got useful path to approach the goal. Navigation operation will intelligently to highlight trick for user; and then, Physically-based manipulation can support design integration intuitively in terms of continuous representation of the objects and actions of interest; physical actions or presses of labeled buttons instead of complex syntax; and rapid incremental reversible operations whose effect on the object of interest is immediately visible [9].

Analytical knowledge-based processes. When intuitive rule-based processes do not provide a satisfactory solution or decision and situation is allowed, the decision action will go upward to the Analytical knowledge-based processes. There are three stages in this process: evaluate explanation, evaluate action, and evaluate plan. In our research, for evaluate explanation phase, we build decision-making tree which represents causal net by using scenario-based design. Scenario-based design (SBD) makes ambiguous and dynamic situations easy to evoke reflection in the design process by embodying concrete design actions [1]. A design scenario, in turn, can be expanded into more detailed sub-scenarios, each of which has a set of associated design moves that the respective process should accomplish. Each design move is associated with each operation procedure of physically-based manipulation as well as constructs a node of decision tree, which can help users to know the solution path of the design decision-making process. In evaluate action phase, we build a generative system which can generate prototype in a show time, so that user can find plausible solutions and judging the validity of solutions relative to the goals and selection among alternatives. Finally, evaluate plan phase will justify decision-making tree by rational design management. The process of designing consists of a long series of design decisions tree so that we can analyze causal-effect history path by rational design management that attempts to capture the rationale behind the design (required decisions, the reasoning behind the choice of final solution(s) and dependencies between decision

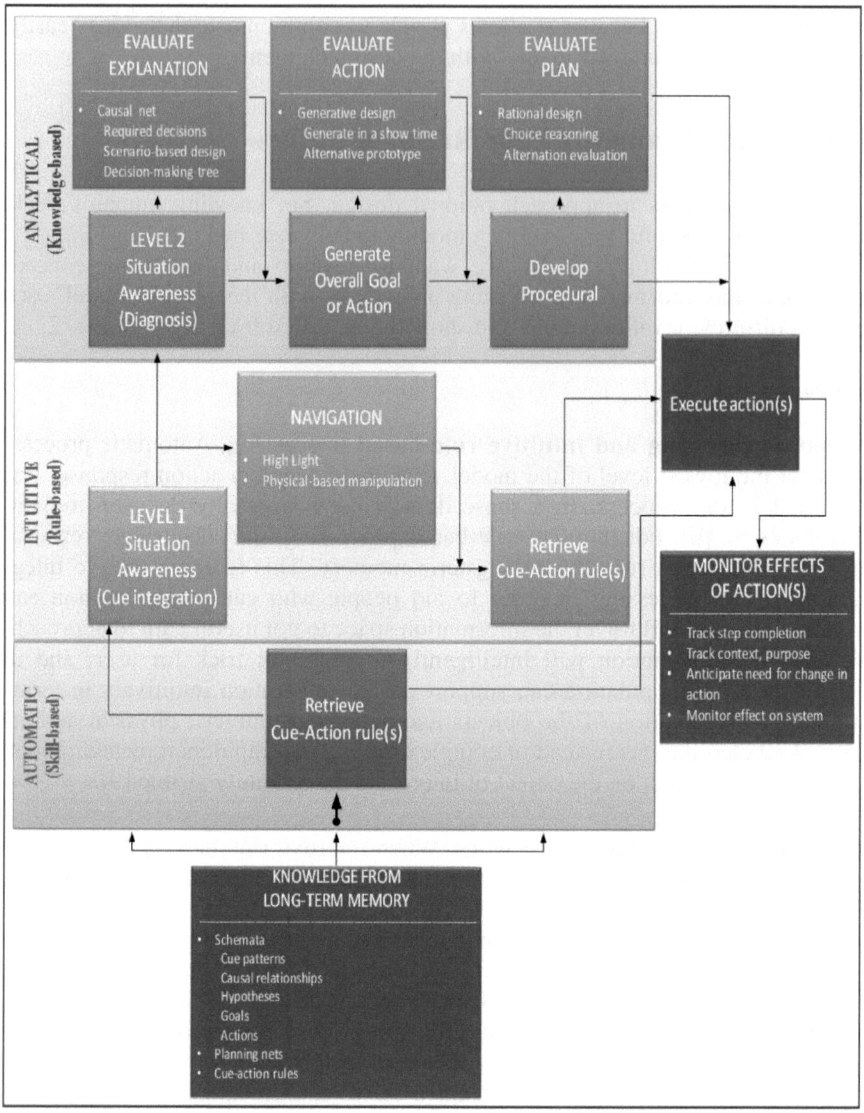

Fig. 3. Modified from the generic information-processing model by Gordon [6]

nodes). The system has an optimization design rationale network which will match with the decision making tree. After that, it can infer the related knowledge and suggest design advices or solutions to user.

3.2 Irrationality Reason and Solution

All of technique place to framework in order to deal with irrationality problem. We mention the cause in section 2. Table 1 is arranged for match the problems and solution.

Table 1. The irrationality reason and solution

irrationality reason	Solution
1. People do not know exactly what they want and are unable to tell us all that they know.	Fuzzy
2. Many of the details are not follow rational design process.	Knowledge based system
3. Human beings are unable to comprehend fully details.	Navigation
4. Subject to change for external reasons.	No
5. Human errors.	Decision-Making tree support to revise. Physically based manipulation.
6. Preconceived design ideas.	Rational design
7. Economic reasons use design component from other project.	Generative design

3.3 Difference from Original Model

Several theoretical positions have been sketched out here. This research has presented a general framework which includes technical method to reduce complex in decision making process. In intuitive rule-based processing, navigation guide people to deal with computer operation for easing off use obstacle. Forward, causal net assist to understand whole situation and encourage consistency in evaluate explanation. Generative design urges to explore different design alternative and show possible outcome immediately in evaluate action. Rational design provides a craftsmanship to support effective judge.

4 Prototype System

4.1 Implementation

The prototype implements various components based on the framework introduced in Section 3: the knowledge base and inference module implemented by commercial software (*fuzzy*TECH), a 3D layout generator implemented using physically based manipulation, a design justification module, and a graphical user interface (GUI) module. Fig. 3 shows the Fuzzy inference using by *fuzzy*Tech.

The 3D layout generator is implemented using Jogl on top of OpenGL as APIs, to visualize and provide an efficient physically based manipulation mechanism to communicate between the user and the system. The design justification represented by a decision tree is implemented using the Java programming language. Finally, we built a GUI implemented by Java to communicate with the fuzzyTECH Java Runtime Library and Jogl for the 3D layout generator, and to seamlessly integrate all the modules with a coherent interface.

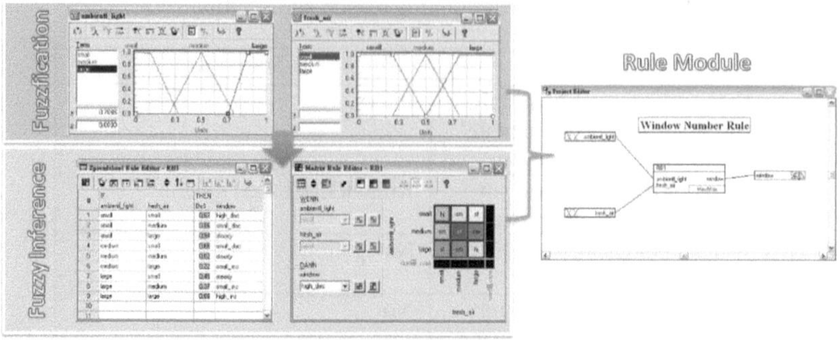

Fig. 3. Fuzzy Inference by *fuzzy*Tech

The GUI of our system includes three major parts: (1) a query window to capture the user requirements and preferences, (2) a 3d decision cone tree window to record design alternatives created by the user manipulation and to support the design justification, and (3) a main window to visualize the 2D and 3D floor plan.

4.2 Result of the Application to Apartment Customization Process

The purpose of this research is helpful to apartment customization process in real world. We provide a scenario to describe practical application. We show the work flow between client and agency in Fig. 4.

When a client is willing to buy a pre-sale apartment, he/she will go to the sale agency to acquire information of unit. If all condition is satisfy except layout, he/she

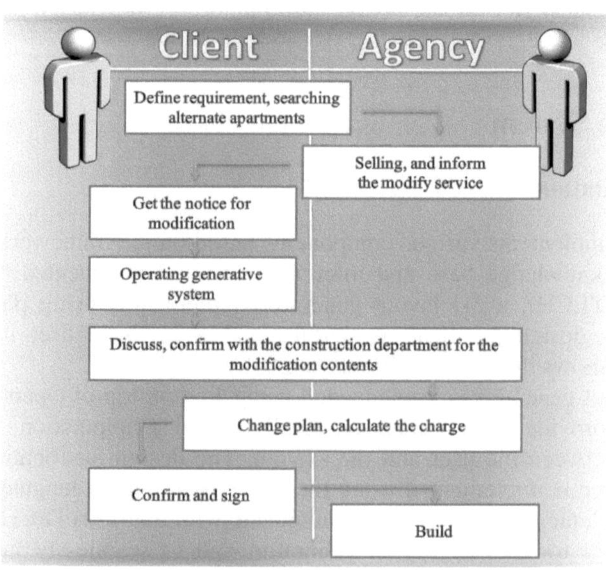

Fig. 4. Describes the process of apartment plan customization between client and agency (revised from Cheng [2])

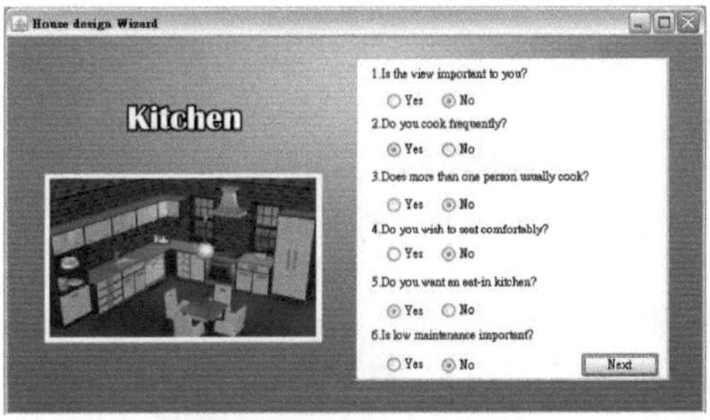

Fig. 5. Query window

could ask the agency that he wants to modify the layout. In this moment, agency can provide the generative system to help client express what kind of layout he/she want.

A client can start by answering a series of design move questions from the Query window (Fig. 5). According to the client's preferences, our system proceeds with the fuzzy inference mechanism.

The knowledge window (Fig. 6) provides related knowledge to the client in the plain language form via the defuzzification mechanism in the right column of the window. The left column of knowledge window shows a 3D decision cone tree. Decision cone tree can help to present a lot of elements in restricted space. It also has another advantage which can let user know forward goal or roll back from error situation. Each hierarchical step represents a corresponding design move, which is selected from the knowledge base. When the client selects an L-D-K configuration node from the decision tree, for example, the right column of the window explains the basic relations between the living room (L), dining room (D), and kitchen (K) to the clients. The main window (Fig.7) displays a 2D floor plan and 3D model. The client easily edits the plan via moving a component or physically selecting it and dragging the edge of the component. The updating immediately propagates to the 3D model.

After client is satisfied, he/she can make a deal with agency to change the layout of unit.

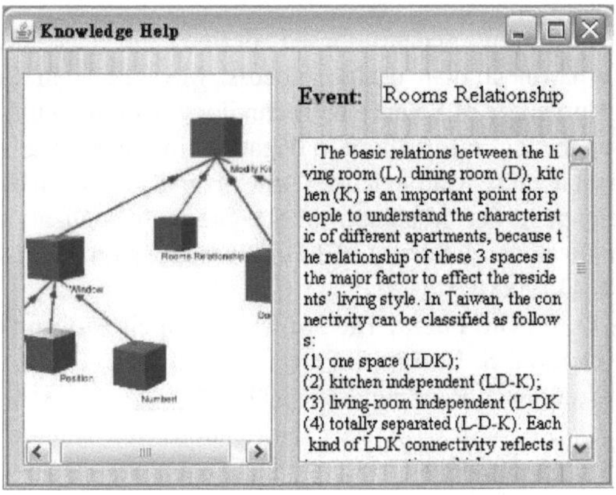

Fig. 6. A 3d decision cone tree window

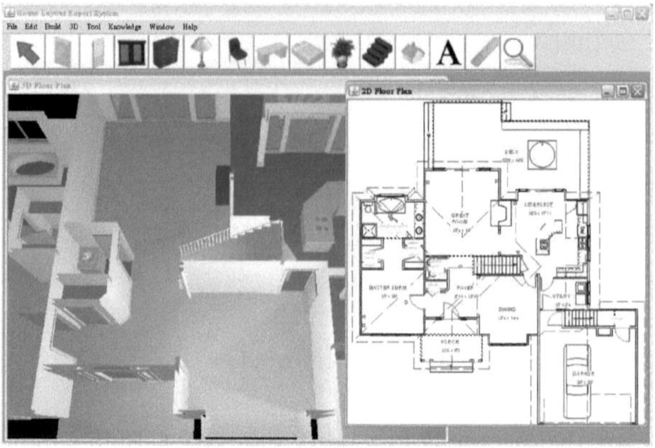

Fig. 7. Main window for 2D & 3D floor plan

5 Conclusion

This paper presents a new rational approach by integrating the decision-making tree with 3D physically based manipulation for the client customization process in apartment plan design. We also provide a research model to validate system ability. This research project forms a comprehensive study, of which only a summarized version of the findings can be presented here. Nonetheless, there is clearly a large number of interesting research issues is raised. For example, intelligent adjustment is a good topic for fuzzy rule base. How does system use rational to assist in problem diagnosis and solution. And, how does system collect people's feedback and evaluation of design alternatives.

References

[1] Carroll, J.M.: Five reasons for scenario-based design. Interacting with computers 13(1), 43–60 (2000)

[2] Cheng, Y.-C.: Integrating scenario-based design and case-based design for user participation in Apartment plan design process, MS Thesis in Computational Design. National Yunlin University of Science & Technology: Douliu, Yunlin, Taiwan (2005)

[3] Chien, S.-F.: Supporting information navigation in generative design systems, PhD Dissertation in Architecture. Carnegie Mellon University: Pittsburgh, PA, USA (1998)

[4] Herman, I., Melancon, G., Marshall, M.S.: Graph Visualization and Navigation in Information Visualization: A Survey. IEEE Transactions on Visualization and Computer Graphics 6(1), 24–43 (2000)

[5] Lin, X.-C., Peng, Q.-F.: Oh! Fuzzy. The 3rd wave, Taipei, Taiwan (1994)

[6] Mickens, C.-D., Gordon, S.-E., Liu, L.: An introduction to Human Factors Engineering. Addision Wesley Longman, Reading, MA (1997)

[7] Papaconstantinou, G.: Screen Space: Navigation and Interactivity. In: eCAADe conference, Greece, pp. 392–398 (2006)

[8] Parnas, D.L., Clements, P.C.: A rational design process:How and why to fake it. IEEE Trans. Softw. Eng. 12(2), 251–257 (1986)

[9] Shneiderman, B.: Direct Manipulation for Comprehensible, Predictable and Controllable User Interfaces. In: 2nd international conference on Intelligent user interface, Orlando, USA, pp. 33–39 (1997)
[10] Terry, J., Standing, C.: The Value of User Participation in E-Commerce Systems Development. Informing Science Journal 7, 31–45 (2004)
[11] Xiao, L., Hao, X.: A Knowledge Base System in the Residential Intelligent CAD System. In: CAADRIA, Shanghai, China, pp. 161–170 (1999)

Text Analysis of Consumer Reviews: The Case of Virtual Travel Firms

Xinran Lehto[1], Jung Kun Park[1], Ounjoung Park[1] and Mark R. Lehto[2]

[1] School of Consumer Science, Purdue University, West Lafayette, IN 47906, USA
{xinran,park4,opark}@purdue.edu
[2] School of Industrial Engineering, Purdue University, West Lafayette, IN 47906, USA
{lehto}@purdue.edu

Abstract. This study uncovered critical domains and themes of compliments and complaints that influence consumers overall satisfaction with using virtual travel agencies. Four domains, namely, "customer Service & Support", "trip schedule change", "product experience" and "firm credibility" were identified as areas where problems arose and extreme dissatisfaction resulted. Three domains emerged as the best predictors for satisfaction: "product experience", Customer Service & Sopecially at managerial level is critical for these virtual travel firms. The outcomes of this research lend insights into how to effectively manage consumer online reviews and turn this domain of valuable resources into knowledge and a strategy tool for the virtual travel segment of the hospitality and tourism industry.

Keywords: Virtual travel agent, e-satisfaction, online product review, complaint, compliment.

1 Introduction

Online product and service review websites allow consumers to express both negative and positive opinions. Typically, complaints are the outcome and expression of consumer dissatisfaction while compliments arise from elevated levels of consumer satisfaction. Consumers are increasingly resorting to these online public channels to publicize their assessment of products or services they have purchased and share their advices for or against using them. In fact, consumers have gone as far as taking the role of protest framing on the web (Ward & Ostrom, 2006), seeking justice. On the other hand, increasingly potential consumers have turned to online consumer reviews for product opinions and other information. In fact, online user evaluation is becoming a dominant factor influencing consumer product decision choices. Because its managerial relevance and potential for enhancing customer relationship management, online consumer product reviews have drawn increasing attention from marketing practitioners and marketing academics (Goetzinger, 2006). The rich textual data from online consumer reviews and feedback have provided a golden mine for insight generation. Researchers have experimented with various methodological approaches including computer-assisted programs to analyze these textual data.

M.J. Smith, G. Salvendy (Eds.): Human Interface, Part I, HCII 2007, LNCS 4557, pp. 490–499, 2007.

As a direct outcome of the internet advances and a newly emerging but evolving segment of the tourism and hospitality industry, virtual travel ventures such as expedia.com and travelocity.com have replaced the traditional travel distribution channels. The characteristics of these business entities being entirely virtual and the intense competition from up and coming new ventures backed by powerful technologies have led to an increasing need to assess their customer satisfaction level by identifying the "good and beautiful" as well as the "the bad and ugly" of this business segment of the hospitality industry. As a result, feedback directly from consumers can bear strategically important implications. Customer Relationship Management is more than implementing technology or software; it is finding new ways of acquiring, retaining, and increasing the profitability of current customers based on better knowledge about the critical domains and determinant factors contributing to consumers' e-satisfaction. To date, however, studies that examine consumers' positive or negative opinions in the form of complaints or compliments within the context of virtual travel intermediaries have been sparse.

Against this backdrop, this study intends to identify critical domains and factors of compliments and complaints that influence consumers overall satisfaction with using virtual travel agencies. It was hoped that the outcome of this research would lend insights into how to effectively manage consumer online reviews and turn this domain of valuable resources into knowledge and a strategy tool for the virtual travel segment of the hospitality and tourism industry.

2 Complaining, Complimenting, and Satisfaction Behavior

The advent of the Internet brought with it a very convenient channel for consumer articulation of company feedback. In contrast to offline consumer complaining behavior, the nature of the Internet allows consumers to both view and post comments that transcend time and physical space. Research has demonstrated that factors driving complaint and compliment behavior do not necessarily converge (Friman & Edvardsson, 2003) and can impact firm customer relationship differently (Rusbult et al. 1991).

Although complaints inherently include negative information about products and services, they can be used by companies to address service recovery opportunities and improve on future services. Consumers have several choices of channels for expressing their complaints or compliments if they choose to do so (Huppertz, 2003; Singh, 1989). Consumers are able to articulate their communication directly to the company, privately to friends, family members or acquaintances, or to third-parties such as the Better Business Bureau or a lawyer.

One of the main goals of customer relationship management is to address and manage consumer complaints and compliments because they are directly related to consumer (dis)satisfaction (Zeithaml, 2000). It is particularly concerning to companies when negative word-of-mouth or complaints are expressed online since they have the potential to be viewed by a potential multitude of people and can remain online indefinitely. On the other hand, online complaints can be viewed as an opportunity if handled correctly. In the case that companies are able to satisfactorily recovery from complaints, they may be able to sustain levels of consumer satisfaction (Bitner & Meuter, 2000; Buttle & Burton, 2001). It may take effort for companies to monitor and respond to complaints not directly expressed to them, but the effort may be well worth it. Complaints, as a general rule, should be welcomed by companies. However, when complaints are expressed online to the public for all to see, other consumers may be impacted by what they are exposed to. If a consumer is not familiar with a particular company that receives criticism, the complaint may be enough to discourage new customers from purchasing that company's products or services. Online third-party complaints are on the rise, and have the potential to create devastating effects for offending companies (Hogarth, English, & Sharma, 2001).

In general, consumers tend not to express their compliments as aggressively as when they need to complaint. Consumer complimenting behavior, however, also plays a role in understanding consumer satisfaction. Unfortunately, research has shown that consumer compliments receive less attention by companies when compared with consumer complaints (Erickson & Eckrich, 2001; Kraft & Martin, 2001). An act as simple as acknowledging a consumer's input is enough to make them feel appreciated and satisfied, whether they have provided negative or positive feedback. No response to any consumer-initiated contact may distance the consumer.

In general, consumers tend to view other consumers' opinions as more credible and reliable when compared with company-generated information. Thus, complaints and compliments that originate from other consumers may be especially persuasive. Moreover, the attributes that lead to dissatisfaction and satisfaction are rarely the same (Friman & Edvardsson, 2003; Oliver, 1997). One particular attribute may lead to dissatisfaction if missing or functioning poorly, but does not necessarily lead to elevated levels of satisfaction if it is present and functioning adequately. The opposite can be true, where the absence of a particular attribute may not be noticed, whereas its presence leads to satisfaction and complimenting behavior.

Because consumer complaints and compliments may not necessarily be motivated by the same attributes, it is especially important to understand the underlying motives for each. However, both consumer complaints and compliments stemming from dissatisfaction or satisfaction should be studied and addressed to ensure customer retention.

3 Dimensions of E-Satisfaction

Purchasing flight ticket and making hotel reservation through an online travel agency are fairly recent phenomenon. Although the majority of online travel agencies are only a few years old, innovative applications of technology have provided efficient customer service tools for travel intermediaries such as Price.com and Travelocity.com (Kim et al., 2005). Earlier research showed that online customers were not moderately satisfied with such hospitality and tourism websites and their services (Jeong et al, 2001). This finding lies in the fact that the intermediaries offer collected information of travel products without purchasing them before they sell them to the travelers but that travelers have to purchase the product before they experience it. Complaints or compliments inherent in this form can happen in searching information, making reservation and transaction, communicating with customer service center, or comparing with the purchased and experienced product (Kim et al, 2005).

Hsu (2006) identified examined five information dimensions on international hotel websites deemed as important to users. The five dimensions are reservation information, facilities information, contact information, surrounding area information, and

Website management. Cai, Card, and Cole (2004) empirically examined web delivery performance of tour operators in the United States and showed a high level of customers' dissatisfaction with the services. Kim et al. (2005) identified six factors contributing to Chinese hotel website users' satisfaction and purchase intention: information needs, service performance and reputation, convenience, price benefits, technological inclination, and safety. Most recently, Oldenburger et al (2007) while examining critical incidents in e-business in general, identified seven frequently mentioned or examined domains for customer e-satisfaction: marketing activities, information provision, transaction (ordering, processing, packing, delivery and returning), product, customer service and support (contact and response), and website (website structure, security, ease of use and technology access). These underlying dimensions, commonly found previous studies on e-transaction, online communities, and customer relationship management can readily serve as a starting point for this research.

4 Research Objectives

Up to date, research on online mega travel intermediaries has been very limited. The existing research on hospitality websites are largely concerned with website functionality and design attributes, and tend to rely on structured survey questionnaires. The objectives of this research are two fold: First, our research was intended to uncover critical and prominent (dis)satisfaction factors for online virtual travel agencies through investigation of textual data of consumer reviews of their purchase and use of services and products provided by these organizations. Second, the researchers attempted to identify underlying structures and domains of these satisfaction and dissatisfaction factors.

5 Data

Consumer feedback comments and reviews from 6 virtual travel companies, namely, expedia.com, orbitz.com, travelocity.com, priceline.com, hotwire.com and cheapticket.com, were collected online from an independent third party website epinions.com between January 2004 and February 2005. This website was designed to provide a platform on the web for consumers to express their feedback on products they have purchased from online sources. Epinion is deemed as unbiased as it is a consumer oriented open forum and does not allow filtering of comments (Goetzinger et al, 2006). Consumers' reviews on the six virtual companies were chosen as they are representative of the mega technology-back-up virtual travel agencies. The sample size included 672 review items including 465 complaining incidents and 216 complimenting incidents.

6 Analysis Procedure

This research attempted to analyze textual data of consumer reviews of virtual travel companies through two a combination of computer assisted analysis using a text mining software TextMiner (Lehto, 2004) and traditional qualitative inductive approach that use expert judgments.

Complaining and complimenting reviews were analyzed separately using TextMiner. Text Mining is a software program that allows analysis of large quantity of textual data and can identify key predictors for an outcome variable. Researchers utilized TextMiner for the purpose of uncovering key words or indigenous concepts. This first order analysis was supplemented with a second order expert judgment process to uncover themes and domains and the interrelationship between those themes. This procedure was deemed as more appropriate than conducting a factor analysis on the identified key words due to the nature of textual data with large missing values and distribution concerns.

While a deductive analysis approach using existing frameworks such as prior classification schemes and domains identified by other researchers was initially contemplated for the second order analysis, the researchers eventually decided to utilize a open and inductive approach due to the unique nature of the virtual travel agencies and the uniqueness of a tourism product, which in essence is an experience based good, that required the make-up of multiple components of the hospitality industry, e.g. hotels, airlines, destinations, car rental firms etc. Once key words/concepts are identified directly from the data with the aid of TextMiner, the researchers went through a further induction process to construct themes and patterns. These researcher-constructed domains was one step removed away from the actually data and can impose new meanings to the original textual data. To combat possible subjectiveness of this second order analysis, three researchers conducted this procedure independently to achieve better reliability.

7 First Order Analysis

First order analysis from TextMiner analysis was intended to identify key indigenous concepts that best predict consumers review outcomes. Results from this procedure revealed 72 indigenous predictive concepts for consumer dissatisfaction and 44 indigenous predictive concepts for consumer satisfaction in using virtual travel companies. These concepts are directly derived from the original texts and have predictive rate of .87 or higher. In other words, 87% of the time, when these concepts appeared in a consumer's review, it results in either a very negative or very positive evaluation of his or her use of the virtual travel companies (Table 1 and Table 2). A few examples of prominent dissatisfaction predictors were "supervisor", "manager", "beware", "USAir" and "horrible". Clearly there is a dissatisfied human interaction element that contributes to customers' extreme negative evaluations. Top predictive concepts for satisfaction included "allow", "efficient", Hawaii", "preferences" and "destinations". There was strong product experience element ("destination", "Hawaii"). Destination, which tends to be the central component of a vacation experience, clearly leads to a very positive outcome for the consumers.

8 Second Order Analysis

In order to gain better knowledge of the contextual and domain information from the data, the researchers conducted a second order analysis based on these identified indigenous concepts with no computer assistance. This further induction process was aimed at uncovering underlying domains or structures of e-(dis)satisfaction.

Table 1. Predictive Indigenous Concepts for Dissatisfaction

Indigenous Concept	Freq.	Prediction Rate	Indigenous Concept	Freq.	Prediction Rate
supervisor	65	1.00000	Warranty	15	0.93750
manager	29	1.00000	response	40	0.93421
beware	19	1.00000	speak	30	0.93333
USAir	18	1.00000	rude	21	0.92308
horrible	16	1.00000	useless	15	0.92308
fix	13	1.00000	upgrade	12	0.92308
terrible	12	1.00000	cancellation	150	0.91997
admitted	10	1.00000	standby	11	0.91667
arguing	9	1.00000	advised	11	0.91667
Bug	9	1.00000	screwed	11	0.91667
Louisville	9	1.00000	refunded	110	0.89610
International	9	1.00000	contacted	30	0.90909
envelope	8	1.00000	claimed	20	0.90179
script	8	1.00000	department	18	0.90000
death	8	1.00000	word	26	0.89899
dirty	8	1.00000	agreement	35	0.89828
unwilling	7	1.00000	rep	42	0.89362
report	7	1.00000	upset	8	0.88889
lied	7	1.00000	wasted	8	0.88889
departs	7	1.00000	indicated	8	0.88889
disconnected	7	1.00000	automated	8	0.88889
incorrect	7	1.00000	policy	45	0.88235
attitude	6	1.00000	mistake	28	0.87500
apology	6	1.00000	circumstances	9	0.87500
moral	6	1.00000	claiming	8	0.87500
ram	6	1.00000	Alexandria	8	0.87500
disaster	6	1.00000	immediate	7	0.87500
recourse	6	1.00000	misleading	7	0.87500
sucks	6	1.00000	fax	7	0.87500
Aaron	6	1.00000	bait	7	0.87500
Canada	6	1.00000	Cleveland	7	0.87500
Shuttle	6	1.00000	property	7	0.87500
reschedule	6	1.00000	ratings	7	0.87500
broker	6	1.00000	deliver	7	0.87500
refused	32	0.96970	answers	6	0.87500
worse	48	0.94816	warning	6	0.87500

Table 2. Predictive Indigenous Concepts for Satisfaction

Indigenous Concept	Freq.	Prediction Rate	Indigenous Concept	Freq.	Prediction Rate
allows	18	1.00000	consis-tently	6	1.00000
efficient	15	1.00000	prompt	6	1.00000
Hawaii	14	1.00000	Timely	6	1.00000
destina-tions	13	1.00000	gas	39	0.97619
prefer-ences	12	1.00000	navigate	29	0.96667
meal	11	1.00000	Packages	20	0.95238
awesome	10	1.00000	feature	55	0.94595
driving	10	1.00000	surprised	14	0.93333
agencies	10	1.00000	content	13	0.92857
painless	8	1.00000	quick	38	0.92683
domestic	8	1.00000	excellent	46	0.92000
Suites	8	1.00000	gallon	32	0.91608
pump	8	1.00000	register	20	0.90909
enjoyed	7	1.00000	cruises	19	0.90476
breakfast	7	1.00000	amazed	9	0.90000
Upscale	7	1.00000	rentals	34	0.89474
terrific	6	1.00000	cheapest	64	0.88889
ranging	6	1.00000	Florida	24	0.88889
resorts	6	1.00000	glad	8	0.88889
sections	6	1.00000	courteous	8	0.88889
Disney	6	1.00000	easy	186	0.88570
Westin	6	1.00000	handy	7	0.87500

Figure 1 demonstrated the analyst-constructed domains that best explain satisfaction and dissatisfaction. Four key complaint domains emerged from the second order analysis: customer services and support (e.g. rude, refuse, unwilling), product experience (e.g. bug, dirty, stranded), firm credibility (e.g. lied, scam) and Trip schedule change (e.g. reschedule, depart). These were areas where problems arose and extreme dissatisfaction resulted. Three key compliment domains were noted from the second order analysis: product experience (e.g. enjoyed, Hawaii, Disney), customer services and support (timely, courteous, prompt), and website (e.g. efficient, easy, painless). It is interesting to note that domains or elements that predict satisfaction and dissatisfaction are not identical. These results concur with Goetzinger (2006) et al's research which detect bivalent (dis)satisfiers (belonging to both satisfaction domain and dissatisfaction domain) and monovalent (dis)satisfiers (belonging exclusively to either satisfaction or dissatisfaction domain). Scheduling change and transaction domains in this case predicts dissatisfaction but not satisfaction, whereas website design related issues contributes strongly to satisfaction is not a prominent issue for dissatisfaction.

Customer service and support contributes to both dissatisfaction and satisfaction but is a more dominant domain predicting dissatisfaction. Product experience, on the other hand, contributes to both satisfaction and dissatisfaction but remains a more dominant predictor for satisfaction. These assessments took into consideration of both the average prediction rate of indigenous concepts under a domain and the frequency of occurrences of those concepts.

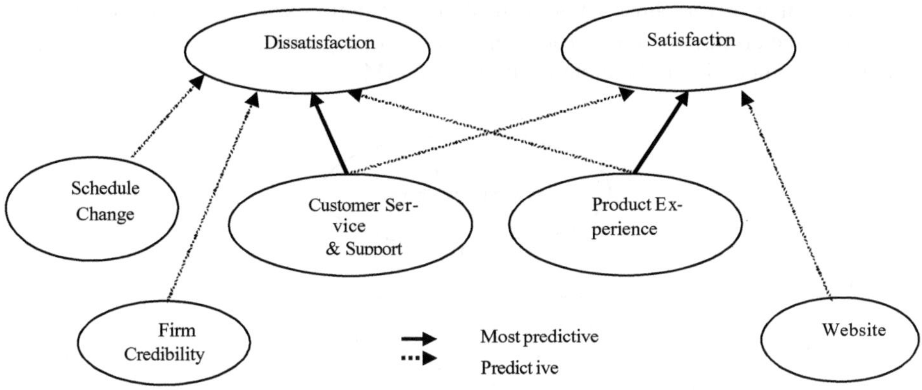

Fig. 1. Overall Predictive Domains

Complaints that were originated from these domains seem to invoke very negative reviews and lead to extreme dissatisfaction. It is apparent that when extremely unhappy with their experience with a company, consumers use very strong negative expressions such as "sucks", "horrible", "disaster" and "worst". These crucifying and flaming concepts certainly serve as strong negative word of mouth (NWOM). Customer service and support domain being the most dominant predictor can be further classified into 4 key themes: customer point of contact, firm employee attitude, firm employee competency and negotiation outcome (Figure 2). Use point of contact as an example, it seems that when a problem occurs, the consumers tend to resort to off line human/customer support. It is most interesting to note that there seems to be a pattern of an escalation of communicated parties. When communication with front line employees (firm representatives) failed to resolve issue at hand, the supervisors or managers were sought after. The ineptitude of managerial personnel could really tick consumers off. Off line support, when with sub-par performances at the managerial level, seems to lead to a landslide of negativity.

Compliments from these domains contribute to profuse eulogy towards the services and product provided such as "awesome" and "terrific". The most dominant predictor was "product experience". This domain can be further classified into two key themes: product features (e.g. Hawaii, Disney, cruises) and experiences (e.g. enjoyed, excellent). It seems a very positive product experience, in this case, trip experience is crucial for evoking elevated level of satisfaction.

9 Conclusion and Implications

This research sought to investigate consumer's complaints and compliments towards virtual travel firms using a grounded theory approach. Key domains and sub-domains were effectively identified for both ends of consumer satisfaction spectrum: aggravated dissatisfaction level which leads to complaints and elevated satisfaction level which leads to compliments. The results demonstrated that a very unhappy customer seems to be one that was very unsatisfied with their interaction with firm off-line support, both during the communication process and with the negotiation outcomes. Extremely unhappy customers do not hesitate to flame (NWOM).

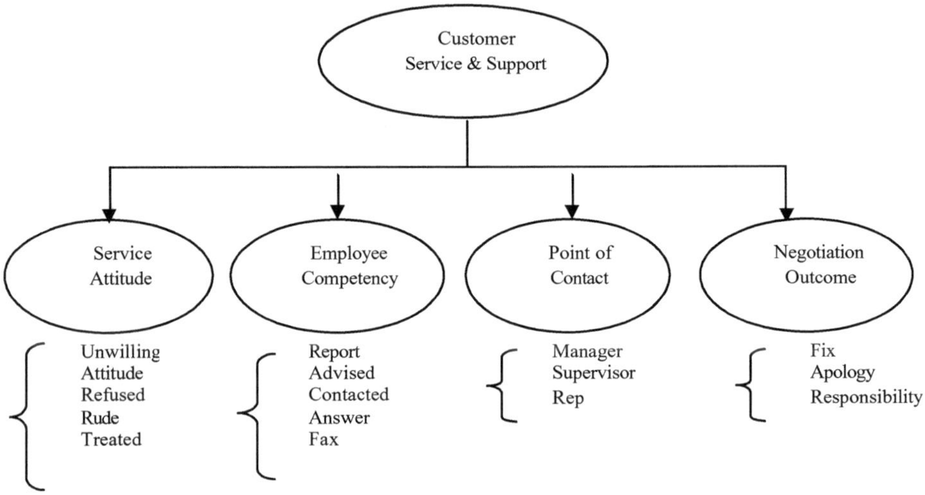

Fig. 2. Themes within Customer Service & Support Domain

This research agrees with the proposition that the key predictors for satisfaction and dissatisfaction are different. While the dissatisfactory areas need to be dealt with carefully, these areas do not necessarily lead to elevated level of satisfaction. This study affirms the proposition that virtual travel companies are unique in their own right. This has been shown through the discovery that the strongest complaints that travelers have are, paradoxically, off-line related. Previous research attributes dissatisfaction to a multitude of website functionality factors, this research, while on all virtual entities, does not reveal this domain as critical in contributing to dissatisfaction. But rather, Off-line customer support is key. Virtual travel firms need to take note of broken linkages between customer online interaction with technology and off-line human communication. When there is a concern or issue with a purchase, supervisors or managers seem to play a pivotal role in resolving the conflict. Otherwise, consumers perceive their ineptitude as very detrimental. Considering that virtual travel agents are already one step removed from human communications, human connection and effective problem solving are ever more important.

The huge textual data sources, straight from the consumer's mind, provides a rich and open data platform to understand consumers' perspectives on product and product use. Systematic assessment of this data can have great potential to identity underlying constructs for a broad range of issues concerning consumer satisfaction. Knowledge of e-satisfaction factors and domains generated directly from consumer online reviews can help define a standard set of criteria that virtual travel intermediaries should meet to achieve the basic thresholds of customer satisfaction for these virtua ventures.

References

1. Bitner, M.J., Meuter, M.L.: Technology infusion in service encounters. Academy of Marketing Science 28(1), 138–149 (2000)
2. Erickson, G.S., Eckrich, D.W.: Consumer affairs responses to unsolicited consumer compliments. Journal of Marketing Management 17, 321–340 (2001)
3. Friman, M., Edvardsson, B.: A content analysis of complaints and compliments. Managing Service Quality 13(1), 20–26 (2003)
4. Goetzinger, L., Park, J.K., Widdows, R.: E-customers' third party complaining and complimenting behavior. International Journal of Service Industry Managementn 17(2), 193–206 (2006)
5. Jeong, M., Oh, H., Gregoire, M.: An Internet marketing strategy study for the lodging industry. American Hotel & Lodging Foundation (2001)
6. Kim, W.G., Ma, X., Kim, D.J.: Determinants of Chinese hotel customers' e-satisfaction and purchase intentions. Tourism Management 27, 890–900 (2006)
7. Oldenburger, K., Lehto, X., Feinberg, R., Lehto, M., Salvendy, G.: Critical purchasing incidents in e-business. Behaviour & Information Technology (In press, 2007)
8. Hogarth, J.M., English, M., Sharma, M.: Consumer complaints and third parties: Determinants of consumer satisfaction with complaint resolution efforts. Journal of Consumer Satisfaction, Dissatisfaction and Complaining Behavior 14, 74–87 (2001)
9. Huppertz, J.W.: An effort model of first-stage complaining behavior. Journal of Consumer Satisfaction, Dissatisfaction and Complaining Behavior 16, 132–144 (2003)
10. Law, R., Hsu, C.H.C.: Importance of Hotel Website Dimensions and Attributes: Perceptions of Online Browsers and Online Purchasers. Journal of Hospitality & Tourism Research 30(3), 295–312 (2006)
11. Oliver, R.L.: Satisfaction. A behavioral perspective on the consumer. McGraw-Hill, New York (1997)
12. Patton, M.Q.: Qualitative Research & Evaluation Methods. Sage Publications, Thousand Oaks, CA (2002)
13. Rusbult, C.E., Verette, J., Whitney, G., Slovik, L., Lipkus, I.: Accommodation Processes in Close Relationships: Theory and Preliminary Research Evidence. Journal of Personality and Social Psychology 60(1), 53–78 (1991)
14. Singh, J.: Determinants of consumers' decisions to seek third party redress: An empirical study of dissatisfied patients. Journal of Consumer Affairs 23(2), 329–363 (1989)
15. Ward, J.C., Ostrom, A.L.: Complaining to the Masses: The Role of Protest Framing in Customer-Created Complaint Web Sites. Journal of Consumer Research 33 (2006)

Computer Classification of Injury Narratives Using a Fuzzy Bayes Approach: Improving the Model

Helen R. Marucci[1,3], Mark R. Lehto[2], and Helen L. Corns[1]

[1] Liberty Mutual Research Institute for Safety, 71 Frankland Road, Hopkinton, MA 01748, USA
[2] School of Industrial Engineering, Purdue University, 1287 Grissom Hall, West Lafayette, In 47907, USA
[3] Department of Work Environment, UMASS Lowell, One University Avenue, Lowell, MA 01854, USA

Abstract. This paper summarizes improvements to an earlier developed Fuzzy Bayes approach for assigning coding categories to injury narratives randomly extracted from a large U.S. insurer. Improvements to the model included: adding sequenced words as predictors and removing common subsets prior to calculation of word strengths. Removing subsets and adding word sequences improved prediction strengths for sequences found frequently in the training dataset, and resulted in more intuitive predictions and increased prediction strengths. Improved accuracy was found for several categories that had proved difficult to code in the past. This study also examined the effectiveness of a two-tiered approach, in which narratives were first categorized at the broad level (such as [falls]), before classification at a more refined level (such as [falls from heights].) The overall sensitivity following a two-tiered approach was 79% for predicting classifications at the broad category level and 66% for the more refined prediction categories.

Keywords: Textmining, Occupational Safety, Workers Compensation, BLS, Textminer.

1 Introduction

Injury narratives can provide useful information for selection of interventions for the prevention of injuries [1], [2] [6]. Using the computer to help classify narratives has the potential for reducing the burden implicit in manually reviewing and classifying large numbers of narratives from administrative injury databases.

A computerized approach based on Fuzzy Bayes logic was introduced in previous studies [3], [4], [5]. This study furthers that investigation by: 1) using separate prediction and training datasets, 2) adding sequence words as predictors, 3) reducing the strength of common words, found in more than one category, by removing common subsets, and 4) determining the value of using this method to assign more specific classifications.

M.J. Smith, G. Salvendy (Eds.): Human Interface, Part I, HCII 2007, LNCS 4557, pp. 500–506, 2007.

2 Methods

Workers' compensation claims case narratives were randomly extracted from a large US insurer. A trained, three person coding panel classified the narratives into two digit event categories (leading to injury) according to the Bureau of Labor Statistics (BLS) Standard Occupational Injury and Illness Classification (SOIIC) coding scheme [8]. Only narratives where at least 2 out of 3 coders agreed on the code were used (n=10,389), these codes were considered as the gold standard for comparing computerized coding. Three thousand narratives were extracted as the prediction data set, while the Fuzzy Bayes classifier "trained' from the probabilities of categories given word, word combinations or word sequences of 7,389 narratives. A previous study [5] found that using multiple word predictors, (the MW model) when compared with using single words alone, improved both the sensitivity and specificity of the computer generated codes. In this study all models were compared with the MW model.

As reported earlier [5] and repeated here using a new classification strategy and narratives, this method is capable of predicting injury event categories with a high level of accuracy (MW model, 78%). The new strategy included adding two word sequences to the wordlist and removing subsets (MW2SEQSUB model), this marginally improved the overall accuracy of the model at the category level (79%). Results for each category (for training and prediction datasets combined and the prediction data set only) are given in Tables 1 and 2 respectively. However, the prediction of the category when including word sequences were determined at higher strengths and were clearly more intuitive (Table 3). The optimistic bias [7] of predicting training and prediction narratives combined as compared with prediction narratives alone ranged from 6 to 10%.

3 Results

If *sequenced* words are found frequently in a category of the training set, the strength of the sequence as a predictor can be much higher than using words not in sequence. The MW2SEQ model correctly predicted 33 additional narratives, which had been poorly predicted in the MW model. The mean prediction strength for those 33 narratives was 7% higher with the sequenced words. Overall the sequence model predicted classifications at higher strengths than word combinations (MW mean strength .82, MW2SEQ mean strength .87). Examples of the improved predictions from the MW model to the MWM2SEQ model are shown in table 3.

There was a larger improvement when adding word sequences at the two digit level with the largest improvement in the contact category 'struck by' (8%). This may be important because the sensitivity of contact categories have been consistently low [5]. Results for two digit categories (with greater than 100 claim narratives) are provided in Table 4. The larger categories maintained a high level of accuracy ('Fall from same level' had sensitivity of 84% and 'highway incident' had sensitivity of 97%).

Finally, although the model with word sequences predicted more specific classifications, some of the categories had too few training narratives (i.e. [fires and explosions], n=42) for an adequate number of unique predictors to be generated.

Table 1. Sensitivity, Specificuity and Positive Predicitive Value of Bureau of Labour Statistics (BLS) Event Categories Comparison to Manual Coding of two Different Fuzzy Bayes Models: Training and Prediction Cases Combined (N=10,389)

		Multiple word model			Multiple & 2 word sequences model (including removing subsets)		
BLS Event Category	N	Sensitivity	Specificity	Positive Predictive	Sensitivity	Specificity	Positive Predictive
Contact with objects & equipment	1,890	0.73	0.98	0.90	0.76	0.98	0.91
Falls	1,617	0.93	0.94	0.76	0.93	0.95	0.78
Bodily Reaction & Exertion	3,315	0.89	0.92	0.84	0.91	0.93	0.85
Exposure to Harmful Substances or Env	1,290	0.90	0.98	0.87	0.90	0.98	0.88
Transportation Accidents	1,351	0.95	0.94	0.71	0.95	0.97	0.82
Fires & Explosions	63	0.25	1.00	0.94	0.30	1.00	0.86
Assaults & Violent Acts	382	0.59	1.00	0.90	0.59	1.00	0.92
Other Events or Exposures	481	0.22	1.00	0.88	0.31	1.00	0.88
Overall	**10,389**	**0.83**	**0.95**	**0.83**	**0.85**	**0.96**	**0.85**

Table 2. Sensitivity, Specificuity and Positive Predicitive Value of Bureau of Labour Statistics (BLS Event Categories Comparison to Manual Coding of two Different Fuzzy Bayes Models: Prediction cases Only (N=3,000)

		Multiple word model			Multiple word and 2 Word Sequences Model (including removing subsets)		
BLS Event Category	N	Sensitivity	Specificity	Positive Predictive	Sensitivity	Specificity	Positive Predictive
Contact with Objects & Equipment	557	0.65	0.97	0.84	0.69	0.97	0.84
Falls	470	0.90	0.92	0.68	0.89	0.93	0.69
Bodily reaction & Exertion	964	0.84	0.90	0.80	0.85	0.91	0.81
Exposure to Harmful Substances or Environment	361	0.88	0.98	0.85	0.87	0.98	0.85
Transportation Accidents	370	0.92	0.96	0.75	0.91	0.96	0.77
Fires & Explosions	21	0.19	1.00	1.00	0.19	1.00	1.00
Assaults & Violent Acts	117	0.49	1.00	0.88	0.47	1.00	0.90
Other Events or Exposures	140	0.17	1.00	0.71	0.25	1.00	0.73
Overall	**3,000**	**0.78**	**0.94**	**0.79**	**0.79**	**0.95**	**0.80**

Table 3. Examples of Improved Predictions for Multiple and 2 Word Sequence Model

		Multiple Word Model (Predicted incorrectly)		Multiple and 2 word Sequences Model (Predicted correctly)	
Claim Narrative	BLS event category	Prediction Strength	Predictor	Prediction Strength	Predictor
Employee was en route to job site and was in a motor vehicle accident and lacerated his face.	Transportation	0.78	LACERATED	0.94	IN-MOTOR
Battery fell on foot.	Contact w Obj & Equip	0.78	FELL&ON	0.92	FELL-ON&ON-FOOT
Hit knee on floor	Contact w Obj & Equip	0.57	FLOOR &HIT&KNEE	0.70	KNEE-ON
On the phone when the phone zapped employee in the ear then computer blew up and started smoking.	Expos to Harmful Sub or Env	0.78	COMPUTER &IN&ON	0.89	IN-EAR
Emp was struck on left foot by a lamp when the lamp fell offthe counter	Contact w Obj & Equip	0.78	FELL&ON	0.88	FELL&STRUCK-ON
Order selecting food products, balancing full. Box fell against his leg.	Contact w Obj & Equip	0.77	FELL	0.83	BOX-FELL
Emp was putting items away above her head on the retail floor causing a strain to her rt shoulder blade	Bodily Reaction & Exertion	0.80	FLOOR &HEAD	0.86	SHOULDER&TO-RT
Emp in back of vehicle man with gun jumped in vehicle emp jumpedout hitting lt knee on steering wheel	Assaults & Violent Acts	0.80	STEERING &WITH	0.86	WITH-GUN
Emp unloading trailer, tried to move bundle of rebar, 1400 lb. Pipe fell on left foot	Contact w Obj & Equip	0.87	FELL&TO &TRIED	0.92	FELL-ON&ON-FOOT
Clt wa putting things on shelf and a co-wprker struck clt with forklift	Transportation	0.86	SHELF &STRUCK	0.91	WITH-FORKLIFT
Trying to prevent student from falling	Bodily Reaction & Exertion	0.75	STUDENT	0.80	FROM-FALLING&TO

Table 4. Sensitivity, Specificuity and Positive Predicitive Value of Bureau of Labour Statistics (BLS) Event Categories Comparison to Manual Coding of two Different Fuzzy Bayes Models: Prediction Cases Only (N=3,000)

		Multiple Word Model				Multiple and 2 word Sequences Model (including removing subsets)			
Two Digit BLS Event Code	N	Sensitivity	Specif icity	Positive Predictive	Negative predictive	Sensitivity	Specif icity	Positive Predictive	Negative predictive
Contact:									
Struck against	163	0.35	0.99	0.67	0.96	0.42	0.99	0.71	0.97
Struck by	272	0.42	0.97	0.56	0.94	0.50	0.97	0.62	0.95
Caught in-between	100	0.79	0.99	0.65	0.99	0.81	0.99	0.67	0.99
Falls:									
Fall to lower level	159	0.57	0.98	0.56	0.98	0.59	0.98	0.63	0.98
Fall on same level	299	0.84	0.93	0.58	0.98	0.84	0.94	0.60	0.98
Bodily reaction & exertion									
Bodily Motion	218	0.36	0.98	0.62	0.95	0.36	0.98	0.60	0.95
Slip,Trip without fall	162	0.36	0.99	0.63	0.96	0.43	0.99	0.64	0.97
Overexertion	460	0.85	0.90	0.61	0.97	0.85	0.91	0.64	0.97
Repetative Motion	98	0.77	0.99	0.68	0.99	0.80	0.99	0.72	0.99
Exposure to Harmful sub or Env									
Exposure to temperature extremes	108	0.86	0.99	0.70	0.99	0.85	0.99	0.71	0.99
Exposure to caustic substances	123	0.79	0.99	0.85	0.99	0.81	0.99	0.79	0.99
Transportation									
Highway accident	205	0.97	0.96	0.64	1.00	0.97	0.96	0.67	1.00
Assaults and Violent Acts									
Assaults & violent acts by persons	111	0.59	0.99	0.77	0.98	0.59	0.99	0.78	0.98
Unclassifiable	140	0.29	0.99	0.69	0.97	0.35	0.99	0.71	0.97
Overall	**3,000**	**0.63**	**0.97**	**0.63**	**0.98**	**0.66**	**0.97**	**0.66**	**0.98**

4 Conclusions

From this study it can be seen that using a Fuzzy Bayes approach, threshold values can be used to filter out more difficult narratives for manual review. Examination of Tables 1 and 2 suggest that the mean strengths of the [falls] category compared with the [contact] category would result in more of the latter narratives filtered out for manual review, a process that would improve the final accuracy substantially. Some additional improvements are being considered for improving the accuracy of smaller categories.

In summary, a more refined computerized approach for classifying narratives using a Fuzzy Bayes approach has been developed. Sequenced words can be stronger predictors, if found sufficiently frequently in the training dataset, and their results use more intuitive predictors. Removing subsets is an important strategy to consider when common single word predictors are found in more than one category.

References

1. Sorock, G., Smith, G., Reeve, G., et al.: Three perspectives on work-related injury surveillance systems. Am J. Ind. Med. 32, 116–128 (1997)
2. Smith, G.S.: Public health approaches to occupational injury prevention: do they work? Inj. Prev. 7(suppl. I), i3–i10 (2001)
3. Lehto, M., Sorock, G.: Machine learning of motor vehicle accident categories from narrative data. Methods Info Med. 35(4-5), 309–316 (1996)
4. Sorock, G., Ranney, T., Lehto, M.: Motor vehicle crashes in roadway construction work zones: an analysis using narrative text from insurance claims. Accid. Anal. Prev. 28, 131–138 (1996)
5. Wellman, H.M., Lehto, M.R., Sorock, G.S.: Computerized coding of injury narrative data from the National Health Interview Survey. Accid Anal. Prev. 36, 165–171 (2004)
6. Lincoln, A.E., Sorock, G.S., Courteney, T.K., Wellman, H.M., Smith, G.S., Amoroso, P.J.: Using narrative text and coded data to develop hazard scenarios for occupational injury interventions. Inj. Prev. 10, 249–254 (2004)
7. Clancy, E.A.: Factors Influencing the Resubstitution Accuracy. Multivariate Classification Analysis: Implications for Study Design in Ergonomics, Ergonomics 40(4), 417–427 (1997)
8. Bureay of Labor Statistics. Occupational injury and illness classification manual. Us Department of Labor, Washington, DC (December 1992)

Involving the User in Semantic Search

Axel-Cyrille Ngonga Ngomo and Frank Schumacher

University of Leipzig, Institute for Computer Sciences, Johannisgasse 26,
04103 Leipzig, Germany
{ngonga,schumacher}@informatik.uni-leipzig.de

Abstract. Retrieval systems have become one of the most used categories of computer tools. Yet the interfaces of modern information retrieval systems fail to address the correlation between the user's context and her information need. Furthermore, they usually do not integrate methods that allow whole communities or groups of users to profit of single retrieval instances. In this paper, we present our vision of an innovative, collaborative information retrieval and presentation approach based on the human factors balance model. Our approach combines automatic natural language processing results with handcrafted knowledge models and integrates implicit retrieval based on intelligent document segmentation and presentation, providing users with contextually relevant information.

Keywords: Information presentation, Intelligent Systems, Knowledge Management.

1 Introduction

The increasing tendency to utilize computer systems in various organization related tasks leads to a steadily growing amount of digitalized information that needs to be retrieved when completing certain tasks. Retrieval systems have therefore become one of the most used categories of computer tools. In the context of companies, IR tools are used to retrieve supplementary information linked to the current task of the user, the primary information needs of the user being modeled in the business process along which she works. Nevertheless, retrieving such information can be of crucial nature for the completion of a task, especially when operating in the information society in which we live. Yet the interfaces of current information retrieval systems present many drawbacks that lead to poor search results and thus to high investments in search with respect to time. Most of them are still very primitive and usually consist of a simple text field for the input and a link list as output. Even more elaborate interfaces fail to integrate the user's context in her retrieval process and do not allow to map the users' knowledge with the content of the searchable corpus, leading primarily to an explorative but uncontrolled query refinement: the user successively reformulate her query to shape the retrieval result without knowledge of the relations between their semantic concepts, the keywords and the documents. The output is usually presented by displaying a short excerpt from the document and a link. The user is thus obliged to evaluate the relevance of each document based on

M.J. Smith, G. Salvendy (Eds.): Human Interface, Part I, HCII 2007, LNCS 4557, pp. 507–516, 2007.

very little knowledge about its real content. Furthermore she is not given the possibility to transfer her insights concerning the relevance of a piece of information for a query to the other members of the organization.

We present our vision of an innovative, collaborative IR and presentation approach based on the human factors balance model applied to individuals using a search engine to retrieve information and on interfaces for IR systems. We first present some related work, then some ergonomic considerations related to user interfaces for retrieval systems. In the fourth and fifth section user-friendly semantic search and an interface modeling this search are presented. The sixth section is concerned with further discussing our results and pointing out some future work.

2 Related Work

User interfaces for IR systems have been studied during the past four decades. The information access process is usually assumed to look as follows: For each information need, a query is formulated and a set of documents retrieved. This information is then analyzed by the user and the results of the analysis are then used to successively alter the input query in such a way that it represents the user needs in a more accurate way [8]. Yet other models such as the Bates' "berrypicking" [1], which takes into consideration the shift in interest of the user consuming retrieved information, are also valid for some users. There are different theories concerning how these research strategies can be supported. According to Hearst [3, p. 265] most of them can be categorized as contrasting browsing, querying, navigating or scanning.

Some modern interfaces try to assist the user during her search process by displaying large information spaces, through which the user can navigate [7]. Yet such approaches are likely to present too much information to the user. Therefore, different viewing operations such as view changes [6], zooming and panning [2], etc. have been defined. The major drawback of these techniques lies in the fact that they demand supplementary interaction from the user during the retrieval process. To achieve similar goals the category-based approach uses taxonomies or similar structures to cluster information and guide the users during their search [4]. They present drawbacks similar to those of the information space approach. Furthermore they potentially lead to a labeling problem, since different categories of users might tend to use different labels for the same categories. A possible solution consists of learning the labels or the meaning of the labels for users or user groups.

Another category of information that could be used to guide users is their context, since the demands to retrieval systems differ depending on the user's current task. Especially in companies, IR is a support process, designed to find information supplementary that complement the input information necessary to efficiently complete a given task. It is a much more controlled process than web retrieval and operates on a small number of data sources. Some interfaces utilize this fact and integrate views on the data sources in which the search should occur. Yet the users who know the content of a data source usually know where to find the relevant data and do not need to search. The fact that they search can be assumed to imply that they do know where the information they need can be found.

None of the methods presented above consider the drawbacks linked to the standard two-fold search process of IR, which consists of searching for relevant documents and subsequently searching for relevant information. Hearst calls these phases search and analysis [3]. We tackle this problem by providing the user with fine granular content and integrating implicit feedback collection in the interaction with the content. The design of the proposed interface is based on considerations on interactions between the user and her working environment, especially the technology she uses and the user himself. As theoretical basis, we use the human factors balance model.

3 Ergonomic Considerations

The interactions between users and their working environment can be modeled using a wide range of theories. We will consider the factors on a user retrieving information using the human factors balance model (HFBM) introduced by Smith and Sainfort [9]. It is a holistic approach to the factors, which potentially influence a user during the completion of tasks in the work context. The HFBM subdivides the influences on a user completing a task in four main points (see Fig 1): environment, task, organization and technology.

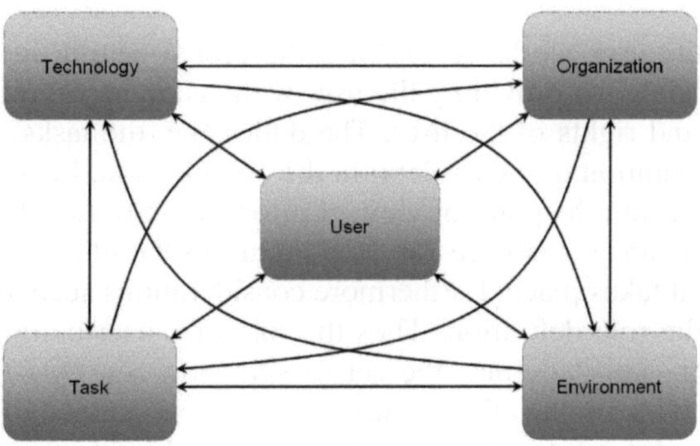

Fig. 1. The elements of a work system according to [9]

3.1 Environment

Meant here is the physical environment of the user. This aspect of the human factors can hardly be integrated in a user interface and will not be considered in our approach. Yet the "semantic environment", i.e. the set of cognitive concepts that are activated in the mind of the user when she retrieves information in a given context will be taken into consideration. Modeling the knowledge of a user is a difficult if not impossible task. Nevertheless, it is possible to model the knowledge contained in the data source(s) the user retrieves information from, making it easier for her to map the concepts activated in her mind and the concepts available in data. The semantic environment predefines the information need of the user. Hence it also alters the relevance she gives to the information she receives from a retrieval engine. It is thus of great importance to give the user the possibility to define the concepts she considers as relevant in a given task. Different knowledge modeling formalisms can be utilized when modeling concepts. In this work, we will focus on the use of terminological ontologies for modeling concepts.

3.2 Task

The user's task is of primordial importance when trying to model her semantic environment since it alters the relevance of information. This is due to the fact that the relevance of information is bound its potential usefulness when trying to achieve the goals set by the current task, on which it thus depends. Also relevant for the successful completion of a task are the input and output information that are related to it. The knowledge about these pieces of information is modeled in the business process model along which the user works. Since in- and output of a process are of static nature for a given process instance, the documents related to them should be aggregated by the process execution engine, which controls the information flow between tasks in processes. Thus it does not belong to the supplementary information that users might need when completing a task. Therefore we will not integrate this information in the user interface proposed.

3.3 Organization

The organization the user works for defines the global setting in which the search occurs. It is responsible for providing the user with accurate information sources and defines the duties and rights of the user. The duties (i.e. the tasks) of employees are condensed in the definition of their role(s) in the organization. Each role equally gives them certain rights. From the point of view of information retrieval, capturing the role of users is essential, since it is necessary for capturing the global scope in which the information retrieval takes place. Furthermore considerations such as data security are directly linked to the role definition. Thus the role of a user defines the view of the user on information, i.e. it defines the set of semantic concepts the user might be interested in at all. This implies that each role should be mapped to a concept model, i.e. an ontology. Considerations such as data sources and rights management are much more of administrative nature and will not be further considered. We will suppose that the user can only be presented with information she can access.

3.4 Technology

The technology utilized to complete a task can produce an information overload or underload [10]. From the point of view of information retrieval systems, both information underload and overload can be observed. The query input interface of modern information retrieval system is usually restricted to a text field, in which the user can give in boolean queries (this also underlines the insufficiency of the model in the background). A direct implication of this interface is that the user must formulate queries using a vocabulary that might differ completely from the one used in the corpus she is searching through. Thus there is a gap between the information available in the system and the knowledge the user has about this information. Another information underload occurs when the user is presented with the result of a retrieval process. The information she receives and out of which she must determine the relevance of a document for her query is usually reduced to the metadata of the document or a short excerpt from the document, generated without any consideration of context. The information overload comes about because the context of the user is not taken into consideration when processing her query. Thus she has to search through a large set of irrelevant information before getting access to the information that might of interest for him. Furthermore the relevance the user assigns to the documents once the retrieval results are presented is not registered by the system. Although explicit relevance feedback methods have been developed to tackle this drawback, they are seldom accepted by the users [5]. The need is thus for implicit methods able to collect the users' feedback.

In the following, we present an implementation of the aspects presented above in a user interface.

4 User Friendly Semantic Search

Required for our approach is a representative text corpus of adequate size containing domain specific documents. The preparation process consists of three steps. In the first step, we create a word similarity graph from the text corpus. The result is a graph, whose nodes represent words that are typical for the domain and whose edges depict the similarity of two adjacent words. We will refer to these words as terms. In the second step we extract domain specific ontologies from the calculated similarity graph. The computed ontologies are then manually revised. Alternatively, we can use existing ontologies and map their concepts with the terms in the similarity graph by computing the relevance value for each term-concept pair. For example, the relevance value between a concept and a term of the same word can be set to 1 by default. Last, we model typical, search intensive tasks executed frequently in this domain. We mark relevant concepts of the ontology for these tasks. With this data (i.e. an ontology with a set of sub ontologies and an underlying similarity graph connected to the concepts of the ontology), we can start our user friendly semantic search.

The three steps that lead the user to an improved IR experience combined with an improved search performance are

1. Configuration of the search environment
2. Query Generation
3. Presentation of IR results

The first step consists in configuring the search environment, i.e. in defining the context in which the user searches, by enabling the user to define her current context and task, which provides additional information that can be used in the next step: query generation. Current search engines do not assist the user in formulating their query, although this is an important point of any retrieval process. The best search algorithm cannot supply satisfying answers, when the user does not utilize relevant terms to describe her information need. In the query generation step, we use the role-dependent ontology and the similarity graph to offer the user an overview of relevant concepts and terms linked with these concepts, thus helping her to find the relevant keywords. The third step is the presentation of the IR results, which are subsequently analyzed by the user. Instead of only presenting links to possibly relevant documents, links which would lead to the user having to search through the document to find relevant pieces of information, user friendly semantic search displays only relevant parts of the document, called chunks. The user is now able to better and faster estimate the relevance of a document compared to only guessing the content from the title.

5 Graphical User Interface for Semantic Search

Our graphical user interface (short GUI, see Figure 2) consists of three frames according to the three steps mentioned above, of which each holds special functionality implementing each step of the search process: The upper right frame, called options frame, provides the user with the parameters necessary to configure her search and context, thus implementing the first step cited above. After setting these, the user can start exploring the ontology for concepts, which may be of interest in her search request. This functionality is realized in the navigation frame. For a quick start, the user can select the start node from within a list of the alphabetically ordered concepts. When the user has finished selecting concepts or words, she can start the search for chunks, i.e. document segments, which will be displayed in the lower frame, the chunk viewer. It is possible to interact with the shown chunks. The underlying framework monitors this interaction and uses it to adapt the ontology. Each of the frames will be presented in more detail in the following subsections.

5.1 The Options Frame

In the options frame, the user can customize the search process. The option frame offers a variety of roles the user can choose from. These roles mostly describe a job, like web designer, programmer, graphics designer and so on. The selection of a role is equivalent to the choice of a domain specific ontology, which is displayed in the navigation frame (see 5.2). The user then selects a specific task, which actually narrows down the ontology by solely activating a sub-ontology. With this method, it is possible to present potentially relevant concepts, from which the user can choose the most relevant for her search. Each ontology is modified during runtime by user behavior, which is collected using implicit feedback techniques. When the user activates a concept of the displayed ontology in the navigation frame, the framework checks the relevance values of terms in relation to the chosen concept. The user is given the possibility of setting upper and lower similarity thresholds. Terms with similarity values higher than the upper threshold value are automatically added to the search query. Likewise, terms are added, when they have a similarity value equal or higher the upper threshold in relation to the just added term. If the similarity value is between the upper and lower threshold value, the term is displayed as being relevant for the chosen concept but not added to the query.

Additionally, the user can influence the display of the frames with check boxes in the option frame. It is possible to deactivate the display of interface elements like the concept list or the search word line to achieve a maximum visibility for the remaining elements, like the ontology.

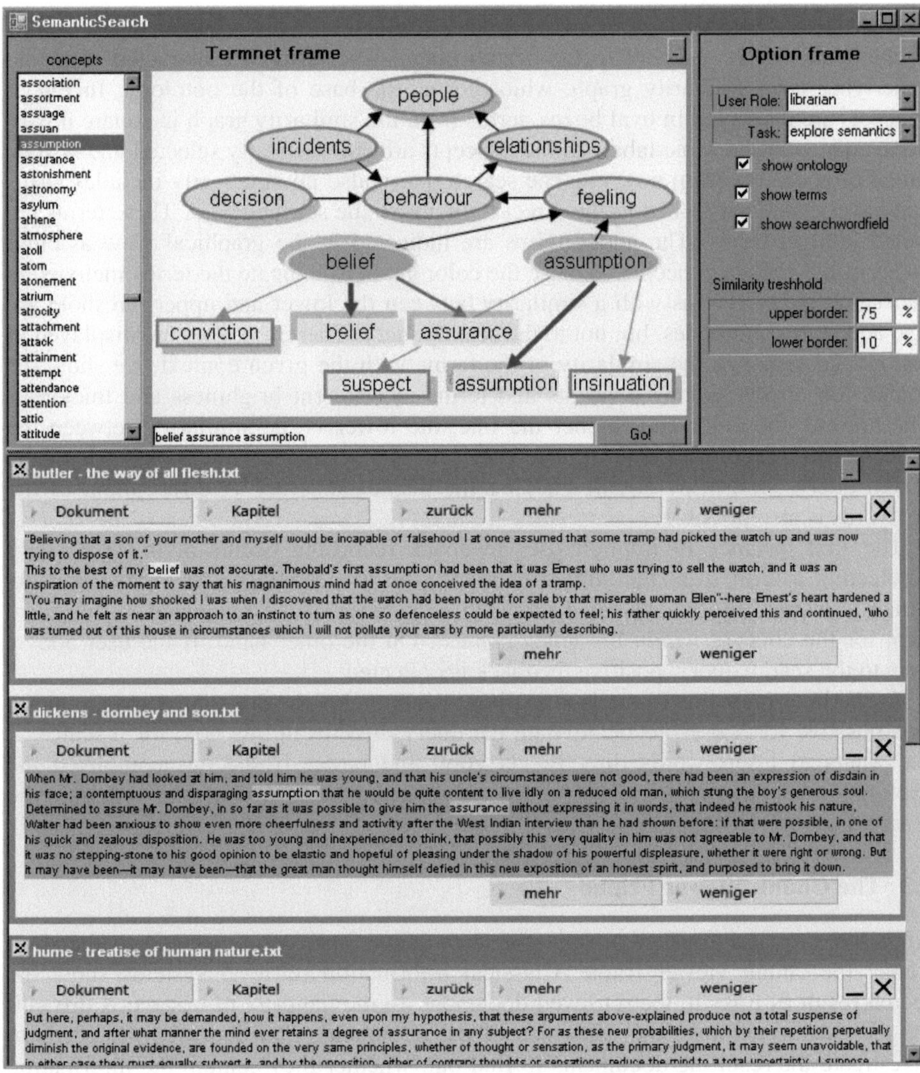

Fig. 2. The Semantic-Search GUI

5.2 The Navigation Frame

The navigation frame displays the part of the ontology that was activated with the selection of the task and user role in the options frame. Since this activated set can be rather large, the user has the possibility to select a first concept from a list box, which lists the concepts contained in the ontology. The window is then focused on this concept. The user can now examine the neighborhood of the chosen concept and can decide on the relevance of other concepts for her query. If the ontology does not fit into the navigation frame, the user can dynamically expand the frame, or drag the ontology inside the frame.

If the user considers a concept adequate, she can add the relevant terms with a double-click on the concept to the search query. The terms are extracted from the underlying term similarity graph, which forms the base of the ontology. In Fig. 2 concepts are displayed in oval boxes, terms from the similarity graph in square boxes. Terms that bear the same labels as the concepts are automatically selected and will be called direct terms from now on. The search query also automatically includes terms that have a similarity value to the direct term above the set threshold. These terms are called indirect terms. The added terms are indicated in the graphical view as child nodes of the chosen concepts. We use the color green to indicate the terms included in the search query. Terms with a similarity between the lower and upper threshold are also added as child nodes, but not to the search query. Therefore, they are displayed in yellow. To visualize the similarity of the terms with the given context, we show the connection arrows among concepts and terms in different brightness and thickness. The brighter the color and thinner the line, the lower is the similarity between the concept and the connected term. The user can now exclude or include words to the search query, simply by double-clicking the displayed terms. A direct input of the keywords is also possible.

The first instance of implicit user feedback recording occurs at this point: By deselecting a term from the similarity graph, the user produces negative feedback, which is learned by the system, leading to a decrease of the relevance of the given term for the concept *within the given context*. On the other hand, if the user adds a term to the search query, positive feedback is collected.

At the bottom of this frame is a text box, which holds the cumulated search query. It is possible for a user, to manipulate the query directly in this box, for instance, if she wants to insert words directly via copy and paste. If she has completed the exploration of the ontology and added all related search words, she can start the search using the Go! button.

5.3 The Chunk Viewer Frame

After the user has finished formulating her query, her focus changes to the third frame: the chunk viewer frame. Classical information retrieval systems show the search result as links to the retrieved document. In the majority of cases it is difficult for the user to estimate the relevance of the document. Consequently she needs to investigate the resulting documents to find out, whether they contain the information she is looking for. Often, this means to start another viewer, like the acrobat PDF-reader and to search through that document.

The chunk viewer frame uses a different approach. It shows the user document parts that most likely answer her question. To achieve this, we segment the documents into small coherent units we call chunks. Instead of showing only links to whole documents, we display sections of documents. This provides more information about the document than only the name of it. Ideally, it already contains the information, the user was looking for. At least she gets an idea about the content of the underlying document and whether it suits her need for information. This reduces the time she needs to investigate in analyzing search results, because she does not need to change her search environment and she does not need to search again through the results to encounter the exact location of the needed information.

The user can now decide to open the whole document (from which she already knows, that it is of high relevance), or she can use the functionality of the chunk viewer frame to get more information. She can e.g. expand the view of the chunk by the previous or next chunks or display the whole chapter in the chunk viewer. If chunks do not match with her search request, she can delete them from the result list. All those interactions are monitored by the system and stored in a database for further computation of the implicit feedback. The implicit feedback of a user's interaction with the concepts of the ontology and the terms of the similarity graph in the navigation frame, as well as the manipulation of the chunks in the chunk frame, result in a continuous re-adjustment of a term's relevance in the context of the user's role and task. Therefore, other users in the same role and task benefit from the re-adjustments generated due to the feedback of other users.

6 Discussion

We presented an approach that integrates human factors into the implementation of user interfaces for retrieval systems. We then elaborated on our vision of the process leading to user-friendly information retrieval, presenting an interface that implements the requirements generated from an analysis of existing systems and the HFBM. The proposed interface is constituted of three frames: the options frames, which allows the parameterization of the search process, the navigation frame, which enables the user to customize the search query and the chunk viewer frame, which display potentially relevant micro-content. The proposed implementation allows addressing some of the main issues discussed in section 2. It allows the user to browse through a representation of the available content, to query the content, to navigate through the retrieval results and eases the scan and analysis of retrieved document by displaying key micro-content directly in the browser. By these means, we implement all forms of interaction suggested by [3]. The complexity of the interface can be adapted to the user, since she can choose not to display certain windows. Our system uses a combination of information space and category-based browsing. A combination of these approaches might lead to an information overload. We address this potential drawback by learning through the user's feedback and thus ensuring that she is presented with a small number of contextually relevant concepts. Our system is based on the standard approach to information retrieval, which assumes an iterative reformulation of the user's query until the desired documents are found. Considering other approaches such as "berrypicking" might lead to some modifications of the interface.

References

1. Bates, M.J.: The design of browsing and berrypicking techniques for the online search interface. Online Review 13(5), 407–424 (1989)
2. Bederson, B.B., Hollan, J.D.: Pad: A zoomable graphical interface. In: Proceedings of ACM Human Factors in Computing Systems Conference Companion (CHI 95) (1995)
3. Hearst, M., Baeza-Yates, R., Ribeiro-Neto, B.: Modern Information Retrieval. ACM Press/Addison-Wesley Longman Publishing Co., Harlow, England

4. Hearst, M.A., Karadi, C.: Cat-a-Cone: an interactive interface for specifying searches and viewing retrieval results using a large category hierarchy. In: Proceedings of SIGIR-97, 20th ACM International Conference on Research and Development in Information Retrieval, Philadelphia, US, pp. 246–255 (1997)
5. Kelly, D., Teevan, J.: Implicit feedback for inferring user preference: a bibliography. SIGIR Forum 37(2), 18–28 (2003)
6. Mäkelä, E., Hyvönen, E., Sidoroff, T.: View-based user interfaces for information retrieval on the semantic web. In: Gil, Y., Motta, E., Benjamins, V.R., Musen, M.A. (eds.) ISWC 2005. LNCS, vol. 3729, Springer, Heidelberg (2005)
7. Robertson, G., Mackinlay, J., Card, S.: Information visualization using 3d interactive animation. In: CHI '91: Proceedings of the SIGCHI conference on Human factors in computing systems, pp. 461–462. ACM Press, New York (1991)
8. Salton, G.: Automatic Text Processing – The Transformation, Analysis, and Retrieval of Information by Computer. Addison–Wesley, Reading, MA (1989)
9. Smith, M.J., Carayon-Sainfort, P.: A balance theory of job design for stress reduction. International Journal of Industrial Ergonomics (4), 67–79 (1989)
10. Smith, M.J., Cohen, B.G.F., Stammerjohn, L.W., Happ, A.: An investigation of health complaints and job stress in video display operations. Human Factors 4(23), 387–400 (1981)

Hybrid Singular Value Decomposition: A Model of Human Text Classification

Amirali Noorinaeini, Mark R. Lehto, and Sze-jung Wu

School of Industrial Engineering, Purdue University,
1287 Grissom Hall, West Lafayette, IN 47907-1287, USA
lehto@purdue.edu

Abstract. This study compared the accuracy of three Singular Value Decomposition (SVD) based models developed for classifying injury narratives. Two SVD-Bayesian models and one SVD-Regression model were developed to classify bodies of free text. Injury narratives and corresponding E-codes assigned by human experts from the 1997 and 1998 US National Health Interview Survey (NHIS) were used on all three models. Using the E-code categories assigned by experts as the basis for comparison all methods were compared. Further experiments showed that the performance of the equidistant Bayes model and regression model improved as more SVD vectors were used for the input. The regression model was compared to a fuzzy Bayes model. It was concluded that all three models are capable of learning from human experts to accurately categorize cause-of-injury codes from injury narratives, with the regression-based model being the strongest, while all were dominated by multiple-word fuzzy Bayes model.

Keywords: accident narratives, bayes, regression, singular value decomposition, statistical modeling, text classification.

1 Introduction

During the course of this study, three SVD based models were developed. Two were based on the Naïve Bayes classifiers and one was based on regression. The models were compared with other SVD models in the literature. They were then compared with each other, and the best performing model was compared with the fuzzy Bayes model developed by Lehto and Sorock [1].

All three models consisted of two phases. The first phase was the SVD dimension reduction which was similar in all the models. The term-document matrix was input to the SVD model and the term-term output matrix was multiplied by the transpose of the original term-document matrix to construct the image of the document set on a lower dimension feature space. This feature space (each column is called a feature as will be discussed later) was used with a different number of features in each model to find the near optimal parameters. The number of features used was 60, 70, 90, and 180 for all the three models. Other numbers of features were also tested on some of the models.

M.J. Smith, G. Salvendy (Eds.): Human Interface, Part I, HCII 2007, LNCS 4557, pp. 517–525, 2007.

Three goals were described for this study, two of them aiming at evaluation of the strength of the models and one aiming at validation of a theoretical characteristic of the SVD based models. Only one of the hypotheses based on these goals could not be verified. The details of the study are discussed in the following sections.

2 The Data

The comparisons among the three models and the fuzzy Bayes model were done using a test bed of 5295 accident narratives from the 1996-1997 National Center for Health Statistics (NCHS) data. The document set first served as a training set, and was then broken into two subsets; "training" and "test". The training set contained 4712 narratives and the test set contained the 583 remaining narratives. All three sets had 13 categories of narratives, matching 11 of the accident categories defined by NCHS experts, one category consisting of all other categories put together and the last category containing all of the non categorized narratives. The breakdown of the narratives within the corpus is shown in Table 1 below:

Table 1. Table of categories, and their distribution over the complete dataset

Category	Description	N_{expert}*	Percentage
1	Motor-vehicle	539	10%
2	Other Transport	83	2%
3	Other unspecified	40	1%
4	All falls	1932	36%
5	Animals	112	2%
6	Foreign Body Eye	56	1%
7	Struck by Caught In	914	17%
8	Machinery	88	2%
9	Cutting Piercing	452	9%
10	Hot corrosive	86	2%
11	Over Exertion	698	13%
12	Other	243	5%
13	Assault	52	1%

* Number of narratives assigned to the category by NCHS experts.

3 The Models

As mentioned in the introduction, two of the models used the Naïve Bayesian classifier as the prediction part of the model. These two models were different from each other in the mapping of the feature vector results onto the Bayes classifier. The columns are usually called the features here because after the SVD technique is applied to the vector space, the resulting matrix is no longer a term-document representation of the data, but rather a representation of different features of the data, each presented in a separate column, resulting in a change in the matrix name change to the "feature vector".

The "equidistant" model used equidistant sub-ranges for each of the vectors' ranges to calculate the probabilities. This means that the coding of the information from the feature vector into the Bayesian apriory was done by assigning a separate value to each equidistant range within each vector's total values range. The "equinumerous" model on the other hand, kept the same number of narratives in each of the sub-ranges in order to exclude the possible bias in probabilities due to non-homogeneous ranges of the equidistant model. The rest of the structure of the models was similar. The SVD-Regression model, on the other hand, used the feature vectors as inputs to a multivariate least squares fit model and mapped them to the categories matrix. The models were tested for three hypotheses. Below is a summary of the results for each of the three hypotheses:

3.1 Hypothesis One

First Hypothesis stated that all of the three models perform better than a random classifier. This was to show that the models are capable of classifying the narratives to begin with. This hypothesis was validated by showing that all of the three models perform better than such a classifier using any of the 60, 70, 90, and 180 feature vectors. To conclude that, one should observe that a random classifier would achieve the best results by classifying all the narratives into the most frequent class. In this case therefore a random classifier would do no better than 36%, which is the percentage of narratives in class 4. Since all the models have performed much better than 36% as shown in Table 2, one can state that the models are performing better than random.

It was also shown that these models' performance on the available corpus is better than the average SVD based models performance. This step did not prove that the present models are better than any of the other models in the literature because the models used very different corpuses. It was not possible to try the models on the same corpus because the other models were not at hand and their corpuses were different from one another as well. SVD models' performances developed by other researchers for the classification task vary greatly depending on the corpus they have used. Wu and Gunoplulos [2] have reported six models on the same dataset with an average sensitivity of over 82% and a high of 92%. Hoffman [3] reports three models and two combination models, with an average sensitivity of less than 45% over four standard test collections. In Hoffman's results, the best sensitivity is 67.5%. Finally, Yang [4] reports one model with three different parameter settings tried over three standard test collections with an average sensitivity of 52% and a best score of 88%. It is really hard to find a baseline for the SVD based model performance in the present study. The best one can do is to take an average of all of these studies' average performances and use that as a rough comparison. If this is done, the average model performances are around 59.7%. This percentage shows that the present model does better than the average SVD model in the literature as shown in Table 2 and therefore it is a promising model.

Note that for each of the models two results are reported. The best guess result indicates the percentage of classification in accordance with expert ratings if only the

Table 2. Performance of all three models for all vector combinations

		60 V.	70 V.	90 V.	140 V.	180 V.
Best Guess	Equidistant	67.7	67.4	69.9	-	77.2
	Equinumerous	66.9	68.5	70.6	76.6	71.1
	Regression	68.9	70.8	72.2	-	76.9
Two Best Guesses	Equidistant	81.7	82.8	85.6	-	91.2
	Equinumerous	79.9	82.2	84.3	90.1	85.7
	Regression	82.5	83.7	85.6	-	89.4

All numbers are percentages.

class with highest probability is reported. The two best guesses result shows the percentage of the narratives that one of the two highest probable classes assigned to them by the model are the same as the class assigned by the human expert. Also note that for the equinumerous model the results for 140 features are reported as well. This result relates to the second hypothesis and will be discussed later.

3.2 Hypothesis Two

The SVD-Regression model was believed to perform as well as the SVD models in the literature because it uses the inputs from the SVD technique, causing its input to be less noisy than if the actual words were used; this would improve the model results as the inputs convey more useful data than they would convey if the original terms were used, not to mention the extreme savings in processing effort. It was also believed that the SVD-Regression model would perform better than the two Bayesian models developed in this study because it does not assume conditional independence between terms, given the categories. The conditional independence assumption has regularly been reported as the cause of the Bayesian classifiers' low performance [5, 6]. Since the SVD-Regression model uses the same input as the Bayes models and does not assume the conditional independence, it was expected that it would perform better than the naïve Bayes classifiers. This belief was expedited through the second hypothesis.

In order to test this hypothesis, first the best performance of the three models in the range between 60 to 180 vectors was searched. This led to finding the best performance of the equinumerous Bayes model using 140 feature vectors. The two other

Table 3. Performance of all models for all vector combinations over the training and test sets

		60 V.	70 V.	90 V.	140 V.	180 V.
Training Set	Equidistant	68.8	70.6	73.2	-	79.0
	Equinumerous	69.3	70.6	72.6	77.5	72.8
	Regression	69.4	71.4	72.6	-	77.8
Test Set	Equidistant	55.4	56.1	55.9	-	56.6
	Equinumerous	47.7	49.2	50.1	56.1	48.0
	Regression	59.3	61.6	67.8	-	71.0

All numbers are percentages.

Table 4. SVD-Regression model using 180 vectors compared with single and multiple-word fuzzy Bayes model

Category	SVD-Regression model			Single-word fuzzy Bayes	
	Sensitivity	Specificity	PPV*	Sensitivity	Specificity
1	94.2%	99.1%	92.2%	93.0%	95.1%
2	79.5%	99.7%	78.6%	26.9%	99.7%
3	15.0%	99.9%	60.0%	22.4%	99.5%
4	88.8%	85.5%	77.9%	92.9%	77.2%
5	77.7%	99.6%	81.3%	79.4%	99.5%
6	80.4%	99.6%	68.2%	28.6%	99.9%
7	71.9%	92.1%	65.4%	46.7%	97.3%
8	35.2%	99.7%	67.4%	19.2%	99.8%
9	72.8%	98.7%	83.9%	70.4%	98.6%
10	24.4%	99.9%	84.0%	60.2%	99.7%
11	73.1%	96.7%	78.9%	65.7%	97.1%
12	39.5%	99.1%	67.1%	25.0%	100.0%
13	1.9%	99.9%	25.0%	67.9%	99.8%
Total	76.9%			71.3%	

* Positive Predictive Value

models showed their highest performance with the highest number of feature vectors (180). The results showed that all the models are comparable in this range. The second step was to test the models over the training and test sets. The results again showed that the equinumerous model performs best with 140 feature vectors while the other two perform best with 180 feature vectors even when the training and test sets are differentiated. It was also concluded that the regression model outperforms the two Bayesian models over the test set, while it requires much less computational power. The regression model performed better than the two other models over the training set as well, although the superiority was marginal. The results are shown below in Table 3:

Further experiments also showed that the regression model is dominated by the multiple-word fuzzy Bayes model. In order to compare the two models, first the results from Wellman et al. [7] were compared to the results from this study. The results are shown in Table 4.

Results with more features showed that the regression model outperforms the fuzzy Bayes model using 500-800 feature vectors. The model had to be checked for over fitting. Since the multiple-word fuzzy Bayes was not tested on a training and test bed separately, new experiments with the fuzzy Bayes model were done on the exact same training and test beds used for the regression model experiments. Results showed that the fuzzy Bayes model performs as well on the test set as it does on the training set. The results are shown in Table 5. The regression model results, on the other hand, showed that the regression model has already reached its best performance using 300 feature vectors (sensitivity reached 78.6%), and that the 500-800 feature vectors' results are high just because of the over fitting of the data. Therefore, it was concluded that the multiple-word fuzzy Bayes model performs better than the regression model. This observation strengthens the assumption that the regression model's better performance in comparison with the naïve Bayes based models is related to the removal of conditional independence assumption. This is because the fuzzy Bayes model does not assume conditional independence.

Table 5. Multiple-word fuzzy Bayes model results for training and test sets

Category	Training set Sensitivity	Test set Sensitivity
1	99.0%	100.0%
2	90.1%	91.7%
3	8.6%	50.0%
4	96.4%	96.9%
5	92.9%	50.0%
6	65.9%	75.0%
7	74.4%	89.6%
8	59.2%	88.2%
9	88.6%	78.7%
10	78.9%	93.3%
11	77.1%	60.5%
12	36.8%	56.0%
13	60.5%	42.9%
Total	84.7%	84.8%

Finally, the second hypothesis was validated, although the regression model did not perform better than the fuzzy Bayes model. It is believed that the better performance of the fuzzy Bayes model is related to its simplicity, causing it to conduct fewer computational steps and data mapping. This higher performance can also be credited to the fuzzy Bayes model's ability to explore the semantic structure of the database because it uses up to four word combinations. This is verified by the fact that the regression model performs better than the single-word fuzzy Bayes model, where, due to using single word predictors, the semantic structure of the text remains mainly unnoticed.

3.3 Hypothesis Three

Theoretically the SVD technique is very useful in the field of Information Retrieval as it filters the noise in the database automatically. This property is mostly credited to the fact that the noise in the database does not follow any pattern, and therefore is not recognized as a major feature of the database. Applying the SVD technique will cause the rank of feature vectors capturing the noise to be very low; therefore, they are hardly ever kept in the reduced space vectors.

One way to investigate the correctness of this assumption is to check the behavior of the classifier models which employ the SVD technique. If these classifiers' results first increase by adding more vectors up to a certain limit, and then decrease, their behavior could be explained using the above theory. The reason would be that the vectors are ordered according to their information content, and including more vectors adds more information, but the amount of information added every time is decreased until a vector is added that contains more noise than information, causing the performance of the model to drop. Such behavior was investigated in the third hypothesis. Results showed that the equidistant model and the regression model don't show this trend for the first 180 vectors; that is, their behavior in the investigated range is asymptotic, while the equinumerous model shows such a behavior. The reason for two of the models not showing this behavior could be the fact that they were not tested over the complete range of 4553 vectors, and that if this was tested, they would show a decreasing performance after some point. This possibility was not further investigated due to computational issues.

Since all three models used the same feature vectors, another question was why the other models do not show the same behavior in the same range. The reason for this difference could be related to how effectively the model uses the data embedded in the SVD vectors. The more of this information a model can extract, the more powerful it will become, and the later the noise embedded in the SVD vectors will affect its overall performance. This explanation was not tested either as it was beyond the scope of this study. Having noted all of the above, the third hypothesis could not be validated.

4 Discussion

The models developed above all have one drawback; that is, they are all using a reduced space in which the term-term relationships are the main concern of the model, and the relationships weights (the SVD diagonal matrix, Σ) are omitted as they are believed to cause the infrequent categories to be treated as noise. Although these actions will cause the infrequent categories to be classified more accurately, theoretically they can cause lower performance over the more frequent categories, thus lowering the overall performance of the model. The extent of this problem has not been explored yet to see how sensitive the performance is to this problem.

The second point to be noticed is that all of these models are based on SVD calculations. This means that when a new document set is going to be introduced to the training set, the intensive SVD calculations may have to be done over the entire training set again. There is no way of appending the new set's SVD calculations to the existing one if the new narratives are significantly different from the original narratives. In practice, this poses a big problem as the new training documents should be added to the model on a daily basis. This causes the re-training of the models to be computationally inefficient compared to models such as the fuzzy Bayes. However, if the new narratives are not significantly different form the original ones, feature extraction can be performed by folding in the new narratives into the space spanned by the original narratives [8]. It should be noted that for the regression part of the model the fitting of data (document set's image on the feature space) should be done again, which does not require heavy computations.

5 Future Work

The results of this study suggest that the optimal performance reached by the equidistant Bayes model should be explored. It was noted that the regression model is relatively inefficient in classifying the narratives belonging to infrequent categories, while the Bayesian based techniques, especially the fuzzy Bayes model, perform relatively better on the infrequent categories, even with less input. It would be interesting to study the possible development of a hybrid of these two models to reach higher performances with fewer vectors used from the SVD technique applied to the database. This is especially important as the speed with which new knowledge is being produced makes it very difficult to keep the search engines' training databases up to date and in the meantime efficient.

In addition to applications on commercial software, internet security and biology science, the hybrid SVD approach for text mining can contribute significantly to improving the efficiency of healthcare informatics and information flow and delivery. Our ambition is to develop a prototype of a text classification tool focusing on the following potential applications. First of all, the proposed hybrid SVD can be a robust model to classify medical narratives in electronic medical records in comparison to other text mining algorithms [9-10]. Another application of SVD would be developing an automatic text entry system for both free text and semi-structured template-driven documentation in electronic medical record systems [11-13]. This application can also resolve the problem of inefficient text entry in the use of small hand-held mobile medical devices. Last but not least, SVD text categorization can be integrated with clinical decision support systems to identify irrelevant and false-positive clinical information [14]. One of the examples would be our continuing development of an information filtering system for computerized clinical reminders to identify false alarms so as to reduce the information overload on healthcare providers [15].

References

1. Lehto, M., Sorock, G.: Machine learning of motor vehicle accident categories from narrative data. Methods Info Med. 35(4-5), 309–316 (1996)
2. Wu, H., Gunopulos, D.: IEEE International Conference on Data Mining (ICDM'02), pp. 713–716. IEEE Computer Society Press, Los Alamitos (2002)
3. Hofmann, T.: Probabilistic latent semantic indexing. 22nd Annual International ACM SIGIR Conference on Research and Development in Information Retrieval, Berkeley, CA, pp. 50–57 (1999)
4. Yang, Y.: Noise reduction in a statistical approach to text categorization. In: Proceedings of the 18th annual international ACM SIGIR conference on Research and development in information retrieval, pp. 256–263. ACM, New York (1995)
5. Deerwester, S., Dumais, S.T., Furnas, G.W., Landauer, T.K., Harshman, R.: Indexing by latent semantic analysis. Journal of the American Society for Information Science 41(6), 391–407 (1990)
6. Berry, M.W., Dumais, S.T., O'Brien, G.W.: Using linear algebra for intelligent information retrieval. SIAM Review 37(4), 573–595 (1995)
7. Wellman, H.M., Lehto, M.R., Sorock, G.S.: Computerized coding of injury narrative data from the National Health Interview Survey. Accid. Anal. Prev. 36, 165–171 (1995)
8. Berry, M.W., Fierro, R.D.: Low rank orthogonal decomposition for information retrieval applications. Numerical Linear Algebra with Applications 3(4), 301–327 (1996)
9. Tange, H.J., Hasman, A., de Vries Robbé, P.F., Schouten, H.C.: Medical narratives in electronic medical records. International Journal of Medical Informatics 46, 7–29 (1997)
10. Mikkelsen, G., Aasly, J.: Manual semantic tagging to improve access to information in narrative electronic medical records. International Journal of Medical Informatics 65(1), 17–29 (2002)
11. van Mulligen, E.M., Stam, H., van Ginneken, A.M.: Clinical Data Entry. In: Chute, C.G. (ed.) AMIA Annual Symposium, Hanley & Belfus, Philadelphia 81 (1998)
12. Harrast, J.J., Koris, M.J., Chen, S.F., Poss, R., Sledge, C.B.: Design and implementation of an automated operative note, M D Computing, vol. 12(6), pp. 559–565 (1995)
13. Stocky, T., Faaborg, A., Lieberman, H.: A commonsense approach to predictive text entry. Proceedings of Conference on Human Factors in Computing Systems, pp. 24–29 (2004)
14. Sona, D., Avesani, P., Moskovitch, R.: Automated multi-classification of clinical guidelines in concept hierarchies, Artificial Intelligence in Medicine, Aberdeen, Scotland, UK (2005)
15. Wu, S.-j., Lehto, M., Yih, Y., Flanagan, M., Zillich, A., Doebbeling, B.: A Logistic Regression Model for Assessing Clinicians' Perceived Usefulness of Computerized Clinical Reminders, the 36th International Conference on Computers and Industrial Engineering, Taipei, Taiwan (June 2006)

A Method for Constructing a Movie-Selection Support System Based on *Kansei* Engineering

Noriaki Sato, Michiko Anse, and Tsutomu Tabe

Department of Industrial and Systems Engineering, Graduate School of Science and Engineering, Aoyama Gakuin University 5-10-1,Fuchinobe,Sagamihara-shi,Kanagawa-ken, 229-0006, Japan
c5606063@cc.aoyama.ac.jp
{anse, tabe}@ise.aoyama.ac.jp

Abstract. When a person requests, for example, "I want to see a bright and exciting movie," the words "bright" and "exciting" are called *Kansei* keywords. With a retrieval system to retrieve recommended movies using these *Kansei* keywords, a viewer will be able to select movies that fit the *Kansei* without actually having to view samples or previews of the movies. The purpose of this research is to clarify a method to construct a support system capable of selecting movies that fit the viewer's *Kansei,* and to verify the effectiveness of this method based on *Kansei* engineering, for the selection of recommended movies. To accomplish this, we extract the features of a movie using factor-analysis from data from a Semantic Differential Gauge questionnaire, then link the viewer's *Kansei* with the features using multiple linear regression analysis. After constructing a prototype · system to verify the effectiveness, ten examinees viewed a movie selected by the prototype·system. "The selected movie fit the Kansei" at a level of about 70 percent.

Keywords: *Kansei*, retrieval system, factor-analysis, multiple linear regression analysis.

1 Introduction

Ordinarily it is easy to access information through searches based on attributes such as "title," "genre," "performer," etc., using movie-retrieval engines inside conventional rental shops or on websites. Viewers retrieve the movies they wish to see by inputting the keywords via the keyboard. When, on the other hand, a person types in a request such as "I want to see a bright and exciting movie," the words "bright" and "excited" so-called *Kansei* keywords, are difficult to handle. With a retrieval system for recommending movies with registered *Kansei* keywords of this type, the movie viewer will be able to select a movie likely to fit the *Kansei* without actually having to preview the movie. Recent research on opinion extraction has focused on methods to extract opinions from text information such as reviews, book reviews, and text on the Web. Liu, Lieverman, and Selker [2] study ways to assess the feelings of readers who read texts. To understand expressions that evince feelings in a text, methods are used to extract the portions of e-mail text that convey feelings,

M.J. Smith, G. Salvendy (Eds.): Human Interface, Part I, HCII 2007, LNCS 4557, pp. 526–534, 2007.

based on categories of feeling defined in advance. Dave, Lawrence, and Pennock [1] classify evaluation expressions from online reviews of products and other online content into the categories of "affirmation" and "denial," and on that basis extract the opinion. Yu and Hatzivassiloglo [6] use Naïve Bayes classifier to classify newspaper articles into "opinion" and "fact," then judge whether to classify similar level of sentences and unit of sentence into opinion or not. Moreover, the research focused on text information related to movies deals with subjective expressions extracted from the text. Turney [5] judges whether film reviews are "thumbs up" or "thumbs down." By keeping track of adjectives from text phrases, they judge whether the phrase is a positive assessment (thumbs up) or negative assessment (thumbs down) based on statistical analysis to quantify commonality. Nakayama and Kando [3] focus on words and phrases from film reviews that indicate feelings, then attempt to understand the characteristic by extracting the "reason" as a related element. The difference of the reason in the same feeling is shown by the writer, and the work of reason is analyzed. Kobayashi [4] focuses on comments from audiences after movie viewings and applies frequency-analysis to connect up with *Kansei* words and phrases categorized as "comments" and "movie genres," in order to extract *Kansei* words for movies in every genre. It is expected that these-extracted results become effective information for movie selection when shown to users in applications in the field of marketing or work retrieval. However, researchers are further away from accomplishing the same thing in studies on retrieval-method for movies based on subjective expression.

Therefore, we have attempted to clarify a method of movie selection by extracting "human *Kansei*," that is, subjective expressions, from text information. By relating human *Kansei* with "text," we construct a retrieval method for movie recommendations that fit two or more *Kansei* requested by viewers. We clarify how to extract human *Kansei* based on the content of a text, and how to statistically analyze commonality in quantitative evaluations of movies and human *Kansei* from an engineering approach. The purposes of this research are therefore to clarify a method for constructing a system for selecting movies that fit a viewer's *Kansei*, and to verify the effectiveness of our method for constructing a prototype system for selecting recommended movies based on *Kansei* engineering. To accomplish these purposes by three methods: first, we extract the features of a movie using a factor-analysis of questionnaire data with a Semantic Differential Gauge made from *Kansei* keywords; second, we quantify the features of a movement by frequency analysis of keywords associated with the features based on_features present in the movie story which was obtainable easily; third, we analyze how a viewer's *Kansei* and the features of a movie connect up using multiple linear regression analysis. Through these methods, we construct a prototype support system for a movie-selection engine to retrieve movies that fit a viewer's *Kansei*, and then test the system by a trial run.

2 Constructing the Support System

2.1 Process Concept

Fig. 1 shows the concept of the support system. Items 1, 2, and 3 in Fig. 1 are the content of the processing.

First, the features of a movie are extracted to clarify the structured factors of the movie.

Second, the features of the movie are quantified by frequency-analysis of the keywords associated with the features present in the movie story, after extracting keywords collected for each feature based on interpretations of factors obtained.

Third, a multiple linear regression formula is drawn up to calculate a user *Kansei*-evaluation value from the questionnaire results.

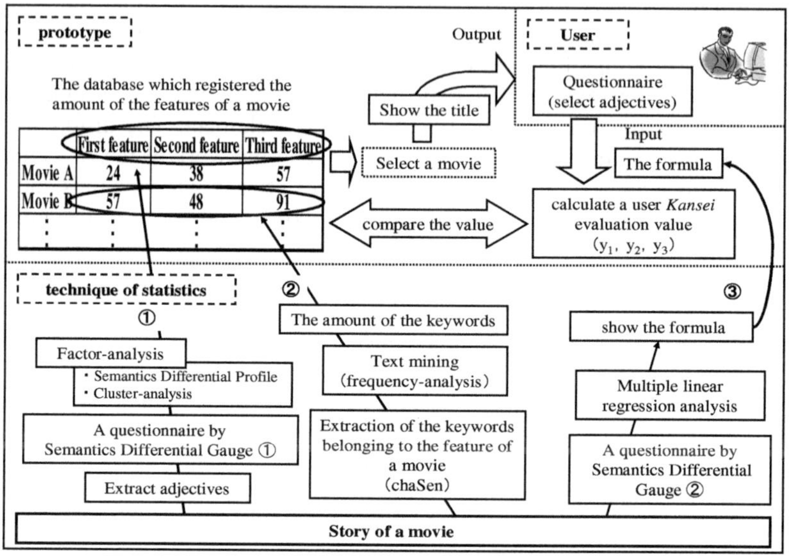

Fig. 1. Concept of the Support System

2.2 Extract the Movie Features

Movie features are collected for the stories of two or more movies, then 20 frequently appearing adjectives (set up uniquely) are extracted from the stories to determine which of the adjectives show the "the *Kansei* feature of the movies." These adjectives are grouped into opposite pairs and are considered in the Semantics Differential Gauge questionnaire in five steps (step 1 to step 5). To investigate the relation between "a movie genre" and "*Kansei* feature" in this case, we have an audience of 20 examinees view a movie preview chosen from among 11 movie genres and then have the examinees answer the above-mentioned questionnaire. Each adjective-pair is estimated by the examinee's *Kansei*. The 11 movie genres chosen are the genres with the highest frequency of appearance in our movie genre investigation.

The factors constituting movies are clarified by a factor analysis of the questionnaire data, then three factors are extracted. Table 1 arranges the adjective pairs based on the factor loads for each in descending order. The feature of the factor is expressed by adjective-pairs over 0.45 points of the absolute value of the amount of

Table 1. Factor Loads

Adjective-pairs		The 1st Factor	The 2nd Factor	The 3rd Factor
very serious	very laughable	0.87	0.14	−0.05
very depressed	very cheerful	0.86	0.27	−0.09
very tragic	very comical	0.85	0.15	−0.03
very dark	very bright	0.84	0.35	−0.15
very sad	very pleasant	0.82	0.18	0.04
very fine	very painful	−0.70	0.15	−0.18
very deep	very light	0.69	−0.23	0.17
very lonely	very busy	0.66	−0.01	0.45
very fearful	very peaceful	0.65	0.41	−0.42
not impressive	very impressive	0.14	0.86	−0.22
not lovely	very lovely	0.20	0.76	−0.21
not heart-warming	very heart-warming	−0.07	0.68	−0.14
very cold	very warm	0.56	0.61	−0.32
very slow	very speedy	0.18	−0.21	0.85
settled down very much	very excited feeling	−0.07	−0.28	0.81
very unreal	very realistic	0.00	0.10	−0.46

each factor-load. The first factor has nine adjective-pairs ("Very laughable," etc.) The second factor has four adjective-pairs ("Very impressive," etc). The third factor has three adjective-pairs ("Very speedy," etc.).

From the above, it turns out that the movie is composed of three obtained factors. Thus, we analyze the meaning of each factor as a key for treating these factors quantitatively. We interpret a factor by two methods: first, we create an SD profile for correlation with the three obtained factors and "movie genre"; second, we reclassify the genres, grouping those that cannot be distinctly divided by the SD profile into "similar impression" categories by cluster analysis (categories that make similar impressions on audiences).

First, we investigate the SD profile by calculating the average value of the questionnaire result for each adjective-pair associated with the three factors for every title. The strength of the factor of a title is taken as the number of adjective-pairs whose average values from five steps are "2 or less and 4 or more." A factor is featured for the highest genre of the rate of the sum total of strength by each of three factors, as each genre has three titles at a time (Refer to Table 2). As a result, the "Comedy" genre belongs to the first factor, the "Romance" genre belongs to the second factor, and the "Horror" genre belongs to the third factor.

Second, we carry out a cluster analysis based on the factor-score obtained by the factor analysis for every genre. As a result, "Horror" from the third factor is positioned in a close relation to "Action" and "Science Fiction/Fantasy." Therefore, it is judged that these movie genres make similar impressions upon viewers. In addition,

Table 2. Factor Strengths

The genre of a movie	The 1st Factor	The 2nd Factor	The 3rd Factor
Animation	18.5	33.3	44.4
Comedy	100.0	33.3	44.4
Action	3.7	0.0	55.6
Youth	3.7	33.3	0.0
Human Drama	3.7	66.7	0.0
Science Fiction/Fintasy	7.4	33.3	55.6
Documentary	0.0	58.3	0.0
Period Drama	0.0	8.3	55.6
Suspense	0.0	0.0	44.4
Romance	29.6	83.3	11.1
Horror	0.0	0.0	66.7

"Youth" is added as a genre of the movie in the second factor. None of the movies in the first factor give impressions of any genre. From these results above, it is determined that "Comedy" belongs to the first factor, "Romance" and "Youth" belong to the second factor, and "Horror," "Action," and "Science Fiction/Fantasy" belonging to the third factor. The interpretation of each factor determines the genre as the first, second, or third feature of the movie.

2.3 Movie Features by Morphological Analysis

Each movie featured is assessed by counting the frequency of keywords belonging to the three abovementioned features in the story of a movie.

2.3.1 Keyword Extraction for Every Feature

The keyword is resolved into morphemes after collecting the stories in six genres belonging to the first, second and third features, in order to extract a keyword for every feature. Next, from the morphemes we extract keywords for three parts of speech which are meaningful as words and phrases, i.e., "Noun," "Noun-General" and "Noun-Adjective verb stem." This leaves words and phrases with high frequency (ten times or more), and those that are common to the six genres are deleted from among them. The extracted words and phrases from the above steps are the keywords for each genre. The first-feature keyword is "Comedy," the second-feature keywords are "Romance" and "Youth," and the third-feature keywords are "Horror," "Action," and "Science Fiction/Fantasy." When we count the number of keywords of each feature, we come up with 120 first-feature keywords, 189 second-feature keywords, and 262 third-feature keywords. We define the "wealth" of the keywords for to each feature keyword category as the number of the keywords extracted in each feature common to all keywords (Refer to Table 3).

Table 3. The wealth of one keyword belonging to each feature

The 1st Feature Comedy	The 2nd Feature Romance/Youth	The 3rd Feature Horror/Action Science Fiction/Fantasy
4.8	2.2	3

2.3.2 Feature of a Movie

The frequency of keywords belonging to the three above-mentioned features in the story of one title is counted using frequency-analysis, one of the text mining techniques. Next, the values for the features for every title are determined by the value obtained by multiplying the frequency of keywords belonging to each feature and the wealth per keyword shown in Table 3. In addition, the length of one story (one title), when printed out, takes up about one sheet of A4 paper.

Table 4. Example movie features

Title name (Genre)	The 1st Feature Comedy	The 2nd Feature Romance Youth	The 3rd Feature Horror/Action Sicence Fiction/Fantasy
A.I. (Sicence Fiction/Fantasy)	86.4	114	154
EYES WIDE SHUT (Human)	72	120	66
ICE AGE (Animation)	144	123	151.8
EXORCIST (Horror)	76.8	45	145.2
SIN・CITY (Action)	14.4	69	110
HITCH	86.4	192	99
MASK2 (Comedy)	268.8	54	140.8

2.4 Linking Human *Kansei* with Movie Features

The same questionnaire described in Section 2.2 is used to link human *Kansei* with movie features. Ten members of an audience (examiners) are asked to fill out a questionnaire made up of nine adjective-pairs with a high factor-load for each factor based on the results of the factor analysis described in Section 2.2. To begin with, by checking strong correlations among the adjective-pairs of the five higher ranks with the high factor loads belonging to the 1st factor, it is judged that there is multicollinearity. For adjustment, we delete four adjective-pairs, e.g., from "very depressed – very cheerful" to "very sad – very pleasant." Moreover, we delete the adjective-pair "not heart-warming – very heart-warming," as an examiner considers these features unconsciously (no person consciously considers whether he or she wants to see a movie which will not remain in his or her heart). Twenty two movies from 11 genres are covered in the questionnaire. Thus, a multiple linear regression analysis is performed by the variable decreasing method using a purpose-variable, the strength of the features for each title, an explanation-variable, and an evaluation of an adjective-pair, in order to show formulas of relations for calculating a

Kansei-evaluation value. If adjustment-R2 becomes 0.5 or more, the accuracy of the regression is good. It is assumed that 0.5 or more base points are fulfilled, and that the model has no variable with multicollinearity. According to the analysis result, all adjustment-R2's for each feature were 0.5 or more. The multiple linear regression formula is shown in (1) ~ (3). In addition, Table 5 shows the adjective-pair of X_1~X_9 used for the explanation-variable

Multiple linear regression formulas obtained with Section 2.4

[Value for the first feature] $y_1=60.1*X_1-42.6*X_2+27.9*X_3+24.8*X_4-45.6*X_5+58.4$ (1)
[Value for the second feature] $y_2=-41.2*X_3+45.9*X_4-34.3*X_6-13.5*X_7+9.2*X_9-22.9$ (2)
[Value for the third feature] $y_3=33.0*X_3+34.5*X_4-37.8*X_6+14.0*X_7-11.0*X_9+26.6$ (3)

Table 5. Adjective-pair used for explanation-variable

	Adjective-pair	
Explanation-value	**smallness**	**largeness**
X_1	**very serious**	**very laughable**
X_2	**very fine**	**very painful**
X_3	**very deep**	**very light**
X_4	**not impressive**	**very impressive**
X_5	**not lovely**	**very lovely**
X_6	**very cold**	**very warm**
X_7	**very slow**	**very speedy**
X_8	**settled down very much**	**very excited feeling**
X_9	**very unreal**	**very realistic**

3 Construction of a Prototype and an Experiment

3.1 Construction of a Prototype and Its Concept

The purpose of constructing a prototype is to verify the effectiveness of a method for constructing the above-mentioned support-systems. The concept of a prototype is shown in Fig. 2, and an outline is described below.

"Input *Kansei* information": A user chooses the grade of an adjective-pair that fits the user's *Kansei* according to the questionnaire described in Section 2.4.

"Calculate the *Kansei*-evaluation value": The three obtained strengths of the features determine "a user's *Kansei*-evaluation value," a value which substitutes the evaluation-values of the respective adjectives input for the multiple linear regression formula gained from Section 2.4.

"Database": A database registering the title of a movie and the strengths of the three features is constructed in advance.

"Selection-processing of a movie": The Euclid distance is calculated by the difference of "a user's *Kansei*-evaluation value" and "the strengths of the features of a movie" after comparing a user's *Kansei*-evaluation value with the database. Next, two movie titles are selected from the smallest value.

"Display movie's name": The selected move title is shown to the user.

In addition, the title of a movie in an impartial genre is selected and registered into the database. To verify the effectiveness and to let the user actually see a movie, a movie already burned onto a DVD (DVD-ized) is adopted.

Moreover, the DVD-movie is already adopted because an examiner has actually seen the movie to verify the effectiveness.

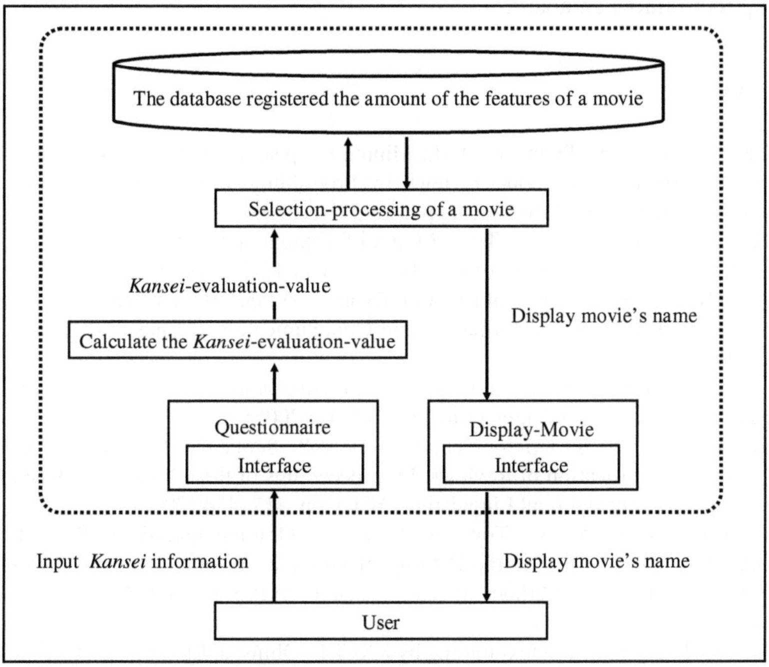

Fig.2. Conceptual Figure of the System

3.2 Verifying the Effectiveness

The experiment is carried out by ten examinees using a prototype. Each examinee chooses an adjective matching the *Kansei* of the movie he or she wants to see. Next, the examinee sees one movie between two other movies selected by the prototype from the results in consideration of the burden of time, that is, the time required to see a movie. Then, when the examinees are asked, in the questionnaire survey, "To what degree (expressed in percent) does the movie you have just seen fit your *Kansei*?," 70% or more of the examinees reply that the selected movie currently fits their *Kansei*.

4 Conclusion

It is clarified that the prototype system built through this research is able to select the movie which most closely fits the *Kansei* a user seeks, among the movies registered in the system database based on verified the effectiveness. Therefore, this research clarifies the method to construct a system capable of selecting movies that fit a viewer's *Kansei*. To raise analytic accuracy in the future, we hope to accomplish the following: first, analyze methods other than correlation with "movie genre" to interpret the factors obtained; second, to examine questionnaire methods that pose less of a burden on examinees, in order to collect more experimental data; third, to automate a series of processes for reading the stories and attributes of movies and registering them into a database.

References

1. Dave, K., Lawrence, S., Pennock, D.M.: Mining the peanut gallery: opinion extraction and semantic classification of product reviews. In: Proceedings of the 12th International World Wide Web Conference (WWW2003), pp. 519–528 (2003)
2. Liu, H., Lieverman, H., Selker, T.: A Model of Textual Affect Sensing using Real-World Knowledge. In: The Proceedings of IUI 2003 (January 12-15, 2003)
3. Norio, N., Noriko, K.: Analysis of Emotion Expression Focusing on Reason. proceedings of the Institute of Electronics, Information and Communication Engineers (IEICE) 105(291), 51–56 (2005)
4. Shinobu, O.: Kansei by Words and Movies Extracting Kansei Using Text Mining. Journal of Japan Society of Kansei Engineering 5(3), 43–47 (2005)
5. Turney, P.D.: Thumbs Up or Thumbs Down? Semantic Orientation Applied to Unsupervised Classification of Reviews. In: proceedings of the 40th Annual Meeting of the Association for Computational Linguistics (ACL), pp. 417–424 (2002)
6. Yu, H., Hatzivassiloglou, V.: Towards Answering Opinion Questions: Separating Facts from Opinions and Identifying the Polarity of Opinion Sentences. In: Proceedings of the Conference on Empirical Methods in Natural Language Processing (EMNLP), pp. 129–136 (2003)
7. Toshikatsu, H.: Approach to text mining by EXCEL. Ohmsha, Ltd., pp. 68–69 (2002)

LensList: Browsing and Navigating Long Linear Information Structures

Hongzhi Song, Yu Qi, Yun Liang, Hongxing Peng, and Liang Zhang

HCI Group, College of Informatics,
South China Agricultural University, Guangzhou, 510642, China
{hz.song,yuliangqi,yliang,xyphx}@scau.edu.cn,
200530740631@stu.scau.edu.cn

Abstract. List is a simple and useful graphical user interface (GUI) component for organizing linearly structured information when it contains only a few elements. However long lists are difficult to use because only a small part of the list is shown each time, to gain an overview or select elements from a long list the user usually needs to scroll the list many times. This problem becomes more serious as the list gets longer. This paper presents a novel solution, LensList, by applying the focus and context technique to the view of a list. LensList dynamically changes the elements to bigger sizes to form a focal area around the mouse cursor while keeping the elements in the peripheral area in smaller sizes as context. This enables it to display a longer list within the same screen area. Therefore it can be more efficient for performing browsing and navigation tasks.

Keywords: Information Navigation, List Navigation, Long List, Focus + Context, Multiple Foci.

1 Introduction

List is a simple yet powerful mechanism for organizing linearly structured information. It is very useful when there are only a small number of elements in the list. However as the quantity of information grows, lists are getting longer and longer. A list sometimes contains tens or even hundreds of elements, such as lists used for selecting monetary exchange rates, countries, character encodings, fonts and favourite web sites etc. Long lists are not as usable as short lists even though they are normally equipped with scrolling mechanism. Because only a small part of the list is shown each time, to gain an overview or select some elements from a long list the user usually needs to scroll the list several times. For example, in an alphabetically ordered list of all countries, how can one choose such three countries as Afghanistan, Saudi Arabia and Yemen. Since the list is long, which contains 266 countries and areas, and also these three countries separate apart from each other in the list, selecting three of them needs more interaction. This navigation problem becomes more serious as the list gets longer.

One solution to the navigation problem mentioned above is to divide the elements in the list into different categories and change the list into a hierarchy. This would be

M.J. Smith, G. Salvendy (Eds.): Human Interface, Part I, HCII 2007, LNCS 4557, pp. 535–543, 2007.

a good solution when the elements can be well classified, which means the categories are intuitive and the user would easily know which element should be in which category. However this approach prolongs the access path to the target elements by providing intermediate elements. Moreover selecting many elements in different categories would be more difficult.

Another solution is to compress the elements into small sizes to form a compact view in order to show more elements at the same time. Although more elements can be shown, the sizes of the elements could be too small to read if the total number of elements is big.

Other solutions are also possible, such as dividing the list into several columns to display, but it does not save any screen space. This paper presents a novel solution, called LensList, by applying the focus and context technique to the view of a list.

2 Related Work

Focus and context (sometimes in the form of focus + context, also often interchangeably referred to as the fisheye distortion technique) has been existing for many years since first introduced by Furnas in 1986 [1]. In that paper the cognitive aspects of how people view and remember information was discussed. The fisheye distortion technique was then applied to a variety of applications [2–5]. Several variations of the fisheye technique have been explored. They have been used in one dimension for word processing [6], access to time [7], and for long lists [8,9]. They have been used in two dimensions for tables [10], graphical maps [11] and space scale diagrams [12]. They have even been used in three dimensions for document browsing [13]. Some applications of fisheye distortion techniques have been carefully evaluated [14], often finding a significant advantage to fisheye views [15–17]. Masui's work [8,9] is more relevant to ours in terms of handling long lists. However they adopted the overview + detail strategy, which is essentially different to the focus and context method.

3 Objectives

Despite the careful investigation of fisheye distortion techniques, and their application to a broad set of complex tasks, fisheye views have never been applied to the widely used lists. By introducing fisheye distortion, LensList becomes fundamentally different from a traditional list, because every element may change size and they are not static anymore. This makes the design of LensList more difficult than traditional lists.

The goal of this work was to alleviate the browsing and navigation difficulty of traditional GUI lists in handling long linear information structures. To achieve this goal, four objectives were set up as below:

1. A GUI component was to be designed which should be compatible with traditional lists in functionality. This is helpful in reducing the users' learning time.

2. An efficient graphical fisheye distortion algorithm should be implemented, which relates the mouse movement to locating the visual focus. Since mouse event listening is not a fast process, an efficient algorithm is necessary to guarantee the smooth rendering of screen scenes.
3. The screen space occupation of this component ought to be fixed in size, no matter the focus located anywhere. This is to ensure there is no overlapping with other on screen components.
4. Multiple foci selection should be supported by this component so that working on multiple elements simultaneously is possible.

4 Methods

In LensList, elements near the focus are displayed at larger sizes, and elements further away from the focus are displayed at smaller sizes. In addition, the interline spaces between elements are also increased in the focal area, and decreased further away from the focal area. In this manner, more elements can fit into the same screen area. The elements are dynamically scaled so that as the cursor moves, a "hump" of readable elements moves with the cursor. The "hump" is like a magnifying lens over the list, and it is thus named as LensList. The effect can beseen in Figure1.

4.1 Locating Focus

LensList changes elements sizes by a distortion function, and the distortion function is based on the location of focus. Thus how to locate the focus becomes an important issue. A simple way is to relate the focus with the mouse cursor. There are two parameters to determine the position of the cursor, *p*, horizontal coordinate p_x and vertical coordinate p_y. P_x does not matter in LensList, because a list displays elements in vertical manner, the horizontal position of an element has nothing to do with distortion. P_y is the one that the designer should be concerned with. An element's position p_e, also has two coordinates p_{ex} and p_{ey}, again it is the p_{ey} that is relevant to distortion. P_{ey} is the parameter to determine the focus, but it has to be acquired through cursor position, p_y. When p_{ey} is equal to p_y, the element at p_e is selected as the focus.

4.2 Degree of Interest

The common approach to implementing fisheye distortion is to compute a "Degree of Interest" (DOI) function for each element to be displayed. The DOI function calculates the element's size. A typical DOI function includes both the distance of an element from the focal point as well as the element's a priori importance [1]. Thus certain landmark elements may be shown at a large size even though they are far from the focal point.

The fisheye view of LensList was controlled by a simple Degree of Interest (DOI) function. The two parameters that determine the degree of the lens effect are the maximum and minimum sizes of the elements. LensList uses a simple DOI function

Fig. 1. LensList displaying a number of countries with three countries being selected

to calculate elements' sizes. The function is shown in Figure2, the horizontal axis represents the distance to focus, and the vertical axis is the font size. It keeps the element at the focal point at the maximum size. Then elements get smaller, one point in font size at a time until the minimum font size is reached at which time, all more distant elements stay at the minimum font size.

The user may change two parameters of the DOI function, i.e. the maximum font size and minimum font size. Since elements away from the dynamic focus get smaller one point in font size at a time, the slope of the function is constant. Increasing or decreasing the maximum font size would also change the distortion distance. The result can be seen in Figure 2(a). The middle line is the initial distortion function. When increasing the maximum font size from 6 to 7, the distortion distance increases by 2, from -4~4 to -5~5. The function changes to the top line. The bottom line is the function after changing the maximum font size from 6 to 5. Changing the minimum font size has the opposite effect. The distortion function will change as shown in Figure 2(b). The middle line is the initial distortion function. When increasing the minimum font size from 2 to 3, the distortion distance decreases by 2, from -4~4 to -3~3. The function changes to the top line. The bottom line is the function after changing the minimum font size from 2 to 1.

The graphical fisheye algorithm uses an array to store the sizes of all the elements. When the focus changes the array is updated accordingly. The elements sizes are retrieved from the array when elements are painted on the screen.

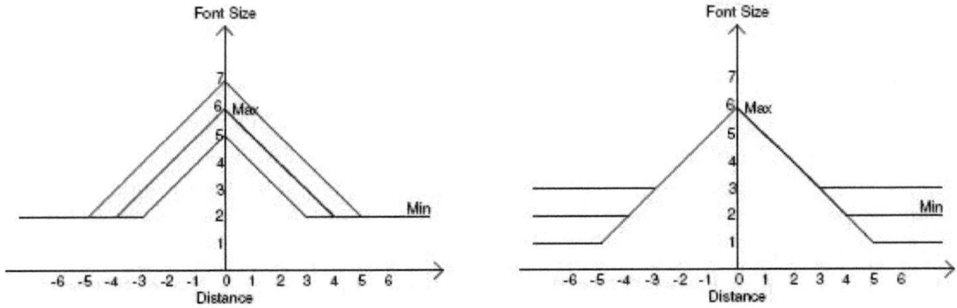

(a) Adjusting max font size. (b) Adjusting min font size.

Fig. 2. LensList DOI function

4.3 Multiple Foci of Selection

Locating foci by the mouse cursor is a practical solution for a single focus situation. However it would be different to achieve multiple foci because there is only one cursor.

The first concern is how to acquire the other foci if the first focus is obtained using the mouse cursor. Since it is not practicable to set predefined foci, the number and location of foci must be chosen by the user interactively. As the first focus and the rest of the foci are acquired differently, two terms are defined to distinguish them. The first focus is acquired through the mouse cursor, and it changes whenever the cursor moves. It is therefore called *dynamic focus* (only one). The other foci are selected by the user, and their positions do not change with cursor movement. They are hence named as *static foci* (may be many).

The second concern is whether all the foci should apply the same distortion function. If they all apply fisheye distortion, more space will be taken to accommodate the magnified elements. One of the objectives of LensList is to increase information density of the traditional lists. If all the foci use the same distortion function as the dynamic focus, this objective would be compromised. To keep high information density only the dynamic focus uses fisheye distortion.

Elements at static foci only change to highlighted color without magnifying the peripheral elements. The effect can be seen in Figure 1. Element *Bahamas* is the current dynamic focus. The three elements in highlighted color are selected as static foci.

When an element is selected by the user, it changes from the dynamic focus to a static focus, and highlighted in different color until it is deselected. Alternatively we can also think of it as a landmark, because it is set at a high a priori importance. In Figure 1, the static focus *Austria* is in the distortion range of the dynamic focus *Bahamas*, but there is no conflict between them because they are emphasized by different mechanism. This solution not only keeps high information density but also avoids conflict between foci.

4.4 Length Compensation

As the focus moving towards one end of the list, the length of the focal area is decreasing. The minimum length, half of the usual length, reaches when the first or last element becomes the focus. In this process the tail of the list will change its position, either moving up or moving down. We call this phenomenon "waving tail" effect for easy understanding. The screen space occupation algorithm compensates the length of the focal area by increasing the elements sizes on the opposite side of the focus moving direction. Hence the screen space that LensList occupies is fixed. The effect can be seen in Figure 3. As the mouse cursor moving from top to bottom, the length of the list stays unchanged.

5 Implementation

LensList was developed in Java. It subclasses the Java *JList* component in the Swing GUI toolkit, so it can be used as a replacement for *JList*. The example used to generate Figure 1 is the list of countries. While the system is up and running, it responds to user interaction immediately. The time lag for computing the distortion and refreshing the screen is unnoticeable.

6 User Test

A couple of users were invited to try the LensList tool, and they were all experienced computer users. The trial was more subjective than objective, and the goal was mainly to find user preference of the new tool comparing to its traditional peers. The task was set to browse the list of countries and a list of web sites.

(a) Focus at the top. (b) Focus in the middle. (c) Focus at the bottom.

Fig. 3. Moving focus on LensList from top to bottom

The users showed more preference towards LensList comparing to the traditional list. Some of them thought the LensList view was "visually appealing". This could be caused by their first image of the fisheye effect in graphical user in-terfaces. Most users thought that it was easier to gain an overview with LensList since it was able to display a longer list. Multiple foci was also thought to be a good feature because more elements in the list can be processed in a batch.

Moreover, some users stated that it was easier for them to concentrate on the foci without being disturbed by other elements.

A problem was also found, which was the difficulty in selecting an element when the minimum size of elements was set too small. This was an important issue in the literature. Because it is a common problem for almost all fisheye distortion based visualization techniques, a good solution is yet to be proposed.

7 Discussion

The goal of this project was reached by fulfilling the four preset objectives, which are discussed as follows:

1. The GUI component, called LensList, was implemented as a subclass of the Java JList component. The functionality of JList is fully supported in LensList, and it can be generally used as a substitute of JList.
2. The graphical fisheye algorithm had been designed twice and modified several times. The current version is O(n) in memory space consumption. The time efficiency is O(1) and dependent on the processing speed the mouse event. It was believed that this algorithm is optimized and can be used on any linear data structures.
3. The length compensation algorithm was designed independently from the graphical fisheye algorithm. This algorithm effectively solved the unexpected "waving tail" effect. It works well in one dimensional fisheye views, and can be extended to two or three dimensions.
4. Multiple foci selection was implemented in LensList. This is a useful feature when the user wants to process many elements altogether.

8 Conclusion and Future Work

Browsing and navigating lists are important and frequent tasks in modern GUIs. Long lists bring challenges to the traditional GUI list component. Fast navigation is becoming more expected with the increasing quantities of information to be processed. LensList, an enhanced GUI list is presented in an attempt to gain better performance in performing this type of tasks. LensList pursues this goal by introducing the focus and context technique, fisheye distortion, into its view. LensList dynamically changes the elements to bigger sizes to form a focal area around the mouse pointer while keeping the elements in the peripheral area in smaller sizes as context. This enables it to display a longer list than the traditional lists within the same screen area. Therefore scrolling is less often used and it can more efficient for performing browsing and navigation tasks.

The merit of this technique had been seen at this stage, therefore it was planned to continue studying this technique by conducting a series of task oriented user evaluations and refinements.

Acknowledgements. This work was jointly sponsored by the National Science Foundation of Guangdong Province of China under grant number 06300433, and Talent Introducing Fund from South China Agricultural University under grant number 2005K099. The feedback and suggestions to improve LensList from voluntary users are appreciated.

References

1. Furnas, G.W.: Generalized fisheye views. In: Proc. ACM CHI '86, Boston, Massachussetts, pp. 16–23. ACM Press, New York (1986)
2. Dill, J., Bartram, L., Ho, A., Henigman, F.: A continuously variable zoom for navigating large hierarchical networks. In: Proc. IEEE International Conference on System, Man and Cybernetics, San Antonio, Texas, pp. 386–390. IEEE SMC Society Press, Los Alamitos (1994)
3. Mitta, D., Gunning, D.: Simplifying graphics-based data: Applying the fisheye lens viewing strategy. Behaviour & Information Technology 12(1), 1–16 (1993)
4. Spence, R., Apperley, M.: Database navigation: An office environment for the professional. Behaviour & Information Technology 1(1), 43–54 (1982)
5. Spenke, M., Beilken, C., Berlage, T.: FOCUS: The interactive table for product comparison and selection. In: Proc. User Interface and Software Technology (UIST), Seattle, Washington, pp. 41–50. ACM Press, New York (1996)
6. Greenberg, S., Gutwin, C., Cockburn, A.: Sharing fisheye views in relaxed-wysiwig groupware applications. In: Proc. Graphics Interface (GI '95), Toronto, Canada, pp. 28–38. Morgan Kaufmann, San Francisco (1995)
7. Koike, Y., Sugiura, A., Koseki, Y.: Timeslider: An interface to time point. In: Proc. User Interface and Software Technology (UIST '97), Banff, Alberta, Canada, pp. 43–44. ACM Press, New York (1997)
8. Masui, T.: LensBar: Visualization for browsing and filtering large lists of data. In: Proc. IEEE Symposium on Information Visualization '98, North Carolina, USA, pp. 113–120. IEEE Computer Society Press, Los Alamitos (1998)
9. Masui, T., Minakuchi, M., Borden, G.R., Kashiwagi, K.: Multiple-view approach for smooth information retrieval. In: Proc. User Interface and Software Technology (UIST '95), Pittsburgh, Pennsylvania, pp. 199–206. ACM Press, New York (1995)
10. Rao, R., Card, S.K.: The table lens: Merging graphical and symbolic representations in an interactive focus + context visualization for tabular information. In: Proc. Human Factors in Computing Systems, CHI '94, Boston, Massachusetts, pp. 318–322. ACM Press, New York (1994)
11. Sarkar, M., Brown, M.H.: Graphical fisheye views of graphs. In: Proc. CHI '92: Human Factors in Computing Systems, Monterey, California, pp. 83–91. ACM Press, New York (1992)

12. Furnas, G.W., Bederson, B.B.: Space-scale diagrams: Understanding multiscale interfaces. In: Proc. ACM CHI '95, Denver, Colorado, USA, pp. 234–241. ACM Press, New York (1995)
13. Robertson, G.G., Card, S.K., Mackinlay, J.D.: The document lens. In: Proc. 1993 ACM User Interface Software and Technology, New Orleans, Louisiana, pp. 101–108. ACM Press, New York (1993)
14. Cockburn, A., Karlson, A., Bederson, B.B.: A review of focus and context interfaces. HCIL Tech Report 2006-09, Department of Computer Science, University of Maryland, College Park, MD 20742, USA (2006)
15. Donskoy, M., Kaptelinin, V.: Windows navigation with and without animation: A comparison of scrollbars, zoom and fisheye view. In: Proc. Extended Abstracts of Human Factors in Computing Systems ACM (CHI '97), Atlanta, Georgia, pp. 279–280. ACM Press, New York (1997)
16. Hollands, J.G., Carey, T.T., Matthews, M.L., McCann, C.A.: Presenting a graphical network: A comparison of performance using fisheye and scrolling views. In: Proc. 3rd International Conference on Human-Computer Interaction, Boston, Massachusetts, pp. 313–320. Elsevier Science Press, Amsterdam (1989)
17. Schaffer, D., Zuo, Z., Bartram, L., Dill, J., Dubs, S., Greenberg, S., Roseman, M.: Comparing fisheye and full-zoom techniques for navigation of hierarchically clustered networks. In: Proc. Graphics Interface (GI' 93), Toronto, Ontario, Canada, Canadian Information Processing Society, pp. 87–96 (1993)

Context-Based Loose Information Structure for Medical Free Text Document

Tadamasa Takemura[1], Kazuya Okamoto[2], Hyogyong Kim[2], Masahiro Hirose[3], Tomohiro Kuroda[1], and Hiroyuki Yoshihara[1]

[1] Department of Medical Informatics, Kyoto University Hospital,
[2] Graduate School of Informatics, Kyoto University
[3] Graduate School of Medicine, Kyoto University
Syogoin-Kawaracho 54, Sakyo-ku, Kyoto City, Kyoto, Japan
{takemura,kazuya,hkim.mhirose,Tomohiro.Kuroda, lob}@kuhp.kyoto-u.ac.jp

Abstract. It is efficient that free text interface is implemented in medical record when Health Care Workers (HCWs) want to write patient detail information. However contents of free text are not be able to evaluate easily. If we evaluate free text, we can use various natural language technique. Especially, knowledge of human recognition to free text is very useful, and we can construct a model of writing medical document. Concretely, this knowledge is "Context based Structure (CBS)" based on Diagnostic-Therapeutic cycle. On this time, we analyzed actual medical document (Incident / Accident Report in hospital) using this CBS and we made navigate system that was able to display something HCWs should write about according to judge contents.

Keywords: medical free text document, context based structure, machine learning.

1 Introduction

Information-oriented society had arrived, and many people thought almost all information treated as numeric data and classified things, because computers are not able to process these data and things. True, early-period information system had numeric and classified data using "Fill in Blank" input method. This fill in blank input method is very useful if we can classify and modeling strictly, but the more we use information with computer systems, the more we want to use ambiguous information semantically. Consequently, current almost all computer systems have free text information interface.

By the way, information system can be enable efficient sharing and utilization of information and innovate medical service quality. Hospital Information system (HIS) was implemented many hospitals in the world. Firstly, HIS was introduced at hospital in order to communicate order information between departments. However Health care workers (HCWs) became to need not only order entry information but also Electronic Medical Records (EMR). Today HIS has also patient information record function. The EMR are demanded not only "fill in blank" information but also ambiguous information as expressed by free text. So recent HIS has free text recorded system in order to record patient information completely.

M.J. Smith, G. Salvendy (Eds.): Human Interface, Part I, HCII 2007, LNCS 4557, pp. 544–548, 2007.

However, HCWs have to record patient information neither too much nor too little but it is difficult that we judge free text information is efficient from the viewpoint of contents semantically.

We think free text information has not to be complete free but to be small restriction and we can explain this restriction based on context which in which HCWs write this information. Therefore, we consider that this knowledge navigates HCWs in order to write free text information efficient, and we can use stored free text data structured by this knowledge in order to alert HCWs about less or more information.

On this time, we targeted "incident/Accident report", is very important medical document in order to prevent medical accident. And we make a navigation tool in order to write this report sufficiently in terms of content.

2 Method

2.1 Situation-Oriented Medical Record

If medical documents written by free text were judged by a machine automatically, it is necessary that a machine must comprehend the semantics of the documents. And if a machine comprehend the semantics, we must implement a knowledge of natural language to the machine. In the research field of automatic natural language comprehension, many researchers have tried to explain the system of human recognizing natural language. In almost human activities we can discern three stages [1]:

1. Observation
2. Reasoning
3. Action

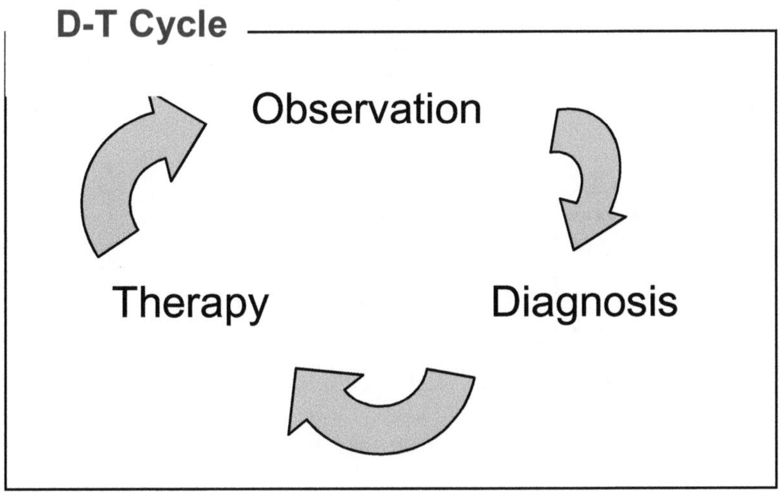

Fig. 1. Diagnostic-Therapeutic Cycle

That means our free text in order to record our activity depends on this three stages, so we considered that we could use this knowledge as explaining human activities. Also in health care the same three stages can be seen in the so-called diagnostic-therapeutic cycle. (D-T cycle). Figure1 is concept of D-T cycle.

Of course, medical records are written by context changes depend on this D-T cycle. Therefore, we can label all sentences with “observation”, “diagnosis” and “therapy” in medical documents.

We called this concept is “situation-oriented medical record” [2]. Most of all medical records can be constructed by this main body (D-T cycle) and specific details of each documentation are demanded [3].

2.2 Incident/Accident Report

Incident/Accident report is very important and expected reuse in order to prevent or decrease medical accident. Recently, this report has been accumulated using HIS subsystem. An incident/accident report is constructed of some blanks and choose items, and free text area. When HCWs encountered incident or accident, they report his or her experience, reason and so on. Figure 2 is an example of Incident/accident reporting system’s user interfaces.

Fig. 2. Incident/Accident Reporting System in Kyoto University Hospital

However we have a problem that the report is not done quality control as represent of incident/accident because free text is difficult to be judged quality automatically.

2.3 Navigation System for Incident/Accident Report

Before we made a navigation system, we made a structure using situation changes and demanding for Incident/Accident report according to analyze free text report. These free text reports, of course, include good and bad reports. But, the good reports have common classes which inform “patient summary”, “process of before accident”, “diagnosis and care”, “treatment”, “underlying cause” and “measure” based on context. We named this “Context Based Structure (CBS) label” for Incident/Accident reports.

Next, we selected actual incident/accident reports written at Kyoto University Hospital classified “falling” category. And all sentences in which the reports included suitable the CBS.

All labeled sentences were as teacher for machine learning, and we tried to use support vector machine (SVM) in order to judge a kind of label of sentences in unknown sentences. A SVM is a statistical learning technique. A SVM finds an optimal hyperplane separating training samples each of which is in vector space and has positive or negative class. We tried to compare result of labeling by human and SVM.

Next, we constructed a “reporting navigation system” using this machine learning results. This system navigates result of machine leaning to support HCWs. Concretely, this system judges contents and gives CBS label and displays something HCWs should write about.

3 Result

Table 1 shows result of performance evaluation of automatic labeling.

Table 1. Result of automatic labeling

CBS label	Correct answer	correct class	precision	recall	Num. of base data
patient summary	76	125	0.61	0.35	216
process of before accident	269	362	0.74	0.60	447
diagnosis and care	244	290	0.84	0.74	330
treatment	149	185	0.81	0.72	208
underlying cause	10	36	0.28	0.10	99
measure	122	180	0.68	0.53	231

Figure 3 shows interface of the navigation system for incident / accident report.

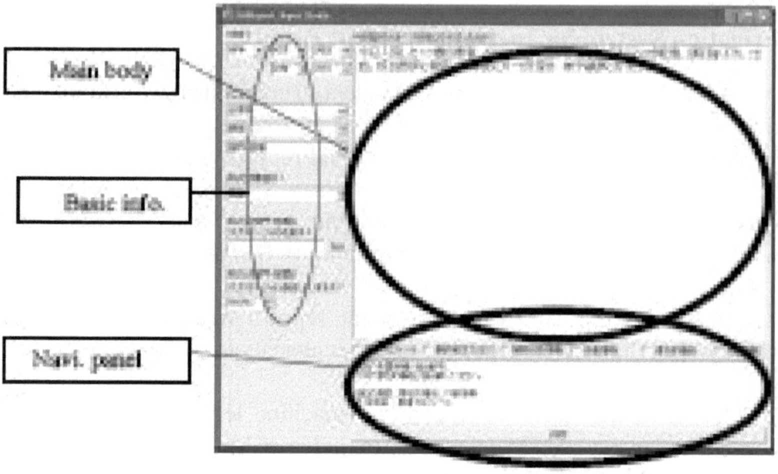

Fig. 3. Interface of the navigation system

4 Conclusion

We constructed new free text navigation system based on the context-based structure. This structure was made by analyzing actual free text reports, therefore the CBS label is suitable for process of writing of HCWs but result of automatic labeling have difference each CBS labels, because some contents of CBS labels is very various and many teacher is necessary in order to judge correctly.

The CBS class is not solid but loose, however HCWs can be navigated naturally by this system because this system matches human cognitive process. In addition, this system is able to alert excess and deficiency according to automatic analysis based on actual examples. This function depends on knowledge of this structure.

We are able to joint flexible of free text and classified data by CBS label.

References

1. Van Bemmel, J.H., Musen, J.H.: Handbook of Medical Informatics. pp. 4–7. Springer, Netherlands (1997)
2. Takemura, T., Ashida, N.: A study of the medical record interface to natural language processing. Journal of Medical Systems 26(2), 79–87 (2002)
3. Okamoto, K., Takemura, T., Kuroda, T., Yoshihara, H.: Context-Based Retrieval System for Similar Medical Documents. In: Proceedings of Asia Pacific Association of Medical Informatics (APAMI), pp. 177–183 (2006)

MyView: Personalized Event Retrieval and Video Compositing from Multi-camera Video Images

Cheng Chris Zhang[1], Sung-Bae Cho[2] and Sidney Fels[1]

[1] Dept. of Electronics and Computer Engineering, University of British Columbia, Vancouver, Canada
[2] Dept. of Computer Science, Yonsei University, Seoul 120-749, Korea
chriszc@ece.ubc.ca, sbcho@cs.yonsei.ac.kr, ssfels@ece.ubc.ca

Abstract. Video retrieval continues to be one of the most exciting research areas in the field of multimedia technology. With the advancement of sensing and tracking technologies it is possible to generate multiple personal video streams during events from different perspectives. This paper presents the first prototype of the MyView system that provides on-demand personalized video streams with multiple cameras and additional sensors. This system captures and stores video streams in an indoor office, extracts high-level events from local positioning system (LPS) tracking information, and provides on-demand video segments for high-level queries by mixing the multi-camera video streams for a stream of the best view.

Keywords: Event retrieval, video compositing, multi-camera.

1 Introduction

Using tracking technology and pervasive video/audio capture technology, it is possible to automatically create personal video programs and diaries of people at events from different perspectives [1]. Suppose that audio, video and positional data are streamed to a server continuously using video cameras, microphones and tracking technology. As people are tracked we can detect when they are in view of each video camera. During capture (possibly with some small lag), the video feeds can be viewed in real-time on demand, and at the end of the day, a personalized video can be automatically created by using the parts of the streamed video where the person is detected and editing them together.

To investigate the potential of this technology, we constructed an in-vitro environment, called MyView, consisting of four video cameras and four local positioning system (LPS) camera tracking units to track people and objects streamed to a multi-media database server. By analyzing the LPS tracking information we annotate the video streams with predefined high-level events, and produce on-demand video segments with the selection of the best views for high-level queries from human users, which is usually difficult because the process typically starts using a high-level query from a person but only low-level features are easily measured and computed.

M.J. Smith, G. Salvendy (Eds.): Human Interface, Part I, HCII 2007, LNCS 4557, pp. 549–558, 2007.

There are many wearable systems to collect personal activities by recording video data [2, 3], which aim to create intelligent recording systems for a user. A system like ID Cam [4] also used ID tags similar to our system, but it required a special camera that supported sub-regions in order to recognize those ID tags quickly. Sumi, *et al.* [5, 6, 7] also designed a video retrieval prototype to make user's favorite video based on events occurred in the scene, where a number of sensors and trackers should be attached on various objects and persons for their experiment.

2 MyView System: Design

The MyView system consists of a capturing and processing block, a query and compositing block and a multimedia database. The capturing and processing block can be divided into a video-camera capturing module, a tag-camera capturing and decoding module, 3D-tag-calculation module and an event-extraction module. The query and compositing block consists of the event interpretation engine, the best view camera selection module and the video compositing and mixing module. The multimedia database stores all the video and data created at the first block and feeds these data for the query block.

2.1 Video-Camera Capturing Module

For video capturing, we set four video cameras on each corner of our 25'x16.7' office environment which covers most of the surveillance region of the room. We envision multiple types of cameras being used in a real system, thus we encapsulate various camera drivers into the system to support different types of video cameras—from the economical USB webcams to professional network cameras. In our system, we integrate two USB webcams and two wireless network cameras to illustrate some of this flexibility. Fig. 1 shows one of the wireless D-Link PTZ IP camera used in our system.

Fig. 1. One of the PTZ camera used for capturing video

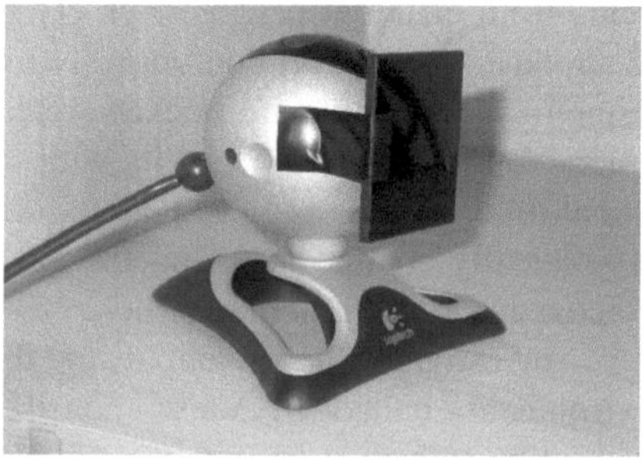

Fig. 2. One of the webcams with an IR pass filter used for a Tracking camera. Note: the camera has its IR filter removed.

2.2 Tracking-Camera Capturing and Decoding Module

We use active tags for our LPS system to detect human motion. The tag is an infrared LED that blinks a unique bit pattern at a known rate, i.e., 30 fps. Tags can be worn by people or fixed on some objects which we are interested in tracking. A camera can be tuned to capture images at the same rate as the blinking pattern. In our experiment, the bit pattern is designed with a 5-bit starting phase and a 3-bit identifying phase (Fig. 3), which means at most 8 tags could be detected at the same time. In this way, the number of tags could be easily extended to 256 only by changing the identifying phase to 8 bits. Error detecting and correcting codes may be added as well to improve robustness in noisy conditions.

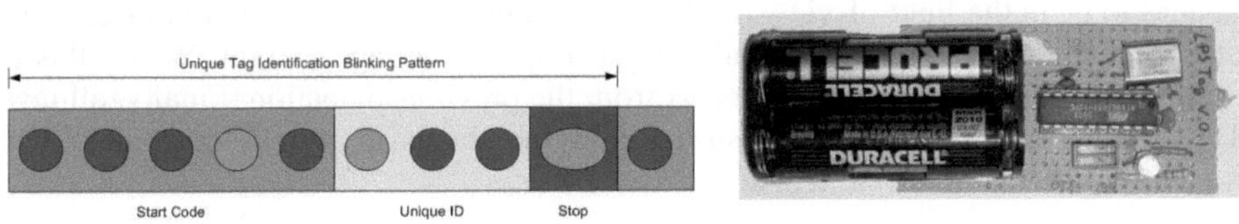

Fig. 3. The tag's ID pattern and a prototype active Tag used in our system

We attach an IR-pass filter in front of each camera to improve detection of the IR LED and reduce interference with visible light providing good quality, simple to decode images. Our tags are designed to work with cheap webcams running at 30fps allowing for large scale deployment. Fig. 2 and Fig. 3 show the tag cameras attached with the filter and the ID tag we used in our system.

A full pattern cycle takes 13/30=0.433 seconds for our current tags, which means decoding the first whole tag's ID takes less then ½ second (i.e., a tag update rate of 2Hz). Due to noise and other optical interference, a tag's blink may not always be captured consistently. However, after analyzing a couple of consecutive cycles, the pattern is always decoded correctly. Finally, each decoded tag's ID and its 2D position in each frame are recorded into a database. When used in real-time, we also perform predictive updates on the tag location, providing, on average, a 15Hz update rate of a tags location.

Usually, a tag is captured by at least two tag cameras. Therefore, we can calculate the tag's 3D location based on normal stereo vision techniques using the 2D coordinates in two cameras [8, 9]. The method can be described as follows.

Suppose the tag's unknown 3D location in world coordinate system, W, is $P:(x,y,z)$. Assume we have already calibrated the two tag cameras, C_1 and C_2, which are located at $S_1:(x_{s1},y_{s1},z_{s1})^T$, $S_2:(x_{s2},y_{s2},z_{s2})^T$ with rotation matrix as $R_{1(3\times3)}$ and $R_{2(3\times3)}$, respectively. Since we also know the 2D coordinates that the point *P* projected onto each camera plane, $(x_1,y_1)^T$ and $(x_2,y_2)^T$, we can calculate the 3D coordinates of point *P* in each camera's coordinate system as $P_{c1}:t_1(x_1,y_1,f_1)^T$ and $P_{c2}:t_2(x_2,y_2,f_2)^T$ respectively, in which f_1 and f_2 are the focal lengths of each camera, and t_1 and t_2 are an unknown scale we need to calculate using equation (1). Actually, P_{c1} and P_{c2} correspond to the same point P although they belong to different coordinate systems. Thus, we can get the transforming equation from camera coordinates to world coordinates shown in equation (1).

$$\begin{cases} P = R_1 t_1 (x_1, y_1, f_1)^T + S_1 \\ P = R_2 t_2 (x_2, y_2, f_2)^T + S_2 \end{cases} \quad (1)$$

Thus, from equation (1), we can get t_1 and t_2, and the 3D location *P*.

When capturing tag video in a real environment, however, in some frames, a tag may only be captured by one tracking camera, either because of occlusion or the tag happens to be in the field-of-view (FOV) of only one tracking camera. In this case, we can still calculate the tag's 3D location by estimating the tag's height. We do this by assuming the tag's height is the same as from the previous detection. Finally, all tags' 3D locations and time are also stored into the database providing context data associated with the video cameras.

2.3 Event Specification and Interpretation

In order to facilitate the high-level query posed useful for people we have designed a list of ***events***, which are significant intervals or moments of activities. We divide events into two categories: primitive and composite. The definition of the primitive events is as follows:

- Enter (*A*), if non-exist (*A*, room, *t*-1) and exist (*A*, gate-area, *t*)
- Leave (*A*), if exist (*A*, gate-area, *t*) and non-exist (*A*, room, *t*+1)
- Stay (*A*, *area*), if exist (*A*, *area*) and speed (*A*) <= *minSpeed*
- Move (*A*, [*x*, *y*]), if speed (*A*) >= *minSpeed*
- Close (*A*, *B*), if exist (*A*, *x*) and exist (*B*, *y*) and $|x - y| <=$ *minDistance*
- Apart (*A*, *B*), if exist (*A*, *x*) and exist (*B*, *y*) and $|x - y| >$ *minDistance*

Based on these primitive events we can define composite events such as Grab (*A*, *object*), Putdown (*A*, *object*), Carry (*A*, *object*), Look (*A*, *object*), Meet (*A*, *B*), and Converse (*A*, *B*).

The interpretation engine is crucial yet difficult to realize in MyView. Based on the database and the interpretation engine, we have implemented query operations to retrieve video pieces requested by users. The queries may be combined with multiple conditions. The engine then decomposes and interprets these conditions into primitive events set which are the elements of the SQL query statement which is finally executed on the database.

2.4 The Multimedia Database

The multimedia database is the base of the whole system. It is responsible for accepting and storing all the video and data created by the capturing and processing block as well as the event interpretation regulations. At the user side, the database provides the query results—data and video, and stores the final composed video.

The database consists of several data tables which mainly include:

- EventsType Table--Recording every event's name, level, etc.

NAME	TYPE	DESCRIPTION
EventID	Int	Primary ID
EventName	Text	Name of Event
EventLevel	Int	Level of Event
ObjectNumber	Int	Number of objects related to this event
HasDuration	Bool	Indicating whether an event has duration

- EventsStream Table– Recording all events occurred in the video stream.

NAME	TYPE	DESCRIPTION
EventName	Text	Event name
StartTime	Int	Start time of event (millisecond)
DurationTime	Int	Duration time of event (millisecond)
Person1	Text	
Person2	Text	
Object	Text	Object related to event
StartLocation	Text	Start location of event (from)
EndLocation	Text	End location of event (to)

- Tag2D_LocationStreamFile—The file recording a single tag's 2D position in each camera, each frame,

Filename: "Tag2D_XXX_YYY.DAT" XXX: TagID, YYY: CameraID

- Tag2D_LocationInfo Table– Record of a single tag's information and the corresponding 2D stream position.

NAME	TYPE	DESCRIPTION
TagID	Int	Tag's ID
CameraID	Int	Belongs to which camera
FrameRate	Int	Tags may have different frame rate
TotalFrame	Int	Total frame number of (millisecond)
DurationTime	Int	Duration time of tag's data (millisecond)
FileName	Text	File name of this 2D stream file
FilePath	Text	File path

- Tag3D_LocationStreamFile– The file recording a single tag's 3D location in each frame

Filename: "Tag3D_XXX.DAT" XXX: TagID

- Tag3D_TrackingInfo Table–Records a single tag's 3D location, direction, speeds, etc, in each frame.

NAME	TYPE	DESCRIPTION
CameraID	Int	
LocationX	Int	Camera's Location
LocationY	Int	
LocationZ	Int	
Direction	Float	Camera's direction (degree: 0-360)
FOV	Float	Camera's FOV (degree)

- BestViewCamera Table—Recording best view camera for each tag in each frame

- CamerasInfo Table – Recording every camera's location and camera features: FOV, direction, etc.

NAME	TYPE	DESCRIPTION
CameraID	Int	
LocationX	Int	Camera's coordinates
LocationY	Int	
LocationZ	Int	
Direction	Float	Camera's direction (degree: 0-360)
FOV	Float	Camera's FOV (degree)

3 MyView System: Implementation

We have developed a first prototype of MyView that captures a scene with four video cameras and four tracking cameras. It extracts events in the scene using the tag information, and retrieves a sequence of video segments for individual users from high-level queries.

3.1 Tag Recognition

As we mentioned before, the tracking cameras only capture the infrared blink signal emitted from the tags. In this way, the tags' position can be easily found by searching for bright spots in the video stream. Every tag's ID is reliably decoded after tracking a few blinking cycles and can be updated at 15Hz once detected.

3.2 3D Location Calculation

Our surveillance space is around 300 inches by 200 inches. The biggest error between the 3D position calculated in this step and the real position is less than 3 inches, which is accurate enough for the tags' events calculation in later steps. The system has two tag positions (blue and pink dots) in our surveillance room. The yellow rectangles are the door area. The four pairs of squares on each corner represent only four video cameras but reflect which camera has the best view for different tag—the bigger the camera's square, the better of view for this tag.

3.3 Event Extraction

Based on the event specification in section 2, we can find every primary event just from the tags' location, speed and direction. Composite events are then extracted using these primary events.

3.4 Final Video Composition

The video compositing step requires creating a final video stream from the sequence of the segments returned from a query. For each segment, we have four video sources possible, however, the simplest approach is to choose the camera with the best view in each time span. The best view camera is determined as follows:

At frame *i*, supposing f_i, given the person A (Tag A)'s direction and position are d and p, the best camera C_k is decided by whether the tag is currently in C_k's FOV and how close the tag and the camera's directions are. The equation is shown as follow:

$$C_k = \arg\max_i f_i^A(d,p) = C_k \middle| p \subset FOV(C_k) \wedge Min(\Delta(d,C_k)) \quad (2)$$

where $\Delta(d,C_k)$ is the angle difference between the camera and the tag.

Finding the best-view camera is implemented during the tag recognition stage. The result is stored into the database once computed. At the moment, we return the best-camera view segment for a query. Occasionally, in one video segment, there may be more than one best view camera, i.e., a "move" event of a person walking around a table may create different best view cameras. In that case, if the segment is too long, another different best view camera—if applicable, may be included as the video source to keep a reasonable camera selection. Other algorithms may be used to determine which camera angle to use when depending upon the desired compositing result for different contexts. This is part of our ongoing research.

4 Experimental Results

Fig. 4 shows a scenario of two persons who enter a room, move, converse each other, and leave the room.

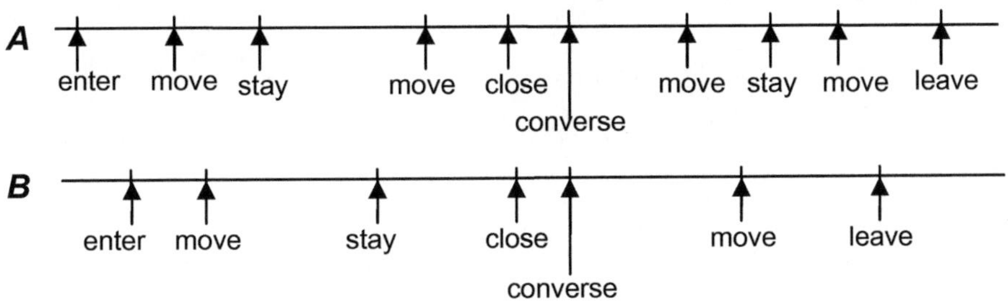

Fig. 4. A scenario of two persons in a room

Fig. 5 shows the user query and compositing interface. Users compose a query by selecting different conditions, i.e., which person, which action, where, when, etc. All these input data are analyzed and combined into the corresponding SQL query and then executed based on the multimedia database. The searched results may consist of several time spans. In this example, the user's query is to find all "move" events of person A. The red blocks show the time spans that satisfy the query. The four video pictures show the videos taken by video cameras. Users can easily play each video piece found by the query or play the whole video in different angles at the same time.

Fig. 5. Video query and mixing

Each query from users is interpreted into SQL statements by the system. The following sections show some example query results showing video query, retrieval and composition.

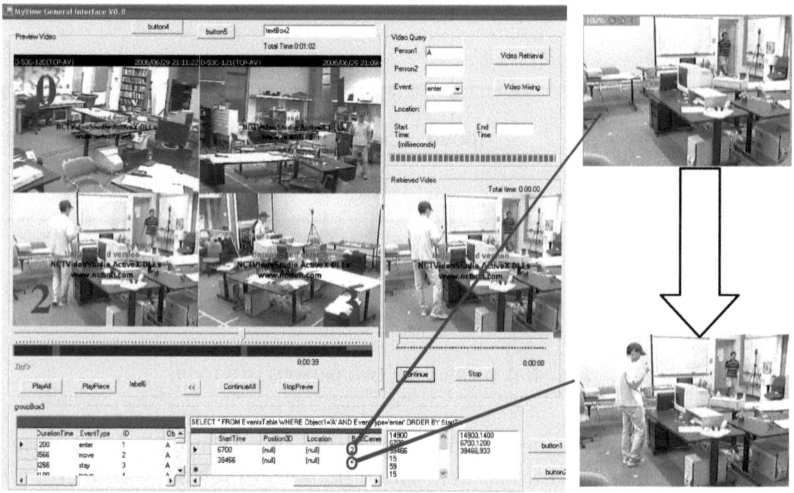

Fig. 6. Query of "Enter" event for person A

4.1 "Enter" Event

In this example, a user wants to search for all person(A)'s "Enter" events. Based on the system described before, the processing can be divided into:

- Making SQL statement. At this time, the statement created is:

```
SELECT * FROM EventsStreamTable WHERE Object1='A' AND EventType='enter'
ORDER BY StartTime ASC, StartTime+DurationTime ASC
```

- Query the database and getting results. Fig. 6 (left) shows two video parts are returned from the database.
- Video composition. Although each result part is actually captured by four video cameras at the same time, we currently select the best view camera as the final video source. In this demo, camera No. 2 is the best camera for both parts and is marked with red lines in Fig. 6.

4.2 "Move" Event

In this example, a user wants to search all the person(B)'s "Move" event.

- SQL statement:

```
SELECT * FROM EventsTable WHERE Object1='B' AND EventType='move' ORDER
BY StartTime ASC, StartTime+DurationTime ASC
```

- Query results. Fig. 7 (left) shows three video parts are found.
- Video composition. The best camera for each part is 2, 1 and 3, respectively.

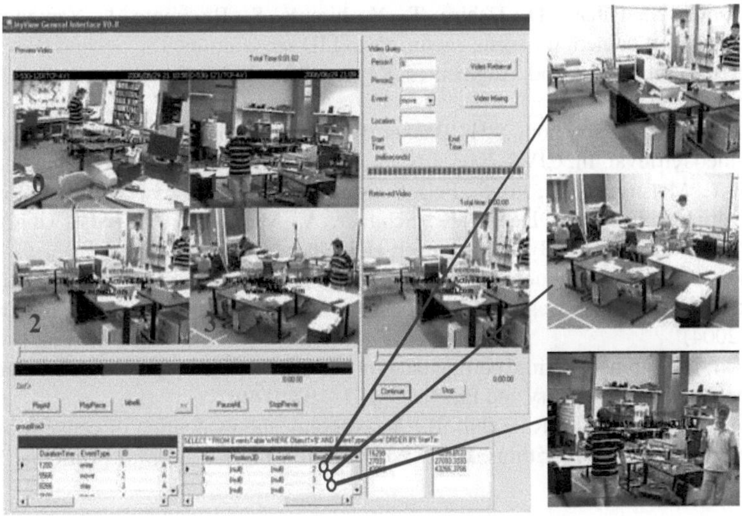

Fig. 7. Query of "Move" event for person B

5 Concluding Remarks

This paper has proposed a multimedia storage and retrieval system with multiple cameras and sensor networks, which provides on-demand personalized video stream for high-level queries. Ultimately, we plan to create the necessary research infrastructure to demonstrate the feasibility of large sensor networks being used to provide a record from multiple perspectives of indoor and outdoor events such as sports arenas, museums, exhibition and entertainment locations. This record can be accessed in real-time, on demand for multi-perspective views. As well, it can be accessed after the activities to create personalized memories or see new perspectives on the event. Both methods for people to access the stored media provide a rich, meaningful and entertaining way to experience life during an event as well as in the future.

Acknowledgements. This work is supported by the NSERC, Bell University Lab (Canada), D-Link (Canada) and MIC (Korea) under ITRC IITA-2005-(C1090-0501-0019).

References

1. Weiser, M.: The computer for the 21st century. Scientific American 265(30), 94–104 (1991)
2. Kawamura, T., Kono, Y., Kidode, M.: Wearable interfaces for a video diary: Towards memory retrieval, exchange, and transportation. Int. Sym. Wearable Computers, pp. 31–38 (2000)
3. Mann, S.: Humanistic intelligence: WearComp as a new framework for intelligence signal processing. Proceedings of the IEEE 86(11), 2123–2125 (1998)
4. Matsushita, N., Hihara, D., Ushiro, T., Yoshimura, S., Rekimoto, J., Yamamoto, Y.: ID CAM: A smart camera for scene capturing and ID recognition. The Second IEEE and ACM Int. Symp. on Mixed and Augmented Reality, 227. IEEE Computer Society Press, Los Alamitos (2003)
5. Sebe, N., Lew, M.S., Zhou, X.H., Bakker, E.M., Huang, T.S.: The state of the art in image and video retrieval. In: CIVR 2003. LNCS, vol. 2728, pp. 1–8. Springer, Heidelberg (2003)
6. Sumi, Y., Matsuguchi, S., Ito, S., Fels, S., Mase, K.: Collaborative capturing of interactions by multiple sensors. In: Dey, A.K., Schmidt, A., McCarthy, J.F. (eds.) UbiComp 2003. LNCS, vol. 2864, pp. 193–194. Springer, Heidelberg (2003)
7. Sumi, Y., Ito, S., Matsuguchi, T., Fels, S., Mase, K.: Collaborative capturing and interpretation of interactions. Pervasive 2004 Workshop on Memory and Sharing of Experiences, pp. 1–7 (2004)
8. Szeliski, R., Zabih, R.: An experimental comparison of stereo algorithms. Int. Workshop on Vision Algorithms: Theory and Practice, pp. 1–19 (1999)
9. Scharstein, D., Szeliski, R., Zabih, R.: A taxonomy and evaluation of cense two-frame stereo correspondence algorithms. IEEE Workshop on Stereo and Multi-Baseline Vision, pp. 131–140 (2001)

Part IV

Development Methods and Techniques

Context-Aware Information Agents for the Automotive Domain Using Bayesian Networks

Markus Ablaßmeier, Tony Poitschke, Stefan Reifinger, and Gerhard Rigoll

Institute for Human-Machine Communication
Technical University of Munich
Arcisstr. 21, 80333 Munich, Germany
phone: +49 89 289-28541
{ablassmeier,poitschke,reifinger,rigoll}@tum.de

Abstract. To reduce the workload of the driver due to the increasing amount of information and functions, intelligent agents represent a promising possibility to filter the immense data sets. The intentions of the driver can be analyzed and tasks can be accomplished autonomously, i.e. without interference of the user. In this contribution, different adaptive agents for the vehicle are realized: For example, the fuel agent determines its decisions by Bayesian Networks and rule-based interpretation of context influences and knowledge. The measured variables which affect the driver, the system, and the environment are analyzed. In the context of a user study the relevance of individual measured variables was evaluated. On this data basis, the agents were developed and the corresponding networks were trained. During the evaluation of the effectiveness of the agents it shows that the implemented system reduces the number of necessary interaction steps and can relieve the driver. The evaluation shows that the intentions are interpreted to a high degree correctly.

1 Introduction

Nowadays, the vehicle is much more than only a mere transport medium. Particularly in cars of the premium segment and increasingly in middle class vehicles, familiar technical systems for comfort have been integrated. For example, for a long time cell phones and navigation systems have been very popular, and even TV, mp3-player and internet are already in the equipment list of some manufactures.

This multiplicity of new multimedia applications must remain operated by the driver. The car manufacturers walk on new paths to accomplish more complex communication between the human and the machine. The used numerous hard keys will become partly replaced by so-called soft keys and well-known control elements are provided with multiple functions (e.g. the wiper lever is added with cruise control). Besides that, also the information flow between driver and automobile changes. In the past, system feedback in cars was only available by small warning lamps. Today, system feedback is given multimodal - thus visually, haptically and/or acoustically (in particular also by speech feedback). Additionally, the driver can give verbal instructions to the system. Also, the instrument cluster is not any longer the only displaying area: it is supplemented by a Head-Up Display (HUD) and a Central Information Display (CID).

M.J. Smith, G. Salvendy (Eds.): Human Interface, Part I, HCII 2007, LNCS 4557, pp. 561–570, 2007.

However, by this variety of new functions and variants in prompting, the driver is often overloaded. This fact causes criticism and skeptical reactions especially by the media [1]. Above all, if the operation is not intuitive and uncommon arranged for the driver, unknown new functions or new ways of presentation (modality) makes it difficult for the user to be handled. This altogether contributes to a substantial cognitive workload as well as the increasing traffic on the roads.

In order to arrange the aforementioned complexity and to make the interaction with the vehicle to a high degree efficient, the use of intelligent procedures is very eligible. Embedded in software agents these procedures support the driver and optimize thereby the human-machine interface. Such intelligent software agents are utilized already in different domains. Since they arrange their tasks mostly in the system background, they are not disturbing and this is an advantages of such agent applications. In particular, these agents are widely spread unnoticed in the internet. For example, depending upon surfing strategies, fitted commercial information are indicated to the user.

2 Basics and Background

In the vehicle domain, agents can reduce the driver's cognitive workload by filtering all data that affects him. Behavior and preferences of the user are to be likewise considered as occurrences while driving. Therefore, agents do not only decide, which information has to be shown on a certain output channel, but also at which point of time. It is surely desirable that an agent recognizes that it should not offer the results of a fuel station search during full braking. As well, the linguistic route guidance should be changed to a visual representation in the HUD during a telephone call. At current time, a navigation system is not capable of calculating the route to the workplace by itself every morning. The agent-based system can recognize regular behavior patterns and fulfills the necessary steps automatically. It even can provide the driver with traffic jam information.

By the cross-linking of the agents with other technologies, they can be used more comprehensively and more efficiently. Sensory inputs are linked and interpreted context-depending. So, a substantially more differentiated evaluation of linguistic inputs is possible for example in connection with a head tracking system. The expression "there is a cold draft..." together with a view to the top will somewhere cause an appropriate agent to close the sun roof.

In this contribution it is examined to which extent agents can be of assistance to the driver. The emphasis is thereby in the adjustment of the agents to the user. Based on their knowledge, learned from the behavior and from the preferences of the driver, they are supposed to filter information and implement inputs independently. Statistic methods offer suitable possibilities of filtering from earlier actions and in this way enclose on the intentions of the user. Here Bayesian Networks are used, because they can draw conclusions by means of conditioned probabilities from data observations.

2.1 Taxonomy of Driving-Tasks

Compared to the automobile domain, in front of a desktop PC, the user can predominantly execute her or his operations in a concentrated way, as there is no dual task competition. Especially in the car domain, often error-prone situations occur regarding human-machine interaction with different in-car applications, as the driver often has a certain mental workload. This basic stress level is due to the execution of so-called primary and secondary tasks, and may be increased by environmental impacts, like the conversation with a co-driver. If the driver interacts, e.g. with a communication and infotainment system in such a stress phase (tertiary task), inattention, distraction, and irritation occur as a consequence of the high workload resulting from a superposition of the tasks mentioned above, which will become manifest in an increased error potential and in erroneous operations of these tertiary systems. [2] and [3] introduce an in-depth classification of driving tasks.

2.1.1 Primary Tasks

Primary tasks only include steering operations. These are segmented into navigation, steering, and stabilization. Choosing the route from departure to destination corresponds to the navigation task. Steering includes, for example, lane changes due to the current traffic situation. User interaction with the car to navigate and steer is called stabilization. It is accomplished by utilizing the steering wheel as well as accelerator and break pedals. These tasks are essential for a safe control of the car, and therefore have highest priority while driving.

2.1.2 Secondary Tasks

Secondary tasks are operations, like reactions to and dependent on driving demands, but they are not essential to keep the vehicle on track. Examples are the turn signal, honking, and turning the headlights up and down. These tasks can again be subdivided into active and reactive actions. Reactive actions happen because of external influences, e.g. windshield wiping. Active operations are initiated with intent to communicate with other traffic participants. This is done by honking or using the turn signal.

2.1.3 Tertiary Tasks

Tasks not concerning the actual driving itself are categorized as tertiary tasks. Besides convenience tasks like adjusting the temperature of the air condition, but communication and entertainment features count in here as well.

2.2 Software Agents

The term "agent" describes a software abstraction, an idea, or a concept. The concept of an agent provides a convenient and powerful way to describe a complex software entity that is capable of acting with a certain degree of autonomy in order to accomplish tasks on behalf of its user.

2.2.1 Definition and Characteristics

In literature there is no uniform definition for agents, so that e.g. Caglayan [4] defines it with the following explanation: a person or a thing, who is authorized to act on behalf of a third. With the two substantial characteristics arising in this explanation that an agent

1. accomplishes tasks
2. on behalf of a person or a object.

Caglayan gives a possible definition for software agents: a software entity, which fulfills delegated tasks from the user autonomously. According to Brenner [5] software agents have one of the following characteristics:

- autonomy
- mobility
- communication/co-operation
- social behavior
- reactivity
- deliberative behavior

Reactive agents are developed according to the stimulus-response principle and represent a rather simple agent structure. According to given behavior rules the agent reacts to conditions or environmental influences. Reactive agents are less complex developed than deliberative agents, but more easily to adapt for new challenges. Their intelligence is due interaction of several agents and therefore is called distributed Artificial Intelligence (AI).

The internal structure of deliberative agents is based on BDI (beliefs, desires and intentions). They possess a model of their environment as knowledge base and a learning as well as a communication module according to the classical AI. The knowledge is constantly updated from sensory data from that intentions are derived. Intentions result from special desire or certain goals. In addition a plan data basis is available for the deliberative agent. The plans selected in each case are called intentions. The conclusion effects from the selection of a plan, consists of several (sub-)goals. On lowest level of this proceeding such a plan is for instance an instruction for the control of an actuator.

2.3 Context and Knowledge

For the development of agents it is important to define the criteria, which are of importance for decision making [6]. Generally, context can be subdivided into user, system and environmental context(see Fig. 1). In the following the relevant data are generally presented. The relevant variables are deposited in several data bases. Besides that, necessary knowledge takes place and allocation of the context sizes in

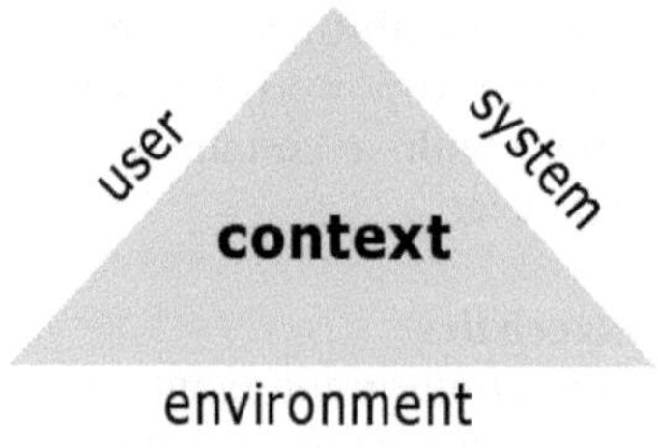

Fig. 1. Context Triangle

user- and vehicle-specific data. User-specific context summarizes data, which contribute to the modeling of the user and therefore are needed to adjust the system to the user and his preferences.

2.4 Knowledge Basis

Besides world and expert knowledge also local relations - e.g. geographical knowledge - rank among the data relevant for the agents (e.g. current time, date, weekday, GPS coordinates, etc.). GPS coordinates are independent of the viewer and its position; thereby through the indication of length, width and height exactly one location on earth is defined. Fuel stations and other destinations are to be stated as examples of the indication of points of interest (POI) by means of GPS coordinates. In our case also physical relationships and/or the relations between physical dimensions are to be mentioned, e.g. the definition of the speed v as quotient from distance s and time t.

2.4.1 Vehicle-Specific Data

Here all information is contained, that refers to the vehicle (e.g. current vehicle position, range readings, required fuel type, consumption and level of fuel in the tank (maximum and cur-rent)).

2.4.2 User-Specific Data

The user is affected by various factors. His preferences provide again a different weighting of user-relevant sizes. These are for example fuel station brand, fuel line price, relative distance, location (own, fuel station, POIs) and the level of fuel in the tank when user refills it. Within the relative distance it is summarized whether the driver wishes as soon as possible a fuel station (and/or POI), whether this is to be on the route or at least in driving direction, and whether he would accept a detour. In addition dates and the contacts deposited in the directory are summarized in this database.

2.4.3 History

In the history decisions and actions of the user are stored. Additionally, also context-information must be deposited, in order to be able to recognize, which sizes of the actions of the user in which extent to affect. Only by the evaluation of a history these conclusions can be drawn. For adaptive systems the history is consequently a crucial element.

Similarly to the knowledge the history is stored however not only explicitly, but finds consideration by the change of variables. Particularly regarding Bayesian networks the state variables in the Conditional Probability Tables (CPT) of the individual knots are occupied by each action of the users with new values, weightings again are computed and assigned.

2.5 Basics of Bayesian Information Processing

The everyday life is often coined from each other dependent events. For example the occurrence of certain circumstances has influence on (or several) decision(s), which we have to meet. Differently than in the Boolean logic, with which a met acceptance is either true (yes or 1) or wrong (no or 0), clearly definable occurrences do not only exist in the reality. Rather indefinite (indistinct) events meet us; instead of saying, “there are a few clouds in the sky, therefore it will rain”, the statement "the sky is covered to 90 %, therefore it will rain with a probability of 60 %” is a lot more precise. Bayesian networks offer now a relatively simple possibility of modeling events and the resulting reactions. They also provide an opportunity of computing final probabilities for their (not -) occurrence. In order to be able to compute the probability for the occurrence and/or non-occurrence of an event under several conditions, a mathematical regulation is needed. The basis for the Bayesian networks is supplied by a theorem from the mathematician and minister Thomas Bayes who lived in England in the 18th century. The Bayesian theorem, to which the Bayesian network owes its name, indicates a computation regulation for conditioned probabilities. For two given events *A* and *B* - whereby A represents the cause and *B* the effect - the Bayesian theorem [7] reads as follows:

$$P(A) = \frac{P(A \mid B) \cdot P(B)}{P(A)} \tag{1}$$

This can be deduced easily from the fundamental law of the conditioned probabilities through the computation of the probabilities of group. From

$$P(A \mid B) \cdot P(B) = P(A, B) \tag{2}$$

$$P(B \mid BA \cdot P(A) = P(A, B) \tag{3}$$

and

$$P(A \mid B) \cdot P(B) = P(B \mid A) \cdot P(A) \tag{4}$$

one receives by equation, hence

$$\sum_{i=1}^{n} P(a_i) = 1 \tag{5}$$

The outcome of rearranging the equations is the Bayesian theorem. Here the variable *A* possesses finally many states (a_1, ..., a_n ($a_i \neq a_j$, *i*, *j*=*1*, ..., *n*)). Similar b_1..., b_n is finite many conditions for the variable *B*. The general probability *P(A)* for the occurrence of the event *A* is called a-priori probability. With *P(B|A)* one receives the probability for the occurrence of *B* under the condition that *A* occurs. To the sum of the probabilities of all possible conditions of *A* applies.

3 Agent Concepts

Adaptive agents are conceivable for an employment in many different domains. If one is limited to the vehicle-specific domain, many interesting possibilities result of using agents for the improvement of man-machine communication. In order to limit this variety, a scenario was sketched. For this, several specific agents could be developed.

From this default the assumption was met that a businessman has to notice a date in far distance. In the vehicle all addresses, dates, and further contacts are stored. So not only the businessman, but also the developed agents can access this information. The search for a fuel station plays a substantial role, at least in this scenario if the distance is larger to the destination than the range dependent on the level of fuel in the tank. A possible necessary fuel stop might cause that the date cannot be achieved in time, this also needs to be addressed in this scenario. Substantial aspects are thereby on the one hand in the adaptation and on the other hand in the autonomous behavior of the agents. They are supposed to learn from the behavior of the user and act without explicit request.

3.1 User Analysis

In order to be able to use Bayesian networks they must be equipped with knowledge. The context data is analyzed for different user profiles. It is possible to consider preferences of individual persons and selection mechanisms. For the collection of data records two programs were developed to collect the data sets of the subjects. By a random process different scenarios are generated and stored in text files. In this way for each subject 25 data records for the tank agent as well as for the contact agent are collected. Eight persons (1 female, 7 male, average age 26.0 years) accomplished the program, so that for the further work per agent 200 data records are finally available.

3.2 Fuel Agent

The fuel agent supports the driver with all aspects connected with the fuel procedure. The special attention is on the adaptation of the agents and mixed-initiative. That means the fuel agent can be activated explicitly by the user or even takes initiative by itself. With consideration of user preferences the agent answers questions in respect to where and when. However the questions are not limited to fuel station search but also consider the dialog with the user. The agent must decide, e.g., when he informs the user and when he probably wants to refuel.

In order to accomplish his task, the agent must be provided with distances to the intermediate and/or final destinations. Of particular importance is also the knowledge about the current level of fuel in the tank and current consumption. This can be calculated from empirical average values formed by several track sections. With an appropriate inquiry the fuel agent gets a preselection of fuel stations. Containing information concerning actual distance, linear distance, fuel station brand, prices, fuel type, address, and the detour what the fuel station would cause. Then it waits for the selection of the driver. The fuel agent supplies the rating of every fuel station transferred to him. Differently expressed the agent makes information available about probability of the user selecting a certain fuel station in the given context. The fuel agent continuously supervises the level of fuel in the tank and adjusts the range with the distance to existing (between-) destinations. If the level of fuel in the tank does not suffice to reach a given destination, then it communicates this to the driver and asks whether a fuel stations should be located. Depending on user preferences the fuel agent will either be initiated immediately or at a certain remaining quantity the driver would like to refuel. This threshold depends on the context, e.g., the kind of road (highway, side street, etc.), the distance to the fuel stations. As default value 80 km are used. However this threshold value can be adapted. If the search for fuel stations was arranged, the agent makes an inquiry to the data base and receives a preselection with ten fuel stations. With the help of the Bayesian networks all fuel stations are rated and these data is provided to the user. If he selects a fuel station at any time, then the tank agent rates the associated data and updates the Bayesian network. Concerning the recognition of the user intention the agent adapts from time to time more exactly.

With the Bayesian network the fuel agent processes the four criteria brand, distance, price, and level of fuel. Each criterion is divided in a number of ranges. The level of fuel in the tank is the available amount for reaching the fuel station. The continuous values are transferred into discrete ranges. Thus, the classifications of the values can be better compared. In the second level the criteria are weighted concerning the user preferences. After the arrival of the preselection, the criteria of the fuel stations are compared and divided in the appropriate ranges. The values for the probability distributions are computed on basis of the user analysis.

3.3 Contact Agent

Similar to the fuel agent, the contact agent has to learn from the behavior of the user. Of special importance is the processing of indistinct inputs, for being able to forecast the user's intentions. Thus, the system calculates on the request "Call Mr. X!" the different probabilities for a call on the mobile, the private or the office number.

If the contact agent is initiated, it has to be provided with different information concerning the current weekday and the current time. Further the agent has to know whether the driver has fixed a date with the communication partner which is deposited. Additionally the status and the kind of date are of importance. The status depends on whether the date will be reached in time, too late or if it was already missed. The kind of date differentiates between the two options in business and private. The contact agent supplies the probabilities for the use of the five possible media: private number, office number, mobile telephone, SMS and e-mail.

The contact agent finally receives still another feedback. The chosen medium (e.g. mobile, private or business phone, etc.) is also used to update the Bayesian Network, which is used for internal processing of the probabilities.

According to the attributes, which were conveyed for the criteria weekday, time, kind of date, date status and user instruction, in this network evidences are assigned. For a criterion a characteristic value does not have to be assigned compellingly; no observation for this condition was done.

If the user wants to contact a dialog partner, he must choose from seven possible instructions. Either he decides explicitly for one of the five media - mobile, private telephone (private), office telephone (business), short message on the mobile (SMS) or e-mail - or it expresses only an indistinct instruction. In the second case there is the option that the driver wishes any of the five methods (generic_contacting) to get in touch with someone.

Another alternative would be to call someone, but did not commit a special medium (generic_call). Still there would be room for contacting someone via SMS/e-mail (written_notice). Since the preliminary evaluation showed that the e-mail is hardly used as contacting medium and the SMS can be selected directly this alternative was not implemented yet.

The time was divided into six ranges, in order to meet the fact that the choice of a medium depends on whether a call happens during the work time, midday, in the avocation, etc. Also the possible knots for the time of day in the Bayesian Network are to be limited by this classification.

4 Evaluation

This chapter presents the results of the usability test that the developed and implemented agents had to face.

4.1 Methods

After this concept for the intelligent, user-adaptive agents was integrated into the simulation, now the efficiency of the assistants was to be examined. Although the provided system covers several agents, only those are examined here for their reliability, where Bayesian networks are used and stand with adaptation in the foreground. Two different behaviors are in special focus. On the one hand it is the extent of the adaptation, i.e. the choice of the agent agrees with the intention of a user. On the other hand the adaptation duration is regarded, that means, how often the user must make a selection until the agent makes right decisions.

4.2 Results

After a learning phase, the fuel agent could recognize the intention of the user in 55 % of the cases and computed the correct fuel station as the most likely. If the user was provided a preselection of ten listed fuel stations, then 81 % of the fuel stations selected by the users were ranked and assortment by the agent at place one to four; without precalculation this was only 44 %. Concerning the number of selection steps (2.725) needed for the selection, this is a saving of 48.1 % (in the median 53.7 %) compared with a representation without pre-sorting by an agent (5.25 steps). With the comparison between users and agent, which medium in given context is to be selected, a success ratio from 57 % is registered using the contact agent. In approximately 15 % of the cases, the medium selected by the user was ranked after the agent's computations at second place.

5 Conclusions

In this contribution, we gave an introduction to the automotive domain in terms of driving task and context and presented context-aware agent systems.

Resulting from this contribution, software agents in the vehicle represent a suitable approach to support the driver. Particularly the recognition of the driver's intentions by intelligent procedures places a kind of electronic secretary functionality to the driver.

This work presents a basic structure of an agent system that was created and examined for negotiability. In order to estimate the potential advantages resulting from the employment of intelligent agents and the integration of Bayesian Networks, an objective evaluation was accomplished.

Long-term studies can verify on the one hand the efficiency of the agents. On the other hand also subjective impressions about agents can be inquired, e.g., whether agents are generally judged as meaningful and for which tasks these are to be applied. Further questions concern the transparency of an agent system: does a user understand, why an agent does its actions, did the user notice the agent's actions at all or is the agent's work disturbing for the user. At worst case agents confuse the user, at best case the agents are judged as helpful.

References

1. Wüst, C.: Irrfahrt durchs Untermenü. Der Spiegel 12, 151 (2005)
2. Rasmussen. Skills, rules and knowledge. In: IEEE Transactions, SMC-13, pp. 257–266 (1983)
3. Donges, E.: Das Prinzip Vorhersehbarkeit als Auslegungskonzept für Maßnahmen zur aktiven Sicherheitsmaßnahmen zur aktiven Sicherheit. In: Das Mensch-Maschine System im Verkehr, VDI-Berichte (1992)
4. Harrison, C.G., Caglayan, A.K.: Intelligente Software-Agenten. Carl Hanser Verlag München, Wien (1998)
5. Wittig, H., Brenner, W., Zarnekow, R.: Intelligente Softwareagenten - Grundlagen und Anwendungen. Springer, Berlin (1998)
6. McTear, M.F.: Spoken dialogue technology: Enabling the conversational user interface. ACM Computing Surveys (CSUR) 34, 90–169 (2002)
7. Jensen, F.V.: Bayesian Networks and Decision Graphs. Springer, New York (2001)

Signposts to Tomorrow's Human-Computer Interaction

Hans-Jörg Bullinger[1], Dieter Spath[2], and Matthias Peissner[2]

[1] Fraunhofer-Gesellschaft zur Förderung der angewandten Forschung e.V.
Hansastraße 27c, 80686 München, Germany
hans-joerg.bullinger@zv.fraunhofer.de
[2] Fraunhofer Institute for Industrial Engineering (IAO),
Nobelstr. 12, 70569 Stuttgart, Germany
{dieter.spath, matthias.peissner}@iao.fraunhofer.de

Abstract. The Fraunhofer-Gesellschaft has selected "human-machine interaction" as one of twelve areas of technology with a particular potential for innovation and market-relevance. The paper gives a brief overview of the goals and research topics of the initiative.

Keywords: Human-machine interaction, Fraunhofer-Gesellschaft, enhanced interaction, contextual user interfaces, user experience engineering.

1 Introduction

The Fraunhofer-Gesellschaft undertakes applied research aimed at promoting the innovative capacity of German industry and thus strengthening Germany's status as an industrial location. Our work involves analyzing markets, developing new products, processes and services, and enhancing existing production plant and organizational structures. We play our part in ensuring that German companies are able to gain a competitive edge in the face of international competition, and can maintain and expand that edge.

To help us find our bearings in the fast-moving current of technological trends around the world, our experts have evaluated numerous foresight studies conducted by other industrial nations and roadmaps drawn up by international corporations, and have discussed and further evolved the results with internal and external experts. A comparison of national and international research trends with the present competencies and strengths of the Fraunhofer-Gesellschaft has revealed twelve areas of technology in which we particularly expect to see market-relevant innovations.

These twelve thematic areas have been collected under the title of "Signposts to tomorrow's markets". They are characterized by their outstanding potential for innovation and their remarkable relevance to the market. The Fraunhofer-Gesellschaft is particularly well equipped to meet the great need for research and development in these areas. The Fraunhofer Institutes have therefore joined forces with partners from industry and are vigorously pressing ahead with the corresponding activities. The twelve thematic areas are:

- Internet of things
- Smart products and environments
- Micro power engineering
- Adaptronics

M.J. Smith, G. Salvendy (Eds.): Human Interface, Part I, HCII 2007, LNCS 4557, pp. 571–574, 2007.

- Simulated reality: Materials, products, processes
- Human-machine interaction
- Grid computing
- Integrated lightweight construction systems
- White biotechnology
- Tailored light
- Polytronics
- Security

2 Fraunhofer Innovation Topic »Human-Machine Interaction«

In the thematic area "human-machine interaction", the competencies and activities of fifteen Fraunhofer-Institutes are bundled in order to completely cover all major

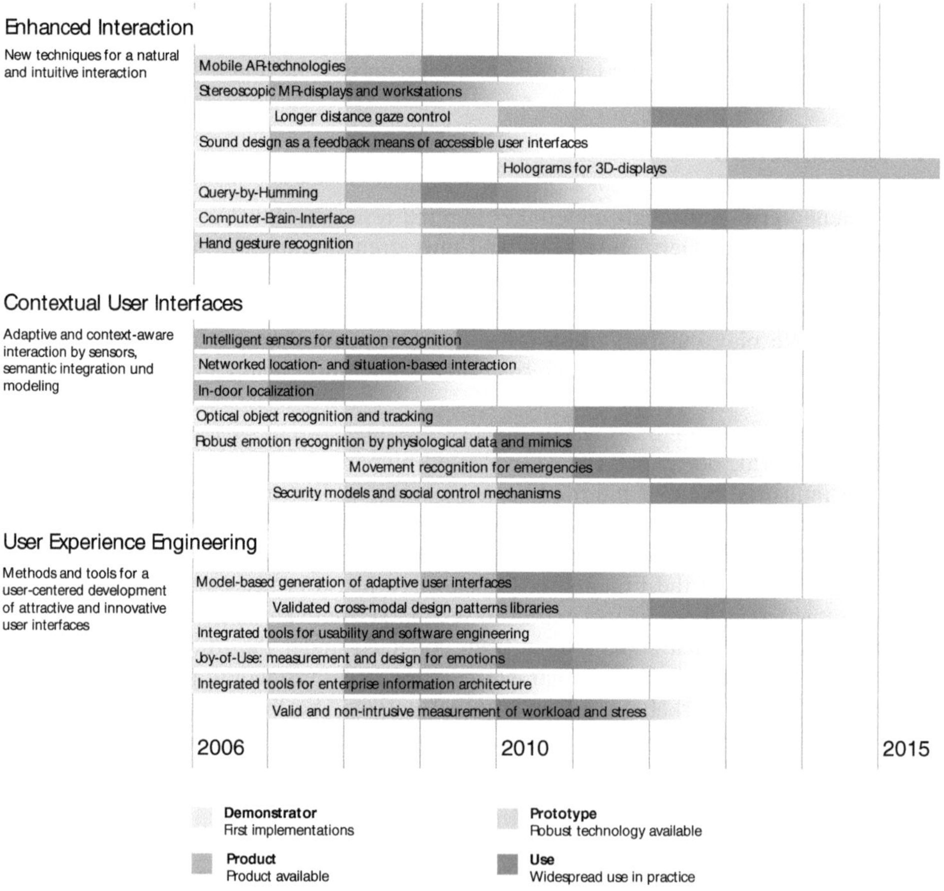

Fig. 1. Extract from the 2006 research roadmap for human-machine interaction at Fraunhofer

aspects of the field. Towards the market and external partners, the consortium enables the participating institutes to act as one strong player with experts from diverse areas. For the partnering institutes, the consortium provides a valuable network that supports multidisciplinary cooperation and knowledge sharing. Fraunhofer experts in the field come from diverse disciplines, including computer science, microelectronics, surface technologies, psychology, production and industrial engineering.

According to the Fraunhofer principle of a direct transfer of new technologies and recent results from research to the market, the institutes are involved in research projects as well as in cooperation with the industries. Fraunhofer researchers are working on the design and development of interactive systems in many fields of application: from ticket vendor machines to service robots, from in-vehicle systems to multimodal telecommunication services and intelligent production environments. Moreover, Fraunhofer has been recognised as a beacon of cutting edge research in human-machine interaction. Fraunhofer developments in the fields of brain-computer interfaces, virtual and augmented reality, 3D-displays and optical recognition systems receive attention all around the world.

3 Research Agenda

As a major activity, the consortium members have identified research topics which will be focused on in the future. They have worked out a roadmap for central technologies and developments. The discussions and workshops with experts of the participating institutes yielded three main research areas (see figure 1 for an excerpt of the roadmap):

- **Enhanced interaction**
  This topic includes the development of new techniques for a natural and more intuitive interaction. This encompasses, for example, the adoption and refinement of recognition technologies, such as continuous speech recognition, optical recognition and eye tracking technology as a means of user input. New developments in the field of mobile technologies for Augmented Reality (AR) and stereoscopic Mixed Reality (MR) displays will lead to a higher level of immersion and will support virtual engineering in an effective manner.
- **Contextual user interfaces**
  In the future, user interfaces will not only respond to explicit user actions but they rather will also incorporate information from the context of use. Intelligent sensors will be able to recognize certain situations. In combination with a localization of the user it will be possible to provide context-aware services and adaptive user interfaces.
- **User Experience Engineering**
  Modern approaches to user-centred design go beyond the traditional concepts of ergonomics and usability. In order to systematically engineer a pleasant user experience, new methods will be needed for considering the "soft" human factors of emotions, branding, trust, etc. while designing and evaluating interactive products. As a link to software engineering, design patterns will help to assure a consistent interaction across various user interfaces and will reanimate the idea of a partly automated generation of ergonomic user interfaces.

4 Conclusion

The importance of research in the field of human-machine interaction will continue to increase in the near future. New technologies require and allow for new forms of interaction. At the moment, we are on the brink of a paradigm shift in interactive systems. The realisation of smart environments will change the way humans interact with technical systems. Besides working on the needed technological groundwork, it will be a major challenge to develop a coherent metaphor for human-machine interaction in smart environments. This metaphor will be necessary in order to concretely communicate the benefits and opportunities of the new technologies to a broader public and to facilitate an intuitive and trustworthy interaction.

Moving Object Contour Detection Based on S-T Characteristics in Surveillance

Yuan-yuan Cao[1], Guang-you Xu[1], and Thomas Riegel[2]

[1] Tsinghua National Lab. On Information Science and Technology, Tsinghua University, Beijing, 100084, P.R. China
caoyy04@mails.tsinghua.edu.cn,xgy-dcs@ tsinghua.edu.cn
[2] Siemens AG, Corporate Technology, Munich, 81730, Germany
Thomas.Riegel@siemens.com

Abstract. We present a method for moving object contours detection based on spatial-temporal characteristics. Using S-T features, the contour of moving object can be well distinguished from background; therefore the moving objects are detected without the need of establishing and updating background models. The detection method can handle situations where the background of the scene suffers from the noises due to the various facts, including the weather condition such as snow or fog and flicker of leafs on trees, and bushes. The algorithm estimates the probability of observing pixel as a contour pixel based on a sample of intensity values for each pixel during a period of time and its local gradient in current frame. The experiments show that this method is sensitive to changes caused by moving objects and is able to avoid the affection of complex background. The paper also shows how to separate multi-person based on the contour detection results using template matching. The approach runs in real-time and achieves sensitive detection.

Keywords: motion detection, contour detection, spatial-temporal characteristics, object classification, visual surveillance.

1 Introduction

Visual surveillance technology has been attracting more and more efforts due to its importance in security systems. An effective surveillance system relies on moving object detection and classification heavily.

A lot of algorithms on motion detection have been proposed, which can be classified mainly into two categories: the region-based and the contour-based. The most popular region-based approach is background modeling with mixture of Gaussians [1], which can handle tough cases like illumination changes and the noise caused by small movement in the background. However, a common problem of background modeling is that it takes somewhat long time to estimate the background models due to the slow cadence of illumination changes and small movement. In the case of contour-based approaches, active contours, such as snake [2], geometric active contour [3], and level sets [4] are widely used. In [5], accurate contour of moving object was extracted by means of integrating color segmentation, motion segmentation, and active contour. Geodesic active

M.J. Smith, G. Salvendy (Eds.): Human Interface, Part I, HCII 2007, LNCS 4557, pp. 575–583, 2007.

contours and level sets were used in [6]. However, these techniques are computationally too expensive for real time applications. Some work combined the information from motion and edge detection to extract the contour of a moving object [7], but this method will run into difficulties when scenes are cluttered.

In this paper, a novel method for detecting moving objects is proposed. It is observed that the intensity of the pixel where the contour of moving object is passing by, will undergo a sharp change either from foreground to background or background to foreground. In contrast, the intensity of pixels on the background will keep stable except the noise. Figure 1 shows how the gray level of a pixel from outdoor scene changes over a short period time (x-axis means frame number and y-axis means intensity value). The locations of four points (A, B, C, D) in original frame images are shown in figure 2. Because the background usually disturbed by some kind of flicker motion, the intensities in background usually vibrate in high frequency with small amplitude. In contrast, the intensity at the pixel where objects are passing by will undergo significant change. The intensity varies of those pixels depend on object moving speed and the intensity gradient along the motion direction. Therefore the contour of object can be detected by modeling the changes of intensity value in a short period of time. Further, in the case that the moving objects are human being, in order to separate human bodies from a group and the shadow, the detected contours are matched with 2D human contour templates hierarchically based on Distance Transform, which can separate multi-person. The templates need not to be established explicitly. The approach can eliminate the shadow of humans and handle partial occlusions.

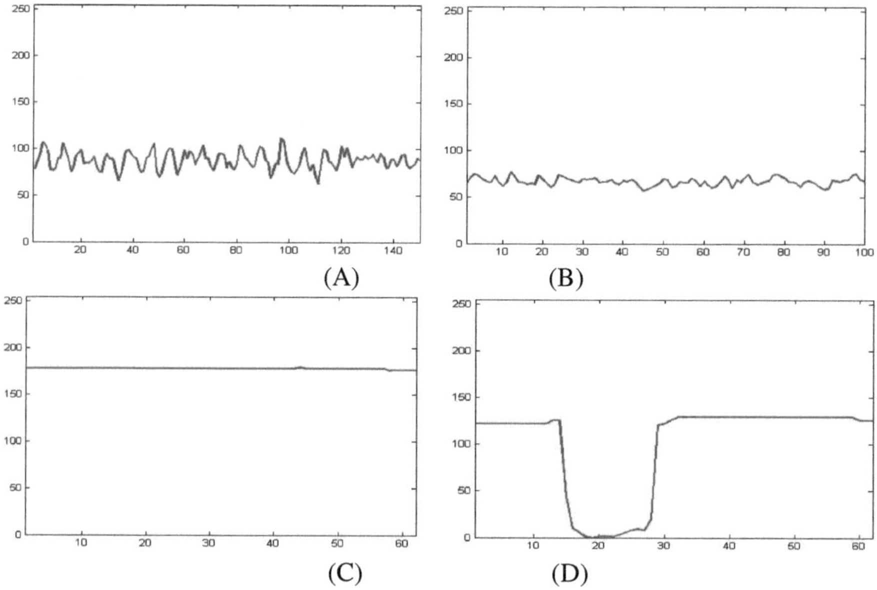

Fig. 1. Intensity value over a period of time (60-140 frames)

Fig. 2. Position of four points in original frame image

The outline of the paper is as follows. Section 2 presents the proposed algorithm of moving object contour detection. Separation of multi-person based on 2D templates matching is described in Section 3. Section 4 lists experimental results. We conclude in Section 5.

2 Moving Object Contour Detection

Let $x_1, x_2, \cdots x_N$ be a recent sample of intensity values for a pixel, that this pixel will have intensity value x_N at time *N*. Using this sample, the probability density function can be estimated using the kernel estimator *K*:

$$\Pr(x_N) = \frac{1}{N}\sum_{i=1}^{N} K(\left|\bar{x} - x_i\right|) \qquad \text{where} \qquad \bar{x} = \frac{1}{n}\sum_{i=1}^{N} x_i \tag{1}$$

If we choose our kernel estimator function, *K*, to be a Normal function $N(0,\Sigma)$, where Σ represents the kernel function bandwidth, then the density can be estimated as:

$$\Pr(x_N) = \frac{1}{N}\sum_{i=1}^{N} \frac{1}{(2\pi)^{\frac{d}{2}}\left|\Sigma\right|^{\frac{1}{2}}} e^{-\frac{1}{2}(\left|\bar{x}-x_i\right|)^T \Sigma^{-1} (\left|\bar{x}-x_i\right|)} \tag{2}$$

The larger $\Pr(x_N)$ is, the more likely this pixel is a contour pixel. However the detected results will be weak when the contour parallels moving direction. An extreme example see figure 3.

To solve this problem, spatial characteristic is combined to measure the probability density by computing the gray gradient of the pixel x_N in position (i, j). The formula is as follows:

$$\begin{cases} \nabla_x I(i,j) = \left[I(i-1,j+1) + 2I(i,j+1) + I(i+1,j+1)\right] - \left[I(i-1,j-1) + 2I(i,j-1) + I(i+1,j-1)\right] \\ \nabla_y I(i,j) = \left[I(i-1,j-1) + 2I(i-1,j) + I(i-1,j+1)\right] - \left[I(i+1,j-1) + 2I(i+1,j) + I(i+1,j+1)\right] \end{cases} \tag{3}$$

$$G(x_N) = \sqrt{\nabla_x I^2 + \nabla_y I^2} \tag{4}$$

$G(x_N)$ is gray gradient of the pixel located in (i, j). Normalized $G(x_N)$ is denoted as $G_n(x_N)$. The sum of the weighted $\Pr(x_N)$ and $G_n(x_N)$ is calculated as probability of a pixel being a contour pixel, which is estimated by the formulation:

$$P(x_N) = \alpha \Pr(x_N) + \beta G_n(x_N) \qquad \text{where} \quad 0 < \alpha, \beta < 1 \tag{5}$$

Using this probability estimate, the pixel is considered as a contour pixel if

$$P(x_N) > th \tag{6}$$

The threshold th is a global threshold over all the images that can be adjusted. The moving object contour detection results are shown in figure 4.

For finer results, morphological transformation is employed to eliminate background noises and fill in tiny holes in the detected contour, which are shown in Figure 5.

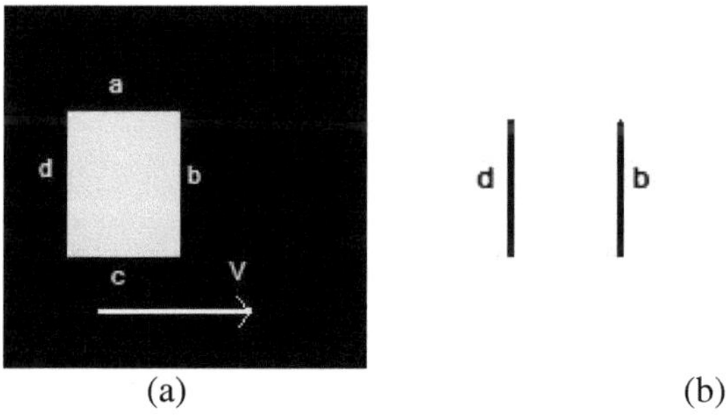

Fig. 3. (a) shows a gray rectangle in a black background is moving in direction which parallels edges a and c. Contour detection result is shown in (b), where a and c are missing for there's no intensity change on the two edges.

Fig. 4. Moving object contour detection results

Fig. 5. Detection results after morphological transformation

This detection method can handle situations when the background of the scene is blurred due to the weather condition, such as snow and fog, and the noises caused by the flicker of tree branches and bushes. We test the approach of background modeling with mixture of Gaussians and the method proposed above in surveillant videos captured by a single static camera with a resolution of 320×240. They show a crowd crossroad in snow weather. The results are compared in figure 6.

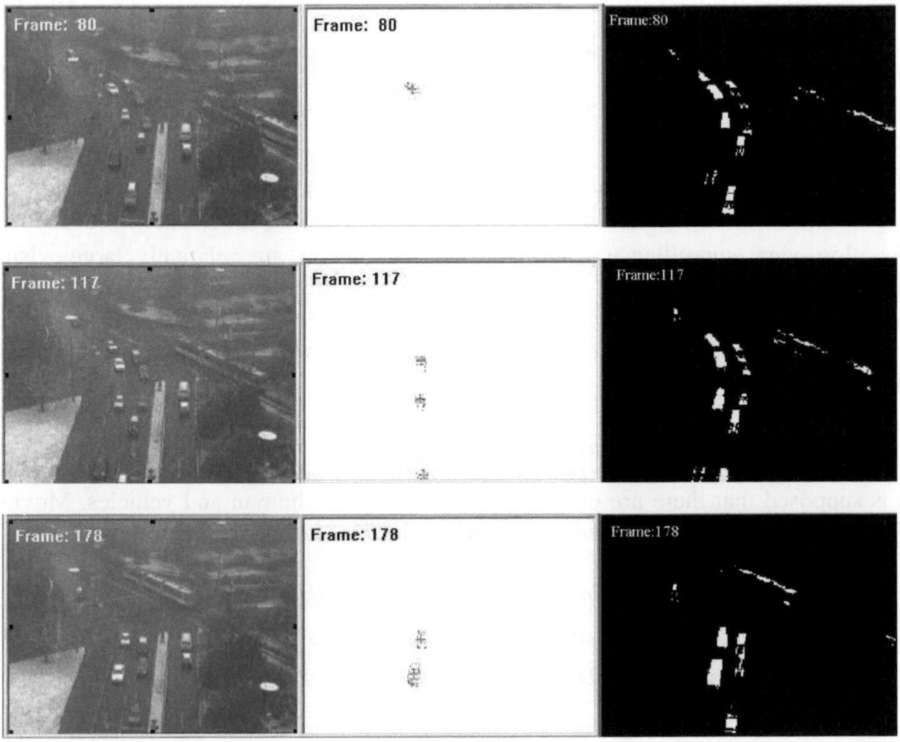

Fig. 6. Images in the first column are original frame samples (frame 80, frame 117 and frame 178); the second column shows detection results of background modeling and the results of approach proposed in this paper are shown in the last column

The experimental results show that background model method is likely to miss the objects whose color is similar to the background. The reason is that in background modeling method, the threshold used to differentiate foreground and background pixels is estimated based on the global statistics, thus can not adapt to different color objects. In our method, the characteristics of relative color change in a local area, instead of absolute difference of color value, are taken into consideration to distinguish moving object contour pixels from others. In the above video viewing from far away, moving objects are relatively small in the image, therefore the detected contour of an object tend to connect as its foreground, However this does not affect the objects detection results. The region bounding algorithm is applied to locate the blob of the detected moving object, results of which are shown in figure 7.

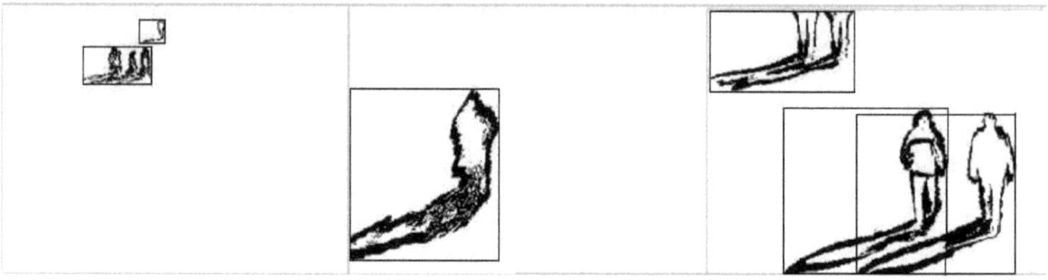

Fig. 7. Detected blob of moving objects

3 Human Detection by Means of Templates Matching

For close-view surveillance where the moving objects are relatively large, visual surveillance usually endures difficult situations like a group of people with partial occlusion and heavy shadow. Based on the contour detection results, 2D template matching is employed to separate a group of human and meanwhile eliminate shadows.

3.1 Contour Blob Classification

It is supposed that there are only two kinds of objects: human and vehicles. Moving objects are first classified based on the knowledge and formulated as follows:

$$ClassID = \begin{cases} a\ single\ human & Dispersedness > \theta_2\ \ area < \theta_1 \\ a\ group\ of\ humans & Dispersedness > \theta_2\ \ area > \theta_3 \\ vehicles & Dispersedness < \theta_2\ \ \theta_1 < area < \theta_3 \end{cases} \tag{7}$$

where $$Dispersedness = \frac{Perimeter^2}{area},\quad Aspect\ Ratio = \frac{height}{width} \tag{8}$$

3.2 Human Contour Template Hierarchy Initialization

A human contour template hierarchy is constructed and object contours are matched in a coarse-to-fine way with templates based on Distance Transform hierarchically. 2D human contour templates are constructed by detecting contours of human with different gait (walking with both legs together, one leg forward, and standing, etc.) from training videos. A template hierarchy is initialized in the way that similar templates are grouped together and represented by a kernel template which holds common properties of this group. Matching is first done with kernel templates rather than with each individual template, a speed-up will then be achieved. See Figure 7 for a general view of template hierarchy which is generally enough to represent human contour in surveillance. Human head and shoulder contour, the common part of all human contour templates, is chosen as the kernel template.

3.3 Human Separation and Shadow Elimination

Distance Transform (DT) [8] based matching is employed to match contour images with the templates. The average distance to the nearest feature, namely chamfer distance, is chosen here as the match measure, which is represented as follows:

$$\mathrm{Dist}_{\mathrm{chamfer}}(\mathrm{M},\mathrm{I}) = \frac{1}{|\mathrm{M}|}\sum_{p\in M} d(I,p) \tag{9}$$

Where |M| denotes the number of contour pixels in template M and $d(I,p)$ denotes the distance between contour pixel p in M and the closed contour pixel in I. A template is considered to be matched at a certain location when the distance measure $\mathrm{Dist}_{\mathrm{chamfer}}(\mathrm{M},\mathrm{I})$ at this position is below a predefined threshold θ.

$$\mathrm{Dist}_{\mathrm{chamfer}}(\mathrm{M},\mathrm{I}) < \theta \tag{10}$$

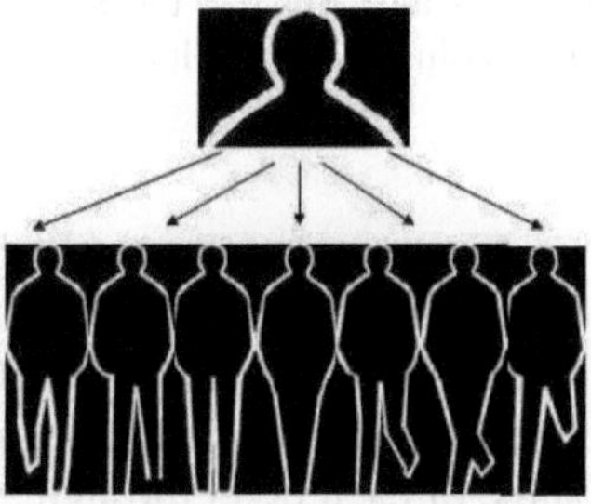

Fig. 8. Human contour template hierarchy

4 Experiment Results

The proposed method for detection and recognition of moving object have been tested with the videos captured by a single static camera in a resolution of 320×240. There are a wide road with shadows of trees and humans casting heavy shadows in the video.

Figure 8 shows some segmentation results,for individuals. The human contour detection results are shown in the second column. The right column shows multi-person separation and shadow elimination results, where humans without being occluded are outlined with blue lines; occluded ones are outlined in the head and shoulder with red lines.

The method, however, encounters some difficulties in certain circumstances. When human or vehicle is far away from the camera, the algorithm fails to detect a clear contour; as mentioned in section 2, as a result the template-based algorithm can not be applied. Another difficulty arises when the head and shoulder of a human is occluded.

Fig. 9. The first column is original frame images; the second column are results of moving object contour detection; the right column are results of multi-person separation and shadow elimination

5 Conclusion

In this paper, a method has been proposed for moving objects detection, multi-person separation and shadow elimination. We employed spatio-temporal characteristics to model the intensity value change of each pixel through a short period time and picked out those lying in contour area. A coarse-to-fine strategy based 2D template matching method is employed to separate multi-person and eliminate shadows.

Acknowledgement. This work was funded under Project 60673189 supported by National Science Foundation of China: Event Detection and Understanding in Dynamic Context for Implicit Interaction.

References

1. Harville, M.: A framework for high-level feedback to adaptive, per-pixel, mixture-of-gaussian background models. In: European Conference on Computer Vision (2002)
2. Seo, K.H., Lee, J.Y., Lee, J.J.: Adaptive color snake tracker using condensation algorithm. In: 5th Asian Control Conference (2004)
3. Huang, F.Z., Su, J.B.: Face contour detection and tracking with complex backgrounds. Proceedings of 2004 International Conference on Machine Learning and Cybernetics (2004)
4. Sethian, J.: Level set methods and fast marching methods. Cambridge Univ. Press, Cambridge (1999)
5. Qiu, L., Li, L.: Contour extraction of moving objects. In: Proc. IEEE Int'l Conf: Pattern Recognition (1998)
6. Paragios, N., Deriche, R.: Geodesic active contours and level sets for the detection and tracking of moving objects. IEEE Trans. Pattern Analysis and Machine Intelligence (2000)
7. Nagao, K.: Detecting contours in image sequences. IEICE Trans. Information and Systems, vol. E76-D(10)
8. Borgefors, G.: Distance Transformations in Digital Images. In: CVGIP (1986)

On Achieving Proportional Loss Differentiation Using Dynamic-MQDDP with Differential Drop Probability

Kyungrae Cho[1], Sangtae Bae[2], Jahwan Koo[1], and Jinwook Chung[1]

[1] School of Information and Communication Engineering , SungKyunKwan University, Chunchun-dong 300, Jangan-gu, Suwon, Gyeonggi-do 440-746, South Korea
{krcho,jhkoo,jwchung}@songgang.skku.ac.kr
http://www.songgang.skku.ac.kr
[2] Dongwon Industry Bldg., 275, Yangjae-dong, Seocho-gu, Seoul, 137-130, Korea
stbae@kistep.re.kr

Abstract. More Recently, researchers have explored to provide a queue management scheme with differentiated loss guarantees for the future Internet. Various types of real time and non-real time traffic with varying requirements are transmitted over the Internet. The sides of a packet drop rate, an each class to differential drop probability on achieving a low delay and high traffic intensity. Improved a queue management scheme to be enhanced to offer a drop probability is desired necessarily. This paper considers multiple random early detection with differential drop probability which is a slightly modified version of the MQDDP model, to get the performance of the best suited, we analyzes its main control parameters (max_{th}, min_{th}, max_p) for achieving the proportional loss differentiation (PLD) model, and gives their setting guidance from the analytic approach. we propose Dynamic-multiple queue management scheme based on differential drop probability, called Dynamic-MQDDP, is proposed to overcome MQDDP's shortcoming as well as supports static max_p parameter setting values for relative and each class proportional loss differentiation. MQDDP is static according to the situation of the network traffic, Network environment is very dynamic situation. Therefore max_p parameter values needs to modify too to the constantly and dynamic. The verification of the guidance is shown with figuring out loss probability using a proposed algorithm under dynamic offered load and is also selection problem of optimal values of parameters for high traffic intensity and show that Dynamic-MQDDP has the better performance in terms of packet drop rate. We also demonstrated using an ns-2 network simulation.

1 Introduction

The goals of differentiated services in IP networks are to manage and provide different types and levels of quality of service (QoS) to satisfy the varying requirements of a diverse set of applications. Many efforts for service differentiation, however, have been relatively limited, because of the additional complexity and the scalability requirements of the continuously growing Internet. As a result, there have been a number of previous works for providing some form of service differentiation. The best known such effort is the proportional differentiated services (PDS) model [3], which

M.J. Smith, G. Salvendy (Eds.): Human Interface, Part I, HCII 2007, LNCS 4557, pp. 584–593, 2007.

attempts to provide a controllable, consistent, and scalable QoS and enables network service providers to have a control knob for the convenient management of services and resources in the next generation network.

In general, the PDS model has two mechanisms: 1) the proportional delay differentiation (PDD) and 2) the proportional loss differentiation (PLD). PDD and PLD can be quantitatively adjusted to be proportional to queuing delay and packet loss in routers, respectively. In this paper, we limit our discussion to PLD for quantifying loss differentiation between different classes of traffic, and forcing the ratios of loss rates of adjacent classes to be proportional.

Dovrolis et al. [5] who developed first the PLD model, claim that several previous mechanisms such as complete buffer partitioning (CBP), partial buffer sharing (PBS), or multi-class random early detection (RED), are not suitable for relative differentiated services. They propose proportional loss rate (PLR) mechanisms that closely approximate the PLD model. Liebeherr et al. [9] propose a novel algorithm called Joint Buffer Management and Scheduling (JoBS) for the integration of buffer management and packet scheduling and compare with PLRs for PLD service. Li et al. [11] propose a novel algorithm by employing Probabilistic Longest Queue First (PLQ) mechanism and claim that its implementation cost is much less than PLRs and even more practical. Zeng et al. [12], [13] propose more enhanced dropping algorithm than PLRs in terms of the packet shortage phenomenon. Aweya et al. [2] propose a scheme for differentially dropping packets in a FIFO queue.

More recently, Even though MQDDP scheme has several benefits such as low complexities and good functionalities, Koo et al. [8] identify that it has some shortcomings such as low throughput, long queuing delays, and selection problem of optimal values of parameters. Unfortunately, most existing schemes supporting the PLD have been implemented in proprietary form. This not only limits the interoperability of routers deployed in many commercial networks, but also reduces the reusability of the prevalent RED scheme recommended by the Internet Engineering Task Force (IETF) for next generation Internet gateways [4]. In this paper, we consider multiple RED with differential drop probability which is capable of easily tuning the original RED, analyze its main control parameters for achieving the PLD, and give their setting guidance from the analytic approach. The optimal setting of RED parameters for the PLD upon their requests helps increase the goals of differentiated services in IP networks.

The rest of this paper is organized as follows. We analyze D-MRED(Dynamic-Multiple RED) queues by deriving drop probability equations, Adjusting drop probabilities for PLD model in section 2, examine the accuracy of the analytic results obtained so far by comparing them with simulation results in section 3, respectively. Finally the paper concludes in Section 4.

2 Dynamic-MRED with Differential Drop Probability

We analyze D-MRED(Dynamic-Multiple RED) queues by deriving drop probability equations using a queuing model [1].

2.1 Deriving Drop Probability Algorithm and Equation

We first derive the drop probability equation from a FIFO tail drop queue, and then we extend it to a single RED queue [2]. Since Dynamic-MRED queues are independent and inherit common properties from a single RED queue, the derived equations of single RED queue are immediately applicable to Dynamic-MRED queues.

For a FIFO tail drop queue with a buffer of size K and a system utilization factor of $\rho = \lambda / \mu$, the probability k packets in the system is $\pi(k) = \left((1-\rho)\rho^k / \left(1-\rho^{k+1}\right)\right)$. As newly arriving packets will be refused to the system and will depart immediately without service in case when K packets are occupied in the system, so a packet drop probability of $P^{TD} = \pi(K) = \left((1-\rho)\rho^k / \left(1-\rho^{k+1}\right)\right)$. With a each class queue *i*'s buffer size of K having TD queue, the steady-state probability of finding k packets in the queue is given by

$$\pi_i(K) = \frac{\prod_{l=0}^{k-1} \rho_i}{\sum_{k=0}^{k} \prod_{l=0}^{k-1} \rho_i} \tag{1}$$

For a single RED queue, however, incoming packets are dropped with a probability that is an increasing function $d(k)$ of the average queue size k . Dynamic-MRED queue, like the original RED, offers three control parameters: maximum drop probability max_p, minimum threshold min_{th}, and maximum threshold max_{th}. It depends on the averaged queue length k with weighted factor w_q to tune RED's dynamics [14]. The average queue size is estimated using an exponential weighted moving average formula, $k = (1-w_q) \cdot k + w_q \cdot \hat{k}$, algorithm for RED scheme are shown in Table 1.

Table 1. Average queue size algorithm for RED

Avg_q = (1-Weight Avg_q)+ Weight * Current queue size;
$0 \le w_q \le 1$
Weight : Weighted parameter of queue, (W_q);
Avg_q: Averaged of queue length, (k);
Current queue size : each time a packet arrivals,($\hat{k}$);

The dropping function of Class *i*, $d_i(k)$, in Dynamic-MRED is defined using three parameters min_{th}, max_{th} and max_p as follows:

$$d_i(k)=\begin{cases}0, & k<\min_{th}\\ \dfrac{\max_{p,i\cdot}(k-\min_{th})}{\max_{th}-\min_{th}}\equiv \max_{p,i\cdot} f(k), & \min_{th}\le k<\max_{th}\\ 1, & k\ge \max_{th}\end{cases} \tag{2}$$

The dropping algorithm Class *i*, $d_i(k)$, in Dynamic-MRED is shown in Table 2.

Table 2. The dropping algorithm of Class *i*, in Dynamic-MRED

Every time t;
if $Avg_q \le \min_{th}$ *then*
enqueue the packet k
if $\min_{th} < Avg_q \le \max_{th}$ *then*
Calculate probability p^{TD} **;**
Drop arriving packet k **with probability** p^{TD} **;**
If $\max_{th} \le Avg_q then$;
Drop arriving packet k **;**

where $max_{p,i}$ is maximum drop probability of Class *i*. Therefore, the drop probability of a packet depending on the dropping function related to each state *K* is defined as follows:

$$\begin{aligned} p_i^{DMR} &= \pi_i^{DMR}(1)d_i(1)+\cdots+\pi_i^{DMR}(k)d_i(K)\\ &= \sum_{k=\min_{th}}^{k} \pi_i^{DMR}(k)d_i(k), \quad \min_{th}\le k<K \end{aligned} \tag{3}$$

Let $\max_{th}$ = K and $\{\lambda_k=\lambda,\mu_k=\mu,\forall k\}$, then the number of packets in the RED queue is actually a birth-death process with the birth rate in state *k* equal to $\lambda_i(1-d_i(k))$ and death rate equal to μ_i. For more details, the formulas for π_i can be referenced in [10]. Accordingly, the steady-state probability of finding *k* packets in the system, $\pi_i^{DMR}(k)$, is derived by modifying Eq.(1) as follows:

$$\pi_i^{DMR}(k)=\frac{\rho_i^k\prod_{l=0}^{k-1}(1-d_i(l))}{\sum_{k=0}^{k}\rho_i^k\prod_{l=0}^{k-1}(1-d_i(l))} \tag{4}$$

2.2 Adjusting Drop Probabilities for PLD Model

For PLD, δ_L, which is given as network service provider's request, is described as

$$\frac{l_{i+1}}{l_i} \quad \delta_L, \qquad 1 \le i \le M \tag{5}$$

where M is service class, l_i is the desired loss probability of class i, δ_L is the differentiation factor between adjacent classes, and loss-priorities are decreasing order with class i.

Then we want to deduce how to set $max_{p,i}$ for given δ_L. From the estimated packet drop probabilities for class i, p_i^{DMR}, in Eq.(3), a suitable $max_{p,i}$ can be retrieved from the following equation by assuming the same ρ_i.

$$\delta_L = \frac{p_{i+1}^{DMR}}{p_i^{DMR}} = \frac{\sum_{k=\min_{th}}^{k} \pi_{i+1}^{DMR}(k) d_{i+1}(k)}{\sum_{k=\min_{th}}^{k} \pi_i^{DMR}(k) d_i(k)}$$

$$= \frac{\max_{p,i+1} \cdot \sum_{k=\min_{th}}^{k} f(k) \prod_{l=0}^{k-1} (1 - \max_{p,i+1} f(l))}{\max_{p,i} \cdot \sum_{k=\min_{th}}^{k} f(k) \prod_{l=0}^{k-1} (1 - \max_{p,i} f(l))} \tag{6}$$

$f(l)$ is an increasing function with queue length l, approaching either one with a heavy queue length (i.e., max_{th}) or a negligible value with a smaller queue length (i.e., min_{th}). Eq. (6) can be approximated by considering only several higher l, terms, as the following:.

$$\delta_L \simeq \frac{\max_{p,i+1} \cdot (1 - \max_{p,i})^n}{\max_{p,i} \cdot (1 - \max_{p,i})^n} \tag{7}$$

where n is the counting number, depending on how many l terms are considered.

From Eq.(7), the value of $max_{p,i}$ for a control knob when δ_L is given as a service request, can be determined. Numerical analysis of Eq.(7) can be evaluated by selecting $max_{p,i}$ for PLD across dynamic offered load patterns presented in Fig. 1. First, we calculate the maximum packet drop probabilities for Dynamic-MRED queue i, $max_{p,i}$ (t) in Fig. 1, which is used during time interval $[t, t + \Delta\tau]$, from the accurate formulation in Eq.(3) when the drop probability in a service class is provided under the known offered load, ρ_i. Then we can perform initial setting of $max_{p,i}$ for the loss probability of referenced class i. Secondly, we determine next value of $max_{p,i+1}$ of adjacent class, i.e. $(i+1)_{th}$ class with given loss differentiation factor δ_L from Eq.(7) for PLD. This process is continued to obtain $max_{p,i+2}$ for next adjacent class and so on.

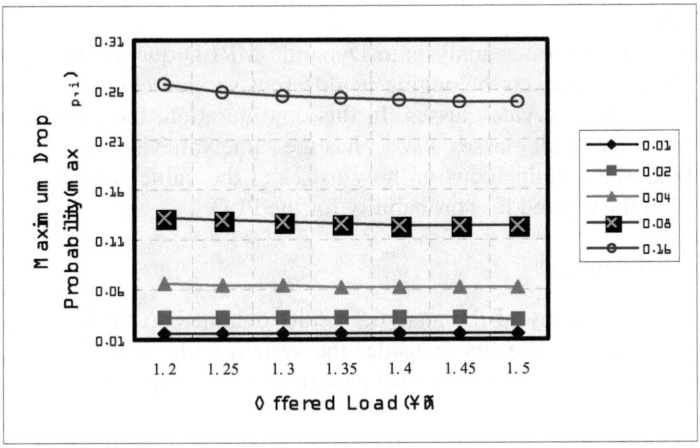

Fig. 1. $max_{p,i}$ versus offered load for loss probability of 0.01, 0.02, 0.04, 0.08, and 0.16

3 Evaluation

3.1 Configuration

Rough guidelines for configuring RED were presented in the original RED paper by Floyd and Jacobson [6].It was suggested that w_q should be set greater than or equal to

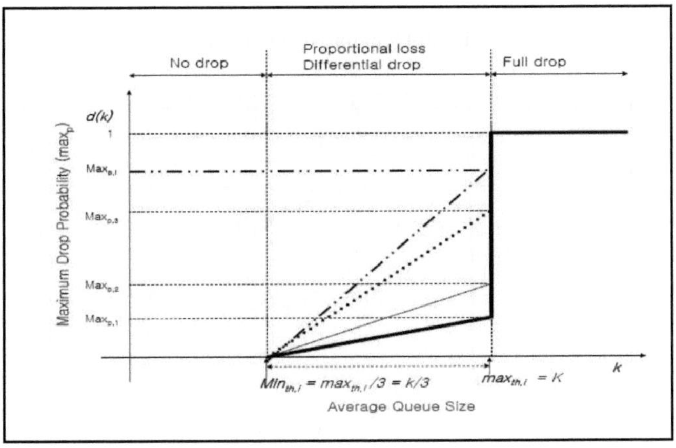

Fig. 2. Dynamic-MRED with differential $max_{p,i}$ according to each class i

0.002 and that max_{th} should be sufficiently large to avoid global synchronization. Also, min_{th} should be set sufficiently large to avoid low utilization of the output link. A more recent set of guidelines is presented in [7] which recommends that max_{th} should be three times min_{th}, max_p should be set to 0.1, and w_q should be set to 0.002. The proposal notes that the optimal setting for min_{th} depends on the tradeoff between low average delay and high link utilization. However, those settings are not considered for the PLD.

Specifically, we limit our analysis to Dynamic-MRED queues with different RED curves for dropping packets belonging to different service classes. Figure 2 depicts one example of three service classes. In this configuration, the parameters of $min_{th,i}$ and $max_{th,i}$ are same in all classes, $max_{th,i}$ has the same value as buffer size K, and $w_{q,i}$ is set to 0.002. We mainly focus on how to select the values of $max_{p,1}$, $max_{p,2}$, and $max_{p,3}$ based on the desired loss probability for the PLD.

3.2 Numerical Analysis

We examine the accuracy of the analytic results obtained so far by comparing them with simulation results. We first consider the system with $K = 120$, $min_{th} = 40$, and $max_{th} = 120$ across dynamic offered load pattern shown in Fig.3, Using Eq. (3),

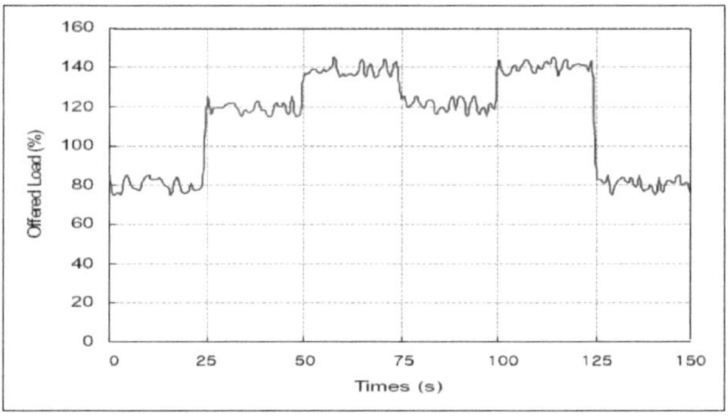

Fig. 3. Dynamic offered load

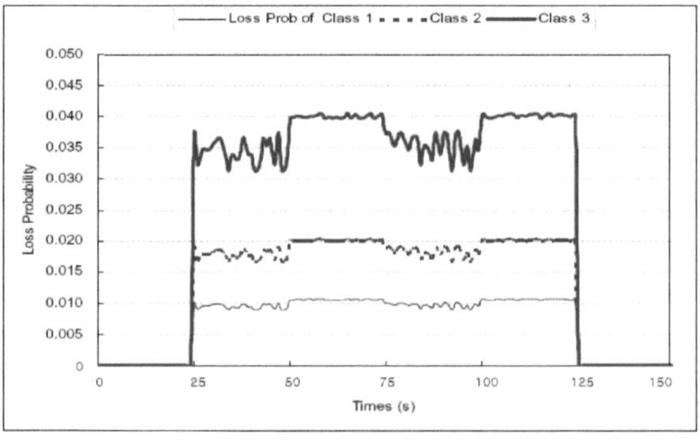

Fig. 4. Numerical result on PLD ($l_1 = 0.01, \delta_L = 2$)

we can calculate the loss probability per service class shown in Fig. 4 and it shows that the desired PLD of δ_L = 2can be achieved by changing $max_{p,i}(t)$ under dynamic offered load.

As a practical guide we can realize the PLD model via our proposed setting of the value of $max_{p,i}$ using Eq. (7).

3.3 Simulation Analysis

We performs network simulation using *ns-2* in order to verify the selection method of $max_{p,i}$ for PLD. The simulation environment consists of a router, a destination, and a set of sources, each source which generates packets with a fixed size of 500 bytes and has the same traffic pattern with a constant bit rate exponential distribution. Each source chooses only one class type among three service classes and is connected to the router node with a link, which bandwidth and delay are set to 10 Mbps and 5 ms, respectively.

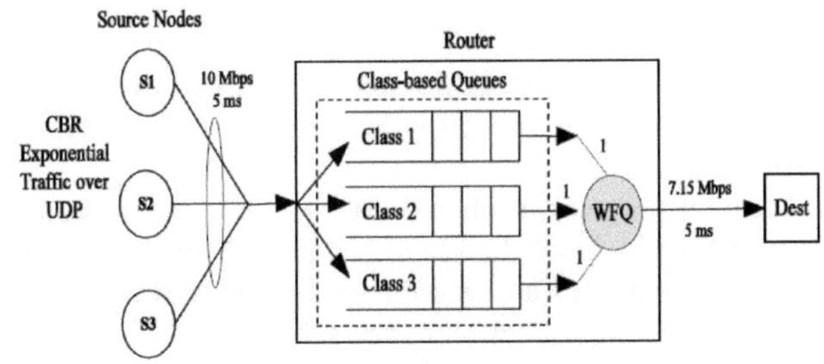

Fig. 5. Using ns-2 Simulation Environment(with Nortel's DiffServ module)

We set the router output link with the capacity of 7.15 Mbps resulting in the offered load as the value of 1.4 and the router has three Dynamic-MRED queues in accordance with three classes of traffic, allowing each queue to have a queue length

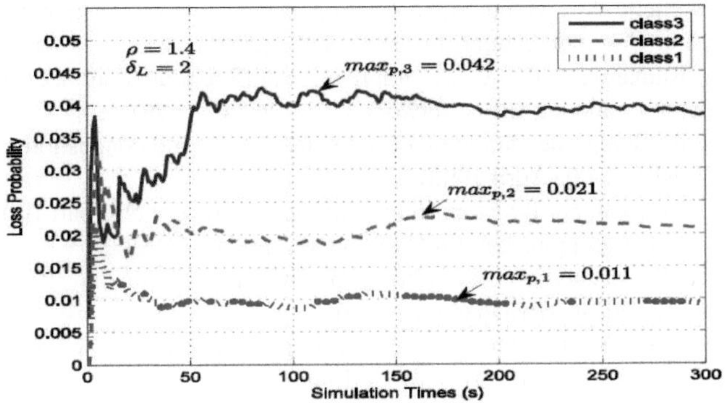

Fig. 6. Simulation result on PLD (l_1:l_2:l_3=1:2:4, n=15)

of 120. Following the guidelines of the previous section, Dynamic-MRED parameters were identically set to $min_{th,i}$ = 40, $max_{th,i}$ = 120, and $w_{q,i}$ = 0.002 except for the main control parameter of ($max_{p,1}$, $max_{p,2}$, $max_{p,3}$) = (0.011, 0.021, 0.042) which is guided by Eq. (7) in the case of δ_L = 2 and n = 15. In addition, a weighted fair queue scheduler is used to allocate service rates equally among classes and the experiment lasts for 300 seconds of simulated time. Simulation results in Fig. 6 demonstrate that the proposed approach is desirable, for adjusting the PLD model among different classes after initial transient periods. This is also well-matched with numerical results, which means Dynamic-MRED queue plays a major role for PLD among different classes.

4 Conclusions

We have performed an analysis of Dynamic-MRED with differential drop probability in order to achieve the PLD model using a queuing model and given some guidance on how to select maximum drop probability in Dynamic-MRED queues. Compared to the analytic results, we have also verified through network simulation that the guidance is suitable for determining an optimum value of the main control parameters. It will be of great assistance in terms of achieving the goals of differentiated services in the future Internet if network service providers are aware exactly of the setting of RED parameters for PLD.

Although conventional protection scheme does provide quick recovery time, it has disadvantage of using up too much bandwidth and lack of ability to find sufficient disjoint paths. This paper proposes a new enhanced path recovery algorithm that overcomes these problems of conventional recovery schemes. The great advantage of the proposed recovery algorithm is that it provides much more recovery path compared to the conventional m:n type recovery method.

References

1. Boland, T., May, M., Bolot, J.C.: Analytic evaluation of RED performance. In: Proc. IEEE INFOCOM, Tel Aviv, Israel, pp. 1415–1424 (March 2000)
2. Aweya, J., Ouellette, M., Montuno, D.Y.: Proportional loss rate differentiation in a FIFO queue, Computer Communications, pp. 1851–1867 (2004)
3. Dovrolis, C., Ramanthan, P.: A case for relative differentiated services and the proportional differentiation model. IEEE Network 13(5), 26–34 (1999)
4. Braden, B. et al.: Recommendation on queue management and congestion avoidance in the Internet, IETF RFC 2309 (April 1998)
5. Dovrolis, C., Ramanathan, P.: Proportional differentiated services, part II: loss rate differentiation and packet dropping. In: Proc. of IWQoS, pp. 52–61 (2000)
6. Floyd, S.: RED: discussions of setting parameters (November 1997) available at http://www.aciri.org/floyd/REDparameters.txt
7. Floyd, S., Jacobson, V.: Random early detection gateways for TCP congestion avoidance. IEEE/ACM Trans.\ Networking 1(4), 397–413 (1993)
8. Koo, J., Shakhov, V.V., Choo, H.: An Enhanced RED-based scheme for differentiated loss guarantees. In: Proc. of 9th Asia-Pacific Network Operations and Management Symposium (2006)

9. Liebeherr, J., Christin, N.: JoBS: joint buffer management and scheduling for differentiated services. In: Proc. of IWQoS, pp. 404–418 (2001)
10. Shakhov, V.V., Koo, J., Choo, H.: On modelling reliability in RED gateways. In: Schärfe, H., Hitzler, P., Øhrstrøm, P. (eds.) ICCS 2006. LNCS (LNAI), vol. 4068, pp. 948–951. Springer, Heidelberg (2006)
11. Li, J.-S., Lai, H.-C.: Providing proportional differentiated services using PLQ. Proc. of Globecom, pp. 2280–2284 (2001)
12. Zeng, J., Ansari, N.: An enhanced dropping scheme for proportional differentiated services. In: Proc. of ICC, pp. 1897–1901 (2003)
13. Li, J.-S., Lai, H.-C.: Providing proportional differentiated services using PLQ. In: Proc. of Globecom, pp. 2280–2284 (2001)
14. Floyd, S., Jacobson, V.: Random Early Detection for Congestion Avoidance, IEEE/ACM (August 1993)

Converting Information Through a Complete and Minimal Unit Transcoder for QoS Adaptation*

Sungmi Chon[1], Dongyeop Ryu[2], and Younghwan Lim[3]

[1] Soongsil University, Information and Media Research Institute
1-1 Sangdo 5-Dong, Seoul, South Korea
smchon@hananet.net
[2] Soongsil University, School of Computer
1-1 Sangdo 5-Dong, Seoul, South Korea
aceryu@naver.com
[3] Soongsil University, School of Media
1-1 Sangdo 5-Dong, Seoul, South Korea
yhlim@computing.ssu.ac.kr

Abstract. MPEG-21's digital item adaptation technology becomes a new way for universal multimedia access. It needs transcoder to change media resource's format and so on according to delivery context. Then, the use of heavy transcoder with various transcoding functions integrated into one altogether is so complicated and difficult in supporting universal multimedia access. Unit transcoder is useful is to resolve this question, in which a transcoder has only one transcoding function. This requires considering how to compose a set of unit transcoders. Thus, given a set for end-to-end different service quality pairs according to the character of application as defined by user, this study suggests how to compose complete unit transcoders that can always create one and more transcoding path(s) for each pair in the set. This method has a question of creating too many transcoding paths for each pair of end-to-end different service quality. Thus, this study also suggests the algorithm that generates minimum unit transcoder sets to support multimedia adaptation with minimum unit transcoder. The algorithm suggested was implemented into multimedia stream engine, and this paper describes the results of experiment for this algorithm.

1 Introduction

MPEG-21 defines an integrated framework for delivering and consuming multimedia contents. In addition, it standardizes a technology called Digital Item Adaptation(DIA), which is a new possible way for Universal Multimedia Access(UMA)[1-3]. To this end, it requires transcoder for multimedia adaptation[4]. Transcoder falls into heavy transcoder and unit transcoder depending on the number of transcoding functions available. Heavy transcoder denotes a transcoder that incorporates a variety of transcoding functions in it. This is too complicated and difficult to support 'Universal Multimedia Access(UMA)'[5]. To solve this problem, there has been a series of methods proposed to use unit transcoder equipped with only one transcoding function.

* This work was supported by grant No.(R01-2004-000-10618-0) from the Basic Research Program of the Korea Science & Engineering Foundation.

M.J. Smith, G. Salvendy (Eds.): Human Interface, Part I, HCII 2007, LNCS 4557, pp. 594–603, 2007.

For this sake, it is necessary to consider how to organize a set of unit transcoders and how to create transcoding path using the set of unit transcoders. In particular, many studies have focused on how to create transcoding path using unit transcoders. The algorithm creating transcoding path based on QoS(Quality of Service) Transition Diagram used queue-based breath first method and brute-force method[6]. CFG-based transcoding path creation algorithm using context-free grammar adopted a method to integrate information on QoS of digital items stored, QoS preferred by users, and a set of unit transcoders[7]. Furthermore, in similar context with the concept of unit transcoder, a study sought to interconnect modules in the order of frame rate, color and resolution in regard to bandwidth, out of transcoding modules for each QoS in heavy transcoder[8]. And a study addressed frame rate based on bit rate and resolution transcoding as well[9]. Another latest study proposed a media adaptation framework using a component comparable to unit transcoder in place of heavy library[5]. On the other hand, there are little studies considering how to organize unit transcoders. Thus, this paper proposes an algorithm to create complete set of unit transcoders, which can create one or more transcoding paths for each pair of elements in a given set for different end-to-end QoS pairs depending on the characteristics of application defined by user. But this algorithm faces a question of creating too many transcoding paths for each pair of different end-to-end QoSs. That is why this paper also proposes another algorithm to create a minimum set of unit transcoders to support multimedia adaptation with the least number of unit transcoders. The algorithms proposed herein were implemented into Transcore, a multimedia stream engine, and the results of corresponding experiments can be outlined. The algorithms as proposed above are available to determine the least number of unit transcoder sets that support UMA in ubiquitous system environment according to given nature of application.

2 Problems and Solutions

2.1 Problems and Solutions of Multimedia Adaptation Using Heavy Transcoder

Transcoding using heavy transcoder evokes a question of only multimedia adaptation supportable according to the ability of converting each heavy transcoder. So it is necessary to seek a solution to enable transcoding regardless of application. As a means of such solutions, a method of using a set of unit transcoders with only one transcoding function in place of heavy transcoder proposed. This method is an application-independent method that supports a variety of transcoding ways required for application by creating transcoding paths of unit transcoders connected.

2.2 Problems and Solutions of Application-Independent Multimedia Adaptation Framework

2.2.1 Architecture of Application-Independent Multimedia Adaptation Framework

The application-independent multimedia adaptation framework complying with QoS at various destinations employs a set of unit transcoders. For instance, as shown in

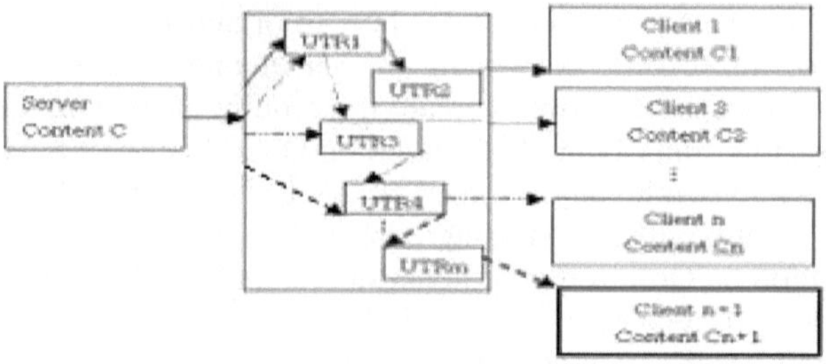

Fig. 1. Application-Independent adaptation using a group of unit transcoders

Fig. 1, Client 1, 2, …, n, n+1 interconnects a set of unit transcoders such as UTR1, (UTR1 UTR3), …, UTR3, (UTR4 UTRm) for the contents of server to meet QoS requirements.

2.2.2 Problems and Solutions

It is sometimes impossible to implement multimedia adaptation using a set of unit transcoders as organized without any strategy, depending on random and different end-to-end QoS requirements. For example, assume that QoS required at source of digital item and that at destination are (CIF, 24-bit color, bmp) and (CIF, 16-bit color, gif). If current system has just unit transcoder for file format conversion from bmp to gif and unit transcoder for the sizing of digital item in bmp type, no set of unit transcoders can meet the needs of clients. To resolve the problem, this paper proposes an application-independent multimedia adaptation framework model that uses a set of strategic unit transcoders to support respective multimedia adaptations for each pair of given end-to-end QoSs. In other words, given a set of end-to-end different QoS pairs according to application and nature defined by user, this paper proposes an algorithm that creates a complete set of unit transcoders so that one or more transcoding paths must be available for each pair. With a complete set of unit transcoders supporting clients, there is an question of system that must lots of unit transcoders supporting given end-to-end different QoS pairs. As a result, there are too many transcoding paths generated. This system is not efficient in practical application. So it is necessary to organize unit transcoders to create transcoding paths for supporting end-to-end QoS pairs depending on the nature of application while reducing the number of unit transcoders. To this end, this paper proposes an algorithm to create least number of transcoder sets.

3. Model of Application-Independent Multimedia Adaptation Framework Considering Unit Transcoder Set

3.1 Components of Model

The model on application-independent multimedia adaptation framework considering a set of unit transcoders proposed consists of three elements as shown in Fig. 2.

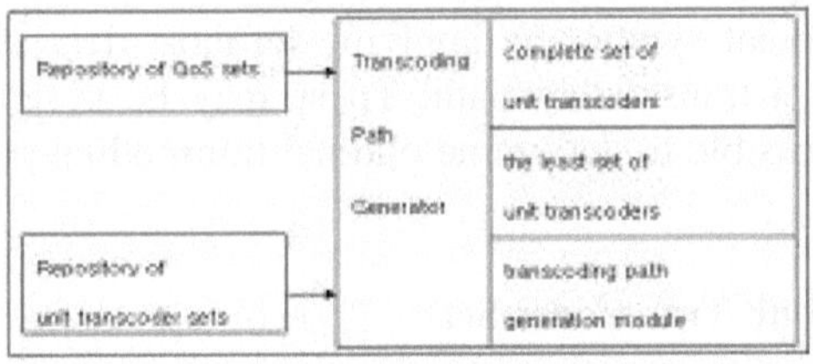

Fig. 2. Application independent adaptation framework model considering Unit transcoder set

- **Repository of QoS sets:** Stores a set of pairs of digital item QoSs at server and a set of QoSs at client to support all QoSs depending on nature of application.
- **Repository of unit transcoder sets:** Stores unit transcoders, and contains their characteristics such as type, I/O attribute, and more. The type encompasses encoder, decoder, color unit transcoder(CT), size unit transcoder(ST), format unit transcoder(FT) and more. I/O attribute is a data type that can be processed or is passed by the unit transcoder.
- **Transcoding path generator:** When end-to-end digital item has different QoSs, this generator has a function to create transcoding paths in connection with unit transcoders for adaptation into digital item required at destination. This paper uses CFG-based transcoding path generator. It is different from other existing CFG-based generator[7] in a sense that the former creates a set of unit transcoders which must be able to generate transcoding path in support of each pair of end-to-end different QoSs, depending on the nature of application.

Based on CFG algorithm, $G_{\text{transcoding path}}$, a grammar to create transcoding path is defined as follows: $G_{\text{transcoding path}}=(S, V_N, V_T, P)$
(S: start symbol, V_T: terminal symbol to express unit transcoder, V_N: non-terminal symbol to express intermediary symbol required for creating paths, P: production rule of creating paths)

Out of elements of context-free grammar, the start production rule signifies an adaptation required for different QoSs needed by clients for digital items at server. For instance, suppose that QoS required at both source and destination of digital item is (CIF, 24-bit color, bmp) and (QCIF, 24-bit color, jpeg) respectively. The information can be integrated into start production rule as follows:

S::=<bmp FT jpg><jpg ST jpg>
|<bmp ST bmp><bmp FT jpg>

Here, non-terminal <bmp FT jpg> represents file format conversion from .bmp to .jpg, and the production rule for enabling such conversion is to use bmp ft jpg or bmp ft gif gif ft jpg in connection. The latter helps generate transcoding paths when there is no bmp ft jpg in system. This can be represented as follows:

<bmp FT jpg> = bmp ft jpg
|bmp ft gif gif ft jpg

Terminal refers to unit transcoder available for multimedia adaptation. So all we have to do is just use the information on unit transcoders of system. If we create a string in form of terminal symbol by applying leftmost derivation from start symbol[10], this string is a transcoding path. There may be several transcoding paths obtained. Here, it is possible to determine optimal transcoding path by selecting least delaying path[11].

3.2 Organization of Unit Transcoder Set

When organizing unit transcoders, it is required to consider completeness and minimum. A complete set of unit transcoders refers to a collection of unit transcoders that support each pair of end-to-end different QoSs depending on the nature of application so that one and more transcoding paths must be available. For each pair of given end-to-end different QoSs, there are too many transcoding paths that can be created in a complete set of unit transcoders; there are many unit transcoders available, resulting in many non-terminal production rules as a connection of these transcoders. So we create a set of unit transcoders, which is made up of least number of unit transcoders without losing completeness in the complete set of unit transcoders. This creation is realized by removing any random unit transcoder out of a complete set of unit transcoders, and replacing the function of unit transcoders removed by a connection of different unit transcoder sets. Chapter 4 describes the possible algorithms to generate a complete set of unit transcoders and possible algorithms to create least number of unit transcoder sets.

4 Algorithm for Creating a Complete Set of Unit Transcoders

The complete set of unit transcoders is created during the generation of CFG components. So it is necessary to make components required for grammar $G_{\text{transcoding path}}$ before. Start production rule refers to QoS item input by user. Here, it is created using pairs of QoS set repository (source QoS, destination QoS) as created according to the nature of application. The creation of non-terminal and related production rules requires selecting a set of non-terminals as not overlapped but contained in start production rule. Then a production set comprising terminals alone should be created for this non-terminal set (See Section 3.1). Terminal is included in production rule for non-terminal, and it is the complete set of unit transcoders proposed herein that excludes anything overlapped in this terminal set. The algorithm for creating a complete set of unit transcoders can be outlined as follows:

<Algorithm for creating a complete set of unit transcoders>
Input: set of end-to-end different QoS pairs according to application.
Output: complete set of unit transcoders and
$CFG_{\text{complete set of unit transcoders}}$
Procedures
1) Create information on QoS set repository: Use a set of end-to-end different QoS pairs according to application.
2) Generate start production rule: Express a combination of unit transcoders required for each pair of 1) as non-terminal.

3) Collect only unredundant non-terminals from 2).
4) Prepare production rule for non-terminals in 3).
5) Replace each production rule of 4) by terminal.
6) the complete set of unit transcoders ← Unredundant terminals of 5).
7) CFG $_{\text{complete set of unit transcoders}}$ ← The results of 2), 4) and 5).
8) Output the complete set of unit transcoders and the CFG $_{\text{complete set of unit transcoders}}$

5 Algorithm for Creating the Least Set of Unit Transcoders

To create a least set of unit transcoders, we can take an approach of eliminating random unit transcoders one by one, till the least number of transcoding paths remains for production rule of non-terminal right-hand side. As a result, final remaining set of unit transcoders indicates the least set of unit transcoders in support of completeness.

The algorithm for creating the least set of unit transcoders can be outlined as follows.

<Algorithm for creating the least set of unit transcoders>
Input: complete set of unit transcoders and
CFG $_{\text{complete set of unit transcoders}}$
Output: least set of unit transcoders and
CFG $_{\text{least set of unit transcoders}}$
Procedures
1) CFG $_{\text{temporariness}}$ ← CFG $_{\text{complete set of unit transcoders}}$
2) Remove random unit transcoders in CFG $_{\text{temporariness}}$.
- Check the production rule of non-terminal right-hand side
- If the rule exists, repeat the Step 2).
- Otherwise, restore current state.
3) Least set of unit transcoders ← A set of unit transcoders left in CFG $_{\text{temporariness}}$
4) CFG $_{\text{least set of unit transcoders}}$ ← CFG $_{\text{temporariness}}$
5) Output least set of unit transcoders and CFG $_{\text{least set of unit transcoders}}$

6 Experiments

6.1 Implementation Environment and Tool

The environment for implementing the model proposed herein can be outlined as follows:

O.S : MS Windows XP Professional
CPU : Intel Pentium IV 3 GHz
Memory : 1024 MB
Tool : Visual C++ 6.0

The formats of original image files used for experiment herein BMP, JPG and GIF. And experiments should be conducted for size and color conversion. Each image has same contents and same resolution of CIF and same color depth of 24-bit color. 4 QoS sets used for experiments are listed in Appendix, and the experiment was repeated for one QoS set under identical environment at 5 times. Then, the mean result was measured as performance of conversion.

The contents and results of this experiment can be summed up as follows:
First, a complete set of unit transcoders was created for end-to-end different QoSs given, and the number and type of complete set were identified as shown in Table 1.

Table 1. Test Result 1

	number of unit transcoders	Generated types of complete unit transcoder
QoS Set 1	6	bmp ft gif bmp ft jpg jpg ft gif jpg ft bmp gif ft jpg gif ft bmp
QoS Set 2	9	bmp ft gif bmp ft jpg jpg ft gif jpg ft bmp gif ft jpg gif ft bmp bmp ct bmp gif ct gif jpg ct jpg
QoS Set 3	9	bmp ft gif bmp ft jpg jpg ft gif jpg ft bmp gif ft jpg gif ft bmp bmp st bmp gif st gif jpg sct jpg
QoS Set 4	12	bmp ft gif bmp ft jpg jpg ft gif jpg ft bmp gif ft jpg gif ft bmp bmp ct bmp gif ct gif jpg ct jpg bmp st bmp gif st gif jpg sct jpg

Second, the least set of unit transcoders was generated for same end- to-end different QoS sets as above, and the number and type of complete set were identified as shown in Table 2.

Table 2. Test Result 2

	number of unit transcoders	generated types of the least set of unit transcoder
QoS Set 1	3	bmp ft gif gif ft jpg jpg ft bmp
QoS Set 2	4	bmp ft gif gif ft jpg jpg ft bmp jpg ct jpg
QoS Set 3	4	bmp ft gif gif ft jpg jpg ft bmp bmp st bmp
QoS Set 4	5	bmp ft gif gif ft jpg jpg ft bmp jpg ct jpg bmp st bmp

Third, the complete set of unit transcoders and the least set of unit transcoders were used for same end-to-end different QoS sets as above respectively to determine time required for calculating transcoding path, mean number of transcoding paths generated, and actual time spent in transcoding, as illustrated in Table 3 and 4.

Table 3. Test Result 3

	QoS Set 1	QoS Set 2	QoS Set 3	QoS Set 4
calculation time for paths (ms)	1.83	7.36	7.37	13.16
mean number of paths	2	16.6	16.6	64
transcoding time(ms)	73.56	105.56	21.56	89.78

Table 4. Test Result 4

	QoS Set 1	QoS Set 2	QoS Set 3	QoS Set 4
calculation time for paths (ms)	1.20	1.58	1.71	3.52
mean number of paths	1	1	1	2
transcoding time(ms)	78.78	105.11	23.56	92.67

As shown in the results of experiment, the complete set shows significant deviations - 2ms to 13ms - in time required for calculating transcoding path. However, the least set doesn't show much significant deviations - 1.2ms to 3.52ms - in the time as compared to the former case, and takes less time to calculate. This is because when user creates desired transcoding path, the use of complete set can help create more paths than that of least set. In addition, actual time spent in trnascoding increased to max. 5ms in case of using least set than in case of using complete set. This is because when creating desired trnascoding paths, users sometimes have to interconnect more unit transcoders with one another due to less number of unit transcoders than complete set. However, in terms of total time as the sum of time required for calculating paths and time spent in trnascoding, it is found that there are less deviations for given QoS set along with less time.

7 Conclusion and Further Research

For the Universal Multimedia Access(UMA) as goal pursued by MPEG-21's Digital Item Adaptation(DIA) technology, this paper suggested application-independent multimedia adaptation framework that encompasses how to organize unit transcoder and how to create transcoding paths.

That is, given a certain set for end-to-end different QoS pairs depending upon the nature of application as defined by user, this study proposed possible algorithms to organize a complete set of unit transcoders, which must be able to generate one or more transcoding paths for each of all pairs in a set.

To this end, this study adopted CFG and sought to create a complete set of unit transcoders during the creation of components. Moreover, it also suggested an algorithm that can create one or more transcoding paths surely and has least number of components in the set of unit transcoder.

In the future, it is expected that this study will be further extended in the aspect of current system availability. That is, the system will contain any mixture of random unit transcoders or heavy transcoders. Here, follow-up studies will focus on investigating how to support adaptations for different QoS pairs in all cases within the nature of application.

References

1. Yang, S.G., Truong, C.T., Ro, Y.M., Nam, J.H., Hong, J.W.: Visual Impairment Description for MPEG-21 Digital Item Adaptation. Journal of Broadcast Engineering 8(4), 352 (2003)
2. Pereira, F., Burnett, I.: Universal Multimedia Experiences for Tomorrow. IEEE Signal Processing Magazine (2003)
3. Vetro, A.: MPEG-21 Digital Item Adaptation: Enabling Universal Multimedia Access. In: Smith, J.R (ed.) IEEE Multimedia, pp. 84–87 (2004)
4. http://www.w3.org/TR/2004/NOTE-di-atdi-20040218
5. Klaus, L., Dietmar, J., Hermann, H.: A Knowledge and Component Based Multimedia Adaptation Framework. IEEE Multimedia Software Engineering Proceedings, pp. 10–17 (2004)
6. Chon, S.M., Lim, Y.H.: An Algorithm Generating All the Playable transcoding paths using the QoS transition diagram for a multimedia presentation requiring different QoS between the source and the destination. Journal of Korea Multimedia Society, 6(2), 208–215 (2003)
7. Chon, S.M., Lim, Y.H.: A CFG Based Automated Search Method of an Optimal Transcoding Path for Application Independent Digital Item Adaptation in Ubiquitous Environment. Journal of Korea Information Processing Society, 12-B(3), 313–322 (2005)
8. Lee, S.J., Lee, H.S., Park, S.Y., Lee, S.W., Jeong, G.D.: Bandwidth Control scheme using Proxy-based Transcoding over Mobile Multimedia Network. Journal of the Korea Information Science Society, 29(2), 157–159 (2002)
9. Kim, J.W., Kim, Y.H., Park, J.H., Choi, B.H., Jung, H.K.: Design and Implementation of Video Transcoding System for the Real-Time Multimedia Service, Workshop for Image Processing and Understanding, pp. 322–327 (2003)
10. Kim, D.S.: Automata and Computational Theory, SaengNeung Publishing Co, pp. 259–261 (1996)
11. Chon, S.M., Lim, Y.H.: A Context Free Grammar based Algorithm for Generating Playable Transcoding Paths of the Multimedia Presentation with Different End-to-End QoS. Journal of Korea Information Processing Society, 9-C(5), 699–708 (2002)

Appendix

QoS Sets for Experiments

	QoS Set 1	QoS Set 2	QoS Set 3	QoS Set 4
1	(bmp, 24bit, CIF), (bmp, 24bit, CIF)	(bmp, 24bit, CIF), (bmp, 24bit, QCIF)	(bmp, 24bit, CIF), (bmp, 8bit, CIF)	(bmp, 24bit, CIF), (bmp, 8bit, QCIF)
2	(bmp, 24bit, CIF), (jpg, 24bit, CIF)	(jpg, 24bit, CIF), (jpg, 24bit, QCIF)	(jpg, 24bit, CIF), (jpg, 8bit, CIF)	(jpg, 24bit, CIF), (jpg, 8bit, QCIF)
3	(bmp, 24bit, CIF), (gif, 24bit, CIF)	(gif, 24bit, CIF), (gif, 24bit, QCIF)	(gif, 24bit, CIF), (gif, 8bit, CIF)	(gif, 24bit, CIF), (gif, 8bit, QCIF)
4	(jpg, 24bit, CIF), (jpg, 24bit, CIF)	(bmp, 24bit, CIF), (jpg, 24bit, QCIF)	(bmp, 24bit, CIF), (jpg, 8bit, CIF)	(bmp, 24bit, CIF), (jpg, 8bit, QCIF)
5	(jpg, 24bit, CIF), (bmp, 24bit, CIF)	(bmp, 24bit, CIF), (gif, 24bit, QCIF)	(bmp, 24bit, CIF), (gif, 8bit, CIF)	(bmp, 24bit, CIF), (gif, 8bit, QCIF)
6	(jpg, 24bit, CIF), (gif, 24bit, CIF)	(jpg, 24bit, CIF), (bmp, 24bit, QCIF)	(jpg, 24bit, CIF), (bmp, 8bit, CIF)	(jpg, 24bit, CIF), (bmp, 8bit, QCIF)
7	(gif, 24bit, CIF), (gif, 24bit, CIF)	(jpg, 24bit, CIF), (gif, 24bit, QCIF)	(jpg, 24bit, CIF), (gif, 8bit, CIF)	(jpg, 24bit, CIF), (gif, 8bit, QCIF)
8	(gif, 24bit, CIF), (bmp, 24bit, CIF)	(gif, 24bit, CIF), (bmp, 24bit, QCIF)	(gif, 24bit, CIF), (bmp, 8bit, CIF)	(gif, 24bit, CIF), (bmp, 8bit, QCIF)
9	(gif, 24bit, CIF), (jpg, 24bit, CIF)	(gif, 24bit, CIF), (jpg, 24bit, QCIF)	(gif, 24bit, CIF), (jpg, 8bit, CIF)	(gif, 24bit, CIF), (jpg, 8bit, QCIF)

Knowledge Management in the Development of Optimization Algorithms

Broderick Crawford[1,2], Carlos Castro[2], and Eric Monfroy[2,3,*]

[1] Pontificia Universidad Católica de Valparaíso, PUCV, Chile
FirstName.Name@ucv.cl
[2] Universidad Técnica Federico Santa María, Valparaíso, Chile
FirstName.Name@inf.utfsm.cl
[3] LINA, Université de Nantes, France
FirstName.Name@univ-nantes.fr

Abstract. This paper captures our experience developing Algorithms to solve Combinatorial Problems using different techniques. Because it is a Software Engineering problem, then to find better ways of developing algorithms, solvers and metaheuristics is our interest too. Here, we fixed some concepts from Knowledge Management and Software Engineering applied in our work.

Keywords: Knowledge Management, Software Engineering, Agile Development, Creativity, Optimization Algorithms.

1 Introduction

Solving Optimization Problems requires more knowledge than any single person can possess. It requires the collaboration of numerous individuals with complementary skills. The necessary resources to solve problems are distributed among the stakeholders and creative solutions emerge out of collaborative work. Creative thinking is an area that has been ignored in the development of Optimization Algorithms. Nevertheless, its successful application in the real world depends on a high degree of creativity and innovation [23]. The development of Optimization Algorithms and Metaheuristics to solve Combinatorial Problems assumes the same connotations it assumes in the field of Software Engineering. Then, the software development life cycle of them might be quite diverse and different models from other fields can be appropriate. This paper captures our experience with valuable concepts from Knowledge Management and Software Engineering applied when we are developing algorithms.

In the last time Optimization Algorithms and Metaheuristics have grown to be an important paradigm in solving large scale combinatorial optimization problems and rapid prototyping of them is an important topic of research today. Clearly, this is a Software Engineering problem, then a vision of the methodologies that improve productivity and quality of software is absolutely necessary to find better ways of developing this kind of solvers.

* The first author has also been partially supported by the project PUCV 209.473/2006. The second author has also been partially supported by the Chilean National Science Fund through the project FONDECYT 1070268.

M.J. Smith, G. Salvendy (Eds.): Human Interface, Part I, HCII 2007, LNCS 4557, pp. 604–612, 2007.

Software Engineering is a creative and knowledge intensive process that includes some aspects of Knowledge Management (KM) in all phases: eliciting requirements, design, construction, testing, implementation, maintenance, and project management. No worker of a development project possess all the knowledge required for fulfilling all activities. This underlies the need for knowledge sharing support to share domain expertise between the customer and the development team [6]. The traditional approaches (often referred to as plan-driven, task-based or Tayloristic), like the waterfall model and its variances, facilitate knowledge sharing primarily through documentation. They also promote usage of role based teams and detailed plans of the entire software development life-cycle. It shifts the focus from individuals and their creative abilities to the processes themselves. In contrary, agile methods emphasise and value individuals and interactions over processes.

There are few studies reported on the importance of creativity in software development. In management and business, researchers have done much work about creativity and obtained evidence that the employees who had appropriate creativity characteristics, worked on complex, challenging jobs, and were supervised in a supportive, noncontrolling fashion, produced more creative work. Since human creativity is thought as the source to resolve complex problem or create innovative products, one possibility to improve the software development process is to design a process which can stimulate the creativity of developers. The agile principles and values have realized the importance of collaboration and interaction in the software development and, by other hand, creative work commonly involves collaboration in some form and it can be understood as an interaction between an individual and a sociocultural context, the study of the potential of techniques to foster creativity in software engineering is a very interesting issue [11].

We believe that in Optimization Algorithms development projects, a better understanding of some valuable and interdisciplinary concepts from Creative Solving Problem [23] and Knowledge Management [18] offers important insights about the use of Software Engineering methodologies.

This paper is organised as follows: Section 2 is dedicated to the presentation of Knowledge Management in Software Engineering. We include a short overview of basic concepts from the area of Knowledge Management in Section 3, presenting the two approaches to KM: Product and Process. A Background on Agile Development Approaches is given in section 4. Section 5 introduces the relevance of Creativity in Software Development. Finally, in Section 6 we conclude the paper and give some perspectives for future research.

2 Knowledge Management in Software Engineering

The main argument to Knowledge Management in software engineering is that it is a creative and knowledge intensive activity. Software development is a process where every person involved has to make a large number of decisions and individual knowledge has to be shared and leveraged at a project and organization level, and this is exactly what KM proposes. People in such groups must collaborate, communicate, and coordinate their work, which makes knowledge management a necessity. In software development one can identify two types of knowledge: Knowledge embedded in the products or artifacts, since they are the result of highly creative activities and Meta-knowledge, that is knowledge about the products and processes. Some of the sources of knowledge (artifacts, objects, components, patterns, templates and containers) are stored in electronic form.

However, the majority of knowledge is tacit, residing in the brains of the employees. A way to address this problem can be to develop a knowledge sharing culture, as well as technology support for knowledge management. There are several reasons to believe that knowledge management for software engineering would be easier to implement than in other organizations: technology is not be intimidating to software engineers and they believe the tools will help them do a better job; all artifacts are already in electronic form and can easily be distributed and shared; and the fact that knowledge sharing between software engineers already does occur to a large degree in many successful software collaborative projects [22].

3 A Framework for Knowledge Management

Knowledge Management focuses on corporate knowledge as a crucial asset of the enterprise and aims at the optimal use and development of this asset, now and in the future. Knowledge Management has been the subject of much discussion over the past decade and different KM life-cycles and strategies have been proposed. One of the most widely accepted approaches to classifying knowledge from a KM perspective is the *Knowledge Matrix* of Nonaka and Takeuchi [18]. This matrix classifies knowledge as either explicit or tacit, and either individual or collective.

Nonaka and Takeuchi also proposes corresponding knowledge processes that transform knowledge from one form to another: socialization (from tacit to tacit, whereby an individual acquires tacit knowledge directly from others through shared experience, observation, imitation and so on); externalization (from tacit to explicit, through articulation of tacit knowledge into explicit concepts); combination (from explicit to explicit, through a systematization of concepts drawing on different bodies of explicit knowledge); and internalization (from explicit to tacit, through a process of learning by doing and through a verbalization and documentation of experiences). Nonaka and Takeuchi model the process of organizational knowledge creation as a spiral in which knowledge is amplified through these four modes of knowledge conversion. It is also considered that the knowledge becomes crystallized within the organization at higher levels moving from the individual through the group to organizational and even inter-organizational levels [4].

To make social creativity a reality, Fisher [7] has explored the externalization supporting social creativity. Externalizations support social creativity in the following ways:

- they cause us to move from vague mental conceptualizations of an idea to a more concrete representation of it
- they provide a means for others to interact with, react to, negotiate around, and build upon an idea
- they allow more voices from other stakeholders to be brought in
- they create a common language of understanding

Externalizations of individual knowledge make it possible to accumulate the knowledge held by a group or community. An important challenge for social creativity is to capture a significant portion of the knowledge generated by work done within a community.

3.1 Two Approaches to KM: Product and Process

Traditional methods of software development use a great amount of documentation for capturing knowledge gained in the activities of a project life-cycle. In contrast, the agile methods suggest that most of the written documentation can be replaced by enhanced informal communications among team members and customers with a stronger emphasis on tacit knowledge rather than explicit knowledge. In the KM market a similar situation exists and two approaches to KM have been mainly employed; we will refer to them as the *Product* and the *Process* approaches. These approaches adopt different perspectives in relation to documentation and interactions between the stakeholders [16].

Knowledge as a product. The product approach implies that knowledge can be located and manipulated as an independent object. Proponents of this approach claim that it is possible to capture, distribute, measure and manage knowledge. This approach mainly focuses on products and artefacts containing and representing knowledge.

Knowledge as a process. The process approach puts emphasis on ways to promote, motivate, encourage, nurture or guide the process of learning, and abolishes the idea of trying to capture and distribute knowledge. This view mainly understands KM as a social communication process, which can be improved by collaboration and cooperation support tools. In this approach, knowledge is closely tied to the person who developed it and is shared mainly through person-to-person contacts. This approach has also been referred to as the *Collaboration* or *Personalization* approach. Choosing one approach or other will be in relation to the characteristics of the organization, the project and the people involved in each case [2].

4 Agile Methods

A new group of software development methodologies has appeared over the last few years. For a while these were known as lightweight methodologies, but now the accepted term is Agile methodologies. The most common of them are: eXtreme Programming, the Crystal Family, Agile Modeling, Adaptive Software Development, Scrum, Feature Driven Development, Dynamic System Development Method [8]. There exist many variations, but all of them share the common principles and core values specified in the Agile Manifesto [5]. Through this work they have come to value individuals and interactions over processes and tools. Working software over comprehensive documentation. Customer collaboration over contract negotiation. Responding to change over following a plan. These new methods attempt a useful compromise between no process and too much process, providing just enough process to gain a reasonable payoff. The result of all of this is that agile methods have some significant differences with the former engineering methods [8]:

Agile methods are adaptive rather than predictive. Engineering methods tend to try to plan out a large part of the software process in great detail for a long span of time, this works well until things change. So their nature is to resist change. Agile methods, however, welcome change. They are processes that try to adapt and thrive on change, even to the point of changing themselves.

Agile methods are people oriented rather than process oriented. The goal of engineering methods is to define a process that will work well whoever happens to be using it. Agile methods assert that no process will ever make up the skill of the development team, so the role of a process is to support the development team in their work.

Most agile methodologies assume that change is inevitable, these methodologies have the ability to address variance and adaptability within the processes. In [12] Highsmith and Cockburn have fixed the role of creativity in agile teams assuming a world view that organizations are complex adaptive systems. A complex adaptive system is one in which decentralized, independent individuals interact in self organizing ways, guided by a set of simple, generative rules, to create innovative, emergent results. Agile methods offer generative rules, a minimum set of things you must do under all situations to generate appropriate practices for special situations. A team that follows generative rules depends on individuals and their creativity to find ways to solve problems as they arise. Creativity, not voluminous written rules, is the only way to manage complex software development problems and diverse situations.

5 Creativity in Software Development

There are many definitions of creativity, we use some ideas from [9]: Creativity is defined as the tendency to generate or recognize ideas, alternatives, or possibilities that may be useful in solving problems, communicating with others, and entertaining ourselves and others. There are three reasons why people are motivated to be creative:

- need for novel, varied, and complex stimulation
- need to communicate ideas and values
- need to solve problems

In order to be creative, you need to be able to view things in new ways or from a different perspective. Among other things, you need to be able to generate new possibilities or new alternatives. Tests of creativity measure not only the number of alternatives that people can generate but the uniqueness of those alternatives. The ability to generate alternatives or to see things uniquely does not occur by change; it is linked to other, more fundamental qualities of thinking, such as flexibility, tolerance of ambiguity or unpredictability, and the enjoyment of things heretofore unknown.

In order to understand creativity in organizations, the use of a creativity management framework may be useful. Amabile [1] had proposed a theory for the development of creativity. In her framework, creativity is hypothesized as a confluence of three kinds of resources:

- creativity-relevant skills (across domains)
- domain-relevant knowledge and skills (domain-specific)
- task motivation

Domain-relevant resources include factual knowledge, technical skills and special talents in the domain. Creativity-relevant resources include appropriate cognitive style, personality trait, conducive work style and knowledge of strategies for generating novel ideas. In specific, the major features of the appropriate cognitive style are the preference of breaking perceptual set and cognitive sets, keeping response options open, suspending judgment, etc. Furthermore, Amabile had proposed that intrinsic motivation was conducive to creativity; whereas extrinsic motivation was detrimental. Concerning the nurturing of intrinsic motivation, she and others highlighted the importance of promoting a playful attitude in the environment. Persons who are able to maintain playfulness, may continue to focus on the interest and enjoyment they derived from the task. They are more likely to keep their intrinsic motivation, even under external constraints.

Then, according to the previous ideas the use of creativity in software development teams is undeniable but requirements engineering is not recognized as a creative process [14]. The importance of creativity has been investigated in all the phases of software development process [10, 11] and focused in the requirements engineering too [21, 15, 17]. Nevertheless, the use of techniques to foster creativity in requirements engineering is still shortly investigated. It is not surprising that the role of communication and interaction is central in many of the creativity techniques. The most popular creativity technique used for requirements identification is the classical brainstorming and more recently, role-playing-based scenarios, storyboard-illustrated scenarios, simulating and visualizing have been applied in an attempt to bring more creativity to requirements elicitation. These techniques try to address the problem of identifying the viewpoints of all the stakeholders [17]. However, in requirements engineering the answers do not arrive by themselves, it is necessary to ask, observe, discover, and increasingly create requirements. If the goal is to build competitive and imaginative products, we must make creativity part of the requirements process. Indeed, the importance of creative thinking is expected to increase over the next decade [13]. The industrial revolution replaced agriculture as the major economic activity, and then information technology replaced industrial production. Now, the information technology will be replaced with a new dominant economic activity focusing on creativity: The Conceptual Age. According to [19] we are moving from High Tech to High Touch and High Concept. The skill of storytelling is now a mandatory business skill. The workers in highest demand will be those with great social skills and a strong drawing portfolio. With the prevalence of search engines, facts are abundant and free, what is in demand now is the ability to put those facts in order and in context. The shift of IT organizations toward the creative sector and companies striving to design innovative products that combine and use existing technologies in unanticipated ways is beginning to justify this prediction.

5.1 Inventing Requirements?

In [21, 20] very interesting open questions are proposed: Is inventing part of the requirements activity? It is if we want to advance. So who does the inventing? We can not rely on the customer to know what to invent. The designer sees his task as deriving the optimal solution to the stated requirements. We can not rely on programmers because they are too far removed from the clients work to understand what needs to be invented.

Requirements analysts are ideally placed to innovate. They understand the business problem, have updated knowledge of the technology, will be blamed if the new product does not please the customer, and know if inventions are appropriate to the work being studied. In short, requirements analysts are the people whose skills and position allows, indeed encourages, creativity.

In [3] the author, a leading authority on cognitive creativity, identifies basic types of creative processes: exploratory creativity explores a possible solution space and discovers new ideas; combinatorial creativity combines two or more ideas that already exist to create new ideas; and transformational creativity changes the solution space to make impossible things possible. Then, most Requirements Engineering activities are exploratory, acquiring and discovering requirements and knowledge about the problem domain. And the Requirements Engineering practitioners have explicitly focused on combinatorial and transformational creativity.

6 Conclusions and Future Directions

Human and social factors have a very strong impact on the success of software development. This paper was focused on some aspects of Knowledge Management and Creativity in the context of Optimization Algorithms development. In our main research topic of interest we are trying to find better ways of developing Optimization Algorithms. Because it is a Software Engineering problem, some ideas, concepts and open issues about it are important in supporting of our work too. The development of Optimization Algorithms is a field well suited for creative studies, since it is a creative activity where the problems often can only be solved through an iterative process faciliting Knowledge Management and exploration of new ideas.

Agile methods emphasis on people, communities of practice, communication, and collaboration in facilitating the practice of sharing tacit knowledge at a team level. They also foster a team culture of knowledge sharing, mutual trust and care. Agile development is not defined by a small set of practices and techniques. Agile development defines a strategic capability, a capability to create and respond to change, a capability to balance flexibility and structure, a capability to draw creativity and innovation out of a development team, and a capability to lead organizations through turbulence and uncertainty. They rough out blueprints (models), but they concentrate on creating working software. They focus on individuals and their skills and on the intense interaction of development team members among themselves and with customers and management.

The agile principles and values have recognized the importance of collaboration and interaction in the software development team. Because creative work commonly involves collaboration the study of techniques to foster creativity in software engineering is very interesting. Agile process to be helpful to generate novel and useful product. On the contrary, the discipline based work are perceived to be useless to produce novel products. The difference between them is that creative work can motivate the generation of something new.

Software development is a creative and knowledge intensive process that involves the integration of a variety of business and technical knowledge, an understanding from a Knowledge Management perspective offers important insights for designing and implementing Optimization Algorithms and Metaheuristics.

References

1. Amabile, T.M.: Creativity in Context: Update to the Social Psychology of Creativity. Westview Press (1996)
2. Apostolou, D., Mentzas, G.: Experiences from knowledge management implementations in companies of the software sector. Business Process Management Journal, 9(3) (2003)
3. Boden, M.: The Creative Mind. Abacus (1990)
4. Bueno, E.: Knowledge management in the emerging strategic business process. Journal of knowledge Management 7(3), 1–25 (2003)
5. Chau, T., Maurer, F.: Knowledge sharing in agile software teams. In: Lenski, W. (ed.) Logic versus Approximation. LNCS, vol. 3075, pp. 173–183. Springer, Heidelberg (2004)
6. Chau, T., Maurer, F., Melnik, G.: Knowledge sharing: Agile methods vs tayloristic methods. In: Twelfth International Workshop on Enabling Technologies: Infrastructure for Collaborative Enterprises, WETICE, May 2003, pp. 302–307. IEEE Computer Society Press, Los Alamitos (2003)
7. Fischer, G.: Social creativity: turning barriers into opportunities for collaborative design. In: PDC 04: Proceedings of the eighth conference on Participatory design, pp. 152–161. ACM Press, New York (2004)
8. Fowler, M.: The new methodology, Available (2001), at http://www.martinfowler.com/articles/newMethodology.html
9. Franken, R.E.: Human Motivation. Thomson Learning College (2002)
10. Glass, R.L.: Software creativity. Prentice-Hall, Englewood Cliffs (1995)
11. Gu, M., Tong, X.: Towards hypotheses on creativity in software development. In: Bomarius, F., Iida, H. (eds.) PROFES 2004. LNCS, vol. 3009, pp. 47–61. Springer, Heidelberg (2004)
12. Highsmith, J., Cockburn, A.: Agile software development: the business of innovation. Computer 34(9), 120–127 (2001)
13. Maiden, N., Gizikis, A.: Where do requirements come from? IEEE Softw. 18(5), 10–12 (2001)
14. Maiden, N., Gizikis, A., Robertson, S.: Provoking creativity: Imagine what your requirements could be like. IEEE Software 21(5), 68–75 (2004)
15. Maiden, N., Robertson, S.: Integrating creativity into requirements processes: Experiences with an air traffic management system. In: 13th IEEE International Conference on Requirements Engineering (RE 2005), Paris, France, 29 August - 2 September 2005, pp. 105–116. IEEE Computer Society Press, Los Alamitos (2005)
16. Mentzas, G.: The two faces of knowledge management. International Consultant's Guide, pp. 10–11 (May 2000) Available at http//imu.iccs.ntua.gr/Papers/O37-icg.pdf
17. Mich, L., Anesi, C., Berry, D.M.: Applying a pragmatics-based creativity fostering technique to requirements elicitation. Requir. Eng. 10(4), 262–275 (2005)
18. Nonaka, I., Takeuchi, H.: The Knowledge Creating Company. Oxford University Press, Oxford (1995)

19. Pink, D.: A Whole New Mind: Moving from the Information Age to the Conceptual Age. Riverhead Hardcover (March 2005)
20. Robertson, J.: Eureka! why analysts should invent requirements. IEEE Softw. 19(4), 20–22 (2002)
21. Robertson, J.: Requirements analysts must also be inventors. Software, IEEE 22(1), 48–50 (2005)
22. Rus, I., Lindvall, M.: Knowledge management in software engineering. IEEE Software 19(3), 26–38 (2002) Available at http://fcmd.umd.edu/mikli/RusLindvallKMSE.pdf
23. Vidal, V.V.: Creativity for operational researchers. Investigacao Operational 25(1), 1–24 (2005)

Research of Model-Driven Interactive Automatic / Semi-automatic Form Building*

Xiuyun Ding[1] and Xueqing Li[2]

School of Computer Science & Technology, Shandong University, Jinan, P. R. China 250061
[1]ddxiuyun@163.com, [2]xqli@sdu.edu.cn

Abstract. Forms are ubiquitous in today's software applications, so automation of form generation is highly desirable. In this paper, we provide improvements on Xforms model including data and event model. At the same time, we give a new method using use case for automatic form-building and the transformations from use case models to form user interfaces.

Keywords: XML, XForms, Form Model, Event Model, Data Model, Use Case, User Interface.

1 Introduction

Forms are often used in user interfaces, which provide a clear and intuitive way for the input/output data. Forms are ubiquitous in today's software applications. Forms facilitate the user input which is critical to business processes. Automation of form generation is highly desirable, as the development of any practical software system requires the creation of one or more forms to gather user input.

Nowadays, many researches have been done on the form styles of user interface. Such as a Functional Programming Technique for Forms in Graphical User Interfaces [1] which puts forward an idea about how to separate the data from presentation by using forms with reference based on the GUI library wxHaskell[2] ; And now one of the most popular research direction includes is the XForms[3]standard for the presentation and collection of form data which has been developed by The World Wide Web Consortium (W3C) as stated in the W3C recommendation. There are also some form-editing tools such as XML Forms Generator[4] from IBM, which can generate rich, working forms quickly by a given description of data to be collected (such as an XML document) and an EMF[5] model of that data.

In the paper, we make some improvements of data and event processing of forms based on the Xforms model. In addition, we provide a model-driven form-building method with the MDA [6] technology, whose inputs are use cases[7] describing the functions of forms. Starting from the forms' functions presented by some use cases, we build a Structured Use Case Model (SUCM) [8] by extending and formalizing them. Next, the SUC models can be transformed into a platform independent form model (PIFM) by using a serial of algorithms and rules immediately. PIFM conforms

---

* Project supported by Shandong Province Foundation for Middle-aged and Young Scientist (2004BS01002).

M.J. Smith, G. Salvendy (Eds.): Human Interface, Part I, HCII 2007, LNCS 4557, pp. 613–622, 2007.

to the XML rules, so we can run them on various platforms, such as web, desktop applications and others like PDA.

2 The Architecture

The architecture includes three parts: Raw Use Case Model （RUCM） [9], Structured Use Case Model and Platform Independent Form Model (PIFM).

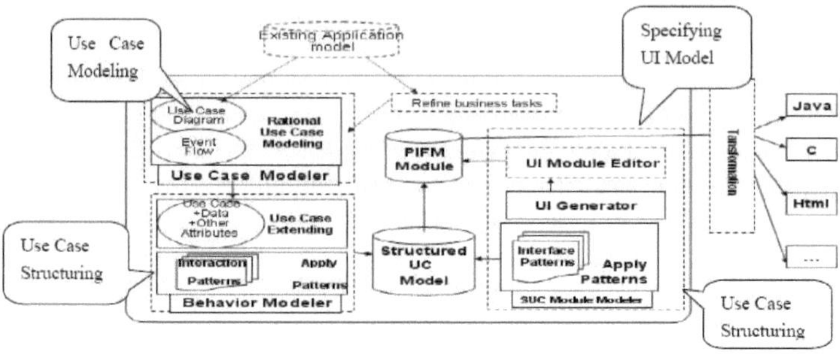

Fig. 1. The architecture

Raw Use Case Model (RUCM): We take RUCM as the input which describes business processes of the forms. The reason of using use cases as input is they provide a simple and intuitive description of functions. However, it is not easy to transform function in use cases to user interfaces components. We need some formalizations and structures for RUCM models and then get Structured Use Case Model (SUCM).

Structured Use Case Model (SUCM): In order to implement the transformation from business process to user interfaces, SUCM models must have some compulsory conditions. Then there are some algorithms and rules for transforming them into concrete use interfaces automatically.

Platform Independent Form Model (PIFM): PIFM is based on Xforms models and we make some improvements which have different data model and event model. And this model can be parsed into different types, and displayed in various platforms and devices.

In the following sections, we will introduce SUCM, PIFM and methods for transformations from SUCM models to PIFM concrete models.

3 Introduction to Models

3.1 Structured Use Case Model （SUCM）

Structured Use Case Model （SUCM） [9] includes Use Case Model and some extensions of Use Case Event (UCEvent). We summarize four events: input, view, modify and command. Each event object (named UCEvent) has an attribute marking the event type. And UCEvents have four attributes: isCommandType (may refer to the logic and background business which will be discussed later), pre_Condtion, post_Condtion, and related data. The followings are some key points:

1. Pre/post conditions are the inputUI/outputUI in the picture. We consider them as data and status of data. Taking an event object of input user information for example, the pre_Condition is a user information data which is null, and the post_Condtion is a concrete data of user information.
2. Which is the next, a UI component, a new event or a new use case after a UCEvent occurs? Here we introduce a Branch_Condition to control the jumping after a command event. Users can set them up in the SUCM models.

In the SUCM models, UCEvents and data are related, i.e. we have built relations between them to get and operate the data, which make all form business processes gathered and related, preparing for continuous interactions in form user interfaces.

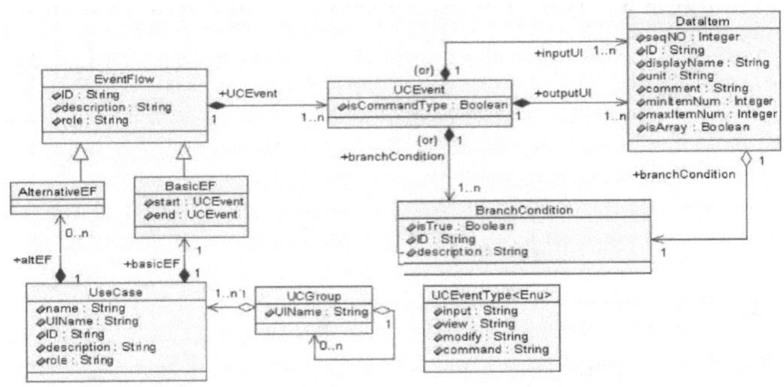

Fig. 2. Structured Use Case Model

3.2 Platform Independent Form Model

PIFM[10][11] model includes two parts: relations and UI Object(UIO).Here we define six basic relations: consistsOf, triggers, composedOf, contains, hasInput, hasOutput, representing relations between UIOs, UIO and data, UI command(UIC)and events. UI Object contains individual UI Objects and Container UI Objects. Each individual UI Object can be connected with related event and input/output data. At the same time, navigators with some Branch_Conditions offer

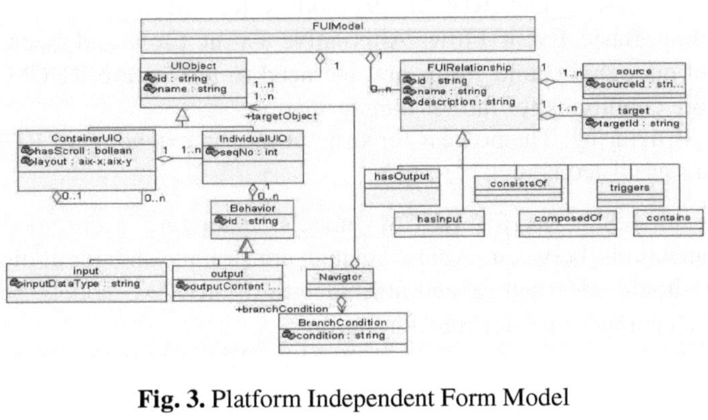

Fig. 3. Platform Independent Form Model

all the interactive functions in forms. As for how to use the model, we will discuss in the form building process.

4 Model-Driven Method for Form Building

Then we provide a model-driven method for the form building, which is based on MDA architecture. The following are the details of the transformation from use cases to PIFM models. We need different processors of transformations from PIFM to PDFM models to display in various platforms or devices which are the transformation between abstract and concrete models. The process is shown in the Fig. 4. Use cases are the input. Output is a forest of some UI objects, with data, events, constraints and so on.

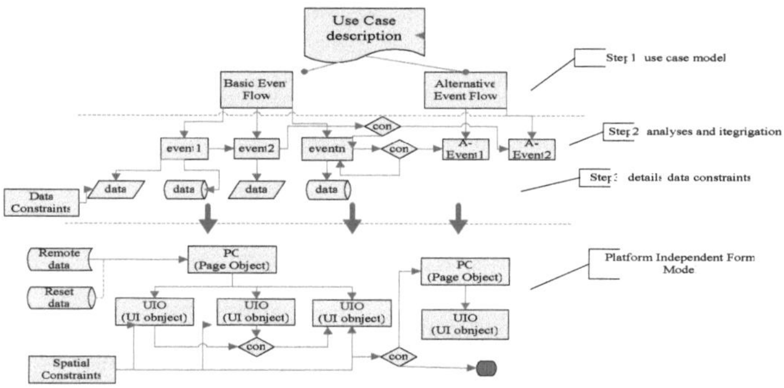

Fig. 4. The process from use cases to platform independent form models

The inputs are some use cases describing the function of forms, which can be given in natural language, and translated into formal ones. There are three steps of transforming use cases to independent form models:

Use Case Modeling: The input is a section of words describing form functions and those words are several use cases. We can formalize them and build formal use cases named Raw Use Case Model (RUCM) [9] models. RUCM models are traditional use cases, including Basic Event Flow, Alternative Event Flow and Scenarios. Those above are not enough to build forms and we need to restructure RUCM models and add some more conditions for farther transformation.

Use Case Structuring: The process for structuring and extending of RUCM models, two key points are listed here.

(1) Event modeling: as for RUCM models, extending event models aims at building connections between events, related constraints and input/output data of events. User should add such related attributes to the RUCM models, so the SUCM models are rich enough for later transformations.

(2) Data modeling: just as described in PIFM models, data are input or output of an event. One point is separation of data from presentation, meaning separation of data

from UI objects here, so that data sources are free. On the other hand, we define constraints between data which we'll give introductions later, such as semantics of data.

Automatic Transformation (SUCM to PIFM): This is a complicated process implementing the transformation from SUCM models to PIFM models, that's to say how to get platform independent form models from SUCM models. We'll describe those in details in next sections.

4.1 Event Processing

In SUCM models, we define several events. Actually, according to influence, events [12] can be abstracted two sorts: events on the whole form and those on the single form element.

The former will influence the change of the whole form, such as submission, save and jumping between pages, which need particular business logics.

In the presentation layer, we can often use a button for such type of event.

```
<button name="submit">submit
<handler name="onclick">//button action process
    parent.inputAccountInfo.remove(); //remove the current page
    parent.resultShow.apply(); //show the result page
</handler>
    </button>
```

Another thing we cannot deny is controls after a command event, generally, there are two or more branches, such as a successful path and a failed path.

Command events are executed by users, deciding the flow of pages and pages' order. When a user clicks a button, he can trigger to another page, where there may be two or more branches: true or false, which is expressed in the use case model.

About branches of a command event, we design them as a Branch_Condition, it can bind a command and a new page, dialog or other object which is triggered by the command event.

Events on form elements involve data. We use some rules for the mapping, deciding what UI objects can be fix for the data with a kind of event, like some showed in table1, and what's more, there may be some constraints logics involved in, such as constraints between several data items and system events at the time when data are changed.

Table 1. Mapping rules for transformation from data into UI components

PIDM Data Type	PIDM Data Type		Human behavior	PIUIM Type
Atom Types	Boolean		select	Default= CheckBox, RadioButton
	String	Maxmum == 1	input	Text , CodeText, PopUpCodeText
		Maxmum >1	input	TextList , ScrollText List
	Number	Maxmum == 1	Input/select	Text , Combo , Spin , Scale
		Maxmum >1	Inoput/select	TextList , ScrollTextList
	Enumeration	Maxmum == 1	select	Combo, RadioGroup , ScrollRadioGroup, ListSelect , ScrollListSelect
		Maxmum > 1	select	CheckGroup , ScrollCheckGroup , TwoListChange
	Date& Time	== Date	Input/select	Text , ThreeCombo , PopUpDate
		== Time	Input/select	Time, Combo
		== DateTime	Input/select	DateTime, Combo
Composite Data Type	hasChildren==0 & Maximum == 1		read/input	RadioGrid , ScrollRadioGrid , MutiValCombo
	hasChildren==0 & Maximum > 1		read/input	CheckGrid, ScrollCheckGrid
	hasChildren==1& Maximum == 1		read/input	Group, ScrollGroup
	hasChildren==1 & Maximum > 1		read/input	GridEdit, ScrollGridEdit
Group Type	Array		select	tabPane,
				Label (for all the read only data type)

4.2 Data Processing

Data that form takes in our form model is either an XML document (or fragment) or another object, that is to say, it doesn't need to be reset firstly, but we can get them in real time.

We design our data sources in the following ways to make it easy to manipulate data and tie that data to a user interface. In particular, we provide:

embedded data directly into an application
receiving data from a remote data source at runtime
receiving data from a web service
about data binding, we follow the Xforms.

Table 2. Syntax and Processing of data source

What's the reference	How to load	Syntax
Embedded	Compilt-time	<df:dataset name="exa_Data"> <df:dataInstance> <!-- ... other XML tags ... --> </df:dataInstance > </df:dataset>
Included	Compile-time	<df:dataset name="exa_Data" src=" exa_Data.xml"/>
HTTP data	Runtime	<df:dataset name=" exa_Data" request="true" type="http" src="exa_Data.xml" />

4.3 Constraint Model

Constraint Model includes Data Constraint and Layout Constraint. Data Constraint combines constraints of single data item and those between data items.

Data Constraints

There are two kinds data constraints: one is constraint of data instance(Static Constraint), another is constraint of method (Dynamic Constraint).

Static Constraint : illustration of single data item which can be a UI object (such as "label") following the UI component of data item. Such as length of a password and etc. Basic rules about data constraint are in the table 3. We define related methods to them. For each attribute of data item, we can setup whether it should be constrained.

Dynamic Constraints : used as constraints between data items and definition of dynamic data: type and express for implementation. Type means how to get the result of the data item (update, available, visible…), and express means what is the result of the data item from parameters (+,-,<,>…).

Table 3. Items of data constraints

Data_ Constraints	comment	minItemNum	maxItemNum	isArray
Description	CommentType data type	Min value	Max value	Decide a UI style like: combox, checkbox, radiogroup…

Table 4. Items of data constraints

Constraints	arrayLength	defVal	instanceVals	typename
Description	Array has a length ,deciding UI Style	Default value	Some values can be selected from	Used for validating

Data operations are often complicated, in order to free users from complex data definitions, we give semantics.

Now we focus on constraints on single data item and those between data items, the latter includes Computing Constraint and Connection Constraint. Computing Constraint defines relations between one data and some other data items which can resolve general computations. Connection constraint means data item or characters that will change when other related data change, i.e. becoming available, visible, changeable. Users only need to set up parameters according to the format and semantics [14] parser will work out logics.

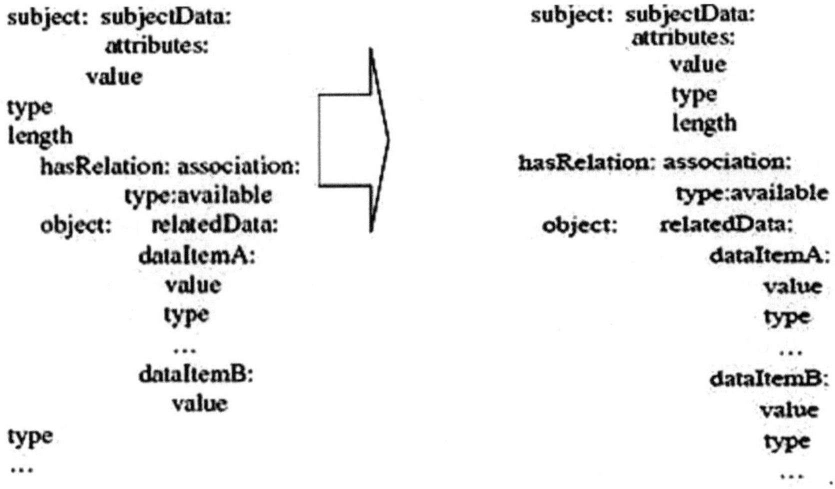

In order to separate data from logics, we'll define logic codes in the parsers. So data and logics can change respectively, disregarding the other.

Layout Constraints

There are two kinds of layout constraints [13][14], one is absolute constraint, another is relative constraint. Absolute constraint means only giving x and y coordinate; Relative one means giving its coordinate in its container. Container UI objects can set the arrangement for its children, vertical or horizontal。 Some mapping rules are listed in the following.

1. We'll provide default width and height for different UI components. At the same time, in the design, users can set up them.
2. As for the position of UI components, we can set them with absolute or relative coordinate. For container UI objects, we can set up an arrangement for its all

children: axis-x, axis-y, and the position will be computed automatically by algorithms.

3. For alignment, we also deal them with default configurations: left or right alignment. Users also can set up them on their own. For complicated situation, we can consider a table for the setup.

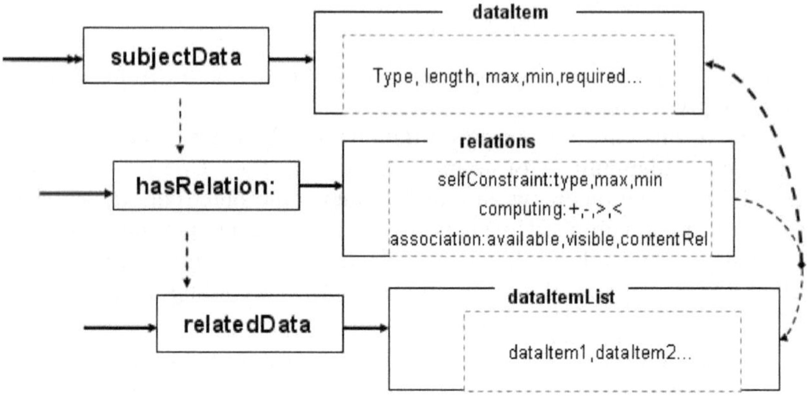

Fig. 5. Data semantics

4.4 Generation of Interface

The input of the system is use case descriptions. The output is form user interface.

Step1: Formalization of traditional use case.

The input is a section words like traditional use cases describing the business process. We build a RUCM model by event flows in use case: Basic Even flow, Alternative Event Flow(event, input/output data).

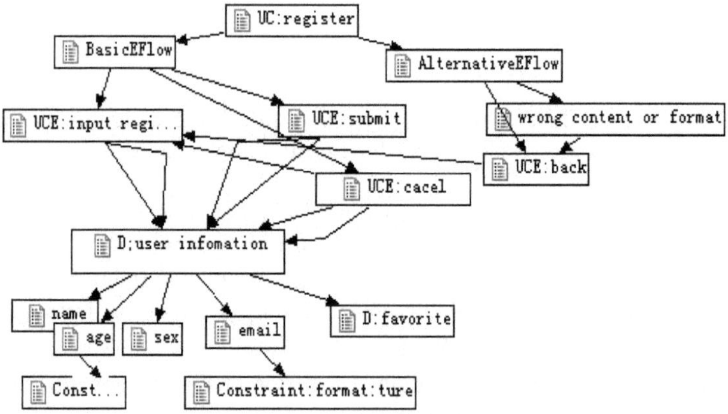

Fig. 6. SUCM model

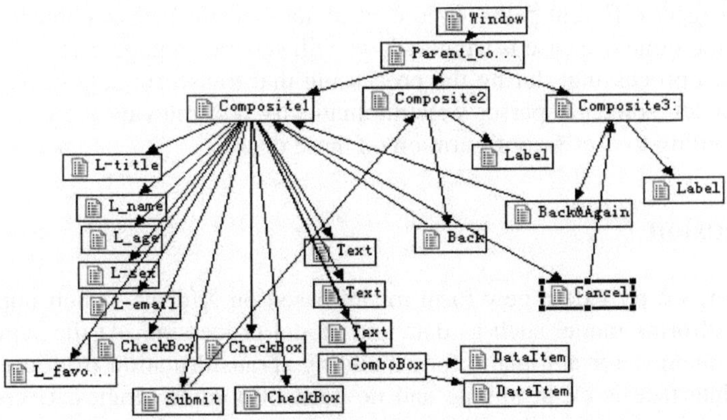

Fig. 7. PIFM model

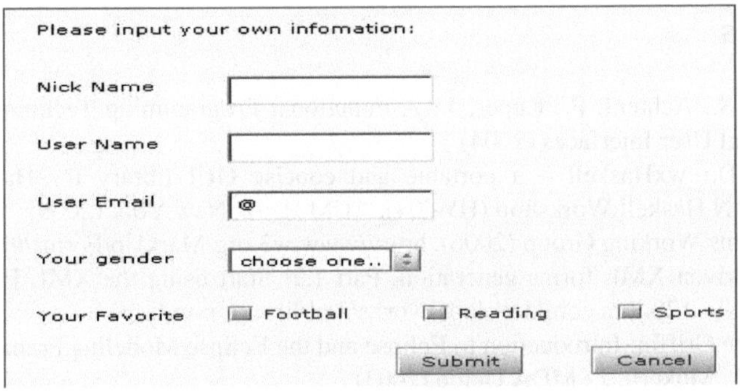

Fig. 8. Web Interface

Step2: Structure of use cases

We extend the RUCM model including type of event, adding data, jumping of event and constraints on event and data. i.e. we add a data structure(including name, gender, age, email) to an event of input user information.

Step3: Transformations from RUCM models to PIFM models

(1) Presentation layer processing: the data and events of SUCM models need to be loaded by UI components in form models. The data can be mapped to UI components through data type and event that related to these data, using the mapping rules listed in table1. For example, data which is String type in SUCM, with an "input" event related, can be mapped to a "TextField" in PIFM. The "Command" event can be rendered to a "Button" component in presentation layer.

(2) Layout and page separation processing: we can assign one container for each of use case and all the PFIM components that mapping from this use case will be lie in this container. We can set up relations between data through constraints, and all related data will be assigned to the same page or the same container. System will arrange the layout automatically according to the number and type of component.

Event will trigger different paths. If content of another use case or content that are not relevant in the same use case is triggered, we will separate a page here.

(3) Data processing: during the processing that transform data of SUCM model to PIFM model, semantic parser will automatically add relevant logic to PIFM data model, according to user's configuration on these data.

5 Conclusion

In this paper, we provide a new form model based on Xforms, which improve some defects of Xforms model such as data and model processing. At the same time, we give a new method for automatic form-building. Transformation from use cases to a form user interface is clear, simple and no need of complex logics. Users know the business process ,then he will build a working form quickly by the steps in the above. Next, we will pay attentions to develop different processors for various platforms.

References

[1] Eversl, S., Achtenl, P., Kuper, J.: A Functional Programming Technique for Forms in Graphical User Interfaces (2004)
[2] Leijen, D.: wxHaskell – a portable and concise GUI library for Haskell. In: ACM SIGPLAN Haskell Workshop (HW'04), ACM Press, New York (2004)
[3] The Forms Working Group (2006), http://www.w3.org/MarkUp/Forms/#implementations
[4] Model-driven XML forms generation, Part 1,2: Start using the XML Forms Generator: http://www-128.ibm.com/developerworks/xml/library/x-mdxfg1/
[5] Catherine Griffin: Introduction to Eclipse and the Eclipse Modeling Framework (2004)
[6] Miller, J., Mukerji, J.: MDA Guide (2003)
[7] Jacobson, I.: Use Cases—Yesterday, Today, and Tomorrow (2002)
[8] Graham McLeod:Beyond Use Cases (2000)
[9] Ksenia Ryndina, IBM Zurich Research Laboratory; Pieter Kritzinger,Department of Computer Science University of Cape Town: Analysis of Structured Use Case Models through Model Checking
[10] M.Sc. Sari A. Laakso: User interface design for business process modeling and visualization (2001)
[11] Université catholique de Louvain (UCL), School of Management (IAG), Information Systems Unit (ISYS), Belgian Lab. of Computer-Human Interaction (BCHI), A MDA-Compliant Environment for Developing User Interfaces of Information Systems
[12] Mogaki, M., Kato, N., Shimada, N., Yamada, Y.: A Layout Improvement Method Based on Constraint Propagation for Analog LSI's
[13] Joung, S., Tanaka, J.: joung: Generating a Visual System with Soft Layout Constraints (2000)
[14] Latva-Koivisto, A.M.: User interface design for business process modeling and visualization (2001)

HEI! – The Human Environment Interaction

José L. Encarnação

INI-GraphicsNet Stiftung
Rundeturmstrasse 10, 64283 Darmstadt, Germany
jle@inigraphics.net

Abstract. As computers are becoming more and more ubiquitous, moving from the desktop into the infrastructure of our everyday life, they begin to influence the way we interact with this environment - the (physical) entities that we operate upon in order to achieve our daily goals. The most important aspect of future human-computer interaction therefore is the way, computers support us in efficiently managing our personal environment. This paper addresses the fundamental components that are involved in the forthcoming human-computer-environment interaction.

Keywords: human computer interaction, ambient intelligence, mobile computing, knowledge management, media management.

1 Introduction

A human being's daily activities – professional or private – are based on a broad range of interactions with numerous external objects: discussing project plans with colleagues, setting up a multimedia presentation in the conference room, editing documents, delegating travel planning to a secretary, driving a car, buying a ticket from a vending machine, visiting an exhibition, controlling the TV at home, etc. As computers are becoming more and more ubiquitous, moving from the desktop into the infrastructure of our everyday life, they begin to influence the way we interact with this environment - the (physical) entities that we operate upon in order to achieve our daily goals. The most important aspect of future human-computer interaction therefore is the way, computers support us in efficiently managing our personal environment. Conventionally, human-computer interaction looks at a process, where only two partners are involved: the human and the computer. However, looking at the computer as a mediator between the user and his environment, we have to acknowledge a more complex communication process. This paper addresses the fundamental components that are involved in this forthcoming human-computer-environment interaction.

2 Human Computer Interaction Today

By the human computer interaction (HCI) the user is in command, he is the operator of the interaction. The user requests actions and services from the environment by

M.J. Smith, G. Salvendy (Eds.): Human Interface, Part I, HCII 2007, LNCS 4557, pp. 623–631, 2007.

means of function calls, by assigning parameters, by activating menus, etc. Combining and interpreting several HCI's we can also implement different types of "computer-supported "Human-Human-Interaction (HHI)" (Fig. 1).

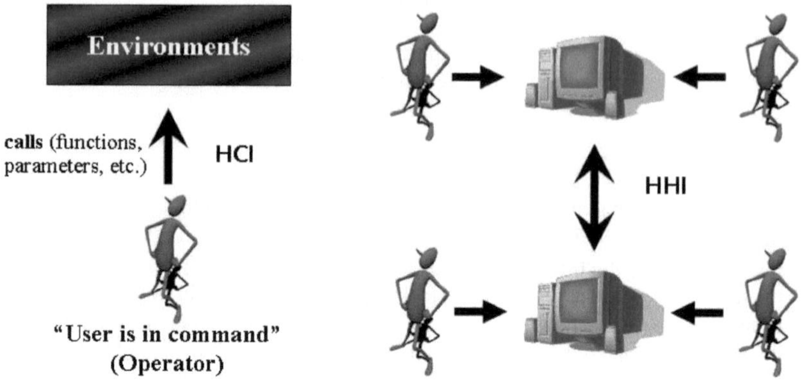

Fig. 1. Today's forms of computer interaction: (left) Human Computer Interaction (HCI) and (right) Computer-based Human-Human Interaction (HHI)

Fig. 2. Marker-based augmented reality (augmented video) applied in marketing: The video shows a car with a marker placed on the front-left felly (left). The user then selects one of the available fellies to augment the video with the virtual felly following the movement of the car in the right perspective position (right). In this way the user get a visual impression of the felly.

Research in this area today includes topics like:

a) integration of multimediality and multimodality
b) integration of computer generated "realities" (virtual reality, augmented reality, mixed reality, simulated reality)
c) making interfaces easy to understand and easy to use

Example applications include:

- marker-based and marker-free augmented visualization of auto-parts (Fig. 2)
- physically-based simulation with virtual objects (Fig. 3)
- interactive exploration of a large set of simulation results (Fig. 4)

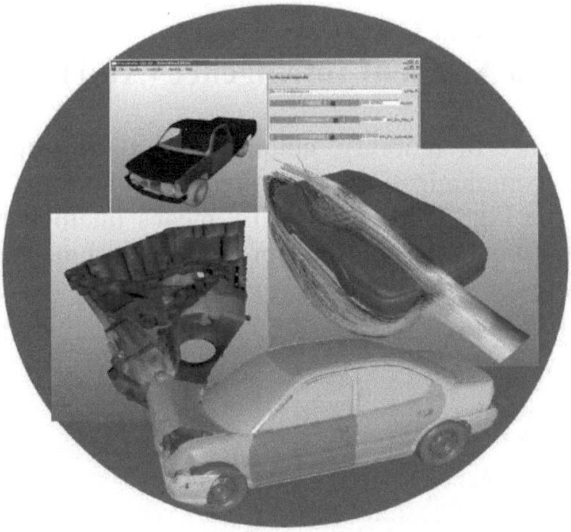

Fig. 3. Examples of simulated reality with virtual objects: The images include physically-based simulation of stress on a gear box, the streamline visualization of a aerodynamic resistance simulation of a car, and a crash test simulation, respectively

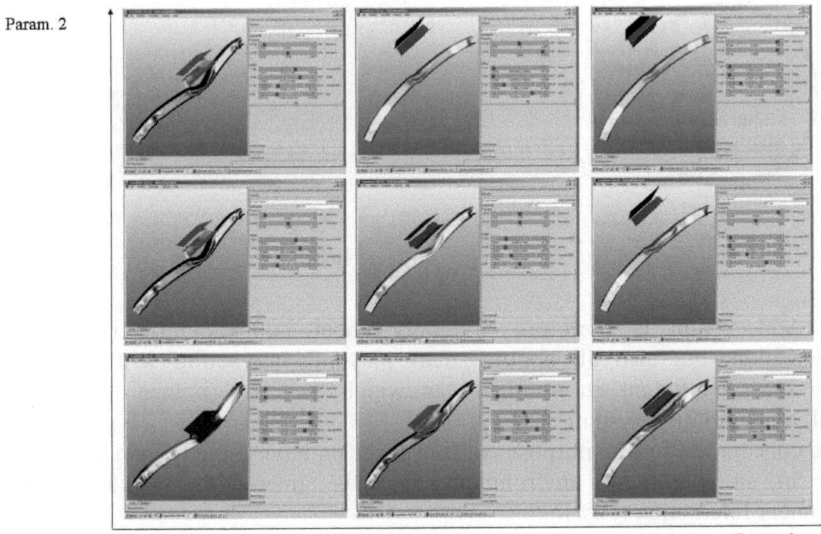

Fig. 4. Design of experiment example: the physical behavior of a component is interactively explored by interpolating between the pre-computed simulation results (here the deformations of the pillar of a car on impact with different parameters like energy of impact and thickness of material)

3 Metaphor Change: From Desktop to Ambient Intelligence (AmI)

Through the convergence of mobility, ubiquity and multimediality/multimodality a new information & communication technology paradigm is emerging: Ambient Intelligence (AmI) (see Fig. 5). AmI basically means intelligent environments with intelligent products serving "smart players" in these environments.

There are many types of "smart players" in AmI:

- humans
- animals
- smart objects
- smart work places
- smart machines
- etc.

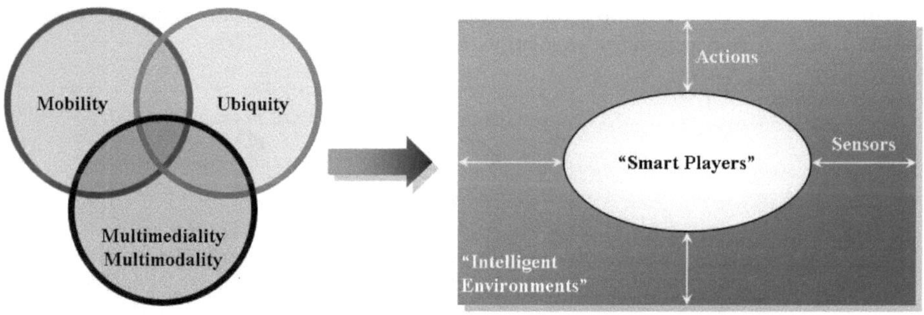

Fig. 5. Metaphor change: Ambient Intelligence (AmI)

The AmI vision is therefore not only about human-centric dialogue and communication, but also all about "smart players" interacting with and getting services from intelligent environments in which they act (home, office, school, hospital, transportation system, factory, etc.).

If we concentrate for the purposes of this paper on having the human as the "smart player" in an intelligent environment, hen the interaction is called Human Environment Interaction (HEI).

4 Human in an Intelligent Environment

In the area of intelligent environments we see the merging and convergence of mobile computing, media management and knowledge management (Fig. 6).

In such a context we have two key enabling technologies in the implementation of the related HEI-interfaces: presence and awareness.

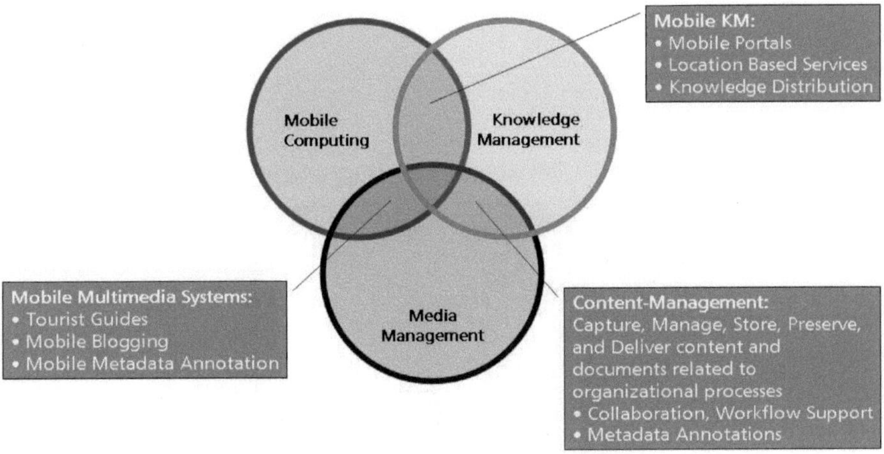

Fig. 6. HEI: Human Environment Interaction

The human is not anymore the operator of the environment, but he is served by the environment. This is implemented based on agent technologies.

A possible HEI architecture is shown in Fig. 7. It is based on the communication and interaction between three different agents:

- user agent
- environment agent
- dialogue agent

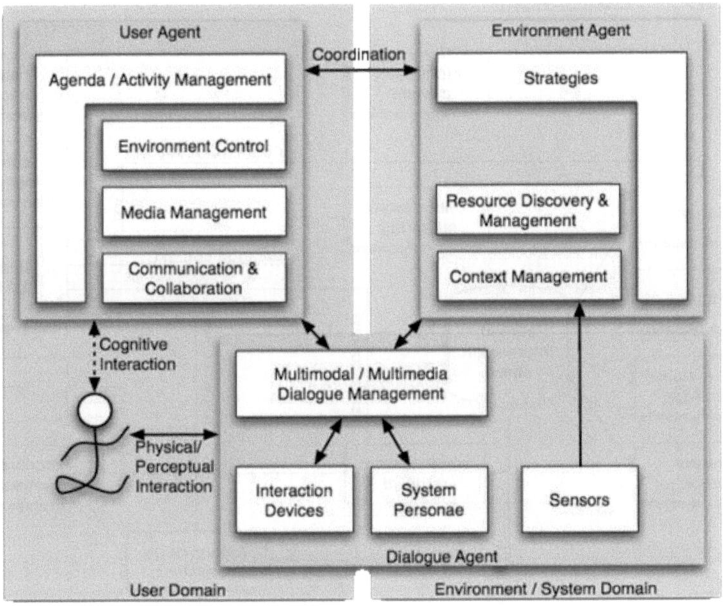

Fig. 7. HEI: System Architecture

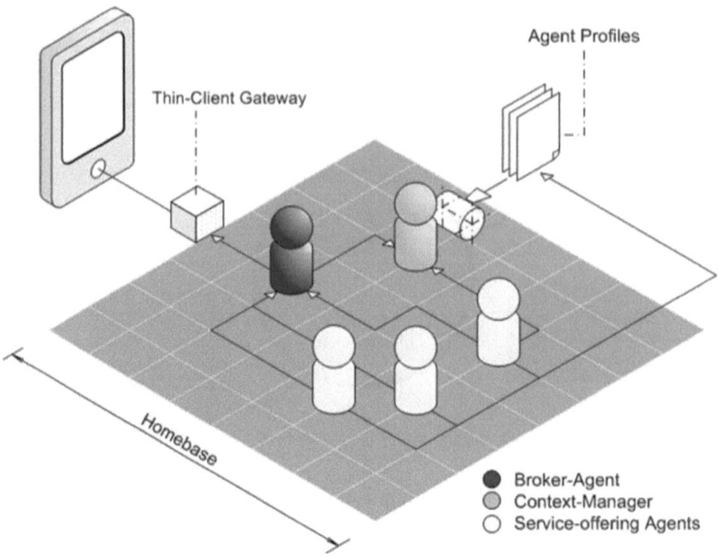

Fig. 8. HEI: key functionalities

From an action (service) point of view there are three other agent functionalities as illustrated in Fig 8:

- broker agent
- context agent (manager)
- service-offering agent

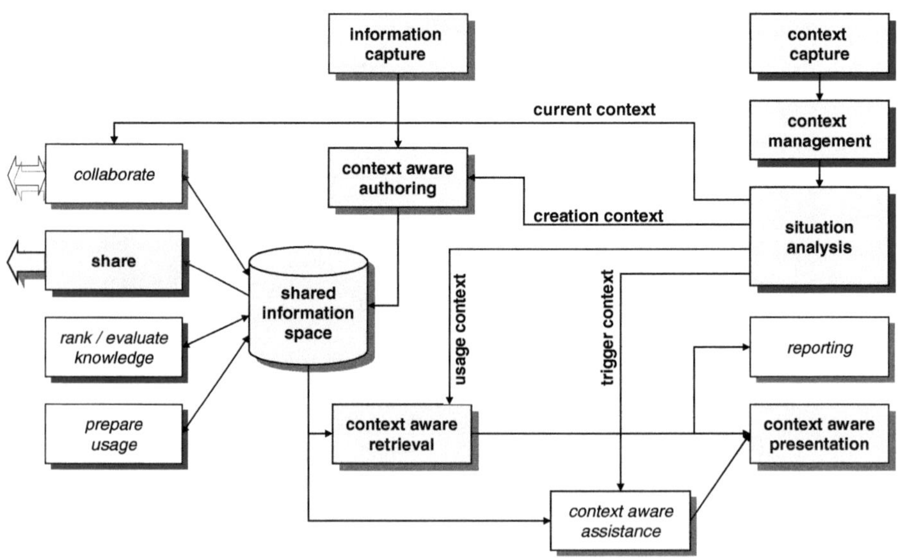

Fig. 9. HEI: context-aware information processing

From a software engineering point of view these agents allow to implement a "context aware information processing" as shown in Fig. 9. This is based on a shared information space, a situation analysis and information capturing (for example using sensors or cameras).

Early prototypes which we developed to test this concept are:

- mobile information and knowledge handling by extending knowledge management to support the authoring, sharing, retrieval and visualization process in mobile work
- just-in-time mobile assistance to speed up the workflow of spatially distributed business processes
- platform for mobile, multilingual services to offer information, communication and orientation for individual experiences in events and performances.

5 HEI Research Issues

The experience in implementing the prototypes mentioned above show the following needs:

- interdisciplinarity
  - integration of many different technologies
  - integration of non-technological disciplines (psychology, social science)
- multiculturality
  - applicability across cultures and specialization towards the specific originalities of individual cultures
- interoperability
  - between components developed by different companies in different nations
  - based on a common reference model and a shared vision

In order to satisfy these needs some advanced research issues arise, like

- interactive screens and multiple interaction tools (Fig. 10) to interact with those screens (examples: laser pointing, magnifying glass and flash light as new metaphors);
- feature-based tracking for indoor and outdoor applications (Fig. 11). 3D geometry models and reference images are the starting point based on which then different approaches (tracking of edges, feature matching, patches, etc.) are being followed;
- natural dialogues that are another important area of research. The goal is to have a guided operation in the intelligent environment in a natural dialog based on new approaches like narrative environments, interactive storytelling or others. These conversational user interfaces also may make use of human modeling for virtual characters and avatars with a realistic appearance and integrating voice, mimic and gestures as special functionalities.

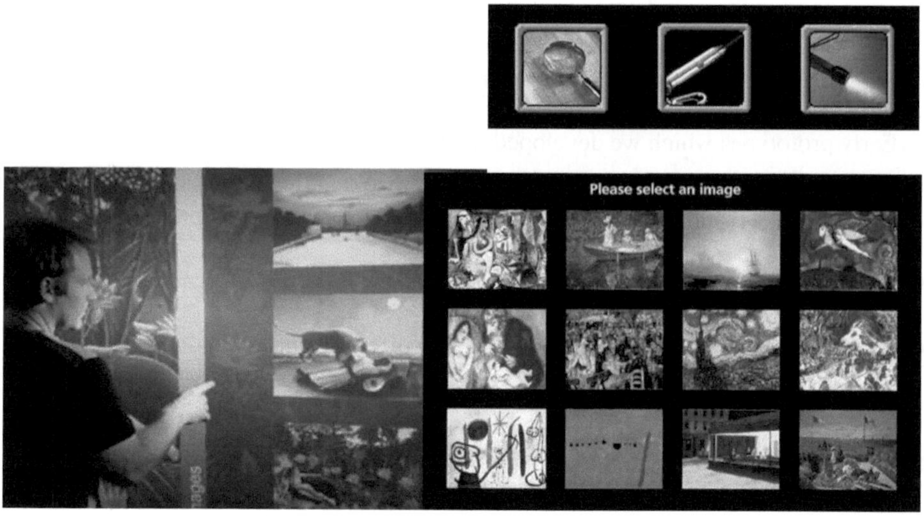

Fig. 10. Interactive screens and interaction tools (top-right)

Fig. 11. Feature-based tracking in partially unknown dynamic scenes

6 Conclusion

A paradigm shift from HCI/HHI to HEI is taking place in advanced user interfaces for dialogues in the context of intelligent environments (AmI). There are two fundamental enabling technologies for this (presence, awareness) and some new key interaction concepts (context aware information processing, service brokering), which have to be researched and further developed in order to be able to implement this paradigm shift. These new forms of interaction and natural dialogues require interdisciplinarity, multiculturality and interoperability.

References

1. Hellenschmidt, M.: Distributed Implementation of a Self Organizing Appliance Middleware. In: Smart Objects and Ambient Intelligence SOC-EUSAI 2005, pp. 201–206 (2005)
2. Encarnação, J.L., Kirste, T.: Ambient Intelligence: Towards Smart Appliance Ensembles. In: Hemmje, M., Niederée, C., Risse, T. (eds.) From Integrated Publication and Information Systems to Information and Knowledge Environments. LNCS, vol. 3379, pp. 261–270. Springer, Heidelberg (2005)
3. Windlinger, L., Grimm, M., Binda, G., Hoffmann, T.: Computer Supported Mobile Work - the Case of Mobile Information Processing Using MUMMY. In: Mobile Work Employs IT – MoWeIT, Prag, Czech Republic, pp. 37–44 (2005)
4. Balfanz, D., Grimm, M., Tazari, M.-R.: A Reference Architecture for Mobile Knowledge Management. In: Dagstuhl-Seminar 05181 2005, Mobile Computing and Ambient Intelligence. Wadern (2005)
5. Tazari, M.-R., Thiergen, S.: Servingo: A Service Portal on the Occasion of the FIFA World Cup 2006. In: Proceedings of the IWWPST '06, Vienna, Austria, pp. 73–93 (2006)
6. Göbel, S., Schneider, O., Iurgel, I., Feix, A., Knöpfle, C., Rettig, A.: Virtual Human: Storytelling & Computer Graphics for a Virtual Human Platform. In: Proceedings of Technologies for Interactive Digital Storytelling and Entertainment, Darmstadt (2004)
7. Hoffmann, A., Göbel, S., Schneider, O., Iurgel, I.: Storytelling-Based Edutainment Applications. In: Tan, L. (ed.) E-Learning and Virtual Science Centers, pp. 190–214. Information Science Publishing (2005)

Mining Attack Correlation Scenarios Based on Multi-agent System

Sisi Huang, Zhitang Li, and Li Wang

Computer Science Department
Huazhong University of Science and Technology, Hubei Wuhan 430074, China
Huangsisi1984@gmail.com, leeying@hust.edu.cn, wtwl@hust.edu.cn

Abstract. Nowadays, one very complicated problem bothering network analysts too much is the redundant data generated by IDS. The objective of our system SATA (Security Alert & Threat Analysis) is trying to solve this problem. Several novel methods using data mining technologies to reconstruct attack scenarios were proposed to predict the next stage of attacks according to the recognition the attackers' high level strategies. The main idea of this paper is to propose a novel idea of mining "complicated" attack scenarios based on multi-agent systems without the limitation of necessity of clear attack specifications and precise rule definitions. We propose SAMP and CAST to mine frequent attack behavior sequences and construct attack scenarios. We perform a series of experiments to validate our method on practical attack network environments of CERNET. The results of experiments show that our approach is valid in multi-agent attack scenario construction and correlation analysis.

Keywords: correlation analysis; attack scenario; frequent attack sequence.

1 Introduction

Nowadays, more and more intrusion detection systems (IDSs) such as intrusion detection system, firewall, anti-virus software, vulnerability scanner and VPN etc are deployed to defend attacks against enterprise networks. Unfortunately, these different security sensors provide not only wealthy information but also a large volume of security data to network administrator. Therefore, it is important to develop a network security correlation system whose functions are reducing the redundancy of alarms, correlating different alerts, constructing attack scenarios, discovering attack strategies and predicting the next step of attacks.

In this paper, we focus on the attack scenario construction module. Attack scenario construction module is a very important component of such systems because it can help forensic analysis, response and recovery, and even prediction of forthcoming attacks. However, there have been several proposals on alert correlation, but most of these proposed approaches are lack of flexibility and depend on complex correlation rules definition and hard-coded domain knowledge that lead to their difficult implementation.

M.J. Smith, G. Salvendy (Eds.): Human Interface, Part I, HCII 2007, LNCS 4557, pp. 632–641, 2007.

Based on the substantial researches about attack behaviors for several years, we discovered the features of attack behaviors in the same attack strategy are: sequence and frequency. We proposed novel methods to discover attack strategies via mining frequent patterns.

1.1 Related Work

The research of the intrusion detection has started for many years, and some researchers have already developed several practical approaches to facilitate analysis of intrusion. In [1], Valdes used a probabilistic method to correlate alerts according to constitute similarity metric between their features. In [2], Wenke Lee and Xinzhou Qin proposed a GCT-based and Bayesian-based correlation approach without the dependence of the prior knowledge of attack transition patterns. Debar,H and Wespi use a consequence mechanism[3] to specify that types of alerts may follow a given alert type. In [4], Templeton and Levitt proposed an attack modeling language based on prerequisite consequence relation called JIGSAW. Ning in [6] also developed a method considered as a variation of JIGSAW. The focuses of these methods are both to uncover attack scenarios based on specification of individual attacks, and the main idea of the approach is also used by Cuppens and Meige in their work of MIRADOR[7].

Although these approaches mentioned above can potentially discover the relationship between alerts, most of these approaches have limited capabilities because they rely on predefined knowledge of attack conditions and consequences. They lack of the capabilities of recognizing a correlation when an attack is new or the relationship is new. The method this paper mentioned of construct attack scenario through attack sequence and frequency pattern mining method is enlightened by Jian Pei, who introduced a data mining method in [8] to mine the sequence and frequency patterns in a database.

1.2 Organization of the Paper

In this work we present the approach we suggest implementing the scenario construction function. The remainder of this paper is organized as follows. Section 1 introduces the main object and related work. Section 2 introduces the overview of our work and the framework of SATA system. Section 3 presents the problem of mining attack sequence patterns and some concepts used in the problem. Section 4 proposes algorithms to construct attack scenario. Section 5 reports our experiments on the branch of CERNET. In section 6 we conclude with a summary and directions for future work.

2 Overall of Our Work

In this section, we present an overview of our system SATA, which aims to provide a platform for integrated network security data management. Figure 1 presents the main principles we suggest to develop a security event management system SATA (Security alerts & Threats analysis) for intrusion detection. There are six main

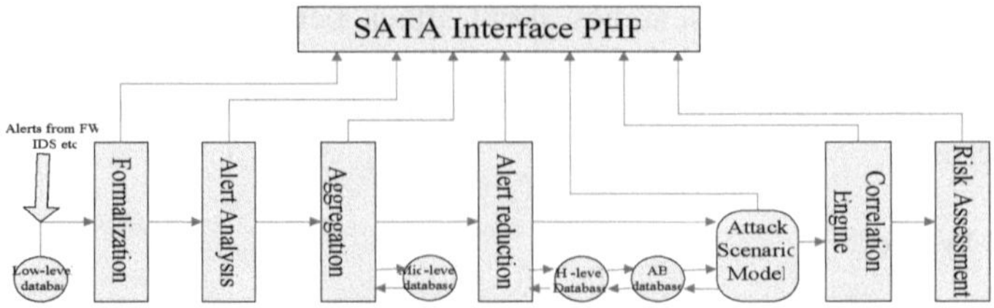

Fig. 1. Framework of SATA

modules in this system: Formalization, alert analysis, aggregation, alert reduction, correlation, and risk assessment. The main target of SATA is to build an integrated and centralized platform of security event management and analysis.

Formalization: SATA receives the alerts generated by different IDS and stores them into the low-level alert database after normalization and format standardization procedures for further analysis.

Alert analysis: provides scores of reliability, priority, asset and a final assessment to the degree of threat for each alert, to indicate how much it poses threat to the current protected network from various aspects.

Aggregation: the modules tries to aggregate the incoming alert with existing alert cluster or single alerts in the alert clusters which are sets of alerts that correspond to the same occurrence of an attack.

Alert reduction: the main component is a filter whose function is to filter the false positives and low-interest alerts.

Correlation engine: contains a construct attack scenarios model that we can recognize the attack strategies by matching attacks' behaviors to attack scenarios.

Risk assessment: the module aims to assess the real-time security state of the protected networks or hosts through a particular risk assessment mechanism.

As mentioned in the above introduction, we will only suggest a specification for the alert management, correlation module. The other modules which are briefly sketched in the conclusion of this paper are not presented.

3 Problem Statement

There are 4 databases in SATA: *low-level* alert database which contains raw alerts, *mid-level* alert database whose contents are the alerts after aggregation process, *and hi-level* alert database which only contains highly improved alerts after alert analysis, reduction and alert aggregation process. The database we are concerned is the *hi-level* alert database, whose alerts are largely less than other database, and also the quality of the alerts in *hi-level* database is more improved. The last database is *AB* database (attack behavior database), which deposits the result of attack behavior analysis from hi-level database.

Each alert record in database *hi-level* consists of the following attributes alert-id (id-number, sensor-id, and signature-id), attack type, timestamp (happen-time, end-time), source (source-IP, source-port), destination (destination-IP, destination-port), risk (reliability, asset, and priority), and protocol.

An itemset is a non-order and non-empty set of items. Without loss of generality, we assume that each alert as an item, mapped to contiguous integers. The itemset in our paper is denoted as a set of attack behaviors reported by different IDS sensors at the same time. We denote these itemsets i by ($i_1, i_2, \ldots i_m$), where i_k is an item. A sequence is an ordered list of itemsets. We denote a sequence s by { $s_1, s_2, \ldots s_m$ }, where s_j is an itemset. We also denote 3 sequences: time sequence T { $t_1, t_2, \ldots t_m$ } where t_i is a time-stamp representing the attack occurring time; alert-id sequence { $a_1, a_2, \ldots a_m$ }, which represent an alert ID. Attack behavior sequence B { $b_1, b_2, \ldots b_m$ }, where b_i is an attack behavior cluster itemset corresponds to different attack behavior types happened in the same time. Attack behavior itemset was denoted as b_i ($k_1, k_2, \ldots k_m$), where k_i is a behavior-id item. In a set of sequences, a sequence S is maximal if S is not contained in any other sequence.

All the attack behavior item in *hi-level* database can be viewed as a global attack behavior sequence, and the sorted attack behavior item can be ordered by the increasing time-stamp $t_1, t_2, \ldots t_m$, which are contained in the time sequence T.

The problem of mining attack behavior correlation scenario is to find the maximal attack sequences among all global attack sequences that have a certain user-specified minimum frequency and a time order. Each such maximal attack sequence represents a frequent *attack sequence*

EXAMPLE Consider the database shown in Table.1. Because of abundant data in database, we focus on 20 of the whole datum which multi-agents generate. This database has been ordered by real attack occurring time sequence T{ $t_1, t_2, \ldots t_m$ }. The alert ID describes attack behaviors in global sequence ordered by attack occurring time. We map the alerts to the integer signature ID because we only concentrate on the attack behavior type attributes. It is more effictive to mine attack correlation scenario and save the cost of string matching.

4 Finding a Sequence Pattern

4.1 Transform Global Attack Items to Sequences

As mentioned in figure 1, attack behaviors after formalization module are arranged in the *hi-level* database by the order of attack occurring timestamp. We found that most of attackers usually complete their attacks in 12 hours based on our long time experience and analysis. Because the time interval between the first step of an attack and the last step is usually in a certain time interval, we divide the global attack sequence according to sliding window gap definition. The range of sliding window gap is denoted as the attack time interval. In other words, attacks during one time interval will probably belong to the same attack correlation scenario. We define time interval attack sequences as *Si*. We also divide one day into 2 parts: active-time from 8:00-20:00 and rest-time from 20:00-8:00(the next day).

Table 1. Ordered alerts segment

TIMESTAMP	Alert ID	Signature ID
2006-02-07-09:23:03	1	1
2006-02-07-10:56:12	2	1□ 3□ 2
2006-02-07-13:47:14	3	1□ 2
2006-02-07-15:15:20	4	8
2006-02-07-18:10:11	5	2□ 4
2006-02-07-23:11:35	6	1□ 8
2006-02-08-06:11:15	7	2
2006-02-08-07:12:22	8	3□ 2
2006-02-08-07:13:21	9	1□ 5
2006-02-09-08:43:21	10	5□ 4
2006-02-09-10:54:12	11	1□ 3
2006-02-09-15:11:43	12	8□ 4
2006-02-09-16:35:28	13	2
2006-02-09-17:13:13	14	3
2006-02-10-09:15:21	15	5
2006-02-10-10:16:11	16	7
2006-02-10-15:51:43	17	1□ 4
2006-02-10-16:42:43	18	2
2006-02-10-17:21:11	19	3
2006-02-10-18:18:21	20	2

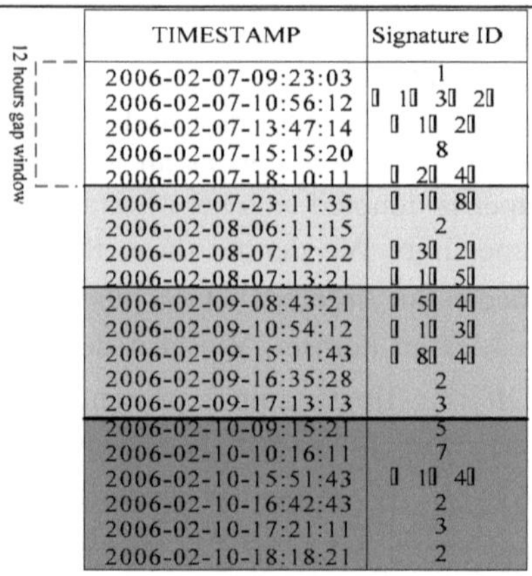

TIMESTAMP	Signature ID
2006-02-07-09:23:03	1
2006-02-07-10:56:12	□ 1□ 3□ 2□
2006-02-07-13:47:14	□ 1□ 2□
2006-02-07-15:15:20	8
2006-02-07-18:10:11	□ 2□ 4□
2006-02-07-23:11:35	□ 1□ 8□
2006-02-08-06:11:15	2
2006-02-08-07:12:22	□ 3□ 2□
2006-02-08-07:13:21	□ 1□ 5□
2006-02-09-08:43:21	□ 5□ 4□
2006-02-09-10:54:12	□ 1□ 3□
2006-02-09-15:11:43	□ 8□ 4□
2006-02-09-16:35:28	2
2006-02-09-17:13:13	3
2006-02-10-09:15:21	5
2006-02-10-10:16:11	7
2006-02-10-15:51:43	□ 1□ 4□
2006-02-10-16:42:43	2
2006-02-10-17:21:11	3
2006-02-10-18:18:21	2

Fig. 2. Transformation time slide window in *hi-level* database for generate sequential parts

Shown as Figure 2, the global attack sequence has been divided by sliding window to generate several sequential parts. We use these sequential parts to represent the attack behaviors occurring in each certain time interval. Each sequential part represents an attack behavior sequence. The problem of mining attack correlation scenario in *hi-level* database transforms to mine the frequent attack sequence from these attack sequential parts in *AB* database.

Terminology from prefixspan[17] The number of instances of items in a sequence is called the **length** of the sequence. A sequence with length l is called an l-sequence. Suppose all the items in an element are listed alphabetically.

Prefix: Given a sequence $\alpha=\{e_1,e_{2,}...e_n\}$, a sequence $\beta=\{e'_1,e'_{2,}...e'_m\}(m\le n)$ is called a prefix of α if and only if (1) $e'_i=e_i$ for $(i\le m-1)$; (2) $e'_m\subseteq e_m$ and all the items in ($e_m-e'_m$) are alphabetically after those in e'_m.

Postfix: $\beta=\{e_1,e_2,...,e_{m-1},e'_m\}(m\le n)$ is α prefix sequence, sequence $\gamma=\{e''_m,e_{m+1},...,e_n\}(m\le n)$ is called postfix of α. Prefix β denoted as $\gamma=\alpha\,|\,\beta$, where $e'_m=(e_m-e'_m)^2$.

4.2 Mining Frequent Attack Sequences

Find length-1 sequential patterns: Scan *AB* database once to find all the frequent items in sequences. Each of these frequent items is a length-1 sequential pattern. One item is frequent if its occurrence in these sequences is greater than a threshold min_frequency. In table 2, the length-1 frequent items are 1, 2, 3, 4, 5, 8.

Search space partition: The complete set of sequential patterns can be partitioned into the following six subsets according to the six prefixes: the ones having prefix 1,2,3,4,5,8. The other parts of the sequence are named postfix subsequence.

Table 2. 1-project sequence pattern

Attack sequence	1-project	2-project	3-project
1,(1,3,2),(1,2),8,(2,4)	(1,3,2),(1,2),8,(2,4)	(1,2),8,(2,4)	(_,2),(1,2),8,(2,4)
(1,8),2,(3,2),(1,5)	(_,8),2,(3,2),(1,5)	(3,2),(1,5)	(_,2),(1,5)
(5,4),(1,3),(8,4)2,3	(_,3),(8,4),2,3	3	(8,4),2,3
5,7,(1,4),2,3,2	(_,4),2,3,2	3,2	2

Table 2 presents the results of the divide sequence patterns by the length-1 1-project sequence pattern.

Find subsets of sequential patterns: After the process of dividing search place, finding all length-2 sequential patterns is our job. We construct corresponding projected databases to mine the subsets of sequential patterns recursively. When getting the length-1 sequential patterns, we can mine the length-2 sequential patterns.

The difference between the real data collected from different sensors to the *AB* database and our examples mentioned above is the amount of real attack behaviors in *AB* database is more than that we mentioned. Notice that the steps of practical attack are mostly more than 3 steps and less than 10 steps, the range of sequential patterns is set as 3 to 10. Differ from the prefixspan, we do not take attention at all sequential pattern which the algorithm could get, but focus on the long length patterns. The projected attack sequential patterns founded in Table 2 are list in Table 4.

Table 3. Attack sequential patterns

prefix	Attack sequential patterns
1	{1,(3,2),1} {1,3,1} {1,3,2} {(1,3),8,2} {1,2,1} {1,2,3} {1,2,2} {1,8,2}
2	NULL
3	{3,8,2}
4	{4,3,2} {4,2,3}
5	{5,1,3} {5,1,2} {5,1,2,3} {5,3,2} {5,2,3} {5,4,2} {5,4,2,3}
8	{8,2,3}

In table 3, we consider that 1-prefix sequential pattern contain some sequences that contain itemset. Take {1,(3,2),1}for example, it means that 2 sensors generate 2 alerts in the same time. To construct attack scenario effectively and accurately, we divide the sequence into 2 sequence: {1,3,1}, {1,2,1}.

SAMP The SAMP (Sequence attack behavior mining based on prefixspan) algorithm is described as follows:

```
Algorithm (SAMP)
Input: a sequence database AB, the min_frequency, window gap
        active-time, window gap rest-time
Output: The frequent attack behavior sequence
Method:
1. Scan the hi-level database, order the attack behaviors by
   time-stamp;
```

```
2. foreach the timestamp, put the attack behaviors itemsets
   among the interval of windows in a sequence labeled by
   sequence id;
3. put attack behavior sequence into AB database;
4. Call PrefixspanSome({},0,S,min,max);
  Subroutine prefixspanSome(b, l, S|b ,min,max)

  Parameters: b: a sequential pattern; l: the length of b;
  S|b: b-projected database, if b≠{}; otherwise, the sequence
database S. max: the maximal length of the patterns; min: the
minimal length of the patterns ,
  Method:
  1. Scan S|b once, find the set of frequent items a such that
     a or {a} can be append to the prefix b to form a
     sequential pattern
  2. Append a to b, to form a new sequential prefix pattern b';
      if (l≠min) JUMP 3;else { scan b'
       if b' contain itemset, output b' and replace the itemset
       to each item in it once;otherwise output b';
  3. Construct prefix b'-projected database S|b'
  4. If l≤max, call prefixSpansom(b',l+1, S|b')
       Otherwise subroutine termination.
```

4.3 Construct Attack Scenario Tree

The process of constructing AS-tree (Attack Scenario tree) can be more effective for recognizing the strategies of attacks. We represent every length-1 frequent item which can be found in attack sequential patterns as a root note. After setting each length-l+1 item as the child of length-l item, the branch of AS-tree is ordered by the frequency of the first embranchment item of each branch of the AS-tree. The character of the AS-tree is the frequency of first embranchment item in right branch is always larger than left, what can make the matching phrase much more effectively.

Algorithm CAST (Construct Attack Scenario Tree) can be described as follows

```
Input: attack sequence patterns
Output: Its sequence and frequent pattern tree, AS-tree
Method: The AS-tree is constructed in the following steps.
1. Scan the attack sequence patterns database once, arrange the
   sequence patterns by the length-1 prefix frequency in order.
2. Create the root of an AS-tree T, and label it as "NULL".
3. Call insert_tree([S/a],T)
4. delete T.

Subroutine Insert_AStree([S/a],T)
Parameters: [S/a]: a is the first element and S is the remaining
              list
1. if T has a child b such that b.name=a.name, then
   b.frequency=a.frequency+b.frequency
2. else create a new node b, make b.frequency=a.frequency, its
   parent link be linked to T as right child of T.
```

```
3. if b have a child c formerly, compare a.frequency and
   c.frequency,
4. if a.frequency ≥ c.frequency exchange a note with b sub-
   branch, jump to step 3,otherwise do not change;
5. if S is nonempty, call insert_tree(S,b)
```

Example: Figure 3 represents the AS-tree attack sequence patterns.

5 Recognize Attack Strategies

After constructing a sequence frequent attack tree from history data, the next step of our work is recognizing attack steps. When AB database receives an alert, we calculate the correlativity and priority to match with certain attack sequence tree easier.

Definition. Cor-degree: $Cor(h_i,h_j)$ $h_i,h_j \in H,(1 \le i,j \le n)$. Alert h_i,h_j is described by p attributes: $x_1,x_2,......x_p$, $y_1,y_2,......y_p$ respectively. The correlativity between h_i,h_j is described as:

$$Cor(h_i,h_j) = \frac{\sum_{i,j=1}^{p} w_{ij} Cor(x_i,y_j)}{\sum_{i,j=1}^{p} w_{ij}}.$$

Definition. pre-degree: $Pre(h_i,h_j)$, $h_i,h_j \in H,(1 \le i,j \le n)$. Alert h_i,h_j is described by p attributes $x_1,x_2,......x_p$, $y_1,y_2,......y_p$ respectively. The correlativity between h_i,h_j is described as:

$$Pre(h_i,h_j) = \frac{\sum_{i,j=1}^{p} k_{ij} Pre(x_i,y_j)}{\sum_{i,j=1}^{p} k_{ij}}$$

The weighted values of w_{ij} and k_{ij} are set empirically and can be tuned in practice. The formulation $Cor(h_i,h_j)$ defines the degree of matching between a new alert and a history alert in a attack scenario and the formulation $Pre(h_{i-1},h_j)$ describe the degree of correlation between the new attack action and the previous attack behavior of the possible-matched alert in the attack scenario. When the value of $Cor(h_i,h_j)$ and $Pre(h_{i-1},h_j)$ are more than certain thresholds, we consider the new attack is belong to a scenario and then predict next possible attack behavior according to the known attack patterns and report the security warning reports to users.

6 Experiments

Experiments 1: To evaluate the effectiveness of our techniques, we conducted the experiments in a branch of CERNET(China Education and Research Network).We ran the SATA system that persisted for 4 weeks to collect history data for SAMP algorithm, and then continue our experiment for 2 weeks to evaluate our method of constructing attack scenarios. During four weeks test, our system received 96300 alerts and then reduced to 2369 after the process of aggregation and verification phase, and last constructed 42 attack scenarios.

The final result of attack scenario construct of alerts in the experiment was presented in Figure4. Newattack number is denoted the amount of the new attacks which IDS generated and the newattack time is denoted three of the latest alerts which has correlation with incidents contained in a attack scenario.

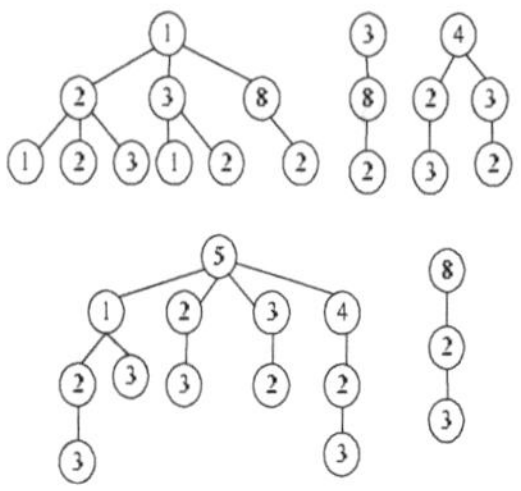

Fig. 3. AS-tree attack sequence

Fig. 4. Attack Scenario correlation interface patterns

Experiment 2: In this section, we simulate several types of practical attacks on attack testing network to test the correlation-ship between attack scenario "BSD 4.2 UNIX mail exploit", "Dos on DNS" with real attack. The attack behaviors with $cor(h_i,h_j)$ and $pre(h_i,h_{i-1})$ more than 0.68 are considered as correlated attack behaviors in the same attack scenario. The valid of our methods can be shown by the results of each attack tests presented in Table 4,5 The $cor(h_i,h_j)$ and $pre(h_i,h_{i-1})$ of each new attack actions and attact scenarios are all up to 0.68. This experiment shows the accuracy and availability of our approach.

Table 4. Attack scenario "BSD 4.2 UNIX mail exploit" and new attacks

BSD 4.2 UNIX mail exploit	Attack actions	$Cor\ (h_i, h_j)$	*pre(hi,hi-1)*
Consist_root_mail	Consist_root_mail	0.84	
Set_root_setuid	Set_root_setuid	0.78	0.76
Touch_file	Touch_file	0.74	0.76
Mail_root	Mail_root	0.78	0.79
Run_root_shell	Run_root_shell	0.68	0.71

Table 5. Attack scenario "DOS on DNS" and new attacks

Dos on DNS	Attack actions	$Cor\ (h_i, h_j)$	*pre(hi,hi-1)*
Lookup_target_DNS	Lookup_target_DNS	0.91	
Ping_DNS	Ping_DNS	0.93	0.92
Nmap_DNS	Nmap_DNS	0.78	0.91
Winnuke_target_run	Winnuke_target_run	0.82	0.78

7 Conclusion and Future Work

In this paper, we proposed the approach we used to design the attack scenario construction module of correlation function within SATA (Security Alert and Threat Analysis) project. Attack scenario construction is conducted based on the Algorithm SAMP and CAST to mine frequent attack sequential patterns from attack sequences database transformed from alerts database and construct attack scenario.

The main points of our novel approach can be mentioned as followed: our approach transforms the alerts database into attack sequences to solve problems of multi-agent and multi-stage attack scenario construction. Our approach gets rid of the predefined knowledge limitation of attack conditions and consequences, which most of approaches must rely on. Our approach can discover new attack relationships as long as the alerts of the attacks have calculable correlation. The results of the experiments conducted on the branch of CERNET can demonstrated the potential of our method in multi-agent attack scenario construction and correlation analysis.

There are several interesting and important future directions. We will continue to study the no-limitation of prerequisites approaches for alert correlation, which is an interesting area to study. To further improve the affectivity and accuracy of the correlation and scenario construction is another important direction of our future work.

References

1. Valdes, A., Skinner, K.: Probabilistic alert correlation. In: Proceedings of the 4th International Symposium on Recent Advances in Intrusion Detection (RAID) (October 2001)
2. Lee, W., Qin, X.: Statistical Causality Analysis of INFOSEC Alert Data. In: RAID (2003)
3. Debar, H., Wespi, A.: Aggregation and correlation of intrusion-detection alerts. In: Recent Advances in Intrusion Detection (2001)
4. Templeton, S.J., Levitt, K.: A requires/provides model for computer attacks. In: Proceedings New Security Paradigm Workshop, Ballycotton, Ireland, vol. 31, ACM, New York (2001)
5. Ning, P., Cui, Y., Reeves, D.S.: Constructing attack scenarios through correlation of intrusion alerts. In: Proceedings of the 9th ACM Conference on Computer and Communications Security, Washington, DC (November 18-22, 2002)
6. Cuppens, F.: Managing alerts in multi-intrusion detection environment. In: Proceedings 17th annual computer security applications conference, New Orleans, pp. 22–31 (2001)
7. Cuppens, F., Miege, A.: Alert correlation in a cooperative intrusion detection framework. In: Proceedings of the 2002 IEEE symposium on security and privacy, p. 202e15 (2002)
8. Pei, J., Han, J., Mortazavi-Asl, B., Pinto, H.: PrefixSpan: Mining Sequential Patterns Efficiently by Prefix-Projected Pattern Growth. In: Proceedings of IEEE Conference on Data Engineering, pp. 215–224 (2001)

A Methodology for Construction Information System for Small Size Organization with Excel/VBA

Hyun Seok Jung and Tae Hoon Kim

Department of Systems and Management Engineering, Dongseo University
San 69-1, Jurye-dong, Sasang-gu, Busan, 617-716, South Korea
hsjung@dongseo.ac.kr

Abstract. In Korea, many small and medium size companies have introduced information system for their effective management. The Korean government have supported them to set this system. But many of these companies failed to utilize this system. One reason is high maintenance and customization cost. In this research, we developed an effective methodology for constructing an information system by themselves. And they can easily renew their system according to the change of their business process. This methology cannot support the internet version, but low cost and easy to maintenance will be the strongest point to set up an early version of information system for small size companies.

Keywords: Information system, Small organization, Excel, VBA.

1 Introduction

According to the development of information technology, many companies have tried to introduce information system to improve the productivity, management quality and market share. The ERP system is mainly introduced. But the owner of a small size company(under 50 employees) has a tendency of hesitating to do so. The main reasons of this hesitation can be summarized as follows. (1) Information system developing cost is usually high for them. (2) There is low reliability of ERP dealers or developers. (3) Annual maintenance fee is about 10% of introduction cost. (4) There is no confidence about the effect of ERP. (5) Small size company suffers the lack of operators. (6) The workers feel a fear to learn a new system.

To cope with this situation, the government has supported them to introducing information system, especially small & medium size company(SMC) in financial and technical aspect. But the governmental efforts are centered to companies that have certain volume. The small size companies that show 99.5% share in total number of Korean companies are suffering many difficult problems. Between 2001 to 2003, "Supporting Project for the 30,000 SMCs to Introduce Information Technology" was carried out by government with 739,000 million Won. As a result, there are several desirable effects such as business process improvement and getting better understanding about information system. But it is said that the effect such as productivity improvement, cost down or management efficiency improvement was insufficient[1]. From the result of a survey research for companies that participated in

M.J. Smith, G. Salvendy (Eds.): Human Interface, Part I, HCII 2007, LNCS 4557, pp. 642–649, 2007.

this government project, the response rate of 'high degree of practical use' was 42.7% and the response rate of 'middle' or 'low' degree of practical use was 33.1% and 22.3%. It is known that small size companies that have under 50 employees cannot use the information system effectively. The reasons of these results are high cost of customization and modification of the information system, lack of excellent users etc.

The target of this research is to show a method for constructing an information system for small size companies that are suffering insufficient fund, lack of manpower. In this research, we use the MS-Excel and Visual Basic for Application tool. These kinds of tools are very familiar to workers of small size companies. It means each company can construct and renew an information system for it with low cost. We applied this method to a small size manufacturer, and we could confirm that this method works well in field.

2 Overview of Information System Application in SMCs

2.1 The Results of Government Project

3 million Korean small and medium size companies have taken a big roll in national economic growth. They shares 86.7% in the number of employee, 42% in the annual amount of export of Korea. But the management circumstances, such as the relationship between maker and vender, domestic market condition, international competition etc, are getting worse.

In case of large companies, the application rate of information system(ERP, SCM, e-business supporting system) are increasing and an amount of investment for information system become larger. But SMCs are in the low level of information system application and also reducing the investment amount for information system. This means that (1) the gap of information system application degree between large companies and SMCs is getting bigger, (2) the cooperation between companies will be getting more difficult[2].

The Korean Government carried out a project that promotes SMCs to introduce an information system for improving its competitiveness(from 2001 to 2003). The supporting fields are (1) pre-consulting for introducing information system, (2) introducing basic information system, (3) introducing ERP(Enterprise Resource Planning), (4) introducing MES(Manufacturing Execution System) and (5) introducing SCM(Supply Chain Management) system. The results are summarized as Table 1. Field (1) : 240 companies, Field (2) : 27,750 companies, Field (3) : 2,592 companies, Field (4) : 39 companies, Field (5) 311 companies.

After this project, a survey for assessing the result and making a basic data for setting a new direction of the supporting project was done. According to the report, (1) the responses to the question of the degree of operation skill for the information system were "high" 42.7%, "middle" 33.1%, "low" 22.3%. This means the degree of application is not so high. In case of small company, this degree will be worse. (2)The degree of PC application is showed in Table 2. Small size company group shows extremely low degree.

Table 1. Result of Government Project to 30,000 SMCs

Applied Field	Pre-consulting	Basic Information System	ERP	MES	SCM	Total
No. of company	240	27,750	2,592	39	311	30,932
Invested Amount (100 million Won)	3.3	270	423	6.7	36	739

Table 2. Degree of PC Application

	No. of Employee	No. of Company	Total	Degree of PC Application (%)
Small	1~49	2,932,789	2,932,786	26.6
Medium	50~99	12,003	2,944,792	99.1
	100~299	6,731	2,951,523	98.0
Large	Over 300	1,601	2,953,124	100

2.2 Suggested Direction for Improvement

90% of participated companies had introduced basic information system. As a result, such companies could take advantage in management aspect, such as management time saving, delivery time keeping etc. However, the policy of supporting the introduction of information system should be changed from quantity-oriented one to quality-oriented one. To help Korea leap to an economically powerful nation, the introduction of information system of small size companies is essential, and to do that, by continuous education, refreshing the mind of owners and workers about information system has to be performed. For those who are well-grounded with the mind, a process of raising the ability to use the basic information system has to be strengthened[3].

This research is for small size companies that are suffering the lack of human and financial resources. The purpose of this research is to suggest a methodology for developing an information system of small size company.

3 Methodology for Developing Information System

Tools that can be used when developing a software are numerous. However, MS Office's Excel VBA(Visual Basic for Application)[4, 5, 6] and Access Data Base are mostly used due to the convenience of development and maintenance. The example company of the development is located in Busan and is a company for assembling the parts of automobiles possessing 32 employees, 4 departments, total annual sales amount of 4500 million Won, 15 kinds of 43 machinery. We developed an inventory management system for this company. The time consumed to develop this system was about 12 days. The developer's knowledge was intermediate level of Excel. Because of not using any other developing tools, we could confirm the convenience of development and maintenance. The program can be developed in a short time period and renewed easily when changing the business process, for VBA is easy to use for any managers.

3.1 The System Structure

In case of introducing the commercial ERP package, a server/client system that needs an expensive DB server is required. This research used a file sharing concept on network. By loading an Access DB in Client PC, there as no need a big DB server. Figure 1 shows the whole hardware structure.

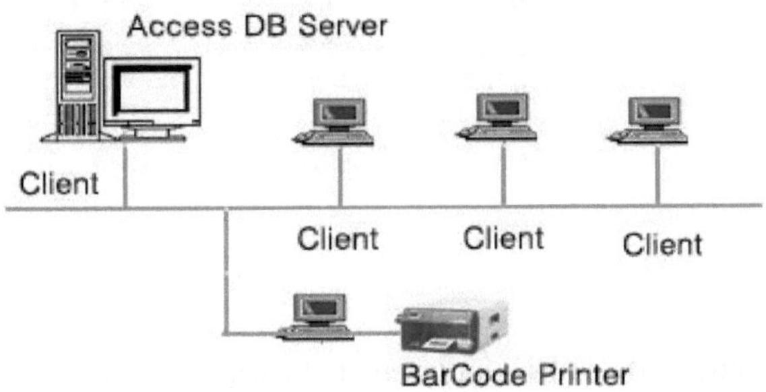

Fig. 1. Hardware Structure

We shortened the time analyzing and designing function model, process model, and data model referring to "The Standard Process Model for Industry[7]" developed by Small and Medium Business Administration. The purpose of this model is to standardize and spread out industry processes of fields with similar production and management in order to propel e-Manufacturing and e-Industry of traditional industry.

(1) Function Model: It is an integrated process for a company to provide products and services with financial resources.
(2) Process Model: A process is a lower part of business function divided into performable unit. Process model is to modeling the processes systematically. This includes the form of input and output for documents or reports needed while performing the business.
(3) Data Model: It is a model systemized in the point of data.

The Figure 2 is the main screen of the inventory management system developed with Excel VBA. Developed system includes several modules. (1) User certification module to make it available for only registered users. (2) User authorizing module is for registering and adjusting users, managing the authorities to use the program. (3) Client management module is concerned about the purchase and sales amount management, outside order management and cooperating company management. In this module, company codes are utilized for the convenience of management. (4) In material management module(Figure 3), the material number consists of client code number and serial number. Selecting a client and clicking the code creation button can generate the material code automatically. Figure 4 also shows an example of automatically generated program code of this screen. This figure shows that users have no need to learn a special program language. Users are just using Excel Macro function.

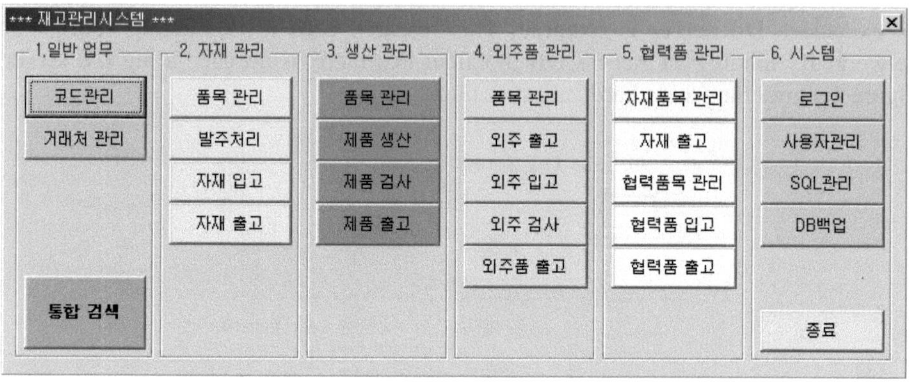

Fig. 2. Main Menu for Inventory Management System

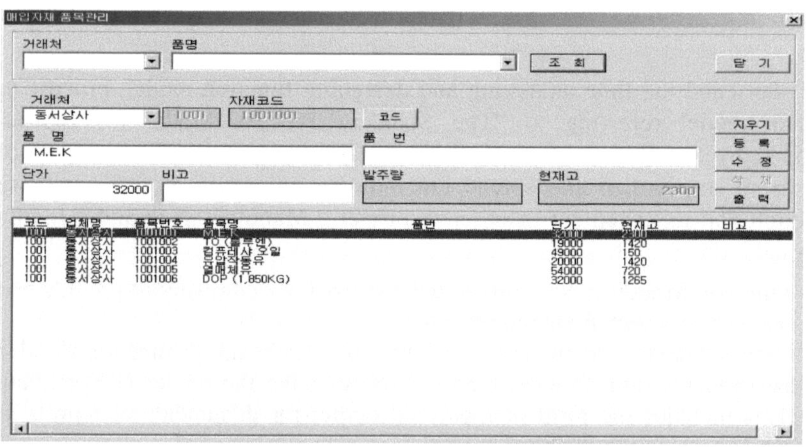

Fig. 3. Material Management Module

(5) Material orders management module processes the order information input and order sheet printing. (6) Warehousing material management module uses the material entry number. This number consists of item number(7 digits) + the date(6 digits) + serial number(2 digits). Totally, 15 digits are used to identify a certain material. It

plays a role of the primary key of entry table. (7) Material handling module controls the issuing of the material from the warehouse. (8) Barcode management module uses the barcode font 39 Code. (9) Order management module controls the outside orders. Figure 5 shows an example of DB connection module to access the DB server.

```
Private Sub btnCode_Click()
Dim MsgStr As String
 If CompCodeText <> "" Then  SQL = "Select max(ItemNo) as a1 from tblInItem " & _
"where ItemNo Like '" & Trim(CompCode.Text) & "*'" Set Rst =
DB.OpenRecordset(SQL)
    If Not Rst.EOF Then
      If IsNull(Rst!a1) = False Then TempItemNo = (Rst!a1) + 1
      Else TempItemNo = CompCode_In.Text & "001"
      End If
      Rst.Close
      txtItemNo.Text = TempItemNo
     End If
 End If
End Sub
```

Fig. 4. Automatically generated Programming codes

```
Public Sub DBConnect()
 Dim strPath As String
 strPath = "z:\DS\DB\Data.MDB"
 If DBState = False Then
   Set DB = DBEngine.Workspaces(0).
               OpenDatabase(strPath)
  DBState = True
 End If
    End Sub
```

Fig. 5. DB Server Access Module

3.2 Checking Points for Developing

Following is the items to check when constructing information system proper to small size companies.

- DB Size : Access DB Max 2G
- End-Users : used by 13-15 persons at the same time
- Safeguarded Users : Backup and user authority management
- Network : Duality of Data Base and Form
- Managing VBA : Simplifying VBA for managing A/S, Development of mutual supplement between employees
- Business Analysis : Maximizing reduced cost by thorough analysis

4 Discussion

4.1 Easiness of Construction

Excel is the most spread software in the whole companies. And Excel has unique language Macro. This software gives many chances to build up various kind of application software easily. On a computer with MS Office, this software can be developed without any extra cost. Even the workers at small size company can construct his management software by himself. The most important strong points of this methodology are (1) low development cost, (2) short development time period, and (3) quick renewal. If workers continuously learn about the Excel and business process remodeling, then he can control this system more effective.

4.2 Easiness of Maintenance

The commercial ERP users do not satisfy with introduced system, even if it is customized to their business process. It is also possible that users suffer the runtime errors. And basically the business processes have to be changed according to the change of circumstances. Or they have to change the system because of the change in a form or a bill etc. These facts are the essential of the information system and management process. With the proposed methodology, users(workers or managers) can renew their system anytime with low cost and easiness.

5 Conclusion

In this paper, we suggested a methodology that can be adopted to manage the small size companies. We developed an information system as an example of a company for confirming the availability of the suggested method. Developed system includes the main module, user certification module, client management module, item management module, order management module, warehousing management module, bill management module, inspection management module. To develop this system, including these modules, we spent 12 days. Developer has studied Excel and VBA for several months. Considering the fact that almost all company workers in Korea are somewhat expert of Excel, they can build this kind of system within the same time period. This means small size companies can be free from the commercial ERP dealer and maintenance cost.

Eventually, small size company can develop a kind of information system by themselves. And they can renew their system with easy if they want. This methodology can not be applied to Web circumstance. But, by using this system, the hesitation about information system could be reduced and the familiarity to it could be stronger. This effect can be said a big progress at the company management.

References

1. Korean Ministry of Commerce, Industry and Energy, The Report for Supporting the 30,000 SMCs to Introduce Information Technology (2004)
2. Ministry of Information and Communication Republic of Korea, The Plan to Promote 1 million SMCs to Introduce Information System, 7 (2004)

3. Korea Technology and Information Promotion Agency for Small and Medium Enterprises, The assessment for the Project for supporting SMCs to Introduce Information System (2006)
4. Byun, J.H.: The Analysis of Business Process with Access VBA (2005)
5. http://www.officeDev.co.kr
6. http://www.officetutor.co.kr
7. Small and Medium Business Administration, The Standard Business Process Model for Industry(Automobile) (2003)

Visual Agent Programming (VAP): An Interactive System to Program Animated Agents

Kamran Khowaja[1] and Sumanta Guha[2]

[1] Department of Computer Science, Isra University, Pakistan
[2] Computer Science & Information Management, Asian Institute of Technology, Thailand
kamran_khowaja@hotmail.com, guha@ait.ac.th

Abstract. An interactive system in which the user can program animated agents visually is introduced: the Visual Agent Programming (VAP) software provides a GUI to program life-like agents. VAP is superior to currently available systems to program such agents in that it has a richer set of features including automatic compilation, generation, commenting and formatting of code, amongst others. Moreover, a rich error feedback system not only helps the expert programmer, but makes the system particularly accessible to the novice user. The VAP software package is available freely online.

1 Introduction

During the last few years animated agents, known also as life-like characters, have become increasingly popular. They emulate humans in application and perform the same tasks that are performed by a human. The objective is to afford the user an experience of human-computer interaction similar to that of human-human interaction. Animated agents are deployed in a variety of applications ranging from E-commerce to electronic learning environments [1], [2], [3], [5], [7], [9], [11]. Fig. 1 shows programmable agents currently available from Microsoft.

Fig. 1. Agents provided by Microsoft

There are a few programming environments – both commercial, as well as freeware – currently available for the scripting of animated agents. A very popular one is the Multimodal Presentation Markup Language (MPML) Visual Editor [6], [10], [12], which allows a content author working with the MPML scripting language to edit a file containing MPML script, as well as to manipulate the graphical representation of the script. However, a significant weakness of this editor is that it provides little by way of feedback or error messages that would enable a user to be able to resolve errors in the MPML script. Neither does it allow access or

M.J. Smith, G. Salvendy (Eds.): Human Interface, Part I, HCII 2007, LNCS 4557, pp. 650–658, 2007.

modification of the finally generated JavaScript code within the editing environment. MPML2.2a [4], DWML [10] and MPML-VR [13] use XSL style sheet to convert MPML script to JavaScript. VHML provides tagging structures for facial and body animation, gesture, emotion, as well as dialogue management [8].

Microsoft Agent Scripting Helper (MASH) provide users an easy to use interface so as they can record and playback their presentations. This is done by dragging & dropping Microsoft Agents on the screen and directing them what to perform next [14]. In addition to the fact that the software is not available for free, the major drawback of MASH is that the generated code is offered for users to edit as well. If user (intentionally or un-intentionally) makes changes in the code there is a possibility that it might generate errors and users are not informed in such situations.

Currently, programming animated agents can be a tedious task, especially for novice-intermediate users who do not have good coding skills. Our goal is to provide an environment that will enable even the inexperienced user to script and deploy agents, thus bringing their utility to a wider audience. This environment, called Visual Agent Programming (VAP), not only provides a graphical user interface (GUI) to program Microsoft Agents, but also enables the user to program these agents manually while simultaneously viewing the underlying script that is generated. An added benefit of VAP's transparency is that a user tends to quickly acquire an understanding of how to use life-like agents to maximum impact in their web application. VAP provides as well a complete and informative debugging environment. VAP statements are automatically compiled when new are added or old are edited, and errors indicated. Microsoft agents currently programmable in VAP are those shown in Fig. 1.

2 VAP Architecture

The architecture of VAP is shown in Fig. 2. There are three main components: the script file, the language settings and the VAP visual editor. The script file contains the script to

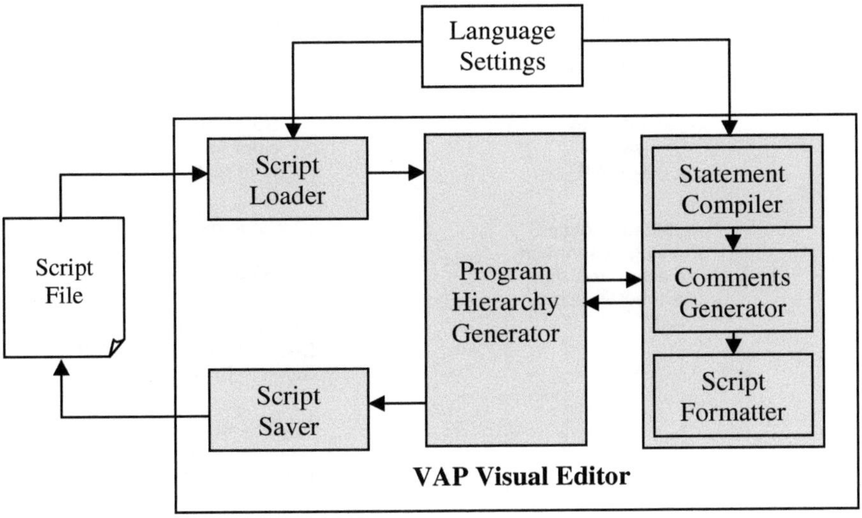

Fig. 2. VAP System Architecture

load characters on a webpage, as well as the actual animation and language-specific script. The language settings contain information related to the languages supported by VAP, their syntax and parameters. The visual editor itself consists of six modules.

- The script loader loads the scripts and passes statements to the program hierarchy generator module.
- The program hierarchy generator creates the tree structure of statements to show parent (function, loop and condition) and child (statements in block) relationship.
- The script saver module saves back the structure created by program hierarchy generator to a script file.
- The statement compiler checks the arguments of each statement and informs the user of errors if any.
- The comments generator module generates comments for each statement.
- The script formatter formats the comments and script generated.

VAP supports both VBScript and JavaScript, unlike most of the competing agent programming environments that support only JavaScript.

3 VAP Features

We describe next the distinguishing features that we believe make VAP a superior choice for programming animated agents. Hopefully, the reader already has had an opportunity to run VAP as described in the previous section. This will make the following easier to appreciate.

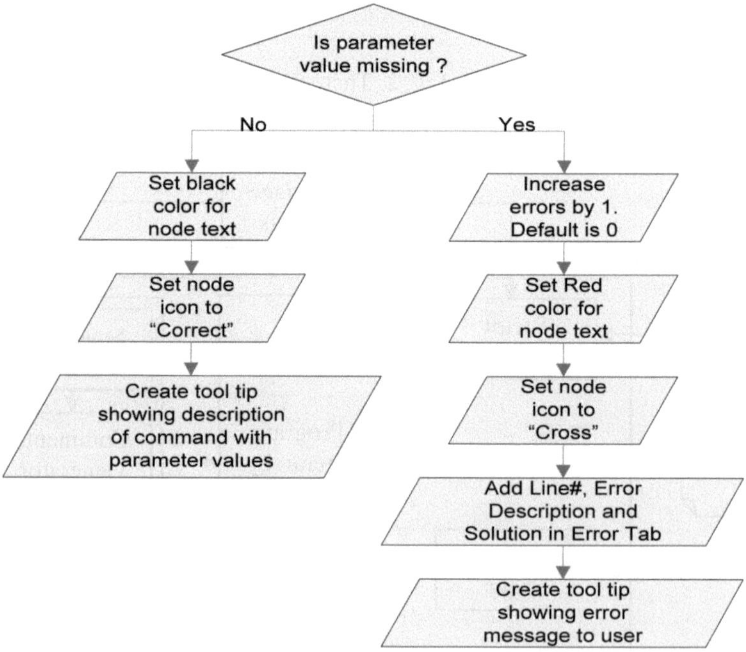

Fig. 3. Flowchart of auto compilation

Stickiness: To make VAP convenient for the novice programmer to use, a feature called stickiness is implemented. Whenever the user either right or left clicks an agent, VAP saves the name of the agent for later use. By default, "Peedy" is the selected agent in VAP. In either case, when the user creates a new command by right clicking on an agent and then selecting command, or right clicking on the program hierarchy and then selecting command, the system will create a new command by automatically adding the name of the agent for whom this command is added. Commands related to a particular agent can be accessed through the agent menu as well.

Auto-compilation: Commands are auto-compiled to show error messages for missing parameters when:

1. A new command is added to the program hierarchy, or
2. Parameters of a command are modified.

Fig. 3 shows the flowchart of automatic compilation. Table 1 shows the commands, their parameters, the types of values expected, and whether the parameter value is optional or not.

Auto-generation: Program code for a command is automatically generated by the system when the user either creates a new command, or modifies an existing one.

Table 1. Command information

Command	Parameter	Data Type	Required/ Optional	Default Value
Speak*	Text	String	Required	Click to modify
Move To*	X	Integer	Required	0
Move To*	Y	Integer	Required	0
Play*	Animation	String	Required	-
Alert	Message	String	Required	Click to modify
Print	Message	String	Required	Click to modify
Loop	From	Integer	Required	0
Loop	To	Integer	Required	0
Loop	Inc/Dec	Integer	Required	0
Function	Name	String	Required	-
Function	Parameters	String	Optional	-
Call	Name	String	Required	-
Call	Parameters	String	Optional	-
Math	Expression	String	Required	-
Condition	Type	String	Required	#

Notes:
- indicates that no default value is provided for this command.
# indicates that the parameter can accept only one of the selected values (If, Else If, or Else).
* indicates that the command is not agent-specific and can be applied only to agents.

Generated script depends very much on the scripting language that is selected. To simplify the interface, script related to loading agents is not shown, as it is not directly relevant to animating the agent. The difference can be seen from fig. 4 and 5, where the former shows only script related to animation, while the latter shows the entire source.

Auto-commenting: The system generates automatic comments for the user and places them before each command and function describing what it will do together with its parameter values. The current version of the system does not allow the user to insert its own comments. Fig. 4 and 5 show example of comments generated by the system.

Auto-formatting: Even if the scripting code in the original file is not properly formatted, the system will re-generate formatted code for the user when the file is saved again. See Fig. 5 and 6, where the latter is unformatted, while the former has been formatted by the system.

Reverse Engineering: This feature of the system allows the user to load scripting code and create program hierarchy. The following steps are performed when the user does reverse engineering.

1. Generate error if the syntax of the command is not correct and set the icon of the node accordingly.
2. Create a tool tip for each command.
3. Create comments for each command.
4. Format comments and commands.

Program code is generated according to the scripting language used. Comments are ignored when reading a file, but generated and inserted when the file is saved back.

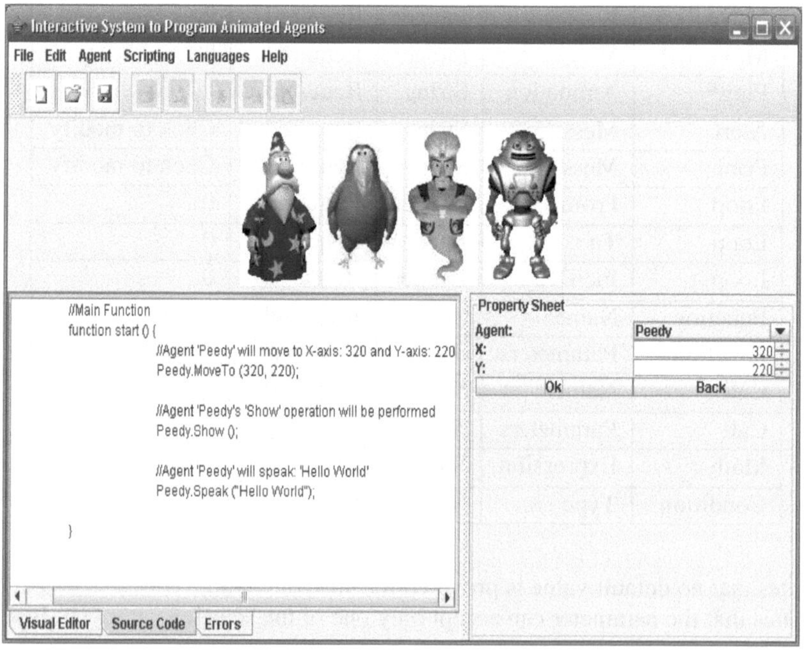

Fig. 4. Partial source code of Hello World

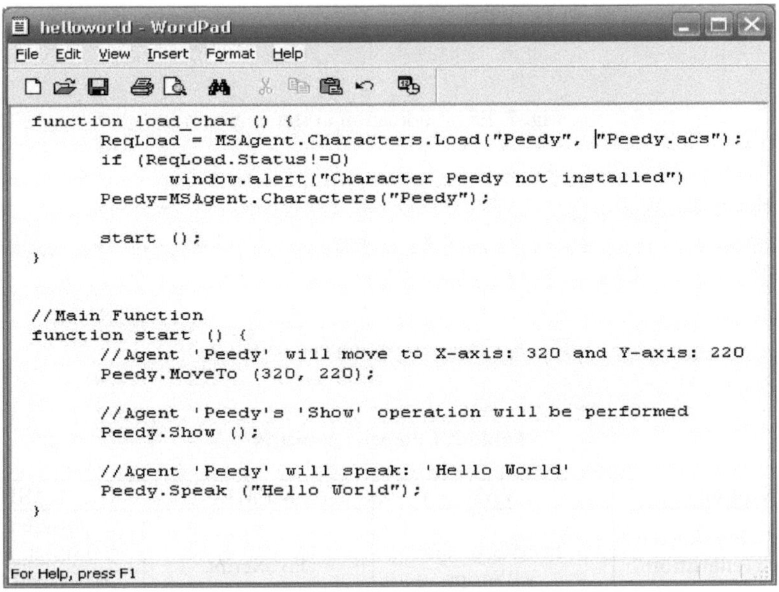

Fig. 5. Complete source code of Hello World

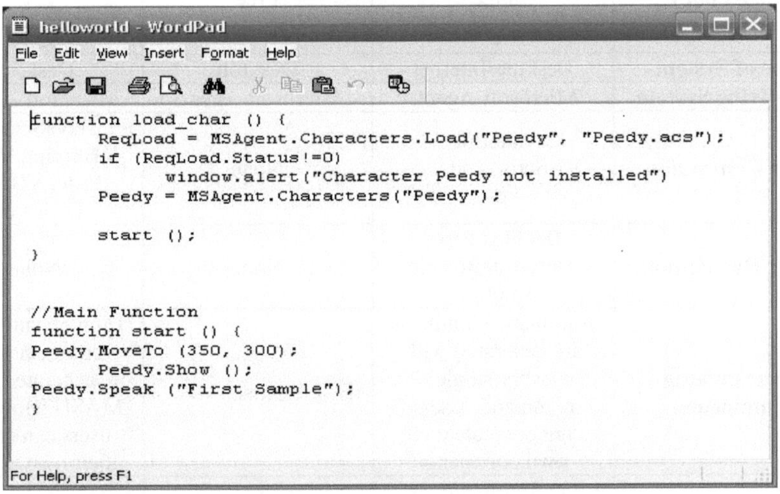

Fig. 6. Source code of Hello World before using the system

Error Message: VAP provides a rich error feedback environment. Fig. 7 shows an example in which an error in a statement is highlighted in red and the icon of the statement turned to cross indicated error. Fig. 8 shows the system telling the user what the error is, where and why it has occurred, and how to resolve it. Consequently, VAP requires little programming experience as the user is constantly guided by error messages.

Fig. 7. Error indication in statement

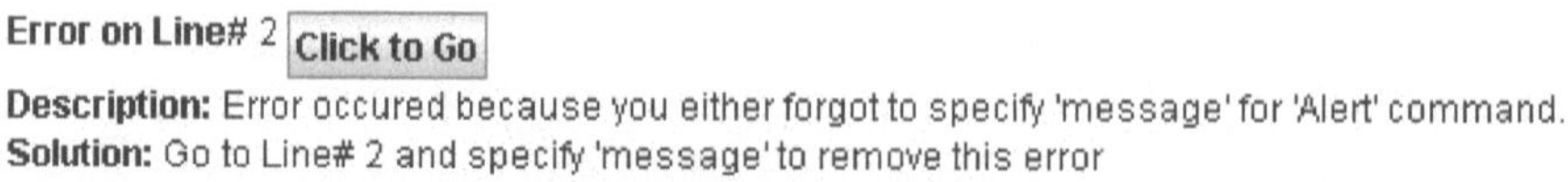

Fig. 8. Error description and solution

Table 2. Comparison matrix

Option	VAP	MPML	MASH
Underlying Source Code Generation Language	JavaScript and VBScript	JavaScript	JavaScript, VBScript, Visual Basic, VBA for Office etc
Usage Policy	Free	Free	Free: 30 days Single User: $25 Educational Site: $250
Type of System	Desktop/Internet	Desktop	Desktop
Underlying System	Microsoft Agents	Microsoft Agents	Microsoft Agents
Code Conversion	JavaScript to VBScript and vice versa	None	JavaScript, VBScript, Visual Basic, VBA for Office etc
Error Description	Detailed Error Description with solution	None	None
Programming Comments	Automatic comments are generated with every single command. Users cannot create their own comments	None	Limited automatic comments are generated by MASH. However, users can create their own as well
Compilation	Automatic compilation of statement when added/edited	Program structure is compiled when saved	None
Character Type	Full Body	Full body	Full body
Audience	Novice – Intermediate	Novice – Intermediate	Novice – Advanced

4 VAP Comparisons

Table 2 shows a comparison between VAP and two currently popular systems. Significant differences can be seen in features such as code conversion, error debugging, automatic compilation, comments generation, etc.

5 Conclusions and Future Work

We believe that the VAP software package, as well as its underlying ideas, will help popularize the deployment of life-like agents in web applications by providing an easy-to-use GUI for their scripting and animation. Planned extensions to VAP include the following:

1. Development of more sophisticated interaction between user and system.
2. Support for concurrent execution of statements.
3. Tagging of statement to overcome problems of absolute locations.
4. Equipping agents with a human voice.
5. Enable speech recognition by agents.

References

1. Andre, E., Rist, T., Mulken, S., van Klesen, M., Baldes, S.: The automated design of believable dialogue for animated presentation teams. In: Cassell, J., Sullivan, J., Prevost, S., Churchill, E. (eds.) Embodied Conversational Agents, pp. 220–255. The MIT Press, Cambridge (2000)
2. Badler, N.I., Allbeck, J., Bindiganavale, R., Schuler, W., Zhao, L., Palmer, M.: Parameterized action representation for virtual human agents. In: Cassell, J., Sullivan, J., Prevost, S., Churchill, E. (eds.) Embodied Conversational Agents, pp. 256–284. The MIT Press, Cambridge (2000)
3. Badler, N.I., Bindiganavale, R., Allbeck, J., Schuler, W., Zhao, L., Palmer, M.: Parameterized action representation for virtual human agents. In: Embodied Conversational Agents, pp. 256–284. The MIT Press, Cambridge, MA (2000)
4. Du, P., Ishizuka, M.: Dynamic web markup language (DWML) for generating animated web pages with character agent and time-control function. In: Proceedings (CD-ROM) IEEE International Conference on Multimedia and Expo (ICME2001) (2001)
5. Huang, Z., Eliens, A., Visser, C.: STEP: a scripting language for embodied agents. In: Prendinger, H. (ed.) Proceedings PRICAI-02 International Workshop on Lifelike Animated Agents. Tools, Affective Functions, and Applications, pp. 46–51 (2002)
6. Ishizuka, M., Tsutsui, T., Saeyor, S., Dohi, H., Zong, Y., Prendinger, H.: MPML: A Multimodal Presentation Markup Language with Character Agent Control Functions (2000)
7. Kitamura, Y., Tsujimoto, H., Yamada, T., Yamamoto, T.: Multiple character-agents interface: An information integration platform where multiple agents and human user collaborate. In: Proceedings First International Joint Conference on Autonomous Agents and Multiagent Systems (AAMAS-02), pp. 790–791. ACM Press, New York (2002)
8. Marriott, A., Stallo, J.: VHML – Uncertainties and problems, A discussion, in: proceedings AAMAD02 Workshop on Embodied Conversational Agents – Let's Specify and Evaluate Them (2002)

9. Marsella, S.C., Johnson, W.L., LaBore, C.: Interactive pedagogical drama. In: Proceedings 4th International Conference on Autonomous Agents (Agents-2000), pp. 301–308. ACM Press, New York (2000)
10. Okazaki, N., Aya, S., Saeyor, S., Ishizuka, M.: A multimodal presentation markup language MPMLVR for a 3D virtual space. In: Proceedings (CD-ROM) of Workshop on Virtual Conversational Characters: Applications, Methods, and Research Challenges (in conj. with HF2002 and OZCHI2002) (2002)
11. Predinger, H., Descamps, S., Ishizuka, M.: Scripting Affective Communication with Life-Like Characters in Web-based Interactions Systems. Applied Artificial Intelligence 16(7-8), 519–553 (2002)
12. Predinger, H., Descamps, S., Ishizuka, M. (eds.): MPML: a markup language for controlling the behavior of life-like characters, Journal of Visual Languages and Computing, January 2004. Elsevier, New York (2004)
13. Saeyor, S.: Multimodal Presentation Markup Language Ver. 2.2a (MPML2.2a), (2003) URL: http://www.miv.t.u-tokyo.ac.jp/~santi/research/mpml2a
14. MASH: http://www.bellcraft.com/
15. VAP: http://www.cs.ait.ac.th/~b101650

The Implementation of Adaptive User Interface Migration Based on Ubiquitous Mobile Agents

Gu Su Kim, Hyun-jin Cho, and Young Ik Eom

School of Information and Communication Engineering, Sungkyunkwan Univ.,
Chunchun-dong 300 Jangan-gu, Suwon, Gyeonggi-do, Korea
{gusukim,hjcho,yieom}@ece.skku.ac.kr

Abstract. MA(Mobile Agent) is active, autonomous, and self-replicable software object containing both computational logic and state information. One advantage of using MA paradigm instead of conventional message passing paradigm is that it can reduce communication traffic among the computing devices in the system. Also, the MA paradigm supports asynchronous interaction of computing devices, enabling more efficient system services. In order to adopt the MA paradigm into the ubiquitous computing technologies, it is necessary to develop a lightweight middleware platform, called MAP(Mobile Agent Platform), that supports agent migration and install it in various devices such as PDAs, hand-held devices, and digital appliances. In this paper, we propose our lightweight MAP, named *KAgentPlatform*, that is developed based on J2ME for ubiquitous environments. Especially, we describe our design and implementation of the *KAgentPlatform*, and show the experiments of adaptive UI migration based on ubiquitous mobile agents.

Keywords: UI migration, mobile agent, ubiquitous, lightweight mobile agent platform.

1 Introduction

Ubiquitous computing and mobile computing are key areas in future computing environments [1]. The concept of ubiquitous computing implies computation in elements that are contained in the environment. However, since these devices have very small displays and limited resources, users cannot deal with large amounts of information with the devices. Users might handle various information by these ubiquitous devices with very small display and limited resource. However, these devices could not provide interface suitable for the user's current environment due to limited device capacities. The migration facilities of user interface based on mobile agent can overcome these limitations.

MA(Mobile Agent) is active and autonomous software object containing both computational logic and state information. An MA running on one host can migrate to another host with its state information and continue execution on that host. The advantages of using MA include low network traffic, asynchronous parallel execution, simple implementation, simple deployment, high reliability, and so on [2].

The adoption of MA paradigm into the ubiquitous computing technologies makes it possible to provide many interesting services such as follow-me service and user

M.J. Smith, G. Salvendy (Eds.): Human Interface, Part I, HCII 2007, LNCS 4557, pp. 659–668, 2007.

preference service to the users by providing migration facilities of adaptive user interfaces [3]. But, in order to adopt the MA paradigm into the ubiquitous computing technologies, it is necessary to develop a lightweight middleware platform, called MAP(Mobile Agent Platform), that supports agent migration and install it in various devices such as PDAs, hand-held devices, and digital appliances.

Due to platform independence of the Java technologies, several MAPs have been implemented with Java technologies. However, most of these platforms were developed for general PC environments on which J2SE(Java2 Standard Edition) can be ported. The development of MAPs based on J2ME(Java2 Micro Edition) for small embedded devices is in its early stage. In this paper, we propose our lightweight MAP, named *KAgentSystem*, that is developed based on J2ME for ubiquitous environments. Especially, we describe our design and implementation of the *KAgentSystem*, and show the experiments of adaptive UI migration based on ubiquitous mobile agents: schedule search agent, mobile MP3 player agent and art gallery guide agent.

The rest of the paper is organized as follows. In Section 2, we describe related work on the Java technology and existing lightweight MAPs. Section 3 describes the architecture and functions of our platform, *KAgentPlatform*. Section 4 shows the some snapshots of the experimental application of *KAgentPlatform*. Finally, Section 5 concludes with a summary.

2 Java and Lightweight Mobile Agent Platform

Java is the most popular development language for MAPs. Until now, several MAPs have been implemented with Java technologies. Due to the property of platform independence of the Java, the MAPs can be ported on various devices without any modification. Sun provides three kinds of Java platforms: J2EE(Java 2 Enterprise Edition), J2SE(Java2 Standard Edition), and J2ME(Java2 Micro Edition). Among the three platforms of Java, J2ME is the Java platform for consumer and embedded devices. J2ME configurations can be classified into CLDC(Connection Limited Device Configuration) and CDC(Connected Device Configuration) according to the resource availability of the device. CLDC is designed for devices with intermittent network connections, slow processor, and limited memory, such as mobile phones, two-way pagers, and smart phones. CDC is designed for devices that have more memory, faster processors, and greater network bandwidth [4,5]. J2ME CLDC doesn't support the essential facilities for the mobility of the MAs, such as reflection, dynamic class loading, and object serialization. So, the existing lightweight MAPs have been implemented on J2ME CDC platforms.

Until now, most of existing MAPs have been developed for desktop-level computer systems and the development of lightweight MAPs for ubiquitous devices is in its early stage. Examples of lightweight MAPs that are experimentally developed for small devices such as PDAs, smart phones, and digital appliances are MAE [6], UbiMAS [7], mP@gent [8], FlipCast [9] and so on.

MAE(Mobile Agent Environment for Resource Limited Devices) [6] is developed at Monash University in Australia and is designed for small wireless devices such as

PDAs or cell phones. MAE is developed to provide wireless agent based applications such as mobile shopping, mobile auction, and so on.

UbiMAS (Ubiquitous Mobile Agent System) [7], developed at Augsburg University in Germany, is a mobile agent system which runs as a service on top of the ubiquitous middleware. The agent system is based on the ubiquitous middleware which uses a JXTA peer-to-peer network as communication infrastructure. In UbiMAS application, mobile agent moves with the user and present location information through embedded devices close to the user.

m-P@gent [8] is a framework developed to run environment-aware applications and to personalize content on resource limited mobile devices. The m-P@gent is developed at Electro-Communication University in Japan. The m-P@gent shows that it is possible to expand the field of ubiquitous computing to include devices that have limited resources. In addition, the framework provides more efficiency in regards to agent migration by providing, more cooperative resource control mechanisms.

Flipcast [9] is developed at Toshiba Corp. in Japan. Flipcast modules can be built in communication devices such as home robots, appliances, and cellular phones. Flipcast consists of platform and scripts. Flipcast platform is the base software which runs on each device and the script consists of operations list to run and acts as a mobile agent.

3 Architecture of KAgentPlatform

In this Section, we introduce the architecture of *KAgentPlatform*, which is a kind of lightweight MAP that can be operated on small ubiquitous devices such as PDAs, hand-held devices, and digital appliances.

3.1 KAgentPlatform Overview

Our *KAgentPlatform* is implemented with J2ME CDC(Connected Device Configuration) / Personal Profile 1.0. Figure 1 shows the architecture of *KAgentPlatform.*

Agent Platform is the key component in *KAgentPlatform.* It consists of 11 sub-components: *KPlatform*, *Information Manager*, *Migration/Immigration Manager*, *Resource Sharer*, *Security Facility*, *Directory Facility*, *Platform Request Handler*, *Communication Manager*, *Platform Plug-in*, *Platform Service*, and *Extended Platform Service*. These components manage the life cycle of MAs and provide the MAs with various service APIs such as migration, authentication, discovery of other MAs, and so on.

In *KAgentPlatform*, three kinds of agents are able to run: *Monitor Agent*, *UI(User Interface) Agent*, and *User Agents*. *Monitor Agent* is a static agent that monitors current states of the *Agent Platform* and *User Agents*. *UI Agent* is also a static agent that provides interface between the users and the platform. The user can create and execute MAs through the interface provided by *UI Agent* and also to control the MAs executing on the *Agent Platform*. *User Agent* means the MA instantiated by user. The *User Agent* can be considered as a Java application program that can migrate from one platform to another in order to perform a job instead of the user. Each *User Agent* has a globally unique ID.

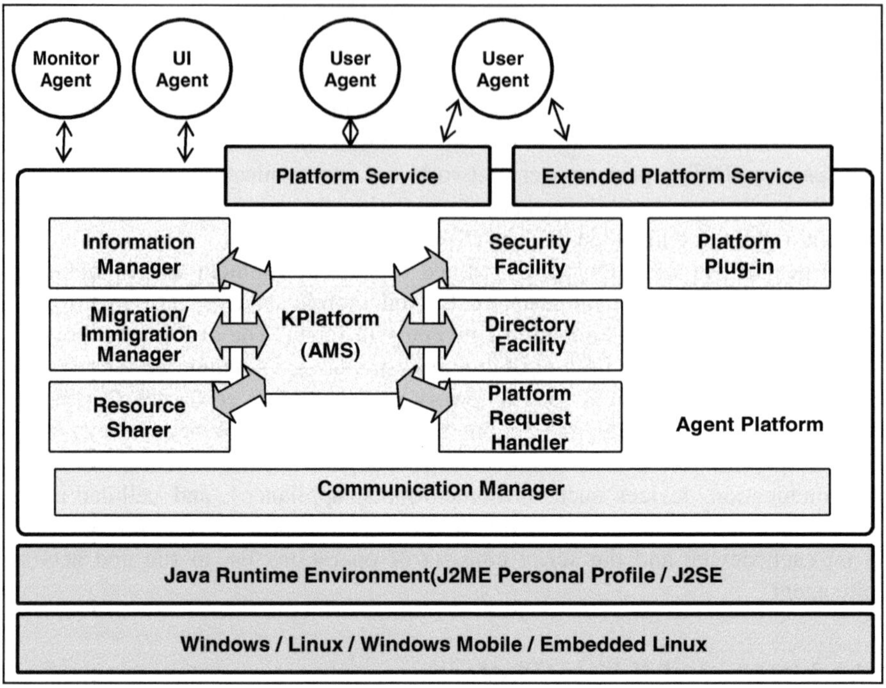

Fig. 1. The architecture of *KAgentPlatform*

3.2 Agent Platform

Agent Platform acts as the middleware for supporting MAs. In this subsection, we describe the main components of the *Agent Platform*.

KPlatform acts as AMS(Agent Management System) and manages the life cycle of the MAs, where it supports creation, cloning, migration, activation, deactivation, and termination of the agents. When a user selects an MA to execute through the *UI Agent*, the pathname of the MA in the file system is passed to the *KPlatform* and it creates a new instance of the MA and registers the MA information into the *Information Manager*. The *Information Manager* manages the information on MAs currently executing on the platform using Java hash table. Our MAs are designed to operate according to the event driven approach. For this approach, *KPlatform* raises events to MA at the specific time associated with the MA's life cycle and invokes the event handler of the MA related to the event.

Migration/Immigration Manager are in charge of MA's migration from one platform to another. When an MA requests migration to another platform, the *Migration Manager* sends the MA's class codes, status data, and other necessary information to the destination platform. At this time, the agent sender converts the MA's class codes into a compressed JAR(Java Archive) file before sending it to the destination. After the destination platform receives the migration request, the *Immigration Manager* receives the MA's class codes, status data, and other necessary information, and registers the MA's class codes into the platform, and recovers the MA for execution. Also, the MA class codes are cached in the destination platform for preventing unnecessary transmission afterwards. The recovered MA is registered into the *Information Manager* of the destination platform and restarts its execution. Figure 2 illustrates the MA migration process.

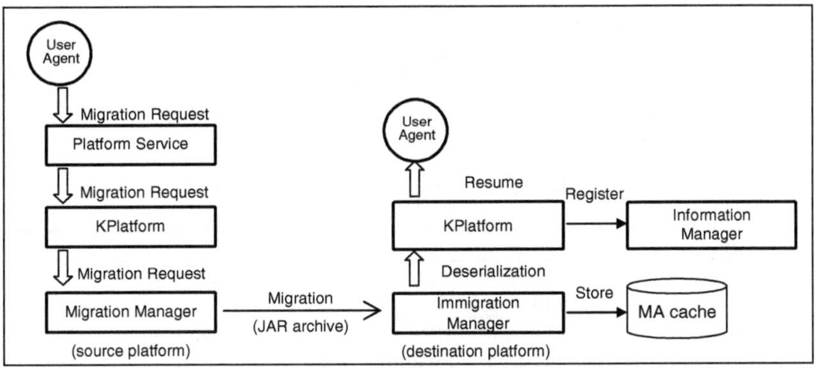

Fig. 2. The migration process of a mobile agent

Security Facility performs the authentication process when a MA enters the platform. We classified the MA authentication process into single-domain authentication and multi-domain authentication. In this paper, we defined a domain as a system area in which each MA can be authenticated by the shared key, which is called single-domain authentication. On the other hand, when an MA comes from other domain, *Security Facility* authenticates the MA by multi-domain authentication process based on the domain's public key. Figure 3 shows the authentication process of *Security Facility*. For the authentication of MA, each MA keeps its own credential, where the credential includes three kinds of authenticator: global authenticator, local authenticator, and domain membership authenticator. Table 1 explains the three authenticators.

Platform Request Handler processes requests such as immigration request, message delivery request, resource sharing request, and so on, that are sent by remote MA. *Platform Request Handler* analyzes the request and invokes the API related for servicing the request.

Communication Manager supports inter-platform and inter-agent communication. It is started as a server daemon thread at the platform initialization stage. It has a thread pool for efficient processing of messages received from remote platforms. Figure 4 shows the architecture of the *Communication Manager*.

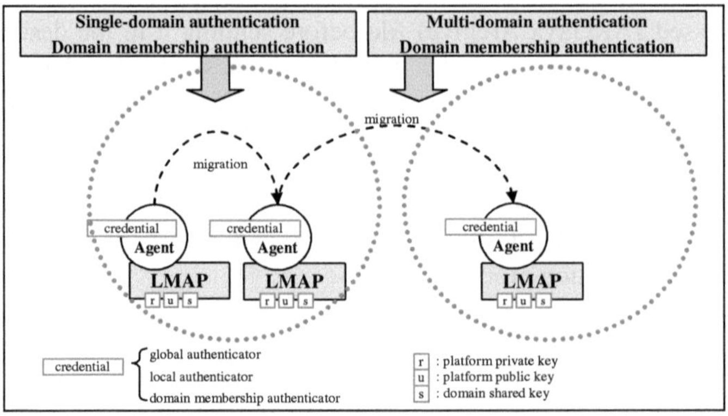

Fig. 3. The authentication process of Security Facility

Table 1. Authenticators in the credential

Authenticator	Description
Global authenticator	- An authenticator for multi-domain authentication - Digitally signed by home agent platform
Local authenticator	- An authenticator for single-domain authentication - Temporary used within a single domain
Domain membership authenticator	- An authenticator that is used for proving the MA's membership of a domain.

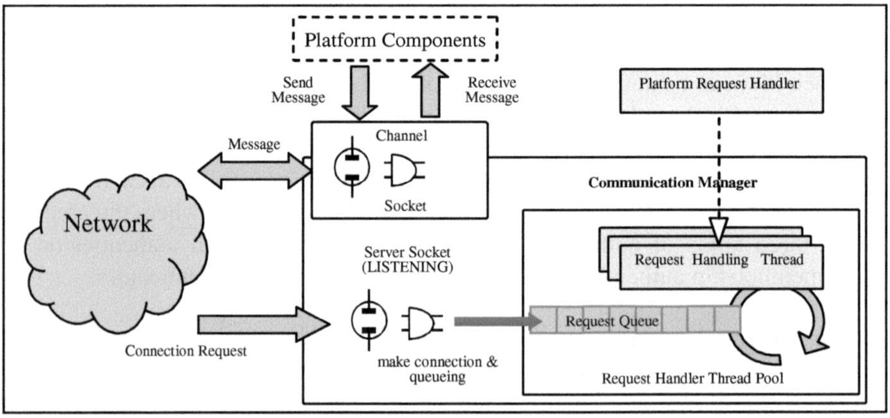

Fig. 4. The architecture of the Communication Manager

Platform Plug-in allows adding new services to the *Agent Platform*. MA developers can use service interface of the Platform Plug-in in order to provide new services to the MA.

4 Experiment

In this Section, we describe our experimental results. We made three experiments: schedule search agent, mobile MP3 player agent as a follow-me service, and art gallery guide agent. Devices used for these experiments are as follows:

- Samsung MITZ PDA, equipped with WinCE and IBM J9
- HP iPAQ PDA, equipped with WinCE and IBM J9
- Sharp Zaurus PDA, equipped with Embedded Linux and IBM J9
- Samsung Web Pad, equipped with Windows XP and Sun J2SE
- Doshiba Notebook, equipped with Windows XP and Sun J2SE

4.1 Schedule Search Agent

In this scenario, we assume that each user has his schedule on his PDA which is equipped with *KAgentPlatform*. When a user wants to make a meeting schedule with his colleagues, and he migrates his ScheduleSearchAgent to his colleagues' PDA to get the meeting time. ScheduleSearchAgent travels and searches the empty time slots in each colleague's schedule, finds the common empty time slot, and reserves the time for meeting. Figure 5 shows the snapshot of ScheduleSearchAgent application.

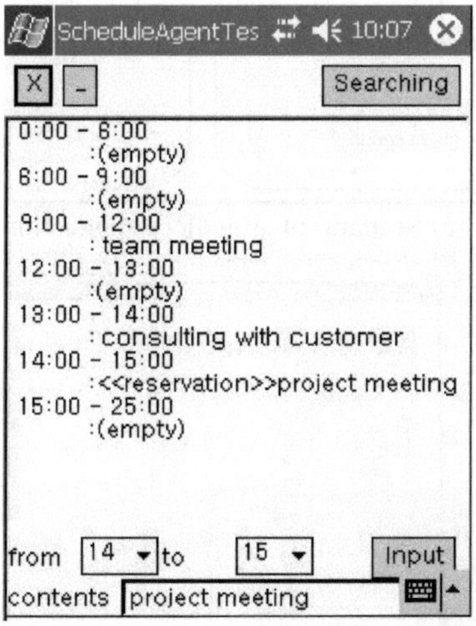

Fig. 5. The snapshot of ScheduleSearchAgent application

4.2 Mobile MP3 Player Agent

In this scenario, an MA plays music while following the user. First, the user executes an MP3 player agent running on his PDA. The MP3 player agent starts playing music and, as the user moves to the living room, it migrates to the audio in the living room. When the user moves to the bed room, the MP3 player agent again finds another audio device near the user and continues playing the music at the new audio device. Figure 6 shows the snapshot of the MP3 player agent application.

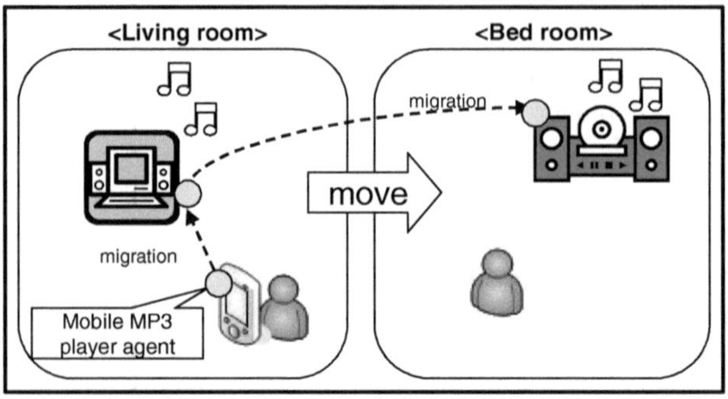

(a) Scenario of playing music

(b) Snapshot of the MP3 player agent

Fig. 6. The mobile MP3 player agent application

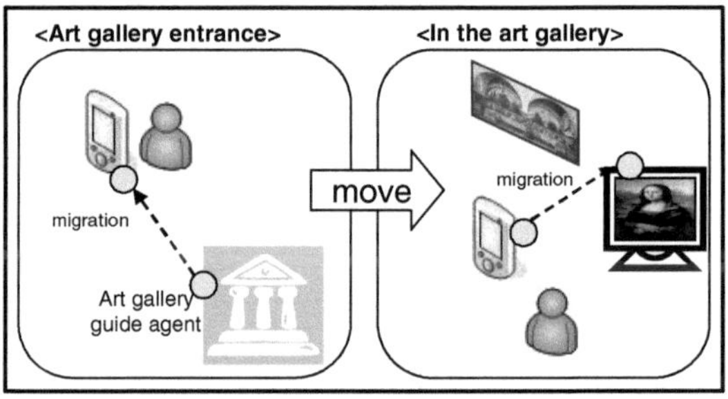

(a) Scenario of art gallery application

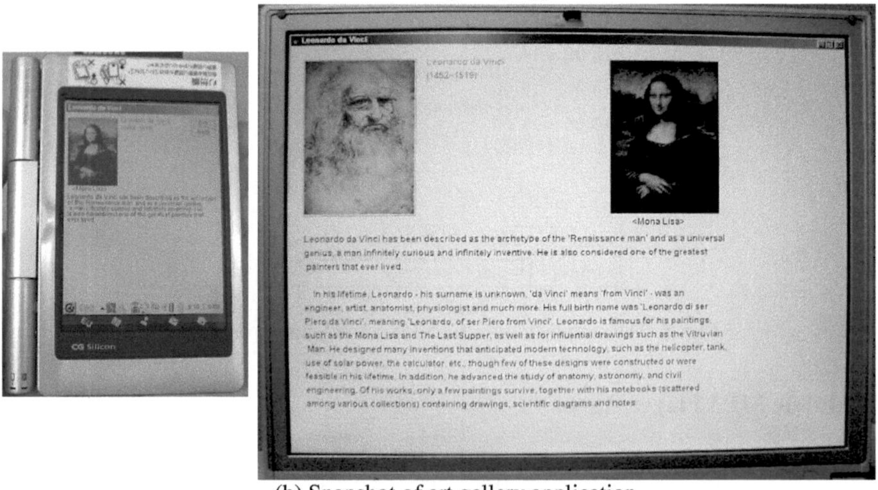

(b) Snapshot of art gallery application

Fig. 7. The art gallery guide agent application

4.3 Art Gallery Guide Agent

In this scenario, an MA plays music while following the user. First, the user executes an MP3 player agent running on his PDA. The MP3 player agent starts playing music and, as the user moves to the living room, it migrates to the audio in the living room. When the user moves to the bed room, the MP3 player agent again finds another audio device near the user and continues playing the music at the new audio device. Figure 7 shows the snapshot of the MP3 player agent application.

5 Conclusion

The MA technology is very useful in distributed and/or ubiquitous environments. But, in order to apply the MA concept to the ubiquitous environments, the MAP should be lightweight and should be designed with the consideration on the characteristics of the ubiquitous devices. In this paper, we presented the architecture of *KAgentPlatform* and described the experimental results of adaptive UI migration based on ubiquitous mobile agents. The *KAgentPlatform* is a Java-based lightweight MAP that runs on PDAs, hand-held devices, and digital appliances equipped with J2ME. The *KAgentPlatform* can provide several user friendly services such as follow-me services and user-preference services, through the adaptive UI migration based on ubiquitous mobile agents over various ubiquitous devices. Also, by using *KAgentPlatform*, we can gain several advantages such as network traffic reduction, asynchronous interaction, and so on.

Acknowledgments. This research was supported by the MIC(Ministry of Information and Communication), Korea, under the ITRC(Information Technology Research Center) support program supervised by the IITA(Institute of Information Technology Assessment), IITA-2006-(C1090-0603-0046).

References

1. Cho, K., Hayashi, H., Hattori, M., Ohsuga, A., Honiden, S.: picoPlangent: An Intelligent Mobile Agent System for Ubiquitous Computing. In: Barley, M.W., Kasabov, N. (eds.) PRIMA 2004. LNCS (LNAI), vol. 3371, Springer, Heidelberg (2005)
2. Aneiba, A., Rees, S.J.: Mobile Agents Technology and Mobility. In: Proc. of the 5th Annual Postgraduate Symposium on the Convergence of Telecommunications, Networking, and Broadcasting (June 2004)
3. Takashio, K., Soeda, G., Tokuda, H.: A Mobile Agent Framework for Follow-Me Applications in Ubiquitous Computing Environment. In: Proc. of 21st International Conference on Distributed Computing System Workshop (April 2001)
4. Ledoux, T., Bouraqadi-Saadani, N.: Adaptability in Mobile Agent Systems using Reflection. In: Proc. of the Workshop on Reflective Middleware (RM2000), New York (April 2000)
5. Mahmoud, Q.H.: Understanding Network Class Loaders, Developer Technical Articles & Tips (October 2004)

6. Mihailescu, P., Binder, W., Kendall, E.: MAE: Mobile Agent Environment for Resource Limited Devices. In: Magnusson, B. (ed.) ECOOP 2002. LNCS, vol. 2374, Springer, Heidelberg (2002)
7. Bagci, F., Petzold, J., Trumler, W., Ungerer, T.: Ubiquitous Mobile Agent System in a P2P-Network. Proc. of the UbiSys-Workshop at the Fifth Annual Conference on Ubiquitous Computing (October 2003)
8. Takashio, K., Mori, M., Tokuda, H.: m-P@gent: A Framework of Environment-Aware Mobile Applications. Proc. of IEEE International Workshop on Networked Appliances (2002)
9. Ueno, K., Kawamura, T., Hasegawa, T., Ohsuga, A., Doi, M.: Cooperation between Robots and Ubiquitous Devices with Network, Script Flipcast. Proc. of Network Robot Systems integrated with environments (IROS 2004 Workshop) (2004)

Construction of Web Application for Cusp Surface Analysis

Yasufumi Kume and Zaw Aung Htwe Maung

Department of Mechanical Engineering, Kinki University,
3-4-1 Kowakae Higashiosaka Osaka, 577-8502, Japan
kume@mech.kindai.ac.jp

Abstract. This paper describes about construction of Web application for cusp surface analysis. Client accesses to the Web server in order to analyze the data for cusp surface analysis, and a user's browser downloads Web page for data input automatically. Client can be offered through Web browsers by using HTML generated dynamically by Java Server Pages (JSP) technology, or Java applets. Tomcat is operated as a plug-in of Apache to start the Java servlet on the server. Tomcat is the servlet container that is used to carry out cusp for Java. Cusp for Java is applied as Java servlet. It is a java class that extends a J2EE-compatible Web server. Cusp for Java receives HTTP (Hypertext Transfer Protocol) request from the browser, and provides views of result data as HTML.

Keywords: web application, cusp surface analysis, server and client, Apache, Tomcat.

1 Introduction

Conventional technological calculation needs complicate calculation ability and calculation time. Standalone computer performs as calculation machine. In this system, it is difficult to investigate usability of developed calculation program. It is impossible to construct the system installed this calculation program by Rotus Notes (group wear). Internet is used to various information system, it is much proper to the general public communication. The system that developed analytical program is exhibited to by web sever on www (world wide web) is investigated. It is important to exhibit the security of system safety. Five seminars were held by participation of advanced system engineer. And Linux as Web server software and Apache as server software and Tomcat is connected with Java servlet container. Therefore, analytic program is used as Java serve let. Java servelet is server side program performed and processed in the servelet container (Tomcat etc.) for input from client. By applying to servelet, it is can be extended the function of web server. Using this servelet, technical calculation system is developed by server and client method that server processes input from client and the result is sent client. Also, technical calculation system connecting Linux, Apache, Tomcat and Java is almost not found. The tool that analyzed unconscious phenomenon included human factors as creative process is proposed using stochastic catastrophe model [1], [2]. This cusp surface analysis is coded by Java language. Cusp for JAVA file formatted by jar file is used as Java

M.J. Smith, G. Salvendy (Eds.): Human Interface, Part I, HCII 2007, LNCS 4557, pp. 669–676, 2007.

application. This paper describes that Cusp for JAVA is safe and smooth behaved on web server in order to make the environment in which more researchers use this Cusp for JAVA.

2 Environment of Web Server

This section describes on the environment of structured web server and the environment for web application carried into practice.

2.1 Linux [3]

Linux is UNIX compatible OS developed based on operating system developed by Linux Torvalds of Graduate school student in Helsinki University on 1991. After that, it is opened as open sauce soft ware and revised by developer of volunteer in the world. Linux is not appropriated by existent OS code and is rewritten quite newly. This soft ware is not only free but also liberty. It is revised and distributed based on license system called GPL. Linux behaves even low performance computer lively comparing with other OS (operating system). Also, it is excellent for network function and security and very stable. Linux is distributed academic organization widely and it is adopted individual Internet server as well as enterprise server. There are three systems, that is, Sack ware system, Red Hat system, Debian system.

2.2 Web server

Www system consists of computer sending information and software having function of sending information. Information of HTML documentation and image are stored on Web server and these information is send by means of network of internet etc. responding to needs of client software that is web browser etc. It run a program responds to requirement and has the function of dynamic page generation by which result is sent to client. Java servelet used Java language and web site increase using JSP, ASP of Microsoft.

2.3 Apache [4]

Apache is web server, which is starting to develop on 1995 based on NCSA httpd 1.3. Apache is exhibited as free software and is developed by volunteer programmer in the world. Any one can revises and distributes. Apache is originally developed as patchwork to revise bag and apply to new performance NCSA httpd developed by super computer center in Illinois University (NCSA). At present, Apache becomes web server behaving by itself and using most often in the world.

2.4 Tomcat [3], [5]

Tomcat is name for Java servlet container and web server created by Jakarta project. This Tomcat equips with Java servelet and Java Server Pages (JSP), and is adapt to send to browser the result of running a program respond to request of user as well as using static HTML page.

2.4.1 Tomcat

Tomcat is name for Java servlet container and web server created by Jakarta project. Tomcat can use as single web server, but add on system to other web server. Which method of use can use the method properly? The license called Apache Software License is exhibited, any one can use, revise and distribute. Tomcat starts from Java Web Server (JWS) which is Java servelet container in Sun Microsystems. After this, Jservlet, Jigsaw of CERN/W3C etc. is proposed as Java servelet container of open source, and after penetrating it graduate, Java Server Development Kit (JSDK) supporting JSP is released on 1997. The next year (1998), after JSDK 2.1 proposed by Sun Microsystems is released by James Duncan Davidson of Sun Microsystems core of JSDK server called Tomcat rewritten into new Java servlet engine and Tomcat3.0 is released as succession on JSDK2.1. Since Tomcat is denoted to the Apache Software Foundation as open source project by intention of developer James in Sun Microsystems, Tomcat is managed by the Apache Software Foundation as Jakarta project.

2.4.2 Connection of Apache to Tomcat

Apache cannot process servelet, and Tomcat can use as web server, but Tomcat lacks in responsibility and security. Therefore, Apache plugs in tomcat to behave generally. Then, four connection modules(Apache, Java2 SDK, Tomcat) must be installed on server. Up-to-date Apache2.0 (http2.0.52.tar.gz) should be downloaded and installed from formal site for Apache (http://www.Apache.org).

Next, Java (TM) 2 SDK, SE (j2sdk-1_4_2_06-linux-i586-rpm.bin) should be downloaded and Java is installed from Sun Microsystems (http://Java.sun.com). At

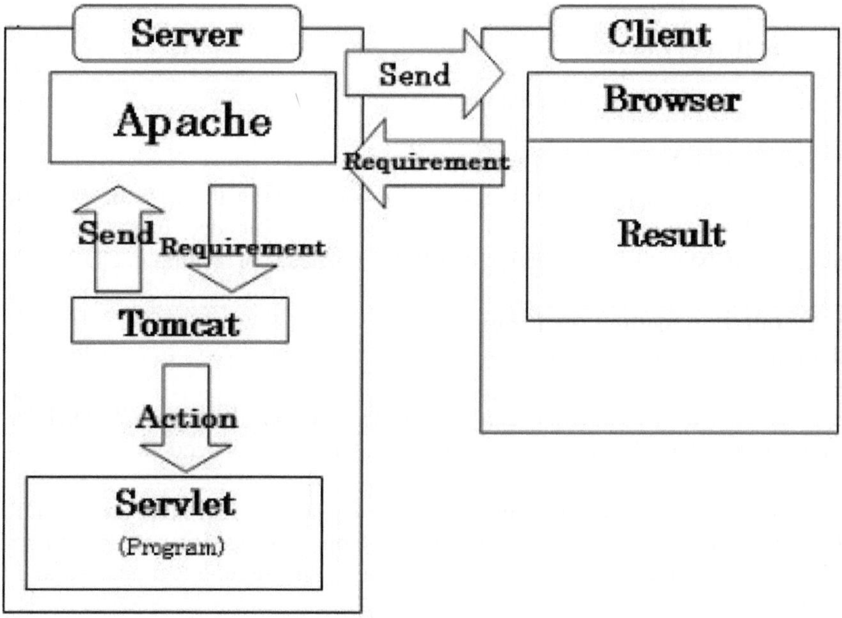

Fig. 1. Connection Behavior Apache and Tomcat

last, from Jakarta Project (http://jakarta.apache.org) on formal site of Apache Tomcat4.1.31 (jakarta-tomcat-4.1.31 trigs) is downloaded and installed. After three installations, it is possible to connect of Apache to Tomcat by means of installation of necessary module (mod_jk2.so) for connection of Apache to Tomcat.

2.5 Java Servelet

Java servelet is Java program that output the result practiced a program and processed to input data from client. Adding servelet, the function of web server can be extended. As servelet is described by Java language, it is possible to operate any web server that equip servelet API without dependence special OS and hardware. Once server program is called, it is permanently stationed at memory. Therefore, it is possible to process speedy different from server side program that is CGI etc. Also, as it is able to treat data permanently, it is shared information between plural users.

2.6 Equipment of Cusp for JAVA to Tomcat

In order to install Java servlet in Tomcat, in the case of installation by default, Tomcat installs in /user/local/tomcat. Tomcat folder in lower layer webapps folder, sarvlet is arranged; the servelet can be started from tomcat.

3 Construction of Web Server

This section describes on the construction method for web server. Starting to install Linux, this section shows installation of server and Java servlet container in detail.

3.1 Installation of Linux

Installation of Linux is describes below.

3.1.1 Installation Method for Linux [3], [4], [6]

Linux is attached as appendix of magazine or book, but in the case of difficulty of acquisition, there is the method that Install file is downloaded from site. The file downloaded can be used as installation CD-ROM by means of writing in CD-ROM etc.

3.1.2 Installation method for Linux

Installation method for Linux is below.

3.1.2.1 CD-ROM. This method can use in the case that there are CD-DRIVE and Red Hat Linux CD-ROM. Boot disk or bootable CD-ROM is needed.

3.1.2.2 Hard disk DRIVE. This method can use in the case that there are CD-DRIVE and Red Hat Linux CD-ROM. Boot disk or bootable CD-ROM is needed. When Red Hat LinuxのISO image is copied on local hard disk, this method can be used and boot disk is needed.

3.1.2.3 NFS image. When ISO image of Red Hat Linux or mirror image is installed from NSF server, this method can be used. In this case, network driver disk is needed.

3.1.2.4 FTP. Installing from FTP server directly, this method is used. Network driver disk is needed.

3.1.2.5 HTTP. Installing from HTTP (Web) server directly, this method is used. Network driver disk is needed.

4 Behavior Confirmation and Discussion

In this section, servlet is booted from client, Behavior confirmation of the servlet for Cusp for Java behave on server and Stand alone that Cusp for JAVA is installed in client compare with the servlet for Cusp for Java behave on server.

4.1 Startup for Cusp for JAVA

Server is accessed to designated page from client, top page of Cusp for JAVA as shown in Fig.2 is emerged on browser. Starting up servlet, input picture as shown is appeared on window and numerical values are input in input form. Pushing start bottom, analysis is started and its result is shown. In Behavior confirmation, sample data used previously is inputted. In stand alone, input form shown in Fig.3 is inputted window information with data.txt and data name analyzed is inputted box inputted as data.txt. Pushing bottom existed in lower window. In behavior confirmation, as function of read in file do not use, it should be performed direct input. In handling operation, number of matrix object data is inputted into the box on upper window and push "New" bottom. Input field is decided. Next, object numerical data shown in Table1 input to column of each variable.

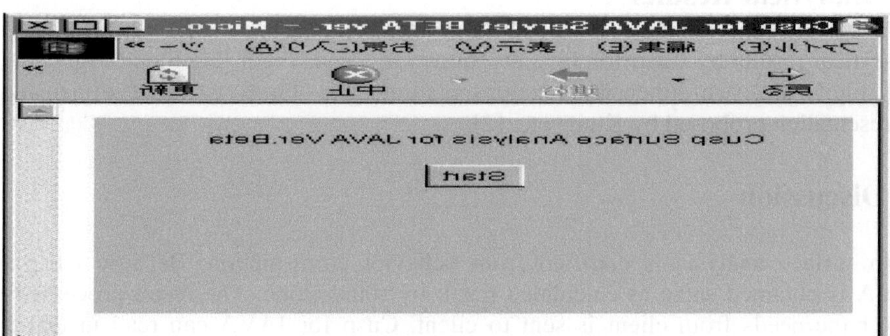

Fig. 2. Top Page for Cusp for Java

Table 1. Sample Data

variable 1	variable 2	variable 3
56.8	56.8	1.81
65.8	69.5	-0.18
71.6	64.2	0.00
61.9	61.1	-2.64
76.8	56.8	-3.23
64.5	53.7	1.29
65.8	69.5	0.53
65.2	60.0	-1.00
58.7	53.7	-3.10
72.3	62.1	1.78
48.4	51.6	4.07
65.2	58.9	0.49
60.0	61.1	-0.10
56.1	64.2	-3.62
61.3	35.8	-0.15
69.0	48.4	-0.97
78.1	61.1	-3.20
78.1	62.1	-0.84
69.0	60.0	-2.11
65.8	60.0	0.43
65.8	60.0	-3.26
69.0	61.1	1.28
65.8	61.1	2.07
71.6	60.0	-0.02
69.0	58.9	-0.60
71.6	58.9	0.17
69.0	61.1	3.36
65.8	60.0	0.42
69.0	61.1	3.50
69.0	60.0	3.81

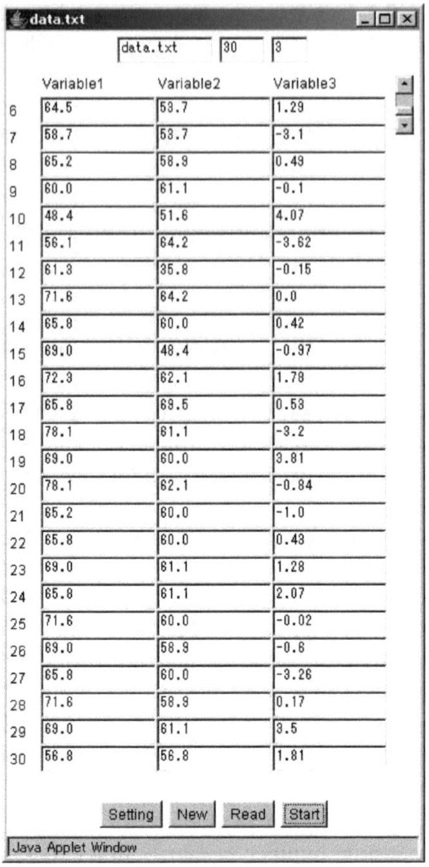

Fig. 3. Input form for Cusp for Java

5 Analytical Results

Analytical results by Cusp for Java are shown in Fig.4, Fig.5, Fig.6 and Fig.7. Fig.4 and Fig.5 are two-dimensional representation, and Fig.6 is three dimensional representation proposed by Kume etc. [2].

6 Discussion

Cusp surface analysis is clarified from behavior confirmation. Because Cusp for JAVA is obtained same as calculated result by stand-alone. The result processed on server the needs from client is sent to client. Cusp for JAVA can read in text file having extension (.txt). When this function for read in text files, the function is excellent reducing handling operation. The function cannot use at present. Client side practiced on server does not interfere. Reason that the function of read in text files does not perform is that the files etc. in client side from Java servlet practiced on server do not interfere using Java servlet.

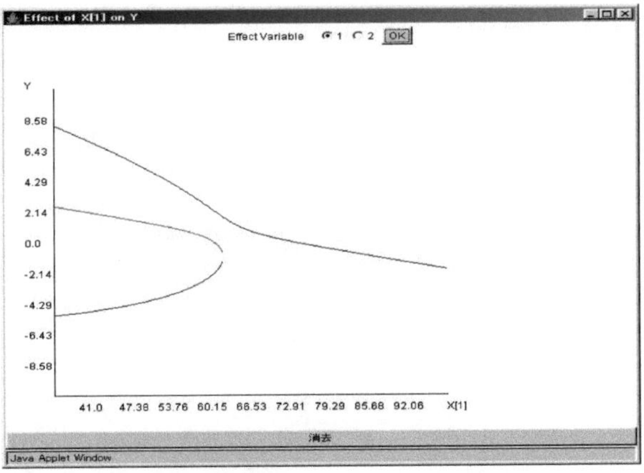

Fig. 4. Relationship between Y and X [1] at X [2]= 59.003

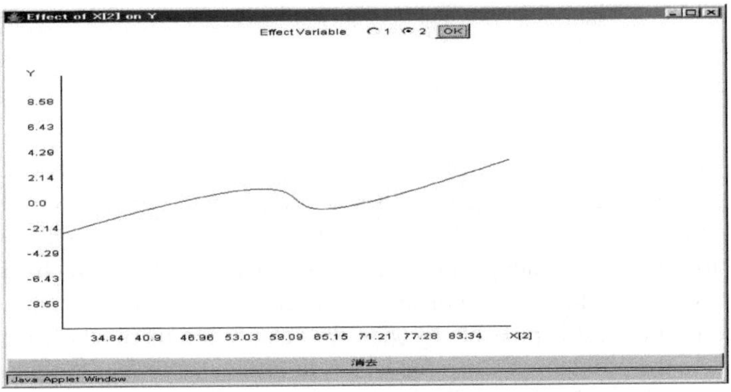

Fig. 5. Relationship between Y and X [2] at X [1]=66.533

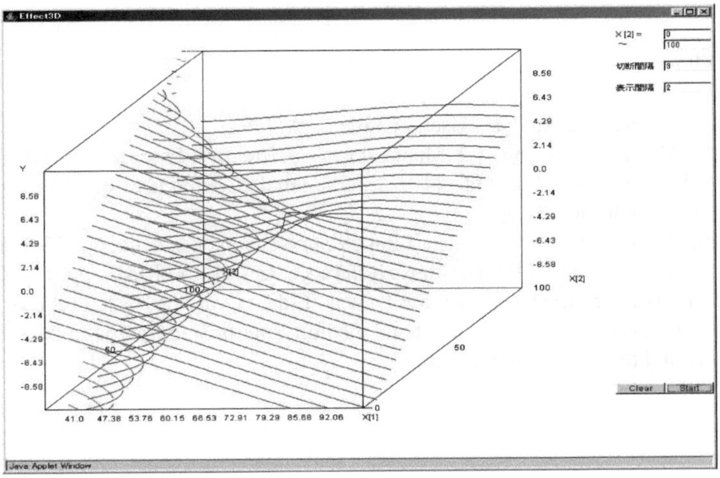

Fig. 6. Effect of X [1] and X [2] to Y

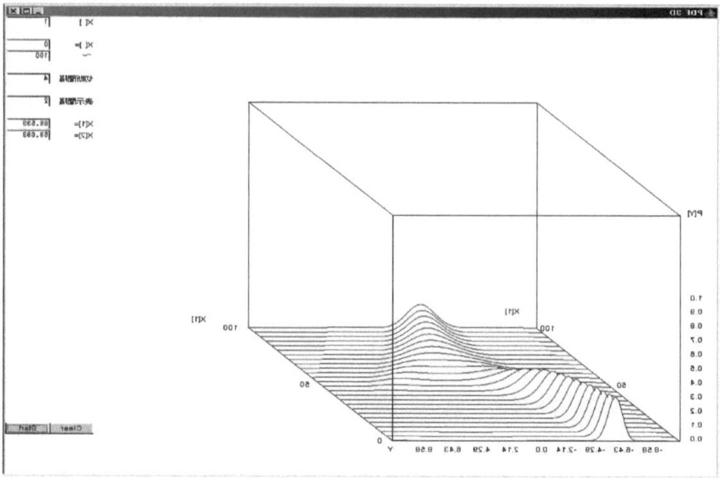

Fig. 7. Transition of Probability density function of Y

7 Conclusion

Handling should operate input of analytical data. But Cusp for JAVA is obtained same analytical result and Apache server receives requirement from client and its requirement transfer to tomcat. Cusp for JAVA carried out in Tomcat start as servlet and send to client. Web application can be constructed. Then, Web exhibition for Cusp for JAVA is able to accomplish this paper. Also, when Java analytical tool is constituted in future, it is possible to open web exhibition using constructed server in this paper. As the problem of Cusp for JAVA in future, server has the function of read in input files and keeping function of output data on server. As the problem of server, it is necessary for servlet user to certify and to customize security.

References

1. Cobb, L.: Cusp Surface Analysis User's Guide, 1–2 (1988)
2. Kume, Y., Okada, K., Cobb, L.: Management of Creative Process Using Cusp Surface Analysis System. In: Proc. of 11th International Conference on Human-Computer Interaction, CD edition (2005)
3. Ttakahara, T.: Introduction to Red Hat Linux 9 Server, Sotech Ltd, co, pp. 77–188 (2003)
4. Ball, B., Pitts, D.: Standard RedHatLinux Reference, Ltd, co INPRESS, pp. 11–205 (2001)
5. Darwin, I.F., Brittain, J.: Hand Book for Tomcat, Ltd.co Orily, pp. 1–282 (2003)
6. SOHO PC-UNIX Lecture for the small and medium enterprise: http://pc-unix.goco.ne.jp/
7. ITNAVI.com: http://www.itnavi.com

Design and Implementation of Enhanced Real Time News Service Using RSS and VoiceXML

Hyeong-Joon Kwon[1], Jeong-Hoon Shin[2], and Kwang-Seok Hong[1]

[1] School of Information and Communication Engineering,
Sungkyunkwan University,
300 Chunchun Dong, Jangan-gu, Suwon, Kyungki-do, 440-746 Korea
katsyuki@skku.edu, kshong@yurim.skku.ac.kr
http://hci.skku.ac.kr

[2] School of Computer and Information Communications Engineering,
Catholic University of Daegu,
300 Geumnak 1-ri, Hayang-eup, Gyeongsan-si, Gyeongsangbuk-do, 712-702 Korea
only4you@cu.ac.kr
http://only4you.or.kr

Abstract. In ubiquitous computing, most people need to track various sources of news using various devices, but it becomes difficult once there are more than a handful of sources. This is the reason for this is that users have to navigate to each page, load it, remember how it's formatted, and find where they last left off in the list. To solve these problems, many service providers provide RDF Site Summary documents. In this paper, we propose a newly designed news service using RSS and VoiceXML. RSS is an XML format that supports the syndication of news stories and similar content. There are several different formats for XML syndication that are referred to as RSS. Since RSS is in XML format, turning it into VoiceXML is easy and the synergy benefits of binding RSS and VXML are great. VoiceXML is a non-proprietary, web-based markup language for creating vocal dialogues between humans and computers. In this paper, we first focus on binding RSS and VXML, and RSS feed parsing. As a result of this research, we implement enhanced real time news service. People can use our service with their wired and wireless phone at any time, at any place. Also, we validate usability by comparing a typical RSS service scenario and typical VoiceXML service scenario, and calculating user's satisfaction.

1 Introduction

The World Wide Web is a huge collection of different sites, and are all updated at different times. The web has quickly grown from a modest hypertext system of interest to computer researchers to a ubiquitous information system that includes virtually all human knowledge. To determine whether or not a site has been updated, a user would have to go to each site and attempt to recollect what information they saw previously, and then find what data, if any, is new. Most users of the web have a collection of sites they frequent to get information. But, nowadays, it is difficult to get the necessary data from the vast scale of web sites. That's because it exists too much information in the web. Users want to get the right data from web sites at the right

M.J. Smith, G. Salvendy (Eds.): Human Interface, Part I, HCII 2007, LNCS 4557, pp. 677–686, 2007.

time with the minimum amount of effort. These kinds of user's needs motivate the change of web sites into a more standardized structure. An increasingly large amount of data is structured, stored, and sent over a network using XML. One such example is Really Simple Syndication (RSS) [1], [2].

RSS has gained popularity thanks to the increased use of web pages and news subscriptions. A site can make its updates available through an RSS feed, a file located within the site that contains the most recently added items in XML format. An RSS feed is written in XML. A feed comprises a channel, which has a title, link, description, etc, followed by a series of items. The real benefit of RSS, apart from the speed of looking at many different sites, is that all the feeds are chosen by the user. No user has the power to set their agenda, and crucially no-one can intervene to send spam. With thousands of sites now RSS-enabled and more on the way, RSS has become perhaps the most visible XML success story to date [2].

In this paper suggests a prototype of a dialog system combining VXML (VoiceXML), which is the W3C's standard XML format for specifying interactive voice dialogues between human and computer, and RSS(RDF Site Summary or Really Simple Syndication), which a representative technology of the semantic web for syndication and subscription of updated web-content. The merits of the proposed system are as follows: 1) It is a new method that recognizes spoken content using wire and wireless telephone networks and then provides content to users via STT(Speech-to-Text), TTS(Text-to-Speech) and visual environment using RSS, 2) It can apply advantages of RSS, where the subscription of updated content is converted to VXML, without modifying traditional methods to provide RSS service, 3) In terms of users, it can reduce restrictions on time-space, in search of content provided by RSS, because it uses wire and wireless telephone networks, not the internet environment. 4) In terms of information provider, it does not need special equipment, design of difficult STT and TTS algorithm, for syndication of new content using speech recognition and synthesis technology. We implemented a news service system using VXML and RSS for performance evaluation of the proposed system. In the experiment results, we estimated the response time and the speech recognition rate in subscription and search of actual content, and confirmed that the proposed system can provide content provided using a RSS Feed.

2 Related Works

In this chapter, we introduce core technologies for our research, that is, RSS and VoiceXML. Then, we introduce basic service architecture for speech applications.

2.1 RSS

Prior to RSS, several similar formats already existed for syndication, but none achieved widespread popularity or are still in common uses today, as most were envisioned to work only with a single service. These originated from push and pull technologies. Two of the earliest examples are "Backweb" and "Pointcast" [3].

Between 1995 and 1997, Ramanathan V. Guha at Apple Computer's Advanced Technology Group developed the Meta Content Framework (MCF). The MCF was a specification for structuring metadata information about web sites and other data, and the basis of Project X (aka Hot Sauce), a 3D flythrough visualizer for the web. When the research project was discontinued, Guha left Apple and went to work at Netscape, where he adapted MCF to use XML and created the first version of the Resource Description Framework (RDF). Then, in 1997 Microsoft created the Channel Definition Format for the Active Channel feature of Internet Explorer 4.0, however, the feature never became popular [2]. Really Simple Syndication (RSS) is a lightweight XML format designed for sharing headlines and other Web content. It can be interpreted as a constantly updated and distributed "What's New" summary for your site. Originated by UserLand in 1997 and subsequently used by Netscape to fill channels for Netcenter, RSS has evolved into a popular means of sharing content between sites (including the BBC, CNET, CNN, Disney, Forbes, Motley Fool, Wired, Red Herring, Salon, Slashdot, ZDNet, and more). RSS solves the myriad problems webmasters commonly face, such as increasing traffic, and gathering and distributing news. RSS can also be the basis for additional content distribution services [2], [3], [4].

2.1.1 Configuration

RSS files are often labeled as XML. RSS version 1.0 is also RDF (any version), which, again, is important only because an RSS file may be labeled as RDF. RSS files (which are also called RSS feeds or channels) simply contain a list of items. Usually, each item contains a title, summary, and a link to a URL (e.g. a web page). Other information, such as the date, creator's name, etc., may also be included. The most common use for RSS files is for news and other reverse-chronologically ordered websites such as blogs. For example, a particular page on Fagan Finder has a change log, which is also available in RSS format. An item's description may contain all of a news article, blog post, etc., or just an extract or summary. The item's link will usually point to the full content (although it may also point to the item linked by the content itself). Figure 1 shows tree types of core elements of RSS 2.0 [2].

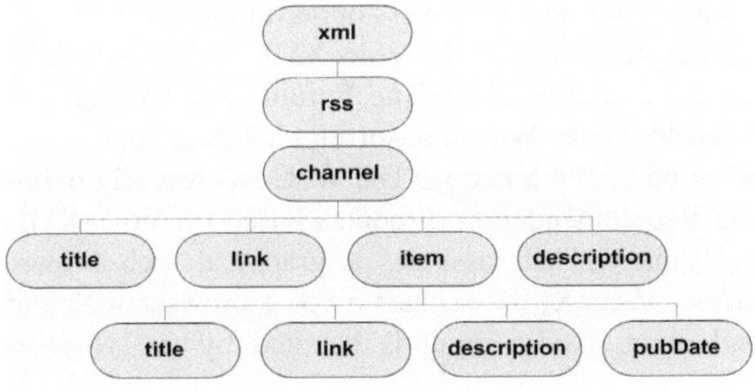

Fig. 1. RSS 2.0 core elements – Tree type

2.1.2 RSS Syntax

RSS is used to share content between websites. With RSS, we can register our content with companies called aggregators. First, create an RSS document and save it with

Table 1. Simplified version of an actual RSS feed

RSS Format in English	Example RSS Feed in English	Example RSS Feed in XML
Title Link	Joe's Breakfast News www.joe.com/news/	<title>Joe's Breakfast News</title> <link>http://www.joe.com/news/</link>
Item Title Link	Item Orange Juice Voted Best Fruit Juice www.joe.com/news/orange-juice.html	<item> <title>Orange Juice Voted Best Fruit Juice</title> <link>http://www.joe.com/news/orange-juice.html</link> </item>
Item Title Link	Item Acme Introduces New Flakes 'n' Nuts Cereal www.joe.com/news/flakes-n-nuts.html	<item> <title>Acme Introduces New Flakes 'n' Nuts Cereal</title> <link>http://www.joe.com/news/flakes-n-nuts.html</link> </item>

an .xml extension. Then, upload the file to your website. Next, register with an RSS aggregator. Each day the aggregator searches the registered websites for RSS documents, verifies the link, and displays information about the feed so clients can link to documents of interest [2], [3].

2.2 VoiceXML

VoiceXML is the HTML of the voice web, the open standard markup language for voice applications. VoiceXML harnesses the massive web infrastructure developed for HTML, to make it easy to create and deploy voice applications. Similar to HTML, VoiceXML has created substantial business opportunities [1].

VoiceXML 1.0 was published by the VoiceXML Forum, which is a consortium of over 500 companies, in March 2000. The Forum then handed over control of the standard to the World Wide Web Consortium (W3C), and now concentrates on conformance, education, and marketing. The W3C has recently published VoiceXML 2.0 as a Candidate Recommendation. Products based on VoiceXML 2.0 are already widely available. While HTML assumes a graphical web browser with display, keyboard, and mouse, VoiceXML assumes a voice browser with audio output, audio input, and keypad input. Audio input is handled by the voice browser's speech recognizer. Audio output consists both of recordings and speech synthesized by the voice browser's text-to-speech system [1].

VoiceXML takes advantage of several trends:

- The growth of the World-Wide Web and its capabilities.
- Improvements in computer-based speech recognition and text-to-speech synthesis.
- The spread of the WWW beyond the desktop computer.

2.3 Service Architecture for Speech Applications

Figure 2 shows service architecture for speech applications using VoiceXML 2.0. In this paper, we use HUVOIS, which contains ASR, TTS and VoiceXML interpreters, provided by Korea Telecom [1], [5].

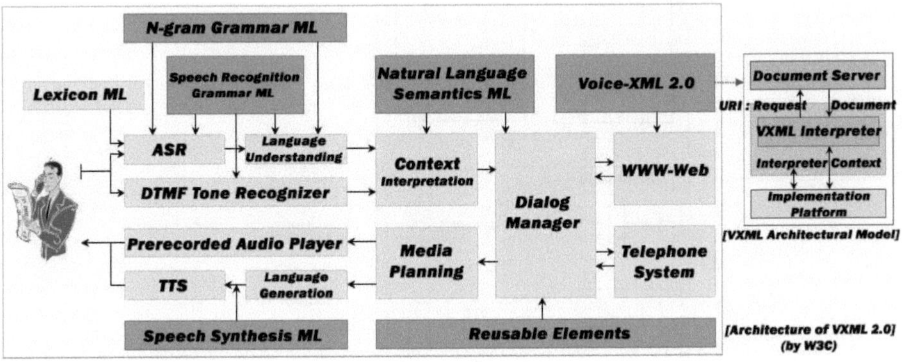

Fig. 2. Service architecture for speech applications using VoiceXML

3 VoiceXML Dialogue System Based on RSS

XSLT is designed for use as part of XSL, which is a style sheet language for XML. In addition to XSLT, XSL includes an XML vocabulary for specifying formatting. XSL specifies the styling of an XML document by using XSLT to describe how the document is transformed into another XML document that uses the formatting vocabulary [6].

XSLT is also designed to be used independently of XSL. That is to say, XSLT could be used as a XML transformation language. However, our proposed system doesn't use XSLT for the purpose of speech applications. For speech applications, we use an HTML parser, the reason for this is that the RSS feeder doesn't include complete content so that its concept is summary, we parses the HTML document that is source of content. To solve these problems, we propose the VoiceXML dialogue system, based on RSS, with a server-side script [6].

3.1 Features

In this chapter, we propose a VoiceXML dialogue system based on RSS. This proposed system provides content which is then provided to the RSS feeder, through a telephony network. The system can provide simultaneously visual RSS service and voice applications using VoiceXML. The system uses a RSS feeder, thus as a result, the proposed system doesn't need an additional DB for VoiceXML service. Figure 3 shows the relationship between components in the proposed system.

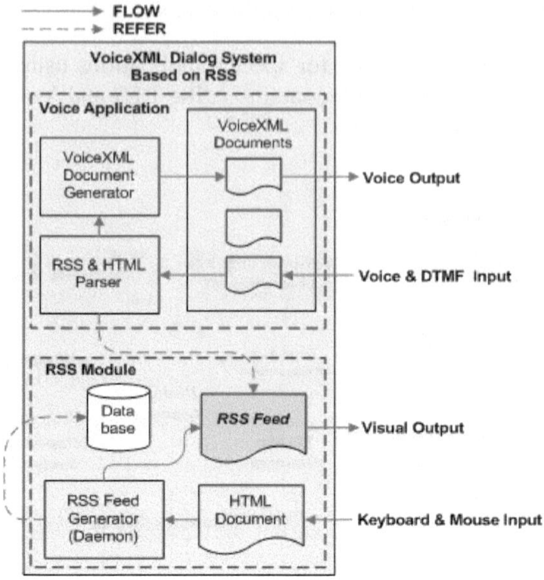

Fig. 3. Components' relationship in the proposed system

3.2 System Components

Our proposed system acts dynamically using scripts language based on Mixed Initiative Forms. In this system, the XML parser and HTML parser are additionally implemented for WEB service using traditional RSS.

3.2.1 Generating RSS Feeds

It is recommended to generate RSS feeds using script languages such as ASP, JSP or PHP [7]. To gather content for the each element in an RSS feed, we use the question and answer method. Then, we record content in the field with elements compatible with RSS version and text format. We use a formatted field for the XML, RSS, channel, title, link and description elements. As shown in Figure 4, in the case of item element, we use the recursive loop algorithm.

```
for (each record in resultSet from database table)
 for (each field in currentRecord)
    if ([column_name] != "pubDate")
     writeline "<[column_name]>", "<[column_content]>", "<[column_name]>"
    endif
 endfor
endfor
```

Fig. 4. Algorithm for generating RSS feeds

The above mentioned algorithm shows, a recursive question and answer method for each element and field.

3.2.2 XML Parser

In this paper, we use the XML parsing algorithm using the DOM. Figure 5 shows the details of the parsing algorithm [8].

```
xmldomVariable <= RSS_FEED_URL
 for (each element in xmlDomVariable)
   if ([element_name] = requiredElements)
     writeline [element_name], [element_content]
   endif
 endfor
```

Fig. 5. Algorithm for XML parsing using the DOM

As shown in Figure 5, our system searches the contents of RSS feed line by line, and finds the necessary element. After that, it records the name of the element and the content of the element. Occasionally, these kinds of algorithms bring about inefficient results. To overcome this inefficient result, we suggest an improved XML parsing algorithm using the DOM. Figure 6 shows the improved XML parsing algorithm using the DOM [8].

```
xmlDomVariable <= RSS_FEED_URL
currentItemCount <= 0
maxItemCount  <= Output Number

 for (each element in xmlDomVariable)
   if ([element_name] = requiredElements)
     if ([element_name] = [item] and
         currentItemCount < maxItemCount)
         currentItemCount ++
         writeline      [element_name]
         writeline      [element_content]
     endif
   endif
 endfor
```

Fig. 6. Improved XML parsing algorithm using the DOM

3.2.3 HTML Parser

Our proposed system has an additional HTML parser for the purpose of using when description elements are incomplete. Generally, description element consists of a part of all content and all content is linked via hypertext. To solve these problems, we implemented in the proposed system the ability to extract necessary content from the hyper linked documents. Figure 7 shows the HTML document parsing algorithm.

```
textString <= Text in html document.
stHeadString <= Insert startpoint string.
stTailString <= Insert endpoint string.

stringArray_1st    Enumerate textString with stHeadString
stringArray_2st    Enumerate stringArray_1st(1) with stTailString
stringResult       Delete HTMLTAG in stringArray_2nd(0)

Return StringResult
```

Fig. 7. HTML document parsing algorithm using annotation

Figure 7 shows the usage of annotation for extracting essential content from the HTML documents.

4 Experimental Results

To verify the performance of proposed system, we implemented a news service using an RSS feed (http://www.chosun.com/rss/rss.xml) compatible with the RSS 2.0 standard. We use the response time and speech recognition rate as indicators.

4.1 Experimental Condition

Table 2 shows the experimental condition and hardware specification. As shown in Table 2, we use an Intel Xeon CPU and Intel Dialogic D/120JCT-LS as a telephony card. The speech recognition and synthesis engine are provided by Korea Telecom (KT HUVOIS).

Table 2. Hardware and software specification

Component	Description
CPU	Intel Xeon 3.0Ghz Dual
RAM	2GB
CTI Board	Intel Dialogic D/120JCT-LS PCI (12CH)
OS	Windows 2000 Professional Service Pack 4
IIS Ver.	5.0
ASR and TTS	KT Huvois

4.2 Performance Evaluation

Ten testee were selected for performance evaluation and learning courses were finished with several kinds of proposed systems. Performance evaluation was conducted 10 times for each subject. The evaluation was based on speech recognition rate, has shown Table 3, and response time.

Table 3. Response Time

Trial Number	Response Time	Spoken Language
1	Within 1 sec	il:b∂̂n
2	Within 1 sec	i:b∂̂n
3	Within 1 sec	sam:b∂̂n
4	Within 1 sec	sa:b∂̂n
5	Within 1 sec	o:b∂̂n
6	Within 2 sec	juk:b∂̂n
7	Within 1 sec	i:b∂̂n
8	Within 1 sec	sam:b∂̂n
9	Within 2 sec	sa:b∂̂n
10	Within 1 sec	o:b∂̂n
11	Within 1 sec	il:b∂̂n
12	Within 1 sec	i:b∂̂n
13	Within 1 sec	sam:b∂̂n
14	Within 1 sec	sa:b∂̂n
15	Within 2 sec	o:b∂̂n

Table 4 shows the speech recognition rate. Testees, designated as testee (A1, A2, …) were people experienced using a voice-enabled service; testees designated as testee (B1, B2, …) were not experienced using voice-enabled service. If voice-enabled service continues to be popular, related research is needed to resolve the deficiencies relating to people's lack of experience of using this kind of system.

Table 4. Speech recognition rate

Category	Testee	False Rcognition	Recognition Rate(%)
Experienced User	A1	1 / 10	90
	A2	2 / 10	80
	A3	1 / 10	90
	A4	1 / 10	90
	A5	2 / 10	80
Non Experienced User	B1	3 / 10	80
	B2	3 / 10	70
	B3	4 / 10	60
	B4	2 / 10	80
	B5	3 / 10	70
Description	Total 10 person	Total 22 / 100	Average 79

5 Conclusions

In this paper, we proposed a prototype of speech application service using VoiceXML and RSS feed. The key benefit of our proposed system is that it doesn't need any change in traditional service architecture. We conducted performance evaluation to verify the usefulness of the proposed system. As shown in the experimental results, our proposed system can provide stable service without delay. Another key benefit of our system is that users could use the provided service on the basis of audio-visual environment. In terms of system architecture, our proposed system provides services with minimum server load. RSS and VoiceXML are some of the most popular technologies. Due to the importance of these technologies, we hope to provide useful reference to researchers of web technology and web architecture.

Acknowledgment. This research was supported by MIC, Korea under ITRC IITA-2006-(C1090-0603-0046).

References

1. McGlashan, S., et al.: Voice Extensible Markup Language Version 2.0 Specification (2004), http://www.w3c.org/TR/voicexml20
2. Winer, D.: RSS 2.0 Specifications, Technology at Harvard Law (2004) http://blogs.law.harvard.edu/tech/rss
3. Finder, F.: (2004) http://www.faganfinder.com/search/rss.php
4. Atomenabled. (2004) http://www.atomenabled.org
5. KT Voice Recognition Service R&D Team.: HUVOIS Solution Manual R1.1, KT Service Develop Laboratory (2003)
6. VOX POPLUI.: (2005) http://www.voiceingov.org/blog?page_id=8
7. Vankayala, R.R., et al.: Dynamic Voice User Interface Using VoiceXML and Active-Server Pages. In: Zhou, X., Li, J., Shen, H.T., Kitsuregawa, M., Zhang, Y. (eds.) APWeb 2006. LNCS, vol. 3841, Springer, Heidelberg (2006)
8. W3C, D.O.M.: interest group.: Document Object Model (2005) http://www.w3.org/DOM/
9. Tsai, M.-J.: The VoiceXML Dialog System for the E-Commerce Ordering Service. In: IEEE Proceedings of the Ninth International Conference (2005)
10. Shin, J.-H., et al.: Simple and Powerful Interface E-Leaning System Using VXML: Design and Implementation of Web and PSTN Linked Efficient Learning System. In: Gavrilova, M., Gervasi, O., Kumar, V., Tan, C.J.K., Taniar, D., Laganà, A., Mun, Y., Choo, H. (eds.) ICCSA 2006. LNCS, vol. 3980, Springer, Heidelberg (2006)
11. De Sutter, R., et al.: Enhancing RSS Feeds: Eliminating Overheard through Binary Encoding. In: IEEE Proceedings of the Third International Conference (2005)
12. Andersson, E.A., et al.: Early Adopter VoiceXML. Wrox (2002)
13. Finder, F.: (2004) http://www.faganfinder.com/search/changes/
14. Finder, F.: (2004) http://www.faganfinder.com/search/changes/?display=rss

Correlation Analysis of Available Bandwidth Estimators for Mobile HCI

Doohyung Lee, Chihoon Lee, Jahwan Koo, and Jinwook Chung

School of Information and Communication Engineering, Sungkyunkwan University
Chunchun-dong 300, Jangan-gu, Suwon 440-746, South Korea
drlee07@skku.edu, chlee10@naver.com
jhkoo@songgang.skku.ac.kr, jwchung@songgang.skku.ac.kr

Abstract. Transport protocol in mobile devices of HCI (human-computer interaction) operates both in wired and wireless heterogeneous network. It has been developed from reactive congestion control schemes to proactive congestion control schemes suitable for both wired and wireless networks environment to overcome the poor wireless environment and restricted mobility. Nevertheless, recently proposed TCP's proactive congestion control schemes could not reflect network environment status in detail, there is also a limit to support new mobile services of HCI users. In this paper, we described issues and problems of rate adjustment metrics used for reducing packet loss and congestion loss in TCP-Jersey's available bandwidth estimators. Also, we presented the guideline of selecting more adequate metrics to improve wireless TCP performance of mobile HCI.

Keywords: Available Bandwidth Estimators, TCP schemes, wireless, Mobile HCI.

1 Introduction

Mobile Devices of HCI consist of five sensing parts (visual, hearing, smell, taste and tactual), middleware part, application part, operating system part and communication part. For example, these devices are mobile PCs, PDAs, cellular phones and all of hybrid devices which support for user mobility. While moving case of use to mobile devices have been presented HCI related issues such as diverse hybrid devices, seamless application roaming [1] . However the mobile wireless environment about several distinguished character of reliability and QoS have not been considered the features of existing focus of transport protocol layer[2]. This is important with aspects of reliability and QoS that accurate delivery to diverse HCI services cope with congestion and packet loss. In early days, in order to reliability and QoS the TCP congestion control have been developed difference between wired and wireless network. However, with increasing user mobility in the wired and wireless heterogeneous network have trended the enabling efficient congestion control of TCP schemes. We summarizes the features to control the congestion detection, notification, metrics of rate adjustment the

M.J. Smith, G. Salvendy (Eds.): Human Interface, Part I, HCII 2007, LNCS 4557, pp. 687–696, 2007.

existing presented schemes and outlines issues, problems centering around the TCP-Jersey recent proposed of proactive congestion control schemes on ABE of respectively metrics. Next this paper focus a selection for more suitable metrics which can reflect the bottleneck link state and its implementation.

The paper is organized as follows. Section 2 describes the existing researches related to the paper. In section 3, the major issues found in existing TCP proactive congestion control schemes. In section 4, the available bandwidth estimator is simulated to make a more accurate estimation for available bandwidth, and the correlation analysis between metrics in determining optimum congestion window is performed. A new metric to estimate available bandwidth is presented also in section 4, and section 5 concludes the paper and presents the future work in research for mobile HCI of TCP scheme.

2 Previous Works for TCP Schemes

Nowadays, mobile HCI devices for seamless application roaming consider easy access to information. for example, a museum or hospital's at indoor/outdoor for local or remote roaming is important to event or guiding services. Because occur packet loss or congestion with this services during roaming, it is well know that TCP's algorithm role has a prevented throughput degrade. Recent congestion control algorithms include the capability of distinguishing from the cause of loss. If the cause is link error, TCP transmission rate is maintained as it is, but if the cause is congestion at a router, the size of congestion window is adjusted through transmission rate control.The congestion control algorithms suited for wired and wireless heterogeneous network environments that have been proposed up to now can be divided into reactive congestion control schemes [4], [5], [6] and [7] and proactive congestion control schemes [8], [9], [10] and [11].

2.1 Reactive Schemes

This method is an algorithm preventing the TCP performance from degrading through early recovery of packet loss such as this of the traditional Reno [3]. they uses the acknowledgements of receiver for the segments sent by the TCP sender and generate timeout or congestion events, and by lowering the value of congestion window, the transmission rate is adjusted.

I-TCP [4], M-TCP [5] and WTCP [6] divide the connection between the sender and receiver into a wired transmission interval and wireless transmission interval. The previous TCP congestion control mechanism is used in the wired network interval between the sender and base station while a congestion control mechanism more suitable for the previous TCP or wireless environment is used in the wireless interval between the base station and receiver. SNOOP [7] adds a TCP-aware module in the link layer (layer 2) which enables the base station to look at the TCP header without having to disconnect the TCP connection. The advantages of such reactive schemes is that the TCP is not affected by the packet loss in the network layer. However, these schemes have a major disadvantage in that they violate end-to-end TCP semantics. Consequently, packets sent by TCP may not match with those arriving at the receiver. The base station must have a TCP layer and there is the burden of having to separately manage the two connections.

2.2 Proactive Schemes

The reactive approach to congestion control estimates the available bandwidth in advance based on the congestion experiences of TCP flow and sends the calculated value of TCP congestion window. This method obtains good goodput by avoiding congestion and lowering the retransmission rate. Based on such fundamental ideas, the congestion window is lowered before congestion arises in the sender. This accounts to the fact that if congestion arises and RTT becomes longer, the TCP transmission efficiency drops since TCP performance is inversely proportional to RTT.

In TCP-Vegas [8], the sender measures RTT (Round Trip Time) at the beginning of connection and sets it as BaseRTT. If a smaller RTT is measured during the packet transmission, the current BaseRTT is renewed to the smaller value. Also, the expected sending rate for the state in which all queues of middle nodes are empty can be obtained by dividing the current window size by BaseRTT, and the sender measures RTT and the number of bytes for packets acknowledged by the sender to obtain the actual sending rate.

TCP-Peach [9] and TCP-Westwood [10] measure the available bandwidth of the current connection and adjust the window size if a packet loss arises, regardless of its cause. TCP-Jersey [11] distinguishes congestion and transmission loss and adjusts the window size. The advantage of these schemes [9],[10],[11] are that the TCP window size is adjusted dynamically, fundamentally preventing packet loss due to congestion. On the other hand, especially in TCP-Vegas, the number of total packets saved in the buffer of middle nodes can be controlled, but the number of packets saved in each queue cannot be controlled.

3 Open Issues of Previous Solutions

The mobile characteristics must be considered sufficiently in the wired and wireless heterogeneous network environment. That is, the cause for packet loss must be clearly identified, and In the case of a congestion, a rate control scheme for more sophisticated control is needed while for a transmission error, the current congestion window (CW) is maintained while using the available bandwidth as much as possible, to display optimal performance.

The most important issue is the need for a rate control scheme which reflects accurate circumstances to bottlenecks. Matters to be considered are described in the following. First issue is Differentiation and detection of causes for packet loss. The second issue is notification of network status. The third issue is selection of metrics for transmission rate adjustment

The problems will be identified and possible measures discussed in the viewpoints of the above-mentioned issues based on the recent literature on TCP-Jersey.

3.1 Differentiation and Detection of Causes for Packet Loss

In previous wired networks (Tahoe, Reno, New-reno, SACK [12]), there was no need to differentiate the cause for packet loss. This is because most of the loss was due to congestion. However, with the increase of wireless network environments, it has become necessary to identify the cause for packet loss. The latest loss identification methods proposed determine the cause for packet loss by a threshold value for a bottleneck of a buffer state. In TCP-Jersey, the threshold value was set at 30 percent of the buffer based on test results. That is, loss under the threshold value at the sender is judged as a transmission error while the value exceeds the threshold as congestion. The problem here is that this causes the waste of buffer resources in router. Another problem is that TCP transmission states are changing frequently in a wired and wireless heterogeneous environment, and this cannot be coped with flexibly by holding a fixed threshold value such as the TCP-Jersey.

These problems can be resolved at a router by setting a dynamic threshold value or accurately identifying the buffer capacity and approximating the actual buffer value.

3.2 Notification of Network Status

As shown in the figure 1, the current method in transmitting the bottleneck situation is to use the ECN (Explicit Congestion Notification) scheme. But, TCP-Jersey uses Congestion Warning(CW) flag instead of ECN. CW operates similarly to ECN and uses the CE bit in the IP header and ECE and CWR in the TCP header. The difference is that ECN uses a maximum and minimum threshold value. If the maximum value is surpassed, the CW flag is marked unconditionally, and if under the minimum value, the flag is not marked. Finally, if the value is between the maximum and minimum values, it is marked by a probability value calculated. On the other hand, TCP-Jersey is simplified by using only one threshold value, and if the value is over the threshold value, the CW flag is marked and sent, but if under, it is not marked.

The problem here is that only 1 bit is used, placing a limit in expressing and sending the various states of congestion in the router. Accordingly, a mechanism which uses the standard ECN while making intelligent judgments at the sender is required.

3.3 Selection of Metrics for TCP's Congestion Control

The rate control is a function performed by the sender. It determines the sending rate based on contextual information of the network. This is a prerequisite for proactive congestion control, and selecting effective metrics to be used determines the success of accurate estimation of available bandwidth.

In the proactive congestion control method, the sender adjusts the congestion window value in precedence based on the feedback information(RTT, ACK) collected in the bottleneck interval. By performing this, the sender can intelligently

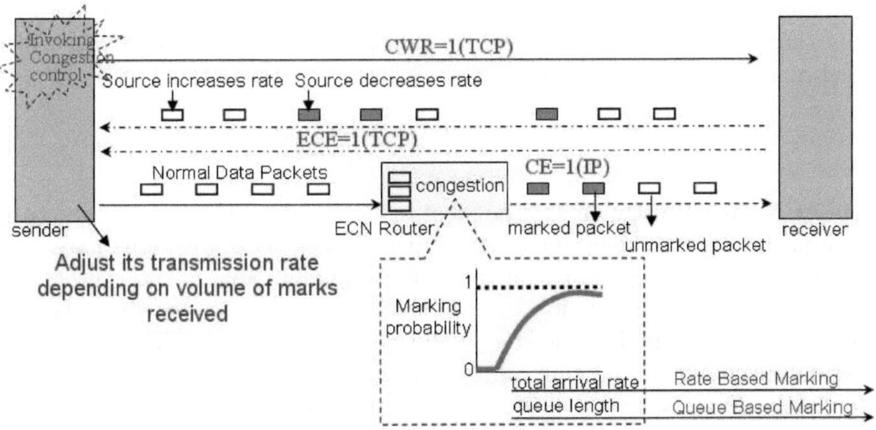

Fig. 1. The Operation of Mark Packets for ECN scheme

cope with the network situation or cause of packet loss. Also, undesirable network conditions(congestion and decrease in unnecessary congestion window) can be prevented in advance. In rate control methods proposed until now, ACK and RTT have been used as major metrics. These metrics have been brought about with the idea of bandwidth delay product [13], and hold a significant importance as they have been used for rate control in the end-to-end approach scheme.

The rate computation in TCP-Jersey was derived from Time Sliding Window(TSW) proposed by David D. Clark et al. [14], and is shown in the following Eq. (1). This equation is applied to the network router and is used to calculate the bandwidth allocated by each individual TCP flow.

$$R_n = \frac{T_w \times R_{n-1} + L_n}{(t_n - t_{n-1}) + T_w} \tag{1}$$

In the above equation, R_n is the calculated bandwidth for packet n when it arrives at time t_n. t_n is the arrival time of the *nth* packet, and t_{n-1} is the previous packet. L_n is the packet size and T_w is a time constant depending on the network situation. R_{n-1} is the previously calculated rate value. TCP-Jersey proposes the following equation in which TW is substituted to RTT. R_n (rate of available bandwidth) in Eq. (2) is multiplied by RTT, the estimation bandwidth and delay, and divided by segment data to obtain the *Ownd* (Optimum Congestion Window) value in Eq. (3). This was proposed to perform efficient TCP rate control for the estimated available bandwidth.

$$R_n = \frac{RTT \times R_{n-1} + L_n}{(t_n - t_{n-1}) + RTT} \tag{2}$$

TCP-Jersey derives the *Ownd* to set *Cwnd* (Congestion Window) and *ssthresh* as Eq (3) with the rate computed in Eq. (2). A contradiction can be seen here as RTT was used to obtain R_n as well as to calculate *Ownd*. That is, the value of RTT is suitable variable to determine $Ownd$ but not for determining the rate. This calls the need for new separate metrics, and this will be verified in section 4 through tests.

$$Ownd = \frac{RTT \times R_n}{seg\_size} \tag{3}$$

The ACK value mostly affects the most in rate control. However, as the transmission path to the receiver and path returning back the ACK value may differ, the delay may differ depending on the buffer queuing value of the router. Also, if an ACK drop of compression occurs, the ACK rate loses its meaning of an adequate sample value. In order to settle this, timestamp should be used or an additional mechanism reflecting only valid ACK values is necessary.

In addition, TCP-Jersey does not perform rate control in the *slowstart* phase. Accordingly, if rate control is performed in the *slowstart* phase, *ssthresh* will generally be formed in a higher point, meaning that the available bandwidth could be reached more quickly. Also, if RTO occurs in the *slowstart* phase, *ssthresh* will not decrease by half as it would in TCP-Reno[3] or TCP-Jersey [11].

4 Verification of Transmission Rate Calculation in TCP-Jersey

In previous section, we have drawn the problems from three aspects of Available Bandwidth Estimator (ABE) in Eq. (2) and Optimum Congestion Window (Ownd) in Eq. (3) of proposed in TCP-Jersey among the proactive congestion control schemes and presented measures to cope with the problems. This section provides the verification through simulation. In particular, the correlation between dependent variables for the rate calculation in Eq. (2) used to compute available bandwidth, and $Ownd$ in Eq. (3) used to compute optimal congestion window have been analyzed to verify the effectiveness of each dependent variable.In addition, verification will be focused on how the R_n and $Ownd$ values change when each independent variable is changed and the correlation with the independent variable using a statistical method.

4.1 Gathering Test Data and Verification Method

Fixed TCP segment data of 600 bytes was generated 500 times/sec for the test data in a topology structure environment shown in Fig. 2 in which there is one input and one output. Fig. 3 shows the value of R_n generated in the test environment. This value has approximated to 1.5 Mbps, similar to that in a TCP-Jersey testing environment. With this, it has been verified to be appropriate data for the test.

The input values RTT, L_n, $t_n - t_{n-1}$ and $RTT * R_{n-1}$ used in Eq. (2) to compute the rate and $RTT * R_n$ in Eq. (3) to compute the optimal congestion window were used to find the correlation with output values R_n and $Ownd$ for the verification. As for the correlation analysis, Pearson's correlation method was used.

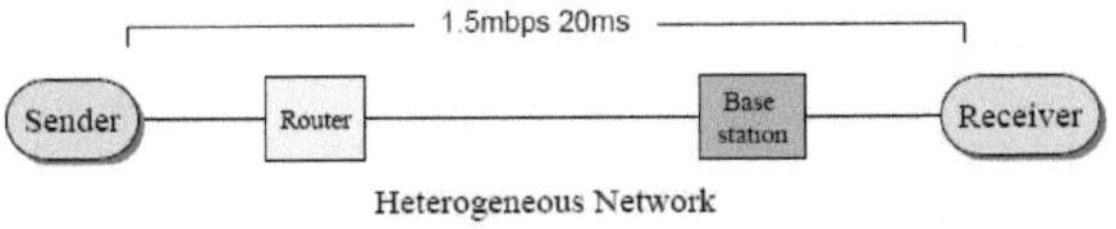

Fig. 2. Experiment environment

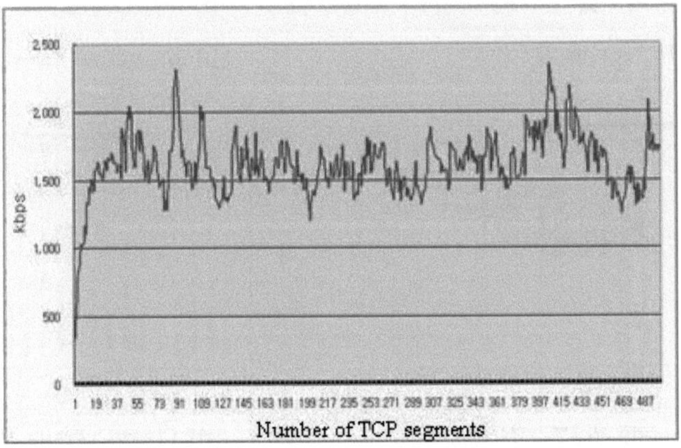

Fig. 3. Raw data gathered from the experiment environment

The significance level was set at P-value 0.05 in this test, That is, if P-value is lower than 0.05 in the 95 percent confidence interval, there is high correlativity, and if higher, then there is low correlativity. + correlativity means positive correlativity and - means negative correlativity. If the correlativity factor comes close to an absolute value of 1, then the correlativity is high, and if it reaches near 0, there is low correlativity. The scatter-plot shows the relation between two quantitative variables visibly.

4.2 Verification of Transmission Rate Adjustment Value Calculation

The correlation between R_n, each metric in Eq. (2) and the available bandwidth estimator for TCP-Jersey will be verified here. Results from analyzing the correlation between R_n and RTT show that P-value is 0.831. Since P-value is higher than 0.05, it is judged to have low correlativity.

Fig. 4(a) shows the scatter-plot between RTT and R_n, and when taking a close look, it can be seen that RTT and R_n do not change in proportion or inverse proportion. Accordingly, this shows that R_n does not change when RTT changes.

The second analysis was performed on the correlativity between R_n and $t_n - t_{n-1}$. P-value turned out to be 0.000, statistically meaning they are significant and are correlative. The correlativity factor was -0.424, showing there was a negative correlation. Fig. 4(b) shows the scatter-plot of R_n and $t_n - t_{n-1}$, and it

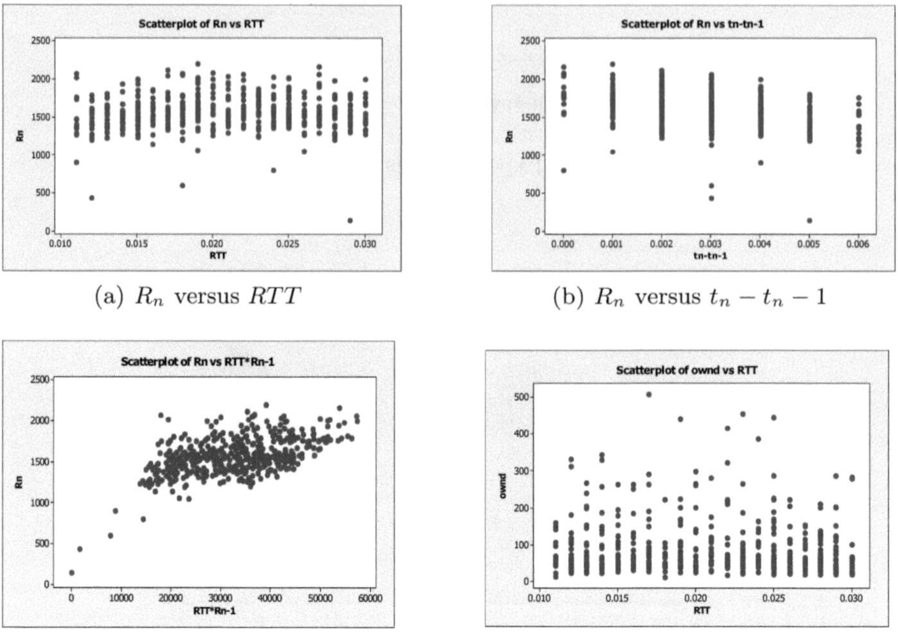

(a) R_n versus RTT (b) R_n versus $t_n - t_n - 1$

(c) R_n versus $RTT * R_{n-1}$ (d) $Ownd$ versus RTT

Fig. 4. Correlation analysis of available bandwidth estimator in TCP-Jersey

can be seen that R_n is inversely proportional to $t_n - t_{n-1}$. The figure too shows the negative correlation.

The third analysis was performed on the correlativity between R_n and $RTT * R_{n-1}$. Just by looking at the scatter-plot in Fig. 4(c), it can bee seen that there is considerable correlation (correlation factor 0.863). This means that there is correlativity with $RTT * R_{n-1}$, but this is not from RTT but by R_{n-1}.

In particular, the scatter-plot in Fig. 4(d) shows a positive correlativity in which R_n makes close to linear changes when $RTT*R_{n-1}$ changes. That is, there is quite a close relation, but the relational factor is based on the previous value of R_n, which is R_{n-1}, rather than RTT. Accordingly, RTT has no correlativity with R_n. It can be seen in the above test that RTT does not have close correlativity in computing the rate value. This means that the rate in estimating available bandwidth is determined by some other variable. This produces counterevidence that RTT should be substituted with a more suitable value.

4.3 Verification of Optimum Congestion Window

The $Ownd$ value is computed by using the available bandwidth estimator R_n, RTT and seg_size. Results from analyzing the correlativity with RTT show that there is a negative correlation (correlation factor -0.024, P-value = 0.004) as can be seen in Fig. 4(d). Last of all, analyzing the correlativity between $Ownd$ and R_n show that a positive correlation (correlation factor 0.206, P-value=0.000) exists. *Ownd* increases in proportion to R_n. Through the above test, it can be seen that *Ownd* has correlation with *RTT* and R_n.

4.4 Discussion and Solution

Combining the test results, we can see that *RTT* is an effective value in producing *Ownd* in Eq. (3) based on Bandwidth Delay Product[13], but not appropriate in obtaining R_n in Eq. (2). Consequently, a new metric in producing the rate is needed. This should be substituted as a constant which considers properly the bottleneck link situation. Accordingly, if the wired/wireless border router can send the buffer value correctly to the sender, the above-mentioned problem can be solved. However, since the only realistic method is the 1 bit ECN, sufficient information cannot be sent.

After all, the solution to this problem is to make an extension for additional information fields by modifying TCP or stochastically marking and sending congestion information to ECN so that the sender can use the statistical information in estimating the buffer value. The advantage of the former is that even though TCP modification is necessary, network congestion state may be transmitted more accurately. And the advantage of the latter is that TCP does not have to be modified but then it will be more difficult to send accurate information. If congestion status of bottlenecklink can be identified accurately through new metrics, two effects are anticipated of proactive TCP congestion control methods for mobile HCI services of seamless application. The first is that if the cause is not due to congestion(CW=0,CE=0), that is, random error, the current rate is maintained and threshold value increased by a reasonable level, enhancing transmission performance. The second is that if it is due to congestion, *Cwnd* is not decreased inconsiderately by half and the buffer threshold value of the router is utilized as mush as possible. This removes resource waste and brings performance improvement in TCP transmission.

5 Conclusions and Future Work Directions

We have described the problems and issues by identifying the cause of packet loss, inspecting the congestion information transmission of existing TCP schemas and testing the validity of metrics in computing the rate of the proactive congestion control method proposed in TCP-Jersey, and presented its solutions. In particular, *RTT* used in TCP-Jersey was found to be useful in computing congestion window but not proper in computing transmission rate through a correlation analysis. Accordingly, we have presented another metric to be used in palace of *RTT* in computing the rate. This makes it possible to send a more accurate buffer capacity from the wired and wireless border router to the sender in order to produce an accurate rate. Through this, proactive congestion control in a wired and wireless heterogeneous environment can be performed by generating an optimum TCP transmission rate for congestion, and if loss is not due to congestion, the current rate may be maintained as it is. Future research in the two measures mentioned above should be done in depth to create an optimum rate computation model and mechanism for the mobile network environment.

References

1. Bardram, J., Nielsen, T.K.C.: Supporting Local Mobility in Healthcare by Application Roaming Among Heterogeneous Devices. In: Chittaro, L. (ed.) Mobile HCI 2003. LNCS, vol. 2795, pp. 161–176. Springer, Heidelberg (2003) 2
2. Sohraby, K.C.W., Daneshmand, M B.L., Hu, Y.: A survey of transport protocols for wireless sensor networks. Network, May-June 2006, vol. 20(3), pp. 34–40. IEEE, New York (2006)
3. Stevens, W.: TCP slow start, congestion avoidance, fast retransmit and fast recovery algorithms. IETF, RFC 2001 (1997)
4. Bakre, A., Badrinath, B.R.: I-TCP: indirect TCP for mobile hosts. In: Proceedings of the 15th International Conference on Distributed Computing Systems, June 1995, pp. 136–143 (1995)
5. Brown, K., Singh, S.: M-TCP: TCP for mobile cellular networks. ACM Computer Communications Review 27, 19–43 (1997)
6. Sinha, P., et al.: WTCP: A reliable transport protocol for wireless wide-area networks. In: Proceedings of ACM Mobicom, pp. 301–316 (1999)
7. Caceres, R., Iftode, L.: Improving the performance of reliable transport protocols in combilecomputing environments. IEEE Journal on Selected Areas in Communications 13(5), 850–857 (1995)
8. Brakmo, L.S., Peterson, L.L.: TCP Vegas: end to end congestion avoidance on a global Internet. IEEE Journal on Selected Areas in Communications 13(8), 1465–1480 (1995)
9. Akylidiz, I.F., Morabito, G., Palazzo, S.: TCP-Peach: a new congestion control scheme for satellite IP networks. IEEE/ACM Transactions on Networking 9(3), 307–321 (2001)
10. Casett, C., et al.: TCP Westwood: bandwidth estimation for enhanced transport over wireless links. In: Proceedings of ACM Mobicom, pp. 287–297 (2001)
11. Xu, K., Tian, Y., Ansari, N.: TCP-Jersey for wireless IP communications. IEEE Journal on Selected Areas in Communications 22(4), 747–756 (2004)
12. Mathis, M. et al.: TCP Selective ACKnowledgment Options, RFC 2018 (October 1996)
13. Chen, K., Xue, Y., Shah, S.H., Nahrstedt, K.: Understanding Bandwidth-Delay Product in Mobile Ad Hoc Networks. Elsevier Computer Communications (ComCom) Journal. Special Issue on Protocol Engineering for Wired and Wireless Networks (2004)
14. Clark, D.D., Fang, W.: Explicit allocation of best effort packet delivery service. IEEE/ACM Trans. Networking 6, 362–373 (1998)

A Bayesian Methodology for Semi-automated Task Analysis

Shu-Chiang Lin and Mark R. Lehto

School of Industrial Engineering, Purdue University, West Lafayette, IN 47906, USA
{slin,lehto}@purdue.edu

Abstract. This research proposes a new task analysis methodology that combines the fuzzy Bayesian model with classic task analysis methods to develop a semi-automated task analysis tool to better help traditional task analysts identify subtasks. We hypothesize that this approach could help task analysts identify activity units performed by the call center agent. The term activity units, in our study, represent the subtasks the agents perform during a remote troubleshooting process. We also investigate whether this tool could help predict the activity units as well. An effort-intensive field-based data collection for the call center's naturalistic decision making's environment was accomplished. A human expert and an additional 18 Purdue students participated in the validation of the assigned subtasks. The machine learning tool's performance was then examined. The preliminary results support our hypotheses that the fuzzy Bayesian based tool is able to learn and predict subtask categories from the agent/customer narrative telephone conversations.

Keywords: Task analysis, subtask, fuzzy Bayesian, machine learning, narratives.

1 Introduction

The purpose of this research is to propose a new task analysis methodology that combines a statistical approach, the fuzzy Bayesian model, with classic task analysis methods, to develop a semi-automated task analysis tool to better help traditional task analysts identify subtasks. We hypothesize that this approach could help task analysts identify activity units performed by the call center agent. The term activity units, in our study, represent the subtasks the agents perform during a remote troubleshooting process. We also investigate whether this tool could help predict the activity units as well [1].

1.1 Task Analysis

Task analysis is one of the oldest and most widely used methods in industrialized systems [2]. It has become a familiar tradition of practice for many human factors specialists. The essence of most task analysis techniques is a systematic identification and decomposition of a user's task into a number of small task components or activity units [3]. Although there are a wide range of available techniques to human factors specialists performing task analysis, identifying and decomposing a user's task into small task components remains as a well known difficult, impractically time-consuming, and expensive process that involves extensive manual effort [4].. Most of

M.J. Smith, G. Salvendy (Eds.): Human Interface, Part I, HCII 2007, LNCS 4557, pp. 697 – 704, 2007.

the task analysis methods also depend on how well a trained analyst understands the context in which he or she is analyzing the task; as a result, an inexperienced analyst will consume more time and more resources performing the analysis. Hence, there remains an essential and practical need for developing task analysis techniques to help practitioners perform task analysis efficiently and effectively.

1.2 Bayesian Methodology

There has been no formal research about utilizing a fuzzy Bayesian model based on semi-automated task analysis tool to help task analysts identify activity units. To learn why a fuzzy Bayesian method of particular is interest in developing an autonomous task analysis tool, we have reviewed some traditional knowledge acquisition strategies in terms of manipulating data. Studies have suggested statistical or probabilistic approaches for eliciting, filtering, parsing, and analyzing data or to classify narrative text into categories to provide decision support to troubleshooters because of their efficiency [5-7]. Among a number of these statistical approaches, applications of Bayesian model are proven to be successful in different disciplines. For example,; Hatakeyama et al. [8] show the Bayesian network model was effective to infer subjects' intention; Chatterjee [9] shows that the fuzzy Bayes model was superior in performance to that of the keyword model by classifying significant amount of free-text accident narratives that the keyword model had failed to classify; Zhou and Huang [10] verifies the efficiency and robustness of the Bayesian framework for hand tracking in high clutter background. Bayesian inference uses Bayes' Rule [11] to combine the various sources of evidence. Two Bayesian inferences are often proposed: classic Bayesian model and fuzzy Bayesian model. If multiple evidences are obtained, classic Bayesian assumes they are conditionally independent given that the hypothesis is true. On the contrary, fuzzy Bayesian makes strong dependence assumptions of the evidences. This approach reflects on the strongest evidence presented with no consideration given to negative evidence while co-occurrence of positive evidence is not aggregated. It is a more relaxed model that produces a relatively simple computation and conservative lower bound. Studies such as Zhu and Lehto [12] propose the Bayesian inference rule as a good statistical model to fit well the contingent relationship between index terms and words in the text. Other research such as Leman and Lehto [13], Qiu and Agogino [14], McCarthy [15], successfully apply fuzzy Bayesian model as a predictive technique to identify and/or predict print defect categories as well as help users diagnose print defects.

The fuzzy Bayesian model is an easily applied approach and could be implemented immediately by a task analyst with a laptop. We are convinced that application of the fuzzy Bayesian approach could lead to the implementation of more complete and systematic use of automated task analysis tools in the future.

2 Methods

Case studies are always vital and have contributed to both theoretical and empirical research in the naturalistic decision making (NDM) environment [16-17]. In this research, we utilize a conventional task analysis tool by recording traditional one-on-one phone conversations at the call center where the customers call to report various printer problems and the knowledge agents help troubleshoot the problems on the phone. The experimental process is conducted in four phases data collection, data manipulation, machine learning environment, and tool evaluation [1].

2.1 Data Collection and Manipulation

With the help of the Remote Print Defect Diagnostics research team, an effort-intensive field-based approach for the call center's naturalistic decision making's environment was accomplished. There were over 120 customer calls collected onsite. 24 experienced agents were observed during their normal working hours.. They were all 25 years of age or older and had at least 4 months or longer and had at least 1500 calls troubleshooting experience. Prior to the data collection, all the knowledge agents were voluntarily participated well informed the purpose of our research and our data collection process, which were the same as what the knowledge agents' mentors and/or their supervisors were constantly doing - monitoring the knowledge agents on the phone on a daily basis to help improve their remote troubleshooting skills as well as communication abilities. More than 60 hours of video/audio data with a total of 150 calls were collected.

In Phase 2 data manipulation, the knowledge agents' daily routine troubleshooting tasks were documented, the recorded audio data were transcribed into 770 pages of text with about 179,000 words, and a task decomposition table for agent with 72 subtasks was created. These subtasks were then tagged to the agent-customer conversation narratives to train and test the machine learning tool.

A significant amount of time and effort was spent in defining and assigning the subtask categories as well as in validating their reliability. There were a total of 20 people, including 2 human experts and 18 Purdue senior or junior students, participated in a four-step validation process to cross validate the assigned subtask categories. All of the students were eighteen years of age or older and had experiences using printers. They all had task analysis background. The author discussed constantly with the participants in validation process the discrepancies and made revisions to the assigned subtasks accordingly. The validation took approximately a year to complete.

2.2 Machine Learning Environment

In our study, we use the fuzzy Bayesian model to classify subtasks into categories using the expression:

$$P(S_i|E) = MAX_j \frac{P(E_j|S_i)P(S_i)}{P(E_j)} \tag{1}$$

where $P(S_i|E)$ is the posterior probability of subtask S_i is true given the evidence E is present, $P(E_j|S_i)$ is the probability of obtaining the evidence E_j given that the subtask S_i is true, $P(S_i)$ is the prior probability of the subtask being true prior to obtaining the evidence E_j, and "MAX is used to assign the maximum value of calculated $P(E_j|S_i)*P(S_i)/P(E_j)$. In our study, we consider the words used by the agent and customer as the sources of evidence, and describe what the agent said when performing subtask A_i by a word vector $WAi = (WAi_1, WAi_2, \ldots ,WAi_q)$ of length q, where WAi_1, WAi_2, … ,WAi_q are the q words in the dialog. Similarly, we describe what the customer said in response to the agent's question by a second word vector WCi = (WCi_1, WCi_2, … ,WCi_q) of length q, where WCi_1, WCi_2, … ,WCi_q are the q words in the dialog. We also consider that A_i is potentially relevant to WAi, WCi-1, and WCi for i greater than 1.

Following this approach we derive equation (2) from equation (1) to calculate the posterior probability of subtask A_i:

$$P(Ai|WAi,WCi,WCi\text{-}1)=MAX[P(WAi|Ai)P(Ai)/P(WAi),\ P(WCi|Ai)P(Ai)/P(WCi),\ P(WCi\text{-}1|Ai)P(Ai)/P(WCi\text{-}1)]$$
$$=MAX\{MAX_j\,[P(WAi_j|Ai)*P(Ai)/P(WAi_j)],\ MAX_j\,[P(WCi_j|Ai)*P(Ai)/P(WCi_j)],\ MAX_j\,[P(WC(i-1)_j|Ai)*P(Ai)/P(WC(i-1)_j)]\}\quad \text{for } j=1,2,\dots.q \qquad (2)$$

In practice, we implement Textminer program [1] as our machine learning tool to carry out the above calculations. A snapshot of the Textminer interface is shown in Figure 1.

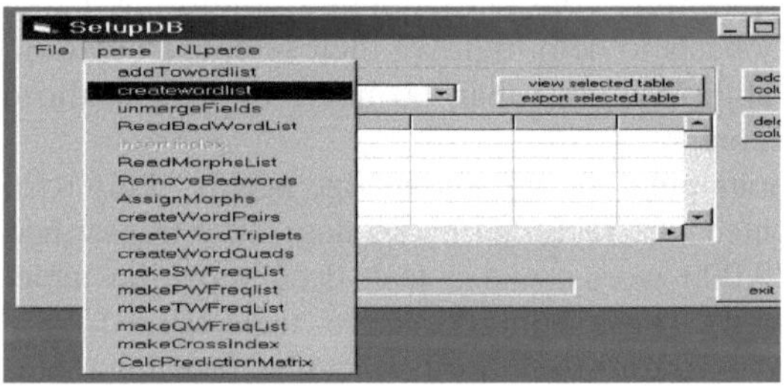

Fig. 1. A snapshot of the machine learning tool – Textminer*

The inputs of the Textminer learning tool are the words said by the agent and the customer. The conversation narratives were parsed and the keywords were elicited. The combinations of words appearing in the narratives were than used as candidates for subtask category predictors. Two-thirds of the 5184 narratives were randomly assigned to the training data set and the other one-third of the narratives was used as testing set. This process was repeated 10 times to allow for cross-validation of the accuracy of the model predictions. The output of each run was the conditional probability of subtask S_i performed by the agent. The preliminary tool performance was evaluated by hit rate and false alarm rates.

3 Results and Discussion

Extensive calculations were carried out and substantial amounts of quantitative and qualitative results, including over 400 tables with contents ranging from 41 to 4.3 million records, were generated during the model development process. The preliminary results summarized below have supported the hypothesis that the tool developed is able to learn and predict subtask categories from the telephone conversation between the customers and the human agents.

* All figures and tables illustrated in this paper are cited from reference [1].

3.1 Model Predictions

A total of thirteen prediction results were obtained and recorded in thirteen prediction tables respectively for fuzzy Bayes model. These tables, each with 5184 prediction results, predicted the subtask categories for various word combinations, ranging from single word alone to single_pair_three_four combination. Table 1 illustrates an example of the prediction table for combinations of single_pair_three_four words. Note that each record reveals the predicted subtask category, the keyword(s) used to make the prediction, and the strength of the subtask category for the particular predictor.

Table 1. A sample list of fuzzy Bayes prediction results based on a combination of single_pair_three_four words predictions*

ID[1]	Category[2]	Prediction[3]	PredAcc[4]	Strength[5]	Predictor[6]
2	211	211	1	1.0000	firstname&your
6	251	251	1	0.8333	Model&of&printer
7	627	626	0	0.6154	back&I'll&right
9	253	253	1	0.9091	on&one&serialnumber
10	254	371	0	0.1149	Right
11	252	621	0	0.3333	Its&said
14	541	541	1	1.0000	bye&day
17	611	521	0	0.5556	csonumber

[1]ID: Identification of a dialog/narrative.
[2]Category: Pre-assigned subtask category.
[3]Prediction: The predicted subtask category based on a combination of words.
[4]PredAcc: Prediction accuracy for that dialog/narrative. 1 means correct hit, 0 if incorrect hit.
[5]Strength: The probability of the subtask category for the particular Predictor.
[6]Predictor: The keyword(s) used to make the prediction.

A contingency table listing three most important and three least important predictors from each subtask category was also derived to illustrate the relationships between the word/word combinations and subtask categories. The word/word combinations are stratified based on subtask categories, strength of the predictions, and types of word combinations. Table 2 illustrates an example of one subtask with a portion of the word combinations.

3.2 Model Performance

The overall hit rates and number of times subtasks were correctly predicted are given in Table 3. Separate values also are given for each of the 10 cross-validation runs. The predicted results are further broken down into those for the training and testing sets. Some variation in the number of hits among the different runs was present, but overall the results are remarkably consistent, as illustrated by Figure 2.

* All figures and tables illustrated in this paper are cited from reference [1].

Table 2. Partial listing of a contingency table for one subtask and word/word combinations*

A[1]	S/W[2]	Single Word[3]	SF[4]	SP[5]	Pair Words[6]	PF[7]	PP[8]
321	S	EIGHTZEROZEROnumber	9	1.0000	EIGHTZEROZEROnumber&Okay	7	1.0000
321	S	dollarTWOPOINTFIVE	6	1.0000	charge&there	6	1.0000
321	W	on	10	0.0085	of&Okay	6	0.0165
321	W	Right	5	0.0084	Okay&on	7	0.0140

[1]A: Pre-assigned subtask category.
[2]S/W: Strong/Weak. Strong means that word or word combination is among the top three predictors with highest conditional probability of the subtask category. Weak means that word or word combination is among the bottom three predictors.
[3]Single Word: Parsed single word from single-word frequency list.
[4]SF: Single word frequency. Number of times that single word was assigned that subtask category.
[5]SP: Strength/conditional probability of that single word for the subtask category.
[6]Pair Words: Parsed two-word combination from two-word frequency list.
[7]PF: Pair words frequency. Number of times that two-word combination was assigned that subtask category.
[8]PP: Strength/conditional probability of that two-word combination for the subtask category.

Table 3. Number of hits based on 10 randomized cross-validation runs*

Run	1	2	3	4	5	6	7	8	9	10	Mean	Std.Dev	Hit
All data	2931	2784	2791	2809	2798	2830	2859	2758	2891	2867	2831.8	54.15	0.5464
Training	2212	2085	2102	2090	2039	2130	2159	2068	2170	2146	2120.2	52.60	0.6106
Testing	719	699	689	719	758	701	701	689	720	722	711.8	20.68	0.4160

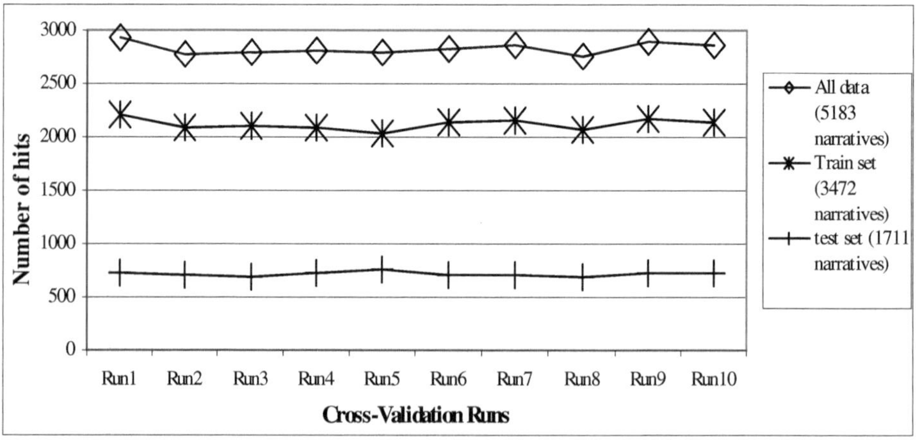

Fig. 2. Number of hits based on 10 randomized cross-validation runs*

* All figures and tables illustrated in this paper are cited from reference [1].

Table 4. ANOVA table to test effect of run on number of hits*

Source of vari.	Sum of squares	DF	Mean squares	F	P-value	F crit
Between Runs	0.0024	9	0.0003	0.0269	0.9999	2.3928
Within Runs	0.1987	20	0.010			
Total	0.2011	29				

ANOVA analysis was conducted to test whether differences between the ten runs statistically significant. The results of this analysis are given in Table 4, and confirm that the differences among the ten runs were not statistically significant ($p = 0.9999$).

The statistical result reveals the tool's robust prediction for our dataset, and supports the conclusion that we can randomly pick up two-thirds of the dataset as a training set and the remaining one-third as a testing dataset and run the tool one time instead of ten times while achieving similar predictions without too much variation. Student's paired t-test of the average hit rates across data sets shows the tool performs more accurately for training set than for testing set ($p=2.409x10^{-10}$ with a two-tailed distribution).

Since the average hit rate 56.55% is significantly greater than the false alarm rate, we conclude that our hypotheses are supported and that the tool is able to learn or predict subtask categories from the agent's and the customer's telephone conversations.

4 Promising Future Studies and Applications

The results presented here serve as a starting point and resource that enables future studies of applications of Bayesian theories to the very important research area of task analysis. With the abundance of both quantitative and qualitative data and results obtained in this work, many advanced analyses such as analyses on coarse subtask categories are being carried out now. The promising results have also provided the resources to explore exciting applications of hybrid Bayes models to many other areas of the naturalistic decision making environment.

References

1. Lin, S.: Machine A Fuzzy Bayesian Model Based Semi-Automated Task Analysis. Unpublished Ph.D. Dissertation. Purdue University, West Lafayette, IN (2006)
2. Luczak, H., Kabel, T., Licht, T.: Task Design and Motivation. In: Salvendy, G. (ed.) Handbook of Human Factors and ergonomics, 3rd edn., Ch. 15, pp. 384–427. Wiley, Hoboken, NJ (2006)
3. Morgeson, F.P., Medsker, G.J., Campion, M.A.: Job and team design. In: Salvendy, G. (ed.) Handbook of Human Factors and ergonomics, 3rd edn., Ch. 16, pp. 428–457. Wiley, Hoboken, NJ (2006)
4. Jeffries, R.: The role of task analysis in design of software. In: Helander, M., Landauer, T.K., Prabhu, P.V. (eds.) Handbook of human-computer interaction, 2nd edn., pp. 347–359. Elsevier Science, Amsterdam (1997)

* All figures and tables illustrated in this paper are cited from reference [1].

5. Yamamoto, H., Sagisaka, Y.: Multi-Class Composite N-gram based on connectiondirection. In: Proceedings IEEE International Conference on Acoustics, Speech & Signal Processing, vol. 1, pp. 533–536 (1999)
6. Salton, G., McGill, M.J.: Introduction to Modem Information Retrieval. McGraw-Hill, New York (1983)
7. Bookstein, A.: Probability and fuzzy-set applications to information retrieval. Annual review conformation science and technology 20, 117–151 (1985)
8. Hatakeyama, N., Furuta, K., Nakata, K.: Model of Intention Inference Using Bayesian Network. In: Stephanidis, C., Jacko, J.A. (eds.) Human-Centred Computing (v2). Human-computer interaction: proceedings of HCI International 2003, pp. 390–394. Lawrence Erlbaum, Mahwah, NJ (2003)
9. Chatterjee, S.: A connectionist approach for classifying accident narratives. Unpublished Ph.D. Dissertation. Purdue University, West Lafayette, IN (1998)
10. Zhou, H., Huang, T.S.: A Bayesian Framework for Real-Time 3D Hand Tracking in High Clutter Background. In: Jacko, J.A., Stephanidis, C. (eds.) Human-Centred Computing (v1). Human-computer interaction: proceedings of HCI International 2003, pp. 1303–1307. Lawrence Erlbaum, Mahwah, NJ (2003)
11. Savage, L.J.: The Foundations of Statistics. Wiley, New York (1954)
12. Zhu, W., Lehto, M.R.: Decision support for indexing and retrieval of information in hypertext system. International Journal of Human Computer Interaction 11, 349–371 (1999)
13. Leman, S., Lehto, M.R.: Interactive decision support system to predict print quality. Ergonomics 46(1-3), 52–67 (2003)
14. Qiu, S., Agogino, A.M.: A Fusion of Bayesian and Fuzzy Analysis for Print Faults Diagnosis. In: Proceedings of The International Society for Computers and Their Application-ISCA 16th International Conference, pp. 229–232 (2001)
15. McCarthy, P.: Machine Learning Applications For Pattern Recognition Within Call Center Data. Unpublished master thesis. Purdue University, West Lafayette, IN (2002)
16. Zsambok, C.E.: Naturalistic Decision Making: Where Are We Now. In: Zsambok, C.E., Klein, G.A. (eds.) Naturalistic decision making, Ch. 1, pp. 3–16. Lawrence Erlbaum, Mahwah, NJ (1997)
17. Hutton, R.J.B., Miller, T.E., Thordsen, M.L.: Decision-Centered Design: Leveraging Cognitive Task Analysis in Design. In: Hollnagel, E. (ed.) Handbook Of Cognitive Task Design, pp. 383–416. Erlbaum, Mahwah, NJ (2003)

Machine Learning and Applications for Brain-Computer Interfacing

K.-R. Müller[1,2], M. Krauledat[1,2], G. Dornhege[2], G. Curio[3], and B. Blankertz[2,⋆]

[1] Technical University Berlin, Str. d. 17. Juni 135, 10 623 Berlin, Germany
[2] Fraunhofer FIRST.IDA, Kekuléstr. 7, 12 489 Berlin, Germany
[3] Dept. of Neurology, Campus Benjamin Franklin, Charité University Medicine Berlin, Hindenburgdamm 30, 12 203 Berlin, Germany
klaus@first.fraunhofer.de

Abstract. This paper discusses machine learning methods and their application to Brain-Computer Interfacing. A particular focus is placed on linear classification methods which can be applied in the BCI context. Finally, we provide an overview on the Berlin-Brain Computer Interface (BBCI).

1 Introduction

Brain-Computer Interfacing is an interesting, active and highly interdisciplinary research topic ([2,3,4,5]) at the interface between medicine, psychology, neurology, rehabilitation engineering, man-machine interaction, machine learning and signal processing. A BCI could, e.g., allow a paralyzed patient to convey her/his intentions to a computer application. From the perspective of man-machine interaction research, the communication channel from a healthy human's brain to a computer has not yet been subject to intensive exploration, however it has potential, e.g., to speed up reaction times, cf. [6] or to supply a better understanding of a human operator's mental states.

Classical BCI technology has been mainly relying on the adaptability of the human brain to biofeedback, i.e., a subject learns the mental states required to be understood by the machines, an endeavour that can take months until it reliably works [7,8].

The Berlin Brain-Computer Interface (BBCI) pursues another objective in this respect, i.e., to impose the main load of the learning task on the 'learning machine', which also holds the potential of adapting to specific tasks and changing environments given that suitable machine learning (e.g. [9]) and adaptive signal processing (e.g. [10]) algorithms are used. Short training times, however, imply the challenge that only few data samples are available for learning to characterize the individual brain states to be distinguished. In particular when dealing with few samples of data (trials of the training session) in a high-dimensional feature space (multi-channel EEG, typically several features per channel), overfitting needs to be avoided. It is in this high dimensional – small sample statistics scenario where modern machine learning can prove its strength.

⋆ The studies were partly supported by the *Bundesministerium für Bildung und Forschung* (BMBF), FKZ 01 IBE 01A/B, and by the IST Programme of the European Community, under the PASCAL Network of Excellence, IST-2002-506778. This paper is based on excerpts of [1].

M.J. Smith, G. Salvendy (Eds.): Human Interface, Part I, HCII 2007, LNCS 4557, pp. 705–714, 2007.

The present paper introduces basic concepts of linear classification (for a discussion of nonlinear methods in the context of BCI, see [11,9,12,13,14]). Finally, we briefly describe our BBCI activities where some of the discussed machine learning ideas come to an application and conclude.

2 Linear Methods for Classification

In BCI research it is very common to use linear classifiers, but although linear classification already uses a very simple model, things *can* still go terribly wrong if the underlying assumptions do not hold, e.g. in the presence of outliers or strong noise which are situations very typically encountered in BCI data analysis. We will discuss these pitfalls and point out ways around them.

Let us first fix the notation and introduce the linear hyperplane classification model upon which we will rely mostly in the following (cf. Fig. 1, see e.g. [15]). In a BCI set-up we measure $k = 1 \dots K$ samples $\mathbf{x}_k$, where $\mathbf{x}$ are some appropriate feature vectors in n dimensional space. In the training data we have a class label, e.g. $y_k \in \{-1,+1\}$ for each sample point $\mathbf{x}_k$. To obtain a linear hyperplane classifier

$$\mathbf{y} = \operatorname{sign}\left(\mathbf{w}^\top \mathbf{x} + b\right) \tag{1}$$

we need to estimate the normal vector of the hyperplane $\mathbf{w}$ and a threshold b from the training data by some optimization technique [15]. On unseen data $\mathbf{x}$, i.e. in a BCI feedback session we compute the projection of the new data sample onto the direction of the normal $\mathbf{w}$ via Eq.(1), thus determining what class label $\mathbf{y}$ should be given to $\mathbf{x}$ according to our linear model.

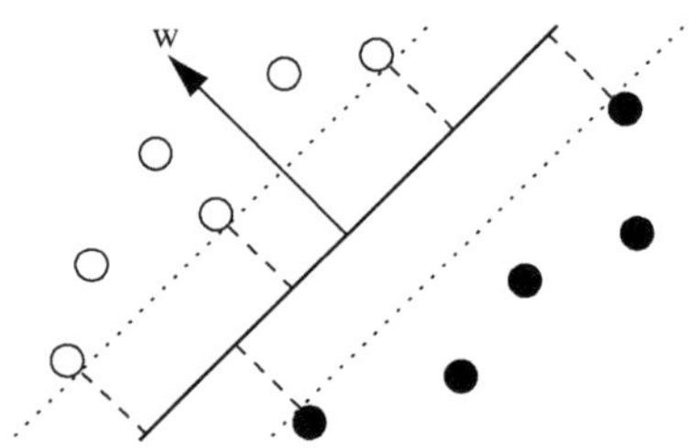

Fig. 1. Linear classifier and margins: A linear classifier is defined by a hyperplane's normal vector $\mathbf{w}$ and an offset b, i.e. the decision boundary is $\{\mathbf{x} \,|\, \mathbf{w}^\top \mathbf{x} + b = 0\}$ (thick line). Each of the two halfspaces defined by this hyperplane corresponds to one class, i.e. $f(\mathbf{x}) = \operatorname{sign}(\mathbf{w}^\top \mathbf{x} + b)$. The *margin* of a linear classifier is the minimal distance of any training point to the hyperplane. In this case it is the distance between the dotted lines and the thick line. From [9].

2.1 Optimal Linear Classification: Large Margins Versus Fisher's Discriminant

Linear methods assume a linear separability of the data. We will see in the following that the optimal separating hyperplane from last section maximizes the *minimal* margin (minmax). In contrast, Fisher's discriminant maximizes the *average* margin, i.e., the margin between the class means.

Large Margin Classification. For linearly separable data there is a vast number of possibilities to determine $(\mathbf{w}, b)$, that all classify correctly on the training set, however that vary in quality on the unseen data (test set). An advantage of the simple hyperplane classifier (in canonical form cf. [16]) is that literature (see e.g. [15,16]) tells us how to select the *optimal* classifier $\mathbf{w}$ for unseen data: it is the classifier with the largest margin $\rho = 1/\|\mathbf{w}\|_2^2$, i.e. of minimal (euclidean) norm $\|\mathbf{w}\|_2$ [16] (see also Fig. 1). Linear Support Vector Machines (SVMs) realize the large margin by determining the normal vector $\mathbf{w}$ according to

$$\min_{\mathbf{w},b,\xi} \; {}^1\!/_2 \|\mathbf{w}\|_2^2 + {}^C\!/_K \|\xi\|_1 \quad \text{subject to} \tag{2}$$
$$y_k(\mathbf{w}^\top \mathbf{x}_k + b) \geqslant 1 - \xi_k \quad \text{and}$$
$$\xi_k \geqslant 0 \qquad \text{for } k = 1, \ldots, K,$$

where $\|\cdot\|_1$ denotes the ℓ_1-norm: $\|\xi\|_1 = \sum |\xi_k|$. Here the elements of vector ξ are slack variables and parameter C controls the size of the margin vs. the complexity of the separation. While the user has not to care about the slack variables, it is essential to select an approppriate value for the free parameter C for each specific data set. The process of choosing C is called model selection, see e.g. [9]. One particular strength of SVMs is that they can be turned in nonlinear classifiers in an elegant and effective way (see e.g. [16,9,12]).

Fisher's Discriminant. Fisher's discriminant computes the projection $\mathbf{w}$ differently. Under the restrictive assumption that the class distributions are (identically distributed) Gaussians of equal covariance, it can be shown to be Bayes optimal. The separability of the data is measured by two quantities: How far are the projected class means apart (should be large) and how big is the variance of the data in this direction (should be small). This can be achieved by maximizing the so-called Rayleigh coefficient of between and within class variance with respect to $\mathbf{w}$ [17,18]. The slightly stronger assumptions have been fulfilled in several of our BCI experiments e.g. in [13,14]. When the optimization to obtain (regularized) Fisher's discriminant is formulated as a mathematical programm, cf. [19,9,20], it resembles the SVM:

$$\min_{\mathbf{w},b,\xi} \; {}^1\!/_2 \|\mathbf{w}\|_2^2 + {}^C\!/_K \|\xi\|_2^2 \quad \text{subject to}$$
$$y_k(\mathbf{w}^\top \mathbf{x}_k + b) = 1 - \xi_k \quad \text{for } k = 1, \ldots, K.$$

2.2 Some Remarks About Regularization and Non-robust Classifiers

Linear classifiers are generally more robust than their nonlinear counterparts, since they have only limited flexibility (less free parameters to tune) and are thus less prone to overfitting. Note however that in the presence of strong noise and outliers *even* linear systems can fail. In the cartoon of Fig.2 one can clearly observe that one outlier or strong noise event can change the decision surface drastically, if the influence of single data points on learning is not limited. Although this effect can yield strongly decreased classification results for linear learning machines, it can be even more devastating for nonlinear methods. A more formal way to control one's mistrust in the available training data, is to use regularization (e.g. [21,22,15]). Regularization helps to limit (a) the influence of outliers or strong noise (e.g. to avoid Fig.2 middle), (b) the complexity of the classifier (e.g. to avoid Fig.2 right) and (c) the raggedness of the decision surface (e.g. to avoid Fig.2 right). No matter whether linear or nonlinear methods are used, one should *always* regularize, – in particular for BCI data!

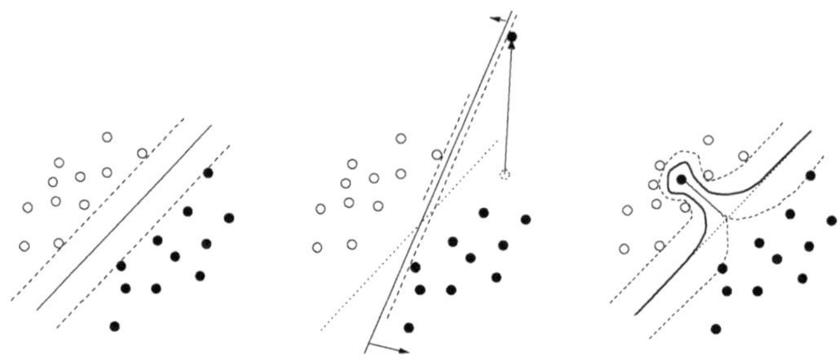

Fig. 2. The problem of finding a maximum margin "hyper-plane" on reliable data (left), data with an outlier (middle) and with a mislabeled pattern (right). The solid line shows the resulting decision line, whereas the dashed line marks the margin area. In the middle and on the right the original decision line is plotted with dots. Illustrated is the noise sensitivity: only one strong noise/outlier pattern can spoil the whole estimation of the decision line. From [23].

3 The Berlin Brain-Computer Interface

The Berlin Brain-Computer Interface is driven by the idea to shift the main burden of the learning task from the human subject to the computer under the motto 'let the machines learn'. To this end, the machine learning methods presented in the previous sections are applied to EEG data from selected BBCI paradigms: selfpaced [13,14] and imagined [24,25,26] experiments.

3.1 Bereitschaftspotential Experiments

In preparation of motor tasks, a negative readiness potential precedes the actual execution. Using multi-channel EEG recordings it has been demonstrated that several brain areas contribute to this negative shift (cf. [27,28]). In unilateral finger or hand movements the negative shift is mainly focussed on the frontal lobe in the area of the corresponding motor cortex, i.e., contralateral to the performing hand. Based on the laterality of the pre-movement potentials it is possible to discriminate multi-channel EEG recordings of upcoming left from right hand movements. Fig. 3 shows the lateralized readiness potential during a 'self-paced' experiment, as it can be revealed here by averaging over 260 trials in one subject.

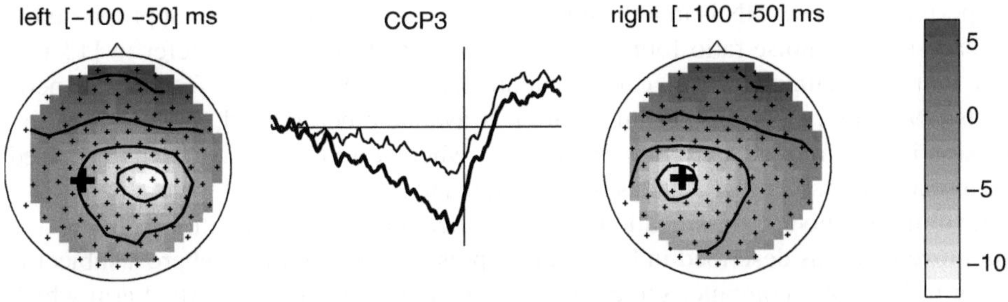

Fig. 3. The scalp plots show the topography of the electrical potentials prior to keypress with the left resp. right index finger. The plot in the middle depicts the event-related potential (ERP) for left (thin line) vs. right (thick line) index finger in the time interval -1000 to -500 ms relative to keypress at electrode position CCP3, which is marked by a bigger cross in the scalp plots. The contralateral negativation (lateralized readiness potential, LRP) is clearly observable. Approx. 260 trials per class have been averaged.

In the 'self-paced' experiments, subjects were sitting in a normal chair with fingers resting in the typing position at the computer keyboard. In a deliberate order and on their own free will (but instructed to keep a pace of approximately 2 seconds), they were pressing keys with their index and little fingers.

EEG data was recorded with 27 up to 120 electrodes, arranged in the positions of the extended 10-20 system, referenced to nasion and sampled at 1000 Hz. The data were downsampled to 100 Hz for further offline analyses. Surface EMG at both forearms was recorded to determine EMG onset. In addition, horizontal and vertical electrooculograms (EOG) were recorded to check for correlated eye movements.

In [6], it has been demonstrated that when analyzing LRP data offline with the methods detailed in the previous sections, classification accuracies of more than 90% can be reached at 110 ms before the keypress, i.e. a point in time where classification on EMG is still at chance level. These findings suggest that it is possible to use a BCI in time critical applications for an early classification and a rapid response.

Table 1 shows the classification results for one subject when comparing different machine learning methods. Clearly regularization and careful model selection are mandatory which can, e.g., be seen by comparing LDA and RLDA. Of course, regularization is of more importance the higher the dimensionality of features is. The reason of the very

Table 1. Test set error (± std) for classification at 110 ms before keystroke; ›mc‹ refers to the 56 channels over (sensori) motor cortex, ›all‹ refers to all 105 channels. The algorithms in question are Linear Discriminant Analysis (LDA), Regularized Linear Discriminant Analysis (RLDA), Linear Programming Machine (LPM), Support Vector Machine with Gaussian RBF Kernel (SVMrbf) and *k*-Nearest Neighbor (*k*-NN).

channels	LDA	RLDA	LPM	SVMrbf	*k*-NN
all	16.9±1.3	8.4±0.6	7.7±0.6	8.6±0.6	28.4±0.9
mc	9.3±0.6	6.3±0.5	7.4±0.7	6.7±0.7	22.0±0.9

bad performance of k-NN is that the underlying Euclidean metric is not appropriate for the bad signal-to-noise ratio found in EEG trials. For further details refer to [13,14]. Note that the accuracy of 90% can be maintained in recent realtime feedback experiments [29]. Here, as no trigger information is available beforehand, the classification decision is split into one classifier that decides whether a movement is being prepared and a second classifier that decides between left and right movement to come.

A similar method of classification can be applied for the discrimination of finger and arm movements, as described in [30]. This is presented in Figure 4, where a subject is sitting in front of a computer screen that shows the movement of a virtual arm which is controlled by a BCI classifier output. The classification is based entirely on pre-movement activity of the EEG, while the subject presses keys with the right index finger or lifts the arm. Scenarios like this can be used for intuitive interaction with a virtual environment.

Fig. 4. The virtual arm in the X-Rooms™virtual reality framework is an example for the applicability of slow cortical potentials in BCI scenarios. See text for details.

3.2 Motor Imagery Experiments

During imagination of a movement, a lateralized attenuation of the μ- and/or central β-rhythm can be observed localized in the corresponding motor and somatosensory cortex. Besides a usual spectral analysis, this effect can be visualized by plotting event-related desynchronization (ERD) curves [31] which show the temporal evolution of the band-power in a specified frequency band. A typical averaged ERD is shown in Fig. 5.

We performed experiments with 6 healthy subjects performing motor imagery. The subjects were sitting comfortably in a chair with their arms in a relaxed position on an arm rest. Two different sessions of data collection were provided: In both a target "L", "R" and "F" (for left, right hand and foot movement) is presented for the duration of 3.5 seconds to the subject on a computer screen. In the first session type this is done by visualizing the letter on the middle of the screen. In the second session the left, right or lower triangle of a moving gray rhomb is colored red. For the whole length of this period, the subjects were instructed to imagine a sensorimotor sensation/movement in

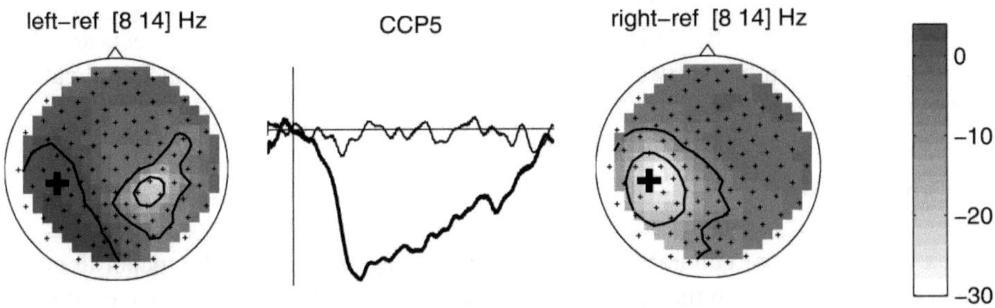

Fig. 5. This scalp plots show the topography of the band power in the frequency band 8–14 Hz relative to a reference period. The plot in the middle shows ERD curves (temporal evolution of band power) at channel CCP5 (mark by a bigger cross in the scalp plots) for left (thin line) and right (thick line) hand motor imagery. The contralateral attenuation of the μ-rhythm during motor imagery is clearly observable. For details on ERD, see [31].

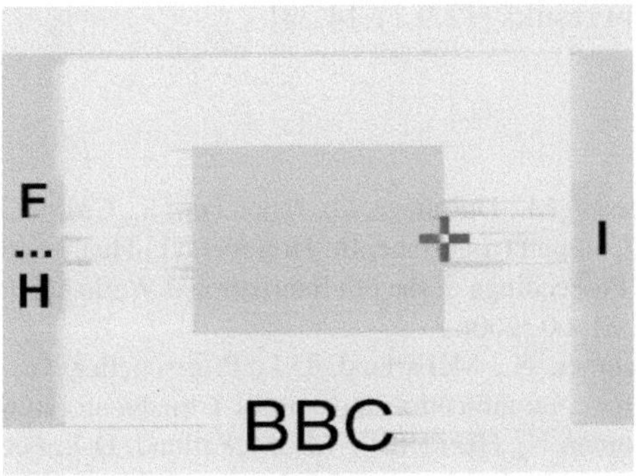

Fig. 6. This screenshot presents an application of the above mentioned 1-D cursor movement. By selecting of one of the targets on the sides of the screen, it is possible to select groups of letters of the alphabet. After the first selection, the choice of letters can be further narrowed by repeated selections, until a single letter remains. This process can then be used as a text input device which is only driven by the BCI.

left hand, right hand resp. one foot. After stimulus presentation, the screen was blank for 1.5 to 2 seconds. In this manner, 35 trials per class per session were recorded. After 25 trials, there was a short break for relaxation. Four sessions (two of each training type) were performed. EEG data was recorded with 128 electrodes together with EMG from both arms and the involved foot, and EOG as described above.

An offline machine learning analysis of the "imagined"-experiments yields again high classification rates (up to 98.9% with the feature combination algorithm PROB [25,24]), which predicts the feasibility of this paradigm for online feedback situations (see also [26]). In fact, our recent online experiments have confirmed this prediction by showing high bitrates for several subjects. These subjects were untrained and had to play video games like 'brain pong', 'basket' and 'controlled 1-D cursor movement'

[32]. Depending on the 'game' scenario the best subjects could achieve information transfer rates of up to 37 Bits/min. With an accuracy of this magnitude, subjects are also able to operate text input software as presented in Figure 6. Details on this application can be found in [33].

4 Conclusion

After a brief review of general linear machine learning techniques, this paper demonstrated their application in the context of real BCI-experiments. Using these techniques, it can be seen that the paradigm shift away from subject training to individualization and adaptation ('let the machines learn') of the signal processing and classification algorithm to the specific brain 'under study' holds the key to the success of the BBCI. Being able to use (B)BCI for untrained subjects dramatically enhances and broadens the spectrum of practical applications in human-computer interfacing.

Acknowledgments. We thank our co-authors from previous publications for letting us use the figures and joint results [12,9,11,14,23].

References

1. Müller, K.-R., Krauledat, M., Dornhege, G., Jähnichen, S., Curio, G., Blankertz, B.: A note on the Berlin Brain-Computer Interface. In: Hommel, G., Huanye, S. (eds.) Human Interaction with Machines: Proceedings of the 6th International Workshop held at the Shanghai Jiao Tong University, pp. 51–60 (2006)
2. Wolpaw, J.R., Birbaumer, N., McFarland, D.J., Pfurtscheller, G., Vaughan, T.M.: Brain-computer interfaces for communication and control. Clin. Neurophysiol 113, 767–791 (2002)
3. Wolpaw, J.R., Birbaumer, N., Heetderks, W.J., McFarland, D.J., Peckham, P.H., Schalk, G., Donchin, E., Quatrano, L.A., Robinson, C.J., Vaughan, T.M.: Brain-computer interface technology: A review of the first international meeting. IEEE Trans. Rehab. Eng. 8(2), 164–173 (2000)
4. Kübler, A., Kotchoubey, B., Kaiser, J., Wolpaw, J., Birbaumer, N.: Brain-computer communication: Unlocking the locked. Psychol. Bull. 127(3), 358–375 (2001)
5. Eleanor, A., Curran, E.A., Stokes, M.J.: Learning to control brain activity: A review of the production and control of EEG components for driving brain-computer interface (BCI) systems. Brain Cogn. 51, 326–336 (2003)
6. Krauledat, M., Dornhege, G., Blankertz, B., Curio, G., Mülle, K.-R.: The Berlin brain-computer interface for rapid response. Biomed. Technik 49(1), 61–62 (2004)
7. Wolpaw, J.R., McFarland, D.J., Vaughan, T.M.: Brain-computer interface research at the Wadsworth Center. IEEE Trans. Rehab. Eng. 8(2), 222–226 (2000)
8. Birbaumer, N., Ghanayim, N., Hinterberger, T., Iversen, I., Kotchoubey, B., Kübler, A., Perelmouter, J., Taub, E., Flor, H.: A spelling device for the paralysed. Nature 398, 297–298 (1999)
9. Müller, K.-R., Mika, S., Rätsch, G., Tsuda, K., Schölkopf, B.: An introduction to kernel-based learning algorithms. IEEE Neural Networks 12(2), 181–201 (2001)
10. Haykin, S.S.: Adaptive Filter Theory. Prentice Hall, Englewood Cliffs (1995)
11. Müller, K.-R., Anderson, C.W., Birch, G.E.: Linear and non-linear methods for brain-computer interfaces. IEEE Trans. Neural Sys. Rehab. Eng. 11(2), 165–169 (2003)

12. Müller, K.-R., Krauledat, M., Dornhege, G., Curio, G., Blankertz, B.: Machine learning techniques for brain-computer interfaces. Biomed. Technik 49(1), 11–22 (2004)
13. Blankertz, B., Curio, G., Müller, K.-R.: Classifying single trial EEG: Towards brain computer interfacing. In: Diettrich, T.G., Becker, S. (eds.) Advances in Neural Inf. Proc. Systems (NIPS 01), vol. 14, pp. 157–164 (2002)
14. Blankertz, B., Dornhege, G., Schäfer, C., Krepki, R., Kohlmorgen, J., Müller, K.-R., Kunzmann, V., Losch, F., Curio, G.: Boosting bit rates and error detection for the classification of fast-paced motor commands based on single-trial EEG analysis. IEEE Trans. Neural Sys. Rehab. Eng. 11(2), 127–131 (2003)
15. Duda, R.O., Hart, P.E., Stork, D.G.: Pattern classification, 2nd edn. John Wiley & Sons, New York (2001)
16. Vapnik, V.N.: The nature of statistical learning theory. Springer, New York (1995)
17. Fisher, R.A.: he use of multiple measurements in taxonomic problems. Annals of Eugenics 7, 179–188 (1936)
18. Fukunaga, K.: Introduction to Statistical Pattern Recognition, 2nd edn. Academic Press, San Diego, London (1990)
19. Mika, S., Rätsch, G., Müller, K.-R.: A mathematical programming approach to the kernel Fisher algorithm. In: Leen, T.K., Dietterich, T.G., Tresp, V. (eds.) Advances in Neural Information Processing Systems, vol. 13, pp. 591–597. MIT Press, Cambridge, MA (2001)
20. Mika, S., Rätsch, G., Weston, J., Schölkopf, B., Smola, A., Müller, K.-R.: Constructing descriptive and discriminative non-linear features: Rayleigh coefficients in kernel feature spaces. IEEE Transaction on Pattern Analysis and Machine Intelligence 25(5), 623–628 (2003)
21. Poggio, T., Girosi, F.: Regularization algorithms for learning that are equivalent to multilayer networks. Science 247, 978–982 (1990)
22. Orr, G.B., Müller, K.-R. (eds.): Neural Networks: Tricks of the Trade. LNCS, vol. 1524. Springer, Heidelberg (1998)
23. Rätsch, G., Onoda, T., Müller, K.-R.: Soft margins for AdaBoost. Machine Learning 42(3), 287–320 (2001) also NeuroCOLT Technical Report NC-TR-1998-021
24. Dornhege, G., Blankertz, B., Curio, G., Müller, K.-R.: Increase information transfer rates in BCI by CSP extension to multi-class. In: Thrun, S., Saul, L., Schölkopf, B. (eds.) Advances in Neural Information Processing Systems, vol. 16, pp. 733–740. MIT Press, Cambridge, MA (2004)
25. Dornhege, G., Blankertz, B., Curio, G., Müller, K.-R.: Boosting bit rates in non-invasive EEG single-trial classifications by feature combination and multi-class paradigms. IEEE Trans. Biomed. Eng. 51(6), 993–1002 (2004)
26. Krauledat, M., Dornhege, G., Blankertz, B., Losch, F., Curio, G., Müller, K.-R.: Improving speed and accuracy of brain-computer interfaces using readiness potential features. In: Proceedings of the 26th Annual International Conference IEEE EMBS on Biomedicine, San Francisco (2004)
27. Cui, R.Q., Huter, D., Lang, W., Deecke, L.: Neuroimage of voluntary movement: topography of the Bereitschaftspotential, a 64-channel DC current source density study. Neuroimage 9(1), 124–134 (1999)
28. Lang, W., Zilch, O., Koska, C., Lindinger, G., Deecke, L.: Negative cortical DC shifts preceding and accompanying simple and complex sequential movements. Exp. Brain Res. 74(1), 99–104 (1989)
29. Krepki, R., Blankertz, B., Curio, G., Müller, K.-R.: The Berlin Brain-Computer Interface (BBCI): towards a new communication channel for online control in gaming applications (invited contribution). In: Journal of Multimedia Tools and Applications (2004)

30. Krepki, R.: Brain-Computer Interfaces: Design and Implementation of an Online BCI System of the Control in Gaming Applications and Virtual Limbs, Ph.D. thesis, Technische Universität Berlin, Fakultät IV – Elektrotechnik und Informatik (2004)
31. Pfurtscheller, G., da Silva, F.H.L.: Event-related EEG/MEG synchronization and desynchronization: basic principles. Clin. Neurophysiol. 110(11), 1842–1857 (1999)
32. Krepki, R., Blankertz, B., Curio, G., Müller, K.-R.: The Berlin Brain-Computer Interface (BBCI): towards a new communication channel for online control of multimedia applications and computer games. In: 9th International Conference on Distributed Multimedia Systems (DMS'03), pp. 237–244 (2003)
33. Krauledat, M., Shenoy, P., Blankertz, B., Rao, R.P.N., Müller, K.-R.: Towards Brain-Computer Interfacing. In: chapter Adaptation in CSP-based BCI systems, MIT Press, Cambridge, MA (2006) (in press)

An Information Filtering Method Based on User's Moods, Situations, and Preferences

Makoto Oka, Hirohiko Mori, and Masaru Saito

1-28-1 Tamadutsumi setagaya-ku, Tokyo, Japan
{oka,mori,saito}@si.musashi-tech.ac.jp

Abstract. We propose a Music Recommendation System (MRS) based on GPS information and user's mood. The system records example automatically when user selects the music, and extracts hypothesis by Inductive Logic Programming (ILP). Then the system calculates recommendation rate by the hypothesis, and recommend the music by using roulette strategy.

Keywords: Recommendation, Moods, Situation, ILP.

1 Introduction

Currently, the common ways to access music are to query by artist or song names or to browse recommendations generated by collaborative filtering. The first approach is that consumers need to know the name of the song or the artist. The second approach is only suitable when a fairly large number of consumers have heard and rated the music.

Recently, portable audio player such as iPod spread into everyone. People can have much music and listen to the music in many moods and situations. Because of mass of music, it is hard to find adequate music in particular situation. When people select the music, we assume that people select similar music in similar situation. Therefore, if some systems can suppose adequate music to user with passed on similar situation, the system can reduce user's selecting cost.

In this paper, we propose music recommendation system in previous idea. Existing recommendation system was often using music attribute or user's favorite for recommendation, but can't be considering user's situation [1]. Even same people, they require different music in different situation.

In recommendation, we add user's mood and GPS information for the situation. The system records example automatically when user selects the music, and extracts hypothesis by Inductive Logic Programming (ILP) [3] [4]. Then the system calculates recommendation rate by the hypothesis, and recommend the music by using roulette strategy.

2 System

2.1 System Structure

The system structure is displayed in figure 1.

M.J. Smith, G. Salvendy (Eds.): Human Interface, Part I, HCII 2007, LNCS 4557, pp. 715–719, 2007.

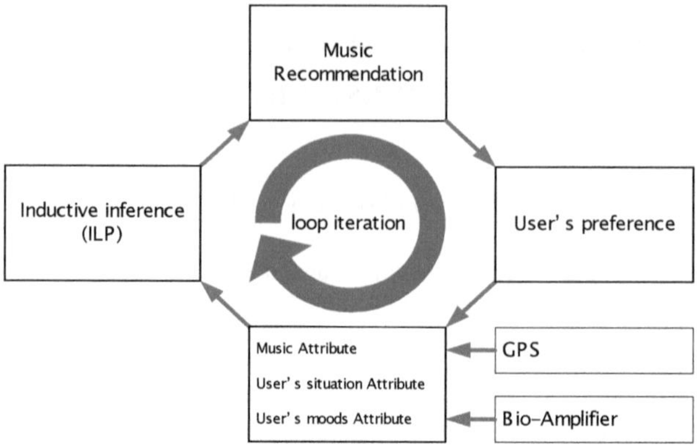

Fig. 1. Music recommendation system using ILP

2.2 Attribute Element

The music attribute are displayed in table 1. The release attribute change elapsed days since release date. The situations attribute are displayed in table 2. The moods attribute are displayed in table 3 [2].

Table 1. The music attribute that used for our system

Attribute name	comment
Title	Song title
Artist	Artist name
Album	Album title
Realese	Release date
Gender	Male of female
Genre	Japanese hip-hop, japanese rock, japanese enka, comedy, hip-hop, or rock
Group	Group or solo
Bpm (beats per minute)	High, normal or low bpm

Table 2. The situations attribute that used for our system

Attribute name	comment
Latitude	North latitude
Longtitude	East longtitude
Landform	Sea, fllat land or mountain land
Area	Redidental area, commercial area or resort area
Time	6 periods of time
Season	spring, summer, autumn or winter

Table 3. The moods attribute that used for our system

Attribute name	comment
Heart rate	Five level
Automatic Nerve	the sympathetic nerve active, the parasympathetic nerve active, or the same activity
Sympathetic nerve(LF/HF)	up trend, down trend, or stable trend

2.3 Knowledge Acquirement

When we construct the system, we need to acquire knowledge and develop knowledge base. One of methods for knowledge acquirement is ILP. Using ILP, we can add the data easily and reduce re-learn cost significantly. When user listen the music, we assume this case is positive example. When user skips to listening to the music, we assume this case is negative example. The system stores the hypothesis by ILP into the knowledge base.

2.4 Target Concept

The system is always watching user's mood and situation. When the system recommends music to user, the system use user's mood and situation as target concept and refers the knowledge base.

2.5 Music Recommendation

To use ILP, the system can judge that user's mood and situation fit the hypothesis. The system calculates recommendation rate, and applies roulette strategy using this rate. Finally, the system determines the recommendation music. By the roulette strategy, always all music can be selected.

3 Evaluation and Experiment

We are now trying to evaluate the system. We would like to describe result found so far.

3.1 Experiment Result

In a certain situation, we evaluated our system. Our system recommends the music for user. When user listen the music, we assume this case is positive example. When user skips to listening to the music, we assume this case is negative example.

Our system recommends the 45 tracks from 217 tracks. We performed an experiment three times. The experiment result is displayed in fig 2. Our system recommends the better music for user by the increased number of case.

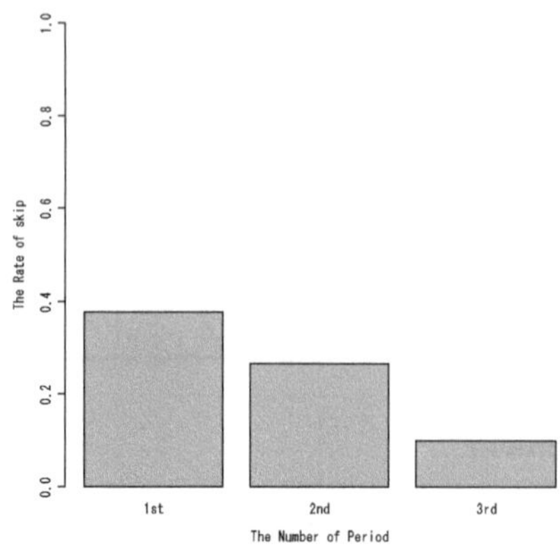

Fig. 2. The rate of skips to listening to the music

3.2 Example of Hypothesis

As an example, we describe few hypotheses that were got from experiments.

```
AutomaticNerve(A, AutomaticNerve_2) :- bpm(A, bpm_1).
```

When subject's autonomic nerve is stable state, subject like to listen the slow tempo music.

```
hr(A, hr_1) :- bpm(A, bpm_0).
```

When subject's heart rate becomes more than 70 beats per minute, subject like to listen the normal tempo music.

```
LatitudeLongtitudeHr(A, Latitude_0, Longtitude_0, Hr_1)
:- title(A, title-1).
```

When subject's heart rate become more than 70 beats per minute at home, subject like to listen the "Ai to Yokubou no Hibi".

4 Conclusion

We described the music recommendation system using user's mood, situation, preference and music attributes. We could make knowledge base by using ILP. Our system can recommend more adequate music than existing systems and reduce user's selecting cost.

References

1. Iwahama, K., Hijikata, Y., Nishida, S.: Content-Based Music Filtering System Using Decision Tree and Its Evaluation. The. Institute of Electronics, Information and Communication Engineers J88-D-I, 642–656 (2005)

2. Mokichi, M., Yokoyama, K., Yoshioka, T., Watanabe, K., Takada, K.: Analysis of R-R Interval Time Series of Automobile Long-Distance Drivers. Society of Automotive Engineers of Japan 27(3), 107–112 (1996)
3. Furukawa, K.: Inductive Logic Programming –Tutorial—. The. Japanese Society for Artificial Intelligence 12(5), 655–664 (1997)
4. Furukawa, K., Ozaki, T., Ueno, K.: Inductive Logic Programming, pp. 161–207 (2001)
5. SWI-Prolog: http://www.swi.psy.uva.nl/projects/SWI-Prolog/

An Adaptive Frame-Based Admission Control for Multimedia Traffic in Wireless LAN

Jinsuk Pak[1], Yongsik Kwon[2], and Kijun Han[1,*]

[1] Department of Computer Engineering, Kyungpook National University
1370, Sankyuk-dong, Puk-gu, Daegu, 702-701, Korea
[2] Advanced Technology Lab. R&D Group, KT2
463-1, Jeonmin-dong, Yusung-gu, Daejeon, P.C 305-811, Korea
jspak@netopia.knu.ac.kr, yongsik@kt.co.kr, kjhan@knu.ac.kr

Abstract. Improving the performance of delay-constrained video streaming over wireless networks is one of the most important issues for various multimedia-related applications, as well as for the efficient overall utilization of the wireless medium. In this paper, we propose a scheme for adaptive admission control and scheduling for frame-based multimedia traffic in the 802.11 wireless local area networks. The conventional scheduler used for IEEE 802.11e is not suitable for VBR traffic such as MPEG-4 because only a single fixed TXOP is allocated from the hybrid Coordinator. In our scheme, we divide an MPEG flow into three different classes of video-frames depending on the user-QoS, and Our scheme is distributed admission control for each divided video-frame types and poll scheduling adaptive admission control for the flow.

Keywords: IEEE 802.11e HCCA, Admission Control.

1 Introduction

IEEE 802.11 Wireless Local Networks (WLANs) have emerged as one of the most successful wireless technologies and are becoming essential in the wireless communications area. As the demand of integrated multimedia services is increasing, high data rates and an efficient Medium Access Control (MAC) scheme become important criteria for the design of WLAN systems.

Recent IEEE 802.11e standard try to provide efficient Quality of Service (QoS) in WLANs. The IEEE 802.11e standard is an extension to the legacy 802.11 standard to provide QoS support to time-sensitive applications [1]. In 802.11e the QoS mechanism is supervised by the Hybrid Coordinator (HC) entity, which implements the Hybrid Coordination Function (HCF). The HC is typically located at the Access Point (AP) and utilizes a combination of a contention-based, called Enhanced Distributed Coordination Access (EDCA), and a polling-based scheme, called HCF Controlled Channel Access (HCCA) to provide QoS-enhanced access to the wireless medium.

In EDCA mode, a traffic class can statistically reduce its transmission delay by making itself a higher prioritized traffic class and utilize an access category (AC) that

* Correspondent author.

M.J. Smith, G. Salvendy (Eds.): Human Interface, Part I, HCII 2007, LNCS 4557, pp. 720–727, 2007.

has higher priority for contending the access channel. However, in heavy load cases, higher prioritized traffic's QoS requirement can easily be breached even though it consumes most of the bandwidth, therefore, lower prioritized traffic can be starved and deteriorates badly both in effectiveness and efficiency [2],[3],[5].

HCCA allows the HC to assign Transmission Opportunities (TXOPs) to the Mobile Stations (STAs) by polling it periodically. A TXOP is a period of time in which a STA or the HC can transmit a burst of contention free data frames. The TXOPs are always initiated and assigned by the HC. The assignment of TXOPs to a STA may be based on a wide set of parameters such as traffic characteristics and current traffic requirements of the STA [4].

The 802.11e standard proposes a simple HCF scheduling mechanism that is efficient for flows with strict CBR characteristics. However, lots of applications such as video conferencing have VBR characteristics and this simple scheduler scheme could not adapt to flows which do not respect exactly their QoS requirements or to flows whose characteristics fluctuate during transmission [8],[9].

In this paper, we propose an admission control and scheduling algorithm for VBR characteristics such as MPEG-4 video traffic based on frame type. We evaluate the capacity of a WLAN system in terms of number of traffic connections that can be supported, given the QoS constraints of voice and video flows. Then, we focus on video traffic and present a technique that exploits the synergies between application layer and MAC layer to improve the system performance.

The remainder of this paper is organized as follows. Section 2 introduces the basis of HCCA in 802.11e WLAN and MPEG-4 video flow. Section 3 discusses a frame-based admission control and scheduling algorithm in the IEEE 802.11e HCCA networks to support real-time multimedia traffic such as MPEG-4, with cross-layer concept. Section 4 describes the simulation step and evaluates the performance of our proposed scheme. Section 5 concludes the paper.

2 Structure of MPEG-4 Video Flow

MPEG-4 provides efficient and scalable video coding. An MPEG encoder generates three types of frames: intra-frames (I-frames), forward predicted frames (P-frames), and bi-directional predicted frames (B-frames). In general, I-frames are encoded as a single image, much like JPEG compression. An I-frame can be decoded independently of other frames and is larger than P frames. P-frames are encoded with a dependency on the closest frame, which can either be an I or a P-frame. Using the past and present frames, motion vectors are computed. B-frames are encoded relative to the past frame, future frame, or both. The name 'bi-directional predicted frame' comes here. Future and past frames are the closest I or P-frames [6], [10].

When compressing a video sequence, typical MPEG encoders use a pre-defined group of picture (GOP) as shown in Fig. 2. A GOP must start with an I-frame so it can be independently decodable. If the MPEG-4 sequence is transmitted over the network, the actual transmission order should be I_0, P_3, B_1, B_2, P_6, B_4, B_5, I_9, B_8, and B_9. The arrows show the dependencies between frames in Fig. 2. For example, frames B_1 and B_2 depend on frames I_0 and P_3. Frames B_4 and B_5 depend on frames P_3 and P_6. Similarly, the whole GOP cannot be decoded properly if the corresponding I frame is not present.

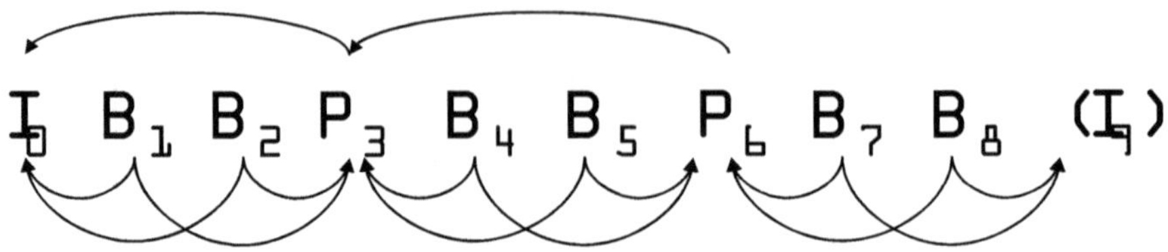

Fig. 1. GOP in MPEG-4

Dependent on the type of frame, the frame loss results in low-quality video picture. The loss leads to perceptible artifacts. The scene of video may become completely unrecognizable or the movement of objects in the video is irregular. From the graph and the explanation above about the dependencies, it is easy to conclude that I-pictures are the most important ones since they contain the actual video content and all other pictures are error-coded based on the I frames.

From the viewpoint of user-perceived quality, the loss of an I or P frame packet causes visible artifacts to appear also during rendering of the dependent frames, while the loss of a B frame packet does not influence the quality of any other frames. This makes the loss of B frames preferable to the loss of P and I frames.

Generally, I-frames take 1-bit/pixel, P-frames take 0.1 bit/pixel, and B-frames take 0.015 bits/pixel, so usually each video frame is composed of 20 to 60 packets. The loss of one packet leads to the loss of a complete frame.

In this paper, we discuss a scheduling algorithm that greatly reduces or simply removes the artifacts that occur during video streaming over a wireless medium.

3 Proposed Scheme

The conventional scheduler used in IEEE 802.11e HCCA is quite inefficient for real-time video streaming applications such as MPEG-4 because it does not reflect unique characteristics of video traffic which varies over time and consists of frames with varying sizes and different decoding priority. In this paper, we propose an admission control and scheduling mechanism to be comfortable with MPEG-4 by focusing on improvement of the overall system utilization as well as the performance of the admitted stations.

3.1 Frame-Based Admission Control

Our admission control is carried out in the frame-based way for video flow such as MPEG-4 which is composed of I, B, and P frames based on the relative priority in terms of the distortion of the decoded video. The frame-based flow mechanism which called by frame-flow is illustrated in Fig. 2.

The frame-flow provides a joint application-MAC optimization that maximizes the number of admitted wireless stations while optimizing the video quality for each admitted station. The application layer enables each type of frame in the video flow to interface with the individual queue of the MAC. Each queue has a different priority which is determined by its type of frame. A frame-flow has its own TSPEC parameters and it allows each admission control independently.

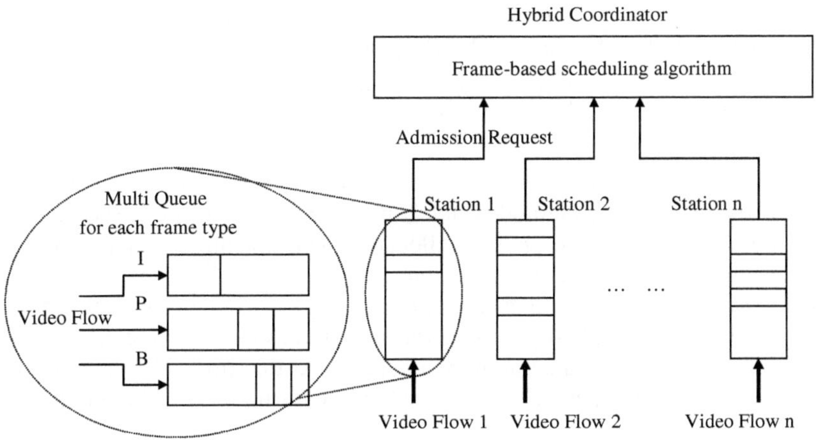

Fig. 2. Frame based admission control and scheduling algorithm

The scheduler calculates the maximum time to transmit N_i MSDUs at minimum PHY rate *R,* and the time to transmit one maximum size MSDU at rate *R* (plus Overheads, *O*) for each TS. From this, TXOP duration is calculated by

$$TXOP_i = \max\left(\frac{N_i \times L_i}{R_i} + O, \frac{M}{R_i} + O\right) \tag{1}$$

We can know the mean data rate for each type of frame by following the trace statistics of the whole transfer data. Therefore, each queue tries to admit its TXOP negotiation with the different frame size. Each method is:

- I frame: peak frame size (or burst frame size: size of the biggest frame)
- P frame: mean frame size (or average frame size of GOP)
- B frame: minimum frame size (or minimum frame size of GOP)

With the method, we can get TXOP duration for each type frame as follows.

$$TXOP_I = \max\left(\frac{Peak\ Frame\ Size}{R_i} + O, \quad \frac{M}{R_i} + O\right) \tag{2-1}$$

$$TXOP_P = \max\left(\frac{Mean\ Frame\ Size}{R_i} + O, \quad \frac{M}{R_i} + O\right) \tag{2-2}$$

$$TXOP_B = \max\left(\frac{Minimum\ Frame\ Size}{R_i} + O, \quad \frac{M}{R_i} + O\right). \tag{2-3}$$

MPEG-4 organizes its GOP units which are composed of I, B, P frames and each type of frame has different mean data rate. But, each frame has fixed interval in one

GOP, also each GOP has fixed interval too. If we denote the TXOP of one GOP which is composed of 15 frames as $TXOP_{GOP}$, we can have

$$TXOP_{GOP} = \max\left(\frac{\rho_I}{R_i} + 4\frac{\rho_P}{R_i} + 10\frac{\rho_B}{R_i} + O, \quad \frac{M}{R_i} + O \right) \tag{3}$$

Where ρ_I, ρ_P and ρ_B are homologous to the mean data rate of I-frame , P-frame and B-frame, respectively. The method which it proposes video traffic I, B and P frames admission control processes especially allocates each transmission queue when each queue it accomplishes its TXOP negotiation.

In this way, frame based admission strategy is more efficient than the flow based one. For example, an MPEG-4 flow generates 25 frames during 1 second and GOP is 12 than each frame is generated at 40ms intervals. At this time, if SI is 40ms than it can allocate more suitableness TXOP for each frame during every SI interval.

3.2 Frame-Based Scheduling Strategy

The priority of the video frame is I-frame > P-frame > B-frame. As previously mentioned, I-frames are the most important ones since they contain the actual video content and one GOP is influenced by the I-frames. Also, P-frame is affected by I-frame or P-frame but it affect B-frames decode. Not only MPEG-4 but also H.263 degrees it has the structure which is similar. In this manner, the HC decides the polling node from the polled-TXOP list which has pending data - different priority frame - in queue when the HC wants to send polled-TXOP. The HC knows the information of pending data in QSTA's queue at the previous frame sent from QSTA.

As S we define the QSTA which is currently selecting the polled-TXOP, the frame which is buffered in QSTA and waiting to be transmitted after received polled-TXOP we mark as W. The list which is admitted QSTA list marks as L, the QSTA which is included polled-TXOP list marks as N. The algorithm in which the HC determines the next polled-TXOP QSTA is shown in Fig. 3.

```
PROCESS select_Poll
BEGIN
WHILE(S is not empty) DO
  FOR (N < L , N ++)
     IF (W is I-frame & didn't polled this SI) THEN
          STORE S with N
  IF ( S is empty) THEN
     FOR (N < L , N ++)
          IF (W is P-frame & didn't polled this SI) THEN
               STORE S with N
  IF ( S is still empty) THEN
     FOR (N < L , N ++)
          IF (W is B-frame & didn't polled this SI) THEN
               STORE S with N
RETURN S
END
```

Fig. 3. The pseudo code for proposed algorithm

4 Simulation

In this section, we evaluate performance of our scheme via a simulation study. We assume that the network consists of several stations and one HC, and each station only has a single VBR traffic flows with the same TSPEC parameters except data rate. PHY and MAC parameters are summarized in Table 1. The size of QoS-ACK and QoS-CF-Poll frame include the MAC header and the CRC overhead.

Table 1. System parameters

Parameter	Value
PHY rate	11Mbps
SIFS	10 μs
QoS-ACK Length	16 octets
QoS-CF-POLL Length	36 octets
MAC header Length	32 octets
MAX MPDU	2304 octets

We consider the dynamic resource management for three different real-time streams represented by real traces. The traces are generated from MPEG-4 streams as well as video sequence coded according to the H.263. The MPEG-4 movie such as "Jurassic Park 1" and "Mr. Bean" have one hour durations as seen in Table 2.

All scheduling decisions were made at the start of a per-flow TXOP for IEEE 802.11e. As the inter-frame time was 40ms for all flows, we observed that the start time of the flows affected the performance.

The results in this section have been produced when the service interval is set to 50 ms and the delay boundary is to 200 ms. First, we examine the importance of the frame-flow concept versus the conventional global flow scheduling. The experiment is carried out with an MPEG-4 "Jurassic Park I" video sequence encoded at 25 frames per second, although similar results have been obtained with a variety of video contents. With the frame-flow scheduling strategy, we show $t_{TXOP} = 2.86$ msec.

Table 2. Trace Statistics of real traffic during 1 hour

Parameter	Jurassic Park I	Mr. Bean
Mean bit rate	770 Kbps	580 Kbps
Peak bit rate	3.3 Mbps	3.1 Mbps
Max frame size	16745 byte	15251 byte
Mean frame size	3.8 Kbyte	2.9 Kbyte
Min frame size	72 byte	93 byte
GOP	12	12

Similarly, with the global flow strategy, we see t_{TXOP} = 4.395 msec. This result is summarized in Table 3 where the t_{TXOP} in the frame-flow strategy is smaller than legacy of IEEE 802.11e. It means that the number of admitted stations can be increased by our frame-flow strategy.

Table 3. TXOP with Optimization Framework

Traffic name	Parameter	Jurassic Park I	Mr. Bean
Global flow	Mean bit rate	4.39ms	3.304ms
I-frame	Max frame size	12.21ms	11.123ms
	Peak bit rate	5.253ms	4.576ms
P-frame	Mean frame size	2.818ms	2.148ms
	Mean bit rate	3.276ms	2.537ms
B-frame	Min frame size	1.707ms	1.707ms
	Minimum bit rate	2.342ms	1.707ms
$\overline{TXOP}_{frame-flow}$	Frame size	2.86ms	2.602ms
	Bit rate	2.818ms	2.154ms

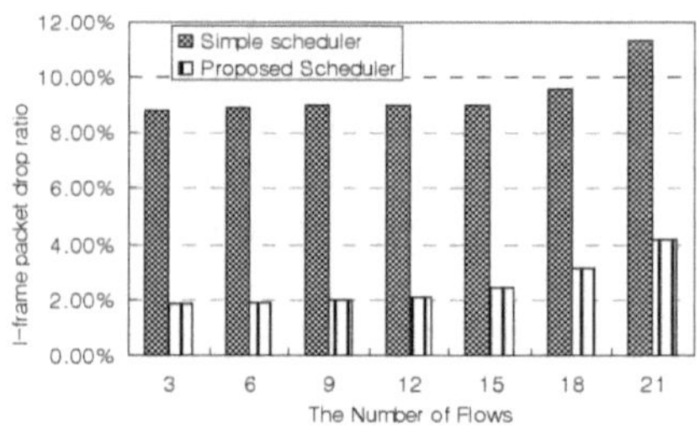

Fig. 4. I-frame packet drop ratio with different flows

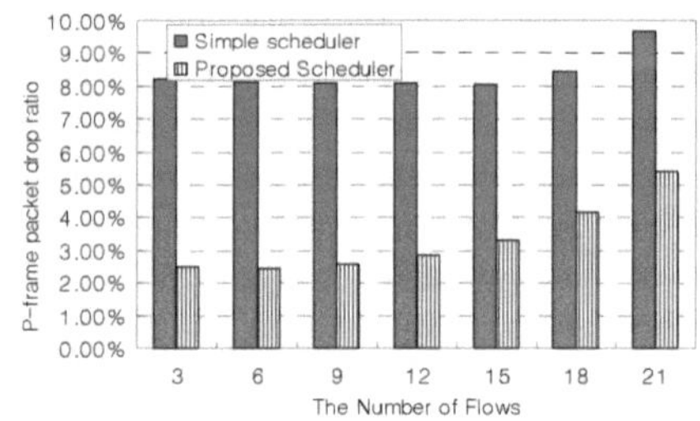

Fig. 5. P-frame packet drop ratio with different flows

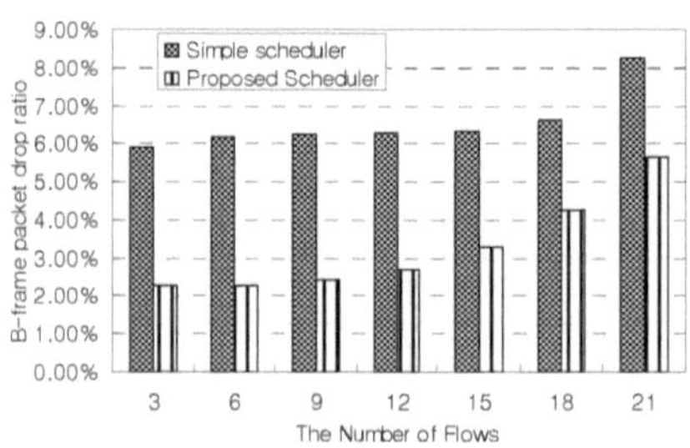

Fig. 6. B-frame packet drop ratio with different flows

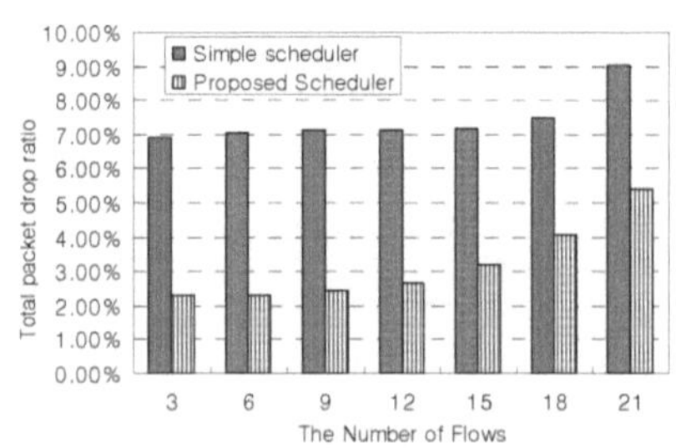

Fig. 7. Total-frame packet drop ratio with different flows

In a collaborative framework, multiple stations may opt to decrease the video frame-rate in order to allow additional stations (or additional video flows) to utilize the wireless medium under HCCA.

From Fig. 4 ~ Fig. 7, we compare the drop ratio by two schemes with different number of traffic flows within 1 hour. We find that our scheme provides a better performance than the conventional scheduling in terms of the packet drop ratio. Each drop of I-frame packet may cause any distortion of displayed content for 480ms. Our scheduler has 2% packet loss which is only 1/3 of the conventional scheduler as shown in Fig. 4. Also we can find out that our scheduler makes a lower packet loss than the conventional scheduler.

5 Conclusion

In this paper, we have proposed an admission control and scheduling strategy for HCCA-based video streaming. Our admission control algorithm can increase the number of admitted stations based on type of frame from one global video flow, each with its own traffic specification. Our scheduling policy effectively allocates the optimal transmission opportunity to each generated frame-flow in order to maximize the utilization of the wireless medium under the HCCA transmission.

In the future, we will investigate the impact of simultaneous scheduling of different flows (such as audio, video, and data having various bit rate and delay requirements) on the admission control mechanism.

References

1. IEEE 802.11 WG, Draft Supplement to Standard for Telecommunication and Information Exchange between Systems – LAN/MAN Specific Requirements – Part II: Wireless Medium Access Control (MAC) and Physical Layer (PHY) Specifications: MAC Enhancements for Quality of Service, IEEE 802.11e Draft 11.0 (October 2004)
2. Choi, S., Prado, J.D., Shankar, S., Mangold, S.: IEEE 802.11e Contention-based Channel Access (EDCF) Performance Evaluation. IEEE ICC'03 2, 1151–1156 (2003)
3. Grilo, A., Nunes, M.: Performance Evaluation of IEEE 802.11e. IEEE PIMRC'02, pp. 511–517 (2002)
4. Lindgren, A., Almquist, A., Schelen, O.: Quality of Service Schemes for IEEE 802.11 Wireless LANs – An Evaluation. Mobile Networks 8, 223–235 (2003)
5. Kong, Z.-N., Tsang, D.H.K., Bensaou, B., Gao, D.: Performance Analysis of IEEE 802.11e Contention-based Channel Access. IEEE JSAC 22(10), 2095–2106 (2004)
6. Kozlov, S., van der Stok, P., Lukkien, J.: Adaptive Scheduling of MPEG Video Frames during Real-Time Wireless Video Streaming. In: WOWMOM 2005, pp. 460–462 (2005)
7. van der Schaar, M., Andreopoulos, Y., Hu, Z.: Optimized Scalable Video Streaming over IEEE 802.11a/e HCCA Wireless Networks under Delay Constraints. IEEE Trans. Mob. Comput. 5(6), 755–768 (2006)
8. Grilo, A., Macedo, M., Nunes, M.: A Scheduling Algorithm for QoS Support in IEEE 802.11e Networks. IEEE Wireless Communications (2003)
9. Ansel, P., Ni, Q., Turletti, T.: FHCF: A Fair Scheduling Scheme for 802.11e WLAN, INRIA Research Report No 4883 (July 2003)
10. Fan, W.F., Tsang, D.H.K., Bensaou, B.: Admission control for variable bit rate traffic using variable service interval in IEEE 802.11e WLANs. In: ICCCN 2004. Proceedings, pp. 447–453 (2004)

A Network Framework on Adaptive Power Management in HCI Mobile Terminals

Hyemee Park[1], Kwangjin Park[2], Tae-Jin Lee[1], and Hyunseung Choo[1,*]

[1] Sungkyunkwan University, Korea
{hyemee,tjlee,choo}@ece.skku.ac.kr
[2] Intelligent HCI Convergence Research Center
kjpark@disys.korea.ac.kr

Abstract. Our goal is to propose the optimal network model for HCI mobile terminals with limited capacity, and to provide ubiquitous services using the minimum cost and to prolong the system lifetime by reducing the energy consumption of mobile devices. In order to achieve these goals, we propose an energy-efficient routing protocol based on clustering scheme for hybrid ad hoc networks. By applying the index-based data broadcasting and selective tuning methods, the infra system performs the major operations related to clustering and routing on behalf of ad hoc nodes. The simulation result presents that our proposed scheme shows up to about 46% of performance improvements in comparison with other schemes.

Keywords: Mobile HCI, Clustering, MANET routing.

1 Introduction

The development of wireless technologies enables to provide high quality Internet access to users using HCI mobile terminals. As Wireless LAN have emerged to fill this growing demand, a structure-based network (*e.g.* cellular network, WiFi network) is deployed with Base Stations (BSs) that are connected to each other via a backbone. It provides an access to a user which is in the physical neighborhood of the BS. Despite its advantages, this technology is still very restrictive, since users should have a direct (single-hop) connection to the BS in order to use it. For the solution to this problem, the mobile terminal has to increase the coverage by adding antennas, or move until it reaches a covered region. The former will require a large amount of transmit power from each node if the base station is far away from the nodes, so quickly drain the battery of the nodes. The latter cannot provide seamless access to the Internet.

Wireless ad hoc networks, which are multi-hop access mode, are being considered as a viable alternative to connect the BS via other users. These two technologies, pure ad hoc networks and fixed infrastructure, can be combined into one to batter satisfy the user needs. In such integrated networks, referred to as *Hybrid Ad hoc Networks* [1], fewer communication infrastructure (*e.g.* access point) can cover a crowded area

* Corresponding author.

M.J. Smith, G. Salvendy (Eds.): Human Interface, Part I, HCII 2007, LNCS 4557, pp. 728–737, 2007.

while the number of fixed antennas is kept relatively small at minimum cost. The energy consumption of the nodes can be reduced because the signal has to cover a smaller distance, and the interference with other nodes diminishes. In addition to having access to the Internet, hybrid ad hoc networks also provide communication between network nodes.

However, to use conventional multi-hop routing protocols [2] cause some data communication problems in hybrid ad hoc networks, such as broadcasting and routing. In these protocols, network nodes unnecessarily consume the energy for receiving and transmitting messages to forward requests from other nodes. Although there are numerous energy-efficient routing mechanisms for ad hoc networks [3], they require additional information, such as link cost, location information, or battery level of nodes. As a result, applying these approaches to dynamic large-scale networks can increase the overhead and costs required to obtain the information.

The goal of this paper is to consider these problems of hybrid ad hoc networks for HCI mobile terminals, and to provide efficient protocol adapted to this kind of networks. We focuses on an energy-efficient protocol based on clustering, reducing the unnecessary energy consumption of nodes. All terminals transmit the data packets after listening to the route information broadcasted from the infra system. We propose the route broadcasting scheme based on indexing to effectively disseminate the route information to the overall network. Each node performs a selective tuning, enabling to become active only when data of interest is being broadcasted. Thus, the proposed scheme can reduce the unnecessary energy consumption and does not require additional cost and delay for path discovery.

The rest of this paper is organized as follows. In Section 2, related works are introduced with some discussion. Section 3 presents an adaptive power management for HCI mobile terminals. The performance of the proposed scheme is evaluated in Section 4. Finally, we conclude in Section 5.

2 Related Works

2.1 Clustering Scheme

Dynamic routing is one of the most important issues in Mobile Ad hoc Networks (MANETs). However, a flat routing based on proactive or reactive protocols cannot perform well in a large dynamic MANET, because of considerable overhead with increased network size. In reactive routing, the disturbing Route Request (RREQ) flooding over the whole network and the considerable route setup delay become intolerable in ad hoc networks. Although proactive routing can provide routes quickly in response to client requests, it periodically incurs route dissemination costs by broadcasting routes throughout a network regardless of the demand for such routes.

Hierarchical routing is essential for achieving basic performance guarantee in a large-scale MANET. In the clustering scheme [4], nodes are organized in a hierarchy, and are thus grouped into clusters. Each cluster has a clusterhead, which coordinates transmission within the cluster, handles inter-cluster traffic, and delivers all packets destined to the cluster. A clustergateway is a non-clusterhead node with inter-cluster links, which can access neighboring clusters and forward information between clusters. A clustermemeber is an ordinary node.

The advantage of clustering is that the generation and spreading of routing information can be restricted in a set of particular nodes. Therefore the number of control overhead transmitted through the network is considerably less than that of traditional flooding methods. However, there are overheads associated with cluster formation and maintenance. To maintain a cluster structure in a dynamic network requires a lot of message exchange between nodes. Frequent information exchange may consume considerable bandwidth and drain nodes' energy quickly. In addition, the clustering scheme may cause the cluster structure to be completely rebuilt over the whole network, when some local changes (*e.g.* re-election of clusterhead) occur. This is called the ripple effect of re-clustering, and greatly affects the performance of systems. Thus, it is a key issue to provide the effectiveness and scalability enhancement of clustering.

2. 2 Energy-Efficient MANET Routing

Although establishing correct routes for communication between nodes is an important research topic in MANETs, energy-efficient routing is the key issue since mobile nodes' battery is the most critical limitation factor. Power failure of a mobile node affects its ability to forward packets of others as well as the node itself, thus the overall network lifetime will be reduced. For this reason, many related works for energy-efficient routing have been proposed, as shown in Table 1.

Table 1. Taxonomy of energy-efficient routing [3]

Approach		Protocols	Goal
Minimize Active Communication Energy	Transmission Power Control	•Flow Argumentation Routing (FAR) •Online Max-Min (OMM) •Power aware Localized Routing (PLR) •Minimum Energy Routing (MER)	Minimize the total transmission energy but avoid low energy nodes.
		•Retransmission-energy Aware Routing (RAR) •Smallest Common Power (COMPOW)	Minimize the total transmission energy while considering retransmission overhead or bi-directionality requirement.
	Load Distribution	•Localized Energy Aware Routing (LEAR) •Conditional Max-Min Battery Capacity Routing (CMMBCR)	Distribute load to energy rich nodes.
Minimize Inactivity Energy	Sleep/Power-Down Mode	•SPAN •Geographic Adaptive Fidelity (GAF) •Prototype Embedded Network (PEN)	Minimize energy consumption during inactivity.

In Transmission Power Control and Load Distribution approaches, each node should know the link cost corresponding to the transmission power required between two end nodes or the residual battery power of intermediate nodes. Thus, it is expensive to obtain the information required for establishing the route. Sleep/power-down mode is to reduce the number of nodes with active state by electing a master node. However, the nodes should perform a particular process to determine and check if they are to become a master, based on the master eligibility rule. It suffers from the master overloading problem and costly master selection procedure. Therefore, these approaches represented in Table 1 are not suitable in dynamic large-scale hybrid ad hoc networks.

3 Proposed Scheme

3.1 System Model

In this paper, we propose energy-efficient protocol for hybrid ad hoc networks. As illustrated in Fig. 1, the wireless ad hoc network can access core networks through the BS. The main goal is to enhance system lifetime by minimizing energy consumption during communications among nodes and Internet access.

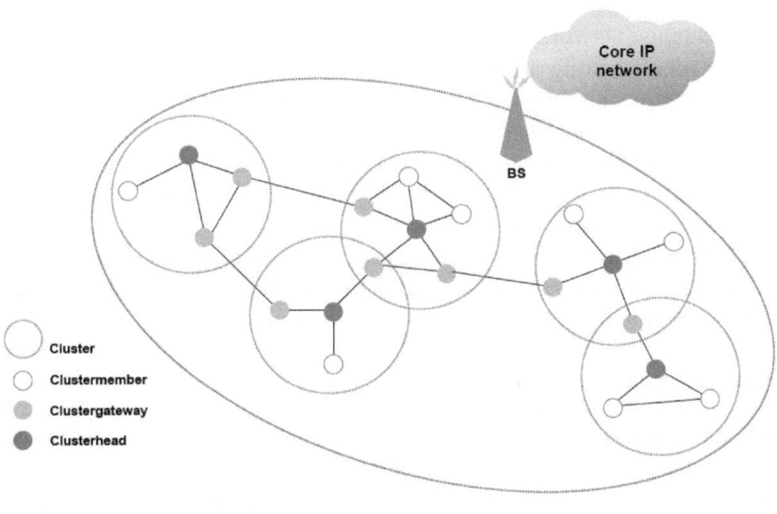

Fig. 1. Network Architecture

The BS is assumed to have high performance and a wide coverage area. As it broadcasts signals to overall networks, the ad hoc nodes within its range can obtain useful information from the system. Each node in the ad hoc network is aware of its location information through use of a Global Positioning System (GPS) device. The network interface of nodes has three energy consumption states: active mode, idle mode, and sleep mode. A mobile node consumes its battery energy not only when it actively sends or receives packets, but also when it stays idle listening on the wireless medium for any possible communication requests from other nodes. The sleep mode has extremely low power consumption, however, the interface can neither transmit nor receive until it is woken up. Therefore, the default mode in ad hoc networks is the idle mode, rather than the sleep mode. Thus, it is important that energy-efficient routing protocols minimize either the active communication energy required to transmit and receive data packets or the energy during inactive periods. We attempt to reduce the number of transmit and receive operations for each message, and save mobile nodes' energy by putting their radio hardware into an idle state using the lower power.

To manage the battery energy of mobile nodes more widely in hybrid ad hoc networks, the proposed scheme uses the clustering-based protocol. The key features of clustering are the efficient utilization of radio channel resources and the reduction of network-layer routing overhead. Thus, energy-efficient clustering enables network lifetime to be greatly prolonged by eliminating unnecessary energy consumption of mobile nodes. The network model is a typical cluster structure as shown in Fig. 1. Network nodes are partitioned into several groups (shown by the dotted circular lines). Each group consists of nodes with a different status and function; clusterhead acting as a local coordinator, clustergateway performing inter-cluster transmission, and clustermember. We consider a network model where all nodes within a cluster are one-hop away from the clusterhead. We make the following assumptions in this paper:

- All nodes in the network are homogeneous and energy-constrained.
- Radio channels among ad hoc nodes are symmetric. In other words, the energy required to transmit messages from A to B is the same as that from B to A for a given SNR (Signal to Noise Ratio).
- Radio channels between ad hoc network and BS are asymmetric. The nodes can receive the signal from the BS directly, but do not have sufficient power for direct communication with the BS.
- The channel between the BS and ad hoc nodes is assigned differently from the communication channel used in the network. Thus, periodical broadcasting by the BS does not affect the data communication among nodes.

3.2 Adaptive Clustering Method

We propose an effective method for clustering using the network model introduced in section 3.1. In the initial phase before a cluster formation, a BS broadcasts its location coordinates to the overall network. The nodes detecting the BS inform their current location to trigger the clustering set-up. When nodes transmit the information to the BS in this phase, they perform directional routing based on the location coordinates of the BS to avoid flooding overhead. The node first selects an estimated node, which is along the path toward the BS. Then, it transmits its location information by piggy-backing the packets received from other nodes. This procedure is recursively performed until the piggy-backed data is forwarded to the BS. If the BS obtains the overall network topology, it elects some nodes to be the clusterhead, and divides network nodes into different virtual groups. The algorithm to solve the NP-hard problem of finding the k optimal clusterhead is out of the scope in this paper. Once the clusters are organized, the BS transmits a message that contains the cluster ID and member nodes' ID to each clusterhead. This is to inform the node selected as the clusterhead, then it performs a role of the clusterhead.

After the clustering set-up, each clusterhead periodically monitors member nodes whether it is or not, by overhearing. When the clustering information in the cache of the clusterhead is changed due to the mobility of the node, the clusterhead reports updated information to the BS. If the member node of its cluster disappears, it transmits an invalid report message to the BS, after deleting the corresponding entry of the cache. The node performing handoff transmits its current location to neighbor node. If the node that receives this message is not a clusterhead, it forwards the message to its clusterhead in order to inform it of a new node. When the clusterhead discovers the node which is not in the cache, it adds the node to its cache and sends the new information to the BS. Then, the BS updates and reconfigures the cluster immediately, so the network can avoid routing error.

By using the infra system with high performance, network nodes do not require additional cost for constructing and maintaining a cluster structure. Conventional clustering schemes require all nodes to advertise cluster-dependent information repeatedly to build and maintain the cluster structure. However, the proposed scheme does not use specific control packets or signals. Once all nodes register their location information in the initial clustering set-up phase, the clusterhead reports the updated node to the BS, whenever an event occurs. Therefore, the scheme can adapt immediately in dynamically changing scenarios without any repair procedure. As a result, our protocol can form and maintain cluster structure without explicitly exchanging the clustering control packets, and thus can completely eliminate the control overhead caused by active clustering.

3.3 Energy-Efficient Routing Protocol

In this section, an energy-efficient routing protocol based on the data broadcasting scheme of the BS is introduced. Since each node obtains a route from the BS, which knows the topology information of the network, the proposed scheme does not require a path discovery procedure by flooding the RREQ message of the typical clustering scheme. When the node wants to transmit the data, it wakes up and forwards the packet along the path obtained by the BS. Thus, it can conserve greater bandwidth, as well as nodes' battery energy than existing schemes

1) **Broadcasting Scheme of Base Station.** The BS periodically broadcasts the route information to the overall network, after clusters are organized. It uses the index-based broadcasting scheme as a way of disseminating information to a massive number of nodes with limited battery [5]. The data is organized by interleaving indexing information among the broadcasting data. The index provides a sequence of pointers that lead to the required data. Data unit consists of cluster information. The node first confirms the arrival time of required data through the index, then it switches to active mode at that time. Since the data is properly organized to provide the needs of nodes, it makes effective use of the low wireless bandwidth, and results in significant improvement in battery utilization. The proposed format of broadcasting data is shown in Fig. 2.

The broadcasting data consists of two segments: *index* segment, and *data* segment. The index segment is divided into *header* and *cluster connectivity* units. The first unit includes <node ID, Cluster ID (CID)> pairs containing information of all nodes in the network, and the arrival time information of each cluster unit (CID unit). The second unit has connectivity information between clusters and the arrival time of the

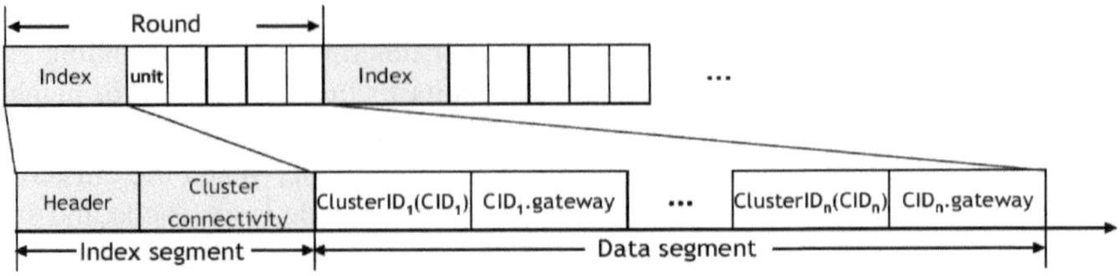

- *Header* unit : (Node ID, Cluster ID) pairs, arrival time of CID data units
- *Cluster connectivity* unit : (CID_i-CID_j) connectivity table, arrival time of CID.gateway data units
- *CID* unit : cluster information
- *CID.gateway* unit : clustergateway information of each CID

Fig. 2. Format of broadcasting data

clustergateway unit (CID.gateway unit). The index segment is located at the beginning point in each round, and all nodes should first listen to this segment for obtaining the subsequent data segment.

The data segment consists of two types of units: *cluster* (CID), and *clustergateway* (*CID.gateway*) units. The CID unit contains cluster information such as node status (*e.g.* clusterhead, clustergateway, and member), and topology of each cluster. The CID.gateway unit consists of two data structures. The first represents the neighboring clusters connected with clustergateways of corresponding CID, enabling access to neighboring clusters, and forwards the data between clusters through the selected clustergateway. The second is an adjacency matrix that represents the connectivity state among clustergateways within a cluster. This information is used to obtain the shortest path from source to destination. In the matrix, 1 of entry indicates that two nodes are connected. In contrast, 0 indicates that they are not connected directly and the routing between two nodes should pass the clusterhead. The node can compute the optimal path among multiple routes without additional information.

2) **Selective Tuning of Node.** The source node listens to the data of route broadcasting from the BS, for obtaining a route to the destination node. To conserve battery energy, nodes require the ability of selective tuning, enabling listen to selected broadcasting data that they want. They can slip into idle mode most of the time and come into active mode only when the data of interest is expected to arrive. The proposed algorithm for selective tuning is represented as follows.

```
Listen the header unit of index segment;
If (source.CID = dest.CID)
        wakeup (own CID unit of data segment); listen;
Else
        listen the clusterconnectivity unit
        wakeup (own CID unit); listen; sleep;
        wakeup (CID.gateway units); listen; sleep;
        wakeup (dest.CID unit); listen;
        compute the optimal path;
Transmit the data using source routing;
```

When the source node wants to send a data, it wakes up and listens to the route broadcasting data from the BS. It first finds cluster IDs of which the source and destination nodes are a member, and confirms the arrival time of their CID units. If the source-destination pair belongs to the same cluster, it slips into idle mode until the corresponding CID unit arrives. After obtaining the routing information within the cluster from data unit, it transmits packets to the destination. In contrast, if their cluster IDs are different, the source node listens continuously to the subsequent cluster connectivity unit. The source node discovers the clusters that should be passed to reach the destination. It also confirms the arrival time of gateway units of clusters. Then, it wakes up when the CID unit of source, CID.gateway units, and CID unit of destination arrive. The rest maintains in the idle mode to save energy.

In this phase, the full path between two end nodes can be obtained. The source node selects proper clustergateways, connected with following clusters in the path, using gateway-cluster connectivity information in CID.gateway units. This may have a route with one or more clustergateways within the same CID, since each cluster can be connected with many neighbor clusters. In this case, the node checks the relationship between clustergateways by referring the adjacency matrix. The proposed scheme can obtain the shortest path among multiple paths. After the full-path is set-up, it forwards the data based on source routing.

4 Performance Evaluation

4.1 Simulation Environment

Simulation is implemented in C. Our simulation model is a wireless network with 25 ~ 100 mobile nodes placed randomly. Simulation traffic is generated by Continuous Bit Rate (CBR) sources spreading the traffic randomly. The packet rate is set to 1 *pkt/s*. Each node has two energy states, idle and active states. We measure the total energy consumptions required to deliver data packets by computing the energy loss of these two states. Our energy consumption model is based on Kasten's measurements [6], represented in Table 2.

Table 2. Energy Model

(mW)

	Model	idle	Normal/active	
CPU	StrongARM SA-1100	0.16	400	
NIC	Digitan 2 Mbps 802.11 Wireless LAN	750	receive	1,550
			transmit	1,900
GPS	μ-blox GPS-MS1E	33	continuous	462
			power save (1s up-date)	165

In this paper, the assumptions are made as follows:

- The node uses a Li-ion battery with a nominal voltage of $10.8V$ and a capacity of $3.2AH$. (This is taken from the specification of the IBM Thinkpad 600 notebook.)
- Each node consumes $0.165W$ by reporting every second in power save mode, since it does not require constant position information.
- The energy for listening to the one unit of broadcasting data is the same as that of the receiving mode ($1.55W$).

4.2 Simulation Results

We compare four schemes: AODV [7] of reactive protocol, CBRP [4] of clustering scheme, flat routing based on route broadcasting scheme, and hierarchical routing based on route broadcasting scheme (the proposed scheme). Fig. 3(a) presents the total power consumption under various network sizes for each scheme. As shown in Fig. 3(a), the proposed scheme has better performance than other schemes as the number of nodes is increasing. In AODV, the path discovery results in considerable energy loss because of RREQ flooding to the overall network. In contrast, the CBRP has lower cost than AODV, since the flooding of this approach is restricted by special nodes. The third is a scheme applying the data broadcasting of the BS to a flat network. Since broadcasting data is not well organized by non-clustering structure, nodes should listen to all data units for finding the routing path. However, as the proposed scheme uses the selective tuning method, depending on cluster structure, it minimizes the total energy loss in large-scale networks. This ultimately affects the network lifetime, as shown in Fig. 3(b). It shows the percentage of battery energy remaining. The result demonstrates that the proposed scheme extends network lifetime considerably; after 300 minutes, 70 ~ 80% of nodes are still alive, while AODV, and CBRP systems are no longer operating.

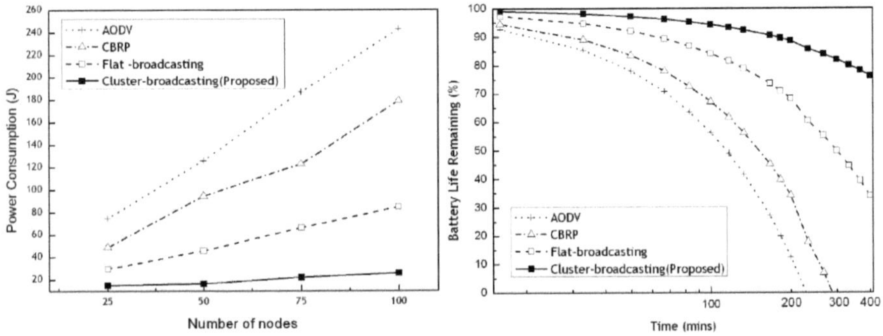

Fig. 3. Simulation results (a) Total system energy, and (b) System lifetime

5 Conclusion

In this paper, we propose an energy-efficient clustering protocol for hybrid ad hoc networks. To solve the limitation of bandwidth and nodes' battery in wireless mobile networks, the proposed scheme uses the infra system with high capacity. As the infra system performs major operations on behalf of nodes with limited ability, the proposed scheme reduces power consumption as well as the cost of path discovery and maintenance, and the delay required to configure the route.

Acknowledgments. This research was supported by the MIC(Ministry of Information and Communication), Korea, under the ITRC(Information Technology Research Center) support program supervised by the IITA(Institute of Information Technology Assessment), IITA-2006-(C1090-0603-0046).

References

1. Ingelrest, F., Simplot-Ryl, D., Stojmenovic, I.: Routing and broadcasting in hybrid ad hoc networks. Report technique-0291, INRIA (2004)
2. Abolhasan, M., Wysocki, T., Dutkiewicz, E.: A Review of Routing Protocols for Mobile Ad Hoc Networks. Ad. Hoc. Networks 2(1), 1–22 (2004)
3. Yu, C., Lee, B., Youn, H.Y.: Energy Efficient Routing Protocols for Mobile Ad Hoc Networks. In: Wireless communications and mobile computing, vol. 3(8), pp. 959–973. Wiley, Chichester (2003)
4. Jiang, M., Li, J., Tay, Y.C.: Cluster Based Routing Protocol (CBRP). Internet-draft, draft-ietf-manet-cbrp-spec-01.txt (1999)
5. Imielinski, T., et al.: Energy Efficient Indexing On Air. In: Proceedings of ACM SIGMOD (1994)
6. Kasten, O.: Energy consumption. ETH-Zurich, Swiss Federal Institute of Technology. Availabl at http://www.inf.ethz.ch/~kasten/research/bathtub/energy_consumption.html
7. Perkins, C.E., et al.: Ad hoc On-demand Destance Vector (AODV) routing. IETF, RFC 3561 (2003)

Real-Time Stereoscopic Conversion with Adaptable Viewing Distance at Personal Stereoscopic Viewing Devices

Ilkwon Park and Hyeran Byun

Department of Computer Science, Yonsei University, Seoul, 120-749, South Korea
{ikheart,hrbyun}@yonsei.ac.kr

Abstract. This paper proposes the real-time stereoscopic conversion method with adaptable viewing distance by the viewer's eye detection at personal stereoscopic viewing devices. The distance between a screen and viewer is often changed by moving a personal display or moving head position. It causes a visual discomfort and low quality stereoscopy. Our proposed method is divided into two parts; eye detection part to acquire viewer's eye positions and stereoscopic generation part in real-time. In the first part, we can measure an interocular distance of a single viewer by the proposed eye detection method and also calculate the distance between a screen and a viewer by using the principle of stereoscopic generation. In the other part, we propose an efficient motion-based approach for conversion from 2D video to stereoscopic video. The disparity between left image and right image is adaptively adjusted by viewer's interocular distance. Adaptation time for a 3D immersion during 3D stereoscopic viewing is remarkably reduced and stereoscopy is personally optimized. Therefore, our proposed system can offer optimized stereoscopic conversion and more freedom of viewer's head considering the distance between a screen and a viewer through the proposed eye detection.

Keywords: Mobile stereoscopic system, Adaptable stereoscopic conversion, Viewer adaptable 3D.

1 Introduction

Within the past decade, many efforts have been made to convert a monoscopic video to a stereoscopic video by the aid of video device and software [1-2]. Generally, stereoscopic video is mainly applied to scientific visualization, simulation and entertainment [3]. However, the stereoscopic displays are getting popular in these days due to low cost by hardware improvements and consumer's visual desire. Moreover, we can find easily personal stereoscopic viewing devices in everywhere according to supply mobile phone that is available with multimedia data, personal digital assistant (PDA), and laptop computer as well as personal computer.

However, stereoscopic contents are still insufficient in spite of many efforts [1-3]. There are two approaches for a stereoscopic video generation. First approach is generated by using stereoscopic camera. This approach generates high quality

M.J. Smith, G. Salvendy (Eds.): Human Interface, Part I, HCII 2007, LNCS 4557, pp. 738–746, 2007.

stereoscopic contents though this camera is too expensive and these stereoscopic contents are only generated by newly recording real scenes without reproducing conventional two dimensional video. The other one is generated by converting a conventional two dimensional video into three dimensional stereoscopic video. Our proposed method is also stereoscopic conversion approach. It has a lot of advantages such as low cost and using a conventional video that is uniquely existed though we have still faced with eyestrain and low-quality stereoscopy as disadvantages [3].

Our proposed system extends fundamentally motion-based stereoscopic conversion system according to motion types. This method converts motion information in conventional 2D video to 3D disparity to generate stereoscopic video. Extracted motion vectors are used to decide motion types and the amount of disparity. Proposed system consists of three motion types such as horizontal, vertical, and static. Motion types follow the decision rule from [3]. For decision of 3D disparity, we adapt roughly the distance between screen and viewer calculated by obtaining user's interocular distance and finely an amount of motion from 2D video.

Based on the principle of stereoscopic generation, the distance between pupils of viewers and the distance between screen and viewer's pupils are important clues to generate optimal 3D stereoscopy. If we generate stereoscopic video by only motion vector without certain information of these distances, viewers must always have kept the approximate distance from the screen uncomfortably as well as struggled for finding the optimal distance each time. To solve this problem, we apply interocular distance based on a viewer's eye positions to generate stereoscopic contents adaptively. This proposed method can make a more freedom of viewer's head and optimal 3D stereoscopy for single viewer.

For detecting viewer's eye positions, we propose robust real-time eye detection method under various illuminations. In our system, we can consider not only personal computer that is fixed but also mobile phone or PDA that is moving in outdoor environments. Furthermore, we can also operate our proposed system in dark circumstance such as night time and home under dim light because of using the infrared illuminators. For real-time processing, we combine the eye detection by using Iris mask for open eye detector with the one by using eye corner filter for closed eye detector [5], [7]. Both methods are very simple and fast. However, each method has relatively low accuracy. Therefore, we highly boost this detection rate by using pattern recognition approach as eye verification step. For real time processing, we configure an efficient formation of the cascaded form of SVM as two level cascaded SVM. This method highly kept the accuracy while speed up the processing time in its test phase.

The remainder of this paper is organized as follows. In Section 2, we describe the proposed system architecture. In Section 3, we represent the proposed eye detection method and the measure of interocular distance. In Section 4, we represent motion detection method in 2D video contents and adaptable stereoscopic conversion method according to viewer's eye position and motion information detected from 2D video contents. Experimental results and the estimation of our system are provided in Section 5, and finally we make a conclusion in Section 6.

2 The Proposed System Architecture

Our proposed system mainly consists of two parts; User Adaptation Engine(UAE) and Video Conversion Engine(VCE). UAE includes the process acquiring viewer's eye position and extracting interocular distance of a single viewer. VCE is the process of motion analysis in 2D video contents and the process of stereoscopic video generation based on viewer's information.

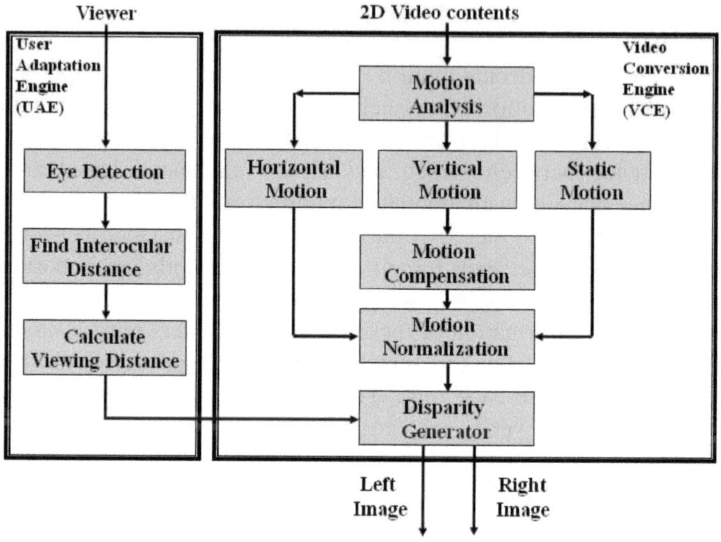

Fig. 1. The overview of the proposed system architecture

Fig.1 shows the overview of the proposed stereoscopic conversion system with viewer adaptable function. We assume that this system is used by only single viewer. The our proposed system is operating on a personal 3D display, such as mobile phone, PDA, and laptop computer as well as personal computer.

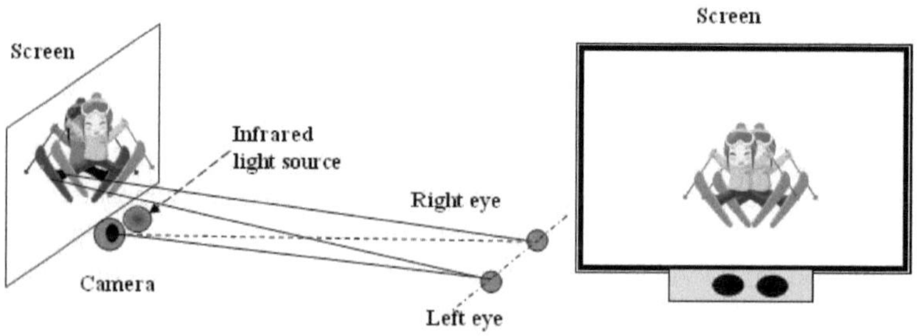

Fig. 2. The installation of personal 3D display and a CMOS camera

UAE is the process module based on computer vision technique. This processing module plays an important role in adaptable stereoscopic generation and it has three parts; eye detection, interocular distance extraction and calculate the distance between the screen and viewer's eye position.

On the other hand, VCE is the processing module that is motion-based stereoscopic converting module with user adaptable modules, while typical systems convert a simply 2D to 3D stereoscopic video based on motion information. Single CMOS camera and infrared light source are installed in front of personal 3D display to acquire viewer's eye position as Fig. 2.

3 User Adaptation Engine

In this paper, UAE extracts user information based on viewer's eye position and applies it to VCE in real-time. Our proposed system must operate not only in bright circumstance but also in dark one. Therefore, infrared illuminators are used with camera as invisible illuminator. To acquire viewer's eye position, we deign efficient eye detection algorithm. This approach consists of three parts; open eye detector, closed eye detector and eye verifier as Fig. 3. This technique gives us fast and accurate performance for finding eye position in spite of combining with a pattern recognition method on eye verification.

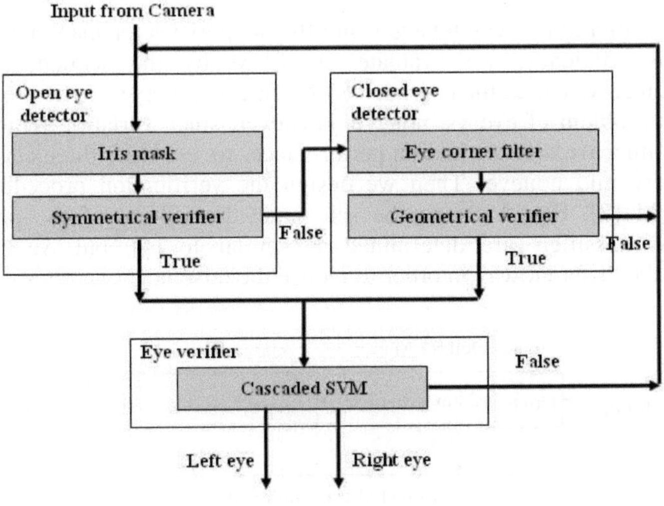

Fig. 3. Proposed eye detection system for real time operation

3.1 Open Eye Detector

Open eye detector is based on geometrical mask as iris shape proposed in [5]. This algorithm is very useful for detecting open eyes or partially open eyes. Through the iris mask, they can directly extract eye candidates in the whole image without face segmentation or limited face region. In [5], search step of similar regions decided eye candidates by using the similarity of both shapes after comparing left eye with right eye. We separate this step as symmetrical verifier. Furthermore, we modified this algorithm to align each candidate patches by the similarity of edge distribution before symmetrical verifier. Although it has a lot of advantages, we can not be satisfied with the results for partially open eyes and closed eyes. Therefore, we use closed eye detector based on eye corner filter for detecting closed eyes and partially open eyes.

3.2 Closed Eye Detector

Eye corner filter is a fast and reliable preprocessing method for eye detection [9]. In this paper, the eye corner filter proposed in [7] is partially used to detect suitable candidates of eye region as closed eyes and partially open eyes. Eye corner filter consists of two shapes: left filter and right filter shapes like eye corners. As input frame is convoluted with the filter, we can detect the positive areas over the threshold. The threshold is determined to take 10% from high rank of the result by a convolution with total eye area. Positive areas by both filters are merged by morphological operation. Each positive area detected by eye corner filter is cropped with 5 pixels margins from the boundary of positive area as eye candidates if each positive area has a reasonable ratio of horizontal length to vertical length. Cropped eye candidates are normalized the size of 41 by 21 for the eye verification.

3.3 Eye Detection by Using Cascaded SVM

We verify the filtered eye candidates using the support vector machine (SVM). For computational efficiency, we evaluate the SVM by the sequential evaluation algorithm [4] that we called the cascaded SVM. In general, eye verification deals with the two-class problem of eye vs. noneye. However, since variation reduction of the eye class can improve the verification performance, we consider three classes of open eye, closed eye and noneye. Then we design the verification procedure with the cascaded SVM [7]. Fig. 4 shows the schematic description of the proposed eye detector. The classifiers are determined as similar to [4], but we choose their thresholds in the different way in order to reduce the false negative rates.

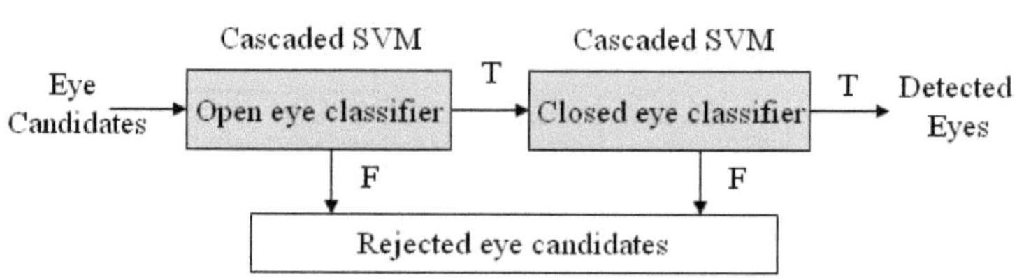

Fig. 4. Schematic description of the proposed eye detector using the cascade SVM

3.4 Calculate Interocular Distance

In the previous section, we found viewer's eye positions by using our proposed method. Based on viewer's eye position, we can calculate Interocular distance in real-time. This is the key idea of this paper that interocular distance reflects not only various personal Interocular distances between two eyes but also the distance between a screen and a viewer. For example, the farther interocular distance is, the closer the distance between a screen and a viewer is. In opposite case, the farther the distance between a screen and a viewer is, the closer interocular distance is. Interocular distance is represented by the number of pixels in the captured image from a camera.

4 Video Conversion Engine

In this section, we represent motion-based 2D to 3D stereoscopic conversion and also we represent how to viewer's information is applied in our proposed system as user adaptable system.

4.1 Motion Detection

In this paper, Optical flow approach that is fast and simple is used for real-time motion estimation. Estimated motion information is represented by orientation and magnitude which is based on the position of each feature point. Each orientation and magnitude is computed by equation 1 and equation 2. Motion vector MV is (MVx, MVy), where MVx and MVy are horizontal and vertical components, respectively.

$$Orientation(\theta) = \tan^{-1}\left(\frac{MV_y}{MV_x}\right). \quad (1)$$

$$Magnitude\ (M) = \left|\sqrt{MV_x^{\ 2} + MV_y^{\ 2}}\right|. \quad (2)$$

To generate a disparity between left and right image from single image, we use only the average motion vector (AMV) of the current frame. If current AMV is larger than threshold (θ) that is experimentally acquired, this frame has enough motion information as horizontal or vertical motion. Otherwise, it classifies static motion. In motion analysis, vertical motion is separated with horizontal motion by empirically acquired threshold and it must be compensated because vertical disparity makes viewer's eye uncomfortable. In the case of static motion, previous AMV is used as the AMV in current frame. Acquired motion magnitude of AMV is too small or large for using stereoscopic disparity. Therefore, it must be normalized within maximum disparity for a viewer. However, our proposed system uses both the disparity by AMV and interocular distance. Therefore, the sum of disparity by AMV and interocular distance is lower than maximum disparity for a viewer. In this paper, the magnitude of AMV is normalized within 10 pixels, this is empirically set.

4.2 Generation of Disparity

To generate disparity between left image and right image from a single image, the magnitude of disparity and shift direction are necessary. Shift direction is represented by only horizontal direction. Disparity assignment is firstly given by viewer's interocular distance. And Magnitude of AMV is applied as a variation from given disparity. Equation 3 represents the disparity assignment with magnitude of AMV and interocular distance. In Eq. 3, *I_disparity* represents the normalized disparity from interocular distance. *nAMV* represents the normalized disparity from the magnitude of AMV and *max_AMV* is a maximum scale of normalized disparity.

$$Disparity = I\_disparity + (nAMV - \max\_AMV/2) \cdot \quad (3)$$

Fig. 5 represents a disparity compensation for the same depth unrelated with the viewing distance. As viewer's eye position is moved from distance 1 to distance 2, depth 1 and converging point 1 are also moved depth 2 and converging point 2. Therefore, viewer watches the video with the different depth although the depth is a physically the same. To solve this problem, the disparity is compensated to be the same converging point as the viewing distance is changed. Finally, for disparity variation, compensated disparity and motion disparity are integrated as Eq. 3.

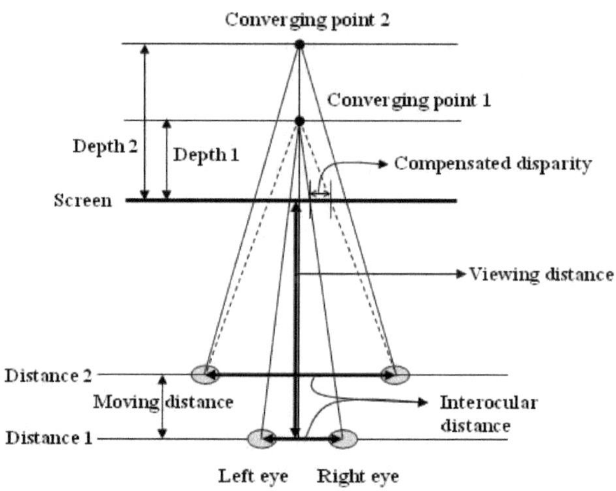

Fig. 5. Disparity compensation for the same depth unrelated with the distance between screen and viewer's eye position

5 Experiments

In this section, we experiments the acquisition process of viewer's information from UAE and the adaptation process of stereoscopic conversion from VCE. To measure viewer's interocular distance, we must define initial distance between screen or camera and viewer's eye position as 70cm in our laptop computer by many experiments and then adjust initial shift distance to watch an optimized stereoscopy.

After initialization for a viewer, the UAE adaptively calculates shift distance followed by viewer's interocular distance. That is, interocular distance reflects the various distances between screen and viewer's eye position. Viewing distance is for minimum 45cm to maximum 120cm for stable watching by characteristics of stereoscopic monitor. Through our experiments, viewing distance is roughly in inverse proportion to the disparity between left image and right image.

Fig. 6 shows the some results of stereoscopic conversion for three video sequences using proposed adaptation methods. In Fig. 6, (a) is converted from typical mpeg 2D test video that called coastguard to stereoscopic video. (b) and (c) is converted from video sequence selected only one view from multiview sequence given by Microsoft Research []. *Disp* represents the disparity between left image and right image. *VD* represents the viewing distance between a screen and a viewer's eye. In Fig. 6, viewing distance is represented as three layers; 100, 75 and 50. Therefore, viewer can watch stereoscopic video with stable depth sense as applying different disparity according to viewing distance

Fig. 6. Some results of stereoscopic conversion adapted by viewing distance

6 Conclusion

This paper proposes the adaptable stereoscopic conversion system by the viewer's eye position on personal stereoscopic viewing devices. For adoptable stereoscopic conversion for a viewer, we propose efficient eye detection method for acquiring viewer eye position. This method is a simple and fast for real-time process as well as accuracy by using cascaded SVM as a verification step. The disparity of left image and right image is continuously adjusted according to the viewing distance represented by acquired viewer's eye position. Therefore, our proposed system can offer personally optimized stereoscopic conversion unrelated with viewing distance.

Acknowledgments. This research was supported by the MIC, Korea, under the ITRC support program supervised by the IITA (IITA-2006-C1090-0603-0017).

References

1. Meesters, L.M.J., IJsselsteijn, W.A., Seuntiëns, P.J.H.: A Survey of Perceptual Evaluations and Requirements of three-dimensional TV. IEEE Transctions on Circuits and systems for Video technology, vol. 14(3) (2004)
2. Battiato, S., Capra, A., Curti, S., La Cascia, M.: 3D Stereoscopic Image pairs by Depth-Map Generation. In: Proceeding of the 2nd International Symposium on 3D Data Processing, Visualization, and Transmission, pp. 124–131 (2004)
3. Kim, M., Park, S., Kim, H., Artem, I.: Automatic conversion of two-dimensional video into stereoscopic video. In: Proceeding of SPIE, November 15, 2005, vol. 6016 (2005)
4. Burges, C.J.C.: Simplified support vector decision rules. Intl. Conf. on Machine Learning 1, 71–77 (1996)
5. D'Orazio, T., Leo, M., Cicirelli, G., Distante, A.: An algorithm for real time eye detection in face images. In: Proc.17th Int. Conference on Pattern Recognition, vol. 3, pp. 278–281 (2004)
6. Liu, X., Xu, F., Fujimura, K.: Real-time eye detection and tracking for driver observation under various light conditions. IEEE Intelligent Vehicle Symposium 2, 344–351 (2002)
7. Park, I., Ahn, J.-H., Byun, H.: Efficient Measurement of Eye Blinking under Various Illumination Conditions for Drowsiness Detection Systems. 18th International Conference on Pattern Recognition, vol. 1, pp. 383–386 (2006)
8. Romdhani, S., Torr, P.H.S., Scholkopf, B., Blake, A.: Computationally efficient face detection. In: Proc. 8th Int. Conf. on Computer Vision, pp. 695–700 (2001)
9. Zhu, J., Yang, J.: Subpixel eye gaze tracking. In: Proc. 5th IEEE International Conference on Automatic Face and Gesture Recognition, pp. 124–129 (2002)

Performance Improvement of SCTP for Heterogeneous Ubiquitous Environment

Doo-Won Seo[1], Hyuncheol Kim[2], Jahwan Koo[1], and Jinwook Jung[1,*]

[1] Information and Communication Engineering, Sungkyunkwan University, 300, Chunchun-dong, Jangan-gu, Suwon, Gyeonggi-do, Korea, 440-746
dwseo@ddc.ac.kr, {jhkoo, jwchung}@songgang.skku.ac.kr
[2] Dept. of Computer Science, Namseoul University, 21 Maeju-ri, Seonghwan-eup, Chonan, Korea, 330-707
hckim@nsu.ac.kr

Abstract. SCTP(Stream Control Transmission Protocol) is a new IETF transport protocol over the IP network. Transport protocol is responsible for reliable delivery of a message form a host to another host. We implemented an efficient file transfer system using the SCTP multiple file transfer and modified STCP congestion control mechanism for heterogeneous ubiquitous environment to improve end-to-end FTP efficiency. The result shows that proposed system reduces transmission latency for huge data and improves the performance.

Keywords: sctp, multifile transfer, congestion control.

1 Introduction

Traditionally, the TCP/IP protocol has UDP(User Datagram Protocol) and TCP(Transmission Control Protocol) as transport layer. Most Internet applications use TCP to ensure end-to-end reliability. SCTP is a new transport layer protocol standardized at the IETF(Internet Engineering Task Force). SCTP was proposed to overcome the limitation of TCP. Like TCP, SCTP provides a full duplex, reliable transmission service to upper applications. In addition to the features of TCP, SCTP supports multi-streaming and multi-homing functions. Due to its attractive features, SCTP has received much attention from the research community and industry [1], [2]. The object of this paper is to implement the efficient file transfer system for heterogeneous ubiquitous environment to improve end-to-end FTP(File Transmission Protocol) efficiency using SCTP.

FTP is a standard application protocol provided by TCP/IP for copying a file from one host to another. FTP uses the service of TCP to provide end-to-end reliability. The FTP control connection is established via TCP port 21 and the FTP data connection is established via TCP port 20. The nonpersistent multiple file transfer on traditional FTP over TCP causes connection setup overhead. And the performance of SCTP is poor when used for bulk transfers in highspeed wide area network. This

* Corresponding author.

M.J. Smith, G. Salvendy (Eds.): Human Interface, Part I, HCII 2007, LNCS 4557, pp. 747–754, 2007.

paper proposed FTP over SCTP-MFCC(SCTP with persistent Multiple File transfer and modified Congestion Control) to overcome these restrictions. The results indicate that FTP over STCP-MFCC outperforms FTP over TCP.

This paper is organized as follows. Section 2 gives the background and related works about SCTP. Section 3 introduces system architecture for FTP over SCTP-MFCC at both server and client. Section 4 presents the experimental results and analyzes the meaning. Conclusions are drawn in Section 5 about pros and cons of the proposed architecture.

2 Background

The basic service offered by SCTP is the reliable transfer of user messages between peer SCTP users. It performs this service within the context of an association between two SCTP endpoints. SCTP is connection-oriented in nature, but the SCTP association is a broader concept than the TCP connection [3]. This section summarizes some of the key features that provides to the ULP(Upper layer protocol).

2.1 Multi-streaming

Multi-streaming allows for multiple streams on an association to send data. If one stream fails, only part of the data is lost and needs to be resent. So SCTP can prevent HOL(Head-Of-Line) blocking. Fig. 1 shows the SCTP multi-streaming concept [4].

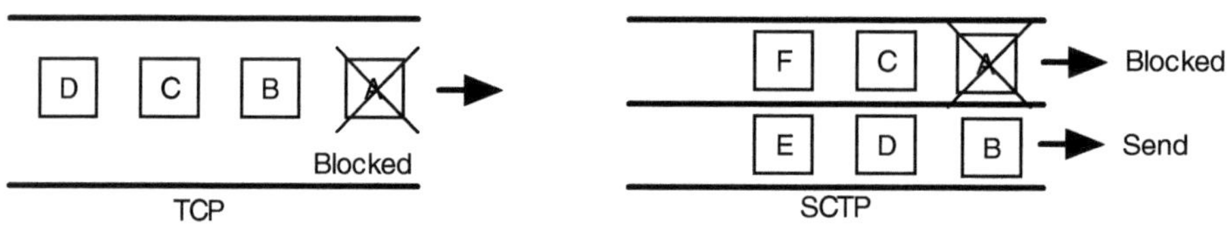

Fig. 1. SCTP multi-streaming - prevention of HOL blocking

2.2 Multi-homing

An SCTP endpoint is considered multi-homed if there are more than one transport address that can be used as a destination address to reach that endpoint. Moreover, the ULP of an endpoint shall select one of the multiple destination addresses of a multi-homed peer endpoint as the primary path. By default, an endpoint should always

Fig. 2. SCTP multi-homing

transmit to the primary path, unless the SCTP user explicitly specifies the destination transport address to use. Fig. 2 shows the SCTP multi-homing concept [5].

2.3 SCTP Congestion Control

Congestion control is one of the fundamental functions in SCTP. For some applications, it may be likely that adequate resources will be allocated to SCTP traffic to assure prompt delivery of time-critical data - thus it would appear to be unlikely, during normal operations, that transmissions encounter severe congestion conditions. The slow start and congestion avoidance algorithms must be used by an endpoint to control the amount of data being injected into the network. The generic congestion control in SCTP is employed in regard to the association, not to an individual stream.

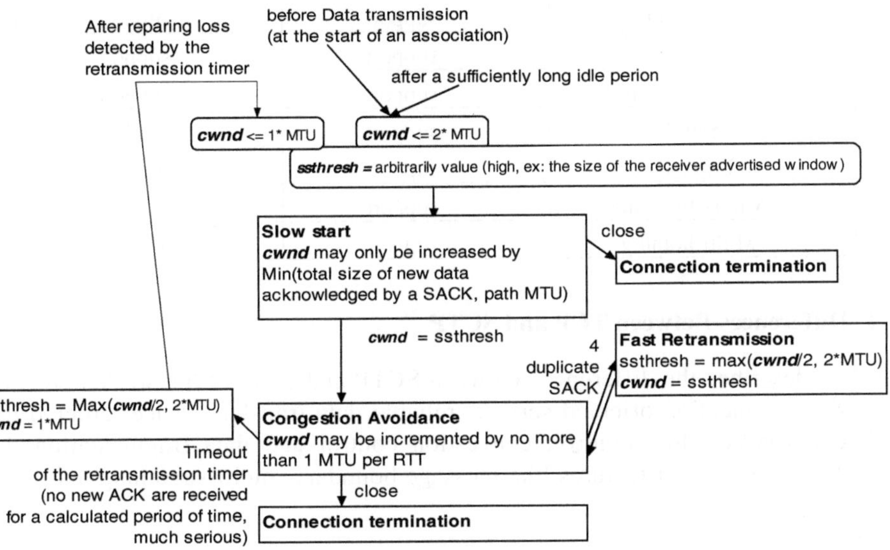

Fig. 3. SCTP congestion control

Upon detection of packet losses from SACK(Selective ACKnowledgement), an endpoint should do the following:

$$ssthresh = \max(cwnd/2, 2*\text{MTU}). \quad (1)$$

$$cwnd = ssthresh. \quad (2)$$

Basically, a packet loss causes *cwnd* to be cut in half.

When the T3-rtx timer expires on an address, SCTP should perform slow start by:

$$ssthresh = \max(cwnd/2, 2*\text{MTU}). \quad (3)$$

$$cwnd = 1*\text{MTU}. \quad (4)$$

Fig. 3 shows the congestion control for SCTP [6].

Table 1. Comparison between SCTP and TCP

Features	SCTP	TCP
Connection Oriented	support	support
Establishment Process	4-way handshake	3-way handshake
Shutdown Process	3-way handshake	4-way handshake
Half-closed connection	not support	support
Reliable data transfer	support	support
Ordered data delivery	support	support
Unordered data delivery	support	not support
Message boundary	boundary preserved	no boundary
Flow control	support	support
Congestion	support	support
ECN Support	support	support
SACK	support	Option
Data Fragmentation	support	support
Multi-streaming	support	not support
Multi-homing	support	not support

2.4 Differences Between TCP and SCTP

Table 1 describes the differences between SCTP and TCP. SCTP works like TCP, providing connection oriented service, reliable data transfer service, ordered data delivery service, flow control, congestion control and data fragmentation. But SCTP have additional features like message boundary, multi-streaming and multi-homing.

2.5 Related Work

Ladha [7] and Gerard [8] were interested in the use of multi-streaming. Ladha [7] proposed modifying FTP to use SCTP and its multi-streaming service. Gerard [8] introduced per-stream priorities as a method of decreasing delays of important data during periods of low bandwidth availability. Ahmed [9] proposed a new congestion control for SCTP by adjusting the congestion window size and proved that performance of SCTP has significantly improved.

3 System Architecture

In this section, the details of proposed file transfer system architecture are explained. It was used for our experimental tests.

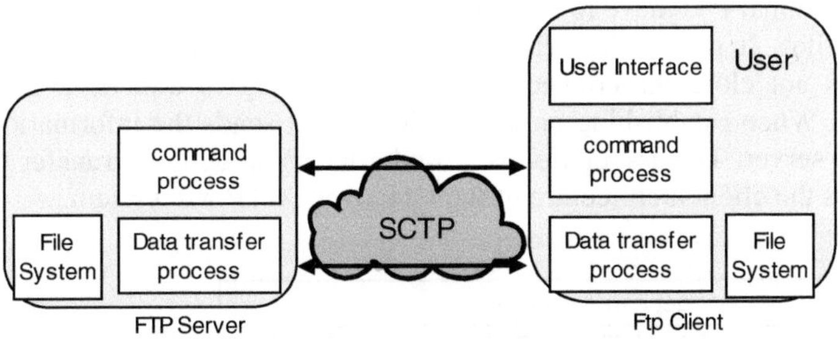

Fig. 4. FTP over SCTP

3.1 FTP over SCTP-MFCC

FTP is used for file backups, source repositories and mirror systems. All these applications require transferring multiple files from one host to another. Fig 4 shows the generic FTP architecture over SCTP. On the FTP over TCP, the number of data connections for multiple file transfer in an FTP session is exactly the number of file transfers performed. This subsection explains FTP over SCTP-MF(persistent Multiple File transfer) which is implemented for the proposal.

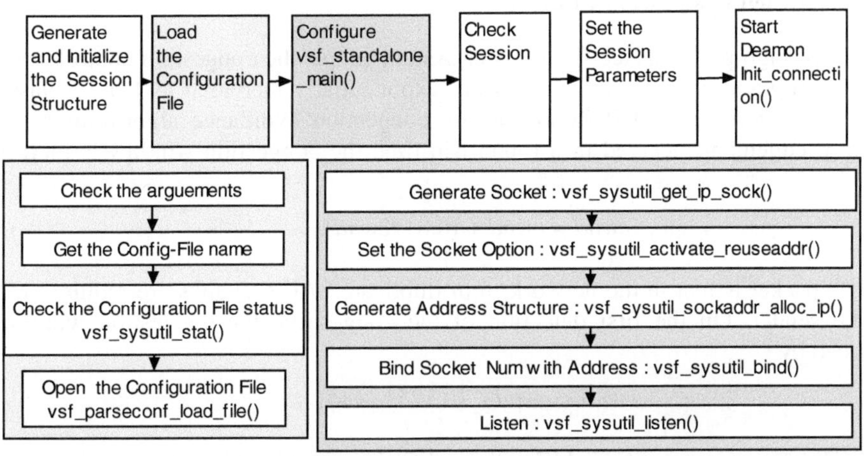

Fig. 5. FTP Server System

Server. Fig. 5 shows the flow of FTP Server system. Using vsf_sysutil_stat() and vsf_parseconf_load_file() function, the configuration file is loaded. Using vsf_sysutil_get_ip_sock(), vsf_sysutil_activate_reuseaddr(), vsf_sysutil_sockaddr _alloc_ip(), vsf_sysutil_bind() and vsf_sysutil_listen() function, configure operation is processed.

Client. Traditional FTP have to establish the data connections as much as the number of transfer files. It makes the performance of transfer system deteriorate. Proposed system does not close the connection after completing to transfer a file and waits another file. When establishing an association, client sends the information about the data files to server. The association is closed after completing to transfer the last file. Fig. 6 shows the client architecture of the FTP over SCTP-MF system.

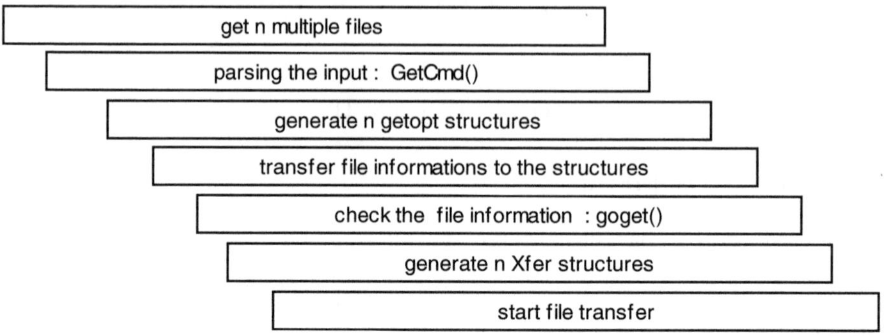

Fig. 6. FTP Client System

3.2 Modified Congestion Control

Proposed high-performance file transfer system adopts the congestion control scheme that is proposed by [9]. The *cwnd* is not exponentially increased as that of classical TCP algorithms like TCP Reno. During congestion avoidance algorithm for each acknowledgement received in a round trip time the congestion window is increased by

$$cwnd = cwnd + [0.01*cwnd]. \tag{5}$$

If a packet is lost in the network, four duplicate SACK chunks are required at the SCTP sender. On the first detection of the congestion, the congestion window is reduced by

$$cwnd = cwnd - [0.125*cwnd]. \tag{6}$$

SCTP-MFCC is the SCTP-MF with modified Congestion Control.

4 Analysis

4.1 Experimental Setup

The simplest network is built to investigate the multi-streaming feature. Two end nodes were used for the measurement process, one for the FTP client and the other for the FTP server. All nodes run Linux. Fig.7 and Table 2 show the testbed to measure the performance of the proposed architecture.

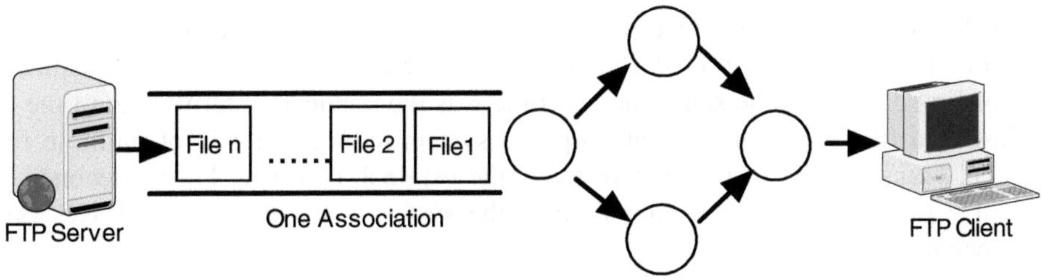

Fig. 7. Experimental Setup

Table 2. Testbed

server host	pentium IV processor	linux (RedHat 9)
client host	pentium IV processor	linux (RedHat 9)
router	cisco 2960	ios

4.2 Result

This section analyzes the experimental results of various file transfer systems. This paper focuses only on experimental results. Performance analysis and comparison are performed on original TCP, original SCTP, SCTP-MF and SCTP-MFCC. SCTP-MF is a file transfer system using persistent multiple file transfer. As stated above, SCTP-MFCC is a file transfer system using persistent multiple file transfer with modified congestion control.

The total transfer times observed for the test of a multiple file transfer are measured. For the systems using SCTP, the starting time was taken as the time after

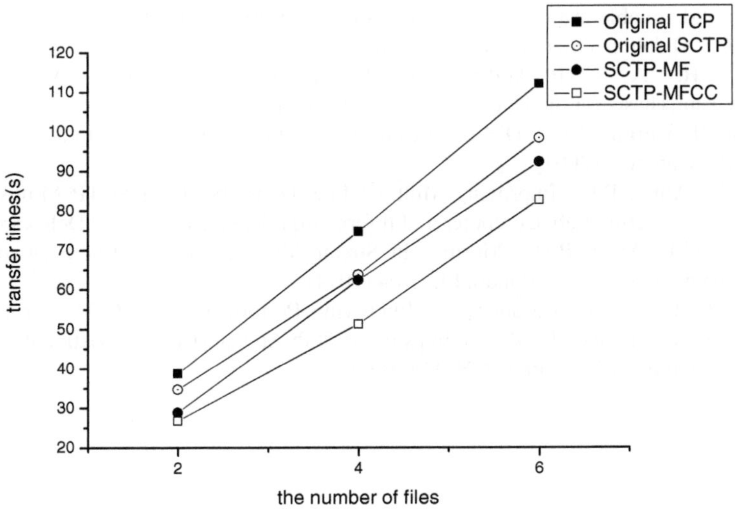

Fig. 8. Performance Analysis

establishment of control association and the end time was the association termination time. For the system using TCP, the starting time was the connection establishment time of first data file to be sent. The end time was the connection termination time of the last data file. Fig 8 shows that proposed system reduces latency of multiple file transfer. Transfer time of the system using the original TCP was 111.98 seconds in case of 6 packet transfer. Transfer time of the system using the SCTP-MFCC took 82.6 seconds.

5 Conclusion

SCTP is an IETF transport layer protocol and its services are at the same layer as TCP and UDP. Proposed file transfer system is running over SCTP instead of TCP. We have modified the FTP over STCP to reduce latency overheads. We implemented the efficient file transfer system using the SCTP multiple file transfer and modified STCP congestion control mechanism for heterogeneous ubiquitous environment to improve end-to-end FTP latency. The result shows that proposed system reduces latency of multiple file transfer and improves the performance.

References

1. Fu, S., Atiquzzaman, M.: SCTP:State of the Art in Research, Products, and Technical Challenges. IEEE Communication Magazine (2003)
2. Stewart, R., Xie, Q.: Stream Control Transmission Protocol (SCTP): A Reference Guide. Addison-Wesley, New York (2002)
3. Stewart, R., Xie, Q., Morneault, K., Sharp, C., Schwarzbauer, H., Taylor, T., Rytina, I., Kalla, M., Zhang, L., Paxson, V.: Stream Control Transmission Protocol RFC 2960 (2000)
4. Tasneem Yunus Kanpurwala: SCTP multi-streaming: Study of transmission of partially reliable and reliable data using dynamic stream priorities, the Degree of MASTER of SCIENCE, Oklahoma State University (2006)
5. Choi, S.-H.: A SCTP Traffic Control Scheduler for Effective Multi-streaming Management,Thesis for the Degree of Master, Korea University (2005)
6. Brennan, R., Curran, T.: SCTP Congestion Control: Initial Simulation Studies,International Teletraffic Congress (2001)
7. Ladha, S., Amer, P.D.: Improving Multiple File Transfers Using SCTP Multistreaming, Protocol Engineering Lab, Computer and Information Science University Delaware (2003)
8. Heinz II, G.J., Amer, P.D.: Priorities In Stream Transmission Control Protocol (SCTP) Multistreaming, SCI '04, Orlando, FL, 7/04 (2004)
9. Ahmed, I., Yasuo, O., Masanori, K.: Improving Performance of SCTP over Broadband High Latency Networks. In: Proceedings of the 28th Annual IEEE International Conference on Local COmputer Network (LCN'03)(2003)

A Suggestion for Analysis of Unexpected Obstacles in Embedded System

Yasufumi Shinyashiki[1], Toshiro Mise[1], Masaaki Hashimoto[2], Keiichi Katamine[2], Naoyasu Ubayashi[2], and Takako Nakatani[3]

[1] Matsushita Electric Works, Ltd., 1048 Kadoma, Kadoma, Osaka, Japan
{yasufumi,tmise}@mail.mew.co.jp
[2] Kyusyu Institute of Technology, 680-4 Kawazu, Iizuka, Fukuoka, Japan
{hasimoto@ai,katamine@ci,ubayashi@ai}.kyutech.ac.jp
[3] University of Tsukuba
nakatani@gssm.otsuka.tsukuba.ac.jp

Abstract. In order to improve the quality of embedded software by finding obstacles unanticipated by software specifications, this paper describes a requirement analysis method based on an Information Flow Diagram and experiments using this method, with discussion. Recently, embedded software has become more large-scale and complicated while the development cycle of the software has typically been shortened. This industry trend compromises the quality of embedded software. In order to improve the quality of the software, unexpected obstacles must be carefully analyzed in the specification and design phases. We propose a diagram-based analysis method to answer this need, and have performed an experiment using the analysis method with an actual product. The results of the experiment made it clear that 1) Novice engineers can construct an Information Flow Diagram under the lead of expert engineers; and 2) The methods by which experts lead novice engineers in the analysis of unexpected obstacles require further refinement.

1 Introduction

Embedded systems like electrical household appliances are used by various kinds of users in various kinds of environments [1, 2]. The operations of these systems must be secure and usable over a long period. For this reason, more than 70% of embedded software is assigned to functions that handle exceptional conditions such as device faults, communication errors, signal noise, operator errors, and bad influences from the environment. Recently, embedded software has become more large-scale and complicated [3], while the development cycle of the software has typically been shortened. Therefore, it has become more difficult for engineers to anticipate exceptional conditions without accidentally leaving out specifications in the design phase of the software development. Furthermore, such work of specification requires knowledge of devices, users, and environments, as well as the software itself [4]. Therefore, engineers sometimes fail to specify certain exceptional conditions. In fact, such specification oversights cause product failure and the delay of development as software must be redesigned. Therefore, if the exceptional conditions are made clear in the specification and design phases, we can expect better quality and productivity

M.J. Smith, G. Salvendy (Eds.): Human Interface, Part I, HCII 2007, LNCS 4557, pp. 755–768, 2007.

in embedded software. In this paper, we define "expected specifications" and "unexpected obstacle specifications" as follows: The former specifies system behaviors usually described in the software operation manual, and explicitly defined at the start of the system architecture design. The latter concerns any deviations from the behaviors determined by the former. The deviations are as follows: the wearing-out and inherent flaws of system hardware, improper operation and overload caused by system users, temperature and radio noise in the system environment, and so on. Generally, only expert engineers can anticipate unexpected obstacles with few oversights. Therefore, we studied a specification method for finding unexpected obstacles by formalizing the method that is tacitly used by experts in embedded software development [5, 6, 7, 8, 9]. One of the experts' methods is that they pay attention to information flow in an embedded system and its operational environment, and find unexpected obstacles by analyzing deviations of the information flow. Then, they apply the ideas of FTA (Fault Tree Analysis) [10], FMEA (Failure Modes and Effects Analysis) [11], and HAZOP (HAZard and OPerability studies) [12], which are used for the analysis of device failures. This paper describes an Information Flow Diagram (IFD) for conceptualizing information flow in an embedded system and its operational environment. One of the main features of our IFD is that the device diagram and process diagram are combined. This paper also describes a method for analyzing unexpected obstacles using the IFD and the concepts of FTA, FMEA, and HAZOP in an integrated manner. In this research area, individual applications of FTA, which applies a top-down approach, and FMEA and HAZOP, which use a bottom-up approach, have been already studied [13, 14, 15]. Misuse cases [16], abuse frames and goal-oriented obstacle handling [17] have also applied the top-down approach. Integrating the ideas of FTA, FMEA, and HAZOP with IFD is effective for preventing the analysis oversight of unexpected obstacles. In addition, this paper describes experiments using the analysis method, and discusses its results. Section 2 outlines the essentials of the requirement analysis method for unexpected obstacles. Section 3 and 4 describes the IFD and the analysis method for unexpected obstacles, respectively. Section 5 explains our experiments, and Section 6 discusses the results.

2 Essentials of a Requirement Analysis Method for Unexpected Obstacles

Before we propose our requirement analysis method for unexpected obstacles in embedded software, we describe the parameters of the method as follows: Business information systems operate using limited kinds of devices such as CPU, memory and hard disk. On the other hand, embedded systems are composed of various kinds of devices such as sensors and controllers. Furthermore, the device types differ according to product. Most unexpected phenomena in embedded systems are caused by exceptional conditions originating from these devices. Moreover, embedded systems are operated not only by users of the systems but also by automatic information triggers such as the light and temperature of the system environment. Exceptional conditions originating in the users and environment also cause unexpected phenomena. Therefore, engineers of embedded systems have to pay attention to users and potential environments as well as devices of the systems.

When a failure occurs in an embedded system, there must be an originating cause or causes. Moreover, there is a causal relation that proceeds from the originating cause(s) to the failure. The devices in which the causes originate are often different from the device in which the failure is found. The links between the originating cause(s) and the observed failure subsist in the deviation of information flowing between the devices. Therefore, engineers of embedded systems must pay attention to the flow of information and processes which create the flow.

It is not difficult to assume a failure caused by a unique serious exception like a device fault. It is more difficult to assume a failure caused by a set of nonserious exceptions such as overload, deterioration of the device, deviation in the user's operation timing, and environmental influences. The failure can occur when the process receives a set of nonserious deviated information as input, then processes it, outputting seriously deviated information. By investigating the set of nonserious exceptions as failure scenarios, expert engineers of embedded software can anticipate failures which are difficult for novice engineers to anticipate. Therefore, failure scenarios representing a set of nonserious exceptions should be introduced into the requirement analysis method.

Expert engineers of embedded software use the concepts of FMEA, FTA and HAZOP to examine, respectively, the failure patterns of devices, failures occurring contrary to the quality specifications of embedded systems, and deviation of the information that flows in embedded systems. These three points of attention correspond to, respectively, the starting point, the terminal point, and the middle point of a failure scenario. Since the points of attention are frequently scattered in failure scenarios, a systematic approach to scenarios is necessary to avoid oversights of unexpected obstacles. Therefore, FMEA, FTA and HAZOP should be applied together to the analysis method.

3 Information Flow Diagram

This section describes the IFD applied to the analysis method of unexpected obstacles. As described in the previous section, it is required for the method that attention should be paid to both devices and processes that are determined by architecture design for implementing expected specifications. Therefore, the IFD is composed of a device diagram (DD) and process diagram (PD). The DD and PD are described in the following subsections:

3.1 Device Diagram

Devices and the connections between them are decided upon in the design phase. The devices and connection are illustrated in a DD. Since, as described in the previous section, attention should be paid to users and the environment of the embedded system, we include the users and environment as objects in the DD. For instance, the sun which sends the light and the space through which the light is transmitted are illustrated as devices in the case of street light systems which determine the time of day using optical sensors.

In the DD, a device is illustrated using a rectangular box, as shown in Figure 1. We divide the rectangular box into three parts, which are used to list the name, attributes and fault patterns of the device, in that order. The fault patterns show the kinds of potential faults such as noise, deterioration and disconnection. The attributes are used to show the properties necessary for the analysis of unexpected obstacles, such as the threshold for noise identification and the time for deterioration.

Information is transmitted between devices by physical media such as electricity, light, and sound. If a fault of the physical media, such as deterioration and disconnection, occurs, the flow of information may be interrupted and a failure may occur. Therefore, it is important in analysing unexpected obstacles to pay attention to physical media that carry information. We call the physical media carrier, and consider that the carrier connects the devices. The DD illustration for a carrier is shown in Figure 2. It uses a diamond box and an arrow. The shaft of the arrow shows the connection between the two devices. The direction of the arrow indicates the direction of information flow. The name of the carrier, its attributes, and fault patterns are described within the diamond box.

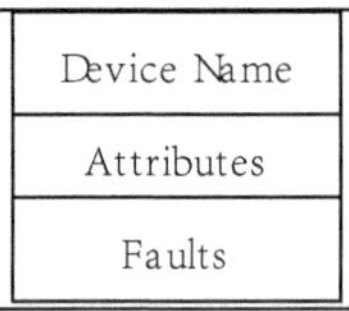

Fig. 1. Device

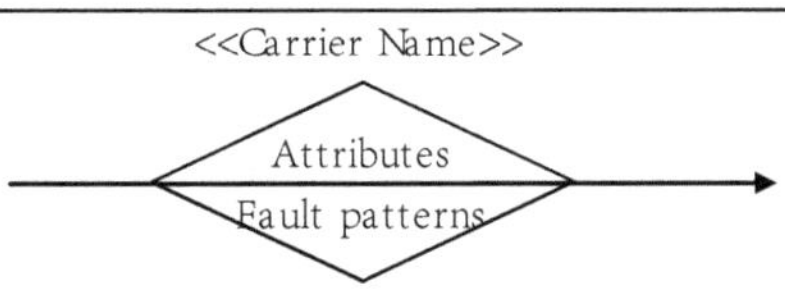

Fig. 2. Carrier

3.2 Process Diagram

We apply the method IDEF0 [18] to statically illustrate the various functions of a process in the PD. In IDEF0, the process is divided into activities. A rectangular box is used to show an activity. The name of the activity is described in the box. If necessary, the input information, output information, control information, and mechanism are specified for the activity using arrows, and their names are given as shown in Figure 3. The activities are connected to each other as part of the information flow. The activity processes input information by the given mechanism, and produces output information. The control information consists of the conditions for the processing.

The PD shows the activities which are processed with the devices shown in the DD, and the information flow between the activities. The relation between the DD and PD is shown by using the connection between the activities and the devices with arrows that indicate the mechanisms. In IDEF0, the process can be described hierarchically in a manner of step-wise refinement. The devices can be also described hierarchically according to subsystems, modules and so on.

The IFD contains not only a representation of an embedded system but also objects for the user and environment of the system. Therefore, the IFD is modelled as a

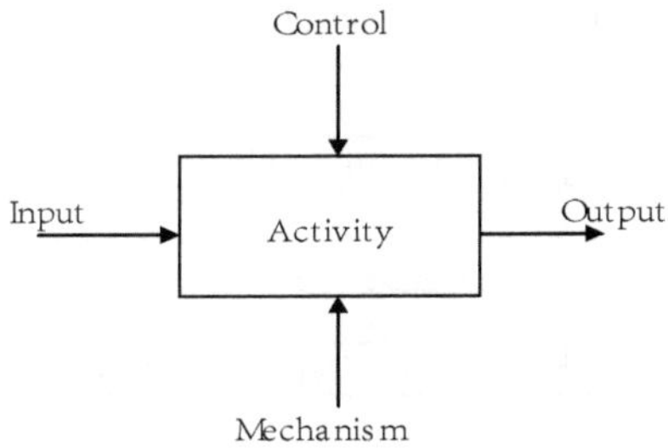

Fig. 3. Process of IDEF0

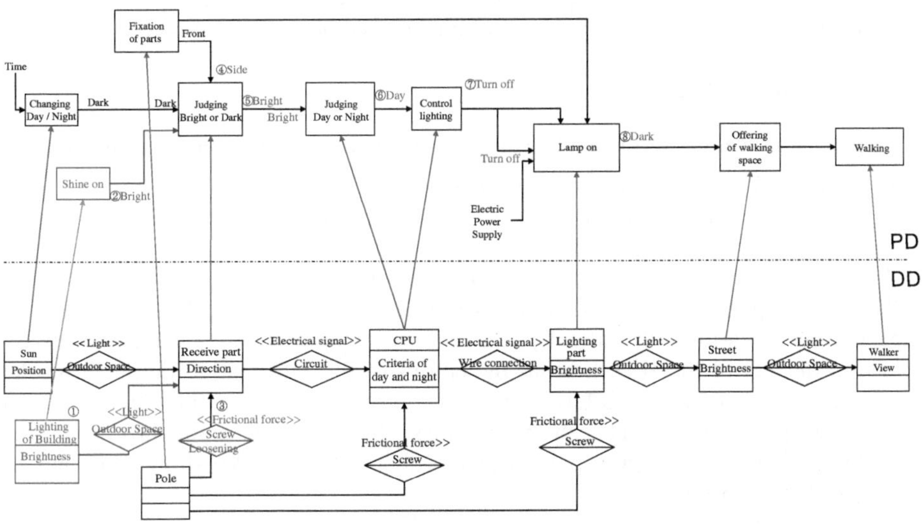

Fig. 4. Example of an information flow diagram

closed world, and information is transferred neither to nor from the IFD. In order to clearly distinguish the information and carrier, we use the term "Information Flow". An example of an IFD is shown in Figure 4.

4 Requirement Analysis Method of Unexpected Obstacles

This section describes an analysis method for unexpected obstacles which integrates the concepts of FTA, FMEA, and HAZOP on the IFD.

4.1 Preconditions of Analysis

The analysis method is assumed to be applicable after the development phase of an embedded system reaches the following situation. First, the function of the system and the failures against the safety requirements of the system are defined as expected

Table 1. Examples of Guide-Words

Type of Influence	Guide-Words
Behavior	Stop, Instability, Fixation, …
Load	Too much, Too few, …
Meaning	Out of range, Undefined, …
Procedure	Misorder, Bad timing, …
Frequency	Continuous, Repetition, Temporary, …

requirement specifications. Then, there are documents describing the hardware architecture and the faults of the devices in the architecture. In addition, there are documents for the software modules. Thus, after the design for expected requirement specifications is concluded, the analysis of unexpected obstacles begins.

4.2 Procedure of Analysis

We show the procedure of the analysis method for unexpected obstacles as follows:

1. *Illustrating the IFD for expected specifications*

   Before applying this method, the IFD of the expected specifications is illustrated on the basis of the architectural design of the system.
2. *Additionally illustrating the failure phenomena to be avoided*

   The failure phenomena to be avoided are illustrated additionally on the IFD of expected specifications. Specifically, the failure phenomena are related to activity boxes on the PD of the IFD. Serious deviations of the information output from the activity boxes related to the failure scenario are also illustrated on the PD, along with exceptional phenomena that can be assumed by applying guide words to the information flow between any pair of processes on the PD. We show examples of guide words in Table 1.
3. *Analyzing exceptional conditions using the IFD*

   We call the conditions for originating exceptional phenomena the exceptional conditions. For each exceptional phenomenon specified in the above-mentioned step (2), its exceptional condition is extracted using the IFD. On the PD, the information flow which has a failure phenomenon is connected to a process which outputs the information. We call this process the "process under examination". The process under examination has its input information, control information, and the device serving as its mechanism. Therefore, the exceptional condition causing the exceptional phenomenon can be classified as follows:

   a) Exceptional condition of the device serving as the mechanism of the process under examination. This condition is furthermore classified into the following two conditions:
      i. Physical fault of the device. This condition is obtained from the design document of the device.
      ii. Logical fault of the device. This condition corresponds to a software bug.
   b) Exceptional condition of the input or control information of the process under examination. This condition is furthermore classified into the following three conditions:

i. Fault of the flow path of the input or control information.
ii. Information received from a disguise device which sends the same kind carrier as the carrier of the input or output information.
iii. Exceptional phenomena of the flow of input or control information.

c) i. combination of two or more of the exceptional conditions described above.

According to the above-mentioned classification, exceptional conditions are searched for the exceptional phenomena under examination. If the exceptional conditions from which the phenomena originate in fact exist, they are adopted. Otherwise, the search for the exceptional conditions causing the phenomena is stopped because the exceptional phenomena have been ruled out. If there is an exceptional condition that is classified into the above-mentioned item b)-iii, the search is repeated for exceptional phenomena newly found from the exceptional condition.

If a disguise device is assumed to appear in the case of the above-mentioned item b)-ii, the disguise device is additionally illustrated. The carrier between the disguise device and the device serving as the mechanism of the process under examination is illustrated. Moreover, a process done on the disguise device is added to the PD. The information flow is also illustrated, from the process on the disguise device to the process under examination.

In this analysis step, the consecutive information flow found in the analysis from an assumed exceptional phenomenon to the fault or disguise device from which the exceptional phenomenon originates is called a failure scenario fragment. It is illustrated on the IFD. Incidentally, it is a top-down approach to first assume an exceptional phenomenon, and then to search for the exceptional conditions from which the phenomenon originates. The search traces the causal relation in reverse.

4. Constructing a failure scenario

In the manner of the bottom-up approach which follows the causal relation, a failure scenario originating from the faults found in the previous step (3) is constructed. The detailed procedure is as follows: First, the influence of the faults for output information of the process is assumed. This procedure is repeated according to the consecutive information flow. If the information representing failure is found, the procedure is finished. The consecutive information flow from the faults to the failure constitutes the failure scenario.

In constructing failure scenarios, it is indispensable to take all of the potential combinations of faults into account. However, the combination is made by the confluence of the information flows. Therefore, the failure scenario is pursued by the following procedures:

a) Examine the process that inputs the information under examination. Then, take all combinations of all possible exceptional phenomena of the input information, control information, and mechanism of the process into account. Repeat this procedure if a combination of the possible exceptional phenomena outputs information of an exceptional phenomenon.
b) Finish this procedure if the process can output no information of an exceptional phenomenon.

c) Reuse the scenario fragment in constructing the failure scenario if the scenario fragment obtained in the above-mentioned step (3) includes the process and its output information under examination mentioned in the previous item a).

This procedure is done for all the faults found in the above-mentioned step (3). The deliverables of the procedure are failure scenarios. In this method, failure scenarios are constructed using the procedure of step (4) after the existence of faults is confirmed using the procedure of step (3). Of course, the other method in which the existence of failures is confirmed prior to faults is also logically applicable. However, expert engineers prefer the method described in this paper.

5 Experiments

This section shows a case study of the analysis method described in the previous section, and describes experiments using an IFD of an actual product.

5.1 Case Study

We use a street light system with an optical sensor as an example. The specifications of the system are to detect the brightness of the environment with the optical sensor, to judge whether it is day or night, and to turn on or off the fluorescent lamp. The system is illustrated in Figure. 5.

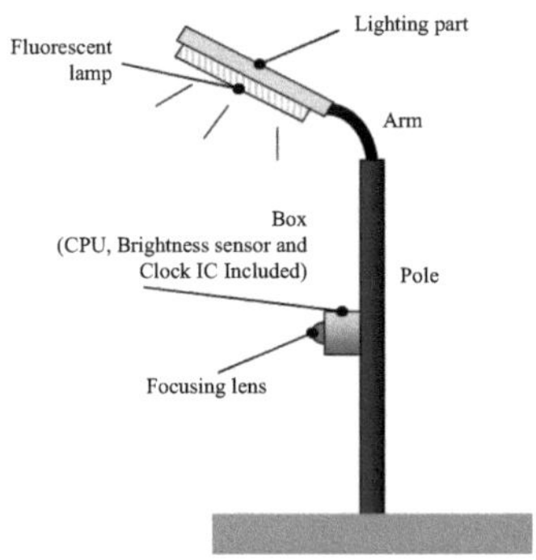

Fig. 5. Street light systems

The obstacles that are necessary to be avoided are shown in Table 2.

Table 2. Obstacles to be avoided in street light systems

Quality	Obstacle	Carrier
Safety	Burning	Heat
	Electrical Shock	Voltage
	...	...

The procedure of unexpected obstacle analysis for the example is described.

1. Making an IFD of expected specifications

The IFD of expected specifications is shown in Figure. 4. In the figure, a failure scenario is also described.

2. Additionally illustrating failure phenomena to be avoided

We describe the obstacles shown in Table 2 on the IFD. The result of the examination shows that the lighting parts can burn out. Therefore, the process "Burn out" is added on the PD, and is related to the lighting part via the mechanism as shown in Figure 6. On the other hand, electrical shocks are disregarded since they can not happen in this example.

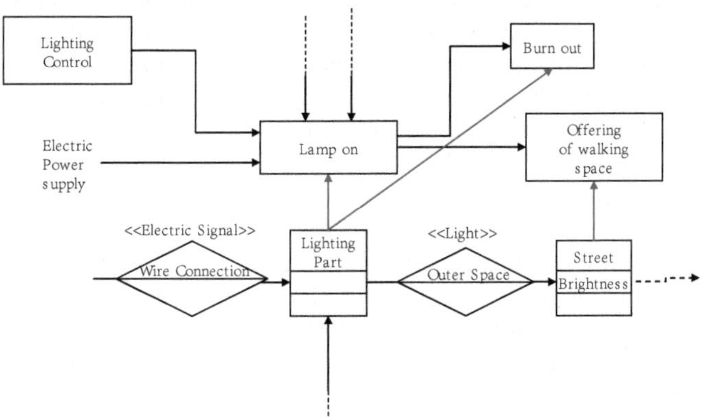

Fig. 6. Addition of process "Burn out"

Moreover, we apply guidewords to each information flow on the PD, and use them to anticipate exceptional phenomena. Examples of exceptional phenomena are shown in Table 3.

Table 3. Example of exceptional phenomena

Process	Output information	Guide-Words	Exceptional phenomena
Judging brightness	"Bright" or "Dark"	Stop	No output information is sent.
		Too few	"Bright" is input; however, the process output "Dark".
		Temporary	The process temporarily outputs] "Bright" during output of "Dark".

3. Analyzing exceptional conditions using the IFD

We analyzed each exceptional phenomenon using a top-down approach, and examined whether exceptional conditions exist for it or not. We show an example of analyzing and examining the exceptional phenomenon in the following: The process "Lamp on" of the mechanism "Lighting part" outputs "Dark" although it is really "Bright".

a) Assumed fault and exceptional phenomenon
   i. Damage, stain, and age deterioration of lighting part
   ii. Interception of input information by disconnection of signal line
   iii. Interception and decrease in power supply
   iv. Exceptional phenomenon of preceding process "Lighting control"
b) Not assumed fault
   i. Irregular direction of lighting due to loosened screws
   ii. Input information deviation by noise overlay of signal line

The assumed exceptional phenomenon iii is analyzed further, according to the classification of the exceptional conditions described in subsection 4.2.

We assumed the exceptional condition "The light of the building disguises itself as the sun" in the analysis. It represents an exceptional phenomenon that would transmit the information "Bright" although it is really night. In addition, we analyzed exceptional conditions for the control information flow "Directing the optical sensor to the front" output from the process "Fixing devices to pole" whose mechanism is a pole. As the result of this analysis, we can assume the fault "Loosening of the screws that fix the optical sensor to the pole".

4. Constructing a failure scenario

We constructed a failure scenario with the bottom-up approach using the fault "Light from a building disguises itself as the sun". Then, on the PD, we traced the flow of information "Bright" output from the process "Lamps on" of the device "Lighting of building". According to the Information flow on the PD, the information reaches the process "Judging brightness". We examine what information will be output from the process by combining the input information, other input information, control information, and exceptional phenomena of the process. In this case, there can be a combination of the input information "Bright" and control information "Directing the optical sensor to the side" because of the fault "Loosening of the screws that fix the optical sensor to the pole". If each of the two pieces of information reaches the process individually, there is no deviation of the output information. However, the combination of the two pieces of information causes the exceptional phenomenon that mistakes "Dark" as "Bright" because the optical sensor receives the light of the building when the loosening of screws causes it to become accidentally turned in the direction of the building. We repeat the procedure for the output information including the exceptional phenomenon mentioned above. Finally, a failure scenario is constructed. We show the scenario in Figure 4 with numbers and comments.

5.2 Description Experiment of IFD

In order for a full-scale experimental testing of the analysis method described in this paper, we have to confirm the effective operation organization of the method. Therefore, we have described an experiment on an IFD using an actual product of embedded software. The source codes of the software are about 50 K steps in size, and are written with C language. In this experiment, 3 students described the IFD under the guidance of an expert engineer of embedded software.

In this experiment, the students first read the documents of specifications and design. Then, they described the IFD of the expected specifications. The PD and DD are divided into 3 layers. The first layer of the DD shows the boundary between the product and its operational environment. The second layer shows the division of the system into the subsystems. The third layer shows the division of subsystems into hardware blocks. Moreover, the 3 layers of the PD are described according to the 3 layers of DD. In describing the IFD of expected specifications, the expert engineer specified only the imperative of describing the 3 layers of the DD. After the students described the IFD individually, we connected the 3 parts of the IFD to each other. Then, a few corrections were made to the IFD. In the next step, the students added the description of failure scenarios which were already found by expert engineers in the product development. In this step, the expert engineers gave the students a simple explanation of the failure scenarios. In this experiment, we found that there were important viewpoints from the device faults and failure phenomena for step-wise refinement of the IFD.

5.3 Discovering Experiment of Exceptional Phenomena

Furthermore, we experimented on the use of guidewords by one student to discover exceptional phenomena. In the experiment, the expert engineer first showed 16 guidewords using the format shown in Table 3. The expert engineer gave no further guidance to the student. The experiment included 46 information flows. The student discovered 76 exceptional phenomena in about 7 hours. Most of the exceptional phenomena not discovered by the student were related to the user or the mechanical parts of the system. The examples which are discovered by the student shown in follows:

- Excessive pressure hangs to the operational panel.
- The program beginning instruction is input in spite of while executing the program.
- A mute instruction is input on the way of the control that increases the volume.

The exceptional phenomena which are not discovered by the student shown in follows:

- The interpretation of the instruction displayed on the screen display is mistaken by the user.
- The guidance voice doesn't synchronize with the screen display.
- The machine had stopped before return to the position.

6 Discussion

The experiment shown in the previous section has proven the following: By applying the analysis procedure to the IFD of expected specifications under the guidance of an expert engineer, non-expert students could describe the IFD of unexpected obstacle specifications including failure scenarios. Therefore, our IFD-based analytic method can be assumed to be effective as a tool by which a small number of engineering experts in embedded software could guide a large number of novice engineers.

Furthermore, we experimented on the possibility of novice engineers discovering exceptional phenomena using only guidewords. However, this was less successful, especially when it came to errors associated with users and the mechanical parts of the system. Therefore, we assume that expert engineers will be required to give more guidance to novice engineers in finding exceptional phenomena.

We have been studying another analysis method applying an analysis matrix other than an IFD [19]. This method constructs failure scenarios by filling in the analysis matrix like a state transition table with exceptional states and events, and by tracing the transitions between exceptional states. This method uses a diagram similar to the DD of the IFD. Of course, the information illustrated in the PD is needed to examine the transitions between exceptional states. In the case of the analysis matrix, however, such information is not explicitly illustrated on the document, but is part of the knowledge base of the expert engineers. Thus, expert skill is required for the method using the analysis matrix. However, the amount of description for the analysis matrix is smaller than that for the PD. Therefore, the method using the analysis matrix is a more efficient method for expert engineers to use than the IFD method.

In the future, we will establish an analysis method using IFD by experimenting with it in real applications. Then, we will integrate the two methods. Furthermore, we will study the effective use of the integrated method by a group composed of a small number of expert engineers and a much larger number of novice engineers. When integrating the methods, we will formalize the integrated method with the qualitative reasoning theory that applies the notion of state transition and constraint conditions. By formalizing the method, we will formalize the experts' knowledge of unexpected obstacles. Then, we will develop a knowledge base according to the formalization of knowledge. The knowledge base will be installed in the CASE tools for analyzing unexpected obstacles. Incidentally, the IFD has the characteristics of a directed graph. Therefore, we will also study the application of the graph method. It will be useful for finding exceptional phenomena such as loss of information flow because of disconnection of a cable, occurrence of new information flow because of electromagnetic induction, and positive or negative feedback on the information flow loop.

7 Conclusion

This paper has described an IFD and an analysis method based on the IFD for analyzing unexpected obstacles in embedded systems. The IFD illustrates the users, environment, and devices in addition to the processes of the systems, all of which sometimes cause unexpected obstacles in embedded systems. In order to analyze unexpected obstacles using both top-down and bottom-up approaches for preventing omissions in analysis, we have jointly applied the concepts of FTA, FMEA and HAZOP. We have confirmed that novice engineers can describe the IFD of expected specifications and the IFD of unexpected obstacle specifications under the guidance of expert engineers in an experiment applying the method to an actual product.

In the future, we will establish the analysis method after further experiments using the IFD-based analysis method. Then, we will integrate the IFD and analysis-matrix-based analysis methods. After that, we will study the quality and efficiency of the new analysis method by considering the possible structure of engineering personnel in a firm using the method. Moreover, we will study the knowledge base of unexpected obstacles and a graph analysis method to use with the IFD.

Acknowledgement. The authors would like to thank Mr.Tanabe, Mr.Tanimoto, Mr.Inoue, and Mr.Kubo for cooperation in the experiment.

References

1. Mise, T., Shinyashiki, Y., Hashimoto, M., Ubayashi, N., Nakatani, T.: A Specification Analysis Method for Unexpected Obstacles in Embedded Software (in Japanese). Proc. of the FOSE2005, Kindai Kagaku Sha. Japan Society for Software Science and Technology, pp. 227–235 (2005)
2. Sumi, T., Hirayama, M., Ubayashi, N.: Analysis of the external environment for embedded systems, IPSJ SIG Technical Reports, 2004-SE-146, pp. 33–40 (2004) (in Japanese)
3. Ministry of Economy, Trade and Industry, editor. Report of actual field survey of embedded software, Edition, Ministry of Economy, Trade and Industry (2005) (in Japanese)
4. Crook, R., Lnce, D., Lin, L., Nuseibeh, B.: Security Requirements Engineering: When Anti-Requirements Hit the Fan. Proc. of the 10th Anniversary Joint IEEE International Requirements Engineering Conference (RE'02), pp. 203–205 (2002)
5. Hatanaka, H., Shinyashiki, Y., Mise, T., Kametani, H., Hashimoto, M., Ubayashi, N., Katamine, K., Nakatani, T.: An Analysis of Information Flow Graph based on Conceptual Model of Exceptions in Embedded Software, Technical Report of IEICE 104-431, pp. 19–24 (2004)
6. Shinyashiki, Y., Mise, T., Eura, Y., Hatanaka, H., Hashimoto, M., Ubayashi, N., Katamine, K., Nakatani, T.: A Conceptual Model of Exceptions in Embedded Software. In: Proceedings of Embedded Software Symposium, pp. 8–11 (2004) (in Japanese)
7. Mise, T., Shinyashiki, Y., Eura, Y., Hatanaka, H., Hashimoto, M., Ubayashi, N., Katamine, K., Nakatani, T.: Exception Analysis Matrix for Embedded System Software Specification. In: Proc. IPSJ/SIGSE Embedded Software Symposium (ESS 2004) (in Japanese) (2004)
8. Mise, T., Shinyashiki, Y., Hashimoto, M., Ubayashi, N., Katamine, K., Nakatani, T.: An Analysis Method with Failure Scenario Matrix for Specifying Unexpected Obstacles in Embedded System. The proceeding of the 12TH Asia-Pacific Software Engineering Conference, pp. 447–454 (2005)
9. Kametani, H., Shinyashiki, Y., Mise, T., Hashimoto, M., Ubayashi, N., Katamine, K., Nakatani, T.: Information Flow Diagram and Analysis Method for Unexpected Obstacle Specification of Embedded Software. Proc. of the Knowledge-Based Software Engineering (JCKBSE'06), pp. 115–124 (2006)
10. Leveson, N.G.: Fault Tree Analysis, Safeware System Safety and Computers, pp. 317–326. Addison-Wesley, Reading, MA (1995)
11. Leveson, N.G.: Failure Modes and Effects Analysis, Safeware System Safety and Computers, pp. 341–344. Addison-Wesley, Reading, MA (1995)
12. Leveson, N.G.: HaZards and Operability Analysis, Safeware System Safety and Computers, pp. 335–341. Addison-Wesley, Reading, MA (1995)
13. Pentti, H., Atte, H.: Failure Mode and Effects Analysis of Software-based Automation Systems, STUK-YTO-TR 190, p. 35 (2002)
14. Dehlinger, J., Lutz, R. R (eds.): Software Fault Tree Analysis for Product Lines. Proceedings of the Eighth IEEE International Symposium on High Assurance Systems Engineering, pp. 12–21 (2004)

15. Redmill, F., Chudleigh, M., Catmur, J.: System Safety: Hazop and Software Hazop, p. 248. John Wiley & Sons Ltd, New York (1999)
16. Alexander, I.: Misuse cases, use cases with hositile intent. IEEE Software 20(1), 55–66 (2003)
17. Lamsweerde, A.V., Letier, E.: Handling Obstacles in Goal-Oriented Requirements Engineering. IEEE Transactions on Software Engineering 26(10), 978–1005 (2000)
18. Bemus, P., Mertins, K., Schmidt, G. (eds.): Handbook on Architectures of Information Systems. Springer, Heidelberg (1998)
19. Mise, T., Hashimoto, M., Katamine, K., Shinyashiki, Y., Ubayashi, N., Nakatani, T.: A Method for Extracting Unexpected Scenarios of Embedded Systems. Proc. of the Knowledge-Based Software Engineering (JCKBSE'06), pp. 41–50 (2006)

Peer-to-Peer File Sharing Communication Detection System Using Network Traffic Mining

Satoshi Togawa[1], Kazuhide Kanenishi[2], and Yoneo Yano[3]

[1] Faculty of Management and Information Science, Shikoku University, 123-1 Furukawa Ojin-cho Tokushima 771-1192, Japan
doors@shikoku-u.ac.jp
[2] Center for Advanced Information Technology, University of Tokushima, 2-1 Minami-Josanjima Tokushima 770-8506, Japan
marukin@cue.tokushima-u.ac.jp
[3] Institute of Technology and Science, University of Tokushima, 2-1 Minami-Josanjima Tokushima 770-8506, Japan
yano@is.tokushima-u.ac.jp

Abstract. In this research, we have built a system for network administrators that visualize the Peer-to-Peer (P2P) file sharing activities of network users. This system monitors network traffic and discerns traffic features using traffic mining. This system visualizes the P2P file sharing traffic activities of an organization by making the processing object not an individual user but a user group. The network administrator can comprehend the P2P sharing activities of the organization by referring to the map. This system extracts a traffic feature from captured IP packets that the users communicated. Afterwards this system creates a traffic model. The features of the traffic model are emphasized by weighting. After that, the traffic model is visualized by a Self-Organizing Map. The network administrator is assisted in understanding users' P2P file sharing communication behavior by this feature map. The administrator can then respond to the situation. As a result, we think we can assist the monitoring operation and network administration.

Keywords: Traffic Mining, Incident Response, Administrator Assistance, Peer-to-Peer Detection.

1 Introduction

Today, Peer-to-Peer (P2P) applications have become on the Internet. It is applied in the field of file sharing, VoIP and groupware. Especially, a lot of file sharing software has been designed on the P2P communication model. If Internet users want to get various kinds of data, the users can easily obtain various files and data using P2P file sharing software. The file content often includes music, movies and so on. However, because most of these files are extracted from music CDs and DVDs protected by copyright law, it is not appropriate to exchange these files. Moreover, popular P2P file sharing applications such as WinMX, Winny and Share need a huge bandwidth because these applications send and receive large amounts of data. As a result, regular communications are obstructed by P2P applications.

M.J. Smith, G. Salvendy (Eds.): Human Interface, Part I, HCII 2007, LNCS 4557, pp. 769–778, 2007.

In addition, a virus that causes data compromise has been generated for the P2P file sharing network. There are many cases of classified information being compromised because of these viruses. These viruses give out sensitive information stored in the user's computer using the publication function on the P2P file sharing application. When an organization's member uses a P2P file sharing application, there is a risk that the organizations security will be compromised. If certain classified information leaks, the data will pass along from place to place in the P2P network.

For example, many data compromise accidents occurred in Japan at 2006. A large amount of military intelligence and investigative information were leaked from Self-Defense Forces and police departments. In addition, a lot of companies leaked customers' private information. These accidents are extremely serious and can result in customers losing trust in an organization.

There are limitation techniques which use packet filters to limit illegal traffic that deviates from the policy established by the company or university. Filter technology, which synchronizes with the packet filter definition, is installed on the firewall, and illegal traffic that does not conform to the site policy are not forwarded to users. However, it is difficult to keep the filter definition perfectly set because the default destination port of each P2P application is different. Moreover, P2P applications such as Winny and Share select the destination port dynamically. This P2P application's traffic limitation can be impossibly strict. Therefore, P2P traffic cannot be limited only by filter technology based on packet filter definitions.

On the other hand, the network administrator can use a specialized firewall system to limit P2P traffic. These P2P traffic limitation techniques are based on the signature information that is extracted from illegal traffic. It is completely blocked when a signature matches a traffic pattern. However, when a signature does not match the traffic, illegal traffic is not restricted. Accordingly, the network administrator has to understand the P2P application's behavior in the organization's network traffic. If the administrator can understand the P2P application's behavior, the administrator can usually ascertain problems at an early stage.

At the present time, if an administrator wants to understand P2P application activities, protocol analysis can be used. However, this method is very labor intensive, and these methods only provide basic information like IP address/port number level classification. The network administrator really wants a result that shows where the P2P file sharing application is used?

For these reasons, we have developed a traffic visualization system for P2P communication detection and administrator assistance. This system provides a feature map of traffic behavior made up from results of network traffic mining. This system assists the monitoring operation of the administrator by showing the feature map that this system presents. As a result, we think that we can assist the monitoring operation of the administrator.

We pay attention to the traffic that the organization users send and receive. These features are extracted from this traffic using traffic mining. The features are the source/destination IP addresses and source/destination TCP (or UDP) port number and TCP flags. In addition, we pay attention to the results of the DNS query from internal clients. We found that when the P2P nodes try to find other nodes, the DNS query amounts are less than normal DNS queries such as Web browsing. We use these DNS query features for discerning P2P application behavior. Moreover, this system acquires the packet occurrence frequency and yield. Consequently, a traffic model is generated from the feature and packet occurrence frequency and a result of the DNS query.

The method of generating the model is the Vector Space Model. The similarity problem between traffic features is replaced with a cosine measure between vectors. Weighting is added to the obtained traffic model to emphasize feature quantity. Afterwards, a feature map is generated by using Self-Organizing Maps (SOM) from the traffic model. This algorithm maps multi-dimensional vectors on a two-dimensional plane.

This map shows an administrator which computer communicated to other computers and the volume of the communication. It expresses not only the summarized traffic amount but also each traffic type and behavior. It can be said that the feature map is a result of traffic mining from the users' traffic, and the administrator is assisted in understanding the organization's traffic behavior by this feature map.

In this paper, we proposed a system framework of traffic visualization for P2P communication detection, and we show a configuration of the prototype system. Next, we show the results of experimental use and examine these. Finally, we describe future study, and we show conclusions.

2 Assisting the Detection of P2P File Sharing Traffic

2.1 Framework of P2P File Sharing Traffic Detection

Fig. 1 shows a framework of administrator assistance for P2P file sharing traffic detection. We assist the monitoring and detecting operation of the network administrator by providing the traffic behavior of the organization users.

We paid attention to traffic between the internal site and the Internet. All users' traffic passes a gateway in the internal site. We collect all IP packets that pass the gateway. In addition, traffic features are extracted from collected IP packets.

In addition, we paid attention to the result of a DNS host queries from internal DNS servers to external DNS servers. Generally, the P2P nodes information is distributed without hostname (Fully Qualified Domain Name). In the result, the DNS host query amounts of the internal P2P nodes are less than normal applications. If the traffic feature of one host has different from other hosts, and that host's DNS host query amount is less than other hosts, it has high probability of the P2P node.

Consequently, a traffic model is generated from extracted features that the users communicated and the results of DNS host queries. The method of generating the traffic model is a Vector Space Model. As a result, the similarity problem between source IP addresses is replaced with a cosine measure between feature vectors. Weighting is added to the obtained traffic model to emphasize feature quantity.

A series of processing described here are traffic mining. Because, the feature related to P2P file sharing communication is extracted from all captured traffic by the

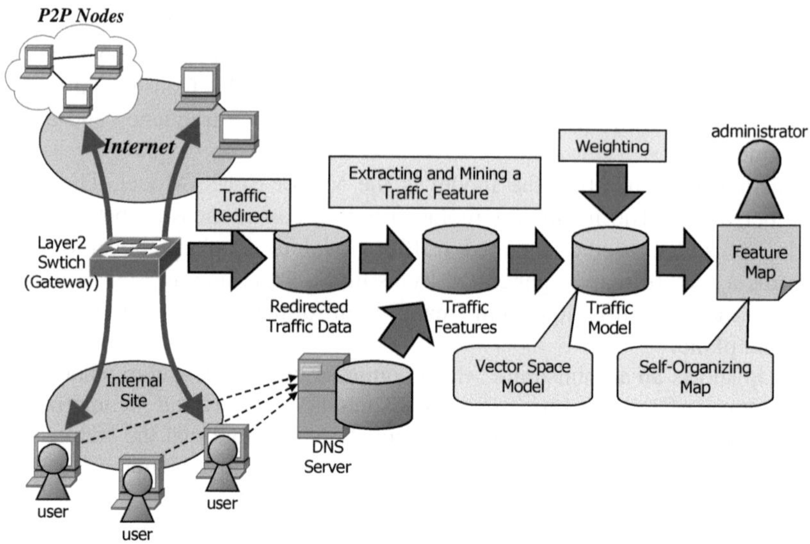

Fig. 1. Framework of administrator assistance for P2P files sharing traffic detection

series of processing. Moreover, extracted and emphasized features are stored to the traffic model. This model adapt to traffic feature visualization.

Afterwards, a feature map is generated by a Self-Organizing Map (SOM). SOM is an algorithm to map multi-dimensional vectors on a two-dimensional plane. As a result, this map expresses the typical source IP addresses that the users communicated.

The administrator gets a bird's-eye view of the organizations communication activities by referring to the map. Therefore, the administrator is assisted in understanding P2P traffic behavior by this feature map.

2.2 P2P Communication Model

Fig.2 is a hybrid type P2P file sharing architecture. Generally, hybrid P2P architecture has a central server which keeps all Meta information such as kind of files and file names. It is directory of user identities and index of resources on the P2P community. If the administrator wants to limit the use of hybrid P2P file sharing communication, it only has to block off the path to the central server.

However, this limitation technique is ineffectual for pure P2P architecture. Fig. 3 is a pure type P2P file sharing architecture. This architecture does not have central server. All information of sharing resources is stored to the some node. In this result, index information of sharing files is distributed on the pure P2P community, it is difficult to block off the path to the resources of sharing information.

Therefore, if an administrator wants to limit pure type P2P file sharing communications, the administrator must keep monitoring users' communication activity of organization.

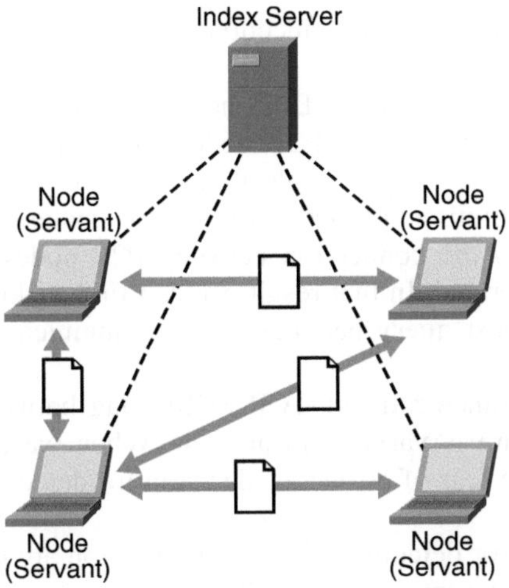

Fig. 2. Hybrid P2P communication model

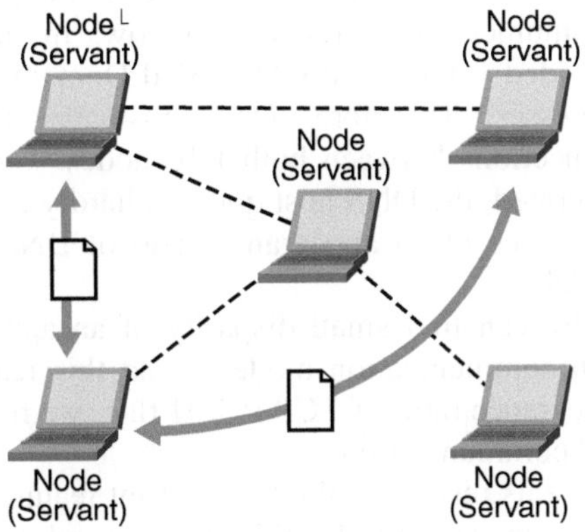

Fig. 3. Pure P2P communication model

2.3 Exploratory Experiment and Result for Traffic Feature Extraction

We made an exploratory experiment to clarify a feature of DNS host query by the P2P node. Especially, we want to clarify a DNS host query feature of the pure P2P nodes.

The P2P file sharing application was installed to experimental computer, and the experimental computer was used for 20 minutes with P2P application. After that, we generated general Web browsing traffic with other computers. Then both traffic was monitored and compared for this exploratory experiment.

Table 1. Measuring Results of pure P2P communication

Amount of sending IP packets	22,235
Amount of destination IP addresses	415
Appearance Ratio of TCP PUSH flag	35.2%
Appearance Ratio of DNS host query	Less than 0.1%

Table 2. Measuring Results of general Web browsing communication

Amount of sending IP packets	6,416
Amount of destination IP addresses	42
Appearance Ratio of TCP PUSH flag	9.2%
Appearance Ratio of DNS host query	91.8%

Table 1 and Table 2 show the measuring results of the exploratory experiment.

First of all, we can find a difference of the amount of sending IP packets. The case of pure P2P communication model much than Web browsing case per same measurement time. It is about 3.5 times larger than Web browsing case. What that means is that P2P communication model makes a lot of connections between internal node and P2P nodes on the Internet.

And then, it is understood that the appearance ratio of DNS host queries by P2P communication is remarkably low. In most situations, P2P nodes information is provided without Fully Qualified Domain Name. In that result, it is provided only IP addresses. Therefore, DNS hostname resolution is not required to make the connections between both P2P nodes. When the connection between P2P nodes is generated, the DNS host query is hardly generated. In that result, we can find striking difference of an appearance ratio of DNS host query between both communication models.

We can find small disparity of an appearance ratio of TCP PUSH flag between both communication models. And this feature is variable in amount. When we use appearance ratio of TCP PUSH flag, we have fear of erroneous decision for detecting P2P communication.

In this result, we think important features for detecting P2P communication are the appearance ratio of DNS host query and the amount of sending IP packets.

3 System Configuration

We show the configuration of proposed system in Fig. 4. This system has 5 modules that includes a “Traffic Collection Module”, “Traffic Analysis Module”, “DNS Query Analysis Module”, “Modeling Module” and “Visualization Module”. A detailed description of each module is provided below.

3.1 Traffic Collection Module

IP packets that users of organization sent and received are redirected by layer2 switch with port mirroring function. Traffic Collection Module accepts the redirected IP packets from layer2 switch. In addition, an Ethernet adapter configuration of this system is set to promiscuous mode. Because, this module have to accept all related IP packets. The accepted IP packets include normal traffic and illegal traffic, and all accepted IP packets are passed to the Traffic Analysis Module.

3.2 Traffic Analysis Module

This module attempts selection for all accepted IP packets. First of all, an administrator gives the IP address information of internal site servers to this module. And source traffic of the internal servers is dropped from all accepted traffic by using the given IP address information. Next, this module attempts to select traffic features from selected IP packets.

This module analyzes a packet field of selected IP packets, and some feature extracted from selected packets. The features are the source/destination IP address, and the source/destination TCP PORT number and TCP flags status. At the same time, each packets occurrence rate is calculated and stored. All extracted and calculated features are passed to the Modeling Module with other features generated from DNS Query Analysis Module.

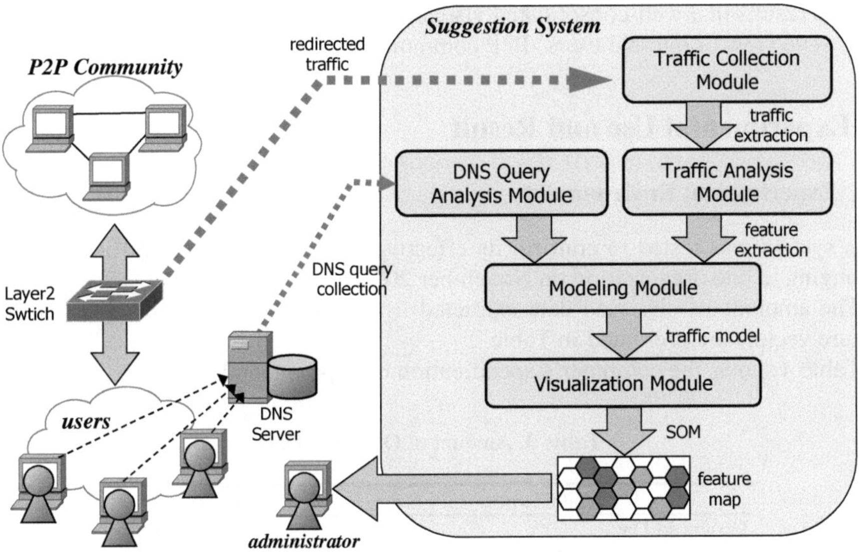

Fig. 4. System Configuration of Proposed System

3.3 DNS Query Analysis Module

The DNS server processes a DNS host resolution requests that was required from internal users. This module collects the results of DNS host resolution and requested client information from DNS server's log. It is selected excluding the incomplete results of DNS request. All extracted complete results of DNS hostname resolution are passed to the Modeling Module.

3.4 Modeling Module

This module generates a traffic model which is defined by the Vector Space Model. One source IP address corresponds to one multi-dimensionally composed vector, and each element of the multi-dimensional vector stores a number of destination IP address and the destination PORT number. We call this multi-dimensional vector a "feature vector". The number of feature vectors is the same as the total number of extracted source IP addresses. The set of these feature vectors becomes the traffic model.

The weighting process done to the feature vectors emphasizes the characteristics of the traffic model according to the occurrence rate with which the source IP address and the PUSH flags appear. As a result, if the module discovers frequency appearing source IP address, it is possible to find the P2P packet spreader host. When the weighting process is finished, the traffic model is passed to the Visualization Module.

3.5 Visualization Module

This module visualizes and making the feature map from the obtained traffic model. The Self-Organizing Map is used as a visualization method in this module. The source IP addresses of the processing object are self-organized by the SOM algorithm.

This results in a well-consolidated visual output that allows the administrator to get bird's-eye view of internal users' P2P communication activities.

4 Experimental Use and Result

4.1 Experimental Environment

This system was tested to confirm its effectiveness. We collected traffic from users belonging to one organization on November 20th 2006.

The amounts of observed data extracted from collected IP packets and generated feature vectors are presented in Table 3.

Table 4 shows the computer's specification of experimental use.

Table 3. Amount of Observed Data

Data Type	Amounts
Observed Data	1,423,592
Generated Feature Vectors	16,356

Table 4. Specification of the Experimental Use

CPU Specification	Intel Pentium4 3.2GHz
System Memory Capacity	1Gbytes
HDD Capacity	300Gbytes
Operating System	Linux (kernel 2.4.18)

4.2 Feature Map

Feature maps were generated once an hour in the experimental period. The traffic data was put into the system. The input packet amount was about 3,200,000 packets and the source IP address count that the system extracted was 720. Each feature map that the system presented had 320 elements, and each element corresponded to a summarized source IP addresses. The number of source IP addresses that the system extracted before clustering was 720. Therefore, the source the source IP addresses appearing in the map where communicated many times related with P2P communication by the computers.

We show the feature map in Fig. 5. This map is one period of the all generated feature maps. All display with application name on the map is marked by the hand for explanation.

We can find two large clusters on the map. These clusters are related to P2P type application's communication. The lower right cluster is related to P2P file sharing application. It is completely detected to P2P file sharing applications traffic. Unfortunately, the other one is not P2P file sharing application. This is related to Skype. However, Skype is based on the P2P type architecture. These application have a same behavior on the Internet, because each application has a lot of nodes on the P2P community.

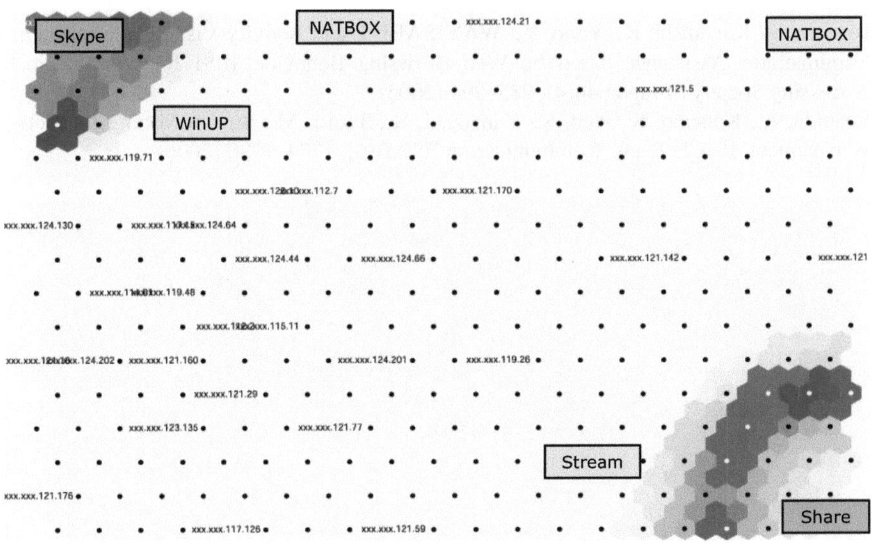

Fig. 5. Feature Map

As a result, we think that we can make complete to detect the P2P file sharing communication. The network administrator is assisted to detect the P2P file sharing traffic using this feature map. We think that the appearance ratio of DNS host query is especially effective for the feature map generating.

5 Conclusion

In this paper, we proposed a traffic visualization system for P2P communication detection. And we explained a configuration of the prototype system. And, we shown the results of experimental use and examine.

This system extracts records of P2P communication activities from the collected IP packets and the collected DNS query results. In addition, this system provides a feature map for the administrator. We developed a prototype system and experimented to confirm its effectiveness. It was shown that an administrator could inspect the results of the feature map.

References

1. WinMX Web site, http://www.winmx.com/
2. Winny Web site, http://www.geocities.co.jp/SiliconValley/2949/
3. BitTorrent Web site, http://bittorrent.com/
4. Skype Web site, http://www.skype.com/
5. Kohonen, T.: Self-Organizing Maps, 3rd edn. Springer, Heidelberg (2001)

6. Togawa, S., Kanenishi, K., Yano, Y.: WAVISABI: Users Activity Visualization System for Administrator Assistance based on Web Browsing Behavior. IPSJ Journal, Information Processing Society of Japan 46(4), 985–994 (2005)
7. Yoshida, K., Katsuno, S., Ano, S., Yamazaki, K., Tsuru, M.: Stream Mining for Network Management. IEICE Trans. Communication E89-B(6), 1774–1780 (2006)

A Method for Rule Extraction by Discernible Vector

E. Xu[1], Shao Liangshan[2], Tong Shaocheng[1], and Ye Baiqing[2]

[1] Department of Computer Science, Liaoning Institute of Technology,
Jinzhou , Liaoning Province, 121001, China
{exu21,jztsc}@163.com
[2] Management School, Liaoning Technical University, Fuxin,
Liaoning Province, 123000, China
{lntushao,baiqing-ye}@163.com

Abstract. To deal with the problem of extracting rules from the information table, a new method was studied and proposed. First, define discernible vector and its addition rule by the indiscernible relation in rough set. Second, scan discernible vectors just only one time by the discernible vector addition rule in order to obtain the core attribute set and the important attributes. Then reduce attribute and attribute value by deleting redundant attributes and attribute values respectively. Finally, a concise rule set was obtained. The illustration and experiment results indicate that the method is effective and efficient for rule extraction.

1 Introduction

Rough set is a new mathematical theory which makes a feature of dealing with imprecise, incomplete and inconsistent data, which don't need some transcendental knowledge or some accessional information but just the data itself. Since rough set was put forward by Professor Pawlak [1], it has been used in many fields such as machine learning, artificial intelligence etc., especially it has been an very efficient method in data mining [2,3,4], and frequently appears in clustering, classification algorithms.

Rule extraction is an important part of rough set, which consists of attribute reduction and attribute value reduction, so many researchers have been studying on it. Many algorithms have been put forward, for example, the attribute reduction based on discernible matrix [5,6], based on entropy [7] and attribute value reduction based on value core[8] and so on. But for now, there isn't an algorithm which can be acknowledged by the researcher in the world. The major problem is that rule extraction is a N-P hard problem [9,10], so in the process of rule extraction, the combination explosion is easily generated.

To extract rules form information table, a new method is proposed in this paper based on discernible vector. Experiment result shows that the method is efficient and effective.

2 Correlative Definition and Theorem

In order to describe attributes reduction, we define some conception as following.

M.J. Smith, G. Salvendy (Eds.): Human Interface, Part I, HCII 2007, LNCS 4557, pp. 779–784, 2007.

Definition 1. In rough set, an information system can be represented as

$$S = (U, A, V, f) \tag{1}$$

Where U is the universe, a finite set of N objects, $U = \{x_1, x_2, ..., x_n\}$, A is a finite set of attributes, $A = C \bigcup D$, where C is the set of condition attributes and D is the set of decision attribute. $V = \bigcup_{q \in A} V_q$, $f : U \times A \rightarrow V$, $q \in A$, $x \in U$.

Definition 2. For given $Q \subseteq A$ and $X \subseteq U$, the Q lower approximation $\underline{Q}X$ of the set X is defined as fellows,

$$\underline{Q}X = \{x \in U : [x]_A \subseteq X\} \tag{2}$$

Definition 3. For given $Q \subseteq A$ and $X \subseteq U$, the Q upper approximation $\overline{Q}X$ of the set X are defined as fellows,

$$\overline{Q}X = \{x \in U : [x]_A \cap X \neq \varnothing\} \tag{3}$$

Definition 4. If an information table is $S = (U, A, V, f)$, the discernible vector was defined as

$$DV = (O, DC, F) \tag{4}$$

Where O is the pair of discernible objects, DC is the discernible attribute set, F is the frequency vector, $F = (f(a_1), f(a_2), ..., f(a_n))$. They were described as below respectively,

$$DC = \begin{cases} \{a_k \mid a_k \in C \wedge a_k(x_i) \neq a_k(x_j)\}, & d(x_i) \neq d(x_j) \\ \varnothing, & d(x_i) = d(x_j) \end{cases}$$

$$f(a_i) = \begin{cases} 1/|DC|, & a_i \in DC \\ 0, & a_i \notin DC \end{cases}$$

Definition 5. If $DV_i = (O_i, DC_i, F_i)$, $DV_j = (O_j, DC_j, F_j)$ are two discernible vectors, then the addition rule for discernable vectors can be defined as

$$DV_i + DV_j = (O_{ij}, DC_{ij}, F_{ij}) \tag{5}$$

Where, $O_{ij} = O_i \cup O_j$,

$$F_{ij} = F_i + F_j = (\max(f_i(a_1), f_j(a_1)), \max(f_i(a_2), f_j(a_2)), ..., \max(f_i(a_n), f_j(a_n)))$$

$$DC_{ij} = \begin{cases} DC_i, & DC_i \subseteq DC_j; \\ DC_j, & DC_j \subseteq DC_i; \\ DC_i \cup DC_j, & (DC_i \not\subset DC_j) \cap (DC_j \not\subset DC_i) \end{cases}$$

Theorem 1. In a discernible vector array, the discernible attribute set of every discernible vector must not be null.

Proof (sketch): Since the discernible vector array represents the discernible state of the information table, every discernible attribute set can distinguish a pair of objects, so assumed $DV_i = (O_i, DC_i, F_i)$ is a discernible vector, where $O_i = (x_m, x_k)$, if $DC_i = \varnothing$, then it indicates that there is no attribute which can distinguish the two objects x_i and x_j ,namely, they are in the same classification. This conflicts with the hypothesis that $DV_i = (O_i, DC_i, F_i)$ is a discernible vector. So, in a discernible vector array, the discernible attribute set of every discernible vector must not be null.

Theorem 2. Given a discernible vector, if one item of its frequency vector is 1, then the corresponding attribute is the core attribute in the initial information table.

Proof (sketch): Assumed $F=(f(a_1),f(a_2),...,f(a_n))$, select an item random from it and let $f(a_i)=1$, since $f(a_i)=\frac{1}{|DC|}=1$, so $|DC|=1$, that is too say, only attribute a_i can distinguish the two objects in discernible set. Therefore, attribute a_i is the core attribute.

Theorem 3. In the discernible vector array DVT, given two discernible vectors, $DV_i \in DV, DV_j \in DV$, and DC_i is the discernible attribute set of DV_i, DC_j is the discernible attribute set of DV_j, and meanwhile, $DC_i \subseteq DC_j$, if add them by the above addition rule for discernible vectors, then taking DC_i as the discernible attribute set of the sum of the two discernible vectors does not change the attribute reduction of the information table.

Proof (sketch): Assumed $DC_j = \{a_1, a_2, ..., a_{m,} a_{m+1}, ..., a_{n-1}, a_n\}$, $DC_i = \{a_1, a_2, ..., a_m\}$, then we can obtain $DC_i = (a_1 \vee a_2 \vee, ..., \vee a_m)$ and $DC_j = (a_1 \vee a_2 \vee ... \vee a_m \vee a_{m+1} \vee ... \vee a_{n-1} \vee a_n)$. And based on the mapping relationship of discernible vector array and information table, we can see that the process of attribute reduction is the logical interaction relationship as following,

$$
\begin{aligned}
DC_i \wedge DC_j &= (a_1 \vee a_2 \vee, ..., \vee a_m) \wedge (a_1 \vee a_2 \vee ... \vee a_m \vee a_{m+1} \vee ... \vee a_{n-1} \vee a_n) \\
&= ((a_1 \vee a_2 \vee, ..., \vee a_m) \wedge (a_1 \vee a_2 \vee ... \vee a_m)) \vee ((a_1 \vee a_2 \vee, ..., \vee a_m) \wedge (a_{m+1} \vee ... \vee a_{n-1} \vee a_n)) \\
&= (a_1 \vee a_2 \vee, ..., \vee a_m) \\
&= DC_i
\end{aligned}
$$

So, the theorem has been proved.

3 Algorithm Description

The algorithm is described as following.

Step1. Establish discernible vector array and initial the scan vector $S' = (C', DF', R')$.

Step2. Generate the final scan vector by scanning the discernible vector based on the addition rule for discernible vectors.

For (i=1; i<=n; i++)

{select $DV_i = (O_i, DC_i, F_i)$ by order from the discernible vector array;

If $f_i(a_k)=1$, then $C=C^{'}\cup\{a_k\}$;

$DF=DF^{'}+DC_i$; $R=R^{'}+F_i$;}

Finally, obtain the final scan vector $S=(C,DF,R)$.

Step3. Initialize attribute set $P=\varnothing$, let $P=P\cup C$, and then reduce the attributes on the scan vector as below.

1) Delete all the items including the attribute a_i from the discernible attribute set and the item $f(a_i)$ from the frequency vector R. Doing like this until the last core attribute, we can obtain a new scan vector.

2) Calculate U/P .

3) If $U/P=D$, then jump to *step4*;

Else select attribute $a_i\in DF$, let $P=P\cup\{a_i\}$, then jump to 2), until the last attribute in the discernible attribute has been done as above.

Step4. Set P is the reduced attribute set of the initial information table.

Step5. Obtain the final decision rules by reducing attribute value as below.

1) Calculate significance of every attribute value and order by it.

2) Delete the minimum important attribute value in order until conflict events generate.

4 Experiment

Now we give an example to explain the process of this algorithm. The initial decision table is denoted by Table 1.

Table 1. The initial decision table

U	a	b	c	d	e
x1	0	0	0	0	0
x2	0	1	0	0	1
x3	0	2	1	1	2
x4	0	0	2	2	2
x5	1	0	0	0	2
x6	2	0	0	0	2

The initial decision table is transformed to discernible vector array as Table 2., every item of which can be described as $DV_i=(O_i,DC_i,F_i)$.

Scan the discernible vectors in the discernible vector array with running the discernible vector addition rule. Since $f_{1,6}(a)=1$, so attribute a is a core attribute, let $P=\{a\}$, calculate $D\neq U/\{a\}=\{\{x_1,x_2,x_3,x_4\},\{x_5\},\{x_6\}\}$. Select another important attribute, since $f_{1,2}(b)=1$, so attribute b is a core attribute and let $P=\{a,b\}$, and calculate $D\neq U/\{a,b\}=\{\{x_1,x_4\},\{x_2\},\{x_3,x_5,x_6\}\}$. In the same way, select another important attribute , let $P=\{a,b,c\}$, obtain $D=U/\{a,b,c\}=\{\{x_1\},\{x_2\},\{x_3,x_4,x_5,x_6\}\}$.

Table 2. The initial decision table

O	DC	F
(x1,x2)	(b)	(0,1,0,0)
(x1,x3)	(b,c,d)	(0, 1/3, 1/3)
(x1,x4)	(c,d)	(0,0, 1/2, 1/2)
(x1,x5)	(a)	(1,0,0,0)
(x1,x6)	(a)	(1,0,0,0)
(x2,x3)	(b,c,d)	(0, 1/3, 1/3)
(x2,x4)	(b,c,d)	(0, 1/3, 1/3)
(x2,x5)	(a,b)	(1/2,1/2,0,0)
(x2,x6)	(a,b)	(1/2,1/2,0,0)

For now, we can obtain the final attribute reduction set $P=\{a,b\}$. So, the final decision table is as Table 3.

Table 3. The final decision table

U	a	b	c	E
x1	0	0	0	0
x2	0	1	0	1
x3	0	2	1	2
x4	0	0	2	2
x5	1	0	0	2
x6	2	0	0	2

In Table.3, reduce the redundant attribute values and obtain the decision rules as fellows,

$Rule1: a_0b_0c_0 \rightarrow e_0$; $Rule2: a_0b_0c_2 \rightarrow e_2$;

$Rule3: b_1 \rightarrow e_1$; $Rule4: b_2 \rightarrow e_2$;

$Rule5: a_1 \rightarrow e_2$; $Rule6: a_2 \rightarrow e_2$;

In order to test the algorithm, we apply it in Iris database and obtain the result as below.

R1: petal-length([1.0,2.5]) $\rightarrow$ class(Iris-setosa)
R2: petal-length([2.5,4.5]) $\rightarrow$ class(Iris-versicolor)
R3: sepal-width([3.9,4.4]) $\rightarrow$ class(Iris-setosa)
R 4: petal-length([4.5,5.0]) AND petal-width([0.1,1.6]) $\rightarrow$ class(Iris-versicolor)
R 5: petal-length([4.5,5.0]) AND sepal-width([3.0,3.9])AND sepal-length([4.3,5.9]) $\rightarrow$ class(Iris-versicolor)
R 6: petal-length([5.0,6.9]) AND petal-width([0.1,1.6]) $\rightarrow$ class(Iris-virginica)
R 7: sepal-length([4.3,5.9]) AND petal-length([5.0,6.9]) $\rightarrow$ class(Iris-virginica)
R 8: sepal-length([4.3,5.9]) AND petal-width ([1.6,1.8]) $\rightarrow$ class(Iris-virginica)

R 9: sepal-length([6.0,6.8]) AND petal-width ([1.6,1.8]) $\rightarrow$ class(Iris-versicolor)
R 10: sepal-length([6.0,6.8]) AND petal-length([5.0,6.9]) $\rightarrow$ class(Iris-virginica)
R 11: sepal-length([6.8,7.9]) AND petal-width ([1.8,2.5]) $\rightarrow$ class(Iris-virginica)
R 12: sepal-length([6.8,7.9]) AND petal-length([4.5,5.0]) $\rightarrow$ class(Iris-versicolor)
R 13: sepal-width([2.0,3.0]) AND petal-width ([1.8,2.5]) $\rightarrow$ class(Iris-virginica)

5 Conclusions

The algorithm defines discernible vector based on discernible relationship to transform the information table to discernible vector array, uses discernible vector addition rule to find the core attribute set, and meanwhile balances the attribute importance by the frequency vector. Additionally, along with the process of finding core attributes, discernible attributes set with less items is generated, which lessen the searching scope of attributes reduction. So the time complexity of the algorithm is much lower than the traditional complexity. Illustration and experiment results indicate that the algorithm of attributes reduction is effective and efficient.

Acknowledgement. The paper is sponsored by Chinese National Foundations (70572070), Chinese Postdoctoral Foundations (2005038319), Chinese 863 Planning (863-511-053, 863-510-031), Chinese Important Project (2005216011, 2005219005), Chinese Chunhui Foundations (Z-1-15007), Chinese National Foundations (60674056) and Liaoning Province Academic Talent Foundations (2005219001).

References

1. Pawlak, Z.: Rough set. International Jjournal of Computer and Information Science (1), 341–356 (1982)
2. Wu, S., Gao, X., Bastian, M: Data Warehouse and Data Mining. Metallurgy press, Beijing, China (2003)
3. Wang, G. (ed.): Rough Set and Knowledge Acquisition. Xi an jiao tong university press, Xi an, China (2003)
4. Zhang, W., Wu, W., Liang, J., et al. (eds.): Rough Set Theory and Methods. Beijing science press, Beijing, China (2001)
5. Hu., X.H., Cercone, N.: Learning in Relational Databases: A Rough Set Approach. Computational Intelligence 11(2), 323–337 (1995)
6. Jelonek, J., Krawiec, K., Slowinski, R.: Rough Set Reduction of Attributes and Their Domains for Neural Networks. Computational Intelligence 11(2), 339–343 (1995)
7. Miao, D., Hu, G.: A Heuristic Algorithm for Reduction of Knowledge. Journal of Computer Research and Development 36(6), 681–684 (1999)
8. Meng, Q.: Information Theory. Xi an jiao tong university press, Xi an, China (1986)
9. Wong, S.K.M., Ziarko, W.: On Optional Decision Rules in Decision Tables. Bulletin of Polish Academy of Science (33), 693–696 (1985)
10. Aleksnder Ohrn.: Discernibility and Rough Seys in Medicine: Tools and Applications. Norwegian University and Science and Technology, Department of Computer and Information Science, N-749 Trondheim, Norway (1999)

The Activation Mechanism for Dynamically Generated Procedures in Hyperlogo

Nobuhito Yamamoto[1] and Tomoyuki Nishioka[2]

[1] Graduate School of Systems and Information Engineering, University of Tsukuba
1-1-1, Tennoudai, Tsukuba, Ibaraki 305, Japan
yamamoto@cs.tsukuba.ac.jp
[2] Department of Industrial Information, Faculty of Industrial Technology, Tsukuba University of Technology
4-3-15, Amakubo, Tsukuba, Ibaraki 305, Japan
nishioka@a.tsukuba-tech.ac.jp

Abstract. The higher-order programming is one of the attractive and powerful ways of expressing our algorithmic/procedural methods of solving in many application fields. It represents our idea in program form naturally. The authors have studied and implemented the function of handling higher-order programming paradigm in Hyperlogo language system. The handling functions are: to make procedure closures, to handle procedure closures in the same way as numerical value and character strings, and to activate generated procedure closures. The third point of the above functions is focused mainly in this paper. Two ways of activating closures in Hyperlogo are presented: (1) add a name to generated closure, and call it up by its name as the need arises, (2) use the assistant procedure which activates a target procedure. The procedure *activate* is introduced to acquire the function.

1 Introduction

Figuring out the mathematical idea to computer programs is an important but rather difficult educational target for information science course students [1]. Hyperlogo is a language developed for them to study some advanced frameworks in the computer science such as the function programming and computational objects. Using the system, students is expected to have clear understandings what is going on during an execution of programs, and how the information is handled under the verious programming environments.

Generating procedures dynamically and activating them is the essential components of the higher-order behavior of these programs.

The higher-order procedures can be divided into three groups.

A) Procedures that accept procedures as their arguments.
B) Procedures that produce procedures as return value.
C) Procedures that behave like both A and B.

The authors made a plan for adding the faculty of handling these procedures to the Hyperlogo language.

In this paper, the way for activating generated procedures is focused mainly so as to get natural coexistence with the usual procedures as possible. An assistant

M.J. Smith, G. Salvendy (Eds.): Human Interface, Part I, HCII 2007, LNCS 4557, pp. 785–792, 2007.

procedure *activate* is introduced to the language for the purpose of running generated procedures, which activates a procedure like that of the funcall in Lisp.

2 Hyperlogo

Hyperlogo has ability of handling lambda functions, mathematical functionality, lexical scoping and so on. The kernel of the language interpreter, which is written using a low level language, is built as small as possible and only has simple basic functions. Other functions are realized/programmed using these simple functions in the same target language [4,5]. The original Logo has two kinds of procedures, commands and reporters [3]. But the Hyperlogo's procedures are integrated into reporters. Each procedure gives back a return value to its caller procedure.

Logo closely resembles the Lisp, but semantics of symbols presented to the interpreter is different. In case of the Lisp, a symbol input to the listener window is interpreted as taking out the value bound to that symbol. When symbols are enclosed in parentheses and form a list, the head symbol expresses a procedure call. On the other hand, a symbol of Hyperlogo is interpreted as a call up of procedure to run. For taking out the bound value, ":" is added in front of a symbol which is a syntactic-sugared expression of procedure *quote*.

The procedure expression of Hyperlogo is written as prefix notation and forms a line of symbols.

<procedure name> <argument 1> <argument 2> --- <argument n>

The number of arguments is fixed at the time when a procedure is defined. The procedures having arguments of arbitrary number can not be used on the system at present.

3 The *Activate* Procedure

There are two ways to launch procedures:

1. add names to procedures, and address them by name.
2. use an assistant procedure for activation..

A clear way for users is employed, which defines procedures using lambda expression and associates them to their procedure names using a binding primitive for the first case above.

The *lambda* procedure makes a closure and returns it as a running result.

lambda [arguments] [procedure body]

After that, the closure is bound with a procedure name and is registered in the some execution environment using the procedure *make*.

make <procedure name> <lambda expression>

Then the defined procedure becomes executable or callable by the name.
e.g.,

```
 =>   make "f lambda [ x ] [ add :x 5 ]
                          /*  defines a procedure f   */
 =>   f 7                 /*  then, runs it with 7    */
   12                     /*  7 + 5   */
```

As described above, the symbol *f* is interpreted as a procedure call and the addition is carried out.

A simple example of Group-A listed in the previous section is shown. The " to ~ end " syntax is the syntactic sugar for lambda form. When input to the listener, the syntactic sugar processor converts it to its equivalent " make ~ lambda ~ " form immediately.

```
to square !x
  mul :x :x
end
```

After running, the procedure closure is stored and registered as the *square* in the running environment in the manner above. The agent procedure *f* accepts some procedure and a value from a caller through the arguments. It means the symbol *a* is bound to the input closure.

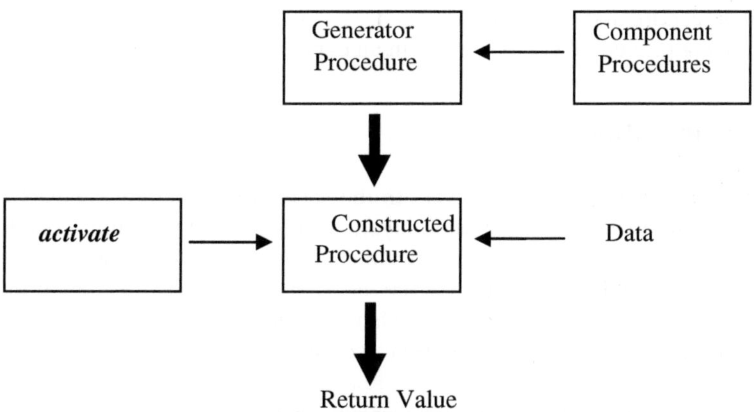

Fig. 1. Processing sequence

```
to f !a !b
     a :b
end
```

Run ***f*** using the *square*,

```
=>   f :square 3
  9                                  /*  3 * 3   */
```

For an example of Group-B, the procedure ***g*** generates a procedure object using *lambda*. The object contains an addition procedure for given value of an argument ***x*** and a variable ***y***.

```
to g !x
  lambda [ y ] [ add :x :y ]
end
```

The above definition is equivalent to the nested lambda expression:

```
make "g lambda [x] [ lambda [y] [ add :x :y ] ]
```

The *g*'s run gives back a closure as its return value.

For another way to launch generated procedures, the *activate* is installed as one-argument procedure and is handled as a special form expression by the Hyperlogo interpreter. It activates the procedure which is generated by the succeeding procedure's execution (Fig. 1).

```
=>  mul activate g 2 3 4
  20                         /*  ( 2 + 3 ) * 4  */
```

The *g* generates a procedure which adds value 2 to its argument ***y***. The *activate* kicks the procedure and (2+3) is performed. The *mul* multiplies the result 5 by 4.

The *activate* procedure can be nested in an expression.

4 Implementation

The Hyperlogo system has three layers in the schematic structure (Fig. 2). The L-abstract machine locates at the middle layer. It has three stacks, the procedure stack (PS), the data stack (DS) and the utility stack (US). Instructions, which are actually the names of procedure, extracted from an input sentence are stored in the PS. Arguments or data are hold in the DS when an instruction is going to be activated. The result produced by an instruction is stored on the top of the DS.

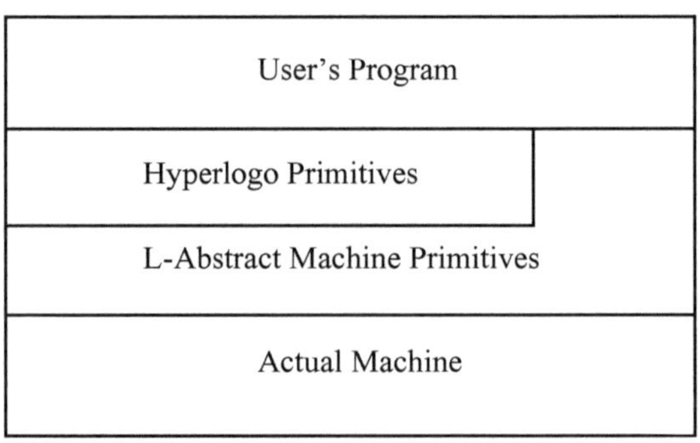

Fig. 2. Hyperlogo system formation

When a datum is stacked in DS, the number of argument of the topmost procedure is decreased by one. And when the number becomes zero, certain numbers of data are popped from the stack and an execution of the procedure begins.

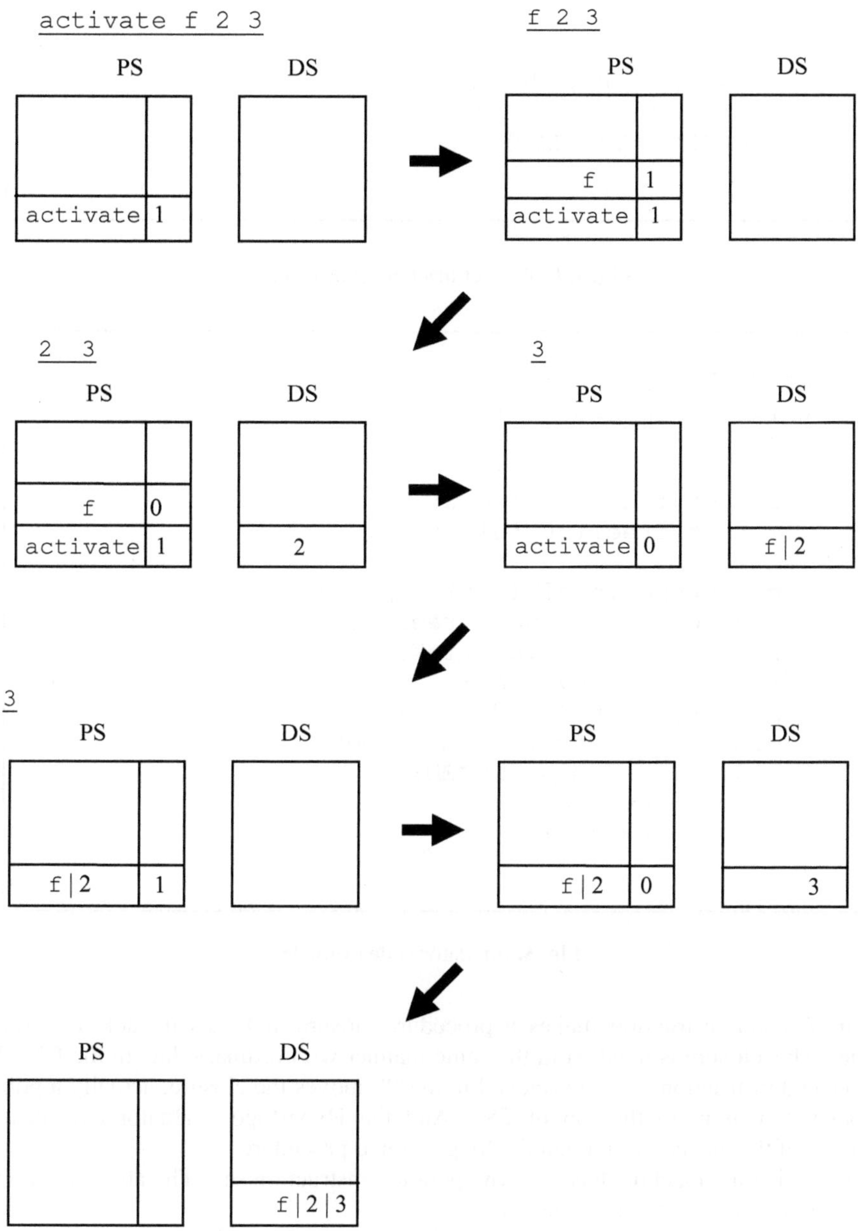

Fig. 3. Input sentence and stacks handling

```
/* EVAL_ACTIVATE */
EVAL_ACTIVATE:
      BRANCH_NOT( SFACTIVATE_P , EVAL_TAG_ERR )
      DUMP_PS_TOP
      POP_DS
      MOVE( GPR , SEQ )
      PUSH_PS
      JUMP( EVAL_LOOP )
/*   */
```

Fig. 4. L-abstract machine code example

```
/* DUMP_PS_TOP */
void dump_ps_top_c()
{
  struct ty_info *p;
  struct freelist *root;
  struct freelist *old_cell;

  p = &type_info[ (int) t_ps ];
  old_cell = (struct freelist *) PSP;
  PSP = PSP -> ps.ps_link;
  root = p -> ty_root;
  p -> ty_root = old_cell;
  old_cell -> free_link = root;
  old_cell -> m = UNUSED;
  ++(p -> ty_ucell);
  --(p -> ty_ncell);
}
```

Fig. 5. Emulator code example

The *lambda* instruction makes a procedure closure and puts it back as a return value. This closure is handled in the same manner with ordinary data in the DS. The introduced instruction *activate* stacked in the PS moves the closure, actually a pointer associated with it, to the top of PS. And the Hyperlogo evaluator executes the contents of the closure as a dynamically generated procedure.

The L-abstract machine has its own operation instruction set. The above action has been programmed using these instructions.

The *activate* is classified to the special form procedure which requires particular handling in the similar way as the *quote* in the evaluator. When the argument number of a procedure becomes zero, the procedure begins its running. The *quote*'s argument number is defined as zero. The processing starts immediately after being read into PS. The next symbol in the input sentence is stored in DS instead of running the associated procedure. The *activate* is defined as a one-argument procedure. It starts its running later when the argument number becomes zero like that of other usual procedures.

Symbols in a Hyperlogo sentence are evaluated from left to right:

```
=>   activate f 2 3
```

Fig. 3 illustrates how an input sentence is processed on the stacks. Let *f* be a one-argument procedure which produces a closure as a result.

Step 1. From the input sentence, *activate* is stacked to PS, The number of needed argument is set to one. It does not run yet.

Step 2. The next symbol *f* and its number of arguments (i.e., one) are stacked to PS. It does not run yet.

Step 3. The numerical value 2 is stored in DS, and the number of necessary argument of *f* at the top is decreased. It becomes zero, which means all arguments are prepared for running.

Step 4. Processing of *f* begins using 2, the content of DS. A procedure closure (*f* | 2) is generated and stored in DS as the running result. The argument number of *activate* is decreased and becomes zero.

Step 5. The *activate* starts running. It pops up a content of DS, i.e. (*f* | 2), and stores in PS with the number of arguments..

Step 6. The numerical value 3 is read into DS and the number of necessary argument of the topmost procedure (*f* | 2) is decreased.

Step 7. The processing of (*f* | 2) begins. It computes using the content of DS and produces a result (*f* | 2 | 3) in DS

Fig. 4 and Fig. 5 are code examples of the L-abstract machine for stacks operations.

5 Conclusion

A mechanism for higher-order handling in programming has been built in Hyperlogo language. The design objective was to get natural coexistence with the usual procedures as possible in the language. The assistant procedure ***activate*** is introduced for the purpose.

Hyperlogo is implemented using the stratified design to obtain flexibility of modification or extension. The procedure ***activate*** was build mainly in the intermediate layer and the extension work was relatively simple. Three groups of the handling have been made possible on the Hyperlogo system. The authors were able to make their language system retain the simple syntax structure and the higher functionality.

It is still remained whether the result is applicable and efficient to other subjects in the higher order programming. The authors would like to study further and show the feasibility of this approach and get more simple and appropriate syntax at the end-user language level.

The authors express their gratitude to Masashi Fujisawa for his assistance to building the system [2]. This work was supported by Grant-in-Aid (No.15500609) for Scientific Research (C) from Ministry of Education, Culture, Sports, Science and Technology (MEXT), Japan.

References

1. Abelson, H., Sussman, G.J., Sussman, J.: Structures and interpretation of computer programs, 2nd edn. MIT Press, Cambridge (1996)
2. Fujisawa, M.: Implementation of procedure activation, Graduation study, University of Tsukuba (2005)
3. Kawano, H.: Introduction to Logo. Baifu-kan, Tokyo (1987)
4. Mizukoshi, Y.: Turtle graphics by the procedure Object generation mechanism, Graduation study, University of Tsukuba (2004)
5. Yamamoto, N., Nishioka, T.: Hyperlogo: A Language Which Provides Visible Structure of Processing Programs. Proceedings of The 11th International Conference on Human-Computer Interaction. LNCS. Springer, Heidelberg (2005)

Part V

Advanced Interaction Technologies and Techniques

The Importance of Human Stance in Reading Machine's Mind (Intention)

Akira Ito and Kazunori Terada

Faculty of Engineering, Gifu University
1-1, Yanagido, Gifu, 501-1193, Japan
{ai,terada}@info.gifu-u.ac.jp

Abstract. Our aim is to implement a mind-reading capability in machines and artifacts. We give a formulation of mind-reading applicable to machines, emphasizing the importance of a human stance to a machine in mind-reading. Next we demonstrate how the stance a human takes can modify the recognition of intention in an object/agent, taking as examples experiments conducted by us, i.e., a moving box experiment, eye-contact robot, gaze-game experiment, and three-card-game experiment. To realize a human-machine communication based on mind-reading, it is essential that a human takes an intentional stance to the machine. We have to use signs of intentionality to trick a human to take an intentional stance. Shared attention is shown to work as a sign of intentionality.

1 Introduction

We are making a research on human-machine (artifact) communication based on mind-reading. We humans interact or communicate with other humans mostly through mind-reading. The research on mind-reading communication has two important roots. One is *the theory of mind* research, first proposed as by Premach and Woodruff[1], and later studied extensively by researchers of various disciplines, especially in relation to autism[2]. Another is *the relevance theory* proposed by Sperber and Wilson[3] as a theory of human communication.

The Theory of Mind is a model of mind a human uses for understanding other's behavior. Many researchers believe that the theory of mind is an innate module in the brain. This module tries to attribute the cause of an action to intention in the actor's mind. The details of mind are not explicitly explained, but rather are taken for granted as everyone knows what his/her mind is. The important point is that mind is what controls the behavior of its owner. Mind-reading means understanding the intention hidden behind observable actions (including utterances) of an agent.

Later researches made clear that human communication (including a verbal communication) is based mostly on mind-reading, i.e., the inference by the receiver (hearer) of the intention of the sender (speaker). The surface meaning of the utterance/sentence is just a hint (though important) to the above inference process. The study of pragmatics tried to repair the discrepancies between semantic (surface) meaning and intended

M.J. Smith, G. Salvendy (Eds.): Human Interface, Part I, HCII 2007, LNCS 4557, pp. 795 – 803, 2007.

meaning (intention) by adding a number of pragmatic rules. But the study of communication, and especially that related to autism, showed that it is very difficult to list up all the pragmatic rules used in communication. Moreover, humans do not seem to use these rules (at least consciously) in ordinary dialogues. Mind-reading ability cannot be replaced by a set of pragmatic rules.

The relevance theory formulated the mind-reading communication as follows: Sender of the information makes some action (utterances, bodily movement etc), which modifies the cognitive environment of a receiver. The receiver perceives the changes of his cognitive environment, and infers first the speaker's communicative intention, and then informative intention. The cognitive environment is what a human can recognize in his environment. Communicative intention is an intention to communicate information, and informative intention is what the sender really want to communicate.

The relevance theory demands that communicated intention should be relevant, i.e., it must have useful meaning to the receiver of the information. If intention is not stated explicitly, the receiver of the information has to search the hidden intention using what information is available to him until he thinks that relevant intention is found. The sender must guarantee that useful information can be found if searched properly.

The reason to endow the ability of mind-reading-based communication to a machine is twofold. One is, of course, a practical one, i.e., to realize better human -machine communication. As we are so much accustomed to mind-reading communication, it is very difficult and stressful to obey the strict semantic rules (like standard programming languages).

The other, which is less mentioned, but more important (we think), is as follows: If humans have come to employ a complex (at least as an algorithm) mechanism of mind-reading for communication, then there must be a substantial reason which makes mind-reading communication superior to the rigid semantics-based communication. These long-range aims in mind, we are developing a machine which can communicate through mind-reading mechanism.

In the following, we give a formulation of mind-reading applicable to machines, emphasizing the importance of a human stance to a machine in mind-reading. Next we demonstrate how the stance a human takes can modify the recognition of intention in an object and vice versa.

2 A Formulation of Mind-Reading Communication

First, we give a few definitions of the terminology which are applicable to both humans and machines.

Intentional agent (hereafter referred to as agent) is an object which has beliefs, and act according to its intentions. As to how action is generated from its beliefs and intentions, we suppose that an agent has some reasoning capability comparable to that of humans. We do not give definitions of beliefs and intentions except that they are some kinds of representations in the mind of an agent.

Designed object is not an intentional agent, but an object which acts by the algorithm (however complex) implemented in it by other agents.

Mind-reading is to infer the beliefs and intentions of an agent from the observation of its behavior (sequence of actions). Actions may include utterances.

Stance is the standpoint from which an agent (human) tries to understand the behavior of an agent. Dennett proposes three stances: physical stance, designed stance, and intentional stance [4].

Physical stance is a stance employed when we try to understand the behavior of a physical object according to the law of nature.

Design stance is a stance employed when we try to understand a designed object.

Intentional stance is a stance employed when we try to understand an intentional agent.

Mind-reading communication is a mechanism to make an intentional agent act at the information-sender's will by controlling the beliefs of the information-receiver. Unfortunately we cannot directly control the inner representation of other agents. The relevance theory explains this mechanism as *by modifying the cognitive environment of the information-receiver.*

Utterances are often used to modify the hearer's cognitive environment. Sometimes non-verbal information is used instead, or in parallel. For us humans, this mechanism is so natural that we regard it a part of a logical reasoning, but it is not true. If you do not know this mechanism, or you do not think that the target object is an intentional agent, you have to employ a more direct method, such as to move the object directly by your hand. This phenomenon is really observed by autistic children (known as a crane phenomenon).

If you argue that every object (including a human) act by the algorithm implemented in it, we must agree. Hence we (and Dennett) used the term *stance*. The definition of an intentional agent may be modified as follows:

Intentional agent (alternative definition) is an agent whose behavior is best understood by taking an intentional stance to the agent.

The definition is circular, but it is not our fault. The circularity is deeply rooted in the mechanism of mind-reading itself. We will discuss this problem in the next section.

3 Paradox of Mind-Reading Mechanism

As we mentioned, the circularity is rooted in the mechanism of mind-reading. We cannot determine beforehand if an object is intentional or not. An easy-going solution may be to first apply an intentional stance, and if failed to predict its behavior, then to switch to other stances. This strategy does not work, or is unpractical at least. Suppose you have to consider the following possibility: If a book dropped from the bookshelf, it may be that the book demanded your attention. If your diary is missing, it may be that the dairy is protesting you against your casual handling. We believe that you are happier to think that the book dropped because it was not placed properly. Mind-reading demands a huge processing cost to the mind-reader.

On the other hand, if you fail to take an intentional stance to an intentional agent, you may miss to see the intentionality behind its behavior. Suppose your friend asked "do you know about X?" and you just answered "No", you might have missed a chance to get valuable information on X. (You should have responded like "What about X?")

A paradox is that you cannot determine if an object is intentional or not before interaction, and the stance you take limits your ability of understanding on that object. Moreover, the stance you take may have some effects on later behaviors of the object itself (if it is an intentional agent). Therefore, you are requested to choose the right stance to an (even unfamiliar) object.

To reduce the information-processing cost, however, a human usually do not take an intentional stance to an unfamiliar object. A human takes an intentional stance only when he finds *a sign of intentionality* in an object. What can be signs of intentionality must be implemented in our brain through the evolution of human race. These signs have been developed through the experiences in a primordial forest, and may not always be applicable for livings in a modern society. If a modern artifact wants a human to take an intentional stance, it must put on some signs of intentionality, or at least mimic them.

In the following sections, we demonstrate the interaction of stances a human takes, and the resultant cognitions and actions.

4 A Moving Box Experiment

A subject was led into a room in which a box is placed (See Fig.1). The box has four wheels and can move automatically. In the experiment the box was controlled by the experimenter in the remote room. When the subject entered the room, the box started to move. It first tried to pass through the gap of an obstacle made by cardboard. The gap was too narrow to pass through. The box next turned around and moved to the subject, and made a forward-backward movement several times in front of the subject. Then the box tried to pass through the gap again. The experiment ended when the subject helped the box passing through by removing the obstacle, or 90 seconds had passed. Among 18 subjects, 12 subjects understood the intention of the box, and 8 subjects actually helped the box. Here the forward-backward movement of the box in front of the subject, which demanded the attention of the subject, worked as a sign of intentionality.

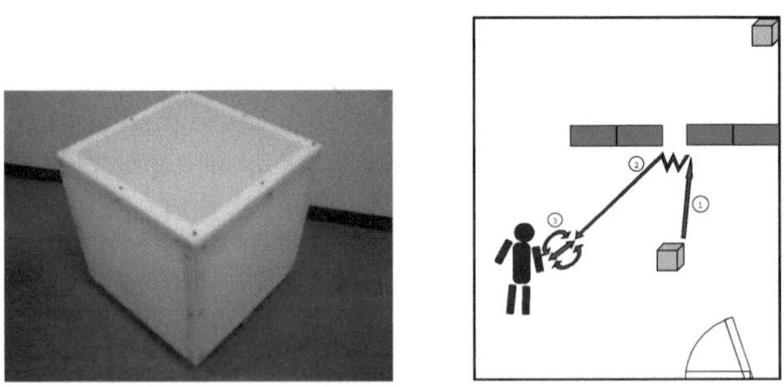

Fig. 1. A moving box and the experimental setup

5 An Eye-Contact Robot and Name-Teaching Experiment

An eye-contact robot *Infanoid* is a head robot with a movable head, eyes and lips [5]. The subject was instructed to teach Infanoid the name of the objects (Fig.2). In that process the eyes of the robot was moved in two conditions, a) to randomly selected direction, b) simulating human eye-contact. Interaction time and attention ratio (the ratio of time the subject was looking at the eyes of the robot) are compared in two conditions. The result is shown in Fig.3. The subject interacted longer, and looked at the eyes of the robot more in simulating human eye-contact conditions.

Fig. 2. Name-teaching experiment

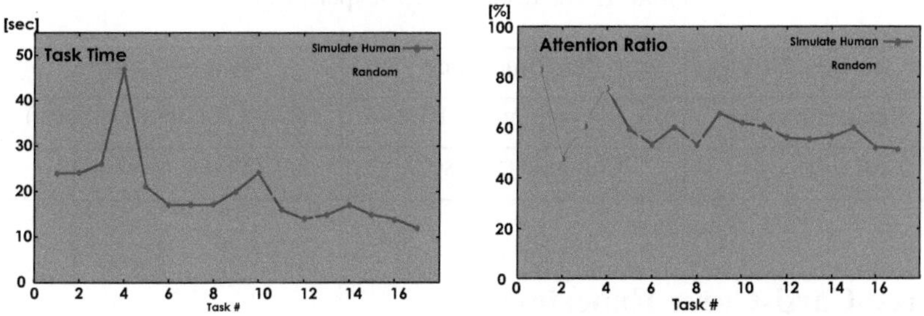

Fig. 3. Task time and attention ratio in the name-teaching experiment

6 The Gaze-Game Experiment

We explained how gaze direction can control the stance a human takes to an object. In this section we show how a stance can modify the recognition of an object.

The eye-contact type dialogue equipment (Fig.4) makes possible eye-contact between subjects in separate rooms [6]. We asked the subjects to quietly keep a gaze on the partner in the display, without expressing any emotions. This is a kind of gaze-game (called *nikamekko* in Japanese) .The experiment confirmed that they cannot keep an eye-contact for long (<10 seconds). They either burst into laughing, or moved his/her gaze away from that of the partner. This is a natural reaction when two humans try to keep a gaze on each other's eye. However, if we (falsely) instructed the subject that the display he was looking was a replay of a pre-recorded video, then the both subject could keep a gaze on the partner's eyes much longer (>30 seconds). On the other hand, if the subject was instructed that video in the display was a real-time production of the partner, they cannot keep a gaze for long even if the video was really a pre-recorded one. The result is summarized in Table.1.

The experiment showed that the instruction to the subject could modify the gaze-pressure of the partner, even if what was displayed is the same. You cannot decide correctly what you are looking at is really an intentional agent, or just a mindless machine (in this case a replay of a pre-recorded video) from the visual information alone.

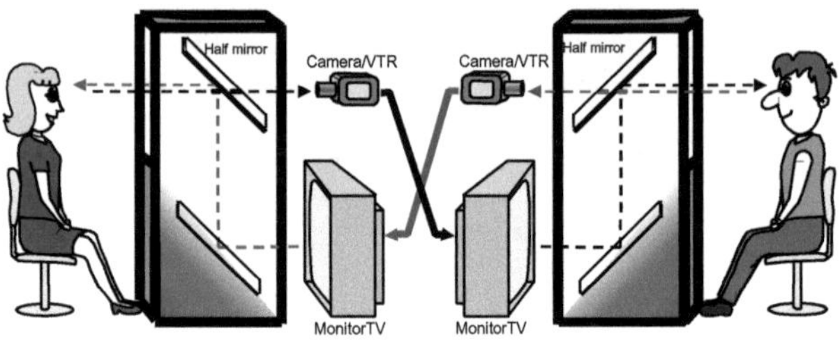

Fig. 4. The eye-contact type dialogue equipment

Table 1. The result of gaze-experiment

Instruction /Reality	Real-time video	Real-time video
Real-time video	Do not continue (<10 sec.)	Do not continue (<15sec.)
Real-time video	Continue (>20 sec.)	Continue (>30 sec.)

7 Three-Card-Game Experiment

The *Thee Card Game* is played by one parent, and two child-players. The each child-player has three cards, King(K), Queen(Q), and Jack(J). The each child-player plays one of the three cards simultaneously, and the reward is paid in coins from the parent depending on the combination of the played cards. The payoff (the number of coins paid) is given by Table.1. For each play, child-player has to pay one coin to the parent. The subjects were asked to play this game 100 times, and the total gain (if exists) was paid in real money. The game was played by PC connected to the network. The subjects were instructed beforehand that the other child-player was also a human subject.

If each player tries his best, or plays randomly, then the average gain is negative. The only way to earn a positive gain is to cooperate with the partner and to win alternatively, i.e., by playing K (to J) and J (to K), resulting to the average gain of 1.5 coins/game (called hereafter as cooperative solution).

Table 2. The payoff of the three card game. That of the left strategy against the top is shown.

	J	Q	K
J	0	1	0
Q	0	0	2
K	3	0	0

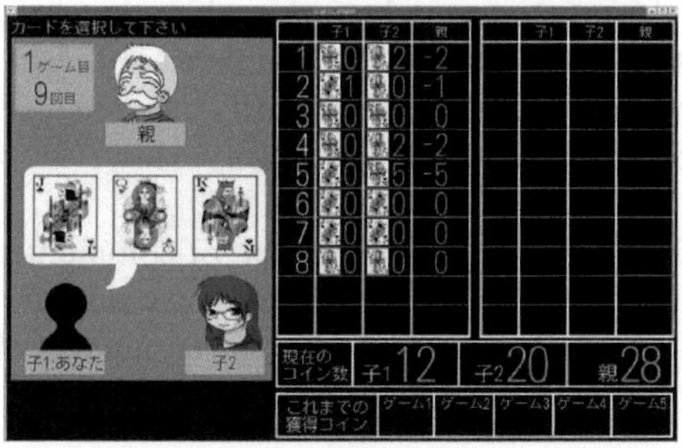

Fig. 5. The Three Card Game Interface Window

Among 5 pairs of subjects, no pair achieved the cooperation. For 2 pairs, however, one of the players really tried to cooperate, and played K and J alternatively. The result was that the cooperative player was exploited by the other player playing Q, K alternatively. In a questionnaire after the experiment, the exploiter player answered that he thought the other player was a PC program. The reason he gave is that it played a very simple pattern. Note that he was instructed beforehand that the other player was also a human subject. Another player in this experiment also thought of the cooperative solution, but did not tried, for he feared (correctly) that the other player would not cooperate.

The experiment demonstrates the difficulty of taking an intentional stance without appropriate communication channels. If he strongly believed the instruction that he was playing with a human, he might have noticed the abnormality of the partner's play, and tried to search a coherent explanation, i.e., the other player must have some intention for doing it. Then he might have thought of the cooperative solution. This is contrasted with a face-to-face version of the experiment, where the parent and the two child-players are seated around the table. In this situation, most players could play cooperative solutions.

The experiment shows another side of the intentional stance. With an appropriate devise we can induce a human to take an intentional stance. On the other hand, without such a device a human cannot keep an intentional stance, and tends to think others as unintentional, which makes him abandon to read the intention behind the behavior of others.

8 Shared Attention and a Sign of Intentionality

The above four examples showed that attention plays an essential role in inducing a human to take an intentional stance to a machine. Careful study shows that not just attention, but shared attention, i.e., the mutual recognition that the target of attention is shared, worked as a sign of intentionality.

This is expected also from the relevance theory. According to the relevance theory, shared attention, or a *mutual manifestness* in its terminology, is what makes possible mind-reading communication. What is interesting is, as we demonstrated by the gaze-game experiment, that shared attention is psychological in nature. If you think that attention is shared, shared attention is realized. If you do not, it does not. This again demonstrates the interplay of stance and mind-reading communication. Shared attention is only realized if an intentional stance is taken, and shared attention induces a human to take an intentional stance. The circularity can be solved if there exists some other device which works as a sign of intentionality. However, we are not sure of this expectation.

The role of shared attention in the development of communication ability in human infant is discussed by Tomasello [7]. How humans solve this circularity is not obvious, but some innate mechanism must be involved, for an autistic child cannot solve this problem.

9 Summary and Discussions

We gave a formulation of mind-reading communication with the help of the theory of mind and the relevance theory. Next we showed with some examples how stance a human takes can modify his cognition. But paradoxically, a human have to depend on the cognition to decide which stance to take for an unfamiliar object. This circularity is deeply rooted in mind-reading mechanism. If you take an intentional stance, you will see an intention in its behavior. If you do not, you fail to see an intention, and may miss an otherwise fruitful communication opportunity.

Another point is that you cannot decide if an object is intentional or not from a pure observation. As the gaze-game experiment and the three-card-game experiment demonstrate, a human is apt to misunderstand/overlook the intention behind the behavior, if communication channel is limited. In face-to-face interaction, there exist abundant information available for the participant to evaluate the intentionality of the partner. For human-machine interaction, or for human-human communication with a limited communication channel, determination of intentionality of the partner is a difficult task.

For a simple mind-reading task (the one in which intention can easily be found) we need not worry much on the stance a human takes. On the other hand, when a machine are involved in a task which demands a deeper search for intentions, the stance a human takes to the machine becomes more important. If you regard the machine as machine-like, you do not want to make a complicated negotiation, you do not want to discuss on ambiguous things, you do not want to give it a favor expecting a return favor, even if these things are mutually beneficent. Do not say that it is too utilitarian. The mind-reading communication is, after all, a mechanism to make others act at one's will by modifying the cognitive environment, as the relevance theory proposes.

References

1. Behavioral and Brain Sciences 4, 515–526 (1978)
2. Frith, U.: AUTISM: Explaining the Enigma, Basil Blackwell Ltd., UK (1989)
3. Sperber, D., Wilson, D.: Relevance: Communication and Cognition. Basil Blackwell, Oxford (1986)
4. Dennett, D.C.: Kinds of Minds, Basic Books (1996)
5. Ito, A.: An approach to the eye contact through the outstaring game niramekko, Proc. RO-MAN2002, pp. 405–410 (2002)
6. Ito, A., Hayakawa, S., Motokata, K., Terada, K.: Can we feel a gaze pressure from a robot? - Development of an eye-contact robot -, RO-MAN 2003, pp. 103–108 (2003)
7. Tomasello, M.: Joint Attention as Social Cognition. In: Moore, C., Dunham, P.J. (eds.) Joint Attention, Its Origins and Role in Development, LEA, pp. 103–130 (1995)

Kansei Analysis for Robotic Motions in Ubiquitous Environments

Janaka Chaminda Balasuriya[1], Chandrajith Ashuboda Marasinghe[2], Keigo Watanabe[1], and Minetada Osano[3]

[1] Department of Advanced Systems Control Engineering
Saga University, 1-Honjomachi, Saga 840-8502, Japan
jcbala@lycos.com, watanabe@me.saga-u.ac.jp
[2] Department of Management and Information Systems Science
Nagaoka University of Technology, 1603-1 Kamitomioka, Nagaoka, Niigata 940-2188, Japan
ashu@kjs.nagaokaut.ac.jp
[3] Software Engineering Lab, Department of Computer Software
University of Aizu, Aizu-Wakamatsu, Fukushima, Japan
osano@u-aizu.ac.jp

Abstract. Human beings are fascinating creatures. Their behavior and appearance cannot be compared with any other living organism in the world. They have two distinct features with compared to any other living being; unique physical nature and emotions / feelings. Anybody who studies on humans or trying to construct human like machines should consider these two vital facts. When robots are interacting with humans and other objects, they certainly have a safe distance between them and the object. But how can this distance be optimized when interacting with humans; will there be any advantages over achieving this; will it help to improve the condition of robots; can it be a mere constant distance; how will the humans react, are some questions arosed. In order to "humanize" robots, they (robots) should also have certain understating of such emotions that we, humans have. In this research project, authors are trying to "teach" one such human understanding, commonly known as "personal space" to autonomous mobile robots.

1 Introduction

As Simmons et al. [1] describe, recent research in mobile robot navigation has utilized autonomous mobile robots in service fields. To operate them in an environment with people, it requires more than just localization and navigation. The robots should recognize and act according to human social behavior to share the resources without conflict [2]. Sometimes, even when humans interact with each other, it leads to resource conflict. At such times humans use social rules to maintain order [3].

The comfort level of the humans for which the robot is working will be very important if the robot is to do its job effectively. Extensive research is being performed in the area of robotics to improve the conditions of the robots, advancing

M.J. Smith, G. Salvendy (Eds.): Human Interface, Part I, HCII 2007, LNCS 4557, pp. 804–813, 2007.

the abilities to do specific tasks, motion planning, etc. But very little work has been performed in trying to understand how people would interact with a robot, how to make them comfortable, factors that make uncomfortable or threatening, methods or ways for robots to indicate their feelings, etc. to analyze the aesthetic qualities of the robots behavior patterns [2].

1.1 Distance from the Robot to the Vicinity

As in the very beginning of the mobile robotic systems, there had been some kind of distance or space between the robots to any other object in the vicinity. This was just a mere distance for safety for easy maneuvering and for collapse avoidance. As Stentz [4] and many others had mentioned, this was just a constant of space. This mechanism was quite acceptable for the systems such as transporting, surveillance and monitoring, etc. In other words, such kind of safe distance was good for non-human interacting purposes. Can the same be applied for human interaction? Although it will give some results, it will not enhance or optimize the real requirement in need, i.e. to build up harmonious relationship with humans.

1.2 Nakauchi Model

When Nakauchi and Simmons [2] studied about personal space and applied it to moving robots, it was some improvement over the "blind" safe distant. In their mechanism, Nakauchi and Simmons had experimented using human subjects for "correct distance" or "personal space" in order to have pleasant feeling towards the two interacting parties.

This experimentally derived average personal space had an approximate shape of an oval that is larger towards the front. Though those results were approximate, it had been aligned with the values that were reported in the cognitive science literature [5].

1.3 Walters Experiments

Another set of experiments were conducted by Walters et al. [6] using adults and children with a robot of mechanistic appearance called $PeopleBot^R$ to find the personal space zones, initial distances between robot and humans, etc. the context of the encounters and the human's perception of the robot as a social being. They had found out that the children showed a dominant response to prefer the "social zone" comparable to distances people adopt when talking to other humans. From the adult studies, they found that, a small number of people preferred the "personal zone" though significant minorities deviate from this pattern.

2 Variation of Personal Space

Although it is possible to find a personal space for a specific instance of environment, it is highly volatile depending on the two interaction parties and not definitely a constant. As Walters et al. [6] suggested, different robot social models, perhaps with very different initial personalities, may be more acceptable to different users (e.g. a discreet servant or even a silent servant, with no obvious initiative or autonomy). They further stated that it probably cannot be assumed that people automatically treat robots socially, apart from simple elements of anthropomorphism as described by Reeves and Nass [7]. A user-friendly robot should automatically refine and adapt its social model (personality) over a longer period of time, depending on information about and feedback from users and the robots own autonomous learning system. For example, adjustments of social distances according to a user's personality trait will be a promising direction (as proposed in [8]) is a promising direction towards a true robot companion that needs to be individualized, personalized and adapt itself to the user [9].

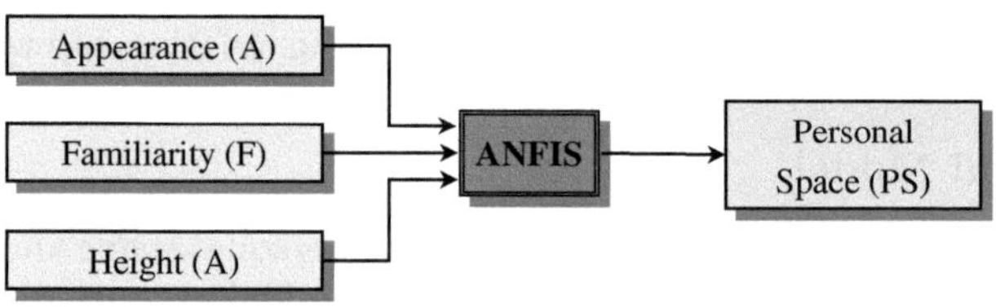

Fig. 1. APS ANFIS block diagram

According to Sack [3] and Malmberg [5], it is reported that the actual size of the personal space at any given instance varies depending on cultural norms and on the task being performed. For a simplified scenario for experimentally analysis; appearance (mainly of the robot), previous acquaintance or familiarity of the either parties, gender, age, height of the bodies, (specially interaction in the standing position), emission of any sound, emotions on the face, carrying objects, etc. were considered to be important. From the list of above, first three: height, appearance and familiarity were considered (as the initial stage for simplicity) to generate an active personal space (APS) determination system and the block diagram is shown in Figure 1.

3 ANFIS for Personal Space Determination

Adaptive Neural Fuzzy Inference System or simply ANFIS can be used as a basis for constructing a set of fuzzy if-then rules with appropriate membership functions to generate the desired input-output combination [10]. It is especially useful when needed to apply a fuzzy inference to already collected input-output data pairs for model building, model following, etc. where there are no predetermined model structures based on characteristics of variables in the system.

3.1 Gathering Data

Considering the procedure as Nakauchi and Simmons [2] or the Walters et al. [6] to obtain a sense of personal space for robot / human interaction, a similar experimental condition was constructed. Here a robot (or a model) is kept at the end of a scaled line in a room and a human is asked to move closer to it.

3.2 Experimental Procedure

As the experiment proceeds, one human subject is instructed to move towards the robots as if he needs to talk with it. The human subject is asked to be along the scaled line and look at the robot face and move closer to it until he feels safe enough to make conversation with it. In the mean time the robotic model was positioned so as to make its face towards the human subject. During the whole time of the experiment, the robot did not do anything and the human subject did all the active tasks of walking, thinking, etc. The robot and the human subject, one person at a time, were supposed to interact at one specific duration of time and it ended once the human subject stops in front of the robot. Then the distance between the two parties was obtained by using a camera or by direct human observer (who reached the two parties once they got stabilized). The human subject had no previous experience with the robot and the authors wanted the human subjects to be curious as well as cautioned about the robot that they are going to meet. In other wards human subjects had no idea what kind of robotic system that they are going to face with or any capabilities that it possesses until they meet the robot.

The robots and the robotic models used in these experiments are 'KooruKun' (PA10 robotic manipulator), 'BoxChyan' (robotic model), and 'BasinKun' (previously known as 'CHRCarry Hospital Robot' re-used with several modifications). The first one was a stationary robot with 200 cm in height and 20 cm average diameter, next was a movable robot model with 100 cm height, 50 cm diameter and around 3 Kg, and the last is also a movable robot with 170 cm height, generalized circular diameter of 60 cm and weight of about 25 Kg. The data gathered are grouped for training, checking and testing for the ANFIS and are described in later.

4 Inputs and Parameters

Out of the three variable parameters considered, height (of the human) and familiarity were two straight-forward measurements while appearance was taken as a collective decision.

4.1 Input Variable 'height (H)'

The height of the human subject is considered in this input variable. The universe of disclose of the input variable 'height (H)' was considered to be 50cm to 200cm, having three membership functions 'big (B)', 'medium (M)', and 'small (S)'. Bell Shape was considered.

4.2 Input Variable 'familiarity (F)'

Previous interactions with the same robot / human pair are considered in this input variable. Keeping in mind that more interactions means more familiar with each other, this variable is set. For the first time of interaction having the value of 1 and 5 times or more having the value of 5 for the input assessment. The universe of disclose of the input variable 'familiarity (F)' was considered as 1 to 5 (without any units), having three membership functions 'big (B)', 'medium (M)', and 'small (S)'. Bell Shape was considered.

4.3 Input Variable 'appearance (A)'

The robots' outer appearance to that of a human is considered in this input variable. Humans are more like to reach one of their own looking rather than to that of very peculiar shaped objects. Human like robot get the more closer feeling with respect to the other crude or rather machine looking robot. The universe of disclose of the input variable 'appearance (A)' was considered to be 1 to 5 (without any units), having three membership functions 'big (B)', 'medium (M)', and 'small (S)'. Bell Shape was considered. Here the appearance value for the 'KooruKun' robot was given as 1, for the 'BoxChyan' as 2, and 'BasinKun' as 5. Although more human like robots were required to get the records, at the time of the experiment, such robots were not available in the laboratory. (Hence the universe of disclose was 1 to 5).

4.4 Output Variable 'personal space (PS)'

Considering the above three inputs that will generate 27 rules in total and each having unity weight for each rule, the output of the ANFIS is obtained using the weighted average defuzzification.

5 ANFIS Architecture

The architecture of the APS determination network is illustrated in Figure 2. Layer (I) to Layer (III) represent the antecedent part of the fuzzy neural network, whereas Layer (V) and Layer (VI) represent the consequence part [11].

As shown in Figure 2, the domain of discourse of height (H) is described by fuzzy variable H with p number of linguistic values ($p = 3$), the domain of discourse of appearance (A) is described by fuzzy variable A with q number of linguistic values ($q = 3$), and the domain of discourse of familiarity (F) is described by fuzzy variable F with r number of linguistic values ($r = 3$). Hence each input variable is unique in the sense of domain of discourse. It is assumed that each node of the same layer has a similar function, as described below. Here we denote the output of the ith node in layer (X) as $O_{(X),i}$.

Layer (I)
Layer (I) consists of three types of nodes; height (H), appearance (A) and familiarity (F). The current value of height (H), i.e. the crisp input to the height

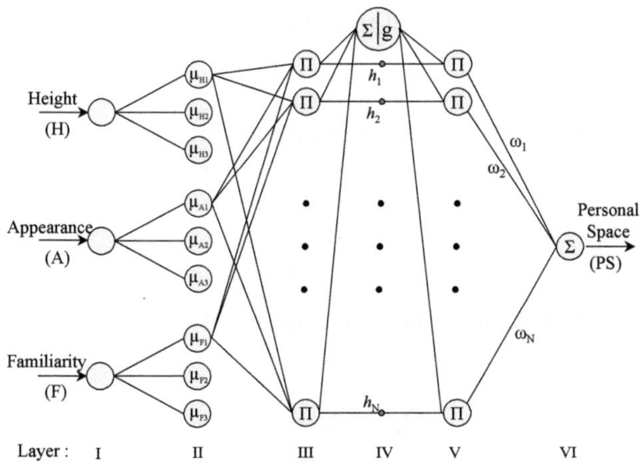

Fig. 2. APS ANFIS architecture

node is H_i, appearance node is A_j and familiarity node is F_k. No computation is carried out at this layer.

Layer (II)
This layer acts as the fuzzification layer of the fuzzy neural network. At this layer, the output of a node connected to the current value of variable input acquires the fuzzy membership value of the universe of discourse. Every node i, where $i = 1, \cdots, p$ (or q or r), in this layer is an adaptive node with a node function

$$O_{II,\,i} = \mu_{X_i}\,(x) \tag{1}$$

where x is the input to node i, and X_i is the linguistic label (big, medium, small, etc.) associated with this node function. In other words, $O_{II,i}$ is the membership function of X_i and it specifies the degree to which the given x satisfied the quantifier X_i. Hence the output from the 2nd layer will be:

$$O_{II,\,p} = \mu_{H_i}\,(H_i) \tag{2}$$

$$O_{II,\,q} = \mu_{A_q}\,(A_j) \tag{3}$$

$$O_{II,\,r} = \mu_{F_r}\,(F_k) \tag{4}$$

for height, appearance and familiarity respectively.

Layer (III)
In this layer, the nodes labeled as Π compute the T-norm of the antecedent part. Thus the rule evaluates the conditions of the inputs and they are continued to the layer (V) for normalization. The output of any node t, where $t = 1, \cdots, N$, where $N = p * q * r$, in this layer is described by the following equation:

$$O_{III,\,t} = h_t = \mu_{H_i}\,(H_i) * \mu_{A_q}\,(A_j) * \mu_{F_r}\,(F_k) \tag{5}$$

where h_t represents the firing strength of the tth rule and there are N such rules as total.

Layer (IV)
The first node of layer (IV) at fuzzy neural network, which has symbols $\sum$ and g, generates the output through the following function:

$$g(x) = \frac{1}{x} \tag{6}$$

with a linear summed input. Then the output of the first node of layer IV is given by

$$O_{IV,1} = \frac{1}{\sum_{t=1}^{N} h_t}. \tag{7}$$

Other nodes just carry forward the outputs of previous nodes to the next layer.

Layer (V)
This layer normalizes the fired rule values. Each node labeled as Π in this layer multiplies the value carried forward by previous node with the output of the first node at Layer (IV). Then the output of any mth node of this layer can be given by the following equation:

$$O_{V,m} = \frac{h_m}{\sum_{t=1}^{N} h_t} \tag{8}$$

Layer (VI)
Layer (VI) is the defuzzification layer of the fuzzy neural network. The node labeled as $\sum$ in this layer calculates the overall output as the summation of all incoming signals. Then the personal distance value for certain input variables is given by:

$$O_{VI} = Personal\_Diatance = \frac{\sum_{m=1}^{N} w_m h_m}{\sum_{n=1}^{N} h_n} \tag{9}$$

where w_m denotes a constant value in the consequence part of the mth rule. The overall output is the weighted mean of w_m with respect to the weight h_m.

The connection weights are trained by applying the hybrid algorithm. The error tolerance was set to zero.

The error is calculated by comparing the output of the expert knowledge with that of fuzzy neural network for the same input data, x. The adaptation of the mth weight, w_m , at the lth time step is given by the following equation:

$$w_m(l+1) = w_m(l) + \gamma [y_d - y_a] \frac{h_m}{\sum_{n=1}^{N} h_n} \tag{10}$$

where γ represents a small positive learning rate, and y_d and y_a represent the desired output and actual output respectively for the personal space value selected for the training.

6 Training, Checking and Testing Data Sets

From the collected data from the experiments, data were rearranged in to 3 groups for training, checking and testing purposes of the APS ANFIS.

6.1 Train Data Set

Train data set was obtained by grouping the input variable 'height (H)'. First, the height was categorized in to 5 groups as:

- 161 cm to 165 cm
- 166 cm to 170 cm
- 171 cm to 175 cm
- 176 cm to 180 cm
- 181 cm to 185 cm

Then under each column, average is taken. The results thus obtained were used as the train data set.

6.2 Check Data Set

In order to optimize the ANFIS and to make a slighter change in the trained data set, following equation was considered when obtaining the check data set. I.e. considering a particular column, average value is subtracted from the maximum value from the gathered data. Then half of that value is added to the minimum value of the gathered data in the same column.

$$ChkDataValue = [\frac{Max.Data - Ave.Data}{2}] + Min.Data \tag{11}$$

6.3 Test Data Set

Considering the original data from the experiments, very close to the training and checking data sets, one data entry is selected for each category. By this selection, another set of data is obtained for testing purpose of the ANFIS.

The trained ANFIS output with train data and check data are shown in Figures 3 and 4.

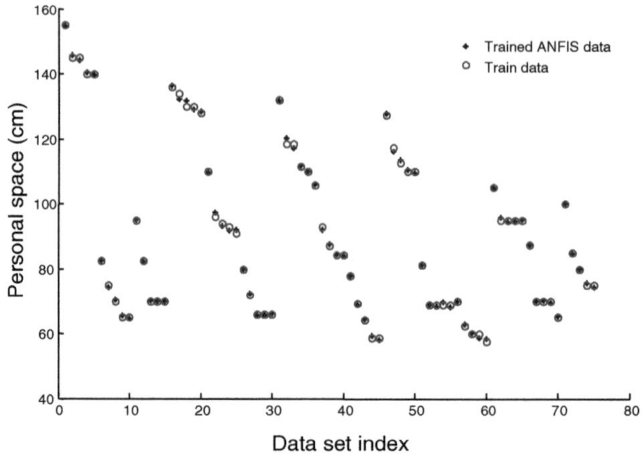

Fig. 3. Trained FIS with train data

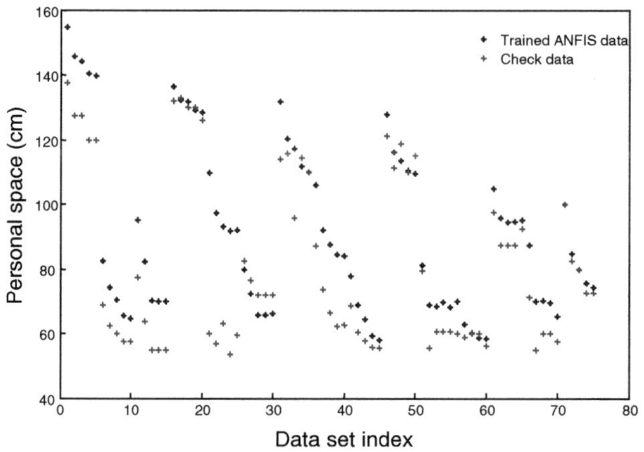

Fig. 4. Trained FIS with check data

7 Summary

In this research project, a determination system of an active personal space has been analyzed. This can be considered as the first step for constructing an active personal space system for any robot in any environment. But for a target as such, many experiments in vast environmental situations should have to be involved. It is a must to obtain similar data with the so-called humanoids to make this experiment complete. Further, more sophisticated supportive equipment such as high speed processing units for human recognition, memory acquisition and manipulation, image processing, etc. should be coupled. This system gave encouraging results in an offline mode with limited facilities. Authors are planning to make the current system more realistic and get the functioning in a real time mode, and are continuously working on it.

References

1. Simmons, R., Goodwin, R., Haigh, K.Z., Koenig, S., O'Sullivan, J.: A layered architechture for office dilivery robots. In: Proc. of Autonomous Agents, pp. 245–252 (1997)
2. Nakauchi, Y., Simmons, R.: A social behavioral robot that stands in line. Autonomous Robots 12, 313–324m (2002)
3. Sack, R.: Human Territory. Cambridge University Press, Cambridge (1986)
4. Stentz, A.: Map-based strategies for robot navigation in unknown environments. In: Proc. of AAAI, pp. 110–116 (1996)
5. Malmberg, M.: Human Territoriality: Survey of behavioural territories in man with preliminary analysis and discussion of meaning. Mouton Publishers (1980)
6. Walters, M.L., Dautenhahn, K., Koay, K.L., Kaouri, C., Boekhorst, R., Nehaniv, C., Werry, I., Lee, D.: Close encounters: spatial distances between people and a robot of mechanistic appearance. In: Proc. of 5th IEEE – RAS Int. Conf. on Humanoid Robots, December 2005, pp. 450–455 (2005)
7. Reeves, B., Nass, C.: The Media Equation: How people treat computers, television and new media like real people and places. Cambridge University Press, Cambridge (1998) ISBN 157586052x
8. Walters, M L, Dautenhahn, K., te Boekhorst, R., Koay, K L, Kaouri, C., Woods, S., Nehaniv, C L, Lee, D., Werry, I.: The influence of subjects' personality traits on personal spatial zones in a human-robot interaction experiment. In: Proc. of 14th Annual Workshop on Robot and Human Interactive Communication (IEEE Ro-man 2005), Tennessee, USA, August 2005, pp. 347–352 (2005)
9. Dautenhahn, K.: Robots we like to live with — A developmental perspective on a personalized life-long robot companion. In: Proc. of 13th Annual Workshop on Robot and Human Interactive Communication (IEEE Ro-Man 2004), Okayama, Japan, September 2004, pp. 17–22 (2004)
10. Jang, J.S.R.: ANFIS: Adaptive-network-based fuzzy inference system. In: Proc. of IEEE SMC, May/June 1993, vol. 23(3), pp. 665–685 (1993)
11. Jang, J.S.R., Sun, C.T.: Neuro-fuzzy modeling and control. In: Proc. of the IEEE, March 1995, vol. 83(3), pp. 378–406 (1995)

The Use of Dynamic Display to Improve Reading Comprehension for the Small Screen of a Wrist Watch

Yu-Hung Chien[1] and Chien-Hsiung Chen[2]

Graduate School of Design, National Taiwan University of Science and Technology
43 Keelung Road, Section 4. Taipei, 106 Taiwan
[1]roland.chien@msa.hinet.net, [2]cchen@mail.ntust.edu.tw

Abstract. This study explored the feasibility of displaying dynamic Chinese text on the screen of a wrist watch. Three design factors (i.e., dynamic display, presentation method, and speed) were examined to investigate their effects on users' reading comprehension under two different conditions of task types. The results of this study indicated the following: (1) There was no significant difference between leading display and RSVP in both single- and dual-task conditions; (2) Presentation method was a significant factor. The participant's reading comprehension was significantly better with the word-by-word format than with the character-by-character one in both single- and dual-task conditions; (3) The participant's reading comprehension was not significantly different among the three different speed settings in the single-task condition. However, participants had significantly higher reading comprehension scores under the slower speed settings of 150 and 250 cpm than under the faster speed setting of 350 cpm in the dual-task condition.

Keywords: Dynamic display, Leading display, Rapid serial visual presentation, Reading comprehension, Small screen, Wrist watch.

1 Introduction

In 2002, Microsoft Chairman Bill Gates introduced the Smart Personal Objects Technology (SPOT) Initiative, aimed at improving the function of everyday objects through the injection of software. SPOT wrist watches are the first SPOT-based application. SPOT turns wrist watches into mobile tools ripe to receive information services and advertisings via radio-frequency links. Recently, users can receive timely information (e.g., traffic information, urgent message, and news) by wrist watches instead of other mobile electronic devices. Fig. 1 shows a snapshot of SPOT wrist watches. Although SPOT wrist watches offer the convenience of portable user interfaces, the display space of wrist watches, however, is at a premium. The amount of information that can be displayed on the screen of wrist watches at one time is very limited. One possibility is to adopt automatic dynamic displays, in that onscreen text moves, to overcome the limitations caused by a small screen [1], [2], [3], [6], [8], [10], [11], [13].

Past studies related to dynamic display were mainly for the reading of English text. However, there are great differences between the reading of dynamically presented Chinese and English. Therefore, further studies are needed.

M.J. Smith, G. Salvendy (Eds.): Human Interface, Part I, HCII 2007, LNCS 4557, pp. 814–823, 2007.

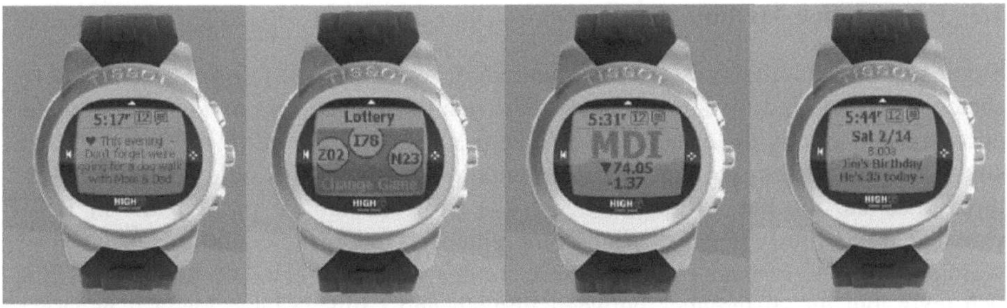

Fig. 1. A snapshot of SPOT wrist watches

1.1 Dynamic Display

The types of dynamic displays widely used in various media include leading display and rapid serial visual presentation (RSVP). In a leading display, text scrolls horizontally from right to left on a one-line display [3], [6], [7], [8], [13]. Leading displays can be divided into the "leaping" or the "smooth-sliding" types. A leaping type of leading display takes the form of the words leaping a specific distance. In contrast, a smooth-sliding type of leading display takes the form of words sliding smoothly by in a pixel-by-pixel manner. Another kind of dynamic display is RSVP, by means of which words are displayed successively, quickly, and continuously in a fixed area of the screen so that users can read words without moving their eyes [1], [2], [3], [6], [7], [8], [10], [11].

The leading display and RSVP have been compared with static display in terms of their effects on the reading of English. As was pointed out in the research, compared with the leading display, the static display had a greater reading efficiency [3]. On the other hand, the comparison between RSVP and static display was believed to have resulted in no significant differences [2], [6].

Meanwhile, studies have also compared directly the different influences of leading display and RSVP on reading English text. A study showed that a higher reading accuracy in RSVP than in the leaping type of leading display [6]. However, another study compared the influences of a smooth-sliding leading display and RSVP on reading comprehension, the result of their study showed that there were no significant differences between the RSVP and the smooth-sliding type of leading display [7].

In addition, there are some studies concerning the dynamic display of Chinese. A study explored the influences of three dynamic displays (i.e., scrolling, leading, and RSVP) on reading comprehension when Chinese characters were presented on different forms of small-screen interfaces. The results revealed that, when a laptop and a mobile phone were involved, users showed no significant differences on comprehension while reading written messages respectively under the conditions of smooth-sliding leading display and RSVP [2]. Another study compared the reading comprehension in the conditions of smooth-sliding leading display and RSVP; the results revealed that the participants had greater recall accuracy with the smooth-sliding leading display than with RSVP [8].

1.2 Presentation Method

Studies compared the presentation method of word-by-word format, sentence-by-sentence format, and the static display. The results showed that no significant differences existed among the effects of the three methods on the reading efficiency [10]. A study also compared another three presentation methods, including word-by-word, three-line, and ten-line formats. The results showed that the participants in RSVP revealed significantly better reading comprehension under the conditions of word-by-word and ten-line formats [1]. The above studies were conducted on the basis of English texts, but the differences between English and Chinese have made it necessary that the exploration of presentation methods for Chinese should be carried out in accordance with Chinese characteristics. Although character in Chinese is the smallest reading unit, it has been pointed out in studies that while reading Chinese characters the eyes usually stare at several characters at the same time. The word is often an important unit in reading Chinese text [4]. The Chinese word consists of one or several Chinese characters with no salient boundary between the words. Research pointed out that reading accuracy would not be affected by any salient boundaries resulting from the insertion of spaces between Chinese words but that the reading time will be shortened [4]. That is to say, spaces between words may help readers separate Chinese characters into words and thus lower the cognitive load and increase their reading speed. Lin and Shieh [8] have investigated the effects of the presentation method with character-by-character and word-by-word formats. It was found that, for the leading display on a single-line screen, the effects on the participants' recall efficiency showed that the word-by-word format scored greater results than did the character-by-character format. The study also investigated the situation with RSVP, and tried to discover the effects of these two presentation methods, i.e., the single-line eight-character format and the word-by-word format. The results of the study revealed that the word-by-word format is significantly superior to the single-line eight-character format.

1.3 Speed

The display speed for the reading of English text is calculated with word as a unit, that is, words per minute, or wpm. An investigation was conducted to reveal the relationships between RSVP and the three display speeds (250, 450 and 650 wpm). Significant differences were found among the different display speeds, with the display speed of 250 wpm producing the best reading comprehension [1]. Juola et al. [6] study the effects of the leading display and RSVP in the reading of English text. The results showed that when the dynamic display was in the manner of leading, participants had the highest reading accuracy when the display speed was 260 wpm; when the dynamic display was in the manner of RSVP, no significant differences were observed for the display speeds of 260 and 171 wpm.

The display speed for the reading of Chinese text is calculated with character as a unit, that is, characters per minute, cpm. A study compared the eye fixation in the reading of English and Chinese texts. The results showed that the eye fixation for each Chinese character was 0.26 second, that is, about 230 cpm (60/0.26), and that the eye fixation for each English word was 0.27 second, that is, about 222 wpm (60/0.27)

[12]. Furthermore, it was pointed out in the research related to dynamic display for Chinese text that when RSVP and the display speed of 140 cpm were employed in the 20-character single-line screen, participants had a significantly higher reading comprehension than in the case of 250 and 350 cpm. When the display was in the manner of leading, the display speed of 195 cpm showed a significantly better reading comprehension [13]. Chen and Chien [2] compared the effects of three different display speeds (171, 250, and 305 cpm) on three different screen sizes. As was revealed by the research, the display speed of 250 cpm was ideal for a laptop screen for reading comprehension. And the display speed of 305 cpm could be adopted for the screens of PDA and mobile phones. However, when Lin and Shieh [8] conducted the research with three display speeds (120, 240, and 500 cpm), they discovered that participants exhibited no significant differences in recalling the materials read when the display speeds were 120 and 240 cpm, though a display speed of 500 cpm greatly affected the participant's recalling rate.

1.4 Task Type

In the past studies concerning the reading in dynamic display, the consideration only focused on participants performing a single task, that is, the exploration of the effects of dynamic-display design factors upon the reading comprehension while concentrating on reading the dynamic text messages [1], [2], [3], [6, [7], [8], [10], [11], [13]. This study, in contrast, explored the effects of design factors on the participant's reading comprehension when they are engaged in one single task or when they are engaged in dual tasks.

2 Methodology

2.1 Participants

Twenty four native Chinese speakers from Taiwan participated in this study. Fourteen were females and ten were males between the ages of 19 to 36 with the average age 23.6 years (SD=3.16). All the participants were pretested to confirm that they had at least 20/25 visual acuity with corrective lenses if required.

2.2 Apparatus and Materials

Testing was conducted on a Pentium IV PC, with all text materials displayed on a 1024 × 768 resolution 17-in. LCD color monitor. Macromedia Flash 8.0 was used to simulate the SPOT watches available in the market. The testing interface simulated in the experimental design was placed right on the center of the monitor's screen.

The Chinese message texts read by participants consisted of 12 passages to be used in the formal experiment and 6 passages for practice. The passages came from the world encyclopedia, with 30 characters for each passage. The determination of Chinese words was based upon the research of Liu et al. [9].

In addition, according to relevant studies, the experimental design for the dual-task operation could be carried out by means of mental arithmetic [14]. The mental arithmetic was conducted via text-to-speech voice.

2.3 Experiment Design

This experiment was designed in order to determine the participant's reading comprehension of the dynamically displayed Chinese under the single- and dual-task conditions, respectively. The three factors likely to affect dynamic display, dynamic-display type, presentation method, and display speed, were explored in this study. The types of dynamic display were divided into two levels: leading display and RSVP. The two levels of presentation methods were character-by-character and word-by-word formats (Fig. 2). Display speeds were set up at 150, 250, and 350 cpm. Participants were tested under two task-type conditions: single- and dual-task conditions. In the dual-task conditions, participants would listen to and then work on a mental-arithmetic task at the same time that they were reading the dynamically displayed texts. There were altogether four variables: three design factors for dynamic display and the different task types among them, which were the dynamic display format, the

Fig. 2. Two presentation methods displayed in two dynamic displays were used in this study, (a) character-by-character format in an RSVP, (b) character-by-character format in a leading display, (c) word-by-word format in an RSVP, and (d) word-by-word format in a leading display

task type belonging to between-subjects design, and display speeds and presentation methods belonging to within-subjects design. Thus, the 24 participants were randomly divided into four groups: leading single-task group (LS), leading dual-task group (LD), RSVP single-task group (RS), and RSVP dual-task group (RD), with six participants in each group.

2.4 Procedure

Participants in the two single-task groups (LS and RS) began with a practice reading of the six short passages, which were presented in different combinations. Participants in the two dual-task groups (LD and RD) would listen to and then would work on the mental-arithmetic task at the same time as when they were practicing reading of the six passages. They would then rest for two minutes before proceeding to the reading of 12 short passages in the LS and RS groups respectively. The simultaneous reading of 12 short passages and calculation of the mental-arithmetic tasks by the LD and RD groups would occur. The short passages, mental-arithmetic tasks, display speeds, and presentation methods were all randomly arranged. Participants in each group would be subjected to six experimental treatments, with each treatment displayed twice. After the participants in the single-task group read a short passage, they would answer two questions according to the content of the passage. Participants in the dual-task group would also fill in the answers for the mental-arithmetic tasks. The whole experiment lasted about 30 minutes.

2.5 Analysis

The measurement of dependent variables for the experiment was conducted through the analysis of variance (ANOVA). The reading comprehension scores and mental-arithmetic scores were counted as dependent variables. The results were tested with $\alpha=0.05$ as the standard for significant difference. If significant differences were observed for independent variables, Tukey HSD (honesty significant difference) would be employed to determine the level differences between the independent variables.

3 Results

3.1 Effects of Dynamic-Display Variables on the Participant's Reading Comprehension in the Single-Task Condition

Table 1 showed the comprehension scores and the standard deviations obtained by the design factors of various dynamic displays from each level when participants were carrying out a single task.

The ANOVA indicated no significant difference between the two dynamic displays. However, there was a significant difference between the two presentation methods (F (1,204) =14.182, p<0.01). The participant's reading comprehension from the word-by-word presentation method was significantly better than that from the

Table 1. The mean of comprehension score under each level of the independent variables for accomplishing a single-task

Independent variables	*n*	Mean (*SD*)	Tukey grouping [a]	
Dynamic text				
Leading	108	0.792 (0.322)		
RSVP	108	0.722 (0.364)		
Presentation method [b]				
Character-by-character	108	0.671 (0.375)	A	
Word-by-word	108	0.843 (0.287)		B
Speed				
150 cpm	72	0.799 (0.343)		
250 cpm	72	0.708 (0.333)		
350 cpm	72	0.764 (0.356)		

[a] values with the different letter are significantly different.
[b] significantly different at $\alpha < 0.01$ level.

character-by-character presentation method. Furthermore, the ANOVA demonstrated no significant differences between the three display speeds. There were also no interactions observed among the design factors of dynamic display.

3.2 Effects of Dynamic-Display Variables on the Participant's Reading Comprehension and Mental-Arithmetic Score in the Dual-Task Condition

Table 2 showed the comprehension scores and the standard deviations obtained by the design factors of various dynamic displays from each level when participants were undertaking dual tasks.

Table 2. The mean of comprehension score under each level of the independent variables for accomplishing a dual task

Independent variables	*n*	Mean (*SD*)	Tukey grouping [a]	
Dynamic text				
Leading	108	0.652 (0.375)		
RSVP	108	0.639 (0.341)		
Presentation method [b]				
Character-by-character	108	0.551 (0.361)	A	
Word-by-word	108	0.713 (0.336)		B
Speed				
150 cpm	72	0.708 (0.333)	A	
250 cpm	72	0.653 (0.342)	A	B
350 cpm	72	0.535 (0.378)		B

[a] values with the different letter are significantly different.
[b] significantly different at $\alpha < 0.01$ level.

The ANOVA revealed no significant difference between the two dynamic displays. However, there was a significant difference between the two presentation methods (F (1,204) =12.243, p=0.001). The participant's reading comprehension from the word-by-word format was significantly better than that from the character-by-character format. Furthermore, the ANOVA revealed significant differences between the three display speeds ($F(1,204)=4.887$, $p=0.008$), the speed group of 150 and 250 cpm registered significantly better results than the speed group of 250 and 300 cpm. There were no interactions observed among the design factors for the dynamic display discussed above.

When participants performed dual tasks, they would listen to and work on mental-arithmetic tasks at the same time as when they read dynamically displayed texts. The ANOVA showed that there were no significant differences between the influences of the dynamic-display design factors upon the participant's mental-arithmetic results. There were also no interactions among these factors.

4 Discussion

There was no significant difference between leading display and RSVP in both single- and dual-task conditions. The results were similar to the studies involved in the comparison of leading display and RSVP [2], [7]. However, research pointed out that the participant's reading accuracy was significantly better in RSVP than in the leading manners [6]. The possible reason for the discrepancy in results might be due to the fact that the leap-type of leading display was adopted for the research. Because when participants were reading the text displayed in the manner of leap-type leading, the position of their eye fixation changed in accordance with the leaping distance of words which were presented in a flashing manner as these words were. On the contrary, the smooth-slide type of leading display involved the smooth movement of words and did not cause flashing. Therefore, a smooth-slide type of leading display was better than a leap-type of leading display. Moreover, Lin and Shieh [8] compared smooth-slide type of leading display and via RSVP on a 7 × 1.2 cm display screen for the participant's reading of written messages. It was found that, when compared with the eight-character single-line display in RSVP, participants achieved greater recall accuracy in the character-by-character and word-by-word formats of smooth-slide leading display. But no significant differences were registered when the comparison was made with word-by-word RSVP display. Thus, it could be concluded that whether the dynamic display was good or not should be determined by judging how the dynamic display factors were combined.

This study showed that the word-by-word presentation method was superior to the character-by-character presentation method for the reading of dynamic text on a wrist watch, no matter whether it was in the single- or dual-task conditions. Lin and Shieh [8] also discovered that the word-by-word presentation method brought about greater recall accuracy, either in the manner of leading or RSVP.

As is shown in this study, when participants conducted the single task of reading dynamic text on a wrist watch, the choice of display speed did not have much effect on the reading comprehension. This result is similar to the related study [2], [8]. However, Wang and Chen [13] pointed out that when dynamic text was employed on the twenty-character single-line screen, display speeds lower than static display were needed to achieve better reading comprehension. In the course of reading, there was a limitation to the number of characters processed at each time of eye fixation (or perceptual span). Inhoff and Liu [5] claimed that the perceptual span in Chinese reading was about 3 to 4 characters on the right of the fixation point and about 1.5 to 2 characters on the left of the fixation point. In the research conducted by Wang and Chen [13], the presentation of twenty characters at a time was larger than the users' perceptual span at each fixation. The wrist-watch interfaces in this study presented a maximum of six Chinese characters, a number that was close to the normal span of each eye fixation in Chinese reading. This lowered the participant's times of eye fixations and saccades in reading. As a result, higher display speeds could be employed for the reading of dynamic texts on the small screen interface of a wrist watch. This also gives a possible reason for the difference between this study and that of Wang and Chen [13].

On the other hand, when participants were performing dual tasks at the same time, display speed was an important factor in affecting reading comprehension. The lower display speeds of 150 and 250 cpm were significantly better than that of 350 cpm. As users were limited in their concentration resources, the simultaneous performing of two tasks or more is likely to cause distraction, which can increase the cost of concurrence for cognitive resources. As a result, lower display speeds should be adopted for dynamic display in dual-task conditions so that the burden on cognitive resources would be lessened.

5 Conclusions and Future Works

The summary was made as follows with regard to the application of various dynamic-design factors to the reading of Chinese texts dynamically displayed on a wrist-watch interface: (1) Both smooth-slide types of leading display and RSVP can be applied to the wrist-watch interface for the presentation of written messages. The design of the interface can employ different dynamic displays in accordance with the different interface configuration requirements. (2) The presentation method of word-by-word format should be adopted for the reading of dynamic text on the wrist watch. (3) When users are involved in a single task, the higher display speed of 350 cpm can be adopted; but when users are performing dual tasks, the lower display speeds of 150-250 cpm are preferable.

References

1. Bernard, M., Chaparro, B.S., Russell, M.C.: Examining Automatic Text Presentation for Small Screens. In: Proc. the Human Factors and Ergonomics Society 45th Annual Meeting, Minneapolis, Minnesota, pp. 637–639 (2001)
2. Chen, C.H., Chien, Y.H: Effect of Dynamic Display and Speed of Display Movement on Reading Chinese Text Presented on a Small Screen. Perceptual and Motor Skills 100, 865–873 (2005)

3. Granaas, M.M., Mckay, T.D., Laham, R.D., Hurt, L.D., Juola, J.F.: Reading Moving Text on a CRT Screen. Human Factors 26(1), 97–104 (1984)
4. Hsu, S.H., Huang, K.C.: Interword Spacing in Chinese Text Layout. Perceptual and Motor Skills 91, 355–365 (2000)
5. Inhoff, A.W., Liu, W.: The Perceptual Span and Oculomotor Activity During the Reading of Chinese Sentences. Journal of Experimental Psychology: Human Perception and Performance 24(1), 20–34 (1998)
6. Juola, J.F., Tiritoglu, A., Pleunis, J.: Reading Text Presented on a Small Display. Applied Ergonomics 26(3), 227–229 (1995)
7. Kang, T.J., Muter, P.: Reading Dynamically Displayed Text. Behaviour and Information Technology 8(1), 33–42 (1989)
8. Lin, Y.C., Shieh, K.K.: Reading a Dynamic Presentation of Chinese Text on a Single-line Display. Displays 27(4-5), 145–152 (2006)
9. Liu, I.M., Chuang, C.J., Wu, J.T.: Guiding Rules for the Computational Analyses of Chinese Words and Propositions. Chinese Journal of Psychology 29(1), 51–61 (1987)
10. Rahman, T., Muter, P.: Designing an Interface to Optimize Reading with Small Display Windows. Human Factors 41(1), 106–117 (1999)
11. Rusell, M.C., Chaparro, B.S.: Exploring Effects of Speed and Font Size with RSVP. In: Proc. the Human Factors and Ergonomics Society 45th Annual Meeting, Minneapolis, Minnesota, pp. 640–643 (2001)
12. Sun, F., Morita, M., Stark, L.W.: Comparative Patterns of Reading Eye Movement in Chinese and English. Perception Psychophysics 37(6), 502–506 (1985)
13. Wang, A.H., Chen, C.H.: Effects of Screen Type, Chinese Typography, Text/Background Colour Combination, Speed, and Jump Length for VDT Leading Display on Users' Reading Performance. International Journal of Industrial Ergonomics 31(4), 249–261 (2003)
14. Wickens, C.D.: Processing Resources and Attention. In: Damos, D.L. (ed.) Multiple-task Performance, Taylor & Francis, pp. 3–34 (1991)

Embodied Communication Between Human and Robot in Route Guidance

Guillermo Enriquez, Yoshifumi Buyo, and Shuji Hashimoto

Science and Engineering, Waseda University
3-4-1 Okubo, Shijuku-ku, Tokyo, Japan 169-8555
{enriquez,buyo,shuji}@shalab.phys.waseda.ac.jp

Abstract. Walking with someone or guiding them to a destination is a simple task that humans do everyday. However, while almost any mobile robot can navigate to a given point while a human walks behind it, these robots do not take into consideration whether the human is following along properly. Our research involves the use of a wireless sensor network for navigation of a guidance robot and a vision-based tracking system for human awareness. The network creates a virtual directional field that provides a directional imperative to the robot, and the vision system allows it to detect if a following human has strayed. We also present a demonstration of our system in a real world application.

Keywords: Human-aware, guidance, tracking, wireless sensor network.

1 Introduction and Objectives

The act of guidance is a simple task that we humans do without much thought. However, in reality, there are many variables to be considered that our brains handle subconsciously. Factors such as path planning, distance from person being guided, and obstacles are all taken into consideration as we move. Robots however are incapable of making these decisions without human programming. While there have been robots that guide the humans in various places such as museum and railway station[1]-[4], these robots did not take into consideration whether or not the humans we actually following. Our system not only moves the robot from its present location to a desired destination, but it also takes into account the present location of the human being guided. Should the human stray, the robot takes measures to collect the human before continuing on to the final destination as described in the flowchart of the robot behavior Fig.1. In such way there exists an embodied communication between human and robot even in simple route guidance. While previous robots utilized vision and maps to navigate the environment our system uses a wireless sensing network to navigate to the destination. In addition, it makes use of a vision tracking system to determine whether or not the human is following the robot properly. In order to test our system, we created a mock-environment to simulate a restaurant.

M.J. Smith, G. Salvendy (Eds.): Human Interface, Part I, HCII 2007, LNCS 4557, pp. 824–829, 2007.

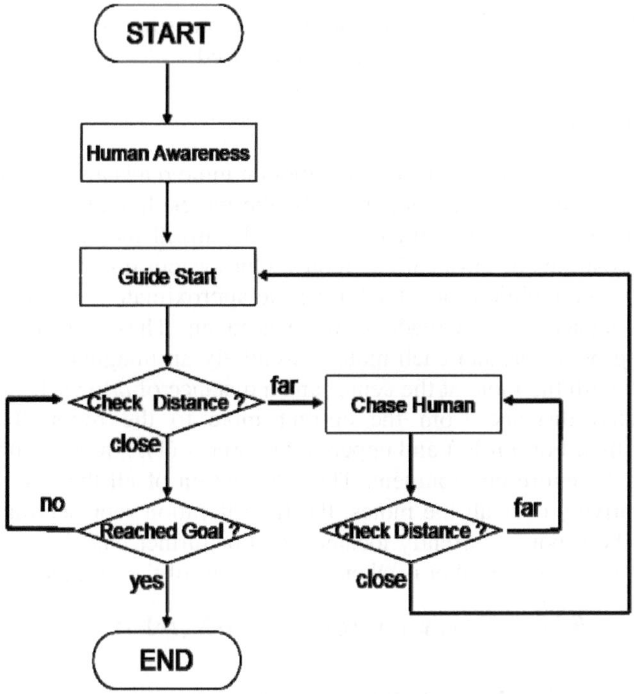

Fig. 1. The flowchart of the guide robot system

2 Navigation Through a Directional Field

Our navigation system has two facets, depending on the state of the robot; the wireless network, and the vision-based tracking system. The wireless sensor network comprised of a number of MICAZ motes and a mobile mote that is mounted on the robot. The network provides general localization and is the basis of the directional field. These motes are pre-programmed with their position within the room. As major static obstacles in restaurants do not often change, pre-programming the map is not costly.

2.1 Wireless Sensor Network

In recent years, interest in wireless sensor networks has grown increasingly due to their wide array of applications. They are ad-hoc, small in size, low in cost, and very scalable. For our system we used MICAZ (MPR2400J) motes and an MIB520 programming board. A total of eight motes were dispersed around the experiment room and pre-programmed with their location and other information. The pre-programming of the motes is not costly in our experiment as restaurants rarely change the position of major obstacles such as walls, partitions, or tables. A ninth mote was used in conjunction with a MIB520 board as an interface between the robot and the sensor network. The mote on the robot, or mobile mote, runs an altered version of the network program allowing localization to be processed.

2.2 General Localization

In order to perform general localization the mobile mote reads a series of signals from the neighboring motes. The signals emitted by the motes, like all radio signals, decay and weaken in strength as they travel. In [5] a localized mobile robot was used to determine the positions of un-localized motes. Our system implements the reverse in order to localize the mobile robot. Each time the approximate position of the robot is calculated, a distance to each readable mote is taken. These distances are used to create bounding boxes around each mote. Essentially an imaginary box around each mote is created with the mote at the center and a distance of *d* to each side, where *d* is the approximated distance from the current mote to the robot. The boxes are represented by their bottom-left and upper-right corners (1). An initial probability box of the robot is the entire environment. The intersection of all the bounding boxes is performed iteratively over all the motes, the final iteration representing the probable area in which the robot is currently located (2). For further calculation purposes, the approximate position of the robot is taken as the center of the probability box (3).

$$B.box = \{(x, y, d) : (blX, blY, trX, trY)\} \quad (1)$$

$$P.box = \{(B.box_i \cap P.box')\} \quad (2)$$

$$approx = \{(P.box) : (x, y)\} \quad (3)$$

2.3 Directional Field

The wireless sensing network performs two duties in our system. Not only is it used in localization, but it provides a directional field that provides basic directional information to the robot. This directional field can be imagined as a set of gradients

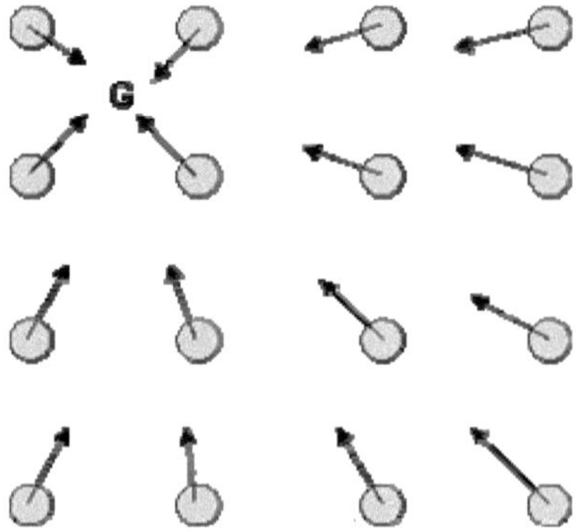

Fig. 2. A visual representation of the Directional Field concept

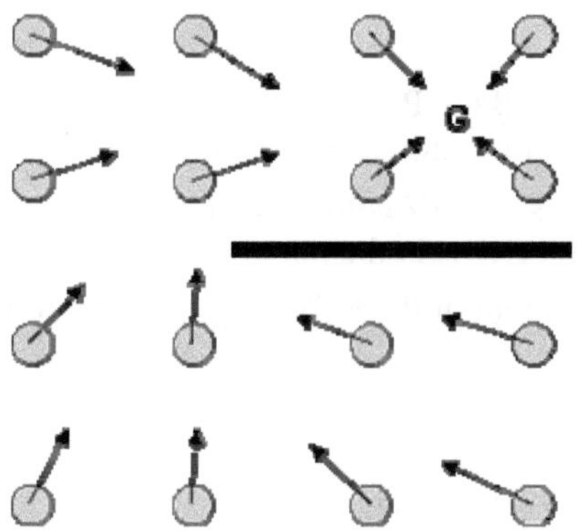

Fig. 3. A visual representation of the Directional Field concept with an obstacle

pointing downwards toward the goal as shown in Fig. 2. The robot examines these gradients and attempts to maintain a downward path. Essentially the robot is rolling downhill on a virtual slope created by the directional field. The directional field can also be used to avoid major obstacles as the directional vectors can be aimed around them as shown in Fig. 3.

3 Human Awareness

Our system provides human awareness through a visual tracking system. This tracking is performed by utilizing a certain difference in luminance value between the background region and the human region of the image. The system extracts human region by focusing on the color feature of the human's clothes.

3.1 Frame Entrance Recognition

The system is started by recognizing when a human appears behind the robot. This is done by frame subtraction method. The robot remains stationary and keeps capturing images until the human enters frame comes within a predefined distance. $I(x,y,t)$ represents the pixel value at (x,y) and at t of the input images. We get a binarized subtraction image $BSI(x,y,t)$ which is computed as

$$BSI(x,y,t)=\begin{cases}255 & I(x,y,t)-I(x,y,t-1)\geq th\\ 0 & otherwise\end{cases} \tag{4}$$

Then, we apply erosion and dilation several times to the BSI in order to eliminate noise or small connected components. After that, we calculate the smallest rectangle which includes all pixels of value 255. When the height of the rectangle is over 90% of the input image, we assume that the human has come sufficiently close to the robot and the system can get the information of the human from the image such as color of clothes and other features. We call this rectangle the initial human region.

3.2 Feature Extraction

We apply the Histogram Back-Projection method to determine the probability distribution image of the human clothes region as follows:

1. Convert the input images to HSV color system.
2. Set the region of interest (ROI) in the initial human region.
3. Calculate H value histogram of the ROI. $\{\hat{q}\}_{u=0\ldots m}$ represents the frequency of the value u of H.
4. Transform pixel value based on the histogram.

$$\left\{ \hat{p}_u = \frac{\hat{q}_u}{\max(\hat{q})} \times \hat{q}_u \right\}_{u=1\ldots m} \tag{5}$$

That is, the histogram values are rescaled from $[0, \max(q)]$ to the new range $[0, 255]$, where pixels with the highest probability of being in the histogram will map as visible intensities in the 2D histogram back-projection image.[6]

Next, we binarize the histogram back-projection image with a threshold which is determined statistically. Then, we apply erosion and dilation to the binarized image and compare the area size of the connected component to determine the human clothes area. In this research, we calculate the approximate distance between the human and the robot by the ratio of extracted human clothes area size to initial human clothes area size.

4 Experiment and Future Works

While this approach has many applications, we simulated a generic restaurant in order to test our system. In many restaurants, customers must wait for a host or hostess to seat them at their table. This is the aspect we chose to replace as it is a common, everyday example of guidance. There are many factors that could affect the guidance and cause it to be interrupted midway.

Using a hallway and several rooms, we dispersed several motes in an array. These motes form the directional field. We placed obstacles such as tables and chairs in order to more closely simulate an actual restaurant.

Under normal conditions, i.e. the human follows the robot properly without straying; we expected our robot to perform properly. In order to test its capability of handling interruptions, we simulated the human stopping, straying, and a combination of both. These cases represent the human stopping perhaps to answer their cell phone or perhaps wandering off to find the bathroom or greet a friend they have spotted in the restaurant.

The details of the experimental results will be shown in our web site with some movie clips soon at http://www.shalab.phys.waseda.ac.jp/index-j.html.

As a future work we are planning to improve the ability of the robot so that it can guide a group of humans. Although we employed wheeled mobile base in our robot at present, we are also considering to use a biped walking system developed in our institute[7] to make the more practical robot that can move on a floor with a difference in level.

Acknowledgments

This research was funded in part by a grant from the Japan Society for the Promotion of Science 21st Century COE Program "Coexistence of Humans and Robotic Technology in the Super-Aging Society" and the Waseda University WABOT-HOUSE Laboratory.

References

1. http://www.cs.cmu.edu/m̃inerva/
2. http://www.informatik.uni-bonn.de/~rhino/
3. http://robotics.naist.jp/research/ikochan/index.html
4. Suzuki, K., Camurri, A., Ferrentino, P., Hashimoto, S.: Intelligent Agent System for Human-Robot Interaction through Artificial Emotion. Proceedings. of SMC, pp. 1055–1060 (1998)
5. Shenoy, S., Tan, J.: Intelligent Robots and Systems, Proceedimgs of IROS, pp. 1636–1641 (2005)
6. Allen, J.G, Xu, R.Y.D, Jin, J.S: Object Tracking Using CamShift Algorithm and Multiple Quantized Feature Spaces. Proceeding of the Pan-Sydney area workshop on Visual information processing, pp. 3–7 (2003)
7. Ogura, Y., Aikawa, H., Shimomura, K., Kondo, H., Morishima, A., Lim, H.-o., Takanishi, A.: Development of a New Humanoid Robot to Realize Various Walking Pattern Using Waist Motions. Proceedings of ROMANSY16 (2006) 16, 279–286 (2006)

A Comparative Study of Brain Activities Engaged in Interface Operations by Means of NIRS Trajectory Map

Miki Fuchigami[1], Akira Okada[1], Hiroshi Tamura[2], Masako Omori[3]

[1] Osaka City University
[2] Tamura Institute for Human Interface
tamura@mobilergo.com
[3] Kobe Women's University

Abstract. This paper studies an elemental human action, hand up-down, and selective button operations, by means of NIRS. In the analysis, we pay attention to NIRS trajectory and the total Hb changes. Our results provided empirical support for the hypothesis that the trajectory of positive inclination indicates brain location actively contributing the task, and negative inclination is the brain locations out of contribution. The locations positively contributing to the task are found most intensively in right and left forebrain area but in middle part of left brain looks less contributing to the task.

Keywords: NIRS, hand up-down, button selection, trajectory, brain map.

1 Introduction

The attempt to apply NIRS to brain activities started to analyze text entry behaviors to mobile phone [1]. The studies are continuing to some of more elemental human actions, like hand up-down, finger shaping, finger matching and, again to actions realistic to human interface, a single button press, and selective button operations.

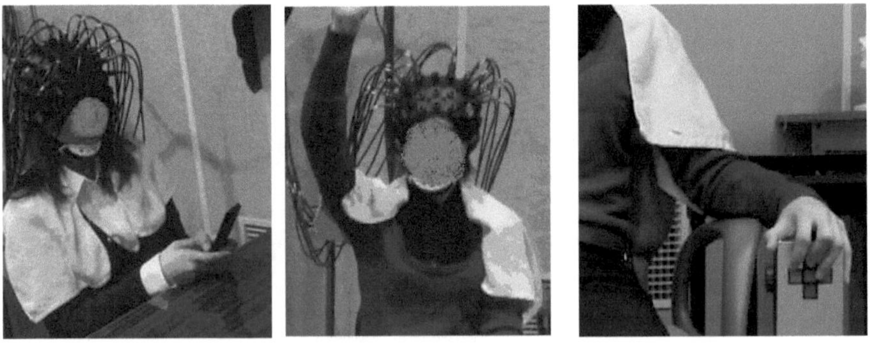

a) mobilephone b) hand up-down c) button selective operation

Fig. 1. Behaviors under experiment by using NIRS

M.J. Smith, G. Salvendy (Eds.): Human Interface, Part I, HCII 2007, LNCS 4557, pp. 830–839, 2007.

In this paper, experimental finding on hand up-down action and on button selective operation are reported.

2 Experimental Systems

NIRS (Near Infra-Red Spectroscopy) facility was a product of Shimadzu co. (OMM3000/12) which is installed in Future ITC Research Center, KARC (Kansai Advanced ICT Research Center, Akashi, Kobe, Japan). The experimenters are honored by Joint Research Treaty with the center to use the NIRS facility. The NIRS had 12 light transmitter and receiver pairs and was capable of processing up to 54 channels of brain activities depending on transmitter and receiver arraignment. The arraignment we adopted in measurement is shown in Fig. 2,

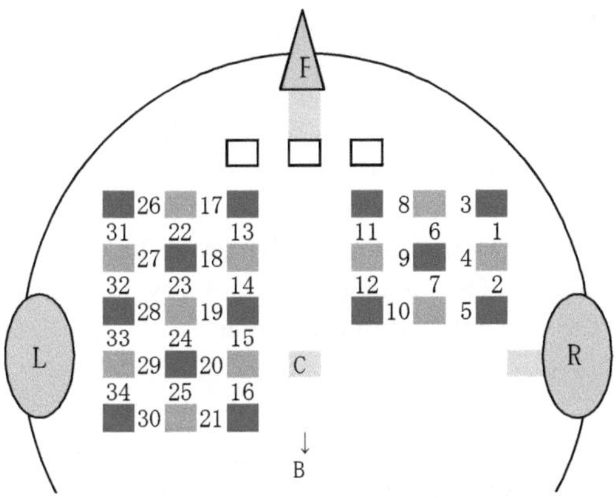

Fig. 2. Transmitter/receiver arraignment

By this arraignment, 34channel brain activities are collected every 0.1second. Beside NIRS data, the machine is capable of 8 external analogue signals, which are used to record start, stop and task completion signals together with NIRS data. The above arraignment was chosen since forehead brain activities are more important to human interface activities and that left brain activities seemed more complex based on preliminary NIRS studies.

Experimental task design and the analysis methods are developed by authors. At this moment, the subject participated in the measurement is graduate student, volunteered with informed consent in accordance with the Institutional Review Board of Kobe Women's University. The purpose of this paper is to show possibilities of NIRS trajectory and the trajectory map from the methodological aspect. Analysis for larger population should be reported elsewhere.

In the previous paper [2], NIRS trajectory and trajectory map are illustrated on one person, one trial base, although they are representing mostly common features of brain activities under similar situations. The analysis in this paper is the attempt to apply the method to multiple trials and take event wise average for equivalent data.

Two experiments in this paper are to examine brain activities related to select hands and fingers proper to the task.

2.1 Hand Selection Task (Hands Up-Down)

Subject sat on a chair with her hands on the knee. Experimenter gave an instruction in a random order, which hand (left, right, both) to raise at the next trial. The subject raised her hand to upright position following a auditory tone. With 2 seconds of time subject raised her hand to the upright position and back to knee position. The hand remained on the knee position another 2 seconds and waited for next instruction.

Two sequences of 9 trials are applied. NIRS data after start tone signal were collected sidewise (left, right, both).

2.2 Finger Selection Task (Button Selection)

Subject rest her hand (left or right) on the button box. Computer generated voice give instruction to choose right, left or up, down button. The subject fingers are allocated at home position, in which middle finger stay in the location of square in the Fig. 3. In case of left or right instruction, index or ring finger is selected depending on the hand in use. In case of up or down instruction, middle finger has to change its location by changing the bending angle. The change in necessary action might be related to different brain activities to be analyzed in this paper.

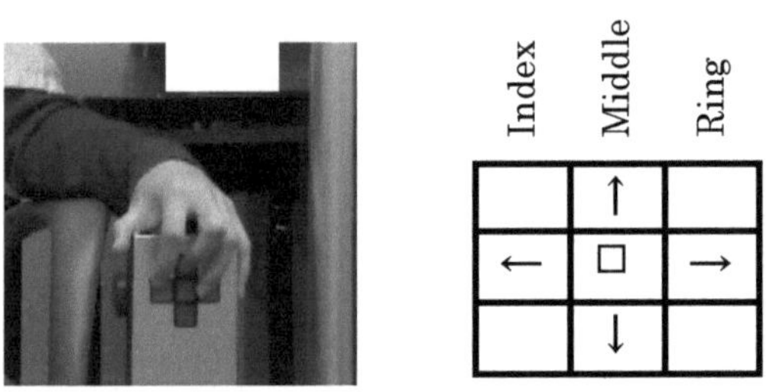

Fig. 3. Button box operated by left hand (photo), and the direction and finger allocation

3 Results

3.1 Hand Selection

NIRS data are zero adjusted at the beginning of each trial, and data from 6 same trials are averaged. Using averaged data of each channel, average NIRS trajectory is drawn. The NIRS trajectory in this case is always starting from the origin of the graph. Average NIRS trajectory for each channel is shown in 34 cells of Fig. 4, where right brain channels are in the right, and the left brain in the left side of the figure. The

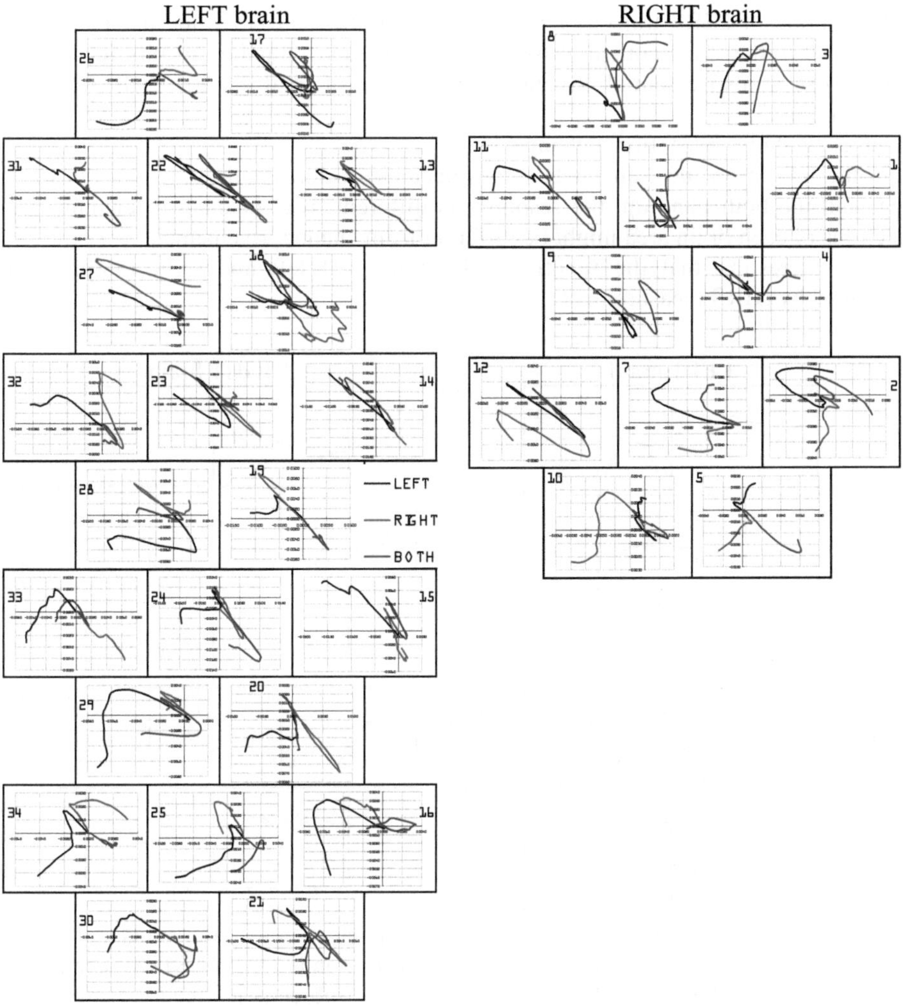

Fig. 4. The brain map of average NIRS trajectory for hand up-down

channel number is indicated in each cell. The blue plot is for the left hand, the red for the right, and the green for the both hands.

If the average trajectory is along -1 inclination line, the total Hb change is slow. If the average trajectory is leaving the origin along +1 inclination, the total Hb change is quick. Our tentative assumption is trajectory of positive inclination indicates brain location actively contributing the task, and negative inclination is the brain locations not contributing the task.

The locations positively contributing to the task are found most intensively in right and left forebrain area but in middle part of left brain looks less contributing to the task.

Each point of NIRS trajectory is corresponding to total Hb value but it does not correspond to the time course of total Hb. For the understanding of physical events, time course is also essential. Average time course of total Hb for each channel is shown in Fig. 5.

Channel 12 shows exceptional change.

In the left brain, most channels shows decrease to left hand-up action, but to the right and the both hand-up, the response is different from channel to channel. The functions of each brain area might be different from channel to channel.

It is interesting that the functional differences of left and right forebrain are confirmed by NIRS measurement.

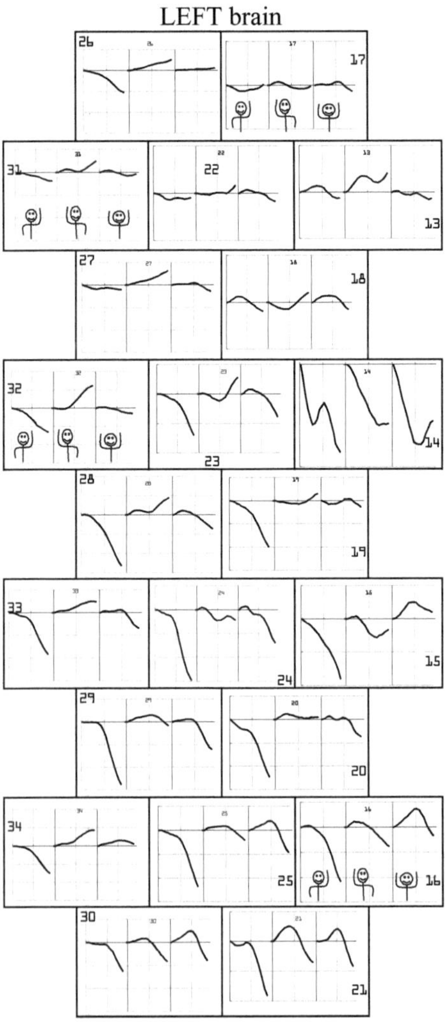

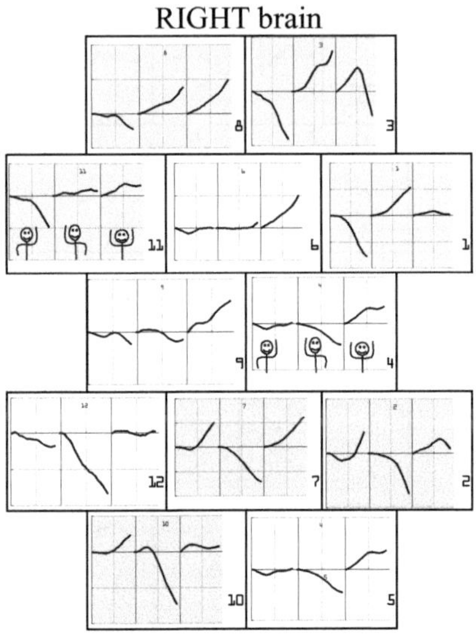

Each cell in this figure corresponds to a NIRS channel. Three sub cells in a cell are showing total Hb change to Left, Right and Both hand-up action, from left to right respectively.

Total Hb changes of right brain are regular in the right brain. In the front two rows, total Hb decreases to left hand-up, increases to right hand-up, and increases to both hand-up. To the contrary, in the channels behind third row, total Hb increases to the left hand-up, and decreases to the right hand-up.

Fig. 5. Average time course of total Hb to hand-up actions

3.2 Finger Selection

NIRS is a relative measuring system. Thus normally at each starting point of trials, the zero adjustment is done. It is also possible to apply the adjustment at each goal point and analyze the data shortly before and after the zero. For the Finger selection trials goal based analysis is introduced. One half of a second before and after the goal point is analyzed.

One set of data of 6 trials taken from a channel is shown in Fig. 6. Fig. 6a is for the case operator pushed the Left direction button with ring finger of Left hand. The action is indicated by LD(ring/LH). Three other actions are indicated in Fig. 6. For each action, 10 trials are included in two sequences of experiment.

Each time course of total Hb change has considerable variation, the average are taken to each cases of each channel. Four average data of 10 channels are shown in Fig. 7 a, b. These channels are selected from 34 channels covering left and right brain.

Some variations among channels can be notified by visual observations. To get quantitative evaluations, correlation analyses are applied to the average data. Since our actions under consideration are four, total 16 correlations are to be considered. But 4 correlations are always 1 (cross hatched in Table 1), and 6 correlations are symmetric to rest of 6 (single hatched). Resultantly, 6 correlations are interesting to study. Criteria to evaluate correlation are set as follows:

- strong (positive, negative) : $p >= 0.8$, $p =< -0.8$
- fair (positive, negative) : $> 0.8 > p >= 0.5$, $-0.5 =< p < -0.8$
- low : $-0.5 < p < 0.5$

Based on the above criteria, strong and fair correlation data are shown in Fig. 7.

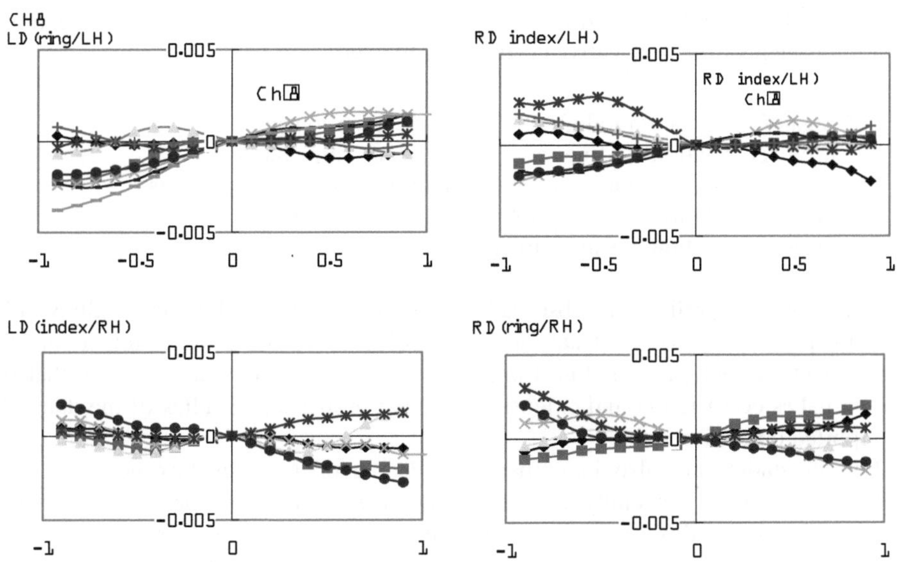

Fig. 6. Total Hb change around the goal (individual responses)

Table 1. Correlations to be considered

	LD(ring/LH)	RD(index/LH)	LD(index/RH)	RD(ring/RH)
RD(ring/RH)	(1)	(2)	(3)	
LD(index/RH)	(4)	(5)		
RD(index/LH)	(6)			
LD(ring/LH)				

If correlations are strong (near +1), or inversely (near -1) correlation in the cells (1)~(6), following remarks can be made:

Selectivity	description
(1) finger	same finger (ring) of different hand to the different direction
(2) direction	same direction by different finger of different hand
(3) hand	different finger of same hand toward different direction
(4) direction	same direction by different finger of different hand
(5) finger	same finger (index) of different hand to the different direction
(6) hand	different finger of same hand toward different direction

Functional features of each channel in Fig. 7 might be described below.

1: negative correlation between index and ring finger of left hand
3: positive correlation to same direction, negative correlation to different direction.
8: strong negative correlation to the same finger of different hand.
10: negative correlation to different direction, strong correlation to same direction
12: strong positive or negative correlations to every action.

16: negative correlation to left, positive correlation to right.
17: strong correlations to fingers of different and same hands
26: high negative correlation to same hand different finger operation.
31: strong correlation to fingers of same and different hands
34: strong correlation to same direction operation

The above are preliminary observation of limited channels. Further results will be explained at the session. Data are also collected from middle finger in this experiment. An action of middle finger is different from index and ring, in that the finger makes constriction and extension to make proper tap. Studies of the data will be reported otherwise.

Further studies are also in preparation. People operate key box not only from orthodox position. Especially for mobile devices, heterodox operations are often noticed. But heterodox operation sometime induces miss operation or unexpected results. Preventing human error at the brain level might be most basic importance in human interface studies.

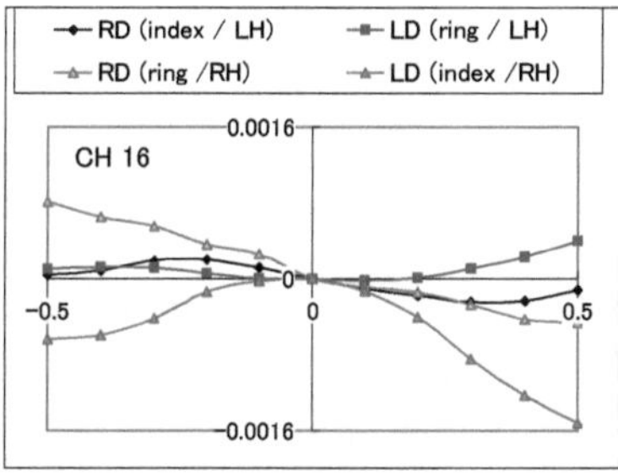

CH 16	LD(ring / LH)	RD(index/ LH)	LD(index / RH)
RD(ring / RH)		**0.81**	
LD(index/ RH)	-0.93	0.55	
RD(index / LH)			

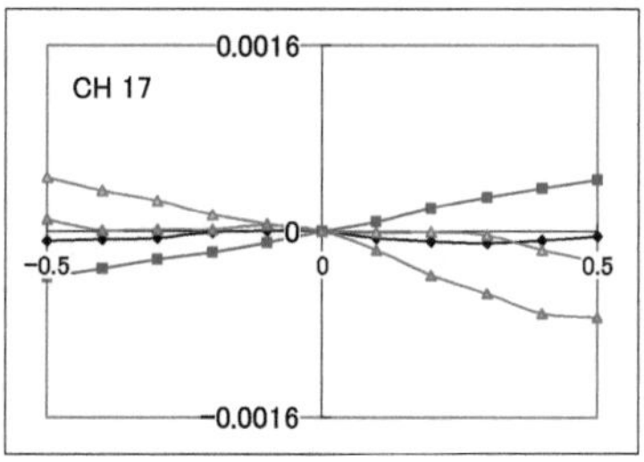

CH 17	LD(ring / LH)	RD(index/ LH)	LD(index / RH)
RD(ring / RH)	-0.95		**0.81**
LD(index/ RH)	-0.93		
RD(index / LH)			

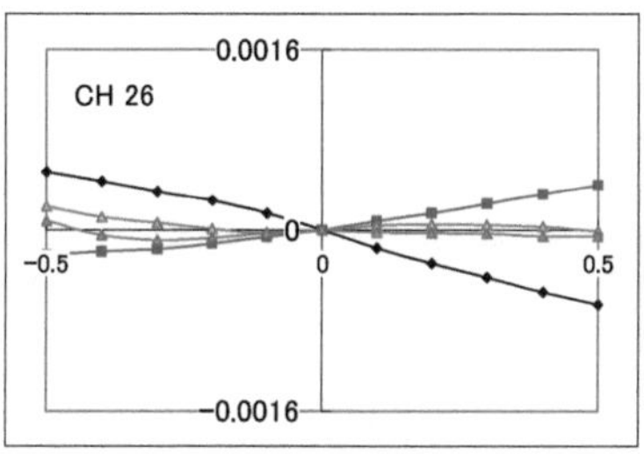

CH 26	LD(ring / LH)	RD(index/ LH)	LD(index / RH)
RD(ring / RH)		0.53	0.60
LD(index/ RH)			
RD(index / LH)	-1.00		

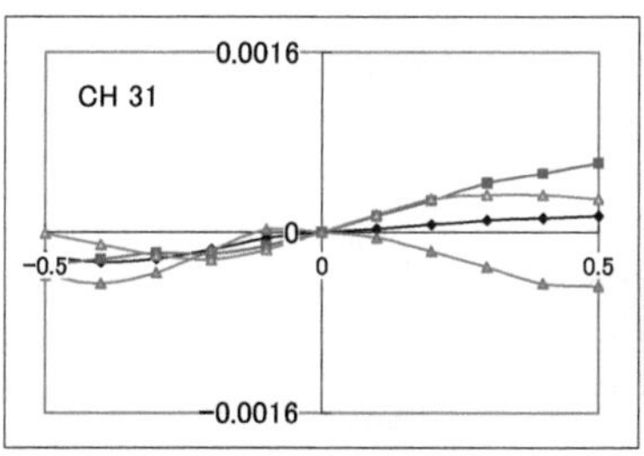

CH31	LD(ring / LH)	RD(index/ LH)	LD(index / RH)
RD(ring / RH)	**0.91**	**0.86**	
LD(index/ RH)			
RD(index / LH)	**0.93**		

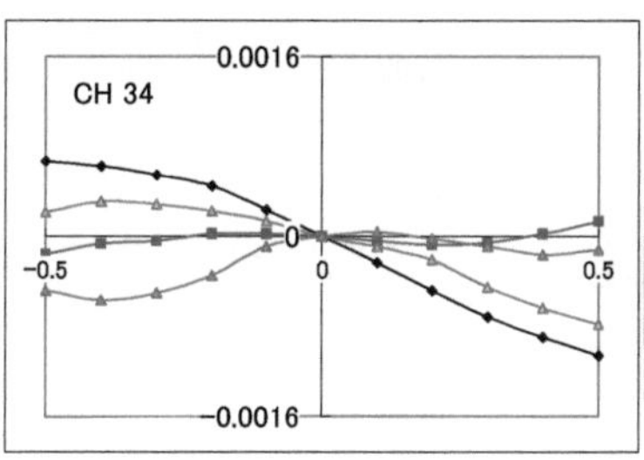

CH34	LD(ring / LH)	RD(index/ LH)	LD(index / RH)
RD(ring / RH)	-0.54	**0.98**	-0.58
LD(index/ RH)		-0.69	
RD(index / LH)	-0.53		

Fig. 7 a). Average total Hb and the correlation for finger selections (left brain)

CH1	LD(ring / LH)	RD(index / LH)	LD(index / RH)
RD(ring / RH)	-0.52		
LD(index / RH)			
RD(index / LH)	-0.97		

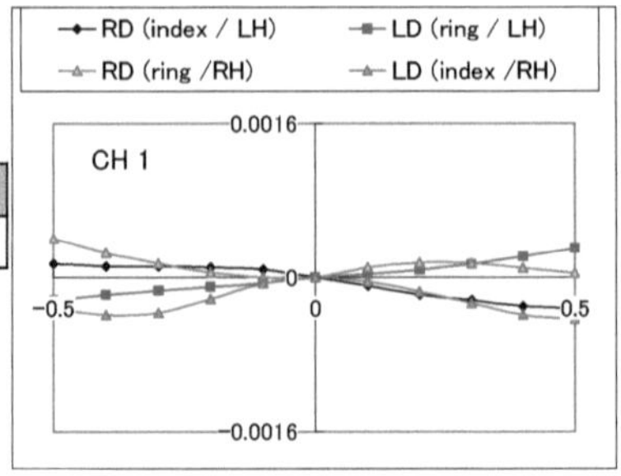

CH 3	LD(ring / LH)	RD(index/ LH)	LD(index / RH)
RD(ring / RH)	-0.90	**0.79**	-0.50
LD(index/ RH)			
RD(index / LH)	-0.95		

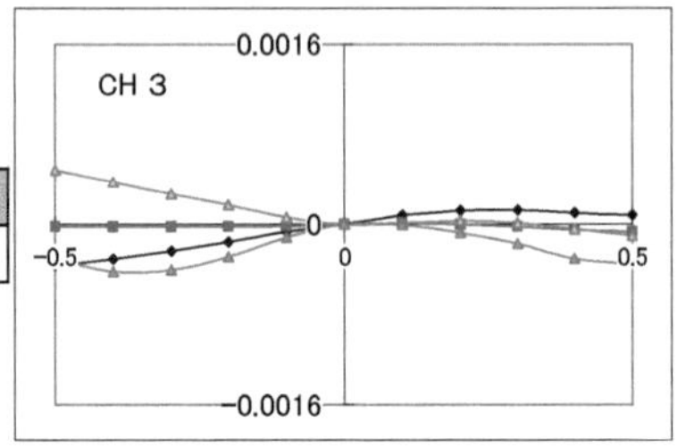

CH 8	LD(ring / LH)	RD(index/ LH)	LD(index / RH)
RD(ring / RH)	-0.61		
LD(index/ RH)		-0.92	
RD(index / LH)			

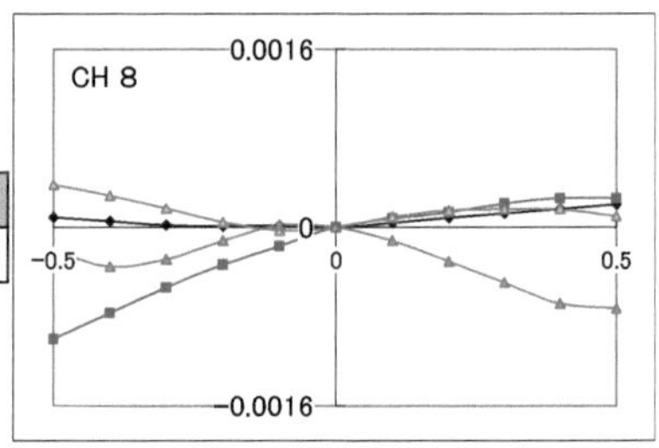

CH 10	LD(ring / LH)	RD(index/ LH)	LD(index / RH)
RD(ring / RH)		**0.74**	-0.66
LD(index/ RH)		-0.96	
RD(index / LH)	-0.58		

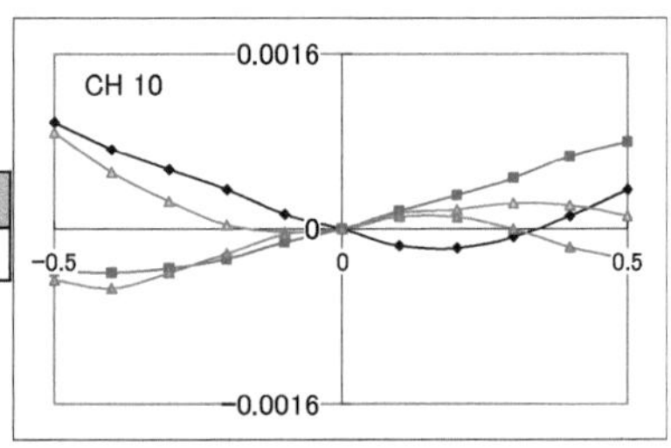

CH 12	LD(ring / LH)	RD(index/ LH)	LD(index / RH)
RD(ring / RH)	-0.95	**0.92**	**0.94**
LD(index/ RH)	-0.87	**0.96**	
RD(index / LH)	-0.85		

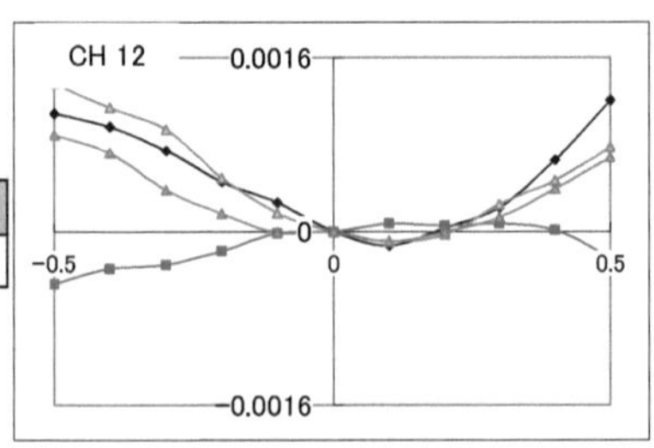

Fig. 7 b). Average total Hb and the correlation for finger selections (right brain)

4 Conclusions

Previously NIRS trajectory is introduced [1], and the trajectory map [2, 3] was proposed to show brain activities performing a task, by allocating trajectories of different part of the brain. Trajectory map study has suggested, the right forebrain is taking more role in some tasks like text entry to mobile phone. And the functional asymmetry forebrain is distinct. NIRS trajectory method and trajectory map are applied to legibility study [4, 5].

Brain function asymmetry is again confirmed in the hand up-down experiment. Finger selection experiments [6, 7] are expected to reveal further knowledge on the brain functional allocation. Past PET studies on tapping are also suggestive to this study.

NIRS measurement will be an essential study tool for human interface.

References

[1] Tamura, H., Omori, M., Choui, M.: Brain activities recorded by NIRS measurement while performing error correction tasks, Preprint of Kansai branch assembly of JES, pp. 35–38 (December 2006) (in Japanese)

[2] Tamura, H., Omori, M., Choui, M.: Brain Activities indicated by two dimensional trajectories of NIRS. Mobile Interactions 3, 9–14 (2007) (in Japanese)

[3] Tamura, H., Omori, M., Choui, M.: NIRS trajectories in Oxy-Deoxy Hb plane and the trajectory map to understand brain activities related to human interface, HCII2007 ID 2590 (2007)

[4] Omori, M., Tamura, H., Choui, M.: A study on brain activities related to visual search. Mobile Interactions 3, 19–22 (2007) (in Japanese)

[5] Omori, M., Hasegawa, S., Miyao, M., Tamura, H., Choui, M.: Brain Functions related to legibility of LCD studied by means of Near Infrared Spectroscopy, HCII2007 ID 2591 (2007)

[6] Fuchigami, M., Okada, A., Yamashita, K.: Cognitive compatibility between direction-of-motion and the human body as a machine control strategy, Preprint of Proceedings of the Human Interface Symposium, pp. 499–504 (2005) (in Japanese)

[7] Fuchigami, M., Okada, A., Yamashita, K.: Influence of body posture on cognitive compatibility between output direction and input direction in manual control, IEA2006 ID 1614 (2006)

[8] Miyamoto, R., Kikuchi, Y., Seno, A.: Neural basis of perception of right and left hand. Japan Society of Physiological Anthoropology 11, 157–162 (2006)

[9] Aoki, T., Tsuda, H., Takasawa, M., Osaki, Y., Oku, N., Hatazawa, J., Kinoshita, H.: The effect of tapping finger and mode differences on cortical and sub cortical activities: a Pet study. Exp. Brain Res 160, 375–383 (2005)

[10] Grafton, S.T., Haseltine, E., Ivry, R.B.: Motor sequence learning. Exp. Brain Res 146, 369–378 (2002)

Interaction Design of a Remote Clinical Robot for Ophthalmology

Kentaro Go[1], Yuki Ito[1], and Kenji Kashiwagi[2]

[1] Interdisciplinary Graduate School of Medicine and Engineering, University of Yamanashi
4-3-11 Takeda, Kofu 400-8511 Japan
go@yamanashi.ac.jp, ito@golab.org
[2] Department of Ophthalmology, Faculty of Medicine, University of Yamanashi
1110 Shimogato, Chuo 409-3898 Japan
kenjik@yamanashi.ac.jp

Abstract. This paper presents an interactive design project of a remote clinical robot for ophthalmology, for use in support of telemedicine. The project is a case study of finding the best organizational structure of a Graphical User Interface (GUI) and Solid User Interface (SUI) in the age of ubiquitous computing. At the early stage of design, we carried out usability tests with paper-based prototypes and obtained requirements from ophthalmologists. Results showed that paper prototyping is applicable for GUI- and SUI-combined products, even they have manual control input and continuously changing visual feedback. We examine lessons learned from the design process and designed product.

Keywords: design method, Graphical User Interface (GUI), interaction design, ophthalmology, paper prototyping, remote clinical robot, Solid User Interface (SUI), telemedicine.

1 Introduction

Design is an essential activity for people. It is fun yet challenging. Especially these days, advancing technology has rapidly increased our knowledge requirements for design. This situation is becoming increasingly complicated because we are in the age of ubiquitous computing [3, 11]. In a ubiquitous computing environment, as we have already experienced it, a combination of many sensors, computers, and network devices seamlessly communicates and configures support of people's activities. Designers must learn and know existing limitations and new opportunities of several specialized areas—it seems impossible for a single designer to design a service or system in ubiquitous computing.

In addition, issues that must be addressed have also been changing. Consumers rapidly shift their emphases from technology-orientation to a convenience-oriented one. The usability issue was not a primary concern 25 years ago. A user-centered view in design has become important in the age of ubiquitous computing because we have many computers, sensors, and networks surrounding us; their collaboration provides us valuable services in a seamless manner. This concept radically changes the design target of computers in human-computer interaction: that target is not a

M.J. Smith, G. Salvendy (Eds.): Human Interface, Part I, HCII 2007, LNCS 4557, pp. 840–849, 2007.

desktop PC with a Graphical User Interface (GUI). It might be a portable device with a communication function, which can consist of numerous Solid User Interfaces (SUIs) [7].

It is difficult to strike an organizational balance between GUI and SUI because some interactions can be represented either way. For example, we can employ a toggle button as a GUI component or as an SUI button. A discussion of GUI and SUI has been conducted in the Japanese science and engineering community [1, 6, 7], but no exhaustive discussions have been put forward. The purpose of this paper is to describe a case study of finding the best organizational structure of GUI and SUI in the age of ubiquitous computing. The case study is an interactive design project of a remote clinical robot for ophthalmological diagnosis.

The organization of the paper is as follows. Section 2 describes the background information and motivation of the robot design project. Section 3 shows practice and lessons learned from the requirements-elicitation process through paper prototyping. Section 4 articulates design activities and designed artifacts of interactive design. Section 5 concludes the paper.

2 Remote Clinical Robot for Ophthalmology

2.1 Current Issues in Ophthalmologic Examination

The number of medical specialists, especially ophthalmology specialists, is decreasing in secluded mountain areas in Japan. Telemedicine has been studied and is expected to be used to ameliorate the situation. In a telemedicine model, a large medical center sets up a network connection to rural health care facilities so that they can perform diagnoses and treatments using a videoconferencing system.

For current ophthalmological telemedicine, general practitioners in a rural health care facility record patients' medical information, typically as still images or moving pictures of eyes, and send it to eye specialists in a large medical center. It does not fully leverage a slit lamp microscope in telemedicine, which is indispensable to eye diagnostics. Consequently, detailed diagnostics cannot be conducted via current telemedicine in ophthalmology.

In ophthalmologic examination, a slit lamp microscope (Fig. 1) is generally used as a fundamental diagnostic device. The microscope consists of a slit lamp unit (the top round shape) and a microscope unit (binocular-shaped). The microscope-unit position is adjusted using the joystick (the bottom stick shape). An eye doctor sits at the front in Fig. 1 and a patient sits at the back. The patient puts her forehead at the forehead rest and chin on the chin rest (the frame at the back).

With the slit lamp microscope, the eye specialist's basic tasks include adjustment of the microscope position with the joystick to find a focal position and to switch slit types with the knobs on the slit lamp unit to set an appropriate diagnostic condition. The eye specialist might use a fronting lens and a blue filter together with the microscope to conduct further examinations. The organization of these special devices requires great skill and experience for microscope control in addition to knowledge of clinical medicine.

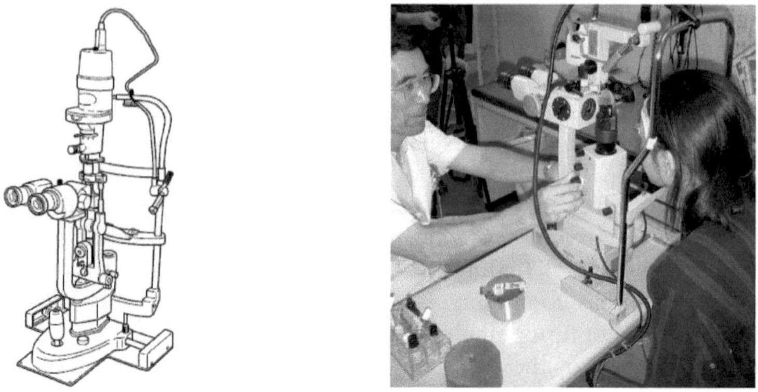

Fig. 1. Overview of slit-lamp microscope. Left: Overview of the slit lamp microscope (Model SM-70N; courtesy of Takagi Seiko Corp., Ltd.) [10]. Right: a slit lamp microscope in use. The eye specialist (left) is almost ready to start an examination of a patient's eye (right).

2.2 A Remote Clinical System for Ophthalmology

As described previously, a conventional and common approach to ophthalmological telemedicine is to capture high-quality still or video images of eyes at the patient's local site and send them to a remote site to ask for a specialists' diagnosis [9]. In this approach, however, the specialist does not use a slit lamp microscope. Consequently, it is difficult to conduct an expert diagnosis and an opportunity to find serious eye problems might be missed. We convert a slit lamp microscope into a remote control robot to take care of this issue. With the robotized slit lamp microscope, an eye specialist can control it from a geographically remote site and directly conduct a diagnosis.

Figure 2 illustrates an overview of our remote clinical system for ophthalmology. The system consists of an input and output (I/O) device at the specialist's site and a clinical robot at the patient's site. We assume that the specialist's site is a central hospital of a region such as a university hospital; the patient's site is a clinic in a remote area. We also assume that, in the remote area clinic, a doctor who is not an eye

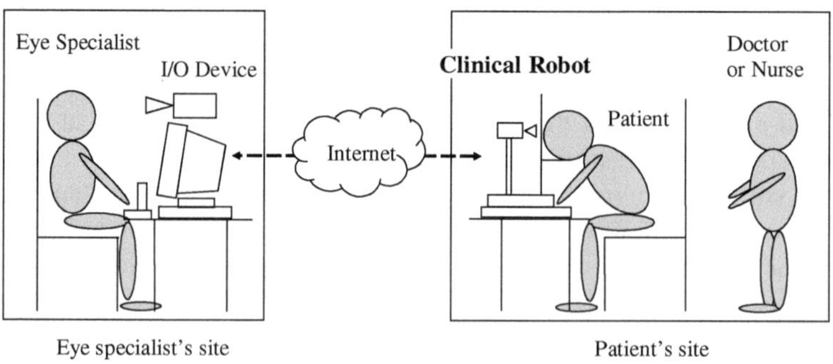

Fig. 2. Overview of the remote clinical system for ophthalmology

specialist, or perhaps a nurse, would provide support for the specialist's robot operation and for the patient.

The eye specialist operates the I/O device and inputs the parameters of the microscope position and slit light. The I/O device transmits the parameters to the robot and the robot changes its appropriate function levels. The robot has a video camera and captures still and moving images; then it transmits them to the I/O device at the specialist's site through the Internet. The eye specialist examines the received images using the I/O device, then conducts a diagnosis.

The following sections explain the prototyping technique enforced for demand acquisition from the ophthalmologist to an interface, and describes the design process and product of an input/output interface.

3 Using Paper Prototyping Techniques

Paper prototyping is a usability testing technique that is conducted using paper-based prototypes [8].

3.1 Basics of Paper Prototyping

At paper prototyping sessions, a user performs tasks assigned by the coordinator using the interactive system represented as a paper prototype. The prototype is paper-based and hand-sketching is usually encouraged. Therefore, a paper prototype can be modified easily if its usability problems are found through the session. Consequently, an iterative process of design and evaluation can be achieved using paper prototyping. Table 1 summarizes the general role assignments of paper prototyping sessions.

Table 1. Role assignment of paper prototyping sessions

Role	*Side*	*Description*
Coordinator	Experimenter	Plans and runs the paper prototyping sessions.
Computer	Experimenter	Manages and operates the paper prototype as the CPU of the computer which has the paper prototype interface. Receives inputs from the user and gives feedback based on pre-defined procedures.
Observers	Experimenter	Observe and record the sessions.
User	Participant	Performs assigned tasks using the computer, which is represented as the paper prototype.

Some exemplary advantages of using paper prototyping are

- much user feedback is obtainable at the early stage of design process;
- a prompt iterative design can be promoted and many design alternatives can be tested; and
- it can encourage participatory design and might improve designers' creativity.

However, potential challenges exist on paper prototyping:

- it is difficult to express the dynamic content of a GUI component, such as a video image, using a paper-based prototype;

- it is difficult to express SUIs such as hardware push buttons and sliders using a paper-based prototype; and
- participants might have negative impressions about low-fidelity prototypes such as paper-based prototypes when running a test session (i.e., it might lack face validity for participants) [5].

In this project, to encourage ophthalmologists' participation in design, we decided to employ paper prototyping as a fundamental tool for design. Before starting the design of the ophthalmologist's interface of the remote clinical robot, we examined paper prototyping itself.

3.2 Pretest for Improving Paper Prototyping

As a pretest, we conducted a paper prototyping session of an automated teller machine (ATM). The ATM we paper-prototyped consists of GUI components and SUI components. The GUI components include a touch screen and several windows and dialogs as visual feedback. The SUI components include a cash insertion slot and hardware push buttons.

Three undergraduate students, three graduate students, and a college professor participated in this pretest. The purpose of the pretest is to conduct a formative evaluation of paper prototyping. Consequently, our intent was to examine how to improve paper prototyping to deal with GUI and SUI design. After paper-prototyping of an ATM interface, participants engaged in a brainstorming session to develop assumptions about paper prototyping for a GUI and SUI combined interface.

Through the pretest, we learned the following lessons.

- Dynamic components can be expressed and processed if we have more than two computers (i.e., more people who have computer roles) at a paper prototyping session.
- Thinking aloud method conducted by test users facilitates linkage of the GUI components and SUI components at a paper-prototyping session.
- Training sessions are important to lower the test users' resistance to low-fidelity prototypes and the thinking aloud method.

These lessons enable us to choose paper prototyping techniques for designing interactions of a remote clinical robot for ophthalmology.

4 Interaction Design of a Remote Clinical Robot for Ophthalmology

We carried out usability testing with a paper prototype in designing the I/O interface for a remote clinical robot for support of ophthalmology. We needed a draft of the I/O interface to conduct paper prototyping. Therefore, we first conducted interviews with eye specialists at a university hospital to elicit basic requirements for the I/O interface for eye specialists to design draft I/O interfaces.

4.1 Requirements of Basic Functions

Table 2 shows the basic requirements obtained from the first interview session. Three kinds of examination are required for telemedicine in ophthalmology with a slit lamp microscope: examination with the slit light control, examination of a foreign body in the eye, and examination of the ocular fundus.

Table 2. Basic required examinations by eye specialists

Required Function	*Description*
Examination with slit light control	This is a fundamental examination with a slit lamp microscope. The eye specialist scans the patient's corneal surface using the slit light and examines its condition.
Examination of a foreign body in the eye	Using a blue light filter, the eye specialist examines the patient's fluorescent-dyed eye to find a foreign body or scratch on the eyeball.
Examination of the ocular fundus	With an optional fronting lens, the eye specialist examines the patient's ocular fundus including the retina, blood vessels, and optic nerve.

To implement the examinations listed in Table 2, the eye specialist needs the following controls: the remote slit-lamp microscope robot requires these functions.

- Moving speed and magnification of the video camera on the robot
- Power source of the slit light
- The slit light width, length and intensity
- Switch to the fronting lens or the blue filter
- Rotation of the fronting lens

At the second interview, we brought three commercial joysticks to elicit concrete ideas about input devices from eye specialists. Participants preferred to use the IP Desktop USB Joystick (HFX Series I, Model 1400 Option 5; CH Products) [2] because it is most similar to that of the slit lamp microscope. Participants said that the other joysticks we brought (one similar to SideWinder by Microsoft Corp. and one similar to Playstation 2 Dualshock 2 Controller by Sony Corp.) were "not business-like."

In addition, they mentioned that the following points.

- The video image of the eye should be as large as possible on the video display terminal (VDT) screen.
- An overview of the remote site should be on the VDT screen; but it can be small.
- The information on the camera position should be visualized on the VDT screen.

Using that information, we started designing the I/O interface of a remote clinical robot for ophthalmology and doing iterative design and evaluation through paper prototyping.

4.2 Conducting Paper Prototyping

Initial Design Proposal. Figure 3 shows the basic design of the I/O interface (a paper-based version). We decided to use a dual display for the interface because of the requirements of the maximum eye images on the VDT. For this design, the left screen (main display) displays the video image transmitted from the slit lamp microscope robot on the remote site. The right screen (sub-display) displays information other than the video image of the eye, such as the camera position and various parameters.

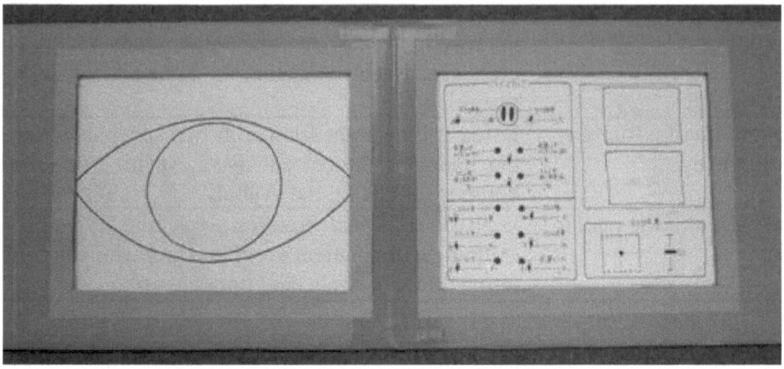

Fig. 3. Dual display design (paper prototype). Left: main display. Right: sub-display.

For the first paper prototyping trial, we prepared three design proposals: (1) an SUI-only design, (2) a GUI-only design, and (3) an SUI and GUI combination design. We offered the design alternatives to provide opportunities to consider a wide design space by the eye specialists. Although they already mentioned the joystick use at the second interview session, we thought that they might not know other technologies of SUIs and GUIs.

The following shows the sub-display of the three designs.

SUI-only Design. On this design, the slit lamp microscope robot is controlled solely using the joystick. Every parameter is assigned to the button on the joystick and the camera position is adjusted using the joystick. Figure 4 shows the sub-display of the SUI-only design. The sub-display is divided into three panels. The left side displays the parameter setting of the joystick. The top right panel shows two overview windows of the remote site. The bottom right panel depicts the camera position.

GUI-only design. Figure 5 shows the sub-display of the GUI-only design. In this case, the sub-display is implemented as a touch screen. Each parameter is assigned to a slider, button, or radio button on the touch screen so that the eye specialist controls the parameters directly on screen with the finger or a stylus. The sub-display has three panels. The top left panel shows two overview windows of the remote site. The right side panel is for the parameter setting. The bottom left panel shows the camera position control area.

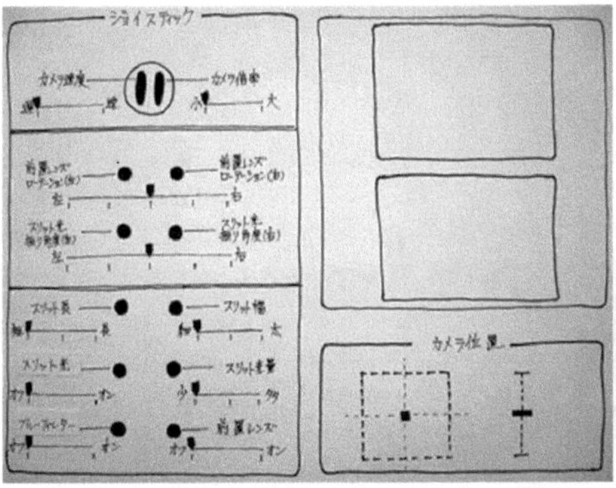

Fig. 4. Sub-display of the SUI-only design

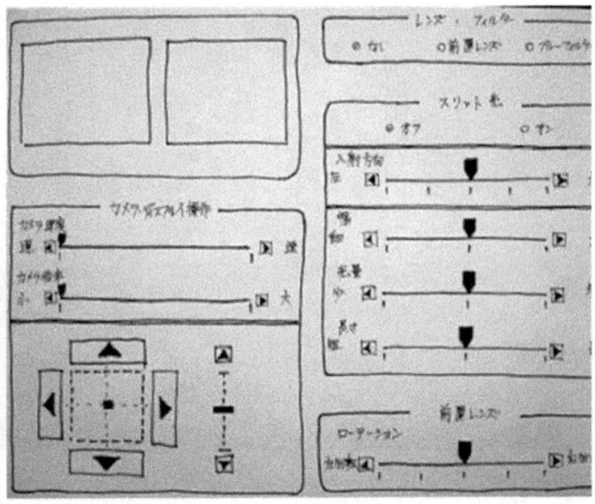

Fig. 5. Sub-display of the GUI-only design

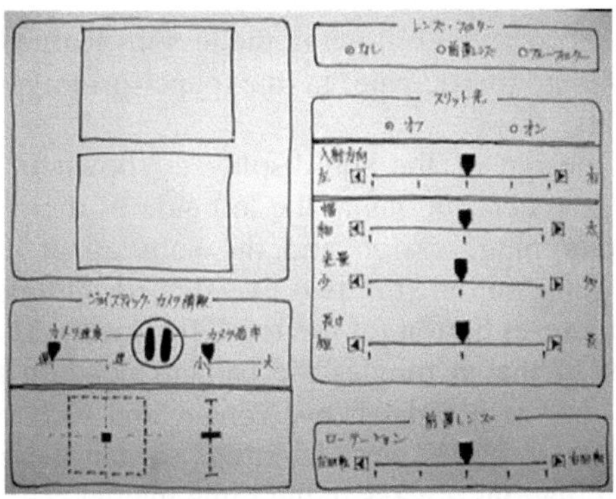

Fig. 6. Sub display of the SUI and GUI combination design

SUI and GUI combination design. Figure 6 shows the sub-display of the GUI-only design. In this case, the sub-display is implemented as a touch screen. Each parameter except for the camera position control is assigned to a slider or button so that the eye specialist controls the parameters directly on screen with the finger or stylus. The camera position is adjusted using the joystick. The basic arrangement of the panels on the sub-display is identical to that of the GUI-only design. However, because it does not adjust the camera position on the screen, the bottom left panel provides the visual feedback of the camera position.

Using the three designs, we conducted a summary evaluation of the designs. We showed them to the eye specialists and asked which one is the best and why. They preferred to use the SUI and GUI combination design. Again, they preferred to use the joystick for the camera control, but they did not want to use the buttons on the joystick because they believed that the touch screen is much better in usability than the buttons on the joystick.

We conducted a paper prototyping session with the SUI and GUI combination design shown in Fig. 6. A scene from the session is shown in Fig. 7, where two computers function mutually, quickly, with simultaneous change of the display image as represented with the sketch. Prior to the session, we conducted a practice session

Fig. 7. Scene from a paper prototyping session. Two computers are working collaboratively.

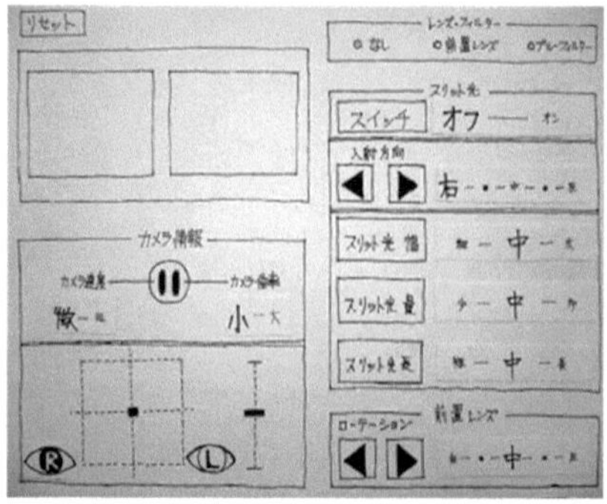

Fig. 8. A revised I/O interface on the sub-display

using an ATM example prototype following the lessons learned from the pretest. The eye specialist was able to participate in the paper-prototyping activities without trouble with face validity.

The final result of design of the sub-display is shown in Fig. 8. It resulted in placing a group of large touch buttons at the left side of the right panel. In addition, through the paper-prototyping session and the subsequent debriefing session, we created the idea of a jump button (the two eye-shaped buttons at the bottom of the bottom left panel) and a reset button (at the top left corner). The jump button sends a command to the robot so that it toggles the camera position from the pre-assigned right eye center to the pre-assigned left eye center, and vice versa. The reset button changes all robot parameters to the initial settings so that the eye specialist can start routine diagnostic examinations always in the same manner.

5 Conclusion

This paper presented discussion of the use of paper prototypes as a technique for clarifying the balance between GUI and SUI to design interactions on a remote clinical robot to be used for ophthalmology. Through the case study, we showed that paper prototyping remains a useful technique, even if the target is not a GUI.

The project on the remote clinical robot for ophthalmology is in the second stage. We are now conducting a formative evaluation of a working version of the robot produced through the design process and are ready to use it at the university hospital's emergency department. We will undertake iterative continuous design through its actual use.

References

1. Black, A., Buur, J.: Making Solid User Interfaces Work. Information Design Journal 8(2), 99–108 (1996)
2. CH Products – Industrial Joystick and Trackballs: (Last accessed: February 23, 2007) http://www.chproducts.com/oem/finger_ip_desktop.html

3. Go, K., Hirasawa, N., Kasai, H., Kurosu, M.: Supporting "Omoiyari": A Design Issue in Post-Ubiquitous Society. In: Proc. the SICE-ICASE International Joint Conference 2006 (2006)
4. Gould, J.D.: How to Design Usable Systems. In: Buxton, W., Grudin, J., Greenberg, S., Baecker, R.M. (eds.) Readings in Human-Computer Interaction: Toward the Year 2000, pp. 93–121. Morgan Kaufmann, San Francisco (2000)
5. Holzinger, A.: Rapid Prototyping for a Virtual Medical Campus Interface. IEEE Software 21(1), 92–99 (2004)
6. Koshizuka, N., Muto, T., Sakamura, K.: Human-Machine Interface Specifications in the Computer Everywhere Age. In: Proc. the 12th TRON Project International Symposium, pp. 87–96. IEEE Computer Society Press, Los Alamitos (1995)
7. Sakamura, K.: TRON Human Interface Guideline. Personal Media, Tokyo (1993)
8. Snyder, C.: Paper Prototyping: The Fast and Simple Techniques for Designing and Refining the User Interface. Morgan Kaufmann, San Francisco (2003)
9. The Medical Information System Development Center: Telemedicine Report—Revised in March, 1997 (Last accessed: January 31, 2007) http://square.umin.ac.jp/ẽnkaku/96/Enkaku-RepSoukatu-nof-eng.html
10. Takagi Seiko Ltd.: Slitlamp microscope, Model SM-70N (Last accessed: January 31, 2007) http://www.takagi-j.com/seihin_e/seihin_e.html
11. Weiser, M.: Computer for the 21st Century, Scientific American, 265, 3 (1991)

Development of Facial Expression Training System

Kyoko Ito[1], Hiroyuki Kurose[2], Ai Takami[2],and Shogo Nishida[2]

[1] Senri Expo Park 1-1, Suita, 5650826 Osaka, Japan
[2] 1-3, Machikaneyama-cho, Toyonaka, 5608531 Osaka, Japan
{ito,kurose,takami,nishida}@nishilab.sys.es.osaka-u.ac.jp

Abstract. The human's facial expression plays an important role as media that visually transmit feelings and the intention. To express the facial expression freely, the action of muscles of facial expression forms the expression is very important. "Facial expression training" is recently paid to attention, and is a training of the face to make a good facial expression. There is a training method by moving the expression muscle of the face. One of the purposes to train the facial expression is to achieve your own ideal facial expression. In this study, it aims at a support of the effective process for facial expression training to achieve the target expression using computer, as a self training system. And, an interface for users to select a target facial expression and a whole development of an effective expression training system is proposed, as a first step toward an effective facial expression training system.

Keywords: Facial expression, Facial expression training system, Muscles of facial expression, Nonverbal communication.

1 Introduction

In the human communication, nonverbal information such as the facial expression, the gesture, the voice plays an important role [1]. Especially, the facial expression is very important media that visually transmit feelings and the intention [2][3]. A study shows that more than 50 percent of the impression of communications is transmitted based on the visual information [4].

As for the expression, the person who expresses a facial expression cannot see the expression directly. It is important to understand your real expression and to image your target facial expression in order to express your own ideal facial expression. The muscles of facial expression play an important role for expressing the human facial expression.

Facial expression training is recently paid to attention to make your facial expression better [5][6][7][8]. Some exercises of the part of the face and the expression muscle are performed for facial expression training. The muscles of facial expression are strengthened by training, and the facial expression is softened, and your own ideal facial expression is expressed. Facial expression training can be utilized effectively by not only the communications scene in daily life but also the scene of the business skill and rehabilitation. Facial expression training has some

M.J. Smith, G. Salvendy (Eds.): Human Interface, Part I, HCII 2007, LNCS 4557, pp. 850–857, 2007.

forms. One is seminar style with a trainer. One of the others is self training using training books or the information on the Internet as a guide. Some seminars are very expensive, and time and space are restricted. On the other hand, when self training is performed, it is difficult to image your target facial expression clearly alone, and to compare the expression the ideal facial expression with your present one.

In this study, it aims at the proposal of an effective expression training system using the computer toward the achievement of your target facial expression. As the first step, an interface to select the target facial expression is proposed. And, the expression training system including the target expression selection interface is developed.

There is a facial expression training system with a virtual mirror as a previous study [9], from the viewpoint of a support system of facial expression training using a computer. Virtual mirror is a facial expression training system that displays your facial expression emphasizing the feature with a virtual mirror. This study is different from virtual mirror from the viewpoint of selecting the target facial expression of actual face.

2 Facial Expression Training System to Achieve Target Facial Expression

2.1 Support Process

In this study, the following processes are considered as a step of facial expression training to achieve the target facial expression.

(1) The target facial expression is imaged.
(2) The current facial expression is expressed, aiming at expressing the target facial expression.
(3) The current facial expression is compared with a target facial expression.
(4) The muscles of facial expression for improvement are trained.

In this study, above processes are supported with a computer. And, it aims at the development of the facial expression training system to achieve the target facial expression. The supporting method with a computer is considered for each process, as follows:

(1) Support to make target facial expression
(2) Support to recognize current expression
(3) Support to compare current facial expression with target facial expression
(4) Support to understand muscles of facial expression for which training is necessary

A main point of this study is (1) support to make target facial expression, and a user interface is proposed. A proposal of the user interface is described as follows.

2.2 Support to Make Target Facial Expression

In this study, it pays attention to the following viewpoints to make the target facial expression.

(a) Your own face is used.
(b) Your target facial expression can be made to correspond to the movement of real human's face, and made freely.

As for (a), it is important to make the target facial expression by using your real face, because human face is different from each one, and your facial expression corresponds to your features of face.

As to (b), it is necessary to make the target expression that can be actually expressed based on anatomic movement using your face. And, it is important to make freely satisfied target facial expression.

From the viewpoints, a user interface to select the target facial expression is considered. And, the following approaches of two stages are considered so that the user may select a target facial expression as satisfactory.

- First stage: Rough expression selection
- Second stage: Detailed expression selection

At the first stage, for the user not to worry about a detailed part, and to select the target expression intuitively, the interface shown in Figure 1 is proposed.

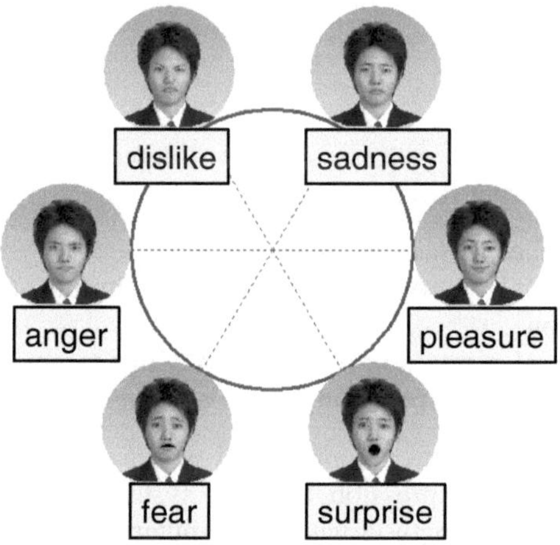

Fig. 1. Interface for selection of target facial expression

As for figure 1, six face images of facial expression made by your face are arranged outside of a big circle at the center [10]. These six images are corresponded to six basic facial expressions of please, surprise, dislike, anger, fear, and sadness. And, the mixture expression of two images and the expression corresponding to strength can be selected by selecting one point on the circle at the center.

In the second stage, a user interface for a user to control a detailed part so that the user is satisfied is considered. In this stage, action unit (AU) n is used from the previous researches on facial expression parts by Ekman et al [11]. The elements of each part corresponding to AU can be set in detail. The parts are eyebrows, eyes, cheeks, mouths, and mandibles. AUs of 3, 5, 2, 10 and 5 are used in the each part. And, 25 kinds of AU are used in total.

3 Design and Development of Facial Expression Training System

A facial expression training system including above mentioned processes and the interface for selection of target facial expression is developed.

First of all, the personal computer camera is used to take a current facial expression of the user.

Figure 2 shows the hardware setup. Visual C++6.0 is used as a software for the development. To fit your face with a wire frame model, FaceFit [12] is used. The facial expression training system has been developed [13][14], and it was named "iFace". The utilization of iFace has the following procedures.

(1) User registration
(2) Selection of target facial expression
(3) Expression of current facial expression
(4) Comparison of current facial expression and target facial expression

Fig. 2. Hardware setup

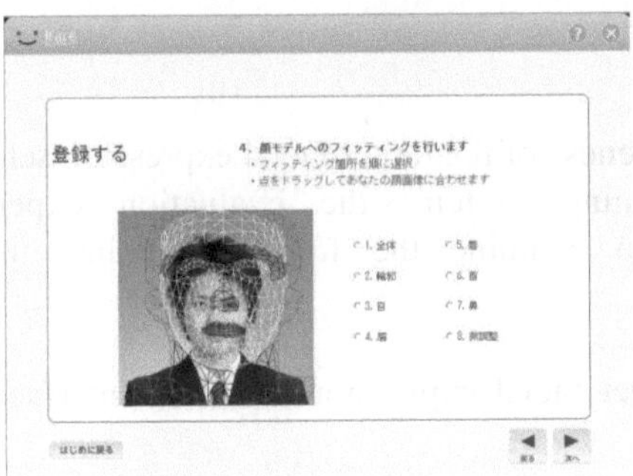

Fig. 3. A screen example of user registration for fitting

Fig. 4. For rough setting

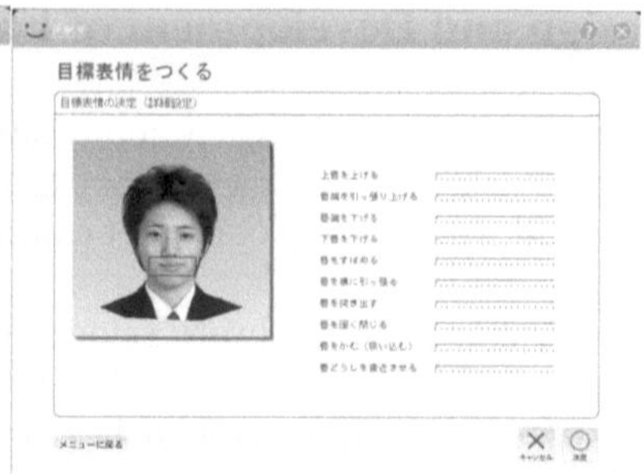

Fig. 5. For detailed setting

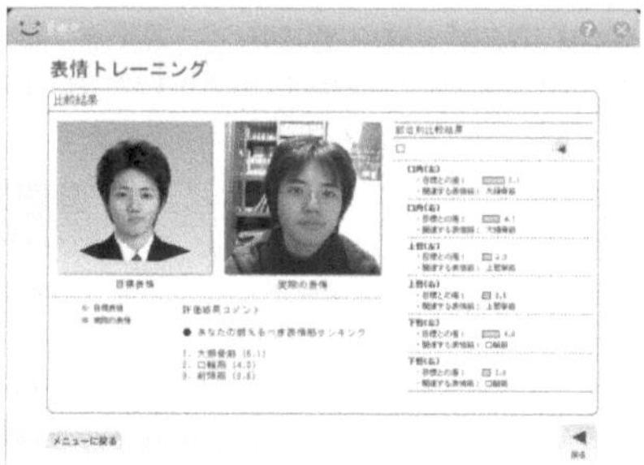

Fig. 6. A screen example of comparison between current and target

Figure 3 shows a screen of the user registration for the fitting. The screens for a rough setting and a detailed setting are shown in figure 4 and figure 5 respectively, for the target facial expression selection. Figure 6 shows the comparison result presentation screen between a current facial expression and target facial expression.

4 Evaluation Experiment of Facial Expression Training System

4.1 Purposes

To examine the effectiveness of the target facial expression selection interface and the facial expression training system, the evaluation experiment is conducted. Concretely, it aims to examine the following points through the evaluation experiment.

- Effectiveness in target facial expression selection interface for facial expression training system
- Possibility of facial expression training system

4.2 Methods

A. Procedure of experiment

(1) The target facial expression is selected by using the facial expression training system by a user.
(2) A user aims at the realization of the target facial expression that the user selected, and a current facial expression of the user is expressed.
(3) The current facial expression is compared with the target facial expression, and the results are presented.

One user makes two target facial expressions. One is the ideal smile of the user, and another the free choice of the user. A current facial expression is expressed three times by the one target. So, six expression of the current facial expression are tried in total.

B. Experiment participant

- 12 female (dentist)

C. Methods of analysis

The results of the questionnaire before and after the experiment are used. In addition, the log data during using the facial expression training system are used.

4.3 Results

- Effectiveness in target facial expression selection interface

Table 1. Results of questionnaire for effectiveness in target facial expression selection interface

#	Item of questionnaire	Avg.
1	Did you satisfactorily come to the target expression? (The first)	-0.3
2	Did you satisfactorily come to the target expression? (The second)	1.7
3	Was the facial expression selection interface intuitive and comprehensible?	2.2
4	Did you feel the synthesized facial expression natural?	-0.3

Table 1 shows the results of the questionnaire for effectiveness in target facial expression selection interface. The questionnaire is seven point scale of +3 to -3. As to the average score of the satisfaction rating, the first time (#1) and the second time (#2) are -0.3 and 1.7. It is shown that the satisfaction rating to the selection of the smile is lower than that of the free choice.

In addition, the time required to select the target facial expression is shown in Figure 7 separately for a rough setting and a detailed setting at the smile selection. In the comparison with a rough setting and a detailed setting, it is shown that there were a lot of people who spend time in a detailed setting.

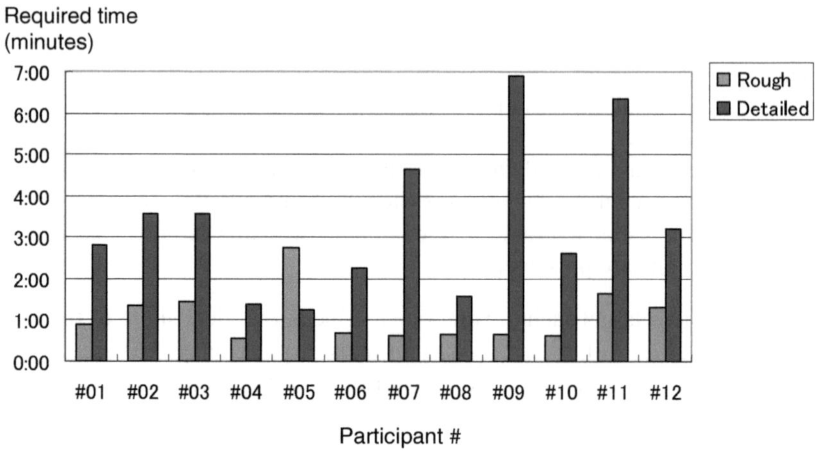

Fig. 7. Required time for rough and detailed selection of target facial expression by each user

- Possibility of facial expression training system

Table 2. Results of questionnaire for possiblity of facial expression training system

#	Item of questionnaire	Avg.
5	Was the training method of deciding the target facial expression concrete and comprehensible?	2.6
6	Was it easy to express your actual expression by comparing the current facial expression with the target facial expression?	2.5
7	Do you daily want to use the facial expression training system?	1.8

Table 2 shows results of the questionnaire for possibility of facial expression training system. The questionnaire is seven point scale of +3 to -3. It is shown that the facial expression training system by this study was positively evaluated.

The following comments were obtained from a free description.

- I think that they can use it to undergo rehabilitation of the face paralytic.
- My motivation for training of the facial expression goes up by making the target facial expression.

5 Conclusion

In this study, a target facial expression selection interface for facial expression training system, and a facial expression training system were proposed and developed. Twelve female dentists used the facial expression training system, and the evaluation and the opinion about the facial expression training system was obtained.

In the future, the improvement of the target facial expression selection interface and the comparison of a current and target facial expression will be tried, and it will aim at the development of an effective facial expression training system that can be actually used for various purposes.

References

[1] Kurokawa, T.: Nonverbal interface. Ohmsha, Ltd, Tokyo (1994) (In Japanese)

[2] Yoshikawa, S.: Facial expression as a media in body and computer, Kyoritsu Shuppan Co., Ltd, pp. 376–388 (2001) (In Japanese)

[3] Uchida, T.: Function of facial expression, Bungeisha, Co., Ltd (2006) (In Japanese)

[4] Mehrabian, A.: Silent messages, Implicit Communication of Emotions and Attitudes, 2nd edn. Wadsworth Pub. Co (1981)

[5] Inudou, F.: Facening Official Site (February 2007) (In Japanese) http://www.facening.com/

[6] Inudou, F.: Facening. Seishun Publishing Co., Ltd, Tokyo (1997) (In Japanese)

[7] COBS ONLINE Business good face by facial muscles training (February 2007) (In Japanese) http://cobs.jp/skillup/face/index.html

[8] Practice of Facial Expression (February 2007) (In Japanese) http://www.nikkeibp.co.jp/style/biz/associe/expression/

[9] Miwa, S., Katayori, H., Inokuchi, M. (eds.): Virtual mirror: Proposal of Facial expression training system by face image data processing. Proc. 43th conference of the institute of system, control and information engineers, pp. 343–344 (1999)

[10] Schlosberg, H.: The description of facial expression in terms of two dimensions. Journal of Experimental Psychology 44 (1952)

[11] Ekman, P., Friesen, W.V.: The Facial Action Coding System. Consulting Psychologists Press (1978)

[12] Galatea Project: (February 2007) (In Japanese) http://hil.t.u-tokyo.ac.jp/~galatea/index-jp.html

[13] Parke, F.I.: Techniques for facial animation, New Trends in Animationand Visualization, pp. 229–241 (1991)

[14] Waters, K.: A Muscle Model for Animating Three dimensional FacialExpression, Computer Graphics. SIGGRAPH'87 2(4), 17–24 (1987)

A Cognitive Approach to Enhancing Human-Robot Interaction for Service Robots

Yo Chan Kim, Wan Chul Yoon, Hyuk Tae Kwon, Young Sik Yoon, and Hyun Joong Kim

IS LAB, Dept. of Industrial Engineering,
Korea Advanced Institute of Science and Technology, Taejon, Korea (305-701)
{nkmissio,wcyoon,jehdeiah,nanhari,yellow145}@kaist.ac.kr

Abstract. As robots become more intelligent and their application fields continue to grow, the decisions and interaction of robots that share work domains with humans become increasingly important. Traditional robots have received only simple commands, and humans' roles have been limited to supervisor. However, for successive task performance, robots' decision making should be approached via collaboration between the human and robot. Interaction also should be regarded as an issue closely associated with joint work plans rather than a simple function. Interaction between the human and robot, moreover, should be systemized in order to decrease the workload of the human and maximize user satisfaction. Accordingly, we developed several cognitive models: a task model, truth-maintenance model, interaction model, and intention rule-base. These models can manage and initiate the interactions based on their tasks and modify robot's activities by using the result of the interaction. We demonstrated the adaptability and usability of the developed models by applying them to the home-service robot, T-Rot.

Keywords: Human-robot interaction, Interaction-based robot behavior, Cognitive model, Task-oriented interaction, Service robot.

1 Introduction

Traditional autonomous robots, such as industrial and professional robots, have performed their tasks by limiting the human's role to merely that of a supervisor [1]. These robots only emphasized robot's autonomy and system development in order to improve efficiency of their performance.

However, currently, studies of personal service robots are brisk with the highest expected growth rate [2]. Personal service robots, such as toy robots, tour-guide robots, and home-care robots, are robots which assist domestic chores or entertain people with recreational activities [3]. The robots in this field have more complex interaction, as well as a higher level of autonomy, since many of these robots are used by people who possess no special skills or training to operate a robot. Therefore, the demands of intimate interaction strategies and collaborative decision making issues increase in environments shared with humans. In this kind of a situation, issues of human-robot interaction have several characteristics.

M.J. Smith, G. Salvendy (Eds.): Human Interface, Part I, HCII 2007, LNCS 4557, pp. 858–867, 2007.

1.1 Characteristic of Human-Robot Interaction Design

Designing human-robot interaction (HRI) is different from design approach of the traditional Artificial intelligence (AI) and Human-computer interaction (HCI). AI models human intelligence directly or loosely, where as HRI involves modeling the linkage that connects human and robot intelligence [4]. The conventional robot's interaction is mostly environment-related and the robot executes a pre-programmed task. Interactions between robots and people merely consist of a command to execute a task, but in our approach to HRI, humans and robots are one of the Joint Cognitive Systems [5]. Both cognitive systems have to attempt to transform and modify tasks each other by mutual interaction.

Compared with AI, HCI and HRI are quite similar-both tries to enhance the quality of interaction and performance between systems based on the analysis of user and system. A computer, however, is not autonomous and is less interactive, where as a robot has more autonomy and behaves dynamically while deeply interacting with the physical environment as well as the human [6]. Fig. 1 explains the differences between ordinary HCI and HRI with an autonomous robot. A computer worked by direct manipulation does not involve parallel activities, but an autonomous robot can monitor, collaborate, give suggestions, and receive directions. Research on HRI should be continued with these facts in mind.

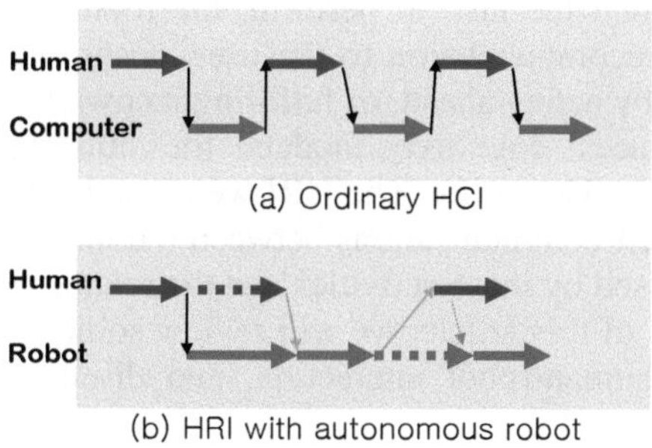

Fig. 1. Task flows of interactions between the human and a computer (a) or a robot (b). Black arrows between two agents mean explicit commands and gray arrows mean implicit interactions.

1.2 Task-Oriented Human-Robot Interaction

Robots sharing same work domain with people, such as service robots, cannot all meet the user's demands perfectly with their own autonomous decision making; therefore, collaboration with humans is essential. Robots perform tasks in conformity with the user's indication, status, or situations. It is very efficient for robots to receive necessary information about task effectuation from the user [6]. Interaction needs to be based on the task since the user's assistance is one of important forms of interaction and should be applied to enhance the performance of the robot. In interaction between the human and robot, the user does not determine the robot's behavior itself, instead there is various information exchange which ranges from abstract to concrete. Therefore, it is critical to make the robot utilize this information to perform the task and affect the user's behavior.

As described above, interaction has to be utilized actively, but, frequent and direct communication increases the human's workload, which can prove frustrating. In related literature, by developing a mixed-initiative system called Lookout, Horvitz reveals that the appropriate timing of services dramatically improves the users' satisfaction[7]. Besides, when interaction occurs untimely, efficiency of problem-solving cannot be maximized. In order to optimize the cognitive systems jointly, it is necessary to elaborate on the interaction's strategies on performing tasks. To build the strategies incrementally, the tasks and related models must be analyzed.

1.3 Interaction Driven Robot's Task Generation

It is crucial to decide when and what kind of tasks a robot will start. In case of a multi-functional robot, this issue becomes even more serious. The robot's ability to decide upon its own task is not only the problem of situation awareness, but also the outset of communications between the robot and user. There are two types of human's behaviors which the robot is responsible for: conversations and movements. In conversations, there are some explicit commands that the robot can follow, but most of conversation is filled with implicit sentences too vague for robot to apply. The user's movements also have ambiguities for the generation of the robot's tasks. Therefore, to decide upon the task to perform, the robot has to be able to initiate interaction or make a recommendation to the user, occasionally inducing the user's intervenient correction by acting ahead, or fulfilling its own task autonomously.

This article introduces cognitive models for human-robot interaction and implemented them to a service robot. The HRI we are consider in this paper involves both implicit and explicit communications between the human and robot and focuses on the robot's tasks caused by these activities and the related process of interaction. In the subsequent sections of this article, we will review some relevant research, present cognitive models for human-robot interaction, and discuss the implementation of these models and future works.

2 Related Work

Recently, various researches on human-robot interaction have been conducted. Attempts to employ interaction with humans as a strategic means of task performance are few of them.

Fong et al. developed a PDA-based teleoperation system called collaborative control[6]. Concerning architecture for interaction, this robot uses each user's profile to configure interaction, adapt the dialogue, and modify how the robot acts, and it manages questions generated from multiple robot modules by QueryManager.

Based on joint intention theory, Alami et al. made a framework and developed high-level symbolic planning for human-robot interaction[8]. The methodology was applied and tested in a tour-guide robot.

Breazeal et al. developed a humanoid robot, called Leo, which has social communication abilities, such as gestures and facial expressions[9]. Leo also is based on joint intention theory as a theoretical framework. This robot represents a goal-oriented and hierarchical task of belief structure, and is able to learn tasks by interaction and collaborate with the human through a turn-taking method.

Hüttenrauch et al. tried a user-centered robot design on a fetch-and-carry robot called Cero[10]. In this research, they created an architecture that controls the robot's activities and several interfaces (gestures, speech recognition, speech synthesis, GUI). Moreover, based on a simulation study involving task analysis and wizard-of-oz framework, they composed the robot's activities and interaction strategies.

Miura et al. developed a robot which can be taught information needed to perform tasks by a user[11]. To do so, they created a task model having a hierarchical task structure, involving several pieces of relevant information and describing the dependency of knowledge. In addition, this robot uses a fill-in process that fills in insufficient information of a task model with interaction.

Despite of the researches discussed above, studies are still in a primary stage that connects the task performance of service robot with interaction jointly and enhances the efficiency of both. Works that strategically utilize interaction itself and optimally apply it to the task are needed.

3 Models for Human-Robot Interaction

Cognitive modeling has a merit of modularizing knowledge, so that it is possible to develop the models cumulatively whenever the province and functions of interaction increases. Basically, each model should deal with related knowledge for its own purpose correctly, but a cognitive model of interaction is also needed to follow structural standards of human's cognition and assume the appropriate form which can be implemented to the robot because cognitive models have to produce interaction like human-human interaction and aid in suitable robot operation. Dynamic and causal relations among models should be linked so that the models are able to handle interaction efficiently by these coherences. We developed a task model, truth maintenance system, interaction model, and intention rule base in order to satisfy these needs.

3.1 Task Model

A task model is the model which offers task information to the truth maintenance system and interaction model and changes the robot's activity using the results of interaction for the purpose of managing task-based interaction. This model taking shape like scripts or event schemas represents activities in tasks which the robot actually performs. This script is distinct from path-finding model or simple procedure description. Interaction between the human and robot has a variety of contents which range from overall needs (ex. "I need something to eat") to specific instances (ex. "Turn on the TV in the room A"). Therefore, the task model assumes the form of a hierarchy and each task between the higher level and lower level has a means-ends relation. When a higher task decomposes into lower tasks, if information must be resolved, the higher node branches two or more candidates with probabilities of ambiguity. Lower ambiguity indicates a more ambiguous sub-task to be performed, and there are two types of candidates which form branches: a verb and object. Verb means what kind of activity the robot performs, while object implies what the robot performs the verb activity with. The candidate creates different tasks in different situations, and the probabilities applied in the candidate become bases in which the robot is capable of making a query. This task model can be applied to the support task manager of the robot as a descriptive model; otherwise, it can generate and modify the robot's tasks as a task planner and manger itself.

3.2 Truth Maintenance System

For resolving ambiguities identified from the task model, a truth maintenance system employs the following strategies. It updates the probabilities of the ambiguities by using information about the user's preferences, some assumption of the user's intention, and situational information. Then, the robot decides whether to make a query or not with updated results. These updated probabilities shape various distribution for each task and situation. If there is no outstanding candidate over the threshold, the system raises questions and answers through the interaction model in order to resolve the ambiguities. After resolution, the final results make the task model modify its own task.

However, probabilities sometimes have biased distributions. If the highest probability exceeds the threshold, direct questions and answers are avoided. In this situation, the user interacts with the robot by monitoring the robot's behavior. In the case that the user approves of the robot's activities, the user takes no action, but in the opposite situation, the user can demand modification of the robot's action. Due to such intervention of the user, the robot changes the assumptions and the probabilities of the related tasks. With the changed result, the robot executes an alternative action.

3.3 Interaction Model

The interaction model manages questions asked to the user and answers obtained from the user. When the truth maintenance system judges the robot should ask to the user, the interaction model generates adequate sentences of question about the candidates. There are two or more candidates for each ambiguity, and the number of candidates and the distribution of each candidate are diverse. Since it is possible that so many candidates exist, it is not recommended to ask about every candidate. According to these distributions, the model is able to formulate yes-no questions (ex. "Do you want -?") or wh- questions (ex. "Which do you want? -, - or -?").

After generating appropriate questions, the robot has to decide when to ask them. The interaction model determines the time point of interaction based on task model and the user's status, such as feelings and busyness.

When the user replies to the robot's question, the interaction model matches this information to the robot's question and then returns the results to the truth maintenance system in order to resolve ambiguities.

3.4 Intention Rule-Base

Although robot's task can be performed by the user's concrete command, the user does not usually transfer all information to the robot. When initiating a task by according to a user's statement, a robot is required to get the user's intention. The user's statement also has ambiguity, so the intention rule base is constructed in order to associate each statement with related intention and to assess the probability of the association. The intention rule base determines the robot's mission by using the truth maintenance system and interaction model.

4 Application to a Robot

We demonstrated the adaptability and usability of the developed models by applying them to a robot, called T-Rot. T-Rot is made by the Korean Ministry of Commerce, Industry and Energy's Intelligent Robot taskforce. It is a home service robot created to assist the elderly and disabled.

T-rot can recognize the voices of users and generate its own voice in order to hold conversations. In addition, it plans and manages various tasks based on the JAM agent [12] so as to fulfill a user's commands. It has two cameras that enable it to recognize people and objects and navigate to target locations. Furthermore, its highly developed sense of touch makes it possible to conduct simple manipulation, such as lifting an object up.

4.1 Target Scenario

There are three scenarios in order to reveal robots which are applied to the cognitive model for interaction.

Respond to a Call: a user calls out, "T-rot!" Then, the robot speaks "Master, what can I do for you?" and is ready for the next order.

Fetch and Carry Beverage: The user says he is thirsty. T-rot suggests something to drink. And With some dialogue, it brings a beverage to her.

Inform to User What There Is with Its Location: a user asks, "What is on the shelf?" T-rot moves to the shelf in the kitchen and comes back to the user's side. Finally, it informs her which objects placed on the shelf.

4.2 Implementation of the Model

We implemented a simple-shaped interaction/task script, disambiguation of TMS, questions and answer management, and intention rule-base. Fig. 2 describes a configuration of the system related to our project.

The user gives the robot multi-modal input, such as voice and touch screen commands. Information that passes through each kind of sensor and interpreter is represented a symbol by the Multi-ModalInterpreter (MMI) and is transferred to the TaskManager. The TaskManager · forwards received information to the CognitiveInteractionModule (CIM).

CIM is made of the state machine in order to operate the IntentionRuleBase, InteractionModel, and TruthMaintenanceSystem (TMS) for each situation. CIM extracts possible user intentions in the IntentionRuleBase. TMS updates the probabilities of these possible intentions by analyzing the user's profile and determines whether or not a query is needed by calculating the entropy value. In the case that a query is needed, the InteractionModel make an interrogative sentence consisting of no more than three candidates according to probability and formulates the question. The user's response is

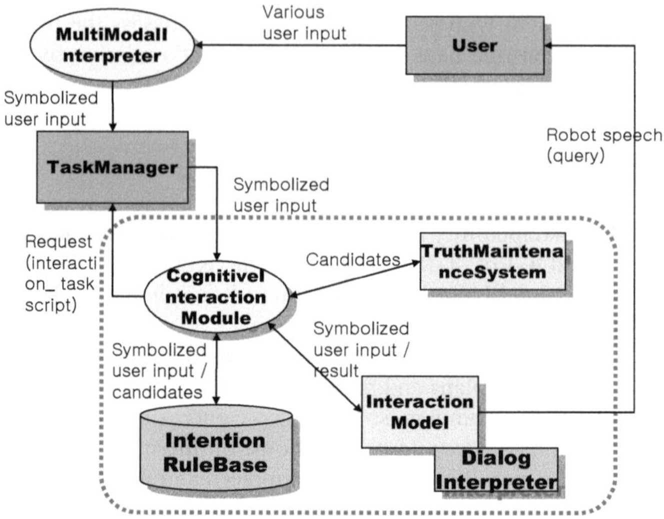

Fig. 2. A system configuration related to the CognitiveInteractionModule

inputted through MMI and is then applied in the DialogInterpreter of the InteractionModel. By the updated probability, as a result of the response, TMS determines the interaction_task script. After that, CIM checks if there is additional ambiguous information to perform the task. Particularly in the fetch-and-carry beverage scenario, ambiguity exists in the target objects which will be transferred to the user. In order to resolve the ambiguity, the robot asks additional questions. The script finally chosen is sent to and is implemented in the TaskManager.

Progress of the robot's activity with its interaction for each scenario is illustrated of discourse chart below.

Table 1. A discourse chart about scenarios - discourses (1), (2), (3), (4) can be changed by situation and interaction

Respond to call	Fetch-and-carry beverage	Inform what there-are
User: "T-rot!" T-rot: (turns toward the user) T-rot: "Master, what can I do for you?"	User: (1)"I'm thirsty" T-rot: "Do you want something to drink?" User: (2)"Yes, I do" T-rot: (3)" do you want a glass of cola?" User: (4)"Something else?" T-rot: "There are orange juice and lemonade." User: "give to me orange juice." T-rot: "OK I'll bring to you orange juice." T-rot: (move to a refrigerator) T-rot: (return) T-rot: (transfer) T-rot: "Here you are."	User: "what is on the shelf?" T-rot: "I will go and see." T-rot: (move to kitchen) T-rot: (perceives objects and their arrangement) T-rot: (return) T-rot: "There are cookies beside coffee and orange juice on the shelf in the kitchen."

In order to assist in understanding variations of activities and conversations, we employed a diagram of Fetch-and carry beverage scenario (Fig. 3). The shaded circle and rounded boxes are the user's statements. The user can initiate the task or interaction by one of the shaded circles. The robot decides upon its course of action, which are depicted with white rectangles and speaks some sentences which are depicted with white rounded boxes. When finishing a set of questions and queries, the robot executes the actions depicted with a dotted lining oval.

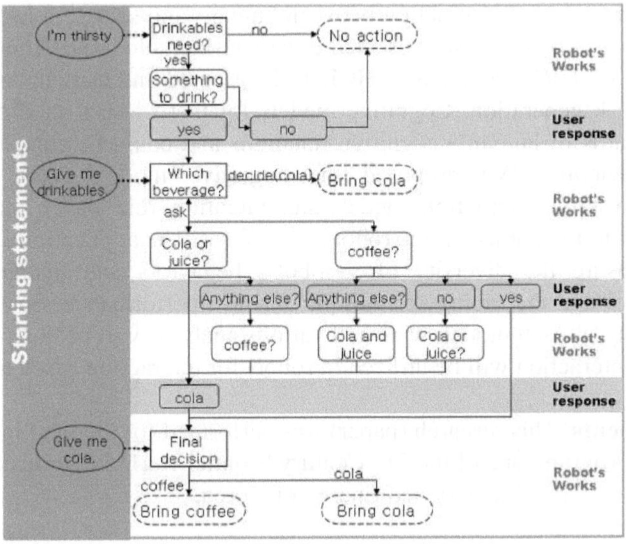

Fig. 3. A diagram for the fetch-and-carry beverage scenario, it shows diverse conversation

5 Future Work

Since developing a robot's intelligence accumulatively, the task model should obtain an adequate amount of procedural knowledge of the user. Several task analyses in various scenarios are necessary to obtain the procedural knowledge. Inference and the learning capability of the knowledge also are needed to create a more flexible task model.

Task planning and interaction with people are interdependent of task execution. Particularly, user intervention is the process that induces a change of plan during task execution and reflects its change to task execution. Therefore, the models for interaction should not be applied in only when the robot plans the task. Every performing process of a task can be a target of interaction, so we must strive to create a system architecture that always allows user's intervention. With the exception of the intervention problem, the robot's architecture should be designed to basically support the interaction of various modalities from humans.

The intention rule base premises an analysis about the user's speech and behaviors. By analysis of the human's task, it is necessary to study the associate primitives of speech and behavior with regard to the user's intention. As a result of that research, the system which can monitor user's action will be constructed. Based on the monitoring system, the robot can understand user's intention and decide things to do on its own initiative.

6 Conclusion

We aim to enable natural interaction between humans and service robots and optimal task performance. To do so, we presented three considerable issues: the differences in characteristics of HRI and AI and HCI, task oriented interaction, and interaction driven robot task generation. Cognitive models for these issues should be complied with the standards of human's cognitive functions and could be applied to the robot the most efficiently. We proposed the cognitive models (task model, truth maintenance system, interaction model, and intention rule base) and applied the partial function to the home-service robot called T-rot. The application shows that it is possible to ensure the diversity of a robot's behaviors and interactions by the cognitive model and that doing so is useful for task performance. For the flexibility and extensibility of each model, wide and various analyses will be done hereafter, and the model for interaction will be utilized in robots for interaction based architecture.

Acknowledgments. This research (paper) was performed for the Intelligent Robotics Development Program, one of the 21st Century Frontier R&D Programs funded by the Ministry of Commerce, Industry and Energy of Korea.

References

1. Breazeal, C.: Social Interactions in HRI: The Robot View. In: IEEE Trans. on Systems, Man, and Cybernetics, Part C, vol. 32(2), pp. 181–186. A publication of the IEEE SMC Society, Los Alamitos (2004)
2. U.N. and I.F.R.R.: United Nations and The International Federation of Robotics: World Robotics 2002. New York and Geneva: United Nations (2002)
3. Thrun, S.: Toward a Framework for Human-Robot Interaction. Human-Computer Interaction 19, 9–24 (2004)
4. Kang, W.L., Kim, H.R., Yoon, W.C., Yoon, Y.S., Kwon, D.S.: Designing a human-robot interaction framework for home service robot. ROMAN 2005. IEEE International Workshop, pp. 286–293 (2005)
5. Hollnagel, E., Woods, D.D.: Joint cognitive systems: Foundations of cognitive systems engineering, pp. 25–47. CRC Press / Taylor & Francis, Boca Raton, FL (2005)
6. Fong, T.: Collaborative control: A robot-centric model for vehicle teleoperation. Ph.D. dissertation, Robotics Inst. Carnegie Mellon Univ, Pittsburgh, PA (2001)
7. Horvitz, E.: Uncertainty, Action, and Interaction: In Pursuit of Mixed-Initiative Computing. IEEE Intelligent Systems 14(5), 17–20 (1999)
8. Alami, R., Clodic, V., Montreuil, E., Sisbot, A., Chatila, R.: Task planning for human-robot interaction. In: Joint sOc-EUSAI conference (October 2005)

9. Breazeal, C., Hoffman, G., Lockerd, A.: Teaching and Working with Robots as a Collaboration. In: Adaptive Agents and Multi-Agent Systems II. LNCS (LNAI), vol. 3394, pp. 1030–1037. Springer, Heidelberg (2005)
10. Huttenrauch, H., Green, A., Norman, M., Oestreicher, L., Eklundh, K.S.: Involving users in the design of a mobile office robot. IEEE Trans. on Systems, Man, and Cybernetics, A publication of the IEEE SMC Society, Part C, vol. 34(2), pp. 113–124 (2004)
11. Miura, J., Iwase, K., Shirai, Y.: Interactive Teaching of a Mobile Robot. In: Proc. IEEE Int. Conf. on Robotics and Automation, Barcelona, Spain, pp. 3389–3394 (2005)
12. Huber, M.J.: JAM Agents in a Nutshell, Version 0.60 + 0.80i

A Study on a Stereoscopic Display System Using a Rotary Disk Type Beam Shutter

Kwang-Hyung Lee and Tae-Jeong Jang

Human Interface Laboratory, Department of Electronic and Telecommunication Eng., Kangwon National University, Chuncheon, 200-701, Korea
{hunterrts,jangtj}@kangwon.ac.kr

Abstract. In this paper, A stereoscopic display system using a rotary disk type beam shutter and two beam projectors is proposed. It is a kind of active stereoscopic display and can be easily converted from a passive stereoscopic system using polarizing filters. If it is possible to synchronize the revolution speed of the beam shutter with the integer multiple of the scanning frequency of the beam projectors, we can obtain a comparable performance with our system to an active stereoscopic system using one expensive high performance beam projector. Further, if we rotate the beam shutter at sufficiently high revolution speed, our active stereoscopic system works regardless of synchronization and thus the system is much easier to implement.

Keywords: stereoscopic display, beam projectors, rotary disk type beam shutter, stereo vision, shutter glasses.

1 Introduction

We obtain various information by seeing and hearing lots of things through a number of media such as TV, video, movie, etc.. But indirect experience under a two-dimensional plane display environment is bound to give us less actual feeling than direct experience. To cope with such an existing two-dimensional plane display environment, studies on three-dimensional stereoscopic displays have been actively in progress and various methods have been proposed. But, while its usefulness and potentiality were limitless, its spread has been restrictive due to a technical limit, complexity of a device and high price.

It can be said that three-dimensional stereoscopic display technology applies to extensive fields such as broadcasting, medical treatment, education, CAD, a game, animation, virtual reality, multimedia, military equipment, etc. and is one of essential technology commonly required in various fields.

As one of the methods of providing stereopsis, this paper proposes a stereoscopic display system using a rotary disk type beam shutter which can additionally implement active-type stereoscopic display even under a passive-type stereoscopic display system environment using two general-purpose beam projectors and polarizing filters. Chapter 2 introduces stereopsis implementation device, and chapter 3 describes the stereoscopic display system using rotary disk type beam shutter, proposed in this paper. And Chapter 4 explains the experiment and result using the proposed device, and Chapter 5 concludes this paper.

M.J. Smith, G. Salvendy (Eds.): Human Interface, Part I, HCII 2007, LNCS 4557, pp. 868–875, 2007.

2 Introduction of Stereoscopic Display Systems

Usually, the stereoscopic feeling a man realizes occurs by comprehensive combination of change of the thickness of the eye lens according to location of the object to be observed, difference of an angle between the both eyes and the object, difference in location and a type of the object seen to the right and left eye, parallax and light and shade which occur motion of the object, and various effects by psychology and memory. Among them, the binocular disparity that occurs by two eyes being located about 6~7cm distant in the horizontal direction can be said to be the most important element of a stereoscopic feeling. That is, if both eyes see one object having some angular difference by binocular disparity, and, due to such difference, respective different two images come into each eye and transmitted to the brain, the brain can feel the original three-dimensional stereoscopic by exactly combining this two information [1-3].

As typical stereoscopic display types, there are a passive method using polarizing filters and an active method using shutter glasses [2]. A passive method (Fig. 1) is one by which a man feels a stereoscopic feeling by projecting a right and a left image on one screen at the same time using projectors having built-in polarizing filters respectively cross perpendicularly and by seeing respective right and left independent images through polarizing filter glasses in the same direction. Such a passive polarizing filter method has a merit that a stereoscopic image is easy to be able to be seen to many people at the same time. But this method has a limit that a stereoscopic feeling can drop in case a filter is used whose polarizing function is low, and a high-priced special screen such as a silver or pearl screen is needed to prevent polarizing interference. Also, there may appear a phenomenon that the brightness of a screen greatly reduces in the process of passing through of a polarizing filter two times, one the filter in front of the beam projector and the other the filter glasses.

An active method (Fig. 2) is one by which one beam projector projects a right and a left image by turns at a certain interval and if an emitter receives this synchronizing signal and retransmits it to a shutter glasses in the infrared-method, the shutter glasses, in response to this synchronizing signal, intercepts view of the user's each eye in turn so that the user may see a right and left image independently and thus

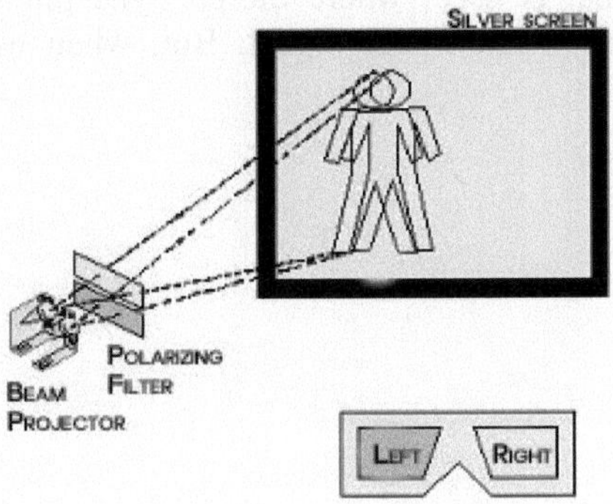

Fig. 1. A passive stereoscopic display system using polarizing filters

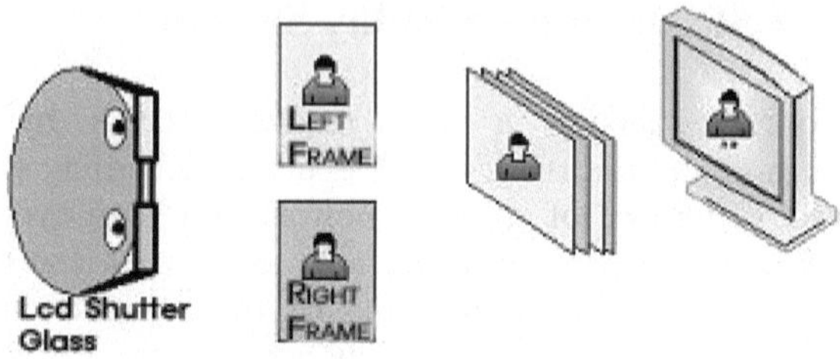

Fig. 2. An active stereoscopic display system using shutter glasses

sense a stereoscopic feeling. This method, compared with a passive method using a polarizing filter, requires graphic card and beam projector which can convert a screen two times as fast. Because vertical synchronization frequency of most commercial beam projectors is 40~90Hz, use of 20~45Hz can not settle the problem of screen flickering, so that it requires a specially manufactured high-priced high speed beam projector. Therefore it costs much money in purchasing the beam projector.

3 Stereoscopic Display System Using Rotary Disk Type Beam Shutter

This paper proposes a display system which implements stereopsis of high quality at the low price using both the merits of a passive and an active method. This system enables stereopsis even with an ordinary screen instead of a high-priced silver screen required only of passive stereoscopic display system and can implement a system at a relatively low price. Also, it completely intercepts a right and a left image, so it can provide virtual environment of actual feeling, giving better picture quality, and, by the nature of composition of this device, the brightness reduction of this system is small compared to that of the system using polarizing filters.

At present, there are many products simultaneously supporting a passive and an active method in a commercialized graphic card enabling stereoscopic display. When making an active method stereo graphic synchronizing signal from such a graphic card, 3PIN DIN terminal is used, where Stereo Sync pin signal repeats high/low according to a right and a left image signal. But, when using a passive method,

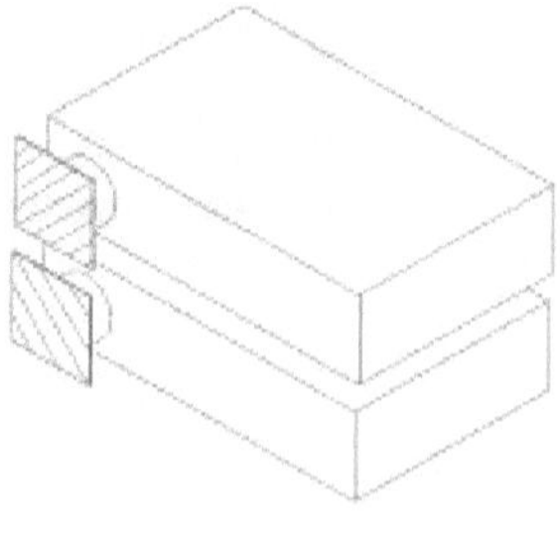

Fig. 3. Stereoscopic display system using a polarizing filter

respective right and left image signals are continuously made through two independent image-making terminals mounted on a graphic card, and the image signals is projected from two beam projectors having built-in two polarizing filters that cross each other perpendicularly, as showed in Fig. 3.

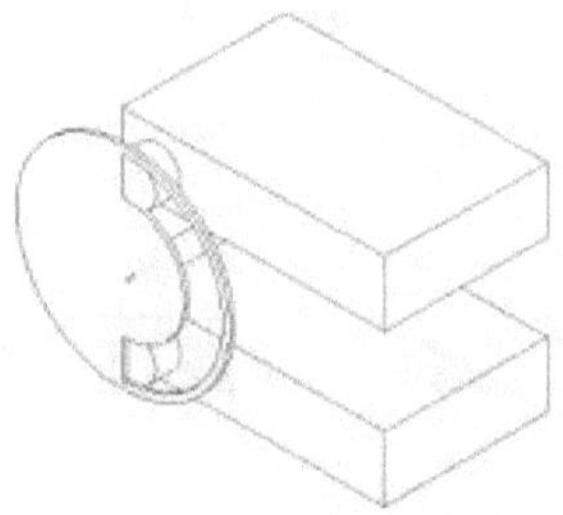

Fig. 4. Stereoscopic display system using a rotary disk type beam shutter

Supposing that, in a passive type stereoscopic system, a polarizing filter is removed and, as shown in Fig 4, rotary disk type beam shutter is installed. In the Fig. 5, we explained how our system works. For each figure, the first and the third line belt represent the frames of the left and the right beam projectors, and the second and the fourth line belt represent the status of the left and the right beam projectors, respectively. As shown in Fig. 5(a), (b), (c) or (d), a right and a left image are intercepted in turn and at the same time shutter glasses can block the right and the left eye in turn, an active-type stereoscopic system using two beam projectors can be configured. The solid line shown in Fig. 5 is a base point of a base for frame start of the each right and left beam projector, the dotted line indicates an emitter synchronization base point by a beam shutter. The white and the black colored part represent 'open' and 'closed' by shutter, respectively.

Fig. 5(a) shows a complete synchronization method by which, during one image cycle, in turn intercepts the half of a right and a left image and have it open. Fig. 5(b) shows complete high speed synchronization method by which, during one image cycle, a process of interception and opening in turn 2 or more integer multiple times as exactly is repeated at the high speed. Fig. 5(c) shows incomplete synchronization method by which every frame the ratio of interception and opening is maintained each at the rate of 50% through phase locking is not consistent. And Fig. 5(d) shows incomplete high speed synchronization method by which the ratio of interception and opening is maintained at the rate of 50% and repeated at the high speed through phase synchronization is not consistent.

In complete synchronization and a complete high speed synchronization, it is necessary to adjust the revolution speed of and the phase locking of a motor as well as an emitter synchronizing signal from VGA output signal on the basis of a vertical Sync signal. In the meantime, incomplete synchronization and incomplete high speed synchronization do not use VGA output vertical Sync signal and require only a system adjusting the synchronization of a beam shutter, an emitter and shutter glasses. The

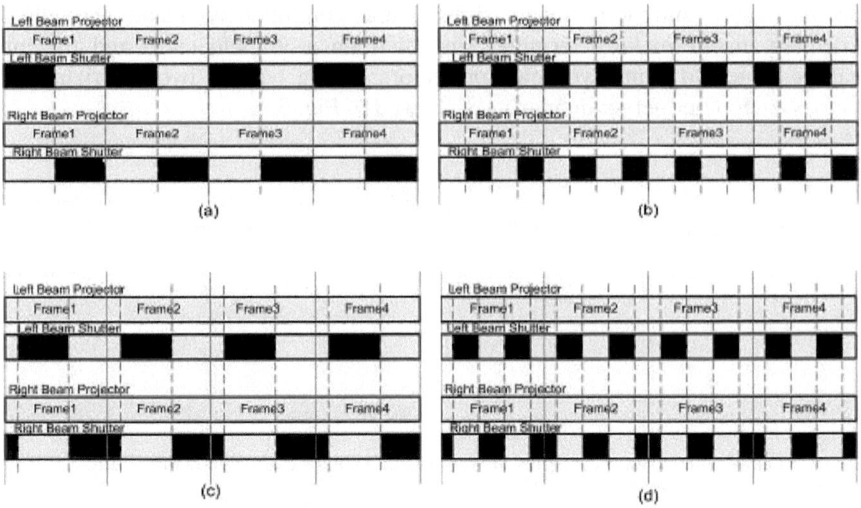

Fig. 5. Timing diagram of the proposed stereoscopic display system using a rotary disk type beam shutter. (a) Perfect sync (b) Perfect high-speed sync (c) Imperfect sync (d) Imperfect high-speed sync.

concept of incomplete synchronization and incomplete high speed synchronization is that if the system detects synchronizing signal by revolution of a beam shutter and transmits it to the emitter and the emitter re-transmits this signal to shutter glasses, the shutter glasses, in response to this synchronizing signal, intercepts in turn the right and left LCD as shown in Fig. 5(c) and 5(d), so that the user can feel a stereoscopic feeling.

The most important in an incomplete (high speed) synchronization system is the structure of the beam shutter for intercepting or opening a beam projector and a method for making an exact emitter synchronizing signal when a beam projector is intercepted or opened by a beam shutter. Fig. 6 shows an example the structure of the disk for several beam shutters and an example of a way an optical sensor is attached which can be installed in such a disk. Like this, it is possible to drive an emitter and shutter glasses with the pulse occurring according to an optic sensor turning ON/OFF by disk revolution as a synchronizing signal. Because with the use of a method like Fig. 6(a), a projection image of a beam projector can be intercepted, a method can be considered by which the location of sensing is changed or index hall is made around the center as shown in Fig. 6(b). In that time, phase locking of a beam shutter is adjusted according to location of an optic sensor for sensing, and if the location of this sensor is unsuitable, a stereoscopic feeling becomes low, so exact adjustment to location of an optic sensor is needed.

In a passive type system mounted with a polarizing filter of Fig 3, about 60% of the light is intercepted in the polarizing filter installed in the upper and lower end and only 40% passes through it. Also, because only the light of the same rate passes in polarizing glasses used by the user, it can be said 16% light as a whole can be

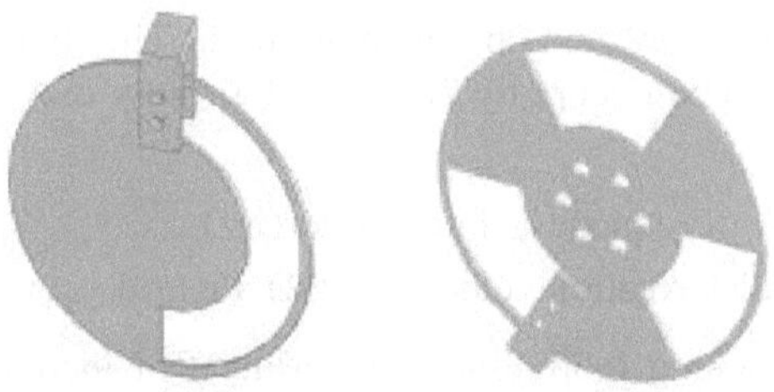

(a) (b)

Fig. 6. Rotary disk type beam shutters. (a) 1 Period per round. (b) 3 Periods per round.

transmitted to the user. In case of an active method system using rotary disk type beam shutter which was made by a passive method being remodeled, as shown in Fig. 4, 50% of the light can be said to pass by a rotary disk, and supposing about 40% of the light passes through shutter glasses, about 20% of the light is transmitted to the user as a whole, so that it can be said the transmission rate of light is a little higher than a passive method system. But, differently from a polarizing filter, in case of shutter glasses, a product whose light transmission is high is likely to be produced, so the whole light transmission rate of shutter glasses is expected to increase. And, differently from a passive type system, the proposed active type stereoscopic display system has merits that ghost phenomenon by incomplete polarization does not occur, can present stereopsis even if an ordinary screen is used instead of a high-priced silver screen, and can implement an active type stereoscopic display system at a relatively low price.

Incomplete high speed synchronization is possible if an emitter synchronizing signal is given to a disk fast rotating using an element such as an optical sensor, etc. The faster the rotation of a motor, the less the necessity of a complete synchronization in which phase locking must be adjusted reduces and the higher the refresh ratio of a screen. In conclusion, even in incomplete high speed synchronization, the user scarcely feels flickering. To have an effect of a beam projector having high scanning frequency in a beam projector having low scanning frequency, the rotation of a motor must be fast or a great number of beam shutter wings have as shown in Fig 6(b). However, the rotation of a motor is too fast, a problem of noise may occur, and if the number of a shutter wing increases, the ratio of the section of transient state of not being completely closed or opened increases, so that a problem of quality deterioration in stereoscopic may occur. Accordingly, the rotation speed of a motor and the number of a beam shutter must be adjusted within a suitable range.

4 Experiment and Result

The center of a lens should be aligned horizontally and vertically in placing two beam projectors, and because, by the principle of implementation, when one shutter opens the other shutter must be closed, so that the wing of a mechanic beam shutter must be divided into 2n (n is an odd natural number) equal parts and compose a rotary disk in turn in an interception part and an opening part. The frequency of an emitter synchronizing signal which occurs by 2n divided a rotary disk can be represented as follows.

$$\text{Frequency of an emitter(Hz)} = \text{disk rotation velocity(RPM)} \times n / 60. \quad (1)$$

In this paper, an opening loop control experiment of incomplete synchronization and incomplete high speed synchronization was conducted for certifying effectiveness of the proposed method. For the experiment, two beam projectors whose vertical synchronization frequency is 60Hz and the 6 equally divided beam shutter were installed as shown in Fig. 7 and an incomplete synchronization method of 60Hz was implemented by rotating the motor at 1200 RPM and an incomplete high speed synchronization method implemented by rotating the motor at 2400 RPM.

In stereopsis perceived by the user through shutter glasses, there is little vibration and the presented stereopsis was bright compared with the existing passive polarizing type stereopsis.

Here, even if opening/interception of 120Hz is implemented in incomplete high speed synchronization method, the actual screen changes into only 60Hz. But if an active method system using one beam projector of the same 60Hz, stereopsis could have been forced to implement at the speed of only each 30Hz right and left. Accordingly, this paper can be said to propose a structure which can implement an active method stereoscopic display system as two times as fast by using two ordinary beam projectors.

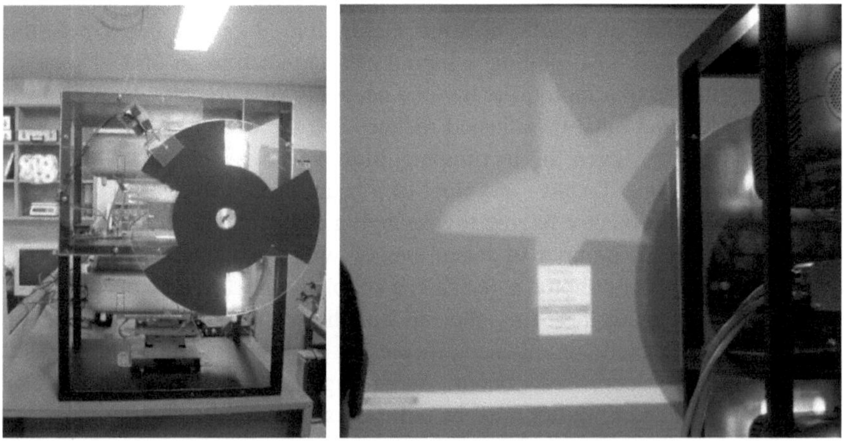

Fig. 7. Rotary disk type stereoscopic display system

5 Conclusion

This paper proposed a stereoscopic display system using rotary disk type beam shutter which can additionally enable an active type stereoscopic display under an passive type stereoscopic display environment using two universal beam projectors and a polarizing filter, by adding a simple device. If the rotation speed of a beam shutter can be adjusted integer multiple times as fast as the scanning frequency of a beam projector and thus synchronization can be adjusted, the rotation speed of a disk increased and thus, with using two low-priced projectors, it was possible to obtain a result similar to the environment in which an active type stereopsis is created using a high-priced projector. A system has been proposed up to now which implemented an active method stereoscopic system using two beam projectors under an electronic shutter method, but no beam shutter method has been proposed using a rotary disk, like the one proposed in this paper, and it seems that this method could be stereoscopic implementation method considerably useful until low-priced high speed beam projector could be generalized.

It is expected that if the proposed method is used, the demerits of an active method requiring much money and a passive method requiring a high-priced silver or a pearl screen whose brightness is greatly reduced could be solved more or less, materializing a better stereopsis environment at a low price. Also, this device costs less than the one using an electronic shutter whose principle is similar, and it has merits that it can completely intercept the right and left image and increase the quality of stereopsis. In addition, if necessary, it can be converted into passive type stereoscopic display system with ease, and thus it can create various stereopsis environment.

References

1. Dhond, U.R, Aggarwal, J.K.: Structure from Stereo - A Review. IEEE Trans. Systems, Man & Cybernetics 19(6), 1489–1510 (1989)
2. Miyake, S. et al.: Basics of 3D Images, NHK Science and Technical Research Laboratories, Ohmsha, Ltd. and Kidary, In: Korean, 2nd edn. (2000) (originally in Japanese)
3. Lipton, L.: The StereoGraphics Developers' Handbook. white paper, Available at http://www.reald.com/3dhandbook_signup.asp

Internal Timing Mechanism for Real-Time Coordination
- Two Types of Control in Synchronized Tapping

Yoshihiro Miyake and Koji Takano

Tokyo Institute of Technology, Dept. of Computational Intelligence and Systems Science
Midori, Yokohama, 226-8502 Japan
miyake@dis.titech.ac.jp

Abstract. We clarify internal timing control mechanism in a synchronization tapping task. Synchronization error (SE) was directly controlled within 450–1800 ms of the inter stimulus-onset interval (ISI) with our experimental method. Two types of internal phase correction mechanism became evident. The first showed a strong negative correlation between SE change and change in inter tap-onset interval (ITI), suggesting a simple negative feedback mechanism. This correlation was observed under all ISI conditions. The second type showed a larger response with a low negative correlation and was observed only with long (1200–1800 ms) ISIs. Only the second mechanism was inhibited in a dual task condition. These results suggest that there are two types of internal timing mechanism.

Keywords: Timing control, Synchronized tapping, Attentional resources.

1 Introduction

Our rhythmic body movements can become synchronized with external dynamics and is widely observed in our face to face communication process. This synchronization phenomenon is thought to be an essential ability for realizing real-time coordination with a dynamic environment and is a kind of interpersonal co-creation process of subjective space and time.

To investigate this kind of subjective timing mechanisms, the synchronization tapping task in which the subject is required to synchronize their taps with periodic stimuli as precisely as possible has been widely used [1–3]. In this experimental task, two error correction mechanisms have been proposed to synchronize the stimulus-onset and the tap-onset [4–9]: phase correction controls the coincidence of tap timing with the stimulus timing, and the period correction synchronizes the tap period with the stimulus period.

These error correction mechanisms have been investigated by varying the inter stimulus-onset interval (ISI) during the synchronization tapping task [7–13]. However, when only the ISI is controlled, there is no clear distinction between phase correction and period correction. To distinguish between these two correction processes, the relationship between the stimulus-onset and the tap-onset must be

M.J. Smith, G. Salvendy (Eds.): Human Interface, Part I, HCII 2007, LNCS 4557, pp. 876–883, 2007.

directly controlled. However, previous experiments have only been conducted with a zero synchronization error (SE) [3]. Our experimental method allows direct control of SE [14,15] and is similar to another recently described procedure [16].

Negative asynchrony (NA) has been observed during the synchronization tapping task and is thought to be involved in phase correction mechanism. This interesting phenomenon shows that the onset of a subject's tap precedes the onset of a stimulus by a few tens of milliseconds, although the subject is unaware of this [1,2,17,18]. This suggests that timing control is not simply a passive reaction to the stimulus, but an active response based on internal anticipation of the timing of the next stimulus. Mates *et al.* has shown that NA is observed when the ISI is between 450 and 3600 ms [17]. Furthermore the dual task method (see Methods section) shows that attentional resources are involved in the generation of NA when the ISI is longer than 1800 ms [19]. These suggest that phase correction does not appear to be a single mechanism, but is hierarchical and depends on the length of ISI.

Our aim was to clarify the internal phase correction mechanisms involved in timing control during the synchronization tapping task. We directly controlled the SE, which allowed us to quantitatively analyse the effects of SE on phase correction. The effects of ISI and attentional resources on timing control were also investigated.

2 Methods

The subject's task was to synchronize their taps with a periodic auditory stimulus as precisely as possible. The SE for each tap was directly and precisely controlled. This involved presenting the auditory stimulus at a fixed time interval following each tap, producing an externally controlled SE. The phase correction mechanism was analysed by measuring the relationship between SE, the controlled parameter, and the inter-tap onset interval (ITI), the observed parameter.

The ITI gradually decreases when the SE is zero [3]. This experimental condition is known as "Pseudo Synchronization" and the results suggest that the auditory stimulus is perceived to precede the onset of the tap. The internal phase difference is not an external SE, but an internal perception that may be controlled in the above SE-control method. If phase correction is a negative feedback mechanism [4,5], our method should elucidate the internal control dynamics in relation to timing perceptions.

Our experiments used several values of ISI to determine if phase correction was affected by the ISI [19]. Each trial consisted of two different parts. The subject tapped to a stimulus with a fixed ISI in the first half of each trial, whereas the SE was controlled during the second half of each trial. Using this hybrid task, the relationship between the phase correction mechanism and ISI was determined.

Each trial consisted of 60 taps, with the stimulus presentation changed after 20 taps (the first half), and with 40 taps in the second half. The ISI was fixed for the first 20 taps as in the typical synchronized tapping task (left side of Fig. 1). This is referred to as the fixed ISI (FISI) condition and was used to control the tapping period of the subject. For the 40 taps in the second half of each trial, stimuli were presented at a fixed time interval following the onset of each tap (right side of Fig. 1). This is

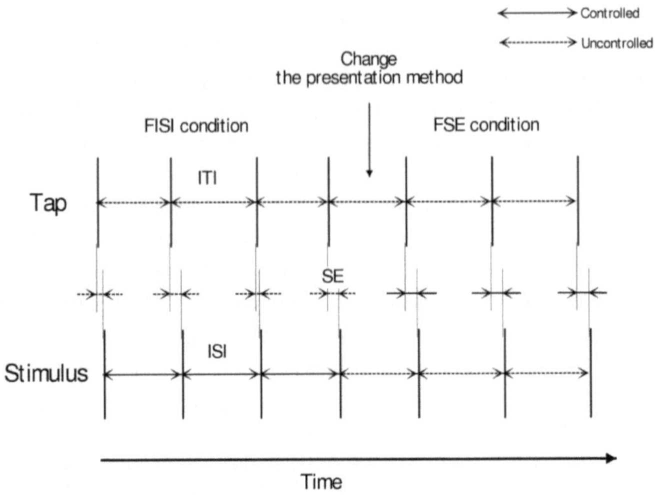

Fig. 1. Experimental procedure

referred to as the fixed SE (FSE) condition. This makes it possible to directly control the SE between the stimulus onset and the tap onset with various ISIs. The subjects were not informed when the stimulus presentation method changed during each trial.

Five different ISIs were used in the FISI condition (450, 600, 900, 1200, and 1800 ms). Seven different SE values (0, –10, –30, –50, –70, –90, and –110 ms) were used in the FSE condition. All SEs were negative i.e., the auditory stimulus was presented after the onset of the subject's tap. Under a combination of these conditions, 35 trials were conducted in a single session for each subject.

The dual task method [20] was used under the same conditions described above to examine the effect of attentional resources on phase correction. The primary task was the synchronized tapping task, and the secondary task was a silent reading task. The primary task alone is referred to as the single task condition, whereas the dual task condition is the simultaneous execution of both the primary and secondary tasks.

The Japanese language version of *The Cathedral and the Bazaar* by Eric S. Raymond [21] was used as the silent reading task. At the end of the experiment, the subject was asked several questions on the content of the text. The subject had to choose between two answers for each question. The mean percentage of correct answers was 85.2% with a range of 76.9–94.7%, showing that the subjects had closely attended to the silent reading task.

Six healthy males (mean age 25.2 years) volunteered for this study. All subjects were right handed and exhibited no hearing abnormalities. They all had experience of the synchronization tapping task, having participated in similar trials in the past. The experiment was conducted in a quiet room with the subject sitting in a chair with their eyes closed. The taps were made using the right index finger. Subjects were forbidden to count using any motion other than their right index finger and to do their tapping with subjective accentuation.

The experimental system was run on a PC (IBM, ThinkPad 560E) with a single-task OS (IBM, PC-DOS2000). Stimuli were transmitted to the subject via headphones connected to the PC. Stimulus duration was 100 ms and the frequency was 500 Hz. The sound pressure was set to an appropriate level to ensure that the auditory stimuli could be clearly heard. The button detecting the tap had a short stroke (< 1 mm). The time measurement was done with a real time clock at a resolution of 1/2048 s. An auditory stimulus was generated at a fixed time interval after each tap, and the timing of taps and stimuli onsets were recorded on the PC.

Each trial consisted of 60 taps with a combination of conditions in the first and second halves of each trial. A session consisted of 35 trials. The subject was given a 5 s break between each trial, and trials were randomly ordered within a session. In the dual task method, a single task condition session was first conducted, followed by a dual task session. However, the order of trials was the same within the session for the single task condition and the dual task condition for the same subject. Each subject participated in 70 trials over the course of two sessions, and the six subjects participated in a total of 420 trials.

The ITI, ISI, and the SE for each tap were calculated. The data from the first five taps at the start of a trial were not used in the analysis to eliminate unstable data.

3 Results and Discussion

Typical time courses of SE and ITI observed with short and long ISIs are shown in Fig. 2. Three types of ITI responses were observed in the FSE condition. The first type was characterized by an increase in the ITI (Fig. 2a), the second type had a relatively constant ITI (Fig. 2b, d), and the third type had a decrease in the ITI (Fig. 2c, e, and f). Fig. 2f shows an especially large decrease in ITI.

We assume that these ITI responses reflect the subject's perception of the temporal order between the tap and the auditory stimulus because the subject was instructed to synchronize the tap with the auditory stimulus. If the ITI is increased in the FSE condition, then the subject's perception may be that the stimulus was delayed after the tap. If the ITI is unchanged, then the tap and the stimulus would seem to be simultaneous, and if the ITI decreased, then the stimulus would appear to precede the tap. However, these ITI responses did not always reflect the physical order between the tap and the stimulus because the stimulus was always presented after the tap in the FSE condition. Therefore, we should interpret these results as evidence for an internal perceptual relationship between the auditory stimulus and the tap response.

We aim to determine the internal phase correction mechanism by controlling the SE. However, SE is constant in the FSE condition and the phase correction itself cannot be observed. Therefore, the following data analysis was performed to extract the internal response to the fixed SE.

An important parameter measured is the SE change (ΔSE). This is the difference between the mean SE for 15 taps in the FISI condition and the fixed SE for the 40 taps in the FSE condition. Our hypothesis is that the ΔSE value is the input to the internal phase correction mechanism. We can express ΔSE as:

$$\Delta SE = (\text{fixed SE in the FSE condition}) - (\text{mean SE in the FISI condition})$$

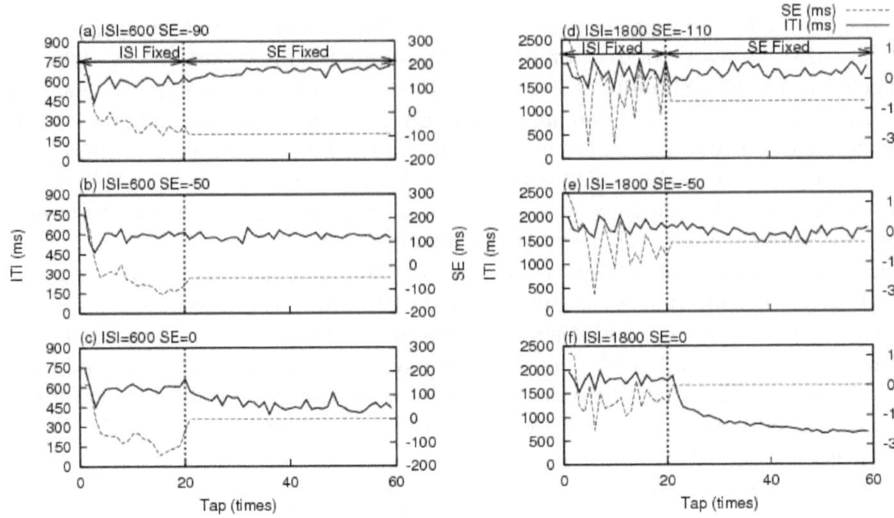

Fig. 2. Typical time course of ITI and SE in ISI = 600 and 1800 ms. The horizontal axis shows the tap and the vertical axis represents the ITI and SE. The period left of the vertical broken line corresponds to the FISI condition. The period to the right corresponds to the FSE condition, and the SEs were fixed. In each ISI condition, across-subject data for mean ITIs in the FISI condition were 450, 596, 899, 1200, and 1799 ms, mean ITIs in the FSE condition were 417, 545, 813, 1065, and 1587 ms, and mean SEs in the FISI condition were –68, –74, –63, –77, and –91 ms.

The other important parameter is the ITI change (ΔITI), which was obtained from the mean ITI for 15 taps in the FISI condition and the mean ITI of 40 taps in the FSE condition. The expression for ΔITI is:

$$\Delta ITI = (\text{mean ITI in the FSE condition}) - (\text{mean ITI in the FISI condition})$$

We assume that the ΔITI reflects the perception of the subject, and can be regarded as an output from the phase correction mechanism. The mean ITI in both conditions and the mean SE in the FISI condition were calculated for each trial.

Fig. 3a–e show the relationships between the ΔSE and ΔITI. These relationships were obtained for all ISIs under FISI conditions. Each point in the graph corresponds to one trial, and each graph consists of 42 trials. These points were clustered by the "nearest neighbour method" [22] to quantitatively analyse their characteristics. There was only one cluster (cluster 1) in the ISI range of 450–900 ms, and its configuration was a straight line that passed through the origin. Moreover, in addition to the clusters with a similar structure observed in Fig. 3a–c, a second, rather different cluster (cluster 2), was observed at ISIs from 1200 to 1800 ms as shown in Fig. 3d and e. Cluster 1 had a strong negative correlation (mean r = –0.73, SD = 0.17) in the ISI range of 450–1800 ms, while cluster 2, in the ISI range of 1200–1800 ms, did not (mean r = –0.25, SD = 0.03). These two clusters are significantly different in their configuration ($p < 0.05$, t-test).

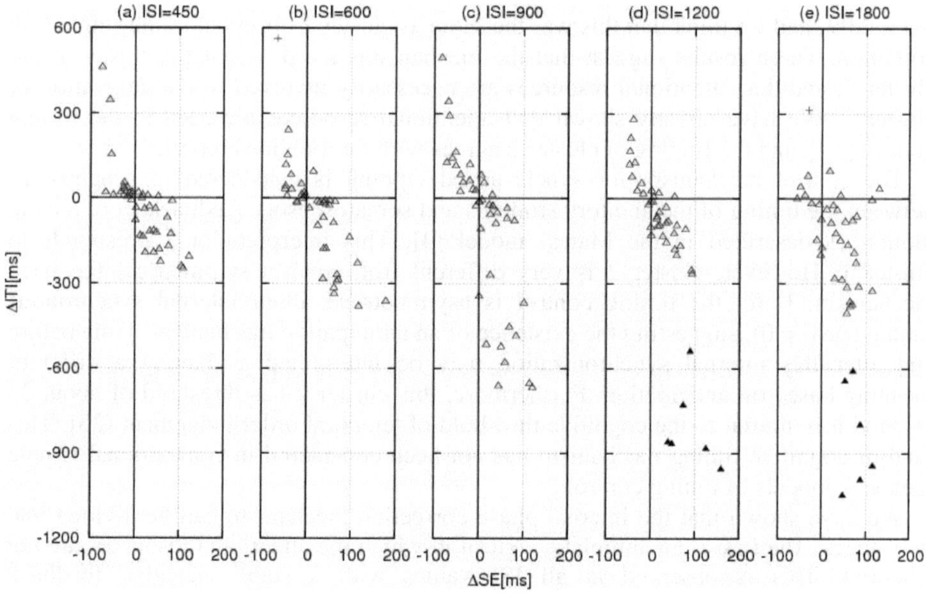

Fig. 3. Relationship between SE change and ITI change under the single task condition. The horizontal axis shows the change of SE and the vertical axis represents the change of ITI. Each point in a graph corresponds to a single trial. These points were clustered by the "nearest neighbour method", and a threshold value of a squared Euclid distance calculated from ΔSE and ΔITI was 200. The clusters were numbered in the order of the cluster with the largest number of points. All numbered clusters had more than one point.

Cluster 1 was present at all ISI values and had a strong negative correlation that sloped downward to the right as shown in Fig. 3. This corresponds to a decrease in ITI as SE increases, and an increase in ITI when SE decreases. This cluster was observed over a wide range of ΔSE and its configuration was stable. This means that the ITI is adjusted to recover a stable SE value when it drifted, suggesting a negative feedback SE mechanism to maintain the internal synchronization of the tap with the auditory stimulus. In previous models of synchronized tapping, the phase correction has been assumed to be a negative feedback for the internal phase difference [4,9], and our experimental results directly support this mechanism.

Cluster 2 was generated under the longer ISI conditions (1200–1800 ms) and showed a weaker correlation and larger ITI change than was seen for cluster 1. Fig. 2f shows an example of such a significant ITI change. This cluster was observed in the region with a large positive ΔSE that exceeded 35 ms. Repp has already reported the overshoot response of ITI and found that this response only appeared when the ISI is substantially changed during synchronized tapping [8,9]. We found this response mainly in the ΔSE positive region where the subject perceives the stimulus as preceding the tap response. This suggests cluster 2 does not reflect a simple negative feedback mechanism for phase correction.

In the dual task condition, only one cluster was found under all ISI conditions. This cluster had a strong negative correlation in all conditions of ISI (mean r = –0.75,

SD = 0.09) and we think that this was the same as cluster 1 observed in the single task condition. These results suggest that the mechanisms are different for cluster 1 and cluster 2, and that attentional resources are necessarily involved in the generation of cluster 2. We have already shown that attentional resources are used in the timing control [19], and the finding of cluster 2 agrees with our previous report.

The control mechanism for synchronized tapping is considered to synchronize between the timing of the auditory stimulus and somatosensory feedback from the tap motion as described in the Mates' model [4]. This interpretation corresponds to cluster 1. However, cluster 2 is very different from such a symmetrical feedback mechanism. Here, the timing control is asymmetrical about internal synchronous timing ($\Delta SE = 0$), suggesting the existence of an anticipation mechanism. Time before and after this internal synchronization may be interpreted as having a different meaning based on anticipation. Furthermore, this cluster 2 has threshold of about 35 ms, and it is similar to the cognitive threshold of temporal order judgement [23]. This kind of cognitive timing mechanism has not been considered in conventional simple feedback models of timing control.

We have shown that the internal phase correction mechanism can be divided into two stages. The first is an automatic control in which the attentional resources are not essential. This is observed at all ISI values with a stable negative feedback mechanism present over a wide range of SE changes. In contrast, the second mechanism requires attentional resources and is observed under the longer ISI conditions. This mechanism appears only when the SE changes are positive and large, suggesting internal anticipation of synchronized timing. We believe the first mechanism is related to the motor control level, while cognition is involved in the second mechanism. We here demonstrate that phase error correction is based on these two mechanisms.

References

[1] Miyake, I.: Researches on rhythmic action, Studies from the Yale Psychology Laboratory, 10, 1–48 (1902)

[2] Dunlap, K.: Reactions to rhythmic stimuli, with attempt to synchronise. Psychol. Rev. 17, 399–416 (1910)

[3] Fraisse, P., Voillaume, C.: Les reperes du sujet dans la synchronisation et dans la pseudo-synchronisation. L'annee Psychologique 71, 359–369 (1971)

[4] Mates, J.: A model of synchronisation of motor acts to a stimulus sequence: I. Timing and error corrections. Biol. Cybern. 70, 463–473 (1994)

[5] Mates, J.: A model of synchronisation of motor acts to a stimulus sequence: II. Stability analysis, error estimation and simulations. Biol. Cybern. 70, 475–484 (1994)

[6] Semjen, A., Vorberg, D., Schulze, H.H.: Getting synchronised with the metronome: Comparisons between phase and period correction. Psychol. Res. 61, 44–55 (1998)

[7] Thaut, M.H., Miller, R.A., Schauer, L.M.: Multiple synchronisation strategies in rhythmic sensorimotor tasks: phase vs period correction. Biol. Cybern. 79, 241–250 (1998)

[8] Repp, B.H.: Phase correction, phase resetting, and phase shifts after subliminal timing perturbations in sensorimotor synchronisation. J. Exp. Psychol. Hum. Percept. Perform. 27, 600–621 (2001)

[9] Repp, B.H.: Processes underlying adaptation to tempo change in sensorimotor synchronisation. Hum. Mov. Sci. 20, 277–312 (2001)

[10] Thaut, M.H., Tian, B., Azimi-Sadjadi, M.R.: Rhythmic finger tapping to cosine-wave modulated metronome sequences: Evidence of subliminal entrainment. Hum. Mov. Sci. 17, 839–863 (1998)

[11] Repp, B.H.: Compensation for subliminal timing perturbations in perceptual-motor synchronisation. Psychol. Res. 63, 106–128 (2000)

[12] Thaut, M.H., Kenyon, G.P.: Rapid motor adaptations to subliminal frequency shifts during syncopated rhythmic sensorimotor synchronisation. Hum. Mov. Sci. 22, 321–338 (2003)

[13] Repp, B.H., Keller, P.E.: Adaptation to tempo changes in sensorimotor synchronisation: Effects of intention, attention, and awareness. Q. J. Exp. Psychol A 57, 499–521 (2004)

[14] Takano, K., Miyake, Y.: Two types of time perception in synchronisation tapping task. In: Proc. of SICE Annual Conference (SICE2003). Fukui, Japan, pp. 3211–3216 (2003)

[15] Takano, K., Miyake, Y.: Two types dynamics in negative asynchrony of synchronisation tapping. In: Proc. of SICE Annual Conference 2004 (SICE2004). Sapporo, Japan, 1792–1796 (2004)

[16] Flach, R.: The transition from synchronisation to continuation tapping. Hum. Mov. Sci. 24, 465–483 (2005)

[17] Mates, J., Radil, T., Müller, U., Pöppel, E.: Temporal integration in sensorimotor synchronisation. J. Cogn. Neurosci. 6(4), 332–340 (1994)

[18] Aschersleben, G.: Temporal control of movements in sensorimotor synchronisation. Brain Cogn. 48, 66–79 (2002)

[19] Miyake, Y., Onishi, Y., Pöppel, E.: Two types of anticipation in synchronisation tapping. Acta Neurobiol. Exp (Warsz) 64(3), 415–426 (2004)

[20] Baddeley, A.: Working Memory. Oxford University Press, New York (1986)

[21] Raymond, E.S.: The Cathedral and the Bazaar (H. Yamagata Trans. to Japanese). Kobosya, Tokyo (2001) (Original work released 1997 on http://www.catb.org/~esr/writings/athedral-bazaar/)

[22] Romesburg, H.C.: Cluster analysis for researchers. Robert E. Krieger Publishing Company, Florida (1989)

[23] Pöppel, E.: Time perception. In: Held, R., Leibowitz, H.W., Teuber, H-L. (eds.) Perception. Handbook of sensory physiology, vol. VIII, pp. 713–729. Springer, Heidelberg (1978)

Measuring Brain Activities Related to Understanding Using Near-Infrared Spectroscopy (NIRS)

Masayoshi Nagai, Nobutaka Endo, and Takatsune Kumada

National Institute of Advanced Industrial Science and Technology (AIST)
AIST Tsukuba Central 6, 1-1-1 Higashi, Tsukuba, Ibaraki, 305-8566, Japan
masayoshi-nagai@aist.go.jp

Abstract. We examined brain activation in the frontal cortex during the contents understanding of comic stories with near-infrared spectroscopy (NIRS). The present results showed that the left prefrontal lobe region is activated when people actively try to understand the comic stories and to memorize their contents for reporting in the future. The activated area could overlap with or be around Broca's area. We discussed how to apply this brain activation related to human understanding to evaluation of contents presented in the human-machine interfaces.

Keywords. contents understanding, left prefrontal region, near-infrared spectroscopy.

1 Introduction

For developing visual presentation systems and human-machine interfaces which are user-friendly, it is very important to establish techniques for measuring how well people understand information or contents presented in a display of the interfaces. Generally in the mobile situation, since visual displays are small, only small amount of information could be presented. Thus, better quality of information should be presented for people to easily understand the content with such a limited presentation source. To assess the quality of presented information or contents, we can measure human understanding processing for the contents. There are two measurements to evaluate people's understanding: subjective report and contents-understanding test. Subjective report is to rate understanding level by oneself (e.g., rating it with a seven points scale between "understood completely" and "did not understand at all"). Although this measure is commonly used in a variety of research areas, we could point out one weakness that it is *not objective* measure. For example, some people may think that they understood the content very well even if they actually did not understand it. On the other hand, others may think that they did not understand it well even if they really understood it. Therefore, it is not reasonable to compare subjective rating scores between individuals if the criterion of rating is different across them. This problem is critical when applying to children or people with cognitive impairment because it should be more difficult for them to monitor their own state of understanding than healthy young adults. Another measurement, tests for the contents-understanding, is for people to ask questions on the contents. This is an

M.J. Smith, G. Salvendy (Eds.): Human Interface, Part I, HCII 2007, LNCS 4557, pp. 884–893, 2007.

objective measure and is better than subjective rating in the sense that the score can be compared between individuals.

These two measures (subjective rating and understanding tests) cannot show *online* processes of understanding while people are actually seeing and/or reading contents to be understood. However, online monitoring of understanding is important in mobile situations researches. Even if subjective rating and understanding tests showed that people completely and easily understood some contents, mental workload to understand this content should be low, but mental workload *at specific moment* could be actually high. This instant high mental workload for understanding cause inattention to important information and might lead to an accident.

In this study we developed to make a new online understanding measure using near-infrared spectroscopy (NIRS). NIRS is an optical method that noninvasively monitors cerebral hemodynamics by measuring changes in the attenuation of near-infrared light passing through tissue. NIRS has proven to be effective in assessing oxygenation changes in response to functional brain activation in many studies, utilizing the tight coupling between neural activity and regional cerebral blood flow. Variety of human cognitive processes including motion illusion [1], Stroop interference [2], and false memory [3] have been studied with NIRS. Using this brain imaging technique, we measured the online brain activity related to content-understanding. Brain activity (increase of oxy-hemoglobin concentration) in some brain locus correlated to participants' understanding the displayed content could be used as online understanding measure or mental workload measure for understanding: the contents eliciting high understanding-related brain activity is evaluated as difficult contents to be understood.

2 Method

2.1 Participants

Thirteen healthy people (9 males and 4 females; 20 - 25 years of age, 22.4 ± 1.4 (mean ± SD)) participated in the present experiment. All the participants were right-handed and had normal or corrected-to-normal vision. They were naïve as to the purpose of the experiment and were paid 1,000 yen (about US$9.00) for their participation.

2.2 Stimuli and Task

Six series of short comic stories were used as stimuli in the experiment. Each comic story consisted of four frames. Four of six comics were the original and complete contents (normal story), while the remaining two comics consisted of four frames which were randomly chosen from different stories (scramble story). Each frame of a story was presented for 5 sec on a 21-inch CRT monitor (Iiyama, MA203DD) and thus the total duration of each story was 20 sec. Participants were required to observe the comic stories in the following three conditions. In the first condition with the normal stories participants were instructed to understand why the stories were funny, and to entitle the stories and to report the contents of them and their funniness after the experiment (the report condition). In the second condition, participants were instructed to just observe the normal stories (the non-report condition). In the third condition, participants were instructed to observe the scramble stories which could not be understood as reasonable ones (the scramble condition). In the first two condition, comic stories were equalized in both their understandability and funniness based on a prior rating by nine people who did not participate in this experiment. Two of four normal stories were randomly assigned to the report and non-report conditions. Each condition was repeated two times in the experiment, and the order of the conditions was counterbalanced across participants. There were 30 sec rests between the conditions.

2.3 Brain Activity Measurement by NIRS

We used a multichannel near-infrared spectroscopy (NIRS) system (ETG-4000, Hitachi Medical Corporation). Near-infrared light with wavelengths of 695 nm and 830 nm was guided by optical-fiber bundles and transmitted into the cranium. The reflection of the infrared light was sampled by receiving probes placed on the scalp, 3.0 cm away from the emitting probes, and it was detected every 0.1 sec by silicon photodiodes. The detected signal was separated into two components that corresponded to the two different wavelengths with lock-in amplifiers [4]. The temporal changes in concentrations of oxy- hemoglobin (oxy-Hb) and deoxy-hemoglobin (deoxy-Hb) were calculated based on the difference in the absorption indices for these two wavelengths. The moving average of 5 sec was used to reduce noise in each Hb data.

We used a pair of probes, each consisting of 4 x 4 arrays. Sixteen optodes (8 emitting and 8 detecting optodes) constituted twenty-four channels in each probe, but we analyzed only seven channels (i.e., the anterior portion of each 4 x 4 arrays) in each hemisphere including F7/F8 of the international 10/20 system because we focused on prefrontal area of the brain (see Fig 1).

Although the raw data of NIRS were originally relative values, and could not be averaged directly across participants or channels, the normalized data such as the z-scores could be averaged regardless of a unit [5, 6, 7]. The z-scores of oxy- and deoxy-Hb (z-Hb) were respectively calculated using the mean value and the standard deviation of oxy- and deoxy-Hb changes during the pre and task periods as follows:

$$\text{z-Hb} = (Hb_{task} - \text{mean}(Hb_{pre}))/\text{std}(Hb_{pre}) \quad (1)$$

Here Hb_{pre} and Hb_{task} represent Hb data in pre-task and task periods, respectively. Previous studies showed that the hemodynamic response typically lags behind the stimulation by a few seconds. Additionally, in this study there should be some task-related brain activity after the end of the comic presentation. Especially in the report condition, it was highly possible that participants tried to remember the content of story during the rest period. Therefore, we defined the comic presentation (20 sec) and the whole rest (30 sec) periods as task period, and the last 10 sec in the rest as pre-task period before the next comic presentation. In the NIRS data analysis, the averaged z-scores of oxy- and deoxy-Hb were calculated in each channel in each condition for individual participant.

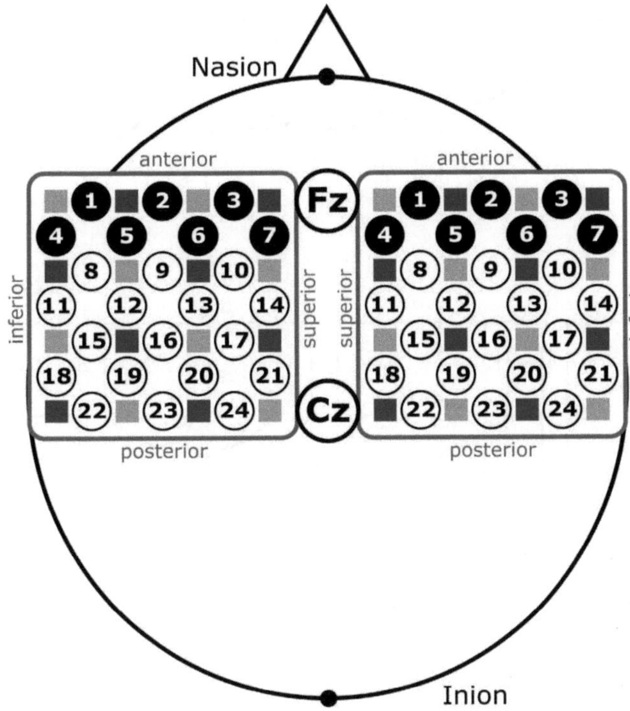

Fig. 1. Location of optodes placed on the prefrontal areas in both hemispheres. The distance between each emitter (indicated by light gray squares) and the corresponding detector (indicated by dark gray squares) was set at 3 cm. NIRS responses at the channels indicated by black circles (Ch-1 to Ch-7) in both hemispheres were mainly analyzed in the present study.

3 Results

3.1 Report Tests

All the participants could reasonably entitle the stories presented in the report condition, and they could also correctly recall their contents. Accordingly, it was confirmed that they properly understood why the stories were funny.

3.2 NIRS Data

Figs 2 and 3 show the time courses for grand averaged z-scores of oxy- and deoxy-Hb changes among the report, the non-report, and the scramble conditions respectively. Figs 4 and 5 show the mean z-scores during the task period of oxy- and deoxy-Hb changes respectively. Overall, oxy-Hb in the report condition tended to increase during and after the comic presentation in the left and the right prefrontal regions. However, no such trends were observed in the other two conditions (Fig 2). In contrast, deoxy-Hb in the report condition tended to decrease during the task period in

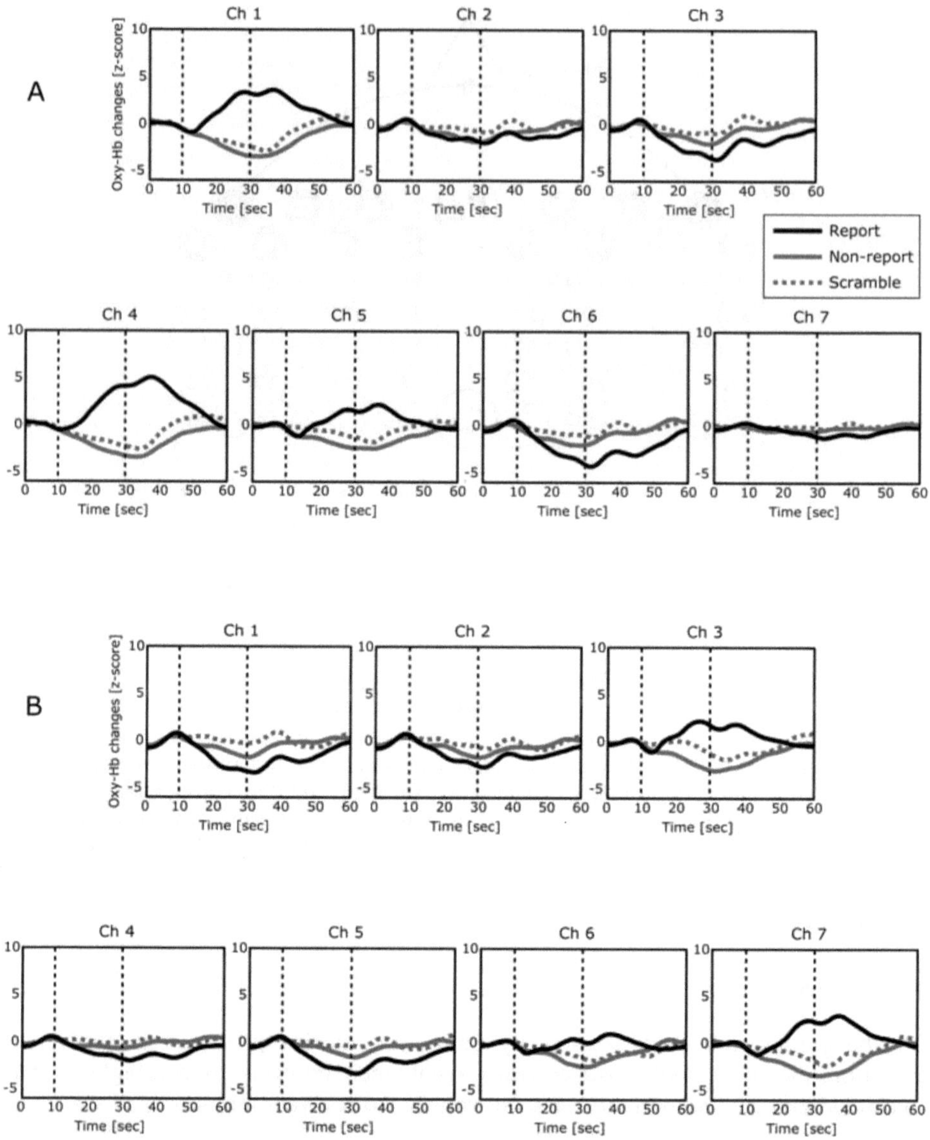

Fig. 2. The time course of the average change of the oxy-Hb concentration during the pre-time and the task periods. Comic stories were presented for the time period between the vertical dashed lines. The results of the left and the right hemispheres are shown in the figures in the upper (A) and the lower (B) panels respectively.

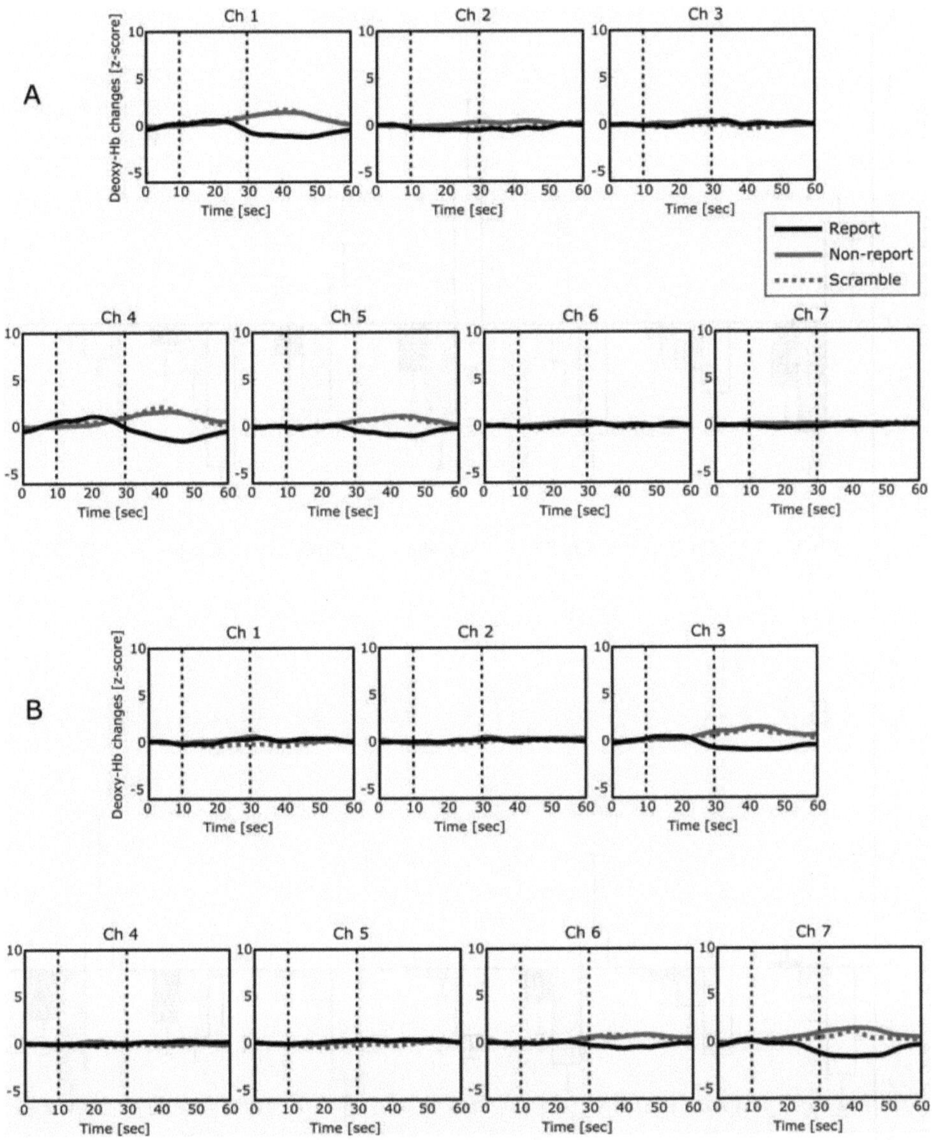

Fig. 3. The time course of the average change of the deoxy-Hb concentration during the pre-time and the task periods. Comic stories were presented for the time period between the vertical dashed lines. The results of the left and the right hemispheres are shown in the figures in the upper (A) and the lower (B) panels respectively.

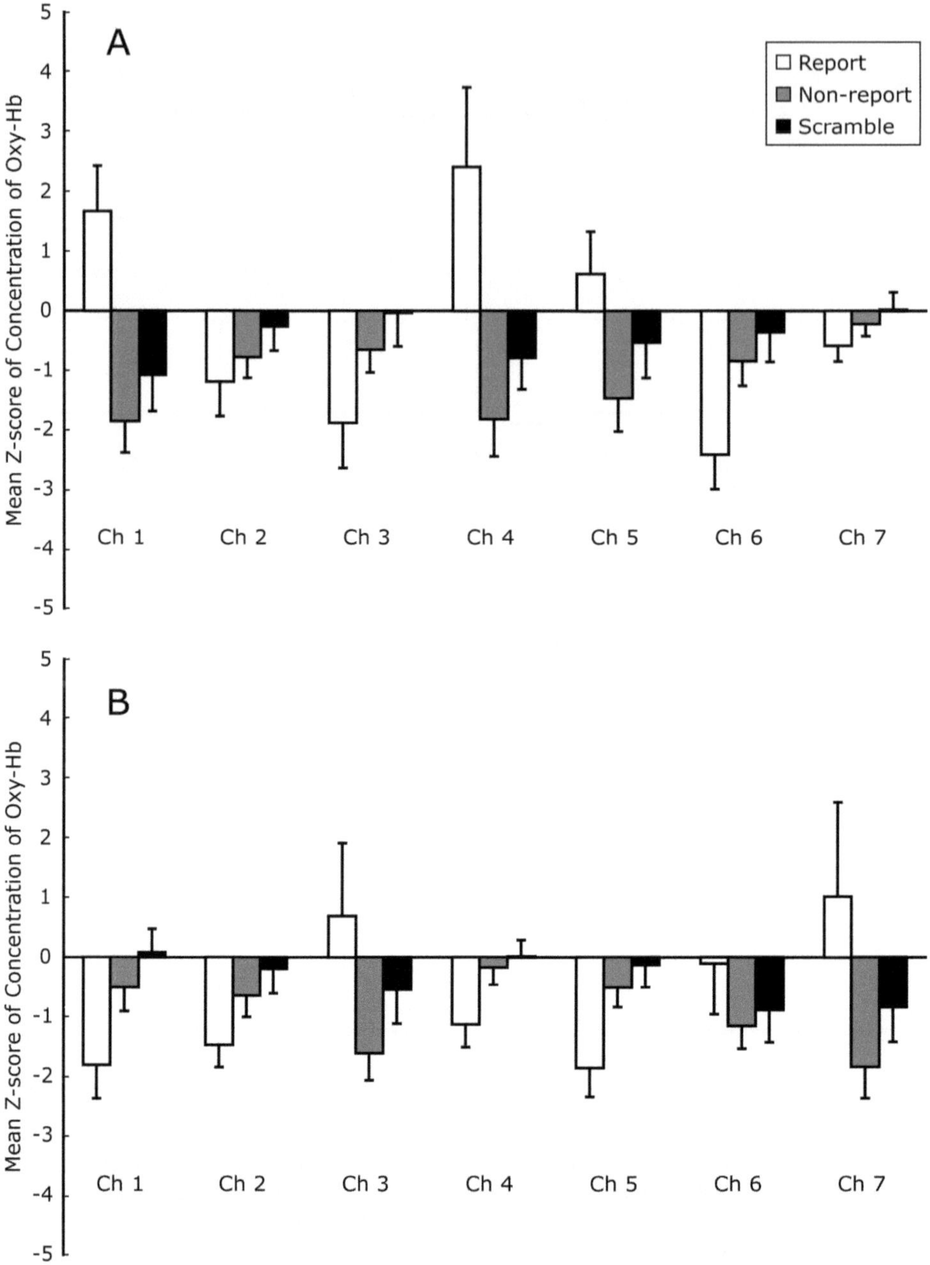

Fig. 4. Mean z-scores of the oxy-Hb concentration in each task during the task period. The results of the left and the right hemispheres are shown in the upper (A) and the lower (B) figures respectively. The vertical bars in the figures represent 1 standard error (SE).

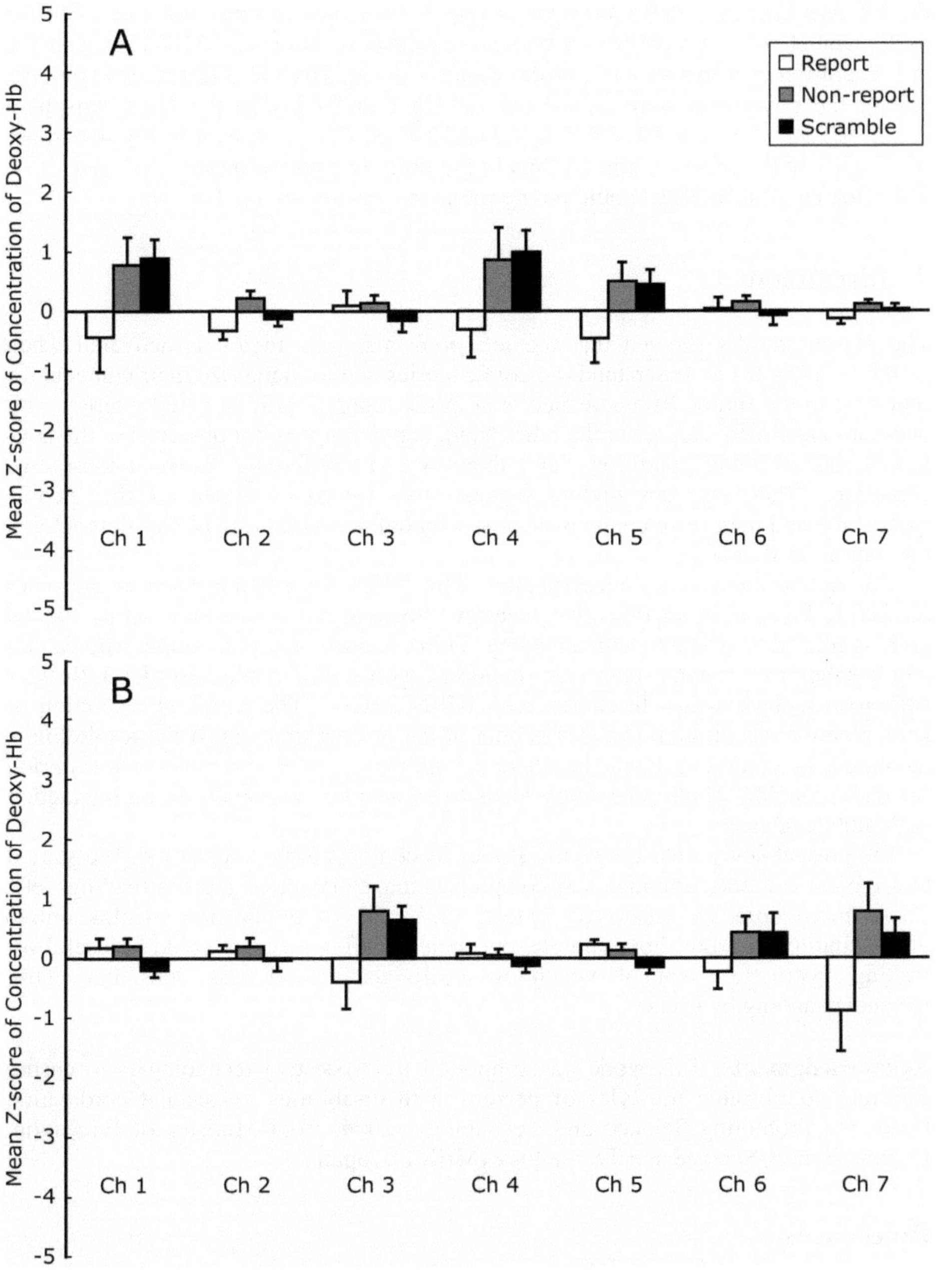

Fig. 5. Mean z-scores of the deoxy-Hb concentration in each task during the task period. The results of the left and the right hemispheres are shown in the upper (A) and the lower (B) figures respectively. The vertical bars in the figures represent 1 standard error (SE).

the left and the right prefrontal regions (Fig 3). Z-scores of oxy- and deoxy-Hb for each channel were subjected to a one-way analysis of variance (ANOVA) with the task condition as a main term. For the changes in oxy-Hb, the differences in z-score among the conditions were significant for Ch-1 and Ch-4 in the left hemisphere (Ch-1: $F(2,24)=8.90$, $p<.01$; Ch-4: $F(2,24)=5.64$, $p<.01$). As shown in Fig 4, z-score was higher in the report condition than in the other two conditions in Ch-1 and Ch-4. The other significant effect could not be found in oxy- and deoxy-Hb data.

4 Discussion

The present results showed that the left prefrontal lobe region is activated when people actively try to understand the comic stories and to memorize their contents for reporting in the future. The activated area in the report condition could overlap with or be around Broca's area. On the other hand, activation was not observed in the non-report and scramble conditions, and there was no difference between these two conditions. Therefore, our finding suggests that we can evaluate whether people understand and memorize contents of stories by online monitoring of oxy-hemoglobin concentration change.

This online measure of understanding with NIRS has advantages over the other measures. First, it is an objective measure. Second, it can monitor online mental process related to contents-understanding. Third, it could apply to people who cannot monitor their own mental states (e.g., children, people with mental disorders). Besides such merits, there was a limitation with NIRS method. The temporal resolution of hemoglobin concentration changes is poor at the present moment. If the resolution is increased, we could check whether there is high mental workload at *very short period* for understanding which potentially leads to an accident especially in an interaction with mobile devices.

The present study used the comic stories as contents to be understood. We started to use other contents: complex text comprehension, story making, proof readings, etc. These more complex tasks can extend the areas of application of this online understanding measure. For example, our understanding measure could be useful for making instruction manual which is easily understood and does not cause misunderstanding by users.

Acknowledgments. This work was supported by "Assistive technologies to ensure safe and comfortable lifestyles of persons with disabilities", Special Coordination Funds for Promoting Science and Technology, 2004, from Ministry of Education, Culture, Sports, Science and Technology (MEXT), Japan.

References

1. Hashimoto, T., Minagawa-Kawai, Y., Kojima, S.: Motion illusion activates the visual motion area of the brain: A near-infrared spectroscopy (NIRS) study. Brain Research 1077, 116–122 (2006)
2. Ehlis, A.C., Herrmann, M.J., Wangener, A., Fallgatter, A.J.: Multi-channel near-infrared spectroscopy detects specific inferior-frontal activation during incongruent Stroop trials. Biological Psychology 69, 315–331 (2005)

3. Kubota, Y., Toichi, M., Shimizu, M., Mason, R.A., Findling, R.L., Yamamoto, K.J.R.: Prefrontal hemodynamic activity predicts false memory—A near-infrared spectroscopy study. Neuroimage 31, 1783–1789 (2006)
4. Maki, A., Yamashita, Y., Ito, Y., Watanabe, E., Mayanagi, Y., Koizumi, H.: Spatial and temporal analysis of human motor activity using noninvasive NIR topography. Medical. Physics. 22, 1997–2005 (1995)
5. Matsuda, G., Hiraki, K.: Sustained decrease in oxygenated hemoglobin during video games in the dorsal prefrontal cortex: A NIRS study of children. NeuroImage 29, 706–711 (2006)
6. Shimada, S., Hiraki, K.: Infant's brain responses to live and televised action. NeuroImage 32, 930–939 (2006)
7. Schroeter, M.L., Zysset, S., Kruggel, F., von Cramon, D.Y.: Age dependency of the hemodynamic response as measured by functional near-infrared spectroscopy. NeuroImage 19, 555–564 (2003)

Brain Activities Related to Legibility of Text, Studied by Means of Near Infrared Spectroscopy

Masako Omori[1], Satoshi Hasegawa[2], Masaru Miyao[3], Masami Choui[4], and Hiroshi Tamura[5]

[1] Kobe Women's University
Faculty of Home Economics, Kobe Women's University, Suma-ku Kobe, Japan
masako@suma.kobe-wu.ac.jp
[2] Department of Information Culture, Nagoya Bunri University, Inazawa Aichi, Japan
hasegawa@nagoya-bunri.ac.jp
[3] Information Technology Center, Nagoya University, Chikusa-ku Nagoya, Japan
miyao@itc.nagoya-u.ac.jp
[4] Hiroshima International University, Hiroshima, Japan
choui@hw.hirokoku-u.ac.jp
[5] Tamura Institute for Human Interface, Kobe, Japan
tamura@mobilergo.com

Abstract. Using NIRS, we investigated brain activity in subjects searching for target characters in given text. Visual stimuli consisted of two different contrasts: positive and negative. Both stimuli were displayed for 30 sec on a cathode-ray tube (CRT: 21 inches). The subjects were instructed to watch a visual display on the CRT. This display included text without meaning. When target characters were detected, subjects were instructed to perform three tasks: (1) count the number of target characters by visual observation, (2) tap a button with the thumb and fingers of their left hand carefully and as fast as possible and (3) operate a mobile phone.

Keywords: brain activities, visual observation, tap a button, operate a mobile phone, human interface, text entry, near infrared spectroscopy.

1 Introduction

Near-infrared spectroscopy (NIRS) is a recently developed neuroimaging methodology. Changes in brain activity have been measured in many studies using positron emission tomography (PET), single photon emission computed tomography (SPECT), and functional magnetic resonance imaging (fMRI), but all have encountered the problem of low temporal resolution. NIRS is a noninvasive [1] and portable technique that can be used in natural settings, which enables examination of male and female subjects over a wide age range, from infancy to old age, while they are performing cognitive tasks. It also has a high temporal resolution that is not affected by the magnetic field produced by TMS (transcranial magnetic stimulation). NIRS can detect changes in the concentration of forms of cerebral blood hemoglobin, such as oxy-hemoglobin (oxy-Hb), deoxy-hemoglobin (deoxy-Hb) and total-hemoglobin (total-Hb), which is the sum of oxy-Hb and deoxy-Hb. The purpose of

M.J. Smith, G. Salvendy (Eds.): Human Interface, Part I, HCII 2007, LNCS 4557, pp. 894–902, 2007.

this study was to explore the possibility of applying neuroimaging techniques to the study of brain activity in humans performing daily communication tasks.

2 Materials and Methods

2.1 Subjects and Materials

Three healthy Japanese adult women (aged 21 years old) were informed of the purpose and safety of experiments in this study, and voluntarily consented to participate. All subjects' had normal or corrected to normal vision, were right-handed, and had no neurological abnormalities. Each subject was seated in front of a table on which a 21-inch cathode-ray tube (CRT) display was placed. Stimuli were presented on the display in the form of text without meaning. This text was made from randomly ordered alphanumeric characters according to ISO 9241-3 Amendment 1 [2], using 240 characters that included 15% spaces as in normal English sentences. The distance from the subject's eye to the display was set at approximately 50 cm to 70 cm. Two different character sizes of L (~2.5 mm height) and S (~1.0 mm) were examined with two different contrasts of W (negative) and X (positive), as shown in Table 1 and Fig. 1.

Table 1. Character sizes and contrast of the graphic text displayed in the experiment

Code	Character	Back Ground
W: Positive	white	black
X: Negative	black	white

Character size	Characters per line	Lines per page	high (mm)	Wide (mm)
L	20	12	2.5	1.3
S	40	6	1.0	0.5

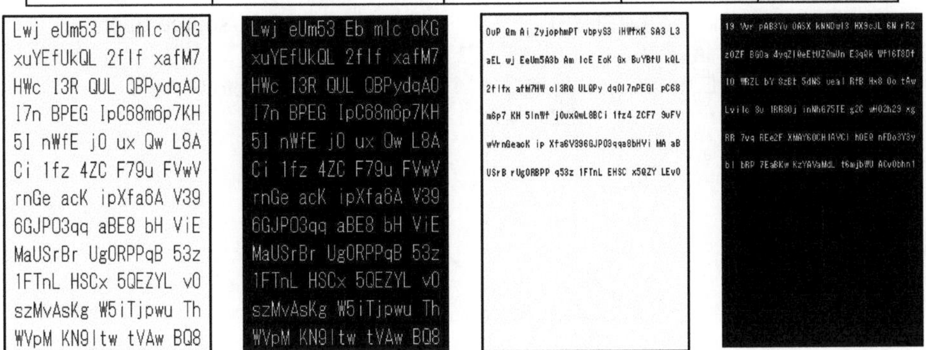

size L W X size S W X

Fig. 1. Samples of graphic characters. They are presented in the JPEG format.

2.2 Task Procedure

The subjects counted the number of times a target letter (A, B, E, respectively) appeared by searching the strings as if they were reading the sentences carefully and as fast as possible. Each subject performed the task, which consisted of a 10-s pretask baseline period, a 30-s task period, and a 10-s post-task. The pretask was display of the target characters. In the post-task period, the number of characters counted by the subjects was recorded, and the subjects conducted subjective evaluations of readability. In the present study, the subjects performed three different tasks of visual evaluation: visual observation, tapping a button with the thumb and fingers on their left hand, and operating the buttons on a cellular phone. In the mobile phone task, the subjects had to input the letter following a target character, as well as the target character itself as a capital letter and small letter, and also numeric characters.

2.3 NIRS Measurements

We measured and recorded brain activation of subjects by NIRS while they were detecting the target character for 30-second periods. Brain activation was recorded using a NIRStation (OMM3000, Shimazu, Kyoto, Japan). We employed a 34-channel NIRS system (OMM3000, Shimazu, Kyoto, Japan) that could detect concentration changes in oxygenated hemoglobin (oxy-Hb), deoxygenated hemoglobin (deoxy-Hb), and their sum (total-Hb) by using three types of near infrared light (wavelengths: 780, 805, and 830 nm). This system was used to measure temporal changes in oxy-Hb, deoxy-Hb, and total-Hb concentrations. We used 34-channel system with 24 optodes, consisting of 13 light-source fibers and 11 detecters.. Each channel consists of one emitter optode and one detector optode located 3 cm away from the emitter optode. The location of the optodes on the skull covering an area in the frontoparietasl region is schematically showen in Fig. 2. Thus, 34 square grids of NIRS topographic map were labeled as channel (CH) 1 to 34.

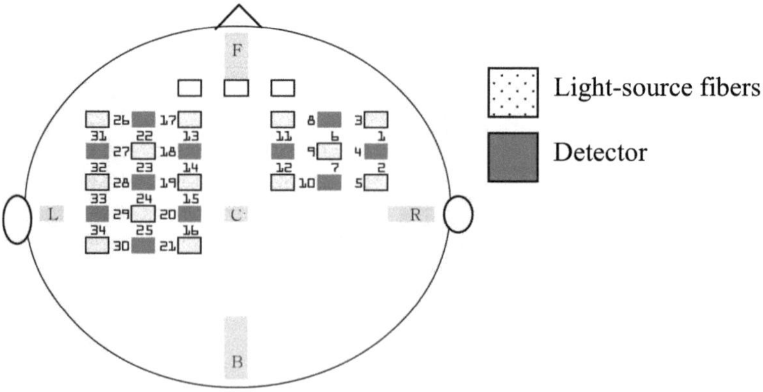

Fig. 2. Locations of optodes and channels. This figure illustrates an overhead view of an optode holder.

2.4 Data Analysis

In this experiment, The result of one subject not wrong is shown in the experiment procedure. It was illustrated the two-dimensional orbital distribution of relationship between oxy-Hb and deoxy-Hb by NIRS data, which the horizontal axis is oxy-Hb and vertical axis is deoxy-Hb [3]. The two-dimensional orbital distribution shows various types of orbits: slope of minus 1 is rest activity, slope of plus 1 is activity, and loop is local activity. The activity in the brain was considered from this orbit feature. It is shown in Fig. 3.

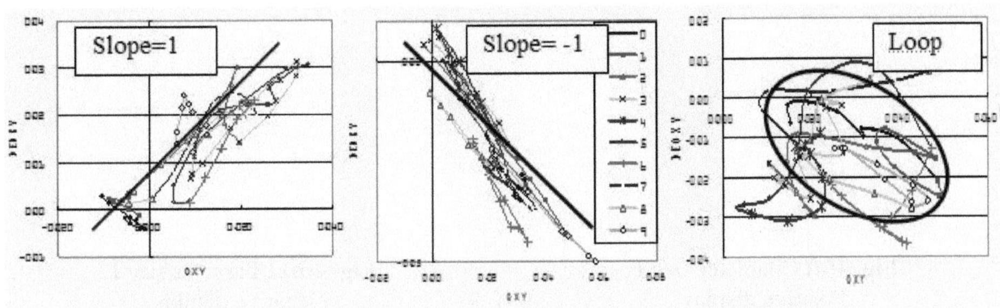

Fig. 3. The plots show the correlation between oxy-Hb and deoxy-Hb

3 Results

3.1 Results for Visual Observation

The two-dimensional orbital distribution during visual observation is shown in Fig. 4. The display number in the graph corresponds to the channel number shown in Fig. 1. Graphs of 2-dimensional orbit are shown for visual observation in Fig. 4, button tapping in Fig. 5, and mobile phone input in Fig. 6. The numbers shown in the graphs correspond to the channel numbers shown in Figure 1.The relation between brain activity and the differences in font size and contrast are described in the following.With negative contrast and character size L, activation was seen at the bilateral frontal region channels (Fig.4 (b), CH1, 2, 18, 22). In the NIRS data, rates of change in total-Hb and oxy-Hb concentrations began to increase.

However, rest was seen with positive contrast and character size L (Fig.4 (a)), and positive and negative contrast and character size S (Fig.4 (c) (d)).

3.2 Result for Button Tap

The two-dimensional orbital distribution when the subjects tapped a button with their thumb is shown in Fig. 5.

Compared with negative contrast and character size L (Fig.5 (b), a wider range of bilateral frontal was seen with positive contrast ((Fig.5 (a), Right: CH 1, 2, 4, 6, 9, 11, Left: CH 13, 17, 22, 26, 27, 31, 32). With negative contrast and character size S, activation was seen at five channels ((Fig.5 (d), CH 3, 5, 16, 20, 22). However, with

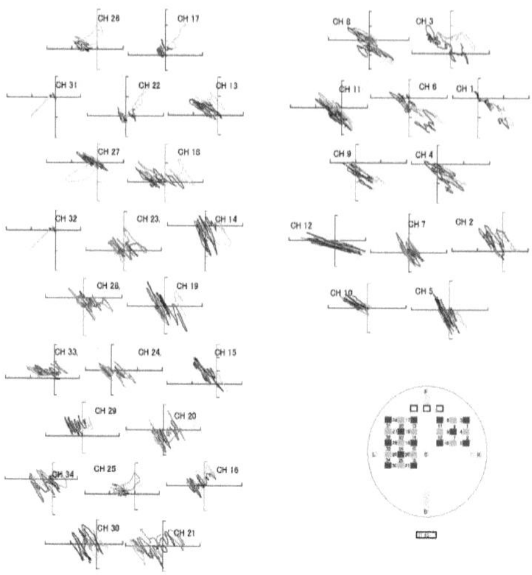

Fig. 4(a) Character size L
Positive display

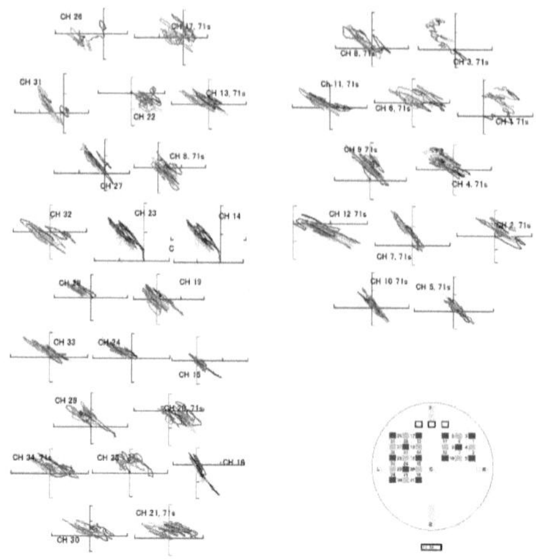

Fig. 4(b) Character size L
Negative display

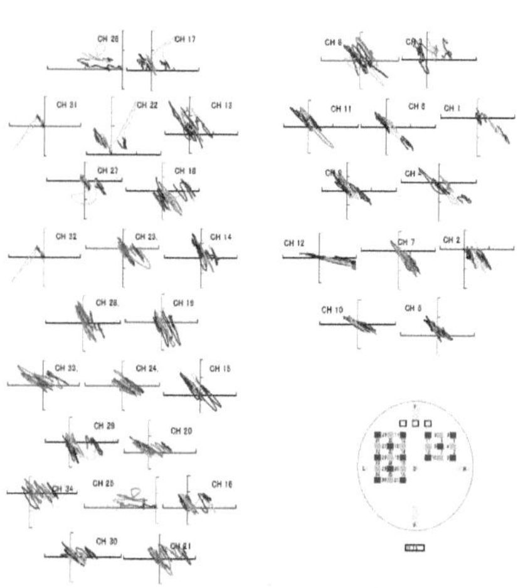

Fig. 4(c) Character size S
Positive display

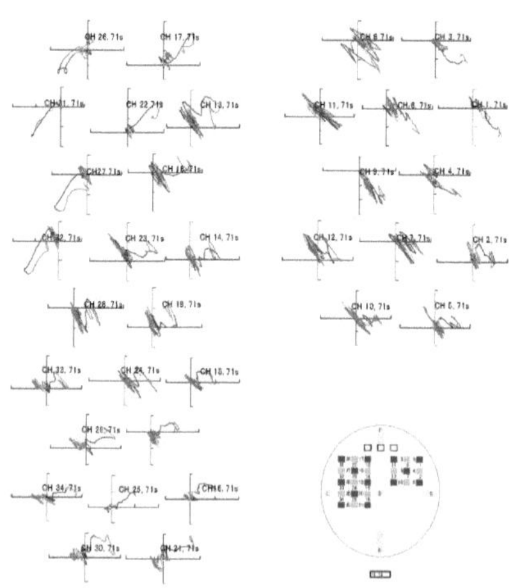

Fig. 4(d) Character size S
Negative display

Fig. 4. The plots show the correlation between oxy-Hb and deoxy-Hb during visual observation

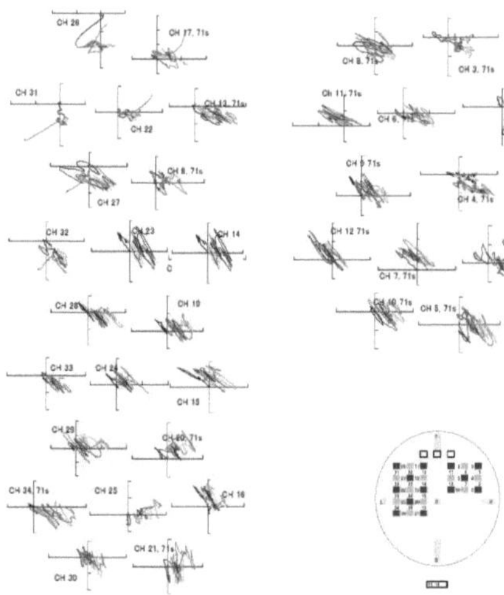

Fig. 5(a) Character size L
Positive display

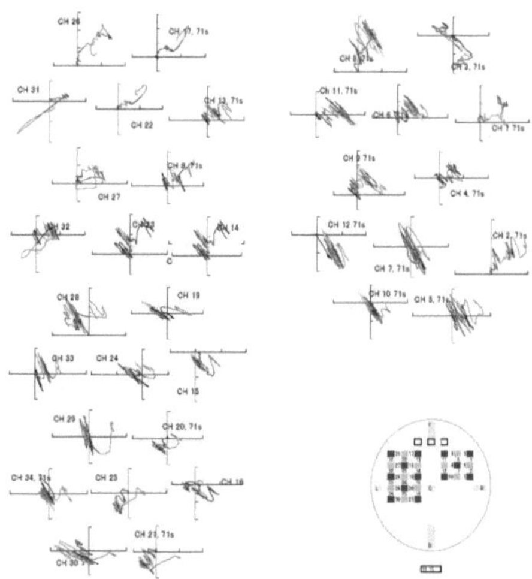

Fig. 5(b) Character size L
Negative display

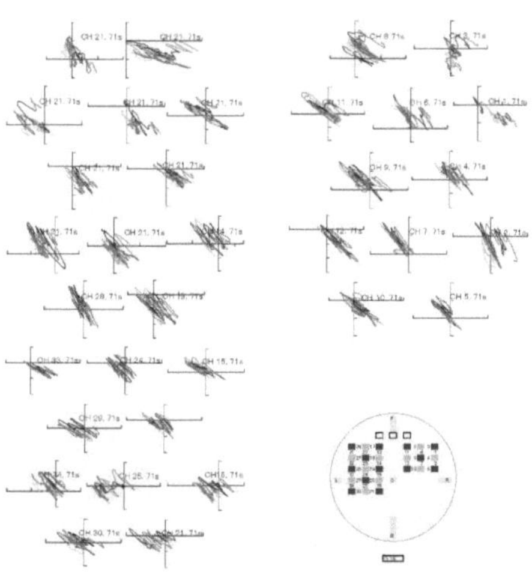

Fig. 5(c) Character size S
Positive display

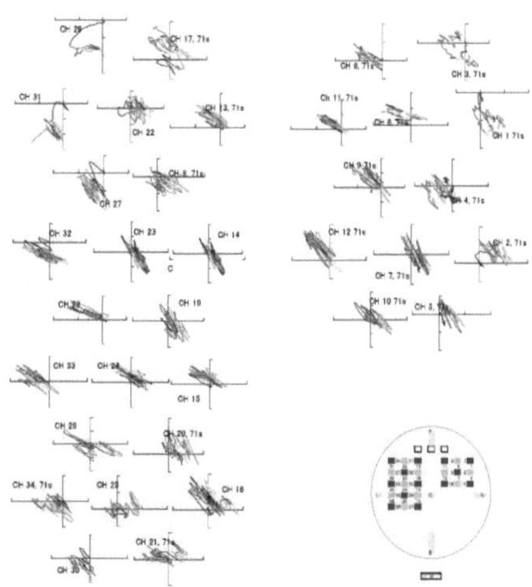

Fig. 5(d) Character size S
Negative display

Fig. 5. The plots show the correlation between oxy-Hb and deoxy-Hb during the button tapping task

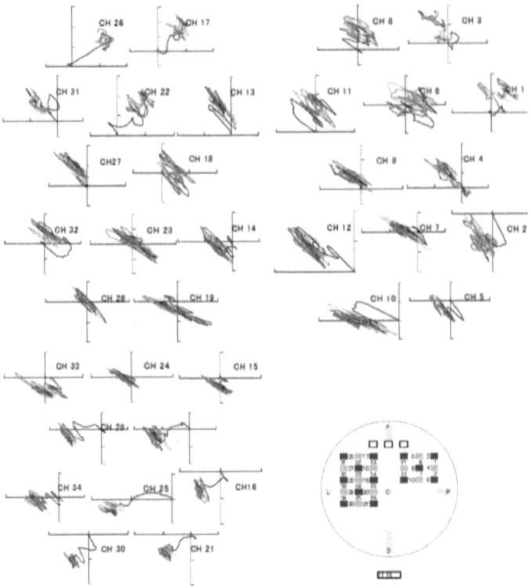

Fig. 6(a) Character size L
Positive display

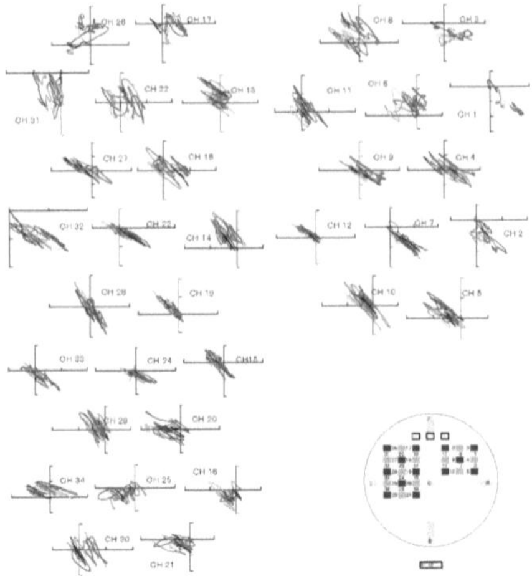

Fig. 6(b) Character size L
Negative display

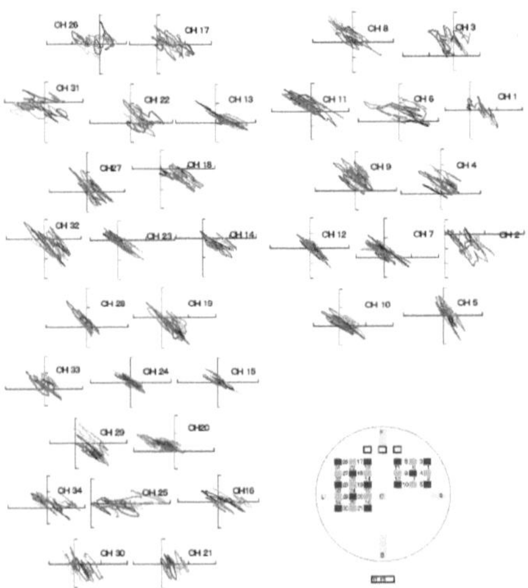

Fig. 6(c) Character size S
Positive display.

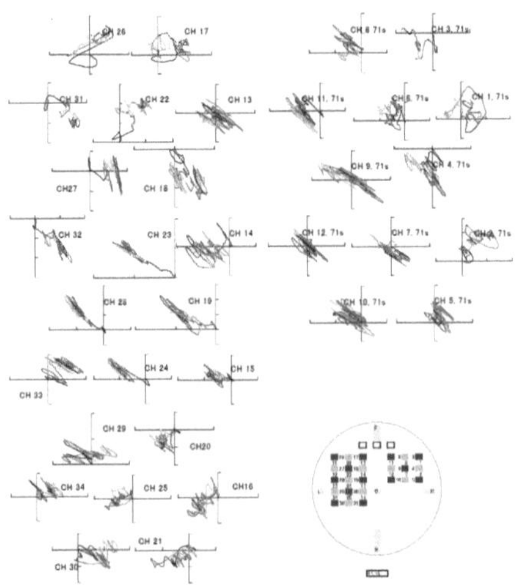

Fig. 6(d) Character size S
Negative display

Fig. 6. The plots show the correlation between oxy-Hb and deoxy-Hb during operation of mobile phone positive contrast activation over a wide area of the bilateral frontal and center regions was seen ((Fig.5 (c), Right: CH 2, 5, 6, 9, 10 Left: CH 13, 14, 15, 16, 17, 18, 19, 20, 21, 22, 24, 25, 28, 29, 30, 34). In the NIRS data, total-Hb and oxy-Hb concentrations began to increase. With negative contrast and character size S, activity was seen at CH 3, 5, 22, 20, and 16, but with positive contrast activity was seen over a wide range in the bilateral frontal and center regions.

3.3 Results for Mobile Phone Input

The two-dimensional orbital distribution with operation of a mobile phone is shown in Fig. 6. With positive contrast activation was seen in the bilateral frontal and center regions (CH 8, 11, 16, 17, 20, 21, 22, 25, 26, 29, 30, 34). It was showed the Fig. 6 (a). In the frontal region, total-Hb and oxy-Hb concentrations began to increase. However, in the central region, total-Hb, oxy-Hb and deoxy-Hb decreased. Activation in the frontal region was seen with both positive and negative contrast. Concentrations of total-Hb and oxy-Hb were increased at frontal region channels CH 2, 5, 13, 14, 17 (Fig. 6 (c)). Total-Hb and oxy-Hb concentrations were increased at central region channels CH 16, 21, 25, 30, 34 (Fig. 6 (a)). In the NIRS data for negative contrast, total-Hb, oxy-Hb and deoxy-Hb increased at CH 2, 17, 26, and total-Hb and oxy-Hb increased at CH1, 27, 31, 32 (Fig. 6 (d)).

4 Conclusion

The purpose of this study was to explore the possibility of applying neuroimaging techniques to the study of brain activity in humans performing daily communication tasks. We investigated brain activity by NIRS while the subjects searched for target characters in text without meaning. Visual stimuli consisted of two different contrasts: positive and negative. The subjects were instructed to perform three tasks when target characters were detected: (1) count the number of target characters by visual observation, (2) tap a button with the thumb and finger on their left hand carefully and as fast as possible, and (3) operate a mobile phone.

The results showed that brain activation differed among tasks. In the visual observation, positive contrast with character size L and positive and negative contrast with character size S showed rest. However, brain activity was seen when subjects tapped the button and operated the mobile phone, especially in the frontal and central region area in the case of positive contrast with character size S.

When positioning the probe in this experiment, we measured the nasal root (Fig. 1:F), posterior internode (Fig. 1:B), and the bilateral preauricular points (Fig. 1:L-C-R) in order to arrive at the derivation area according to the international 10/20 method. We then positioned the probe so that Cz was at the center (Fig. 1: C). If these derivation areas and the functional distribution of the brain are made to correspond, 2-dimensional orbits of activity were seen in the channel locations corresponding to motor area and somatosensory area in the left brain button tapping response, and to the somatosensory area in the mobile phone operation. In watching also, wide-ranging activity was seen at the channel locations corresponding to motor and somatosensory areas.

Activity was seen in the motor and somatosensory areas in relation to the button pressing task, and this result is similar to reports of increased blood flow in primary motor area and first somatosensory area with a simple task of repeatedly pressing a spring upon instruction [4].

However, a resting state was seen in the secondary orbit corresponding to motor areas with the mobile phone operation task, and action was observed only in the frontal region and somatosensory area. While button tapping and mobile phone operation are similar finger movement tasks, mobile phone operation is thought to be affected by the complexity of the finger manipulation needed for character conversion of upper and lower case letters and numerals in which the preceding and subsequent letters are input after the target is found.

In the future we would like to increase the number of subjects for a more detailed investigation of the relation between 2-dimensional orbit mas and NIRS waveforms from the waveforms of hemoglobin oxygenation (Oxy) and deoxygenation (Deoxy).

Acknowledgements. This work was supported by NIRS (Near Infra-Red Spectroscopy) facility was a product of Shimadzu co. (OMM3000/12) which is installed in Future ITC Research Center, KARC (Kansai Advanced ICT Research Center, Akashi, Kobe, Japan). The authors are honored by Joint Research Treaty with the center to use the NIRS facility.

References

1. Ito, Y., Kennan, R.P., Watanabe, E., Koizumi, H.: Assessment of heating effects in skin during continuous wave near infrared spectroscopy. J. Biomed. Opt. 5, 383–390 (2000)
2. ISO 9241-3. Amendment 1 (2000). ISO 9241-3. Ergonomic requirements for office work with visual display terminals (VDTs) –Part 3: Visual display requirements. International Standard (1992)
3. Tamura, H., Omori, M., Choui, M.: Brain Activities indicated by two dimensional trajectories of NIRS. Mobile Interactions 3, 9–14 (2007) (in Japanese)
4. Roland, P.E., Larsen, B., Lassen, N.A., Skinhoj, E.: Supplementary motor area and other cortical areas in organization of voluntary movements in man. J.Neurophysiol. 43, 118 (1980)

A Mobile Terminal User Interface for Intelligent Robots

Ji-Hwan Park[1], Gi-Oh Kim[2], Pham Dai Xuan [1], Key Ho Kwon[1], Soon-Hyuk Hong[2], and Jae Wook Jeon[1]

[1] School of Information and Communication Engineering, Sungkyunkwan University Chunchun-Dong, Jangan-Gu, Suwon City, 440-746 Korea
jwjeon@yurim.skku.ac.kr
[2] Telecommunication R&D Center Samsung Electronics
Maetan-Dong, Yeongtong-Gu, Suwon City, 442-600 Korea

Abstract. Intelligent robots can obtain environmental information by processing sensor signals, understand contextual information, and perform necessary tasks autonomously. They can also perform tasks directed by commands from a user. They can be used for housekeeping, entertaining, and educational purposes. In order to use intelligent robots efficiently, a convenient user interface is needed to control them. In this study, we analyze how to organize such a user interface according to the function and level of autonomy of intelligent robots and develop a user interface for controlling them remotely via a PDA (Personal Digital Assistant) or mobile phone.

Keywords: robot, level of autonomy, user interface, mobile device.

1 Introduction

Previous generations of robots mainly performed simple and repetitive tasks for industry, such as assembling, welding, and paint spraying. Nowadays, however, more advanced robots are available which are capable of having a more intimate relationship with humans and, as such, can be used in the fields of medicine, welfare, and even administration. Intelligent robots can perform various services such as housekeeping, education, and entertainment through their interaction with human beings. Their construction brings together a number of difference specialties, including electronics, information, semiconductors, artificial intelligence, medical engineering, bionics, mechanical engineering, advanced materials, and MEMS technologies. In existing industrial robots, the position and velocity controls were mainly developed to precisely perform repetitive tasks and a simple UI (user interface) was developed in the form of a sequential program [1]. However, since the functions and LOA (Level of Autonomy) of intelligent robots have been improved, a different UI needs to be developed for them [2]. Also, in order to use intelligent robots more efficiently, a method needs to be developed which intelligent robots can use for delivering the information obtained by self learning and sensor signal processing. In particular, in the case where mobile terminals such as PDAs and mobile phones are used to control remote intelligent robots, the availability of a UI for intelligent robots becomes even more important.

M.J. Smith, G. Salvendy (Eds.): Human Interface, Part I, HCII 2007, LNCS 4557, pp. 903 – 911, 2007.

In this study, we propose and implement a UI to control various intelligent robots efficiently using mobile terminals. In section 2, intelligent robots are analysed based on their UI functions. In section 3, we analyse the manner in which each function of a mobile terminal can be used to develop a UI for controlling intelligent robots. In section 4, a UI for intelligent robots is developed using mobile terminals such as PDAs and mobile phones. Finally, our conclusions are presented in section 5.

2 Intelligent Robot Interface

In contrast to industrial robots that perform repetitive tasks in a structured environment, intelligent robots should be able to move and determine their actions autonomously in response to the changes that occur in and their continuous interaction with an unstructured environment which inevitably includes humans. The interfaces of intelligent robots can be classified as shown in Fig. 1 according to the type of interaction: robot to environment, robot to robot/information device, and robot to human.

The interface between intelligent robots and their environment represents the method of obtaining information by processing sensor signals. That is, intelligent robots can use various sensors and cameras to collect information from their environment and use this information to perform tasks such as collision avoidance and self localization.

Since information devices can communicate with each other through a network, intelligent robots can communicate with other robots or information devices. Intelligent robots can communicate with a home network server to control home digital appliances using IrDA, Bluetooth, and wireless LAN.

The interface between a human being and an intelligent robot is called an HRI (Human Robot Interaction) and it includes technology to deliver commands from the human being to the robot and to transfer the information concerning the robot and task environment from the robot to the human being. The HRI can be operated using a remote controller, a mobile terminal, or human natural senses such as sight, auditory, and tactile senses. A remote controller can be used to give simple commands to the robot over a short distance. A mobile terminal can be used to exchange various type of information such as voice and vision with robots over a short or long distance. Since intelligent robots can recognize human voices or faces, human beings can communicate with robots using their voice or facial expression, with no equipment being required between the human and the robots.

Robots controlled by a remote controller have built-in programs related to their tasks. They can only perform predetermined tasks corresponding to simple commands from a remote controller. Education robots, such as maze robots and line tracing robots, whose purpose is to help students to understand the basic concepts of robotics, household robots such as cleaning and lawn mower robots and toy robots, are examples of robots controlled by a remote controller [3-9]. For these robots, the interface with their environment is considered to be more important than that with human beings. Therefore, most of these robots use infra-red sensors and supersonic sensors to perform tasks in a given environment.

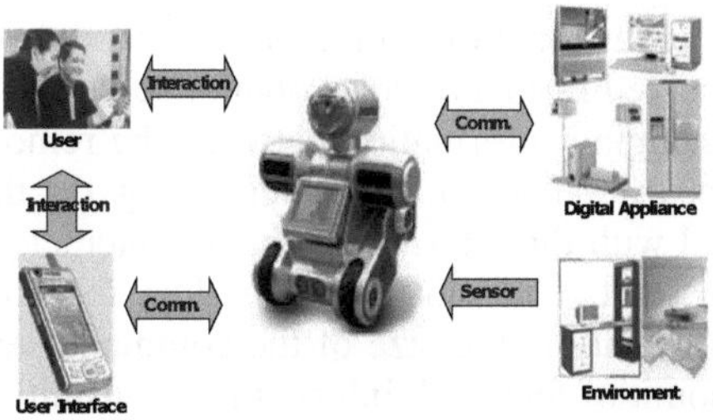

Fig. 1. Interface of intelligent robots

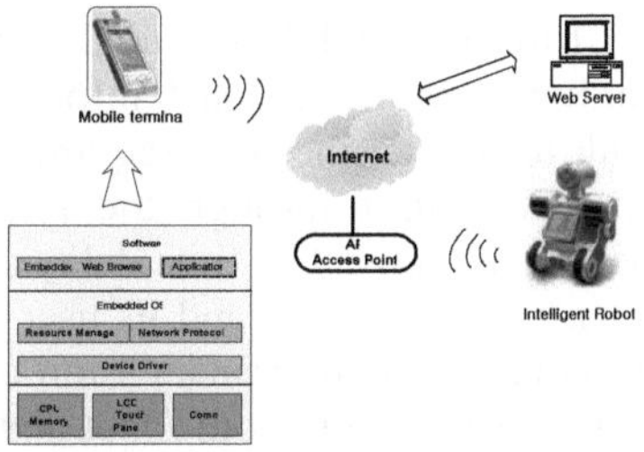

Fig. 2. Robot control using a mobile terminal

Robots controlled by a mobile terminal over a long distance have wireless communication capability and can send information about a remote location to human beings using cameras or sensors [7, 10, 11]. Humans can control remote intelligent robots easily using the multimedia functions provided by a mobile terminal. As shown in Fig. 2, a remote robot can be controlled by a mobile terminal through the internet. In this case, the robot is connected to the server with a wireless LAN and the server supplies the necessary information to the human to control the robot [12-14]. The images obtained using cameras connected to the robot or server can be used as feedback information, together with the information concerning the robot and its surrounding environment obtained using other sensors [15-18]. When exchanging information through the internet, the time required to transfer the data cannot be estimated exactly and, therefore, the remote control of the robot may be difficult to achieve. Therefore, the amount of information exchanged should be kept to a minimum, in order to reduce the uncertainty caused by the time delay associated with the exchange of information [12]. In order to reduce amount of information exchanged, the LOA of the robot should be increased, so that the commands and feedback information can be made more simple.

Robots controlled by humans' natural senses have various functions incorporated into them so that they can be controlled easily without any special devices being required [19-24]. Technologies designed to recognize the sound of clapping or simple words and sentences, or to recognize humans or objects, have been utilized in these robots. Touch screens and microphones attached to robots are used as the input devices, while speakers, LEDs (Light-Emitting Diodes), and LCDs (Liquid Crystal Displays) are used as the output devices. Future robots will recognize the intention of human beings in a more advanced way, in order to perform a desired task and provide the latter with the result.

3 User Interface for Intelligent Robots Using a Mobile Terminal

In contrast to a PC (Personal Computer), a mobile terminal has a small size display and a limited number of I/O devices. Therefore, this restriction should be taken into consideration when developing a UI for intelligent robots using a mobile terminal [16].

A UI using a mobile terminal consists of input functions for user commands and output functions for feedback information received from a remote robot or server. It is difficult to predict the time required for exchanging information between the remote robot and its user through the internet, because this can vary depending on the ambient conditions. The time delay associated with the internet makes the interaction between a remote robot and its user more difficult. Thus, in order to reduce the time delay required for the transmission of the information, the size of the command sent from the user to the remote robot and the amount of feedback information transmitted from the remote robot to the user should be reduced.

3.1 User Command Input

Since a mobile terminal includes several input devices such as external buttons, a touch screen, microphone, and camera, these devices can be used for controlling a remote robot.

The external buttons on a mobile terminal can be used by the user to directly control the motion of low LOA robots. Using input devices such as wheels or joysticks makes it easier to control the direction and speed of the robot. By using previously defined function keys, it is possible to make high LOA robots perform specific tasks. The user can give the desired commands to the robot using the touch screen of the terminal either by pressing a button, using the virtual keyboard, performing drag and drop or designating some area on the screen. He or she can provide voice or images using a microphone or camera, respectively. Also, voice and images can be used to control a remote robot or to authenticate the user.

If the LOA of the robot is increased and supervisory control can be exercised for the robot, then the user does not have to control the robot in full and this will decrease the amount and frequency of the data exchanged to control the robot, thereby reducing the time delay [25].

3.2 Output of Feedback Information

Users have to receive feedback information from a remote robot in order to be aware of the robot's status, task status, and environment during the control of the robot. Image information is usually used for feedback information and most of these images are transferred using the JPEG (Joint Photographic Experts Group) image type. Using JPEG images decreases the transmission time compared with using the raw images. However, in some cases, the time required to send even JPEG images is too long. To solve this problem, JPEG image information should only be used to check the result of execution of a given task, and other, simpler types of information should be sent for the purpose of verifying the execution result [26-28]. Such simpler types of information include numbers or characters that can be easily transmitted, sounds such as voice or beeps, and the vibration of the terminal.

The LCD device of a mobile terminal can show images, characters, and graphs of numerical data to represent the information received from the remote robot or the server. By processing this information, the LCD can display the robot on a map or a in a 3 dimensional virtual environment [29-30].

If a remote robot encounters an unexpected situation while performing its tasks, it should inform the user of the situation, and using sound and vibration is an efficient way of accomplishing this. The information on the robot's status can easily be represented using previously recorded voice and beep sounds, or by using the vibration of the terminal. This method provides an efficient way of sending information to the user, when he or she is not paying attention to the motion of the remote robot.

3.3 Classification of Mobile Terminal UIs

The ideal solution would be to develop a customized UI for each robot. However, this is not a viable solution and, therefore, it is better to develop an integrated UI according to the LOA and functions of a given type of robot. Table 1 shows the classification of UI functions in a mobile terminal according to the LOA, motion, and task of a robot.

Movement Control classifies the UI according to the type of robot movement. *Direct Control* is a method of UI that uses external buttons or buttons on the display device of a mobile terminal to control a low LOA robot. *Destination Assignment* is a method of controlling a robot using the display device of a mobile terminal. In this case, the user can watch the image and then designate a certain point toward which the robot should move. This can be accomplished by pressing a destination point on the touch screen or entering a set of coordinates using a virtual keyboard. *Tracking* means that a robot uses sensor information to track a moving object and therefore a simple UI is needed. *Autonomous Motion* means that a robot can decide

Table 1. Classification of UI Functions

User Interface		***Intelligent Robot***
Movement Control	Direct Control	Moves under manual control.
	Destination Assignment	Moves to a destination.
	Tracking	Follows a moving object.
	Autonomous Motion	Decides the context and performs an appropriate task autonomously.
Task Control	Modularization of robot functions	Performs an appropriate task according to the function
Sensor Based Information Output	Numbers and Characters	Provides information using numbers and characters.
	Image	Provides information using images from a camera
	Map	Displays the location of a robot on a previously defined or sensor based map
	Virtual Environment	Displays a 3 dimensional virtual environment
Supervisory Control	Request for additional user commands	Requests additional commands from the user when a robot cannot make a decision autonomously

Fig. 3. A mobile terminal for controlling a robot

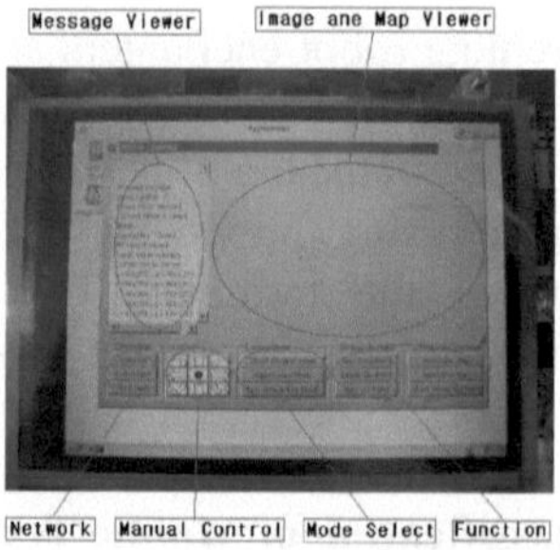

Fig. 4. Image based user interface

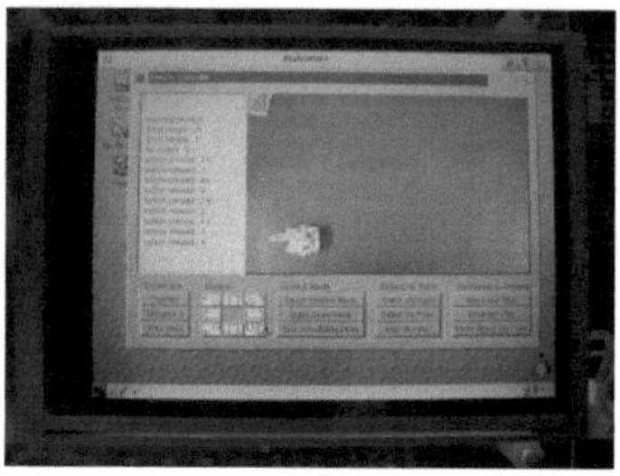

Fig. 5. Direct Control of robot

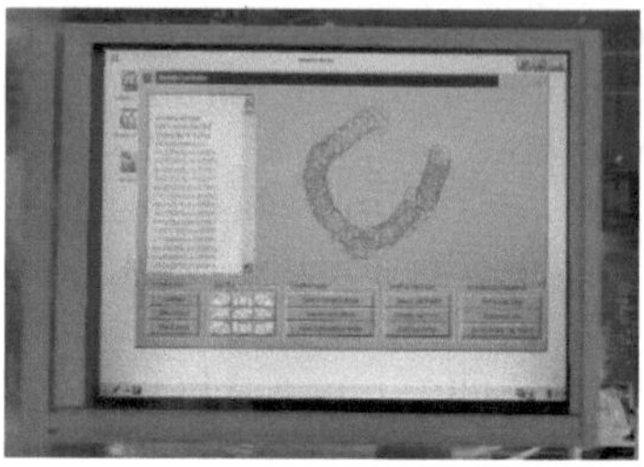

Fig. 6. Display of robot locations

autonomously based on its understanding of the environmental context and execute an appropriate task and therefore the minimum UI is needed.

Task Control refers to the modularization of a UI according to the functions of a robot. Since the function of the robot can determine its UI, it needs to be modularized, so that the necessary buttons can be added for each function that it has to perform. For example, the UI for a housekeeping robot should include cleaning functions.

Sensor Based Information Output refers to the feedback from the robot designed to give its user the necessary environmental information, including information on the robot itself, by using the robot sensors. *Number and Characters* is a method of providing this environmental information using numbers and characters. *Image* is a method of providing the environmental information using images obtained by robot cameras or server cameras. *Map* is a method of supplying the environmental information such that the robot location is displayed on a simple map of the environment obtained using the robot sensors. The robot location can also be displayed on a predefined map. *Virtual Environment* is a method of supplying the environmental information using a 3 dimensional virtual environment that is similar to the real environment of the robot.

Supervisory Control is a method of providing additional user commands when the robot is in need of input from the user, because it cannot make the decision required to perform the next task autonomously under a given condition. User commands are given to the robot only when it requests them.

The functions listed in Table I are combined as necessary to develop a UI for a robot depending on its LOA and purpose.

4 Development of User Interface

As described above, we can develop one UI function that is based on the image information on a mobile terminal to control an intelligent robot with other UI functions. Also, we can develop another UI function that is based on map information in order to control it with other UI functions. In some cases, these two UI functions can be combined to control a robot.

4.1 Image Information Based User Interface

The mobile terminal shown in Fig. 3 was developed for the purpose of controlling robots. It consists of an Intel XScale PXA255, an LCD to show the feedback information, an input touch panel, and CF card and PCMCIA card slots. A wireless LAN card or other devices can be connected to these slots. Embedded Linux is used as the operating system of the mobile terminal. The Qt/Embedded GUI toolkit is used for developing the GUI.

Fig. 4 shows the UI that is based on image information to control a robot. It contains *network buttons* to connect the mobile terminal to the server, *manual control buttons* to directly control the robot, *mode select buttons* to choose the control mode, and *function buttons* to execute the functions of the robot. It also contains an *image and map viewer* to show the feedback information, and a *message viewer* to show the character information received from the robot or server.

In order to allow the user to directly control the robot, the UI can display the image of the robot's surroundings on the mobile terminal transmitted from the camera of server, as shown in Fig. 5. In order to reduce the amount of information transmitted, the server can calculate the current location of the robot from the camera images and send the location data to the mobile terminal. Then, the UI can display the location of the robot in the form of a circle on the mobile terminal as shown in Fig. 6.

4.2 Map Information Based User Interface

To develop a UI that is based on the map information, a Samsung Electronics MITs PDA was used. This PDA consists of a 200Mhz S3C24A0ARM9 processor, an LCD to show the feedback information, and an input touch panel. It has IEEE 802.11b functions, its operating system is Embedded Linux and the Qt/Embedded GUI toolkit is used for developing the GUI.

Fig. 7 shows the UI that is based on map information using the Qt/Embedded toolkit. It contains a *robot position part* to show the coordinates of the current robot position, a *robot move button* to send robot movement commands to the server, *map zoom in/out buttons* to magnify and reduce the size of the map, and a *map viewer* to show the robot's location. In order to make the robot move to a destination point, the

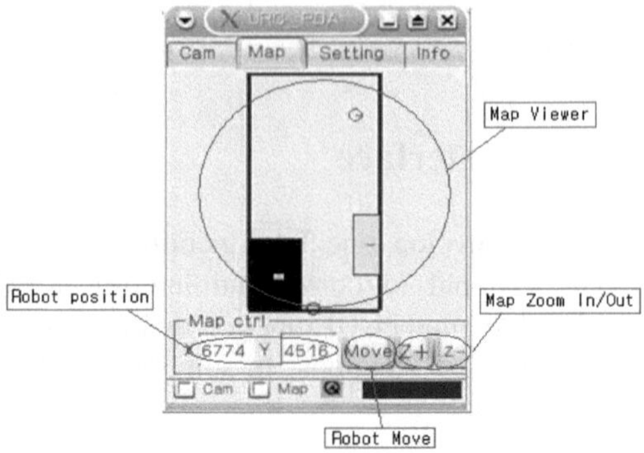

Fig. 7. Map based user interface

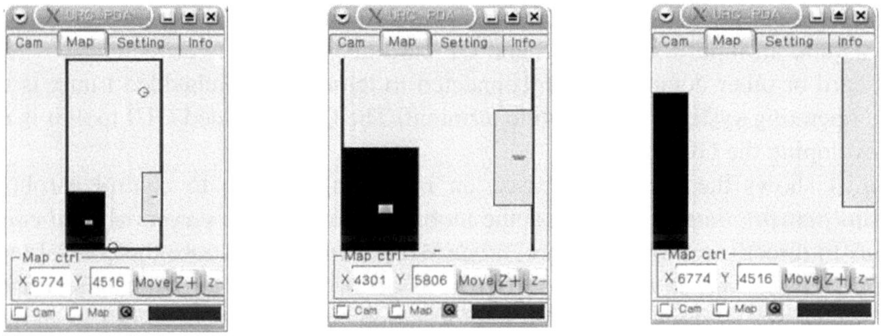

(a) Original size (b) Double size (c) Quadruple size

Fig. 8. Magnification of a map

destination point of the robot can be designated by simply pressing a specific point on the touch screen of the LCD and its coordinates are sent to the server by pressing the *robot move button*. Fig. 8 shows the double sized and four times magnified maps obtained using the *map zoom in/out buttons*.

5 Conclusion

It is not viable to develop a dedicated UI for each intelligent robot. Therefore, it is necessary to make a standard UI and to integrate the necessary UI functionalities depending on the LOA of the intelligent robot and its functions. Thus, each component of the UI should be modularized, in order to make it easier to add and remove each component of the UI. In this way, it becomes possible to make an appropriate UI for a given robot efficiently. Furthermore, it will also be possible to modify the UI depending on the changes made to the robot functions.

Acknowledgement. This research was supported by the MIC, Korea under ITRC, IITA-2006-(C1090-0603-0046).

References

1. Fu, K.S., Gonzalez, R.C., Lee, C.S.G.: Robotics: control, sensing, vision and intelligence. New York, McGraw-Hill (1987)
2. Guglielmelli, E., et al.: A modular and distributed supervisory system for a semi-autonomous personal robot for household applications. Proc. of the 8th International Conf. ICAR Robotics (July 7-9, 1997)
3. Microrobot, http://www.microrobot.com
4. Izirobotics, http://www.izirobotics.co.kr/
5. Sharper image, http://www.sharperimage.com
6. Friendly robotics, http://www.friendlyrobotics.com
7. iRobot, http://www.irobot.com
8. ElectroLux, http://trilobite.electrolux.com
9. Karcher, http://www.robocleaner.de
10. Personal Robotics, http://www.personalrobots.com
11. Joy mecha, http://www.joymecha.com
12. Luo, R.C., et al.: Network intelligent robot through the Internet. In: Proc of the IEEE, March 2003, vol. 91(3), pp. 371–382 (2003)
13. Yan, L.: A Study of teleoperation of robotic system via the Internet, University of Alberta at Canada (2001)
14. Chanitnan, Internet-based control, MS Thesis, University of Texas of Arlington (May 2002)
15. Baruch, J.E.F., Cox, M.J.: Remote control and robots: An Internet solution. IEEE Comput. Control Eng. 7(1), 39–44 (1996)
16. Lundberg, C., et al.: PDA interface for a field robot. International Conference on Intelligent Robots and System, pp. 27–31 (October 2003)
17. Ok, J.S., et al.: The Remote Control of Mobile Robots on the Web. The. Korean Society of Power System Engineering 4(4), 78–83 (2000) (in Korean)
18. Fong, T., Thorpe, C., Baur, C.: Multi-Robot Remote Driving With Collaborative Control. IEEE Transactions on Industrial Electronics 50(4), 699–704 (2003)
19. Tomy, http://www.tomy.co.jp
20. ROBOTIS, http://www.robotis.com
21. Tiger Electronics, http://www.tigerelectronics.com
22. Digitek, http://www.digitek.co.kr
23. SAIT, http://www.hcilab.com/robot
24. Yujin robotics, http://www.yujinrobot.com
25. Sheridan, T.B.: Telerobotics, automation, and human supervisory control. MIT Press, Cambridge, MA (2000)
26. Luo, R.C., Chen, T.M.: Remote supervisory control of a sensor based mobile robot via Internet. In: Proc. of the Conf. IEEE/RSJ International, September 1997, vol. 2, pp. 7–11 (1997)
27. Lee, J., Hsu, P.: Remote supervisory control of the Human-in-the-Loop System by Using Petri Nets and Java. IEEE Transactions on industrial electronics 50(3), 431–439 (2003)
28. Bruemmer, D.J., et al.: Mixed-Initiative Control for Remote Characterization of Hazardous Environments. In: Proc. of the 36th Hawaii International Conf. on System Sciences (2002)
29. Jo, J., et al.: Virtual Testing of Agile Manufacturing Software Using 3D Graphical Simulation. In: Proc. of the IEEE International Conf. on Robotics and Automation, Albuquerque, April 1997, pp. 1223–1228 (1997)
30. Safaric, R., et al.: Control of Robot Arm with Virtual Environment via the Internet. In: Proc. of the 2003 IEEE, March 2003, vol. 91(3), pp. 422–429 (2003)

A Modular User Interface of Robots

Ji hwan Park, Tae Houn Song, Key Ho Kwon, and Jae Wook Jeon

School of Information and Communication Engineering, Sungkyunkwan University
Chunchun-dong 300, Jangan-gu, Suwon 440-746, South Korea
fellens@gmail.com, thsong@ece.skku.ac.kr,
{khkwon, jwjeon}@yurim.skku.ac.kr

Abstract. This paper discusses development of a control system for home robots. When a user requests a user interface for home robot, the home robot sends this robot's user interface to the user's PDA. Then, the robot can be efficiently controlled from this user interface on the PDA. The home server distributes the service of robots, and manages the transmission of service interface modules. When this procedure is used, home network traffic is reduced, and the user can simply access a service of a robot.

Keywords: user interface, robot, UI, interface module, robot interface.

1 Introduction

As robots evolve, robots are increasing becoming move widely used for a variety of functions. This increased usage makes robots' interface programs very important. A robot generally is in need of special controllers or application programs. When a robot is controlled by using a specific controller, the mobility and expandability of the robot can be attenuated. When a user installs an application program for robot control from a PC, the resource of the PC is wasted. In many situations of robot control, the user must re-install the application program when the PC changes. Also, A PC is not mobile whereas a PDA is. The user interface programs used for robots are not necessarily similar for each robot. So a user must study several new user interfaces when robots change. In previous work, the robot control method via internet is showed good performance [1-2].

Most robots operate and think independently of other robots. However, robot technology and computer networks are developing rapidly, and many research organizations have attempted to integrate robot technology and communication networks. This research assumes that robots interact with each other or a computer network, and can perform a large number of functions [3].

In this paper, we propose a PDA based user interface module management system for a home service robot. The user interface is divided into service modules and standard interface modules. These modules are made from XML. The robot language to control and manage robots is many. XML based RobotML exist of them [4]. RobotML is suitability and useful language at internet environment. And XML is simple grammar and very easy. Also data can be expressed intuition [5].

M.J. Smith, G. Salvendy (Eds.): Human Interface, Part I, HCII 2007, LNCS 4557, pp. 912–921, 2007.

A home server manages the service robot via a home network. The home server has a service list for the service robot which is connected to the home server through a network. When a user connects to the home server by means of the user's PDA and then requests any service, the home server searches for the robot which has the requested service. The home server then transfers this robot's user interface module and service list to the user's PDA. We assume that the robot has a powerful CPU, large data storage, and a service interface module for this proposal. Also we assume that the command protocol is standardized.

Then, when we use this system and connect to the network, we can control any robot anywhere on the network. When a user requests a service, the home server transfers the service interface module of the robot which has the requested service. So a user can control the robot anywhere he or she has access to the network. Also resources of the home server are saved because it is necessary that a user interface of service robots be installed.

Section 2 describes the interface transmission system. Section 3 describes the service interface module and standard interface module. Section 4 describes a result of the user interface program; Section 5 provides a conclusion and presents some directions for future work.

2 Interface Transmission System

We now describe the user interface modules management system for a robot. The robot interface program consists of many different interface modules. A security robot consists of a remote supervising computer, a security module, application modules, a RF interface, and GSM controller modules. These components of the robot system manage the robot function [6]. In our work described in this paper, the modules are grouped into standard interface modules and service interface modules and then managed. The standard interface modules have the common function of the service robot and are stored in a user's PDA.

In this paper, we assume that the command protocol is standardized. So the standard interface module can control all robots. The service interface module controls special functions. The service interface modules are stored in the service robots or on the home server. When the service interface module is requested, the service interface modules are transferred to the user's PDA.

2.1 Home Server and Service Robots

Classically, a home server controls the electronic products in a home that are connected to the home server [7, 8]. In our work, the home server controls home service robots also. A user's PDA is connected to the home server through a home network. So, the role of home server is that of a bridge between the user and the service robots [7, 8]. When the service robot requests registration, the home server gives an ID to the robot and requests a service list from the service robot. When the home server receives the service list, the home server registers and renews the service list of the home server. Then the home server has a service list for all of the robots.

When a user selects any service, the home server searches the service list of the home server for the requested service. The home server then links the user with the robot which has the requested service and higher priority.

A service robot supports services for a user. The service robots have a service list which is list of the supported services for the robots. The service list is transferred from the service robot to the home server. The service robot has a service interface module which is adequate to support the service list. This module is implemented using XML and has command protocols and a layout of a user interface.

2.2 Personal Digital Assistant (PDA)

The user must select the controller which controls the service robot. A PDA is widely used and has more mobility than other possible control devices. A PDA can connect to a network to using a wireless LAN. Because of this possibility, a PDA can be a good controller. When a user connects to a home server, the home server transfers a service list from the home server to the user's PDA. A user can select services that are indicated in a black color, and not select services that are listed in a gray color. The gray color indicating unavailability of a service means that another is using that service.

When a user selects a service, the user's PDA requests the selected service. After, the home server searches for a robot which supports this requested service. The home server then transfers a service list and service interface module for the robot to the requesting user's PDA. Then the user's PDA that received these components builds a user interface program with the service list, the service interface module, and the standard interface module.

2.3 Interface Transmission

Figure 1 shows a flow chart of the transfer of the service interface module to the user's PDA. First, when a service robot is powered on, the service robot checks the home network periodically. When a home server is found, the service robot connects to the network and requests registration on the home server. To control the robot, a user must connect to the home server. The service robot is also connected to the home server and has one service or more. When the service robot requests a service registration on the home server, the home server gives an ID to the robot. Then, the home server demands a service list of the service robot. The service list has all the possible functions for the service robot. When a service robot transfers its service list to the home server, the home server registers and renews the service list of the home server. The home server decides priorities of all the robots.

To control any service robot, a user must connect to the home server. When a user's PDA is connected to a home server, the home server can transfer a requested service list. Then a user will select which service to use. The PDA requests service to the home server. The home server then requests a service search for a robot which is supported and has a highest priority. If the service robot is found, the home server requests this service robot's service list and the service interface module of this service robot. Then, the service robot transfers a service list and a service interface module to the home server. The home server which received the service list and

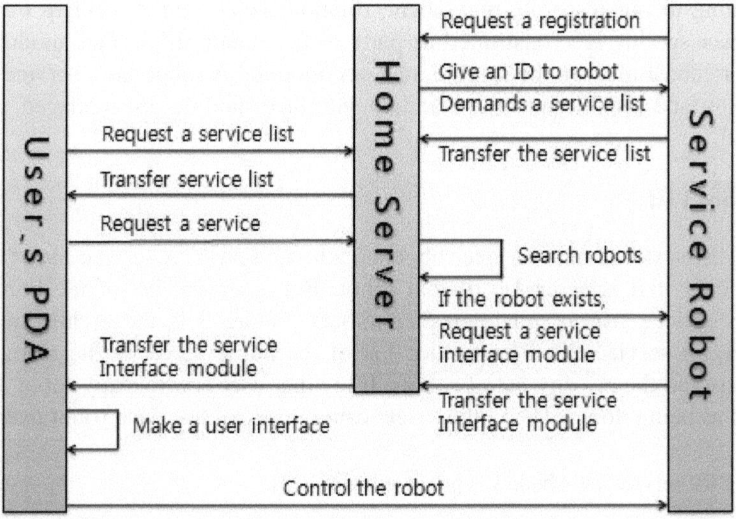

Fig. 1. A flow chart for the interface transmission system

service interface module then transfers these components of the service robot to the requesting user's PDA. This user's PDA, upon receiving these components, builds a user interface program into these components and standard interface module.

Even if a user requests the same service at another time, a new user interface is built as it is may not the same as a prior time, because the robot which is now selected may be different.

3 Interface Modules

In our work described in this paper, a user interface is in the user's PDA. This user interface is divided into service modules and standard interface modules. The standard interface module is a common part of the interface. The service interface module is a module transferred to a service robot. Table 1 show that many robots may have common services. A large number of robots have movement, rotation, voice output, camera output, and a range sensor service [9-18].

Table 1. The common function of robots

	Movement	*Voice Output*	*Camera Output*	*Range sensor*
Patrol robot(5)	5	3	5	5
Cleaning robot(6)	6	1	1	3
Housework robot(5)	5	4	5	5
Toy robot(14)	12	3	3	6
Military robot(10)	10	3	8	7
Total(40)	38	14	22	26

According to our research, movement, rotation, voice output, camera output, and range sensor service are constituted as parts of the standard interface module. Other services are constituted in the service interface module. A robot has a service list. The service interface module and the standard interface module are managed using the service list.

3.1 Service List

A service list is stored on a service robot and a home server. A service list of the robot is stored only if it is a service of that robot. But a service list of the home server represents service lists of all connected robots. Figure 3 is the architecture of the home server's service list. The service list of the home server is the stored service robots' lists and the priority list of robots. If another user controls any robot, the robot is marked as being down. Then other users cannot control a service robot overlap.

Example of the service list which was produced with XML

```
<server>
    <service>
        <robot name="Patrol robot">
            <ID>2</ID>
        </robot>
        <robot name="Cleaning robot">
            <ID>1</ID>
        </robot>
        <service_list>
            <service_rot name= "movement"/>
                <ID="2" priority="10"/>
                <ID="1" priority="9"/>
            </service_rot>
            <service_rot name= "range sensor"/>
                <ID="2" priority="10"/>
                <ID="1" priority="3"/>
            </service_rot>
        </service_list>
    </service>
</server>
```

An example is the service list that constitutes XML. The service list of the service robot is different from the service list of the home server. The service list of the home server is stored as a command protocol of all functions of the robot and the supported service. For example, we have camera output, navigation, and patrol service.

3.2 Interface Modules

Many robots have common functions. Generally, a robot has functions of movement and rotation. Recently robots support both a voice output and camera output. We define a standard interface module if that module is made with common functions.

When any robot uses standard protocol which is suitable for a standard interface module, a user can control the robot by using the standard interface module. The standard interface module is contained in the user's PDA.

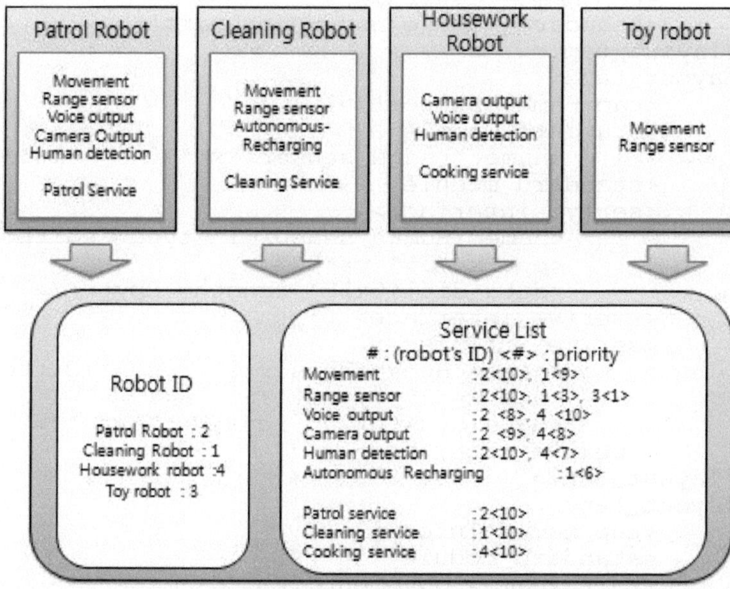

Fig. 2. The architecture of the Service List in a Home server

A service interface module is a special interface module. It supports functions without a standard interface module. Each robot has a different service interface module. The service interface module is implemented using XML. When a home server requests a service interface module, the service interface module is transferred from the robot to the home server. It can use special functions which are not included in the standard interface module. It defines a layout for the user interface. For example, some examples of special functions are the emotion of toy robots or a mine detector of military robots.

3.3 XML Based User Interface Programs

The service interface module and the service list are implemented in XML. The service interface module is stored protocols and a layout of a user interface. When the service interface module and the service list is transferred to a user's PDA, a PDA builds a user interface program with the service interface module and standard interface module. When a robot is changed, a robot interface program will be changed. But these are similar interface programs. So, it is easy to learn how to use the slightly changed program.

Example of a Service Interface Module

```
<ui_map>
    <layout_tap1>
          <tap name="Control"/>
          <standard_module name="movement"/>
    </layout_tap1>
    <layout_tap2>
          <tap name="Sensor"/>
          <standard_module>
                <name="range_sensor" x="10" y="10"/>
          <standard_module>
          <serive_interface>
                <name="Human_sensor 1" type="editbox"/>
                <pos x="35" y="20"/>
                <data did="0x1F" length="2byte"/>
          </serive_interface>
          <serive_interface>
                <name="Human_sensor 2" type="editbox"/>
                <pos x="35" y="10"/>
                <data did="0x20" length="2byte"/>
          </serive_interface>
    </layout_tap2>
    <layout_tap3>
          <tap_name="Info"/>
          <standard_module>
                <name="movement"/>
          </standard_module>
    </layout_tap3>
</ui_map>
```

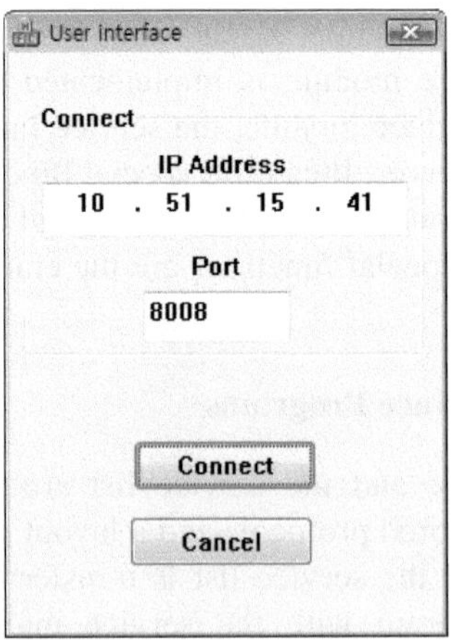

Fig. 3. The home server connection screen of the user interface

4 Results

Figure 4 is a screen showing attempts to connect to a home server. When a user enters the IP of the home server in the edit window, the user interface program tries to connect to the home server. When a user connects to the home server, a user interface program requests the home server to transfer a service list from the home server.

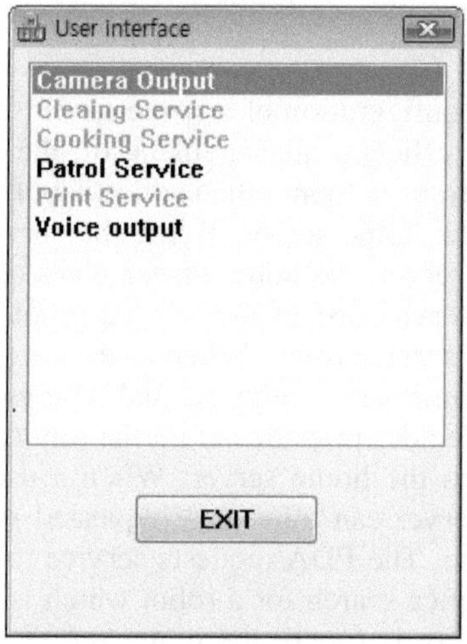

Fig. 4. The service list menu

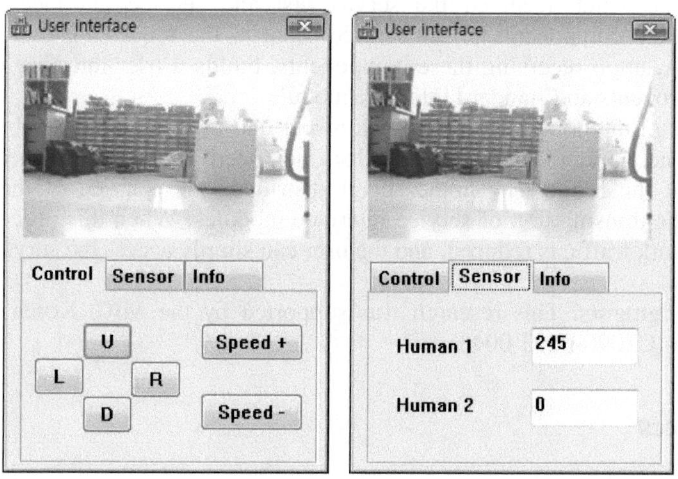

Fig. 5. The user interface program of the robot which is supported the patrol service

When the home server transfers the service list to the user's PDA, the user interface program in the PDA shows a service list screen. Figure 5 show that list screen. Currently, a user only selects the patrol service. The other services – cooking, print and the cleaning services – are used by other user. So these services are shown in a gray font.

When a user selects the patrol service, the user interface program in a PDA requests a service interface module and service list of the service robot. Figure 6 is a user interface program which builds a user interface program with these and the standard interface module.

5 Conclusion

This paper proposed the configuration of a home server, home network, and interface transmission system for efficient management of the home robot by the home network. When a home server is found, the service robot connects to the network and requests registration on the home server. When the service robot requests a service registration on the home server, the home server gives an ID to the robot. Then, the home server demands a service list of the service robot. The service list has all the possible functions for the service robot. When a service robot transfers its service list to the home server, the home server registers and renews the service list of the home server. The home server decides priorities of all the robots.

A user must connect to the home server. When a user's PDA is connected to a home server, the home server can transfer a requested service list. Then a user will select which service to use. The PDA requests service to the home server. The home server then requests a service search for a robot which is supported and has a highest priority. If the service robot is found, the home server requests this service robot's service list and the service interface module of this service robot. Then, the service robot transfers a service list and a service interface module to the home server. The home server which received the service list and service interface module then transfers these components of the service robot to the requesting user's PDA. This user's PDA, upon receiving these components, builds a user interface program into these components and standard interface module.

The role of the home server is to link robots and user. When new robot is added to the home network, the home server does not need to install the special interface program of the robot. The home server distributes the service of the robots, and manages the transmission of service interface modules. When this procedure is used, home network traffic is reduced, and the user can simply access the service of a robot.

Acknowledgments. This research was supported by the MIC, Korea under ITRC, IITA-2006-(C1090-0603-0046).

References

1. Suzuki, T., Fujii, T., Yokota, K., Asama, H., Kaetsu, H., Endo, I.: Teleoperation of Multiple Robots through the Internet. Proc. of IEEE Int. Workshop on Robot and Human Communication, pp. 84–89 (1996)
2. Taylor, K., Dalton, B., Trevelyan, J.: Web-Based Telerobotics. Robotica 17, 49–57 (1999)

3. Jung, B.-c., Park, J.-h., Choi, D.-s., Kim, H.-m.: A study on intelligent robot based on home network. The. Korea Society of Mechanical Engineers autumn collected papers, pp. 792–798 (2003)
4. Makatchev, M., Tso, S.K.: Human-robot interface using agents communicating in an XML-based markup language. Robot and Human Interactive Communication, 2000. RO-MAN 2000. Proceedings. 9th IEEE International Workshop, September 27-29, 2000, pp. 270–275 (2000)
5. Young, M.J.: Step by Step XML. Microsoft.
6. Luo, R.C., Hsu, T.Y., Lin, T.Y., Su, K.L.: The development of intelligent home security robot. ICM '05. IEEE International Conference on Publication, July 10-12, 2005, pp. 422–427 (2005)
7. Nakajima, T., Satoh, I.: Personal home server: enabling personalized and seamless ubiquitous computing environments. Pervasive Computing and Communications, pp. 341–345 (2004)
8. Bae, C., Yoo, J., Kang, K., Choe, Y., Lee, J.: Home server for home digital service environments Consumer Electronics. In: ICCE. 2003 IEEE International Conference, June 17-19, 2003, pp. 382–383 (2003)
9. http://www.cybermotion.com/datasheets/sr3esp.pdf
10. http://www.joymecha.com
11. http://imoyo.com/
12. http://www.roombavac.com
13. http://www.roboblock.co.kr
14. http://www.aibotown.co.kr
15. http://www.airobot.com
16. http://irobot.com
17. http://packbot.com
18. http://www.honda.co.jp/ASIMO/

Intuitive Human-Machine-Interaction and Implementation on a Household Robot Companion[*,**]

Christopher Parlitz, Winfried Baum,
Ulrich Reiser, and Martin Hägele

Fraunhofer IPA, Nobelstr. 12, D-70569 Stuttgart, Germany
parlitz@ipa.fraunhofer.de

Abstract. The increasing capabilities of experimental household robot platforms require more and more sophisticated methods of interaction. While there are many developments in all directions of Human-Machine-Interaction, the integration and combination of several modalities into one robot system require some effort. To ease the development of applications supporting several types of interaction, Fraunhofer IPA has developed a framework named "Go". Within this framework we have integrated different kinds of interaction methods into one robot platform "Care-O-bot 3", a mobile service robot for accomplishing daily tasks. This framework and its interaction methods are presented here.

1 Introduction

Service robots or robot companions that interact with humans require a special form of Human-Machine-Interaction (HMI). This form of interaction, commonly known as Human-Robot-Interaction (HRI) is manifold. It can contain active interaction by the robot, recognizing what a person is doing and possibly why he is doing it, but also simple passive interaction like waiting for a button to be pressed. Thus HRI incorporates methods from traditional Human-Computer-Interaction, but also, due to the embodied nature of the machine, observations from human-human interaction. Based on a typical household scenario, we have developed a framework for integrating several human-robot modalities into one robot system. The robot system is our future robotic household companion Care-O-bot 3. In the second chapter, we will present our scenario in which we choose the required human-robot interfaces. The next chapter will explain our approach how we integrate all components including the HRI of the robot into one system. The fourth chapter will explain details of the implementation of HRI components derived from the scenario. The final chapter shows implementation and conclusions.

[*] The work described in this paper was partially conducted within the EU Integrated Project COGNIRON ('The Cognitive Robot Companion' - www.cogniron.org) and funded by the European Commission Division FP6-IST Future and Emerging Technologies under Contract FP6-002020.

[**] This work was also partly funded as part of the research project DESIRE by the German Federal Ministry of Education and Research (BMBF) under grant no. 01IME01A.

M.J. Smith, G. Salvendy (Eds.): Human Interface, Part I, HCII 2007, LNCS 4557, pp. 922–929, 2007.

2 Scenario

We envisage our robot being useful in typical household scenarios. For more detailed analysis of, amongst others, the HRI components needed we have chosen a household application area. The gist of the experiment is to demonstrate to the robot how to lay a table, i. e. a human picks up objects from a storage area and places them in a typical place setting on a table. The robot observes the human during his movements and acquires the information necessary for repeating this task. After the initial learning phase the robot is asked to repeat the task several times for a complete table of e. g. six place settings. By analysing this scenario we have chosen the following HRI components for implementation and integration into our robot system: Speech Input / Output, Gesture Recognition and Remote Control. Additional components are the "traditional" functionalities of mobile robots like navigation, manipulation, object recognition and the like. While a remote control is not necessarily needed for the scenario above, our vision of a robot integrated into a household environment also e. g. incorporates control of the house itself. Additional visual feedback or supervision of the robot when it is not in close proximity requires some form of remote control.

We chose reliable, working implementations of the HRI components listed, adapted to our special needs of service robotics. Our focus is on the development of an integration framework for the components allowing simple integration and development of scenarios beyond the one described above.

3 Integration Framework "Go"

A seamless integration of robot components into one system is an ongoing topic in many robotic developments. This is due to near countless hard- and software components (Sensors, Actuators, diverse OSes, etc.) and requirements (bandwidth, real-time, etc.) that need to be integrated into one system. Several solutions already exist (e. g. ORCA [1], GenoM [2], MCA2 [3]), but are usually only suitable for some areas of applications or require a high effort of installation and maintenance. During the development of several robot systems the following key issues for a framework are indispensable:

- Modularity: A loose coupling of components is an advantage if changing setups, typical for service robot developments occur.
- Platform independence: Diverse hardware components need to be able to communicate.
- Inter-process communication: Different kinds of components and their data require diverse ways of communication, depending e. g. on bandwidth and response times.
- Resource management: Processes which use common sensors or actuators need to be coordinated, so that no conflict occurs.
- Configuration of the system: A flexible configuration of the control should separated and take place before the actual execution.
- Robustness: The framework should have a supervising functionality over the individual components.

- Tools for testing and debugging: Primarily during development, but also during run-time, tools for monitoring the processes are necessary.
- Performance analysis: If timing problems in the system's control occur, tools for performance analysis are necessary.

Our framework "Go" covers these requirements. Go is a high-level control framework implemented in Python [4], integrating and controlling lower component architectures. It was implemented as a module and a Go application is written as a Python programme using this module. This means, that all of Python's functionalities are available and therefore all advantages of a dynamic object-oriented script programming language (e. g. step-wise execution, object-orientated programming, consoles).

The extension of Python by Go allows simple and intuitive definition of hierarchical activities of individual system components. An activity is the smallest entity of a component. The activities can be executed synchronous or asynchronous to each other. Cyclic or one-shot executions are possible. Figure 1 shows some possible forms of activities.

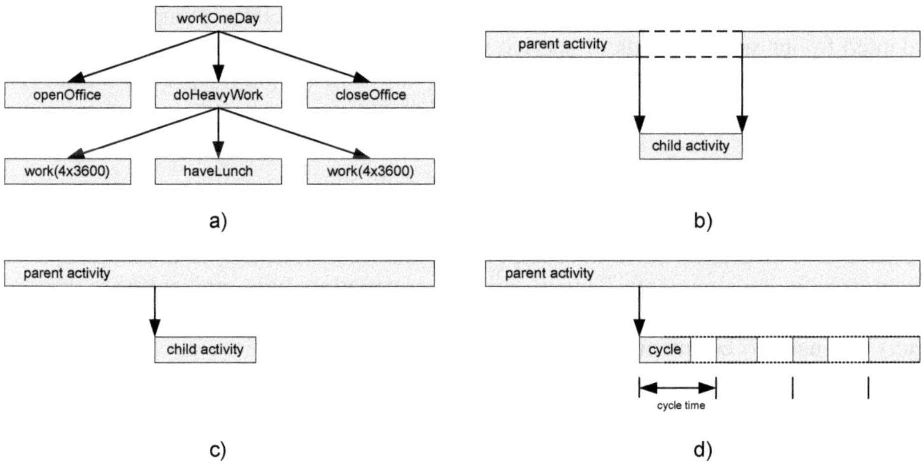

Fig. 1. Typical forms of activities: (a) hierarchical, (b) synchronous, (c) asynchronous, (d) cyclic

Therefore individual functions of the components, e. g. speech output can be addressed as an action, combined with other actions and joined together as one overall robot control operation.

A synchronization method guarantees that activities do not drift apart. The synchronisation is achieved by different kinds of wait commands in the Go module. These wait commands allow some activities to be suspend until other activities taking place in parallel have ended.

For process control across hardware components "GoCo" was developed, another Python module for activity communication across TCP/IP sockets. By using GoCo, activities on different computers can be treated as if they were implemented on one machine during control development. Processes that require an exchange of a lot of data, like images or movies in GUIs can be combined with the help of "GoSmaP". This is another module implemented with the help of smart pointers, which use shared memory to reduce copying time if possible. They also make sure that memory is released, if the data is no longer required by any activity. An error messaging system based on exception handling which extends through all application levels can be used for debugging and error handling. This debugging information can also be supervised remotely.

One disadvantage of Go is the lack of real-time functionality of script programming languages. While this is not required for most HRI components, it is as yet necessary for some applications like two-handed manipulation. If two components need real-time communication, this is done at a lower level. Go then causes the initialization of this communication at a higher level.

4 Details on HRI Components

Each component below has been implemented as a module and offers one or more Go activities, which are integrated into the overall robot control. The creation of modules from existing components for usage in Go is either done manually, or can be done by an automated interface generator like Swig [5]. Figure 2 shows how the components are joined within the Go framework.

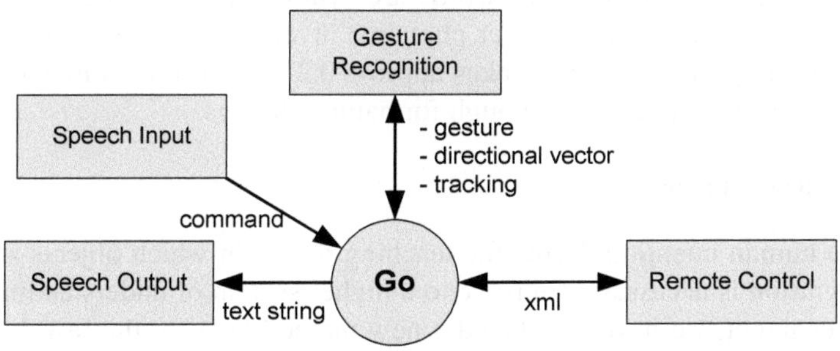

Fig. 2. Go and its HRI components with data flow

4.1 Speech I/O

Although speech is only a small part [6] of human-human interaction, reliable speech I/O is a feature of service robots that seems to be expected by most people. We were able to deduce from several installed service robots (e. g. Communication Museum Berlin). We use speech I/O for simple one word commands to the robot in different phases of the scenario and more elaborate status output by the robot. Speech I/O has several advantages. No special training is required compared to e. g. a keyboard and screen. The user has both his hands free and does not need to pay visual attention to the robot. For input we have selected a commercial application [7] based on the following conditions: Speaker independent, no training required, sufficient recognition rate and highly configurable. The key aspect for achieving a reliable speech input is the method of perception. While microphone and microphone arrays mounted on the robot are sufficient for lab and home environment, public shows make the use of speech very difficult. This problem is commonly called the "cocktail party effect" [8] and although current results are promising [9, 10] a common Bluetooth headset delivers a more reliable performance.

For speech output the most widely used system is the concatenative speech synthesis, where recorded natural utterances are combined to form new naturally sounding speech output. By using the speech synthesis program Festival [11] and the FestVox module [12], a voice based on unit selection [13] was created. While most concatenative systems use phonemes or diphones with fixed length the unit selection has variable length of the individual speech units reaching from a single phoneme to complete sentences. The speech corpus of Care-O-bot 3 created from sound files spoken by a professional narrator covers current and future scenarios of the robot. Additionally, all 48 phonemes of the German language are included, so that although we only have a limited domain of 379 different words, unknown words can be synthesized as well. If sentences are made up only from words known in the dictionary speech quality is very good. Occasional minimal variations of speech rate and prosodic problems in interrogative sentences occur. These problems are due to the relatively small speech corpus and the never quite perfect concatenative syntheses. The quality of the system declines in parts clearly, if unknown words are synthesized. This was expected and can only be overcome by a much larger phoneme inventory. The synthesis of a "normal" statement of approx. 10 words takes approx. 0.2 s on current computer systems. Therefore the speech output is fast enough for natural dialogs.

4.2 Gesture Recognition

To detect the human intention during the teaching phase, in which objects are placed, gesture recognition is necessary. En route to a higher system of understanding and detection of user intent, we have developed a new method [14] for the fast detection of body-parts, based on a modern range imaging sensor [15] and scale-space theory [16]. This approach has advantages over other published methods. While most methods are robust for fixed distances, certain environmental conditions or based on color (e. g. skin color [17]) or grey-level images for detection they have some simple drawbacks. If e. g. the hand is covered by a glove, there is no skin color or a poster showing a human might be falsely detected.

The idea behind our approach is to detect body-parts in 3D as extreme points over several scales as convex structure in the scale-space of range images. Since we currently only detect head and hands with this method, no complete kinematical setting of the human can be derived. Nevertheless with this information we were able to implement a demonstrator at the CeBIT 2006 computer fair and the AUTOMATICA 2006 in Munich as shown in figure 3. In this set-up a pneumatic robot torso mimicked

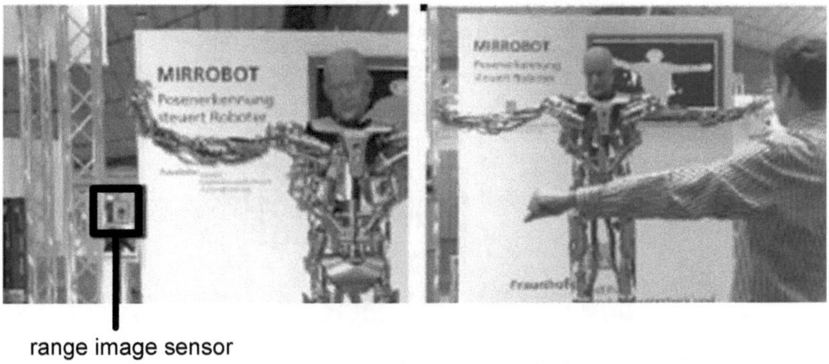

Fig. 3. The set-up at the fairs: A robot mimics the movements of a visitor

the movements of a visitor by detecting the head and hand poses via the 3D sensor SwissRanger SR 3000.

For a more accurate pose detection, the information gathered can be used to initialize a tracking algorithm to capture the complete human motion. Tracking algorithms can be based on color or again on using range information from a 3D sensor. The method we have currently implemented is an iterative-closest-point (ICP) algorithm presented in [18]. This combined method allows the interpretation of first simple gestures and the use of deictic information by the robot. We use it in the scenario for defining the position of the additional place settings to be laid out by the robot.

4.3 Remote Control

While a graphical user interface (GUI) is not necessarily needed for the scenario described above, a concept for controlling a household robot companion was developed as part of the multimodal user interface. Typical elements of a GUI were combined, which allow a wide range of control based on content and context. The concept was developed within a usability engineering process following a user centred scenario based design by Rosson and Carroll [19]. Following the design guide, an analysis of the major factors was undertaken. The usage context analysis was divided into three important areas: the typical user, the task and the surroundings. For the analysis of the tasks, the user centred approach had to give way to a feasibility study, since potential users overestimated the possibilities of a household robot of today. Based on typical scenarios a user interface was realized in Java. The first prototype contained different kinds of interaction ideas, which were later used in user studies. In the following usability tests the interface was refined and some problems eliminated. A proof of concept was achieved by showing that the users were able to command the robot via the interface unknown to the users, to fulfil tasks which were chosen by us. The current state of the interface is shown in figure 4. The GUI is implemented on a tablet computer which was chosen based on criteria like size, weight and resolution.

The actual current display status of the GUI is configurable via XML and is controlled by Go. In this case Go and GoCo only see that the XML strings are sent between the control on the robot and the tablet computer.

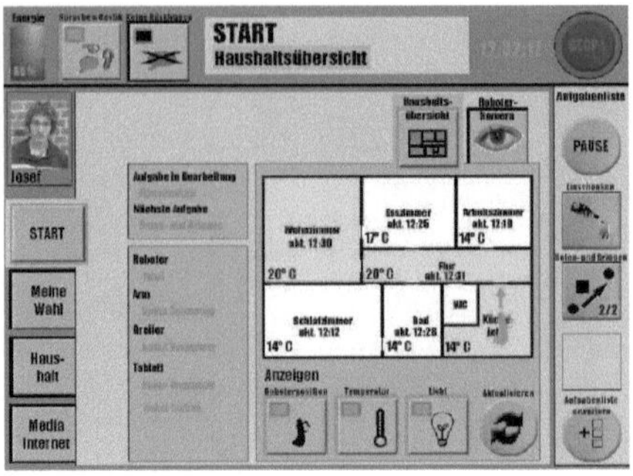

Fig. 4. Design of a GUI for a robot household companion

5 Implementation and Conclusion

Every interface has been individually tested on appropriate hardware and was designed to allow implementation on Care-O-bot 3. It is created for helping people in a household in an everyday environment. The preliminary design can be seen in figure 5a. The HRI interfaces described above are currently being implemented on a test setup (figure 5b) of Care-O-bot 3. Since the HRI components have not been fully integrated into the system, no results on performance or efficiency can be yet given. First versions of Go have already been used on several service robots (Care-O-bot generations, museum robots, etc. [20]).

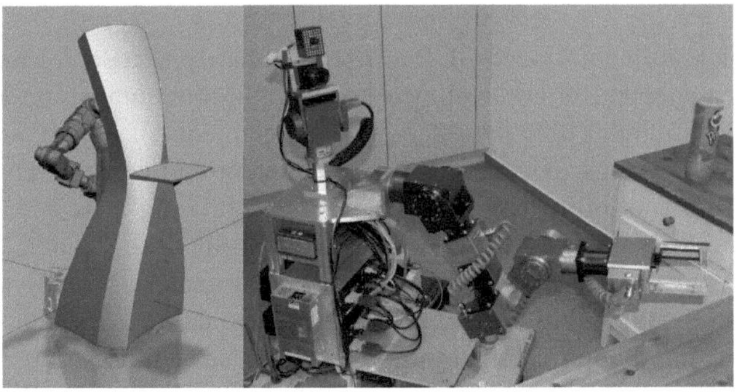

Fig. 5. Care-O-bot 3 in an early design phase (a) and test setup (b)

References

1. (Retrieved November 2006) http://orca-robotics.sourceforge.net/
2. (Retrieved November 2006) http://softs.laas.fr/openrobots/tools/genom.php
3. (Retrieved November 2006) http://mca2.sourceforge.net/
4. (Retrieved November 2006) http://www.python.org/
5. (Retrieved November 2006) http://www.swig.org/

6. Delhees, K.: Soziale Kommunikation. Psychologische Grundlage für das Miteinander in der modernen Gesellschaft. Westdeutscher Verlag (1994)
7. (Retrieved November 2006) http://www.sympalog.de
8. Handel, S.: Listening: An Introduction to the Perception of Auditory Events. MIT Press, Cambridge (1989)
9. Rodeman, T., et al.: Real-time Sound Localization With a Binaural Head-system Using a Biologically-inspired Cue-triple Mapping. In: Proceedings of the 2006 IROS, Beijing, China (2006)
10. Tamai, Y., Sasaki, Y., Kagami, S., Mizoguchi, H.: Three ring microphone array for 3D sound localization and separation for mobile robot audition. In: Intelligent Robots and Systems (2005)
11. (Retrieved November 2006) http://www.cstr.ed.ac.uk/projects/festival/
12. (Retrieved November 2006) http://festvox.org/
13. Möbius, B.: Corpus-based speech synthesis: methods and challenges. In: Arbeitspapier des Instituts für Maschinelle Sprachverarbeitung (Univ. Stuttgart) AIMS (2000)
14. Kubacki, J., Reiser, U.: Mirrobot: Fast Detection of Body-Parts in the Scale-Space of Range Images with an Application to Human Motion Copying, RO-MAN, University of Hertfordshire, Hatfield, United Kingdom, pp. 552–557 (2006)
15. Oggier, T., Büttgen, B., Lustenberger, F.: SwissRanger SR3000 and first experiences based on miniaturised 3D-TOF cameras. In: Proc. Of the First Range Imaging Research Day, pp. 97–108 (2005)
16. Lindeberg, T.: Scale–space: A framework for handling image structures at multiple scales. Technical Report CVAP-TN15, Royal Instute of Technology (1996)
17. Bretzner, L., Laptev, I., Lindeberg, T.: Hand gesture recognition using multi-scale colour features, hierarchical models and particle filtering. In: Face and Gesture, Washington, DC, pp. 423–428 (2002)
18. Knoop, S., Vacek, S., Dillmann, R.: Modeling joint constraints for an articulated 3D human body model with artificial correspondences in ICP. In: International Conference on Humanoid Robots (Humanoids 2005); Tsukuba, Japan (2005)
19. Rosson, M., Carroll, J.: Usability Engineering. Academic Press, San Diego, CA (2002)
20. (Retrieved November 2006) http://www.care-o-bot.de/

KANSEI Information Processing of Human Body Movement

Mamiko Sakata[1] and Kozaburo Hachimura[2]

[1] Faculty of Culture and Information Science, Doshisha University
Tatara Miyakodani 1-3 Kyotanabe, Kyoto, Japan
msakata@mail.doshisha.ac.jp
[2] College of Information Science and Engineering, Ritsumeikan University
Noji Higashi 1-1-1 Kusatsu, Shiga, Japan

Abstract. To quantitatively clarify the cognitive mechanism of KANSEI information conveyed by body movement, we conducted evaluation experiments and movement analysis by optical motion capture. We attempted to develop multiple regression models to investigate the effect of physical movement on subjective evaluation. The results show the relation between subjective feature of movements and the physical measurement. Thus, we expect that our research is effective to the guidance of dancing and acting, because it is possible to paraphrase sensitive image (KANSEI information) in another word (movement feature) which has strong relation with physical movement. Furthermore, this relational model which can replace human sensibility by digital data is useful for human-computer interaction.

Keywords: Body movement, KANSEI information, Motion capture.

1 Introduction

In the field of non-verbal communication research, the facial expression has been focused on, because it was considered that the body movement played only a secondary role(Ekman & Friesen, 1975). However, in the research by authors, people can perceive KANSEI information(emotion, impression and image etc.) from body movement, even without facial expressions, and the body does not only have additional or secondary function, but also plays an important role as independent media in human communications(Sakata et al., 2004). Recently, there are some researches aiming to find the relation between KANSEI information and body movement, we don't very often find the researches with a quantitative underpinning. In effective communications among humans, and also between human and robots, to quantitatively examine KANSEI information conveyed by body movement is very important.

So far, it was tried to demonstrate the relationship between body motion and the impression arising from the motion. When doing this they extracted the physical values of movement through analysis of silhouette images from video image taken

M.J. Smith, G. Salvendy (Eds.): Human Interface, Part I, HCII 2007, LNCS 4557, pp. 930–939, 2007.

from directly in front of the performer (Camurri et al., 1999, 2002, Inoue et al., 2005). However, this method discards the three dimensional information.

Therefore, we use optical motion capture system to extract the characteristics of body movement, because a time series 3-D coordinates value are obtained.
In this paper, we aim to analyse the relation between the subjective evaluation of body movement and the physical measurement, and to quantitatively clarify the cognitive mechanism of KANSEI information conveyed by body movement.

2 Target Movements

In this research, Matsumoto's 7 Motives (Matsumoto, 1985) was subjected to study. It empirically verified the connection between images and movements. The 7 Motives are the seven dance movements that Matsumoto set up to be the minimum unit of expressions, in which each Motive contains clearly distinguishable emotional value. Matsumoto added the arms raising/lowering and closing/opening movements to the ordinary walking movement, and then set up the movements that involve the entire body, including exhalation and inhalation, as the "basic pattern" and called it the Natural Mv. Based on the Natural Mv., she varied the time, energy and design factors of movements, while retaining the characteristics of the Natural Mv. to the extent possible, and created three movements at both extremes. She named them as Happy Mv., Lonely Mv., Sharp Mv., Solemn Mv., Dynamic Mv., and Flowing Mv. Mv. is the abbreviation of "Motive" used by Matsumoto (see Table 1).

In the two experiments (evaluation experiment and motion capture), a skilled female dancer who is a 26-year-old majoring in dance with 15 years of training in modern dance, classical ballet, and contemporary dance performed 7 Motives. Before starting the experiments, she was briefed on the experiment, which was followed by

Table 1. Target Movement; Matsumoto's 7 Motives

	Time	Energy	Design	Movement
Happy Mv.	M.M. ♩=120 ♪♬♪♬♪♬	High	Skip Turn	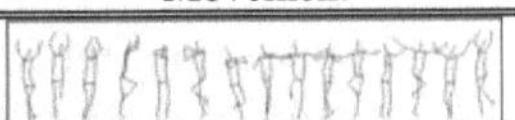
Sharp Mv.	M.M. ♩=100 ♬♬♪♩.♪♩	High	Straight	
Lonely Mv.	M.M. ♩=36 𝅝 𝅝	Low	Up & Down, Asymmetrical	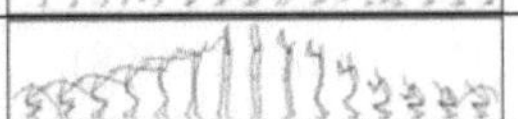
Natural Mv.	M.M. ♩=96 ♩♩♩♩♩♩♩♩	Normal	Symmetrical, Balance	
Flowing Mv.	M.M. ♩=92 ♩♩♩♩♩♩	Normal	Smooth -faced turn	
Dynamic Mv.	M.M. ♩=192 ♩♩𝅗𝅥.♩𝅗𝅥.♩	High	Jump	
Solemn Mv.	M.M. ♩=15 ♩♩𝅗𝅥	Low	Symmetrical, Balance	

several weeks rehearsal. During the rehearsal, the authors familiar with the 7 Motives were present and the performer rehearsed until she was able to represent the 7 Motives in a loyal manner.

3 Extraction of KANSEI Information from Body Movement

In this chapter, to extract KANSEI information from the body movement, we conducted an evaluation experiment using 7 Motives as the stimulus. We researched what the kinds of KANSEI information perceived from body movements and how the raters catch the feature of movement, and developed relation models to investigate the effect of movement feature on KANSEI information conveyed through the body movement.

Fig. 1. Mosaic

Forty-eight raters (39 men, 9women, with average age of 22.4, SD=2.05) participated in this experiment. To allow the raters to sufficiently observe the stimuli, movement image were projected on a large display.

Matsumoto's 7 Motives were newly acted out by the foregoing dancer and was recorded on a video-tape, which was used as the displayed stimulus. The face of the dancer in the image was obscured by mosaic to exclude the influence of the expression (see Figure 1). The name of each motive was not taught the raters.

After the raters were asked to glance over the two lists of rating terms, they were shown the display stimulus showing the motives one by one. After the same motive was shown, the vision changed to an interval. During the interval, the raters were asked to fill in the answer sheets with the following questions.

i. How much of the seven KANSEI information (happy, sharp, lonely, natural, flowing, dynamic and solemn) were expressed in each Motive: (Rated in the 5-level scales ranging from 1.'not expressed' through 5. 'very clearly expressed')
ii. What kind of motion each motive showed: (Rated in the 5-level extreme scales by semantic differential method, see Table 2) This rating terms are selected in terms of time, energy and design; the elements that composes movement (Matsumoto, 1968).

Table 2. The Rating Terms for Movement Feature

Time	T1: Speedy—Slow T2: Accentual—Smooth T3:Irregular—Regular
Energy	E1: Strong—Weak E2: Heavy—Light E3: Variable—Constant
Design	D1: Linear—Curved D2: Expansive—Shrinking D3: Balanced—Imbalance

3.1 The Subjective Features of Body Movement

Rating on KANSEI Information of the 7 Motives

The rating results using the 5-level scales for the seven KANSEI information (happy, sharp, lonely, natural, flowing, dynamic and solemn) were converted into numerical values (from 1 to 5 points). The mean value were obtained and tabulated in Table 3.

Table 3. Mean Value of KANSEI Information

		7 Motives						
		Happy Mv.	*Sharp Mv.*	*Lonely Mv.*	*Natural Mv.*	*Flowing Mv.*	*Dynamic Mv.*	*Solemn Mv.*
KANSEI	Happy	4.56	1.98	1.79	2.19	4.08	4.33	1.58
	Sharp	2.04	4.67	1.40	1.60	1.81	2.92	1.54
	Lonely	1.29	2.25	3.81	2.54	1.73	1.40	3.67
	Natural	1.94	1.56	2.75	3.27	2.42	1.71	2.54
	Flowing	4.06	1.75	3.35	4.02	4.56	3.54	3.23
	Dynamic	4.33	3.06	1.81	1.96	3.77	4.83	1.85
	Solemn	1.43	3.63	2.96	2.25	1.85	1.65	3.90

Rating on Movement Features of the 7 Motives

Table 4 shows the results of rating on movement features of the 7 Motives. "It applies" was made a score as one point to the left term of Table 2, and "It applies" was made a score as five points to the right term, and "There is not both" was made a score as three points.

Table 4. Mean Value of Movement Features

	Happy Mv.	*Sharp Mv.*	*Lonely Mv.*	*Natural Mv.*	*Flowing Mv.*	*Dynamic Mv.*	*Solemn Mv.*
T1	1.77	2.19	4.85	3.88	3.17	1.25	4.81
T2	2.88	1.25	4.63	4.08	3.79	1.81	4.42
T3	3.75	4.50	3.06	4.54	3.63	3.40	3.98
E1	2.58	1.42	3.69	3.33	2.75	2.08	2.96
E2	4.33	2.42	2.52	3.27	3.98	4.13	1.88
E3	3.50	2.15	3.98	4.25	3.52	2.31	4.25
D1	3.85	1.38	4.60	3.15	4.52	2.21	3.85
D2	2.58	2.75	2.79	2.79	2.04	1.69	2.69
D3	2.58	2.17	3.00	1.77	2.48	2.67	2.17

3.2 The Relation Between KANSEI Information and Movement Feature

In order to examine the relation between the perceived KANSEI Information and movement feature, we conducted multiple regression analyses using the nine evaluated scores of each movement feature (mean value of Table 4) as explanation variables, and the evaluated scores of the seven KANSEI Information (mean value of Table 3) as destination variables (by stepwise way). Table 5 shows the standardized coefficients ($p<0.05$) and the scores of adjusted R^2.

As shown by Table 5, a regression model with high significance was obtained for all of the KANSEI information, except for “flowing” (significance level p<0.05). In the author's earlier study, a multiple regression model for the flowing impression could not be obtained, too. From the values of the coefficient concerned with each model, we can find that “light” motion contribute to happy impression, while it is “strong” motion for sharp, “heavy” and “slow” for lonely, “balanced” and “weak” motion for natural, “speedy”, “curved”, “Expanding” and “heavy” motion for dynamic, and “heavy”, “strong” motion for solemn. It seems that KANSEI information perceived from each motive are greatly affected by the energy factor of movement.

The relation between KANSEI information and movement feature is shown in the Figure 2.

Table 5. Relation between KANSEI Information and Movement Feature

	T1	T2	T3	D3	E1	E2	adjusted R^2
Happy**	—	—	—	—	—	0.952**	0.887
Sharp**	—	—	—	—	-0.927**	—	0.831
Lonely**	0.590 *	—	—	—	—	-0.466*	0.924
Natural**	—	—	—	-0.427***	0.946***	—	0.952
Flowing	—	—	—	—	—	—	—
Dynamic**	-1.178**	0.468**	-0.287**	—	—	-0.213*	0.998
Solemn***	—	—	—	—	-0.180*	-1.006 ***	0.991

*...p<0.05, **...p<0.01, ***...p<0.001.

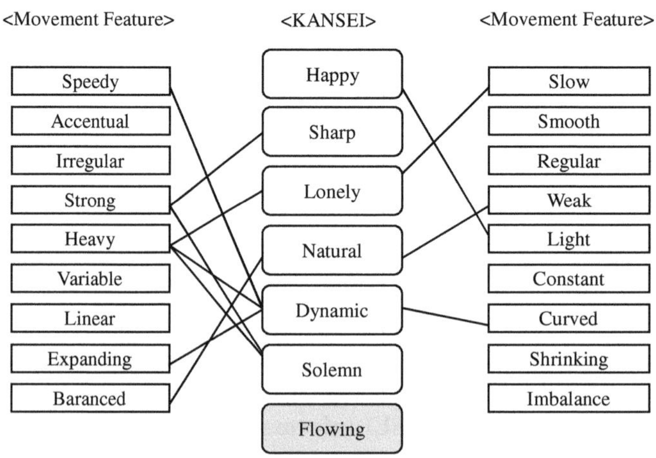

Fig. 2. Relation between the KANSEI Information and the Movement Feature

4 Measurement of Body Movements

In the preceding chapter, we examined the relation between KANSEI information perceived from body and the movement feature. It was clarified that KANSEI information perceived from each motive were greatly affected by the energy factor of movement. However, the feature of movement is only a subjective factor. It is necessary to examine the relation to the above-mentioned movement feature by the rater and a physical actual one. Therefore, we measured the dancer's movements of 7 Motives by using the optical motion capture.

4.1 Optical Motion Capture

The optical motion capture system employs video camera to track the motion of reflective markers attached to joints of the actor's body. The system uses LED's mounted around the camera lens. In this system, the centre of the marker images are matched from the various camera views using triangulation to compute their frame-to-frame positions in 3D space. Several problems often occur during the tracking process, including of markers, noisy or missing data and false reflections.

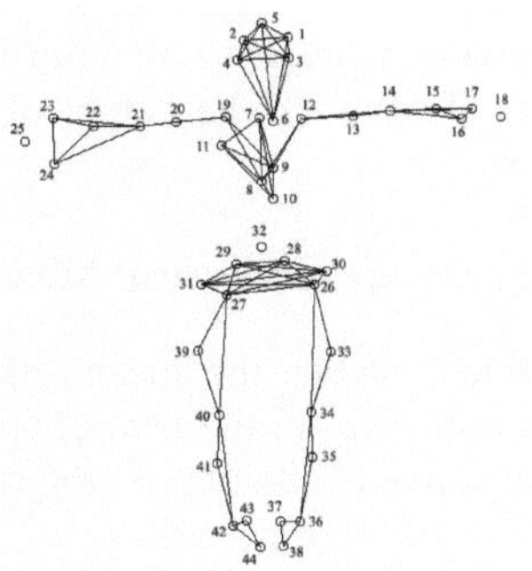

Fig. 3. Positions of Markers

In this experiment, 44 markers were put on the same dancer's body to the way of Figure 3, and the movement was measured with ten LED's cameras. The acquired data is obtained as a time series coordinates value(x, y, z) of each marker position by each frame (frame rate 1/60sec.).

4.2 Physical Measurement

When the motion capture is used, the enormous quantity of data by each part is obtained. Then, it becomes important to discover a significant parameter which influences KANSEI information.

In this paper, the following 16 measurements are calculated as the physical parameters that seem to influence KANSEI information; mean velocity of hand, SD velocity of hand, mean acceleration of hand, SD acceleration of hand, Mean velocity

<Physical Measurements>
P1: Mean velocity of hand
P2: SD velocity of hand
P3: Mean acceleration of hand
P4: SD acceleration of hand
P5: Mean velocity of foot
P6: SD velocity of foot
P7: Mean acceleration of foot
P8: SD acceleration of foot
P9: Mean velocity of root
P10: SD velocity of root
P11: Mean acceleration of root
P12: SD acceleration of root
P13: Mean dimensions of opened body
P14: SD dimensions of opened body
P15: Mean height of root
P16: SD height of root

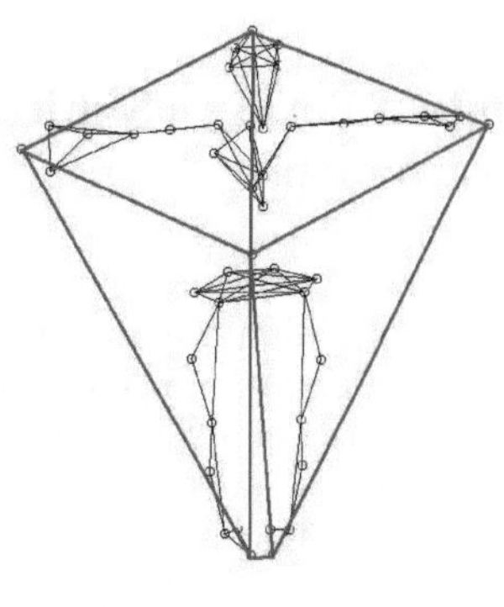

Fig. 4. Dimensions of Opened Body

of foot, SD velocity of foot, Mean acceleration of foot, SD acceleration of foot, Mean velocity of root, SD velocity of root, Mean acceleration of root, SD acceleration of root, Mean dimensions of opened body, SD dimensions of opened body, mean height of root, SD height of root. Because 7 Motives are basically symmetric movements, only the right were adopted as for measurements of both hands and feet. The dimensions of opened body are the sum total of the triangular area made when drawing a straight line from five end parts (head top, both hands, both ankles) to the root (Figure 4).The height of root was based on the absolute origin of the capturing area.

4.3 Result of Physical Measurements

Table 7 shows the mean value and standard deviation of physical measurement of each motive. In the next chapter, we examine the relation between this result and the subjective evaluation by the raters.

Table 7. Result of Physical Measurement.

		Happy Mv.	*Sharp Mv.*	*Lonely Mv.*	*Natural Mv.*	*Flowin g Mv.*	*Dynami c Mv.*	*Solemn Mv.*
P1	(cm/sec)	40.82	26.04	5.13	7.27	34.49	51.16	3.34
P2		19.32	21.89	3.66	3.28	17.40	24.01	2.42
P3	(cm/sec^2)	6.94	5.00	1.56	1.70	4.66	9.97	1.26
P4		5.63	3.94	2.16	1.74	4.52	7.46	1.98
P5	(cm/sec)	24.61	22.57	2.76	15.57	17.05	47.96	2.47
P6		14.89	24.17	3.83	7.72	10.31	22.98	4.29
P7	(cm/sec^2)	7.52	4.86	1.07	2.11	3.00	10.50	0.55
P8		8.29	5.64	1.85	1.66	2.31	8.97	0.68
P9	(cm/sec)	19.35	11.29	1.84	13.41	10.46	44.54	1.89
P10		7.68	6.13	1.35	2.31	5.28	7.90	2.74
P11	(cm/sec^2)	3.58	1.11	0.40	0.66	1.00	3.69	0.29
P12		3.57	0.98	0.65	0.36	0.62	2.29	0.33
P13	(cm^2)	7579.24	7146.98	6315.32	7641.95	8030.88	9277.37	8464.56
P14		2867.52	3425.65	2061.86	2554.98	2442.39	2547.81	2799.85
P15	(cm)	1.44	-0.92	-13.24	-2.04	-1.08	7.20	-10.88
P16		6.19	5.64	13.40	0.77	8.93	13.98	9.31

5 Relation Between Subjective Evaluation and Physical Measurement

Section 3.5 above, we attempted to develop a multiple regression model of the seven KANSEI information using movement features. However, both of them are only subjective evaluation. In this chapter, we made sure the relation between the subjective evaluation by the rater and the physical movement by motion capture.

Table 8 shows the correlation matrix of subjective evaluation and physical measurements. Many of physical measurements correlated with "speed-slow", because P1~P12 were velocity and acceleration. All of them correlated with at least one subjective term except for <mean dimensions of opened body>.

Next, we conducted multiple regression analysis using the nine evaluated scores of each movements features (Table 4) as explanation variables, and the physical measurement (Table 7) as destination variables (by stepwise way). Table 9 shows standardized coefficients($p < 0.05$) and the scores of adjusted R2.

As the result, we obtained a multiple regression model for all subjective features (significance level $p<0.05$). When each feature was observed, the physical measurement that contributes to "Speedy-Slow" is <SD velocity of root>. Similarly, <mean and SD velocity of foot> contribute to "Accentual-Smooth", whereas <SD height of root> is the factors of "Irregular-Regular". <SD velocity of foot> contributes for "Strong-Weak", whereas it was <mean velocity of hand> for "Heavy-Light" and "Variable-Constant". <SD velocity of hand and foot> contribute for "Linear-Curved", and <SD acceleration of hand and foot>, <mean velocity of root> are the factors of "Expanding-Shrinking". <Mean dimensions of the opened body>

Table 8. Correlation Matrix of Subjective Evaluation and Physical Measurements

		Movement Features								
		T1	T2	T3	E1	E2	E3	D1	D2	D3
Physical Measurements	P1	-0.932**	-0.711	-0.229	-0.633	0.811*	-0.718	-0.325	-0.783*	0.318
	P2	-0.941**	-0.880**	-0.050	-0.839*	0.602	-0.889**	-0.532	-0.626	0.230
	P3	-0.945**	-0.773*	-0.233	-0.661	0.728	-0.775*	-0.450	-0.764*	0.326
	P4	-0.908**	-0.706	-0.323	-0.614	0.749	-0.729	-0.341	-0.807*	0.400
	P5	-0.920**	-0.787*	-0.053	-0.646	0.687	-0.767*	-0.605	-0.744	0.108
	P6	-0.909**	-0.994***	0.201	-0.927**	0.370	-0.961**	-0.832*	-0.449	0.003
	P7	-0.947**	-0.781*	-0.161	-0.630	0.708	-0.743	-0.514	-0.662	0.263
	P8	-0.939**	-0.807*	-0.140	-0.656	0.636	-0.749	-0.515	-0.493	0.309
	P9	-0.821*	-0.640	-0.161	-0.475	0.687	-0.626	-0.497	-0.769*	0.152
	P10	-0.963***	-0.816*	-0.025	-0.779*	0.675	-0.761*	-0.469	-0.619	0.166
	P11	-0.873*	-0.606	-0.251	-0.454	0.776*	-0.538	-0.291	-0.589	0.329
	P12	-0.764*	-0.497	-0.250	-0.356	0.690	-0.395	-0.142	-0.324	0.384
	P13	0.059	0.028	-0.273	-0.008	-0.066	-0.122	-0.114	-0.639	0.046
	P14	-0.454	-0.690	0.714	-0.810*	-0.174	-0.525	-0.717	0.201	-0.501
	P15	-0.932**	-0.750	0.154	-0.647	0.777*	-0.673	-0.555	-0.672	-0.101
	P16	-0.010	0.005	-0.878**	0.034	0.003	-0.231	0.206	-0.518	0.838*

Table 9. Relationship between subjective evaluation and physical Measurements

	P1	P2	P4	P5	P6	P8	P9	P10	P13	P16	adjusted R^2
T1***	—	—	—	—	—	—	—	-0.963	—	—	0.913
T2***	—	—	—	0.173	-1.14	—	—	—	—	—	0.995
T3**	—	—	—	—	—	—	—	—	—	-0.878	0.724
E1**	—	—	—	—	-0.927	—	—	—	—	—	0.83
E2*	0.811	—	—	—	—	—	—	—	—	—	0.588
E3**	—	—	—	—	-0.961	—	—	—	—		0.908
D1***	—	1.285	—	—	-2	—	—	—	—	—	0.972
D2***	—	—	-1.575	—	—	1.328	-0.495	—	—	—	0.992
D3**	—	—	—	—	—	—	—	—	-0.572	1.146	0.901

*...$p<0.05$, **...$p<0.01$, ***...$p<0.001$ ※The table shows only statistically-meaningful items of physical feature.

and <SD height of root> contribute for "Balanced-Imbalance". From these results, it is assumed that the feature of movement that the raters subjectively felt can be replaced with the physical measurement.

6 Discussion

When the results of this research are brought together, it is schematised as shown in the following Figure 5.

The left lines show the relation between KANSEI information and movement feature in the raters cognition. On the right side, the relation between subjective value (movement feature) and physical value are connected. Put simply, the subjective and the objective feature of movement conveyed KANSEI information was revealed. For example, lonely motion is slow and heavy, and it is the actual movement that the hand is slowly moved and the difference of fast and slow of root (waist) is not applied. This result is effective to the guidance of dancing and acting, because it is possible to paraphrase sensitive image (KANSEI information) in another word (movement feature) which has strong relation with physical movement.

The reason of the failure to develop a multiple regression model on "flowing" image is attributed to the fact that it is difficult to replace "flowing" with another word representing motion. It seems that "flowing" is not KANSEI information, but movement feature.

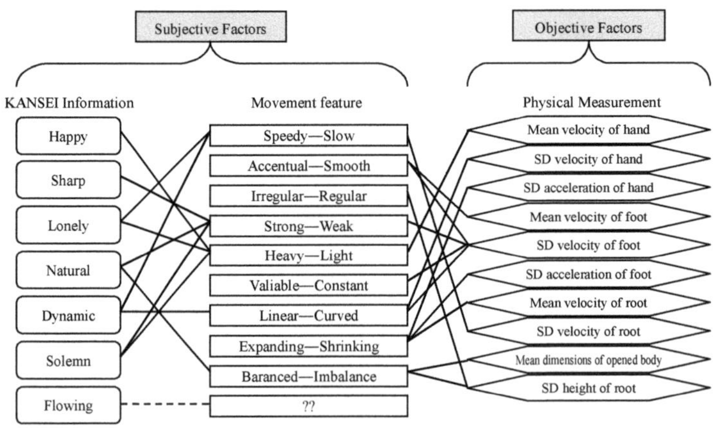

Fig. 5. Relation among KANSEI Information and Movement Feature and Physical Measurement

7 Conclusion and Future Work

To quantitatively clarify the cognitive mechanism of KANSEI information conveyed by body movement, we conducted evaluation experiments and movement analysis by optical motion capture. We attempted to develop multiple regression models to investigate the effect of physical movement on subjective evaluation. The results show the relation between subjective feature of movements and the physical measurement. Thus, we expect that our research is effective to the guidance of dancing and acting, because it is possible to paraphrase sensitive image (KANSEI information) in another word (movement feature) which has strong relation with physical movement. Furthermore, this relational model which can replace human sensibility by digital data is useful for human-computer interaction.

In the future, we will examine the all kinds of movement, not apply only to 7 Motives. Furthermore, we will verify the results of this paper and construct a more exquisite relational model that various movements are covered.

References

1. Camurri, A., Coletta, P., Mazzarino, B., Trocca, R., Volpe, G.: Improving the man-machine interface through the analysis of expressiveness in human movement. In: Proc. 2002 IEEE ROMAN conf., pp. 417–421 (2002)
2. Camurri, A., Hashimoto, S., Suzuki, K., Trocca, R.: KANSEI Analysis of Dance Performance. IEEE International Conference on System, Man and Cybernetics, vol. IV, pp. 327–332 (1999)
3. Ekman, P., Friesen, W.V.: Unmasking the face. Prentice-Hall, Inc., NJ (1975)
4. Inoue, M., Iwadate, Y., Shiba, M., Auzuki, M.: Kansei Feature Extraction Related to Human body-Movements. The Journal of The Virtual Reality Society of Japan 1(10), 31–40 (2005) (in Japanese)
5. Matsumoto, C.: Dance research: Problem situation and learning of problem solving II - Qualities of movements and feeling values. Lecture at the Tenth Congress of the international Association of Physical Education and Sport for Girls and Women. Lerwick (1985)
6. Matsumoto, C.: An introduction of kinesiology. Taishukan Inc., Tokyo (1968)
7. Sakata, M., Shiba, M., Maiya, K., Tadenuma, M.: Human Body as the Medium in Dance Movement. International Journal of Human- Computer Interaction 17(3), 427–444 (2004)

A Japanese Text Input Interface Using On-Line Writing-Box-Free Handwriting Recognition and Kana-to-Kanji Conversion

Takeshi Sakurada, Yoichi Hagiwara, Hideto Oda, and Masaki Nakagawa

Tokyo University of Agriculture and Technology,
2-24-16, Naka-cho, Koganei-shi, Tokyo, Japan
take-s@cc.tuat.ac.jp

Abstract. People often forget how to write Kanji characters even though they are able to read those characters. If one cannot write a Kanji character, one cannot input it with current handwriting recognition engines. Therefore, we propose to combine Kana (phonetic characters) to Kanji conversion with a handwriting recognition engine. Moreover, we remove writing boxes to write characters one by one, since they are unnatural for many applications. This is similar to writing on a plain paper with a pen. When, handwritten characters are misrecognized, the user corrects them using correction gestures. When the user cannot write Kanji characters, the user writes in Kana and invokes the Kana-to-Kanji conversion. In reality, people often write Kana characters instead of Kanji characters when they cannot write those characters or it takes more time than writing Kana characters. Kana-to-Kanji conversation is common in keyboard interface. This paper describes the design of a pen interface by which the user can write text without writing boxes, correct misrecognitions and specify Kana-to Kanji conversion. When a user writes Kana characters instead of Kanji characters, it would be desirable to translate them to Kanji characters automatically.

Keywords: pen interface, free-format handwriting recognition, Kana-to-Kanji conversation.

1 Introduction

Pen-based interfaces are attracting people's attention again due to the wide spread of mobile PDAs (Personal Digital Assistant) with pen input, expanding sales of interactive electronic whiteboards, shipping of Tablet PC of Microsoft Corp., etc. Especially, Nintendo DS, which is a portable game machine with pen, is opening a new market in Japan.

Pen input has several benefits. It does not require large space. People can use it without training. Pen input on interactive whiteboards can easily attract the attention of an audience. People can express or annotate their ideas without being bothered to wonder how to use a pen. Thinking is not interrupted by the actions for writing. However, there are still some problems with existing handwritten character recognition.

M.J. Smith, G. Salvendy (Eds.): Human Interface, Part I, HCII 2007, LNCS 4557, pp. 940–949, 2007.

One problem is that handwriting recognition engines might make misrecognitions. Another problem is that many systems require people to write characters one by one into so-called writing boxes in order to avoid segmentation errors. However, a user likes as few restrictions as possible, when writing. There is also another problem. People often forget how to write Kanji characters even though they are able to read those characters. It takes time to write complex Kanji characters. If one cannot write a Kanji character, one cannot input it with current handwriting recognition engines.

Therefore, we propose to combine Kana (phonetic characters) to Kanji conversion with a handwriting recognition engine. Moreover, we remove writing boxes to write characters one by one, since they are unnatural for many applications. This is similar to writing on a plain paper with a pen.

2 Lazy Recognition and Correction by Gestures

We have been working on on-line writing-box-free handwriting recognition [1][2]. A user can write text without using writing boxes, as if writing on a plain paper. Even with the latest character recognition engine, however, the correct recognition rate is 93% and incorrect recognitions occur and the correction work is surely needed. Some of systems have adopted the method of recognizing and showing a recognition result immediately after a character is written [3]. This uses the real-time nature of on-line input. However, it forces a user to verify a recognition result and correct misrecognition when necessary, thus interrupts user's thinking. Therefore, we have proposed the lazy recognition [4], which delay the display of recognition results of handwritten character patterns until needed.

In handwriting recognition, misrecognition is unavoidable. Therefore, recognition error correction interface is necessary. Error correction by pen gestures is effective. We have already proposed a set of pen gestures for writing-box-free handwriting input [5][6].

This section summarizes the state of the art and our previous studies.

2.1 A Character Recognition with Writing-Boxes

The character recognition with writing-boxes is a system, which is generally used by the present PDA etc. and recognizes by filling in one character at a time to one frame (Fig.1). The recognized result may be mistaken, although character segmentations become right since the character recognition with a frame recognizes a character for every frame. Therefore, what is necessary is to consider only the interface, which corrects the mistake of the recognized result. For example, the candidate of a recognition result is taken out with carrying out the tap of the inside of the frame which wrote in the character with an electronic pen, and a user is made to choose the right result from there. Since the character recognition with writing-boxes has the right character segmentations, it can raise the rate of recognition. However, the burden that a character must be written to a frame is given to a user.

Fig. 1. Character recognition with writing-boxes

2.2 A Writing-Box-Free Handwriting Recognition

Writing-box-free character recognition recognizes that which the user took notes of freely as a character string, and there is no input frame in every character (Fig.2).

Fig. 2. A writing-box-free handwriting recognition

Writing-box-free character recognition may incorrect-recognize a character simple substance like recognition with a writing-box. Furthermore, since writing-box-free character recognition does not have a frame for every character, character segmentations may be incorrect-recognized. When character segmentations have been incorrect-recognized, though natural, it becomes what also mistook the last result (Fig.3).

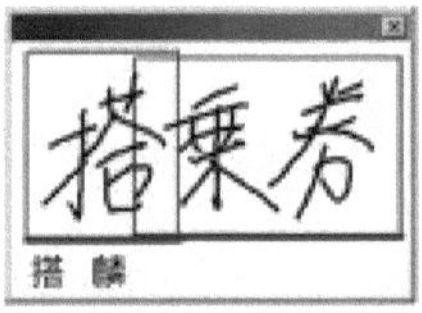

Fig. 3. Incorrect-recognized character segmentations

2.3 The Display of the Recognition Result of Character Segmentations

As mentioned above, in writing-box-free character recognition, there is a problem that character segmentations are sometimes incorrect-recognize. Therefore, in order to correct character segmentations, it is necessary to indicate how the character recognition system has recognized character segmentations to a user. The following two methods are mentioned as a method of displaying character segmentations.

1. The color of a stroke is changed and segmentations are displayed (Fig.4).
2. A rectangle (frame) is displayed and segmentations are displayed (Fig.5).

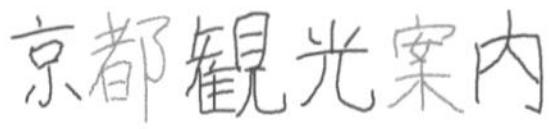

Fig. 4. Character segmentations are expressed with the color of strokes

Fig. 5. Character segmentations are expressed using a rectangle

In this paper, the method of displaying character segmentations using a rectangle is adopted. It is because there may be a case of monochrome environment by PDA etc. or a color may be used by emphasis of a stroke etc. also in a color environment.

2.4 Correction of Character Segmentations

Since character segmentations were accidentally detected by writing-box-free character recognition, the incorrect-recognized example is shown in Fig.6 and Fig.7. The example shown in Fig.6 is an example, which has incorrect-recognized the pattern which recognition engine should recognize as two characters as one character. In this case, a user performs the gesture, which draws a line and divides a rectangle in a place to divide a pattern (Fig.6-b). Then, a character is divided in the corrected place and a new recognition result is displayed (Fig.6-c). By the gesture that divides this rectangle, both the starting point and the ending point of a division line are a rectangular outside. Moreover, when the place, which should be recognized as one character, has been divided like the example of Fig.7-a, the gesture surrounding a stroke to set to one is carried out (Fig.7-b). Then, a character recognition system corrects the boundary of a character, and displays a new recognition result (Fig.7-c).

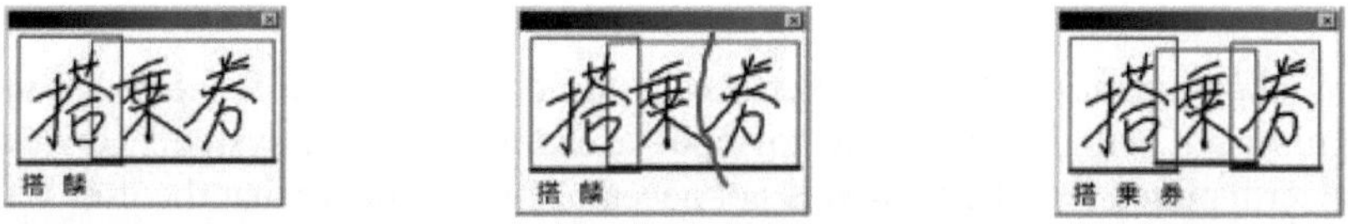

(a)Segmentations incorrect (b)Correction of segmentations (c)New recognition result

Fig. 6. Incorrect recognition of character segmentations (it is intension about excessive strokes)

(a)Segmentations incorrect (b)Correction of segmentations (c)New recognition result

Fig. 7. Incorrect recognition of character segmentations (when dividing too much)

2.4.1 The Evaluation Experiment of the Correction Gesture of Character Segmentations

In order to confirm the validity of the above-mentioned character segmentations correction gesture, we experimented in preliminary evaluation. Although connection gesture (Fig.8) was used in former study [7] in order to have made the divided character into one character, we experimented this time by adding the gesture that was defined above in addition to it.

The evaluation experiment was conducted on ten university students who are skilled in computer operation and pen operation using Tablet PC. Before beginning an evaluation experiment, explanation of a display interface and a correction interface is given, and we operated the system to them freely and had them get used.

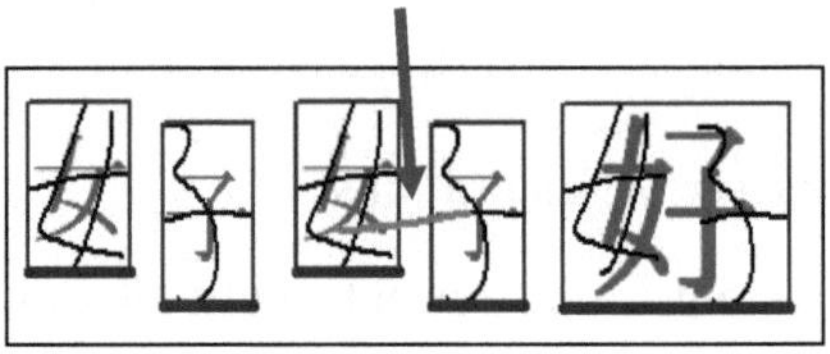

Fig. 8. Combination of the character by connection gesture (former system)

In the input of about 20 characters, both the gesture to divide and the gesture to surround were used about 1.5 averages as a result of the experiment. However, connection gesture was not used once. Since the gesture to surround is more nearly intuitive than connection gesture, it is thought that the gesture to surround was used. From this experiment, it supposes that connection gesture will not be adopted from now on, and this gesture is used for the below-mentioned conversion-of-kana-into-kanji input gesture.

3 Conversion-of-Kana-into-Kanji Input Gesture

If the above-mentioned user interface is used, while a user corrects incorrect recognition, it will become easy to input a character into a system. However, the character inputted into the system is not necessarily a character that the user wanted to input truly. As a result of our cooperating in the experiment for program work of NHK (Japan Broadcasting Corporation) before, it recognizes that the character, which can be read and which can be written are not the same. In the input of Japanese by the keyboard, the conversion of kana into kanji is performed and the kanji character is inputted. In this case, if reading is known even when memory of a character is ambiguous, reading is inputted and the target kanji character can be inputted by the conversion of kana into kanji. However, since in handwriting input it cannot write if the kanji character is not known in detail even if reading is known, a user cannot input a character into a system. Since the writing order is sharply different or a line and a point are lacking, the kanji character currently regarded as the ability of a user to write may not be exact in fact, either. Therefore, in character recognition, the kanji character, which the user desired, is not displayed as a recognition result. Therefore, a user inputs kana characters unavoidably instead of a character to input first. In order to solve this situation, the interface for using the conversion of kana into kanji in a handwriting input is designed.

3.1 Combined Use of Incorrect Recognition Correction and Character Segmentations Correction

It is possible to change the mode like character segmentation correction mode, incorrect recognition correction mode, conversion-of-kana-into-kanji mode, and then to use conversion-of-kana-into-kanji input gesture at the time of conversion-of-kana-into-kanji mode. In this case, a user always needs to grasp in which mode it is working now, and has the time-consuming problem of a mode change.

In this paper, the conversion-of-kana-into-kanji input gesture, which can perform correction of character segmentations and revision of incorrect recognition, is proposed, without changing the mode.

3.2 An Addition of a Conversion-of-Kana-into-Kanji Candidate

Before describing conversion-of-kana-into-kanji input gesture, the incorrect recognition correction gesture used by the conventional character recognition with a frame [5][6] is described. In the conventional incorrect recognition correction, when the user did the tap of the inside of the rectangle to which the recognized character was written as shown in Fig. 9, the recognition candidate list was displayed and the user had chosen the right thing out of it (Fig.9-d).

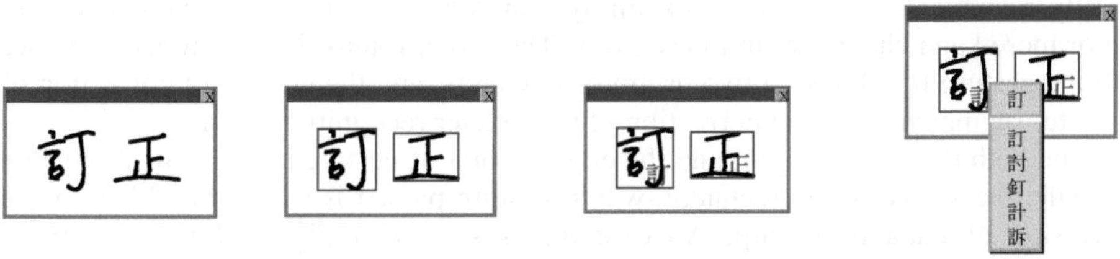

(a)Stroke input (b)Character segmentation recognition result (c)Character recognition result (d)Correction of a recognition result

Fig. 9. The conventional incorrect recognition correction

By this system, it extends by adding a conversion-of-kana-into-kanji result to this conventional incorrect recognition correction gesture. First, the example in one character is shown to Fig.10. The stroke input which the user inputted is recognized (Fig.10-b,c). Usually, character segmentation recognition result and character recognition result are displayed almost simultaneous. If a user does the tap of the inside of the rectangle to which the recognized character was written, the candidate list of recognition results and a conversion-of-kana-into-kanji result will be

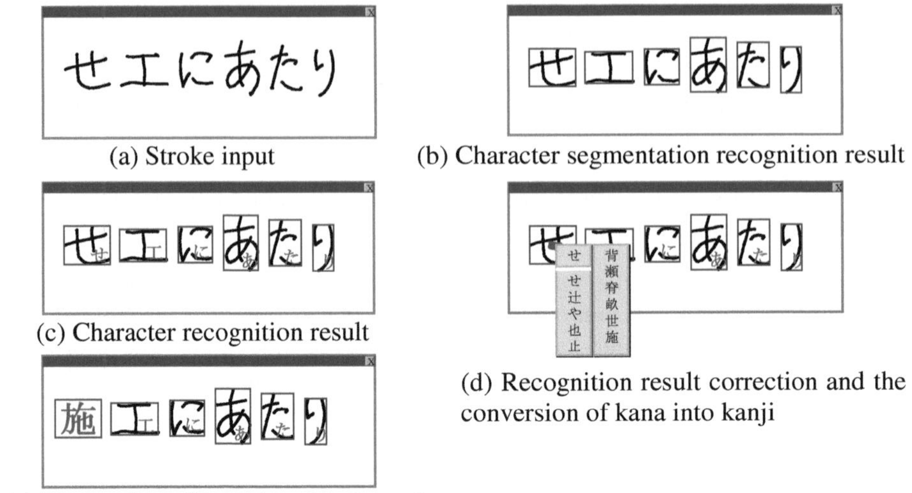

(a) Stroke input

(b) Character segmentation recognition result

(c) Character recognition result

(d) Recognition result correction and the conversion of kana into kanji

(e) Conversion-of-kana-into-kanji result

Fig. 10. The addition of the conversion of kana into kanji (one character)

displayed, and the thing which a user wants to input out of them will be chosen (Fig.10-d,e).

3.3 The Addition of the Gesture for the Conversion of Kana into Kanji

Although the above-mentioned thing converts one kana character to a kanji character, actually a conversion-of-kana-into-kanji system converts it to kanji characters from two or more kana characters in many cases. Therefore, it must be able to use together with the gesture for choosing two or more characters, and the gesture of correction of character segmentations, and correction of a character recognition result.

When both the starting point and the ending point of gesture are contained in some rectangle, the stroke in the rectangle which gesture passed is set as the object of the conversion of kana into kanji. An example is shown in Fig.11. First, the stroke (Fig.11-a), which the user inputted, is recognized (Fig.11-b,c). Usually, the recognition result of character segmentations and character recognitions result are displayed continuously. If a user does the tap of the inside of the rectangle surrounding the recognized character, correction of a character recognition result and selection of a simple sentence character conversion-of-kana-into-kanji result are possible. Correction of the above-mentioned character segmentations can be performed by the gesture, which surrounds strokes or divides a rectangle at this time. When the starting point and the ending point of gesture are contained in the respectively different rectangle (Fig.11-d), the portion which gesture passed is converted to kanji characters from the kana characters (Fig.11-e). Finally, the result of the conversion of kana into kanji is chosen and displayed (Fig.11-f). However, since it surrounds for correcting character segmentations when the starting point and the ending point of gesture are in the same rectangle, correction of the character segmentations instead of the conversion of kana into kanji is performed.

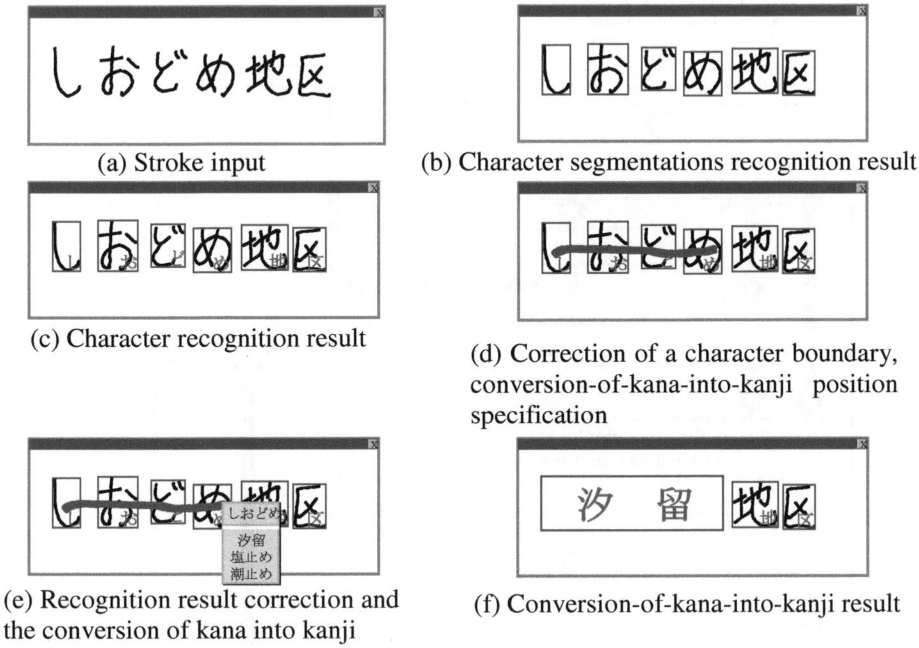

(a) Stroke input

(b) Character segmentations recognition result

(c) Character recognition result

(d) Correction of a character boundary, conversion-of-kana-into-kanji position specification

(e) Recognition result correction and the conversion of kana into kanji

(f) Conversion-of-kana-into-kanji result

Fig. 11. The addition of the conversion of kana into kanji (Two or more characters)

The flowchart of the interface, which added the gesture for the conversion of kana into kanji, is shown in Fig.12.

A user applies recognition processing to arbitrary timing to the inputted stroke to write. Recognition engine presumes character segmentations first, performs character recognition, and displays the character string of a result. If a user inputs the character segmentations correction gesture, character segmentations will be presumed again and a character recognition result will be displayed. When character recognition results differ, character recognition result correction gesture, which carries out the tap of the circumscription rectangle of a character, is performed. Then, below the next candidate of character recognition result and the result of the simple sentence character conversion of kana into kanji are displayed. A user chooses from what was displayed and performs correction of a character recognition result, and the simple sentence character conversion of kana into kanji. When a user inputs the gesture, which performs the conversion of kana into kanji, a conversion-of-kana-into-kanji candidate is displayed from two or more selected Kana characters, and a user chooses. Character recognition and a conversion-of-kana-into-kanji result are decided by the above-mentioned procedure.

By the method of proposing in this paper, character recognition and the conversion of kana into kanji can be performed to a user's arbitrary timing. For this reason, the following becomes possible. When it is in a hurry, or when there is no necessity of carrying out character recognition, it saves with the handwritten stroke. When there is time, or when it is necessary to recognize and to make it a printing type, character recognition can be carried out using the saved stroke, and a result can be used. Since

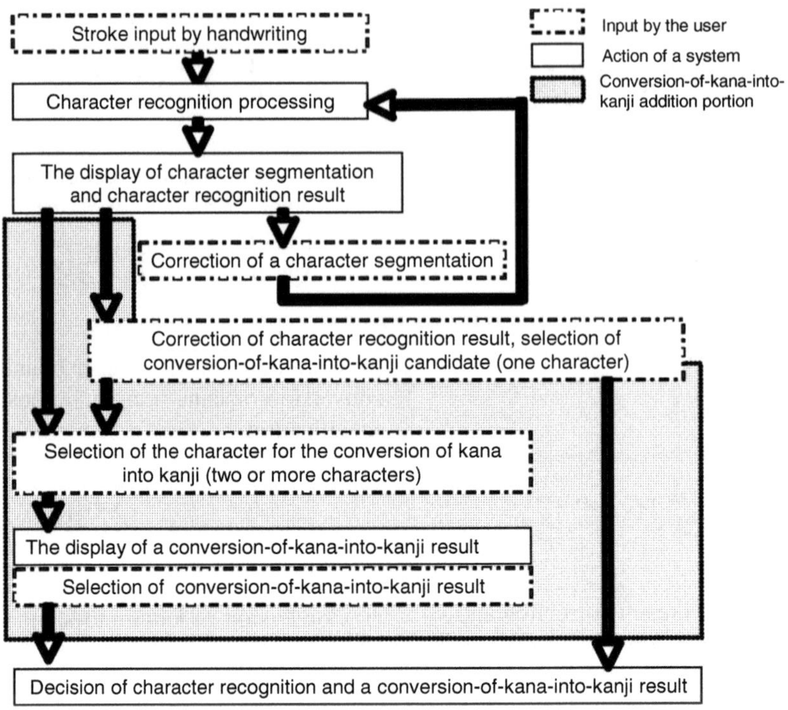

Fig. 12. This systems flowchart

it can carry out by the gesture which neither correction of recognition nor the conversion of kana into kanji overlaps, a user does not need to change mode and needs to perform neither correction nor conversion.

4 Extendibility

As extension of this design, improvement using a context analysis in recognition accuracy and improvement using a prediction input in input speed can be considered. The system independent of client environment can be built by processing character recognition and the conversion of kana into kanji by the server installed on the network. Furthermore, collection of a handwriting character pattern and collection of the character string, which carried out the conversion of kana into kanji, become possible on a server. Although we have already started for construction of a character recognition server [8], it is necessary to realize in the form incorporating a conversion-of-kana-into-kanji server from now on.

5 Conclusion

This paper described the design of the input interface using on-line writing-box-free handwriting recognition and Kana-to-Kanji conversion. During a handwriting input, if this interface is used, when a kanji character is not known, it writes in kana characters and the conversion of kana into kanji can be carried out later. A free input is attained without barring thinking in frame-less handwriting input environment by using the interface described in this paper.

References

1. Fukushima, T., Nakagawa, M.: On-line Writing Box Free Recognition of Handwritten Text based on Stochastic Models. IEICE Technical Report, PRMU 98-139(489), 25–30 (1998)
2. Oki, T., Nakagawa, M.: Implementation and Evaluation of a Combined Recognition System using Normalization of Likelihood Values. Technical Report for IEICE Japan, PRMU2002-196 102(555), 43–48 (2003)
3. Fukushima, T., Yamada, H.: A Predictive Pen-Based Japanese Text Input Method and Its Evaluation. Trans. of IPS Japan 37(1), 23–30 (1996)
4. Souya, T., Fukushima, H., Takahashi, N., Nakagawa, M.: A Basic Design of Handwriting User Interface with Lazy Recognition. Trans. of IPS Japan 34(1), 158–166 (1993)
5. Bandoh, H., Fukushima, T., Kato, N., Nakagawa, M.: User interfaces for correcting errors in writing-box-free recognition of handwritten text. IPSJ SIG Notes, 2000-HI-89, pp. 81–88 (2000)
6. Bandoh, H., Fukushima, T., Kato, N., Nakagawa, M.: User interfaces for correcting errors in writing-box-free recognition of handwritten text. Trans. of IPS Japan 43(6), 1996–2005 (2002)
7. Nakagawa, M., Yorifuji, M.: Error Correction Gestures for Free-format Text Input by Pen Interfaces. In: Proc. Human Interface Symposium 2005, vol. 1, pp. 43–46 (2005)
8. Sakurada, T., Yorifuji, M., Onuma, M., Nakagawa, M.: Web-based Applications Using Pen-Based Interfaces and Network-Based on-line Handwriting Recognition. In: Human-Computer Interaction 10th International Conference (HCI International 2003), vol. 2, pp. 268–272 (2003)

Tasting Robot with an Optical Tongue: Real Time Examining and Advice Giving on Food and Drink

Hideo Shimazu[1], Kaori Kobayashi[1], Atsushi Hashimoto[2], and Takaharu Kameoka[2]

[1] NEC System Technologies Ltd.
8916-47, Takayama, Ikoma, Nara, Japan, 630-0101
{shimazu-hxa,kobayashi-kxd}@necst.nec.co.jp
[2] Mie University,
1577 Kurimamachiya-cho, Tsu, Mie, Japan, 514-8507
hasimoto@bio.mie-u.ac.jp, kameoka@mie-u.ac.jp

Abstract. This paper describes the tasting robot with an artificial sense of taste. Its optical tongue examines the taste of food and drink within several seconds and gives the name as well as its ingredients without crashing them and without opening paper or plastic packages. It is the integration of infrared (IR) spectroscopic technologies with pattern recognition technologies. The robot is the world's first tasting robot which gives advices on the food and on health issues based on the information gathered. It sees if foods are best to eat or if they are good for health from the point of the content of sugar and fat. The robot was first exhibited in EXPO 2005 AICHI, Japan.

Keywords: Optical tongue, Sense of taste, Robot, Infrared sensor, Food recognition.

1 Introduction

This paper describes an artificial sense of taste and the world's first tasting robot with it. Using its optical tongue, the robot is capable of examining the taste of food and drink and giving the name as well as its ingredients. Furthermore, it can give advices on the food and on health issues based on the information gathered.

Specifically we have developed an optical tongue with an infrared ray (IR) sensor. It is attached to an arm of a table-top robot called the tasting robot. The optical tongue is the integration of IR spectroscopic technologies with pattern recognition technologies. It generates an IR spectrum of a food sample, analyzes the quantities inside the food components, and identifies the name of the food. As these results are transmitted to the body of the tasting robot, the robot promptly gives various advices on health and meal to a user in natural language with graphical images.

Aimed for the realization of a robot close to human, the five senses of human are drawing attention. Compared to sight, hearing, and touch with which the physical sensor techniques have developed from the early stage, taste and smell have been left behind. But, checking with the health boom in these later days, environmental problems, rising consciousness for the safety of food (additives and tractability), and care and problems of nursing at home in the aging society, the demand for the intellectual machines with five senses clearly seems to be growing.

M.J. Smith, G. Salvendy (Eds.): Human Interface, Part I, HCII 2007, LNCS 4557, pp. 950–957, 2007.

Therefore, the robot with the sense of taste has been developed. In this paper, its design decisions, capabilities, architecture and the prototype robot exhibited at EXPO 2005 AICHI are described.

2 Design Decisions

This section describes the major decisions made in designing the tasting robot.

2.1 Definition of Tasting

There are at least three major abilities a robot is expected to have if it has a sense of taste.

The first ability is to identify the ingredients in a food sample like sugar and fat and to estimate their quantities. Because a human doesn't have it, a robot with this ability can give useful advices to a human in various situations, like "If you eat this sweet, you can't eat anything more because the accepted calorie will be over the limit", or "This apple is good for eating now".

The second ability is to generate literal expressions of taste for a food sample. Humans naturally have various vocabularies to express their tastes and enjoy talking about food with each other. A robot with this ability can become a critique of food and a human can enjoy discussion on food and drink with the robot, like "This wine is very dry and a little salty. I am sure you like this and it matches well with your sushi meal".

The third ability is to identify the names of food and drink. Humans need trainings to acquire this skill. For example, only wine sommeliers can taste a glass of wine and identify the name among various wines. If a robot has this ability, a human can have his/her own sommelier and gets useful and interesting advices. In addition, because main ingredients and their average quantities of major food products are already investigated and its report is published in [6, 7], if the name of a food product is identified, its main ingredients and average quantities are predicted by referring to such literatures.

Because the first and the third ability are difficult for a human to have, we decided to develop these two abilities for the tasting robot.

2.2 Basic Approaches to Tasting

There are two approaches to analyze the proportion of the inclusive elements for the tastes. One is the use of chemical sensors which simulate a human's sense mechanism of taste directly [13]. The approach is effective to realize the second and third ability defined in the previous subsection. The size of the sensor is now similar to a desktop PC.

The other is the use of physical sensors which analyze IR spectra and is called chemometrics [5, 8, 9, 11]. By analyzing the spectral characteristics of a food sample, the ingredients and their quantities are measured. Though it is not as accurate as the chromatography method [1] which takes several hours, it has an advantage of analysis time length because it takes only several seconds to minutes. It is already used widely in food factories. This approach is effective to realize the first and third ability defined in the previous subsection. The size of an IR sensor is getting small. Because hand-held sensors are already commercialized, it is now possible to attach an IR sensor to a table-top robot.

2.3 The Architecture of the Tasting Robot

The tasting robot architecture has consists of three modules as shown in Figure 1.

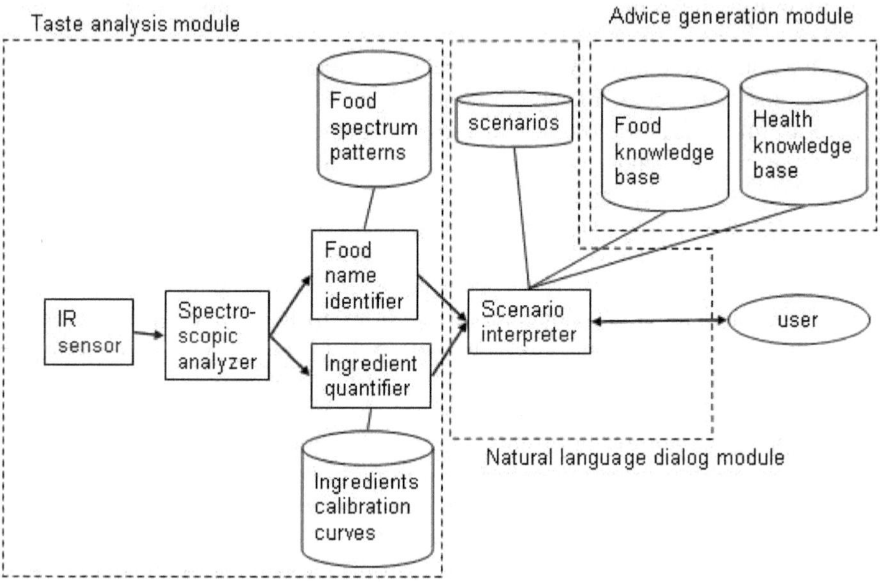

Fig. 1. The architecture of the tasting robot

- Natural language dialog module has a speech recognition sub-module, a speech generation sub module, and a scenario interpretation sub-module.
- Taste analysis module has an IR sensor and its controller, the ingredient quantification module, and the food name identification module.
- Advice generation module has a knowledge base of foods and health, and a reasoning engine to generate various advices.

We used a commercial robot software development tool, RoboStudio [12] which provides the natural language dialog module. By using RoboStudio, a human designer only defines a set of scenarios which describes a sequence of conversation dialogs between a human and a robot. RoboStudio scenario interpreter selects a scenario and performs the sequence in the scenario step by step.

2.4 Appearance of the Tasting Robot

We assumed that the tasting robot would be placed at a dining kitchen or on a dining table. Its size must be similar to a thermos bottle. PaPeRo [4] is a human partner robot

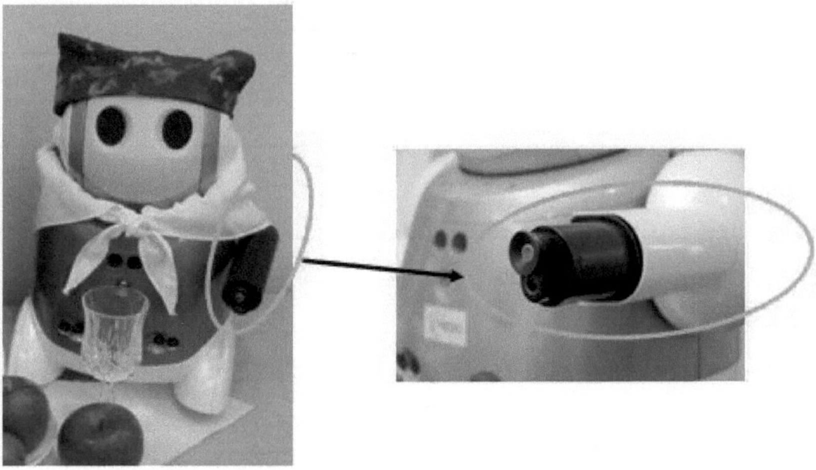

Fig. 2. The appearance of the tasting robot

developed at NEC Corporation and is about 50 cm tall with a speech recognition and generation functions. Because it satisfies our requirements, we used PaPeRo by attaching an arm with an IR sensor as shown in Figure 2. The lower-side of the edge of the arm is an IR transmitter and its upper-side is an IR receiver. During the analysis, IR ray continues to be transmitted for a few seconds. If a spectrum pattern is not generated by some reasons, IR ray is again transmitted for another few seconds until the spectrum pattern is correctly generated.

3 Examining Food Ingredients

3.1 Spectroscopic Analysis on Food

It is known that each food has different IR spectral characteristics [8, 9]. Figure 3 shows the spectra of varieties of cheeses including Gouda, Edam, Parmigiano Reggiano, and Camembert. Its Y-axis is absorbance ratio; $\log_{10}$ (Input light / Output light). Its X-axis is wave number, the number of waves within 1 cm. According to the Lambert-Beer's law, the absorbance ratio at a set of special wave numbers is in proportion to the concentration of a sample. Therefore, a concentration of a food sample is estimated by measuring its absorbance ratio.

There are two independent methods for analyzing a food spectrum as shown in Figure 4. One is the quantification analysis of food components using calibration curves. The other is the food name identification analysis using food pattern recognition. Both of them are stored in the taste analysis module shown in Figure 1.

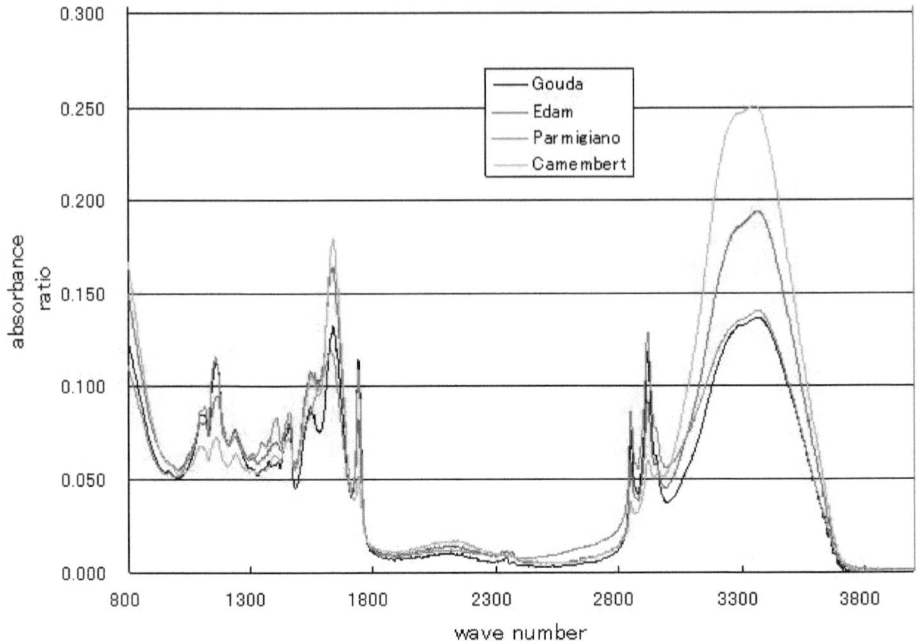

Fig. 3. Spectra of varieties of cheeses

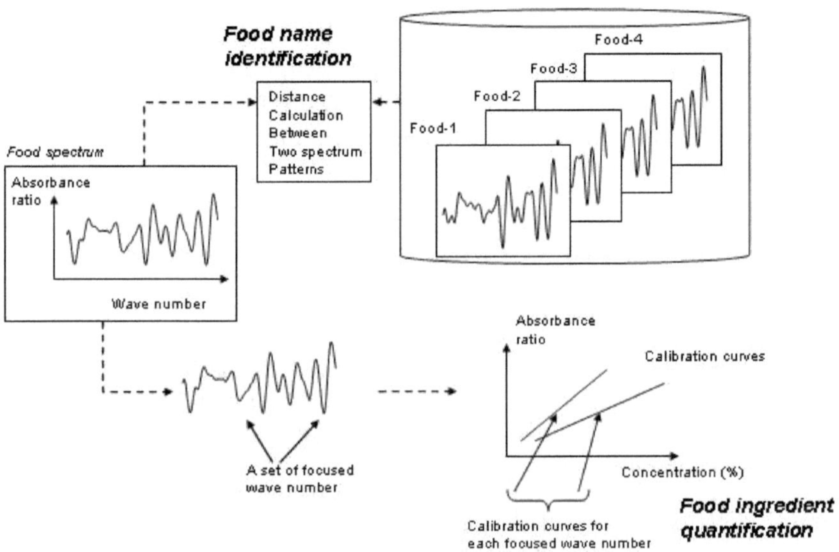

Fig. 4. Two methods for food analysis

3.2 Quantification Analysis of Food Components Using Calibration Curves

Calibration curves of food components are generated based on precise experiments by spectroscopic experts. Figure 5 shows the linear calibration curves of glucose, sucrose and fructose. The wave numbers specified in Figure 5 are known to be directly proportional to the concentrations of sugars.

The quantification analysis is effective if the composition of a food sample is simple and homogeneous. But, in case of sophisticated food samples like cheese, its ingredients are various and varieties of molecules influence the absorbance ratio of a specific wave number. The quantification analysis of such a food sample is less effective. The tasting robot can only analyze the quantification of simple foods like fruits such as apples, bananas and peaches. In case of fruits, the robot discriminates a sample fruit into five categories, “very unripe”, “unripe”, “good for eating”, “too ripe”, or “rotten”. In this level of categorization, it works correctly.

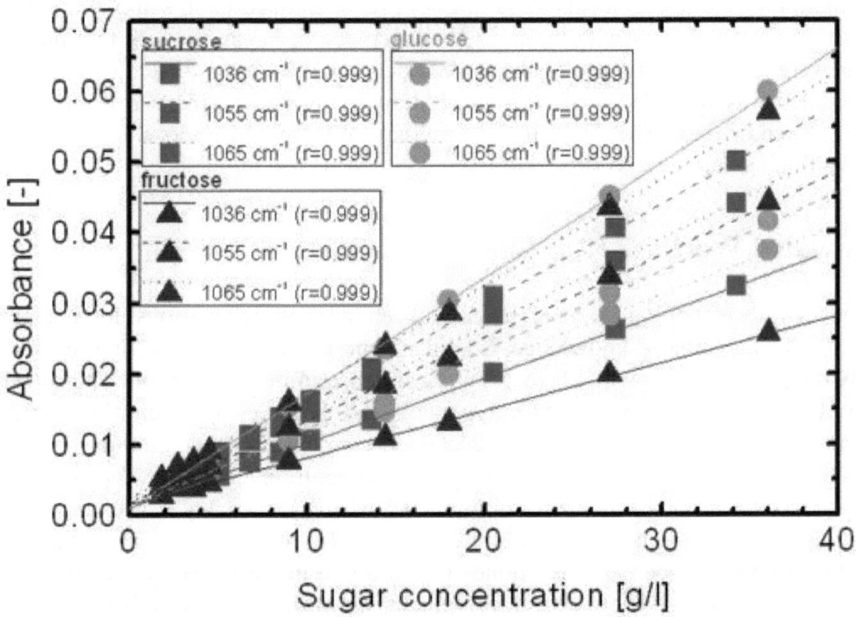

Fig. 5. Linear calibration curves of glucose, sucrose and fructose

3.3 Food Name Identification Analysis Using Food Pattern Recognition

Pattern recognition techniques were introduced to identify food names. Varieties of food samples were measured their spectrum patterns and these patterns were stored with their food names in the IR spectrum pattern database. If a new food sample is given, its spectrum pattern is compared with stored spectrum patterns and calculated the distances. A food name with a closest distance is assumed to be identical to the given food sample.

The basic algorithm of the food name identification analysis consists of two steps. Step 1 selects a set of wave numbers which are effective to calculate the distances between a food sample's spectrum and stored IR spectrum patterns. Because each spectrum pattern spreads over a wide range of wave numbers from 800 to 4000 (1/cm), measuring distances between two spectrum patterns at all wave numbers is very time-consuming and ineffective. Therefore, only 20 to 30 wave number points are selected of which the degree of dispersion of all spectrum patterns in the pattern database is high. These points are used to measure the distances between a food sample's spectrum and spectrum patterns in the pattern database. Step 2 uses canonical correlation analysis [3] where its input variables are the absorbance ratio at selected wave numbers of a spectrum, and its output variable is a food name. Various pairs of a food name and absorbance ratio values of selected wave numbers of the food are used to generate the discrimination equations of canonical correlation analysis.

Because distances among spectra of different cheeses were large, its average accuracy to discriminate one among seven or eight different cheeses was about 80%. Discrimination of processed meats including loin ham, smoked ham, sausage and bacon got similar results. Teas without sugar, with a piece of sugar, or with two pieces of sugar were also discriminated with similar accuracy. The most difficult food we experimented was wine. With the first generation IR sensor, a wine was discriminated among three or four wines with lower accuracy. We are still improving both pattern recognition techniques and IR sensors.

4 Related Work

Spectroscopic technologies have recently been applied for various food analyses [5, 8, 9]. To the authors' knowledge, Edelmann and Lendl [2] was the first research to apply the technology to wine. It reported a wine discrimination method. Their method is to identify wines by measuring polyphenols and tannin using calibration curves. It is similar to our quantification analysis of food components using calibration curves. On the other hand, they haven't applied pattern recognition techniques for food identification. Keller reported a spectra library [10] where a company's medical products are registered. Although the library was limited to a number of medical products, if it covers varieties of foods, the techniques we proposed in our paper can be applied and will be widely used. As for robot development, the authors believe this is the world's first tasting robot.

5 Summary

This paper reported the tasting robot, its design decisions and architecture, its capabilities and performances. The robot was first exhibited in EXPO 2005 AICHI, Japan.

In the future, this kind of robot will possibly be applied to the estimation of materials, support for the seasoning plan of food, early problem analysis in the food market. Also, it is expected to develop into various fields like a refrigerator robot with the sensor to detect the deteriorated food, foods fit to eat, and the smell of mold, a robot for the prevention and cure with the diet for adult diseases, the nursing robot which have artificial tongue and nose, and helper robot.

References

[1] 1993 Annual Book of Astm Standards: Section 14: General Methods and Instrumentation: Vol.14.01: Analytical Methods-Spectroscopy; Chromatograph (1993)
[2] Edelmann, A., Lendl, B.: Characterization of Red Wine Polyphenols by FT-IR Spectroscopy, 2001 Second Symposium o. In: Vino analytica Scientia 2001 (2001)
[3] Fisher, R.A.: The use of multiple measurements in taxonomic problems. Annals of Eugenics 7, 179–188 (1936)
[4] Fujita, Y.: Personal Robot PaPeRo. Journal of Robotics and Mechatronics 14(1), 60–63 (2002)
[5] Hashimoto, A., Mori, H., Kanou, M., Yamanaka, A., Kameoka, T.: Mid-Infrared Spectroscopic Analysis on Brewed Coffee Characteristics, Proceedings of 10th Asian Pacific Confederation of Chemical, Engineering Conference (CD-ROM version), 3P-01-068 (2004)
[6] Gijutsu, K., Chousa-kai, S., Kagawa, Y. (eds.): Standard Tables of Food Composition in Japan, revised 5th edn. Kagawa Nutrition University Press, Kagawa Nutrition (2005)
[7] Kagawa, Y. (ed.): Shokuhin 80 Kilo Calorie Guidebook. Kagawa Nutrition University Press (2005)
[8] Kameoka, T., Okuda, T., Hashimoto, A., Noro, A., Shiinoki, Y., Ito, K.: FT-IR/ATR Analysis of Sugars in Aqueous Solution Using ATR Method. Nippon Shokuhin Kagaku Kogaku Kaishi 45(3), 192–198 (1998)
[9] Kameoka, T., Okuda, T., Hashimoto, A., Noro, A., Shiinoki, Y., Ito, K.: A Rapid FT-IR/ATR Method for Sugar Content Determination in Food. Nippon Shokuhin Kagaku Kogaku Kaishi 45(3), 199–204 (1998)
[10] Keller, R.: THE SIGMA LIBRARY of FT-IR SPECTRA, Sigma Chemical Company, St. Louis (1986)
[11] Ogawa, C., Wakaki, M.: Hikari Kogaku Nyumon - Hikari no Kisochishiki no subete, Jikkyo Shuppan Co (2001)
[12] RoboStudio Web page, http://www.necst.co.jp/product/robot/
[13] Toko, K. (ed.): Kansei Bio Sensor Asakura Shoten (2001)

Concept-Based Question Answering System

Seung-Eun Shin[1], Yu-Hwan Kang[2], and Young-Hoon Seo[2]

[1] Chungbuk National University BK21 Chungbuk Information Technology Center
Cheongju, Chungbuk, 361-763, Korea
seshin@nlp.chungbuk.ac.kr
[2] School of Electrical & Computer Engineering, Chungbuk National University
Cheongju,Chungbuk, 361-763, Korea
eric@nlp.chungbuk.ac.kr, yhseo@chungbuk.ac.kr

Abstract. This paper describes a concept-based approach for question answering system in which concept rather than keyword makes an important role on question analysis, document retrieval, and answer extraction. Our idea is that we can extract correct answer from various paragraphs with different structures when we use well-defined concepts because concepts occurred in questions of same answer type are similar. We defined a concept rule for each answer type. The concept rule contains core concepts occurred in questions of that answer type. Question analysis module extracts concepts from user's question and determines the answer type. Document retrieval module retrieves more relevant documents using extracted concepts. Answer extraction module extracts a probable answer from retrieved documents using concepts. Empirical results show that our concept-based approach can retrieve more relevant documents and extract more accurate answer than any other conventional approach. Also, our approach has additional merits that it is language universal model, and can be combined with arbitrary conventional approaches.

Keywords: Concept, Question Answering System, Document Retrieval, Information Retrieval, Concept-based Question Answering System.

1 Introduction

Information retrieval techniques used to find information fast and exactly from tremendous documents have been rapidly developed with the growth and commercial application of the Internet. However, we can often find that high ranked documents retrieved from a general web search engine may be far from a user intension. Therefore, effective retrieval and rank techniques are needed to provide more relevant documents to users or question answering systems are demanded for user's convenience.

Question answering systems return answers rather than documents containing answers, in response to a question. The task of a question answering system is to provide direct, succinct responses to natural language questions posed by a user. Functionally, most question answering systems today can be decomposed into three major components: question analysis, document retrieval, and answer extraction [1]. The question analysis component classifies user questions by expected semantic types of answers, e.g., the expected answer type of "Who wrote Hamlet?" is a person. Document retrieval component retrieves relevant documents using conventional

M.J. Smith, G. Salvendy (Eds.): Human Interface, Part I, HCII 2007, LNCS 4557, pp. 958–967, 2007.

information retrieval methods. Finally, answer extraction component extracts named entity related with answer type from relevant documents as answer. However, general question answering systems cannot accurately analyze question, retrieve relevant documents, and extract answers, because they do not reflect the meaning of sentences and use a statistical approach using only keywords.

The studies for question analysis mostly extract answer type corresponding to question type using named entities, answer type taxonomies, and ontologies such as WordNet [2], [3], [4], [5], [6], [7]. The number of answer types varies widely from single digits to a few thousands. The subdivided classification of answer type helps to extract more accurate answer by reducing the number of answer candidates in phase of answer extraction.

The studies for answer extraction are largely divided into keyword approaches and language analysis approaches. Keyword approaches [2], [3], [5] use keywords' frequency and position in passages, an expansion of keywords, and so on. These approaches are efficient in speed, but it is difficult to find correct answer because only keywords and corresponding limited knowledge are used. Language analysis approaches [4], [6], [8], [9] use natural language processing technologies such as syntax analysis. Approaches using linguistic knowledge can extract more accurate answer than keyword approaches.

Approaches using syntactic patterns for answer extraction have a limit to extract answer because there are many answers which can not be found by syntactic information obtained from question, and because syntactic patterns for answer extraction can hardly be derived. Therefore, we need semantic information as well as syntactic information for more accurate question analysis, document retrieval, and answer extraction.

This paper describes a concept-based question answering system in which concept rather than keyword makes an important role in question answer system. We define concepts commonly occurred in the same type of questions, and use those concepts on question analysis, document retrieval, and answer extraction to retrieve more relevant documents and extract more accurate answer.

2 Concept-Based Question Analysis

Concept-based question analysis applies NLP techniques on a user's natural language questions and analyzes them semantically for an efficient document ranking. For each answer type, we define a concept rule which contains core concepts occurred in questions of that answer type. Concept is a well-defined semantic component for each answer type.

Let's consider following questions whose answer types are authors.

(Question 1) Who wrote "Hamlet"?

(Question 2) Who was the author of the book about computer hackers called "The Cuckoo's Egg: Tracking a Spy Through the Maze of Computer Espionage"?

(Question 3) Who wrote the hymn "Amazing Grace"?

(Question 4) What composer wrote "Die Gotterdammerung"?

(Question 5) Writer of "Huckleberry Finn"?

We can see concepts to be used commonly to represent the information demand of users in questions (Question 1-5). They are titles ("Hamlet", "The Cuckoo's Egg: Tracking a Spy Through the Maze of Computer Espionage", "Amazing Grace", "Die Gotterdammerung", "Huckleberry Finn"), an interrogative pronoun (who, what), nouns to express the author (author, composer, writer), verbs to express the author (write), and nouns to express the genre (book, hymn).

The statistical model extracts index terms (Who, write, Hamlet) from Question 1 and index terms (What, composer, wrote, Die Gotterdammerung) from Question 4. Then it generates queries using index terms and retrieves relevant documents by the query-document similarity. Therefore, it will retrieve the document which includes simply many index terms to more relevant document than the really relevant document which includes sentences such as "Shakespeare is the author of Hamlet". To solve this problem, we determine answer types and subtypes of answers, extract concepts from user's questions, and use them for accurate document retrieval and answer extraction.

< Results of concept-based question analysis >

(Question 1) Who wrote Hamlet?

- Answer type : Person
- Subtype of the answer : Author
- Book Title : Hamlet

(Question 4) What composer wrote "Die Gotterdammerung"?

- Answer type : Person
- Subtype of the answer : Author
- Book Title : Die Gotterdammerung

As shown above, we determine answer types and subtypes of answers and extract concepts from user's questions. A concept is not simple meaning of a word but the semantic role of a word or phrase in sentences and is used to represent the user's intention. We defined subtypes of the answer and concepts from 643 questions whose answer type is a person in TREC QA Data and Web, and we constructed a concept dictionary by tagging concepts manually on those questions and expanded it using a synonym dictionary.

Table 1 shows a sample of subtypes of the answer and concepts. Subtypes of the answer type 'person' are classified by 24 categories such as author, family, prizewinner, politician, developer, inventor, scholar, entertainer, player etc. We are currently defining 125 concepts.

Table 1. Sample of subtypes of the answer and concepts

Subtype of the answer	Concepts
Common	Nationality, Time, Sex, Person, …
Author	Book Title, Author Noun, Author Verb, …
Family	Relationship, Base Person, Relationship Info, …
Prizewinner	Prize, Prize Noun, Prize Verb, Ceremony/Place, …
Politician	Position, Event, Organization, Election Noun, …

We defined a concept rule manually for each answer type to extract concepts from a user's questions. Such a concept rule is represented as concepts and grammatical morphemes in order to consider semantic and syntactic structure of the user's questions. Fig. 1 shows the BNF notation of the concept rule and characteristics of Korean considered in our discussion. The concept rule consists of a list of <word information> which consists of concepts and grammatical morphemes according to the answer type.

```
<Concept Rule> ::= <Word List>
<Word List> ::= <Word Information> | <Word List><Word Information>
<Word Information> ::= '(' <Concept> ')' | '(' <Concept> <Grammatical Morpheme> ')'
<Concept> ::= '@Who'|'@Person'|'@Author_N'|'@Author_V'|'@Genre'|
                                  '@Relationship'|'#Base_Person'|'#Book_Title'|...
<Grammatical Morpheme> ::= 'jc'|'jx'|'jm'|'etm'|'co'|'ef'|'oj'|'co+etm'
```

jc : case particle	co : copula
jx : auxiliary particle	etm : adnominal transition ending
jm : adnominal case particle	ef : sentence ending
oj : objective case particle	

Fig. 1. BNF notation for the concept rule

Tag for each concept consists of 'Property' + 'Concept'. Properties are divided into three properties such as '%', '@', '#'. We are currently constructing the concept dictionary which includes 2,039 vocabularies.

Concepts with '%' property are placed relatively free in sentence, and so they are inappropriate to handle by rules. Some '%' property concepts such as '%Adverb' may be extracted from concept dictionary, and others such as '%Time' from rules. These concepts are used importantly in document ranking though they are not used in concept rule matching.

Concepts with '@' property are extracted from concept dictionary. That is, we tag a word or phrase to '@' property concept if it is found in concept dictionary. Concepts with '#' property are ones to be extracted only from concept rules. Proper nouns such as title of the book and person name belong to '#' property concepts. '#' property concepts are extracted from question after a concept rule is selected using '@' property concepts.

The following examples are a concept rule for 'author' that is represented as an extended BNF.

- Example of the concept rule for the author

  ……

  4. (#Book_Title jc) (@Author_V etm) (@Person|@Author_N jx?) (@Who)?
  5. (#Book_Title jm) (@Author_N jx?) (@Who)?

  ……

We tag concepts to morphological analysis result of a user's question and construct the concept list. The concept list is matched by each concept rule. If there is a matched concept rule, we determine answer type and extract concepts of '#' property using that concept rule. We select the longest rule when we have several matched concept rules.

If there is no concept rule to be matched to a user's questions, we extract concepts from the user's questions by following manner. We classified questions whose answer type is a person whether it contains a verb or not. The question which includes a verb consists of Event_V (verb to represent an event), Person (noun to represent a person), Property_N (noun to represent the property of a person such as doctor, author, and so on), and Who (interrogative). The question which does not include a verb consists of Property_N, NP (noun phrase), and Who. Therefore, we designed the common concept rule according to the syntactic structure of Korean natural language questions as follow.

- Common concept rule for the question which includes a verb
  1. (#NP jc) (#Event_V etm) (@Person|@Property_N) (@Who)?
  2. (#NP jx) (@Who) (#Event_V ef)
  3. (@Who) (#NP jc) (#Event_V ef)
- Common concept rule for the question which does not include a verb
  1. (#NP jm) (@Property_N) (@Who)?
  2. (#NP jc) (@Property_N) (@Who)?
  3. (@Property_N) (@Who)?

Although we cannot determine the subtype of the answer, we can extract concepts by using common concept rules from a user's question which has not the concept rule to be applied.

3 Concept-Based Document Retrieval

We generate queries that reflect various syntactic structures to represent the answer and utilize them for accurate document retrieval. We can retrieve a document that include generated queries as a relevant document and improve the precision of document retrieval. Other approaches that do not analyze the user's question semantically can hardly obtain queries whose syntactic structure is different from the user's question, but our approach can generate queries using the answer type and concepts as results of concept-based question analysis [10].

We designed query generation concept rules in order to generate queries, and it is made of concepts and grammatical morphemes. The query generation concept rules have concepts and syntactic structures used to represent an answer. We can generate queries using the query generation concept rules, concepts, concept dictionary, and synonym dictionary. Query generation concept rules for an author are examples of query generation concept rules.

- Query generation concept rules for an author
  1. (Book_Title oj) (Author_V)
  2. (Book_Title jm) (Author_N)
  ……

We can generate queries using query generation concept rules and results of the concept-based question analysis of (Question 1) as below.

< Results of the query generation >

•Query generation concept rule 1 : (Book_Title oj) (Author_V)

▪Generated queries : Hamlet-eul jeo-sul-hal|jeo-jak-hal|......|sseu (write|wrote|......|to compose Hamlet)

•Query generation concept rule 2 : (Book_Title jm) (Author_N)

▪Generated queries : Hamlet-ui jeo-ja|jak-ga|......|geul-sseun-i (writer|author of Hamlet)

Our approach can generate queries which have the same meaning as the original question, but have different structures such as above example. Generated queries reflect the syntactic structures of almost all phrases which have answers for that question. We retrieve a document that includes generated queries as a relevant document because they consist of concepts and syntactic structures used to represent an answer.

Formula (1) is a transformation of the cosine coefficient to determine query-document similarity when our approach is combined with the vector model [11]. We calculate the query-document similarity by using formula (1) to retrieve a document which includes generated queries as a relevant document.

$$\text{if } \vec{d}_i \cdot \vec{q}_{ge} \neq 0 \text{ then } sim(d_i, q_u) = \vec{d}_i \cdot \vec{q}_{ge}$$
$$\text{else } sim(d_i, q_u) = \frac{\vec{d}_i \cdot \vec{q}_u}{|\vec{d}_i||\vec{q}_u|} \quad (1)$$

d_i : document, q_u : user's question, $\vec{d}_i$: document vector

$\vec{q}_{ge}$: generated query vector, $\vec{q}_u$: query vector by query expansion

We can rank documents by formula (1) because queries are generated by concepts and syntactic structures which are used to represent an answer. We can increase the precision at document retrieval by ranking documents which include a generated query in the high position.

4 Concept-Based Answer Extraction

We designed answer extraction concept rules that have concepts and syntactic structures used to represent an answer, and it is made of answer, concepts and grammatical morphemes. Following concept rules are examples of answer extraction concept rules for an author and answer extraction common concept rules. We apply answer extraction common concept rules to questions which are analyzed by common concept rules.

• Example of answer extraction concept rules for an author
1. ($Book oj) (@Author_V etm) (Answer)
2. (Answer j_) ($Book oj) (@Author_V)
3. ($Book oj) (@Author_V etm) (@Author_N|@Person j_) (Answer co)
4. (Answer j_) ($Book oj) (@Author_V) (@Author_N|@Person co)
……

• Example of answer extraction common concept rules
1. (!NP jc) (!Event_V etm) (Answer)
2. (Answer j_) (!NP jc) (!Event_V)
3. (!NP jc) (!Event_V etm) (@Person|!Property_N j_) (Answer co)
4. (Answer j_) (!NP jc) (!Event_V etm) (@Person|!Property_N co)
……

'$' is a property symbol whose concept is tagged by a question analysis result, '@' is a property symbol whose concept is tagged by a concept dictionary, and '!' is a property symbol whose concept is tagged by a synonym dictionary and question analysis result. 'Answer' is named entity that corresponds to answer type.

After morphological analysis for relevant documents, we fills an answer extraction concept rule with a concept tagged word or phrase. Following example shows an answer extraction process.

< Result of the answer extraction >
(Question 1) Nu-ga Hamlet-eul sseot-na-yo? (Who wrote Hamlet?)

• Answer type : Person
• Subtype of the answer : Author
• Book_Title : Hamlet
• Sentence which include an answer : Shakespeare-neun Hamlet-ui jeo-ja-i-da. (Shakespeare is the author of Hamlet.)
• Morphological analysis result : Shakespeare/nh+neun/jx Hamlet/nh+ui/jm jeo-ja/nc+i/co+da/ef+./s
• Concept tagging result : Shakespeare/Answer+neun/jx Hamlet/Book+ui/jm jeo-ja/ Author_N +i/co
• Answer extraction concept rule for an author : (Answer j_) ($Book jm) (@Author_N co)

Answer	j_	$Book	jm	@Author_N	co
Shakespeare	neun	Hamlet	ui	jeo-ja	i

We rank answer candidates using the precedence of applied concept rules and frequency of them.

5 Experimental Results

We randomly selected 100 questions as a test set from natural language questions whose answer type is a person. They were questions which were used actually for IR in the Web. We measured the precision at N documents. In our experiments, we used

Table 2. Accuracy of concept-based question analysis

	Question which concept rule is applied	Question which common co ncept rule is applied
The number of questions	69	31
Accuracy of answer type determination	1.000	0.903
Accuracy of subtype determination	1.000	We cannot determine the su btype of the answer
Accuracy of concept extraction	0.918	0.845

Table 3. Precision at N documents of the document retrieval result

Micro Averaging Precision				
N	Google	Google+ Our approach	Yahoo	Yahoo+ Our approach
At 3 docs	0.584	0.803(+0.219)	0.580	0.804(+0.224)
At 5 docs	0.585	0.770(+0.185)	0.557	0.743(+0.186)
At 10 docs	0.548	0.646(+0.098)	0.511	0.604(+0.093)
At 15 docs	0.523	0.591(+0.068)	0.478	0.541(+0.063)
At 20 docs	0.489	0.541(+0.052)	0.458	0.506(+0.048)

Google and Yahoo as the IR systems and used only top 30 results of such systems for the precision at N documents. Table 2 shows the accuracy of concept-based question analysis and Table 3 shows the precision at N documents of the document retrieval result.

Accuracy of concept extraction is (the number of concepts which are extracted correctly)/(the total number of concepts). In case that the concept-based document retrieval is applied to Google and Yahoo, the test of precision at N documents was improved by +0.2215(N=3), +0.1850(N=5) and +0.0955 (N=10). If our approach is applied to more documents, the precision at N documents can be improved more than that of table 3. In addition, we found that it is possible to make document retrieval more efficient by analyzing questions based on concepts which are comparatively short but fully expressing a user's intentions.

In answer extraction experiment, we evaluated the accuracy of answer extraction. Table 4 shows the evaluation result for concept-based answer extraction.

Table 4. Accuracy of answer extraction

	Concept-based Ans wer Extraction	TREC 2005 Highest	TREC 2005 Medium
Accuracy of answ er extraction	0.749	0.713	0.152
Questions	Our test set	TREC 2005 factoid questions	

Accuracy of concept-based answer extraction is 0.749 and our approach extracts more accurate answer than any other conventional approach (TREC 2005 Highest 0.713) [12]. Accuracy of concept-based answer extraction is difficult to compare directly with TREC 2005 Highest approach because our result is from the restricted domain (questions whose answer type is a person). But if concept-based approach is combined with arbitrary conventional approaches, we can expect a notable improvement in the accuracy of answer extraction.

6 Conclusion and Future Work

In this paper, we proposed a concept-based approach for question answering system. For each answer type, we defined a concept rule which is composed of well-defined concepts commonly occurred in that type of question. A concept rule is represented as concept components and grammatical morphemes in order to consider the semantic and syntactic structure of questions and answer containing paragraphs.

Concept-based question analysis extracts concepts from morphological analysis result for a user's question and determines answer type. And then, we use those concepts to extract relevant documents and to extract an accurate answer. Document retrieval and answer extraction can be merged into one if relevant document retrieval is not needed.

The accuracy of concept-based answer extraction was 0.749. And we applied our concept-based approach to document retrieval system, Google and Yahoo, and obtained a notable improvement (+0.2215, N=3) in the precision at N documents. Although we make an experiment in the restricted domain in which questions require a person name as its answer, our concept-based approach can retrieve more relevant documents and extract more accurate answer than any other conventional approach. Also, our approach has additional merits that it is a language universal model, and can be combined with arbitrary conventional approaches how concept-based approach is used when the given question can be analyzed to one of the defined concept rules, and other approach is used otherwise.

We plan to expand concept rules for various domains, and expect incremental performance improvement.

Acknowledgments. This research was supported by the MIC(Ministry of Information and Communication), Korea, under the ITRC(Information Technology Research Center) support program supervised by the IITA(Institute of Information Technology Assessment).

References

1. Bilotti, M.W., Katz, B., Lin, J.: What Works Better for Question Answering: Stemming or Morphological Query Expansion? IR4QA: Information Retrieval for Question Answering. A SIGIR 2004 Workshop (2004)
2. Abney, S., Colins, M., Singhal, A.: Answer Extraction. In: 6th Applied Natural Language Processing Conference (2000)

3. Ittycheriah, A., Franz, M., Zhu, W., Ratnaparkhi, A.: IBM's Statistical Question Answering System. In: 9th Text Retrieval Conference, pp. 229–334 (2000)
4. Haragagiu, S., Pasca, M., Maiorano, S.: Experiments with open-domain with open-domain textual question answering. In: COLLING-2000, pp. 292–298 (2000)
5. Lee, G.G, Seo, J., Lee, S., Jung, H., Cho, B., Lee, C., Kwak, B., Cha, J., Kim, D., Ann, J., Kim, H., Kim, K.: SiteQ:Engineering High Performance QA System Using Lexico-Semantic Pattern Matching and Shallow NLP. In: 10th Text Retrieval Conference, pp. 437–446 (2001)
6. Cardie, C.: Examining the Role of Statistical and Linguistic Knowledge Sources in a General-Knowledge Question-Answering System. In: 6th Applied Natural Language Processing Conference, pp. 180–187 (2000)
7. Pasca, M., Harabagui, S.: High Performance Question / Answer. In: 24th Annual International ACM SIGIR Conference on Research and Development in Information Retrieval, pp. 366–374 (2001)
8. Buchholz, S., Daelemans, W.: Complex Answers: a case study using a WWW question answering system. Natural Language Engineering 7, 301–323 (2001)
9. Kesel, V.: Question Answering using Unification-based Grammar. In: Stroulia, E., Matwin, S. (eds.) Advances in Artificial Intelligence. LNCS (LNAI), vol. 2056, Springer, Heidelberg (2001)
10. Shin, S.-E., Seo, Y.-H.: Seung-Eun Shin and Young-Hoon Seo: Query Generation using Semantic Features. In: Sugimoto, S., Hunter, J., Rauber, A., Morishima, A. (eds.) ICADL 2006. LNCS, vol. 4312, pp. 234–243. Springer, Heidelberg (2006)
11. Salton, G.: Automatic Text Processing. Addison-Wesley, Reading, MA (1989)
12. Voorhees, E.M., Dang, H.T.: Overview of the TREC 2005 Question Answering Track, TREC 2005 (2005)

Real IT: Information Technology in Real Space

Ronald Sidharta, Tomohiro Tanikawa, and Michitaka Hirose

RCAST, The University of Tokyo
4-6-1 Komaba, Meguro-Ku, Tokyo, Japan
{ronalds,tani,hirose}@cyber.t.u-tokyo.ac.jp

Abstract. There are needs recently to push information technology beyond the desktop environment. It is because even decades after the proliferation of information technology, it still, by large, confine the users in its own domain. Conventional IT confined users to the desktop and virtual reality technology confined the users onto specialiced devices such as the HMD, large screen display room, or the CAVE. Our vision is to push for the ubiquity of IT beyond the mobile cell phone media. This paper discusses several requirements for the real IT vision. Spefically we are interested in the characteristics of display technologies that suit our vision. We will describe several key technologies that have been developed within this requirements, and our most recent work in this field.

Keywords: information technology, display technology, virtual reality, augmented reality, volumetric display.

1 Introduction

There has been a great deal of research and development both in the conventional IT field and also in the field of virtual reality. Many new algorithms and technologies have been invented to better support the users in the respective field. However, despite the development, we feel that the users are still confined to the domain of the field.

Conventional IT usually confined its users to the desktop environment. Its users must sit in front of a terminal and interact with the terminal in one-to-one basis. There is little support for multi-users. Even if there are several users that want to conduct a collaborative task, they have to crowd in front of one terminal, in which there can only be one person that operates the system. It is difficult for the users to see all the available information while crowding in front of the terminal and it is also difficult for them to convey their idea to the operator. This kind of setting bottlenecks the input and output of communication from and to the users.

Many researches in virtual reality technology have been focused to address this collaborative problem in conventional desktop environment. However, since VR technology usually put their users in its HMD environment, the CAVE, or large screen display room, we feel that the users are still disconnected from the real space, and from their peers. HMD and CAVE displays completely put users into a virtual reality space in which they are separated from the real space [3]. Usually in this type

M.J. Smith, G. Salvendy (Eds.): Human Interface, Part I, HCII 2007, LNCS 4557, pp. 968–974, 2007.

of projection system, co-location collaborative work is not supported due to the stereo rendering technique used. Collaboration with other users is usually done with a virtual representation called avatars. Furthermore, there are still no unifying interface models that support collaboration and interaction in such environments. On the other hand, large screen displays solve the problem of viewing information in a collaborative setting. The users are not crowding in front of a terminal, and information disseminated to the users is well displayed. However, we feel that there are situations when large screen displays are not enough. For example, the situations when the users are mobile, when they occupy a wide area larger than the display, or if the information that needs to be displayed are three dimensional.

1.1 Requirements for Real IT

Considering the situation described in previous section, we have developed a vision for IT technology to penetrate real spaces. The characteristic of a real space and the requirement of technology needed have been address previously by Eitoku et al. [5]. Based of such requirement, we have set the following guidelines in achieving our vision of real IT:

1. Information exists physically within the real world.
2. Not isolate users from their peers and support collaboration.
3. Not encumbering the users with special apparatus.
4. Support the need to naturally display three dimensional image

The main goal of our vision is to realize information penetration into the real space; it must co-exist within the real space in harmony. This means, ideally, the users shall not be confined to the desktop environment or to the specialized VR devices. Current technology that allows information to exist within the real world usually uses HMD as a medium to display the information [2] [8]. This breaks requirement two and three. HMD encumbered the user with its weight and cables; furthermore it isolates its user from the environment, disadvantageous for collaborative work.

2 Related Work

2.1 Projected Reality

There have been several techniques proposed to integrate information into the real world without encumbering the users with devices that disconnect them from the environment and their peers. Projection based technology has been the focus of research in this field. Using an image projector, various techniques have been developed to augment information efficiently into the real world. Raskar et al. [10] described a technique that uses projectors to augment textural information onto a real non-planar object in the real world. Furthermore, textural information could be animated and combined with a projected animation of a moving background to create an illusion of moving objects. A video demo demonstrated a situation where a physical mockup of a car was projected with textures of the painting and rotating tires. Combined with projected images of a moving background, it gave the illusion that the car is moving.

Ehnes et al. [4] took further the concept of projected reality and enabling it to project to not only static objects, but also moving objects. Using several rotatable projectors with marker based tracking system; marked static object, as well as moving objects in the environment can be augmented with information (see Figure 1). This is very useful in a situation where a wide area and several objects are available for augmentation, such as in a museum.

Projected reality is a suitable solution to augment textual or other 2D information onto the real space. However, it does not support the presentation of three dimensional images naturally, breaking requirement four that we have discussed previously. In order to display 3D data, stereoscopic display techniques using polarized glasses are required. This requires the user to wear special device. Non-glasses techniques such as using lenticular sheets have a limited field of view, thus limiting the number of participants. Furthermore, displaying 3D data on a flat 2D screen introduces accommodation and convergence problem that could result in visual fatigue and nausea. [7]

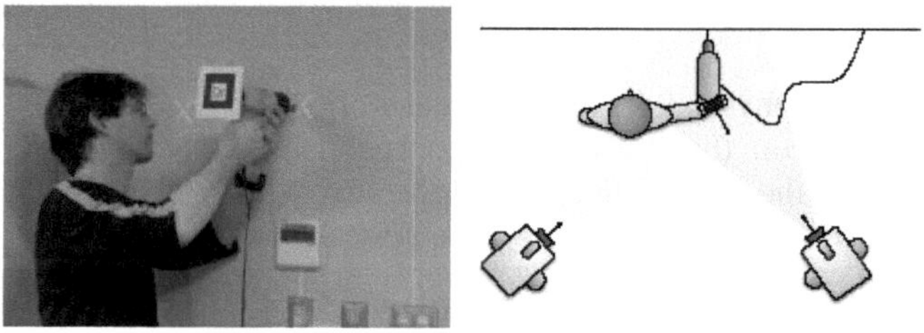

Fig. 1. Using multiple projectors and selection algorithm to augment information to the real space

2.2 Volumetric Displays

One way to display three dimensional data naturally, without accommodation and convergence mismatch, is to use volumetric display. There are two well-known types of display that are able to display information with real depth, thus eliminating the accommodation and convergence problem. Falavora et al. [6] developed a swept-volume type volumetric display called the Perspecta Spatial 3D Display. A swept volume display uses a spinning projection screen to project images at different depths. By spinning the projection screen fast enough, 3D volumetric image can be perceived by the eyes. Tanikawa and Suzuki et al. [9] developed similar system with much larger size using two plasma displays. With this size, it was possible to display human-sized 3D data; furthermore, this type of displays supports presentation to multi-users really well because it could display 3D information in 360 degrees, and special apparatus are not required.

Rotational screen displays begs the users to reach in and interact with the 3D information being displayed. However, it was not possible because the display system itself is in the way. Eitoku et al. [5] addressed this need by developing a volumetric display based on projection onto water particles thus it is possible to reach in into the display. Arrays of water drops are released in a timely fashion so that it creates vertical planar surfaces. A projector from below projects onto these planar surfaces, timed carefully so that each planar surfaces displays a cross-section of a three dimensional objects.

3 Volumetric Augmented Reality Display

As we have described previously, the first requirement of real IT is to enable the presentation of information within the real space. Projected reality system fulfilled this requirement suitably, especially in a situation where information must be displayed directly on a physical object in the real space. Although current volumetric display addressed the need of displaying 3D data naturally so it could co-exist within the real space, it did not address the need of augmenting real object. In other words, the displayed information is separate from the real object. You couldn't for example, display floating textual information over a real printer.

In 2001, Bimber et al [1] developed a mixed reality system for museum exhibition that uses half-mirror reflection technique called the Pepper's Ghost setup with active stereo projection to show real historical artifacts and the virtual representation at the same time. This system consisted of a convex assembly of half-silvered mirror and a BARCO display. The observers could see the stereoscopic virtual objects reflected off the display in the half-silvered mirror, coexisting with the physical object located inside the showcase. Because the system displayed in stereo, the virtual objects could be perceived to have a varying depth. However because the system uses active stereo projection technique, it requires the user to wear a stereo glass. Furthermore, it still suffers from accommodation and convergence mismatch problem.

We have developed an augmented reality display that is able to display volumetric 3D data augmenting the physical object.

3.1 Display System Setup

Our system setup is illustrated in figure 2. To enable physical object and 3D data to co-exist in the same space, we use the Pepper's Ghost setup. In order to display with real varying depths, we used layers of Polymer Dispersed Liquid Crystals (PDLC) film placed on the horizontal plane. A PDLC film can be turned on/off; in off condition, the liquid crystals are in relaxed state causing the layer to have opacity thus it is possible to project image onto it. We specified in our system so that at any given time, there could only be one liquid crystal layer that was in opaque state. This layer was projected with images coming from a projector. To control the depth of the virtual object's appearance, we controlled which PDLC layer was turned off. To minimize the height of the system, we placed a projector on the lower horizontal plane and we used a mirror to reflect the projection upward onto the PDLC layers. A half-silvered mirror was placed at a forty five degrees angle, while the diorama was placed behind the half-silvered mirror. With this setup, the observer would see the images from the PDLC layers inside the half-screen mirror. Since the PDLC layers were stacked physically, when it was reflected, the depth of resulting image corresponded directly on the depth of the PDLC layer. To complete the setup, we

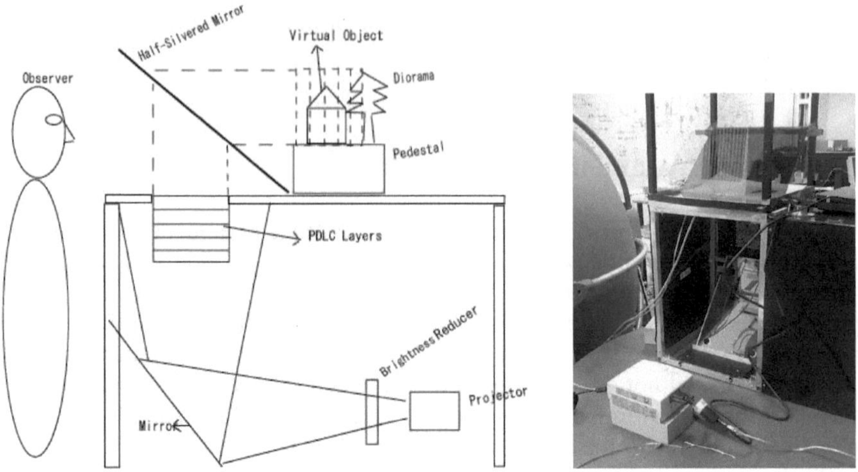

Fig. 2. General System Setup

lighted the diorama with study lamp in order for it to be visible; it was because the projector was really bright. Additionally we placed a dark film in front of the projector to further reduce the light that is coming into the user's eyes, and to prevent the diorama from being washed out due to the projector's bright light.

3.2 Liquid Crystal Layers

We used six layers of PDLC film as our projection screen. PDLC is a relatively new material that is commonly used in an interior design as a digitally switchable window. It is an array of liquid crystal droplets that can be excited with electricity. In relaxed state, the liquid crystals are randomly dispersed, and it prevents light from transmitting, thus it is opaque. When electricity is applied, the liquid crystals become excited, and they lined up properly allowing light to go through, thus it becomes transparent. In Japan, this sheet is sold under the brand UMU films (NSG UMU Products Co., Ltd). We used a particular model that when 100 volts of current is

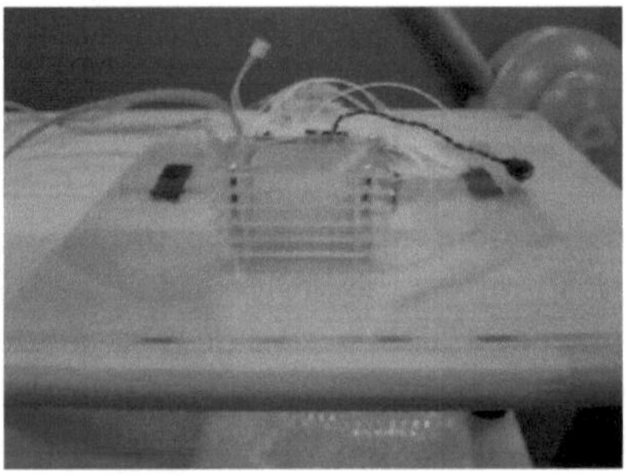

Fig. 3. UMU Film layers sized at 40mmx40mm; each spaced at 7 mm

applied, opaque state is reached with 93 percent haze ratio, and when there is 0 volts of current, it became clear with 12 percent haze ratio.

The PDLC takes 0.78 millisecond to turn from opaque to clear, and from clear to opaque it takes 6 millisecond. We project images onto the film when it turned opaque, and because we are limited by the slowest speed, which is 6 milliseconds.

For our prototype, we used UMU films that are sized at 40 millimeters by 40 millimeters. We placed the UMU film onto a clear acrylic panel that was 2 mm thick, and we set it so that each layer is spaced at 7 mm distance (including the acrylic thickness) from each other (See figure 3), and we used 6 layers altogether.

3.3 Display Rate Consideration and Rendering

There are several things to consider when calculating how fast a display can perform. In our case, we have six layers of PDLC in which each layer's flicker takes 6 milliseconds. Because we are developing a solid stack volumetric display, one frame is defined when all layers within the display have finished rendering; thus, theoretically, it takes 36 milliseconds to complete a volumetric frame. That means in one second, our display could theoretically display 27.77 frames.

However, we must also consider the display rate of the projector used in the system. Currently off-the-shelf DLP projector could display at a max of 85Hz, which it takes 11 ms to draw a frame. So with six layers in the system, the system could still display at 15 frame per second (FPS), which is still within the minimum requirement for a real time display system.

We use OpenGL graphic library to do the rendering. We specify a 3 dimensional frustum with size and resolution that is identical to our physical display which was 40mm(width)x40mm(height)x42mm(depth). Whenever a layer is opaque, we send command to the OpenGL library to draw a cross-section of the frustum.

4 Summary and Discussion

We have presented four specifications that guide us in developing display systems for realizing our vision of real IT. The most of the important aspect of this vision is that we, as the developers, must not separate our users from their environment and their peers when using they are using our system. This means that the information being displayed must be in real space, instead of in the virtual space, for example, in the CAVE or HMD; furthermore, the users must not be burdened with equipments that could increase their cognitive load when interacting or communicating with their peers. In addition, when displaying 3D data, it is important for the users to be able to experience them without focus and convergence mismatch, and preferably the 3D data could exist in the same space with the physical object for information augmentation purpose.

Our recent work of volumetric augmented reality display is in progress to address the need of displaying three dimensional data for augmenting physical objects.

We realize that in order for the vision of real IT to fully be realized, there is really crucial need to address the issue of interface. This means that we need to research a set of interface beyond mouse and keyboard that could be used in a real space along with display technologies that we have defined previously. Input and output system must be defined in order for a system to be fully functional. As part of our progress, we are planning to investigate interaction method suitable for our volumetric augmented reality display. We still start by identifying several requirements that support our vision of real IT and implementing the system according to that guideline. Ideally we would not want to use special purpose hardware that might burden the users, and the paradigm should support three dimensional interaction.

References

1. Bimber, O., Fröhlich, B., Schmalstieg, D., Encarnação, L.M: The Virtual Showcase. IEEE Computer Graphics & Applications 21(6), 48–55 (2001)
2. Billinghurst, M., Kato, H.: Collaborative Mixed Reality. In: ISMAR 1999, Mixed Reality - Merging Real and Virtual Worlds1, pp. 261–284 (1999)
3. Cruz-Neira, C., Sandin, D.J., DeFanti, T.A.: Surround Screen Projection Based Virtual Reality: The Design and implementation of the CAVE. In: Computer Graphic (SIGGRAPH 93), pp. 135–142 (1993)
4. Ehnes, J., Hirota, K., Hirose, M.: Projected Augmentation ii - a scalable architecture for multi projector based AR-Systems based on Projected Aplications in ISMAR, pp. 190–191 (2002)
5. Eitoku, S., Tanikawa, T., Hirose, M., et al.: Display Composed of Water Drops for Filling Space with Materialized Virtual Three-Dimensional Objects. IEEE VR 2006, pp. 159–166 (2006)
6. Falavora, G.E., et al.: 100 Million-voxels Volumetric Displays, Cockpit Display IX: Displays for Defense Application. In: Proc. SPIE. SPIE, vol. 4712, pp. 300–312 (2002)
7. Howarth, P.A.: Empirical Studies of Accommodation, Convergence, and HMD Use. In: Proceedings of the Hoso-Bunka Foundation Symposium, Tokyo (December 3, 1996)
8. Sidharta, R., Oliver, J., Sannier, A.: Augmented Reality Tangible Interface for Distributed Design Review. In: International Conference on Computer Graphics, Imaging and Visualisation, pp. 464-470 (2006)
9. Tanikawa, T., Suzuki, Y., Hirota, K., Hirose, M.: Real World Video Avatar: Real-time and Real-size Transmission and Presentation of Human Figure, ICAT (2005)
10. Raskar, R., Welch, G., Low, K., Bandyopadhyay, D.: Shader Lamps, Eurographics Workshop on Rendering, London (2001)

New Approaches to Intuitive Auditory User Interfaces

Dieter Spath, Matthias Peissner, Lorenz Hagenmeyer, and Brigitte Ringbauer

Fraunhofer Institute for Industrial Engineering (IAO),
Nobelstr. 12, 70569 Stuttgart, German
{dieter.spath,matthias.peissner,lorenz.hagenmeyer,
brigitte.ringbauer}@iao.fraunhofer.de

Abstract. This paper gives an overview of recent projects done at Fraunhofer IAO which take innovative approaches to the use of non-speech audio in human-computer interfaces. The examples illustrate the benefits that non-speech sound - alone and in combination with other modalities - can provide for supporting an effective interaction.

Keywords: Auditory user interface, sound design, acoustic feedback, navigation, vigilance.

1 Introduction

The use of non-speech audio in human-computer interfaces offers significant benefits. The most obvious advantage of sound as an interaction technique is that it can help to make computer interfaces more usable for vision impaired users. But sound is beneficial also for a broader user group, especially in situations and during tasks that do not allow a permanent visual scanning of a graphical user interface. This applies, for example, to many user tasks in industrial contexts, to vehicle driving, and to most situations of using a mobile device. Sound helps to reduce the load on the visual channel. It grabs the user's attention no matter where he is currently looking at. Therefore, it can be used effectively for alerts and reminders and in order to direct the user's visual attention to the graphical user interface for more detailed information. Another issue with mobile devices is their small visual displays. A reasonable complement of graphical and auditory elements for the presentation of information can help to avoid cluttered screens. Moreover, sound is capable of conveying information which can hardly be displayed on a screen, for instance, information about internal mechanisms of complex objects (Gaver, 1989). Finally, a combination of sound and vision can improve the usability, e.g. by "sonically enhanced widgets" (Brewster, 1998), and contribute to an enriched user experience.

Many of the above mentioned benefits also apply to speech output. But non-speech audio offers significant advantages over speech in some concerns and for some types of information, such as, for example, feedback on user actions, system states and transitions. Non-speech audio can transmit even complex information within very short messages. While speech is always presented in a sequential and therefore time-consuming mode, non-speech audio provides the possibility to encode multiple bits of information in a parallel manner, for example, by using the different attributes of

M.J. Smith, G. Salvendy (Eds.): Human Interface, Part I, HCII 2007, LNCS 4557, pp. 975–984, 2007.

sound, such as pitch, rhythm, loudness and timbre (cf. Blattner et al., 1989). As another advantage, non-speech sound is much less disruptive than speech. This is especially important for giving continuous status feedback about ongoing processes. While habituation is difficult to achieve with speech, continuous sound can fade into the background of consciousness and come to the foreground again only when it changes or stops (cf. Brewster, 2002).

In this paper, we give an overview of some selected projects in which we are investigating innovative approaches to the use of non-speech audio in human-computer interfaces. In the first contribution, we present a universal, multimodal human-machine interface (HMI) for hypovigilance management systems which feature different acoustic stimuli for warnings and vigilance maintenance. The second project deals with the design of intuitive sounds for generic user interface events such as affirmative and negative feedback, alerts and requests for clarification or confirmation. Following an empirical approach, we show how paraverbal utterances commonly used in human-to-human conversations, for instance "m-mh" and "mmh?", can serve as an effective design metaphor. The third project describes the development of a concept for an orientation and navigation guide for blind people and visually impaired people using audio cues.

2 Development of a Universal, Multimodal HMI for Hypovigilance Management Systems

Today's "24 hour society" seems to pressure many individuals to sacrifice sleep in favor of other activities, without realizing the negative effects. In addition, as much as 10% of the general population suffers to some degree from sleep disorders, such as: hypersomnia, narcolepsy, and sleep apnea (Backhaus & Riemann, 1999) eventually leading to extreme sleepiness.

Sleepiness reduces alertness and with this, vigilance, and attention, so that the ability to perform is impaired. The speed at which information is processed and the quality of decision-making is also affected (Reissman, 1996). Sleepiness or hypovigilance, therefore, are among the key causes of serious accidents: According to NASA Aviation Safety Reporting System, approximately 21% of the aviation incidents are fatigue-related. In a similar way, most nuclear accidents (among them Chernobyl and Three-mile Island) have clear fatigue-related causes (Mitler et al, 1988). Moreover, more than 40% of automotive accidents on US-highways can be related to hypovigilance (Garder, 1998).

Obviously, the support of technical systems, so called Hypovigilance Management Systems or HVMSs, in states of hypovigilance is desirable. In principles, HVMS measure the state of vigilance of the respective user and take according measures if a critical state of vigilance with respect to the work task is reached: The user is warned and informed about his/her state of vigilance, and the system might try to keep the user awake for a restricted time (if at all possible) in order to enable the user to finish dangerous actions (e.g. to safely drive to the next highway exit). However, the systems known so far are mostly in a research state of development. They mainly base on single-sensor approaches and lack reliability. Moreover, most of these systems focus on the sensor system but do not feature an effective HMI. Finally, the systems present are restricted to single use contexts such as the car. As hypovigilance is a problem in many different use contexts, a universal system is needed.

Therefore, Fraunhofer IAO developed a new universal, multimodal HMI for HVMSs. It includes both the communication strategy (i.e. the warning strategy) and its implementation, the actual physical HMI-elements. In order to create a universally applicable interface, first, a representative set of application scenarios with respect to the dynamic and the mobility of the work system was chosen. The car driver, for example, is part of a work system of high system dynamic as he/she needs to quickly react to his/her surrounding but is immobile with respect to the work place, i.e., the car.

The requirements for the construction of the HMI were gathered by means of a thorough review of the relevant human factors literature as well as a questionnaire survey in the pilot sites chosen (Löher & Hagenmeyer, 2006). A cluster analysis was performed on these requirements with the aim to identify a universally applicable HMI-configuration. As a result, three functional modes were introduced:

Upon activation, the system starts in normal mode in which a status indicator informs the user about the hypovigilance state monitored and indicates that the system is working properly. Such continuous information should be presented in a way that it is not disturbing the user but should be recallable at any time; a traffic light like status indicator was chosen for this purpose.

The warning mode is activated when the user reaches a critical level of hypovigilance. Within this mode, depending on the current hypovigilance state of the user, either a cautionary or an imminent warning is displayed with the aim to stop the user conducting the work task and, by doing so, avoid accidents.

According to Wickens et al. (1998), an effective warning must first draw attention and then inform about the hazard, the potential consequences and give recommendations on how to react. Optionally, a feedback might be required from the addressee in order to confirm the reception of the information. Therefore, for cautionary warnings, the attention of the user is drawn by a complex sound according to DIN ISO 7731 (2002) in combination with a haptic vibration pulse and a bright flash. Then, more complex information about the hazard, its potential consequences and according countermeasures are presented by means of speech messages. These were preferred vs. text displays because of their higher compatibility with most work tasks in respect of the mobility of the user and interference with the work task itself. Finally, feedback from the user is demanded which is to be given via a push-button. In case no feedback is given, the warning is repeated. If still no feedback is received or if the user reaches a more critical hypovigilance level, an imminent alarm is given which follows the same structure. However, the stimuli to draw attention are intensified by an increased volume, higher pitch and vibration frequency. Moreover, the speech messages are adapted, emphasizing the urgency to quickly react to the warning.

After an imminent warning was given and confirmed, the system switches into the vigilance maintaining mode aiming to support the user safely stopping his/her work task, e.g., safely driving to the next exit. The user is informed about the activation of the vigilance maintaining system (VMS) as well as its restricted, short term usefulness by a speech message. The VMS then stimulates the user by means of so called “Landström Sounds”: Landström et al. (1998) showed that specific disharmonious sequences of sounds, each lasting around 4 seconds, that are presented in irregular intervals of up to 5 minutes significantly enhance wakefulness. Moreover, a high level of acceptance by the participants could be observed.

The above warning strategy was implemented in SWITCHBOARD, a software package developed at Fraunhofer IAO with various I/O-capabilities. The status indicator, visual master alarm and feedback button were integrated in a wrist worn device (compare to fig. 1; for presentation purposes, all LEDs are activated). Haptic stimuli were presented by a belt mounted vibration device and acoustic stimuli were presented by a standard ear phone. All of the HMI elements were connected to the central computer system via a body area/local area network enabling the user to move freely.

Fig. 1. Wrist worn HMI-device

3 Intuitive Auditory Cues for Basic Interface Events

According to Buxton et al. (1994), non-speech audio mainly fulfils one of the following three basic functions: (1) alarming and warning the user, (2) providing status feedback and information about ongoing tasks, and (3) encoding numerical data. Whereas the latter often applies to highly specialized tasks, for example, monitoring continuous biomedical data, the other two functions play a major role in nearly all types of user interfaces. These two basic functions are focused also in the project here described.

The project aims at developing empirically founded design guidelines for intuitive sounds that can be adopted for generic interaction events in different application contexts. In order to achieve guidelines that can be applied to diverse situations and domains, the events to be sonified are addressed on a general and abstract level. In our project, we were focusing the following nine events:

1. Affirmative feedback when the user successfully triggers a function
2. Affirmative feedback when the system has successfully completed a given task
3. Negative feedback when the user tries to trigger a function currently not available
4. Negative feedback when the system has failed to complete a given task
5. Request for confirmation before executing an irreversible system action
6. Request for clarification in case of ambiguous or underspecified user inputs
7. Critical system state – medium priority
8. Critical system state – high priority
9. Alert/reminder

The above listed events resemble situations that often arise also in human-to-human conversations. In human dialogs, some of these or similar messages are simply conveyed by a short paraverbal utterance such as “m-mh” or “mmh?”. This led us to the idea that paraverbal human utterances might be an effective design metaphor for auditory cues in human-computer interfaces. This approach corresponds with Bill Gaver’s notion of *auditory icons* (Gaver, 1989), that rely on conveying information about computer events by analogy with every-day sound producing events.

3.1 Recording and Evaluating Paraverbal Utterances

As a first step, we recorded paraverbal human utterances which were systematically stimulated in an experimental setting[1]. This should help us to identify the needed pattern later used as a model for sound design. Eight students of speech training courses at the State University of Music and Performing Arts Stuttgart participated in our study. We had translated the above listed events to every-day situations that served as a setting for the recordings. In a scenario-based role play, the participants were instructed to react vocally – but without using speech - to certain events, and actions or questions by the experimenter. The participants' utterances were recorded. As the result of this first study we collected a total of 72 human voice recordings, eight for each of the nine everyday-life situations that were derived from the generic user interface events above.

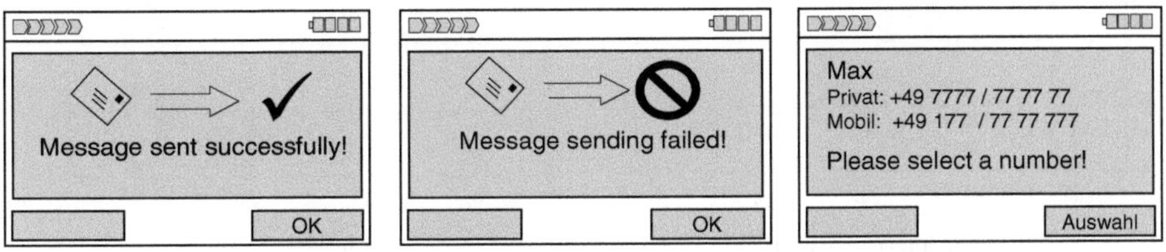

Fig. 2. Schematic screen views (examples) used for empirical evaluations

We conducted a second study in order to evaluate how effective the recorded sounds were in communicating a clear message to a naive listener. For this study, we translated the generic interface events to schematic screen views as possibly displayed on a mobile phone or a washing machine (see figure 2). 48 participants were instructed to assign each of the 72 recordings to the one of the concrete screen views which they felt suited best. The participants were randomly and in equal shares assigned to one of the two conditions of mobile phone screens and washing machine screens. For none of the presented sounds, the results from both test conditions differed significantly. This allows for a conjoint analysis of the total data set, and suggests that the findings might be generalized also to other similar applications. For 65 of the 72 recorded sounds, the participants' matching behavior differed significantly from an equal distribution over the nine events. It seems that most of the recorded sounds convey information that is related somehow to the presented interface situations.

For each interface event, we selected the recorded sounds, with which the event was most frequently associated. We found that in many cases, the listeners of a sound were not able to discriminate between two semantically similar events, such as the two affirmative feedback events (1 and 2 in the list above), the two negative feedback events (3 and 4), the two requests (5 and 6), and the two critical states (7 and 8). Sounds which were strongly associated with one event of such a pair often ranked high also for the other event of the same pair. For this reason, we decided to consider

[1] Sounds downloadable at http://www.hci.iao.fraunhofer.de/de/projekte/sounddesign

classification rates on basis of semantic pairs rather than single events. Doing so, we obtained highly accumulated classification patterns with peaks at 60 to 80%. One single exception is the reminder event. No recorded sound was clearly associated with the reminder screen. Figure 3 (a) illustrates the classification distributions over the paired events for some example sounds. In summary, it shows that the selected recordings can be effectively used for the discussed interface events. However, the chosen design approach seems to be applicable only to a restricted set of messages.

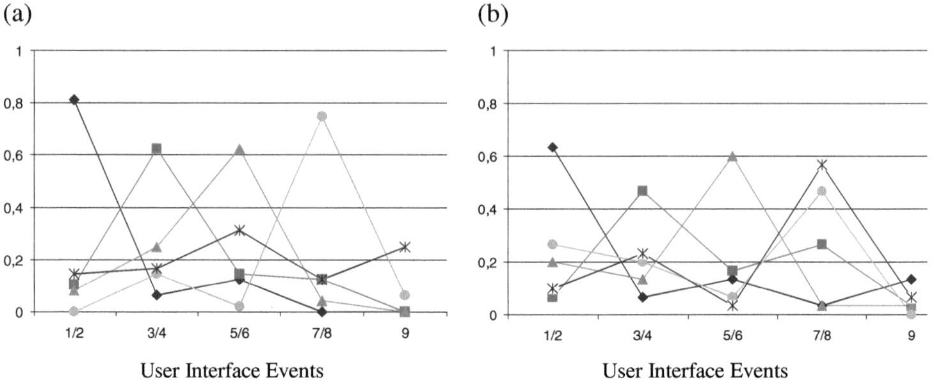

Fig. 3. Assigning sounds to user interface events: empirical distributions (relative frequencies) for selected paraverbal utterances (a) and respective synthesized sounds (b)

3.2 Generating and Evaluating Synthesized Sounds

Despite the evident effectiveness of the recorded human utterance, they will hardly be acceptable as interface sounds for a concrete product, as they do not meet essential requirements: The auditory user interface plays an important role for emphasizing certain product features, for example, valuable materials, and for contributing to an aesthetic and coherent over-all impression. Moreover, sound is more and more recognized as an important arena for branding.

Therefore, we need basic design guidelines which concentrate only on the sound features that are most substantial for conveying the intended message and leave undefined all the other sound features that can then be used for differentiation. Another related requirement of the envisioned guidelines is their succinctness that shall make the synthetic sound reproduction as economic and easy as possible.

In a first step, we generated synthesized sounds that imitate the selected recordings in a reduced and simplified manner[2]. (1) Rhythm was copied unchanged, i.e. the number and duration of the sound segments of the original sound were directly adopted. (2) We conserved the dynamics of the recorded sounds. The intensity of each sound segment was kept constant at the average intensity of the respective segment in the original recording. Variations within segments were eliminated. (3) Pitch variations in the recorded sounds that can be interpreted as involuntary were

[2] Sounds downloadable at http://www.hci.iao.fraunhofer.de/de/projekte/sounddesign

eliminated by setting the pitch of each segment to the average fundamental frequency (f_0) of the respective segment. We identified also obviously intended pitch variations within segments. Most of them were upward slides of about an octave with a doubling of frequency from the beginning to the end point of a segment. These segments were modelled by linear octave slides with the same average f_0 than the original segment. (4) Timbre was eliminated by using sinus waves.

The synthesized sounds were evaluated with 30 participants in the same fashion as reported for the recorded ones. The results are illustrated in figure 3 (b). Although some sounds obtained a significantly different assignment distribution for their synthesized versions than for the recorded ones, the results suggest that most of the relevant information was kept during the described process of synthesizing and simplifying the sounds.

To sum up, the described study supports the effectiveness of paraverbal utterances as a metaphor for the design of auditory interface cues. Further investigations will be needed in order to yield a deeper understanding of the sound features relevant for communicating the intended message. The presented procedure resulted in concrete design guidelines which can be written down in musical notation and easily adopted for reproduction. In future studies, we will investigate the effect of diverse timbres on the comprehensibility of the sounds.

4 Pedestrian Navigation for Blind and Visually Impaired People

4.1 Motivation and Related Work

The wayfinding requirements for blind people differ significantly from the ones using their visual senses to orientate. Blind pedestrians can not see obstacles on the pavement or spot the church in the distance for orientation. For micro-navigation, i.e. navigation through the immediate environment, the long cane, a guiding dog and additional electronic mobility aids help navigation. For macro-navigation, i.e. navigation through the more distant environment like the church example, mobility support is much more complicated (Petrie 1995).

There have been several projects that tried to meet this challenge. The MoBIC project (Strothotte 1996; Petrie 1997) used a preparation system to plan a walk in an urban area combining touch tablet maps with speech output for guiding exploration. The selected route can transferred to the MoBIC outdoor system (MoODS) that guides the user outdoor on his way. The MoODS takes the journey plan and assists travellers finding their way in by matching GIS information against travellers' positions (GPS). Directions, warnings, and other information is given in speech or Braille.

Loomis et al. 1994 aimed at developing a portable, self-contained system that allows visually impaired individuals to travel through familiar and unfamiliar environments without the assistance of guides. The system used GPS, a Geographic Information System comprising a database of the environment and functions for automatic route planning and for selecting the database information. The system provides a virtual acoustic environment for travellers, guiding them to buildings and other objects in the environment by sound signals which appear to come from the direction of the respective object.

There are also several navigation aid systems on the market (for an overview see http://www.tiresias.org/equipment/eb24.htm). An interesting recent product is Trekker (www.humanware.ca/web/en/p_DA_Trekker.asp). It is a navigation aid for blind and visually impaired people and uses GPS and digital maps to help blind persons find their way in urban and rural areas. It complements existing aids (white canes and guide dogs) and provides information through synthesized speech output.

4.2 Pedestrian Navigation for Blind in ASK-IT Project

ASK-IT[3] is a project that aims at supporting mobility impaired people in a mobile world. The prototype that is developed in this context will support blind people in wayfinding based on acoustic interface elements. It is researched in more detail how the provision of sound elements can be used to guide navigation.

Sound offers broad possibilities for unobtrusive information provision. This might help blind people to know that they are on the right way, without annoying spoken information that is currently not as important. Also privacy plays a major role here. The idea is to give blind people the power to choose what they attend to (Strothotte 1996, p. 3). The requirements analysis in the project revealed that it is especially important to give the control on the level of detail of guidance to the blind person. Some users require detailed guidance whereas others simply need reassurance that they are where they expect to be and are headed towards their destination.

We propose that acoustic icons and earcons for orientation-related and wayfinding information. A close connection to the prototypes on acoustic user interface elements and the speech user interface is seeked, e.g. to clarify ambiguous situations in wayfinding.

In December 2006 interviews with six severely visually impaired and blind people have been performed to learn about essential information to be presented by an electronic travel aid and nice-to-have information and events. The acceptance of the sound-based concept was very high although concerns on the challenges of sound design and concept development were raised as well. The sounds should e.g. pleasant and not annoying.

The concept can be divided in provision of orientation-related information and navigation-related information. The first class of information has the aim to orientate the person in the environment without a planned route – similar to the visual scanning of the physical surroundings of an able-bodied person. The second class of information deals with navigation-related information i.e. guide a person on a pre-planned route.

Those three categories of information need to be considered for **orientation**:

1. Information for orientation: Essential information like street names or directions,,detailed information like public buildings, shopping malls, and transport related information like stations, bus stops

---

[3] The project ASK-IT is funded by the EC in the 6th framework program, priority IST-2002-2.3.2.10 eInclusion. Contract no.: IST-2003-511298.

2. Safety-related information: Critical situations like street crossings and stairs and information on safe pathways
3. Temporary barriers like construction sites

For **navigation support** information like "You are on the right / wrong way", "You have to turn right / left soon", and "Now you have to turn right / left" needs to be given.

It turned out that the overall concept for navigation for blind people should not only contain sounds, but also some speech elements will be needed for convenient interaction. Examples include the commands: Where am I? Am I on the right way? What is the next orientation point? Additionally it will be researched, if some information (like "turn right now") should be better given as a sound or in voice.

Currently the design of sounds is under development. Not only sounds have to be designed that are easy to learn and to distinguish, but also a concept for sound families needs to be developed to indicate the information categories and to allow the chunking of information. The user tests to evaluate the produced sounds are planned for April 2007. They will then be optimized and evaluated in the real world context in field tests two months later.

The developed sounds for orienting and guiding blind people will be also used to enhance graphical user interfaces for navigation. Even though orientation for visually able people is mainly supported by the visual sense, the developed sounds for navigation support will be useful: being on the move it is more comfortable to get a sound feedback if the selected path is still the right one than having to constantly monitor the PDA GUI to find out about wrong turns.

With this approach we do not only aim at developing a pedestrian navigation system especially suited to the needs of visually impaired people. We also aim to collect some important knowledge on sound families, knowledge on how complex and rich an audio based information system without visual feedback can be and get an idea about the potential of an audio-based display in real-world mobile situations.

5 Concluding Remarks

We expect an increasing need for effective auditory user interfaces in the near future. One the one hand, this will be due to the persistent growth in the use of mobile devices and their expanding functionalities. On the other hand, advances in the field of intelligent systems and environments will lead to a higher level of automation and to the need of keeping the users informed about current system actions and states. The vision of an "implicit human-computer interaction" (cf. Schmidt, 2000) will result in vanishing user interfaces. Human behavior and actions that are not explicitly directed towards a computer interface will be recognized and interpreted as a trigger of system reactions. Auditory cues and feedback will be needed in order to support user orientation and to enable a transparent and trustable human-computer interaction in smart environments.

References

1. Backhaus, J., Riemann, D.: Schlafstörungen. Fortschritte in der Psychotherapie. Hogrefe, Göttingen (1999)
2. Blattner, M., Sumikawa, D., Greenberg, R.: Earcons and icons: Their structure and common design principles. Human Computer Interaction 4(1), 11–44 (1989)
3. Brewster, S.A.: The design of sonically-enhanced widgets. Interacting with Computers 11(2), 211–235 (1998)
4. Brewster, S.A.: Non-speech auditory output (Chap 12). In: Jacko, J.A., Sears, A. (eds.) Human-Computer Interaction Handbook, pp. 220–239. Lawrence Erlbaum Associates, Mahwah, NJ (2002)
5. Buxton, W., Gaver, W., Bly, S.: Auditory Interfaces: The Use of Non-Speech Audio at the Interface. Unfinished book (1994) (retrieved February 2007) manuscript at: http://www.billbuxton.com/Audio.TOC.html
6. DIN ISO 7731, Gefahrensignale für öffentliche Bereiche und Arbeitsstätten. Akustische Gefahrensignale (2002)
7. Garder, P.: Continuous Shoulder Rumble-Strips – A safety evaluation. In: CSRS Seminar, Linkoeping (August 1998)
8. Gaver, W.: SonicFinder: An interface that uses auditory icons. Human Computer Interaction 4(1), 67–94 (1989)
9. Landström, U., Englund, B., Nordström, B., Aström, A.: Laboratory Studies of a Sound System that Maintains Wakefulness. Perceptual and Motor Skills 86(1), 147–161 (1998)
10. Loomis, J.M., Golledge, R.G., Klatzky, R.L., Speigle, J.M., Tietz, J.: Personal guidance system for the visually impaired. In: Proceedings of the First Annual ACM Conference on Assistive Technologies. Assets '94, pp. 85–91. ACM Press, New York (1994)
11. Löher, L., Hagenmeyer, L.: Gestaltungsrichtlinien für die Nutzerschnittstelle von Hypovigilanz-Management-Systemen. In: GfA (Edt.): Innovationen für Arbeit und Organisation, pp. 39–42. GfA-Press, Dortmund (2006)
12. Petrie, H.: User requirements for a GPS-based travel aid for blind people. In: Gill, J.M., Petrie, H. (eds.) Proceedings of the conference on orientation and navigation systems for blind persons, Hatfield, UK, Royal National Institute for the Blind, London (1995)
13. Petrie, H., Johnson, V., Strothotte, T., Raab, A., Michel, R., Reichert, L., Schalt, A.: MoBIC: An Aid to Increase the Independent Mobility of Blind Travellers. The. British Journal of Visual Impairment 15(2), 63–66 (1997)
14. Reissman, C.J.: The Alert Driver. A Trucker's Guide to Sleep, Fatigue, and Rest in our 24-Hour Society. American Trucking Associations, Alexandria (1996)
15. Schmidt, A.: Implicit Human Computer Interaction Through Context. Personal Technologies, vol. 4(2&3), pp. 191–199. Springer, Heidelberg (2000)
16. Strothotte, T., Fritz, S., Michel, R., Raab, A., Petrie, H., Johnson, V., Reichert, L., Schalt, A.: Development of dialogue systems for a mobility aid for blind people: initial design and usability testing. In: Proceedings of the Second Annual ACM Conference on Assistive Technologies. Assets '96, pp. 139–144. ACM Press, New York (1996)
17. Wickens, C.D., Gordon, S.E., Liu, Y.: An introduction to human factors engineering. Addison Wesley Longman, New York (1998)

A Study on Haptic Interaction and Simulation of Motion and Deformation of Elastic Object

Kazuyoshi Tagawa, Koichi Hirota, and Michitaka Hirose

Graduate School of Information Science and Technology,
University of Tokyo
tagawa@cyber.t.u-tokyo.ac.jp, hirota@media.k.u-tokyo.ac.jp,
hirose@cyber.rcast.u-tokyo.ac.jp
http://www.tagawa.info

Abstract. An approach to haptic interaction and simulation of motion and deformation of elastic object is proposed. In our approach, the motion of an object is defined by changes of kinetic momentum and angular momentum after impulse force is applied to each degree of freedom, and the deformation of the object is defined by a set of temporal deformation patterns after impulse force is applied to each degree of freedom, which we call extended impulse response deformation model. The time complexity of computing interaction force is independent of the complexity of the model. This feature is advantageous for time-critical applications. Also, the time complexity of computing object deformation is linearly proportional to the complexity of the model. Through implementation of a prototype environment and evaluation of its performance, the feasibility of the proposed approach is demonstrated.

Keywords: Elastic objects, deformation model, dynamic motion, record and retrieval approach.

1 Introduction

Simulation of deformable object is one of the greatest interests in the field of Virtual Reality. Escpecially in the field of haptics, presentation of force in the moving and deforming interaction is a challenging topic. A difficulty of the problem is the amount of computation to simulate the behavior of deformable objects.

One way to avoid the problem of the computational complexity of such simulations is to employ the record and retrieval (or record reproduction) approach [7,4,5,2], where the behavior of an object under interaction is measured on a real object or pre-computed off line using a precise computation algorithm and then the data is used to generate a reaction of the virtual object in a real-time application. In a previous paper, we proposed an impulse response deformation model using this record and retrieval approach [6]. This model is a potential candidate because of the low computational complexity of calculation of interaction force, however, this approach cannot take into account the motion of the object.

In this paper, we propose an extending method for the impulse response deformation model to solve the problem of considering the object's motion. In our

M.J. Smith, G. Salvendy (Eds.): Human Interface, Part I, HCII 2007, LNCS 4557, pp. 985–993, 2007.

approach, advantages of the impulse response deformation model are taken over, moreover motion of the object is considered.

Our approach enables haptic interaction with dynamic deformable and movable object that is too complex to be solved by a common Finite Element Method (FEM). Also, an analysis on computational complexity suggests that our approach becomes more advantageous than FEM as the complexity of the model increases.

In the next section, the concept of our approach is discussed and an outline of the implementation is described. Experimental systems and results will be presented in Section 3.

2 Extended Impulse Response Deformation Model

In previous paper, we proposed the impulse response deformation model[6]. In our approach, the behavior of an object is defined by a set of temporal deformation patterns after impulse force is applied to each degree of freedom. Deformation resulting from interaction is obtained by computing convolution of the model and the history of interaction force.

The idea of the impulse response deformation model is based on the premise that the model is linear, which means that the influences caused by impulse forces on different degrees of freedom or at different times are independent of each other, and the resulting deformation is computed as the sum total of the influences. This assumption makes it possible to perform haptic interaction with dynamically deformable and complex object in real-time. However, part of the object must be fixed to a floor, a wall, etc., in order to avoid computational complexity of changing boundary conditions and nonlinerity resulting from rotary motion.

In this study, to solve these problems, the following a novel assumption is introduced.

2.1 Separating Impulse Response into Component of Deformation and Motion

Suppose an object is remainning stationary under gravity free conditions. When a unit impulse force is applied to some degree of freedom on the object, definite amounts of kinetic momentum and angular momentum are produced. Although component of the object's translation is proportional to the input force, component of the object's rotation has nonlinearity resulting from change of inertia matrix and centrifugal force. If the angular rate of the object is not large, effect of the nonlinearity is able to assume negligible small. At the same time, this assumption makes it possible to improve degrees of freedom of interaction dramatically.

In our apprpach, this assumption is introduced. In this assumption, when a unit impulse force is applied to some degree of freedom on the object, definite amounts of acceleration $\ddot{x}$ and angular acceleration $\ddot{\theta}$ are produced. Using this assumption, deformations after the impulse force is applied are separated into components of deformation and motion.

2.2 Deformation Model

In our approach, the following deformation model based on a idea of impulse response is used.

One Dimensional Model. In case of a one-dimensional continuous system, the relationship between the force $f(t)$ and displacement $u(t)$ is described using impulse response $r(s)$ by

$$u(t) = \int_0^\infty r(s)f(t-s)ds. \tag{1}$$

In case when $f(t)$ is a delta function at $t = 0$, it becomes identical to $r(t)$.

With the convenience for digital processing, we use a temporary discrete model. In the following explanation, we use s and t as indexes of discrete time, and the time for each value is indicated in a superscript with square brackets. We also confine the duration of the impulse response in a limited time T. Then the one-dimensional model is described as follows:

$$u^{[t]} = \sum_{s=0}^{T-1} r^{[s]} f^{[t-s]}. \tag{2}$$

In a precise sense, if we consider acceleration and angular acceleration of the object after the impulse force is applied, equation (2) becomes as follows:

$$u^{[t]} = \sum_{s=0}^{T-1} (r^{[s]} + \ddot{x}^{[s]} \cdot \Delta t^2 + l^{[s]}\ddot{\theta}^{[s]}\Delta t^2) f^{[t-s]}, \tag{3}$$

where $l^{[s]}$ is distance between the center of gravity of the object and point of application of the force, and Δt is time steps of the simulation. In usual haptic interaction, Δt becomes very small, therefore terms of $\ddot{x}^{[s]} \cdot \Delta t^2$ and $l^{[s]}\ddot{\theta}^{[s]}\Delta t^2$ are negligible. So, we used the approximate model described by equation (2).

In the interaction using a haptic interface device, or more precisely in the case when position-input/force-output device is used, the interface point (i.e. a representative point through which user interacts with object) of the device causes a displacement boundary condition on the deformation model. In other words, $u^{[t]}$ is given and $f^{[t]}$ is unknown. The unknown force is obtained by solving:

$$u^{[t]} = r^{[0]} f^{[t]} + \tilde{u}^{[t]}, \tag{4}$$

where $\tilde{u}^{[t]}$ is the current displacement derived from the past force and it is formulated by:

$$\tilde{u}^{[t]} = \sum_{s=1}^{T-1} r^{[s]} f^{[t-s]}. \tag{5}$$

Since all past force is known at the present time, the value of $\tilde{u}^{[t]}$ can be computed.

2.3 Multi Degrees of Freedom Model

Suppose we have a n degrees of freedom system. We describe the force and displacement on all degrees of freedom by vectors $F^{[t]}$ and $U^{[t]}$ $(n \times 1)$, respectively. Also, we describe the impulse response by a matrix $R^{[s]}$ $(n \times n)$, because the impulse response is defined for all combinations of n input impulses and n output displacements. Then their relationship is written by:

$$\begin{aligned} U^{[t]} &= \sum_{s=0}^{T-1} R^{[s]} F^{[t-s]} \\ &= R^{[0]} F^{[t]} + \tilde{U}^{[t]}, \end{aligned} \tag{6}$$

where

$$\tilde{U}^{[t]} = \sum_{s=1}^{T-1} R^{[s]} F^{[t-s]}. \tag{7}$$

In a usual deformation operation, the displacement boundary condition is given to some degrees of freedom, on which current displacement is given and the force is unknown. On other degrees of freedom, the current displacement is unknown and the force is not applied (or 0). To formulate this relationship, we introduce a description separated by the boundary conditions, as follows:

$$\begin{pmatrix} U_o^{[t]} \\ U_c^{[t]} \end{pmatrix} = \begin{pmatrix} R_{oo}^{[0]} & R_{oc}^{[0]} \\ R_{co}^{[0]} & R_{cc}^{[0]} \end{pmatrix} \begin{pmatrix} F_o^{[t]} \\ F_c^{[t]} \end{pmatrix} + \begin{pmatrix} \tilde{U}_o^{[t]} \\ \tilde{U}_c^{[t]} \end{pmatrix}, \tag{8}$$

where suffix c and o indicates the degrees of freedom on which the displacement boundary condition is applied and not applied, respectively. By solving the equation, the unknown values of $F_c^{[t]}$ and $U_o^{[t]}$ are obtained by

$$F_c^{[t]} = (R_{cc}^{[0]})^{-1}(U_c^{[t]} - \tilde{U}_c^{[t]}), \tag{9}$$
$$U_o^{[t]} = R_{co}^{[0]} F_c^{[t]} + \tilde{U}_o^{[t]}. \tag{10}$$

In the experiments discussed below, the proposed approach is applied to an object model consisting of triangular patches, and interaction with one or two interface points or one plane is investigated. In point interaction, since each node has three degrees of freedom, one or two interface points interacting with one or two nodes cause displacement boundary condition on those nodes, respectively (i.e. $n_c = 3$ or 6). In the case when the interface points are interacting with triangle patches, rather than with nodes, we approximately determine the interaction force that is distributed to the vertex nodes of the triangle patches. We employed an interpolation approach that is similar to those used by Hirota et al[?], where interpolated interaction force is computed by repeatedly using the formula of one- and two-node interaction.

2.4 Motion Model

In our approach, motion of the object is described by Euler method, as follows:

$$\dot{x}^{[t+1]} = \dot{x}^{[t]} + \ddot{x}^{[t]} \cdot \Delta t \tag{11}$$

$$x^{[t+1]} = x^{[t]} + \dot{x}^{[t+1]} \cdot \Delta t \tag{12}$$

$$\dot{\theta}^{[t+1]} = \dot{\theta}^{[t]} + \ddot{\theta}^{[t]} \cdot \Delta t \tag{13}$$

$$\theta^{[t+1]} = \theta^{[t]} + \dot{\theta}^{[t+1]} \cdot \Delta t. \tag{14}$$

2.5 Computational Complexity

The computation of solving the equation of the deformation model becomes easy as the degree of freedom of displacement boundary condition (n_c) is small through time. Since only n_c factors of $F^{[t]}$ have non-zero values for any given t, the values of $\tilde{U}_c$ and $\tilde{U}_o$ are computed by $O(n_c^2 \cdot T)$ and $O(n \cdot n_c \cdot T)$, respectively. $R_{cc}^{[0]}$ becomes a matrix of $n_c \times n_c$, hence the Equation 9 is solved by $O({n_c}^3)$ even if it is solved by Gauss elimination method. Most memory resource is consumed to store the impulse response matrix $R^{[s]}$, which requires $O(n^2 \cdot T)$, and some more area is needed for storing past forces, $O(n_c \cdot T)$.

The computation of solving the equation of the motion model requires $O(N)$, where N is the number of the objects. This is negligible small, under the condition that the number of the objects is small.

3 Experiment

3.1 Pre-computation of Deformation Model

Pre-computation is a process where the measurement of an impulse response is performed by simulation. A unit force is applied for a unit time on each degree of freedom and displacements of all degrees of freedom are recorded. By repeating this process on all degrees of freedom, the impulse response matrix $R^{[s]}$ is obtained. The dynamic behavior of the object was simulated using a FEM model consisting of tetrahedral elements.

We employed a *cube* model, as shown in Figure 1. The complexity of the model is listed in Table 1. Also, Young's modulus $E = 2000\text{N/m}^2$, Poisson's ratio $\nu = 0.49$, density $\rho = 110\text{kg/m}^3$. The length of a side of the model is 10cm. All through the experiments described in this paper, the sampling frequency of the impulse response was set to 500Hz and the duration of response 1s, hence we obtained 500 samples for each element of an impulse response matrix (or

Table 1. Complexity of model

free nodes (n)	866
triangle patches	1728
entire nodes	2197
tetrahedral elements	8640
pre-computation time per d.o.f. (s)	92.8
data size per d.o.f. (MB)	9.9

$T = 500$). In the pre-computation process, however, to improve accuracy in relatively quick deformation followed immediately after the affection of force, the time step for the simulation was set to 1ms, and then averaged them for every 2ms. The computation time to obtain impulse response matrix for each model is also listed in the table, where the computation was performed by a PC (CPU:Itanium2 1.4GHz×4).

Figure 2 is an example of impulse response in the case when a downward impulse force by a point is applied to a node on the top surface. Elastic wave caused by the impulse force is propagating from the point where the impulse is applied, and in this example, it is taking approximately 16ms before entire nodes in the model start causing displacement. Also, Figure 3 is an example of impulse response in the case when downward impulse forces by a plane are applied to all nodes on the top surface.

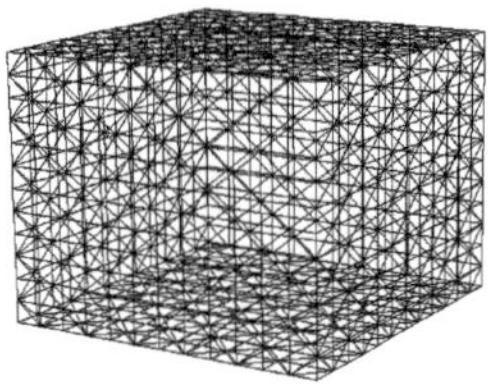

Fig. 1. Experimental model

3.2 Experimental System

The experimental system consists of two PCs and two PHANToM devices[3], as shown in Figure 4. All of the computation related to the model is performed by PC1 (CPU:Itanium2 1.4GHz×4, memory:16GB, OS:Linux); the computation of force and deformation are executed as separate processes using a thread mechanism.

In the *force* process, the following tasks are performed: the information of interface points of the devices is received from the *haptic update* process on PC2 through an Ethernet; collision between the interface points and the object surface is detected and the displacement cause by those points is determined; the impulse

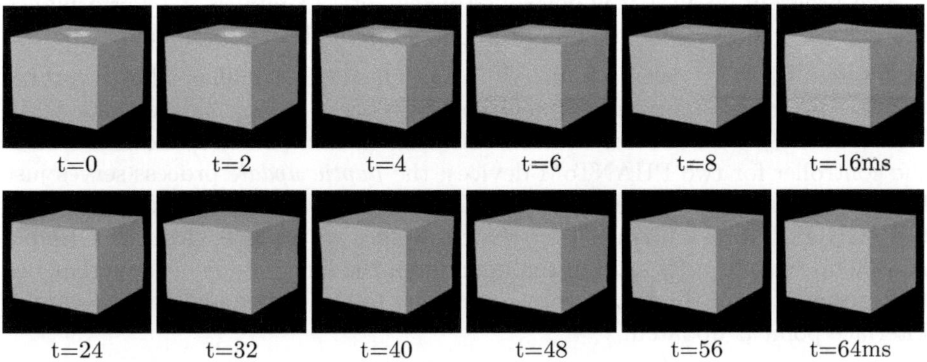

Fig. 2. Examples of impulse response (in the case when a downward impulse force by a point is applied to a node on the top surface)

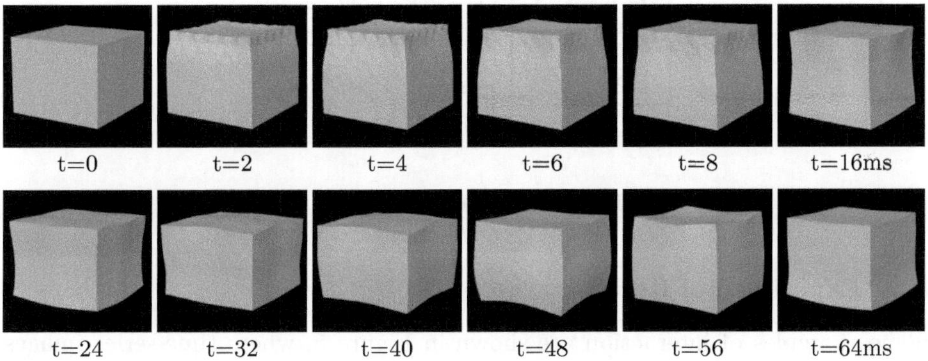

Fig. 3. Examples of impulse response (in the case when downward impulse forces by a plane are applied to all nodes on the top surface)

force (or force boundary condition at each time) and feedback force is computed and output to the haptic update process; the history of impulse forces is updated. The *force* process repeats this computation every 2ms. The *deformation* process computes the deformation of the entire model using the history of impulse forces, which is passed from the *force* process through a shared memory. The process runs asynchronously from the *force* process except for minimum exclusive access control while the *force* process is writing to the shared memory.

Computation of collision was carried out by using the God-object Method[8]. This algorithm is advantageous for our purpose in that it is capable of providing the position of the collision (or ideal interface point) and that the displacement on the point can be computed as the relative position of the interface point from the ideal interface point.

The program on PC1 was developed using Intel Compiler and Performance Libraries [1]; the process of computing the current deformation caused by the past forces (i.e. $\tilde{\boldsymbol{U}}_c^{[t]}$) was parallelized using OpenMP Compiler allotting three CPUs. Also, the computation of *deformation* process was optimized using Math Kernel Library. PC2 (CPU:Pentium3 500MHz×2, OS:Windows) is considered as the controller for two PHANToM devices; the *haptic update* process serves just for converting the interface from GHOST API to the TCP/IP connection interface. GHOST library was used for basic interface with the device; the GHOST main loop runs at 1kHz, and in the main loop, the latest value of force that has been received from the force process is passed to the device and the position of interface point is updated.

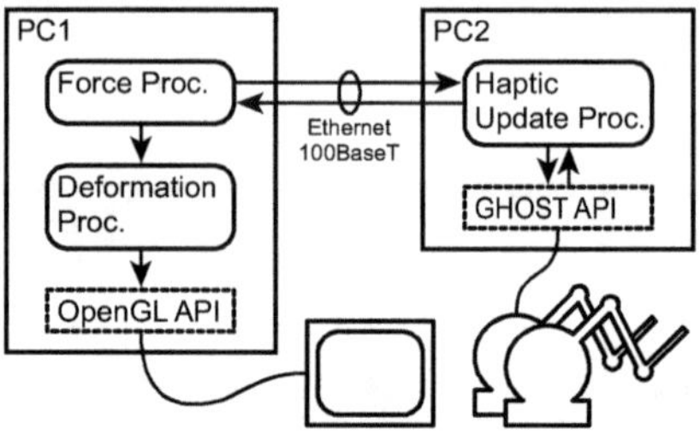

Fig. 4. Experimental system

3.3 Experimental Results

Some examples of interaction are shown in Figure 5, where time-series images of dynamic deformation are listed. Since it is difficult to store every frame of screen image in real-time, these images were generated off line based on the log of the sequence of the interaction force.

Figure 5(a) shows the case where the *cube* model is colliding and bouncing on a floor after free falling. Motion and deformation resulting from collision and inertia force is represented and relatively fast motion of the restoring process is presented.

An example of interaction cooperatively using two interface points is shown by Figure 5(b), where the user is holding both the left and the right side of *cube* model by both the right and the left interface points, and the user is shaking the model at 2Hz from side to side under gravity free conditions. Displacement on the left and right interface point causes motion and deformation of the *cube* model and consequently changes the forces on the left and right interface points.

The computation on interaction force required at most approximately 0.5ms, which is sufficiently fast for usual force feedback application. The entire *force* process including collision detection and communication was completed within each cycle-time of 2ms.

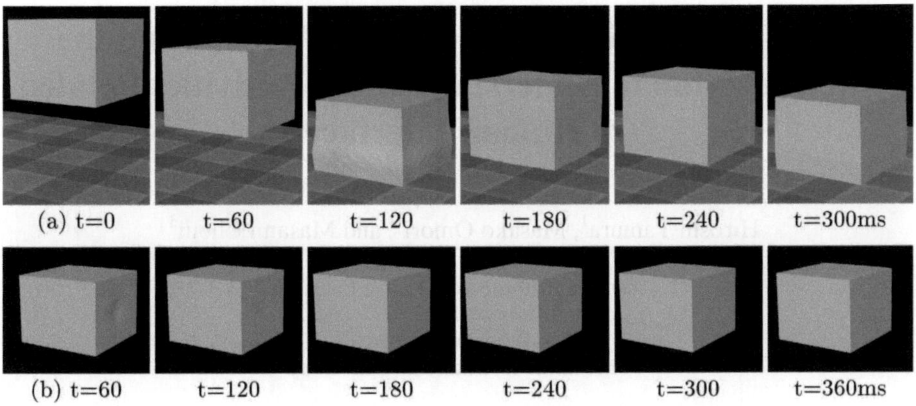

Fig. 5. Examples of dynamic motion and deformation

4 Conclusion

We presented the concept of an extended impulse response deformation model. The model provides an approach to simulating motion and deformation of elastic objects and computes the interaction force at haptic rate, which was barely attained by other methods such as FEM. We implemented a prototype interaction system and demonstrated the feasibility and performance of the proposed approach.

References

1. Intel software development products. pages http://www.intel.com/cd/software/products/asmo–na/eng/index.htm
2. Kuchenbecker, K.J., Fiene, J., Niemeyer, G.: Event-based haptics and acceleration matching: Portraying and assessing the realism of contact. In: Proc. WHC 2005, pp. 381–387 (2005)
3. Massie, T.H.: Initial Haptic Explorations with the Phantom: Virtual Touch Through Point Interaction. Master Thiese at M.I.T (1996)
4. Okamura, A.M., Dennerlein, J.T., Howe, R.D.: Vibration feedback models for virtual environments. Proc. IEEE ICRA 3, 2485–2490 (1998)
5. Okamura, A.M., Hage, M.W., Cutkosky, M.R., Dennerlein, J.T.: Improving reality-based models for vibration feedback. Proc. ASME DSCD DSC-69-2, 1117–1124 (2000)
6. Tagawa, K., Hirota, K., Hirose, M.: Impulse response deformation model: an approach to haptic interaction with dynamically deformable object. In: Proc. IEEE Haptics Symposium 06, pp. 209–215 (2006)
7. Wellman, P., Howe, R.D.: Towards realistic vibrotactile display in virtual environments. Proc. ASME DSCD DSC-57-2, 713–718 (1995)
8. Zilles, C., Salisbury, K.: A constraint-based god object method for haptic display. Proc. IROS '95, pp. 145–151 (1995)

NIRS Trajectories in Oxy-Deoxy Hb Plane and the Trajectory Map to Understand Brain Activities Related to Human Interface

Hiroshi Tamura[1], Masako Omori[2], and Masami Choui[3]

[1] Tamura Institute for Human Interface
tamura@mobilergo.com
[2] Kobe Women's University,
[3] Hiroshima International University

Abstract. NIRS trajectory method is proposed to understand brain activities engaged in mobile device operation. The trajectory is the plot of oxy- and deoxy- Hb (hemoglobin concentration) data to the Oxy-Deoxy Hb plane. Three basic trajectories and some complex trajectories are explained. Total brain activities are represented by a set of NIRS trajectories derived from NIRS channels located on the head surface. Trajectory maps of text entry to mobile phone are shown. The trajectory map increases its complexity with the increase of work task complexity, which we defined by normal and reverse order hiragana and kanji conversion entry of Japanese text. The results suggest the right brain is taking dominant roles in text and other finger activities. NIRS trajectory method can be applied further to find centers of brain activities located on the surface area of the brain. Kindly use color version in CD-ROM for better understanding of this paper.

Keywords: brain activities, mobile device, human interface, text entry, near infrared spectroscopy, NIRS trajectory, trajectory map.

1 Introduction

Studies of human brain functions are greatly advanced by introducing PET, fMRI, MEG, and recently NIRS. Near Infrared spectroscopy (NIRS) has great advantages compared to others in that it can be applied to intact human brain at work.

Near Infrared light of certain wave length can penetrate human skin and skull, and go through surface area of brain tissue. The light scattered in the brain tissue are reflected and comes out through skull and surface skin. Finally the out coming light is guided through light fiber to light sensitive optical multiplier. NIRS transmitter contains two/three light components, whose light absorption rate of oxy and deoxy Hb (hemoglobin) is different. By the differences of reflection rate of each wavelength light, relative concentrations of Oxy-, Deoxy and Total Hb are calculated.

NIRS data consist of three components: oxy-hemoglobin (Hb), deoxy-Hb and total Hb of brain tissue shortly under the skull.

M.J. Smith, G. Salvendy (Eds.): Human Interface, Part I, HCII 2007, LNCS 4557, pp. 994–1003, 2007.

Current applications of NIRS are medical, psychological and welfare examination fields. But since the systems are non-intrusive, it can be applied to people in working environment. It is promising to be applied to human interface design and evaluation.

Since the introduction of mobile devices (mobile phone, car navigation), considerable social attention are paid to brain work load related to visual and cognitive load, and finger operations on small size keypad, under moving or unstable operating conditions.

In many countries, people are using email based on mobile phone, and a car navigation device is equipped on most of the cars in Japan.

We have started the study on brain functions of people using the mobile phone together with basic responses of human activities.

Some scientists are focused on the activation of brain activities which is indicated by oxy-Hb concentration. On the other hand, fMRI studies use deoxy-Hb concentration for indication of brain activities. Some NIRS scientists believe total Hb is most reliable index of brain activities. For the thorough understanding of brain functions, combined analysis of oxy- and deoxy- Hb, involved in total Hb change are essential. In the next chapter, author introduce NIRS trajectory method to grasp the brain activities by taking account jointly of oxy- and deoxy- Hb. Further in Chapter 3, trajectory map is introduced to represent activities of the different parts of brain participating in a task.

2 NIRS Trajectory

The NIRS trajectory is the plot of oxy-Hb and deoxy-Hb data in NIRS plane, in which the horizontal axis is taken for oxy-Hb and vertical axis for deoxy-Hb. By connecting data points next to each other, a NIRS trajectory is shown by a curve. Time axis is not shown on the NIRS trajectory plane, but each dot of a plot represent time change in a fixed length. If the distance between two dots is large, the data is changing fast in time. If the plots are aggregated in some area, the changes are small and slow.

In general, total Hb is in relation to oxy and deoxy-Hb.

$$\text{TTL} = \text{Oxy} + \text{Deoxy}. \tag{1}$$

Assuming TTL=0, equation (2)

$$\text{Deoxy} = -\,\text{Oxy} \tag{3}$$

is derived.

Total Hb is not represented explicitly on NIRS plane. But it is also shown by the distance from a data point to the line of -1 declination, which goes through the origin of the plane (pail blue line in Fig. 1). If the data point is above the line, total Hb is positive, and if it is below the line, total Hb is negative. Total Hb is a good measure to understand brain activities, but since it is effected jointly by oxy- and deoxy- Hb, it is essential to look into NIRS trajectory.

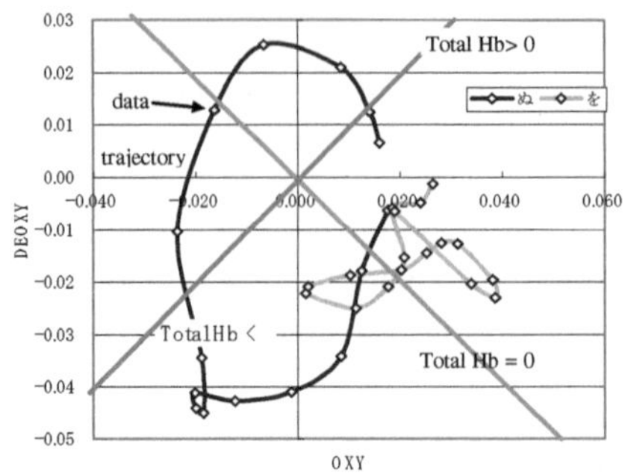

Fig. 1. NIRS trajectory in relation to time and total Hb

2.1 Basic NIRS Trajectories

Three trajectories (Fig. 2, 3, and 4) are basic in NIRS analysis.

The lower wave form of Fig.2 is showing a set of NIRS data, when oxy and deoxy Hb is in opposite phase. Reverse phase may happen when some amount of Hb increase in the local tissue, equivalent amount of deoxy- Hb decrease, so that total Hb remains constant.

If total Hb remains constant, oxy- and deoxy-Hb should be in the opposite phase.

The NIRS trajectory is the plot of oxy-Hb in horizontal and deoxy-Hb in vertical axis of a plane. Fig. 2b) is the NIRS trajectory of Fig.2a).The trajectory is representing the trace of about 3 seconds. The trajectory is composed of three parts. The second part has some divergence in trajectory, while the 3rd part is almost along the line of -1 declination. The dark blue part is the trajectory and the light blue line is the line of -1 slope. Empirically we understand the trajectory along the line of -1 slope is not contributing the task.

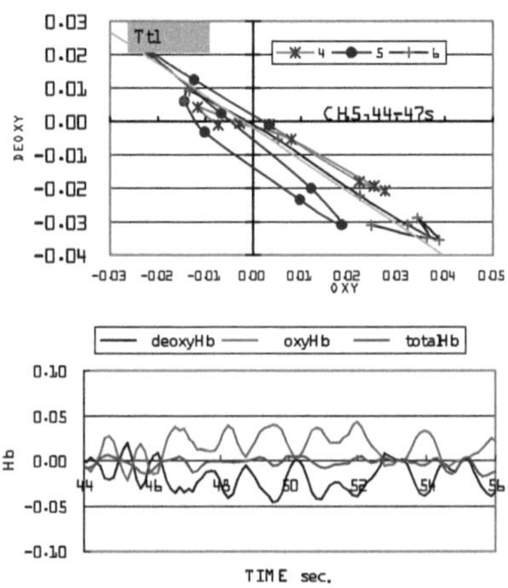

Fig. 2. Opposite phase Oxy- and deoxy- Hb

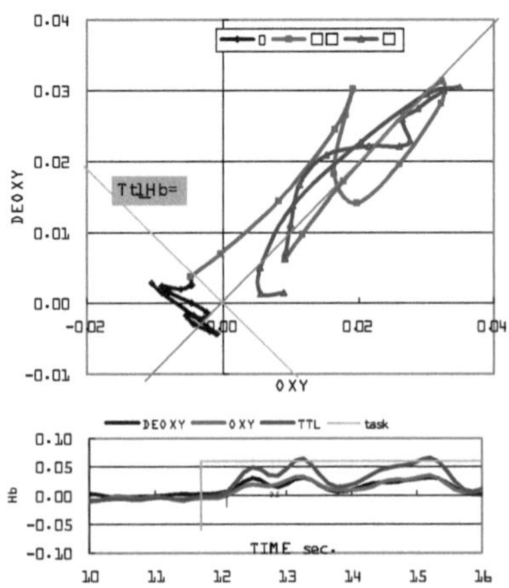

Fig. 3. In phase Oxy- and deoxy- Hb

In the wave form shown in Fig.3a), the oxy- and deoxy- Hb are changing in phase. Resultantly, the change of oxy and deoxy Hb is doubled in total Hb. The NIRS trajectory for this wave form is along the line of +1 slope (in orange).

From the above relation, as far as fluctuation in oxy- and deoxy- Hb are in opposite phase, it is clearly understood that TTL remains in constant level. Since the fluctuations in Oxy and Deoxy are of considerable in amplitude, if the fluctuations change in phase, total-Hb might increase or decrease quickly.

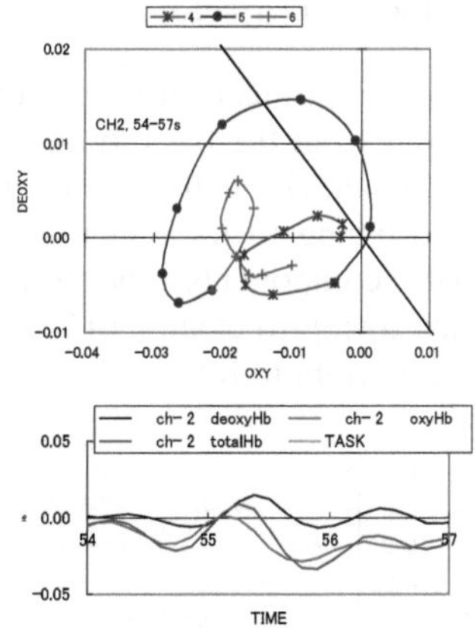

Fig. 4. Trajectory for variable phase Hb

Time change of oxy Hb in local tissue might be complex and dependent on incoming oxy Hb from fine arterial and oxygen consumption in the local tissue. Also time change of deoxy Hb is complex and dependent on production of deoxy Hb in local tissue and outgoing of deoxy Hb to fine vein. In cases the incoming and outgoing blood flow is not much, the oxy-deoxy plot might reflect diagram typical to oxygen consumption, and deoxy Hb production in the local tissue. The diagrams often show round cocoon like shapes, fork, etc. The tissues with negative declination of -1, are supposed to be out of function temporarily, and tissues with increasing slope are facilitating activation. The tissue showing cocoon shapes are supposed to perform data processing at the moment specific to the location.

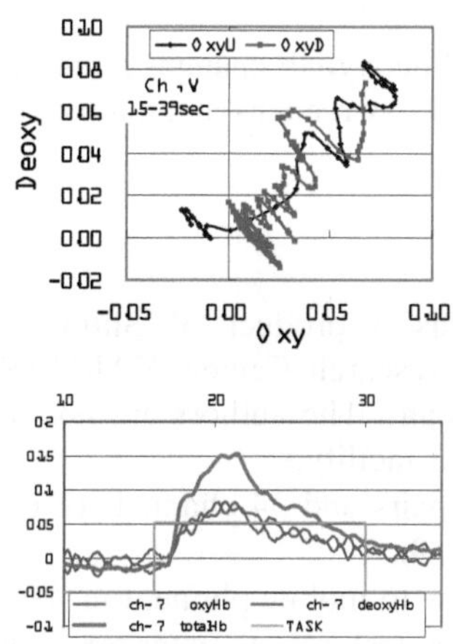

Fig. 5. Change of total Hb with fluctuation, by joint change of Oxy- and deoxy Hb

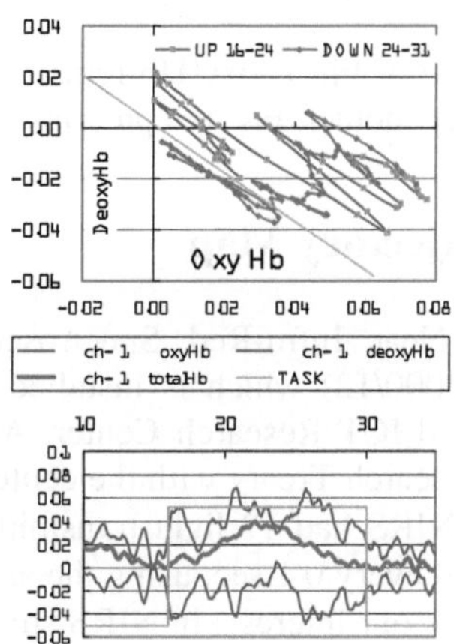

Fig. 6. Total Hb change by change of Oxy-Hb

2.2 Complex Trajectory

More complex trajectories are found when the task level is higher. But many of them are derived from the basic trajectory. For example Fig 5, and 6 are those found from different channels at the same time performing a task. In case of Fig. 5, total Hb increased owing to joint increase of oxy- and deoxy Hb, but due to noise (or high frequency components) the trajectory is fluctuated. In case of Fig. 6, total Hb increased owing to change of oxy-Hb. Thus the trajectory extended from left to right, and back to right to left.

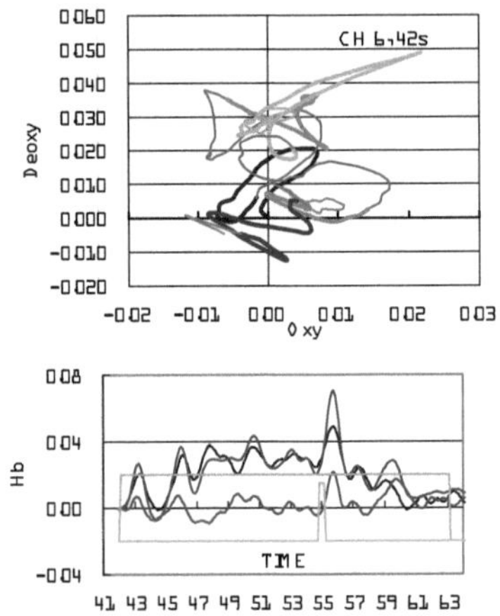

Fig. 7. Multiple loop trajectories along the vertical axis

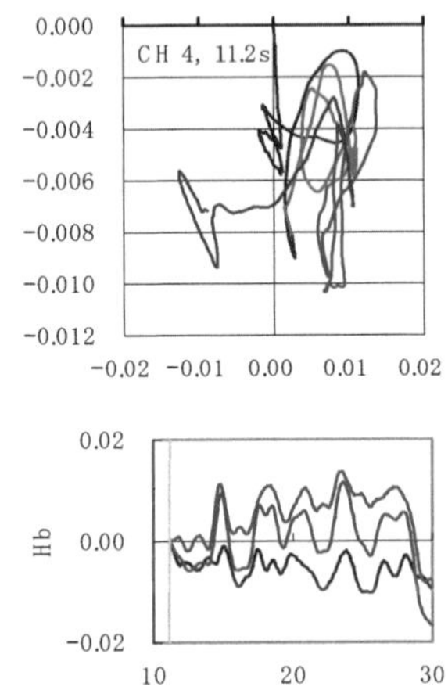

Fig. 8. Multiple loop trajectories around a fixed focus

In case of Fig.7, oxy-Hb remains around zero and deoxy-Hb changes representing a loop after another. Fig. 8 represents the case, deoxy and oxy remains in a forcus area.

3 Trajectory Map

NIRS (Near Infra-Red Spectroscopy) facility was a product of Shimadzu co. (OMM3000/12) which is installed in Future ITC Research Center, KARC (Kansai Advanced ICT Research Center, Akashi, Kobe, Japan). The authors are honored by Joint Research Treaty with the center to use the NIRS facility.

The NIRS had 12 light transmitter and receiver pairs and 34 channels of data are collected every 0.1 sec. using the arraignment in Fig. 9.

One of our interests in NIRS study is the text entry to mobile phone. Fig. 10 shows time wave forms of various channels at one trial. The task was to enter text, which is presented on screen while the square wave form (in orange) is high. The text consisted of 12 Japanese hirakana (phonetic letter) with and without meaning. When he/she completed entry, he /she was required to push a response button to acknowledge. The completion response is also recorded by the green trigger in Fig.10. Wave forms of every channels are zero adjusted at start point of the task. The subjects used multitap method for text entry.

The NIRS trajectory for each channel are drawn and arraigned in the proper location of brain, which is named trajectory map as shown in Fig.11 & 12. Fig.11 is for the first task in fig.10, and Fig.12 is for second task in Fig. 10.

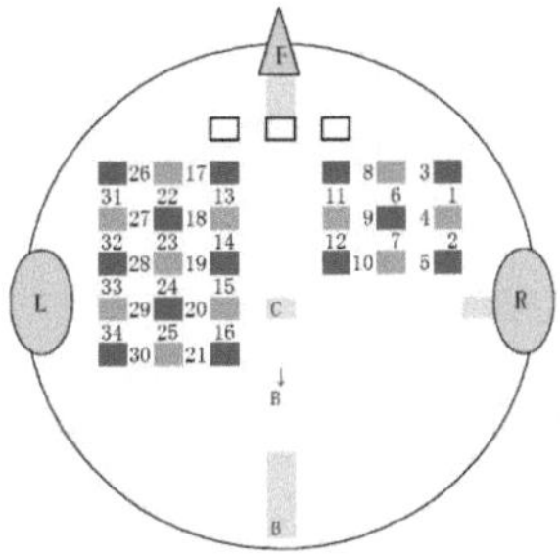

Fig. 9. Transmitter/receiver arraignment for NIRS

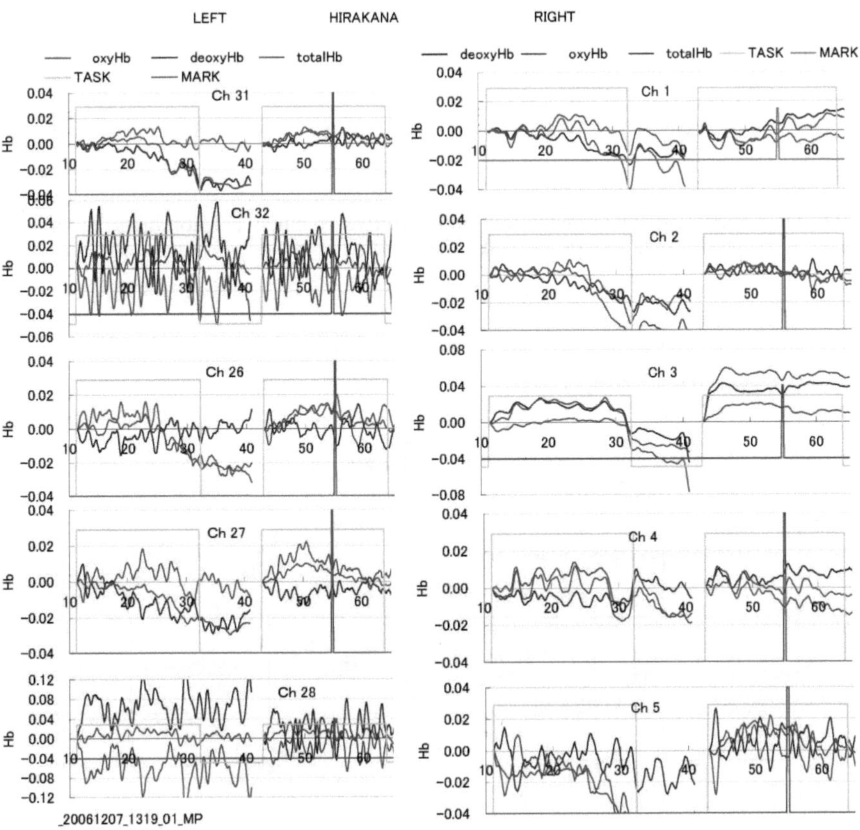

Fig. 10. NIRS time wave form of text entry. Five records from left and right brain are shown. Orange color square mark shows task, and green vertical mark indicate completion of text entry.

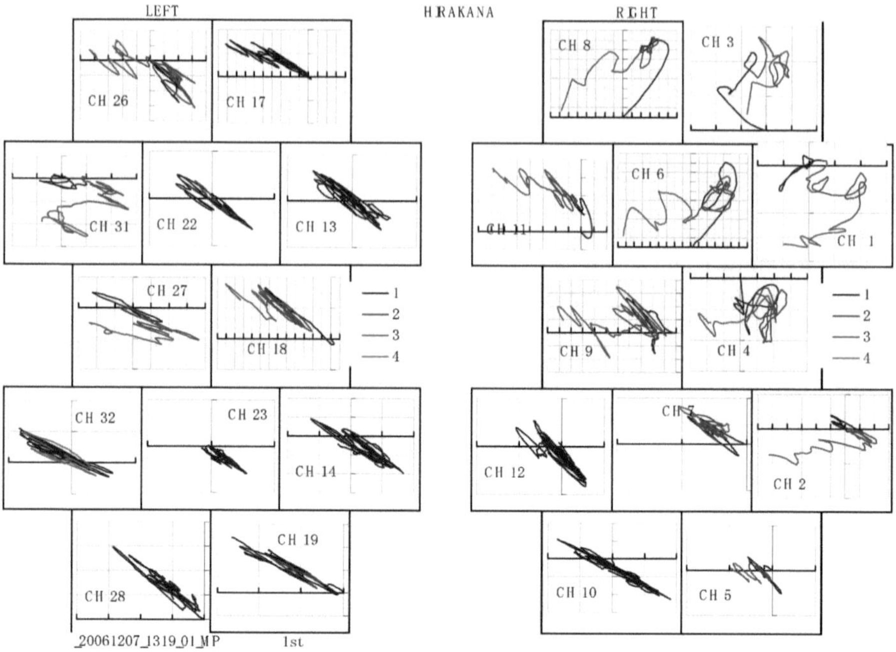

Fig. 11. Trajectory map for hiragana text entry. Active channels are dominant in right brain, while many channels are in rest (blue) in the left.

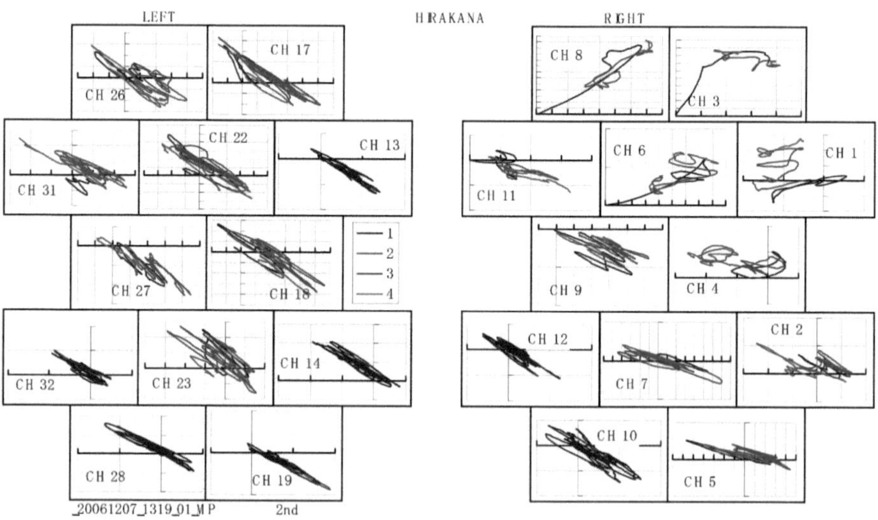

Fig. 12. Trajectory map for hiragana text entry

In Fig. 11, and 12 a trajectory is colored by 5 second segment. Each task length is 20 seconds, but the subject completed a task after a little over 10 seconds. So roughly, during the first (blue) and second (red) segments text is entered, and while in the green segment the task is completed. As you see in Fig. 10, subject missed to push completion button. The fourth (brown) segment is after completion trajectory.

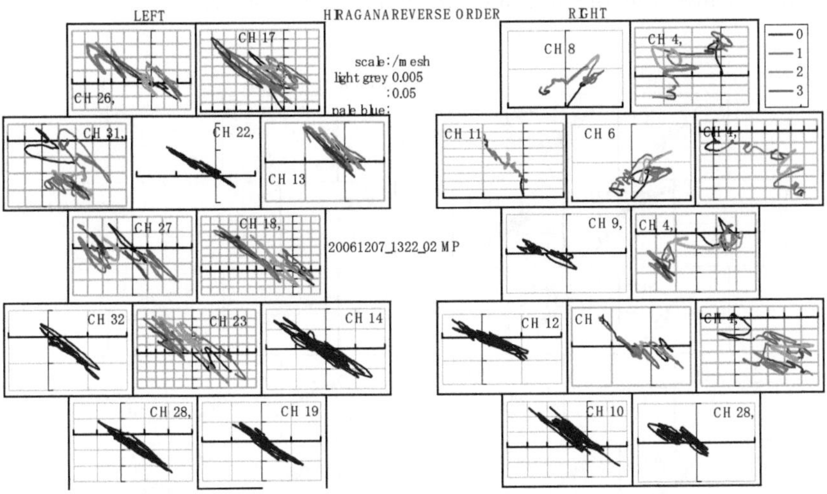

Fig. 13. Trajectory map for reverse order hiragana text. The dominant channels are almost equal to normal ordered text.

For those channels all the trajectory segments are in rest, whole trajectories are colored in blue so that it is easy to identify the channels in rest. Many channels are in rest in left brain, while most of the channels are active in the right. In channel 3, 6, 8, of Fig. 12, strong activation of positive inclination appeared at the first segment. Cocoon like segments are seen in channel 4, and 6. rest mode with drifting oxy-Hb increase or decrease are frequently seen in right brain, but a few in the left. Rest channels with small drift of oxy or deoxy-Hb are seen in channels, 18, 22, 23, 21, 26, 31 of Fig 12.

Entering Text in Reverse Order

People feel it difficult to enter text in reverse order. It may be true, but the time to enter text in normal and reverse order is almost same [6]. It is not the time that people feel it difficult. Fig. 13 is the trajectory map of the previous subject entering a text in reverse order. Active channels are mostly in right brain as in fig.11 and 12, and some channels in left are rest. But the trajectories in Fig. 13 are more complex. This may be the indications of people feeling difficult.

Also in the left brain, channels are apparently activated. Trajectory in channel 17, 18, 26, 27, 23 are expanding horizontally, implicating oxy- Hb increase and decrease during performing the task.

Kanji Conversion Input

Kanji conversion input is higher in task load. The trajectory map for kanji conversion input is derived as in Fig. 14. In the left brain, the trajectories are more complex wild, except for channel 10, and 12. Those two channels are working for other functions.

In the left brain, most of the channels remain as in Fig. 13. They are not responding to the increased task load, except for channel 31, 32, 23. The exceptional channels are supposed to contributing the language processing in the forebrain.

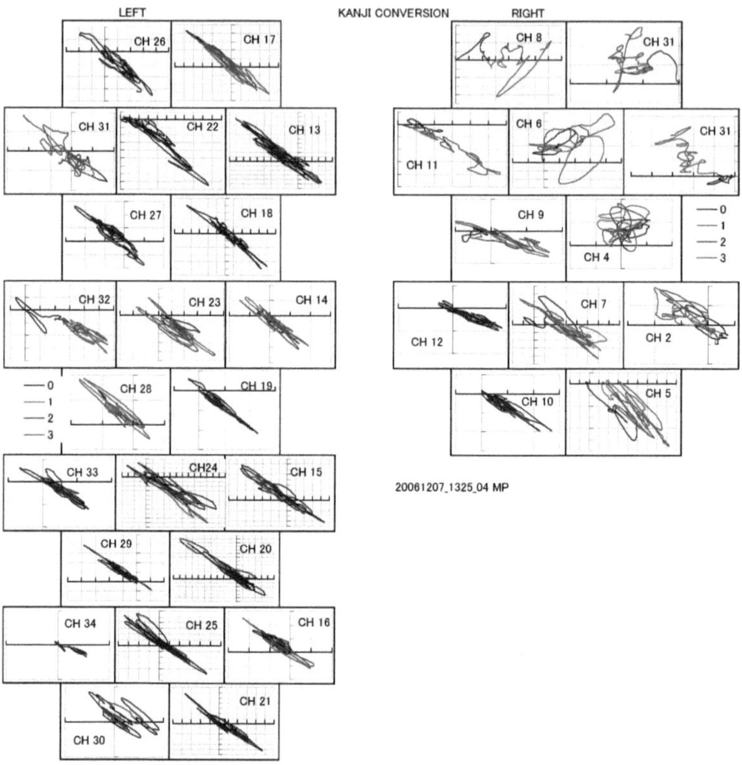

Fig. 14. Trajectory map for Kanji conversion entry

The above results suggest the forebrain are organized to have great asymmetry, and hand and finger actions are mostly under control of right forebrain.

4 Conclusion

Trajectory method is introduced to NIRS analysis and the brain activity while performing a task, is represented by trajectory map. The attempt was applied to text entry tasks. Many channels of right brain and some channel in the left (17, 18, 31) change responses to the change of work load. The studies are continuing to some of more elemental human actions, in one side and evaluation of realistic interface operations on the other side.

References

[1] Tamura, H., Omori, M., Choui, M.: Brain activities recorded by NIRS measurement while performing error correction tasks, Preprint of Kansai branch assembly of JES December 2006, pp. 35–38 (in Japanese) (2006)

[2] Tamura, H., Omori, M., Choui, M.: Brain Activities indicated by two dimensional trajectories of NIRS. Mobile Interactions 3, 9–14 (in Japanese) (2007)

[3] Tamura, H., Omori, M., Choui, M.: NIRS trajectories in Oxy-Deoxy Hb plane and the trajectory map to understand brain activities related to human interface, HCII2007 ID: 2590 (2007)

[4] Omori, M., Tamura, H., Choui, M.: A study on brain activities related to visual search. Mobile Interactions 3, 19–22 (in Japanese) (2007)

[5] Omori, M., Hasegawa, S., Miyao, M., Tamura, H., Choui, M.: Brain Functions related to legibility of LCD studied by means of Near Infrared Spectroscopy, HCII2007 ID: 2591 (2007)

[6] Tamura, H., Choi, S.: Japanese and Hangul Text Entry Time on mobile phone, in comparison to reading and writing, ESK-JES Joint International Symposium on Ergonomics, in The Japanese Journal of Ergonomics, pp. 58–61 (2004)

Brain Computer Interface Via Stereoscopic Images in CAVE

Hideaki Touyama and Michitaka Hirose

Graduate School of Information Science and Technology, University of Tokyo
tou@cyber.rcast.u-tokyo.ac.jp

Abstract. The electroencephalogram signals of steady-state visual evoked potentials were recorded for three subjects in immersive virtual environment. A machine learning technique of support vector machine with single trial EEG data for 1.0 seconds resulted in 92.1 % of averaged recognition rate in selecting a virtual button among two. The online demonstrations in CAVE showed good performance in position control of a simple 3D object.

1 Introduction

It is so exciting to consider that we could operate a variety of devices only by brain activities. The brain computer interface (BCI) is a communication channel with which the computer or machine can be operated [1]. In recent, the BCI has been investigated in immersive virtual environment [2]. The novel works on walking from thought via motor imagery tasks was performed there. The immersive projection technology in virtual reality was expected to be suitable to evaluate in advance the BCI developed in the real world.

The purpose of this paper is to show the possibility to detect the brain activity of steady-state visual evoked potential (VEP) [3] induced by two virtual stimuli in immersive virtual environment for future applications. In later section, only by using the VEP, the position of a virtual object will be briefly controlled in real time.

2 Experimental Method

The virtual cubes were used as visual stimuli in our experiment. The visual stimuli were presented in CABIN, which is an extended multi-screen display system from CAVE [4] to have a five-screen configuration (Fig. 1). Two virtual cubes illuminated with constant frequencies of 4.80 Hz and 6.86 Hz, respectively. Note that, in novel studies of BCI based on the VEP, the visual stimuli were virtual buttons with a few degrees of visual angles on a computer monitor [5, 6].

The subject was sitting on a luxury arm chair about 2.0 m in front of the front screen, wearing a LCD shutter glasses to obtain stereoscopic images. Adopting non-invasive electroencephalogram (EEG) technique, three electrodes (O_1, O_2 and O_z) were attached over visual cortex. The EEG signals were amplified and were band pass filtered between 0.5 and 30 Hz. The analogue data was sampled at 100 Hz in an A/D converter to be stored in a PC.

M.J. Smith, G. Salvendy (Eds.): Human Interface, Part I, HCII 2007, LNCS 4557, pp. 1004–1007, 2007.

Fig. 1. The overview of immersive virtual environment

For all three subjects, two types of experimental task were imposed. The task 1 was to gaze and focus attention to a visual stimulus of 6.86 Hz illumination, and task 2 was of 4.80 Hz illumination. The task 1 and 2 were performed by turns. One trial lasted 30 seconds, and after that the subjects took about one minute rest.

3 Offline Analyses

The recorded EEG data was analyzed in offline in the following procedures; (1) The extraction of features of steady-state VEP, (2) The machine learning, and (3) The estimation of recognition rates.

The features were extracted from the linear combination of voltage value of three electrodes on visual cortex, expecting the reduction of environmental noise. The result of estimated power spectral density is shown in Fig. 2 for one subject. The steady-state VEP was observed and the power spectral densities from 3 Hz to 25 Hz, with 1.0 second of analyzing time period, were used to construct a feature vector.

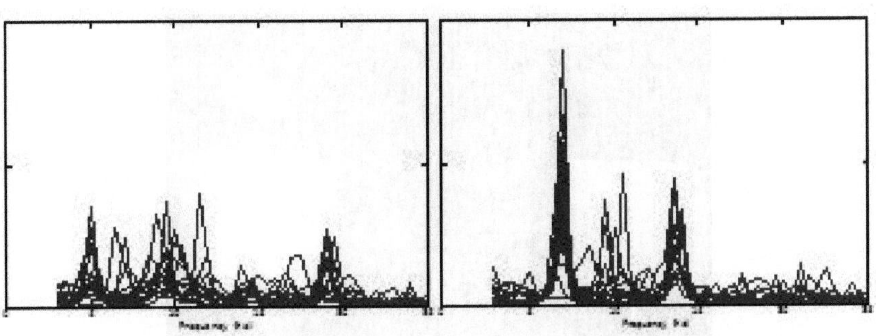

Fig. 2. The left and right figure show the estimated power spectral density of single trial EEG data, for a visual stimulus of 4.80 Hz and 6.86 Hz, respectively. The 15 densities are superposed in each figure, dividing one trial sequence data for 30.0 seconds into 2.0-seconds segments.

In the process of machine learning, we adopted support vector machine (SVM) [7]. To estimate the recognition rate, a leave-one-out method was applied. By using the mentioned feature vectors, the averaged rate was found to be 96.7 %, 89.7 %, and 90.0 % for three subjects, respectively. The average was 92.1 %. These high recognition rates would be promising for online experiments in the following section. We expected that the VEP enabled the user to control the position of a simple 3D virtual object.

4 Online Control

We demonstrated a preliminary experiment on online control of a simple virtual object floating in CAVE. The online control system consists of three modules; (1) The EEG data acquisition module, (2) The signal processing module, and (3) The image rendering module. The data acquisition module was the same with that in the offline analyses. After the data acquisition, the digitized EEG data was transmitted to a signal processing server through the Ethernet. At the server, the SVM classification was performed, and the discrimination result was transmitted to the workstation of CAVE for real-time image renderings.

In the online experiments, the visual stimulus A and B might have a role to be a virtual button to control the position of a specific virtual object C (Fig. 3). The user was instructed to gaze and focus attention to a red-coloured fixed point. The fixed point changed its position every 10 seconds between the fixed point on the center of stimulus A and B. It was expected that the steady-state VEP induced by the virtual button would be adequately processed to yield the command to control the position of the cube C. During the online session, the user obtained the visual feedback of the stereo images with LCD shutter glasses.

It was found that the position control of a virtual cube C was successfully performed reflecting the user's intention to select the virtual button A or B, while the time delay and the error control was observed.

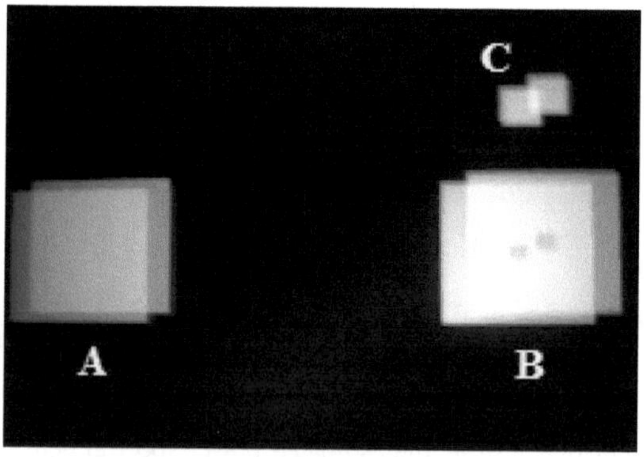

Fig. 3. The virtual button A, B and controlled object C in online experiments

5 Summary

We investigated the steady-state VEP in immersive virtual environment. The EEG signals, induced by visual stimulus of stereoscopic virtual object, were recorded and analyzed. It was found that, in button selecting tasks, the recognition rate was 92.1 % averaged for all three subjects, and the position of a 3D object could be controlled only via the steady-state VEP in real time. This technology of brain CAVE interfacing would enable us to realize intuitive and non-muscular manipulations of objects. We expect that non-invasive BCI in immersive virtual environment will be one of the research paradigms of the next generation of virtual reality, that is, "Virtual Reality 2.0" for people who enjoy it.

Acknowledgement

This research is partly supported by The Telecommunications Advancement Foundation, Japan.

References

[1] Wolpaw, J.R., et al.: Brain-computer interfaces for communication and control. Clinical Neurophysiology 113, 767–791 (2002)
[2] Pfurtscheller, G., et al.: Walking from thought. Brain Research 1071, 145–152 (2006)
[3] Kuroiwa, Y., et al.: Visual Evoked Potentials with hemifield pattern stimulation, Their use in the diagnosis of retrochiasmatic lesions. Arch. Neurol. 38, 86–90 (1981)
[4] Cruz-Neira, C. et al.: Surround-screen projection-based virtual reality: The design and implementation of the CAVE, ACM SIGGRAPH'93 Proc., pp. 135–142
[5] Middendorf, M., et al.: Brain-Computer Interfaces Based on the Steady-State Visual-Evoked Response. IEEE Transactions on Rehabilitation Engineering 8(2), 211–214 (2000)
[6] Cheng, M., et al.: Design and Implementation of a Brain-Computer Interface With High Transfer Rates. IEEE Transactions on Biomedical Engineering 49(10), 1181–1186 (2002)
[7] Vapnik, V.N.: The nature of statistical learning theory, Statistics for Engineering and Information Science. Springer, New York (2000)

Human-Entrained E-COSMIC: Embodied Communication System for Mind Connection

Tomio Watanabe

Faculty of Computer Science and System Engineering, Okayama Prefectural University
111 Kuboki, Soja, Okayama, 719-1197 Japan
watanabe@cse.oka-pu.ac.jp

Abstract. The embodied communication system for mind connection (E-COSMIC) has been developed by applying the entrainment mechanism of the embodied rhythms of nodding and body movements to physical robots and CG characters. E-COSMIC comprises an embodied virtual communication system for human interaction analysis by synthesis and a speech-driven embodied interaction system for supporting essential human interaction and communication based on the analysis that uses the embodied virtual communication system. Human-entrained embodied interaction and communication technology for an advanced media society is introduced through some applications of E-COSMIC.

Keywords: Human Interface, Human Communication, Human Interaction, Enbodied Communication, Enbodied Interaction, Enbodied Interface.

1 Introduction

In a human face-to-face conversation, embodied rhythms between speech and body motions such as nodding are mutually synchronized not only between talkers but also in a talker. The phenomenon is observed in mother-infant interaction as a primitive form of communication [1], [2]. This synchrony of embodied rhythms in communication, referred to as entrainment, generates the sharing of embodiment in human interaction, which plays an important role in human interaction and communication. Entrainment in communication is also observed in physiological indices such as respiration and heart rate variability [3]. This embodied communication closely related to behavioral and physiological entrainment is an essential form of communication that forms the relation of interaction between talkers through mutual embodiment. Hence, the introduction of this mechanism to human interface is indispensable to the realization of human-centered essential interaction and communication systems.

In this paper, by focusing on the embodied entrainment, the embodied interaction and communication technology through the development of the embodied communication system for mind connection (E-COSMIC) is proposed for supporting human interaction and communication [4]. E-COSMIC mainly comprises an embodied virtual face-to-face communication system and a speech-driven embodied

M.J. Smith, G. Salvendy (Eds.): Human Interface, Part I, HCII 2007, LNCS 4557, pp. 1008 – 1016, 2007.

interaction system, as shown in Fig. 1. The former is developed for human interaction analysis by synthesis and the latter, for supporting human interaction and communication based on the analysis that uses the former. The effectiveness of the system is demonstrated by some actual applications on CG/robot and human interactive communications.

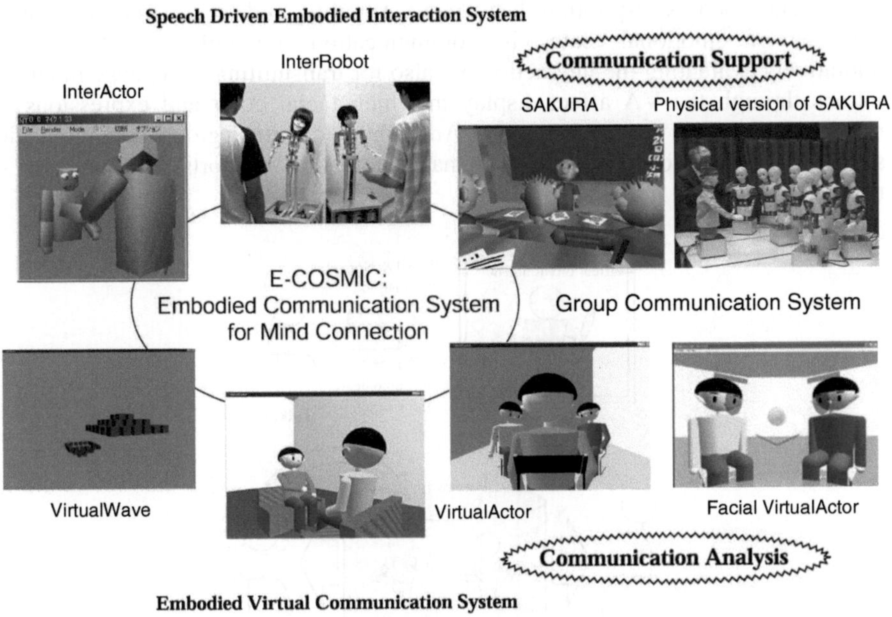

Fig. 1. E-COSMIC

2 Embodied Virtual Communication System

The concept of an embodied virtual face-to-face communication system is illustrated in Fig. 2. The figure presents a VirtualActor (VA), an interactive avatar, that represents the talker's interactive behavior such as gesture, nodding, blinking, facial color and expressions, paralanguage, respiration, etc., based on one's verbal and nonverbal information as well as physiological information in a virtual face-to-face communication environment. Fig. 3 provides an example of a virtual face-to-face scene with two VAs from the diagonal backward viewpoint of one's own VA. The motions of the head, arms, and body for each VA are represented based on the positions and angles measured by four magnetic sensors that are placed on the top of the talker's head, both wrists, and the back of the body [5]. Two remote talkers can communicate through their VAs and become aware of the interaction through the embodied interaction of VAs in the same virtual communication environment from any viewpoint. The analysis by synthesis for interaction in communication is performed by processing the behavior of VAs, such as cutting or delaying the motion and voice of VAs in various conditions of their spatial relations and positions. For example, to examine the effects of only nodding on interaction, it is possible for a VA to represent just nodding without body motion even if the talker nods with body motion. Thus, the characteristics and relations of the embodied interaction between talkers are systematically clarified through the analysis by synthesis of interaction in communication by using the system in which talkers are the observers of interactions as well as the operators of interaction through their VAs. Further, physiological measurements such as respiration, heart rate variability, and facial skin temperature, as indices of emotional states in communication are utilized not only for quantitatively evaluating the interaction but also for transmitting the change in talkers' emotions through the VA affect display in which facial color and expressions are synthesized based on the measurement. An embodied virtual group communication system is also developed for three human interaction supports and analyses by synthesis, as indicated in Fig. 1 [6].

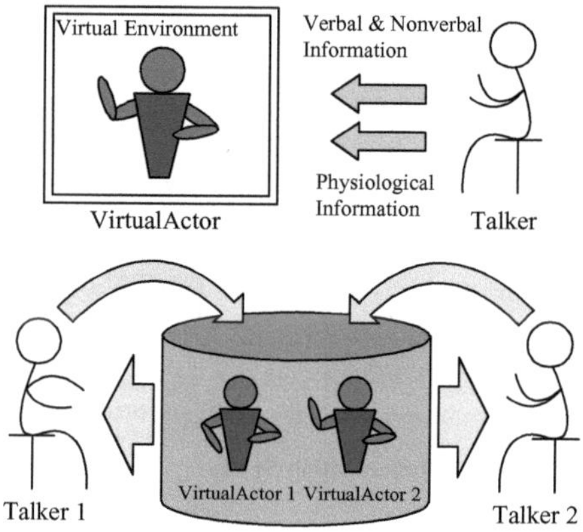

Fig. 2. Concept of the embodied virtual communication system

Fig. 3. Example of a virtual face-to-face scene with two VirtualActors representing the talker's own self and his/her partner

3 Speech-Driven Embodied Interaction System

Based on the human interaction analysis that uses the embodied virtual communication system, a speech-driven embodied interaction system is developed for supporting human interaction by generating the communicative motions of a CG character known as InterActor or a physical robot referred to as InterRobot; these communicative motions are coherently related to speech input [7]. The concept and system are presented in Fig. 4. The system comprises two InterActors (InterRobots) that have both the functions of a speaker and a listener based on speech input. When Talker 1 speaks to InterActor 2, InterActor 2 responds to Talker 1's utterance with an appropriate timing through its entire body motions, including nodding, blinking, and actions, in a manner similar to the body motions of a listener. Thus, Talker 1 can talk smoothly and naturally. Subsequently, the speech is transmitted via a network to the remote InterActor 1. InterActor 1 can effectively transmit Talker 1's message to Talker 2 by generating the body motions similar to those of the speaker based on the time series of the speech and by simultaneously presenting both the speech and the entrained body motions. This time, Talker 2 in the role of a speaker achieves communication in the same way by transmitting his/her speech via InterActor 1 as a listener and InterActor 2 as the one talking to Talker 1. Thus, in this manner, two remote talkers enjoy a conversation via InterActors. The information transmitted and received by this system is only through speech. Of significance is the fact that it is a human who transmits and receives the information; the InterActor/InterRobot merely generates the entrained communicative movements and actions based on speech input and supports the sharing of mutual embodiment in communication.

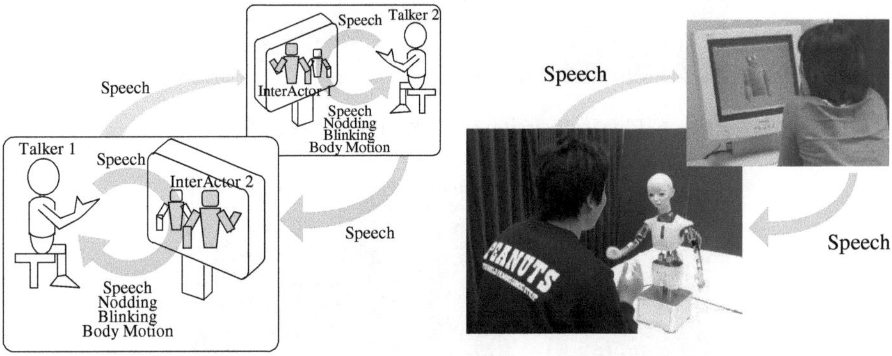

Fig. 4. Concept and system of the speech-driven embodied interaction system

Fig. 5 illustrates a speech-driven embodied group-entrained communication system referred to as SAKURA [8]. SAKURA activates group communication in which InterActors are entrained to one another as a teacher and some students in the same virtual classroom. By using SAKURA, talkers can communicate with a sense of unity through the entrained InterActors by using only speech input via the network. Fig. 6 depicts a physical version of SAKURA with InterRobots. Their entrained movements and actions based on speech can activate and assist human embodied interaction and

Fig. 5. SAKURA: The speech-driven embodied group-entrained communication system

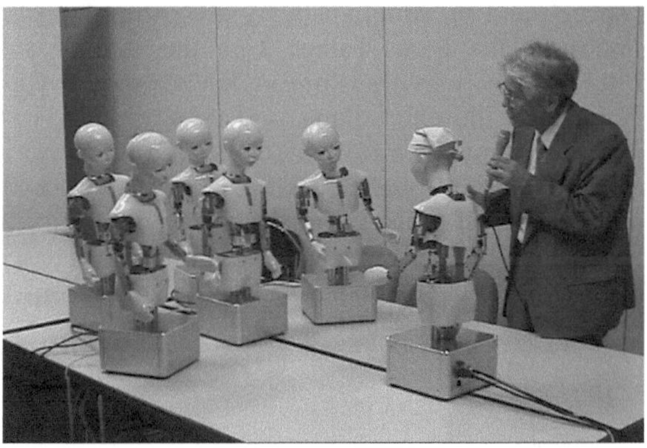

Fig. 6. Physical version of SAKURA with InterRobots

Fig. 7. Speech-driven embodied interaction system with InterRobots and an InterActor in the National Museum of Emerging Science and Innovation

communication. Fig. 7 indicates another physical version of SAKURA with four InterRobots and one InterActor, which is exhibited in the National Museum of Emerging Science and Innovation where visitors can enjoy a dynamic experience of embodied communication. They perceive the effects of group-entrained communication environment intuitively and recognize the importance of embodied communication.

4 Embodied Interaction and Communication Technology

In this section, some actual applications of InterActor/InterRobot to human interface are introduced. Fig. 8 depicts an interaction scene between an InterRobot and children in a kindergarten. This InterRobot is commercialized and marketed for kindergarten. Children enjoy and are excited about having conversations with the InterRobot, while the teacher standing behind the InterRobot enjoys talking and encouraging children in a new communication mode from a completely different standpoint, just changing to a friend and so forth. By focusing on an animal character that is most preferred by children, an animal-type InterActor/InterRobot known as InterAnimal is developed in

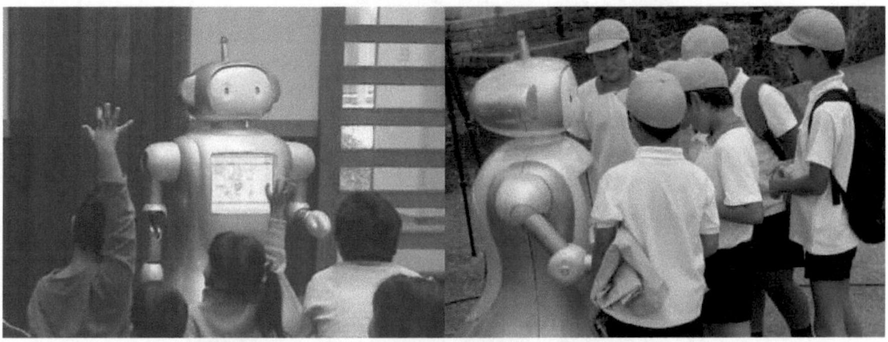

Fig. 8. Interaction scene between an InterRobot and children in a kindergarten

Fig. 9. InterAnimal: Animal-type InterActor

order to encourage and cheer up children, as depicted in Fig. 9. The bear-type InterAnimal shown in Fig. 10 was popular in the 2005 exhibit EXPO. Fig. 11 illustrates a toy version of InterRobot with the function of a listener, which generates the listener's actions of nodding, tilting his/her head, and moving his/her arms up and down, based on speech input. The stuffed toy bear is eager to listen without ever uttering a word. It is commercialized and marketed under the name of Unazukikun.

Fig. 10. InterAnimal in EXPO 2005

Fig. 11. Toy version of InterRobot

The InterActor, as indicated in Fig. 12, is also commercialized under the name of InterCaster through which news and media contents are effectively and cordially transmitted in a commercial program. By superimposing InterActors as listeners on the video images of a lecture such as an education program, the InterActor-superimposed learning support system is developed, as illustrated in Fig. 13 [9]. The system provides group-entrained interaction effects for audiences who watch the video, in which two InterActors at the bottom of reduced images are entrained with the lecturer's speech. The InterActor is a speech-driven CG-embodied interaction character that can generate communicative movements and actions for an entrained interaction. An InterPuppet, on the other hand, is an embodied interaction character that is driven by both speech input, similar to the InterActor, and hand motion input, like a puppet. Therefore, humans can use the InterPuppet to communicate effectively by using deliberate body movements as well as natural communicative movements and actions. An advanced InterPuppet with a cellular phone-type device is developed as indicated in Fig. 14. This interaction character can be used in a mobile environment.

Fig. 12. InterCaster

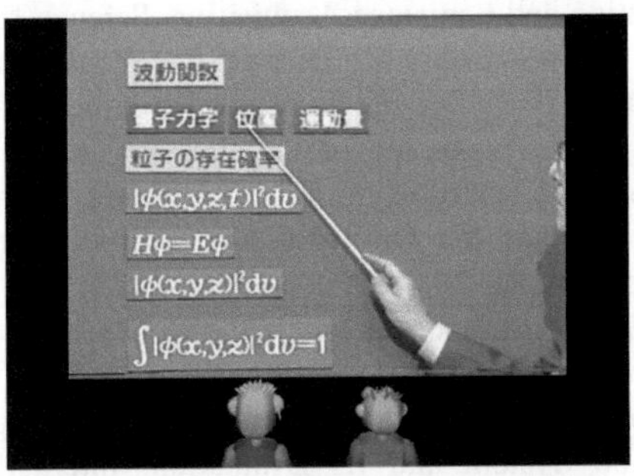

Fig. 13. InterActor-superimposed learning support system

Fig. 14. InterPuppet with a cellular phone-type device

5 Conclusion

The human-entrained embodied interaction and communication technology for an advanced media society was proposed through the development of E-COSMIC for supporting essential human interactive communication based on the entrainment mechanism of the embodied rhythms between speech and body movements such as nodding. Some actual applications pertaining to CG/robot and human interactive communication were also demonstrated. In particular, the speech-driven embodied interaction system, such as InterActor and InterRobot, is a robust and practical communication support system for everyday living, which activates embodied interaction and communication in a new communication mode by using only speech input. The speech-driven entrainment technology for enhancing interaction and communication would be expected to form the foundation of mediated communication technologies as well as the methodology for the analysis and understanding of human interaction and communication, and to develop a new embodied communication industry for supporting essential human interactive communication.

Acknowledgments. This work has been supported by Core Research for Evolutional Science and Technology (CREST) of Japan Science and Technology Agency (JST).

References

[1] Condon, W.S., Sander, L.W.: Neonate Movement is Synchronized with Adult Speech: Interactional Participation and Language Acquisition. Science 183, 99–101 (1974)
[2] Kobayashi, N., Ishii, T., Watanabe, T.: Quantitative Evaluation of Infant Behavior and Mother-Infant Interaction. Early Development and Parenting 1, 23–31 (1992)
[3] Watanabe, T., Okubo, M.: Evaluation of the Entrainment Between a Speaker's Burst-Pause of Speech and Respiration and a Listener's Respiration in Face-to-Face Communication. In: Proc. of the 6th IEEE International Workshop on Robot-Human Interactive Communication (RO-MAN'97), pp. 392–397 (1997)
[4] Watanabe, T.: E-COSMIC: Embodied Communication System for Mind Connection? In: Proc. of RO-MAN 2004, pp. 1–6 (2004)
[5] Watanabe, T., Ogikubo, M., Ishii, Y.: Visualization of Respiration in the Embodied Virtual Communication System and Its Evaluation. International Journal of Human-Computer Interaction 17(1), 89–102 (2004)
[6] Shintoku, T., Watanabe, T.: An Embodied Virtual Communication System for Three Human Interaction Support and Analysis by Synthesis. In: Proc. of the 5th IEEE International Symposium on Computational Intelligence in Robotics and Automation (CIRA2003), pp. 211–216 (2003)
[7] Watanabe, T., Okubo, M., Nakashige, M., Danbara, R.: InterActor: Speech-Driven Embodied Interactive Actor. International Journal of Human-Computer Interaction 17(1), 43–60 (2004)
[8] Watanabe, T., Okubo, M.: SUKURA: Voice-Driven Embodied Group-Entrained Communication System. In: Proc. of HCI International 2003, vol. 2, pp. 558–562 (2003)
[9] Watanabe, T., Yamamoto, M.: An Embodied Entrainment System with InterActors Superimposed on Images. In: Proc. of HCI International 2005, vol. 4(2045), pp. 1–6(2005)

Development of an Embodied Image Telecasting Method Via a Robot with Speech-Driven Nodding Response

Michiya Yamamoto and Tomio Watanabe

Okayama Prefectural University, Faculty of Computer Science and System Engineering
111 Kuboki, Soja, Okayama 719-1197, Japan
{yamamoto,watanabe}@cse.oka-pu.ac.jp

Abstract. InterRobot is an embodied interaction robot driven by InterRobot technology (iRT) in order to make communicative motions and actions based only on speech input. In this paper, the concept of an embodied image telecasting method is proposed for activating embodied interaction and communication via such a robot. The prototype systems are developed: a new InterRobot equipped with a camera and a projector, and an embodied visual effect on telecasted images. The sensory evaluation demonstrates that the visual effect is effective for realizing presence and a sense of unity.

Keywords: human communication, human interaction, embodied interaction, entrainment.

1 Introduction

In recent years, robot technology has been developed with the aim of developing a collaborative robot, such as a communication robot, service robot, etc., such that it can communicate with humans irrespective of their age (from children to old people). For realizing these robots, remote supervisory and remote control functions as well as autonomous behavior are required [1],[2]. In addition, the functions for fluent communication via the robot itself are essential for such partner robots so that they can provide various network services [3],[4].

In human face-to-face communication, not only verbal messages but also nonverbal behavior such as nodding and body actions are rhythmically related and mutually synchronized between speakers. The authors have developed InterRobot technology (iRT) to generate communicative motions and actions such as nodding, blinking, and the actions of the head, arms, and waist based on only speech input. The effectiveness of iRT was demonstrated by using a speech-driven embodied interactive robot called "InterRobot" with both the functions of a speaker and listener [5].

In this study, we propose a new form of communication via a robot with a speech-driven nodding response, which can activate embodied communication and interaction for mind connection. In this study, we first propose the concept of an embodied image telecasting method via such a robot. Next, we develop the embodied image telecasting method via the robot with the speech-driven nodding response: a new InterRobot equipped with a camera and a projector for creating an embodied

M.J. Smith, G. Salvendy (Eds.): Human Interface, Part I, HCII 2007, LNCS 4557, pp. 1017–1025, 2007.

visual effect on telecasted images. Then, we create a sample telecasting image of a TV news program and evaluate the system by sensory evaluation. The results demonstrate the effectiveness of the system.

2 Concept

2.1 A Robot with Speech-Driven Nodding Response

In this study, we present a partner robot through which a human can communicate in daily life. Fig. 1 shows the concept of the communication. The robot plays a central role in the communication environment by generating communicative motions and actions, such as nodding, blinking, and the actions of the head, arms, and waist based on only speech input, and by activating human communication. In addition, the embodied entrainment of the motions and actions of the robot enables remote users to take part in the communication via the robot and enjoy communication. The robot performs both the functions of InterRobot as a communicative partner robot and embodied media for telecasting presence.

Fig. 1. Communication via a robot with a speech-driven nodding response

2.2 Embodied Image Telecasting

In order to realize the aforementioned concept, it is necessary to develop a new interface in order to smoothly take part in the communication from a remote location. The interface should not be businesslike wherein users can only watch a customer who they are talking with but an interface in which users can enjoy interaction and communication via the robot.

In this study, we propose the embodied image telecasting method via the robot with the speech-driven nodding response. Here, we introduce InterRobot equipped with the camera and projector. Fig. 2 shows the concept of embodied image telecasting. The robot as a listener fits into the scene with a natural entrained response and transmits the image by using the onboard camera. The robot can also watch the telecasted image at the receiving side by using a TV or the onboard projector. The users at both sides can feel a sense of unity and presence as if they are listening together.

By using this telecasting method, robots can create content by recording the scene; further, they can enliven lectures and meetings with the best embodied response and telecast the image.

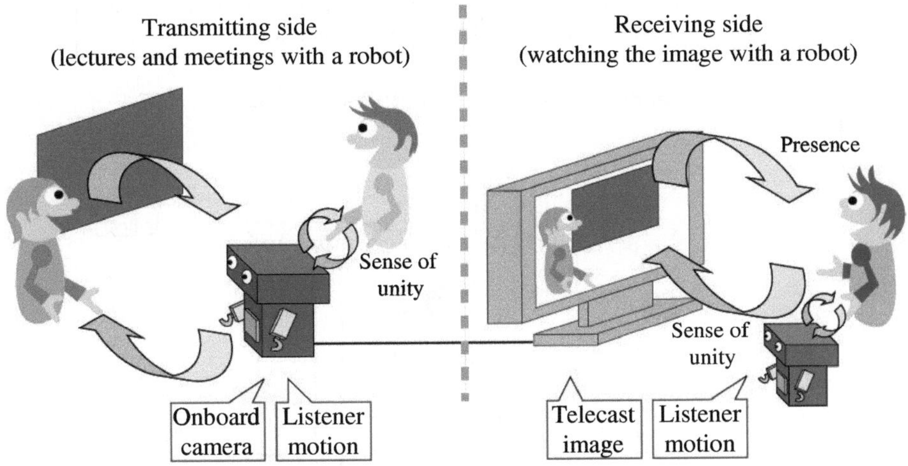

Fig. 2. Concept of embodied image telecasting

3 Development of Embodied Image Telecasting

3.1 Robot with Speech-Driven Nodding Response

In order to realize the concept described in section 2, we developed a prototype of the robot with the speech-driven nodding response. Fig. 3 is an image of the robot. It is made of wood and has the camera (SONY, DCR-TRV70) and projector (PLUS, V-1100) inside the head. Its dimensions are 315 mm (L) × 360 mm (W) × 480 mm (H). The nodding motion is in the range of –5° to +15°.

Fig. 4 shows the system configuration. The angle of the neck is controlled by a servo motor (Kondo, KRS-2350HV) by a control board (Kondo, RCB-1 HV) to adjust the nodding angle. The nodding time is 0.5 s when it is driven at DC 10.8 V. The nodding response is estimated by the MA (moving average) model as shown in equation (1).

$$y(i) = \sum_{j=1}^{J} a(j)x(i-j) + w(i) \tag{1}$$

where $a(j)$ is the linear prediction coefficient and $w(i)$ is the noise [5].

Fig. 5 shows an example of the telecast by the robot with the speech-driven nodding response. When the robot is listening to the speaker and responding via the nodding motion of its head, the image moves up and then down.

Dimensions	315 mm (L) x 360 mm (W) x 480 mm (H)
Weight	5.9 kg
Drive	DC servo motor (PC control)
Functions	·Embodied interaction ·Camera / projector
Operations	·Speech-driven nodding response ·Image telecast

Fig. 3. Robot with a speech-driven nodding response

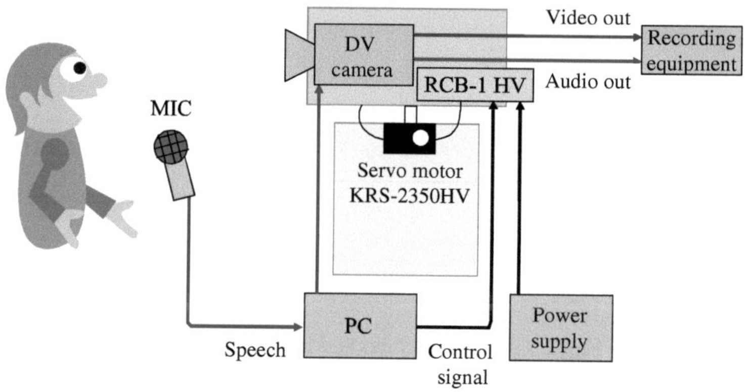

Fig. 4. System configuration

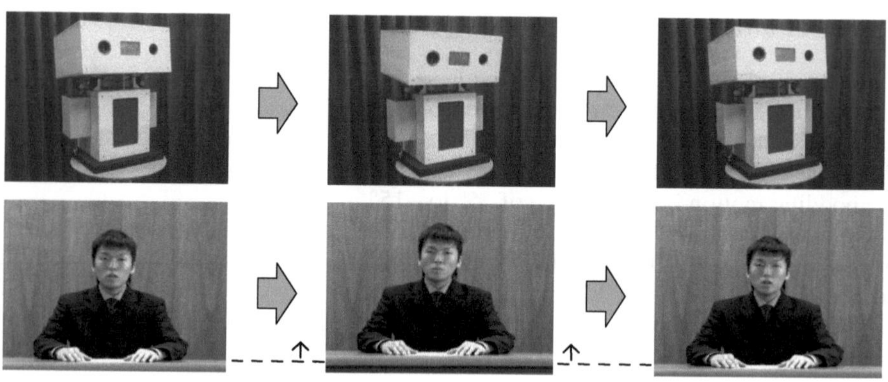

Fig. 5. Telecast by the robot with the speech-driven nodding response

3.2 Speech-Driven Embodied Visual Effect

If we use two robots with the speech-driven nodding response with synchronous head motion, the object positions in the telecasted image will not change. This is because, when the robots nod, the onboard camera image at the transmitting side moves up while the projected image at the receiving side simultaneously moves down. Focusing on this telecasting mechanism, we propose another image telecasting method: an embodied visual effect—the superimposition of a moving black frame on the still image. The black frames move down during a nodding response and then move up.

We developed a prototype system as shown in Fig. 6. From the audio signal of the source image, the embodied visual effect can be realized. We used DirectX 9.0 SDK to capture the source image from the DV port on a PC and superimposed the black frames in real time. The response time of the frame is 0.5 s, which is the same as that of the robot with the speech-driven nodding response. The transparency and moving amount can be adjusted by the PC. The nodding frame at a transparency of 0% corresponds to the telecasted image by the robot with the speech-driven nodding response. Here, we also introduced the MA model for nodding estimation.

Fig. 7 shows an example of the speech-driven embodied visual effect. The black frames at the top and bottom of the image move down against the source image during nodding and then the frames move up.

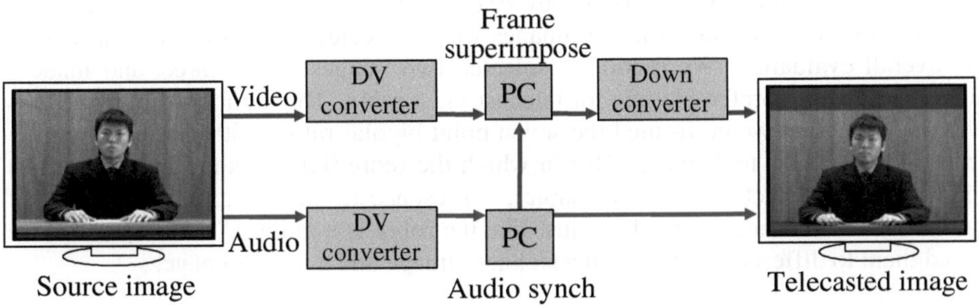

Fig. 6. System configuration

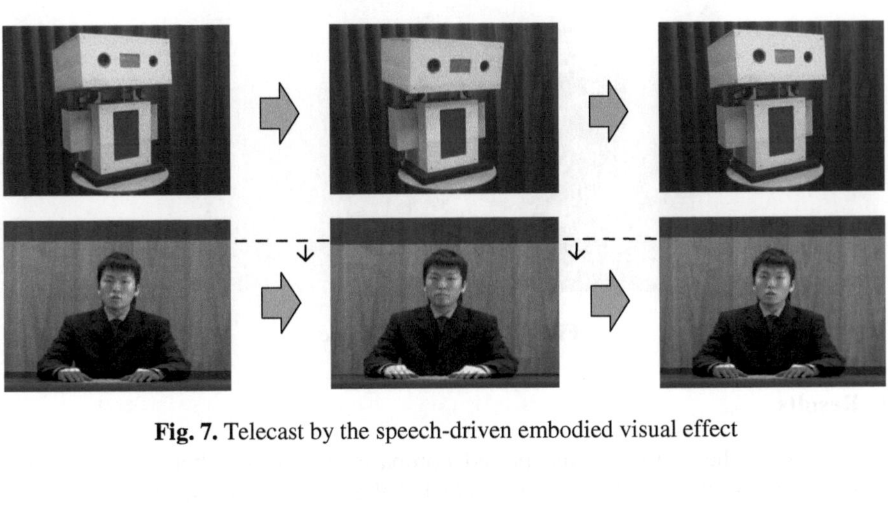

Fig. 7. Telecast by the speech-driven embodied visual effect

4 Telecasting Experiment

4.1 Image Production

For the telecasting experiment, we created three types of sample telecasting images of a TV news program by setting the robot with the speech-driven nodding response and a conventional static video camera side by side (about 10 cm): (A) a conventional static telecasting image (conventional), (B) an image shot by the robot with the speech-driven nodding response (nodding camera), and (C) an image in which the speech-driven embodied visual effect is superimposed (nodding TV). The content of the program is about how a bio-venture beaten the international project in genetic analysis, and it continues for about 2 min. In Fig. 5 and Fig. 7, examples of the images are shown. The amount of motion is 1 cm on a 20 inch display.

4.2 Experimental Setup

By using the images, we performed an evaluation experiment. Fig. 8 is an example of the experimental scene. A 20 inch LCD display (Sharp, LC-20B1-S) is used to show the image. The experiment was performed by 40 (20 male, 20 female) Japanese students.

In this experiment, three kinds of images were subjected to a paired comparison for the overall evaluation. We randomly selected two images out of three, and made a subject to select a preferable one for telecast via a robot. The number of comparisons was 6 (${}_3C_2$). Next, we performed the seven point bipolar rating with the scale ranging from –3 (not at all) to 3 (extremely) in which the score 0 denotes moderate from the viewpoints of "watch (easily watchable)," "presence (a subject can feel a sense of presence)," and "sense of nodding (just like the robot is nodding)." Subsequently, we asked them to differentiate between a sickness image and a usable image.

Fig. 8. Experimental scene

4.3 Results

Table 1 shows the results of the paired comparison. The number in each column represents the number of subjects that selected the pattern against the pattern in the row. Here, the Bradley-Terry model was fitted to the result in order to analyze quantitatively as shown in the following equations. π_i is the intensity of the preference for mode i, and P_{ij} is the probability that i is better than j.

$$P_{ij} = \frac{\pi_i}{\pi_i + \pi_j}$$
$$\sum \pi_i = const.\ (100) \tag{2}$$

Fig. 9 shows the intensity of the preference π solved by the maximum likelihood method. The matching of the model was validated by the goodness-of-fit test and the likelihood ratio test. As a result, nodding TV, conventional, and nodding camera, in this order, were selected as the preferable telecasting method via a robot.

Fig. 10 shows the result of the seven point bipolar rating. The result of the Friedman test is also shown in the figure. The nodding TV was rated the highest in terms of "presence" and "nodding." The effectiveness of the proposed method was thus demonstrated.

Fig. 11 shows the result of the interview. Half the subjects answered that the nodding camera was a sickness image; however, the rest answered that there was no sickness image. On the other hand, the nodding TV was evaluated as the most usable image.

Table 1. Result of paired comparison

	A	B	C	Total
A: Conventional		66	37	103
B: Nodding camera	14		7	21
C: Nodding TV	43	73		116

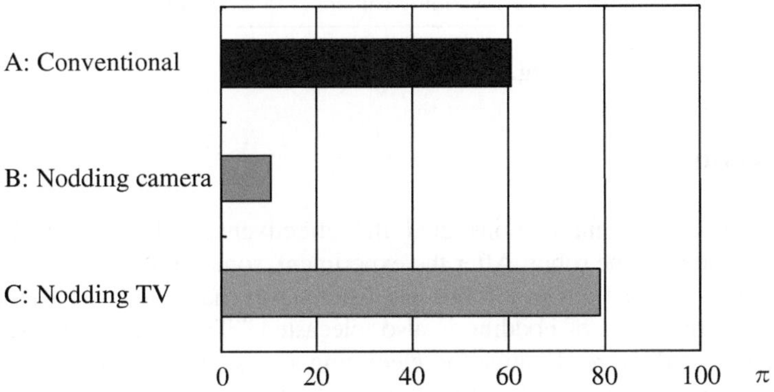

Fig. 9. Comparison of π

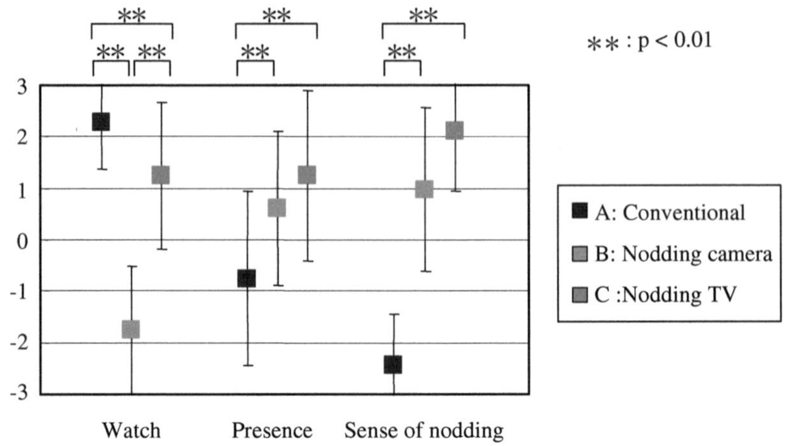

Fig. 10. Result of the seven point bipolar rating

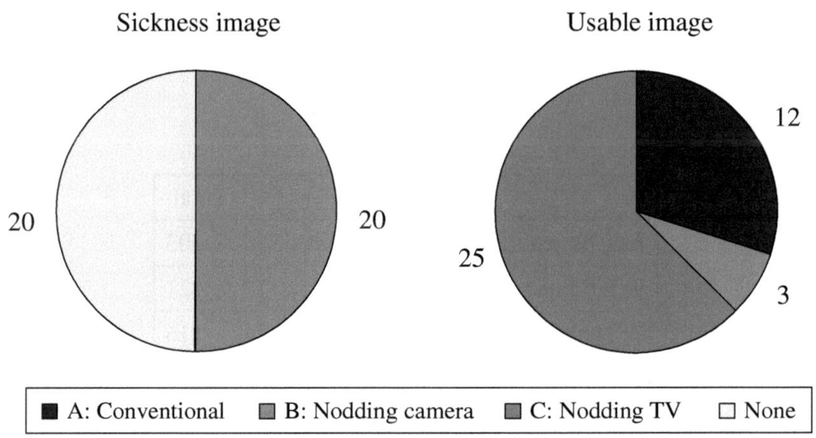

Fig. 11. Result of the interview

5 Discussion

The evaluation experiment demonstrated the effectiveness of the embodied image telecasting method via the robot. After the experiment, some of the subjects commented as follows: "I feel as if I am in a robot and listening to the caster" or "I feel sense of reality, because the sense of nodding is also telecasted." These comments suggest that the proposed methods were effective for telecasting a sense of unity and presence.

We presented the initial results of both the embodied image telecasting method and its effect on users. Several improvements to our prototype need to be carried out by eliminating the slight sickness images and the unclear telecasted images caused by the NTSC quality of the camera.

6 Conclusion

The concept of an embodied image telecasting method was proposed for activating embodied interaction and communication via a speech-driven embodied robot. The prototype of the system was developed: a new InterRobot equipped with a camera and a projector for creating a visual effect on telecasted images. The sensory evaluation demonstrated that the visual effect was effective for realizing presence and a sense of unity.

Acknowledgments

This work under our project "Generation and Control Technology of Human-entrained Embodied Media" has been supported by CREST (Core Research for Evolution Science and Technology) of JST (Japan Science and Technology Agency).

References

1. Finke, M., Koay, K., Dautenhahn, K., Nehaniv, C.L., Walters, M.L., Saunders, J.: Hey, I'm over here - How can a robot attract people's attention? CDROM of the 2005 IEEE International Workshop on Robots and Human Interactive Communication, pp. 7–12 (2005)
2. Kozima, H., Nakagawa, C., Yano, H.: Attention coupling as a prerequisite for social interaction. In: Proceedings of the 12th IEEE Workshop on Robot and Human Interactive Communication (RO-MAN 2003), pp. 109–114 (2003)
3. Yamamoto, M., Watanabe, T.: Timing control effects of utterance to communicative actions on embodied interaction with a robot. In: Proceedings of the 13th IEEE International Workshop on Robot and Human Interactive Communication (RO-MAN 2004), pp. 467–472 (2004)
4. Yamamoto, M., Watanabe, T.: Time lag effects of utterance to communicative actions on CG character-human greeting interaction. In: Proceedings, R.O.-M.A.N. (ed.) 2006 - the 15th IEEE International Symposium on Robot and Human Interactive Communication, pp. 629–634 (2006)
5. Watanabe, T., Okubo, M., Nakashige, M., Danbara, R.: InterActor: speech-driven embodied interactive actor. International Journal of Human-Computer Interaction 17(1), 43–60 (2004)

Author Index

Lecture Notes in Computer Science

For information about Vols. 1–4507

please contact your bookseller or Springer

Vol. 4611: J. Indulska, J. Ma, L.T. Yang, T. Ungerer, J. Cao (Eds.), Ubiquitous Intelligence and Computing. XXIII, 1257 pages. 2007.

Vol. 4610: B. Xiao, L.T. Yang, J. Ma, C. Muller-Schloer, Y. Hua (Eds.), Autonomic and Trusted Computing. XVIII, 571 pages. 2007.

Vol. 4605: D. Papadias, D. Zhang, G. Kollios (Eds.), Advances in Spatial and Temporal Databases. X, 479 pages. 2007.

Vol. 4602: S. Barker, G.-J. Ahn (Eds.), Data and Applications Security XXI. X, 291 pages. 2007.

Vol. 4600: H. Comon-Lundh, C. Kirchner, H. Kirchner, Rewriting, Computation and Proof. XVI, 273 pages. 2007.

Vol. 4598: G. Lin (Ed.), Computing and Combinatorics. XII, 570 pages. 2007.

Vol. 4597: P. Perner (Ed.), Advances in Data Mining. XI, 353 pages. 2007. (Sublibrary LNAI).

Vol. 4596: L. Arge, C. Cachin, T. Jurdzinski, A. Tarlecki (Eds.), Automata, Languages and Programming. XVII, 953 pages. 2007.

Vol. 4595: D. Bošnački, S. Edelkamp (Eds.), Model Checking Software. X, 285 pages. 2007.

Vol. 4594: R. Bellazzi, A. Abu-Hanna, J. Hunter (Eds.), Artificial Intelligence in Medicine. XVI, 509 pages. 2007. (Sublibrary LNAI).

Vol. 4592: Z. Kedad, N. Lammari, E. Métais, F. Meziane, Y. Rezgui (Eds.), Natural Language Processing and Information Systems. XIV, 442 pages. 2007.

Vol. 4591: J. Davies, J. Gibbons (Eds.), Integrated Formal Methods. IX, 660 pages. 2007.

Vol. 4590: W. Damm, H. Hermanns (Eds.), Computer Aided Verification. XV, 562 pages. 2007.

Vol. 4589: J. Münch, P. Abrahamsson (Eds.), Product-Focused Software Process Improvement. XII, 414 pages. 2007.

Vol. 4588: T. Harju, J. Karhumäki, A. Lepistö (Eds.), Developments in Language Theory. XI, 423 pages. 2007.

Vol. 4587: R. Cooper, J. Kennedy (Eds.), Data Management. XIII, 259 pages. 2007.

Vol. 4586: J. Pieprzyk, H. Ghodosi, E. Dawson (Eds.), Information Security and Privacy. XIV, 476 pages. 2007.

Vol. 4585: M. Kryszkiewicz, J.F. Peters, H. Rybinski, A. Skowron (Eds.), Rough Sets and Intelligent Systems Paradigms. XIX, 836 pages. 2007. (Sublibrary LNAI).

Vol. 4584: N. Karssemeijer, B. Lelieveldt (Eds.), Information Processing in Medical Imaging. XX, 777 pages. 2007.

Vol. 4583: S.R. Della Rocca (Ed.), Typed Lambda Calculi and Applications. X, 397 pages. 2007.

Vol. 4582: J. Lopez, P. Samarati, J.L. Ferrer (Eds.), Public Key Infrastructure. XI, 375 pages. 2007.

Vol. 4581: A. Petrenko, M. Veanes, J. Tretmans, W. Grieskamp (Eds.), Testing of Software and Communicating Systems. XII, 379 pages. 2007.

Vol. 4580: B. Ma, K. Zhang (Eds.), Combinatorial Pattern Matching. XII, 366 pages. 2007.

Vol. 4578: F. Masulli, S. Mitra, G. Pasi (Eds.), Applications of Fuzzy Sets Theory. XVIII, 693 pages. 2007. (Sublibrary LNAI).

Vol. 4577: N. Sebe, Y. Liu, Y. Zhuang (Eds.), Multimedia Content Analysis and Mining. XIII, 513 pages. 2007.

Vol. 4576: D. Leivant, R. de Queiroz (Eds.), Logic, Language, Information, and Computation. X, 363 pages. 2007.

Vol. 4575: T. Takagi, T. Okamoto, E. Okamoto, T. Okamoto (Eds.), Pairing-Based Cryptography. XI, 408 pages. 2007.

Vol. 4574: J. Derrick, J. Vain (Eds.), Formal Techniques for Networked and Distributed Systems – FORTE 2007. XI, 375 pages. 2007.

Vol. 4573: M. Kauers, M. Kerber, R. Miner, W. Windsteiger (Eds.), Towards Mechanized Mathematical Assistants. XIII, 407 pages. 2007. (Sublibrary LNAI).

Vol. 4572: F. Stajano, C. Meadows, S. Capkun, T. Moore (Eds.), Security and Privacy in Ad-hoc and Sensor Networks. X, 247 pages. 2007.

Vol. 4570: H.G. Okuno, M. Ali (Eds.), New Trends in Applied Artificial Intelligence. XXI, 1194 pages. 2007. (Sublibrary LNAI).

Vol. 4569: A. Butz, B. Fisher, A. Krüger, P. Olivier, S. Owada (Eds.), Smart Graphics. IX, 237 pages. 2007.

Vol. 4566: M.J. Dainoff (Ed.), Ergonomics and Health Aspects of Work with Computers. XVIII, 390 pages. 2007.

Vol. 4565: D.D. Schmorrow, L.M. Reeves (Eds.), Foundations of Augmented Cognition. XIX, 450 pages. 2007. (Sublibrary LNAI).

Vol. 4564: D. Schuler (Ed.), Online Communities and Social Computing. XVII, 520 pages. 2007.

Vol. 4563: R. Shumaker (Ed.), Virtual Reality. XXII, 762 pages. 2007.

Vol. 4562: D. Harris (Ed.), Engineering Psychology and Cognitive Ergonomics. XXIII, 879 pages. 2007. (Sublibrary LNAI).

Vol. 4561: V.G. Duffy (Ed.), Digital Human Modeling. XXIII, 1068 pages. 2007.

Vol. 4560: N. Aykin (Ed.), Usability and Internationalization, Part II. XVIII, 576 pages. 2007.

Vol. 4559: N. Aykin (Ed.), Usability and Internationalization, Part I. XVIII, 661 pages. 2007.

Vol. 4558: M.J. Smith, G. Salvendy (Eds.), Human Interface and the Management of Information, Part II. XXIII, 1162 pages. 2007.

Vol. 4557: M.J. Smith, G. Salvendy (Eds.), Human Interface and the Management of Information, Part I. XXII, 1030 pages. 2007.

Vol. 4553: J.A. Jacko (Ed.), Human-Computer Interaction, Part IV. XXIV, 1225 pages. 2007.

Vol. 4552: J.A. Jacko (Ed.), Human-Computer Interaction, Part III. XXI, 1038 pages. 2007.

Vol. 4551: J.A. Jacko (Ed.), Human-Computer Interaction, Part II. XXIII, 1253 pages. 2007.

Vol. 4550: J.A. Jacko (Ed.), Human-Computer Interaction, Part I. XXIII, 1240 pages. 2007.

Vol. 4549: J. Aspnes, C. Scheideler, A. Arora, S. Madden (Eds.), Distributed Computing in Sensor Systems. XIII, 417 pages. 2007.

Vol. 4548: N. Olivetti (Ed.), Automated Reasoning with Analytic Tableaux and Related Methods. X, 245 pages. 2007. (Sublibrary LNAI).

Vol. 4547: C. Carlet, B. Sunar (Eds.), Arithmetic of Finite Fields. XI, 355 pages. 2007.

Vol. 4546: J. Kleijn, A. Yakovlev (Eds.), Petri Nets and Other Models of Concurrency – ICATPN 2007. XI, 515 pages. 2007.

Vol. 4545: H. Anai, K. Horimoto, T. Kutsia (Eds.), Algebraic Biology. XIII, 379 pages. 2007.

Vol. 4544: S. Cohen-Boulakia, V. Tannen (Eds.), Data Integration in the Life Sciences. XI, 282 pages. 2007. (Sublibrary LNBI).

Vol. 4543: A.K. Bandara, M. Burgess (Eds.), Inter-Domain Management. XII, 237 pages. 2007.

Vol. 4542: P. Sawyer, B. Paech, P. Heymans (Eds.), Requirements Engineering: Foundation for Software Quality. IX, 384 pages. 2007.

Vol. 4541: T. Okadome, T. Yamazaki, M. Makhtari (Eds.), Pervasive Computing for Quality of Life Enhancement. IX, 248 pages. 2007.

Vol. 4539: N.H. Bshouty, C. Gentile (Eds.), Learning Theory. XII, 634 pages. 2007. (Sublibrary LNAI).

Vol. 4538: F. Escolano, M. Vento (Eds.), Graph-Based Representations in Pattern Recognition. XII, 416 pages. 2007.

Vol. 4537: K.C.-C. Chang, W. Wang, L. Chen, C.A. Ellis, C.-H. Hsu, A.C. Tsoi, H. Wang (Eds.), Advances in Web and Network Technologies, and Information Management. XXIII, 707 pages. 2007.

Vol. 4536: G. Concas, E. Damiani, M. Scotto, G. Succi (Eds.), Agile Processes in Software Engineering and Extreme Programming. XV, 276 pages. 2007.

Vol. 4534: I. Tomkos, F. Neri, J. Solé Pareta, X. Masip Bruin, S. Sánchez Lopez (Eds.), Optical Network Design and Modeling. XI, 460 pages. 2007.

Vol. 4531: J. Indulska, K. Raymond (Eds.), Distributed Applications and Interoperable Systems. XI, 337 pages. 2007.

Vol. 4530: D.H. Akehurst, R. Vogel, R.F. Paige (Eds.), Model Driven Architecture- Foundations and Applications. X, 219 pages. 2007.

Vol. 4529: P. Melin, O. Castillo, L.T. Aguilar, J. Kacprzyk, W. Pedrycz (Eds.), Foundations of Fuzzy Logic and Soft Computing. XIX, 830 pages. 2007. (Sublibrary LNAI).

Vol. 4528: J. Mira, J.R. Álvarez (Eds.), Nature Inspired Problem-Solving Methods in Knowledge Engineering, Part II. XXII, 650 pages. 2007.

Vol. 4527: J. Mira, J.R. Álvarez (Eds.), Bio-inspired Modeling of Cognitive Tasks, Part I. XXII, 630 pages. 2007.

Vol. 4526: M. Malek, M. Reitenspieß, A. van Moorsel (Eds.), Service Availability. X, 155 pages. 2007.

Vol. 4525: C. Demetrescu (Ed.), Experimental Algorithms. XIII, 448 pages. 2007.

Vol. 4524: M. Marchiori, J.Z. Pan, C.d.S. Marie (Eds.), Web Reasoning and Rule Systems. XI, 382 pages. 2007.

Vol. 4523: Y.-H. Lee, H.-N. Kim, J. Kim, Y. Park, L.T. Yang, S.W. Kim (Eds.), Embedded Software and Systems. XIX, 829 pages. 2007.

Vol. 4522: B.K. Ersbøll, K.S. Pedersen (Eds.), Image Analysis. XVIII, 989 pages. 2007.

Vol. 4521: J. Katz, M. Yung (Eds.), Applied Cryptography and Network Security. XIII, 498 pages. 2007.

Vol. 4519: E. Franconi, M. Kifer, W. May (Eds.), The Semantic Web: Research and Applications. XVIII, 830 pages. 2007.

Vol. 4517: F. Boavida, E. Monteiro, S. Mascolo, Y. Koucheryavy (Eds.), Wired/Wireless Internet Communications. XIV, 382 pages. 2007.

Vol. 4516: L. Mason, T. Drwiega, J. Yan (Eds.), Managing Traffic Performance in Converged Networks. XXIII, 1191 pages. 2007.

Vol. 4515: M. Naor (Ed.), Advances in Cryptology - EUROCRYPT 2007. XIII, 591 pages. 2007.

Vol. 4514: S.N. Artemov, A. Nerode (Eds.), Logical Foundations of Computer Science. XI, 513 pages. 2007.

Vol. 4513: M. Fischetti, D.P. Williamson (Eds.), Integer Programming and Combinatorial Optimization. IX, 500 pages. 2007.

Vol. 4511: C. Conati, K. McCoy, G. Paliouras (Eds.), User Modeling 2007. XVI, 487 pages. 2007. (Sublibrary LNAI).

Vol. 4510: P. Van Hentenryck, L. Wolsey (Eds.), Integration of AI and OR Techniques in Constraint Programming for Combinatorial Optimization Problems. X, 391 pages. 2007.

Vol. 4509: Z. Kobti, D. Wu (Eds.), Advances in Artificial Intelligence. XII, 552 pages. 2007. (Sublibrary LNAI).

Vol. 4508: M.-Y. Kao, X.-Y. Li (Eds.), Algorithmic Aspects in Information and Management. VIII, 428 pages. 2007.

MIX
[illegible]
FSC® C105338

[illegible]
ProductSafety@springernature.com
[illegible]
[illegible] Customer Service Center GmbH [illegible]
Europaplatz 3, 69115 Heidelberg, Germany
Printed by [illegible]
in Hamburg, Germany

MIX
Papier aus verantwortungsvollen Quellen
Paper from responsible sources
FSC® C105338

If you have any concerns about our products,
you can contact us on
ProductSafety@springernature.com

In case Publisher is established outside the EU,
the EU authorized representative is:
Springer Nature Customer Service Center GmbH
Europaplatz 3, 69115 Heidelberg, Germany

Printed by Libri Plureos GmbH
in Hamburg, Germany